Teacher, Student, and Parent One-Stop Internet Resources

ONLINE STUDY TOOLS

- Section Self-Check Quizzes
- Interactive Tutor
- Chapter Review Tests
- Standardized Test Practice
- Vocabulary PuzzleMaker

ONLINE RESEARCH

- WebQuest Projects
- Prescreened Web Links
- Career Links
- Microscopy Image Links
- Internet Labs

INTERACTIVE ONLINE STUDENT EDITION

- Complete Interactive Student Edition available at mhln.com

FOR TEACHERS

- Teacher Bulletin Board
- Teaching Today—Professional Development

SAFETY SYMBOLS

SAFETY SYMBOLS	HAZARD	EXAMPLES	PRECAUTION	REMEDY
DISPOSAL	Special disposal procedures need to be followed.	certain chemicals, living organisms	Do not dispose of these materials in the sink or trash can.	Dispose of wastes as directed by your teacher.
BIOLOGICAL	Organisms or other biological materials that might be harmful to humans	bacteria, fungi, blood, unpreserved tissues, plant materials	Avoid skin contact with these materials. Wear mask or gloves.	Notify your teacher if you suspect contact with material. Wash hands thoroughly.
EXTREME TEMPERATURE	Objects that can burn skin by being too cold or too hot	boiling liquids, hot plates, dry ice, liquid nitrogen	Use proper protection when handling.	Go to your teacher for first aid.
SHARP OBJECT	Use of tools or glassware that can easily puncture or slice skin	razor blades, pins, scalpels, pointed tools, dissecting probes, broken glass	Practice common-sense behavior and follow guidelines for use of the tool.	Go to your teacher for first aid.
FUME	Possible danger to respiratory tract from fumes	ammonia, acetone, nail polish remover, heated sulfur, moth balls	Make sure there is good ventilation. Never smell fumes directly. Wear a mask.	Leave foul area and notify your teacher immediately.
ELECTRICAL	Possible danger from electrical shock or burn	improper grounding, liquid spills, short circuits, exposed wires	Double-check setup with teacher. Check condition of wires and apparatus.	Do not attempt to fix electrical problems. Notify your teacher immediately.
IRRITANT	Substances that can irritate the skin or mucous membranes of the respiratory tract	pollen, moth balls, steel wool, fiberglass, potassium permanganate	Wear dust mask and gloves. Practice extra care when handling these materials.	Go to your teacher for first aid.
CHEMICAL	Chemicals can react with and destroy tissue and other materials	bleaches such as hydrogen peroxide; acids such as sulfuric acid, hydrochloric acid; bases such as ammonia, sodium hydroxide	Wear goggles, gloves, and an apron.	Immediately flush the affected area with water and notify your teacher.
TOXIC	Substance may be poisonous if touched, inhaled, or swallowed.	mercury, many metal compounds, iodine, poinsettia plant parts	Follow your teacher's instructions.	Always wash hands thoroughly after use. Go to your teacher for first aid.
FLAMMABLE	Flammable chemicals may be ignited by open flame, spark, or exposed heat.	alcohol, kerosene, potassium permanganate	Avoid open flames and heat when using flammable chemicals.	Notify your teacher immediately. Use fire safety equipment if applicable.
OPEN FLAME	Open flame in use, may cause fire.	hair, clothing, paper, synthetic materials	Tie back hair and loose clothing. Follow teacher's instruction on lighting and extinguishing flames.	Notify your teacher immediately. Use fire safety equipment if applicable.

Eye Safety
Proper eye protection should be worn at all times by anyone performing or observing science activities.

Clothing Protection
This symbol appears when substances could stain or burn clothing.

Animal Safety
This symbol appears when safety of animals and students must be ensured.

Handwashing
After the lab, wash hands with soap and water before removing goggles.

New York, New York Columbus, Ohio Chicago, Illinois Woodland Hills, California

Glencoe Science

The American bald eagle can only be found in North America. Bald eagles are predators and mainly eat fish. Female bald eagles lay one to three eggs each spring in a nest called an aerie. Both the male and female of a pair share the work of feeding and protecting their young. Parents shred pieces of meat to feed the eaglets.

The McGraw·Hill Companies

Send all inquiries to:
Glencoe/McGraw-Hill
8787 Orion Place
Columbus, OH 43240-4027

ISBN: 978-0-07-877800-1
MHID: 0-07-877800-X

Printed in the United States of America.

3 4 5 6 7 8 9 10 027/055 09 08 07

Contents In Brief

Authors

NATIONAL GEOGRAPHIC
Education Division
Washington, D.C.

Alton Biggs
Retired Biology Teacher
Allen High School
Allen, TX

Lucy Daniel, EdD
Teacher/Consultant
Rutherford County Schools
Rutherfordton, NC

Edward Ortleb
Science Consultant
St. Louis, MO

Peter Rillero, PhD
Associate Professor of
Science Education
Arizona State University West
Phoenix, AZ

Dinah Zike
Educational Consultant
Dinah-Might Activities, Inc.
San Antonio, TX

Science Consultants

CONTENT

Sandra K. Enger, PhD
Associate Director,
Associate Professor
UAH Institute for Science Education
Huntsville, AL

Michael A. Hoggarth, PhD
Department of Life and Earth
Sciences
Otterbein College
Westerville, OH

Jerome A. Jackson, PhD
Whitaker Eminent Scholar in Science
Program Director
Center for Science, Mathematics,
and Technology Education
Florida Gulf Coast University
Fort Meyers, FL

Connie Rizzo, MD, PhD
Depatment of Science/Math
Marymount Manhattan College
New York, NY

Dominic Salinas, PhD
Middle School Science Supervisor
Caddo Parish Schools
Shreveport, LA

Series Consultants

MATH

Michael Hopper, DEng
Manager of Aircraft Certification
L-3 Communications
Greenville, TX

Teri Willard, EdD
Mathematics Curriculum Writer
Belgrade, MT

READING

Elizabeth Babich
Special Education Teacher
Mashpee Public Schools
Mashpee, MA

Barry Barto
Special Education Teacher
John F. Kennedy Elementary
Manistee, MI

Carol A. Senf, PhD
School of Literature,
Communication, and Culture
Georgia Institute of Technology
Atlanta, GA

Rachel Swaters-Kissinger
Science Teacher
John Boise Middle School
Warsaw, MO

SAFETY

Aileen Duc, PhD
Science 8 Teacher
Hendrick Middle School, Plano ISD
Plano, TX

Sandra West, PhD
Department of Biology
Texas State University-San Marcos
San Marcos, TX

ACTIVITY TESTERS

Nerma Coats Henderson
Pickerington Lakeview Jr. High School
Pickerington, OH

Mary Helen Mariscal-Cholka
William D. Slider Middle School
El Paso, TX

Science Kit and Boreal Laboratories
Tonawanda, NY

Series Reviewers

Deidre Adams
West Vigo Middle School
West Terre Haute, IN

Sharla Adams
IPC Teacher
Allen High School
Allen, TX

Maureen Barrett
Thomas E. Harrington Middle School
Mt. Laurel, NJ

John Barry
Seeger Jr.-Sr. High School
West Lebanon, IN

Desiree Bishop
Environmental Studies Center
Mobile County Public Schools
Mobile, AL

William Blair
Retired Teacher
J. Marshall Middle School
Billerica, MA

Tom Bright
Concord High School
Charlotte, NC

Lois Burdette
Green Bank Elementary-Middle School
Green Bank, WV

Marcia Chackan
Pine Crest School
Boca Raton, FL

Obioma Chukwu
J.H. Rose High School
Greenville, NC

Nerma Coats Henderson
Pickerington Lakeview Jr. High School
Pickerington, OH

Karen Curry
East Wake Middle School
Raleigh, NC

Inga Dainton
Merrilvills High School
Merrilville, IN

Joanne Davis
Murphy High School
Murphy, NC

Robin Dillon
Hanover Central High School
Cedar Lake, IN

Anthony J. DiSipio, Jr.
8th Grade Science
Octorana Middle School
Atglen, PA

Annette D'Urso Garcia
Kearney Middle School
Commerce City, CO

Dwight Dutton
East Chapel Hill High School
Chapel Hill, NC

Carolyn Elliott
South Iredell High School
Statesville, NC

Sueanne Esposito
Tipton High School
Tipton, IN

Sandra Everhart
Dauphin/Enterprise Jr. High Schools
Enterprise, AL

Mary Ferneau
Westview Middle School
Goose Creek, SC

Cory Fish
Burkholder Middle School
Henderson, NV

Linda V. Forsyth
Retired Teacher
Merrill Middle School
Denver, CO

George Gabb
Great Bridge Middle School
Chesapeake Public Schools
Chesapeake, VA

Judith Helton
RS Middle School
Rutherfordton, NC

Lynne Huskey
Chase Middle School
Forest City, NC

Maria E. Kelly
Principal
Nativity School
Catholic Diocese of Arlington
Burke, VA

Michael Mansour
Board Member
National Middle Level Science
Teacher's Association
John Page Middle School
Madison Heights, MI

Mary Helen Mariscal-Cholka
William D. Slider Middle School
El Paso, TX

Michelle Mazeika
Whiting Middle School
Whiting, IN

Joe McConnell
Speedway Jr. High School
Indianapolis, IN

Sharon Mitchell
William D. Slider Middle School
El Paso, TX

Amy Morgan
Berry Middle School
Hoover, AL

Norma Neely, EdD
Associate Director for Regional
Projects
Texas Rural Systemic Initiative
Austin, TX

Annette Parrott
Lakeside High School
Atlanta, GA

Nora M. Prestinari Burchett
Saint Luke School
McLean, VA

Mark Sailer
Pioneer Jr.-Sr. High School
Royal Center, IN

Vicki Snell
Monroe Central Jr.-Sr. High School
Parker City, IN

Joanne Stickney
Monticello Middle School
Monticello, NY

Dee Stout
Penn State University
University Park, PA

Darcy Vetro-Ravndal
Hillsborough High School
Tampa, FL

Karen Watkins
Perry Meridian Middle School
Indianapolis, IN

Clabe Webb
Permian High School
Ector County ISD
Odessa, TX

Alison Welch
William D. Slider Middle School
El Paso, TX

Kim Wimpey
North Gwinnett High School
Suwanee, GA

Michael Wolter
Muncie Central High School
Muncie, IN

Kate Ziegler
Durant Road Middle School
Raleigh, NC

Teacher Advisory Board

The Teacher Advisory Board gave the authors, editorial staff, and design team feedback on the content and design of the Student Edition. They provided valuable input in the development of the 2008 edition of ***Glencoe Life Science.***

John Gonzales
Challenger Middle School
Tucson, AZ

Rachel Shively
Aptakisic Jr. High School
Buffalo Grove, IL

Roger Pratt
Manistique High School
Manistique, MI

Kirtina Hile
Northmor Jr. High/High School
Galion, OH

Marie Renner
Diley Middle School
Pickerington, OH

Nelson Farrier
Hamlin Middle School
Springfield, OR

Jeff Remington
Palmyra Middle School
Palmyra, PA

Erin Peters
Williamsburg Middle School
Arlington, VA

Rubidel Peoples
Meacham Middle School
Fort Worth, TX

Kristi Ramsey
Navasota Jr. High School
Navasota, TX

Student Advisory Board

The Student Advisory Board gave the authors, editorial staff, and design team feedback on the design of the Student Edition. We thank these students for their hard work and creative suggestions in making the 2008 edition of ***Glencoe Life Science*** student friendly.

Jack Andrews
Reynoldsburg Jr. High School
Reynoldsburg, OH

Peter Arnold
Hastings Middle School
Upper Arlington, OH

Emily Barbe
Perry Middle School
Worthington, OH

Kirsty Bateman
Hilliard Heritage Middle School
Hilliard, OH

Andre Brown
Spanish Emersion Academy
Columbus, OH

Chris Dundon
Heritage Middle School
Westerville, OH

Ryan Manafee
Monroe Middle School
Columbus, OH

Addison Owen
Davis Middle School
Dublin, OH

Teriana Patrick
Eastmoor Middle School
Columbus, OH

Ashley Ruz
Karrer Middle School
Dublin, OH

The Glencoe middle school science Student Advisory Board taking a timeout at COSI, a science museum in Columbus, Ohio.

HOW TO...

Use Your Science Book

Why do I need my science book?

Have you ever been in class and not understood all of what was presented? Or, you understood everything in class, but at home, got stuck on how to answer a question? Maybe you just wondered when you were ever going to use this stuff?

These next few pages are designed to help you understand everything your science book can be used for . . . besides a paperweight!

Before You Read

- **Chapter Opener** Science is occurring all around you, and the opening photo of each chapter will preview the science you will be learning about. The **Chapter Preview** will give you an idea of what you will be learning about, and you can try the **Launch Lab** to help get your brain headed in the right direction. The **Foldables** exercise is a fun way to keep you organized.
- **Section Opener** Chapters are divided into two to four sections. The **As You Read** in the margin of the first page of each section will let you know what is most important in the section. It is divided into four parts. **What You'll Learn** will tell you the major topics you will be covering. **Why It's Important** will remind you why you are studying this in the first place! The **Review Vocabulary** word is a word you already know, either from your science studies or your prior knowledge. The **New Vocabulary** words are words that you need to learn to understand this section. These words will be in **boldfaced** print and highlighted in the section. Make a note to yourself to recognize these words as you are reading the section.

As You Read

- **Headings** Each section has a title in large red letters, and is further divided into blue titles and small red titles at the beginnings of some paragraphs. To help you study, make an outline of the headings and subheadings.
- **Margins** In the margins of your text, you will find many helpful resources. The **Science Online** exercises and **Integrate** activities help you explore the topics you are studying. **MiniLabs** reinforce the science concepts you have learned.
- **Building Skills** You also will find an **Applying Math** or **Applying Science** activity in each chapter. This gives you extra practice using your new knowledge, and helps prepare you for standardized tests.
- **Student Resources** At the end of the book you will find **Student Resources** to help you throughout your studies. These include **Science, Technology,** and **Math Skill Handbooks,** an **English/Spanish Glossary,** and an **Index.** Also, use your **Foldables** as a resource. It will help you organize information, and review before a test.
- **In Class** Remember, you can always ask your teacher to explain anything you don't understand.

FOLDABLES™ Study Organizer

Science Vocabulary Make the following Foldable to help you understand the vocabulary terms in this chapter.

STEP 1 **Fold** a vertical sheet of notebook paper from side to side.

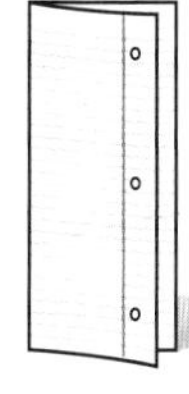

STEP 2 **Cut** along every third line of only the top layer to form tabs.

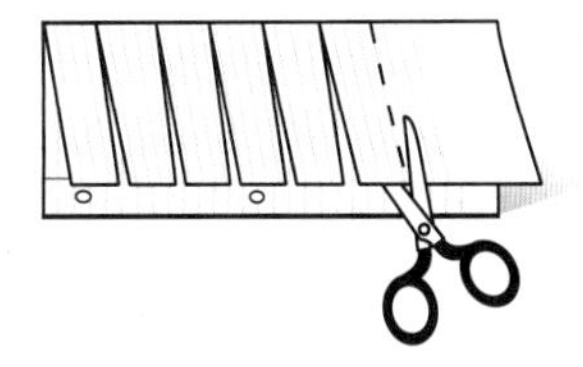

STEP 3 **Label** each tab with a vocabulary word from the chapter.

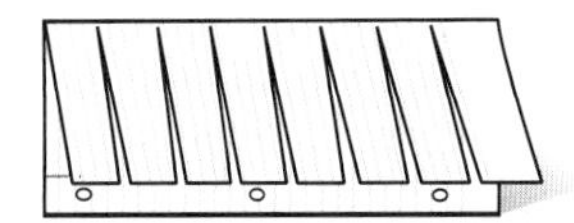

Build Vocabulary As you read the chapter, list the vocabulary words on the tabs. As you learn the definitions, write them under the tab for each vocabulary word.

Look For...
FOLDABLES™
At the beginning of every section.

In Lab

Working in the laboratory is one of the best ways to understand the concepts you are studying. Your book will be your guide through your laboratory experiences, and help you begin to think like a scientist. In it, you not only will find the steps necessary to follow the investigations, but you also will find helpful tips to make the most of your time.

- Each lab provides you with a **Real-World Question** to remind you that science is something you use every day, not just in class. This may lead to many more questions about how things happen in your world.
- Remember, experiments do not always produce the result you expect. Scientists have made many discoveries based on investigations with unexpected results. You can try the experiment again to make sure your results were accurate, or perhaps form a new hypothesis to test.
- Keeping a **Science Journal** is how scientists keep accurate records of observations and data. In your journal, you also can write any questions that may arise during your investigation. This is a great method of reminding yourself to find the answers later.

Look For...
Launch Labs start every chapter.
MiniLabs in the margin of each chapter.
Two Full-Period Labs in every chapter.
EXTRA Try at Home Labs at the end of your book.
the Web site with laboratory demonstrations.

Before a Test

Admit it! You don't like to take tests! However, there *are* ways to review that make them less painful. Your book will help you be more successful taking tests if you use the resources provided to you.

- Review all of the **New Vocabulary** words and be sure you understand their definitions.
- Review the notes you've taken on your **Foldables,** in class, and in lab. Write down any question that you still need answered.
- Review the **Summaries** and **Self Check questions** at the end of each section.
- Study the concepts presented in the chapter by reading the **Study Guide** and answering the questions in the **Chapter Review.**

a or b?

?

T or F?

Look For...

- Reading Checks and caption questions throughout the text.
- the summaries and self check questions at the end of each section.
- the Study Guide and Review at the end of each chapter.
- the Standardized Test Practice after each chapter.

Let's Get Started

To help you find the information you need quickly, use the Scavenger Hunt below to learn where things are located in Chapter 1.

1. What is the title of this chapter?
2. What will you learn in Section 1?
3. Sometimes you may ask, "Why am I learning this?" State a reason why the concepts from Section 2 are important.
4. What is the main topic presented in Section 2?
5. How many reading checks are in Section 1?
6. What is the Web address where you can find extra information?
7. What is the main heading above the sixth paragraph in Section 2?
8. There is an integration with another subject mentioned in one of the margins of the chapter. What subject is it?
9. List the new vocabulary words presented in Section 2.
10. List the safety symbols presented in the first Lab.
11. Where would you find a Self Check to be sure you understand the section?
12. Suppose you're doing the Self Check and you have a question about concept mapping. Where could you find help?
13. On what pages are the Chapter Study Guide and Chapter Review?
14. Look in the Table of Contents to find out on which page Section 2 of the chapter begins.
15. You complete the Chapter Review to study for your chapter test. Where could you find another quiz for more practice?

Life's Structure and Function—2

Chapter 1 Exploring and Classifying Life —4

Chapter 2 Cells—36

Chapter 3 Cell Processes—64

In each chapter, look for these opportunities for review and assessment:
- Reading Checks
- Caption Questions
- Section Review
- Chapter Study Guide
- Chapter Review
- Standardized Test Practice
- Online practice at **life.msscience.com**

Get Ready to Read Strategies

Contents

From Bacteria to Plants—182

In each chapter, look for these opportunities for review and assessment:
- Reading Checks
- Caption Questions
- Section Review
- Chapter Study Guide
- Chapter Review
- Standardized Test Practice
- Online practice at **life.msscience.com**

Get Ready to Read Strategies

Contents

Plant Processes—300

Animal Diversity—326

Introduction to Animals—328

Mollusks, Worms, Arthropods, Echinoderms—358

In each chapter, look for these opportunities for review and assessment:
- Reading Checks
- Caption Questions
- Section Review
- Chapter Study Guide
- Chapter Review
- Standardized Test Practice
- Online practice at **life.msscience.com**

Contents

unit 4 Human Body Systems—480

In each chapter, look for these opportunities for review and assessment:
- Reading Checks
- Caption Questions
- Section Review
- Chapter Study Guide
- Chapter Review
- Standardized Test Practice
- Online practice at **life.msscience.com**

Contents

In each chapter, look for these opportunities for review and assessment:
- Reading Checks
- Caption Questions
- Section Review
- Chapter Study Guide
- Chapter Review
- Standardized Test Practice
- Online practice at life.msscience.com

Get Ready to Read Strategies

Student Resources—800

Cross-Curricular Readings

NATIONAL GEOGRAPHIC Unit Openers

NATIONAL GEOGRAPHIC VISUALIZING

Content Details

TIME Science and Society

TIME Science and History

Oops! Accidents in Science

Science and Language Arts

Science Stats

DVD available as a video lab on DVD

Mini LAB

Content Details

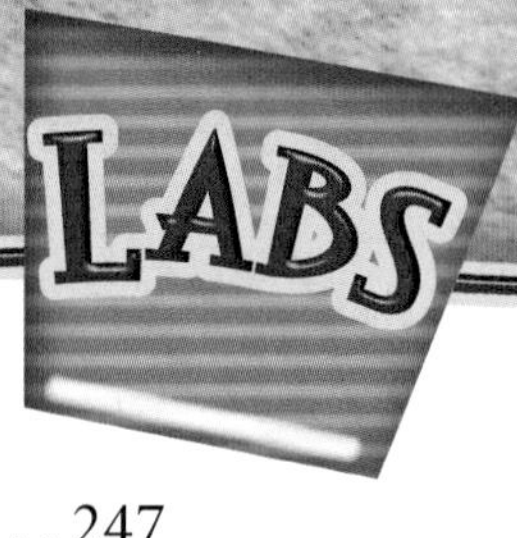

Mini LAB Try at Home

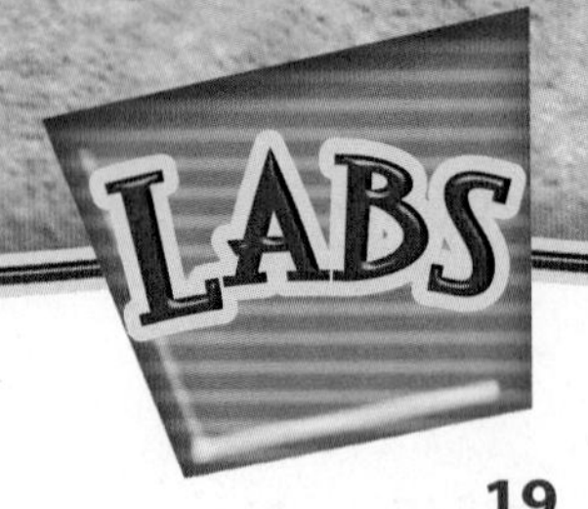

available as a video lab on DVD

One-Page Labs

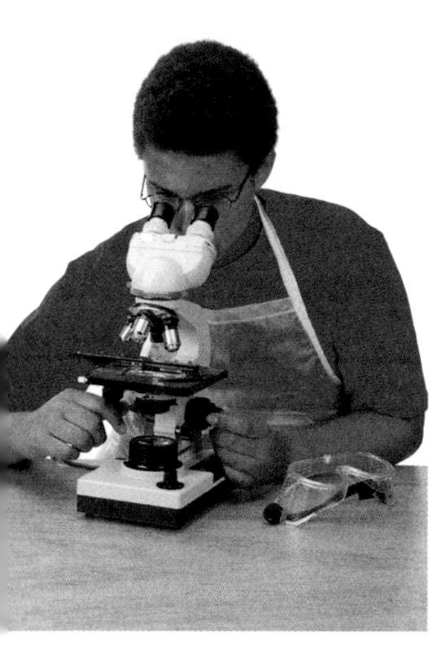

Two-Page Labs

Design Your Own Labs

Model and Invent Labs

Use the Internet Labs

Share your data with other students at **life.msscience.com/internet_lab**.

Activities

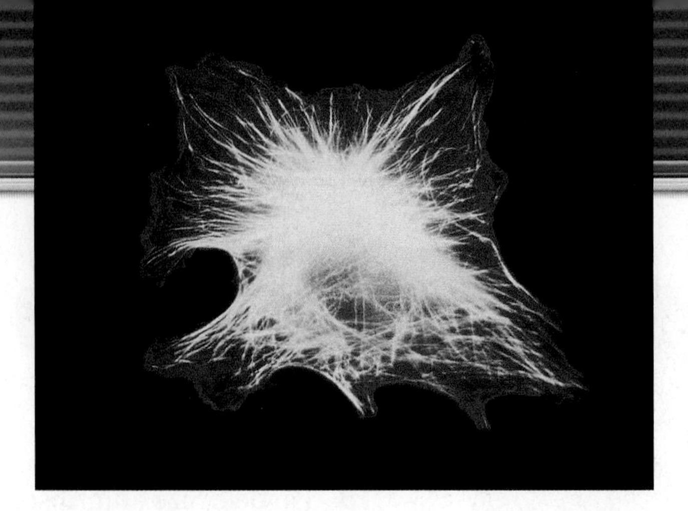

Applying Math

Applying Science

INTEGRATE

Astronomy: 429, 606
Career: 50, 83, 97, 137, 228, 303, 497, 634, 717, 755
Chemistry: 105, 109, 338, 465, 466, 499, 598, 633, 779
Earth Science: 21, 167, 190, 363, 569, 727, 750
Environment: 44, 142, 284, 529, 669
Health: 77, 217, 255, 402, 457, 782
History: 242, 278, 417, 466, 554, 595, 699
Language Arts: 158, 331
Physics: 312, 545, 640
Social Studies: 17, 196, 373, 431, 519, 581, 658, 773

Science Online

8, 15, 23, 53, 54, 70, 84, 113, 115, 127, 135, 156, 165, 189, 197, 214, 223, 248, 259, 274, 282, 306, 316, 334, 340, 368, 382, 409, 413, 432, 441, 459, 468, 486, 491, 514, 526, 547, 551, 571, 574, 599, 601, 609, 629, 637, 654, 663, 686, 692, 717, 725, 741, 757, 780, 790

Standardized Test Practice

34–35, 62–63, 92–93, 122–123, 150–151, 180–181, 206–207, 236–237, 268–269, 298–299, 324–325, 356–357, 390–391, 424–425, 452–453, 478–479, 508–509, 536–537, 564–565, 590–591, 618–619, 648–649, 678–679, 708–709, 736–737, 766–767, 798–799

How Are Seaweed & Cell Cultures Connected?

In the 1800s, many biologists were interested in studying one-celled microorganisms. But to study them, the researchers needed to grow, or culture, large numbers of these cells. And to culture them properly, they needed a solid substance on which the cells could grow. One scientist tried using nutrient-enriched gelatin, but the gelatin had drawbacks. It melted at relatively low temperatures—and some microorganisms digested it. Fannie Eilshemius Hesse came up with a better option. She had been solidifying her homemade jellies using a substance called agar, which is derived from red seaweed (such as the one seen in the background here). It turned out that nutrient-enriched agar worked perfectly as a substance on which to culture cells. On the two types of agar in the dishes below, so many cells have grown that, together, they form dots and lines.

unit projects

Visit **life.msscience.com/unit_project** to find project ideas and resources.
Projects include:

- **Career** Brainstorm a list of questions for a health professional about cell reproduction, or bacteria and virus resistance to drugs.
- **Technology** Design both a chart and a graph that present information on cell reproduction rates during specific time intervals.
- **Model** Construct a thumb flip book that models mitosis. Complete a second book for meiosis to analyze and compare the two processes.

WebQuest *New Research on Cells* is a Web-based investigation of current research involving different types of cells, and why their structures vary according to their location.

chapter 1

Exploring and Classifying Life

The BIG Idea

Life science includes the study of living and once-living things.

SECTION 1
What is science?
Main Idea Science is an organized way of studying things and finding answers to questions.

SECTION 2
Living Things
Main Idea Living things have certain characteristics in common.

SECTION 3
Where does life come from?
Main Idea There are many hypotheses about the origins of life.

SECTION 4
How are living things classified?
Main Idea Classification systems show relationships among living things.

Life Under the Sea

This picture contains many living things—including living coral. These living things have both common characteristics and differences. Scientists classify life according to similarities.

Science Journal List three characteristics that you would use to classify underwater life.

Start-Up Activities

Classify Organisms

Life scientists discover, describe, and name hundreds of organisms every year. How do they decide if a certain plant belongs to the iris or orchid family of flowering plants, or if an insect is more like a grasshopper or a beetle?

1. Observe the organisms on the opposite page or in an insect collection in your class.
2. Decide which feature could be used to separate the organisms into two groups, then sort the organisms into the two groups.
3. Continue to make new groups using different features until each organism is in a category by itself.
4. **Think Critically** How do you think scientists classify living things? List your ideas in your Science Journal.

Preview this chapter's content and activities at life.msscience.com

Vocabulary Make the following Foldable to help you understand the vocabulary terms in this chapter.

STEP 1 **Fold** a vertical sheet of notebook paper from side to side.

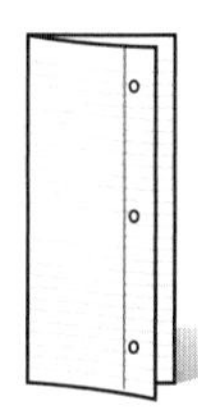

STEP 2 **Cut** along every third line of only the top layer to form tabs.

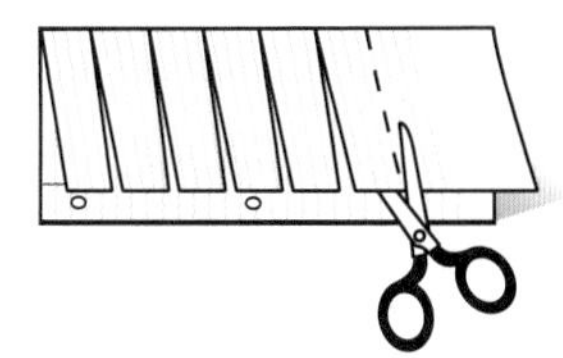

STEP 3 **Label** each tab.

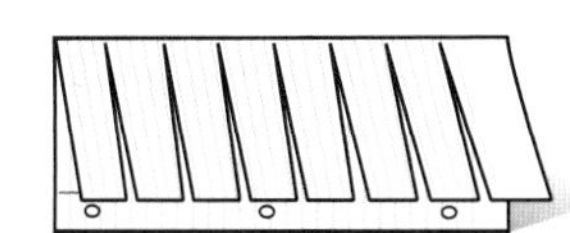

Build Vocabulary As you read the chapter, write the vocabulary words on the tabs. As you learn the definitions, write them under the tab for each vocabulary word.

Get Ready to Read

Preview

1 Learn It! If you know what to expect before reading, it will be easier to understand ideas and relationships presented in the text. Follow these steps to preview your reading assignments.

1. Look at the title and any illustrations that are included.
2. Read the headings, subheadings, and anything in bold letters.
3. Skim over the passage to see how it is organized. Is it divided into many parts?
4. Look at the graphics—pictures, maps, or diagrams. Read their titles, labels, and captions.
5. Set a purpose for your reading. Are you reading to learn something new? Are you reading to find specific information?

2 Practice It! Take some time to preview this chapter. Skim all the main headings and subheadings. With a partner, discuss your answers to these questions.

- Which part of this chapter looks most interesting to you?
- Are there any words in the headings that are unfamiliar to you?
- Choose one of the lesson review questions to discuss with a partner.

3 Apply It! Now that you have skimmed the chapter, write a short paragraph describing one thing you want to learn from this chapter.

Reading Tip

As you preview this chapter, be sure to scan the illustrations, tables, and graphs. Skim the captions.

Target Your Reading

Use this to focus on the main ideas as you read the chapter.

① **Before you read** the chapter, respond to the statements below on your worksheet or on a numbered sheet of paper.

- Write an **A** if you **agree** with the statement.
- Write a **D** if you **disagree** with the statement.

② **After you read** the chapter, look back to this page to see if you've changed your mind about any of the statements.

- If any of your answers changed, explain why.
- Change any false statements into true statements.
- Use your revised statements as a study guide.

Before You Read A or D	Statement	After You Read A or D
	1 If not supported by evidence collected over time, scientists reject a theory.	
	2 Some living things do not require water to survive.	
	3 There is just one way to approach a scientific problem.	
	4 Some organisms grow by enlarging cells.	
	5 Following safety rules in lab not only protects you but your classmates as well.	
	6 All living things use energy.	
	7 Living things can grow spontaneously from nonliving things.	
	8 Scientists rarely repeat experiments.	
	9 An organism's classification can change with the discovery of new information.	
	10 All living things reproduce.	

Science Online
Print out a worksheet of this page at life.msscience.com

section 1

What is science?

as you read

What You'll Learn

- **Apply** scientific methods to problem solving.
- **Demonstrate** how to measure using scientific units.

Why It's Important

Learning to use scientific methods will help you solve ordinary problems in your life.

Review Vocabulary

experiment: using controlled conditions to test a hypothesis

New Vocabulary

- scientific methods
- hypothesis
- control
- variable
- theory
- law

The Work of Science

Movies and popcorn seem to go together. So before you and your friends watch a movie, sometimes you pop some corn in a microwave oven. When the popping stops, you take out the bag and open it carefully. You smell the mouthwatering, freshly popped corn and avoid hot steam that escapes from the bag. What makes the popcorn pop? How do microwaves work and make things hot? By the way, what are microwaves anyway?

Asking questions like these is one way scientists find out about anything in the world and the universe. Science is often described as an organized way of studying things and finding answers to questions.

Types of Science Many types of science exist. Each is given a name to describe what is being studied. For example, energy and matter have a relationship. That's a topic for physics. A physicist could answer most questions about microwaves.

On the other hand, a life scientist might study any of the millions of different animals, plants, and other living things on Earth. Look at the objects in **Figure 1.** What do they look like to you? A life scientist could tell you that some of the objects are living plants and some are just rocks. Life scientists who study plants are botanists, and those who study animals are zoologists. What do you suppose a bacteriologist studies?

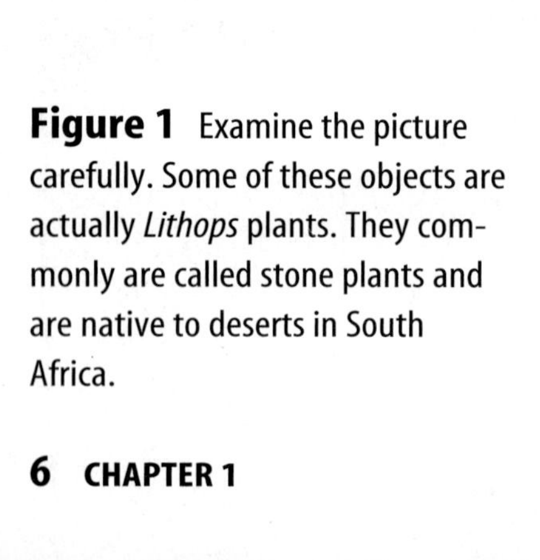

Figure 1 Examine the picture carefully. Some of these objects are actually *Lithops* plants. They commonly are called stone plants and are native to deserts in South Africa.

Critical Thinking

Whether or not you become a trained scientist, you are going to solve problems all your life. You probably solve many problems every day when you sort out ideas about what will or won't work. Suppose your CD player stops playing music. To figure out what happened, you have to think about it. That's called critical thinking, and it's the way you use skills to solve problems.

If you know that the CD player does not run on batteries and must be plugged in to work, that's the first thing you check to solve the problem. You check and the player is plugged in so you eliminate that possible solution. You separate important information from unimportant information—that's a skill. Could there be something wrong with the first outlet? You plug the player into a different outlet, and your CD starts playing. You now know that it's the first outlet that doesn't work. Identifying the problem is another skill you have.

Solving Problems

Scientists use the same types of skills that you do to solve problems and answer questions. Although scientists don't always find the answers to their questions, they always use critical thinking in their search. Besides critical thinking, solving a problem requires organization. In science, this organization often takes the form of a series of procedures called **scientific methods. Figure 2** shows one way that scientific methods might be used to solve a problem.

Figure 2 The series of procedures shown below is one way to use scientific methods to solve a problem.

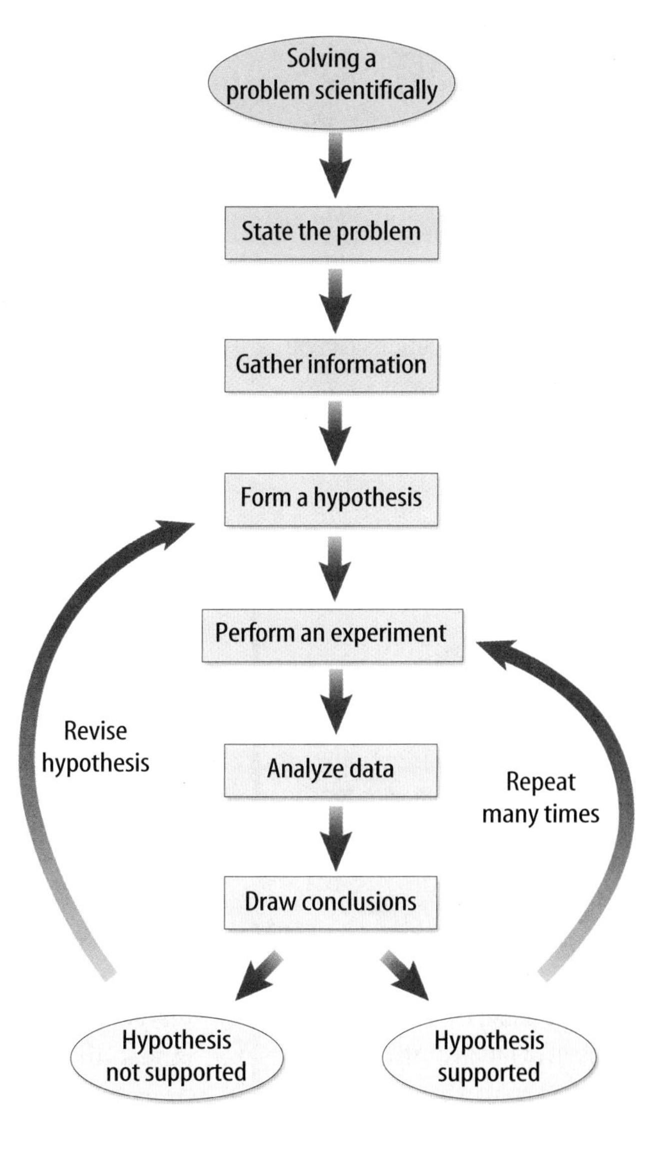

State the Problem Suppose a veterinary technician wanted to find out whether different types of cat litter cause irritation to cats' skin. What would she do first? The technician begins by observing something she cannot explain. A pet owner brings his four cats to the clinic to be boarded while he travels. He leaves his cell phone number so he can be contacted if any problems arise. When they first arrive, the four cats seem healthy. The next day however, the technician notices that two of the cats are scratching and chewing at their skin. By the third day, these same two cats have bare patches of skin with red sores. The technician decides that something in the cats' surroundings or their food might be irritating their skin.

Laboratory investigations

Computer models

Fieldwork

Figure 3 Observations can be made in many different settings.
List *three other places where scientific observations can be made.*

Gather Information Laboratory observations and experiments are ways to collect information. Some data also are gathered from fieldwork. Fieldwork includes observations or experiments that are done outside of the laboratory. For example, the best way to find out how a bird builds a nest is to go outside and watch it. **Figure 3** shows some ways data can be gathered.

The technician gathers information about the problem by watching the cats closely for the next two days. She knows that cats sometimes change their behavior when they are in a new place. She wants to see if the behavior of the cats with the skin sores seems different from that of the other two cats. Other than the scratching and chewing at their skin, all four cats' behavior seems to be the same.

The technician calls the owner and tells him about the problem. She asks him what brand of cat food he feeds his cats. Because his brand is the same one used at the clinic, she decides that food is not the cause of the skin irritation. She decides that the cats probably are reacting to something in their surroundings. There are many things in the clinic that the cats might react to. How does she decide what it is?

During her observations she notices that the cats seem to scratch and chew themselves most after using their litter boxes. The cat litter used by the clinic contains a deodorant. The technician calls the owner and finds out that the cat litter he buys does not contain a deodorant.

Topic: Controlled Experiments
Visit life.msscience.com for Web links to information about how scientists use controlled experiments.

Activity List the problem, hypothesis, and how the hypothesis was tested for a recently performed controlled experiment.

Form a Hypothesis Based on this information, the next thing the veterinary technician does is form a hypothesis. A **hypothesis** is an explanation that can be tested. After discussing her observations with the clinic veterinarian, she hypothesizes that something in the cat litter is irritating the cats' skin.

Test the Hypothesis with an Experiment The technician gets the owner's permission to test her hypothesis by performing an experiment. In an experiment, the hypothesis is tested using controlled conditions. The technician reads the labels on two brands of cat litter and finds that the ingredients of each are the same except that one contains a deodorant.

Controls The technician separates the cats with sores from the other two cats. She puts each of the cats with sores in a cage by itself. One cat is called the experimental cat. This cat is given a litter box containing the cat litter without deodorant. The other cat is given a litter box that contains cat litter with deodorant. The cat with deodorant cat litter is the control.

A **control** is the standard to which the outcome of a test is compared. At the end of the experiment, the control cat will be compared with the experimental cat. Whether or not the cat litter contains deodorant is the variable. A **variable** is something in an experiment that can change. An experiment should have only one variable. Other than the difference in the cat litter, the technician treats both cats the same.

How many variables should an experiment have?

Analyze Data The veterinary technician observes both cats for one week. During this time, she collects data on how often and when the cats scratch or chew, as shown in **Figure 4.** These data are recorded in a journal. The data show that the control cat scratches and chews more often than the experimental cat does. The sores on the skin of the experimental cat begin to heal, but those on the control cat do not.

Draw Conclusions The technician then draws the conclusion—a logical answer to a question based on data and observation—that the deodorant in the cat litter probably irritated the skin of the two cats. To accept or reject the hypothesis is the next step. In this case, the technician accepts the hypothesis. If she had rejected it, new experiments would have been necessary.

Although the technician decides to accept her hypothesis, she realizes that to be surer of her results she should continue her experiment. She should switch the experimental cat with the control cat to see what the results are a second time. If she did this, the healed cat might develop new sores. She makes an ethical decision and chooses not to continue the experiment. Ethical decisions, like this one, are important in deciding what science should be done.

Analyzing Data

Procedure

1. Obtain a **pan balance.** Follow your teacher's instructions for using it.
2. Record all data in your **Science Journal.**
3. Measure and record the mass of a **dry sponge.**
4. Soak this sponge in **water.** Measure and record its mass.
5. Calculate how much water your sponge absorbed.
6. Combine the class data and calculate the average amount of water absorbed.

Analysis

What other information about the sponges might be important when analyzing the data from the entire class?

Figure 4 Collecting and analyzing data is part of scientific methods.

Report Results When using scientific methods, it is important to share information. The veterinary technician calls the cats' owner and tells him the results of her experiment. She tells him she has stopped using the deodorant cat litter.

The technician also writes a story for the clinic's newsletter that describes her experiment and shares her conclusions. She reports the limits of her experiment and explains that her results are not final. In science it is important to explain how an experiment can be made better if it is done again.

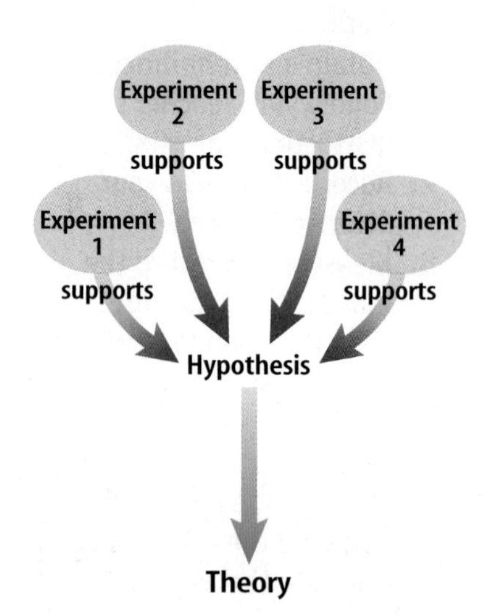

Figure 5 If data collected from several experiments over a period of time all support the hypothesis, it finally can be called a theory.

Developing Theories

After scientists report the results of experiments supporting their hypotheses, the results can be used to propose a scientific theory. When you watch a magician do a trick you might decide you have an idea or "theory" about how the trick works. Is your idea just a hunch or a scientific theory? A scientific **theory** is an explanation of things or events based on scientific knowledge that is the result of many observations and experiments. It is not a guess or someone's opinion. Many scientists repeat the experiment. If the results always support the hypothesis, the hypothesis can be called a theory, as shown in **Figure 5.**

Reading Check *What is a theory based on?*

A theory usually explains many hypotheses. For example, an important theory in life sciences is the cell theory. Scientists made observations of cells and experimented for more than 100 years before enough information was collected to propose a theory. Hypotheses about cells in plants and animals are combined in the cell theory.

A valid theory raises many new questions. Data or information from new experiments might change conclusions and theories can change. Later in this chapter you will read about the theory of spontaneous generation and how this theory changed as scientists used experiments to study new hypotheses.

Laws A scientific **law** is a statement about how things work in nature that seems to be true all the time. Although laws can be modified as more information becomes known, they are less likely to change than theories. Laws tell you what will happen under certain conditions but do not necessarily explain why it happened. For example, in life science you might learn about laws of heredity. These laws explain how genes are inherited but do not explain how genes work. Due to the great variety of living things, laws that describe them are few. It is unlikely that a law about how all cells work will ever be developed.

Scientific Methods Help Answer Questions You can use scientific methods to answer all sorts of questions. Your questions may be as simple as "Where did I leave my house key?" or as complex as "Will global warming cause the polar ice caps to melt?" You probably have had to find the answer to the first question. Someday you might try to find the answer to the second question. Using these scientific methods does not guarantee that you will get an answer. Often scientific methods just lead to more questions and more experiments. That's what science is about—continuing to look for the best answers to your questions.

Applying Science

Does temperature affect the rate of bacterial reproduction?

Some bacteria make you sick. Other bacteria, however, are used to produce foods like cheese and yogurt. Understanding how quickly bacteria reproduce can help you avoid harmful bacteria and use helpful bacteria. It's important to know things that affect how quickly bacteria reproduce. How do you think temperature will affect the rate of bacterial reproduction? A student makes the hypothesis that bacteria will reproduce more quickly as the temperature increases.

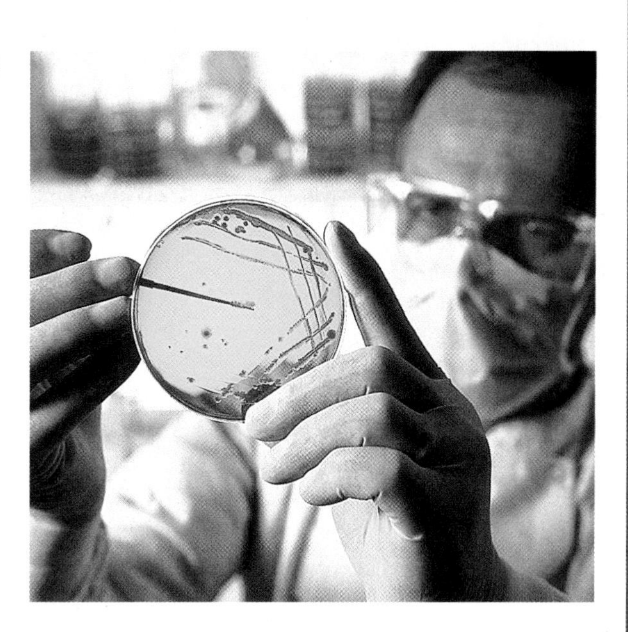

Identifying the Problem

The table below lists the reproduction-doubling rates at specific temperatures for one type of bacteria. A rate of 2.0 means that the number of bacteria doubled two times that hour (e.g., 100 to 200 to 400).

Bacterial Reproductive Rates

Temperature (°C)	Doubling Rate per Hour
20.5	2.0
30.5	3.0
36.0	2.5
39.2	1.2

Look at the table. What conclusions can you draw from the data?

Solving the Problem

1. Do the data in the table support the student's hypothesis?
2. How would you write a hypothesis about the relationship between bacterial reproduction and temperature?
3. Make a list of other factors that might have influenced the results in the table.
4. Are you satisfied with these data? List other things that you wish you knew.
5. Describe an experiment that would help you test these other ideas.

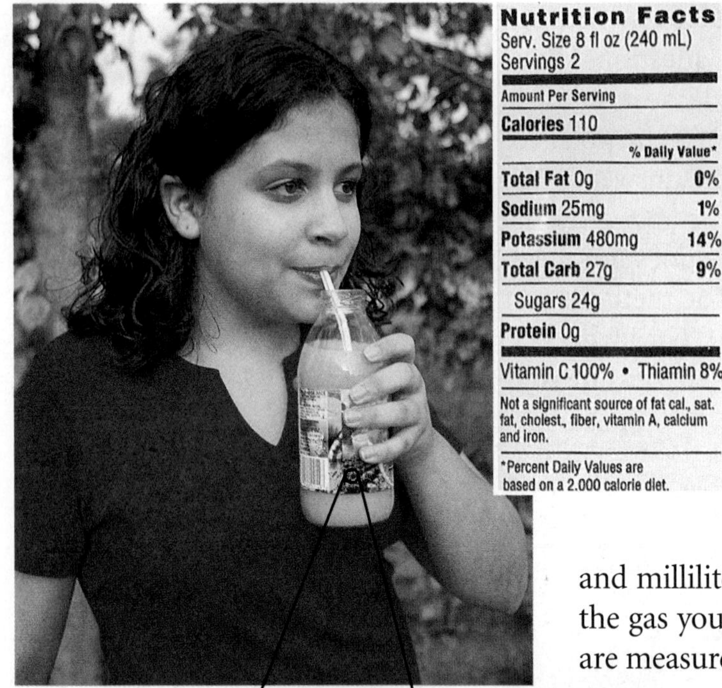

The label of this juice bottle shows you that it contains 473 mL of juice.

Figure 6 Your food often is measured in metric units. Nutritional information on the label is listed in grams or milligrams.

Measuring with Scientific Units

An important part of most scientific investigations is making accurate measurements. Think about things you use every day that are measured. Ingredients in your hamburger, hot dog, potato chips, or soft drink are measured in units such as grams and milliliters, as shown in **Figure 6.** The water you drink, the gas you use, and the electricity needed for a CD player are measured, too.

Reading Check *Why is it important to make accurate measurements?*

In your classroom or laboratory this year, you will use the same standard system of measurement scientists use to communicate and understand each other's research and results. This system is called the International System of Units, or SI. For example, you may need to calculate the distance a bird flies in kilometers. Perhaps you will be asked to measure the amount of air your lungs can hold in liters or the mass of an automobile in kilograms. Some of the SI units are shown in **Table 1.**

Table 1 Common SI Measurements

Measurement	Unit	Symbol	Fractions and Multiples
Distance	Meter	m	1/1,000 m = 1 millimeter (mm) 1/100 m = 1 centimeter (cm) 1000 m = 1 kilometer (km)
Mass	Kilogram	kg	1/1,000 g = 1 milligram (mg) 1/1,000 kg = 1 gram (g) 1000 kg = 1 tonne (t) (metric ton)
Time	Second	s	60 s = 1 minute (min) 60 min = 1 hour (h)

Safety First

Doing science is usually much more interesting than just reading about it. Some of the scientific equipment that you will use in your classroom or laboratory is the same as what scientists use. Laboratory safety is important. In many states, a student can participate in a laboratory class only when wearing proper eye protection. Don't forget to wash your hands after handling materials. Following safety rules, as shown in **Figure 7,** will protect you and others from injury during your lab experiences. Symbols used throughout your text will alert you to situations that require special attention. Some of these symbols are shown below. A description of each symbol is in the Safety Symbols chart at the front of this book.

Figure 7 Proper eye protection should be worn whenever you see this safety symbol.
Predict *what might happen if you do not wear eye protection in the lab.*

section 1 review

Summary

The Work of Science

- Science is an organized way of studying things and finding answers to questions.

Solving Problems and Developing Theories

- Scientific methods are procedures used to solve problems and answer questions.
- A theory is an explanation based on many scientific observations.

Measuring with Scientific Units

- Scientists use the SI system for measurements.

Safety First

- Follow safety rules in the lab.

Self Check

1. **Describe** scientific methods.
2. **Infer** why it is important to test only one variable at a time during an experiment.
3. **Identify** the SI unit you would use to measure the width of your classroom.
4. **Compare and contrast** a theory with a hypothesis.
5. **Think Critically** Can the veterinary technician in this section be sure that deodorant caused the cats' skin problems? How could she improve her experiment?

Applying Skills

6. **Write a paper** that explains what the veterinary technician discovered from her experiment.

section 2

Living Things

as you read

What You'll Learn

- **Distinguish** between living and nonliving things.
- **Identify** what living things need to survive.

Why It's Important

All living things, including you, have many of the same traits.

Review Vocabulary

raw materials: substances needed by organisms to make other necessary substances

New Vocabulary

- organism
- cell
- homeostasis

What are living things like?

What does it mean to be alive? If you walked down your street after a thunderstorm, you'd probably see earthworms on the sidewalk, birds flying, clouds moving across the sky, and puddles of water. You'd see living and nonliving things that are alike in some ways. For example, birds and clouds move. Earthworms and water feel wet when they are touched. Yet, clouds and water are nonliving things, and birds and earthworms are living things. Any living thing is called an **organism.**

Organisms vary in size from the microscopic bacteria in mud puddles to gigantic oak trees and are found just about everywhere. They have different behaviors and food needs. In spite of these differences, all organisms have similar traits. These traits determine what it means to be alive.

Living Things Are Organized If you were to look at almost any part of an organism, like a plant leaf or your skin, under a microscope, you would see that it is made up of small units called cells. A **cell** is the smallest unit of an organism that carries on the functions of life. Some organisms are composed of just one cell while others are composed of many cells. Cells take in materials from their surroundings and use them in complex ways. Each cell has an orderly structure and contains hereditary material. The hereditary material contains instructions for cellular organization and function. **Figure 8** shows some organisms that are made of many cells. All the things that these organisms can do are possible because of what their cells can do.

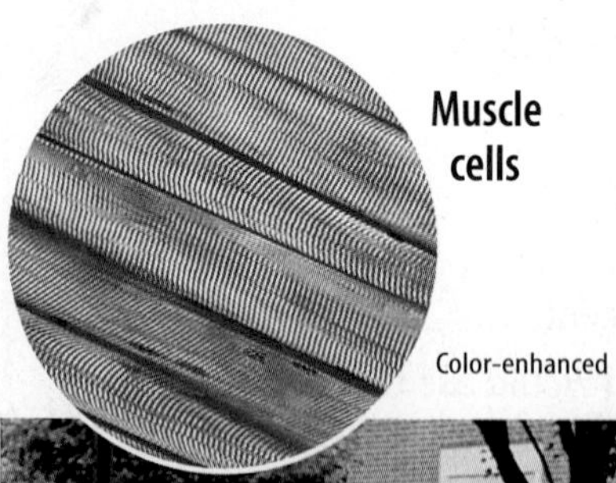

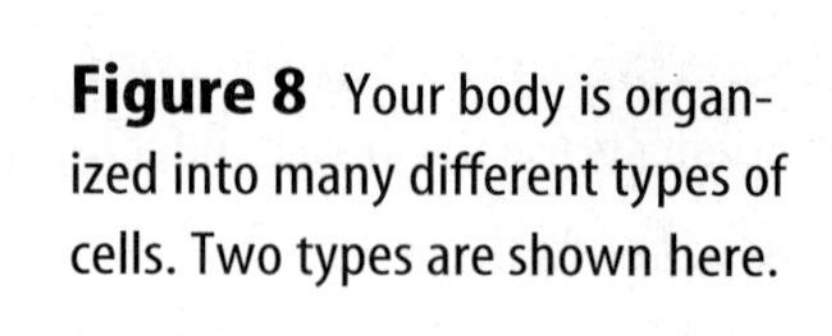

Figure 8 Your body is organized into many different types of cells. Two types are shown here.

Living Things Respond Living things interact with their surroundings. Watch your cat when you use your electric can opener. Does your cat come running to find out what's happening even when you're not opening a can of cat food? The cat in **Figure 9** ran in response to a stimulus—the sound of the can opener. Anything that causes some change in an organism is a stimulus (plural, *stimuli*). The reaction to a stimulus is a response. Often that response results in movement, such as when the cat runs toward the sound of the can opener. To carry on its daily activity and to survive, an organism must respond to stimuli.

Living things also respond to stimuli that occur inside them. For example, water or food levels in organisms' cells can increase or decrease. The organisms then make internal changes to keep the right amounts of water and food in their cells. Their temperature also must be within a certain range. An organism's ability to keep the proper conditions inside no matter what is going on outside the organism is called **homeostasis.** Homeostasis is a trait of all living things.

Figure 9 Some cats respond to a food stimulus even when they are not hungry.
Infer *why a cat comes running when it hears a can opener.*

What are some internal stimuli living things respond to?

Living Things Use Energy Staying organized and carrying on activities like homeostasis require energy. The energy used by most organisms comes either directly or indirectly from the Sun. Plants and some other organisms use the energy in sunlight and the raw materials carbon dioxide and water to make food. You and most other organisms can't use the energy in sunlight directly. Instead, you take in and use food as a source of energy. You get food by eating plants or other organisms that ate plants. Most organisms, including plants, also must take in oxygen in order to release the energy of foods.

Some bacteria live at the bottom of the oceans and in other areas where sunlight cannot reach. They can't use the energy in sunlight to produce food. Instead, the bacteria use energy stored in some chemical compounds and the raw material carbon dioxide to make food. Unlike most other organisms, many of these bacteria do not need oxygen to release the energy that is found in their food.

Topic: Homeostasis
Visit life.msscience.com for Web links to information about homeostasis.

Activity Describe the external stimuli and the corresponding internal changes for three different situations.

Living Things Grow and Develop When a puppy is born, it might be small enough to hold in one hand. After the same dog is fully grown, you might not be able to hold it at all. How does this happen? The puppy grows by taking in raw materials, like milk from its female parent, and making more cells. Growth of many-celled organisms, such as the puppy, is mostly due to an increase in the number of cells. In one-celled organisms, growth is due to an increase in the size of the cell.

Organisms change as they grow. Puppies can't see or walk when they are born. In eight or nine days, their eyes open, and their legs become strong enough to hold them up. All of the changes that take place during the life of an organism are called development. **Figure 10** shows how four different organisms changed as they grew.

The length of time an organism is expected to live is its life span. A dog can live for 20 years and a cat for 25 years. Some organisms have a short life span. Mayflies live only one day, but land tortoises can live for more than 180 years. Some bristlecone pine trees have been alive for more than 4,600 years. Your life span is about 80 years.

Figure 10 Complete development of an organism can take a few days or several years. The pictures below show the development of a dog, a human, a pea plant, and a butterfly.

Figure 11 Living things reproduce themselves in many different ways. A *Paramecium* reproduces by dividing into two. Beetles, like most insects, reproduce by laying eggs. Every spore released by the puffballs can grow into a new fungus.

Beetle

Puffballs

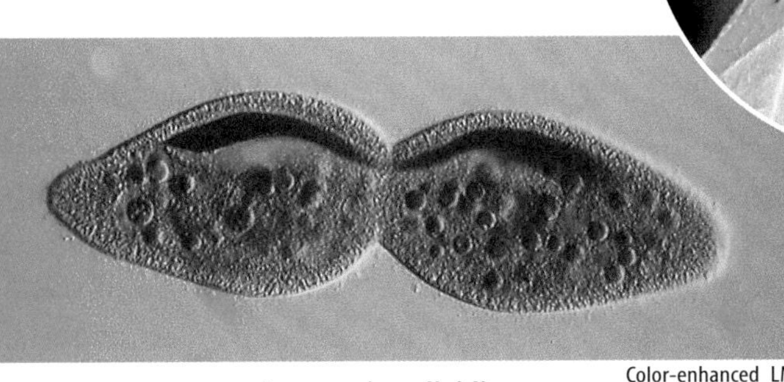

Paramecium dividing

Color-enhanced LM
Magnification: 400×

Living Things Reproduce Cats, dogs, alligators, fish, birds, bees, and trees eventually reproduce. They make more of their own kind. Some bacteria reproduce every 20 minutes while it might take a pine tree two years to produce seeds. **Figure 11** shows some ways organisms reproduce.

Without reproduction, living things would not exist to replace those individuals that die. An individual cat can live its entire life without reproducing. However, if cats never reproduced, all cats soon would disappear.

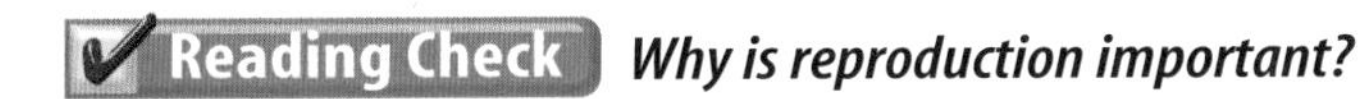

Why is reproduction important?

What do living things need?

What do you need to live? Do you have any needs that are different from those of other living things? To survive, all living things need a place to live and raw materials. The raw materials that they require and the exact place where they live can vary.

A Place to Live The environment limits where organisms can live. Not many kinds of organisms can live in extremely hot or extremely cold environments. Most cannot live at the bottom of the ocean or on the tops of mountains. All organisms also need living space in their surroundings. For example, thousands of penguins build their nests on an island. When the island becomes too crowded, the penguins fight for space and some may not find space to build nests. An organism's surroundings must provide for all of its needs.

Social Development Human infants quickly develop their first year of life. Research to find out how infants interact socially at different stages of development. Make a chart that shows changes from birth to one year old.

Figure 12 You and a corn plant each take in and give off about 2 L of water in a day. Most of the water you take in is from water you drink or from foods you eat. **Infer** *where plants get water to transport materials.*

Raw Materials Water is important for all living things. Plants and animals take in and give off large amounts of water each day, as shown in **Figure 12.** Organisms use homeostasis to balance the amounts of water lost with the amounts taken in. Most organisms are composed of more than 50 percent water. You are made of 60 to 70 percent water. Organisms use water for many things. For example, blood, which is about 50 percent water, transports digested food and wastes in animals. Plants have a watery sap that transports materials between roots and leaves.

Living things are made up of substances such as proteins, fats, and sugars. Animals take in most of these substances from the foods they eat. Plants and some bacteria make them using raw materials from their surroundings. These important substances are used over and over again. When organisms die, substances in their bodies are broken down and released into the soil or air. The substances can then be used again by other living organisms. Some of the substances in your body might once have been part of a butterfly or an apple tree.

At the beginning of this section, you learned that things such as clouds, sidewalks, and puddles of water are not living things. Now do you understand why? Clouds, sidewalks, and water do not reproduce, use energy, or have other traits of living things.

section 2 review

Summary

What are living things like?

- A cell is the smallest unit of an organism that carries on the functions of life.
- Anything that causes some change in an organism is a stimulus.
- Organisms use energy to stay organized and perform activities like homeostasis.
- All of the changes that take place during an organism's life are called development.

What do living things need?

- Living things need a place to live, water, and food.

Self Check

1. **Identify** the source of energy for most organisms.
2. **List** five traits that most organisms have.
3. **Infer** why you would expect to see cells if you looked at a section of a mushroom cap under a microscope.
4. **Determine** what most organisms need to survive.
5. **Think Critically** Why is homeostasis important to organisms?

Applying Skills

6. **Use a Database** Use references to find the life span of ten animals. Use your computer to make a database. Then, graph the life spans from shortest to longest.

Science Online life.msscience.com/self_check_quiz

section 3

Where does life come from?

Life Comes from Life

You've probably seen a fish tank, like the one in **Figure 13,** that is full of algae. How did the algae get there? Before the seventeenth century, some people thought that insects and fish came from mud, that earthworms fell from the sky when it rained, and that mice came from grain. These were logical conclusions at that time, based on repeated personal experiences. The idea that living things come from nonliving things is known as **spontaneous generation.** This idea became a theory that was accepted for several hundred years. When scientists began to use controlled experiments to test this theory, the theory changed.

as you read

What You'll Learn

- **Describe** experiments about spontaneous generation.
- **Explain** how scientific methods led to the idea of biogenesis.

Why It's Important

You can use scientific methods to try to find out about events that happened long ago or just last week. You can even use them to predict how something will behave in the future.

Review Vocabulary

contaminate: to make impure by coming into contact with an unwanted substance

New Vocabulary

- spontaneous generation
- biogenesis

Reading Check *Why did the theory of spontaneous generation change?*

Spontaneous Generation and Biogenesis From the late seventeenth century through the middle of the eighteenth century, experiments were done to test the theory of spontaneous generation. Although these experiments showed that spontaneous generation did not occur in most cases, they did not disprove it entirely.

It was not until the mid-1800s that the work of Louis Pasteur, a French chemist, provided enough evidence to disprove the theory of spontaneous generation. It was replaced with **biogenesis** (bi oh JE nuh suss), which is the theory that living things come only from other living things.

Figure 13 The sides of this tank were clean and the water was clear when the aquarium was set up. Algal cells, which were not visible on plants and fish, reproduced in the tank. So many algal cells are present now that the water is cloudy.

NATIONAL GEOGRAPHIC

VISUALIZING THE ORIGINS OF LIFE

Figure 14

For centuries scientists have theorized about the origins of life. As shown on this timeline, some examined spontaneous generation—the idea that nonliving material can produce life. More recently, scientists have proposed theories about the origins of life on Earth by testing hypotheses about conditions on early Earth.

1668 Francesco Redi put decaying meat in some jars, then covered half of them. When fly maggots appeared only on the uncovered meat (see below, left), Redi concluded that they had hatched from fly eggs and had not come from the meat.

1745 John Needham heated broth in sealed flasks. When the broth became cloudy with microorganisms, he mistakenly concluded that they developed spontaneously from the broth.

1768 Lazzaro Spallanzani boiled broth in sealed flasks for a longer time than Needham did. Only the ones he opened became cloudy with contamination.

Not contaminated

Contaminated

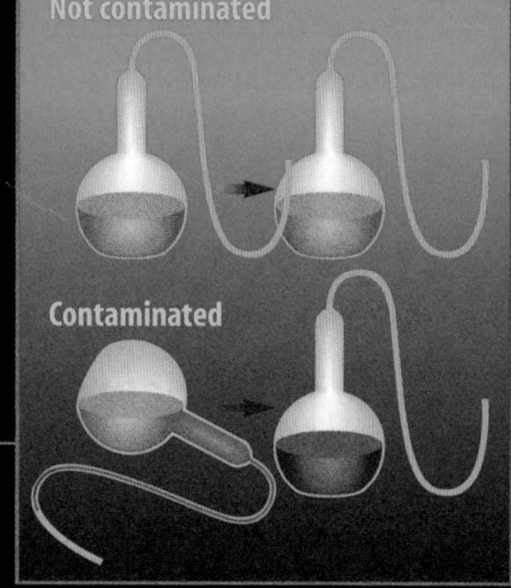

1859 Louis Pasteur disproved spontaneous generation by boiling broth in S-necked flasks that were open to the air. The broth became cloudy (see above, bottom right) only when a flask was tilted and the broth was exposed to dust in the S-neck.

Gases of Earth's early atmosphere

Electric current

Oceanlike mixture forms

Cools

Materials in present-day cells

1924 Alexander Oparin hypothesized that energy from the Sun, lightning, and Earth's heat triggered chemical reactions early in Earth's history. The newly-formed molecules washed into Earth's ancient oceans and became a part of what is often called the primordial soup.

1953 Stanley Miller and Harold Urey sent electric currents through a mixture of gases like those thought to be in Earth's early atmosphere. When the gases cooled, they condensed to form an oceanlike liquid that contained materials such as amino acids, found in present-day cells.

Life's Origins

If living things can come only from other living things, how did life on Earth begin? Some scientists hypothesize that about 5 billion years ago, Earth's solar system was a whirling mass of gas and dust. They hypothesize that the Sun and planets were formed from this mass. It is estimated that Earth is about 4.6 billion years old. Rocks found in Australia that are more than 3.5 billion years old contain fossils of once-living organisms. Where did these living organisms come from?

Oparin's Hypothesis In 1924, a Russian scientist named Alexander I. Oparin suggested that Earth's early atmosphere had no oxygen but was made up of the gases ammonia, hydrogen, methane, and water vapor. Oparin hypothesized that these gases could have combined to form the more complex compounds found in living things.

Using gases and conditions that Oparin described, American scientists Stanley L. Miller and Harold Urey set up an experiment to test Oparin's hypothesis in 1953. Although the Miller-Urey experiment showed that chemicals found in living things could be produced, it did not prove that life began in this way.

For many centuries, scientists have tried to find the origins of life, as shown in **Figure 14.** Although questions about spontaneous generation have been answered, some scientists still are investigating ideas about life's origins.

Oceans Scientists hypothesize that Earth's oceans originally formed when water vapor was released into the atmosphere from many volcanic eruptions. Once it cooled, rain fell and filled Earth's lowland areas. Identify five lowland areas on Earth that are now filled with water. Record your answer in your Science Journal.

section 3 review

Summary

Life Comes from Life

- Spontaneous generation is the idea that living things come from nonliving things.
- The work of Louis Pasteur in 1859 disproved the theory of spontaneous generation.
- Biogenesis is the theory that living things come only from other living things.

Life's Origins

- Alexander I. Oparin hypothesized about the origin of life.
- The Miller-Urey experiment did not prove that Oparin's hypothesis was correct.

Self Check

1. **Compare and contrast** spontaneous generation with biogenesis.
2. **Describe** three controlled experiments that helped disprove the theory of spontaneous generation and led to the theory of biogenesis.
3. **Summarize** the results of the Miller-Urey experiment.
4. **Think Critically** How do you think life on Earth began?

Applying Skills

5. **Draw Conclusions** Where could the organisms have come from in the 1768 broth experiment described in **Figure 14?**

section 4

How are living things classified?

as you read

What You'll Learn

- **Describe** how early scientists classified living things.
- **Explain** how similarities are used to classify organisms.
- **Explain** the system of binomial nomenclature.
- **Demonstrate** how to use a dichotomous key.

Why It's Important

Knowing how living things are classified will help you understand the relationships that exist among all living things.

Review Vocabulary

common name: a nonscientific term that may vary from region to region

New Vocabulary

- phylogeny
- kingdom
- binomial nomenclature
- genus

Classification

If you go to a library to find a book about the life of Louis Pasteur, where do you look? Do you look for it among the mystery or sports books? You expect to find a book about Pasteur's life with other biography books. Libraries group similar types of books together. When you place similar items together, you classify them. Organisms also are classified into groups.

History of Classification When did people begin to group similar organisms together? Early classifications included grouping plants that were used in medicines. Animals were often classified by human traits such as courageous—for lions—or wise—for owls.

More than 2,000 years ago, a Greek named Aristotle observed living things. He decided that any organism could be classified as either a plant or an animal. Then he broke these two groups into smaller groups. For example, animal categories included hair or no hair, four legs or fewer legs, and blood or no blood. **Figure 15** shows some of the organisms Aristotle would have grouped together. For hundreds of years after Aristotle, no one way of classifying was accepted by everyone.

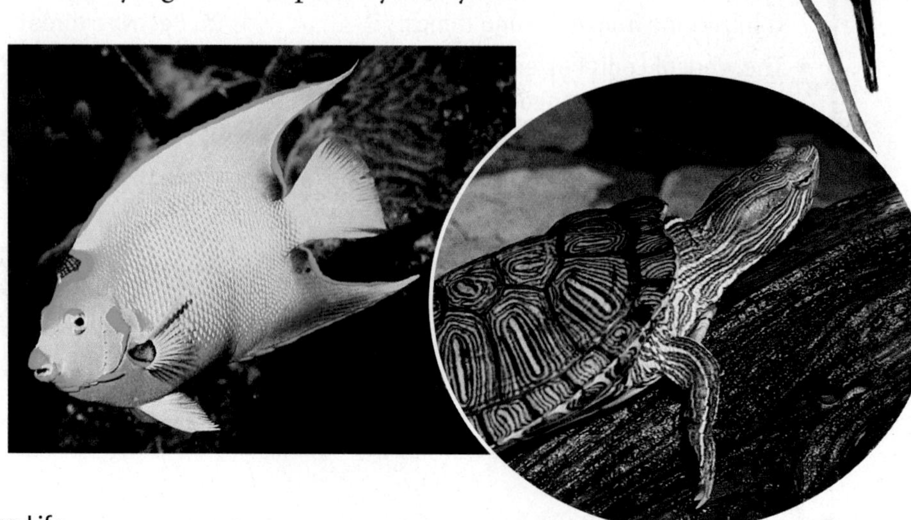

Figure 15 Using Aristotle's classification system, all animals without hair would be grouped together.
List *other animals without hair that Aristotle would have put in this group.*

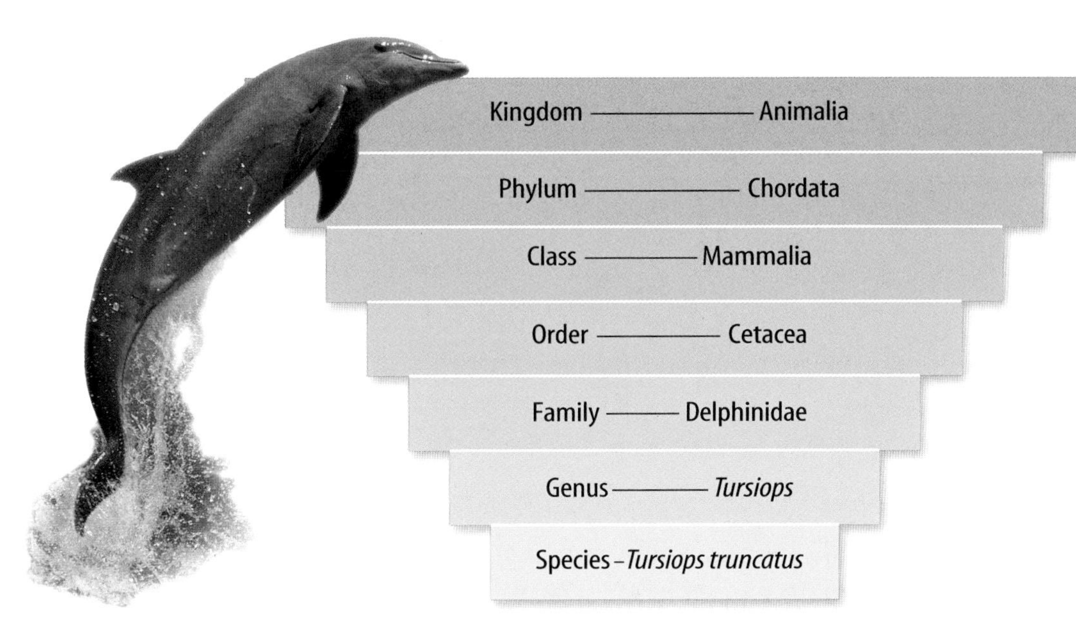

Figure 16 The classification of the bottle-nosed dolphin shows that it is in the order Cetacea. This order includes whales and porpoises.

Linnaeus In the late eighteenth century, Carolus Linnaeus, a Swedish naturalist, developed a new system of grouping organisms. His classification system was based on looking for organisms with similar structures. For example, plants that had similar flower structure were grouped together. Linnaeus's system eventually was accepted and used by most other scientists.

Modern Classification Like Linnaeus, modern scientists use similarities in structure to classify organisms. They also use similiarities in both external and internal features. Specific characteristics at the cellular level, such as the number of chromosomes, can be used to infer the degree of relatedness among organisms. In addition, scientists study fossils, hereditary information, and early stages of development. They use all of this information to determine an organism's phylogeny. **Phylogeny** (fi LAH juh nee) is the evolutionary history of an organism, or how it has changed over time. Today, it is the basis for the classification of many organisms.

What information would a scientist use to determine an organism's phylogeny?

Six Kingdoms A classification system commonly used today groups organisms into six kingdoms. A **kingdom** is the first and largest category. Organisms are placed into kingdoms based on various characteristics. Kingdoms can be divided into smaller groups. The smallest classification category is a species. Organisms that belong to the same species can mate and produce fertile offspring. To understand how an organism is classified, look at the classification of the bottle-nosed dolphin in **Figure 16.** Some scientists propose that before organisms are grouped into kingdoms, they should be placed in larger groups called domains. One proposed system groups all organisms into three domains.

Topic: Domains
Visit life.msscience.com for Web links to information about domains.

Activity List all the domains and give examples of organisms that are grouped in each domain.

Scientific Names

Using common names can cause confusion. Suppose that Diego is visiting Jamaal. Jamaal asks Diego if he would like a soda. Diego is confused until Jamaal hands him a soft drink. At Diego's house, a soft drink is called pop. Jamaal's grandmother, listening from the living room, thought that Jamaal was offering Diego an ice-cream soda.

What would happen if life scientists used only common names of organisms when they communicated with other scientists? Many misunderstandings would occur, and sometimes health and safety are involved. In **Figure 17,** you see examples of animals with common names that can be misleading. A naming system developed by Linnaeus helped solve this problem. It gave each species a unique, two-word scientific name.

Figure 17 Common names can be misleading.

Sea lions are more closely related to seals than to lions. **Identify** *another misleading common name.*

Jellyfish are neither fish nor jelly.

Binomial Nomenclature The two-word naming system that Linnaeus used to name the various species is called **binomial nomenclature** (bi NOH mee ul • NOH mun klay chur). It is the system used by modern scientists to name organisms. The first word of the two-word name identifies the genus of the organism. A **genus** is a group of similar species. The second word of the name might tell you something about the organism—what it looks like, where it is found, or who discovered it.

In this system, the tree species commonly known as red maple has been given the name *Acer rubrum.* The maple genus is *Acer.* The word *rubrum* is Latin for red, which is the color of a red maple's leaves in the fall. The scientific name of another maple is *Acer saccharum.* The Latin word for sugar is *saccharum.* In the spring, the sap of this tree is sweet.

Figure 18 These two lizards have the same common name, iguana, but are two different species.

Uses of Scientific Names Two-word scientific names are used for four reasons. First, they help avoid mistakes. Both of the lizards shown in **Figure 18** have the name *iguana.* Using binomial nomenclature, the green iguana is named *Iguana iguana.* Someone who studied this *iguana,* shown in the left photo, would not be confused by information he or she read about *Dispsosaurus dorsalis,* the desert iguana, shown in the right photo. Second, organisms with similar evolutionary histories are classified together. Because of this, you know that organisms in the same genus are related. Third, scientific names give descriptive information about the species, like the maples mentioned earlier. Fourth, scientific names allow information about organisms to be organized easily and efficiently. Such information may be found in a book or a pamphlet that lists related organisms and gives their scientific names.

Reading Check *What are four functions of scientific names?*

Tools for Identifying Organisms

Tools used to identify organisms include field guides and dichotomous (di KAH tuh mus) keys. Using these tools is one way you and scientists solve problems scientifically.

Many different field guides are available. Most have illustrations or photographs of organisms, information about where each organism lives, and general descriptions of each species. You can identify species from around the world using the appropriate field guide.

Communicating Ideas

Procedure

1. Find a **magazine picture of a piece of furniture** that can be used as a place to sit and to lie down.
2. Show the picture to ten people and ask them to tell you what word they use for this piece of furniture.
3. Keep a record of the answers in your **Science Journal.**

Analysis

1. In your Science Journal, infer how using common names can be confusing.
2. How do scientific names make communication among scientists easier?

Try at Home

Dichotomous Keys A dichotomous key is a detailed list of identifying characteristics that includes scientific names. Dichotomous keys are arranged in steps with two descriptive statements at each step. If you learn how to use a dichotomous key, you can identify and name a species.

Did you know many types of mice exist? You can use **Table 2** to find out what type of mouse is pictured to the left. Start by choosing between the first pair of descriptions. The mouse has hair on its tail, so you go to 2. The ears of the mouse are small, so you go on to 3. The tail of the mouse is less that 25 mm. What is the name of this mouse according to the key?

Table 2 Key to Some Mice of North America

1. Tail hair	**a.** no hair on tail; scales show plainly; house mouse, *Mus musculus* **b.** hair on tail, go to 2
2. Ear size	**a.** ears small and nearly hidden in fur, go to 3 **b.** ears large and not hidden in fur, go to 4
3. Tail length	**a.** less than 25 mm; woodland vole, *Microtus pinetorum* **b.** more than 25 mm; prairie vole, *Microtus ochrogaster*
4. Tail coloration	**a.** sharply bicolor, white beneath and dark above; deer mouse, *Peromyscus maniculatus* **b.** darker above than below but not sharply bicolor; white-footed mouse, *Peromyscus leucopus*

section 4 review

Summary

Classification

- Organisms are classified into groups based on their similarities.
- Scientists today classify organisms into six kingdoms.
- Species is the smallest classification category.

Scientific Names

- Binomial nomenclature is the two-word naming system that gives organisms their scientific names.

Tools for Identifying Organisms

- Field guides and dichotomous keys are used to identify organisms.

Self Check

1. **State** Aristotle's and Linnaeus' contributions to classifying living things.
2. **Identify** a specific characteristic used to classify organisms.
3. **Describe** what each of the two words identifies in binomial nomenclature.
4. **Think Critically** Would you expect a field guide to have common names as well as scientific names? Why or why not?

Applying Skills

5. **Classify** Create a dichotomous key that identifies types of cars.

Classifying Seeds

Scientists use classification systems to show how organisms are related. How do they determine which features to use to classify organisms? In this lab, you will observe seeds and use their features to classify them.

Real-World Question

How can the features of seeds be used to develop a dichotomous key to identify the seed?

Goals

- **Observe** the seeds and notice their features.
- **Classify** seeds using these features.

Materials

packets of seeds (10 different kinds)
magnifying lens
metric ruler

Safety Precautions

WARNING: *Some seeds may have been treated with chemicals. Do not put them in your mouth.*

Procedure

1. Copy the following data table in your Science Journal and record the features of each seed. Your table will have a column for each different type of seed you observe.

Seed Data

Feature	Type of Seed		
Color			
Length (mm)	Do not write in this book.		
Shape			
Texture			

2. Use the features to develop a dichotomous key.
3. Exchange keys with another group. Can you use their key to identify seeds?

Conclude and Apply

1. **Determine** how different seeds can be classified.
2. **Explain** how you would classify a seed you had not seen before using your data table.
3. **Explain** why it is an advantage for scientists to use a standardized system to classify organisms. What observations did you make to support your answer?

Communicating Your Data

Compare your conclusions with those of other students in your class. **For more help, refer to the** Science Skill Handbook.

USING SCIENTIFIC METHODS

Goals

- **Design** and carry out an experiment using scientific methods to infer why brine shrimp live in the ocean.
- **Observe** the jars for one week and notice whether the brine shrimp eggs hatch.

Possible Materials

500-mL, widemouthed containers (3)
brine shrimp eggs
small, plastic spoon
distilled water (500 mL)
weak salt solution (500 mL)
strong salt solution (500 mL)
labels (3)
magnifying lens

Safety Precautions

WARNING: *Protect eyes and clothing. Be careful when working with live organisms.*

Real-World Question

Brine shrimp are relatives of lobsters, crabs, crayfish, and the shrimp eaten by humans. They are often raised as a live food source in aquariums. In nature, they live in the oceans where fish feed on them. They can hatch from eggs that have been stored in a dry condition for many years. How can you use scientific methods to determine whether salt affects the hatching and growth of brine shrimp?

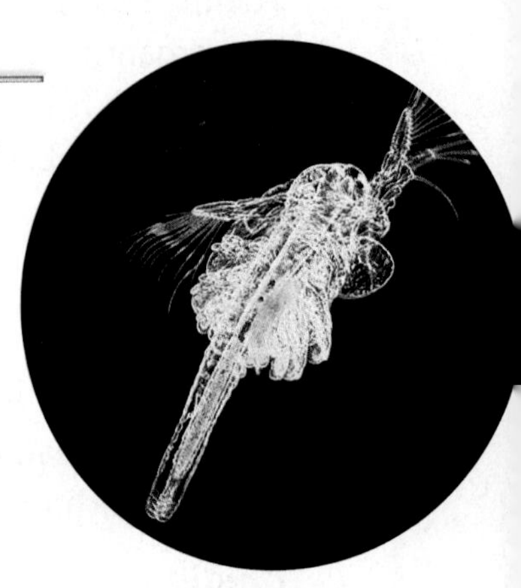

Brine shrimp

Form a Hypothesis

Based on your observations, form a hypothesis to explain how salt affects the hatching and growth of brine shrimp.

Test Your Hypothesis

Make a Plan

1. As a group, agree upon the hypothesis and decide how you will test it. Identify what results will confirm the hypothesis.

2. **List** steps that you need to test your hypothesis. Be specific. Describe exactly what you will do at each step.
3. **List** your materials.
4. **Prepare** a data table in your Science Journal to record your data.
5. Read over your entire experiment to make sure that all planned steps are in logical order.
6. **Identify** any constants, variables, and controls of the experiment.

Follow Your Plan

1. Make sure your teacher approves your plan before you start.
2. Carry out the experiment as planned by your group.
3. While doing the experiment, record any observations and complete the data table in your Science Journal.
4. Use a bar graph to plot your results.

Analyze Your Data

1. **Describe** the contents of each jar after one week.
2. **Identify** your control in this experiment.
3. **Identify** your variable in this experiment.

Conclude and Apply

1. **Explain** whether or not the results support your hypothesis.
2. **Predict** the effect that increasing the amount of salt in the water would have on the brine shrimp eggs.
3. **Compare** your results with those of other groups.

Communicating Your Data

Prepare a set of instructions on how to hatch brine shrimp to use to feed fish. Include diagrams and a step-by-step procedure.

TIME SCIENCE AND Society

SCIENCE ISSUES THAT AFFECT YOU!

Acari marmoset

Monkey BUSINESS

Manicore marmoset

In 2000, a scientist from Brazil's Amazon National Research Institute came across two squirrel-sized monkeys in a remote and isolated corner of the rain forest, about 2,575 km from Rio de Janeiro.

It turns out that the monkeys had never been seen before, or even known to exist.

Acari marmoset

The new species were spotted by a scientist who named them after two nearby rivers the Manicore and the Acari, where the animals were discovered. Both animals are marmosets, which is a type of monkey found only in Central and South America. Marmosets have claws instead of nails, live in trees, and use their extraordinarily long tail like an extra arm or leg. Small and light, both marmosets measure about 23 cm in length with a 38 cm tail, and weigh no more than 0.4 kg.

The Manicore marmoset has a silvery-white upper body, a light-gray cap on its head, a yellow-orange underbody, and a black tail.

The Acari marmoset's upper body is snowy white, its gray back sports a stripe running to the knee, and its black tail flashes a bright-orange tip.

Amazin' Amazon

The Amazon Basin is a treasure trove of unique species. The Amazon River is Earth's largest body of freshwater, with 1,100 smaller tributaries. And more than half of the world's plant and animal species live in its rain forest ecosystems.

Research and Report Working in small groups, find out more about the Amazon rain forest. Which plants and animals live there? What products come from the rain forest? How does what happens in the Amazon rain forest affect you? Prepare a multimedia presentation.

For more information, visit life.msscience.com/time

Reviewing Main Ideas

Section 1 What is science?

1. Scientists use problem-solving methods to investigate observations about living and nonliving things.
2. Scientists use SI measurements to gather measurable data.
3. Safe laboratory practices help you learn more about science.

Section 2 Living Things

1. Organisms are made of cells, use energy, reproduce, respond, grow, and develop.
2. Organisms need energy, water, food, and a place to live.

Section 3 Where does life come from?

1. Controlled experiments finally disproved the theory of spontaneous generation.
2. Pasteur's experiment proved biogenesis.

Section 4 How are living things classified?

1. Classification is the grouping of ideas, information, or objects based on their similar characteristics.
2. Scientists today use phylogeny to group organisms into six kingdoms.
3. All organisms are given a two-word scientific name using binomial nomenclature.

Visualizing Main Ideas

Copy and complete this events-chain concept map that shows the order in which you might use a scientific method. Use these terms: analyze data, perform an experiment, *and* form a hypothesis.

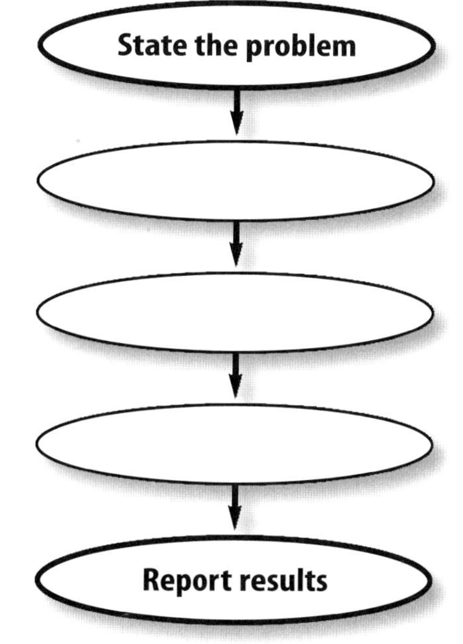

chapter 1 Review

Using Vocabulary

Explain the differences in the vocabulary words in each pair below. Then explain how they are related.

1. control—variable
2. law—theory
3. biogenesis—spontaneous generation
4. binomial nomenclature—phylogeny
5. organism—cell
6. kingdom—phylogeny
7. hypothesis—scientific methods
8. organism—homeostasis
9. kingdom—genus
10. theory—hypothesis

Checking Concepts

Choose the word or phrase that best answers the question.

11. What category of organisms can mate and produce fertile offspring?
A) family **C)** genus
B) class **D)** species

12. What is the closest relative of *Canis lupus?*
A) *Quercus alba* **C)** *Felis tigris*
B) *Equus zebra* **D)** *Canis familiaris*

13. What is the source of energy for plants?
A) the Sun **C)** water
B) carbon dioxide **D)** oxygen

14. What makes up more than 50 percent of all living things?
A) oxygen **C)** minerals
B) carbon dioxide **D)** water

15. Who finally disproved the theory of spontaneous generation?
A) Oparin **C)** Pasteur
B) Aristotle **D)** Miller

16. What gas do some scientists think was missing from Earth's early atmosphere?
A) ammonia **C)** methane
B) hydrogen **D)** oxygen

17. What is the length of time called that an organism is expected to live?
A) life span **C)** homeostasis
B) stimulus **D)** theory

18. What is the part of an experiment that can be changed called?
A) conclusion **C)** control
B) variable **D)** data

19. What does the first word in a two-word name of an organism identify?
A) kingdom **C)** phylum
B) species **D)** genus

Use the photo below to answer question 20.

20. What SI unit could you use to measure the mass of the fish shown above?
A) meter **C)** gram
B) liter **D)** degree

Science Online life.msscience.com/vocabulary_puzzlemaker

Thinking Critically

21. **Predict** what *Lathyrus odoratus,* the scientific name for a sweet pea plant, tells you about one of its characteristics.

Use the photo below to answer question 22.

22. **Determine** what problem-solving techniques this scientist would use to find how dolphins learn.

Use the graph below to answer question 23.

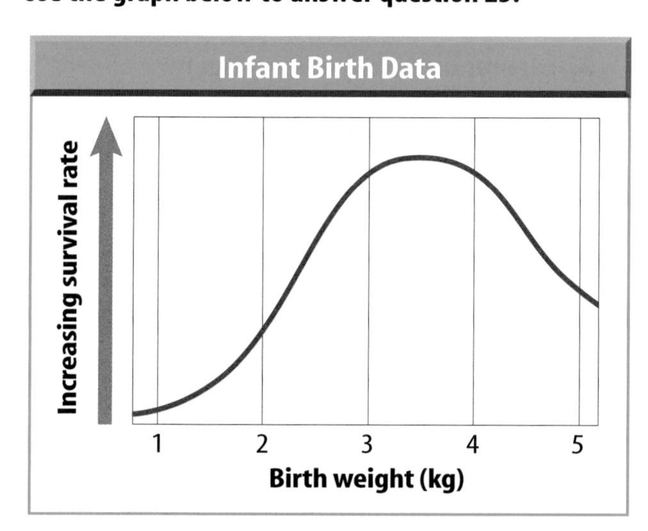

23. **Interpret Data** Do the data in the graph above support the hypothesis that babies with a birth weight of 2.5 kg have the best chance of survival? Explain.

24. **List** advantages of using SI units.

25. **Form a Hypothesis** A lima bean plant is placed under green light, another is placed under red light, and a third under blue light. Their growth is measured for four weeks to determine which light is best for plant growth. What are the variables in this experiment? State a hypothesis for this experiment.

Performance Activities

26. **Bulletin Board** Interview people in your community whose jobs require a knowledge of life science. Make a Life Science Careers bulletin board. Summarize each person's job and what he or she had to study to prepare for that job.

Applying Math

27. **Body Temperature** Normal human body temperature is 98.6°F. What is this temperature in degrees Celsius? Use the following expression, 5/9(°F−32), to find degrees Celsius.

Use the graph below to answer question 28.

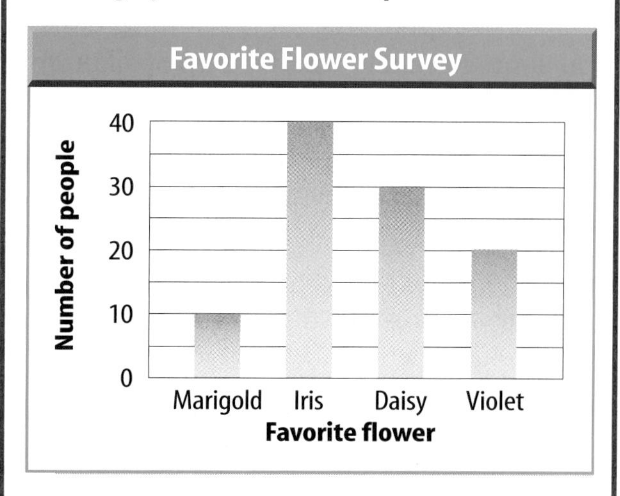

28. **Favorite Flower** The graph above shows how many people selected a certain type of flower as their favorite. According to the graph, what percentage of the people picked daisy as their favorite?

Part 1 Multiple Choice

Record your answers on the answer sheet provided by your teacher or on a sheet of paper.

1. A prediction that can be tested is a
   A. conclusion. **C.** hypothesis.
   B. variable. **D.** theory.

2. Which of the following units would a scientist likely use when measuring the length of a mouse's tail?
   A. kilometers **C.** grams
   B. millimeters **D.** milliliters

Use the illustrations below to answer questions 3 and 4.

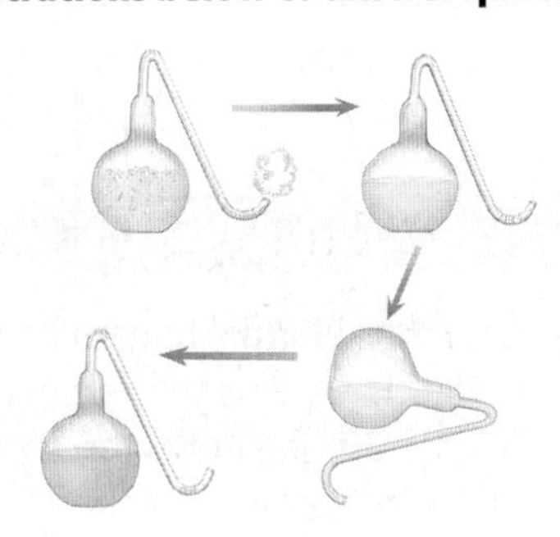

3. What scientist used the flasks pictured above to support the theory of biogenesis?
   A. John Needham **C.** Lazzaro Spallanzani
   B. Louis Pasteur **D.** Francesco Redi

4. Why did only the broth in the flask that was tilted become cloudy and contaminated?
   A. The broth was not boiled.
   B. Flies contaminated the broth.
   C. The broth was exposed to dust in the neck of the flask.
   D. Decaying meat caused the broth to be contaminated.

Test-Taking Tip

Practice Skills Remember that test-taking skills can improve with practice. If possible, take at least one practice test and familiarize yourself with the test format and instructions.

Use the photos below to answer questions 5 and 6.

5. The dog pictured above has increased in size. How did most of this size increase take place?
   A. an increase in cell size
   B. an increase in the number of cells
   C. an increase in cell water
   D. an increase in cell energy

6. What characteristic of life is illustrated by the change in the dog?
   A. reproduction
   B. homeostasis
   C. growth and development
   D. response to stimulus

7. What gas must most organisms take in to release the energy of foods?
   A. oxygen **C.** water vapor
   B. carbon dioxide **D.** hydrogen

8. What characteristic of living things is represented by a puffball releasing millions of spores?
   A. reproduction **C.** organization
   B. development **D.** use of energy

9. When using scientific methods to solve a problem, which of the following is a scientist most likely to do after forming a hypothesis?
   A. analyze data
   B. draw conclusions
   C. state a problem
   D. perform an experiment

10. What are the smallest units that make up your body called?
    A. cells **C.** muscles
    B. organisms **D.** fibers

Part 2 | Short Response/Grid In

Record your answers on the answer sheet provided by your teacher or on a sheet of paper.

11. From where do bacteria that live in areas where there is no sunlight obtain energy?

12. Organisms take in and give off large amounts of water each day. What process do they use to balance the amount of water lost with the amount taken in?

13. After a rain storm, earthworms may be seen crawling on the sidewalk or road. How would the theory of spontaneous generation explain the origin of the worms?

14. List three things modern scientists study when they classify organisms.

Use the illustrations below to answer questions 15 and 16.

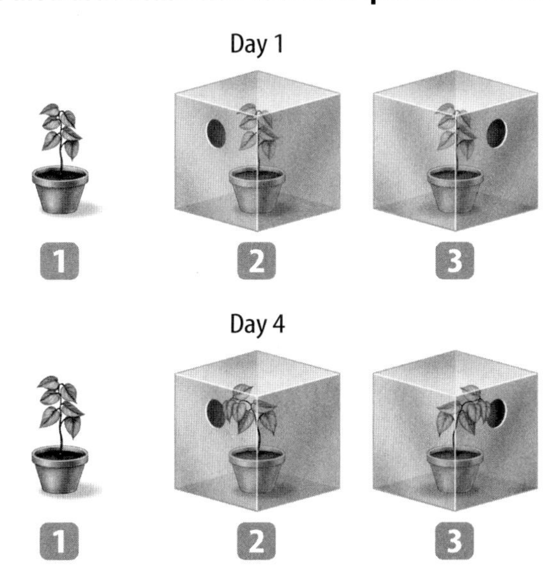

15. A science class set up the experiment above to study the response of plants to the stimulus of light. What hypothesis is likely being tested by this experiment?

16. After day 4, Fatima wanted to find out how plant 2 and plant 3 would grow in normal light. What did she have to do to find out?

Part 3 | Open Ended

Record your answers on a sheet of paper.

17. Describe several different ways scientists gather information. Which of these ways would likely be used to collect data about which foods wild alligators eat in Florida?

18. Some scientists think that lightning may have caused chemicals in the Earth's early atmosphere to combine to begin the origin of life. Explain how the experiment of Miller and Urey does not prove this hypothesis.

Use the photo below to answer questions 19 and 20.

19. Both of these living things use energy. Describe the difference between the source of energy for each. In what similar ways would each of these organisms use energy?

20. How are the needs of the two organisms alike? Explain why the plant is raw material for the beetle. When the beetle dies, how could it be raw material for the plant?

21. Explain stimulus and response. How is response to a stimulus related to homeostasis?

22. What information would you need to write a field guide used to identify garden plants? What other information would you need if the guide included a dichotomous key?

23. Explain the difference between a kingdom and a species in the classification system commonly used.

chapter 2

Cells

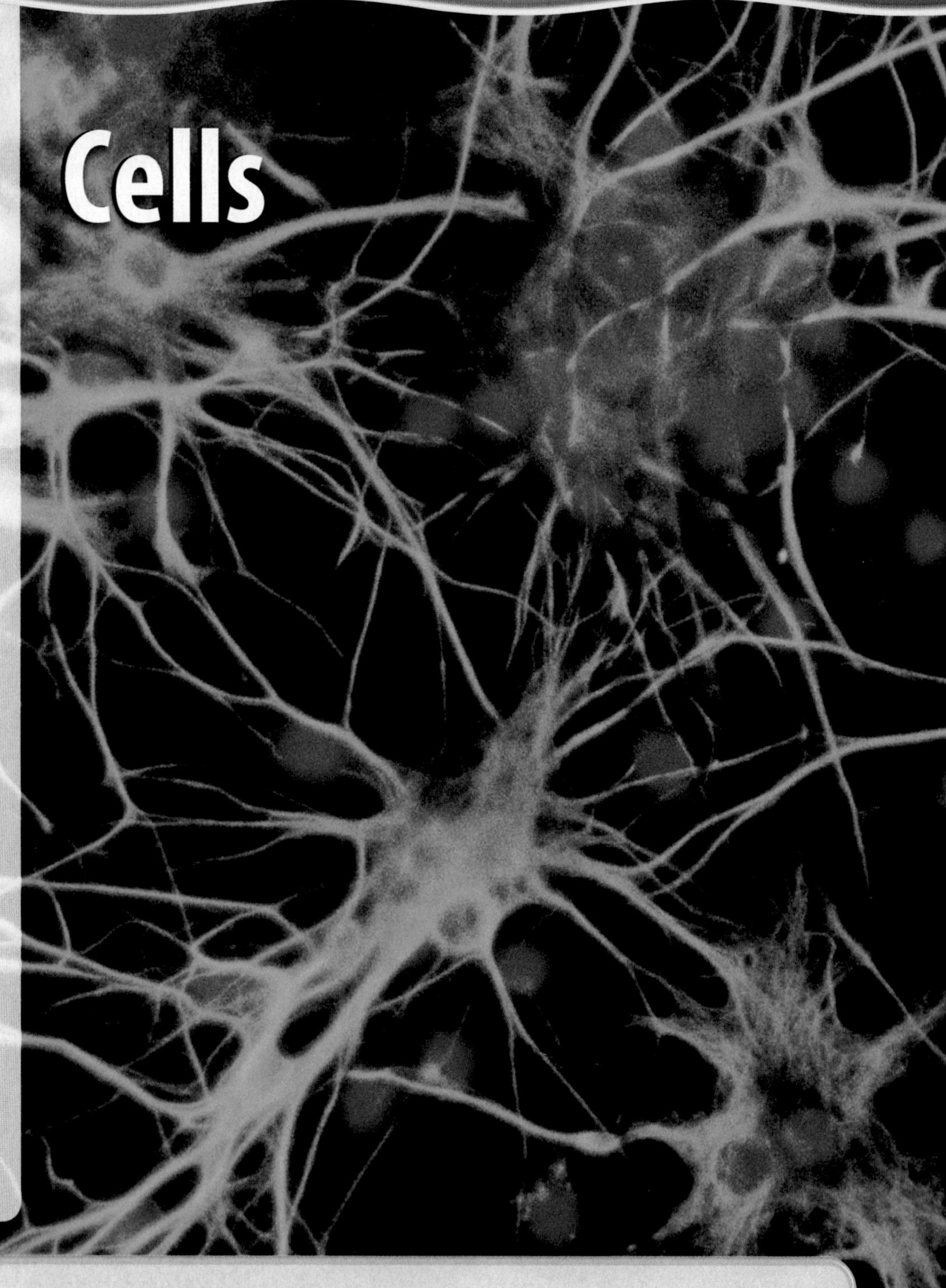

The BIG Idea

The different structures in a cell work together to ensure the cell's survival.

SECTION 1
Cell Structure

Main Idea Different cell types can have different structures, but some cell structures are common to all cells.

SECTION 2
Viewing Cells

Main Idea Scientists can study living things too small to be seen with only the human eye by using microscopes.

SECTION 3
Viruses

Main Idea Although viruses are not considered living things, they can affect all living things.

Too Small To Be Seen

The world around you is filled with organisms that you could overlook, or even be unable to see. Some of these organisms are one-celled and some are many-celled. You can study these organisms and the cells of other organisms by using microscopes.

Science Journal Write three questions that you would ask a scientist researching cancer cells.

Start-Up Activities

Magnifying Cells

If you look around your classroom, you can see many things of all sizes. Using a magnifying lens, you can see more details. You might examine a speck of dust and discover that it is a living or dead insect. In the following lab, use a magnifying lens to search for the smallest thing you can find in the classroom.

1. Obtain a magnifying lens from your teacher. Note its power (the number followed by ×, shown somewhere on the lens frame or handle).
2. Using the magnifying lens, look around the room for the smallest object that you can find.
3. Measure the size of the image as you see it with the magnifying lens. To estimate the real size of the object, divide that number by the power. For example, if it looks 2 cm long and the power is 10×, the real length is about 0.2 cm.
4. **Think Critically** Write a paragraph that describes what you observed. Did the details become clearer? Explain.

Cells Make the following Foldable to help you illustrate the main parts of cells.

STEP 1 **Fold** a vertical sheet of paper in half from top to bottom.

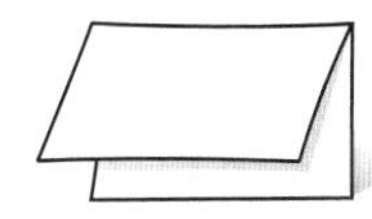

STEP 2 **Fold** in half from side to side with the fold at the top.

STEP 3 **Unfold** the paper once. **Cut** only the fold of the top flap to make two tabs.

STEP 4 **Turn** the paper vertically and **write** on the front tabs as shown.

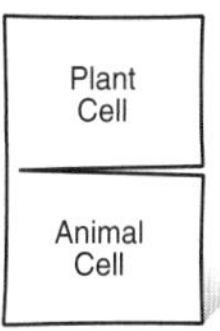

Illustrate and Label As you read the chapter, draw and identify the parts of plant and animal cells under the appropriate tab.

Preview this chapter's content and activities at life.msscience.com

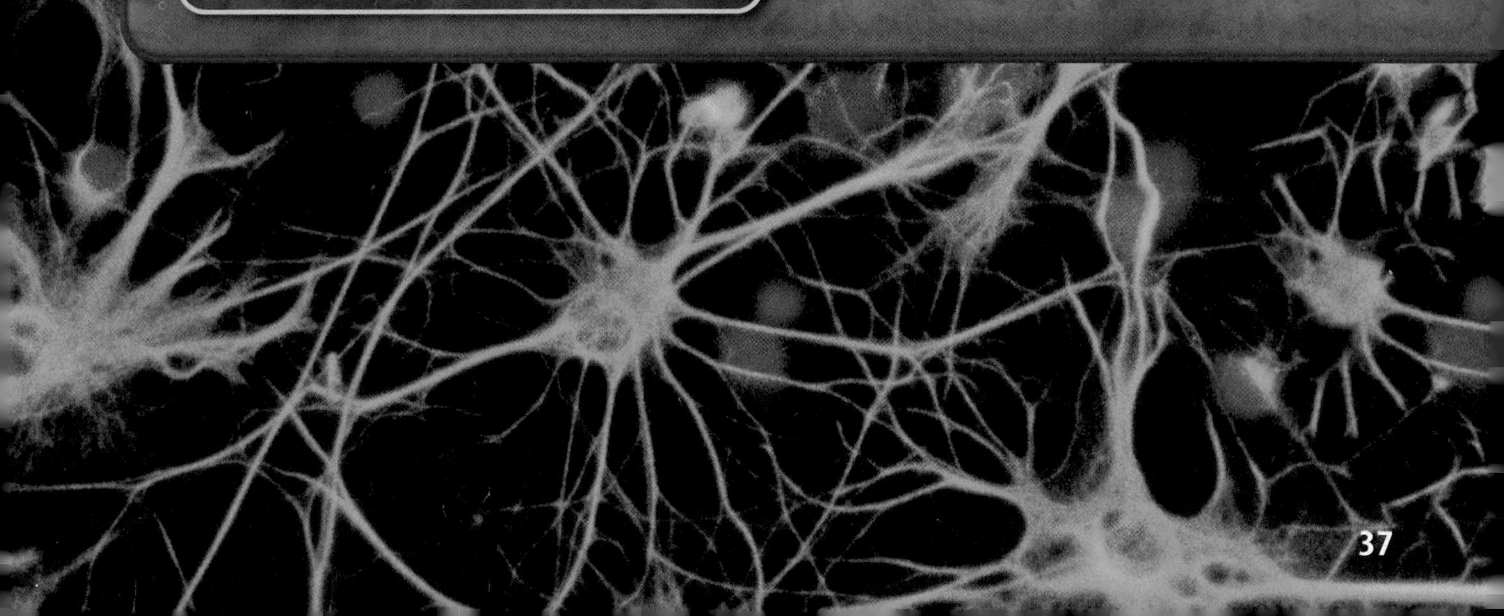

Get Ready to Read

Identify the Main Idea

1 Learn It! Main ideas are the most important ideas in a paragraph, lesson, or chapter. Supporting details are facts or examples that explain the main idea. Understanding the main idea allows you to grasp the whole picture.

2 Picture It! Read the following paragraph. Draw a graphic organizer like the one below to show the main idea and supporting details.

> Things that are too small to be seen with other microscopes can be viewed with an electron microscope. Instead of using lenses to direct beams of light, an electron microscope uses a magnetic field in a vacuum to direct beams of electrons. Some electron microscopes can magnify images up to one million times. To see electron microscope images, they must be photographed or electronically produced.
>
> *—from page 50*

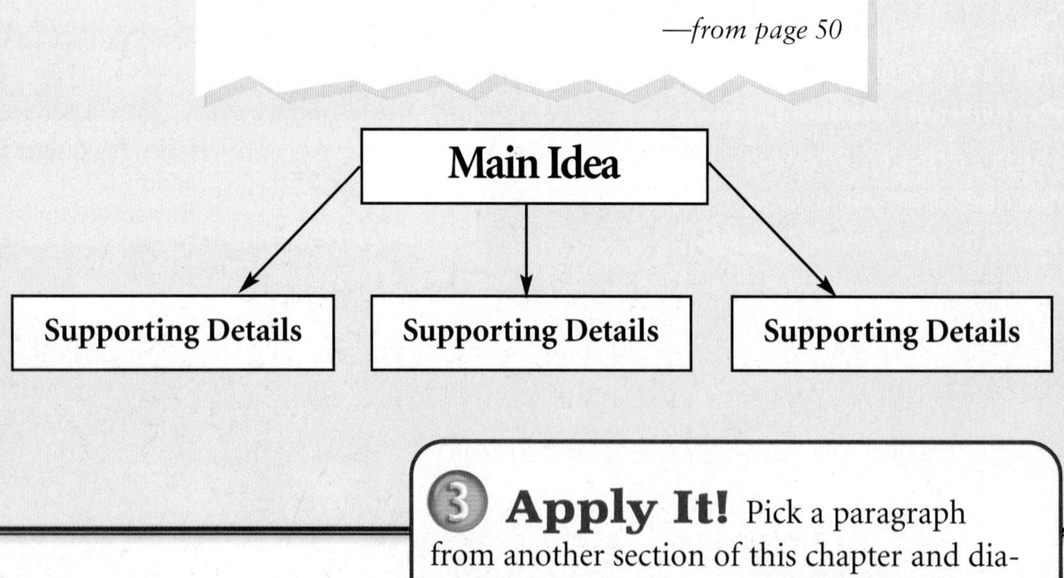

3 Apply It! Pick a paragraph from another section of this chapter and diagram the main ideas as you did above.

The main idea is often the first sentence in a paragraph but not always.

Target Your Reading

Use this to focus on the main ideas as you read the chapter.

1. **Before you read** the chapter, respond to the statements below on your worksheet or on a numbered sheet of paper.
   - Write an **A** if you **agree** with the statement.
   - Write a **D** if you **disagree** with the statement.

2. **After you read** the chapter, look back to this page to see if you've changed your mind about any of the statements.
   - If any of your answers changed, explain why.
   - Change any false statements into true statements.
   - Use your revised statements as a study guide.

Before You Read A or D	Statement	After You Read A or D
	1 All new cells come from preexisting cells.	
	2 You must use a microscope to see most cells.	
	3 A flexible cell membrane surrounds every cell.	
	4 Chromosomes are in the nucleus of every cell.	
	5 A bacterium is larger than an animal cell.	
	6 A cell wall and cytoplasm control the shape of each cell.	
	7 Tissues are groups of similar types of cells that work together to perform a function.	
	8 The most powerful microscopes create images by focusing light through two or more lenses.	
	9 A cell's mitochondria transform light energy into chemical energy.	
	10 Viruses are harmful and never beneficial.	

Science Online
Print out a worksheet of this page at life.msscience.com

section 1

Cell Structure

as you read

What You'll Learn

- **Identify** names and functions of each part of a cell.
- **Explain** how important a nucleus is in a cell.
- **Compare** tissues, organs, and organ systems.

Why It's Important

If you know how organelles function, it's easier to understand how cells survive.

Review Vocabulary
photosynthesis: process by which most plants, some protists, and many types of bacteria make their own food

New Vocabulary
- cell membrane
- cytoplasm
- cell wall
- organelle
- nucleus
- chloroplast
- mitochondrion
- ribosome
- endoplasmic reticulum
- Golgi body
- tissue
- organ

Common Cell Traits

Living cells are dynamic and have several things in common. A cell is the smallest unit that is capable of performing life functions. All cells have an outer covering called a **cell membrane.** A living membrane is made of one or more layers of linked molecules. Inside every cell is a gelatinlike material called **cytoplasm** (SI tuh pla zum). In the cytoplasm of every cell is hereditary material that controls the life of the cell.

Comparing Cells Cells come in many sizes. A nerve cell in your leg could be a meter long. A human egg cell is no bigger than the dot on this *i*. A human red blood cell is about one-tenth the size of a human egg cell. A bacterium is even smaller—8,000 of the smallest bacteria can fit inside one of your red blood cells.

A cell's shape might tell you something about its function. The nerve cell in **Figure 1** has extensions that send impulses to or receive impulses from other cells. A nerve cell cannot change shape but muscle cells and some blood cells can. Some plant stems have long, hollow cells with openings at their ends. These cells carry food and water throughout the plant.

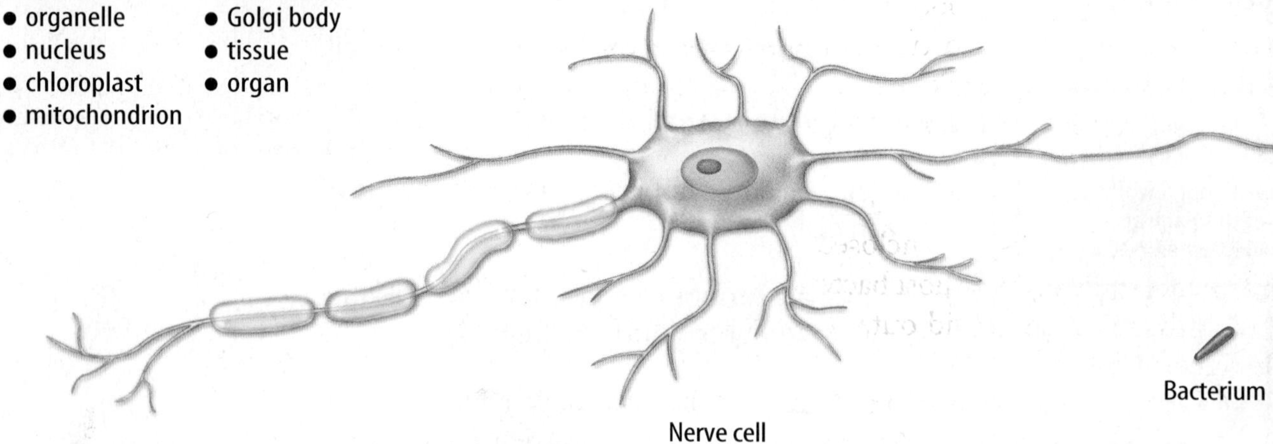

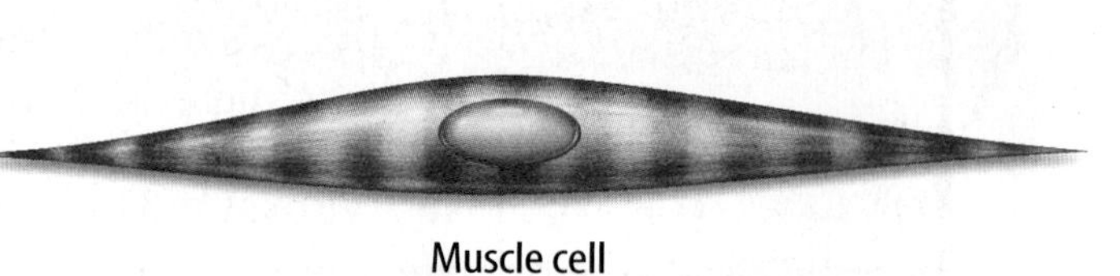

Figure 1 The shape of the cell can tell you something about its function. These cells are drawn 700 times their actual size.

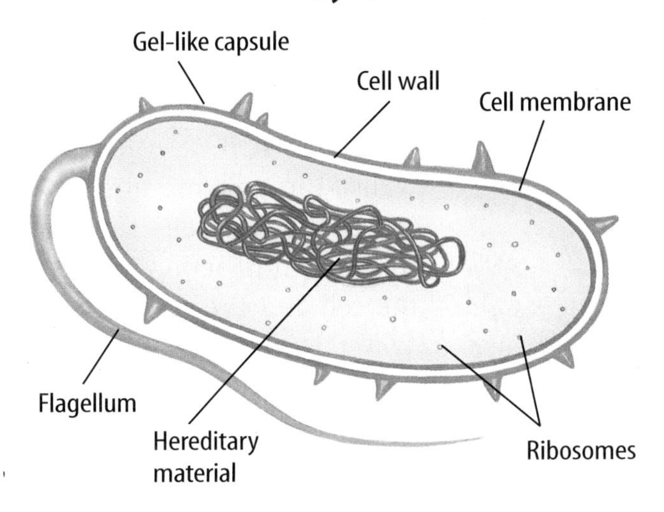

Prokaryotic cell

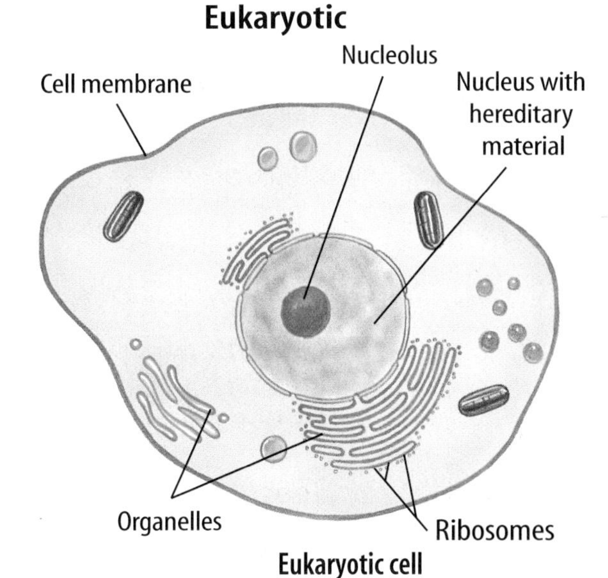

Eukaryotic cell

Cell Types Scientists have found that cells can be separated into two groups. One group has no membrane-bound structures inside the cell and the other group does, as shown in **Figure 2.** Cells without membrane-bound structures are called prokaryotic (proh KAYR ee yah tihk) cells. Cells with membrane-bound structures are called eukaryotic (yew KAYR ee yah tihk) cells.

Figure 2 Examine these drawings of cells. Prokaryotic cells are only found in one-celled organisms, such as bacteria. Protists, fungi, plants, and animals are made of eukaryotic cells. **Describe** *differences you see between them.*

Reading Check *Into what two groups can cells be separated?*

Cell Organization

Each cell in your body has a specific function. You might compare a cell to a busy delicatessen that is open 24 hours every day. Raw materials for the sandwiches are brought in often. Some food is eaten in the store, and some customers take their food with them. Sometimes food is prepared ahead of time for quick sale. Wastes are put into trash bags for removal or recycling. Similarly, your cells are taking in nutrients, secreting and storing chemicals, and breaking down substances 24 hours every day.

Cell Wall Just like a deli that is located inside the walls of a building, some cells are enclosed in a cell wall. The cells of plants, algae, fungi, and most bacteria are enclosed in a cell wall. **Cell walls** are tough, rigid outer coverings that protect the cell and give it shape.

A plant cell wall, as shown in **Figure 3,** mostly is made up of a substance called cellulose. The long, threadlike fibers of cellulose form a thick mesh that allows water and dissolved materials to pass through it. Cell walls also can contain pectin, which is used in jam and jelly, and lignin, which is a compound that makes cell walls rigid. Plant cells responsible for support have a lot of lignin in their walls.

Figure 3 The protective cell wall of a plant cell is outside the cell membrane.

Color-enhanced TEM Magnification: 9000×

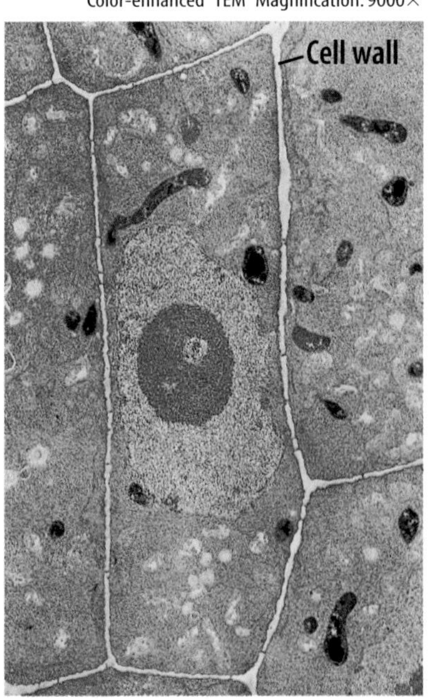

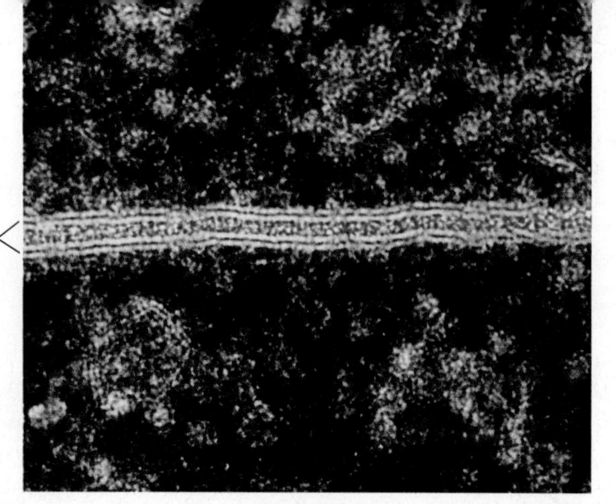

Color-enhanced TEM Magnification: 125000×

Figure 4 A cell membrane is made up of a double layer of fatlike molecules.

Cell Membrane The protective layer around all cells is the cell membrane, as shown in **Figure 4.** If cells have cell walls, the cell membrane is inside of it. The cell membrane regulates interactions between the cell and the environment. Water is able to move freely into and out of the cell through the cell membrane. Food particles and some molecules enter and waste products leave through the cell membrane.

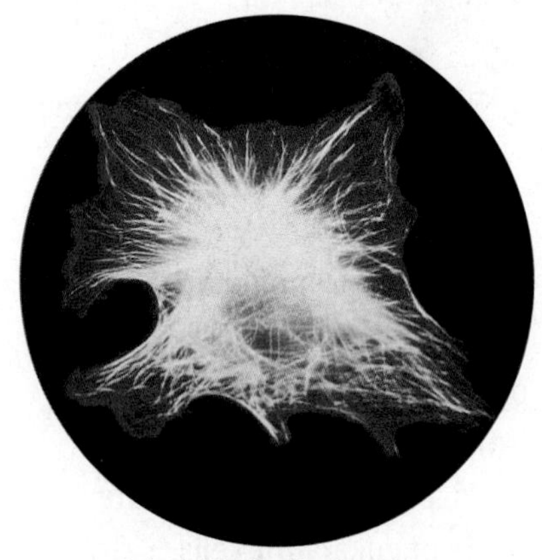

Stained LM Magnification: 700×

Figure 5 Cytoskeleton, a network of fibers in the cytoplasm, gives cells structure and helps them maintain shape.

Cytoplasm Cells are filled with a gelatinlike substance called cytoplasm that constantly flows inside the cell membrane. Many important chemical reactions occur within the cytoplasm.

Throughout the cytoplasm is a framework called the cytoskeleton, which helps the cell maintain or change its shape. Cytoskeletons enable some cells to move. An amoeba, a one-celled eukaryotic organism, moves by stretching and contracting its cytoskeleton. A cytoskeleton is made up of thin, hollow tubes of protein and thin, solid protein fibers, as shown in **Figure 5.** Proteins are organic molecules made up of amino acids.

Reading Check ***What is the function of the cytoskeleton?***

Most of a cell's life processes occur in the cytoplasm. Within the cytoplasm of eukaryotic cells are structures called **organelles.** Some organelles process energy and others manufacture substances needed by the cell or other cells. Certain organelles move materials, while others act as storage sites. Most organelles are surrounded by membranes. The nucleus is usually the largest organelle in a cell.

Nucleus The nucleus is like the deli manager who directs the store's daily operations and passes on information to employees. The **nucleus,** shown in **Figure 6,** directs all cell activities and is separated from the cytoplasm by a membrane. Materials enter and leave the nucleus through openings in this membrane.

The nucleus contains the instructions for everything the cell does. These instructions are found on long, threadlike, hereditary material made of DNA. DNA is the chemical that contains the code for the cell's structure and activities. During cell division, the hereditary material coils tightly around proteins to form structures called chromosomes. A structure called a nucleolus also is found in the nucleus.

Modeling Cytoplasm

Procedure

1. Add 100 mL of **water** to a **clear container.**
2. Add **unflavored gelatin** and stir.
3. Shine a **flashlight** through the solution.

Analysis

1. Describe what you see.
2. How does a model help you understand what cytoplasm might be like?

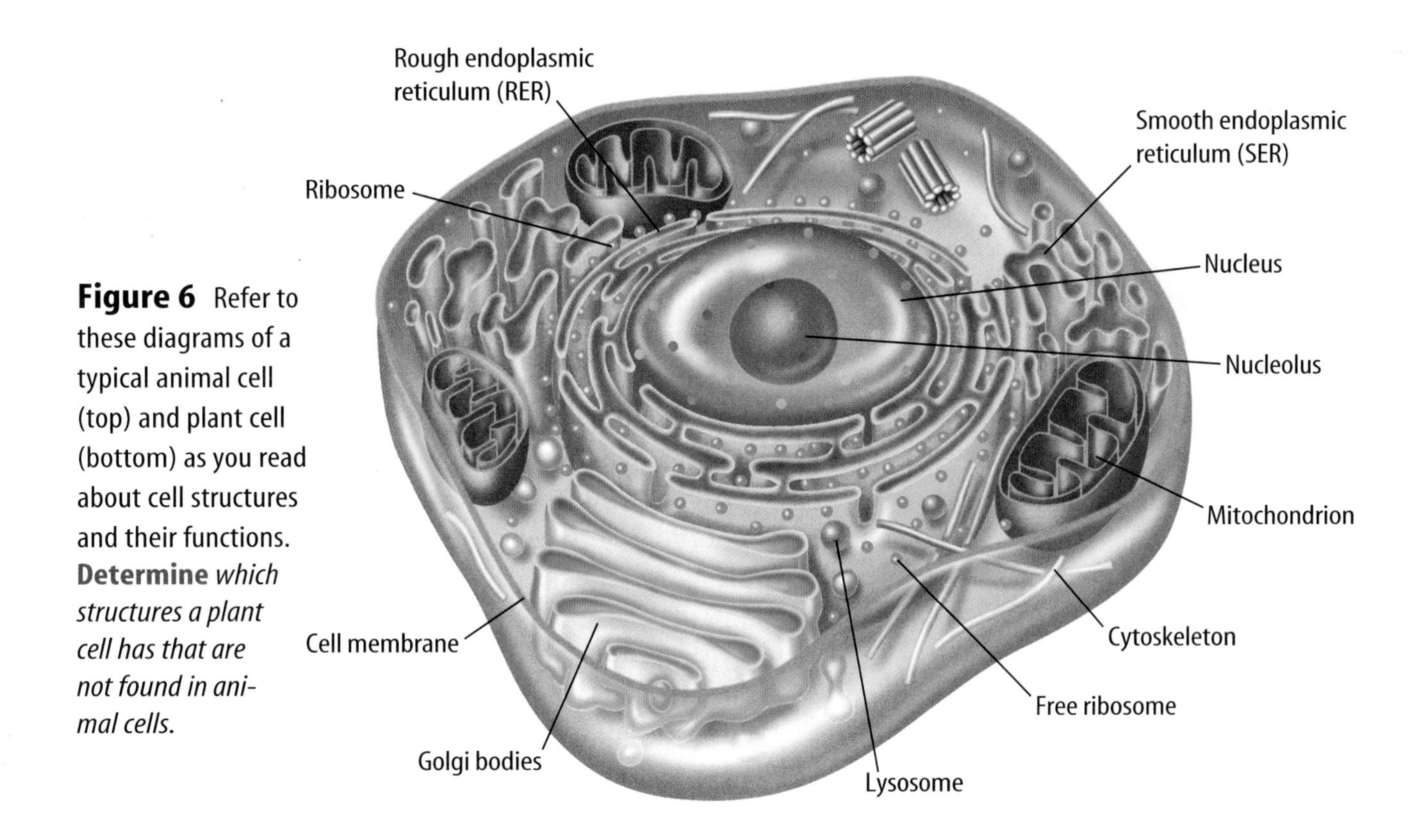

Figure 6 Refer to these diagrams of a typical animal cell (top) and plant cell (bottom) as you read about cell structures and their functions. **Determine** *which structures a plant cell has that are not found in animal cells.*

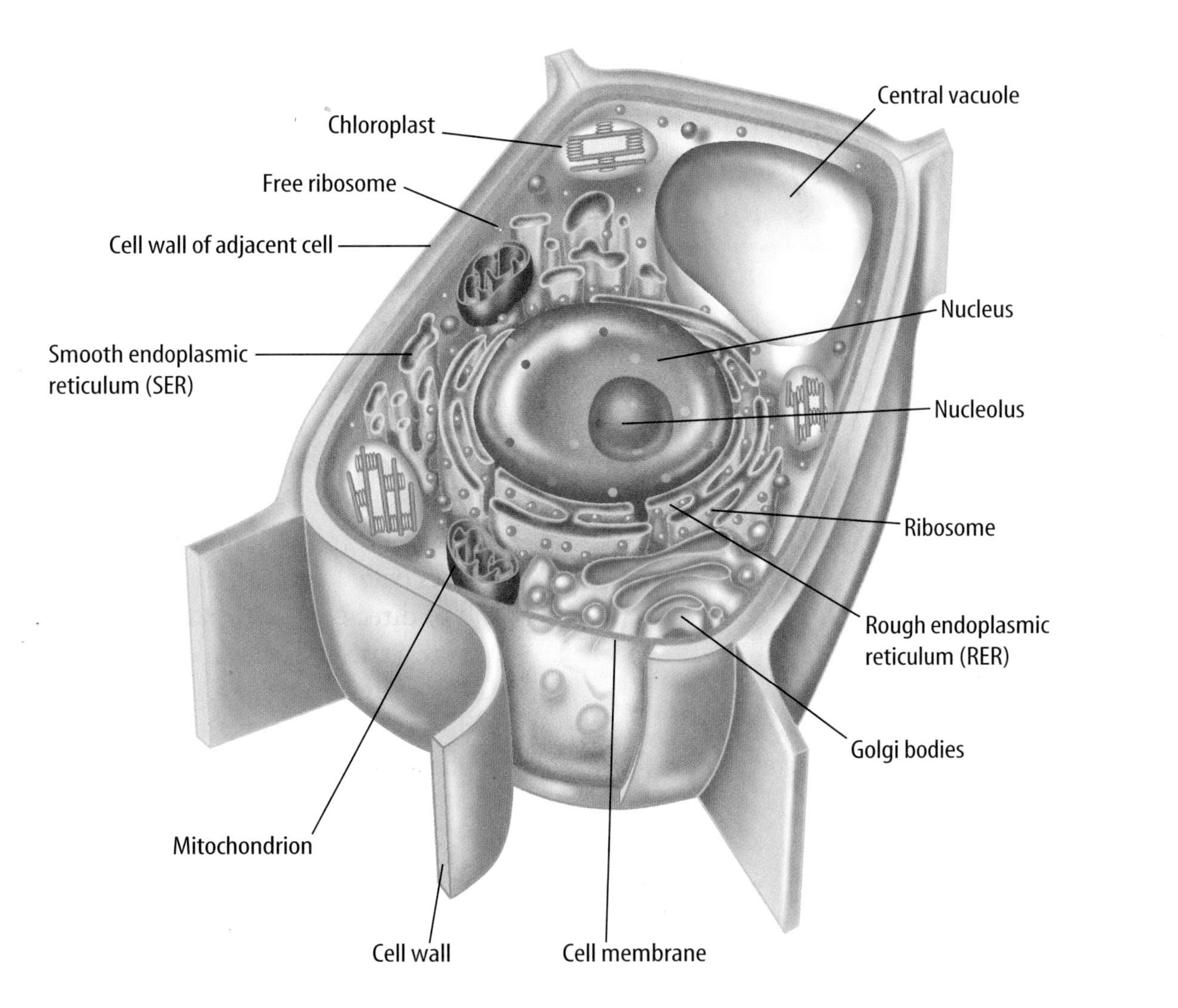

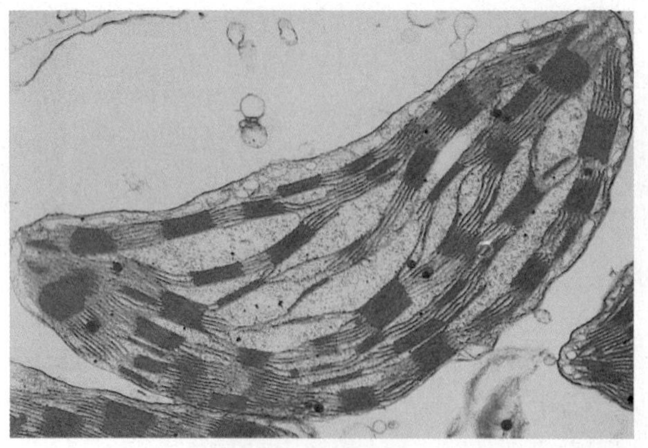
Color-enhanced TEM Magnification: 37000×

Figure 7 Chloroplasts are organelles that use light energy and make sugar from carbon dioxide and water.

Energy-Processing Organelles Cells require a continuous supply of energy to process food, make new substances, eliminate wastes, and communicate with each other. In some plant cells, food is made in green organelles in the cytoplasm called **chloroplasts** (KLOR uh plasts), as shown in **Figure 7.** Chloroplasts contain the green pigment chlorophyll, which gives many leaves and stems their green color. Chlorophyll captures light energy that is used to make a sugar called glucose. Glucose molecules store the captured light energy as chemical energy. Many cells, including animal cells, do not have chloroplasts for making food. They must get food from their environment.

The energy in food is stored until it is released by the mitochondria. **Mitochondria** (mi tuh KAHN dree uh) (singular, *mitochondrion*), such as the one shown in **Figure 8,** are organelles where energy is released from the breakdown of food into carbon dioxide and water. Just as the gas or electric company supplies fuel for the deli, a mitochondrion releases energy for use by the cell. Some types of cells, such as muscle cells, are more active than other cells. These cells have large numbers of mitochondria. Why would active cells have more or larger mitochondria?

Figure 8 Mitochondria are known as the powerhouses of the cell because they release energy that is needed by the cell from food.
Name *the cell types that might contain many mitochondria.*

Manufacturing Organelles One substance that takes part in nearly every cell activity is protein. Proteins are part of cell membranes. Other proteins are needed for chemical reactions that take place in the cytoplasm. Cells make their own proteins on small structures called **ribosomes.** Even though ribosomes are considered organelles, they are not membrane bound. Some ribosomes float freely in the cytoplasm; others are attached to the endoplasmic reticulum. Ribosomes are made in the nucleolus and move out into the cytoplasm. Ribosomes receive directions from the hereditary material in the nucleus on how, when, and in what order to make specific proteins.

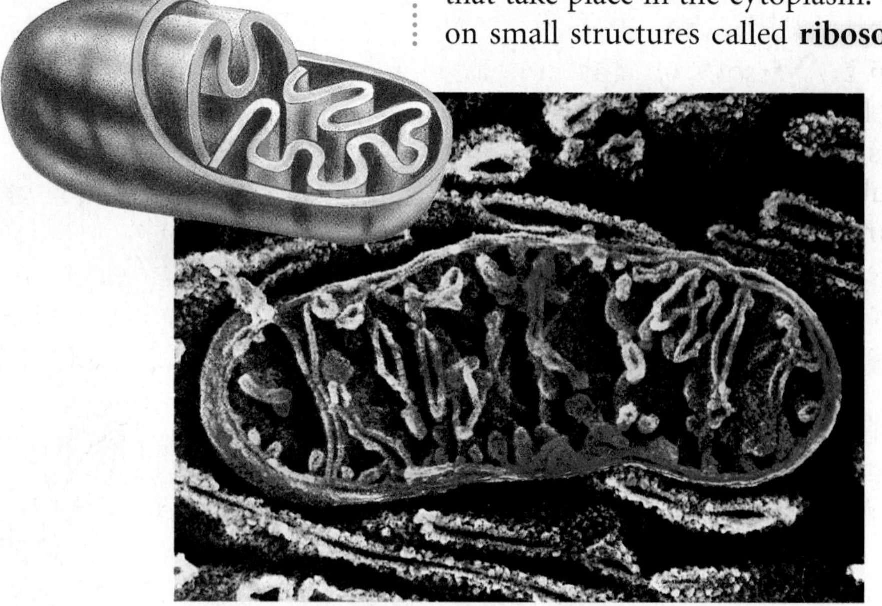
Color-enhanced SEM Magnification: 48000×

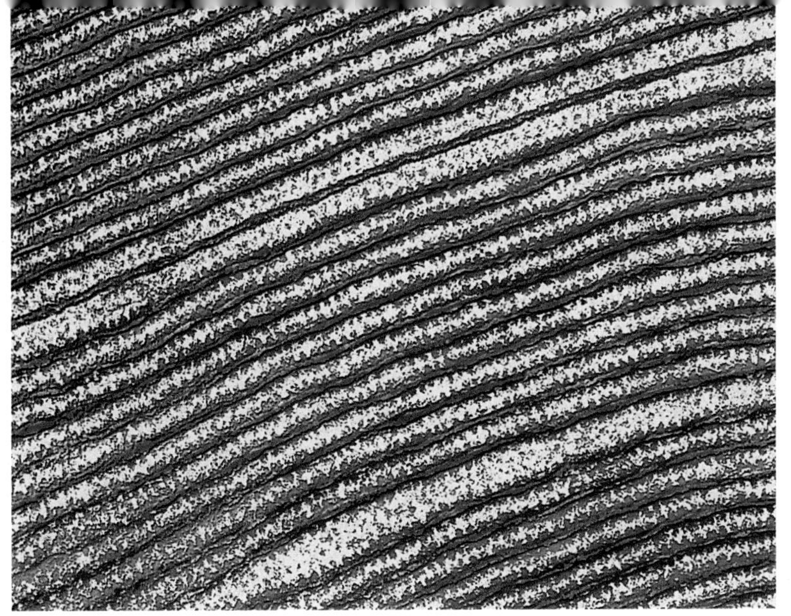
Color-enhanced TEM Magnification: 65000×

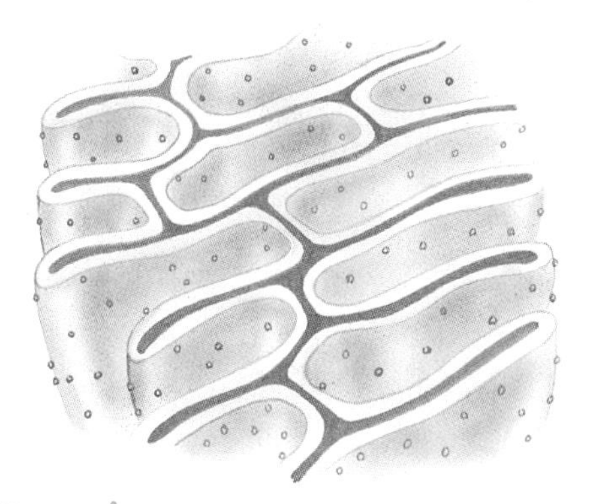

Figure 9 Endoplasmic reticulum (ER) is a complex series of membranes in the cytoplasm of the cell.
Infer *what smooth ER would look like.*

Processing, Transporting, and Storing Organelles

The **endoplasmic reticulum** (en duh PLAZ mihk • rih TIHK yuh lum) or ER, as shown in **Figure 9,** extends from the nucleus to the cell membrane. It is a series of folded membranes in which materials can be processed and moved around inside of the cell. The ER takes up a lot of space in some cells.

The endoplasmic reticulum may be "rough" or "smooth." ER that has no attached ribosomes is called smooth endoplasmic reticulum. This type of ER processes other cellular substances such as lipids that store energy. Ribsomes are attached to areas on the rough ER. There they carry out their job of making proteins that are moved out of the cell or used within the cell.

What is the difference between rough ER and smooth ER?

After proteins are made in a cell, they are transferred to another type of cell organelle called the Golgi (GAWL jee) bodies. The **Golgi bodies,** as shown ion **Figure 10,** are stacked, flattened membranes. The Golgi bodies sort proteins and other cellular substances and package them into membrane-bound structures called vesicles. The vesicles deliver cellular substances to areas inside the cell. They also carry cellular substances to the cell membrane where they are released to the outside of the cell.

Just as a deli has refrigerators for temporary storage of some of its foods and ingredients, cells have membrane-bound spaces called vacuoles for the temporary storage of materials. A vacuole can store water, waste products, food, and other cellular materials. In plant cells, the vacuole may make up most of the cell's volume.

Figure 10 The Golgi body packages materials and moves them to the outside of the cell.
Explain *why materials are removed from the cell.*

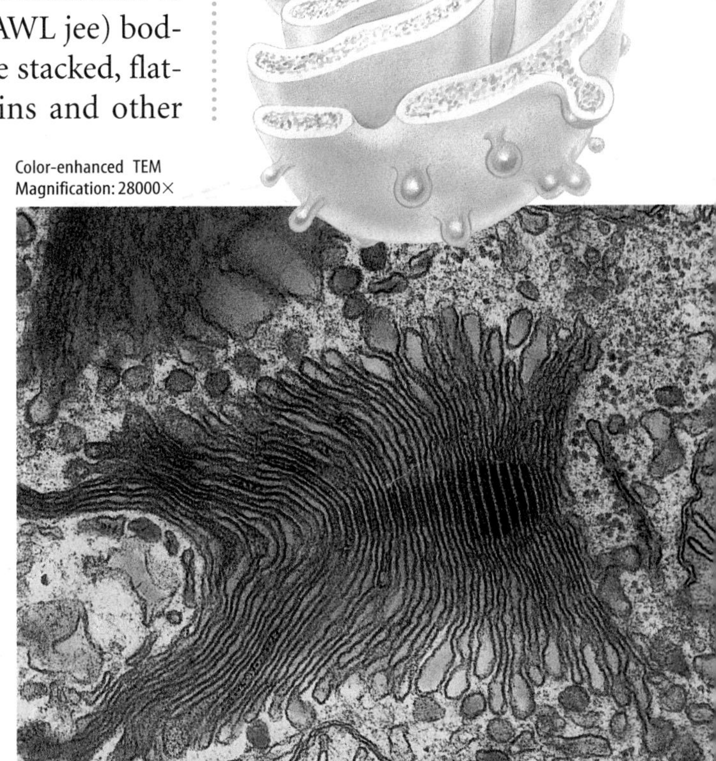
Color-enhanced TEM Magnification: 28000×

Recycling Just like a cell, you can recycle materials. Paper, plastics, aluminum, and glass are materials that can be recycled into usable items. Make a promotional poster to encourage others to recycle.

Recycling Organelles Active cells break down and recycle substances. Organelles called lysosomes (LI suh sohmz) contain digestive chemicals that help break down food molecules, cell wastes, and worn-out cell parts. In a healthy cell, chemicals are released into vacuoles only when needed. The lysosome's membrane prevents the digestive chemicals inside from leaking into the cytoplasm and destroying the cell. When a cell dies, a lysosome's membrane disintegrates. This releases digestive chemicals that allow the quick breakdown of the cell's contents.

What is the function of the lysosome's membrane?

Applying Math — Calculate a Ratio

CELL RATIO Assume that a cell is like a cube with six equal sides. Find the ratio of surface area to volume for a cube that is 4 cm high.

Solution

1. *This is what you know:* A cube has 6 equal sides of 4 cm × 4 cm.
2. *This is what you need to find out:* What is the ratio (R) of surface area to volume for the cube?
3. *These are the equations you use:*
   - surface area (A) = width × length × 6
   - volume (V) = length × width × height
   - $R = A/V$
4. *This is the procedure you need to use:*
   - Substitute in known values and solve the equations

   $A = 4\text{ cm} \times 4\text{ cm} \times 6 = 96\text{ cm}^2$

   $V = 4\text{ cm} \times 4\text{ cm} \times 4\text{ cm} = 64\text{ cm}^3$

   $R = 96\text{ cm}^2/64\text{ cm}^3 = 1.5\text{ cm}^2/\text{cm}^3$
5. *Check your answer:* Multiply the ratio by the volume. Did you calculate the surface area?

Practice Problems

1. Calculate the ratio of surface area to volume for a cube that is 2 cm high. What happens to this ratio as the size of the cube decreases?
2. If a 4-cm cube doubled just one of its dimensions, what would happen to the ratio of surface area to volume?

For more practice, visit life.msscience.com/math_practice

From Cell to Organism

Many one-celled organisms perform all their life functions by themselves. Cells in a many-celled organism, however, do not work alone. Each cell carries on its own life functions while depending in some way on other cells in the organism.

In **Figure 11,** you can see cardiac muscle cells grouped together to form a tissue. A **tissue** is a group of similar cells that work together to do one job. Each cell in a tissue does its part to keep the tissue alive.

Tissues are organized into organs. An **organ** is a structure made up of two or more different types of tissues that work together. Your heart is an organ made up of cardiac muscle tissue, nerve tissue, and blood tissues. The cardiac muscle tissue contracts, making the heart pump. The nerve tissue brings messages from the brain that tell the heart how fast to beat. The blood tissue is carried from the heart to other organs of the body.

What types of tissues make up your heart?

A group of organs working together to perform a certain function is an organ system. Your heart, arteries, veins, and capillaries make up your cardiovascular system. In a many-celled organism, several systems work together in order to perform life functions efficiently. Your nervous, circulatory, respiratory, muscular, and other systems work together to keep you alive.

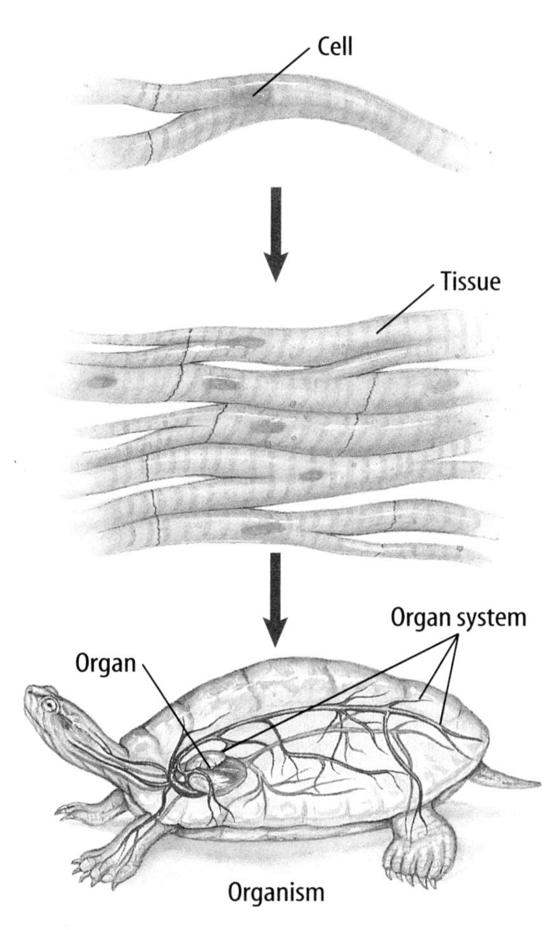

Figure 11 In a many-celled organism, cells are organized into tissues, tissues into organs, organs into systems, and systems into an organism.

section 1 review

Summary

Common Cell Traits

- All cells have an outer covering called a cell membrane.
- Cells can be classified as prokaryotic or eukaryotic.

Cell Organization

- Each cell in your body has a specific function.
- Most of a cell's life processes occur in the cytoplasm.

From Cell to Organism

- In a many-celled organism, several systems work together to perform life functions.

Self Check

1. **Explain** why the nucleus is important in the life of a cell.
2. **Determine** why digestive enzymes in a cell are enclosed in a membrane-bound organelle.
3. **Discuss** how cells, tissues, organs, and organ systems are related.
4. **Think Critically** How is the cell of a one-celled organism different from the cells in many-celled organisms?

Applying Skills

5. **Interpret Scientific Illustrations** Examine **Figure 6.** Make a list of differences and similarities between the animal cell and the plant cell.

Comparing Cells

If you compared a goldfish to a rose, you would find them unlike each other. Are their individual cells different also?

Real-World Question

How do human cheek cells and plant cells compare?

Goals

- **Compare and contrast** an animal cell and a plant cell.

Materials

microscope
microscope slide
coverslip
forceps
tap water
dropper
Elodea plant
prepared slide of human cheek cells

Safety Precautions

Procedure

1. Copy the data table in your Science Journal. Check off the cell parts as you observe them.

Cell Observations		
Cell Part	**Cheek**	***Elodea***
Cytoplasm		
Nucleus		
Chloroplasts	Do not write in this book.	
Cell wall		
Cell membrane		

2. Using forceps, make a wet-mount slide of a young leaf from the tip of an *Elodea* plant.
3. **Observe** the leaf on low power. Focus on the top layer of cells.
4. Switch to high power and focus on one cell. In the center of the cell is a membrane-bound organelle called the central vacuole. Observe the chloroplasts—the green, disk-shaped objects moving around the central vacuole. Try to find the cell nucleus. It looks like a clear ball.
5. **Draw** the *Elodea* cell. Label the cell wall, cytoplasm, chloroplasts, central vacuole, and nucleus. Return to low power and remove the slide. Properly dispose of the slide.
6. **Observe** the prepared slide of cheek cells under low power.
7. Switch to high power and observe the cell nucleus. Draw and label the cell membrane, cytoplasm, and nucleus. Return to low power and remove the slide.

Conclude and Apply

1. **Compare and contrast** the shapes of the cheek cell and the *Elodea* cell.
2. **Draw conclusions** about the differences between plant and animal cells.

Communicating Your Data

Draw the two kinds of cells on one sheet of paper. Use a green pencil to label the organelles found only in plants, a red pencil to label the organelles found only in animals, and a blue pencil to label the organelles found in both. **For more help, refer to the** Science Skill Handbook.

section 2

Viewing Cells

Magnifying Cells

The number of living things in your environment that you can't see is much greater than the number that you can see. Many of the things that you cannot see are only one cell in size. To see most cells, you need to use a microscope.

Trying to see separate cells in a leaf, like the ones in **Figure 12,** is like trying to see individual photos in a photo mosaic picture that is on the wall across the room. As you walk toward the wall, it becomes easier to see the individual photos. When you get right up to the wall, you can see details of each small photo. A microscope has one or more lenses that enlarge the image of an object as though you are walking closer to it. Seen through these lenses, the leaf appears much closer to you, and you can see the individual cells that carry on life processes.

as you read

What You'll Learn

- **Compare** the differences between the compound light microscope and the electron microscope.
- **Summarize** the discoveries that led to the development of the cell theory.

Why It's Important

Humans are like other living things because they are made of cells.

Review Vocabulary
magnify: to increase the size of something

New Vocabulary
- cell theory

Early Microscopes In the late 1500s, the first microscope was made by a Dutch maker of reading glasses. He put two magnifying glasses together in a tube and could see an image that was larger than that made by either lens alone.

In the mid 1600s, Antonie van Leeuwenhoek, a Dutch fabric merchant, made a simple microscope with a tiny glass bead for a lens, as shown in **Figure 13.** With it, he reported seeing things in pond water that no one had ever imagined. His microscope could magnify up to 270 times. Another way to say this is that his microscope could make the object appear 270 times larger than its actual size. Today you would say his lens had a power of 270×. Early compound microscopes were crude by today's standards. The lenses would make a larger image, but it wasn't always sharp or clear.

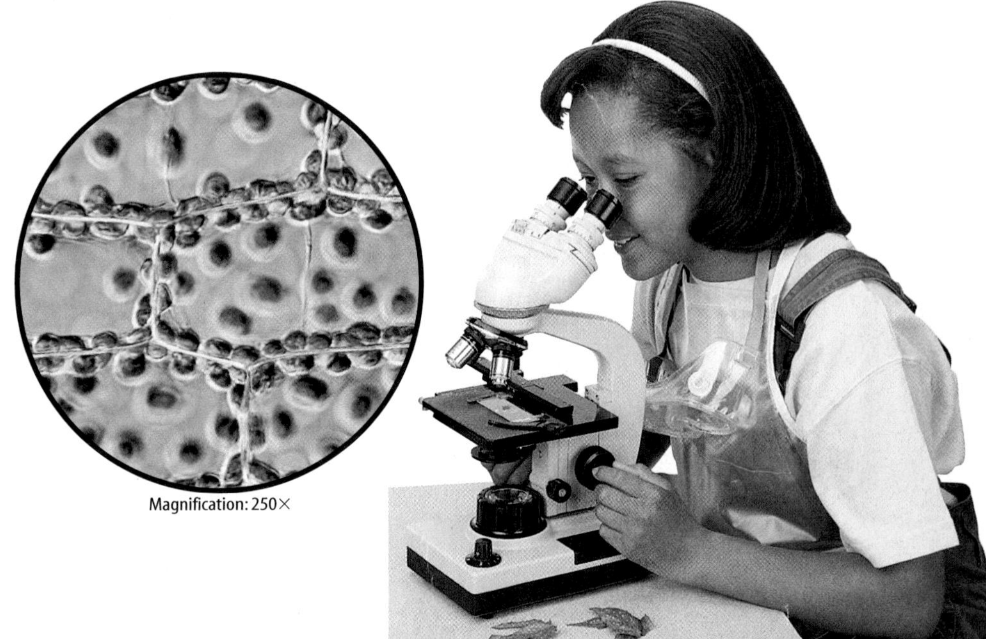

Figure 12 Individual cells become visible when a plant leaf is viewed using a microscope with enough magnifying power.

NATIONAL GEOGRAPHIC VISUALIZING MICROSCOPES

Figure 13

Microscopes give us a glimpse into a previously invisible world. Improvements have vastly increased their range of visibility, allowing researchers to study life at the molecular level. A selection of these powerful tools—and their magnification power—is shown here.

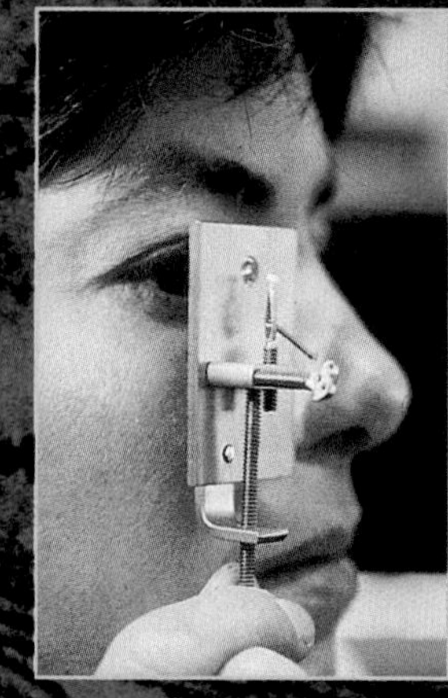

▶ **Up to 250×** LEEUWENHOEK MICROSCOPE Held by a modern researcher, this historic microscope allowed Leeuwenhoek to see clear images of tiny freshwater organisms that he called "beasties."

▼ **Up to 2,000×** BRIGHTFIELD / DARKFIELD MICROSCOPE The light microscope is often called the brightfield microscope because the image is viewed against a bright background. A brightfield microscope is the tool most often used in laboratories to study cells. Placing a thin metal disc beneath the stage, between the light source and the objective lenses, converts a brightfield microscope to a darkfield microscope. The image seen using a darkfield microscope is bright against a dark background. This makes details more visible than with a brightfield microscope. Below are images of a *Paramecium* as seen using both processes.

▲ **Up to 1,500×** FLUORESCENCE MICROSCOPE This type of microscope requires that the specimen be treated with special fluorescent stains. When viewed through this microscope, certain cell structures or types of substances glow, as seen in the image of a *Paramecium* above.

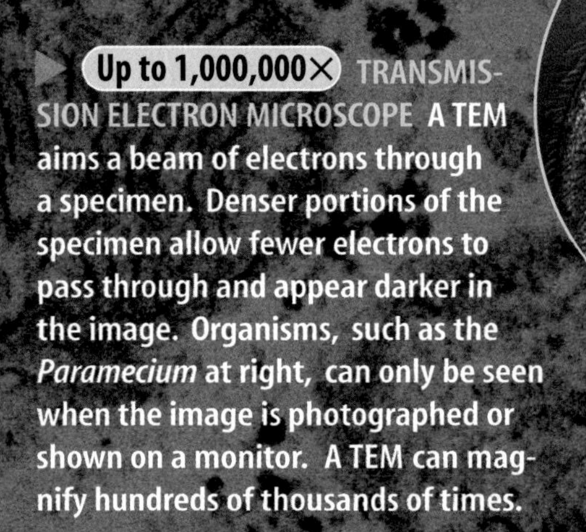

Up to 1,000,000× TRANSMISSION ELECTRON MICROSCOPE A TEM aims a beam of electrons through a specimen. Denser portions of the specimen allow fewer electrons to pass through and appear darker in the image. Organisms, such as the *Paramecium* at right, can only be seen when the image is photographed or shown on a monitor. A TEM can magnify hundreds of thousands of times.

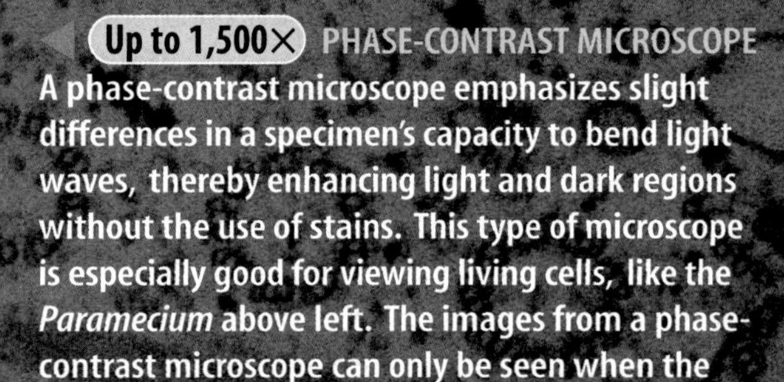

Up to 1,500× PHASE-CONTRAST MICROSCOPE A phase-contrast microscope emphasizes slight differences in a specimen's capacity to bend light waves, thereby enhancing light and dark regions without the use of stains. This type of microscope is especially good for viewing living cells, like the *Paramecium* above left. The images from a phase-contrast microscope can only be seen when the specimen is photographed or shown on a monitor.

Up to 200,000× SCANNING ELECTRON MICROSCOPE An SEM sweeps a beam of electrons over a specimen's surface, causing other electrons to be emitted from the specimen. SEMs produce realistic, three-dimensional images, which can only be viewed as photographs or on a monitor, as in the image of the *Paramecium* at right. Here a researcher compares an SEM picture to a computer monitor showing an enhanced image.

Observing Magnified Objects

Procedure

1. Look at a **newspaper** through the curved side and through the flat bottom of an **empty, clear glass.**
2. Look at the newspaper through a **clear glass bowl filled with water** and then with a **magnifying lens.**

Analysis

In your Science Journal, compare how well you can see the newspaper through each of the objects.

Try at Home

Cell Biologist Microscopes are important tools for cell biologists as they research diseases. In your Science Journal, make a list of diseases for which you think cell biologists are trying to find effective drugs.

Modern Microscopes Scientists use different microscopes to study organisms, cells, and cell parts that are too small to be seen with just the human eye. Depending on how many lenses a microscope contains, it is called simple or compound. A simple microscope is similar to a magnifying lens. It has only one lens. A microscope's lens makes an enlarged image of an object and directs light toward your eye. The change in apparent size produced by a microscope is called magnification. Microscopes vary in powers of magnification. Some microscopes can make images of individual atoms.

The microscope you probably will use to study life science is a compound light microscope, similar to the one in the Reference Handbook at the back of this book. The compound light microscope has two sets of lenses—eyepiece lenses and objective lenses. The eyepiece lenses are mounted in one or two tubelike structures. Images of objects viewed through two eyepieces, or stereomicroscopes, are three-dimensional. Images of objects viewed through one eyepiece are not. Compound light microscopes usually have two to four movable objective lenses.

Magnification The powers of the eyepiece and objective lenses determine the total magnifications of a microscope. If the eyepiece lens has a power of 10× and the objective lens has a power of 43×, then the total magnification is 430× (10× times 43×). Some compound microscopes, like those in **Figure 13,** have more powerful lenses that can magnify an object up to 2,000 times its original size.

Electron Microscopes Things that are too small to be seen with other microscopes can be viewed with an electron microscope. Instead of using lenses to direct beams of light, an electron microscope uses a magnetic field in a vacuum to direct beams of electrons. Some electron microscopes can magnify images up to one million times. To see electron microscope images, they must be photographed or electronically produced.

Several kinds of electron microscopes have been invented, as shown in **Figure 13.** Scanning electron microscopes (SEM) produce a realistic, three-dimensional image. Only the surface of the specimen can be observed using an SEM. Transmission electron microscopes (TEM) produce a two-dimensional image of a thinly-sliced specimen. Details of cell parts can be examined using a TEM. Scanning tunneling microscopes (STM) are able to show the arrangement of atoms on the surface of a molecule. A metal probe is placed near the surface of the specimen and electrons flow from the tip. The hills and valleys of the specimen's surface are mapped.

Cell Theory

During the seventeenth century, scientists used their new invention, the microscope, to explore the newly discovered microscopic world. They examined drops of blood, scrapings from their own teeth, and other small things. Cells weren't discovered until the microscope was improved. In 1665, Robert Hooke cut a thin slice of cork and looked at it under his microscope. To Hooke, the cork seemed to be made up of empty little boxes, which he named cells.

Table 1 The Cell Theory

All organisms are made up of one or more cells.	An organism can be one cell or many cells like most plants and animals.
The cell is the basic unit of organization in organisms.	Even in complex organisms, the cell is the basic unit of structure and function.
All cells come from cells.	Most cells can divide to form two new, identical cells.

In the 1830s, Matthias Schleiden used a microscope to study plants and concluded that all plants are made of cells. Theodor Schwann, after observing different animal cells, concluded that all animals are made up of cells. Eventually, they combined their ideas and became convinced that all living things are made of cells.

Several years later, Rudolf Virchow hypothesized that cells divide to form new cells. Virchow proposed that every cell came from a cell that already existed. His observations and conclusions and those of others are summarized in the **cell theory,** as described in **Table 1.**

Reading Check *Who first concluded that all animals are made of cells?*

section 2 review

Summary

Magnifying Cells

- The powers of the eyepiece and objective lenses determine the total magnification of a microscope.
- An electron microscope uses a magnetic field in a vacuum to direct beams of electrons.

Development of the Cell Theory

- In 1665, Robert Hooke looked at a piece of cork under his microscope and called what he saw cells.
- The conclusions of Rudolf Virchow and those of others are summarized in the cell theory.

Self Check

1. **Determine** why the invention of the microscope was important in the study of cells.
2. **State** the cell theory.
3. **Compare** a simple and a compound light microscope.
4. **Explain** Virchow's contribution to the cell theory.
5. **Think Critically** Why would it be better to look at living cells than at dead cells?

Applying Math

6. **Solve One-Step Equations** Calculate the magnifications of a microscope that has an 8× eyepiece and 10× and 40× objectives.

section 3

Viruses

as you read

What You'll Learn

- **Explain** how a virus makes copies of itself.
- **Identify** the benefits of vaccines.
- **Investigate** some uses of viruses.

Why It's Important

Viruses infect nearly all organisms, usually affecting them negatively yet sometimes affecting them positively.

Review Vocabulary

disease: a condition that results from the disruption in function of one or more of an organism's normal processes

New Vocabulary

- virus
- host cell

What are viruses?

Cold sores, measles, chicken pox, colds, the flu, and AIDS are diseases caused by nonliving particles called viruses. A **virus** is a strand of hereditary material surrounded by a protein coating. Viruses don't have a nucleus or other organelles. They also lack a cell membrane. Viruses, as shown in **Figure 14,** have a variety of shapes. Because they are too small to be seen with a light microscope, they were discovered only after the electron microscope was invented. Before that time, scientists only hypothesized about viruses.

How do viruses multiply?

All viruses can do is make copies of themselves. However, they can't do that without the help of a living cell called a **host cell.** Crystalized forms of some viruses can be stored for years. Then, if they enter an organism, they can multiply quickly.

Once a virus is inside of a host cell, the virus can act in two ways. It can either be active or it can become latent, which is an inactive stage.

Figure 14 Viruses come in a variety of shapes.

Color-enhanced TEM Magnification: 160000×

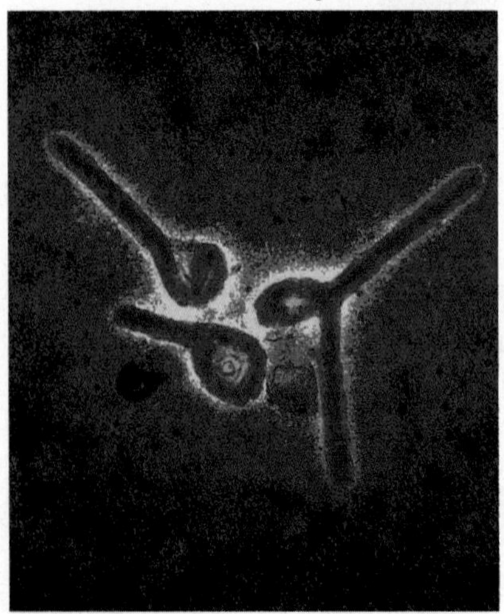

Filoviruses do not have uniform shapes. Some of these *Ebola* viruses have a loop at one end.

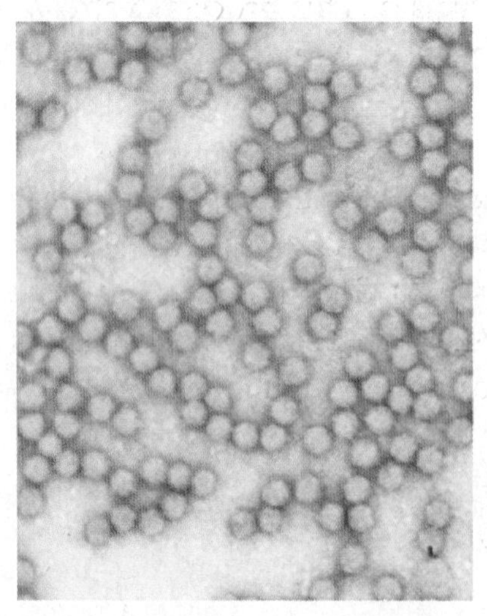

The potato leafroll virus, *Polervirus,* damages potato crops worldwide.

Color-enhanced SEM Magnification: 140000×

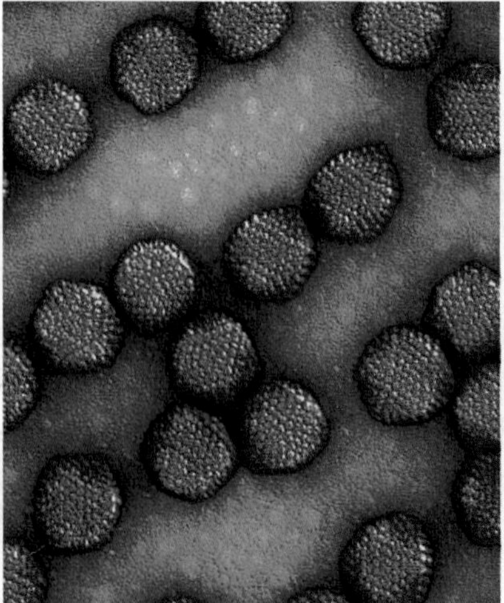

This is just one of the many adenoviruses that can cause the common cold.

Figure 15 An active virus multiplies and destroys the host cell.

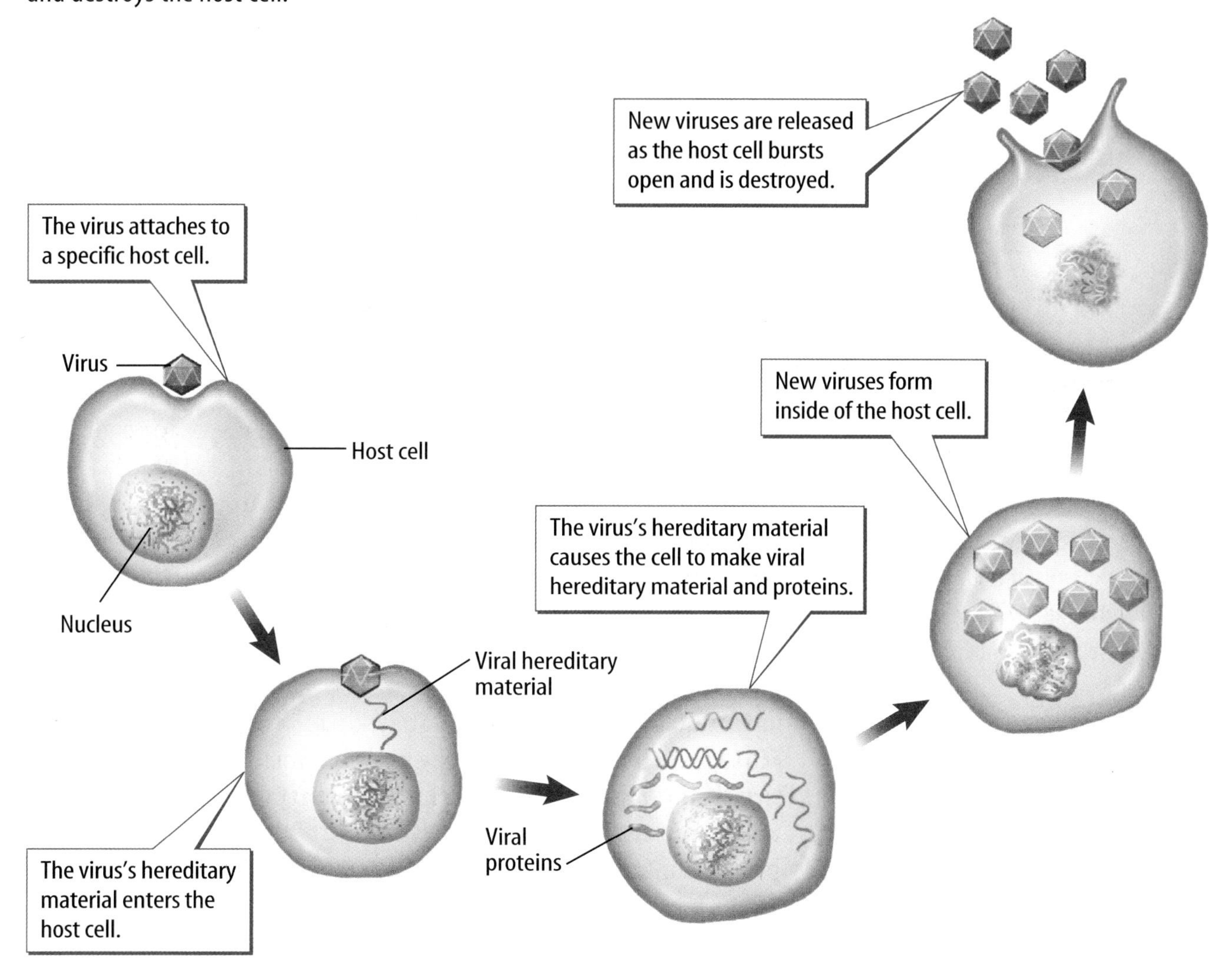

Active Viruses When a virus enters a cell and is active, it causes the host cell to make new viruses. This process destroys the host cell. Follow the steps in **Figure 15** to see one way that an active virus functions inside a cell.

Latent Viruses Some viruses can be latent. That means that after the virus enters a cell, its hereditary material can become part of the cell's hereditary material. It does not immediately make new viruses or destroy the cell. As the host cell reproduces, the viral DNA is copied. A virus can be latent for many years. Then, certain conditions, either inside or outside your body, cause the latent virus to become an active virus.

If you have had a cold sore on your lip, a latent virus in your body has become active. The cold sore is a sign that the virus is active and destroying cells in your lip. When the cold sore disappears, the virus has become latent again. The virus is still in your body's cells, but it is hiding and doing no apparent harm.

Topic: Virus Reactivation
Visit life.msscience.com for Web links to information about viruses.

Activity In your Science Journal, list five stimuli that might activate a latent virus.

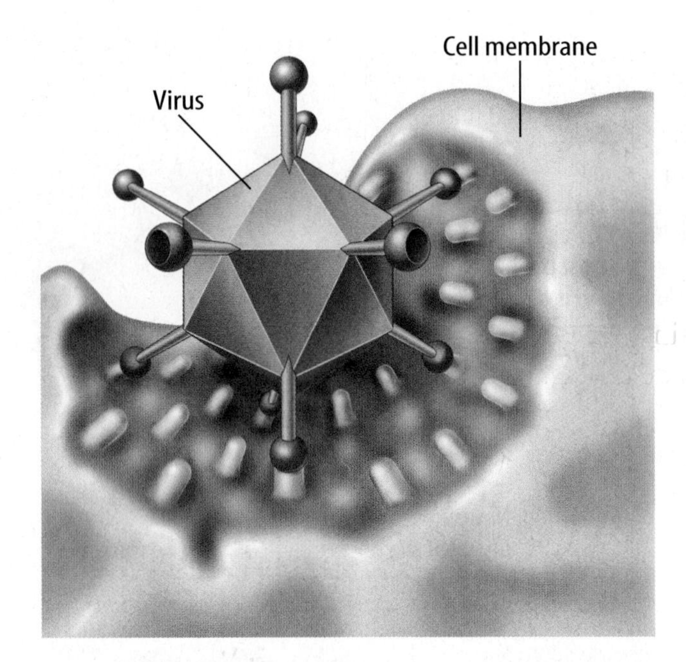

Figure 16 Viruses and the attachment sites of the host cell must match exactly. That's why most viruses infect only one kind of host cell.
Identify *diseases caused by viruses.*

How do viruses affect organisms?

Viruses attack animals, plants, fungi, protists, and bacteria. Some viruses can infect only specific kinds of cells. For instance, many viruses, such as the potato leafroll virus, are limited to one host species or to one type of tissue within that species. A few viruses affect a broad range of hosts. An example of this is the rabies virus. Rabies can infect humans and many other animal hosts.

A virus cannot move by itself, but it can reach a host's body in several ways. For example, it can be carried onto a plant's surface by the wind or it can be inhaled by an animal. In a viral infection, the virus first attaches to the surface of the host cell. The virus and the place where it attaches must fit together exactly, as shown in **Figure 16.** Because of this, most viruses attack only one kind of host cell.

Viruses that infect bacteria are called bacteriophages (bak TIHR ee uh fay jihz). They differ from other kinds of viruses in the way that they enter bacteria and release their hereditary material. Bacteriophages attach to a bacterium and inject their hereditary material. The entire cycle takes about 20 min, and each virus-infected cell releases an average of 100 viruses.

Fighting Viruses

Vaccines are used to prevent disease. A vaccine is made from weakened virus particles that can't cause disease anymore. Vaccines have been made to prevent many diseases, including measles, mumps, smallpox, chicken pox, polio, and rabies.

Reading Check *What is a vaccine?*

The First Vaccine Edward Jenner is credited with developing the first vaccine in 1796. He developed a vaccine for smallpox, a disease that was still feared in the early twentieth century. Jenner noticed that people who got a disease called cowpox didn't get smallpox. He prepared a vaccine from the sores of people who had cowpox. When injected into healthy people, the cowpox vaccine protected them from smallpox. Jenner didn't know he was fighting a virus. At that time, no one understood what caused disease or how the body fought disease.

Science Online

Topic: Filoviruses
Visit life.msscience.com for Web links to information about the virus family *Filoviridae.*

Activity Make a table that displays the virus name, location, and year of the initial outbreaks associated with the *Filoviridae* family.

Treating Viral Diseases Antibiotics treat bacterial infections but are not effective against viral diseases. One way your body can stop viral infections is by making interferons. Interferons are proteins that are produced rapidly by virus-infected cells and move to noninfected cells in the host. They cause the noninfected cells to produce protective substances.

Antiviral drugs can be given to infected patients to help fight a virus. A few drugs show some effectiveness against viruses but some have limited use because of their adverse side effects.

Preventing Viral Diseases Public health measures for preventing viral diseases include vaccinating people, improving sanitary conditions, quarantining patients, and controlling animals that spread disease. For example, annual rabies vaccinations of pets and farm animals protect them and humans from infection. To control the spread of rabies in wild animals such as coyotes and wolves, wildlife workers place bait containing an oral rabies vaccine, as shown in **Figure 17,** where wild animals will find it.

Figure 17 This oral rabies bait is being prepared for an aerial drop by the Texas Department of Health as part of their Oral Rabies Vaccination Program. This five-year program has prevented the expansion of rabies into Texas.

Research with Viruses

You might think viruses are always harmful. However, through research, scientists are discovering helpful uses for some viruses. One use, called gene transfer, substitutes normal hereditary material for a cell's defective hereditary material. The normal material is enclosed in viruses that "infect" targeted cells. The new hereditary material enters the cells and replaces the defective hereditary material. Using gene therapy, scientists hope to help people with genetic disorders and find a cure for cancer.

section 3 review

Summary

What are viruses?

- A virus is a strand of hereditary material surrounded by a protein coating.

How do viruses multiply?

- An active virus immediately destroys the host cell but a latent virus does not.

Fighting Viruses and Research with Viruses

- Antiviral drugs can be given to infected patients to help fight a virus.
- Scientists are discovering helpful uses for some viruses.

Self Check

1. **Describe** how viruses multiply.
2. **Explain** how vaccines are beneficial.
3. **Determine** how some viruses might be helpful.
4. **Discuss** how viral diseases might be prevented.
5. **Think Critically** Explain why a doctor might not give you any medication if you have a viral disease.

Applying Skills

6. **Concept Map** Make an events-chain concept map to show what happens when a latent virus becomes active.

Comparing Light Microscopes

Real-World Question

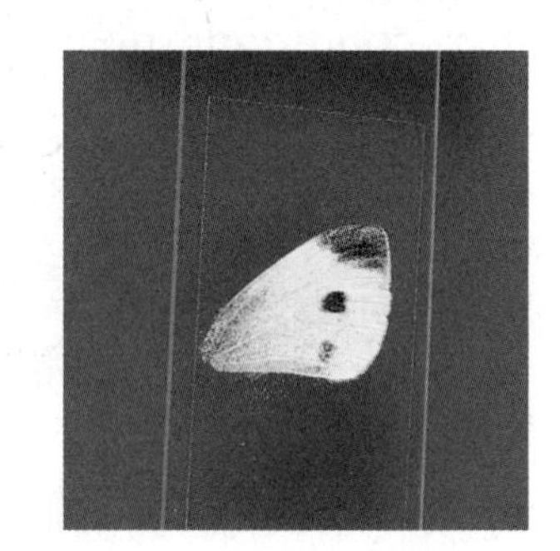

You're a technician in a police forensic laboratory. You use a stereomicroscope and a compound light microscope in the laboratory. A detective just returned from a crime scene with bags of evidence. You must examine each piece of evidence under a microscope. How do you decide which microscope is the best tool to use? Will all of the evidence that you've collected be viewable through both microscopes?

Form a Hypothesis

Compare the items to be examined under the microscopes. Form a hypothesis to predict which microscope will be used for each item and explain why.

Goals

- **Learn** how to correctly use a stereomicroscope and a compound light microscope.
- **Compare** the uses of the stereomicroscope and compound light microscope.

Possible Materials

compound light microscope
stereomicroscope
items from the classroom—include some living or once-living items (8)
microscope slides and coverslips
plastic petri dishes
distilled water
dropper

Safety Precautions

Test Your Hypothesis

Make a Plan

1. As a group, decide how you will test your hypothesis.
2. **Describe** how you will carry out this experiment using a series of specific steps. Make sure the steps are in a logical order. Remember that you must place an item in the bottom of a plastic petri dish to examine it under the stereomicroscope and you must make a wet mount of any item to be examined under the compound light microscope. For more help, see the Reference Handbook.
3. If you need a data table or an observation table, design one in your Science Journal.

Follow Your Plan

1. Make sure your teacher approves the objects you'll examine, your plan, and your data table before you start.
2. Carry out the experiment.
3. While doing the experiment, record your observations and complete the data table.

Analyze Your Data

1. **Compare** the items you examined with those of your classmates.
2. **Classify** the eight items you observed based on this experiment.

Conclude and Apply

1. **Infer** which microscope a scientist might use to examine a blood sample, fibers, and live snails.
2. **List** five careers that require people to use a stereomicroscope. List five careers that require people to use a compound light microscope. Enter the lists in your Science Journal.
3. **Infer** how the images would differ if you examined an item under a compound light microscope and a stereomicroscope.
4. **Determine** which microscope is better for looking at large, or possibly live, items.

Communicating Your Data

In your Science Journal, **write** a short description of an imaginary crime scene and the evidence found there. Sort the evidence into two lists—items to be examined under a stereomicroscope and items to be examined under a compound light microscope. **For more help, refer to the Science Skill Handbook.**

TIME SCIENCE AND HISTORY

SCIENCE CAN CHANGE THE COURSE OF HISTORY!

Cobb Against Cancer

This colored scanning electron micrograph (SEM) shows two breast cancer cells in the final stage of cell division.

Jewel Plummer Cobb is a cell biologist who did important background research on the use of drugs against cancer in the 1950s. She removed cells from cancerous tumors and cultured them in the lab. Then, in a controlled study, she tried a series of different drugs against batches of the same cells. Her goal was to find the right drug to cure each patient's particular cancer. Cobb never met that goal, but her research laid the groundwork for modern chemotherapy—the use of chemicals to treat cancer.

Jewel Cobb also influenced science in another way. She was a role model, especially in her role as dean or president of several universities. Cobb promoted equal opportunity for students of all backgrounds, especially in the sciences.

Light Up a Cure

Vancouver, British Columbia 2000. While Cobb herself was only able to infer what was going on inside a cell from its reactions to various drugs, her work has helped others go further. Building on Cobb's work, Professor Julia Levy and her research team at the University of British Columbia actually go inside cells, and even organelles, to work against cancer. One technique they are pioneering is the use of light to guide cancer drugs to the right cells. First, the patient is given a chemotherapy drug that reacts to light. Then, a fiber optic tube is inserted into the tumor. Finally, laser light is passed through the tube, which activates the light-sensitive drug—but only in the tumor itself. This will hopefully provide a technique to keep healthy cells healthy while killing sick cells.

Write Report on Cobb's experiments on cancer cells. What were her dependent and independent variables? What would she have used as a control? What sources of error did she have to guard against? Answer the same questions about Levy's work.

For more information, visit life.msscience.com/time

Reviewing Main Ideas

Section 1 Cell Structure

1. Prokaryotic and eukaryotic are the two cell types.
2. The DNA in the nucleus controls cell functions.
3. Organelles such as mitochondria and chloroplasts process energy.
4. Most many-celled organisms are organized into tissues, organs, and organ systems.

Section 2 Viewing Cells

1. A simple microscope has just one lens. A compound light microscope has an eyepiece and objective lenses.
2. To calculate the magnification of a microscope, multiply the power of the eyepiece by the power of the objective lens.
3. According to the cell theory, the cell is the basic unit of life. Organisms are made of one or more cells, and all cells come from other cells.

Section 3 Viruses

1. A virus is a structure containing hereditary material surrounded by a protein coating.
2. A virus can make copies of itself only when it is inside a living host cell.

Visualizing Main Ideas

Copy and complete the following concept map of the basic units of life.

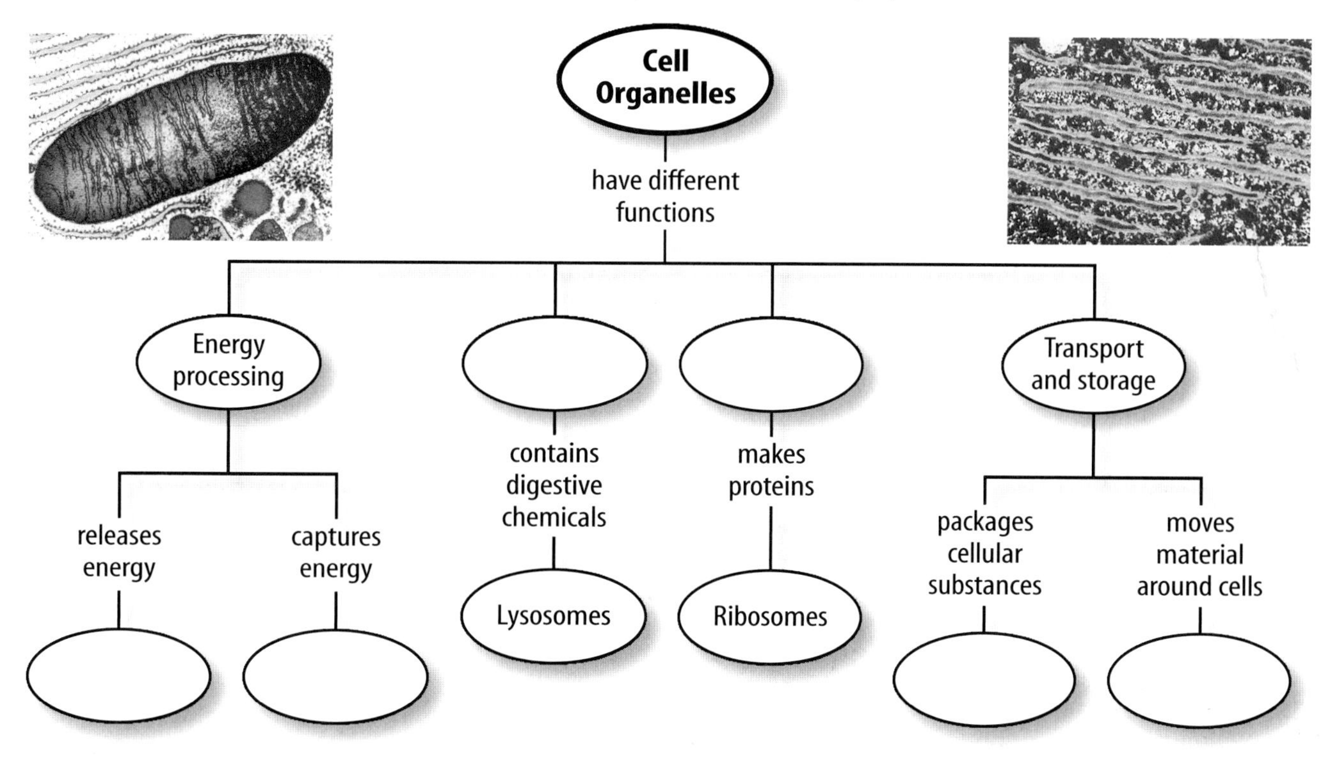

Using Vocabulary

cell membrane p. 38
cell theory p. 51
cell wall p. 39
chloroplast p. 42
cytoplasm p. 38
endoplasmic reticulum p. 43
Golgi body p. 43
host cell p. 52
mitochondrion p. 42
nucleus p. 40
organ p. 45
organelle p. 40
ribosome p. 42
tissue p. 45
virus p. 52

Using the vocabulary words, give an example of each of the following.

1. found in every organ
2. smaller than one cell
3. a plant-cell organelle
4. part of every cell
5. powerhouse of a cell
6. used by biologists
7. contains hereditary material
8. a structure that surrounds the cell
9. can be damaged by a virus
10. made up of cells

Checking Concepts

Choose the word or phrase that best answers the question.

11. What structure allows only certain things to pass in and out of the cell?
 A) cytoplasm **C)** ribosomes
 B) cell membrane **D)** Golgi body

12. What is the organelle to the right?

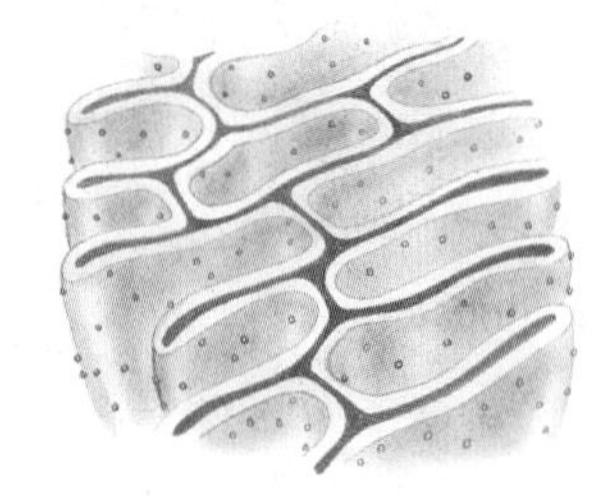

 A) nucleus
 B) cytoplasm
 C) Golgi body
 D) endoplasmic reticulum

Use the illustration below to answer question 13.

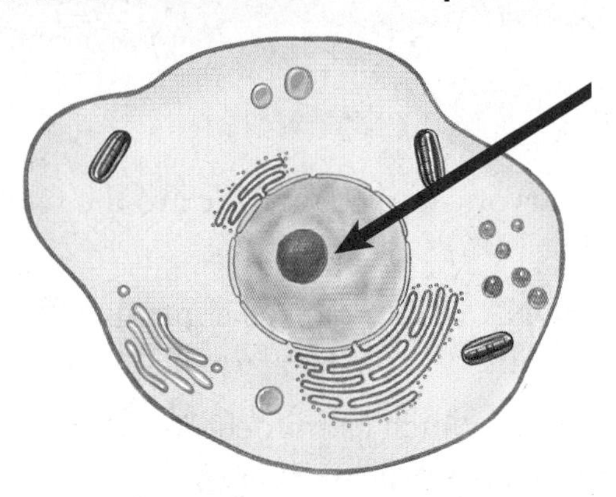

13. In the figure above, what is the function of the structure that the arrow is pointing to?
 A) recycles old cell parts
 B) controls cell activities
 C) protection
 D) releases energy

14. Which scientist gave the name *cells* to structures he viewed?
 A) Hooke **C)** Schleiden
 B) Schwann **D)** Virchow

15. Which of the following is a viral disease?
 A) tuberculosis **C)** smallpox
 B) anthrax **D)** tetanus

16. Which microscope can magnify up to a million times?
 A) compound light microscope
 B) stereomicroscope
 C) transmission electron microscope
 D) atomic force microscope

17. Which of the following is part of a bacterial cell?
 A) a cell wall **C)** mitochondria
 B) lysosomes **D)** a nucleus

18. Which of the following do groups of different tissues form?
 A) organ **C)** organ system
 B) organelle **D)** organism

Science Online life.msscience.com/vocabulary_puzzlemaker

Thinking Critically

19. **Infer** why it is difficult to treat a viral disease.
20. **Explain** which type of microscope would be best to view a piece of moldy bread.
21. **Predict** what would happen to a plant cell that suddenly lost its chloroplasts.
22. **Predict** what would happen if the animal cell shown to the right didn't have ribosomes.

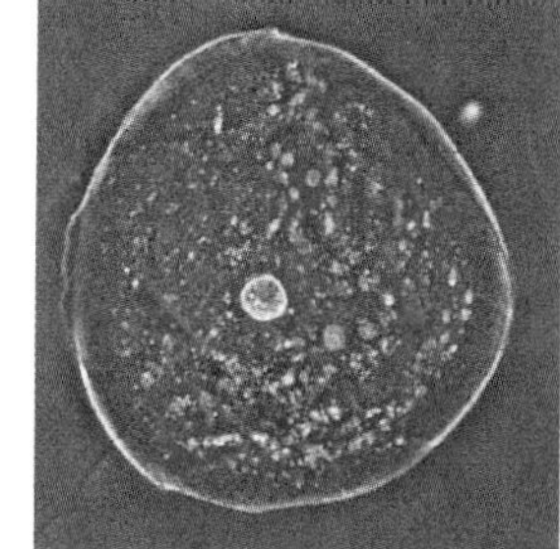

23. **Determine** how you would decide whether an unknown cell was an animal cell, a plant cell, or a bacterial cell.
24. **Concept Map** Make an events-chain concept map of the following from simple to complex: *small intestine, circular muscle cell, human,* and *digestive system.*
25. **Interpret Scientific Illustrations** Use the illustrations in **Figure 1** to describe how the shape of a cell is related to its function.

Use the table below to answer question 26.

Cell Structures

Structure	Prokaryotic Cell	Eukaryotic Cell
Cell membrane		Yes
Cytoplasm	Yes	
Nucleus		Yes
Endoplasmic reticulum		
Golgi bodies		

26. **Compare and Contrast** Copy and complete the table above.
27. **Make a Model** Make and illustrate a time line about the development of the cell theory. Begin with the development of the microscope and end with Virchow. Include the contributions of Leeuwenhoek, Hooke, Schleiden, and Schwann.

Performance Activities

28. **Model** Use materials that resemble cell parts or represent their functions to make a model of a plant cell or an animal cell. Include a cell-parts key.
29. **Poster** Make a poster about the history of vaccinations. Contact your local Health Department for current information.

Applying Math

Use the illustration below to answer question 30.

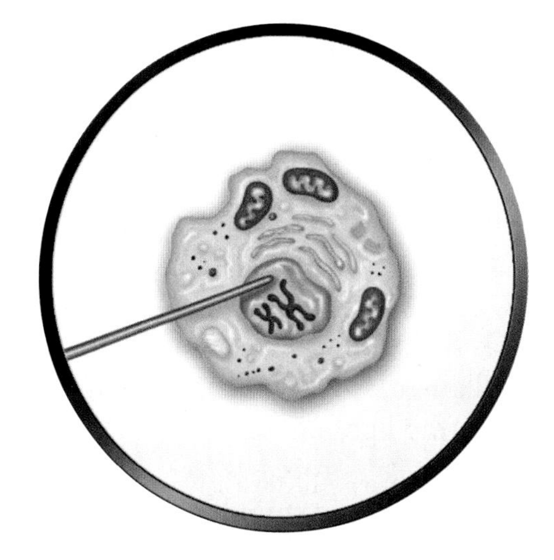

30. **Cell Width** If the pointer shown above with the cell is 10 micrometers (μm) in length, then about how wide is this cell?

    A) 20 μm
    B) 10 μm
    C) 5 μm
    D) 0.1 μm

31. **Magnification** Calculate the magnification of a microscope with a 20× eyepiece and a 40× objective.

Part 1 Multiple Choice

Record your answers on the answer sheet provided by your teacher or on a sheet of paper.

1. What do a bacterial cell, a plant cell, and a nerve cell have in common?
   A. cell wall and nucleus
   B. cytoplasm and cell membrane
   C. endoplasmic reticulum
   D. flagella

2. Which is not a function of an organelle?
   A. cell shape and movement
   B. energy release
   C. chemical transfer
   D. chemical storage

Use the images below to answer question 3.

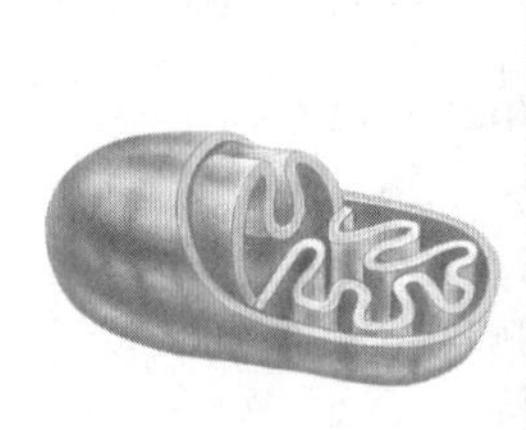
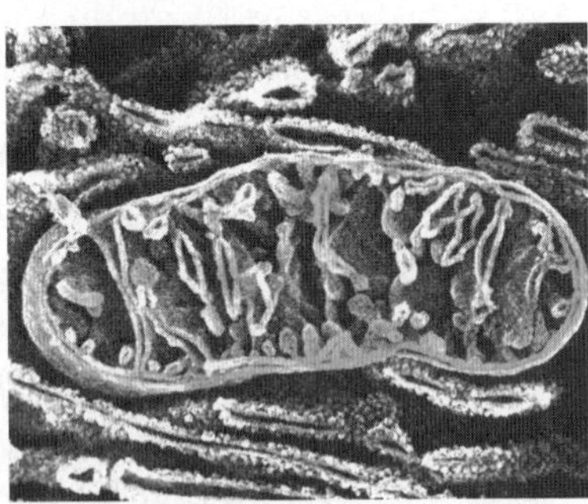

3. What is the primary function of this organelle?
   A. capturing light energy
   B. directing cell processes
   C. releasing energy stored in food
   D. making proteins

4. Which organelles receive the directions from the DNA in the nucleus about which proteins to make?
   A. ribosomes
   B. endoplasmic reticulum
   C. Golgi bodies
   D. cell wall

5. Why is a virus not considered a living cell?
   A. It has a cell wall.
   B. It has hereditary material.
   C. It has no organelles.
   D. It cannot multiply.

Use the illustration below to answer questions 6 and 7.

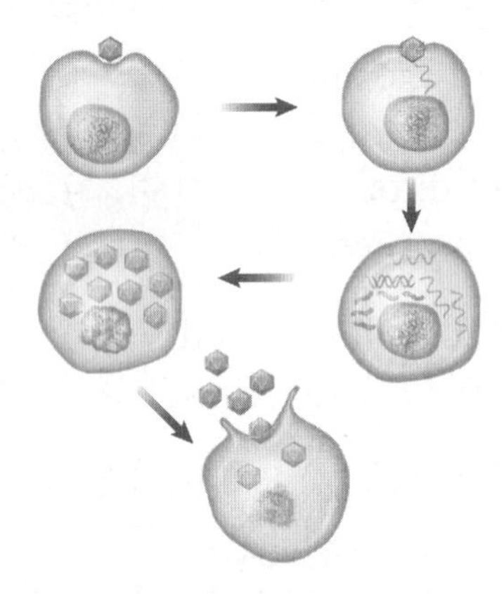

6. What does the diagram above represent?
   A. cell reproduction
   B. bacterial reproduction
   C. active virus multiplication
   D. vaccination

7. What does the largest circular structure represent?
   A. a host cell **C.** a vacuole
   B. a ribosome **D.** the nucleus

8. Where do most of a cell's life processes occur?
   A. nucleus **C.** organ
   B. cell wall **D.** cytoplasm

9. What is a group of similar cells that work together?
   A. tissue **C.** organ system
   B. organ **D.** organism

Test-Taking Tip

Read Carefully Read each question carefully for full understanding.

Part 2 Short Response/Grid In

Record your answers on the answer sheet provided by your teacher or on a sheet of paper.

10. Compare and contrast the cell wall and the cell membrane.

11. How would a cell destroy or breakdown a harmful chemical which entered the cytoplasm?

12. How does your body stop viral infections? What are other ways of protection against viral infections?

13. Where is cellulose found in a cell and what is its function?

Use the following table to answer question 14.

Organelle	Function
	Directs all cellular activities
Mitochondria	
	Captures light energy to make glucose
Ribosomes	

14. Copy and complete the table above with the appropriate information.

15. How are Golgi bodies similar to a packaging plant?

16. Why does a virus need a host cell?

17. Give an example of an organ system and list the organs in it.

18. Compare and contrast the energy processing organelles.

19. Describe the structure of viruses.

20. How do ribosomes differ from other cell structures found in the cytoplasm?

21. What kind of microscope uses a series of lenses to magnify?

Part 3 Open Ended

Record your answers on a sheet of paper.

22. Name three different types of microscopes and give uses for each.

23. Some viruses, like the common cold, only make the host organism sick, but other viruses, like *Ebola,* are deadly to the host organism. Which of these strategies is more effective for replication and transmission of the virus to new host organisms? Which type of virus would be easier to study and develop a vaccine against?

24. Discuss the importance of the cytoplasm.

25. Explain how Hooke, Schleiden, and Schwann contributed to the cell theory.

Use the illustration below to answer question 26.

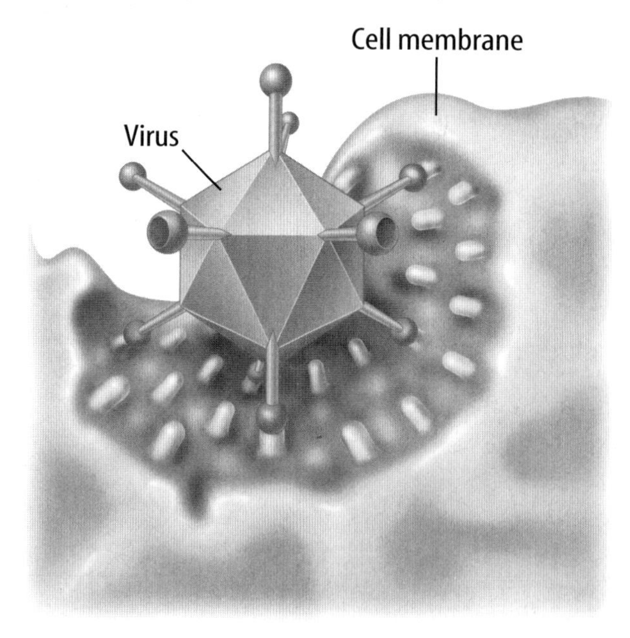

26. What interaction is taking place in the illustration above? What are two possible outcomes of this interaction?

27. Describe how the first vaccine was developed.

chapter 3

Cell Processes

The BIG Idea

Each cell undergoes processes that ensure its survival and often, the survival of other organisms.

SECTION 1
Chemistry of Life

Main Idea All organisms require certain elements that combine and form countless substances needed for life.

SECTION 2
Moving Cellular Material

Main Idea A cell can survive only if substances can move within the cell and pass through its cell membrane.

SECTION 3
Energy for Life

Main Idea All cells require and use energy.

The Science of Gardening

Growing a garden is hard work for both you and the plants. Like you, plants need water and food for energy. How plants get food and water is different from you. Understanding how living things get the energy they need to survive will make a garden seem like much more than just plants and dirt.

Science Journal Describe two ways in which you think plants get food for energy.

Start-Up Activities

Why does water enter and leave plant cells?

If you forget to water a plant, it will wilt. After you water the plant, it probably will straighten up and look healthier. In the following lab, find out how water causes a plant to wilt and straighten.

1. Label a small bowl *Salt Water.* Pour 250 mL of water into the bowl. Then add 15 g of salt to the water and stir.
2. Pour 250 mL of water into another small bowl.
3. Place two carrot sticks into each bowl. Also, place two carrot sticks on the lab table.
4. After 30 min, remove the carrot sticks from the bowls and keep them next to the bowl they came from. Examine all six carrot sticks, then describe them in your Science Journal.
5. **Think Critically** Write a paragraph in your Science Journal that describes what would happen if you moved the carrot sticks from the plain water to the lab table, the ones from the salt water into the plain water, and the ones from the lab table into the salt water for 30 min. Now move the carrot sticks as described and write the results in your Science Journal.

How Living Things Survive Make the following vocabulary Foldable to help you understand the chemistry of living things and how energy is obtained for life.

STEP 1 **Fold** a vertical sheet of notebook paper from side to side.

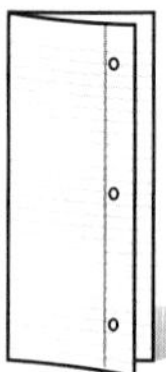

STEP 2 **Cut** along every third line of only the top layer to form tabs.

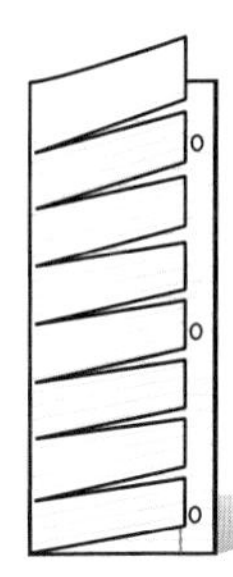

Build Vocabulary As you read this chapter, list the vocabulary words about cell processes on the tabs. As you learn the definitions, write them under the tab for each vocabulary word. Write a sentence about one of the cell processes using the vocabulary word on the tab.

Preview this chapter's content and activities at life.msscience.com

Get Ready to Read

New Vocabulary

① Learn It! What should you do if you find a word you don't know or understand? Here are some suggested strategies:

1. Use context clues (from the sentence or the paragraph) to help you define it.
2. Look for prefixes, suffixes, or root words that you already know.
3. Write it down and ask for help with the meaning.
4. Guess at its meaning.
5. Look it up in the glossary or a dictionary.

② Practice It! Look at the term *inorganic compounds* in the following passage. See how context clues can help you understand its meaning.

Context Clue
Inorganic compounds contain elements other than carbon and have fewer atoms than organic compounds.

Context Clue
Many elements needed for life come from inorganic compounds.

Context Clue
Water is one of the most important inorganic compounds for living things.

Most **inorganic compounds** are made from elements other than carbon. Generally, inorganic molecules contain fewer atoms than organic molecules. Inorganic compounds are the sources for many elements needed by living things. For example, plants take up inorganic compounds from the soil. These inorganic compounds can contain the elements nitrogen, phosphorus, and sulfur. Many foods that you eat contain inorganic compounds. **Table 3** shows some of the inorganic compounds that are important to you. One of the most important inorganic compounds for living things is water.

—from page 71

③ Apply It! Make a vocabulary bookmark with a strip of paper. As you read, keep track of words you do not know or want to learn more about.

Reading Tip

Read a paragraph containing a vocabulary word from beginning to end. Then, go back to determine the meaning of the word.

Target Your Reading

Use this to focus on the main ideas as you read the chapter.

1. **Before you read** the chapter, respond to the statements below on your worksheet or on a numbered sheet of paper.
   - Write an **A** if you **agree** with the statement.
   - Write a **D** if you **disagree** with the statement.

2. **After you read** the chapter, look back to this page to see if you've changed your mind about any of the statements.
   - If any of your answers changed, explain why.
   - Change any false statements into true statements.
   - Use your revised statements as a study guide.

Before You Read A or D	Statement	After You Read A or D
	1 Osmosis is the movement of water into and out of a cell.	
	2 All substances can easily pass through a cell's cell membrane.	
	3 Photosynthesis produces oxygen and a sugar.	
	4 Proteins are organic compounds that are important for storing energy.	
	5 Ions play an important role in many life processes.	
	6 Diffusion continues until equilibrium occurs.	
	7 Matter is anything that has mass and takes up space.	
	8 Only plant cells can transform energy.	
	9 Water is the most abundant compound in a cell.	
	10 Cellular respiration requires oxygen and releases energy for a cell.	

Science Online
Print out a worksheet of this page at life.msscience.com

section 1

Chemistry of Life

as you read

What You'll Learn

- **List** the differences among atoms, elements, molecules, and compounds.
- **Explain** the relationship between chemistry and life science.
- **Discuss** how organic compounds are different from inorganic compounds.

Why It's Important

You grow because of chemical reactions in your body.

Review Vocabulary
cell: the smallest unit of a living thing that can perform the functions of life

New Vocabulary
- mixture
- organic compound
- enzyme
- inorganic compound

The Nature of Matter

Think about everything that surrounds you—chairs, books, clothing, other students, and air. What are all these things made up of? You're right if you answer "matter and energy." Matter is anything that has mass and takes up space. Energy is anything that brings about change. Everything in your environment, including you, is made of matter. Energy can hold matter together or break it apart. For example, the food you eat is matter that is held together by chemical energy. When food is cooked, energy in the form of thermal energy can break some of the bonds holding the matter in food together.

Atoms Whether it is solid, liquid, or gas, matter is made of atoms. **Figure 1** shows a model of an oxygen atom. At the center of an atom is a nucleus that contains protons and neutrons. Although they have nearly equal masses, a proton has a positive charge and a neutron has no charge. Outside the nucleus are electrons, each of which has a negative charge. It takes about 1,837 electrons to equal the mass of one proton. Electrons are important because they are the part of the atom that is involved in chemical reactions. Look at **Figure 1** again and you will see that an atom is mostly empty space. Energy holds the parts of an atom together.

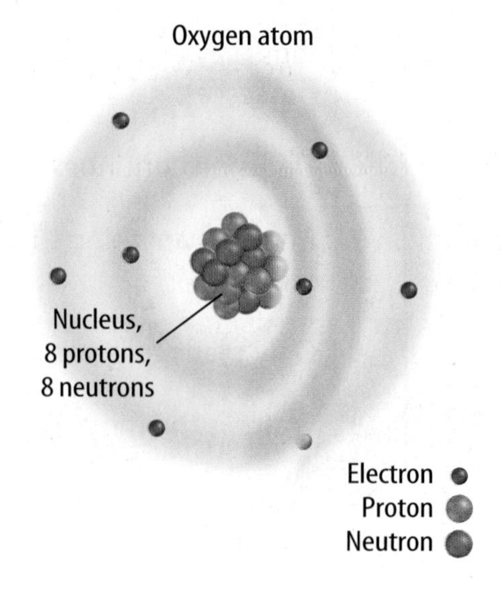

Figure 1 An oxygen atom model shows the placement of electrons, protons, and neutrons.

Table 1 Elements in the Human Body

Symbol	Element	Percent
O	Oxygen	65.0
C	Carbon	18.5
H	Hydrogen	9.5
N	Nitrogen	3.2
Ca	Calcium	1.5
P	Phosphorus	1.0
K	Potassium	0.4
S	Sulfur	0.3
Na	Sodium	0.2
Cl	Chlorine	0.2
Mg	Magnesium	0.1
	Other elements	0.1

Elements When something is made up of only one kind of atom, it is called an element. An element can't be broken down into a simpler form by chemical reactions. The element oxygen is made up of only oxygen atoms, and hydrogen is made up of only hydrogen atoms. Scientists have given each element its own one- or two-letter symbol.

All elements are arranged in a chart known as the periodic table of elements. You can find this table at the back of this book. The table provides information about each element including its mass, how many protons it has, and its symbol.

Everything is made up of elements. Most things, including all living things, are made up of a combination of elements. Few things exist as pure elements. **Table 1** lists elements that are in the human body. What two elements make up most of your body?

Six of the elements listed in the table are important because they make up about 99 percent of living matter. The symbols for these elements are S, P, O, N, C, and H. Use **Table 1** to find the names of these elements.

What types of things are made up of elements?

Figure 2 The words *atoms, molecules,* and *compounds* are used to describe substances.
Explain *how these terms are related to each other.*

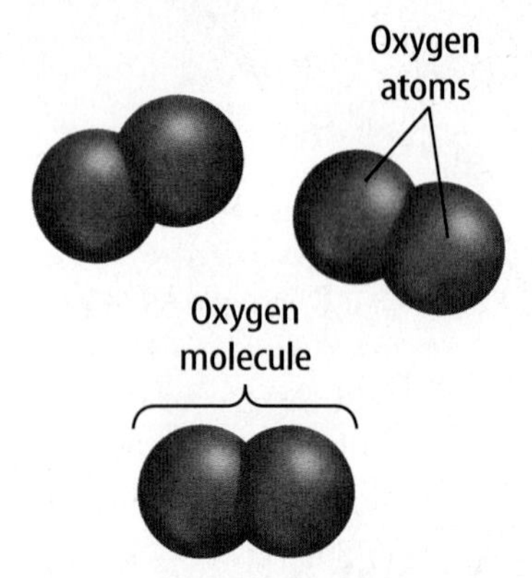

A Some elements, like oxygen, occur as molecules. These molecules contain atoms of the same element bonded together.

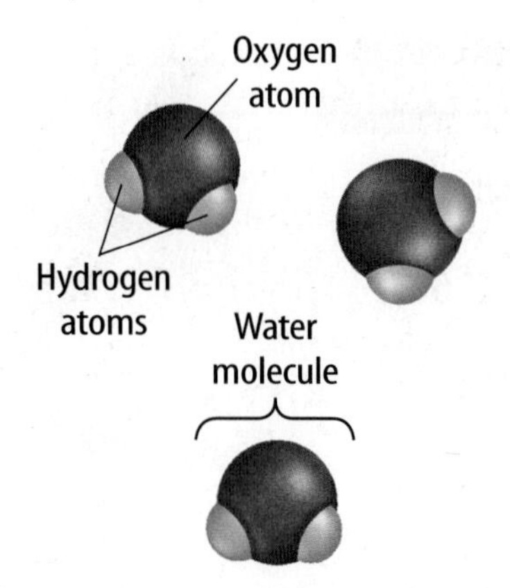

B Compounds also are composed of molecules. Molecules of compounds contain atoms of two or more different elements bonded together, as shown by these water molecules.

Compounds and Molecules

Suppose you make a pitcher of lemonade using a powdered mix and water. The water and the lemonade mix, which is mostly sugar, contain the elements oxygen and hydrogen. Yet, in one, they are part of a nearly tasteless liquid—water. In the other they are part of a sweet solid—sugar. How can the same elements be part of two materials that are so different? Water and sugar are compounds. Compounds are made up of two or more elements in exact proportions. For example, pure water, whether one milliliter of it or one million liters, is always made up of hydrogen atoms bonded to oxygen atoms in a ratio of two hydrogen atoms to one oxygen atom. Compounds have properties different from the elements they are made of. There are two types of compounds—molecular compounds and ionic compounds.

Molecular Compounds The smallest part of a molecular compound is a molecule. A molecule is a group of atoms held together by the energy of chemical bonds, as shown in **Figure 2.** When chemical reactions occur, chemical bonds break, atoms are rearranged, and new bonds form. The molecules produced are different from those that began the chemical reaction.

Molecular compounds form when different atoms share their outermost electrons. For example, two atoms of hydrogen each can share one electron on one atom of oxygen to form one molecule of water, as shown in **Figure 2B.** Water does not have the same properties as oxygen and hydrogen. Under normal conditions on Earth, oxygen and hydrogen are gases. Yet, water can be a liquid, a solid, or a gas. When hydrogen and oxygen combine, changes occur and a new substance forms.

Ions Atoms also combine because they've become positively or negatively charged. Atoms are usually neutral—they have no overall electric charge. When an atom loses an electron, it has more protons than electrons, then it is positively charged. When an atom gains an electron, it has more electrons than protons, then it is negatively charged. Electrically charged atoms—positive or negative—are called ions.

Ionic Compounds Ions of opposite charges attract one another to form electrically neutral compounds called ionic compounds. Table salt is made of sodium (Na) and chlorine (Cl) ions, as shown in **Figure 3B.** When they combine, a chlorine atom gains an electron from a sodium atom. The chlorine atom becomes a negatively charged ion, and the sodium atom becomes a positively charged ion. These oppositely charged ions attract each other and form the ionic compound sodium chloride, NaCl.

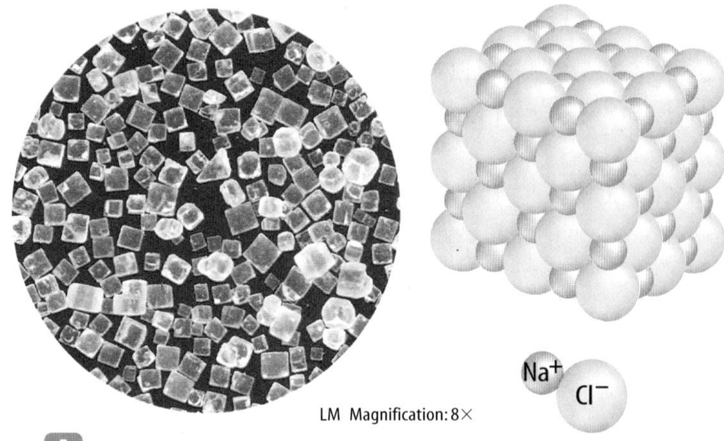

A Magnified crystals of salt look like this.

B A salt crystal is held together by the attractions between sodium ions and chlorine ions.

Figure 3 Table salt crystals are held together by ionic bonds.

Ions are important in many life processes that take place in your body and in other organisms. For example, messages are sent along your nerves as potassium and sodium ions move in and out of nerve cells. Calcium ions are important in causing your muscles to contract. Ions also are involved in the transport of oxygen by your blood. The movement of some substances into and out of a cell would not be possible without ions.

Mixtures

Some substances, such as a combination of sugar and salt, can't change each other or combine chemically. A **mixture** is a combination of substances in which individual substances retain their own properties. Mixtures can be solids, liquids, gases, or any combination of them.

Reading Check *Why is a combination of sugar and salt said to be a mixture?*

Most chemical reactions in living organisms take place in mixtures called solutions. You've probably noticed the taste of salt when you perspire. Sweat is a solution of salt and water. In a solution, two or more substances are mixed evenly. A cell's cytoplasm is a solution of dissolved molecules and ions.

Living things also contain mixtures called suspensions. A suspension is formed when a liquid or a gas has another substance evenly spread throughout it. Unlike solutions, the substances in a suspension eventually sink to the bottom. If blood, shown in **Figure 4,** is left undisturbed, the red blood cells and white blood cells will sink gradually to the bottom. However, the pumping action of your heart constantly moves your blood and the blood cells remain suspended.

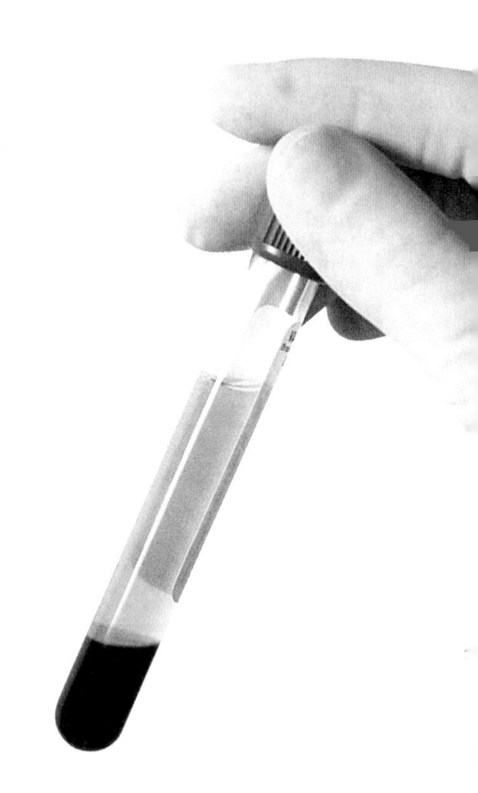

Figure 4 When a test tube of whole blood is left standing, the blood cells sink in the watery plasma.

Table 2 Organic Compounds Found in Living Things

	Carbohydrates	Lipids	Proteins	Nucleic Acids
Elements	carbon, hydrogen, and oxygen	carbon, oxygen, hydrogen, and phosphorus	carbon, oxygen, hydrogen, nitrogen, and sulfur	carbon, oxygen, hydrogen, nitrogen, and phosphorus
Examples	sugars, starch, and cellulose	fats, oils, waxes, phospholipids, and cholesterol	enzymes, skin, and hair	DNA and RNA
Function	supply energy for cell processes; form plant structures; short-term energy storage	store large amounts of energy long term; form boundaries around cells	regulate cell processes and build cell structures	carry hereditary information; used to make proteins

Organic Compounds

You and all living things are made up of compounds that are classified as organic or inorganic. Rocks and other nonliving things contain inorganic compounds, but most do not contain large amounts of organic compounds. **Organic compounds** always contain carbon and hydrogen and usually are associated with living things. One exception would be nonliving things that are products of living things. For example, coal contains organic compounds because it was formed from dead and decaying plants. Organic molecules can contain hundreds or even thousands of atoms that can be arranged in many ways. **Table 2** compares the four groups of organic compounds that make up all living things—carbohydrates, lipids, proteins, and nucleic acids.

Topic: Air Quality
Visit life.msscience.com for Web links to information about air quality.

Activity Organic compounds such as soot, smoke, and ash can affect air quality. Look up the air quality forecast for today. List three locations where the air quality forecast is good, and three locations where it is unhealthy.

Carbohydrates Carbohydrates are organic molecules that supply energy for cell processes. Sugars and starches are carbohydrates that cells use for energy. Some carbohydrates also are important parts of cell structures. For example, a carbohydrate called cellulose is an important part of plant cells.

Lipids Another type of organic compound found in living things is a lipid. Lipids do not mix with water. Lipids such as fats and oils store and release even larger amounts of energy than carbohydrates do. One type of lipid, the phospholipid, is a major part of cell membranes.

What are three types of lipids?

Proteins Organic compounds called proteins have many important functions in living organisms. They are made up of smaller molecules called amino acids. Proteins are the building blocks of many structures in organisms. Your muscles contain large amounts of protein. Proteins are scattered throughout cell membranes. Certain proteins called **enzymes** regulate nearly all chemical reactions in cells.

Nucleic Acids Large organic molecules that store important coded information in cells are called nucleic acids. One nucleic acid, deoxyribonucleic (dee AHK sih ri boh noo klee ihk) acid, or DNA is the genetic material found in all cells at some point in their lives. It carries information that directs each cell's activities. Another nucleic acid, ribonucleic (ri boh noo klee ihk) acid, or RNA, is needed to make enzymes and other proteins.

Inorganic Compounds

Most **inorganic compounds** are made from elements other than carbon. Generally, inorganic molecules contain fewer atoms than organic molecules. Inorganic compounds are the source for many elements needed by living things. For example, plants take up inorganic compounds from the soil. These inorganic compounds can contain the elements nitrogen, phosphorus, and sulfur. Many foods that you eat contain inorganic compounds. **Table 3** shows some of the inorganic compounds that are important to you. One of the most important inorganic compounds for living things is water.

Observing How Enzymes Work

Procedure

1. Get two small cups of **prepared gelatin** from your teacher. Do not eat or drink anything in lab.
2. On the gelatin in one of the cups, place a piece of **fresh pineapple.**
3. Let both cups stand undisturbed overnight.
4. Observe what happens to the gelatin.

Analysis

1. What effect did the piece of fresh pineapple have on the gelatin?
2. What does fresh pineapple contain that caused it to have the effect on the gelatin you observed?
3. Why do the preparation directions on a box of gelatin dessert tell you not to mix it with fresh pineapple?

Table 3 Some Inorganic Compounds Important in Humans

Compound	Use in Body
Water	makes up most of the blood; most chemical reactions occur in water
Calcium phosphate	gives strength to bones
Hydrochloric acid	helps break down foods in the stomach
Sodium bicarbonate	helps the digestion of food to occur
Salts containing sodium, chlorine, and potassium	important in sending messages along nerves

Importance of Water Some scientists hypothesize that life began in the water of Earth's ancient oceans. Chemical reactions might have occurred that produced organic molecules. Similar chemical reactions can take place in cells in your body.

Living things are composed of more than 50 percent water and depend on water to survive. You can live for weeks without food but only for a few days without water. **Figure 5** shows where water is found in your body. Although seeds and spores of plants, fungi, and bacteria can exist without water, they must have water if they are to grow and reproduce. All the chemical reactions in living things take place in water solutions, and most organisms use water to transport materials through their bodies. For example, many animals have blood that is mostly water and moves materials. Plants use water to move minerals and sugars between the roots and leaves.

Applying Math — Solve an Equation

CALCULATE THE IMPORTANCE OF WATER All life on Earth depends on water for survival. Water is the most vital part of humans and other animals. It is required for all of the chemical processes that keep us alive. At least 60 percent of an adult human body consists of water. If an adult man weighs 90 kg, how many kilograms of water does his body contain?

Solution

1. *This is what you know:*
   - adult human body = 60% water
   - man = 90 kg
2. *This is what you need to find:* How many kilograms of water does the adult man have?
3. *This is the procedure you need to use:*
   - Set up the ratio: $60/100 = x/90$.
   - Solve the equation for x: $(60 \times 90)/100$.
   - The adult man has 54 kg of water.
4. *Check your answer:* Divide your answer by 90, then multiply by 100. You should get 60%.

Practice Problems

1. A human body at birth consists of 78 percent water. This gradually decreases to 60 percent in an adult. Assume a baby weighed 3.2 kg at birth and grew into an adult weighing 95 kg. Calculate the approximate number of kilograms of water the human gained.
2. Assume an adult woman weighs 65 kg and an adult man weighs 90 kg. Calculate how much more water, in kilograms, the man has compared to the woman.

For more practice, visit life.msscience.com/math_practice

INTEGRATE Physics

Characteristics of Water

The atoms of a water molecule are arranged in such a way that the molecule has areas with different charges. Water molecules are like magnets. The negative part of a water molecule is attracted to the positive part of another water molecule just like the north pole of a magnet is attracted to the south pole of another magnet. This attraction, or force, between water molecules is why a film forms on the surface of water. The film is strong enough to support small insects because the forces between water molecules are stronger than the force of gravity on the insect.

When thermal energy is added to any substance, its molecules begin to move faster. Because water molecules are so strongly attracted to each other, the temperature of water changes slowly. The large percentage of water in living things acts like an insulator. The water in a cell helps keep its temperature constant, which allows life-sustaining chemical reactions to take place.

You've seen ice floating on water. When water freezes, ice crystals form. In the crystals, each water molecule is spaced at a certain distance from all the others. Because this distance is greater in frozen water than in liquid water, ice floats on water. Bodies of water freeze from the top down. The floating ice provides insulation from extremely cold temperatures and allows living things to survive in the cold water under the ice.

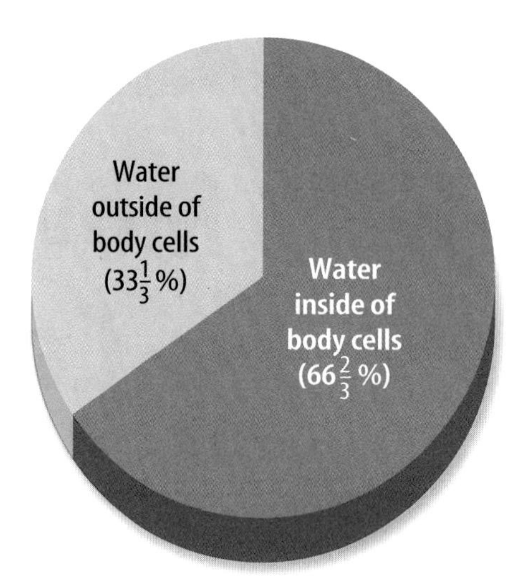

Figure 5 About two-thirds of your body's water is located within your body's cells. Water helps maintain the cells' shapes and sizes. One-third of your body's water is outside of your body's cells.

section 1 review

Summary

The Nature of Matter

- Atoms are made up of protons, neutrons, and electrons.
- Elements are made up of only one kind of atom.
- Compounds are made up of two or more elements.

Mixtures

- Solutions are made of two or more substances and are mixed evenly, whereas substances in suspension eventually will sink to the bottom.

Organic Compounds

- All living things contain organic compounds.

Inorganic Compounds

- Water is one of the most important inorganic compounds for living things.

Self Check

1. **Compare and contrast** atoms and molecules.
2. **Describe** the differences between an organic and an inorganic compound. Given an example of each type of compound.
3. **List** the four types of organic compounds found in all living things.
4. **Infer** why life as we know it depends on water.
5. **Think Critically** If you mix salt, sand, and sugar with water in a small jar, will the resulting mixture be a suspension, a solution, or both?

Applying Skills

6. **Interpret** Carefully observe **Figure 1** and determine how many protons, neutrons, and electrons an atom of oxygen has.

section 2

Moving Cellular Materials

as you read

What You'll Learn

- **Describe** the function of a selectively permeable membrane.
- **Explain** how the processes of diffusion and osmosis move molecules in living cells.
- **Explain** how passive transport and active transport differ.

Why It's Important

Cell membranes control the substances that enter and leave the cells in your body.

Review Vocabulary

cytoplasm: constantly moving gel-like mixture inside the cell membrane that contains hereditary material and is the location of most of a cell's life process

New Vocabulary

- passive transport
- diffusion
- equilibrium
- osmosis
- active transport
- endocytosis
- exocytosis

Passive Transport

"Close that window. Do you want to let in all the bugs and leaves?" How do you prevent unwanted things from coming through the window? As seen in **Figure 6,** a window screen provides the protection needed to keep unwanted things outside. It also allows some things to pass into or out of the room like air, unpleasant odors, or smoke.

Cells take in food, oxygen, and other substances from their environments. They also release waste materials into their environments. A cell has a membrane around it that works for a cell like a window screen does for a room. A cell's membrane is selectively permeable (PUR mee uh bul). It allows some things to enter or leave the cell while keeping other things outside or inside the cell. The window screen also is selectively permeable based on the size of its openings.

Things can move through a cell membrane in several ways. Which way things move depends on the size of the molecules or particles, the path taken through the membrane, and whether or not energy is used. The movement of substances through the cell membrane without the input of energy is called **passive transport.** Three types of passive transport can occur. The type depends on what is moving through the cell membrane.

Figure 6 A cell membrane, like a screen, will let some things through more easily than others. Air gets through a screen, but insects are kept out.

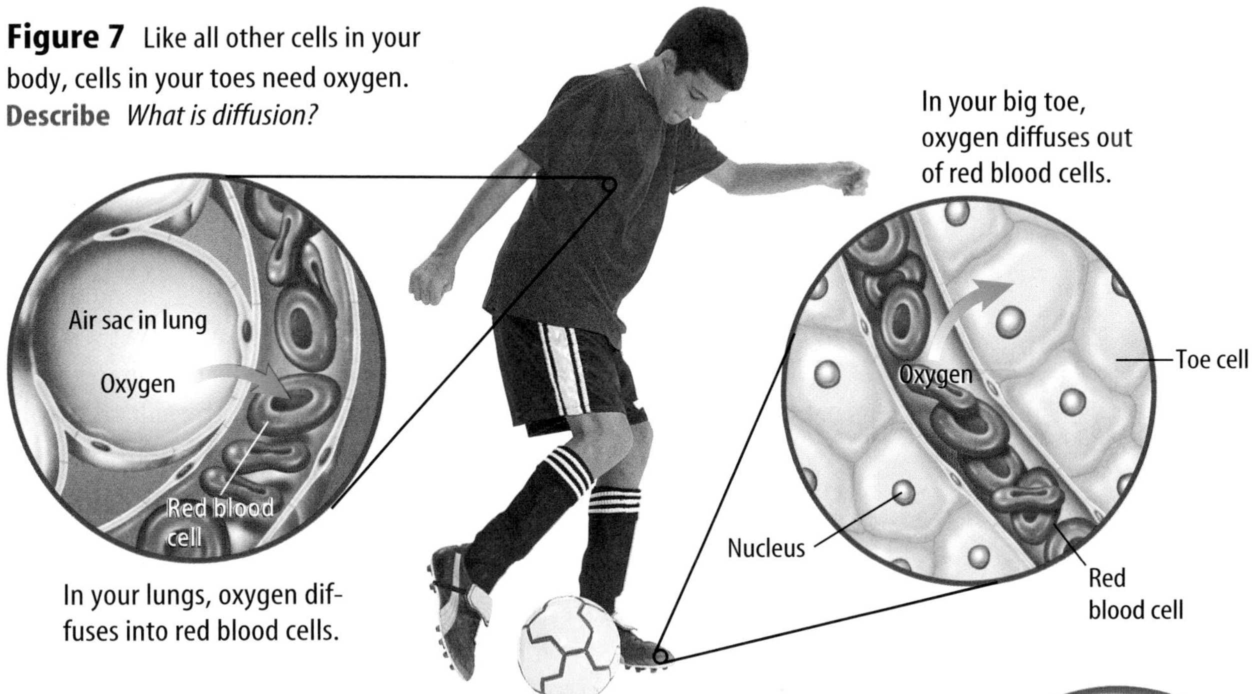

Figure 7 Like all other cells in your body, cells in your toes need oxygen.
Describe *What is diffusion?*

Diffusion Molecules in solids, liquids, and gases move constantly and randomly. You might smell perfume when you sit near or as you walk past someone who is wearing it. This is because perfume molecules randomly move throughout the air. This random movement of molecules from an area where there is relatively more of them into an area where there is relatively fewer of them is called **diffusion.** Diffusion is one type of cellular passive transport. Molecules of a substance will continue to move from one area into another until the relative number of these molecules is equal in the two areas. When this occurs, **equilibrium** is reached and diffusion stops. After equilibrium occurs, it is maintained because molecules continue to move.

Reading Check *What is equilibrium?*

Every cell in your body uses oxygen. When you breathe, how does oxygen get from your lungs to cells in your big toe? Oxygen is carried throughout your body in your blood by the red blood cells. When your blood is pumped from your heart to your lungs, your red blood cells do not contain much oxygen. However, your lungs have more oxygen molecules than your red blood cells do, so the oxygen molecules diffuse into your red blood cells from your lungs, as shown in **Figure 7.** When the blood reaches your big toe, there are more oxygen molecules in your red blood cells than in your big toe cells. The oxygen diffuses from your red blood cells and into your big toe cells, as shown also in **Figure 7.**

Mini LAB

Observing Molecule Movement

Procedure

1. Use two clean glasses of equal size. Label one *Hot,* then fill it until half full with **very warm water.** Label the other *Cold,* then fill it until half full with **cold water. WARNING:** *Do not use boiling hot water.*
2. Add one drop of **food coloring** to each glass. Carefully release the drop just at the water's surface to avoid splashing the water.
3. Immediately observe the water in the glasses. Record your observations and again after 15 min.

Analysis

What is the relationship between temperature and molecule movement?

Try at Home

Osmosis—The Diffusion of Water Remember that water makes up a large part of living matter. Cells contain water and are surrounded by water. Water molecules move by diffusion into and out of cells. The diffusion of water through a cell membrane is called **osmosis.**

If cells weren't surrounded by water that contains few dissolved substances, water inside of cells would diffuse out of them. This is why water left the carrot cells in this chapter's Launch Lab. Because there were relatively fewer water molecules in the salt solution around the carrot cells than in the carrot cells, water moved out of the cells and into the salt solution.

Losing water from a plant cell causes its cell membrane to come away from its cell wall, as shown on the left in **Figure 8.** This reduces pressure against its cell wall, and a plant cell becomes limp. If the carrot sticks were taken out of salt water and put in pure water, the water around the cells would move into them and they would fill with water. Their cell membranes would press against their cell walls, as shown on the right in **Figure 8,** pressure would increase, and the cells would become firm. That is why the carrot sticks would be crisp again.

Reading Check *Why do carrots in salt water become limp?*

Osmosis also takes place in animal cells. If animal cells were placed in pure water, they too would swell up. However, animal cells are different from plant cells. Just like an overfilled water balloon, animal cells will burst if too much water enters the cell.

Figure 8 Cells respond to differences between the amount of water inside and outside the cell.
Define *What is osmosis?*

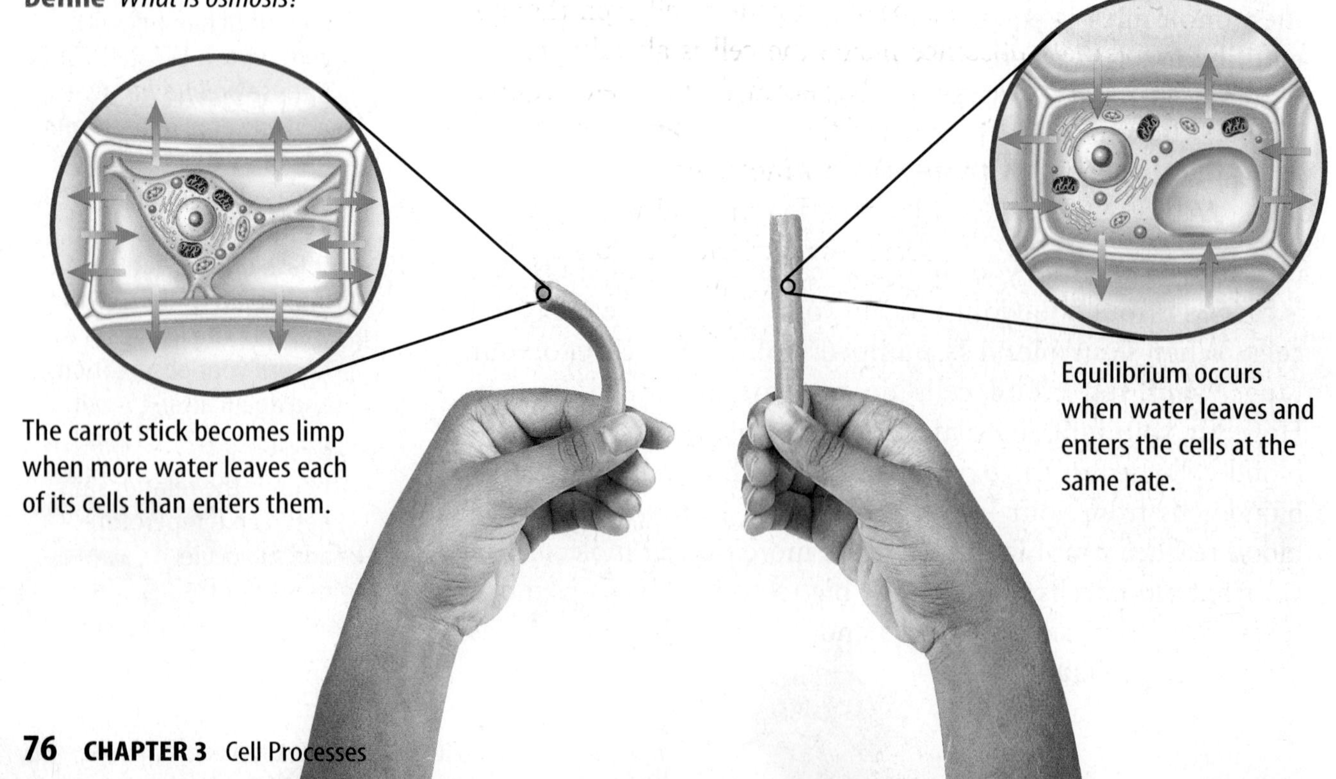

The carrot stick becomes limp when more water leaves each of its cells than enters them.

Equilibrium occurs when water leaves and enters the cells at the same rate.

Facilitated Diffusion Cells take in many substances. Some substances pass easily through the cell membrane by diffusion. Other substances, such as sugar molecules, are so large that they can enter the cell only with the help of molecules in the cell membrane called transport proteins. This process, a type of passive transport, is known as facilitated diffusion. Have you ever used the drive through at a fast-food restaurant to get your meal? The transport proteins in the cell membrane are like the drive-through window at the restaurant. The window lets you get food out of the restaurant and put money into the restaurant. Similarly, transport proteins are used to move substances into and out of the cell.

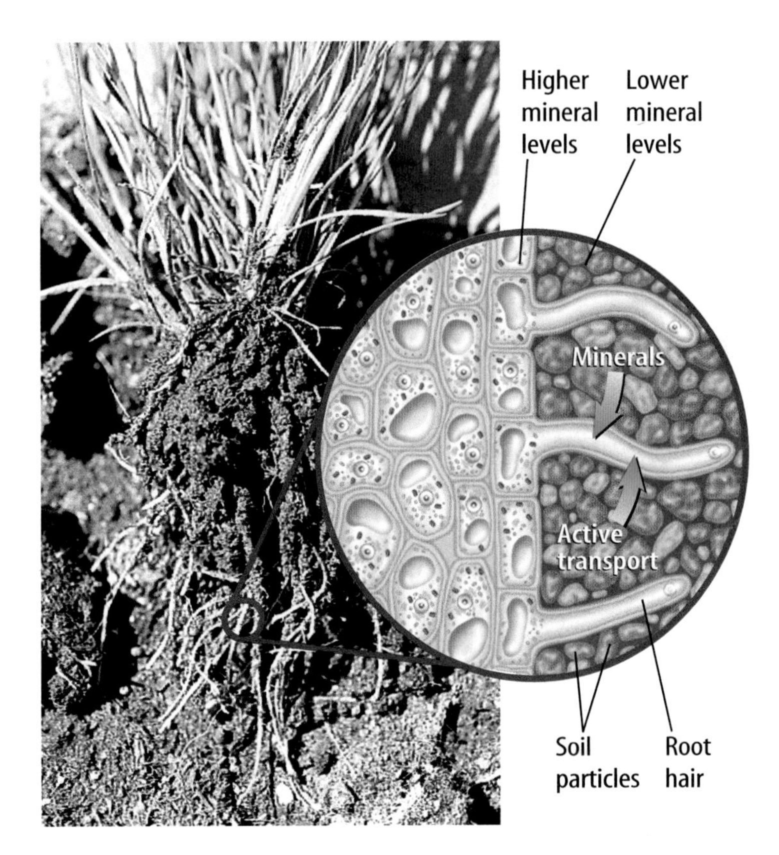

Figure 9 Some root cells have extensions called root hairs that may be 5 mm to 8 mm long. Minerals are taken in by active transport through the cell membranes of root hairs.

Active Transport

Imagine that a football game is over and you leave the stadium. As soon as you get outside of the stadium, you remember that you left your jacket on your seat. Now you have to move against the crowd coming out of the stadium to get back in to get your jacket. Which required more energy—leaving the stadium with the crowd or going back to get your jacket? Something similar to this happens in cells.

Sometimes, a substance is needed inside a cell even though the amount of that substance inside the cell is already greater than the amount outside the cell. For example, root cells require minerals from the soil. The roots of the plant in **Figure 9** already might contain more of those mineral molecules than the surrounding soil does. The tendency is for mineral molecules to move out of the root by diffusion or facilitated diffusion. But they need to move back across the cell membrane and into the cell just like you had to move back into the stadium. When an input of energy is required to move materials through a cell membrane, **active transport** takes place.

Active transport involves transport proteins, just as facilitated diffusion does. In active transport, a transport protein binds with the needed particle and cellular energy is used to move it through the cell membrane. When the particle is released, the transport protein can move another needed particle through the membrane.

Transport Proteins Your health depends on transport proteins. Sometimes transport proteins are missing or do not function correctly. What would happen if proteins that transport cholesterol across membranes were missing? Cholesterol is an important lipid used by your cells. Write your ideas in your Science Journal.

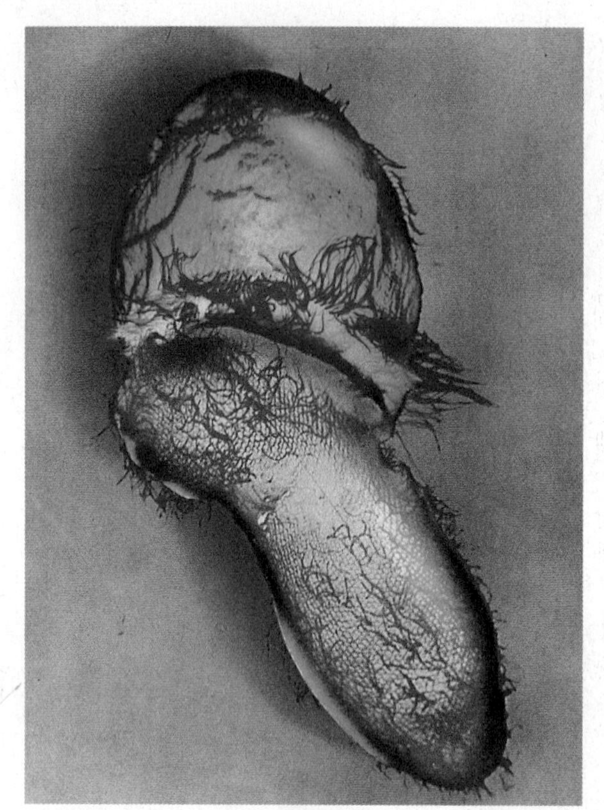

Color-enhanced TEM Magnification: 1,400×

Figure 10 One-celled organisms like this egg-shaped one can take in other one-celled organisms using endocytosis.

Endocytosis and Exocytosis

Some molecules and particles are too large to move by diffusion or to use the cell membrane's transport proteins. Large protein molecules and bacteria, for example, can enter a cell when they are surrounded by the cell membrane. The cell membrane folds in on itself, enclosing the item in a sphere called a vesicle. Vesicles are transport and storage structures in a cell's cytoplasm. The sphere pinches off, and the resulting vesicle enters the cytoplasm. A similar thing happens when you poke your finger into a partially inflated balloon. Your finger is surrounded by the balloon in much the same way that the protein molecule is surrounded by the cell membrane. This process of taking substances into a cell by surrounding it with the cell membrane is called **endocytosis** (en duh si TOH sus). Some one-celled organisms, as shown in **Figure 10,** take in food this way.

The contents of a vesicle can be released by a cell using the process called **exocytosis** (ek soh si TOH sus). Exocytosis occurs in the opposite way that endocytosis does. A vesicle's membrane fuses with a cell's membrane, and the vesicle's contents are released. Cells in your stomach use this process to release chemicals that help digest food. The ways that materials can enter or leave a cell are summarized in **Figure 11.**

section 2 review

Summary

Passive Transport

- Cells take in substances and release waste through their cell membranes.
- Facilitated diffusion and osmosis are types of passive transport.

Active Transport

- Transport proteins are involved in active transport.
- Transport proteins can be reused many times.

Endocytosis and Exocytosis

- Vesicles are formed when a cell takes in a substance by endocytosis.
- Contents of a vesicle are released to the outside of a cell by exocytosis.

Self Check

1. **Describe** how cell membranes are selectively permeable.
2. **Compare and contrast** the processes of osmosis and diffusion.
3. **Infer** why endocytosis and exocytosis are important processes to cells.
4. **Think Critically** Why are fresh fruits and vegetables sprinkled with water at produce markets?

Applying Skills

5. **Communicate** Seawater is saltier than tap water. Explain why drinking large amounts of seawater would be dangerous for humans.

life.msscience.com/self_check_quiz

NATIONAL GEOGRAPHIC

VISUALIZING CELL MEMBRANE TRANSPORT

Figure 11

A flexible yet strong layer, the cell membrane is built of two layers of lipids (gold) pierced by protein "passageways" (purple). Molecules can enter or exit the cell by slipping between the lipids or through the protein passageways. Substances that cannot enter or exit the cell in these ways may be surrounded by the membrane and drawn into or expelled from the cell.

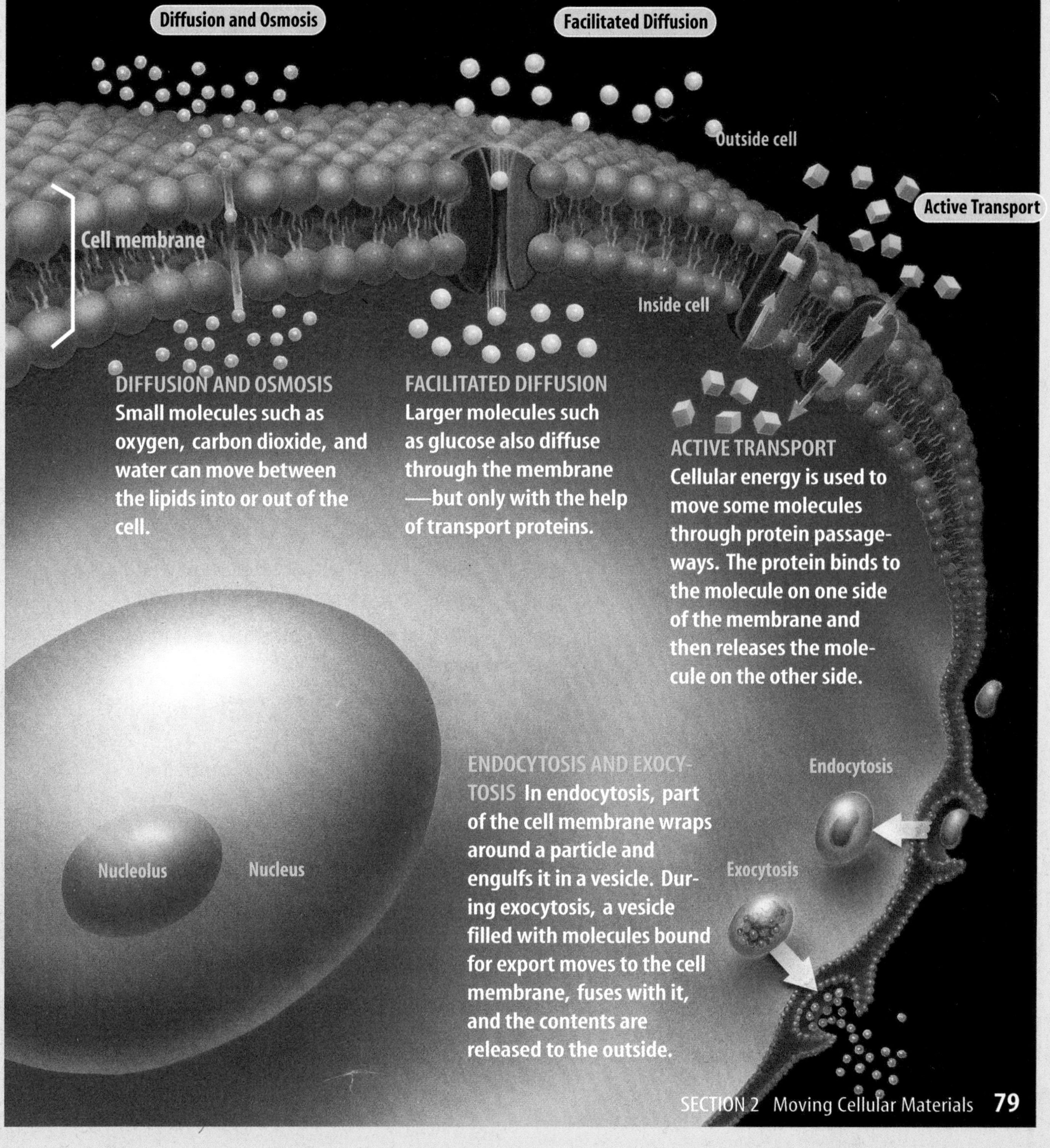

DIFFUSION AND OSMOSIS Small molecules such as oxygen, carbon dioxide, and water can move between the lipids into or out of the cell.

FACILITATED DIFFUSION Larger molecules such as glucose also diffuse through the membrane —but only with the help of transport proteins.

ACTIVE TRANSPORT Cellular energy is used to move some molecules through protein passageways. The protein binds to the molecule on one side of the membrane and then releases the molecule on the other side.

ENDOCYTOSIS AND EXOCYTOSIS In endocytosis, part of the cell membrane wraps around a particle and engulfs it in a vesicle. During exocytosis, a vesicle filled with molecules bound for export moves to the cell membrane, fuses with it, and the contents are released to the outside.

Observing Osmosis

It is difficult to observe osmosis in cells because most cells are so small. However, a few cells can be seen without the aid of a microscope. Try this lab to observe osmosis.

Real-World Question

How does osmosis occur in an egg cell?

Materials

unshelled egg*
balance
spoon
distilled water (250 mL)
light corn syrup (250 mL)
500-mL container

**an egg whose shell has been dissolved by vinegar*

Goals

- **Observe** osmosis in an egg cell.
- **Determine** what affects osmosis.

Safety Precautions

WARNING: *Eggs may contain bacteria. Avoid touching your face.*

Procedure

1. Copy the table below into your Science Journal and use it to record your data.

Egg Mass Data

	Beginning Egg Mass	Egg Mass After Two Days
Distilled water	Do not write in this book.	
Corn syrup		

2. Obtain an unshelled egg from your teacher. Handle the egg gently. Use a balance to find the egg's mass and record it in the table.
3. Place the egg in the container and add enough distilled water to cover it.
4. **Observe** the egg after 30 min, one day, and two days. After each observation, record the egg's appearance in your Science Journal.
5. After day two, remove the egg with a spoon and allow it to drain. Find the egg's mass and record it in the table.
6. Empty the container, then put the egg back in. Now add enough corn syrup to cover it. Repeat steps 4 and 5.

Conclude and Apply

1. **Explain** the difference between what happened to the egg in water and in corn syrup.
2. **Calculate** the mass of water that moved into and out of the egg.
3. **Hypothesize** why you used an unshelled egg for this investigation.
4. **Infer** what part of the egg controlled water's movement into and out of the egg.

Communicating Your Data

Compare your conclusions with those of other students in your class. **For more help, refer to the Science Skill Handbook.**

section 3

Energy for Life

Trapping and Using Energy

Think of all the energy that players use in a soccer game. Where does the energy come from? The simplest answer is "from the food they eat." The chemical energy stored in food molecules is changed inside of cells into forms needed to perform all the activities necessary for life. In every cell, these changes involve chemical reactions. All of the activities of an organism involve chemical reactions in some way. The total of all chemical reactions in an organism is called **metabolism.**

The chemical reactions of metabolism need enzymes. What do enzymes do? Suppose you are hungry and decide to open a can of spaghetti. You use a can opener to open the can. Without a can opener, the spaghetti is unusable. The can of spaghetti changes because of the can opener, but the can opener does not change. The can opener can be used again later to open more cans of spaghetti. Enzymes in cells work something like can openers. The enzyme, like the can opener, causes a change, but the enzyme is not changed and can be used again, as shown in **Figure 12.** Unlike the can opener, which can only cause things to come apart, enzymes also can cause molecules to join. Without the right enzyme, a chemical reaction in a cell cannot take place. Each chemical reaction in a cell requires a specific enzyme.

as you read

What **You'll Learn**

- **List** the differences between producers and consumers.
- **Explain** how the processes of photosynthesis and cellular respiration store and release energy.
- **Describe** how cells get energy from glucose through fermentation.

Why **It's Important**

Because of photosynthesis and cellular respiration, you use the Sun's energy.

Review Vocabulary

mitochondrion: cell organelle that breaks down lipids and carbohydrates and releases energy

New Vocabulary

- metabolism
- photosynthesis
- cellular respiration
- fermentation

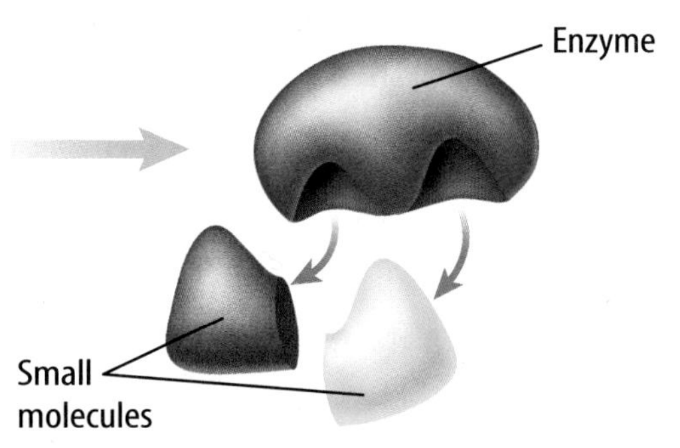

The enzyme attaches to the large molecule it will help change.

The enzyme causes the larger molecule to break down into two smaller molecules. The enzyme is not changed and can be used again.

Figure 12 Enzymes are needed for most chemical reactions that take place in cells.
Determine *What is the sum of all chemical reactions in an organism called?*

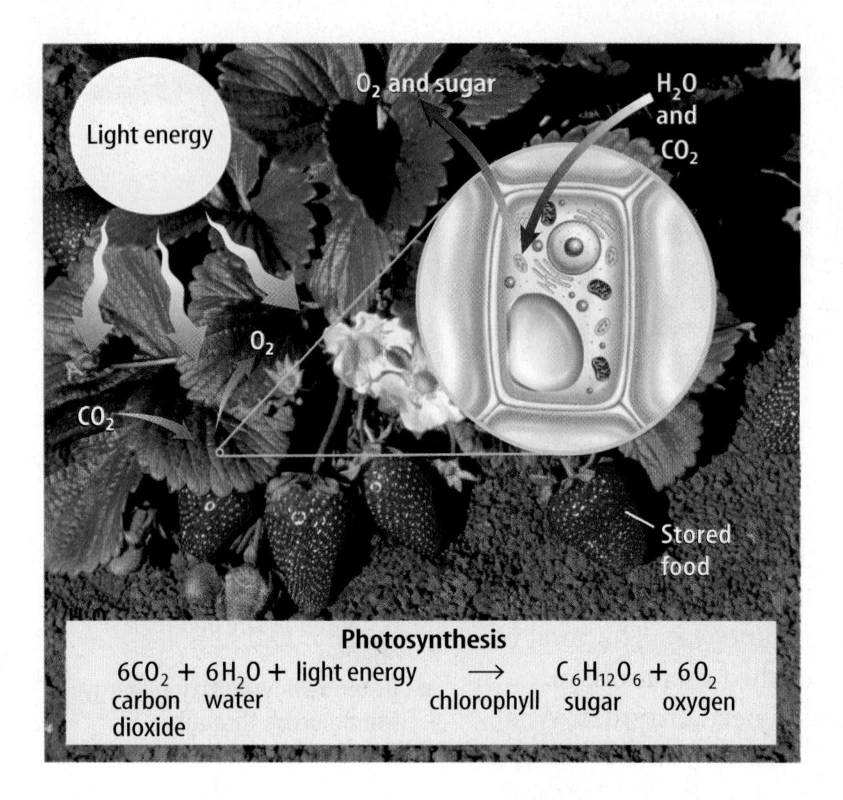

Figure 13 Plants use photosynthesis to make food.
Determine *According to the chemical equation, what raw materials would the plant in the photo need for photosynthesis?*

Photosynthesis Living things are divided into two groups—producers and consumers—based on how they obtain their food. Organisms that make their own food, such as plants, are called producers. Organisms that cannot make their own food are called consumers.

If you have ever walked barefoot across a sidewalk on a sunny summer day, you probably moved quickly because the sidewalk was hot. Sunlight energy was converted into thermal energy and heated the sidewalk. Plants and many other producers can convert light energy into another kind of energy—chemical energy. The process they use is called photosynthesis. During **photosynthesis,** producers use light energy and make sugars, which can be used as food.

Producing Carbohydrates Producers that use photosynthesis are usually green because they contain a green pigment called chlorophyll (KLOR uh fihl). Chlorophyll and other pigments are used in photosynthesis to capture light energy. In plant cells, these pigments are found in chloroplasts.

The captured light energy powers chemical reactions that produce sugar and oxygen from the raw materials, carbon dioxide and water. For plants, the raw materials come from air and soil. Some of the captured light energy is stored in the chemical bonds that hold the sugar molecules together. **Figure 13** shows what happens during photosynthesis in a plant. Enzymes also are needed before these reactions can occur.

Storing Carbohydrates Plants make more sugar during photosynthesis than they need for survival. Excess sugar is changed and stored as starches or used to make other carbohydrates. Plants use these carbohydrates as food for growth, maintenance, and reproduction.

Why is photosynthesis important to consumers? Do you eat apples? Apple trees use photosynthesis to produce apples. Do you like cheese? Some cheese comes from milk, which is produced by cows that eat plants. Consumers take in food by eating producers or other consumers. No matter what you eat, photosynthesis was involved directly or indirectly in its production.

Cellular Respiration Imagine that you get up late for school. You dress quickly, then run three blocks to school. When you get to school, you feel hot and are breathing fast. Why? Your muscle cells use a lot of energy when you run. To get this energy, muscle cells break down food. Some of the energy is used when you move and some of it becomes thermal energy, which is why you feel warm or hot. Most cells also need oxygen to break down food. You were breathing fast because your body was working to get oxygen to your muscles. Your muscle cells were using the oxygen for the process of cellular respiration. During **cellular respiration,** chemical reactions occur that break down food molecules into simpler substances and release their stored energy. Just as in photosynthesis, enzymes are needed for the chemical reactions of cellular respiration.

What must happen to food molecules for respiration to take place?

Microbiologist Dr. Harold Amos is a microbiologist who has studied cell processes in bacteria and mammals. He has a medical degree and a doctorate in bacteriology and immunology. He has also received many awards for his scientific work and his contributions to the careers of other scientists. Research microbiology careers, and write what you find in your Science Journal.

Breaking Down Carbohydrates The food molecules most easily broken down by cells are carbohydrates. Cellular respiration of carbohydrates begins in the cytoplasm of the cell. The carbohydrates are broken down into glucose molecules. Each glucose molecule is broken down further into two simpler molecules. As the glucose molecules are broken down, energy is released.

The two simpler molecules are broken down again. This breakdown occurs in the mitochondria of the cells of plants, animals, fungi, and many other organisms. This process uses oxygen, releases much more energy, and produces carbon dioxide and water as wastes. When you exhale, you breathe out carbon dioxide and some of the water.

Cellular respiration occurs in the cells of many living things. **Figure 14** shows how it occurs in one consumer. As you are reading this section of the chapter, millions of cells in your body are breaking down glucose, releasing energy, and producing carbon dioxide and water.

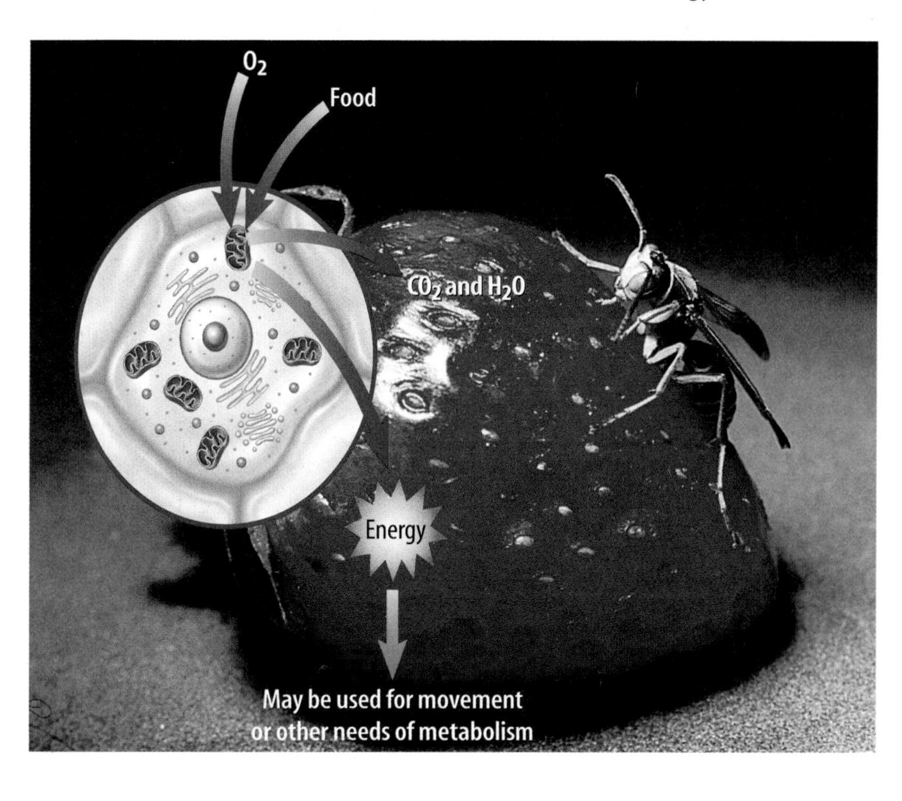

Figure 14 Producers and consumers carry on cellular respiration that releases energy from foods.

Fermentation Remember imagining you were late and had to run to school? During your run, your muscle cells might not have received enough oxygen, even though you were breathing rapidly. When cells do not have enough oxygen for cellular respiration, they use a process called **fermentation** to release some of the energy stored in glucose molecules.

Like cellular respiration, fermentation begins in the cytoplasm. Again, as the glucose molecules are broken down, energy is released. But the simple molecules from the breakdown of glucose do not move into the mitochondria. Instead, more chemical reactions occur in the cytoplasm. These reactions release some energy and produce wastes. Depending on the type of cell, the wastes may be lactic acid or alcohol and carbon dioxide, as shown in **Figure 15.** Your muscle cells can use fermentation to change the simple molecules into lactic acid while releasing energy. The presence of lactic acid is why your muscles might feel stiff and sore after exercising.

Science Online

Topic: Beneficial Microorganisms

Visit life.msscience.com for Web links to information about how microorganisms are used to produce many useful products.

Activity Find three other ways that microorganisms are beneficial.

Reading Check *Where in a cell does fermentation take place?*

Some microscopic organisms, such as bacteria, carry out fermentation and make lactic acid. Some of these organisms are used to produce yogurt and some cheeses. These organisms break down a sugar in milk and release energy. The lactic acid produced causes the milk to become more solid and gives these foods some of their flavor.

Have you ever used yeast to make bread? Yeasts are one-celled living organisms. Yeast cells use fermentation and break down sugar in bread dough. They produce alcohol and carbon dioxide as wastes. The carbon dioxide waste is a gas that makes bread dough rise before it is baked. The alcohol is lost as the bread bakes.

Figure 15 Organisms that use fermentation produce several different wastes.

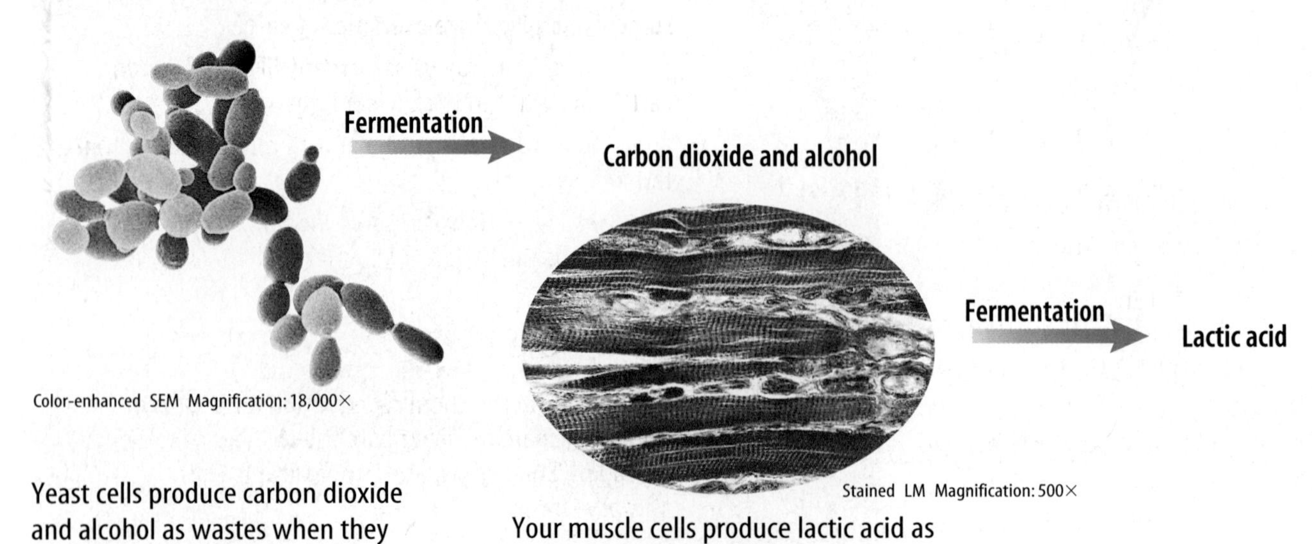

Yeast cells produce carbon dioxide and alcohol as wastes when they undergo fermentation.

Your muscle cells produce lactic acid as a waste when they undergo fermentation.

CO_2, H_2O

Sugars, O_2

Figure 16 The chemical reactions of photosynthesis and cellular respiration could not take place without each other.

Related Processes How are photosynthesis, cellular respiration, and fermentation related? Some producers use photosynthesis to make food. All living things use respiration or fermentation to release energy stored in food. If you think carefully about what happens during photosynthesis and cellular respiration, you will see that what is produced in one is used in the other, as shown in **Figure 16.** These two processes are almost the opposite of each other. Photosynthesis produces sugars and oxygen, and cellular respiration uses these products. The carbon dioxide and water produced during cellular respiration are used during photosynthesis. Most life would not be possible without these important chemical reactions.

section 3 review

Summary

Trapping and Using Energy

- Metabolism is the total of all chemical reactions in an organism.
- During photosynthesis, light energy is transformed into chemical energy.
- Chlorophyll and other pigments capture light energy.
- Consumers take in energy by eating producers and other consumers.
- Living cells use oxygen and break down glucose that releases energy. This is called cellular respiration.
- Fermentation releases energy without oxygen.
- Without photosynthesis and cellular respiration and fermentation, most life would not be possible.

Self Check

1. **Explain** the difference between producers and consumers and give three examples of each.
2. **Infer** how the energy used by many living things on Earth can be traced back to sunlight.
3. **Compare and contrast** cellular respiration and fermentation.
4. **Think Critically** How can some indoor plants help to improve the quality of air in a room?

Applying Math

5. **Solve** Refer to the chemical equation for photosynthesis. Calculate then compare the number of carbon, hydrogen, and oxygen atoms before and after photosynthesis.

Photosynthesis and Cellular Respiration

Goals

- **Observe** green water plants in the light and dark.
- **Determine** whether plants carry on photosynthesis and cellular respiration.

Materials

16-mm test tubes (3)
150-mm test tubes with stoppers (4)
small, clear-glass baby food jars with lids (4)
test-tube rack
stirring rod
scissors
carbonated water (5 mL)
bromthymol blue solution in dropper bottle
aged tap water (20 mL)
distilled water (20 mL)
sprigs of *Elodea* (2)
other water plants
Alternate materials

Safety Precautions

WARNING: *Wear splash-proof safety goggles to protect eyes from hazardous chemicals.*

Real-World Question

Every living cell carries on many chemical processes. Two important chemical processes are cellular respiration and photosynthesis. All cells, including the ones in your body, carry on cellular respiration. However, some plant cells can carry on both processes. In this experiment you will investigate when these processes occur in plant cells. How could you find out when plants were using these processes? Are the products of photosynthesis and cellular respiration the same? When do plants carry on photosynthesis and cellular respiration?

Procedure

1. In your Science Journal, copy and complete the test-tube data table as you perform this lab.

Test-Tube Data

Test Tube	Color at Start	Color After 30 Minutes
1		
2	Do not write in this book.	
3		
4		

2. Label each test tube using the numbers *1, 2, 3,* and *4.* Pour 5 mL of aged tap water into each test tube.
3. Add 10 drops of carbonated water to test tubes *1* and *2.*
4. Add 10 drops of bromthymol blue to all of the test tubes. Bromthymol blue turns green to yellow in the presence of an acid.
5. Cut two 10-cm sprigs of *Elodea.* Place one sprig in test tube *1* and one sprig in test tube *3.* Stopper all test tubes.
6. Place test tubes *1* and *2* in bright light. Place tubes *3* and *4* in the dark. Observe the test tubes for 45 min or until the color changes. Record the color of each of the four test tubes.

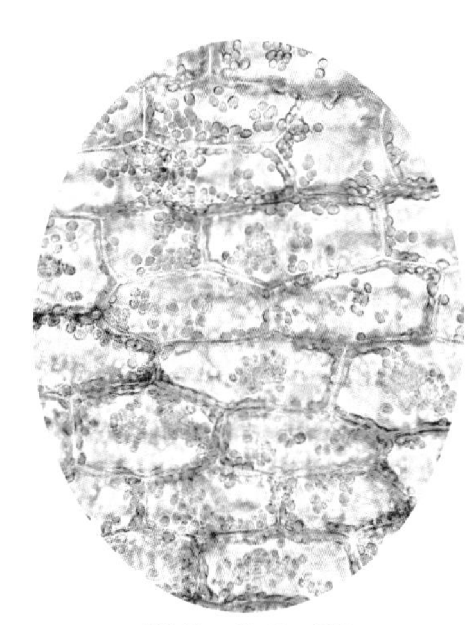

LM Magnification: 225×

Analyze Your Data

1. **Identify** what is indicated by the color of the water in all four test tubes at the start of the activity.
2. **Infer** what process occurred in the test tube or tubes that changed color after 30 min.

Conclude and Apply

1. **Describe** the purpose of test tubes *2* and *4* in this experiment.
2. **Explain** whether or not the results of this experiment show that photosynthesis and cellular respiration occur in plants.

Choose one of the following activities to **communicate** your data. Prepare an oral presentation that explains how the experiment showed the differences between products of photosynthesis and cellular respiration. Draw a cartoon strip to **explain** what you did in this experiment. Use each panel to show a different step. **For more help, refer to the Science Skill Handbook.**

from "Tulip"

by Penny Harter

I watched its first green push
through bare dirt, where the builders
had dropped boards, shingles,
plaster—
killing everything.
I could not recall what grew there,
what returned each spring,
but the leaves looked tulip,
and one morning it arrived,
a scarlet slash against the aluminum siding.
Mornings, on the way to my car,
I bow to the still bell
of its closed petals; evenings,
it greets me, light ringing
at the end of my driveway.
Sometimes I kneel
to stare into the yellow throat . . . .
It opens and closes my days.
It has made me weak with love . . . .

Understanding Literature

Personification Using human traits or emotions to describe an idea, animal, or inanimate object is called personification. When the poet writes that the tulip has a "yellow throat," she uses personification. Where else does the poet use personification?

Respond to the Reading

1. Why do you suppose the tulip survived the builders' abuse?
2. What is the yellow throat that the narrator is staring into?
3. **Linking Science and Writing** Keep a gardener's journal of a plant for a month, describing weekly the plant's condition, size, health, color, and other physical qualities.

Because most chemical reactions in plants take place in water, plants must have water in order to grow. The water carries nutrients and minerals from the soil into the plant. The process of active transport allows needed nutrients to enter the roots. The cell membranes of root cells contain proteins that bind with the needed nutrients. Cellular energy is used to move these nutrients through the cell membrane.

Reviewing Main Ideas

Section 1 Chemistry of Life

1. Matter is anything that has mass and takes up space.
2. Energy in matter is in the chemical bonds that hold matter together.
3. All organic compounds contain the elements hydrogen and carbon. The organic compounds in living things are carbohydrates, lipids, proteins, and nucleic acids.
4. Organic and inorganic compounds are important to living things.

Section 2 Moving Cellular Materials

1. The selectively permeable cell membrane controls which molecules can pass into and out of the cell.
2. In diffusion, molecules move from areas where there are relatively more of them to areas where there are relatively fewer of them.
3. Osmosis is the diffusion of water through a cell membrane.
4. Cells use energy to move molecules by active transport but do not use energy for passive transport.
5. Cells move large particles through cell membranes by endocytosis and exocytosis.

Section 3 Energy for Life

1. Photosynthesis is the process by which some producers change light energy into chemical energy.
2. Cellular respiration uses oxygen, releases the energy in food molecules, and produces waste carbon dioxide and water.
3. Some one-celled organisms and cells that lack oxygen use fermentation and release small amounts of energy from glucose. Wastes such as alcohol, carbon dioxide, and lactic acid are produced.

Visualizing Main Ideas

Copy and complete the following table on energy processes.

Energy Processes

	Photosynthesis	Cellular Respiration	Fermentation
Energy source		food (glucose)	food (glucose)
In plant and animal cells, occurs in		Do not write in this book.	
Reactants are			
Products are			

chapter 3 Review

Using Vocabulary

active transport p. 77
cellular respiration p. 83
diffusion p. 75
endocytosis p. 78
enzyme p. 71
equilibrium p. 75
exocytosis p. 78
fermentation p. 84
inorganic compound p. 71
metabolism p. 81
mixture p. 69
organic compound p. 70
osmosis p. 76
passive transport p. 74
photosynthesis p. 82

Use what you know about the vocabulary words to answer the following questions.

1. What is the diffusion of water called?
2. What type of protein regulates nearly all chemical reactions in cells?
3. How do large food particles enter an amoeba?
4. What type of compound is water?
5. What process is used by some producers to convert light energy into chemical energy?
6. What type of compounds always contain carbon and hydrogen?
7. What process uses oxygen to break down glucose?
8. What is the total of all chemical reactions in an organism called?

Checking Concepts

Choose the word or phrase that best answers the question.

9. What is it called when cells use energy to move molecules?
   A) diffusion **C)** active transport
   B) osmosis **D)** passive transport

Use the photo below to answer question 10.

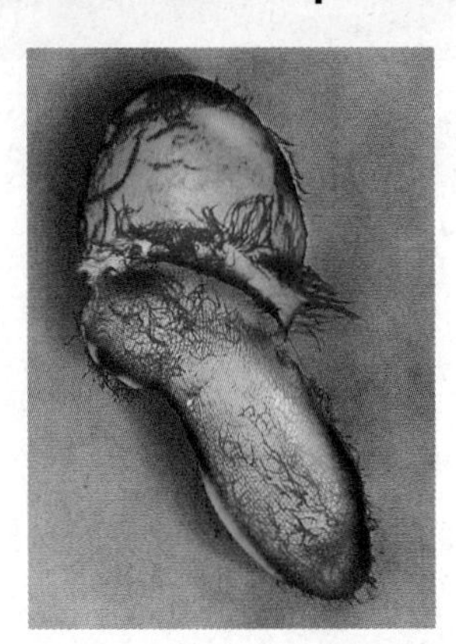

10. What cell process is occurring in the photo?
    A) osmosis **C)** exocytosis
    B) endocytosis **D)** diffusion
11. What occurs when the number of molecules of a substance is equal in two areas?
    A) equilibrium **C)** fermentation
    B) metabolism **D)** cellular respiration
12. Which substance is an example of a carbohydrate?
    A) enzyme **C)** wax
    B) sugar **D)** DNA
13. What is RNA an example of?
    A) carbon dioxide **C)** lipid
    B) water **D)** nucleic acid
14. What organic molecule stores the greatest amount of energy?
    A) carbohydrate **C)** lipid
    B) water **D)** nucleic acid
15. Which formula is an example of an organic compound?
    A) $C_6H_{12}O_6$ **C)** H_2O
    B) NO_2 **D)** O_2
16. What are organisms that cannot make their own food called?
    A) biodegradables **C)** consumers
    B) producers **D)** enzymes

Science Online life.msscience.com/vocabulary_puzzlemaker

Thinking Critically

17. **Concept Map** Copy and complete the events-chain concept map to sequence the following parts of matter from smallest to largest: *atom, electron,* and *compound.*

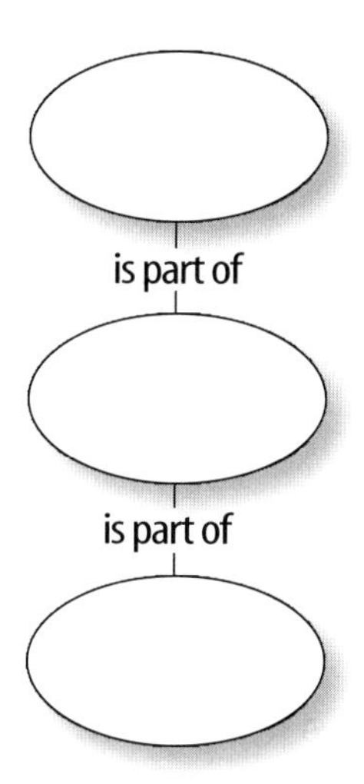

Use the table below to answer question 18.

Photosynthesis in Water Plants

Beaker Number	Distance from Light (cm)	Bubbles per Minute
1	10	45
2	30	30
3	50	19
4	70	6
5	100	1

18. **Interpret Data** Water plants were placed at different distances from a light source. Bubbles coming from the plants were counted to measure the rate of photosynthesis. What can you say about how the distance from the light affected the rate?

19. **Infer** why, in snowy places, salt is used to melt ice on the roads. Explain what could happen to many roadside plants as a result.

20. **Draw a conclusion** about why sugar dissolves faster in hot tea than in iced tea.

21. **Predict** what would happen to the consumers in a lake if all the producers died.

22. **Explain** how meat tenderizers affect meat.

23. **Form a hypothesis** about what will happen to wilted celery when placed in a glass of plain water.

Performance Activities

24. **Puzzle** Make a crossword puzzle with words describing ways substances are transported across cell membranes. Use the following words in your puzzle: *diffusion, osmosis, facilitated diffusion, active transport, endocytosis,* and *exocytosis.* Make sure your clues give good descriptions of each transport method.

Applying Math

25. **Light and Photosynthesis** Using the data from question 18, make a line graph that shows the relationship between the rate of photosynthesis and the distance from light.

26. **Importance of Water** Assume the brain is 70% water. If the average adult human brain weighs 1.4 kg, how many kilograms of water does it contain?

Use the equation below to answer question 27.

Photosynthesis

$$6CO_2 + 6H_2O + \text{light energy} \xrightarrow{\text{chlorophyll}} C_6H_{12}O_6 + 6O_2$$

carbon dioxide, water → sugar, oxygen

27. **Photosynthesis** Refer to the chemical equation above. If 18 CO_2 molecules and 18 H_2O molecules are used with light energy to make sugar, how many sugar molecules will be produced? How many oxygen molecules will be produced?

Part 1 Multiple Choice

Record your answers on the answer sheet provided by your teacher or on a sheet of paper.

1. Which describes a substance that is made up of only one kind of atom and cannot be broken down by chemical reactions?
   A. electron **C.** element
   B. carbohydrate **D.** molecule

Use the illustration below to answer questions 2 and 3.

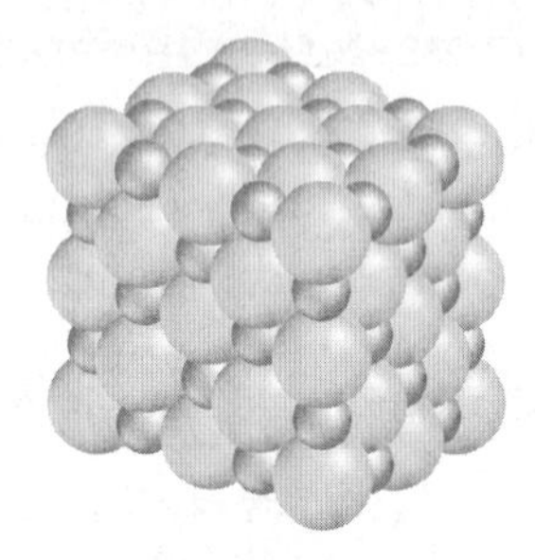

2. What kind of chemical compound do salt and water form?
   A. covalent
   B. ionic
   C. solution
   D. lipid
3. Salt is very important in the human body. What kind of compound is salt?
   A. organic
   B. carbohydrate
   C. protein
   D. inorganic
4. A cell that contains 40% water is placed in a solution that is 20% water. The cell and the solution will reach equilibrium when they both contain how much water?
   A. 30% **C.** 60%
   B. 40% **D.** 20%
5. All chemical reactions in living things take place in what kind of a solution?
   A. protein **C.** gas
   B. water **D.** solid
6. What is the sum of all the chemical reactions in an organism?
   A. cellular respiration **C.** fermentation
   B. metabolism **D.** endocytosis
7. What is needed for all chemical reactions in cells?
   A. enzymes **C.** DNA
   B. lipids **D.** cell membrane
8. What process produces the carbon dioxide that you exhale?
   A. osmosis
   B. DNA synthesis
   C. photosynthesis
   D. respiration
9. Which is needed to hold matter together or break it apart?
   A. gas
   B. liquid
   C. energy
   D. temperature

Use the table below to answer question 10.

Cell Substances		
Organic Compound	**Flexibility**	**Found in**
Keratin	Not very flexible	Hair and skin of mammals
Collagen	Not very flexible	Skin, bones, and tendons of mammals
Chitin	Very rigid	Tough outer shell of insects and crabs
Cellulose	Very flexible	Plant cell walls

10. According to this information, which organic compound is the least flexible?
    A. keratin
    B. collagen
    C. chitin
    D. cellulose

Part 2 | Short Response/Grid In

Record your answers on the answer sheet provided by your teacher or on a sheet of paper.

11. Explain the structure of an atom.
12. How does chewing food affect your body's ability to release the chemical energy of the food?
13. Ice fishing is a popular sport in the winter. What properties of water is this sport based on?
14. Explain where the starch in a potato comes from.
15. Does fermentation or cellular respiration release more energy for an athlete's muscles? Which process would be responsible for making muscles sore?

Use the table below to answer question 16.

Classification of Compounds			
Compound	**Organic**	**Inorganic**	**Type of organic compound**
Salt			
Fat			
Skin			
DNA			
Sugar			
Water			
Potassium			

16. Copy and complete the table above. Identify each item as inorganic or organic. If the item is an organic compound further classify it as a protein, carbohydrate, lipid or nucleic acid.
17. Define selectively permeable and discuss why it is important for the cell membrane.
18. What is the source of energy for the photosynthesis reactions and where do they take place in a cell?

Part 3 | Open Ended

Record your answers on a sheet of paper.

19. Give examples of each of the four types of organic molecules and why they are needed in a plant cell.
20. Trace the path of how oxygen molecules are produced in a plant cell to how they are used in human cells.
21. Describe four ways a large or small molecule can cross the cell membrane.
22. Discuss how water is bonded together and the unique properties that result from the bonds.

Use the illustration below to answer question 23.

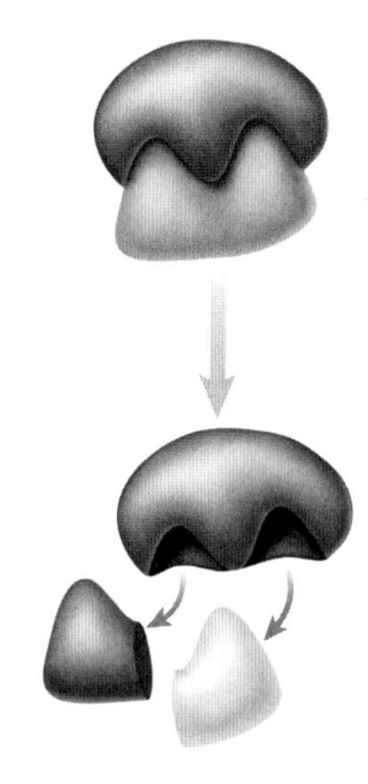

23. Describe in detail what process is taking place in this diagram and its significance for a cell.
24. How do plants use carbon dioxide? Why would plants need oxygen?

Test-Taking Tip

Diagrams Study a diagram carefully, being sure to read all labels and captions.

Cell Reproduction

The BIG Idea

Reproduction must occur for species to survive.

SECTION 1
Cell Division and Mitosis

Main Idea Different organisms can grow, repair damaged cells, and reproduce because of cell division and mitosis.

SECTION 2
Sexual Reproduction and Meiosis

Main Idea Sexual reproduction and meiosis ensure the preservation of species and diversity of life.

SECTION 3
DNA

Main Idea DNA contains the instructions for all life.

Why a turtle, not a chicken?

Several new sweet potato plants can be grown from just one potato, but turtles and most other animals need to have two parents. A cut on your finger heals. How do these things happen? In this chapter, you will find answers to these questions as you learn about cell reproduction.

Science Journal Write three things that you know about how and why cells reproduce.

Start-Up Activities

Infer About Seed Growth

Most flower and vegetable seeds sprout and grow into entire plants in just a few weeks. Although all of the cells in a seed have information and instructions to produce a new plant, only some of the cells in the seed use the information. Where are these cells in seeds? Do the following lab to find out.

1. Carefully split open two bean seeds that have soaked in water overnight.
2. Observe both halves and record your observations.
3. Wrap all four halves in a moist paper towel. Then put them into a self-sealing, plastic bag and seal the bag.
4. Make observations every day for a few days.
5. **Think Critically** Write a paragraph that describes what you observe. Hypothesize which cells in seeds use information about how plants grow.

Preview this chapter's content and activities at life.msscience.com

How and Why Cells Divide Make the following Foldable to help you organize information from the chapter about cell reproduction.

STEP 1 **Draw** a mark at the midpoint of a vertical sheet of paper along the side edge.

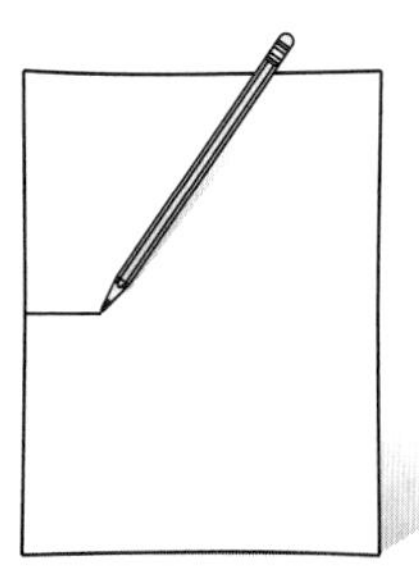

STEP 2 **Turn** the paper horizontally and **fold** the outside edges in to touch at the midpoint mark.

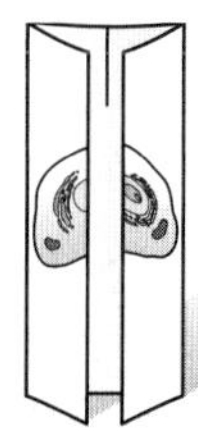

STEP 3 **Use** a pencil to draw a cell on the front of your Foldable as shown.

Analyze As you read the chapter, write under the flaps how cells divide. In the middle section, list why cells divide.

Get Ready to Read

Monitor

① Learn It! An important strategy to help you improve your reading is monitoring, or finding your reading strengths and weaknesses. As you read, monitor yourself to make sure the text makes sense. Discover different monitoring techniques you can use at different times, depending on the type of test and situation.

② Practice It! The paragraph below appears in Section 1. Read the passage and answer the questions that follow. Discuss your answers with other students to see how they monitor their reading.

> Reproduction is the process by which an organism produces others of its same kind. Among living organisms, there are two types of reproduction—sexual and asexual. Sexual reproduction usually requires two organisms. In **asexual reproduction,** a new organism (sometimes more than one) is produced from one organism. The new organism will have hereditary material identical to the hereditary material of the parent organism.
>
> —*from page 101*

- What questions do you still have after reading?
- Do you understand all of the words in the passage?
- Did you have to stop reading often? Is the reading level appropriate for you?

③ Apply It! Identify one paragraph that is difficult to understand. Discuss it with a partner to improve your understanding.

Monitor your reading by slowing down or speeding up depending on your understanding of the text.

Target Your Reading

Use this to focus on the main ideas as you read the chapter.

1. **Before you read** the chapter, respond to the statements below on your worksheet or on a numbered sheet of paper.
   - Write an **A** if you **agree** with the statement.
   - Write a **D** if you **disagree** with the statement.

2. **After you read** the chapter, look back to this page to see if you've changed your mind about any of the statements.
   - If any of your answers changed, explain why.
   - Change any false statements into true statements.
   - Use your revised statements as a study guide.

Before You Read A or D		Statement	After You Read A or D
	1	All cell cycles last the same amount of time.	
	2	Interphase lasts longer than other phases of a cell's cycle.	
	3	Asexual reproduction requires two parents.	
	4	Cell division and mitosis is the same in all organisms.	
	5	Meiosis always happens before fertilization.	
	6	A zygote is the cell formed when an egg and sperm join.	
	7	Diploid cells have pairs of similar chromosomes.	
	8	The exact structure of DNA is unknown.	
	9	A gene is a section of DNA on a chromosome.	
	10	Mistakes in copying DNA result in mutations.	
	11	Budding and regeneration can occur in most organisims.	

Science Online
Print out a worksheet of this page at life.msscience.com

section 1

Cell Division and Mitosis

as you read

What You'll Learn

- **Explain** why mitosis is important.
- **Examine** the steps of mitosis.
- **Compare** mitosis in plant and animal cells.
- **List** two examples of asexual reproduction.

Why It's Important

Your growth, like that of many organisms, depends on cell division.

Review Vocabulary

nucleus: organelle that controls all the activities of a cell and contains hereditary material made of proteins and DNA

New Vocabulary

- mitosis
- chromosome
- asexual reproduction

Why is cell division important?

What do you, an octopus, and an oak tree have in common? You share many characteristics, but an important one is that you are all made of cells—trillions of cells. Where did all of those cells come from? As amazing as it might seem, many organisms start as just one cell. That cell divides and becomes two, two become four, four become eight, and so on. Many-celled organisms, including you, grow because cell division increases the total number of cells in an organism. Even after growth stops, cell division is still important. Every day, billions of red blood cells in your body wear out and are replaced. During the few seconds it takes you to read this sentence, your bone marrow produced about six million red blood cells. Cell division is important to one-celled organisms, too—it's how they reproduce themselves, as shown in **Figure 1.** Cell division isn't as simple as just cutting the cell in half, so how do cells divide?

The Cell Cycle

A living organism has a life cycle. A life cycle begins with the organism's formation, is followed by growth and development, and finally ends in death. Right now, you are in a stage of your life cycle called adolescence, which is a period of active growth and development. Individual cells also have life cycles.

Figure 1 All organisms use cell division. Many-celled organisms, such as this octopus, grow by increasing the numbers of their cells.

Like this dividing amoeba, a one-celled organism reaches a certain size and then reproduces.

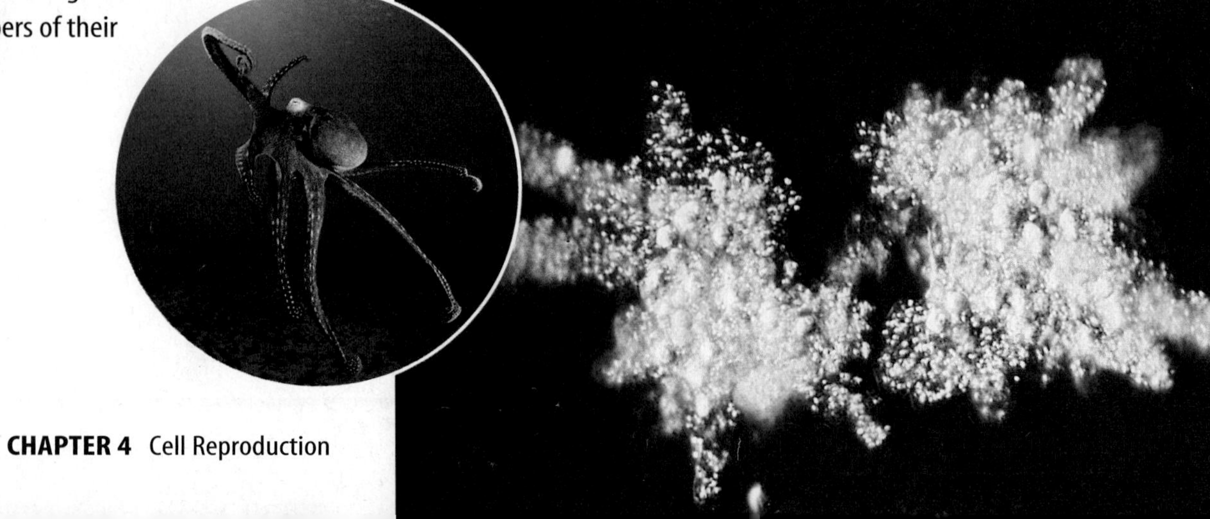

Figure 2 Interphase is the longest part of the cell cycle.
Identify *When do chromosomes duplicate?*

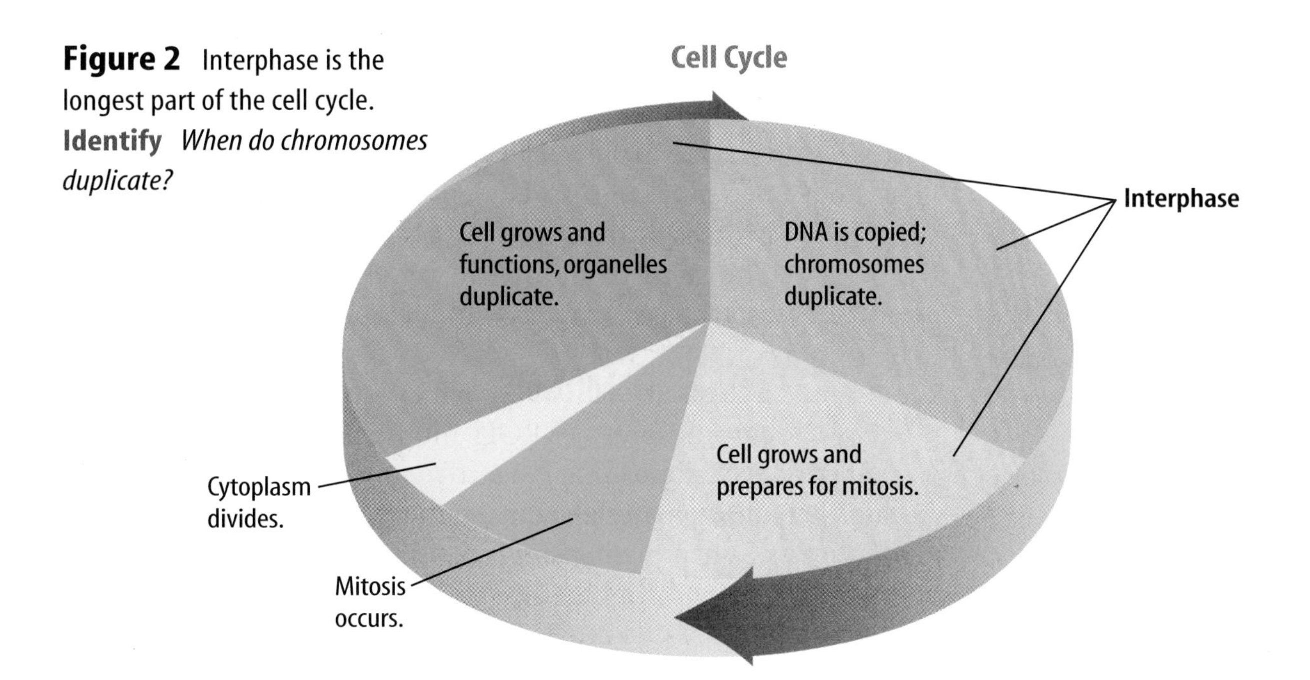

Length of Cycle The cell cycle, as shown in **Figure 2,** is a series of events that takes place from one cell division to the next. The time it takes to complete a cell cycle is not the same in all cells. For example, the cycle for cells in some bean plants takes about 19 h to complete. Cells in animal embryos divide rapidly and can complete their cycles in less than 20 min. In some human cells, the cell cycle takes about 16 h. Cells in humans that are needed for repair, growth, or replacement, like skin and bone cells, constantly repeat the cycle.

Interphase Most of the life of any eukaryotic cell—a cell with a nucleus—is spent in a period of growth and development called interphase. Cells in your body that no longer divide, such as nerve and muscle cells, are always in interphase. An actively dividing cell, such as a skin cell, copies its hereditary material and prepares for cell division during interphase.

Why is it important for a cell to copy its hereditary information before dividing? Imagine that you have a part in a play and the director has one complete copy of the script. If the director gave only one page to each person in the play, no one would have the entire script. Instead the director makes a complete, separate copy of the script for each member of the cast so that each one can learn his or her part. Before a cell divides, a copy of the hereditary material must be made so that each of the two new cells will get a complete copy. Just as the actors in the play need the entire script, each cell needs a complete set of hereditary material to carry out life functions.

After interphase, cell division begins. The nucleus divides, and then the cytoplasm separates to form two new cells.

Oncologist In most cells, the cell cycle is well controlled. Cancer cells, however, have uncontrolled cell division. Doctors who diagnose, study, and treat cancer are called oncologists. Someone wanting to become an oncologist must first complete medical school before training in oncology. Research the subspecialities of oncology. List and describe them in your Science Journal.

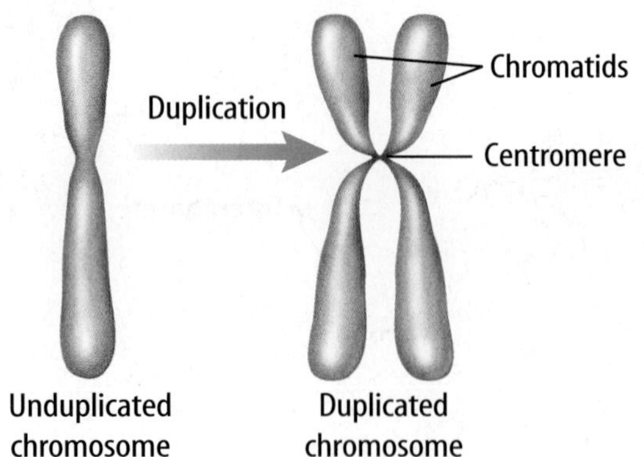

Figure 3 DNA is copied during interphase. An unduplicated chromosome has one strand of DNA. A duplicated chromosome has two identical DNA strands, called chromatids, that are held together at a region called the centromere.

Mitosis and Cell Division

The process in which the nucleus divides to form two identical nuclei is **mitosis** (mi TOH sus). Each new nucleus also is identical to the original nucleus. Mitosis is described as a series of phases, or steps. The steps of mitosis in order are named prophase, metaphase, anaphase, and telophase.

Steps of Mitosis When any nucleus divides, the chromosomes (KROH muh sohmz) play the important part. A **chromosome** is a structure in the nucleus that contains hereditary material. During interphase, each chromosome duplicates. When the nucleus is ready to divide, each duplicated chromosome coils tightly into two thickened, identical strands called chromatids, as shown in **Figure 3.**

Reading Check *How are chromosomes and chromatids related?*

During prophase, the pairs of chromatids are fully visible when viewed under a microscope. The nucleolus and the nuclear membrane disintegrate. Two small structures called centrioles (SEN tree olz) move to opposite ends of the cell. Between the centrioles, threadlike spindle fibers begin to stretch across the cell. Plant cells also form spindle fibers during mitosis but do not have centrioles.

In metaphase, the pairs of chromatids line up across the center of the cell. The centromere of each pair usually becomes attached to two spindle fibers—one from each side of the cell.

In anaphase, each centromere divides and the spindle fibers shorten. Each pair of chromatids separates, and chromatids begin to move to opposite ends of the cell. The separated chromatids are now called chromosomes. In the final step, telophase, spindle fibers start to disappear, the chromosomes start to uncoil, and two new nuclei form.

Figure 4 The cell plate shown in this plant cell appears when the cytoplasm is being divided. **Identify** *what phase of mitosis will be next.*

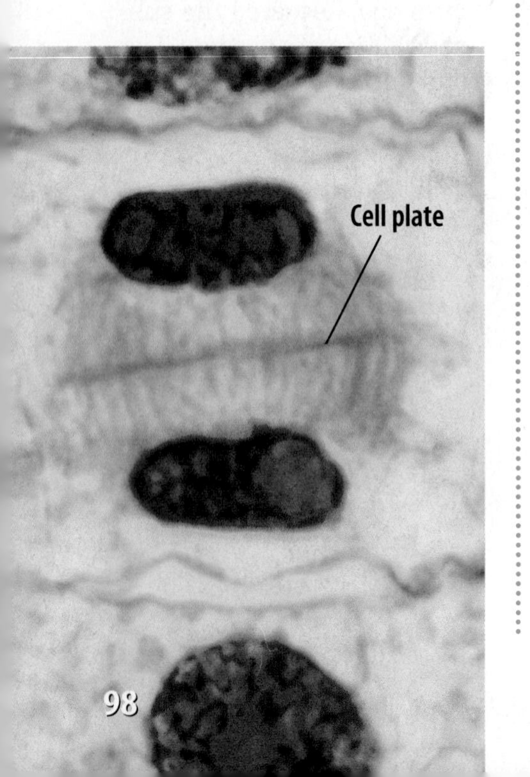

Cell Division For most cells, after the nucleus divides, the cytoplasm separates and two new cells form. In animal cells, the cell membrane pinches in the middle, like a balloon with a string tightened around it, and the cytoplasm divides. In plant cells, the appearance of a cell plate, as shown in **Figure 4,** tells you that the cytoplasm is being divided. New cell membranes form from the cell plate, and new cell walls develop from molecules released by the cell membranes. Following division of the cytoplasm, most new cells begin the period of growth, or interphase. Review cell division for an animal cell using the illustrations in **Figure 5.**

Figure 5 Cell division for an animal cell is shown here. Each micrograph shown in this figure is magnified 600 times.

Centrioles

Nucleus

Nucleolus

Interphase
During interphase, the cell's chromosomes duplicate. The nucleolus is clearly visible in the nucleus.

Mitosis begins

Spindle fibers

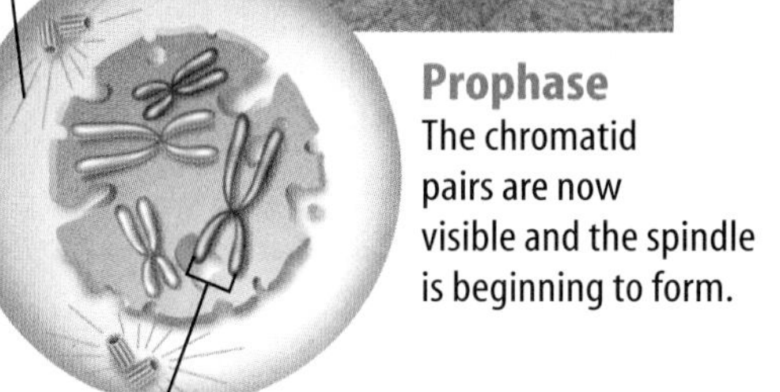

Prophase
The chromatid pairs are now visible and the spindle is beginning to form.

Duplicated chromosome (2 chromatids)

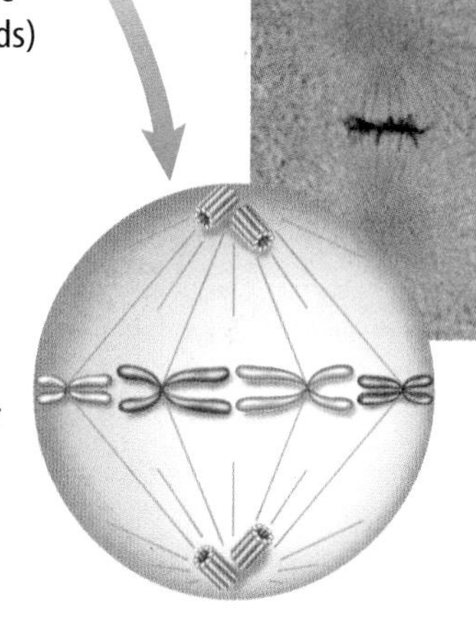

Metaphase
Chromatid pairs are lined up in the center of the cell.

Anaphase
The chromatids separate.

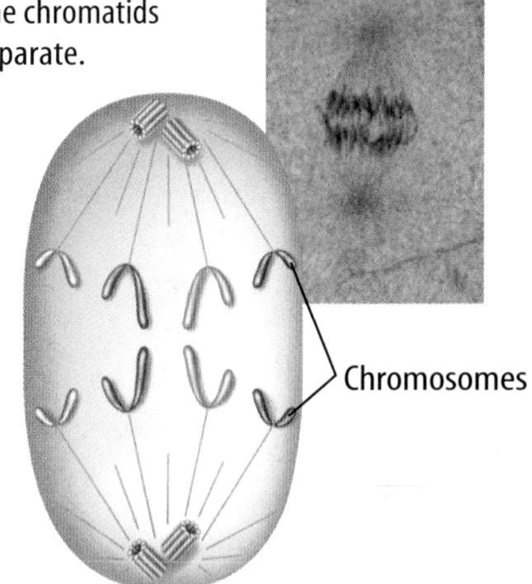

Chromosomes

Telophase
In the final step, the cytoplasm is beginning to separate.

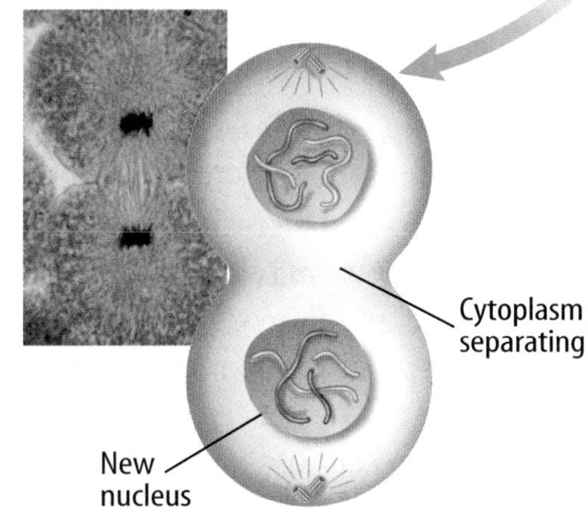

Cytoplasm separating

New nucleus

Mitosis ends

The two new cells enter interphase and cell division usually begins.

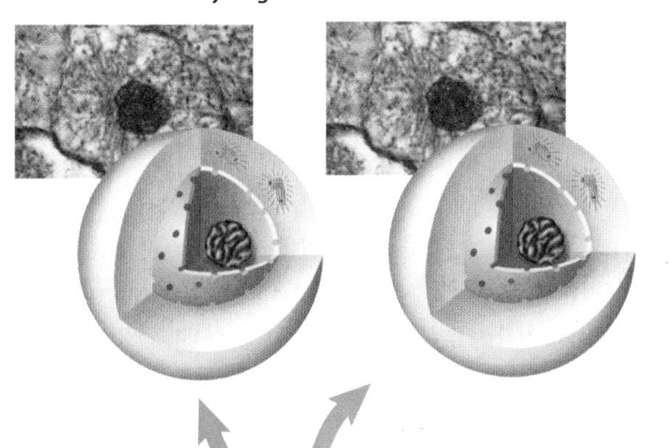

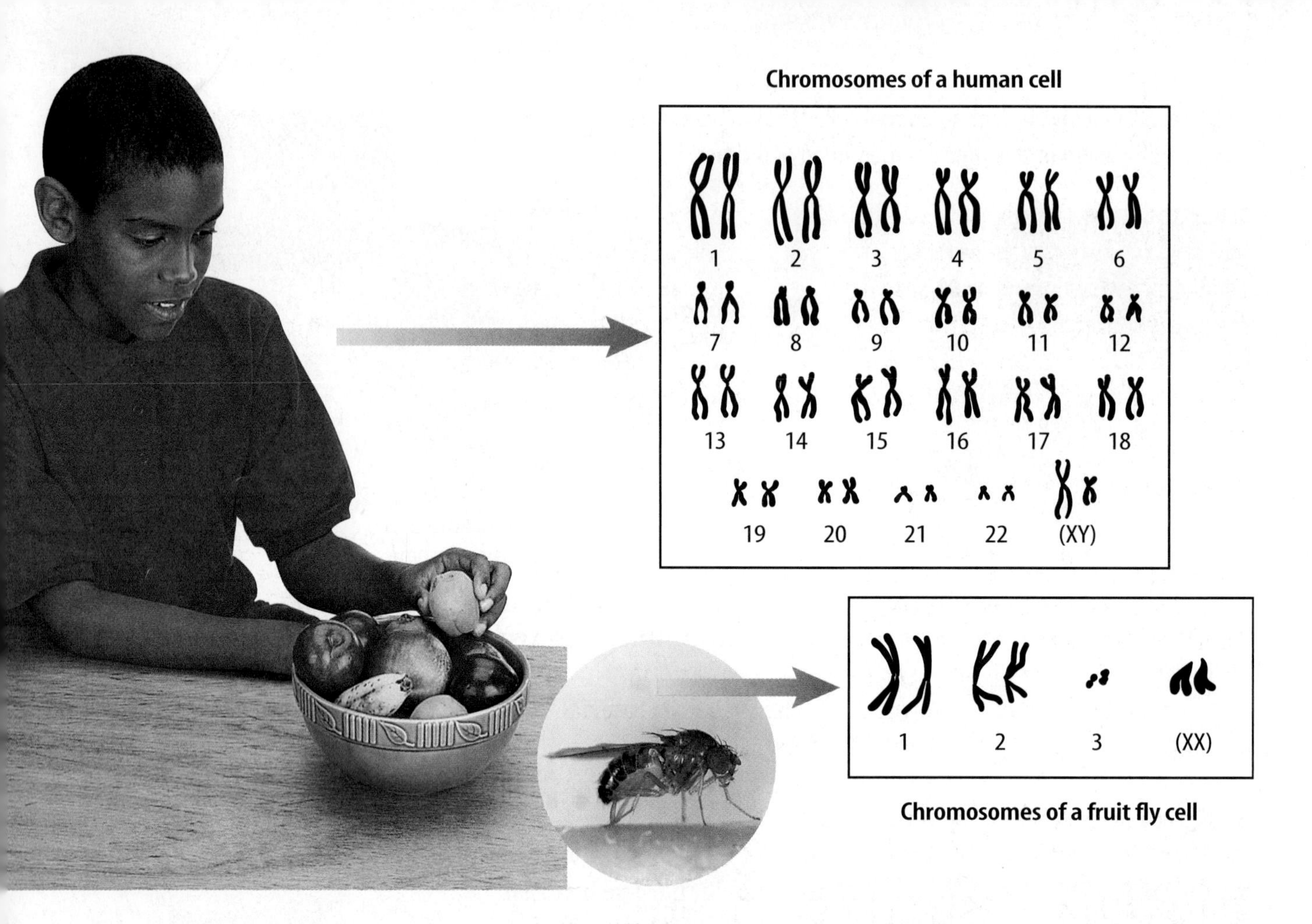

Figure 6 Pairs of chromosomes are found in the nucleus of most cells. All chromosomes shown here are in their duplicated form. Most human cells have 23 pairs of chromosomes including one pair of chromosomes that help determine a person's gender such as the XY pair above. Most fruit fly cells have four pairs of chromosomes.
Infer *What do you think the XX pair in fruit flies helps determine?*

Results of Mitosis and Cell Division There are three important things to remember about mitosis and cell division. First, mitosis is the division of a nucleus. Second, it produces two new nuclei that are identical to each other and the original nucleus. Each new nucleus has the same number and type of chromosomes. Each of the trillions of cells in your body, except sex cells, has a nucleus with a copy of the same 46 chromosomes, because you began as one cell with 46 chromosomes in its nucleus. In the same way, each cell in a fruit fly has eight chromosomes and each new cell produced by mitosis and cell division has a copy of those eight chromosomes, as shown in **Figure 6.** Third, the original cell no longer exists.

You probably know that not all of your cells are the same. Just as all actors in a play use the same script to learn different roles, cells use copies of the same hereditary material and become different cell types with specific functions. Certain cells only used that part of the heredity material with information needed to become a specific cell type.

Cell division allows growth and replaces worn out or damaged cells. You are much larger and have more cells than a baby mainly because of cell division. If you cut yourself, the wound heals because cell division replaces damaged cells. Another way some organisms use cell division is to produce new organisms.

Asexual Reproduction

Reproduction is the process by which an organism produces others of its same kind. Among living organisms, there are two types of reproduction—sexual and asexual. Sexual reproduction usually requires two organisms. In **asexual reproduction,** a new organism (sometimes more than one) is produced from one organism. The new organism will have hereditary material identical to the hereditary material of the parent organism.

How many organisms are needed for asexual reproduction?

Cellular Asexual Reproduction Organisms with eukaryotic cells asexually reproduce by mitosis and cell division. A sweet potato growing in a jar of water is an example of asexual reproduction. All the new stems, leaves, and roots that grow from the sweet potato have the same hereditary material. New strawberry plants can be reproduced asexually from horizontal stems called runners. **Figure 7** shows asexual reproduction in a potato and a strawberry plant.

Recall that mitosis is the division of a nucleus. However, a bacterium does not have a nucleus so it can't use mitosis. Instead, bacteria reproduce asexually by fission. During fission, the one-celled bacterium without a nucleus copies its genetic material and then divides into two identical organisms.

Modeling Mitosis

Procedure

1. Make models of cell division using **materials supplied by your teacher.**
2. Use four chromosomes in your model.
3. When finished, arrange the models in the order in which mitosis occurs.

Analysis

1. In which steps is the nucleus visible?
2. How many cells does a dividing cell form?

Figure 7 Many plants can reproduce asexually.

A new potato plant can grow from each sprout on this potato.

Infer *how the genetic material in the small strawberry plant above compares to the genetic material in the large strawberry plant.*

Figure 8 Some organisms use cell division for budding and regeneration.

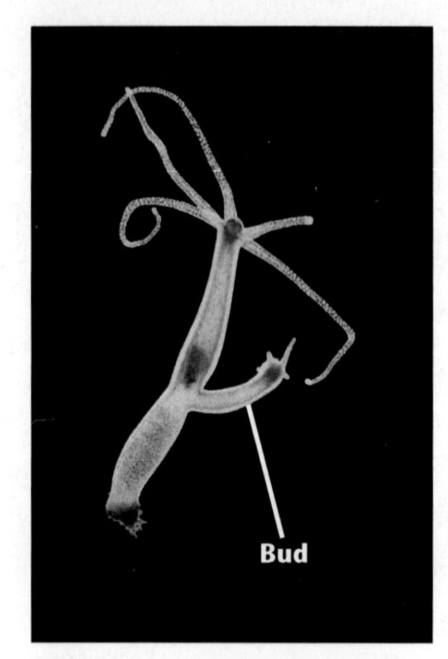

A Hydra, a freshwater animal, can reproduce asexually by budding. The bud is a small exact copy of the adult.

B This sea star is regenerating four new arms.

Budding and Regeneration Look at **Figure 8A.** A new organism is growing from the body of the parent organism. This organism, called a hydra, is reproducing by budding. Budding is a type of asexual reproduction made possible because of mitosis and cell division. When the bud on the adult becomes large enough, it breaks away to live on its own.

Some organisms can regrow damaged or lost body parts, as shown in **Figure 8B.** Regeneration is the process that uses mitosis and cell division to regrow body parts. Sponges, planaria, sea stars, and some other organisms can also use regeneration for asexual reproduction. If these organisms break into pieces, a whole new organism can grow from each piece. Because sea stars eat oysters, oyster farmers dislike them. What would happen if an oyster farmer collected sea stars, cut them into pieces, and threw them back into the ocean?

section 1 review

Summary

The Cell Cycle

- The cell cycle is a series of events from one cell division to the next.
- Most of a eukaryotic cell's life is interphase.

Mitosis

- Mitosis is a series of four phases or steps.
- Each new nucleus formed by mitosis has the same number and type of chromosomes.

Asexual Reproduction

- In asexual reproduction, a new organism is produced from one organism.
- Fission, budding, and regeneration are forms of asexual reproduction.

Self Check

1. **Define** mitosis. How does it differ in plants and animals?
2. **Identify** two examples of asexual reproduction in many-celled organisms.
3. **Describe** what happens to chromosomes before mitosis.
4. **Compare and contrast** the two new cells formed after mitosis and cell division.
5. **Think Critically** Why is it important for the nuclear membrane to disintegrate during mitosis?

Applying Math

6. **Solve One-Step Equations** If a cell undergoes cell division every 5 min, how many cells will there be after 1 h?

Science Online life.msscience.com/self_check_quiz

Mitosis in Plant Cells

Reproduction of most cells in plants and animals uses mitosis and cell division. In this lab, you will study mitosis in plant cells by examining prepared slides of onion root-tip cells.

Real-World Question

How can plant cells in different stages of mitosis be distinguished from each other?

Goals

- **Compare** cells in different stages of mitosis and observe the location of their chromosomes.
- **Observe** what stage of mitosis is most common in onion root tips.

Materials

prepared slide of an onion root tip
microscope

Safety Precautions

Procedure

1. Copy the data table in your Science Journal.

Number of Root-Tip Cells Observed

Stage of Mitosis	Number of Cells Observed	Percent of Cells Observed
Prophase		
Metaphase		
Anaphase	Do not write in this book.	
Telophase		
Total		

2. **Obtain** a prepared slide of cells from an onion root tip.

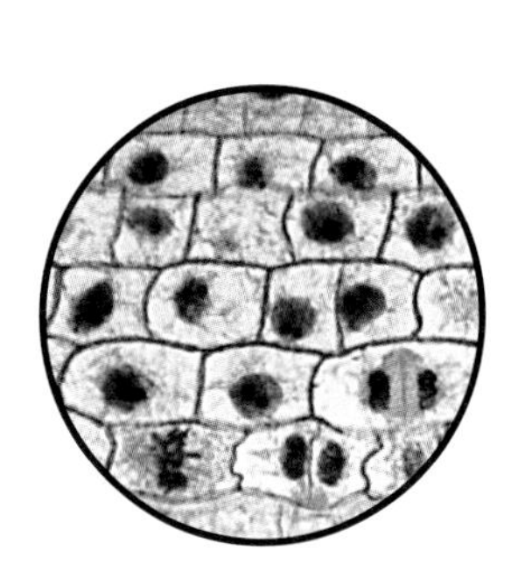

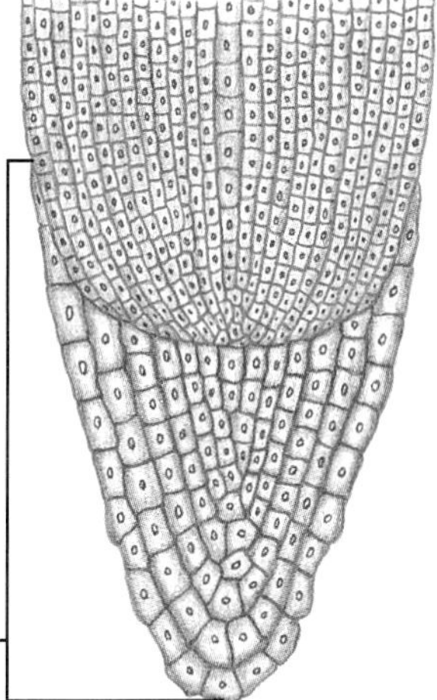

3. Set your microscope on low power and examine the slide. The large, round cells at the root tip are called the root cap. Move the slide until you see the cells just behind the root cap. Turn to the high-power objective.
4. Find an area where you can see the most stages of mitosis. Count and record how many cells you see in each stage.
5. Return the nosepiece to low power. Remove the onion root-tip slide.

Conclude and Apply

1. **Compare** the cells in the region behind the root cap to those in the root cap.
2. **Calculate** the percent of cells found in each stage of mitosis. Infer which stage of mitosis takes the longest period of time.

Communicating Your Data

Write and illustrate a story as if you were a cell undergoing mitosis. Share your story with your class. **For more help, refer to the Science Skill Handbook.**

section 2

Sexual Reproduction and Meiosis

as you read

What You'll Learn

- **Describe** the stages of meiosis and how sex cells are produced.
- **Explain** why meiosis is needed for sexual reproduction.
- **Name** the cells that are involved in fertilization.
- **Explain** how fertilization occurs in sexual reproduction.

Why It's Important

Meiosis and sexual reproduction are the reasons why no one else is exactly like you.

Review Vocabulary

organism: any living thing; uses energy, is made of cells, reproduces, responds, grows, and develops

New Vocabulary

- sexual reproduction
- sperm
- egg
- fertilization
- zygote
- diploid
- haploid
- meiosis

Sexual Reproduction

Another way that a new organism can be produced is by sexual reproduction. During **sexual reproduction,** two sex cells, sometimes called an egg and a sperm, come together. Sex cells, like those in **Figure 9,** are formed from cells in reproductive organs. **Sperm** are formed in the male reproductive organs. **Eggs** are formed in the female reproductive organs. The joining of an egg and a sperm is called **fertilization,** and the cell that forms is called a **zygote** (ZI goht). Generally, the egg and the sperm come from two different organisms of the same species. Following fertilization, mitosis and cell division begins. A new organism with a unique identity develops.

Diploid Cells Your body forms two types of cells—body cells and sex cells. Body cells far outnumber sex cells. Your brain, skin, bones, and other tissues and organs are formed from body cells. Recall that a typical human body cell has 46 chromosomes. Each chromosome has a mate that is similar to it in size and shape and has similar DNA. Human body cells have 23 pairs of chromosomes. When cells have pairs of similar chromosomes, they are said to be **diploid** (DIH ployd).

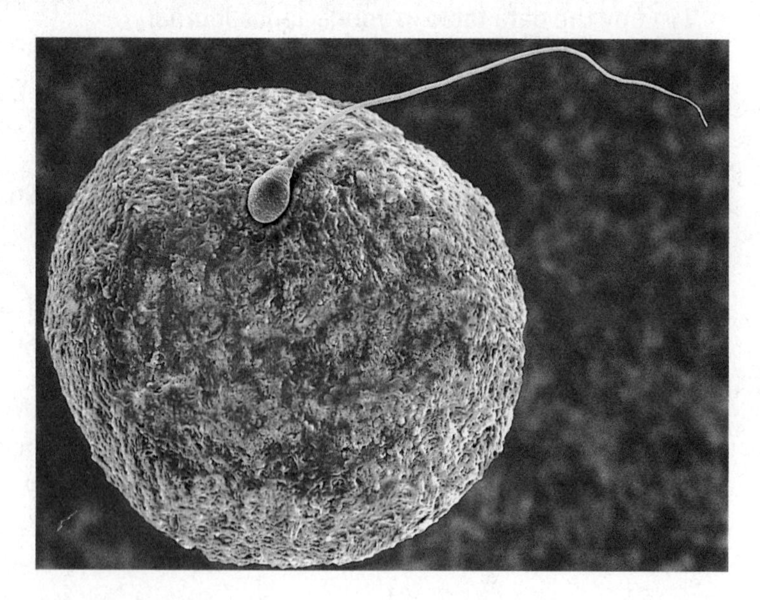

Figure 9 A human egg and a human sperm at fertilization.

Haploid Cells Because sex cells do not have pairs of chromosomes, they are said to be **haploid** (HA ployd). They have only half the number of chromosomes as body cells. *Haploid* means "single form." Human sex cells have only 23 chromosomes—one from each of the 23 pairs of similar chromosomes. Compare the number of chromosomes found in a human sex cell to the full set of human chromosomes seen in **Figure 6.**

Reading Check *How many chromosomes are usually in each human sperm?*

Meiosis and Sex Cells

A process called **meiosis** (mi OH sus) produces haploid sex cells. What would happen in sexual reproduction if two diploid cells combined? The offspring would have twice as many chromosomes as its parent. Although plants with twice the number of chromosomes as the parent plants are often produced, most animals do not survive with a double number of chromosomes. Meiosis ensures that the offspring will have the same diploid number as its parent, as shown in **Figure 10.** After two haploid sex cells combine, a diploid zygote is produced that develops into a new diploid organism.

During meiosis, two divisions of the nucleus occur. These divisions are called meiosis I and meiosis II. The steps of each division have names like those in mitosis and are numbered for the division in which they occur.

Diploid Zygote The human egg releases a chemical into the surrounding fluid that attracts sperm. Usually, only one sperm fertilizes the egg. After the sperm nucleus enters the egg, the cell membrane of the egg changes in a way that prevents other sperm from entering. What adaptation in this process guarantees that the zygote will be diploid? Write a paragraph describing your ideas in your Science Journal.

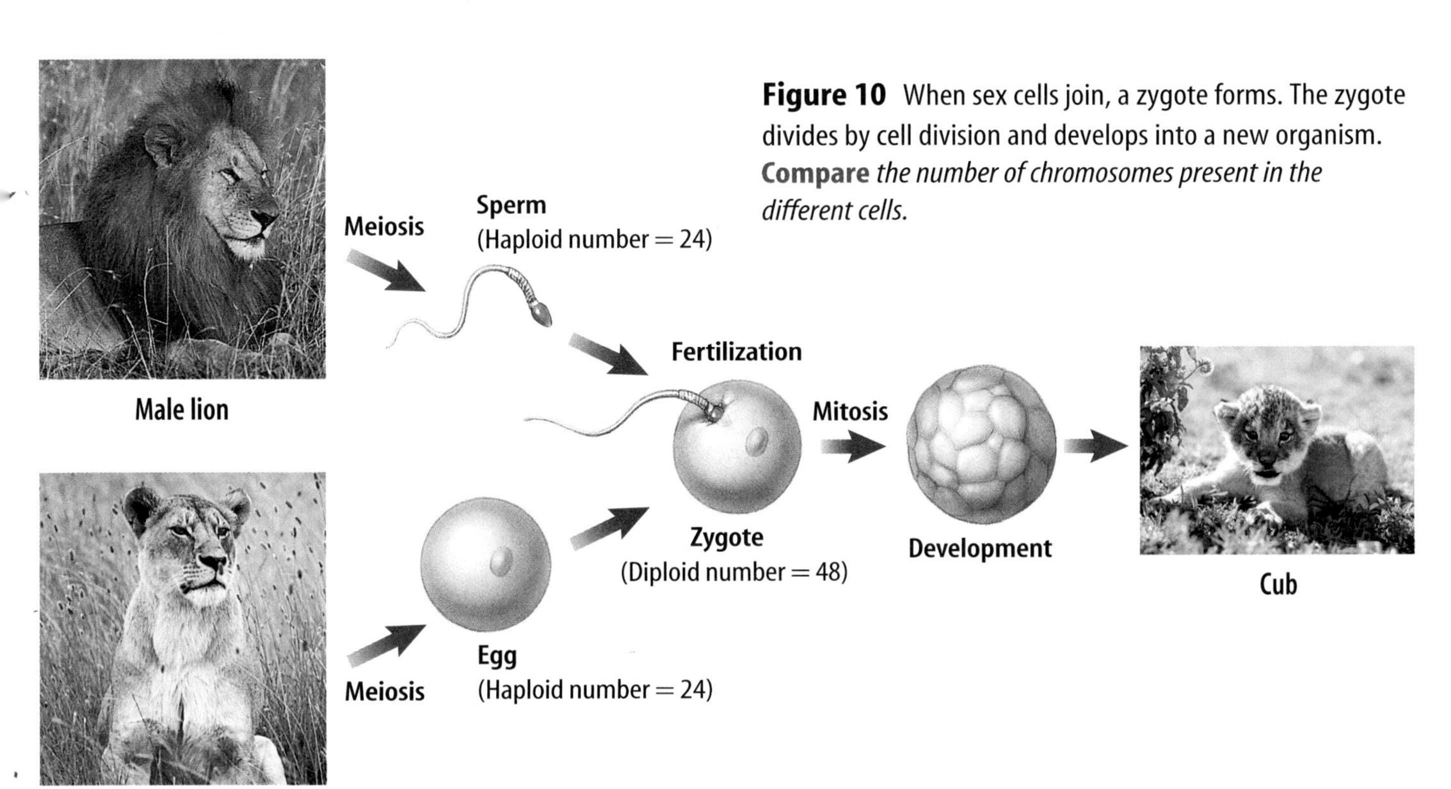

Figure 10 When sex cells join, a zygote forms. The zygote divides by cell division and develops into a new organism. **Compare** *the number of chromosomes present in the different cells.*

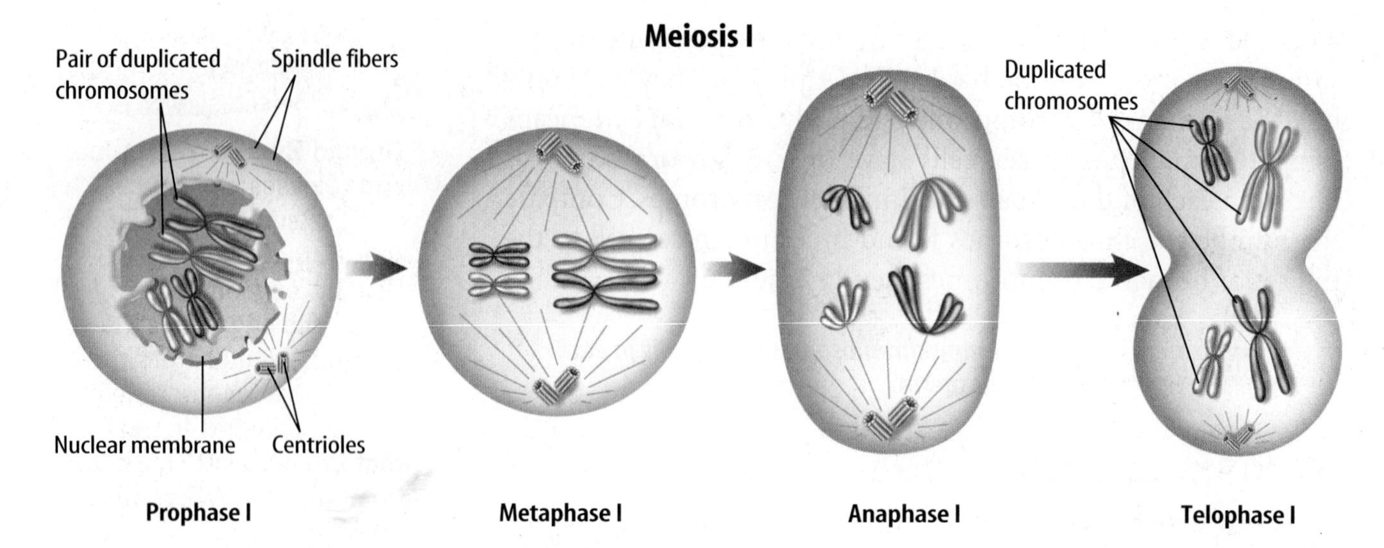

Figure 11 Meiosis has two divisions of the nucleus—meiosis I and meiosis II.
Determine *how many sex cells are finally formed after both divisions are completed.*

Meiosis I Before meiosis begins, each chromosome is duplicated, just as in mitosis. When the cell is ready for meiosis, each duplicated chromosome is visible under the microscope as two chromatids. As shown in **Figure 11,** the events of prophase I are similar to those of prophase in mitosis. In meiosis, each duplicated chromosome comes near its similar duplicated mate. In mitosis they do not come near each other.

In metaphase I, the pairs of duplicated chromosomes line up in the center of the cell. The centromere of each chromatid pair becomes attached to one spindle fiber, so the chromatids do not separate in anaphase I. The two pairs of chromatids of each similar pair move away from each other to opposite ends of the cell. Each duplicated chromosome still has two chromatids. Then, in telophase I, the cytoplasm divides, and two new cells form. Each new cell has one duplicated chromosome from each similar pair.

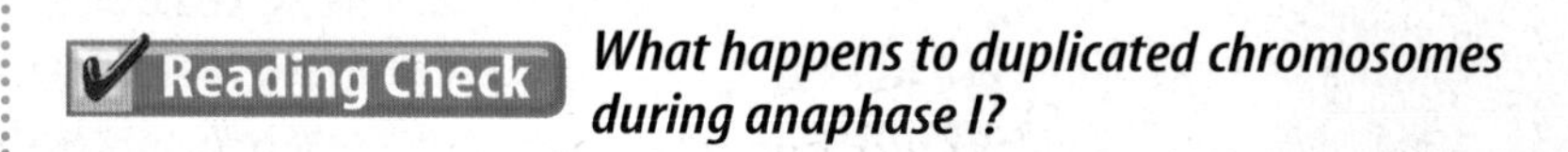

What happens to duplicated chromosomes during anaphase I?

Meiosis II The two cells formed during meiosis I now begin meiosis II. The chromatids of each duplicated chromosome will be separated during this division. In prophase II, the duplicated chromosomes and spindle fibers reappear in each new cell. Then in metaphase II, the duplicated chromosomes move to the center of the cell. Unlike what occurs in metaphase I, each centromere now attaches to two spindle fibers instead of one. The centromere divides during anaphase II. The chromatids separate and move to opposite ends of the cell. Each chromatid now is an individual chromosome. As telophase II begins, the spindle fibers disappear, and a nuclear membrane forms around each set of chromosomes. When meiosis II is finished, the cytoplasm divides.

Meiosis II

Unduplicated chromosomes

Prophase II | Metaphase II | Anaphase II | Telophase II

Summary of Meiosis Two cells form during meiosis I. In meiosis II, both of these cells form two cells. The two divisions of the nucleus result in four sex cells. Each has one-half the number of chromosomes in its nucleus that was in the original nucleus. From a human cell with 46 paired chromosomes, meiosis produces four sex cells each with 23 unpaired chromosomes.

Applying Science

How can chromosome numbers be predicted?

Offspring get half of their chromosomes from one parent and half from the other. What happens if each parent has a different diploid number of chromosomes?

Identifying the Problem

A Grevy's zebra and a donkey can mate to produce a zonkey, as shown below.

Solving the Problem

1. How many chromosomes would the zonkey receive from each parent?
2. What is the chromosome number of the zonkey?
3. What would happen when meiosis occurs in the zonkey's reproductive organs?
4. Predict why zonkeys are usually sterile.

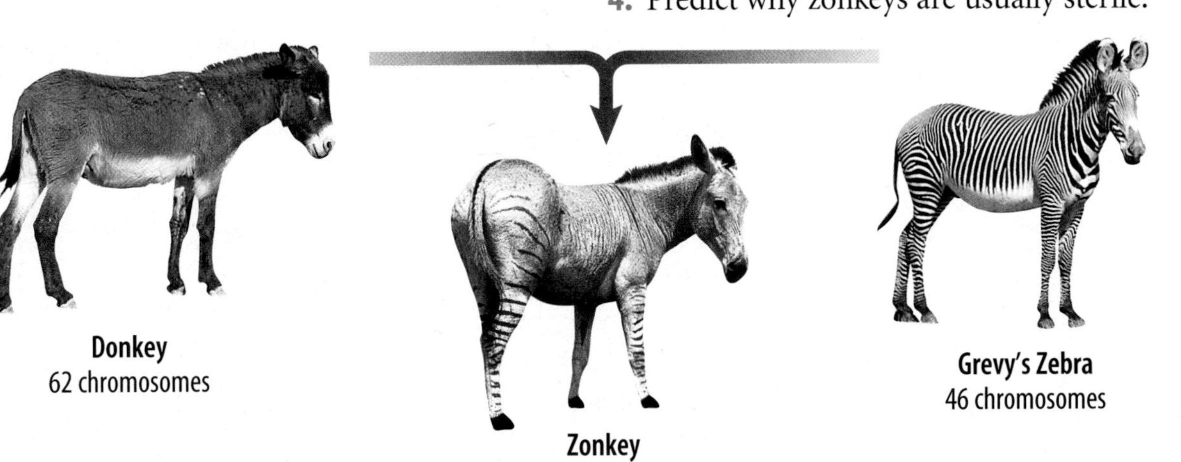

NATIONAL GEOGRAPHIC

VISUALIZING POLYPLOIDY IN PLANTS

Figure 12

You received a haploid (n) set of chromosomes from each of your parents, making you a diploid (2n) organism. In nature, however, many plants are polyploid—they have three (3n), four (4n), or more sets of chromosomes. We depend on some of these plants for food.

▲ TRIPLOID Bright yellow bananas typically come from triploid (3n) banana plants. Plants with an odd number of chromosome sets usually cannot reproduce sexually and have very small seeds or none at all.

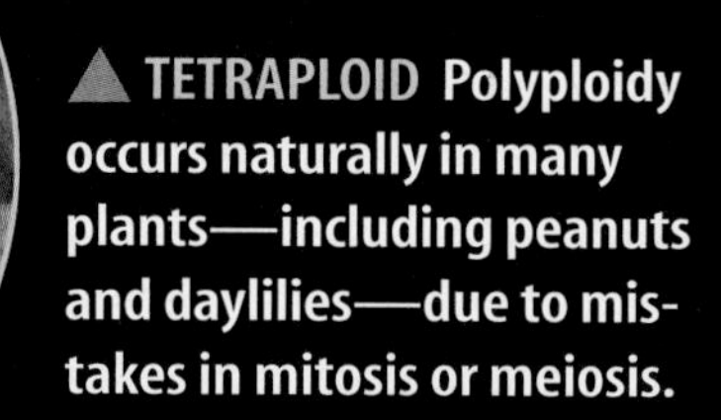

▲ TETRAPLOID Polyploidy occurs naturally in many plants—including peanuts and daylilies—due to mistakes in mitosis or meiosis.

▼ HEXAPLOID Modern cultivated strains of oats have six sets of chromosomes, making them hexaploid (6n) plants.

▲ OCTAPLOID Polyploid plants often are bigger than nonpolyploid plants and may have especially large leaves, flowers, or fruits. Strawberries are an example of octaploid (8n) plants.

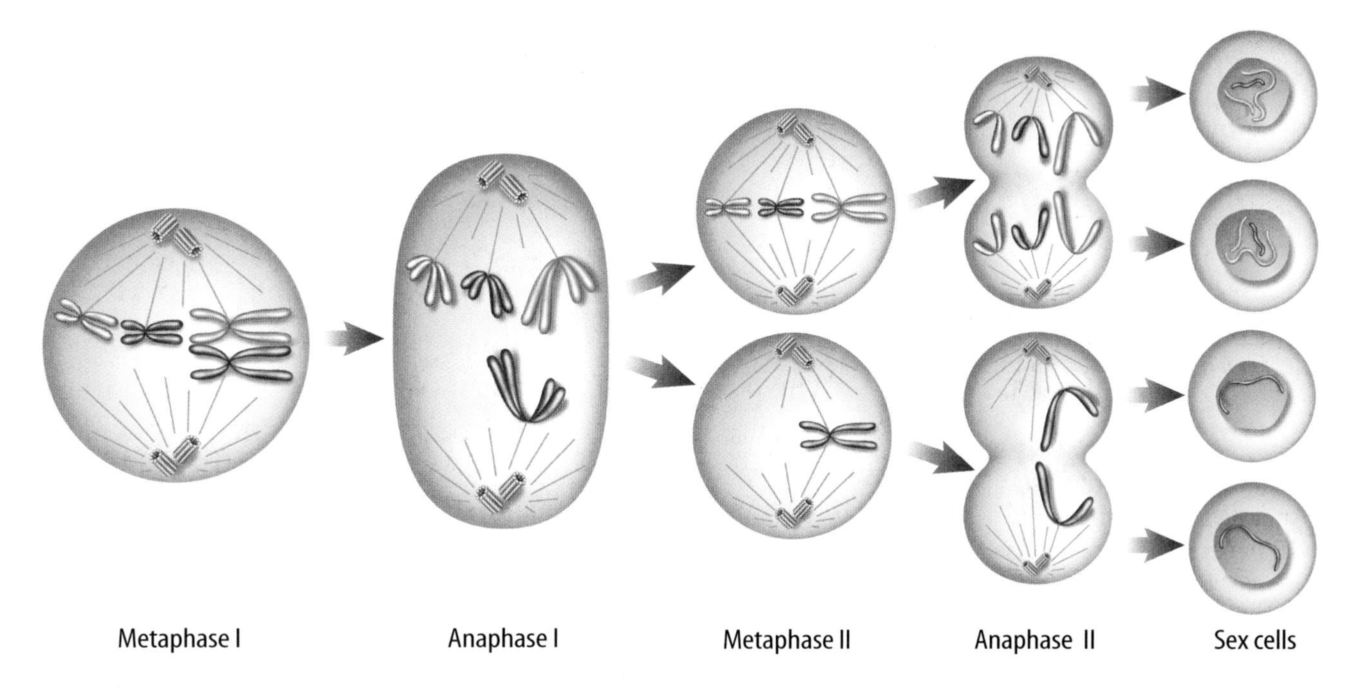

Mistakes in Meiosis Meiosis occurs many times in reproductive organs. Although mistakes in plants, as shown in **Figure 12,** are common, mistakes are less common in animals. These mistakes can produce sex cells with too many or too few chromosomes, as shown in **Figure 13.** Sometimes, zygotes produced from these sex cells die. If the zygote lives, every cell in the organism that grows from that zygote usually will have the wrong number of chromosomes. Organisms with the wrong number of chromosomes may not grow normally.

Figure 13 This diploid cell has four chromosomes. During anaphase I, one pair of duplicated chromosomes did not separate. **Infer** *how many chromosomes each sex cell usually has.*

section 2 review

Summary

Sexual Reproduction

- During sexual reproduction, two sex cells come together.
- Mitosis and cell division begin after fertilization.
- A typical human body cell has 46 chromosomes, and a human sex cell has 23 chromosomes.

Meiosis and Sex Cells

- Each chromosome is duplicated before meiosis, then two divisions of the nucleus occur.
- During meiosis I, duplicated pairs of chromosomes are separated into new cells.
- Chromatids separate during meiosis II.
- Meiosis I and meiosis II result in four sex cells.

Self Check

1. **Describe** a zygote and how it is formed.
2. **Explain** where sex cells form.
3. **Compare** what happens to chromosomes during anaphase I and anaphase II.
4. **Think Critically** Plants grown from runners and leaf cuttings have the same traits as the parent plant. Plants grown from seeds can vary from the parent plants in many ways. Why can this happen?

Applying Skills

5. **Make and use a table** to compare mitosis and meiosis in humans. Vertical headings should include: *What Type of Cell (Body or Sex), Beginning Cell (Haploid or Diploid), Number of Cells Produced, End-Product Cell (Haploid or Diploid),* and *Number of Chromosomes in New Cells.*

section 3

DNA

as you read

What You'll Learn

- **Identify** the parts of a DNA molecule and its structure.
- **Explain** how DNA copies itself.
- **Describe** the structure and function of each kind of RNA.

Why It's Important

DNA helps determine nearly everything your body is and does.

Review Vocabulary
protein: large organic molecule made of amino acid bases

New Vocabulary
- DNA
- gene
- RNA
- mutation

What is DNA?

Why was the alphabet one of the first things you learned when you started school? Letters are a code that you need to know before you learn to read. A cell also uses a code that is stored in its hereditary material. The code is a chemical called deoxyribonucleic (dee AHK sih ri boh noo klay ihk) acid, or **DNA.** It contains information for an organism's growth and function. **Figure 14** shows how DNA is stored in cells that have a nucleus. When a cell divides, the DNA code is copied and passed to the new cells. In this way, new cells receive the same coded information that was in the original cell. Every cell that has ever been formed in your body or in any other organism contains DNA.

INTEGRATE Chemistry

Discovering DNA Since the mid-1800s, scientists have known that the nuclei of cells contain large molecules called nucleic acids. By 1950, chemists had learned what the nucleic acid DNA was made of, but they didn't understand how the parts of DNA were arranged.

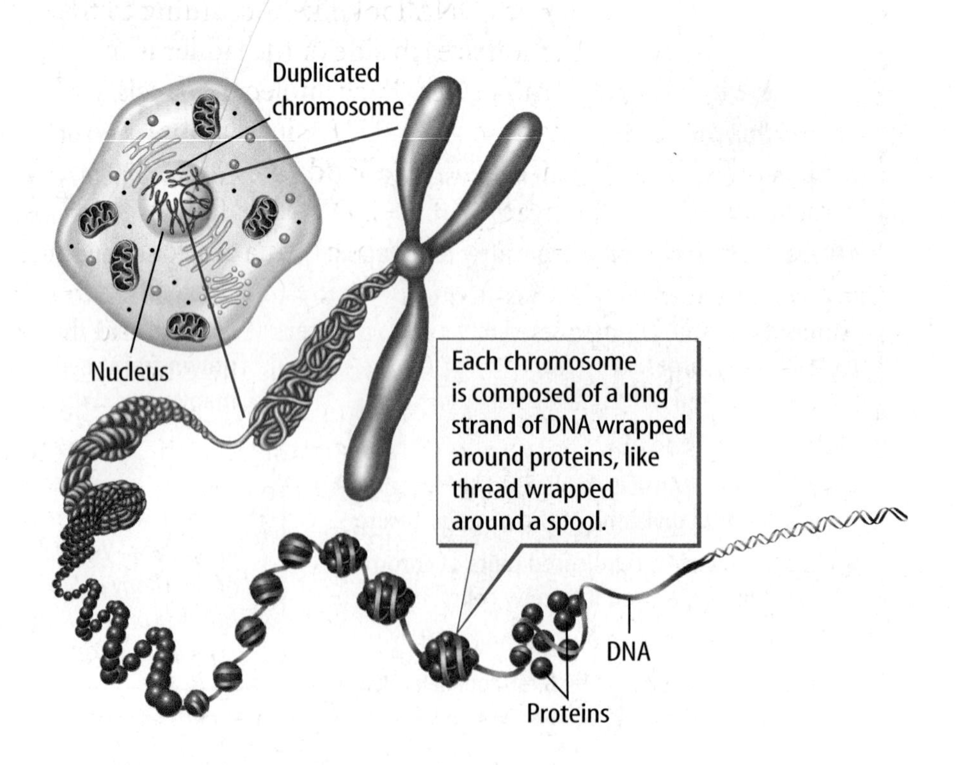

Figure 14 DNA is part of the chromosomes found in a cell's nucleus.

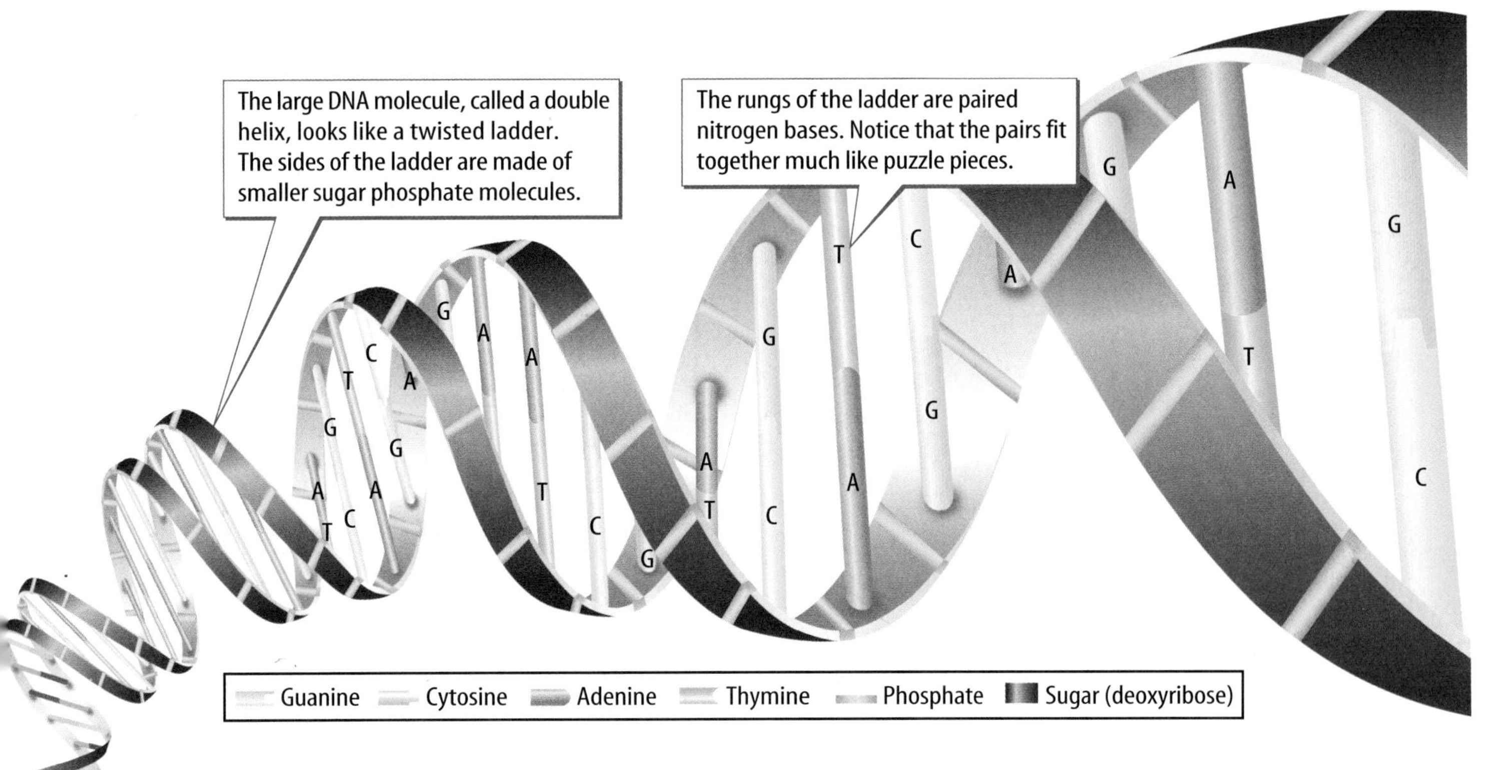

The large DNA molecule, called a double helix, looks like a twisted ladder. The sides of the ladder are made of smaller sugar phosphate molecules.
The rungs of the ladder are paired nitrogen bases. Notice that the pairs fit together much like puzzle pieces.
Guanine
Cytosine
Adenine
Thymine
Phosphate
Sugar (deoxyribose)

DNA's Structure In 1952, scientist Rosalind Franklin discovered that DNA is two chains of molecules in a spiral form. By using an X-ray technique, Dr. Franklin showed that the large spiral was probably made up of two spirals. As it turned out, the structure of DNA is similar to a twisted ladder. In 1953, using the work of Franklin and others, scientists James Watson and Francis Crick made a model of a DNA molecule.

A DNA Model What does DNA look like? According to the Watson and Crick DNA model, each side of the ladder is made up of sugar-phosphate molecules. Each molecule consists of the sugar called deoxyribose (dee AHK sih ri bohs) and a phosphate group. The rungs of the ladder are made up of other molecules called nitrogen bases. DNA has four kinds of nitrogen bases—adenine (A duh neen), guanine (GWAH neen), cytosine (SI tuh seen), and thymine (THI meen). The nitrogen bases are represented by the letters A, G, C, and T. The amount of cytosine in cells always equals the amount of guanine, and the amount of adenine always equals the amount of thymine. This led to the hypothesis that the nitrogen bases occur as pairs in DNA. **Figure 14** shows that adenine always pairs with thymine, and guanine always pairs with cytosine. Like interlocking pieces of a puzzle, each base bonds only with its correct partner.

What are the nitrogen base pairs in a DNA molecule?

Modeling DNA Replication

Procedure

1. Suppose you have a segment of DNA that is six nitrogen base pairs in length. On **paper,** using the letters A, T, C, and G, write a combination of six pairs, remembering that A and T are always a pair and C and G are always a pair.
2. Duplicate your segment of DNA. On paper, diagram how this happens and show the new DNA segments.

Analysis

Compare the order of bases of the original DNA to the new DNA molecules.

Try at Home

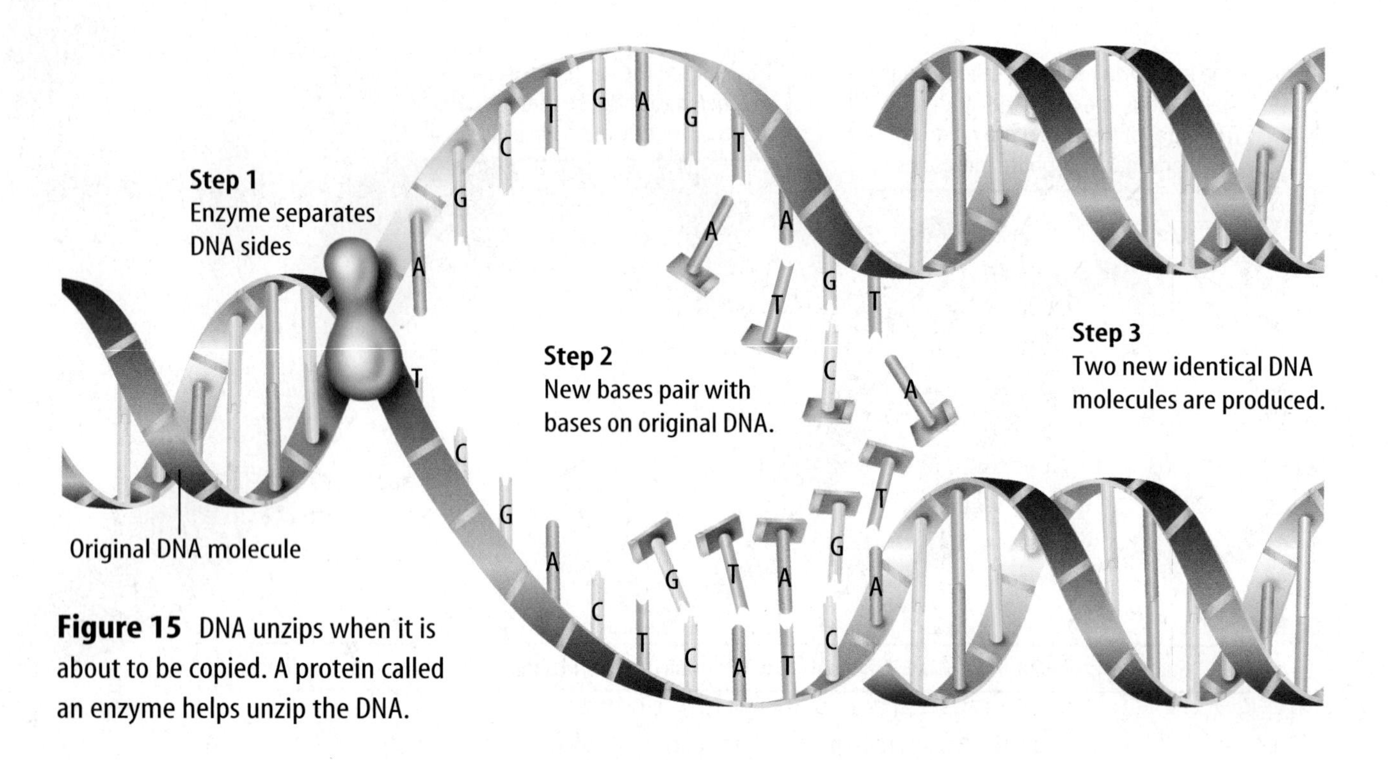

Figure 15 DNA unzips when it is about to be copied. A protein called an enzyme helps unzip the DNA.

Copying DNA When chromosomes are duplicated before mitosis or meiosis, the amount of DNA in the nucleus is doubled. The Watson and Crick model shows how this takes place. The two sides of DNA unwind and separate. Each side then becomes a pattern on which a new side forms, as shown in **Figure 15.** The new DNA has bases that are identical to those of the original DNA and are in the same order.

Genes

Most of your characteristics, such as the color of your hair, your height, and even how things taste to you, depend on the kinds of proteins your cells make. DNA in your cells stores the instructions for making these proteins.

Proteins build cells and tissues or work as enzymes. The instructions for making a specific protein are found in a **gene** which is a section of DNA on a chromosome. As shown in **Figure 16,** each chromosome contains hundreds of genes. Proteins are made of chains of hundreds or thousands of amino acids. The gene determines the order of amino acids in a protein. Changing the order of the amino acids makes a different protein. What might occur if an important protein couldn't be made or if the wrong protein was made in your cells?

Figure 16 This diagram shows just a few of the genes that have been identified on human chromosome 7. The bold print is the name that has been given to each gene.

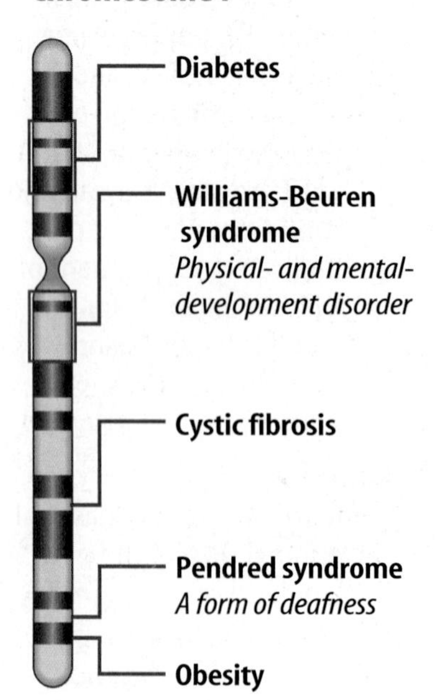

Making Proteins Genes are found in the nucleus, but proteins are made on ribosomes in cytoplasm. The codes for making proteins are carried from the nucleus to the ribosomes by another type of nucleic acid called ribonucleic acid, or **RNA.**

Ribonucleic Acid RNA is made in the nucleus on a DNA pattern but is different from DNA. If DNA is like a ladder, RNA is like a ladder that has all its rungs cut in half. Compare the DNA molecule in **Figure 14** to the RNA molecule in **Figure 17.** RNA has the nitrogen bases A, G, and C like DNA but has the nitrogen base uracil (U) instead of thymine (T). The sugar-phosphate molecules in RNA contain the sugar ribose, not deoxyribose.

The three main kinds of RNA made from DNA in a cell's nucleus are messenger RNA (mRNA), ribosomal RNA (rRNA), and transfer RNA (tRNA). Protein production begins when mRNA moves into the cytoplasm. There, ribosomes attach to it. Ribosomes are made of rRNA. Transfer RNA molecules in the cytoplasm bring amino acids to these ribosomes. Inside the ribosomes, three nitrogen bases on the mRNA temporarily match with three nitrogen bases on the tRNA. The same thing happens for the mRNA and another tRNA molecule, as shown in **Figure 17.** The amino acids that are attached to the two tRNA molecules bond. This is the beginning of a protein.

The code carried on the mRNA directs the order in which the amino acids bond. After a tRNA molecule has lost its amino acid, it can move about the cytoplasm and pick up another amino acid just like the first one. The ribosome moves along the mRNA. New tRNA molecules with amino acids match up and add amino acids to the protein molecule.

Topic: The Human Genome Project

Visit life.msscience.com for Web links to information about the Human Genome Project.

Activity Find out when chromosomes 5, 16, 29, 21, and 22 were completely sequenced. Write about what scientists learned about each of these chromosomes.

Figure 17 Cells need DNA, RNA, and amino acids to make proteins.

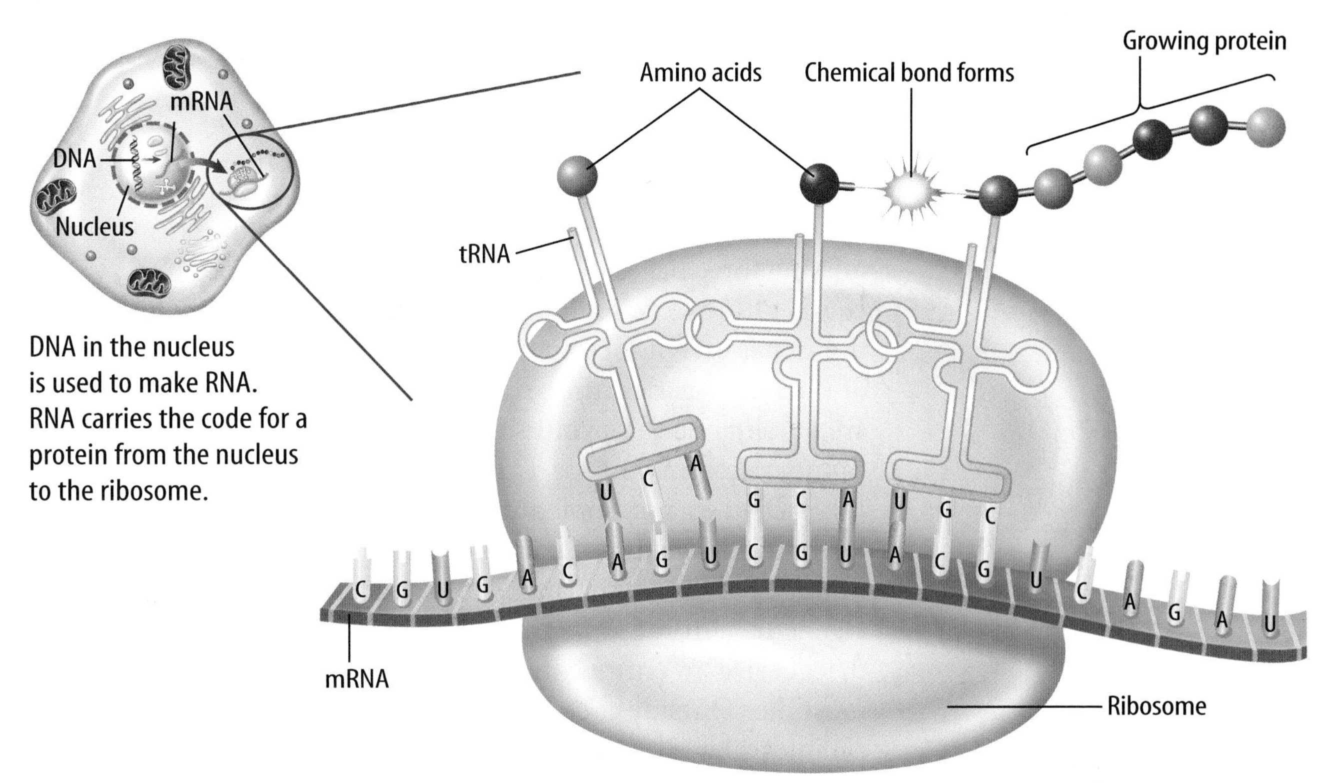

DNA in the nucleus is used to make RNA. RNA carries the code for a protein from the nucleus to the ribosome.

At the ribosome, the RNA's message is translated into a specific protein.

Figure 18 Each cell in the body produces only the proteins that are necessary to do its job.

Controlling Genes You might think that because most cells in an organism have exactly the same chromosomes and the same genes, they would make the same proteins, but they don't. In many-celled organisms like you, each cell uses only some of the thousands of genes that it has to make proteins. Just as each actor uses only the lines from the script for his or her role, each cell uses only the genes that direct the making of proteins that it needs. For example, muscle proteins are made in muscle cells, as represented in **Figure 18,** but not in nerve cells.

Cells must be able to control genes by turning some genes off and turning other genes on. They do this in many different ways. Sometimes the DNA is twisted so tightly that no RNA can be made. Other times, chemicals bind to the DNA so that it cannot be used. If the incorrect proteins are produced, the organism cannot function properly.

Mutations

Sometimes mistakes happen when DNA is being copied. Imagine that the copy of the script the director gave you was missing three pages. You use your copy to learn your lines. When you begin rehearsing for the play, everyone is ready for one of the scenes except for you. What happened? You check your copy of the script against the original and find that three of the pages are missing. Because your script is different from the others, you cannot perform your part correctly.

If DNA is not copied exactly, the proteins made from the instructions might not be made correctly. These mistakes, called **mutations,** are any permanent change in the DNA sequence of a gene or chromosome of a cell. Some mutations include cells that receive an entire extra chromosome or are missing a chromosome. Outside factors such as X rays, sunlight, and some chemicals have been known to cause mutations.

When are mutations likely to occur?

Figure 19 Because of a defect on chromosome 2, the mutant fruit fly has short wings and cannot fly.
Predict *Could this defect be transferred to the mutant's offspring? Explain.*

Results of a Mutation Genes control the traits you inherit. Without correctly coded proteins, an organism can't grow, repair, or maintain itself. A change in a gene or chromosome can change the traits of an organism, as illustrated in **Figure 19.**

If the mutation occurs in a body cell, it might or might not be life threatening to the organism. However, if a mutation occurs in a sex cell, then all the cells that are formed from that sex cell will have that mutation. Mutations add variety to a species when the organism reproduces. Many mutations are harmful to organisms, often causing their death. Some mutations do not appear to have any effect on the organism, and some can even be beneficial. For example, a mutation to a plant might cause it to produce a chemical that certain insects avoid. If these insects normally eat the plant, the mutation will help the plant survive.

Topic: Fruit Fly Genes
Visit life.msscience.com for Web links to information about what genes are present on the chromosomes of a fruit fly.

Activity Draw a picture of one of the chromosomes of a fruit fly and label some of its genes.

section 3 review

Summary

What is DNA?

- Each side of the DNA ladder is made up of sugar-phosphate molecules, and the rungs of the ladder are made up of nitrogen bases.
- When DNA is copied, the new DNA has bases that are identical to those of the original DNA.

Genes

- The instructions for making a specific protein are found in genes in the cell nucleus. Proteins are made on ribosomes in the cytoplasm.
- There are three main kinds of RNA—mRNA, rRNA, and tRNA.

Mutations

- If DNA is not copied exactly, the resulting mutations may cause proteins to be made incorrectly.

Self Check

1. **Describe** how DNA makes a copy of itself.
2. **Explain** how the codes for proteins are carried from the nucleus to the ribosomes.
3. **Apply** A strand of DNA has the bases AGTAAC. Using letters, show a matching DNA strand.
4. **Determine** how tRNA is used when cells build proteins.
5. **Think Critically** You begin as one cell. Compare the DNA in your brain cells to the DNA in your heart cells.

Applying Skills

6. **Concept Map** Using a Venn diagram, compare and contrast DNA and RNA.
7. **Use a word processor** to make an outline of the events that led up to the discovery of DNA. Use library resources to find this information.

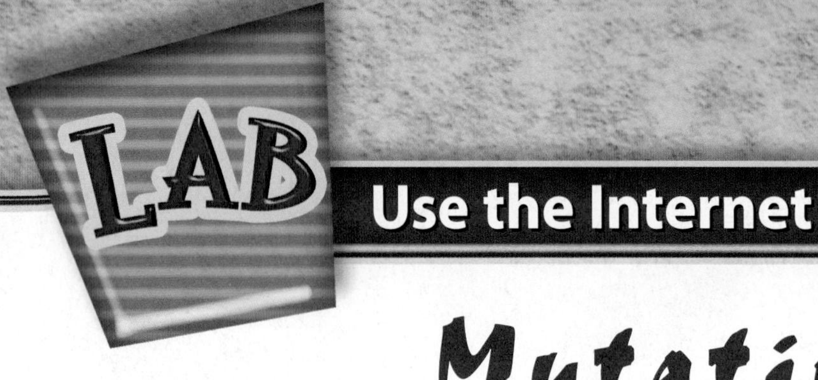

Mutations

Fantail pigeon

Real-World Question

Mutations can result in dominant or recessive genes. A recessive characteristic can appear only if an organism has two recessive genes for that characteristic. However, a dominant characteristic can appear if an organism has one or two dominant genes for that characteristic. Why do some mutations result in more common traits while others do not? Form a hypothesis about how a mutation can become a common trait.

Make a Plan

1. **Observe** common traits in various animals, such as household pets or animals you might see in a zoo.
2. **Learn** what genes carry these traits in each animal.
3. **Research** the traits to discover which ones are results of mutations. Are all mutations dominant? Are any of these mutations beneficial?

Goals

- **Observe** traits of various animals.
- **Research** how mutations become traits.
- Gather data about mutations.
- Make a frequency table of your findings and communicate them to other students.

Data Source

Science Online

Visit life.msscience.com/internet_lab for more information on common genetic traits in different animals, recessive and dominant genes, and data from other students.

White tiger

Follow Your Plan

1. Make sure your teacher approves your plan before you start.
2. Visit the link shown below to access different Web sites for information about mutations and genetics.
3. **Decide** if a mutation is beneficial, harmful, or neither. Record your data in your Science Journal.

Analyze Your Data

1. **Record** in your Science Journal a list of traits that are results of mutations.
2. **Describe** an animal, such as a pet or an animal you've seen in the zoo. Point out which traits are known to be the result of a mutation.
3. **Make** a chart that compares recessive mutations to dominant mutations. Which are more common?
4. **Share** your data with other students by posting it at the link shown below.

Siberian Husky's eyes

Conclude and Apply

1. **Compare** your findings to those of your classmates and other data at the link shown below. What were some of the traits your classmates found that you did not? Which were the most common?
2. Look at your chart of mutations. Are all mutations beneficial? When might a mutation be harmful to an organism?
3. **Predict** how your data would be affected if you had performed this lab when one of these common mutations first appeared. Do you think you would see more or less animals with this trait?
4. Mutations occur every day but we only see a few of them. Infer how many mutations over millions of years can lead to a new species.

Communicating Your Data

Find this lab using the link below. **Post** your data in the table provided. Combine your data with that of other students and make a chart that shows all of the data.

life.msscience.com/internet_lab

Oops! Accidents in SCIENCE

SOMETIMES GREAT DISCOVERIES HAPPEN BY ACCIDENT!

A Tangled Tale

How did a scientist get chromosomes to separate?

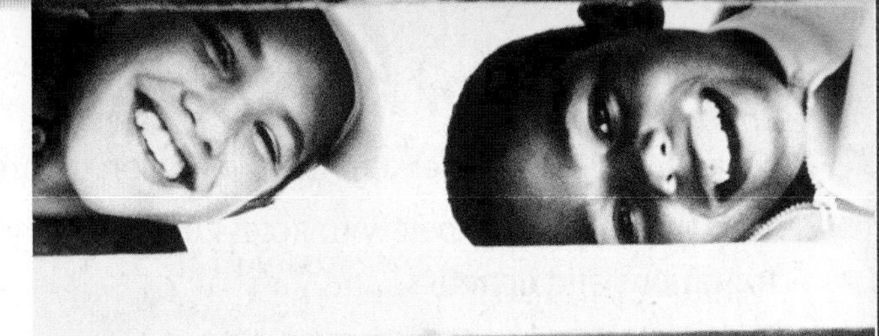

Thanks to chromosomes, each of us is unique!

Viewed under the microscope, chromosomes in cells sometimes look a lot like spaghetti. That's why scientists had such a hard time figuring out how many chromosomes are in each human cell. Imagine, then, how Dr. Tao-Chiuh Hsu (dow shew•SEW) must have felt when he looked into a microscope and saw "beautifully scattered chromosomes." The problem was, Hsu didn't know what he had done to separate the chromosomes into countable strands.

"I tried to study those slides and set up some more cultures to repeat the miracle," Hsu explained. "But nothing happened."

These chromosomes are magnified 500 times.

For three months Hsu tried changing every variable he could think of to make the chromosomes separate again. In April 1952, his efforts were finally rewarded. Hsu quickly realized that the chromosomes separated because of osmosis.

Osmosis is the movement of water molecules through cell membranes. This movement occurs in predictable ways. The water molecules move from areas with higher concentrations of water to areas with lower concentrations of water. In Hsu's case, the solution he used to prepare the cells had a higher concentration of water then the cell did. So water moved from the solution into the cell and the cell swelled until it finally exploded. The chromosomes suddenly were visible as separate strands.

What made the cells swell the first time? Apparently a technician had mixed the solution incorrectly. "Since nearly four months had elapsed, there was no way to trace who actually had prepared that particular [solution]," Hsu noted. "Therefore, this heroine must remain anonymous."

Research What developments led scientists to conclude that the human cell has 46 chromosomes? Visit the link shown to the right to get started.

For more information, visit life.msscience.com/oops

Reviewing Main Ideas

Section 1 Cell Division and Mitosis

1. The life cycle of a cell has two parts—growth and development, and cell division.
2. In mitosis, the nucleus divides to form two identical nuclei. Mitosis occurs in four continuous steps, or phases—prophase, metaphase, anaphase, and telophase.
3. Cell division in animal cells and plant cells is similar, but plant cells do not have centrioles and animal cells do not form cell walls.
4. Organisms use mitosis and cell division to grow, to replace cells, and for asexual reproduction. Asexual reproduction produces organisms with DNA identical to the parent's DNA. Fission, budding, and regeneration can be used for asexual reproduction.

Section 2 Sexual Reproduction and Meiosis

1. Sexual reproduction results when an egg and sperm join. This event is called fertilization, and the cell that forms is called the zygote.
2. Meiosis occurs in the reproductive organs, producing four haploid sex cells.
3. During meiosis, two divisions of the nucleus occur.
4. Meiosis ensures that offspring produced by fertilization have the same number of chromosomes as their parents.

Section 3 DNA

1. DNA is a large molecule made up of two twisted strands of sugar-phosphate molecules and nitrogen bases.
2. All cells contain DNA. The section of DNA on a chromosome that directs the making of a specific protein is a gene.
3. DNA can copy itself and is the pattern from which RNA is made. Messenger RNA, ribosomal RNA, and transfer RNA are used to make proteins.
4. Permanent changes in DNA are called mutations.

Visualizing Main Ideas

Copy and complete the spider diagram below about how organisms can use mitosis and cell division.

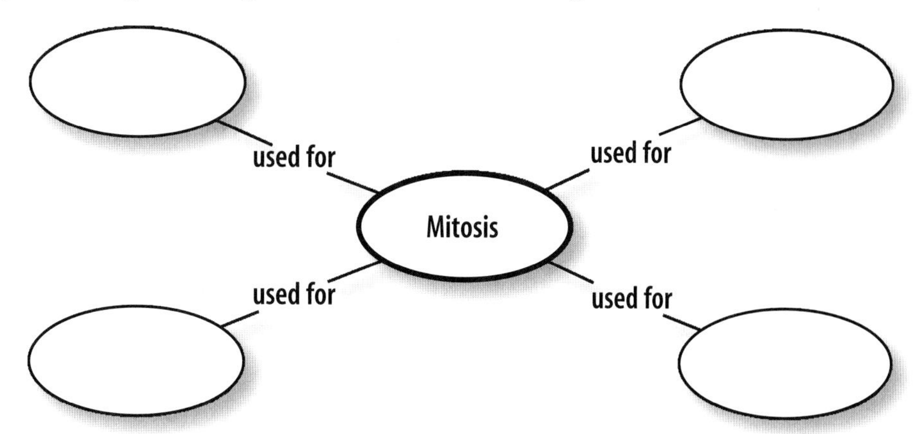

chapter 4 Review

Using Vocabulary

asexual reproduction p. 101	haploid p. 105
chromosome p. 98	meiosis p. 105
diploid p. 104	mitosis p. 98
DNA p. 110	mutation p. 114
egg p. 104	RNA p. 112
fertilization p. 104	sexual reproduction p. 104
gene p. 112	sperm p. 104
	zygote p. 104

Fill in the blanks with the correct vocabulary word or words.

1. _________ and _________ cells are sex cells.
2. _________ produces two identical cells.
3. An example of a nucleic acid is _________.
4. A(n) _________ is the code for a protein.
5. A(n) _________ sperm is formed during meiosis.
6. Budding is a type of _________.
7. A(n) _________ is a structure in the nucleus that contains hereditary material.
8. _________ produces four sex cells.
9. As a result of _________, a new organism develops that has its own unique identity.
10. An error made during the copying of DNA is called a(n) _________.

Checking Concepts

Choose the word or phrase that best answers the question.

11. Which of the following is a double spiral molecule with pairs of nitrogen bases?
    A) RNA **C)** protein
    B) amino acid **D)** DNA

12. What is in RNA but not in DNA?
    A) thymine **C)** adenine
    B) thyroid **D)** uracil

13. If a diploid tomato cell has 24 chromosomes, how many chromosomes will the tomato's sex cells have?
    A) 6 **C)** 24
    B) 12 **D)** 48

14. During a cell's life cycle, when do chromosomes duplicate?
    A) anaphase **C)** interphase
    B) metaphase **D)** telophase

15. When do chromatids separate during mitosis?
    A) anaphase **C)** metaphase
    B) prophase **D)** telophase

16. How is the hydra shown in the picture reproducing?
    A) asexually, by budding
    B) sexually, by budding
    C) asexually, by fission
    D) sexually, by fission

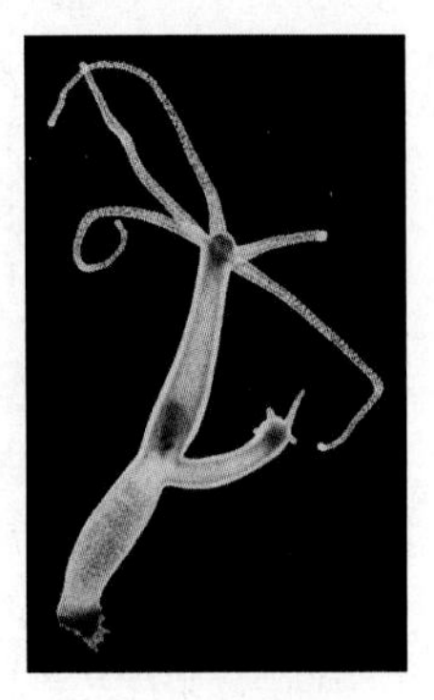

17. What is any permanent change in a gene or a chromosome called?
    A) fission **C)** replication
    B) reproduction **D)** mutation

18. What does meiosis produce?
    A) cells with the diploid chromosome number
    B) cells with identical chromosomes
    C) sex cells
    D) a zygote

19. What type of nucleic acid carries the codes for making proteins from the nucleus to the ribosome?
    A) DNA **C)** protein
    B) RNA **D)** genes

Science Online life.msscience.com/vocabulary_puzzlemaker

Thinking Critically

20. **List** the base sequence of a strand of RNA made using the DNA pattern ATCCGTC. Look at **Figure 14** for a hint.

21. **Predict** whether a mutation in a human skin cell can be passed on to the person's offspring. Explain.

22. **Explain** how a zygote could end up with an extra chromosome.

23. **Classify** Copy and complete this table about DNA and RNA.

DNA and RNA

	DNA	RNA
Number of strands		
Type of sugar	Do not write in this book.	
Letter names of bases		
Where found		

24. **Concept Map** Make an events-chain concept map of what occurs from interphase in the parent cell to the formation of the zygote. Tell whether the chromosome's number at each stage is haploid or diploid.

25. **Concept Map** Copy and complete the events-chain concept map of DNA synthesis.

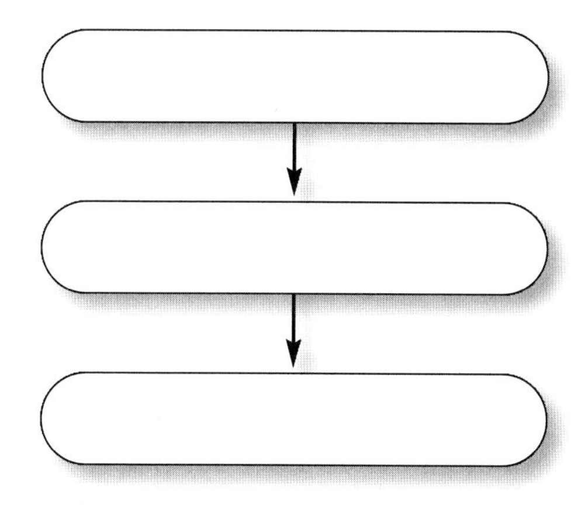

26. **Compare and Contrast** Meiosis is two divisions of a reproductive cell's nucleus. It occurs in a continuous series of steps. Compare and contrast the steps of meiosis I to the steps of meiosis II.

27. **Describe** what occurs in mitosis that gives the new cells identical DNA.

28. **Form a hypothesis** about the effect of an incorrect mitotic division on the new cells produced.

29. **Determine** how many chromosomes are in the original cell compared to those in the new cells formed by cell division. Explain.

Performance Activities

30. **Flash Cards** Make a set of 11 flash cards with drawings of a cell that show the different stages of meiosis. Shuffle your cards and then put them in the correct order. Give them to another student in the class to try.

Applying Math

31. **Cell Cycle** Assume an average human cell has a cell cycle of 20 hours. Calculate how many cells there would be after 80 hours.

Use the diagram below to answer question 32.

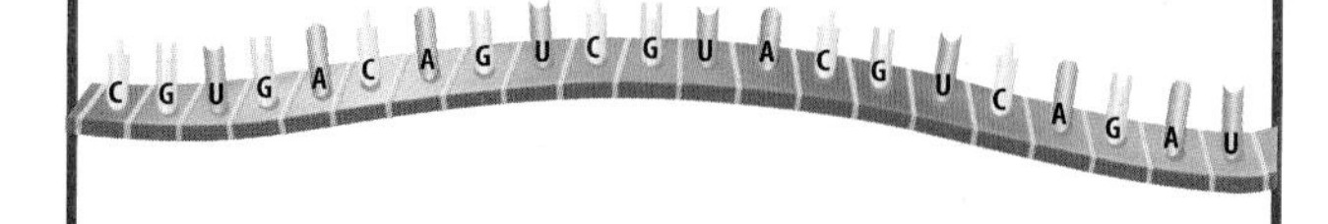

32. **Amino Acids** Sets of three nitrogen bases code for an amino acid. How many amino acids will make up the protein molecule that is coded for by the mRNA molecule above?

chapter 

Standardized Test Practice

Part 1 Multiple Choice

Record your answers on the answer sheet provided by your teacher or on a sheet of paper.

1. What stage of the cell cycle involves growth and function?
   A. prophase
   B. interphase
   C. mitosis
   D. cytoplasmic division

2. During interphase, which structure of a cell is duplicated?
   A. cell plate
   B. mitochondrion
   C. chromosome
   D. chloroplast

Use the figure below to answer questions 3 and 4.

3. What form of asexual reproduction is shown here?
   A. regeneration
   B. cell division
   C. sprouting
   D. meiosis

4. How does the genetic material of the new organism above compare to that of the parent organism?
   A. It is exactly the same.
   B. It is a little different.
   C. It is completely different.
   D. It is haploid.

5. Organisms with three or more sets of chromosomes are called
   A. monoploid.
   B. diploid.
   C. haploid.
   D. polyploid.

6. If a sex cell has eight chromosomes, how many chromosomes will there be after fertilization?
   A. 8
   B. 16
   C. 32
   D. 64

Use the diagram below to answer questions 7 and 8.

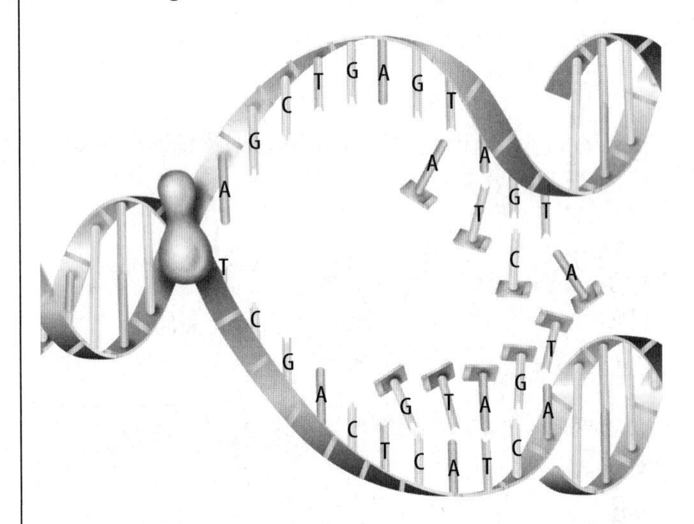

7. What does this diagram illustrate?
   A. DNA duplication
   B. RNA
   C. cell reproduction
   D. RNA synthesis

8. When does the process shown occur in the cell cycle?
   A. prophase
   B. metaphase
   C. interphase
   D. anaphase

9. Proteins are made of
   A. genes
   B. bases
   C. amino acids
   D. chromosomes

Test-Taking Tip

Prepare Avoid rushing on test day. Prepare your clothes and test supplies the night before. Wake up early and arrive at school on time on test day.

Part 2 Short Response/Grid In

Record your answers on the answer sheet provided by your teacher or on a sheet of paper.

10. In the human body, which cells are constantly dividing? Why is this important? How can this be potentially harmful?

11. Arrange the following terms in the correct order: *fertilization, sex cells, meiosis, zygote, mitosis.*

12. What are the three types of RNA used during protein synthesis? What is the function of each type of RNA?

13. Describe the relationship between gene, protein, DNA and chromosome.

Use the table below to answer question 14.

Phase of Cell Cycle	Action within the Cell
	Chromosomes duplicate
Prophase	
Metaphase	
	Chromosomes have separated
Telophase	

14. Fill in the blanks in the table with the appropriate term or definition.

15. What types of cells would constantly be in interphase?

16. Why is regeneration important for some organisms? In what way could regeneration of nerve cells be beneficial for humans?

17. What types of organisms are polyploidy? Why are they important?

18. What happens to chromosomes in meiosis I and meiosis II?

19. Describe several different ways that organisms can reproduce.

Part 3 Open Ended

Record your answers on a sheet of paper.

Use the photo below to answer question 20.

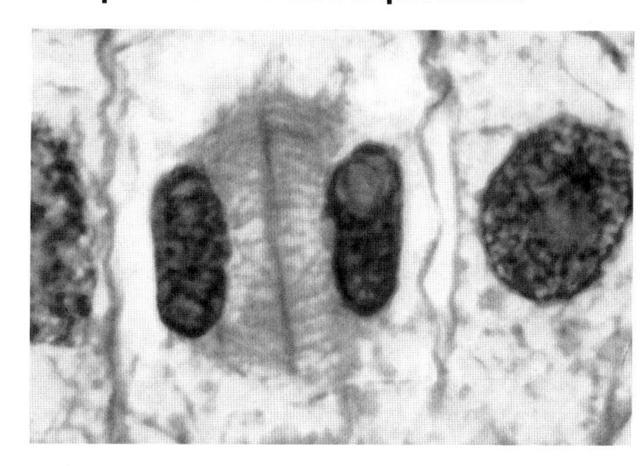

20. Is this a plant or an animal cell? Compare and contrast animal and plant cell division.

21. Describe in detail the structure of DNA.

Use the diagram below to answer question 22.

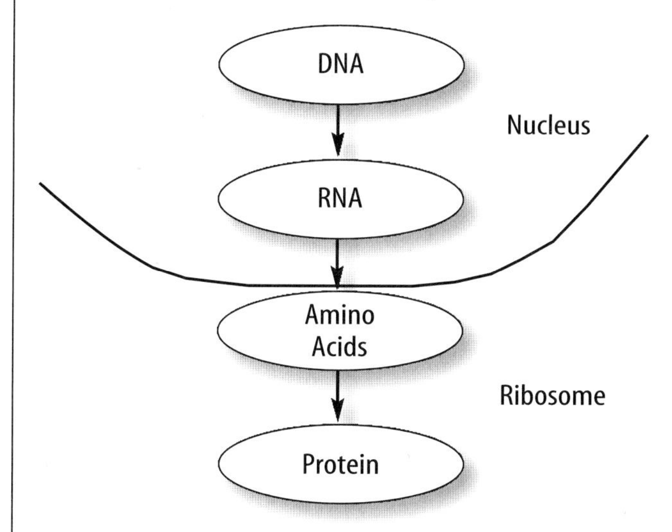

22. Discuss in detail what is taking place at each step of protein synthesis diagrammed above.

23. If a skin cell and a stomach cell have the same DNA then why are they so different?

24. What is mutation? Give examples where mutations could be harmful, beneficial or neutral.

chapter

Heredity

The BIG Idea

Inherited genes determine an organism's traits.

Section 1
Genetics

Main Idea Using scientific methods, Gregor Mendel discovered the basic principles of genetics.

Section 2
Genetics Since Mendel

Main Idea It is now known that interactions among alleles, genes, and the environment determine an organism's traits.

Section 3
Advances in Genetics

Main Idea Through genetic engineering, scientists can change the DNA of organisms to improve them, increase resistance to insects and diseases, or produce medicines.

Why do people look different?

People have different skin colors, different kinds of hair, and different heights. Knowing how these differences are determined will help you predict when certain traits might appear. This will help you understand what causes hereditary disorders and how these are passed from generation to generation.

Science Journal Write three traits that you have and how you would determine how those traits were passed to you.

Start-Up Activities

Who around you has dimples?

You and your best friend enjoy the same sports, like the same food, and even have similar haircuts. But, there are noticeable differences between your appearances. Most of these differences are controlled by the genes you inherited from your parents. In the following lab, you will observe one of these differences.

1. Notice the two students in the photographs. One student has dimples when she smiles, and the other student doesn't have dimples.
2. Ask your classmates to smile naturally. In your Science Journal, record the name of each classmate and whether each one has dimples.
3. **Think Critically** In your Science Journal, calculate the percentage of students who have dimples. Are facial dimples a common feature among your classmates?

Classify Characteristics As you read this chapter about heredity, you can use the following Foldable to help you classify characteristics as inherited or not inherited.

STEP 1 **Fold** the top of a vertical piece of paper down and the bottom up to divide the paper into thirds.

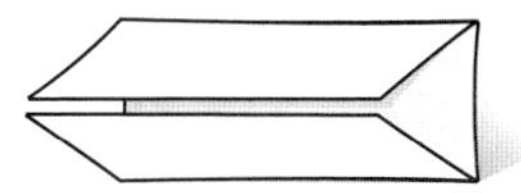

STEP 2 **Turn** the paper horizontally; **unfold and label** the three columns as shown.

Personal Characteristics | Inherited | Not Inherited
eyes
hair
dimples

Read for Main Ideas Before you read the chapter, list personal characteristics and predict which are inherited or not inherited. As you read the chapter, check and change your list.

Preview this chapter's content and activities at life.msscience.com

Get Ready to Read

Visualize

1 Learn It! Visualize by forming mental images of the text as you read. Imagine how the text descriptions look, sound, feel, smell, or taste. Look for any pictures or diagrams on the page that may help you add to your understanding.

2 Practice It! Read the following paragraph. As you read, use the underlined details to form a picture in your mind.

> In a Punnett square for predicting one trait, the letters representing the two alleles from one parent are written along the top of the grid, one letter per section. Those of the second parent are placed down the side of the grid, one letter per section. Each square of the grid is filled in with one allele donated by each parent. The letters that you use to fill in each of the squares represent the genotypes of possible offspring that the parents could produce.
>
> —*from page 131*

Based on the description above, try to visualize a Punnett square. Now look at the *Applying Math* feature on page 131.

- How closely do these Punnett squares match your mental picture?
- Reread the passage and look at the picture again. Did your ideas change?
- Compare your image with what others in your class visualized.

3 Apply It! Read the chapter and list three subjects you were able to visualize. Make a rough sketch showing what you visualized.

Reading Tip

Forming your own mental images will help you remember what you read.

Target Your Reading

Use this to focus on the main ideas as you read the chapter.

1. **Before you read** the chapter, respond to the statements below on your worksheet or on a numbered sheet of paper.
   - Write an **A** if you **agree** with the statement.
   - Write a **D** if you **disagree** with the statement.

2. **After you read** the chapter, look back to this page to see if you've changed your mind about any of the statements.
   - If any of your answers changed, explain why.
   - Change any false statements into true statements.
   - Use your revised statements as a study guide.

Before You Read A or D	Statement	After You Read A or D
	1 The two alleles of a gene can be the same or different.	
	2 Alleles are either dominant or recessive.	
	3 An organism's phenotype determines its genotype.	
	4 A Punnett square shows the actual genetics of offspring from two parents.	
	5 Traits are determined by more than one gene.	
	6 Some organisms inherit extra chromosomes.	
	7 A pedigrees chart can show the inheritance of a trait within a family.	
	8 The female parent determines whether an offspring will be male or female.	
	9 Genetically engineered organisms can produce medicines.	
	10 Sex-linked disorders are more common in females than in males.	

Science Online

Print out a worksheet of this page at life.msscience.com

section 1

Genetics

as you read

What You'll Learn

- **Explain** how traits are inherited.
- **Identify** Mendel's role in the history of genetics.
- **Use** a Punnett square to predict the results of crosses.
- **Compare and contrast** the difference between an individual's genotype and phenotype.

Why It's Important

Heredity and genetics help explain why people are different.

Review Vocabulary

meiosis: reproductive process that produces four haploid sex cells from one diploid cell

New Vocabulary

- heredity
- allele
- genetics
- hybrid
- dominant
- recessive
- Punnett square
- genotype
- phenotype
- homozygous
- heterozygous

Inheriting Traits

Do you look more like one parent or grandparent? Do you have your father's eyes? What about Aunt Isabella's cheekbones? Eye color, nose shape, and many other physical features are some of the traits that are inherited from parents, as **Figure 1** shows. An organism is a collection of traits, all inherited from its parents. **Heredity** (huh REH duh tee) is the passing of traits from parent to offspring. What controls these traits?

What is genetics? Generally, genes on chromosomes control an organism's form and function or traits. The different forms of a trait that make up a gene pair are called **alleles** (uh LEELZ). When a pair of chromosomes separates during meiosis (mi OH sus), alleles for each trait also separate into different sex cells. As a result, every sex cell has one allele for each trait. In **Figure 2,** the allele in one sex cell controls one form of the trait for having facial dimples. The allele in the other sex cell controls a different form of the trait—not having dimples. The study of how traits are inherited through the interactions of alleles is the science of **genetics** (juh NE tihks).

Figure 1 Note the strong family resemblance among these four generations.

Figure 2 An allele is one form of a gene. Alleles separate into separate sex cells during meiosis. In this example, the alleles that control the trait for dimples include *D,* the presence of dimples, and *d,* the absence of dimples.

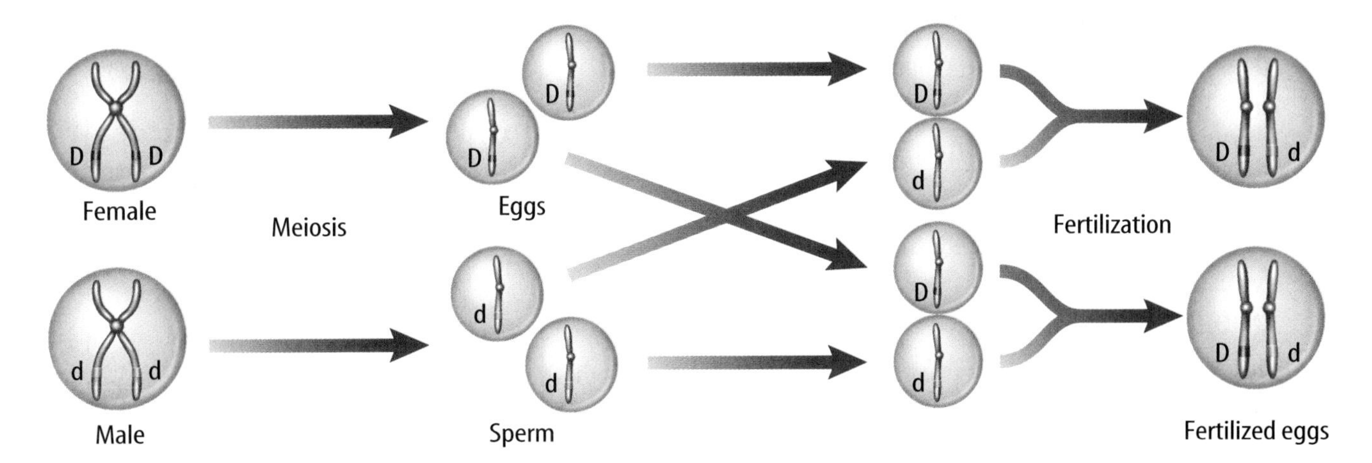

The alleles that control a trait are located on each duplicated chromosome.

During meiosis, duplicated chromosomes separate.

During fertilization, each parent donates one chromosome. This results in two alleles for the trait of dimples in the new individual formed.

Mendel—The Father of Genetics

Did you know that an experiment with pea plants helped scientists understand why your eyes are the color that they are? Gregor Mendel was an Austrian monk who studied mathematics and science but became a gardener in a monastery. His interest in plants began as a boy in his father's orchard where he could predict the possible types of flowers and fruits that would result from crossbreeding two plants. Curiosity about the connection between the color of a pea flower and the type of seed that same plant produced inspired him to begin experimenting with garden peas in 1856. Mendel made careful use of scientific methods, which resulted in the first recorded study of how traits pass from one generation to the next. After eight years, Mendel presented his results with pea plants to scientists.

Before Mendel, scientists mostly relied on observation and description, and often studied many traits at one time. Mendel was the first to trace one trait through several generations. He was also the first to use the mathematics of probability to explain heredity. The use of math in plant science was a new concept and not widely accepted then. Mendel's work was forgotten for a long time. In 1900, three plant scientists, working separately, reached the same conclusions as Mendel. Each plant scientist had discovered Mendel's writings while doing his own research. Since then, Mendel has been known as the father of genetics.

Topic: Genetics

Visit life.msscience.com for Web links to information about early genetics experiments.

Activity List two other scientists who studied genetics, and what organism they used in their research.

Table 1 Traits Compared by Mendel

Traits	Shape of Seeds	Color of Seeds	Color of Pods	Shape of Pods	Plant Height	Position of Flowers	Flower Color
Dominant trait	Round	Yellow	Green	Full	Tall	At leaf junctions	Purple
Recessive trait	Wrinkled	Green	Yellow	Flat, constricted	short	At tips of branches	White

Genetics in a Garden

Each time Mendel studied a trait, he crossed two plants with different expressions of the trait and found that the new plants all looked like one of the two parents. He called these new plants **hybrids** (HI brudz) because they received different genetic information, or different alleles, for a trait from each parent. The results of these studies made Mendel even more curious about how traits are inherited.

Garden peas are easy to breed for pure traits. An organism that always produces the same traits generation after generation is called a purebred. For example, tall plants that always produce seeds that produce tall plants are purebred for the trait of tall height. **Table 1** shows other pea plant traits that Mendel studied.

✔ Reading Check *Why might farmers plant purebred crop seeds?*

Dominant and Recessive Factors In nature, insects randomly pollinate plants as they move from flower to flower. In his experiments, Mendel used pollen from the flowers of purebred tall plants to pollinate by hand the flowers of purebred short plants. This process is called cross-pollination. He found that tall plants crossed with short plants produced seeds that produced all tall plants. Whatever caused the plants to be short had disappeared. Mendel called the tall form the **dominant** (DAH muh nunt) factor because it dominated, or covered up, the short form. He called the form that seemed to disappear the **recessive** (rih SE sihv) factor. Today, these are called dominant alleles and recessive alleles. What happened to the recessive form? **Figure 3** answers this question.

Comparing Common Traits

Procedure

1. Safely survey as many **dogs** in your neighborhood as you can for the presence of a solid color or spotted coat, short or long hair, and floppy or upright ears.
2. Make a data table that lists each of the traits. Record your data in the data table.

Analysis

1. Compare the number of dogs that have one form of a trait with those that have the other form.
2. What can you conclude about the variations you noticed in the dogs?

Try at Home

VISUALIZING MENDEL'S EXPERIMENTS

Figure 3

Gregor Mendel discovered that the experiments he carried out on garden plants provided an understanding of heredity. For eight years he crossed plants that had different characteristics and recorded how those characteristics were passed from generation to generation. One such characteristic, or trait, was the color of pea pods. The results of Mendel's experiment on pea pod color are shown below.

Parents

1st Generation

A One of the so-called "parent plants" in Mendel's experiment had pods that were green, a dominant trait. The other parent plant had pods that were yellow, a recessive trait.

B Mendel discovered that the two "parents" produced a generation of plants with green pods. The recessive color—yellow—did not appear in any of the pods.

2nd Generation

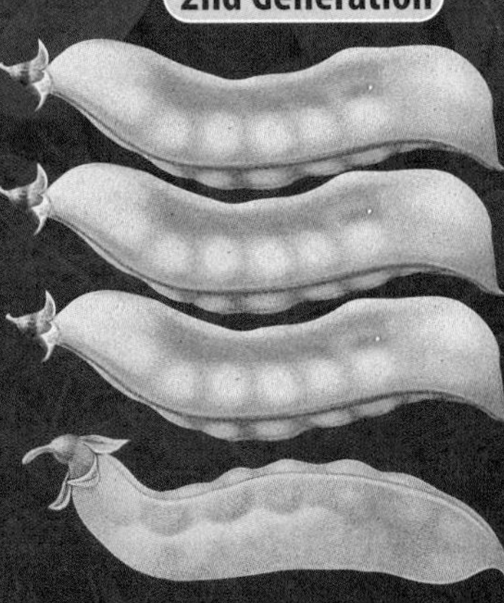

C Next, Mendel collected seeds from the first-generation plants and raised a second generation. He discovered that these second-generation plants produced plants with either green or yellow pods in a ratio of about three plants with green pods for every one plant with yellow pods. The recessive trait had reappeared. This 3:1 ratio proved remarkably consistent in hundreds of similar crosses, allowing Mendel to accurately predict the ratio of pod color in second-generation plants.

Using Probability to Make Predictions If you and your sister can't agree on what movie to see, you could solve the problem by tossing a coin. When you toss a coin, you're dealing with probabilities. Probability is a branch of mathematics that helps you predict the chance that something will happen. If your sister chooses tails while the coin is in the air, what is the probability that the coin will land tail-side up? Because a coin has two sides, there are two possible outcomes, heads or tails. Therefore, the probability of tails is one out of two, or 50 percent.

Mendel also dealt with probabilities. One of the things that made his predictions accurate was that he worked with large numbers of plants. He studied almost 30,000 pea plants over a period of eight years. By doing so, Mendel increased his chances of seeing a repeatable pattern. Valid scientific conclusions need to be based on results that can be duplicated.

Figure 4 This snapdragon's phenotype is red.
Determine *Can you tell what the flower's genotype for color is? Explain your answer.*

Punnett Squares Suppose you wanted to know what colors of pea plant flowers you would get if you pollinated white flowers on one pea plant with pollen from purple flowers on a different plant. How could you predict what the offspring would look like without making the cross? A handy tool used to predict results in Mendelian genetics is the **Punnett** (PUH nut) **square.** In a Punnett square, letters represent dominant and recessive alleles. An uppercase letter stands for a dominant allele. A lowercase letter stands for a recessive allele. The letters are a form of code. They show the **genotype** (JEE nuh tipe), or genetic makeup, of an organism. Once you understand what the letters mean, you can tell a lot about the inheritance of a trait in an organism.

The way an organism looks and behaves as a result of its genotype is its **phenotype** (FEE nuh tipe), as shown in **Figure 4.** If you have brown hair, then the phenotype for your hair color is brown.

Alleles Determine Traits Most cells in your body have at least two alleles for every trait. These alleles are located on similar pairs of chromosomes within the nucleus of cells. An organism with two alleles that are the same is called **homozygous** (hoh muh ZI gus) for that trait. For Mendel's peas, this would be written as *TT* (homozygous for the tall-dominant trait) or *tt* (homozygous for the short-recessive trait). An organism that has two different alleles is called **heterozygous** (he tuh roh ZI gus) for that trait. The hybrid plants Mendel produced were all heterozygous for height, *Tt.*

What is the difference between homozygous and heterozygous organisms?

Making a Punnett Square In a Punnett square for predicting one trait, the letters representing the two alleles from one parent are written along the top of the grid, one letter per section. Those of the second parent are placed down the side of the grid, one letter per section. Each square of the grid is filled in with one allele donated by each parent. The letters that you use to fill in each of the squares represent the genotypes of possible offspring that the parents could produce.

Applying Math — Calculate Percentages

PUNNET SQUARE One dog carries heterozygous, black-fur traits (*Bb*), and its mate carries homogeneous, blond-fur traits (*bb*). Use a Punnett square to determine the probability of one of their puppies having black fur.

Solution

1 *This is what you know:*

- dominant allele is represented by *B*
- recessive allele is represented by *b*

2 *This is what you need to find out:*

What is the probability of a puppy's fur color being black?

3 *This is the procedure you need to use:*

- Complete the Punnett square.
- There are two *Bb* genotypes and four possible outcomes.
- %(black fur) = $\frac{\text{number of ways to get black fur}}{\text{number of possible outcomes}}$

$$= \frac{2}{4} = \frac{1}{2} = 50\%$$

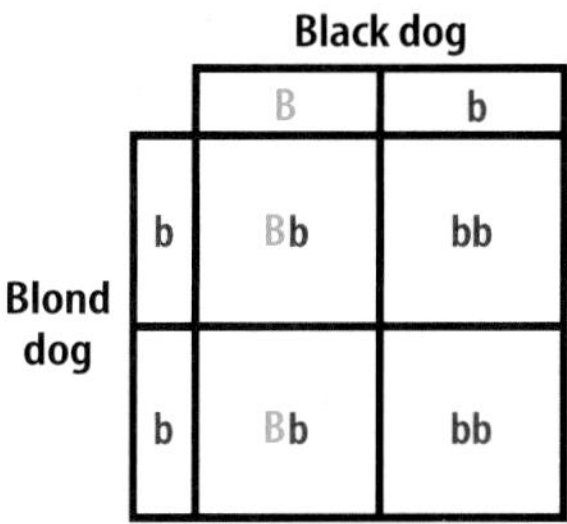

Genotypes of offspring: 2Bb, 2bb
Phenotypes of offspring: 2 black, 2 blond

4 *Check your answer:*

$\frac{1}{2}$ of 4 is 2, which is the number of black dogs.

Practice Problems

1. In peas, the color yellow (*Y*) is dominant to the color green (*y*). According to the Punnett square, what is the probability of an offspring being yellow?
2. What is the probability of an offspring having the *yy* genotype?

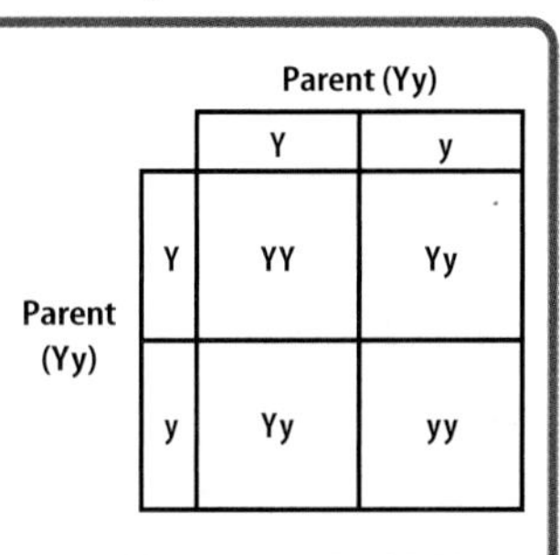

Science Online — For more practice, visit life.msscience.com/math_practice

Principles of Heredity Even though Gregor Mendel didn't know anything about DNA, genes, or chromosomes, he succeeded in beginning to describe and mathematically represent how inherited traits are passed from parents to offspring. He realized that some factor in the pea plant produced certain traits. Mendel also concluded that these factors separated when the pea plant reproduced. Mendel arrived at his conclusions after years of detailed observation, careful analysis, and repeated experimentation. **Table 2** summarizes Mendel's principles of heredity.

Table 2 Principles of Heredity

1	Traits are controlled by alleles on chromosomes.
2	An allele's effect is dominant or recessive.
3	When a pair of chromosomes separates during meiosis, the different alleles for a trait move into separate sex cells.

section 1 review

Summary

Inheriting Traits

- Heredity is the passing of traits from parent to offspring.
- Genetics is the study of how traits are inherited through the interactions of alleles.

Mendel—The Father of Genetics

- In 1856, Mendel began experimenting with garden peas, using careful scientific methods.
- Mendel was the first to trace one trait through several generations.
- In 1900, three plant scientists separately reached the same conclusions as Mendel.

Genetics in a Garden

- Hybrids receive different genetic information for a trait from each parent.
- Genetics involves dominant and recessive factors.
- Punnett squares can be used to predict the results of a cross.
- Mendel's conclusions led to the principles of heredity.

Self Check

1. **Contrast** Alleles are described as being dominant or recessive. What is the difference between a dominant and a recessive allele?
2. **Describe** how dominant and recessive alleles are represented in a Punnett square.
3. **Explain** the difference between genotype and phenotype. Give examples.
4. **Infer** Gregor Mendel, an Austrian monk who lived in the 1800s, is known as the father of genetics. Explain why Mendel has been given this title.
5. **Think Critically** If an organism expresses a recessive phenotype, can you tell the genotype? Explain your answer by giving an example.

Applying Math

6. **Use Percentages** One fruit fly is heterozygous for long wings, and another fruit fly is homozygous for short wings. Long wings are dominant to short wings. Use a Punnett square to find the expected percent of offspring with short wings.

Science Online life.msscience.com/self_check_quiz

Predicting Results

Could you predict how many brown rabbits would result from crossing two heterozygous black rabbits? Try this investigation to find out. Brown color is a recessive trait for hair color in rabbits.

Real-World Question

How does chance affect combinations of genes?

Goals

- **Model** chance events in heredity.
- **Compare and contrast** predicted and actual results.

Materials

paper bags (2)
red beans (100)
white beans (100)

Safety Precautions

WARNING: *Do not taste, eat, or drink any materials used in the lab.*

Procedure

1. Make a Punnett square for a cross between two heterozygous black rabbits, ($Bb \subseteq Bb$). *B* represents the black allele and b represents the brown allele.
2. Model the above cross by placing 50 red beans and 50 white beans in one paper bag and 50 red beans and 50 white beans in a second bag. Red beans represent black alleles and white beans represent brown alleles.
3. Label one of the bags *Female* for the female parent. Label the other bag *Male* for the male parent.
4. Use a data table to record the combination each time you remove two beans. Your table will need to accommodate 50 picks.
5. Without looking, remove one bean from each bag and record the results on your data table. Return the beans to their bags.
6. Repeat step five 49 more times.
7. **Count and record** the total numbers for each of the three combinations in your data table.
8. **Compile and record** the class totals.

Conclude and Apply

1. **Name** the combination that occurred most often.
2. **Calculate** the ratio of red/red to red/white to white/white. What hair color in rabbits do these combinations represent?
3. **Compare** your predicted (expected) results with your observed (actual) results.
4. **Hypothesize** how you could get predicted results to be closer to actual results.

Gene Combinations

Rabbits	Red/ Red	Red/ White	White/ White
Your total	Do not write in this book.		
Class total			

Communicating Your Data

Write a paragraph that clearly describes your results. Have another student read your paragraph. Ask if he or she could understand what happened. If not, rewrite your paragraph and have the other student read it again. **For more help, refer to the Science Skill Handbook.**

section 2

Genetics Since Mendel

as you read

What You'll Learn

- **Explain** how traits are inherited by incomplete dominance.
- **Compare** multiple alleles and polygenic inheritance, and give examples of each.
- **Describe** two human genetic disorders and how they are inherited.
- **Explain** how sex-linked traits are passed to offspring.

Why It's Important

Most of your inherited traits involve more complex patterns of inheritance than Mendel discovered.

Review Vocabulary
gene: section of DNA on a chromosome that contains instructions for making specific proteins

New Vocabulary
- incomplete dominance
- polygenic inheritance
- sex-linked gene

Incomplete Dominance

Not even in science do things remain the same. After Mendel's work was rediscovered in 1900, scientists repeated his experiments. For some plants, such as peas, Mendel's results proved true. However, when different plants were crossed, the results were sometimes different. One scientist crossed purebred red-flowered four-o'clock plants with purebred white-flowered four-o'clock plants. He expected to get all red flowers, but they were pink. Neither allele for flower color seemed dominant. Had the colors become blended like paint colors? He crossed the pink-flowered plants with each other, and red, pink, and white flowers were produced. The red and white alleles had not become blended. Instead, when the allele for white flowers and the allele for red flowers combined, the result was an intermediate phenotype—a pink flower.

When the offspring of two homozygous parents show an intermediate phenotype, this inheritance is called **incomplete dominance.** Other examples of incomplete dominance include the flower color of some plant breeds and the coat color of some horse breeds, as shown in **Figure 5.**

Chestnut horse

Cremello horse

Figure 5 When a chestnut horse is bred with a cremello horse, all offspring will be palomino. The Punnett square shown on the opposite page can be used to predict this result. **Explain** *how the color of the palomino horse shows that the coat color of horses may be inherited by incomplete dominance.*

Multiple Alleles Mendel studied traits in peas that were controlled by just two alleles. However, many traits are controlled by more than two alleles. A trait that is controlled by more than two alleles is said to be controlled by multiple alleles. Traits controlled by multiple alleles produce more than three phenotypes of that trait.

Imagine that only three types of coins are made—nickels, dimes, and quarters. If every person can have only two coins, six different combinations are possible. In this problem, the coins represent alleles of a trait. The sum of each two-coin combination represents the phenotype. Can you name the six different phenotypes possible with two coins?

Blood type in humans is an example of multiple alleles that produce only four phenotypes. The alleles for blood types are called A, B, and O. The O allele is recessive to both the A and B alleles. When a person inherits one A allele and one B allele for blood type, both are expressed—phenotype AB. A person with phenotype A blood has the genetic makeup, or genotype—AA or AO. Someone with phenotype B blood has the genotype BB or BO. Finally, a person with phenotype O blood has the genotype OO.

What are the six different blood type genotypes?

Science Online

Topic: Blood Types

Visit life.msscience.com for Web links to information about the importance of blood types in blood transfusions.

Activity Make a chart showing which blood types can be used for transfusions into people with A, B, AB, or O blood phenotypes.

Palomino horse

Punnett square

	Chestnut horse (CC): C	C
Cremello horse (C'C'): C'	CC'	CC'
C'	CC'	CC'

Genotypes: All CC'
Phenotypes: All palomino horses

Interpreting Polygenic Inheritance

Procedure

1. Measure the hand spans of your classmates.
2. Using a **ruler,** measure from the tip of the thumb to the tip of the little finger when the hand is stretched out. Read the measurement to the nearest centimeter.
3. Record the name and hand-span measurement of each person in a data table.

Analysis

1. What range of hand spans did you find?
2. What are the mean, median, and mode for your class's data?
3. Are hand spans inherited as a simple Mendelian pattern or as a polygenic or incomplete dominance pattern? Explain.

Polygenic Inheritance

Eye color is an example of a trait that is produced by a combination of many genes. **Polygenic** (pah lih JEH nihk) **inheritance** occurs when a group of gene pairs acts together to produce a trait. The effects of many alleles produces a wide variety of phenotypes. For this reason, it may be hard to classify all the different shades of eye color.

Your height and the color of your eyes and skin are just some of the many human traits controlled by polygenic inheritance. It is estimated that three to six gene pairs control your skin color. Even more gene pairs might control the color of your hair and eyes. The environment also plays an important role in the expression of traits controlled by polygenic inheritance. Polygenic inheritance is common and includes such traits as grain color in wheat and milk production in cows. Egg production in chickens is also a polygenic trait.

Impact of the Environment Your environment plays a role in how some of your genes are expressed or whether they are expressed at all, as shown in **Figure 6.** Environmental influences can be internal or external. For example, most male birds are more brightly colored than females. Chemicals in their bodies determine whether the gene for brightly colored feathers is expressed.

Although genes determine many of your traits, you might be able to influence their expression by the decisions you make. Some people have genes that make them at risk for developing certain cancers. Whether they get cancer might depend on external environmental factors. For instance, if some people at risk for skin cancer limit their exposure to the Sun and take care of their skin, they might never develop cancer.

Reading Check *What environmental factors might affect the size of leaves on a tree?*

Figure 6 Himalayan rabbits have alleles for dark-colored fur. However, this allele is able to express itself only at lower temperatures. Only the areas located farthest from the rabbit's main body heat (ears, nose, feet, tail) have dark-colored fur.

Human Genes and Mutations

Sometimes a gene undergoes a change that results in a trait that is expressed differently. Occasionally errors occur in the DNA when it is copied inside of a cell. Such changes and errors are called mutations. Not all mutations are harmful. They might be helpful or have no effect on an organism.

Certain chemicals are known to produce mutations in plants or animals, including humans. X rays and radioactive substances are other causes of some mutations. Mutations are changes in genes.

Chromosome Disorders In addition to individual mutations, problems can occur if the incorrect number of chromosomes is inherited. Every organism has a specific number of chromosomes. However, mistakes in the process of meiosis can result in a new organism with more or fewer chromosomes than normal. A change in the total number of human chromosomes is usually fatal to the unborn embryo or fetus, or the baby may die soon after birth.

Look at the photo of human chromosome 21 in **Figure 7.** If three copies of this chromosome are in the fertilized human egg, Down syndrome results. Individuals with Down syndrome can be short, exhibit learning disabilities, and have heart problems. Such individuals can lead normal lives if they have no severe health complications.

Genetic Counselor Testing for genetic disorders may allow many affected individuals to seek treatment and cope with their diseases. Genetic counselors are trained to analyze a family's history to determine a person's health risk. Research what a genetic counselor does and how to become a genetic counselor. Record what you learn in your Science Journal.

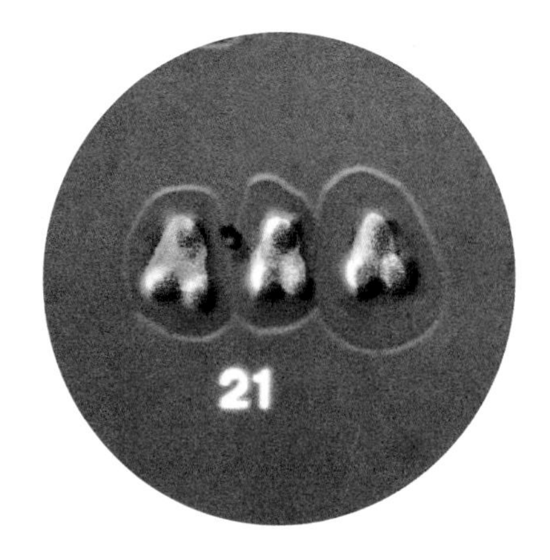

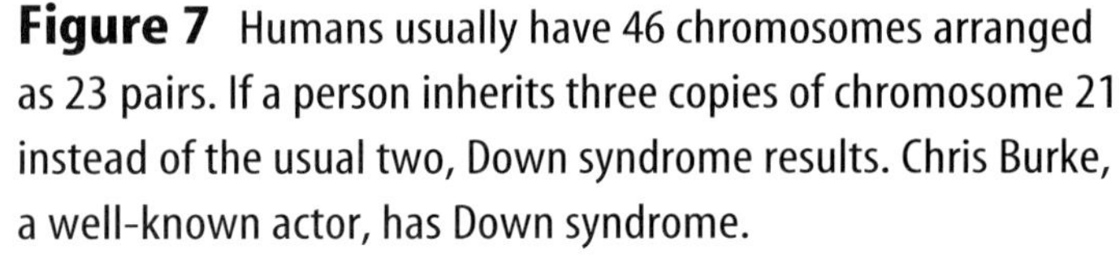

Figure 7 Humans usually have 46 chromosomes arranged as 23 pairs. If a person inherits three copies of chromosome 21 instead of the usual two, Down syndrome results. Chris Burke, a well-known actor, has Down syndrome.

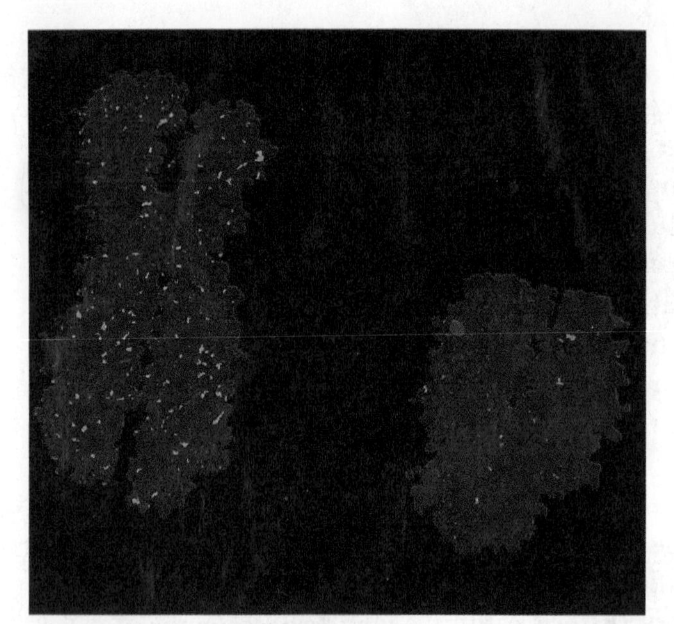

Color-enhanced SEM Magnification: 16000×

Figure 8 Sex in many organisms is determined by X and Y chromosomes.
Observe *How do the X (left) and Y (right) chromosomes differ from one another in shape and size?*

Recessive Genetic Disorders

Many human genetic disorders, such as cystic fibrosis, are caused by recessive genes. Some recessive genes are the result of a mutation within the gene. Many of these alleles are rare. Such genetic disorders occur when both parents have a recessive allele for this disorder. Because the parents are heterozygous, they don't show any symptoms. However, if each parent passes the recessive allele to the child, the child inherits both recessive alleles and will have a recessive genetic disorder.

✓ **Reading Check** *How is cystic fibrosis inherited?*

Cystic fibrosis is a homozygous recessive disorder. It is the most common genetic disorder that can lead to death among Caucasian Americans. In most people, a thin fluid is produced that lubricates the lungs and intestinal tract. People with cystic fibrosis produce thick mucus instead of this thin fluid. The thick mucus builds up in the lungs and makes it hard to breathe. This buildup often results in repeated bacterial respiratory infections. The thick mucus also reduces or prevents the flow of substances necessary for digesting food. Physical therapy, special diets, and new drug therapies have increased the life spans of patients with cystic fibrosis.

Gender Determination

What determines the gender or sex of an individual? Much information on gender inheritance came from studies of fruit flies. Fruit flies have only four pairs of chromosomes. Because the chromosomes are large and few in number, they are easy to study. Scientists identified one pair that contains genes that determine the sex of the organism. They labeled the pair XX in females and XY in males. Geneticists use these labels when studying organisms, including humans. You can see human X and Y chromosomes in **Figure 8.**

Each egg produced by a female normally contains one X chromosome. Males produce sperm that normally have either an X or a Y chromosome. When a sperm with an X chromosome fertilizes an egg, the offspring is a female, XX. A male offspring, XY, is the result of a Y-containing sperm fertilizing an egg. What pair of sex chromosomes is in each of your cells? Sometimes chromosomes do not separate during meiosis. When this occurs, an individual can inherit an abnormal number of sex chromosomes.

Sex-Linked Disorders

A **sex-linked gene** is an allele on a sex chromosome. Some conditions that result from inheriting a sex-linked gene are called sex-linked disorders. Red-green color blindness in humans is a sex-linked disorder because the related genes are on the X chromosome. People who inherit this disorder have difficulty seeing the difference between green and red, and sometimes, yellow. This condition is a recessive sex-linked disorder. A female is color-blind when each of her X chromosomes has the recessive allele. A male has only one X chromosome and, if it has the recessive allele, he will be color-blind.

Dominant sex-linked disorders are rare and result when a person inherits at least one dominant sex-linked allele. Vitamin D-resistant rickets is an X-linked dominant disorder. The kidneys of an affected person cannot absorb adequate amounts of phosphorus. The person might have low blood-phosphorus levels, soft bones, and poor teeth formation.

Pedigrees Trace Traits

How can you trace a trait through a family? A pedigree is a visual tool for following a trait through generations of a family. Males are represented by squares and females by circles. A completely filled circle or square shows that the trait is seen in that person. Half-colored circles or squares indicate carriers. A carrier is heterozygous for the trait, and it is not seen. People represented by empty circles or squares do not have the trait and are not carriers. The pedigree in **Figure 9** shows how the trait for color blindness is carried through a family.

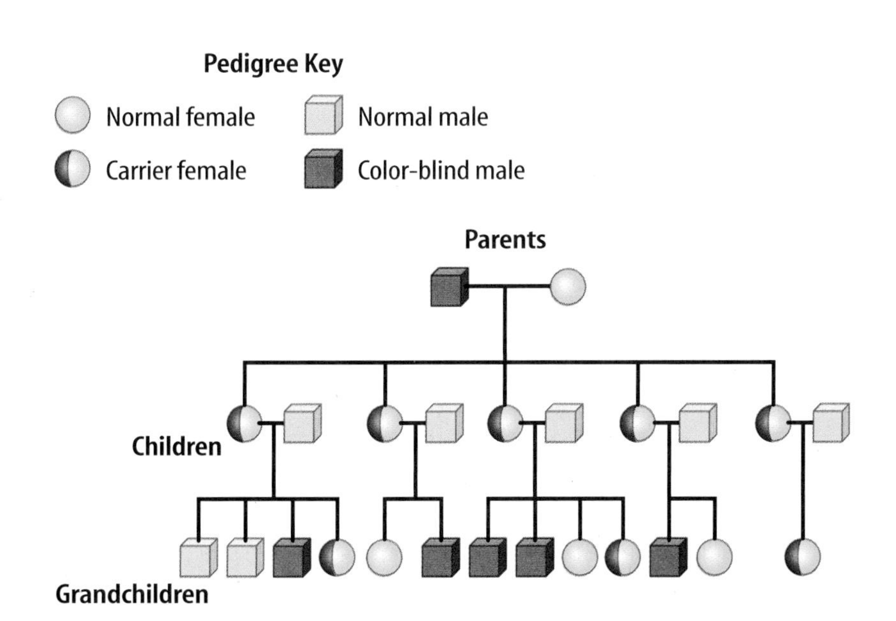

Figure 9 The symbols in this pedigree's key mean the same thing on all pedigree charts. The grandfather in this family was color-blind and married to a woman who was not a carrier of the color-blind allele.
Infer *why no women in this family are color-blind.*

Shih Tzu

Black Labrador

Figure 10
A variety of traits are considered when breeding dogs.

Using Pedigrees A pedigree is a useful tool for a geneticist. Sometimes a geneticist needs to understand who has had a trait in a family over several generations to determine its pattern of inheritance. A geneticist determines if a trait is recessive, dominant, sex-linked, or has some other pattern of inheritance. When geneticists understand how a trait is inherited, they can predict the probability that a baby will be born with a specific trait.

Pedigrees also are important in breeding animals or plants. Because livestock and plant crops are used as sources of food, these organisms are bred to increase their yield and nutritional content. Breeders of pets and show animals, like the dogs pictured in **Figure 10,** also examine pedigrees carefully for possible desirable physical and ability traits. Issues concerning health also are considered when researching pedigrees.

section 2 review

Summary

Incomplete Dominance

- Incomplete dominance is when a dominant and recessive allele for a trait show an intermediate phenotype.
- Many traits are controlled by more than two alleles.
- A wide variety of phenotypes is produced by polygenic inheritance.

Human Genes and Mutations

- Errors can occur when DNA is copied.
- Mistakes in meiosis can result in an unequal number of chromosomes in sex cells.
- Recessive genes control many human genetic disorders.

Sex Determination

- An allele inherited on a sex chromosome is called a sex-linked gene.
- Pedigrees are visual tools to trace a trait through generations of a family.

Self Check

1. **Compare** how inheritance by multiple alleles and polygenic inheritance are similar.
2. **Explain** why a trait inherited by incomplete dominance is not a blend of two alleles.
3. **Discuss** Choose two genetic disorders and discuss how they are inherited.
4. **Apply** Using a Punnett square, explain why males are affected more often than females by sex-linked genetic disorders.
5. **Think Critically** Why wouldn't a horse breeder mate male and female palominos to get palomino colts?

Applying Skills

6. **Predict** A man with blood type B marries a woman with blood type A. Their first child has blood type O. Use a Punnett square to predict what other blood types are possible for their offspring.
7. **Communicate** In your Science Journal, explain why offspring may or may not resemble either parent.

Science Online life.msscience.com/self_check_quiz

section 3

Advances in Genetics

Why is genetics important?

If Mendel were to pick up a daily newspaper in any country today, he'd probably be surprised. News articles about developments in genetic research appear almost daily. The term *gene* has become a common word. The principles of heredity are being used to change the world.

Genetic Engineering

You might recall that chromosomes are made of DNA and are in the nucleus of a cell. Sections of DNA in chromosomes that direct cell activities are called genes. Through **genetic engineering,** scientists are experimenting with biological and chemical methods to change the arrangement of DNA that makes up a gene. Genetic engineering already is used to help produce large volumes of medicine. Genes also can be inserted into cells to change how those cells perform their normal functions, as shown in **Figure 11.** Other research is being done to find new ways to improve crop production and quality, including the development of plants that are resistant to disease.

as you read

What You'll Learn

- **Evaluate** the importance of advances in genetics.
- **Sequence** the steps in making genetically engineered organisms.

Why It's Important

Advances in genetics can affect your health, the foods that you eat, and your environment.

Review Vocabulary
DNA: deoxyribonucleic acid; the genetic material of all organisms

New Vocabulary
- genetic engineering

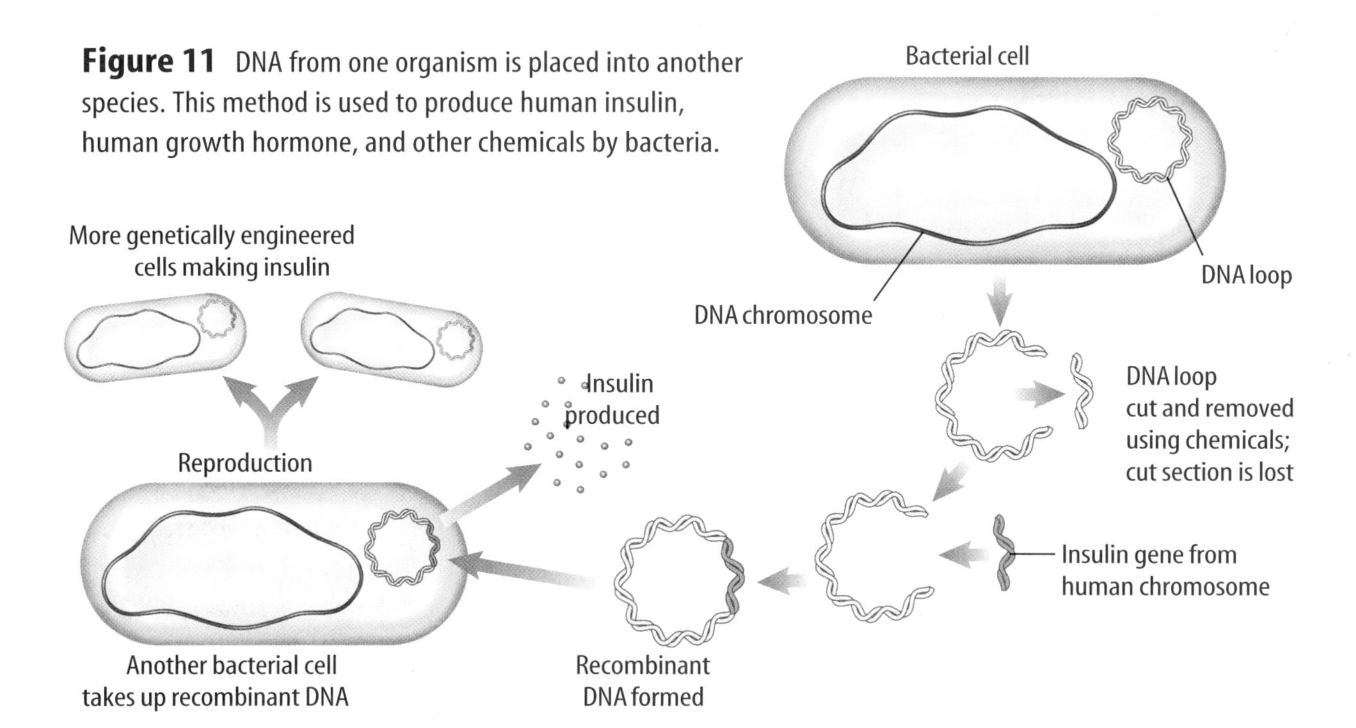

Figure 11 DNA from one organism is placed into another species. This method is used to produce human insulin, human growth hormone, and other chemicals by bacteria.

Genetically Engineered Crops Crop plants are now being genetically engineered to produce chemicals that kill specific pests that feed on them. Some of the pollen from pesticide-resistant canola crops is capable of spreading up to 8 km from the plant, while corn and potato pollen can spread up to 1 km. What might be the effects of pollen landing on other plants?

Recombinant DNA Making recombinant DNA is one method of genetic engineering. Recombinant DNA is made by inserting a useful segment of DNA from one organism into a bacterium, as illustrated in **Figure 11.** Large quantities of human insulin are made by some genetically engineered organisms. People with Type 1 diabetes need this insulin because their pancreases produce too little or no insulin. Other uses include the production of growth hormone to treat dwarfism and chemicals to treat cancer.

Gene Transfer Another application of genetic-engineering is gene transfer. A goal of this experimental procedure is to replace abnormal genetic material with normal genetic material. First, normal DNA or RNA is placed in a virus. Then the virus delivers the normal DNA or RNA to target cells, as shown in **Figure 12.** Gene transfer, also known as gene therapy, might help correct genetic disorders such as cystic fibrosis. It also is being studied as a possible treatment for cancer, heart disease, and certain infectious diseases.

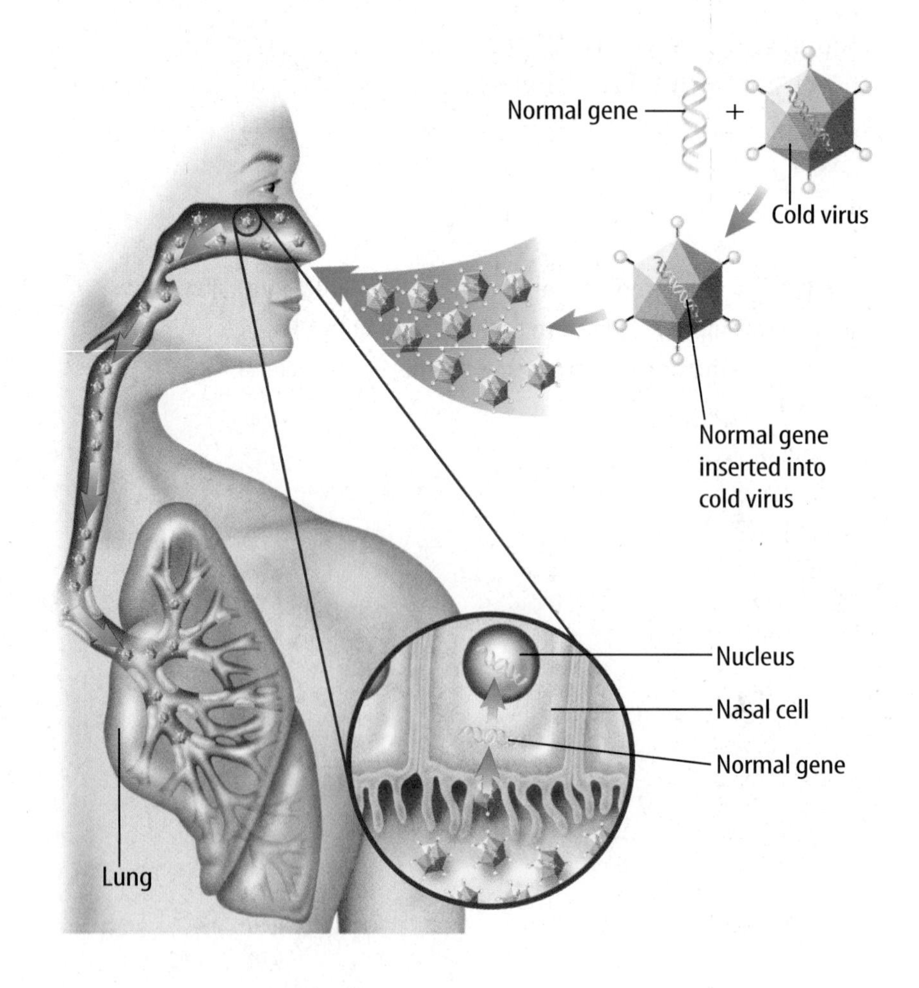

Figure 12 Gene transfer involves placing normal genetic material into a cell with abnormal genetic material. When the normal genetic material begins to function, the abnormal condition is corrected.

Genetically Engineered Plants For thousands of years people have improved the plants they use for food and clothing even without the knowledge of genotypes. Until recently, these improvements were the results of selecting plants with the most desired traits to breed for the next generation. This process is called selective breeding. Recent advances in genetics have not replaced selective breeding. Although a plant can be bred for a particular phenotype, the genotype and pedigree of the plants also are considered.

Genetic engineering can produce improvements in crop plants, such as corn, wheat, and rice. One type of genetic engineering involves finding the genes that produce desired traits in one plant and then inserting those genes into a different plant. Scientists recently have made genetically engineered tomatoes with a gene that allows tomatoes to be picked green and transported great distances before they ripen completely. Ripe, firm tomatoes are then available in the local market. In the future, additional food crops may be genetically engineered so that they are not desirable food for insects.

Figure 13 Genetically engineered produce is sometimes labeled. This allows consumers to make informed choices about their foods.

What other types of traits would be considered desirable in plants?

Because some people might prefer foods that are not changed genetically, some stores label such produce, as shown in **Figure 13.** The long-term effects of consuming genetically engineered plants are unknown.

section 3 review

Summary

Why is genetics important?

- Developments in genetic research appear in newspapers almost daily.
- The world is being changed by the principles of heredity.

Genetic Engineering

- Scientists work with biological and chemical methods to change the arrangement of DNA that makes up a gene.
- One method of genetic engineering is making recombinant DNA.
- A normal allele is replaced in a virus and then delivers the normal allele when it infects its target cell.

Self Check

1. **Apply** Give examples of areas in which advances in genetics are important.
2. **Compare and contrast** the technologies of using recombinant DNA and gene therapy.
3. **Infer** What are some benefits of genetically engineered crops?
4. **Describe** how selective breeding differs from genetic engineering.
5. **Think Critically** Why might some people be opposed to genetically engineered plants?

Applying Skills

6. **Concept Map** Make an events-chain concept map of the steps used in making recombinant DNA.

Design Your Own

Tests for Color Blindness

Goals

- **Design** an experiment that tests for a specific type of color blindness in males and females.
- **Calculate** the percentage of males and females with the disorder.

Possible Materials

white paper or poster board
colored markers: red, orange, yellow, bright green, dark green, blue
**computer and color printer*
**Alternate materials*

Real-World Question

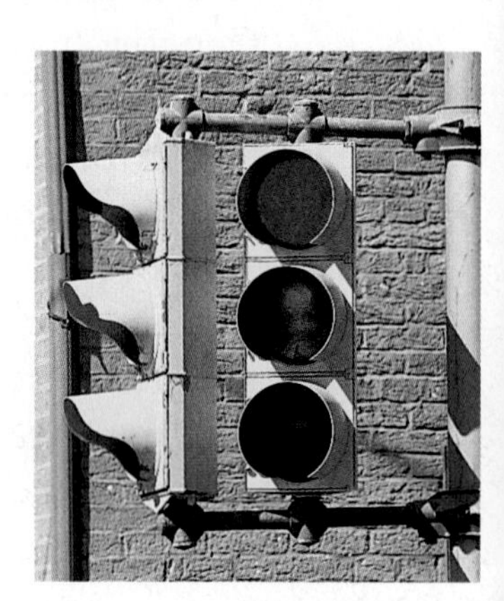

What colors do color-blind people see? That depends on the type of color blindness that they inherit. The most common type is red-green color blindness in which people have difficulty seeing any difference between red and green. People with another inherited type cannot distinguish between blue and yellow. In rare instances, a person can inherit a type of color blindness where the only colors seen are shades of gray. What percentages of males and females in your school are color-blind?

Form a Hypothesis

Based on your reading and your own experiences, form a hypothesis about how common color blindness is among males and females.

Test Your Hypothesis

Make a Plan

1. Decide what type of color blindness you will test for—the common green-red color blindness or the more rare green-blue color blindness.
2. **List** the materials you will need and describe how you will create test pictures. Tests for color blindness use many circles of red, orange, and yellow as a background, with circles of dark and light green to make a picture or number. List the steps you will take to test your hypothesis.
3. Prepare a data table in your Science Journal to record your test results.

4. **Examine** your experiment to make sure all steps are in logical order.
5. **Identify** which pictures you will use as a control and which pictures you will use as variables.

Follow Your Plan

1. Make sure your teacher approves your plan before you start.
2. **Draw** the pictures that you will use to test for color blindness.
3. Carry out your experiment as planned and record your results in your data table.

Analyze Your Data

1. **Calculate** the percentage of males and females that tested positive for color blindness.
2. **Compare** the frequency of color blindness in males with the frequency of color blindness in females.

Conclude and Apply

1. **Explain** whether or not the results supported your hypothesis.
2. **Explain** why color blindness is called a sex-linked disorder.
3. **Infer** how common the color-blind disorder is in the general population.
4. **Predict** your results if you were to test a larger number of people.

Communicating Your Data

Using a word processor, write a short article for the advice column of a fashion magazine about how a color-blind person can avoid wearing outfits with clashing colors. **For more help, refer to the Science Skill Handbook.**

SCIENCE Stats

The Human Genome

Did you know...

. . . The biggest advance in genetics in years took place in February 2001. Scientists successfully mapped the human genome. There are 30,000 to 40,000 genes in the human genome. Genes are in the nucleus of each of the several trillion cells in your body.

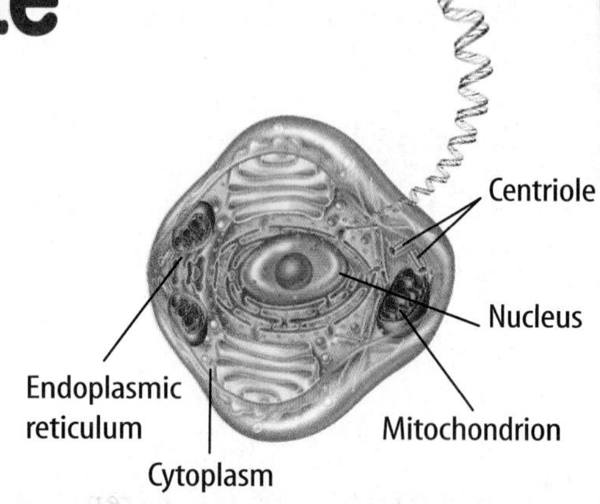

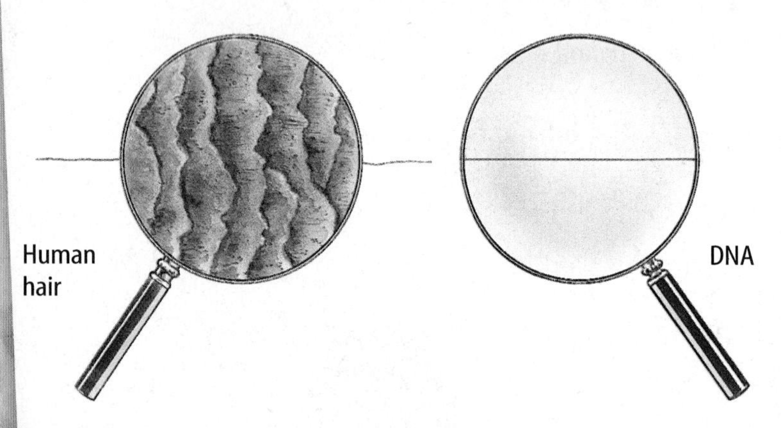

. . . The strands of DNA in the human genome, if unwound and connected end to end, would be more than 1.5 m long—but only about 130 trillionths of a centimeter wide. Even an average human hair is as much as 200,000 times wider than that.

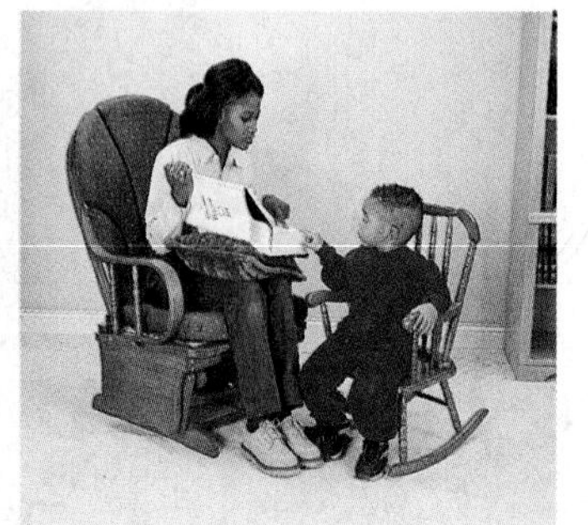

. . . It would take about nine and one-half years to read aloud without stopping the 3 billion bits of instructions (called base pairs) in your genome.

Applying Math If one million base pairs of DNA take up 1 megabyte of storage space on a computer, how many gigabytes (1,024 megabytes) would the whole genome fill?

Find Out About It

Human genome scientists hope to identify the location of disease-causing genes. Visit life.msscience.com/science_stats to research a genetic disease and share your results with your class.

chapter 5 Study Guide

Reviewing Main Ideas

Section 1 Genetics

1. Genetics is the study of how traits are inherited. Gregor Mendel determined the basic laws of genetics.
2. Traits are controlled by alleles on chromosomes.
3. Some alleles can be dominant or recessive.
4. When a pair of chromosomes separates during meiosis, the different alleles move into separate sex cells. Mendel found that he could predict the outcome of genetic crosses.

Section 2 Genetics Since Mendel

1. Inheritance patterns studied since Mendel include incomplete dominance, multiple alleles, and polygenic inheritance.
2. These inheritance patterns allow a variety of phenotypes to be produced.
3. Some disorders are the results of inheritance and can be harmful and even deadly.
4. Pedigree charts help reveal patterns of the inheritance of a trait in a family. Pedigrees show that sex-linked traits are expressed more often in males than in females.

Section 3 Advances in Genetics

1. Genetic engineering uses biological and chemical methods to change genes.
2. Recombinant DNA is one method of genetic engineering to make useful chemicals, including hormones.
3. Gene transfer shows promise for correcting many human genetic disorders, cancer, and other diseases.
4. Breakthroughs in the field of genetic engineering are allowing scientists to do many things, such as producing plants that are resistant to disease.

Visualizing Main Ideas

Examine the following pedigree for diabetes and explain the inheritance pattern.

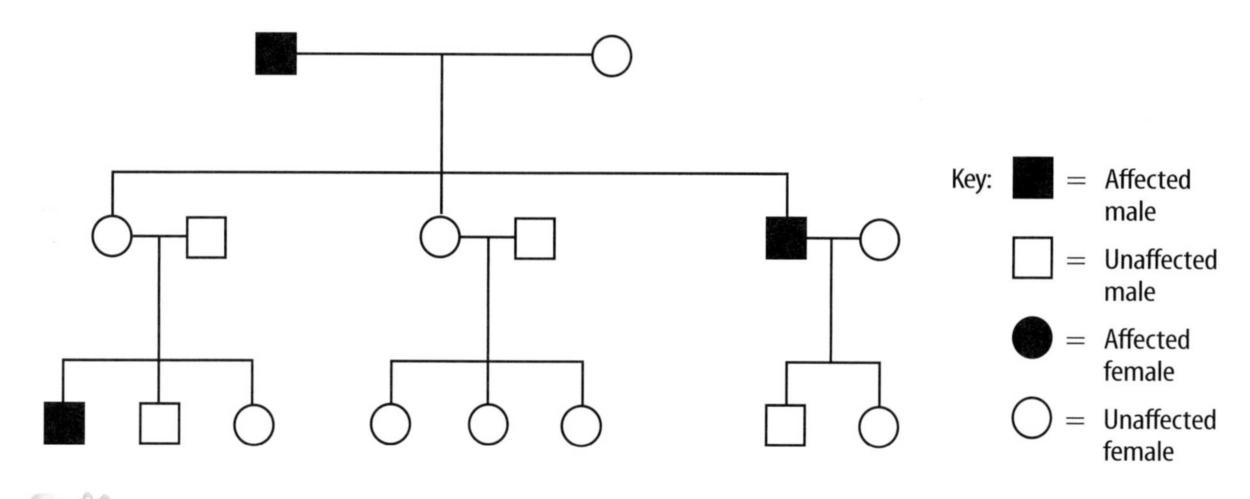

chapter 5 Review

Using Vocabulary

allele p. 126
dominant p. 128
genetic engineering p. 141
genetics p. 126
genotype p. 130
heredity p. 126
heterozygous p. 130
homozygous p. 130
hybrid p. 128
incomplete dominance p. 134
phenotype p. 130
polygenic inheritance p. 136
Punnett square p. 130
recessive p. 128
sex-linked gene p. 139

Fill in the blanks with the correct word.

1. Alternate forms of a gene are called ________.

2. The outward appearance of a trait is a(n) ________.

3. Human height, eye color, and skin color are all traits controlled by ________.

4. An allele that produces a trait in the heterozygous condition is ________.

5. ________ is the science that deals with the study of heredity.

6. The actual combination of alleles of an organism is its ________.

7. ________ is moving fragments of DNA from one organism and inserting them into another organism.

8. A(n) ________ is a helpful device for predicting the probabilities of possible genotypes.

9. ________ is the passing of traits from parents to offspring.

10. Red-green color blindness is a human genetic disorder caused by a(n) ________.

Checking Concepts

Choose the word or phrase that best answers the question.

11. Which describes the allele that causes color blindness?
A) dominant
B) carried on the Y chromosome
C) carried on the X chromosome
D) present only in males

12. What is it called when the presence of two different alleles results in an intermediate phenotype?
A) incomplete dominance
B) polygenic inheritance
C) multiple alleles
D) sex-linked genes

13. What separates during meiosis?
A) proteins
B) phenotypes
C) alleles
D) pedigrees

14. What controls traits in organisms?
A) cell membrane
B) cell wall
C) genes
D) Punnett squares

15. What term describes the inheritance of cystic fibrosis?
A) polygenic inheritance
B) multiple alleles
C) incomplete dominance
D) recessive genes

16. What phenotype will the offspring represented in the Punnett square have?
A) all recessive
B) all dominant
C) half recessive, half dominant
D) Each will have a different phenotype.

	F	f
F	FF	Ff
F	FF	Ff

Science Online life.msscience.com/vocabulary_puzzlemaker

Thinking Critically

17. **Explain** the relationship between DNA, genes, alleles, and chromosomes.

18. **Classify** these inheritance patterns:
    a. many different phenotypes produced by one pair of alleles
    b. many phenotypes produced by more than one pair of alleles; two phenotypes from two alleles; three phenotypes from two alleles.

Use the illustration below to answer question 19.

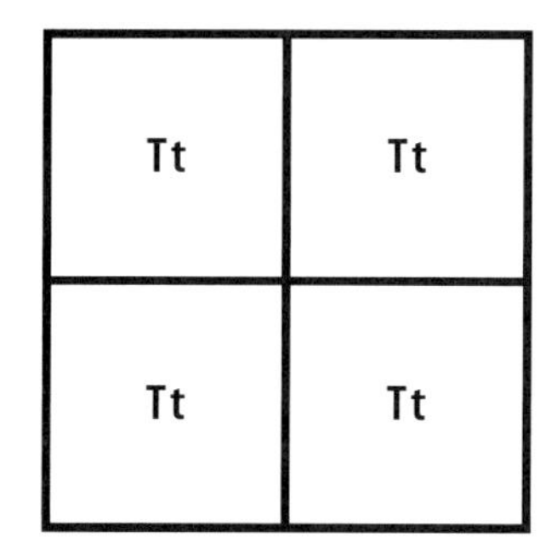

19. **Interpret Scientific Illustrations** What were the genotypes of the parents that produced the Punnett Square shown above?

20. **Explain** why two rabbits with the same genes might not be colored the same if one is raised in northern Maine and one is raised in southern Texas.

21. **Apply** Can a person with a genetic disorder that has been corrected by gene transfer pass the corrected condition to his or her children? Explain.

22. **Predict** Two organisms were found to have different genotypes but the same phenotype. Predict what these phenotypes might be. Explain.

23. **Compare and contrast** Mendelian inheritance with incomplete dominance.

Performance Activities

24. **Newspaper Article** Write a newspaper article to announce a new, genetically engineered plant. Include the method of developing the plant, the characteristic changed, and the terms that you would expect to see. Read your article to the class.

25. **Predict** In humans, the widow's peak allele is dominant, and the straight hairline allele is recessive. Predict how both parents with widow's peaks could have a child without a widow's peak hairline.

26. **Use a word processor** or program to write predictions about how advances in genetics might affect your life in the next ten years.

Applying Math

27. **Human Genome** If you wrote the genetic information for each gene in the human genome on a separate sheet of 0.2-mm-thick paper and stacked the sheets, how tall would the stack be?

Use the table below to answer question 28.

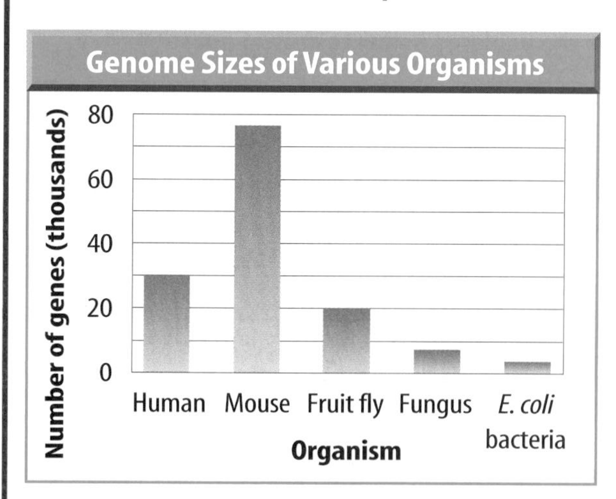

28. **Genes** Consult the graph above. How many more genes are in the human genome than the genome of the fruit fly?

Standardized Test Practice

Part 1 Multiple Choice

Record your answers on the answer sheet provided by your teacher or on a sheet of paper.

1. Heredity includes all of the following except
A. traits.
B. chromosomes.
C. nutrients.
D. phenotype.

2. What is a mutation?
A. A change in a gene which is harmful, beneficial, or has no effect at all.
B. A change in a gene which is only beneficial.
C. A change in a gene which is only harmful.
D. No change in a gene.

3. Sex of the offspring is determined by
A. only the mother, because she has two X chromosomes.
B. only the father, because he has one X and one Y chromosome.
C. an X chromosome from the mother and either an X or Y chromosome from the father.
D. mutations.

Use the pedigree below to answer questions 4–6.

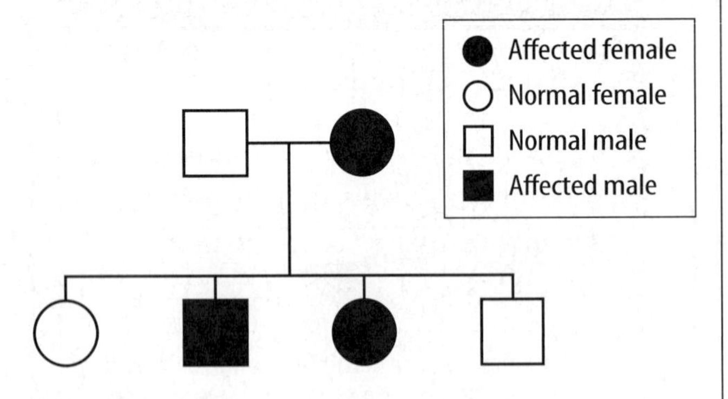

Huntington disease has a dominant (DD or Dd) inheritance pattern.

4. What is the genotype of the father?
A. DD
B. Dd
C. dd
D. D

5. What is the genotype of the mother?
A. DD
B. Dd
C. dd
D. D

6. The genotype of the unaffected children is
A. DD.
B. Dd.
C. dd.
D. D.

7. Manipulating the arrangement of DNA that makes up a gene is called
A. genetic engineering.
B. chromosomal migration.
C. viral reproduction.
D. cross breeding.

Use the Punnett square below to answer question 8.

	A	O
A	AA	AO
B	AB	BO

8. How many phenotypes would result from the following Punnett square?
A. 1
B. 2
C. 3
D. 4

9. Down syndrome is an example of
A. incomplete dominance.
B. genetic engineering.
C. a chromosome disorder.
D. a sex linked disorder.

Test-Taking Tip

Complete Charts Write directly on complex charts such as a Punnett square.

Question 10 Draw a Punnett square to answer all parts of the question.

Part 2 Short Response/Grid In

Record your answers on the answer sheet provided by your teacher or on a sheet of paper.

Use the table below to answer questions 10–11.

Some Traits Compared by Mendel			
Traits	Shape of Seeds	Shape of Pods	Flower Color
Dominant Trait	Round	Full	Purple
Recessive Trait	Wrinkled	Flat, constricted	White

10. Create a Punnett square using the *Shape of Pods* trait crossing heterozygous parents. What percentage of the offspring will be heterozygous? What percentage of the offspring will be homozygous? What percentage of the offspring will have the same phenotype as the parents?

11. Gregor Mendel studied traits in pea plants that were controlled by single genes. Explain what would have happened if the alleles for flower color were an example of incomplete dominance. What phenotypes would he have observed?

12. Why are heterozygous individuals called carriers for non-sex-linked and X-linked recessive patterns of inheritance?

13. How many alleles does a body cell have for each trait? What happens to the alleles during meiosis?

Part 3 Open Ended

Record your answers on a sheet of paper.

14. Genetic counseling helps individuals determine the genetic risk or probability a disorder will be passed to offspring. Why would a pedigree be a very important tool for the counselors? Which patterns of inheritance (dominant, recessive, x-linked) would be the easiest to detect?

15. Explain the process of gene transfer. What types of disorders would this be best suited? How might gene transfer help patients with cystic fibrosis?

Refer to the figure below to answer question 16.

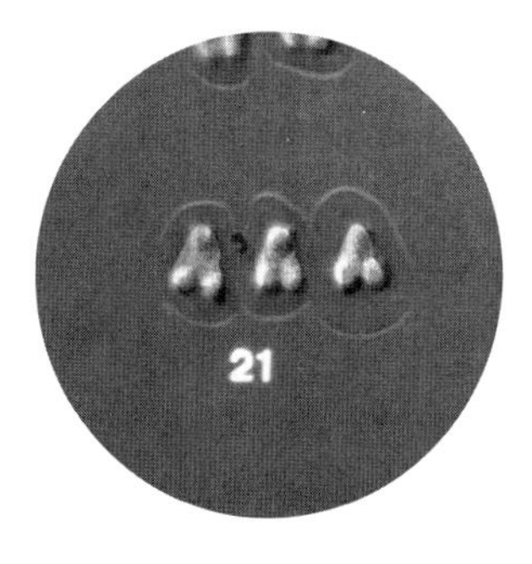

16. What is the disorder associated with the karyotype shown above? How does this condition occur? What are the characteristics of someone with this disorder?

17. Explain why the parents of someone with cystic fibrosis do not show any symptoms. How are the alleles for cystic fibrosis passed from parents to offspring?

18. What is recombinant DNA and how is it used to help someone with Type I diabetes?

19. If each kernel on an ear of corn represents a separate genetic cross, would corn be a good plant to use to study genetics? Why or why not? What process could be used to control pollination?

chapter 6

Adaptations over Time

The BIG Idea

Life-forms have changed over time.

SECTION 1
Ideas About Evolution
Main Idea Charles Darwin and other scientists observed that species change over time by different methods.

SECTION 2
Clues About Evolution
Main Idea Scientists find clues about evolution by studying fossils, development of embryos, structures of organisms, and DNA.

SECTION 3
The Evolution of Primates
Main Idea Evidence indicates that the ancient ancestor of present-day humans appeared on Earth for 4–6 million years ago.

Adaptation? No problem.

Cockroaches have existed for millions of years, yet they are still adapted to their environment. Since they first appeared, many species have disappeared, and other well-adapted species have evolved.

Science Journal Pick a favorite plant or animal and list in your Science Journal all the ways it is well-suited to its environment.

Start-Up Activities

Adaptation for a Hunter

The cheetah is nature's fastest hunter, but it can run swiftly for only short distances. Its fur blends in with tall grass, making it almost invisible as it hides and waits for prey. Then the cheetah pounces, capturing the prey before it can run away.

1. Spread a sheet of newspaper classified ads on the floor.
2. Using a hole puncher, make 100 circles from each of the following types of paper: white paper, black paper, and classified ads.
3. Scatter all the circles on the newspaper on the floor. For 10 s, pick up as many circles as possible, one at a time. Have a partner time you.
4. Count the number of each kind of paper circle that you picked up. Record your results in your Science Journal.
5. **Think Critically** Which paper circles were most difficult to find? What can you infer about a cheetah's coloring from this activity? Enter your responses to these questions in your Science Journal.

FOLDABLES™ Study Organizer

Principles of Natural Selection Make the following Foldable to help you understand the process of natural selection.

STEP 1 **Fold** a sheet of paper in half lengthwise.

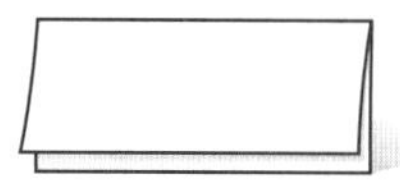

STEP 2 **Fold** paper down 2.5 cm from the top. (Hint: From the tip of your index finger to your middle knuckle is about 2.5 cm.)

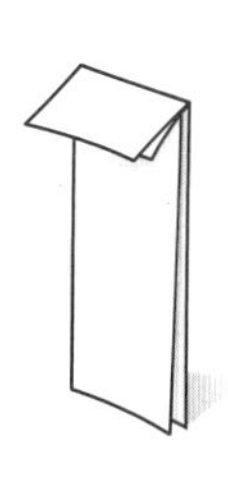

STEP 3 **Open and draw** lines along the 2.5-cm fold and the center fold. **Label** as shown.

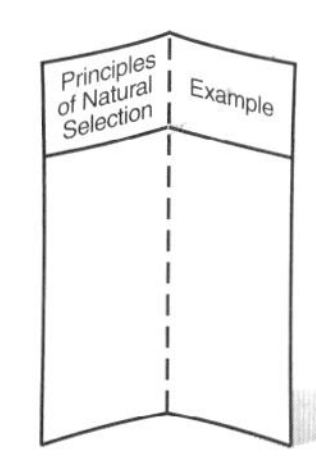

Summarize in a Table As you read, list the five principles of natural selection in the left-hand column. In the right-hand column, briefly write an example for each principle.

Preview this chapter's content and activities at life.msscience.com

Get Ready to Read

Questioning

① Learn It! Asking questions helps you to understand what you read. As you read, think about the questions you'd like answered. Often you can find the answer in the next paragraph or lesson. Learn to ask good questions by asking who, what, when, where, why, and how.

② Practice It! Read the following passage from Section 2.

> One way to find the approximate age of fossils within a rock layer is relative dating. Relative dating is based on the idea that, in undisturbed areas, younger rock layers are deposited on top of older rock layers, as shown in **Figure 10.** Relative dating provides only an estimate of a fossil's age. The estimate is made by comparing the ages of rock layers found above and below the fossil layer. For example, suppose a 50 million-year-old rock layer lies below a fossil, and a 35-million-year-old layer lies above it. According to relative dating, the fossil is probably between 35 million and 50 million years old.
>
> *—from page 165*

Here are some questions you might ask about this paragraph:

- How do the ages of rock layers help determine the age of a fossil?
- What must be true of the area where rock layers are used for relative dating?
- Does relative dating determine the actual fossil age or an estimate of a fossil's age?

③ Apply It! As you read the chapter, look for answers to questions that are part of the text.

Target Your Reading

Use this to focus on the main ideas as you read the chapter.

1. **Before you read** the chapter, respond to the statements below on your worksheet or on a numbered sheet of paper.
   - Write an **A** if you **agree** with the statement.
   - Write a **D** if you **disagree** with the statement.

2. **After you read** the chapter, look back to this page to see if you've changed your mind about any of the statements.
   - If any of your answers changed, explain why.
   - Change any false statements into true statements.
   - Use your revised statements as a study guide.

Reading Tip

Test yourself. Create questions and then read to find answers to your own questions.

Before You Read A or D		Statement	After You Read A or D
	1	Darwin's observations in the Galápagos Islands helped him develop his theory of evolution by natural selection.	
	2	When some geographic barrier, such as mountains, separate members of a species, each group remains unchanged over time.	
	3	One principle of natural selection is that organisms best able to survive in an environment are more likely to reproduce and pass their traits to future generations.	
	4	Variation makes one member of a species different from other members of the same species.	
	5	Fossils can be the actual remains of an organism.	
	6	Evolution only happens slowly over time.	
	7	Present-day organisms can provide clues about evolution.	
	8	Plants were the first forms of life to evolve.	

Science Online
Print out a worksheet of this page at life.msscience.com

section 1

Ideas About Evolution

as you read

What You'll Learn

- **Describe** Lamarck's hypothesis of acquired characteristics and Darwin's theory of natural selection.
- **Identify** why variations in organisms are important.
- **Compare and contrast** gradualism and punctuated equilibrium.

Why It's Important

The theory of evolution suggests why there are so many different living things.

Review Vocabulary
hypothesis: an explanation that can be tested

New Vocabulary
- species
- evolution
- natural selection
- variation
- adaptation
- gradualism
- punctuated equilibrium

Early Models of Evolution

Millions of species of plants, animals, and other organisms live on Earth today. Do you suppose they are exactly the same as they were when they first appeared—or have any of them changed? A **species** is a group of organisms that share similar characteristics and can reproduce among themselves to produce fertile offspring. Many characteristics of a species are inherited when they pass from parent to offspring. Change in these inherited characteristics over time is **evolution. Figure 1** shows how the characteristics of the camel have changed over time.

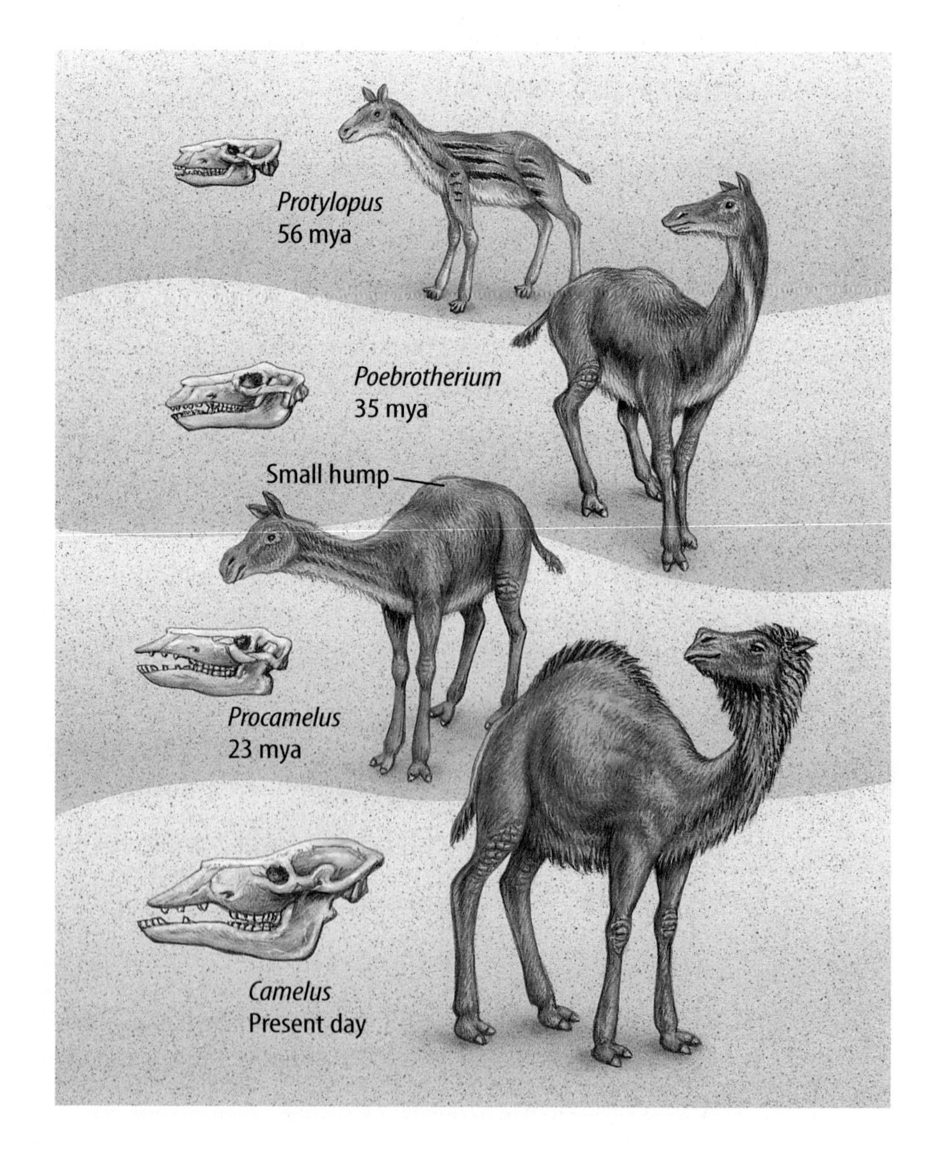

Figure 1 By studying fossils, scientists have traced the hypothesized evolution of the camel.
Discuss *the changes you observe in camels over time.*

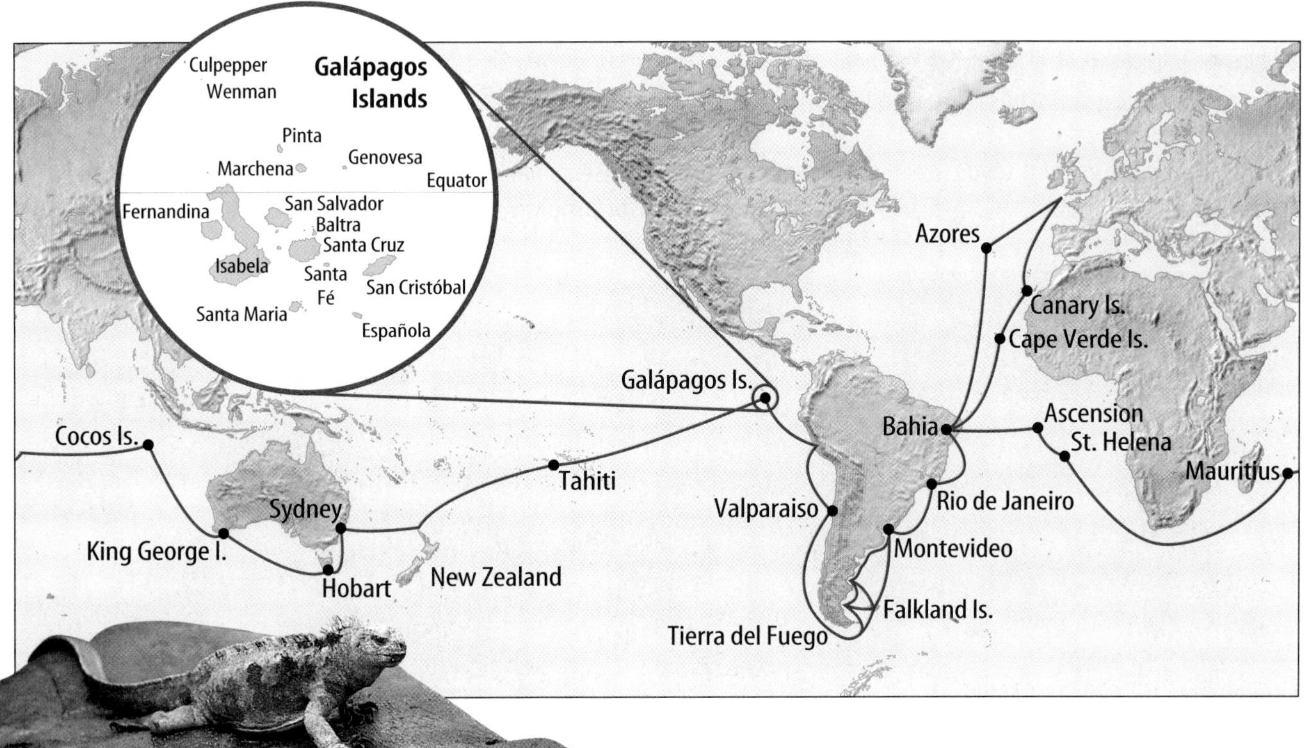

Figure 2 This map shows the route of Darwin's voyage on the HMS *Beagle*. Darwin noticed many species on the Galápagos Islands that he had not seen along the coast of South America, including the marine iguana. This species is the only lizard in the world known to enter the ocean and feed on seaweed.

Hypothesis of Acquired Characteristics In 1809, Jean Baptiste de Lamarck proposed a hypothesis to explain how species change over time. He suggested that characteristics, or traits, developed during a parent organism's lifetime are inherited by its offspring. His hypothesis is called the inheritance of acquired characteristics. Scientists collected data on traits that are passed from parents to offspring. The data showed that traits developed during a parent's lifetime, such as large muscles built by hard work or exercise, are not passed on to offspring. The evidence did not support Lamarck's hypothesis.

✔ Reading Check *What was Lamarck's explanation of evolution?*

Darwin's Model of Evolution

In December 1831, the HMS *Beagle* sailed from England on a journey to explore the South American coast. On board was a young naturalist named Charles Darwin. During the journey, Darwin recorded observations about the plants and animals he saw. He was amazed by the variety of life on the Galápagos Islands, which are about 1,000 km from the coast of Ecuador. Darwin hypothesized that the plants and animals on the Galápagos Islands originally came from Central and South America. But the islands were home to many species he had not seen in South America, including giant cactus trees, huge land tortoises, and the iguana shown in **Figure 2.**

Figure 3 Darwin observed that the beak shape of each species of Galápagos finch is related to its eating habits.

Finches that eat nuts and seeds have short, strong beaks for breaking hard shells.

Finches that feed on insects have long, slender beaks for probing beneath tree bark.

Finches with medium-sized beaks eat a variety of foods including seeds and insects.

Topic: Darwin's Finches
Visit life.msscience.com for Web links to information about the finches Darwin observed.

Activity In your Science Journal, describe the similarities and differences of any two species of Galápagos finches.

Darwin's Observations Darwin observed 13 species of finches on the Galápagos Islands. He noticed that all 13 species were similar, except for differences in body size, beak shape, and eating habits, as shown in **Figure 3.** He also noticed that all the Galápagos finch species were similar to one finch species he had seen on the South American coast.

Darwin reasoned that the Galápagos finches must have had to compete for food. Finches with beak shapes that allowed them to eat available food survived longer and produced more offspring than finches without those beak shapes. After many generations, these groups of finches became separate species.

How did Darwin explain the evolution of the different species of Galápagos finches?

Natural Selection

After the voyage, Charles Darwin returned to England and continued to think about his observations. He collected more evidence on inherited traits by breeding racing pigeons. He also studied breeds of dogs and varieties of flowers. In the mid 1800s, Darwin developed a theory of evolution that is accepted by most scientists today. He described his ideas in a book called *On the Origin of Species,* which was published in 1859.

Darwin's Theory Darwin's observations led many other scientists to conduct experiments on inherited characteristics. After many years, Darwin's ideas became known as the theory of evolution by natural selection. **Natural selection** means that organisms with traits best suited to their environment are more likely to survive and reproduce. Their traits are passed to more offspring. All living organisms produce more offspring than survive. Galápagos finches lay several eggs every few months. Darwin realized that in just a few years, several pairs of finches could produce a large population. A population is all of the individuals of a species living in the same area. Members of a large population compete for living space, food, and other resources. Those that are best able to survive are more likely to reproduce and pass on their traits to the next generation.

The principles that describe how natural selection works are listed in **Table 1.** Over time, as new data was gathered and reported, changes were made to Darwin's original ideas about evolution by natural selection. His theory remains one of the most important ideas in the study of life science.

Table 1 The Principles of Natural Selection

1. Organisms produce more offspring than can survive.
2. Differences, or variations, occur among individuals of a species.
3. Some variations are passed to offspring.
4. Some variations are helpful. Individuals with helpful variations survive and reproduce better than those without these variations.
5. Over time, the offspring of individuals with helpful variations make up more of a population and eventually may become a separate species.

Applying Science

Does natural selection take place in a fish tank?

Alejandro raises tropical fish as a hobby. Could the observations that he makes over several weeks illustrate the principles of natural selection?

Identifying the Problem

Alejandro keeps a detailed journal of his observations, some of which are given in the table to the right.

Fish Tank Observations

Date	Observation
June 6	6 fish are placed in aquarium tank.
July 22	16 new young appear.
July 24	3 young have short or missing tail fins. 13 young have normal tail fins.
July 28	Young with short or missing tail fins die.
August 1	2 normal fish die—from overcrowding?
August 12	30 new young appear.
August 15	5 young have short or missing tail fins. 25 young have normal tail fins.
August 18	Young with short or missing tail fins die.
August 20	Tank is overcrowded. Fish are divided equally into two tanks.

Solving the Problem

Refer to **Table 1** and match each of Alejandro's journal entries with the principle(s) it demonstrates. Here's a hint: *Some entries may not match any of the principles of natural selection. Some entries may match more than one principle.*

Evolution of English
If someone from Shakespeare's time were to speak to you today, you probably would not understand her. Languages, like species, change over time. In your Science Journal, discuss some words or phrases that you use that your parents or teachers do not use correctly.

Variation and Adaptation

Darwin's theory of evolution by natural selection emphasizes the differences among individuals of a species. These differences are called variations. A **variation** is an inherited trait that makes an individual different from other members of its species. Variations result from permanent changes, or mutations, in an organism's genes. Some gene changes produce small variations, such as differences in the shape of human hairlines. Other gene changes produce large variations, such as an albino squirrel in a population of gray squirrels or fruit without seeds. Over time, more and more individuals of the species might inherit these variations. If individuals with these variations continue to survive and reproduce over many generations, a new species can evolve. It might take hundreds, thousands, or millions of generations for a new species to evolve.

Some variations are more helpful than others. An **adaptation** is any variation that makes an organism better suited to its environment. The variations that result in an adaptation can involve an organism's color, shape, behavior, or chemical makeup. Camouflage (KA muh flahj) is an adaptation. A camouflaged organism, like the one shown in **Figure 4,** blends into its environment and is more likely to survive and reproduce.

Figure 4 Variations that provide an advantage tend to increase in a population over time. Variations that result in a disadvantage tend to decrease in a population over time.

Albinism can prevent an organism from blending into its environment.
Infer *what might happen to an albino lemur in its natural environment.*

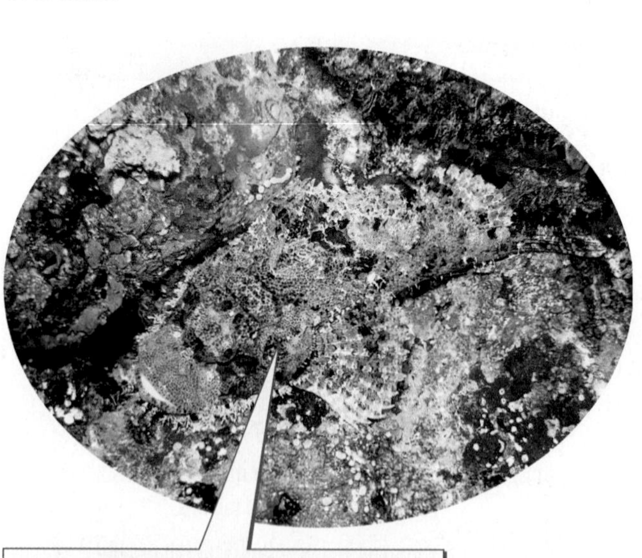

Camouflage allows organisms to blend into their environments.
Infer *how its coloration gives this scorpion fish a survival advantage.*

Figure 5 During the last ice age, Virginia white-tailed deer moved south ahead of an advancing ice sheet. When ice sheets melted worldwide about 4,000–10,000 years ago, ocean levels rose. Some deer were isolated on a chain of islands and evolved into a new subspecies, the Key deer. Key deer live only on approximately 30 islands in the subtropical lower keys of Florida.

A Virginia white-tailed deer can be 0.9 m to 1.1 m tall at the shoulder.

A Key deer can be 0.6 m to 0.7 m tall at the shoulder.
Infer *why Key deer are smaller than Virginia white-tailed deer.*

Changes in the Sources of Genes Over time, the genetic makeup of a species might change its appearance. For example, as the genetic makeup of a species of seed-eating Galápagos finch changed, so did the size and shape of its beak. Many kinds of environmental factors help bring about changes. When individuals of the same species move into or out of an area, they might bring in or remove genes and variations. Suppose a family from another country moves to your neighborhood. They might bring different foods, customs, and ways of speaking with them. In a similar way, when new individuals enter an existing population, they can bring in different genes and variations.

Geographic Isolation Sometimes mountains, lakes, or other geologic features isolate a small number of individuals from the rest of a population. Over several generations, variations that do not exist in the larger population might begin to be more common in the isolated population. Also, gene mutations can occur that add variations to populations. Over time, the two populations can become so different that they no longer can breed with each other. Key deer, like the one shown in **Figure 5,** evolved because of geographic isolation about 4,000–6,000 years ago.

Modeling Evolution

Procedure

1. On a piece of **paper,** print the word *train.*
2. Add, subtract, or change one letter to make a new word.
3. Repeat step 2 with the new word.
4. Repeat steps 2 and 3 two more times.
5. Make a "family tree" that shows how your first word changed over time.

Analysis

Compare your tree to those of other people. How is this process similar to evolution by natural selection?

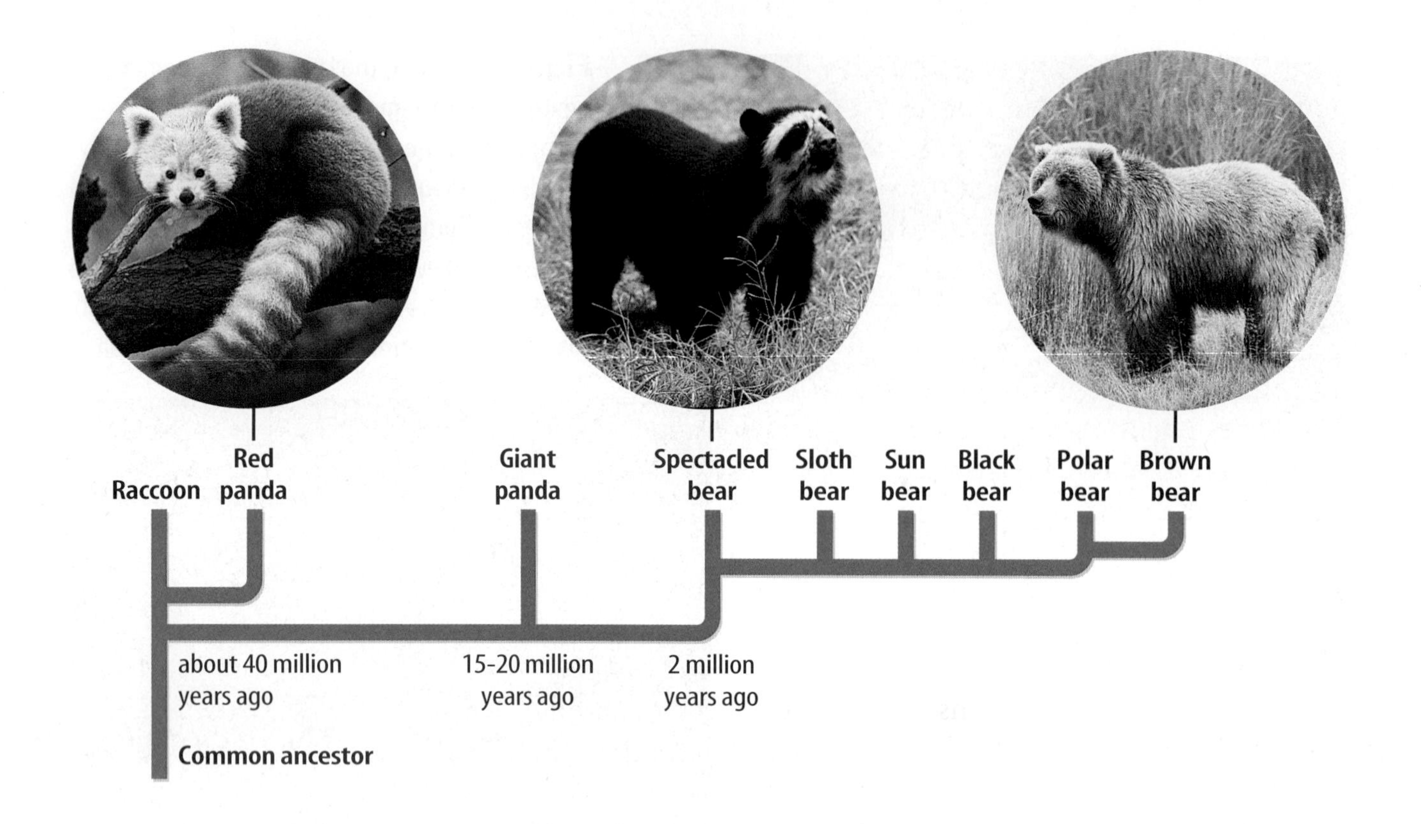

Figure 6 The hypothesized evolution of bears illustrates the punctuated equilibrium model of evolution.
Discuss *how the six species on the far right are explained better by punctuated equlibrium.*

The Speed of Evolution

Scientists do not agree on how quickly evolution occurs. Many scientists hypothesize that evolution occurs slowly, perhaps over tens or hundreds of millions of years. Other scientists hypothesize that evolution can occur quickly. Most scientists agree that evidence supports both of these models.

Gradualism Darwin hypothesized that evolution takes place slowly. The model that describes evolution as a slow, ongoing process by which one species changes to a new species is known as **gradualism.** According to the gradualism model, a continuing series of mutations and variations over time will result in a new species. Look back at **Figure 1,** which shows the evolution of the camel over tens of millions of years. Fossil evidence shows a series of intermediate forms that indicate a gradual change from the earliest camel species to today's species.

Punctuated Equilibrium Gradualism doesn't explain the evolution of all species. For some species, the fossil record shows few intermediate forms—one species suddenly changes to another. According to the **punctuated equilibrium** model, rapid evolution comes about when the mutation of a few genes results in the appearance of a new species over a relatively short period of time. The fossil record gives examples of this type of evolution, as you can see in **Figure 6.**

Punctuated Equilibrium Today Evolution by the punctuated equilibrium model can occur over a few thousand or million years, and sometimes even faster. For example, many bacteria have changed in a few decades. The antibiotic penicillin originally came from the fungus shown in **Figure 7.** But many bacteria species that were once easily killed by penicillin no longer are harmed by it. These bacteria have developed resistance to the drug. Penicillin has been in use since 1943. Just four years later, in 1947, a species of bacteria that causes pneumonia and other infections already had developed resistance to the drug. By the 1990s, several disease-producing bacteria had become resistant to penicillin and many other antibiotics.

How did penicillin-resistant bacteria evolve so quickly? As in any population, some organisms have variations that allow them to survive unfavorable living conditions when other organisms cannot. When penicillin was used to kill bacteria, those with the penicillin-resistant variation survived, reproduced, and passed this trait to their offspring. Over a period of time, this bacteria population became penicillin-resistant.

Figure 7 The fungus growing in this petri dish is *Penicillium,* the original source of penicillin. It produces an antibiotic substance that prevents the growth of certain bacteria.

section 1 review

Summary

Early Models of Evolution

- Evolution is change in the characteristics of a species over time.
- Lamarck proposed the hypothesis of inherited acquired characteristics.

Natural Selection

- Darwin proposed evolution by natural selection, a process by which organisms best suited to their environments are most likely to survive and reproduce.
- Organisms have more offspring than can survive, individuals of a species vary, and many of these variations are passed to offspring.

Variation and Adaptation

- Adaptations are variations that help an organism survive or reproduce in its environment.
- Mutations are the source of new variations.

The Speed of Evolution

- Evolution may be a slow or fast process depending on the species under study.

Self Check

1. **Compare** Lamarck's and Darwin's ideas about how evolution takes place.
2. **Explain** why variations are important to understanding change in a population over time.
3. **Discuss** how the gradualism model of evolution differs from the punctuated equilibrium model of evolution.
4. **Describe** how geographic isolation contributes to evolution.
5. **Think Critically** What adaptations would be helpful for an animal species that was moved to the Arctic?
6. **Concept Map** Use information given in **Figure 6** to make a map that shows how raccoons, red pandas, giant pandas, polar bears, and black bears are related to a common ancestor.

Applying Math

7. **Use Percentages** The evolution of the camel can be traced back at least 56 million years. Use **Figure 1** to estimate the percent of this time that the modern camel has existed.

Hidden Frogs

Through natural selection, animals become adapted for survival in their environment. Adaptations include shapes, colors, and even textures that help an animal blend into its surroundings. These adaptations are called camouflage. The red-eyed tree frog's mint green body blends in with tropical forest vegetation as shown in the photo on the right. Could you design camouflage for a desert frog? A temperate forest frog?

Real-World Question

What type of camouflage would best suit a frog living in a particular habitat?

Goals

- **Create** a frog model camouflaged to blend in with its surroundings.

Materials (for each group)

cardboard form of a frog
colored markers
crayons
colored pencils
glue
beads
sequins
modeling clay

Safety Precautions

Procedure

1. Choose one of the following habitats for your frog model: muddy shore of a pond, orchid flowers in a tropical rain forest, multicolored clay in a desert, or the leaves and branches of trees in a temperate forest.
2. **List** the features of your chosen habitat that will determine the camouflage your frog model will need.
3. **Brainstorm** with your group the body shape, coloring, and skin texture that would make the best camouflage for your model. Record your ideas in your Science Journal.
4. **Draw** in your Science Journal samples of colors, patterns, texture, and other features your frog model might have.
5. Show your design ideas to your teacher and ask for further input.
6. **Construct** your frog model.

Conclude and Apply

1. **Explain** how the characteristics of the habitat helped you decide on the specific frog features you chose.
2. **Infer** how the color patterns and other physical features of real frogs develop in nature.
3. **Explain** why it might be harmful to release a frog into a habitat for which it is not adapted.

Communicating Your Data

Create a poster or other visual display that represents the habitat you chose for this activity. Use your display to show classmates how your design helps camouflage your frog model. **For more help, refer to the Science Skill Handbook.**

section 2

Clues About Evolution

Clues from Fossils

Imagine going on a fossil hunt in Wyoming. Your companions are paleontologists—scientists who study the past by collecting and examining fossils. As you climb a low hill, you notice a curved piece of stone jutting out of the sandy soil. One of the paleontologists carefully brushes the soil away and congratulates you on your find. You've discovered part of the fossilized shell of a turtle like the one shown in **Figure 8.**

The Green River Formation covers parts of Wyoming, Utah, and Colorado. On your fossil hunt, you learn that about 50 million years ago, during the Eocene Epoch, this region was covered by lakes. The water was home to fish, crocodiles, lizards, and turtles. Palms, fig trees, willows, and cattails grew on the lakeshores. Insects and birds flew through the air. How do scientists know all this? After many of the plants and animals of that time died, they were covered with silt and mud. Over millions of years, they became the fossils that have made the Green River Formation one of the richest fossil deposits in the world.

as you read

What You'll Learn

- **Identify** the importance of fossils as evidence of evolution.
- **Explain** how relative and radiometric dating are used to estimate the age of fossils.
- **List** examples of five types of evidence for evolution.

Why It's Important

The scientific evidence for evolution helps you understand why this theory is so important to the study of biology.

Review Vocabulary

epoch: next-smaller division of geological time after a period; is characterized by differences in life-forms that may vary regionally

New Vocabulary

- sedimentary rock
- radioactive element
- embryology
- homologous
- vestigial structure

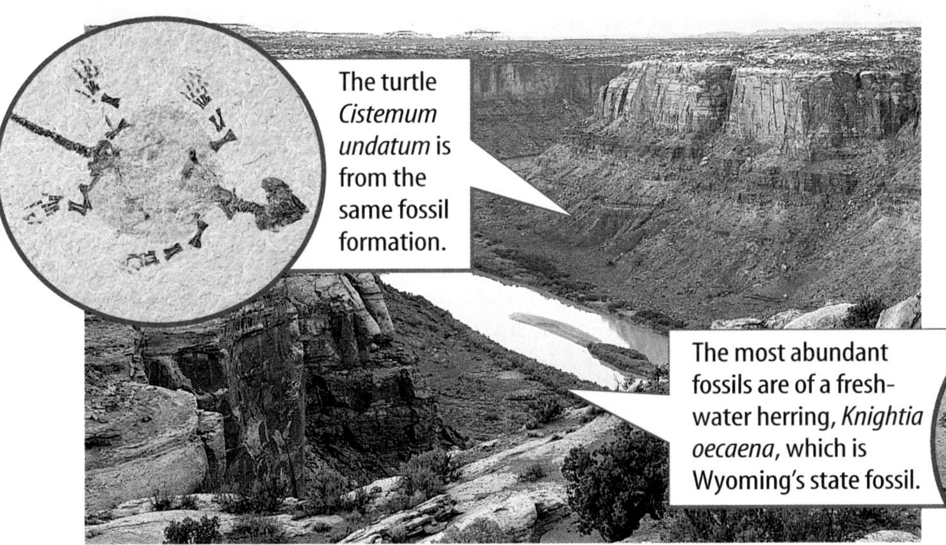

The turtle *Cistemum undatum* is from the same fossil formation.

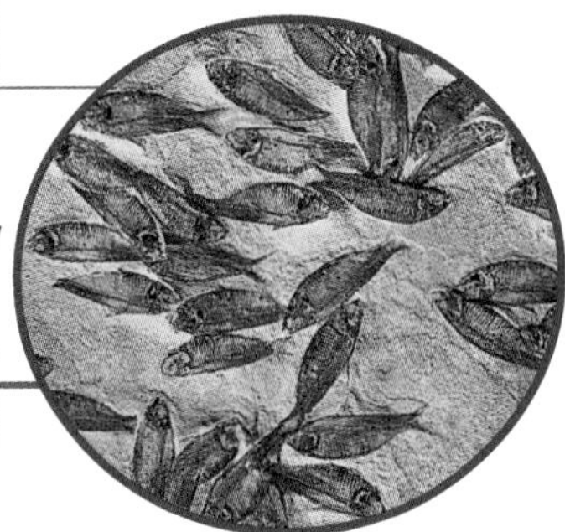

The most abundant fossils are of a freshwater herring, *Knightia oecaena*, which is Wyoming's state fossil.

Figure 8 The desert of the Green River Formation is home to pronghorn antelope, elks, coyotes, and eagles. Fossil evidence shows that about 50 million years ago the environment was much warmer and wetter than it is today.

Figure 9 Examples of several different types of fossils are shown here.
Infer *which of these would most likely be found in a layer of sedimentary rock.*

Mineralized fossils Minerals can replace wood or bone to create a piece of petrified wood as shown to the left or a mineralized bone fossil.

Imprint fossils A leaf, feather, bones, or even the entire body of an organism can leave an imprint on sediment that later hardens to become rock.

Frozen fossils The remains of organisms like this mammoth can be trapped in ice that remains frozen for thousands of years.

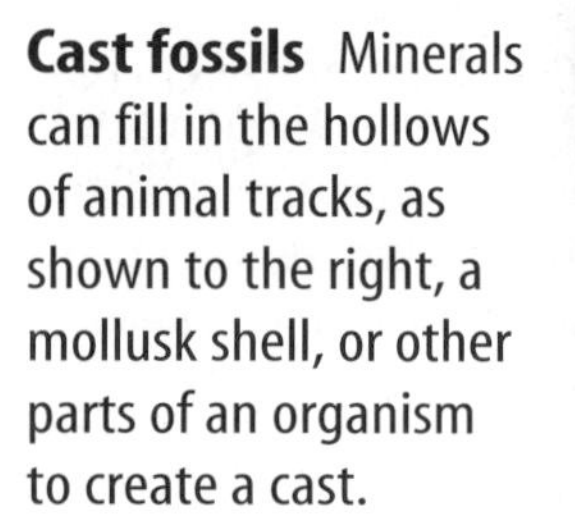

Cast fossils Minerals can fill in the hollows of animal tracks, as shown to the right, a mollusk shell, or other parts of an organism to create a cast.

Fossils in amber When the sticky resin of certain cone-bearing plants hardens over time, amber forms. It can contain the remains of trapped insects.

Types of Fossils

Most of the evidence for evolution comes from fossils. A fossil is the remains, an imprint, or a trace of a prehistoric organism. Several types of fossils are shown in **Figure 9.** Most fossils are found in sedimentary rock. **Sedimentary rock** is formed when layers of sand, silt, clay, or mud are compacted and cemented together, or when minerals are deposited from a solution. Limestone, sandstone, and shale are all examples of sedimentary rock. Fossils are found more often in limestone than in any other kind of sedimentary rock. The fossil record provides evidence that living things have evolved.

Determining a Fossil's Age

Paleontologists use detective skills to determine the age of dinosaur fossils or the remains of other ancient organisms. They can use clues provided by unique rock layers and the fossils they contain. The clues provide information about the geology, weather, and life-forms that must have been present during each geologic time period. Two basic methods—relative dating and radiometric dating—can be used, alone or together, to estimate the ages of rocks and fossils.

Topic: Fossil Finds

Visit life.msscience.com for Web links to information about recent fossil discoveries.

Activity Prepare a newspaper article describing how one of these discoveries was made, what it reveals about past life on Earth, and how it has impacted our understanding of what the past environments of Earth were like.

Relative Dating One way to find the approximate age of fossils found within a rock layer is relative dating. Relative dating is based on the idea that in undisturbed areas, younger rock layers are deposited on top of older rock layers, as shown in **Figure 10.** Relative dating provides only an estimate of a fossil's age. The estimate is made by comparing the ages of rock layers found above and below the fossil layer. For example, suppose a 50-million-year-old rock layer lies below a fossil, and a 35-million-year-old layer lies above it. According to relative dating, the fossil is between 35 million and 50 million years old.

Why can relative dating be used only to estimate the age of a fossil?

Radiometric Dating Scientists can obtain a more accurate estimate of the age of a rock layer by using radioactive elements. A **radioactive element** gives off a steady amount of radiation as it slowly changes to a nonradioactive element. Each radioactive element gives off radiation at a different rate. Scientists can estimate the age of the rock by comparing the amount of radioactive element with the amount of nonradioactive element in the rock. This method of dating does not always produce exact results, because the original amount of radioactive element in the rock can never be determined for certain.

Figure 10 In Bryce Canyon, erosion by water and wind has cut through the sedimentary rock, exposing the layers.
Infer *the relative age of rocks in the lowest layers compared to the top layer.*

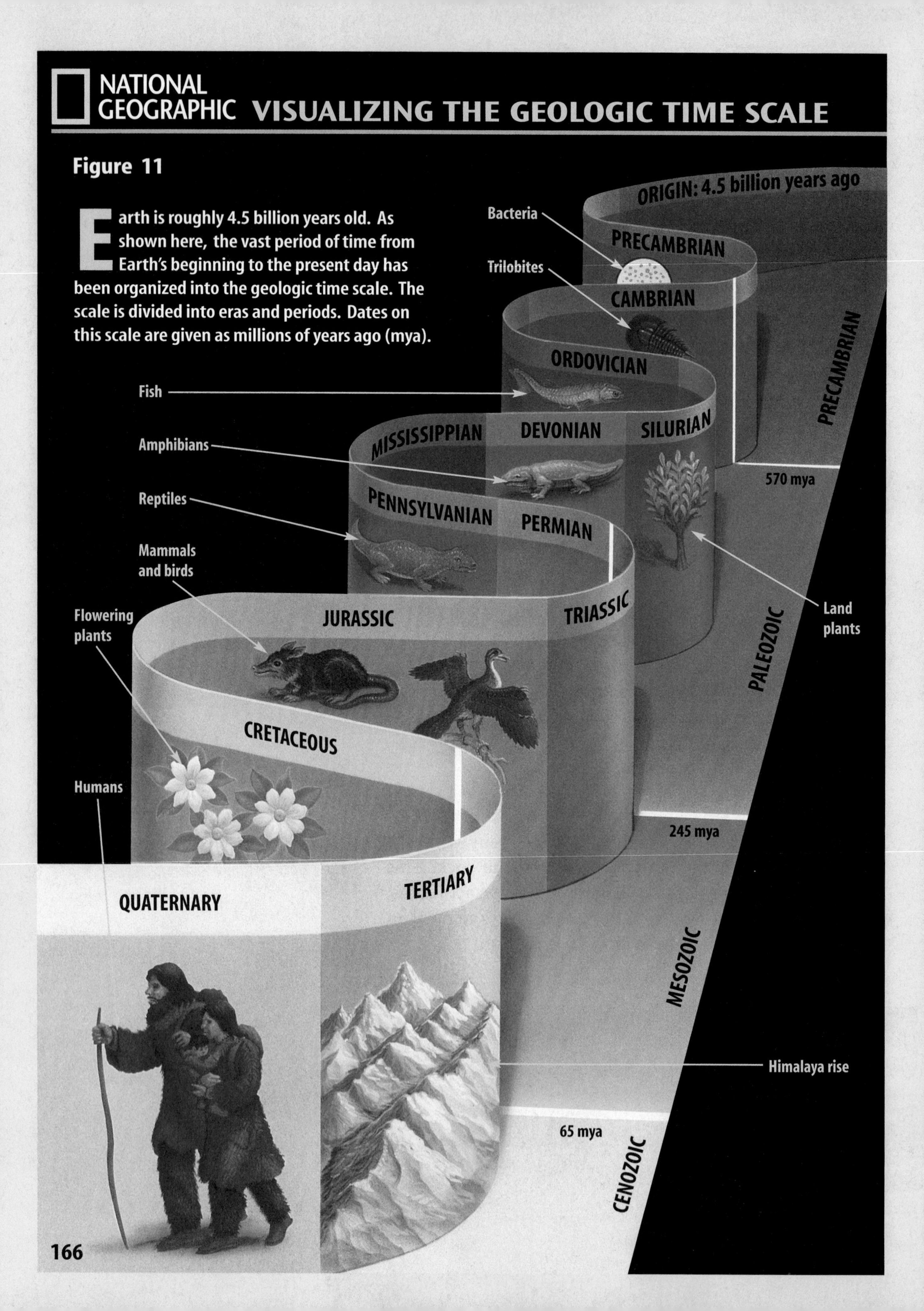
NATIONAL GEOGRAPHIC
VISUALIZING THE GEOLOGIC TIME SCALE
Figure 11
Earth is roughly 4.5 billion years old. As shown here, the vast period of time from Earth's beginning to the present day has been organized into the geologic time scale. The scale is divided into eras and periods. Dates on this scale are given as millions of years ago (mya).
ORIGIN: 4.5 billion years ago
Bacteria
PRECAMBRIAN
Trilobites
CAMBRIAN
ORDOVICIAN
Fish
MISSISSIPPIAN
DEVONIAN
SILURIAN
Amphibians
Reptiles
PENNSYLVANIAN
PERMIAN
Mammals and birds
Flowering plants
JURASSIC
TRIASSIC
CRETACEOUS
Humans
QUATERNARY
TERTIARY
PRECAMBRIAN
570 mya
PALEOZOIC
Land plants
245 mya
MESOZOIC
Himalaya rise
65 mya
CENOZOIC

Fossils and Evolution

Fossils provide a record of organisms that lived in the past. However, the fossil record is incomplete, or has gaps, much like a book with missing pages. The gaps exist because most organisms do not become fossils. By looking at fossils, scientists conclude that many simpler forms of life existed earlier in Earth's history, and more complex forms of life appeared later, as shown in **Figure 11.** Fossils provide indirect evidence that evolution has occurred on Earth.

Almost every week, fossil discoveries are made somewhere in the world. When fossils are found, they are used to help scientists understand the past. Scientists can use fossils to make models that show what the organisms might have looked like. From fossils, scientists can sometimes determine whether the organisms lived in family groups or alone, what types of food they ate, what kind of environment they lived in, and many other things about them. Most fossils represent extinct organisms. From a study of the fossil record, scientists have concluded that more than 99 percent of all organisms that have ever existed on Earth are now extinct.

Evolution in Fossils Many organisms have a history that has been preserved in sedimentary rock. Fossils show that the bones of animals such as horses and whales have become reduced in size or number over geologic time, as the species has evolved. In your Science Journal, explain what information can be gathered from changes in structures that occur over time.

More Clues About Evolution

Besides fossils, what other clues exist about evolution? Sometimes, evolution can be observed directly. Plant breeders observe evolution when they use cross-breeding to produce genetic changes in plants. The development of antibiotic resistance in bacteria is another direct observation of evolution. Entomologists have noted similar rapid evolution of pesticide-resistant insect species. These observations provide direct evidence that evolution occurs. Also, many examples of indirect evidence for evolution exist. They include similarities in embryo structures, the chemical makeup of organisms including DNA, and the way organisms develop into adults. Indirect evidence does not provide proof of evolution, but it does support the idea that evolution takes place over time.

Embryology The study of embryos and their development is called **embryology** (em bree AH luh jee). An embryo is the earliest growth stage of an organism. A tail and pharyngeal pouches are found at some point in the embryos of fish, reptiles, birds, and mammals, as **Figure 12** shows. Fish develop gills, but the other organisms develop other structures as their development continues. Fish, birds, and reptiles keep their tails, but many mammals lose theirs. These similarities suggest an evolutionary relationship among all vertebrate species.

Figure 12 Similarities in the embryos of fish, chickens, and rabbits show evidence of evolution. **Evaluate** *these embryos as evidence for evolution.*

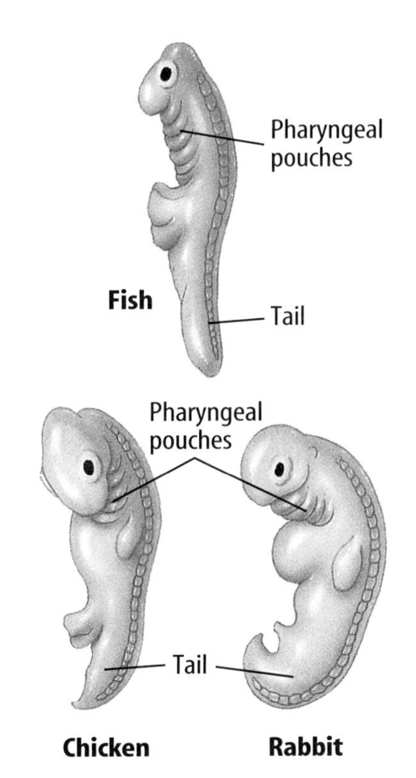

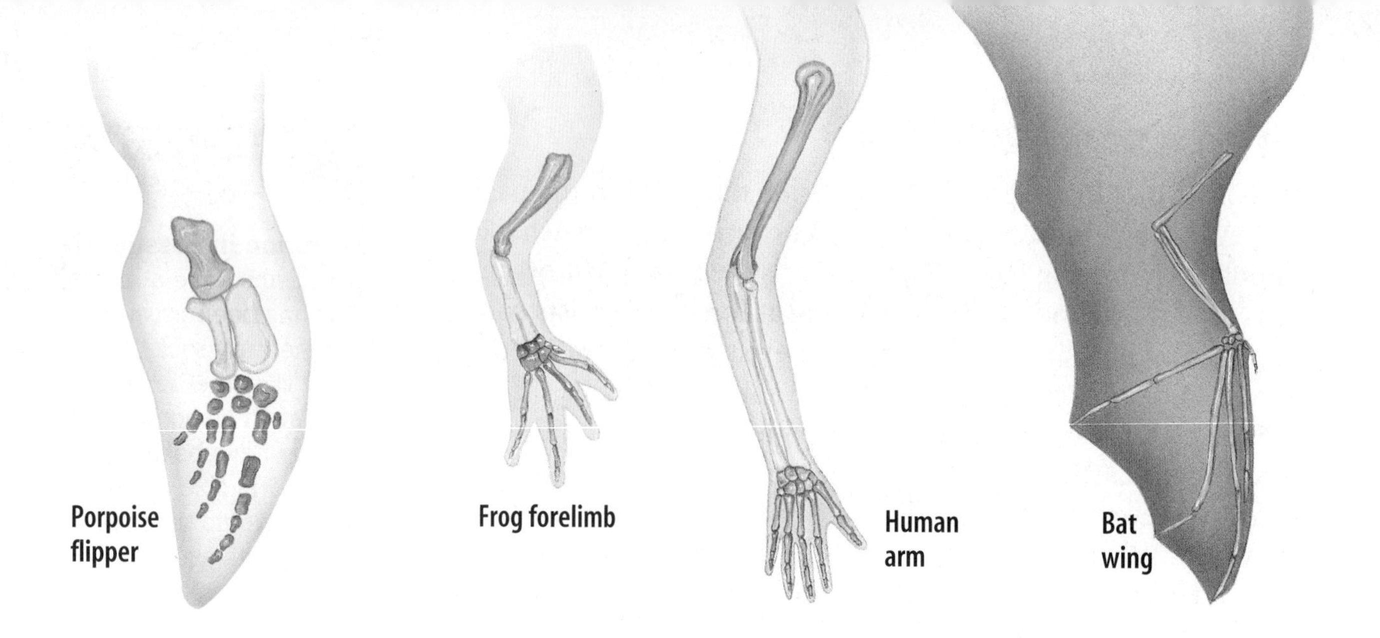

Figure 13 A porpoise flipper, frog forelimb, human arm, and bat wing are homologous. These structures show different arrangements and shapes of the bones of the forelimb. They have the same number of bones, muscles, and blood vessels, and they developed from similar tissues.

Homologous Structures What do the structures shown in **Figure 13** have in common? Although they have different functions, each of these structures is made up of the same kind of bones. Body parts that are similar in origin and structure are called **homologous** (hoh MAH luh gus). Homologous structures also can be similar in function. They often indicate that two or more species share common ancestors.

Reading Check *What do homologous structures indicate?*

Vestigial Structures The bodies of some organisms include **vestigial** (veh STIH jee ul) **structures**—structures that don't seem to have a function. Vestigial structures also provide evidence for evolution. For example, manatees, snakes, and whales no longer have back legs, but, like all animals with legs, they still have pelvic bones. The human appendix is a vestigial structure. The appendix appears to be a small version of the cecum, which is an important part of the digestive tract of many mammals. Scientists hypothesize that vestigial structures, like those shown in **Figure 14,** are body parts that once functioned in an ancestor.

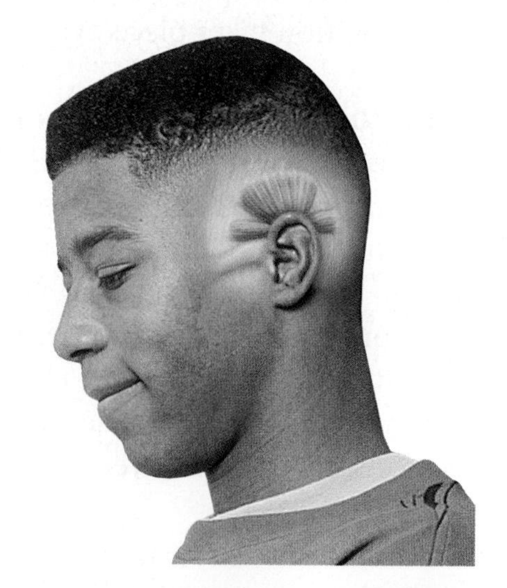

Figure 14 Humans have three small muscles around each ear that are vestigial. In some mammals, such as horses, these muscles are large. They allow a horse to turn its ears toward the source of a sound. Humans cannot rotate their ears, but some people can wiggle their ears.

Figure 15 Gorillas have DNA and proteins that are similar to humans and other primates.

DNA If you enjoy science fiction, you have read books or seen movies in which scientists re-create dinosaurs and other extinct organisms from DNA taken from fossils. DNA is the molecule that controls heredity and directs the development of every organism. In a cell with a nucleus, DNA is found in genes that make up the chromosomes. Scientists compare DNA from living organisms to identify similarities among species. Examinations of ancient DNA often provide additional evidence of how some species evolved from their extinct ancestors. By looking at DNA, scientists also can determine how closely related organisms are. For example, DNA studies indicate that dogs are the closest relatives of bears.

Similar DNA also can suggest common ancestry. Apes such as the gorillas shown in **Figure 15,** chimpanzees, and orangutans have 24 pairs of chromosomes. Humans have 23 pairs. When two of an ape's chromosomes are laid end to end, a match for human chromosome number 2 is formed. Also, similar proteins such as hemoglobin—the oxygen-carrying protein in red blood cells—are found in many primates. This can be further evidence that primates have a common ancestor.

section 2 review

Summary

Clues from Fossils

- Scientists learn about past life by studying fossils.

Determining a Fossil's Age

- The relative date of a fossil can be estimated from the ages of rocks in nearby layers.
- Radiometric dating using radioactive elements gives more accurate dates for fossils.

Fossils and Evolution

- The fossil record has gaps which may yet be filled with later discoveries.

More Clues About Evolution

- Homologous structures, similar embryos, or vestigial structures can show evolutionary relationships.
- Evolutionary relationships among organisms can be inferred from DNA comparisons.

Self Check

1. **Compare and contrast** relative dating and radiometric dating.
2. **Discuss** the importance of fossils as evidence of evolution and describe five different kinds of fossils.
3. **Explain** how DNA can provide some evidence of evolution.
4. **List** three examples of direct evidence for evolution.
5. **Interpret Scientific Illustrations** According to data in **Figure 11,** what was the longest geologic era? What was the shortest era? In what period did mammals appear?
6. **Think Critically** Compare and contrast the five types of evidence that support the theory of evolution.

Applying Math

7. **Use Percentages** The Cenozoic Era represents about 65 million years. Approximately what percent of Earth's 4.5-billion-year history does this era represent?

section 3

The Evolution of Primates

as you read

What **You'll Learn**

- **Describe** the differences among living primates.
- **Identify** the adaptations of primates.
- **Discuss** the evolutionary history of modern primates.

Why **It's Important**

Studying primate evolution will help you appreciate the differences among primates.

Review Vocabulary
opposable: can be placed against another digit of a hand or foot

New Vocabulary
- primate
- hominid
- *Homo sapiens*

Primates

Humans, monkeys, and apes belong to the group of mammals known as the **primates.** All primates have opposable thumbs, binocular vision, and flexible shoulders that allow the arms to rotate. These shared characteristics indicate that all primates may have evolved from a common ancestor.

Having an opposable thumb allows you to cross your thumb over your palm and touch your fingers. This means that you can grasp and hold things with your hands. An opposable thumb allows tree-dwelling primates to hold on to branches.

Binocular vision permits you to judge depth or distance with your eyes. In a similar way, it allows tree-dwelling primates to judge the distances as they move between branches. Flexible shoulders and rotating forelimbs also help tree-dwelling primates move from branch to branch. They also allow humans to do the backstroke, as shown in **Figure 16.**

Primates are divided into two major groups. The first group, the strepsirhines (STREP suh rines), includes lemurs and tarsiers like those shown in **Figure 17.** The second group, haplorhines (HAP luh rines), includes monkeys, apes, and humans.

Figure 16 The ability to rotate the shoulder in a complete circle allows humans to swim through water and tree-dwelling primates to travel through treetops.

Figure 17 Tarsiers and lemurs are active at night. Tarsiers are commonly found in the rain forests of Southeast Asia. Lemurs live on Madagascar and other nearby islands.
List *the traits that distinguish these animals as primates.*

Hominids About 4 million to 6 million years ago, humanlike primates appeared that were different from the other primates. These ancestors, called **hominids,** ate both meat and plants and walked upright on two legs. Hominids shared some characteristics with gorillas, orangutans, and chimpanzees, but a larger brain separated them from the apes.

African Origins In the early 1920s, a fossil skull was discovered in a quarry in South Africa. The skull had a small space for the brain, but it had a humanlike jaw and teeth. The fossil, named *Australopithecus,* was one of the oldest hominids discovered. An almost-complete skeleton of *Australopithecus* was found in northern Africa in 1974. This hominid fossil, shown in **Figure 18,** was called Lucy and had a small brain but is thought to have walked upright. This fossil indicates that modern hominids might have evolved from similar ancestors.

Figure 18 The fossil remains of Lucy are estimated to be 2.9 million to 3.4 million years old.

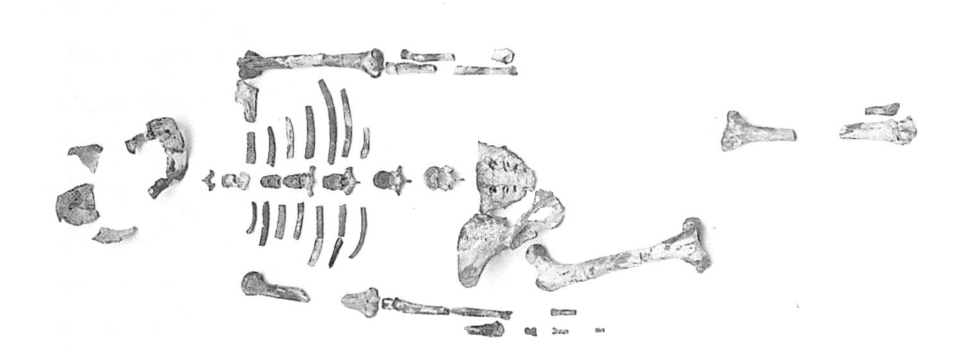

Living Without Thumbs

Procedure

1. Using **tape,** fasten down each of your thumbs next to the palm of each hand.
2. Leave your thumbs taped down for at least 1 h. During this time, do the following activities: eat a meal, change clothes, and brush your teeth. Be careful not to try anything that could be dangerous.
3. Untape your thumbs, then write about your experiences in your **Science Journal.**

Analysis

1. Did not having use of your thumbs significantly affect the way you did anything? Explain.
2. Infer how having opposable thumbs could have influenced primate evolution.

Try at Home

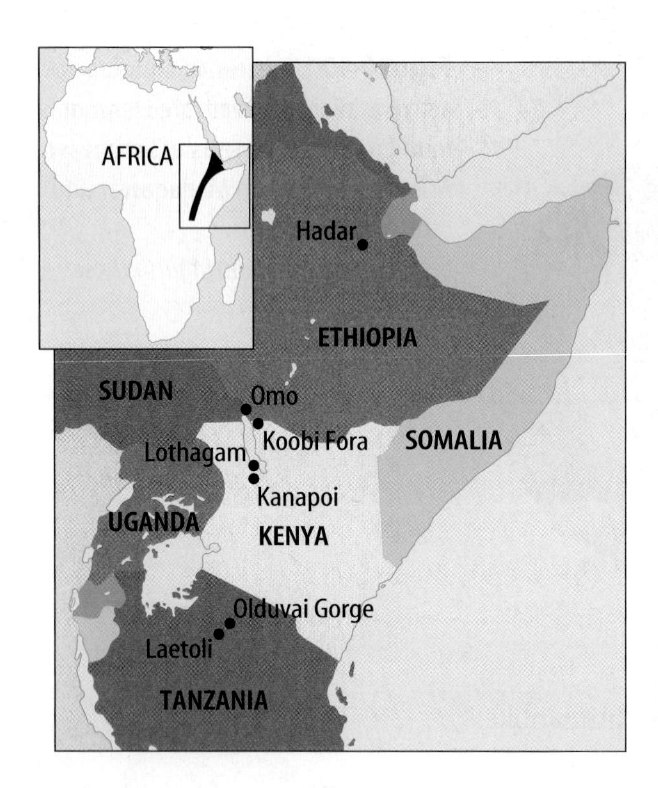

Figure 19 Many of the oldest humanlike skeletons have been found in this area of east Africa.

Early Humans In the 1960s in the region of Africa shown in **Figure 19,** a hominid fossil, which was more like present-day humans than *Australopithecus,* was discovered. The hominid was named *Homo habilis,* meaning "handy man," because simple stone tools were found near him. *Homo habilis* is estimated to be 1.5 million to 2 million years old. Based upon many fossil comparisons, scientists have suggested that *Homo habilis* gave rise to another species, *Homo erectus,* about 1.6 million years ago. This hominid had a larger brain than *Homo habilis. Homo erectus* traveled from Africa to Southeast Asia, China, and possibly Europe. *Homo habilis* and *Homo erectus* are thought to be ancestors of humans because they had larger brains and more humanlike features than *Australopithecus.*

Why was Homo habilis *given that name?*

Humans

The fossil record indicates that ***Homo sapiens*** evolved about 400,000 years ago. By about 125,000 years ago, two early human groups, Neanderthals (nee AN dur tawlz) and Cro-Magnon humans, as shown in **Figure 20,** probably lived at the same time in parts of Africa and Europe.

Neanderthals Short, heavy bodies with thick bones, small chins, and heavy browridges were physical characteristics of Neanderthals. Family groups lived in caves and used well-made stone tools to hunt large animals. Neanderthals disappeared from the fossil record about 30,000 years ago. They probably are not direct ancestors of modern humans, but represent a side branch of human evolution.

Figure 20 Compare the skull of a Neanderthal with the skull of a Cro-Magnon. **Describe** *what differences you can see between these two skulls.*

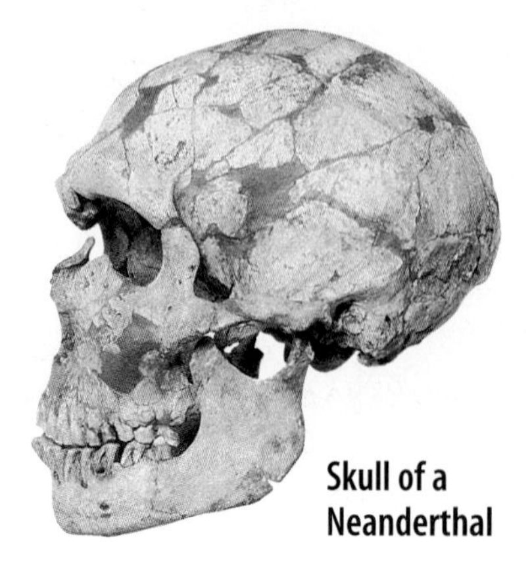

Skull of a Neanderthal

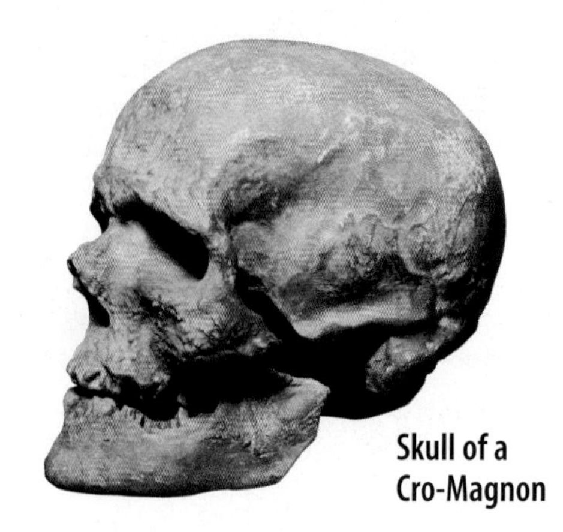

Skull of a Cro-Magnon

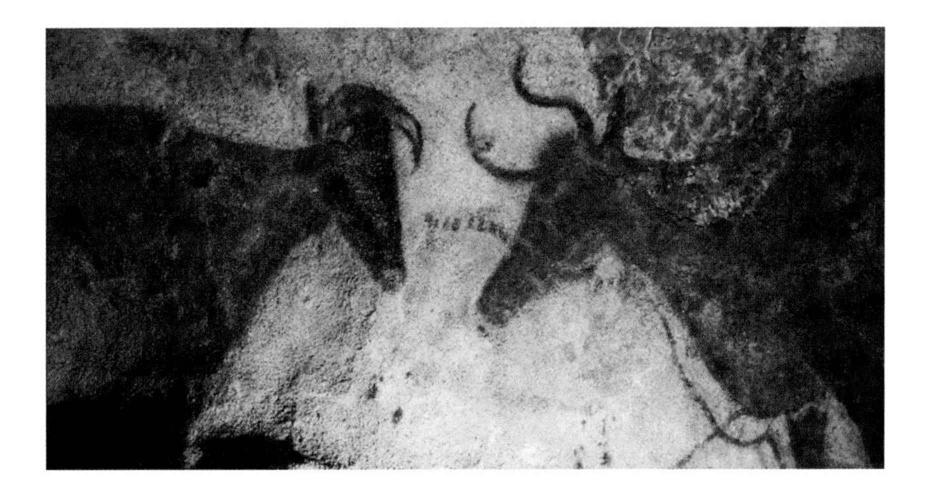

Figure 21 Paintings on cave walls have led scientists to hypothesize that Cro-Magnon humans had a well-developed culture.

Cro-Magnon Humans Cro-Magnon fossils have been found in Europe, Asia, and Australia and date from 10,000 to about 40,000 years in age. Standing about 1.6 m to 1.7 m tall, the physical appearance of Cro-Magnon people was almost the same as that of modern humans. They lived in caves, made stone carvings, and buried their dead. As shown in **Figure 21,** the oldest recorded art has been found on the walls of caves in France, where Cro-Magnon humans first painted bison, horses, and people carrying spears. Cro-Magnon humans are thought to be direct ancestors of early humans, *Homo sapiens,* which means "wise human." Evidence indicates that modern humans, *Homo sapiens sapiens,* evolved from *Homo sapiens.*

section 3 review

Summary

Primates

- Primates are an order of mammals characterized by opposable thumbs, binocular vision, and flexible shoulder joints.
- Primates are divided into strepsirrhines and haplorhines.
- Hominids are human ancestors that first appeared in Africa 4–6 million years ago.
- Hominids in the genus *Homo* first used tools and had larger brains than previous primates.

Humans

- *Homo sapiens* first appeared about 400,000 years ago.
- Cro-Magnon humans and Neanderthals coexisted in many places until Neanderthals disappeared about 30,000 years ago.
- *Homo sapiens* looked like modern humans and are believed to be our direct ancestors.

Self Check

1. **Describe** three kinds of evidence suggesting that all primates might have shared a common ancestor.
2. **Discuss** the importance of *Australopithecus.*
3. **Compare and contrast** Neanderthals, Cro-Magnon humans, and early humans.
4. **Identify** three groups most scientists consider to be direct ancestors of modern humans.
5. **Think Critically** Propose a hypothesis to explain why teeth are the most abundant fossil of hominids.

Applying Skills

6. **Concept Map** Make a concept map to show in what sequence hominids appeared. Use the following: *Homo sapiens sapiens,* Neanderthal, *Homo habilis, Australopithecus, Homo sapiens,* and Cro-Magnon human.
7. **Write** a story in your Science Journal about what life might have been like when both Neanderthals and Cro-Magnon humans were alive.

Recognizing Variation in a Population

Goals

- **Design** an experiment that will allow you to collect data about variation in a population.
- **Observe, measure, and analyze** variations in a population.

Possible Materials

fruit and seeds from one plant species
metric ruler
magnifying lens
graph paper

Safety Precautions

WARNING: *Do not put any fruit or seeds in your mouth.*

Real-World Question

When you first observe a flock of pigeons, you might think all the birds look alike. However, if you look closer, you will notice minor differences, or variations, among the individuals. Different pigeons might have different color markings, or some might be smaller or larger than others. Individuals of the same species—whether they're birds, plants, or worms—might look alike at first, but some variations undoubtedly exist. According to the principles of natural selection, evolution could not occur without variations. What kinds of variations have you noticed among species of plants or animals? How can you measure variation in a plant or animal population?

Form a Hypothesis

Make a hypothesis about the amount of variation in the fruit and seeds of one species of plant.

Using Scientific Methods

Test Your Hypothesis

Make a Plan

1. As a group, agree upon and write out the prediction.
2. **List** the steps you need to take to test your prediction. Be specific. Describe exactly what you will do at each step. List your materials.
3. **Decide** what characteristic of fruit and seeds you will study. For example, you could measure the length of fruit and seeds or count the number of seeds per fruit.
4. **Design** a data table in your Science Journal to collect data about two variations. Use the table to record the data your group collects.
5. **Identify** any constants, variables, and controls of the experiment.
6. How many fruit and seeds will you examine? Will your data be more accurate if you examine larger numbers?
7. **Summarize** your data in a graph or chart.

Follow Your Plan

1. Make sure your teacher approves your plan before you start.
2. Carry out the experiment as planned.
3. While the experiment is going on, write down any observations you make and complete the data table in your Science Journal.

Analyze Your Data

1. **Calculate** the mean and range of variation in your experiment. The range is the difference between the largest and the smallest measurements. The mean is the sum of all the data divided by the sample size.
2. **Graph** your group's results by making a line graph for the variations you measured. Place the range of variation on the *x*-axis and the number of organisms that had that measurement on the *y*-axis.

Conclude and Apply

1. **Explain** your results in terms of natural selection.
2. **Discuss** the factors you used to determine the amount of variation present.
3. **Infer** why one or more of the variations you observed in this activity might be helpful to the survival of the individual.

Create a poster or other exhibit that illustrates the variations you and your classmates observed.

TIME SCIENCE AND HISTORY

SCIENCE CAN CHANGE THE COURSE OF HISTORY!

Fighting HIV

The first cases of AIDS, or acquired immune deficiency syndrome, in humans were reported in the early 1980s. AIDS is caused by the human immunodeficiency virus, or HIV.

A major problem in AIDS research is the rapid evolution of HIV. When HIV multiplies inside a host cell, new versions of the virus are produced as well as identical copies of the virus that invaded the cell. New versions of the virus soon can outnumber the original version. A treatment that works against today's HIV might not work against tomorrow's version.

These rapid changes in HIV also mean that different strains of the virus exist in different places around the world. Treatments developed in the United States work only for people who contracted the virus in the United States. This leaves people in some parts of the world without effective treatments. So, researchers such as geneticist Flossie Wong-Staal at the University of California, San Diego, must look for new ways to fight the evolving virus.

Working Backwards

Flossie Wong-Staal is taking a new approach. First, her team identifies the parts of a human cell that HIV depends on and the parts of the human cell that HIV needs but the human cell doesn't need. Then the team looks for a way to remove—or inactivate—those unneeded parts. This technique limits the virus's ability to multiply.

Wong-Staal's research combines three important aspects of science—a deep understanding of how cells and genes operate, great skill in the techniques of genetics, and great ideas. Understanding, skill, and great ideas are the best weapons so far in the fight to conquer HIV.

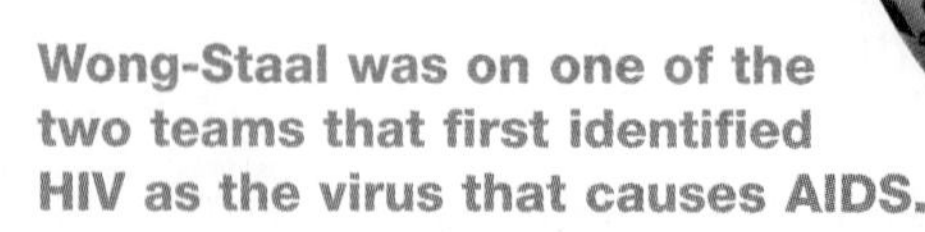

Wong-Staal was on one of the two teams that first identified HIV as the virus that causes AIDS.

Research Use the link to the right and other sources to determine which nations have the highest rates of HIV infection. Which nation has the highest rate? Where does the U.S. rank? Next, find data from ten years ago. Have the rankings changed?

Science Online

For more information, visit life.msscience.com/time

Reviewing Main Ideas

Section 1 Ideas About Evolution

1. Evolution is one of the central ideas of biology. It explains how living things have changed in the past and is a basis for predicting how they might change in the future.
2. Charles Darwin developed the theory of evolution by natural selection to explain how evolutionary changes account for the diversity of organisms on Earth.
3. Natural selection includes concepts of variation, overproduction, and competition.
4. According to natural selection, organisms with traits best suited to their environment are more likely to survive and reproduce.

Section 2 Clues About Evolution

1. Fossils provide evidence for evolution.
2. Relative dating and radiometric dating can be used to estimate the age of fossils.
3. The evolution of antibiotic-resistant bacteria, pesticide-resistant insects, and rapid genetic changes in plant species provides direct evidence that evolution occurs.
4. Homologous structures, vestigial structures, comparative embryology, and similarities in DNA provide indirect evidence of evolution.

Section 3 The Evolution of Primates

1. Primates include monkeys, apes, and humans. Hominids are humanlike primates.
2. The earliest known hominid fossil is *Australopithecus.*
3. *Homo sapiens* are thought to have evolved from Cro-Magnon humans about 400,000 years ago.

Visualizing Main Ideas

Copy and complete the following spider map on evolution.

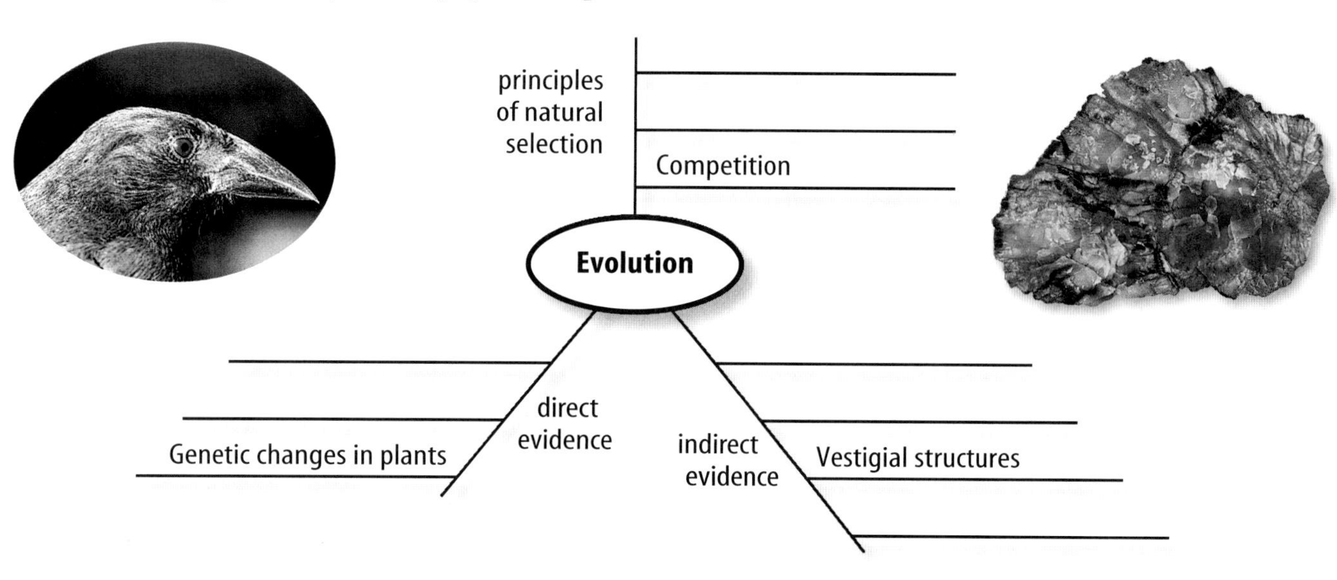

chapter 6 Review

Using Vocabulary

adaptation p. 158
embryology p. 167
evolution p. 154
gradualism p. 160
hominid p. 171
Homo sapiens p. 172
homologous p. 168
natural selection p. 157
primate p. 170
punctuated equilibrium p. 160
radioactive element p. 165
sedimentary rock p. 164
species p. 154
variation p. 158
vestigial structure p. 168

Fill in the blanks with the correct vocabulary word or words.

1. _________ contains many different kinds of fossils.
2. The muscles that move the human ear appear to be _________.
3. Forelimbs of bats, humans, and seals are _________.
4. Opposable thumbs are a characteristic of _________.
5. The study of _________ can provide evidence of evolution.
6. The principles of _________ include variation and competition.
7. _________ likely evolved directly from Cro-Magnons.

Checking Concepts

Choose the word or phrase that best answers the question.

8. What is an example of adaptation?
   - **A)** a fossil
   - **B)** gradualism
   - **C)** camouflage
   - **D)** embryo
9. What method provides the most accurate estimate of a fossil's age?
   - **A)** natural selection
   - **B)** radiometric dating
   - **C)** relative dating
   - **D)** camouflage
10. What do homologous structures, vestigial structures, and fossils provide evidence of?
    - **A)** gradualism
    - **B)** food choice
    - **C)** populations
    - **D)** evolution
11. Which model of evolution shows change over a relatively short period of time?
    - **A)** embryology
    - **B)** adaptation
    - **C)** gradualism
    - **D)** punctuated equilibrium
12. What might a series of helpful variations in a species result in?
    - **A)** adaptation
    - **B)** fossils
    - **C)** embryology
    - **D)** climate change

Use the following chart to answer question 13.

Homo habilis
↓
Homo erectus

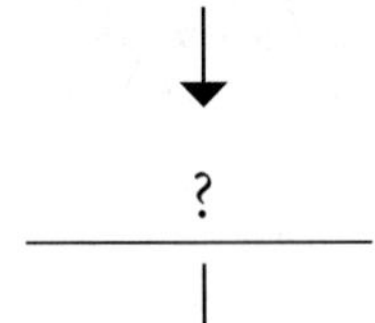

↓
Homo sapiens

13. Which correctly fills the gap in the line of descent from *Homo habilis?*
    - **A)** Neanderthal
    - **B)** *Australopithecus*
    - **C)** Cro-Magnon human
    - **D)** chimpanzee
14. What is the study of an organism's early development called?
    - **A)** adaptation
    - **B)** relative dating
    - **C)** natural selection
    - **D)** embryology

Science Online life.msscience.com/vocabulary_puzzlemaker

Thinking Critically

15. **Predict** what type of bird the foot pictured at right would belong to. Explain your reasoning.

16. **Discuss** how Lamarck and Darwin would have explained the large eyes of an owl.

17. **Explain,** using an example, how a new species of organism could evolve.

18. **Identify** how the color-changing ability of chameleons is an adaptation.

19. **Form a hypothesis** as to why ponds are not overpopulated by frogs in summer. Use the concept of natural selection to help you.

20. **Sequence** Make an events-chain concept map of the events that led Charles Darwin to his theory of evolution by natural selection.

Use the table below to answer question 21.

Chemicals Present in Bacteria	
Species 1	A, G, T, C, L, E, S, H
Species 2	A, G, T, C, L, D, H
Species 3	A,G, T, C, L, D, P, U, S, R, I, V
Species 4	A, G, T, C, L, D, H

21. **Interpret Data** Each letter above represents a chemical found in a species of bacteria. Which species are most closely related?

22. **Discuss** the evidence you would use to determine whether the evolution of a group were best explained by gradualism. How would this differ from a group that followed a punctuated equilibrium model?

23. **Describe** the processes a scientist would use to figure out the age of a fossil.

24. **Evaluate** the possibility for each of the five types of fossils in **Figure 9** to yield a DNA sample. Remember that only biological tissue will contain DNA.

Performance Activities

25. **Collection** With permission, collect fossils from your area and identify them. Show your collection to your class.

26. **Brochure** Assume that you are head of an advertising company. Develop a brochure to explain Darwin's theory of evolution by natural selection.

Applying Math

27. **Relative Age** The rate of radioactive decay is measured in half-lives—the amount of time it takes for one half of a radioactive element to decay. Determine the relative age of a fossil given the following information:
    - Rock layers are undisturbed.
    - The layer below the fossil has potassium-40 with a half-life of 1 million years and only one half of the original potassium is left.
    - The layer above the fossil has carbon-14 with a half-life of 5,730 years and one-sixteenth of the carbon isotope remains.

Use the graph below to answer question 28.

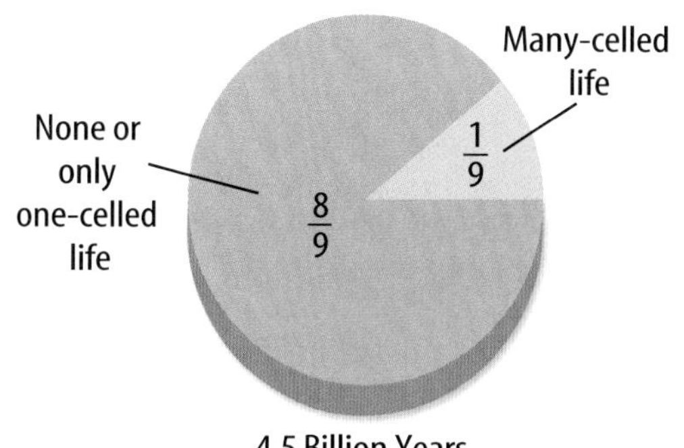

28. **First Appearances** If Earth is 4.5 billion years old, how long ago did the first many-celled life-forms appear?

Part 1 Multiple Choice

Record your answers on the answer sheet provided by your teacher or on a sheet of paper.

1. A species is a group of organisms
   - **A.** that lives together with similar characteristics.
   - **B.** that shares similar characteristics and can reproduce among themselves to produce fertile offspring.
   - **C.** across a wide area that cannot reproduce.
   - **D.** that chooses mates from among themselves.

2. Which of the following is considered an important factor in natural selection?
   - **A.** limited reproduction
   - **B.** competition for resources
   - **C.** no variations within a population
   - **D.** plentiful food and other resources

3. The marine iguana of the Galápagos Islands enters the ocean and feeds on seaweed. What is this an example of?
   - **A.** adaptation
   - **B.** gradualism
   - **C.** survival of the fittest
   - **D.** acquired characteristic

Use the illustration below to answer question 4.

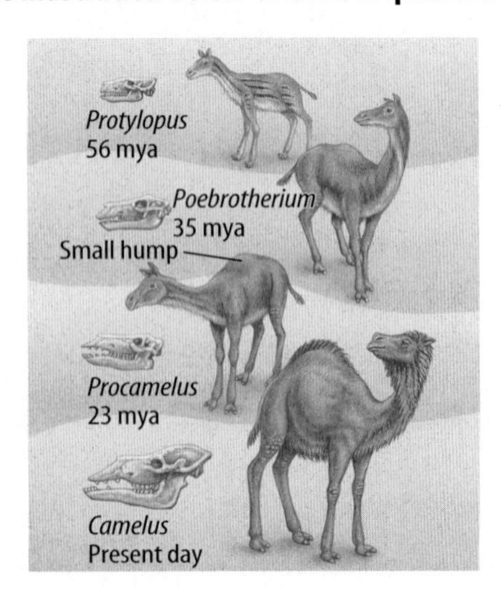

4. According to Lamarck's hypothesis of acquired characteristics, which statement best explains the changes in the camel over time?
   - **A.** All characteristics developed during an individual's lifetime are passed on to offspring.
   - **B.** Characteristics that do not help the animal survive are passed to offspring.
   - **C.** Variation of the species leads to adaptation.
   - **D.** Individuals moving from one area to another carry with them new characteristics.

Use the illustrations below to answer question 5.

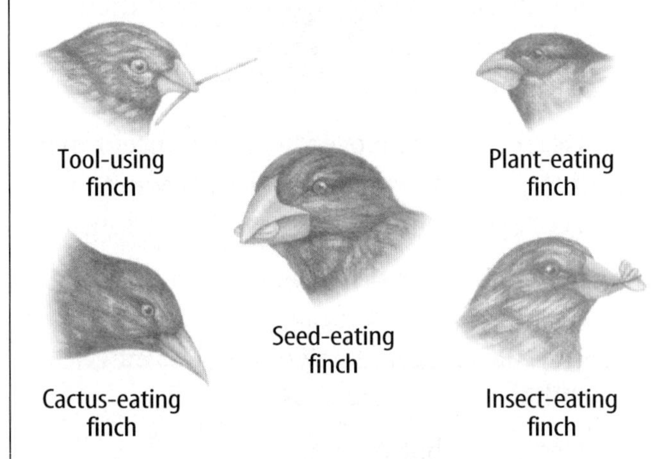

5. What, besides competition for food, contributed to the evolution of the species of Darwin's finches?
   - **A.** predation
   - **B.** natural disaster
   - **C.** DNA
   - **D.** variation in beak shapes

6. Some harmless species imitate or mimic a poisonous species as a means for increased survival. What is this an example of?
   - **A.** acquired characteristics
   - **B.** adaptation
   - **C.** variation
   - **D.** geographic isolation

Part 2 Short Response/Grid In

Record your answers on the answer sheet provided by your teacher or on a sheet of paper.

7. How does camouflage benefit a species?

Use the photo below to answer question 8.

8. Describe an environment where the albino lemur would not be at a disadvantage.
9. Variation between members of a species plays an important role in Darwin's theory of evolution. What happens to variation in endangered species where the number of individuals is very low?
10. Describe what happens to an endangered species if a variation provides an advantage for the species. What would happen if the variation resulted in a disadvantage?
11. Using the theory of natural selection, hypothesize why the Cro-Magnon humans survived and the Neanderthals disappeared.

Test-Taking Tip

Never Leave Any Answer Blank Answer each question as best you can. You can receive partial credit for partially correct answers.

Question 16 If you cannot remember all primate characteristics, list as many as you can.

Part 3 Open Ended

Record your answers on a sheet of paper.

12. What are the two groups of early humans that lived about 125,000 years ago in Africa and Europe? Describe their general appearance and characteristics. Compare these characteristics to modern humans.
13. Explain how bacterial resistance to antibiotics is an example of punctuated equilibrium.
14. Why are radioactive elements useful in dating fossils? Does this method improve accuracy over relative dating?

Use the illustrations below to answer question 15.

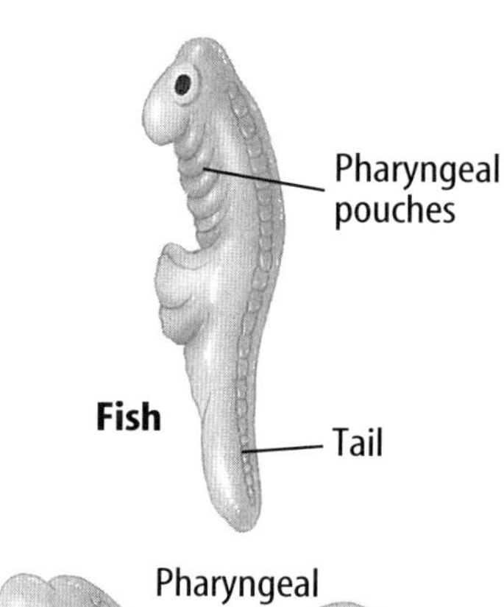

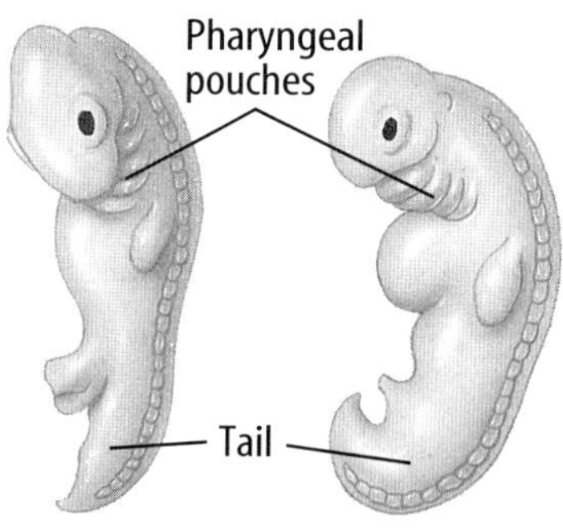

15. Why would scientists study embryos? What features of these three embryos support evolution?
16. How does DNA evidence provide support that primates have a common ancestor?

unit 2 From Bacteria to Plants

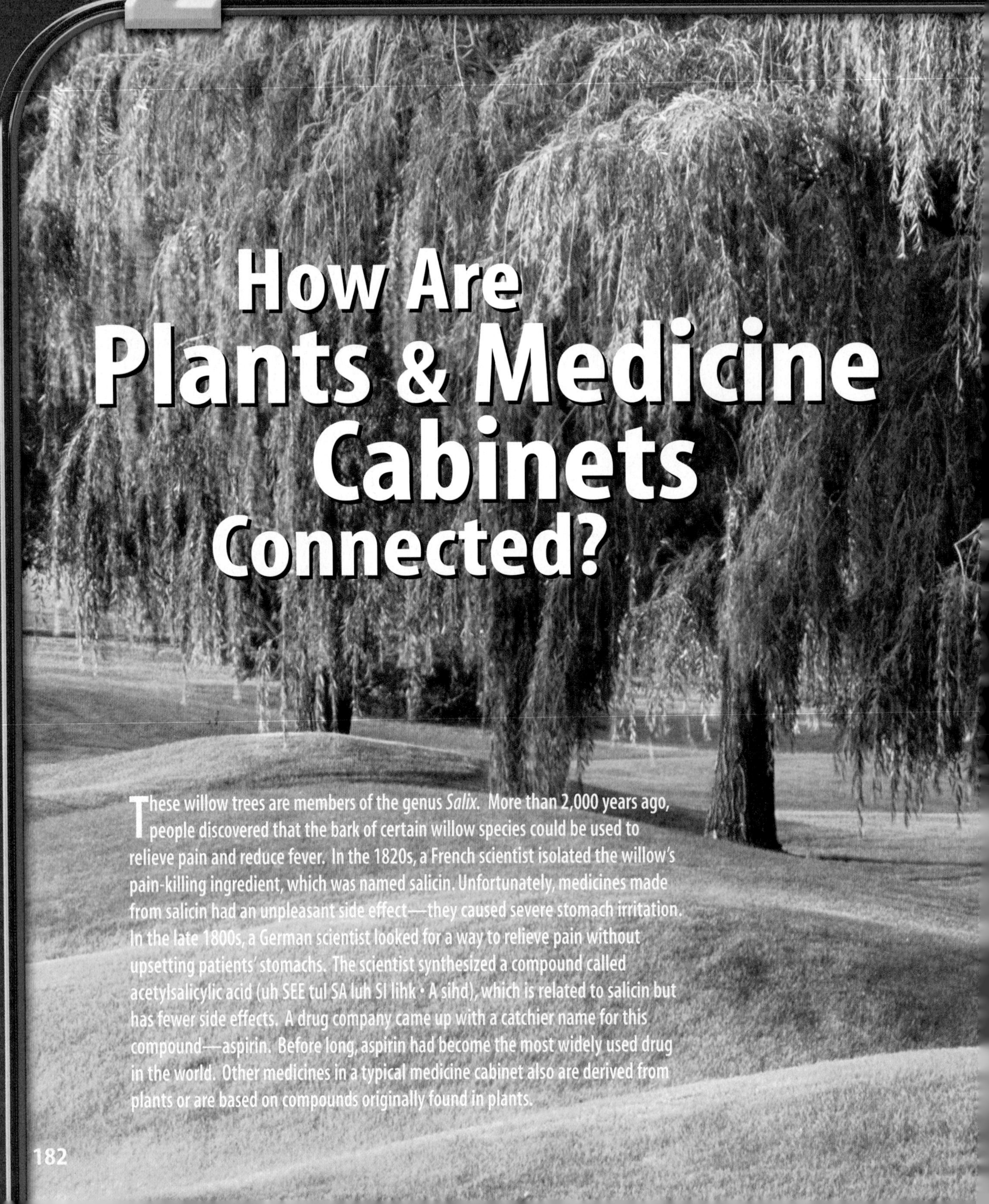

How Are Plants & Medicine Cabinets Connected?

These willow trees are members of the genus *Salix*. More than 2,000 years ago, people discovered that the bark of certain willow species could be used to relieve pain and reduce fever. In the 1820s, a French scientist isolated the willow's pain-killing ingredient, which was named salicin. Unfortunately, medicines made from salicin had an unpleasant side effect—they caused severe stomach irritation. In the late 1800s, a German scientist looked for a way to relieve pain without upsetting patients' stomachs. The scientist synthesized a compound called acetylsalicylic acid (uh SEE tul SA luh SI lihk • A sihd), which is related to salicin but has fewer side effects. A drug company came up with a catchier name for this compound—aspirin. Before long, aspirin had become the most widely used drug in the world. Other medicines in a typical medicine cabinet also are derived from plants or are based on compounds originally found in plants.

unit projects

Visit **life.msscience.com/unit_project** to find project ideas and resources. Projects include:

- **History** Design a slide show to present information on medicines derived from plants and where these plants grow.
- **Technology** Make your own giant jigsaw puzzle illustrating the five systems of a seed plant, including labels and functions of each plant part.
- **Model** Construct a review game demonstrating knowledge of nitrogen and oxygen cycles. The game and instructions should be assembled in an eco-friendly box.

WebQuest Discover *Phytochemicals and a Healthy Diet.* Compare your diet with the suggested diet that helps prevent cancer and heart disease.

chapter 7

Bacteria

The BIG Idea

Bacteria are the smallest organisms but have important roles in the environment.

SECTION 1
What are bacteria?

Main Idea Bacteria are prokaryotic cells with unique structures and functions.

SECTION 2
Bacteria in Your Life

Main Idea Bacteria are both beneficial and harmful to humans.

The Microcosmos of Yogurt

Have you ever eaten yogurt? Yogurt has been a food source for about 4,000 years. Bacteria provide yogurt's tangy flavor and creamy texture. Bacteria also are required for making sauerkraut, cheese, buttermilk, and vinegar.

Science Journal List ways that bacteria can be harmful and ways bacteria can be beneficial. Which list is longer? Why do think that is?

Start-Up Activities

Model a Bacterium's Slime Layer

Bacterial cells have a gelatinlike, protective coating on the outside of their cell walls. In some cases, the coating is thin and is referred to as a slime layer. A slime layer can help a bacterium attach to other surfaces. Dental plaque forms when bacteria with slime layers stick to teeth and multiply there. A slime layer also can reduce water loss from a bacterium. In this lab you will make a model of a bacterium's slime layer.

1. Cut two 2-cm-wide strips from the long side of a synthetic kitchen sponge.
2. Soak both strips in water. Remove them from the water and squeeze out the excess water. Both strips should be damp.
3. Completely coat one strip with hair-styling gel. Do not coat the other strip.
4. Place both strips on a plate (not paper) and leave them overnight.
5. **Think Critically** Record your observations of the two sponge strips in your Science Journal. Infer how a slime layer protects a bacterial cell from drying out. What environmental conditions are best for survival of bacteria?

Archaebacteria and Eubacteria Make the following Foldable to compare and contrast the characteristics of bacteria.

STEP 1 **Fold** one sheet of paper lengthwise.

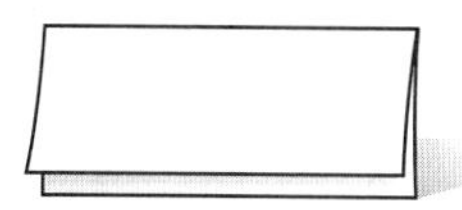

STEP 2 **Fold** into thirds.

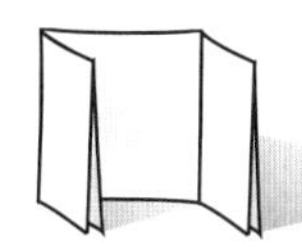

STEP 3 **Unfold and draw** overlapping ovals. Cut the top sheet along the folds.

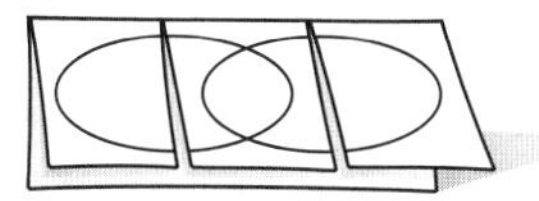

STEP 4 **Label** the ovals as shown.

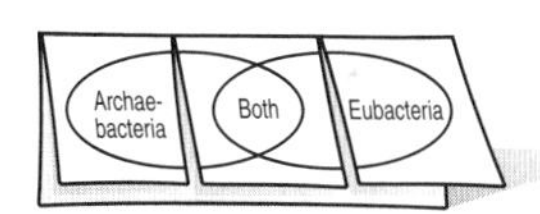

Construct a Venn Diagram As you read the chapter, list the characteristics unique to archaebacteria under the left tab, those unique to eubacteria under the right tab, and those characteristics common to both under the middle tab.

Preview this chapter's content and activities at life.msscience.com

Get Ready to Read

Make Predictions

1 Learn It! A prediction is an educated guess based on what you already know. One way to predict while reading is to guess what you believe the author will tell you next. As you are reading, each new topic should make sense because it is related to the previous paragraph or passage.

2 Practice It! Read the excerpt below from Section 1. Based on what you have read, make predictions about what you will read in the rest of the lesson. After you read Section 1, go back to your predictions to see if they were correct.

Predict how cyanobacteria can cause problems for aquatic life.

Predict what a bloom is.

Predict what can happen when available resources in the water are used up quickly and the cyanobacteria die.

Cyanobacteria also can cause problems for aquatic life. Have you ever seen a pond covered with smelly, green, bubbly slime? When large amounts of nutrients enter a pond, cyanobacteria increase in number. Eventually the population grows so large that a bloom is produced. A bloom looks like a mat of bubbly green slime on the surface of the water. Available resources in the water are used up quickly and the cyanobacteria die. Other bacteria that are aerobic consumers feed on dead cyanobacteria and use up all the oxygen in the water. As a result of the reduced oxygen in the water, fish and other organisms die.

—*from page 190*

3 Apply It! Before you read, skim the questions in the Chapter Review. Choose three questions and predict the answers.

As you read, check the predictions you made to see if they were correct.

Target Your Reading

Use this to focus on the main ideas as you read the chapter.

1. **Before you read** the chapter, respond to the statements below on your worksheet or on a numbered sheet of paper.
   - Write an **A** if you **agree** with the statement.
   - Write a **D** if you **disagree** with the statement.

2. **After you read** the chapter, look back to this page to see if you've changed your mind about any of the statements.
   - If any of your answers changed, explain why.
   - Change any false statements into true statements.
   - Use your revised statements as a study guide.

Before You Read A or D		Statement	After You Read A or D
	1	Bacteria can live almost everywhere on Earth.	
	2	A bacterium has a nucleus and other organelles.	
	3	Bacteria reproduce by fission, a type of asexual reproduction.	
	4	Most bacteria use cellular respiration.	
	5	Cyanobacteria are producers that use light energy to make their food.	
	6	Bacteria cannot live in extremely hot environments, such as the hot springs in Yellowstone National Park.	
	7	Bacteria that produce methane gas cannot survive when oxygen is present in their environment.	
	8	Bacteria are not beneficial for humans.	
	9	Pasteurization kills all bacteria in milk.	
	10	Vaccines can prevent some diseases caused by bacteria.	

Science Online
Print out a worksheet of this page at life.msscience.com

section 1

What are bacteria?

as you read

What You'll Learn

- Identify the characteristics of bacterial cells.
- Compare and contrast aerobic and anaerobic organisms.

Why It's Important

Bacteria are found almost everywhere and affect all living things.

Review Vocabulary
prokaryotic: cells without membrane-bound organelles

New Vocabulary
- flagella
- fission
- aerobe
- anaerobe

Characteristics of Bacteria

For thousands of years people did not understand what caused disease. They did not understand the process of decomposition or what happened when food spoiled. It wasn't until the latter half of the seventeenth century that Antonie van Leeuwenhoek, a Dutch merchant, discovered the world of bacteria. Leeuwenhoek observed scrapings from his teeth using his simple microscope. Although he didn't know it at that time, some of the tiny swimming organisms he observed were bacteria. After Leeuwenhoek's discovery, it was another hundred years before bacteria were proven to be living cells that carry on all of the processes of life.

Where do bacteria live? Bacteria are almost everywhere—in the air, in foods that you eat and drink, and on the surfaces of things you touch. They are even found thousands of meters underground and at great ocean depths. A shovelful of soil contains billions of them. Your skin has about 100,000 bacteria per square centimeter, and millions of other bacteria live in your body. Some types of bacteria live in extreme environments where few other organisms can survive. Some heat-loving bacteria live in hot springs or hydrothermal vents—places where water temperature exceeds 100°C. Others can live in cold water or soil at 0°C. Some bacteria live in very salty water, like that of the Dead Sea. One type of bacteria lives in water that drains from coal mines, which is extremely acidic at a pH of 1.

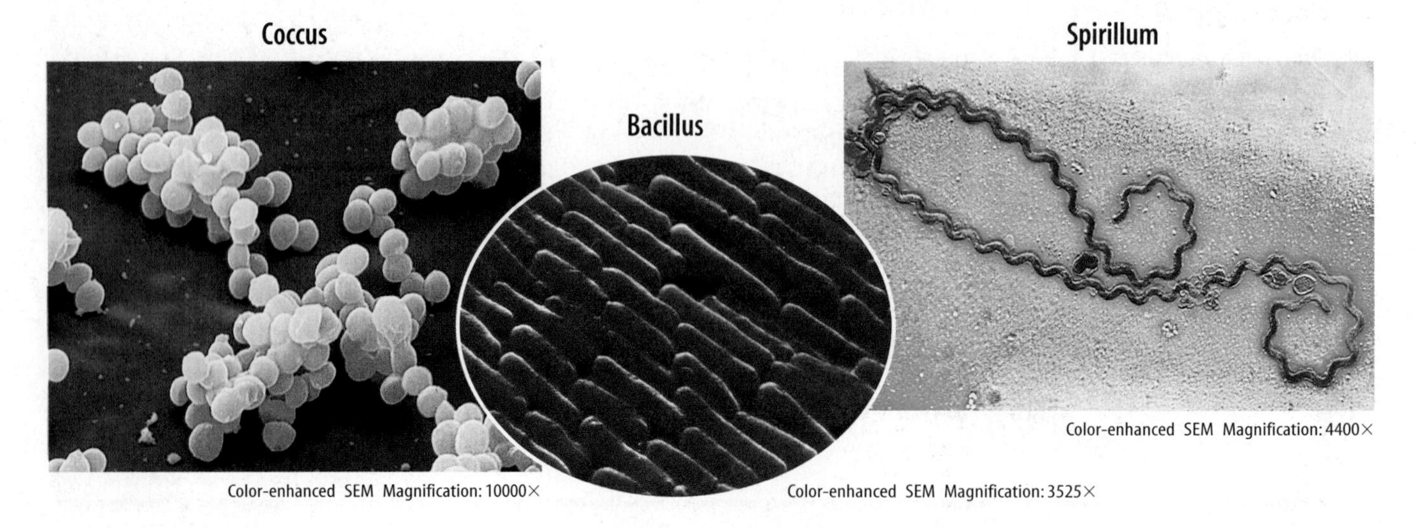

Figure 1 Bacteria can be found in almost any environment.
List *common terms that could be used to describe these cell shapes.*

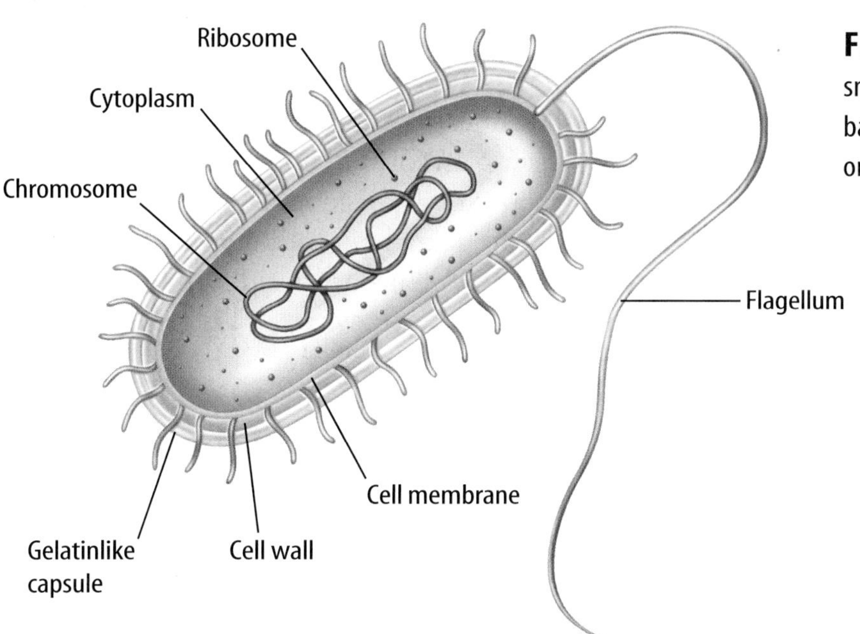

Figure 2 Bacterial cells are much smaller than eukaryotic cells. Most bacteria are about the size of some organelles found inside eukaryotic cells.

Structure of Bacterial Cells Bacteria normally have three basic shapes—spheres, rods, and spirals, as shown in **Figure 1.** Sphere-shaped bacteria are called cocci (KAHK si) (singular, *coccus*), rod-shaped bacteria are called bacilli (buh SIH li) (singular, *bacillus*), and spiral-shaped bacteria are called spirilla (spi RIH luh) (singular, *spirillum*). Bacteria are smaller than plant or animal cells. They are one-celled organisms that occur alone or in chains or groups.

A typical bacterial cell contains cytoplasm surrounded by a cell membrane and a cell wall, as shown in **Figure 2.** Bacterial cells are classified as prokaryotic because they do not contain a membrane-bound nucleus or other membrane-bound internal structures called organelles. Most of the genetic material of a bacterial cell is in its one circular chromosome found in the cytoplasm. Many bacteria also have a smaller circular piece of DNA called a plasmid. Ribosomes also are found in a bacterial cell's cytoplasm.

Special Features Some bacteria, like the type that causes pneumonia, have a thick, gelatinlike capsule around the cell wall. A capsule can help protect the bacterium from other cells that try to destroy it. The capsule, along with hairlike projections found on the surface of many bacteria, also can help them stick to surfaces. Some bacteria also have an outer coating called a slime layer. Like a capsule, a slime layer enables a bacterium to stick to surfaces and reduces water loss. Many bacteria that live in moist conditions also have whiplike tails called **flagella** to help them move.

How do bacteria use flagella?

Modeling Bacteria Size

Procedure
1. One human hair is about 0.1 mm wide. Use a **meterstick** to measure a piece of **yarn or string** that is 10 m long. This yarn represents the width of your hair.
2. One type of bacteria is 2 micrometers long (1 micrometer = 0.000001 m). Measure another piece of yarn or string that is 20 cm long. This piece represents the length of the bacterium.
3. Find a large area where you can lay the two pieces of yarn or string next to each other and compare them.

Analysis
1. Calculate how much smaller the bacterium is than the width of your hair.
2. In your **Science Journal,** describe why a model is helpful to understand how small bacteria are.

Try at Home

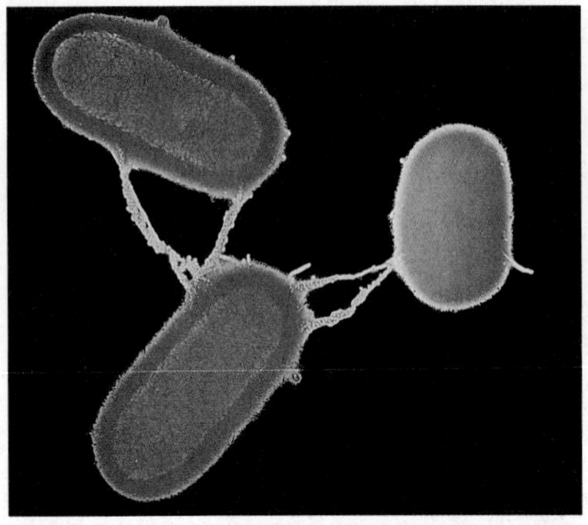

Figure 3 Before dividing, these bacteria are exchanging DNA through the tubes that join them. This process is called conjugation.

Reproduction Bacteria usually reproduce by fission. **Fission** is a process that produces two new cells with genetic material identical to each other and that of the original cell. It is the simplest form of asexual reproduction.

Some bacteria exchange genetic material through a process similar to sexual reproduction, as shown in **Figure 3.** Two bacteria line up beside each other and exchange DNA through a fine tube. This results in cells with different combinations of genetic material than they had before the exchange. This can result in bacteria with variations that give them an advantage for survival.

How Bacteria Obtain Food and Energy Bacteria obtain food in a variety of ways. Some make their food and others get it from the environment. Bacteria that contain chlorophyll or other pigments make their own food using light energy from the Sun. Other bacteria use energy from chemical reactions to make food. Bacteria and other organisms that can make their own food are called producers.

Most bacteria are consumers. They do not make their own food. Some break down dead organisms to obtain food. Others live in or on living organisms and absorb nutrients from their host.

Most organisms need oxygen for the breakdown of food that releases energy within a cell. An organism that uses oxygen when breaking down food is called an **aerobe** (AY rohb). You are an aerobic organism and so are most bacteria. In contrast, an organism that is adapted to live without oxygen is called an **anaerobe** (AN uh rohb). Several kinds of anaerobic bacteria live in the intestinal tract of humans. Some bacteria cannot survive in areas with oxygen.

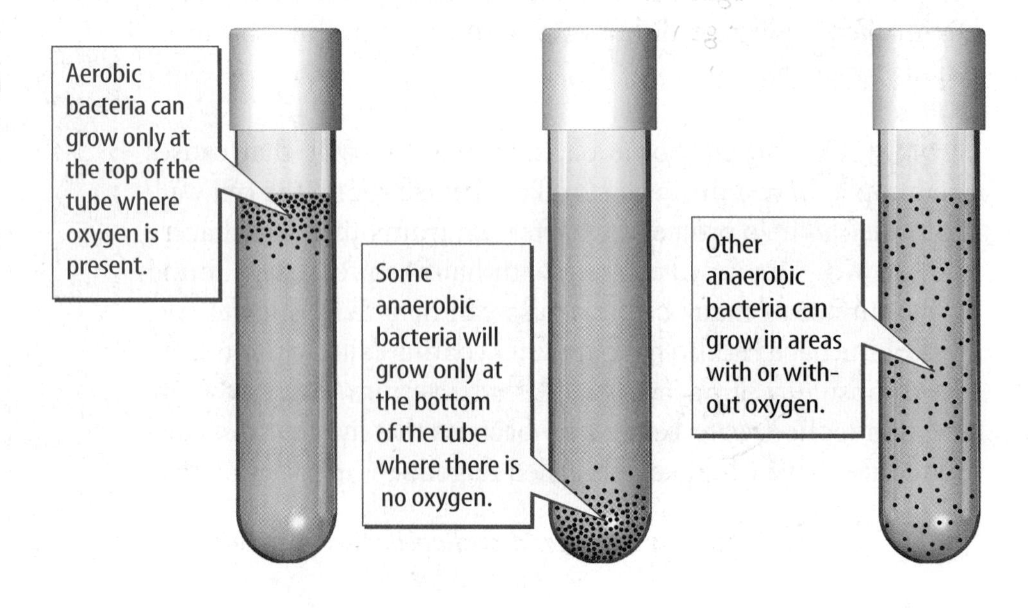

Aerobic bacteria can grow only at the top of the tube where oxygen is present.
Some anaerobic bacteria will grow only at the bottom of the tube where there is no oxygen.
Other anaerobic bacteria can grow in areas with or without oxygen.

Figure 4 Observing where bacteria can grow in tubes of a nutrient mixture shows you how oxygen affects different types of bacteria.

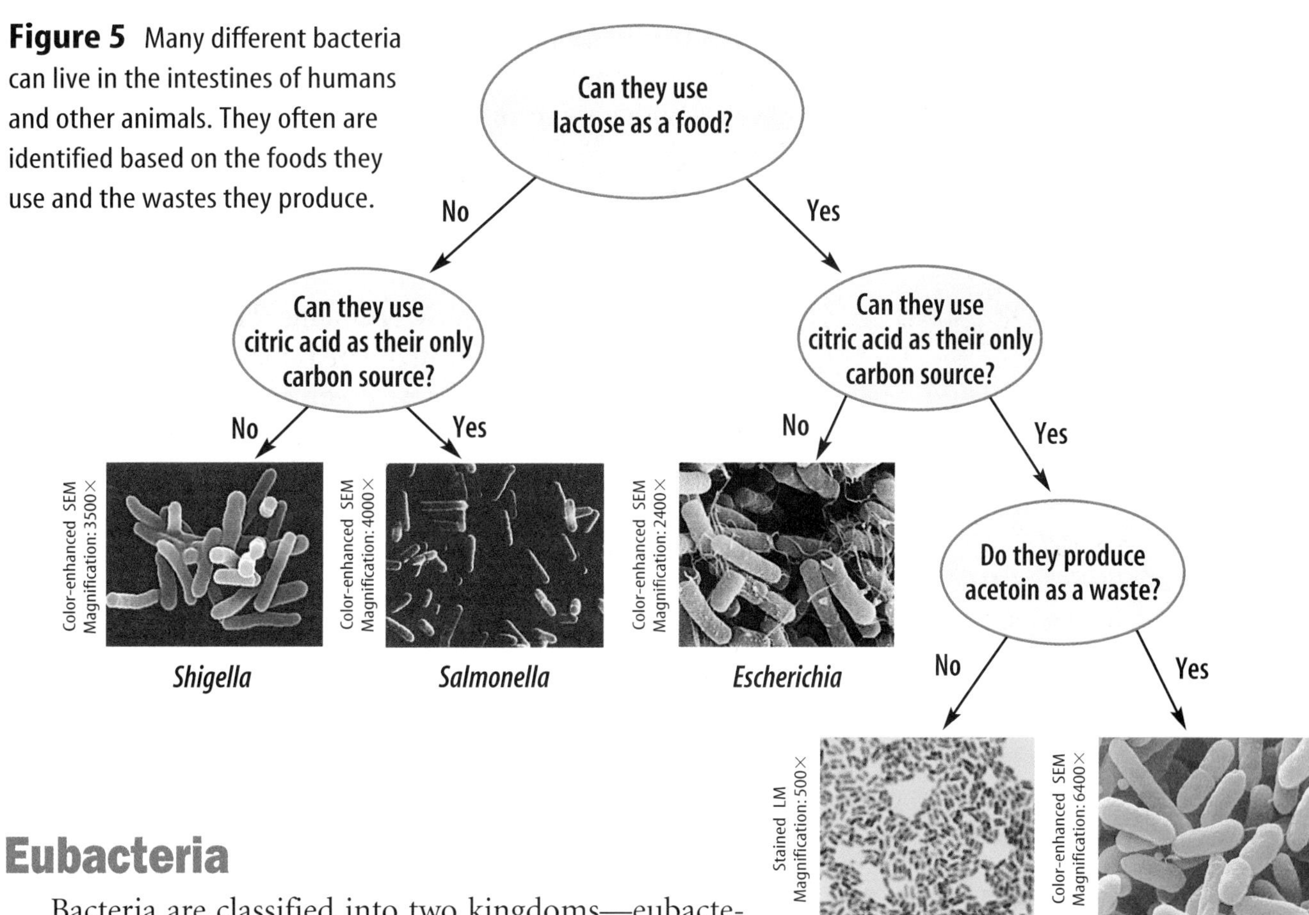

Figure 5 Many different bacteria can live in the intestines of humans and other animals. They often are identified based on the foods they use and the wastes they produce.

Eubacteria

Bacteria are classified into two kingdoms—eubacteria (yew bak TIHR ee uh) and archaebacteria (ar kee bak TIHR ee uh). Eubacteria is the larger of the two kingdoms. Scientists must study many characteristics in order to classify eubacteria into smaller groups. Most eubacteria are grouped according to their cell shape and structure, the way they obtain food, the type of food they consume, and the wastes they produce, as shown in **Figure 5.** Other characteristics used to group eubacteria include the method used for cell movement and whether the organism is an aerobe or anaerobe. New information about their genetic material is changing how scientists classify this kingdom.

Producer Eubacteria One important group of producer eubacteria is the cyanobacteria (si an oh bak TIHR ee uh). They make their own food using carbon dioxide, water, and energy from sunlight. They also produce oxygen as a waste. Cyanobacteria contain chlorophyll and another pigment that is blue. This pigment combination gives cyanobacteria their common name—blue-green bacteria. However, some cyanobacteria are yellow, black, or red. The Red Sea gets its name from red cyanobacteria.

Why are cyanobacteria classified as producers?

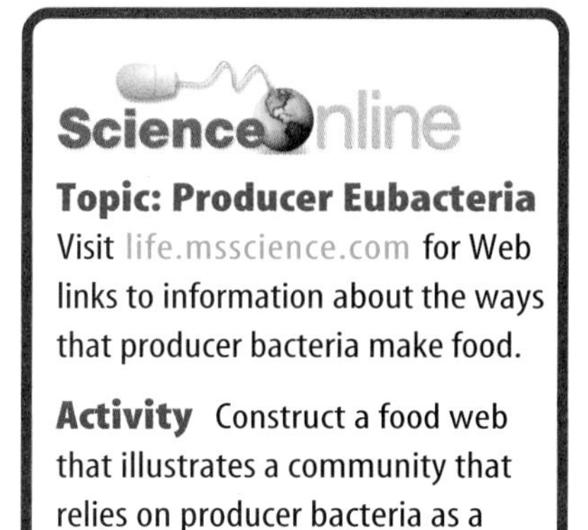

Science Online

Topic: Producer Eubacteria
Visit life.msscience.com for Web links to information about the ways that producer bacteria make food.

Activity Construct a food web that illustrates a community that relies on producer bacteria as a source of energy.

LM Magnification: 100×

Figure 6 These colonies of the cyanobacteria *Oscillatoria* can move by twisting like a screw.

Importance of Cyanobacteria Some cyanobacteria live together in long chains or filaments, as shown in **Figure 6.** Many are covered with a gelatinlike substance. This adaptation enables cyanobacteria to live in groups called colonies. They are an important source of food for some organisms in lakes, ponds, and oceans. The oxygen produced by cyanobacteria is used by other aquatic organisms.

Cyanobacteria also can cause problems for aquatic life. Have you ever seen a pond covered with smelly, green, bubbly slime? When large amounts of nutrients enter a pond, cyanobacteria increase in number. Eventually the population grows so large that a bloom is produced. A bloom looks like a mat of bubbly green slime on the surface of the water. Available resources in the water are used up quickly and the cyanobacteria die. Other bacteria that are aerobic consumers feed on dead cyanobacteria and use up the oxygen in the water. As a result of the reduced oxygen in the water, fish and other organisms die.

Ocean Vents Geysers on the floor of the ocean are called ocean vents. Research to find out how ocean vents form and what conditions are like at an ocean vent. In your Science Journal, describe organisms that have been found living around ocean vents.

Consumer Eubacteria Most consumer eubacteria are grouped into one of two categories based on the results of the Gram's stain. These results can be seen under a microscope after the bacteria are treated with certain chemicals that are called stains. As shown in **Figure 7,** gram-positive cells stain purple because they have thicker cell walls. Gram-negative cells stain pink because they have thinner cell walls.

The composition of the cell wall also can affect how a bacterium is affected by medicines given to treat an infection. Some antibiotics (an ti bi AH tihks) will be more effective against gram-negative bacteria than they will be against gram-positive bacteria.

One group of eubacteria is unique because they do not produce cell walls. This allows them to change their shape. They are not described as coccus, bacillus, or spirillum. One type of bacteria in this group, *Mycoplasma pneumoniae,* causes a type of pneumonia in humans.

Stained LM Magnification: 315×

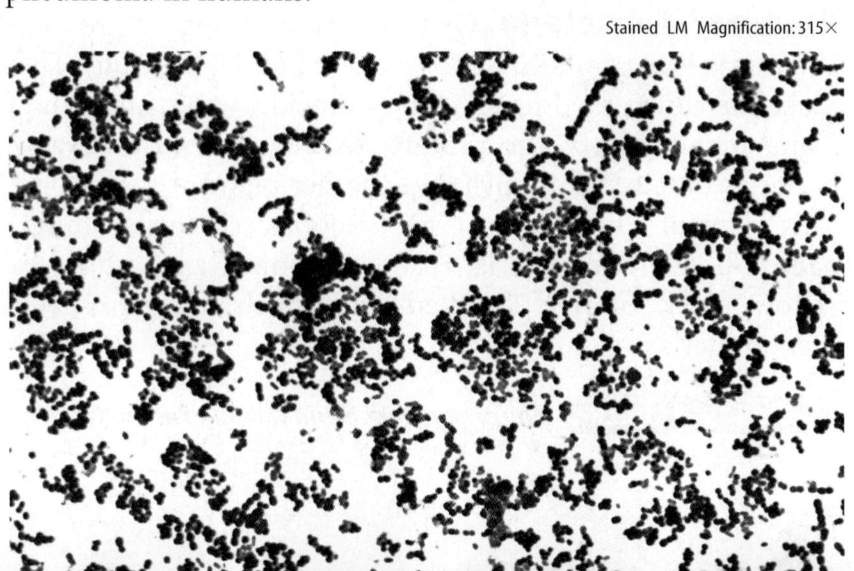

Figure 7 When stained with certain chemicals, bacteria with thin cell walls appear pink when viewed under a microscope. Those with thicker walls appear purple.

Archaebacteria

Kingdom Archaebacteria contains certain kinds of bacteria that often are found in extreme conditions, such as hot springs. The conditions in which some archaebacteria live today are similar to conditions found on Earth during its early history. Archaebacteria are divided into groups based on where they live or how they get energy.

Salt-, Heat-, and Acid-Lovers One group of archaebacteria lives in salty environments such as the Great Salt Lake in Utah and the Dead Sea. Some of them require a habitat ten times saltier than seawater to grow.

Other groups of archaebacteria include those that live in acidic or hot environments. Some of these bacteria live near deep ocean vents or in hot springs where the temperature of the water is above 100°C.

Methane Producers Bacteria in this group of archaebacteria —the methane producers—are anaerobic. They live in muddy swamps, the intestines of cattle, and even in you. Methane producers, as shown in **Figure 8,** use the energy holding molecules of carbon dioxide together and release methane gas as a waste. Sometimes methane produced by these bacteria bubbles up out of swamps and marshes. These archaebacteria also are used in the process of sewage treatment. In an oxygen-free tank, the bacteria break down wastes that have been filtered from sewage water.

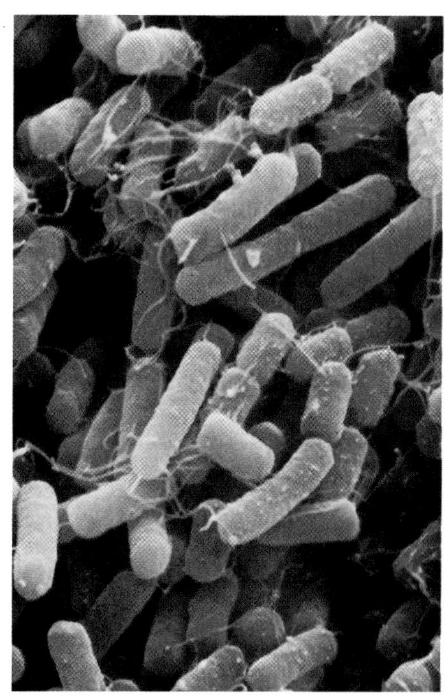

Color-enhanced SEM Magnification: 6000×

Figure 8 Some methane-producing bacteria live in the digestive tracts of cattle. They help digest the plants that cattle eat.

section 1 review

Summary

Characteristics of Bacteria

- Bacteria live almost everywhere and usually are one of three basic shapes.
- A bacterium lacks a nucleus, most bacteria reproduce asexually, and they can be aerobes or anaerobes.

Eubacteria

- Eubacteria are grouped by cell shape and structure, how they obtain food, and whether they are gram-positive or gram-negative.

Archaebacteria

- Archaebacteria can be found in extreme environments.
- Some break down sewage and produce methane.

Self Check

1. **List** three shapes of bacteria cells.
2. **Compare and contrast** aerobic organisms and anaerobic organisms.
3. **Explain** how most bacteria reproduce.
4. **Identify** who is given credit for first discovering bacteria.
5. **Think Critically** A pond is surrounded by recently fertilized farm fields. What effect would rainwater runoff from the fields have on the organisms in the pond?

Applying Math

6. **Solve One-Step Equations** Some bacteria reproduce every 20 min. Suppose that you have one bacterium. How long would it take for the number of bacteria to increase to more than 1 million?

Science Online life.msscience.com/self_check_quiz

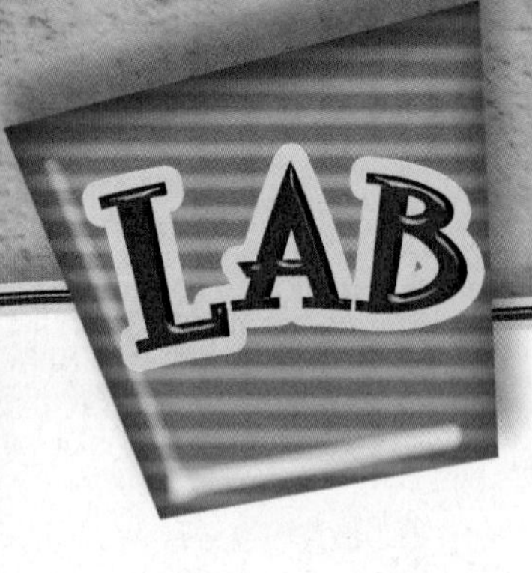

Observing Cyanobacteria

You can obtain many species of cyanobacteria from ponds. When you look at these organisms under a microscope, you will find that they have similarities and differences. In this lab, compare and contrast species of cyanobacteria.

Real-World Question

What do cyanobacteria look like?

Goals

- **Observe** several species of cyanobacteria.
- **Describe** the structure and function of cyanobacteria.

Materials

micrograph photos of *Oscillatoria* and *Nostoc*
**prepared slides of* Oscillatoria *and* Nostoc
prepared slides of *Gloeocapsa* and *Anabaena*
**micrograph photos of* Anabaena *and* Gloeocapsa
microscope
**Alternate materials*

Safety Precautions

Procedure

1. Copy the data table in your Science Journal. As you observe each cyanobacterium, record the presence or absence of each characteristic in the data table.
2. **Observe** prepared slides of *Gloeocapsa* and *Anabaena* under low and high power of the microscope. Notice the difference in the arrangement of the cells. In your Science Journal, draw and label a few cells of each.
3. **Observe** photos of *Nostoc* and *Oscillatoria.* In your Science Journal, draw and label a few cells of each.

Conclude and Apply

1. **Infer** what the color of each cyanobacterium means.
2. **Explain** how you can tell by observing that a cyanobacterium is a eubacterium.

Communicating Your Data

Compare your data table with those of other students in your class. **For more help, refer to the Science Skill Handbook.**

Cyanobacteria Observations

Structure	*Anabaena*	*Gloeocapsa*	*Nostoc*	*Oscillatoria*
Filament or colony		Do not write in this book.		
Nucleus				
Chlorophyll				
Gel-like layer				

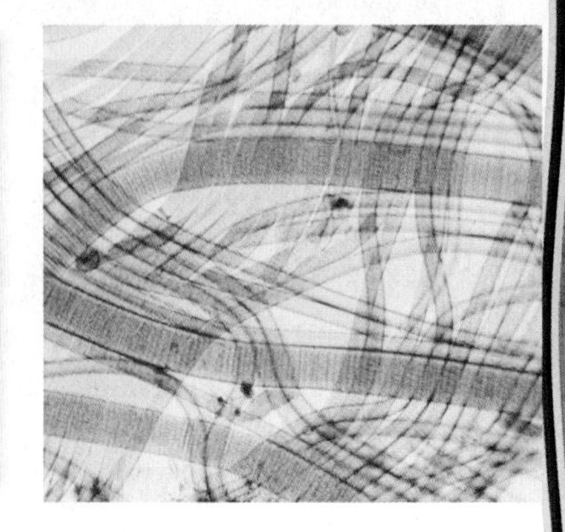

section 2

Bacteria in Your Life

as you read

What You'll Learn

- **Identify** some ways bacteria are helpful.
- **Determine** the importance of nitrogen-fixing bacteria.
- **Explain** how some bacteria can cause human disease.

Why It's Important

Discovering the ways bacteria affect your life can help you understand biological processes.

Review Vocabulary

disease: a condition with symptoms that interferes with normal body functions

New Vocabulary

- antibiotic
- saprophyte
- nitrogen-fixing bacteria
- pathogen
- toxin
- endospore
- vaccine

Beneficial Bacteria

When you hear the word *bacteria,* you probably associate it with sore throats or other illnesses. However, few bacteria cause illness. Most are important for other reasons. The benefits of most bacteria far outweigh the harmful effects of a few.

Bacteria That Help You Without bacteria, you would not be healthy for long. Bacteria, like those in **Figure 9,** are found inside your digestive system. These bacteria are found in particularly high numbers in your large intestine. Most are harmless to you, and they help you stay healthy. For example, some bacteria in your intestines are responsible for producing vitamin K, which is necessary for normal blood clot formation.

Some bacteria produce chemicals called **antibiotics** that limit the growth of other bacteria. For example, one type of bacteria that is commonly found living in soil produces the antibiotic streptomycin. Another kind of bacteria, *Bacillus,* produces the antibiotic found in many nonprescription antiseptic ointments. Many bacterial diseases in humans and animals can be treated with antibiotics.

Lactobacillus LM Magnification: 250×

Klebsiella Color-enhanced TEM Magnification: 11000×

E. coli Color-enhanced SEM Magnification: 3200×

Fusobacterium Color-enhanced TEM Magnification: 3000×

Figure 9 Many types of bacteria live naturally in your large intestine. They help you digest food and produce essential vitamins.

Figure 10 Air is bubbled through the sewage in this aeration tank so that bacteria can break down much of the sewage wastes. **Determine** *whether the bacteria that live in this tank are aerobes or anaerobes.*

Bacteria and the Environment Without bacteria, there would be layers of dead material all over Earth deeper than you are tall. Consumer bacteria called **saprophytes** (SAP ruh fites) help maintain nature's balance. A saprophyte is any organism that uses dead organisms as food and energy sources. Saprophytic bacteria help recycle nutrients. These nutrients become available for use by other organisms. As shown in **Figure 10,** most sewage-treatment plants use saprophytic aerobic bacteria to break down wastes into carbon dioxide and water.

Reading Check *What is a saprophyte?*

Plants and animals must take in nitrogen to make needed proteins and nucleic acids. Animals can eat plants or other animals that contain nitrogen, but plants need to take nitrogen from the soil or air. Although air is about 78 percent nitrogen, neither animals nor plants can use this nitrogen directly. **Nitrogen-fixing bacteria** change nitrogen from the air into forms that plants and animals can use. The roots of some plants, such as peanuts and peas, develop structures called nodules that contain nitrogen-fixing bacteria, as shown in **Figure 11.** It is estimated that nitrogen-fixing bacteria save U.S. farmers millions of dollars in fertilizer costs every year. Many cyanobacteria fix nitrogen and provide usable nitrogen to aquatic organisms.

Bioremediation Using organisms to help clean up or remove environmental pollutants is called bioremediation. One type of bioremediation uses bacteria to break down wastes and pollutants into simpler harmless compounds. Other bacteria use certain pollutants as a food source. Every year about five percent to ten percent of all wastes produced by industry, agriculture, and cities are treated by bioremediation. Sometimes bioremediation is used at the site where chemicals, such as oil, have been spilled. Research continues on ways to make bioremediation a faster process.

Observing Bacterial Growth

Procedure

1. Obtain two or three **dried beans.**
2. Carefully break them into halves and place the halves into 10 mL of **distilled water** in a **glass beaker.**
3. Observe how many days it takes for the water to become cloudy and develop an unpleasant odor.

Analysis

1. How long did it take for the water to become cloudy?
2. What do you think the bacteria were using as a food source?

NATIONAL GEOGRAPHIC

VISUALIZING NITROGEN-FIXING BACTERIA

Figure 11

Although 78 percent of Earth's atmosphere is nitrogen gas (N_2), most living things are unable to use nitrogen in this form. Some bacteria, however, convert N_2 into the ammonium ion (NH_4^+) that organisms can use. This process is called nitrogen fixation. Nitrogen-fixing bacteria in soil can enter the roots of plants, such as beans, peanuts, alfalfa, and peas, as shown in the background photo. The bacteria and the plant form a relationship that benefits both of them.

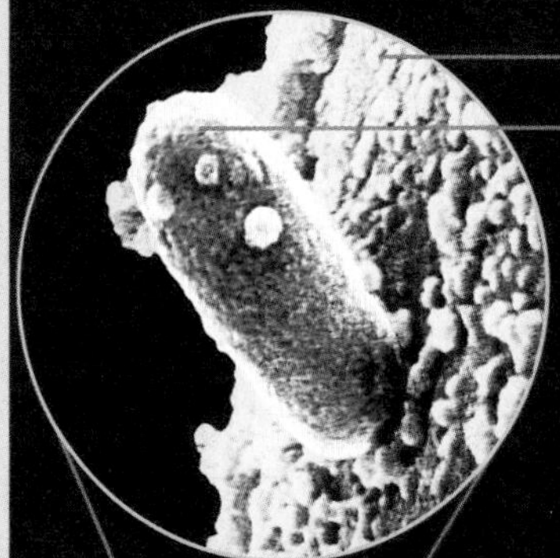

Infection thread

Root hair

Bacterium

Root hair

◀ Nitrogen-fixing bacteria typically enter a plant through root hairs—thin-walled cells on a root's outer surface.

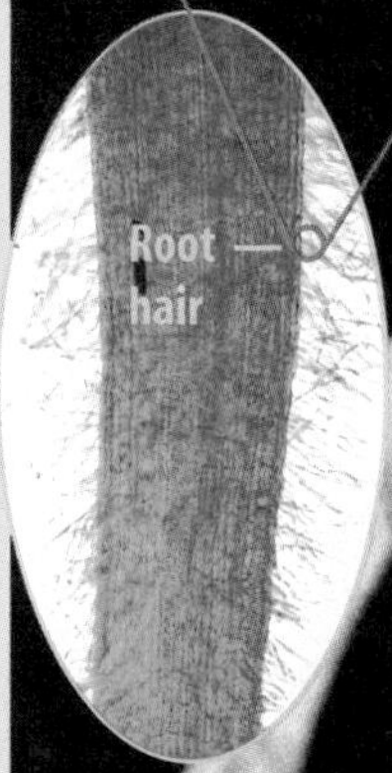

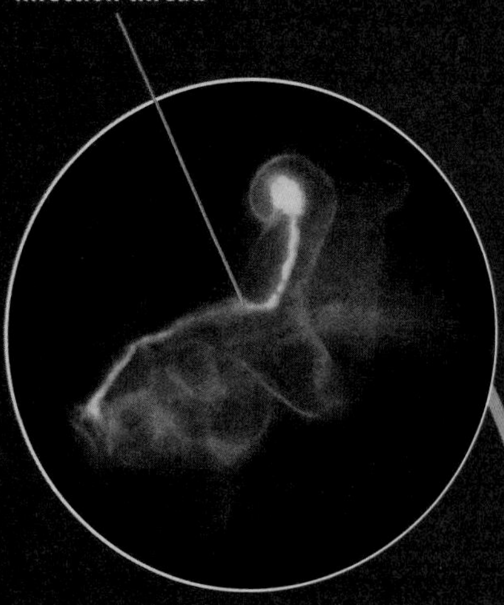

▲ Once inside the root hair, the bacteria enlarge and cause the plant to produce a sort of tube called an infection thread. The bacteria move through the thread to reach cells deeper inside the root.

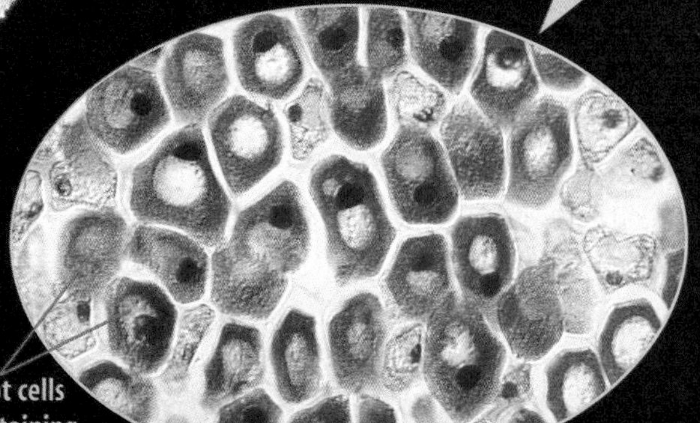

Root cells containing nitrogen-fixing bacteria

Beadlike nodules full of bacteria cover the roots of a pea plant.

▲ The bacteria rapidly divide in the root cells, which in turn divide repeatedly to form tumorlike nodules on the roots. Once established, the bacteria (purple) fix nitrogen for use by the host plant. In return, the plant supplies the bacteria with sugars and other vital nutrients.

Bioreactor Landfills As Earth's population grows and produces more waste, traditional landfills, which take 30 to 100 years to decompose waste, no longer fulfill the need for solid-waste disposal. Bioreactor landfills, which take 5 to 10 years to decompose waste, are beginning to be used instead. Bioreactor landfills can use aerobic or anaerobic bacteria, or a combination of the two, for rapid degradation of wastes.

Bacteria and Food Have you had any bacteria for lunch lately? Even before people understood that bacteria were involved, they were used in the production of foods. One of the first uses of bacteria was for making yogurt, a milk-based food that has been made in Europe and Asia for hundreds of years. Bacteria break down substances in milk to make many dairy products. Cheeses and buttermilk also can be produced with the aid of bacteria. Cheese making is shown in **Figure 12.**

Other foods you might have eaten also are made using bacteria. Sauerkraut, for example, is made with cabbage and a bacterial culture. Vinegar, pickles, olives, and soy sauce also are produced with the help of bacteria.

Bacteria in Industry Many industries rely on bacteria to make many products. Bacteria are grown in large containers called bioreactors. Conditions inside bioreactors are carefully controlled and monitored to allow for the growth of the bacteria. Medicines, enzymes, cleansers, and adhesives are some of the products that are made using bacteria.

Methane gas that is released as a waste by certain bacteria can be used as a fuel for heating, cooking, and industry. In landfills, methane-producing bacteria break down plant and animal material. The quantity of methane gas released by these bacteria is so large that some cities collect and burn it, as shown in **Figure 13.** Using bacteria to digest wastes and then produce methane gas could supply large amounts of fuel worldwide.

✔ **Reading Check** ***What waste gas produced by some bacteria can be used as a fuel?***

Figure 12 When bacteria such as *Streptococcus lactis* are added to milk, it causes the milk to separate into curds (solids) and whey (liquids). Other bacteria are added to the curds, which ripen into cheese. The type of cheese made depends on the bacterial species added to the curds.

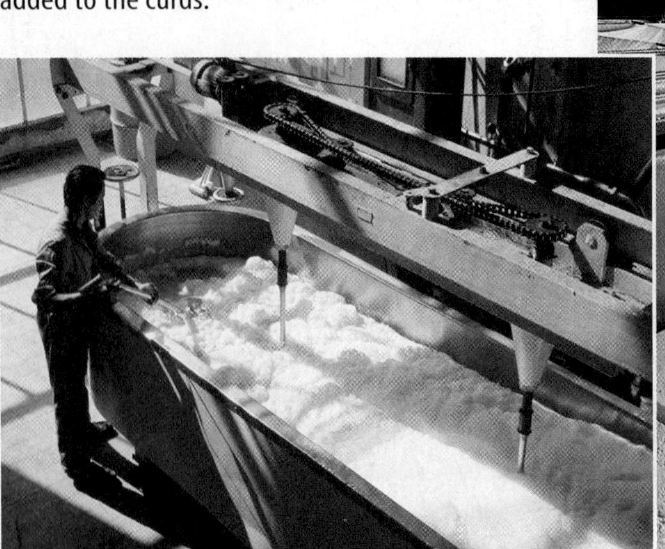

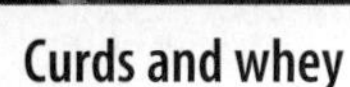
Curds and whey

Curds

Figure 13 Methane gas produced by bacteria in this landfill is burning at the top of these collection tubes.

Harmful Bacteria

Not all bacteria are beneficial. Some bacteria are known as pathogens. A **pathogen** is any organism that causes disease. If you have ever had strep throat, you have had firsthand experience with a bacterial pathogen. Other pathogenic bacteria cause diphtheria, tetanus, and whooping cough in humans, as well as anthrax in humans and livestock.

How Pathogens Make You Sick Bacterial pathogens can cause illness and disease by several different methods. They can enter your body through a cut in the skin, you can inhale them, or they can enter in other ways. Once inside your body, they can multiply, damage normal cells, and cause illness and disease.

Some bacterial pathogens produce poisonous substances known as **toxins.** Botulism—a type of food poisoning that can result in paralysis and death—is caused by a toxin-producing bacterium. Botulism-causing bacteria are able to grow and produce toxins inside sealed cans of food. However, when growing conditions are unfavorable for their survival, some bacteria, like those that cause botulism, can produce thick-walled structures called **endospores.** Endospores, shown in **Figure 14,** can exist for hundreds of years before they resume growth. If the endospores of the botulism-causing bacteria are in canned food, they can grow and develop into regular bacterial cells and produce toxins again. Commercially canned foods undergo a process that uses steam under high pressure, which kills bacteria and most endospores.

Science Online
Topic: Pathogens
Visit life.msscience.com for Web links to information about pathogenic bacteria and antibiotics.

Activity Compile a list of common antibiotics and the bacterial pathogens they are used to treat.

Figure 14 Bacterial endospores can survive harsh winters, dry conditions, and heat.
Describe *possible ways endospores can be destroyed.*

LM Magnification: 600×

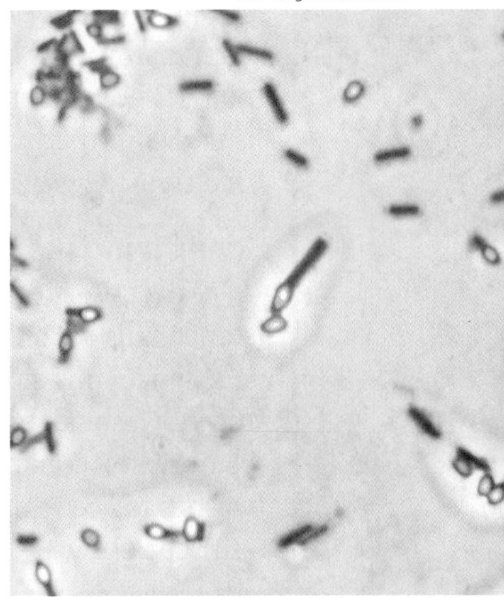

Figure 15 Pasteurization lowers the amount of bacteria in foods. Products, such as juice, ice cream, and yogurt, are pasteurized.

Pasteurization Unless it has been sterilized, all food contains bacteria. But heating food to sterilizing temperatures can change its taste. Pasteurization is a process of heating food to a temperature that kills most harmful bacteria but causes little change to the taste of the food. You are probably most familiar with pasteurized milk, but some fruit juices and other foods, as shown in **Figure 15,** also are pasteurized.

Applying Science

Controlling Bacterial Growth

Bacteria can be controlled by slowing or preventing their growth, or killing them. When trying to control bacteria that affect humans, it is often desirable just to slow their growth because substances that kill bacteria or prevent them from growing can harm humans. For example, bleach often is used to kill bacteria in bathrooms or on kitchen surfaces, but it is poisonous if swallowed. *Antiseptic* is the word used to describe substances that slow the growth of bacteria.

Identifying the Problem

Advertisers often claim that a substance kills bacteria, when in fact the substance only slows its growth. Many mouthwash advertisements make this claim. How could you test three mouthwashes to see which one is the best antiseptic?

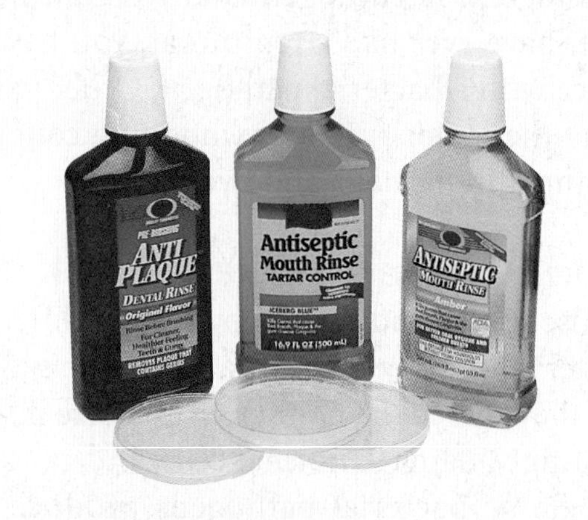

Solving the Problem

1. Describe an experiment that you could do that would test which of three mouthwash products is the most effective antiseptic.
2. Identify the control in your experiment.
3. Read the ingredients labels on bottles of mouthwash. List the ingredients in the mouthwash. What ingredient do you think is the antiseptic? Explain.

Figure 16 Each of these paper disks contains a different antibiotic. Clear areas where no bacteria are growing can be seen around four of the disks.
Infer *which one of these disks contains an antibiotic that is most effective against the bacteria growing on the plate.*

INTEGRATE Health

Treating Bacterial Diseases Bacterial diseases in humans and animals usually are treated effectively with antibiotics. Penicillin, a well-known antibiotic, works by preventing bacteria from making cell walls. Without cell walls, certain bacteria cannot survive. **Figure 16** shows antibiotics at work.

Vaccines can prevent some bacterial diseases. A **vaccine** can be made from damaged particles taken from bacterial cell walls or from killed bacteria. Once the vaccine is injected, white blood cells in the blood recognize that type of bacteria. If the same type of bacteria enters the body at a later time, the white blood cells immediately attack them. Vaccines have been produced that are effective against many bacterial diseases.

section 2 review

Summary

Beneficial Bacteria

- Many types of bacteria help you stay healthy.
- Antibiotics are produced by some bacteria.
- Bacteria decompose dead material.
- Certain bacteria change nitrogen in the air to forms that other organisms can use.
- Some bacteria are used to remove pollutants.
- Bacteria help to produce some foods.

Harmful Bacteria

- Some bacteria cause disease.
- Some bacteria have endospores that enable them to adapt to harsh environments.

Self Check

1. **Explain** why saprophytic bacteria are helpful.
2. **Summarize** how nitrogen-fixing bacteria benefit plants and animals.
3. **List** three uses of bacteria in food production and other industry.
4. **Describe** how some bacteria cause disease.
5. **Think Critically** Why is botulism associated with canned foods and not fresh foods?

Applying Skills

6. **Measure in SI** Air can have more than 3,500 bacteria per cubic meter. How many bacteria might be in your classroom?

Composting

Goals

- **Predict** which of several items will decompose in a compost pile and which will not.
- **Demonstrate** the decomposition, or lack thereof, of several items.
- **Compare and contrast** the speed at which various items break down.

Possible Materials

widemouthed, clear-glass jars (at least 4)
soil
water
watering can
banana peel
apple core
scrap of newspaper
leaf
plastic candy wrapper
scrap of aluminum foil

Safety Precautions

Real-World Question

Over time, landfills fill up and new places to dump trash become more difficult to find. One way to reduce the amount of trash that must be dumped in a landfill is to recycle. Composting is a form of recycling that changes plant wastes into reusable, nutrient-rich compost. How do plant wastes become compost? What types of organisms can assist in the process? What types of items can be composted and what types cannot?

Form a Hypothesis

Based on readings or prior knowledge, form a hypothesis about what types of items will decompose in a compost pile and which will not.

Test Your Hypothesis

Make a Plan

1. **Decide** what items you are going to test. Choose some items that you think will decompose and some that you think will not.
2. **Predict** which of the items you chose will or will not decompose. Of the items that will, which do you think will decompose fastest? Slowest?
3. **Decide** how you will test whether or not the items decompose. How will you see the items? You may need to research composting in books, magazines, or on the Internet.
4. **Prepare** a data table in your Science Journal to record your observations.
5. **Identify** all constants, variables, and controls of the experiment.

Follow Your Plan

1. Make sure your teacher approves of your plan and your data table before you start.
2. **Observe** Set up your experiment and collect data as planned.
3. **Record Data** While doing the experiment, record your observations and complete your data tables in your Science Journal.

Analyze Your Data

1. **Describe** your results. Did all of the items decompose? If not, which did and which did not?
2. Were your predictions correct? Explain.
3. **Compare** how fast each item decomposed. Which items decomposed fastest and which took longer?

Conclude and Apply

1. What general statement(s) can you make about which items can be composted and which cannot? What about the speed of decomposition?
2. **Determine** whether your results support your hypothesis.
3. **Explain** what might happen to your compost pile if antibiotics were added to it.
4. **Describe** what you think happens in a landfill to items similar to those that you tested.

Point of View Write a letter to the editor of a local newspaper describing what you have learned about composting and encouraging more community composting.

Science Stats

Unusual Bacteria

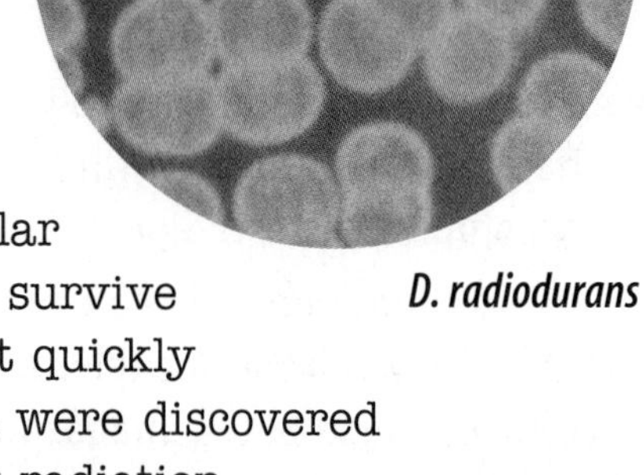
Color-enhanced TEM
Magnification: 4000×

D. radiodurans

Did you know...

...The hardiest bacteria, *Deinococcus radiodurans* (DE no KO kus·RA de oh DOOR anz), has a nasty odor, which has been described as similar to rotten cabbage. It might have an odor, but it can survive 3,000 times more radiation than humans because it quickly repairs damage to its DNA molecule. These bacteria were discovered in canned meat when they survived sterilization by radiation.

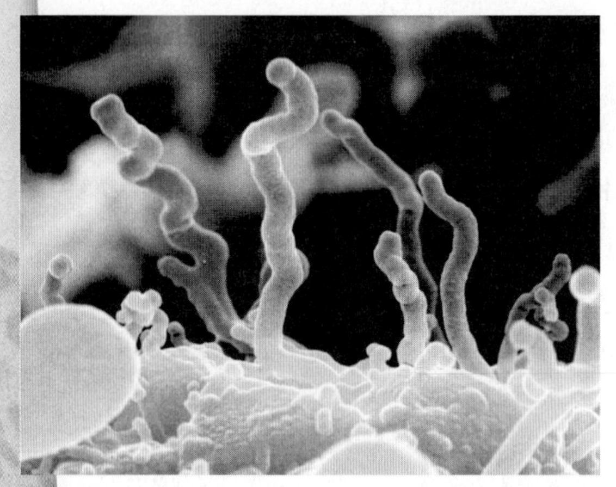
Nanobes

...The smallest bacteria, nanobes (NA nobes), are Earth's smallest living things. They have been found 5 km beneath the ocean floor near Australia. These tiny cells are 20 to 150 nanometers long. That means, depending on their size, it would take about 6,500,000 to 50,000,000 nanobes lined up to equal 1 m!

Applying Math What is the difference in size between the largest nanobe and the smallest nanobe?

...Earth's oldest living bacteria are thought to be 250 million years old. These ancient bacteria were revived from a crystal of rock salt buried 579 m below the desert floor in New Mexico.

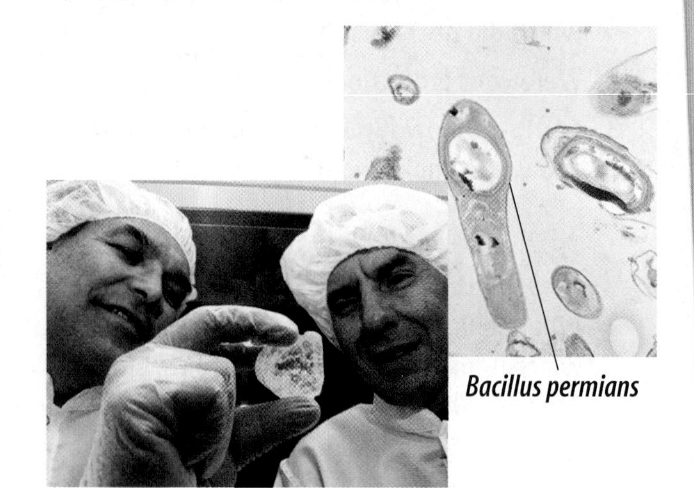
Bacillus permians

Find Out About It

Do research about halophiles, the bacteria that can live in highly salty environments. What is the maximum salt concentration in which extreme halophiles can survive? How does this compare to the maximum salt concentration at which nonhalophilic bacteria can survive? Visit life.msscience.com/science_stats to learn more.

Reviewing Main Ideas

Section 1 What are bacteria?

1. Bacteria can be found almost everywhere. They have one of three basic shapes—coccus, bacillus, or spirillum.
2. Bacteria are prokaryotic cells that usually reproduce by fission. All bacteria contain DNA, ribosomes, and cytoplasm but lack a membrane-bound nucleus.
3. Most bacteria are consumers, but some can make their own food. Anaeroic bacteria live without oxygen, but aerobic bacteria need oxygen to survive.
4. Cell shape and structure, how they get food, if they use oxygen, and their waste products can be used to classify eubacteria.
5. Cyanobacteria are producer eubacteria. They are an important source of food and oxygen for some aquatic organisms.
6. Archaebacteria are bacteria that often exist in extreme conditions, such as near ocean vents or in hot springs.

Section 2 Bacteria in Your Life

1. Most bacteria are helpful. They aid in recycling nutrients, fixing nitrogen, or helping in food production. They even can be used to break down pollutants.
2. Some bacteria that live in your body help you stay healthy and survive.
3. Other bacteria are harmful because they can cause disease in organisms.
4. Pasteurization can prevent the growth of harmful bacteria in food.

Visualizing Main Ideas

Copy and complete the following concept map on how bacteria affect the environment.

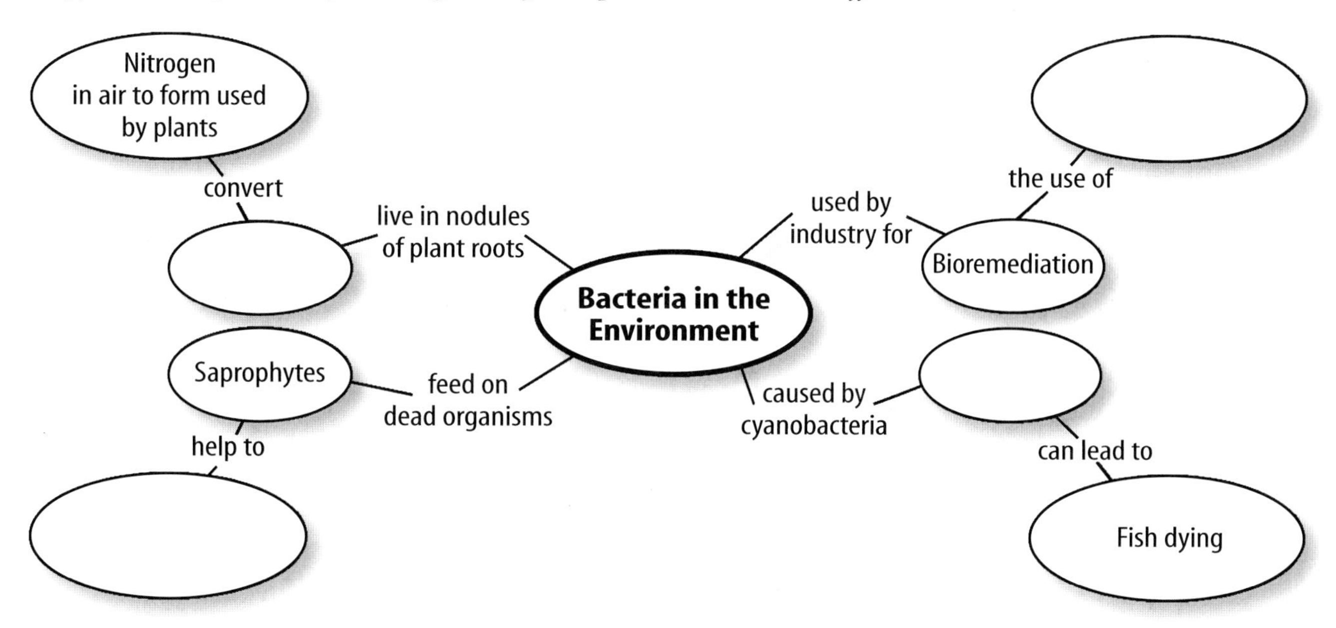

chapter 7 Review

Using Vocabulary

aerobe p. 188
anaerobe p. 188
antibiotic p. 193
endospore p. 197
fission p. 188
nitrogen-fixing bacteria p. 194
pathogen p. 197
saprophyte p. 194
toxin p. 197
vaccine p. 199

Fill in the blanks with the correct word or words.

1. A(n) ________ uses dead organisms as a food source.
2. A(n) ________ can prevent some bacterial diseases.
3. A(n) ________ causes disease.
4. A bacterium that needs oxygen to carry out cellular respiration is a(n) ________.
5. Bacteria reproduce using ________.
6. ________ are bacteria that convert nitrogen in the air to a form that plants can use.
7. A(n) ________ can live without oxygen.

Checking Concepts

Choose the word or phrase that best answers the question.

8. What is a way of cleaning up an ecosystem using bacteria to break down harmful compounds?
   A) landfill **C)** toxic waste dumps
   B) waste storage **D)** bioremediation
9. What pigment do cyanobacteria need to make food?
   A) chlorophyll **C)** plasmids
   B) chromosomes **D)** ribosomes
10. Which term describes most bacteria?
    A) anaerobic **C)** many-celled
    B) pathogens **D)** beneficial
11. What is the name for rod-shaped bacteria?
    A) bacilli **C)** spirilla
    B) cocci **D)** colonies
12. What structure allows bacteria to stick to surfaces?
    A) capsule **C)** chromosome
    B) flagella **D)** cell wall
13. What organisms can grow as blooms in ponds?
    A) archaebacteria **C)** cocci
    B) cyanobacteria **D)** viruses
14. Which organisms are recyclers in the environment?
    A) producers **C)** saprophytes
    B) flagella **D)** pathogens
15. Which results from processes of a pathogenic bacterium?
    A) an antibiotic **C)** nitrogen fixation
    B) cheese **D)** strep throat

Use the photo below to answer questions 16 and 17.

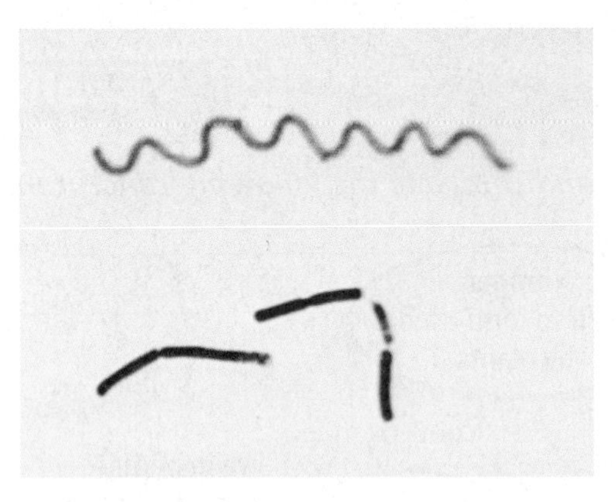

16. What shape are the gram-positive bacteria in the above photograph?
    A) coccus **C)** spirillum
    B) bacillus **D)** ovoid
17. What shape are the gram-negative bacteria in the above photograph?
    A) coccus **C)** spirillum
    B) bacillus **D)** ovoid

Thinking Critically

18. **Infer** what would happen if nitrogen-fixing bacteria could no longer live on the roots of some plants.

19. **Explain** why bacteria are capable of surviving in almost all environments of the world.

20. **Draw a conclusion** as to why farmers often rotate crops such as beans, peas, and peanuts with other crops such as corn, wheat, and cotton.

21. **Describe** One organism that causes bacterial pneumonia is called pneumococcus. What is its shape?

22. **List** the precautions that can be taken to prevent food poisoning.

23. **Concept Map** Copy and complete the following events-chain concept map about the events surrounding a cyanobacteria bloom.

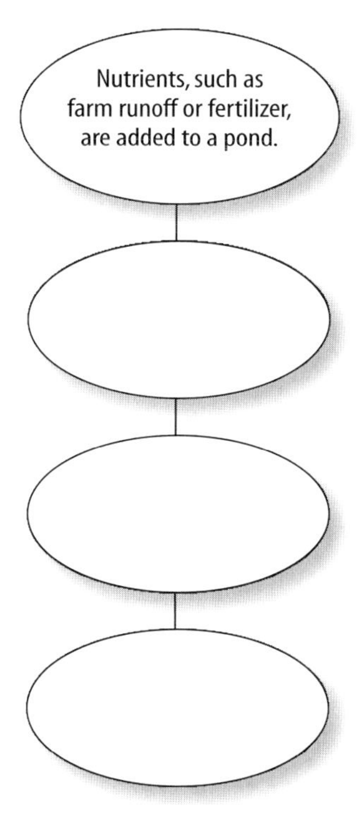

24. **Design an experiment** to decide if a kind of bacteria could grow anaerobically.

25. **Describe** the nitrogen-fixing process in your own words, using numbered steps. You will probably have more than four steps.

26. **Infer** the shape of *Spirillum minus* bacteria.

Performance Activities

27. **Poster** Create a poster that illustrates the effects of bacteria. Use photos from magazines and your own drawings.

28. **Poem** Write a poem that demonstrates your knowledge of the importance of bacteria to human health.

Applying Math

Use the table below to answer questions 29 and 30.

Bacterial Reproduction Rates

Temperature (°C)	Doubling Rate Per Hour
20.5	2.0
30.5	3.0
36.0	2.5
39.2	1.2

29. **Doubling Rate** Graph the data from the table above. Using the graph, determine where the doubling rate would be at 20°C. Where would the doubling rate be at 40°C?

30. **Bacterial Reproduction** Bacteria can reproduce rapidly. At 30.5°C, some species of bacteria can double their numbers in 3.0 hours. A biologist places a single bacterium on growth medium at 6:00 A.M. and incubates the bacteria until 4:00 P.M. the same afternoon. How many bacterium will there be?

chapter 7 Standardized Test Practice

Part 1 Multiple Choice

Record your answers on the answer sheet provided by your teacher or on a sheet of paper.

1. Most pathogenic bacteria are consumer eubacteria and are grouped according to what characteristic?
   A. chlorophyll **C.** cell wall
   B. ribosomes **D.** plasmids

2. Which of the following cannot be found in a bacterial cell?
   A. ribosomes **C.** chromosome
   B. nucleus **D.** cytoplasm

Use the photo below to answer questions 3 and 4.

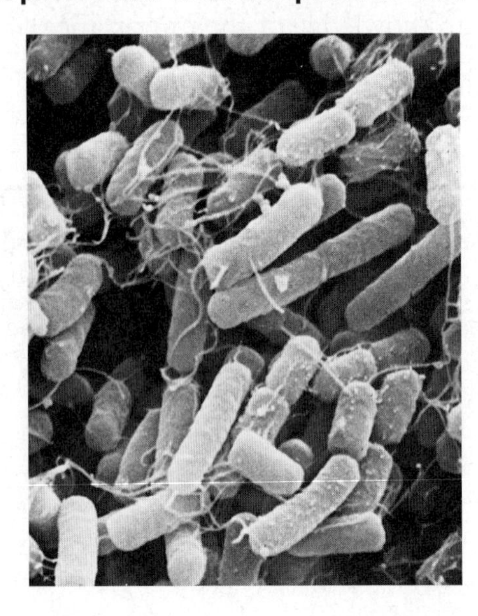

3. What shape are the bacterial cells shown above?
   A. bacillus **C.** spirillum
   B. coccus **D.** tubular

4. These bacteria are methane producers. Which of the following statements is true of these bacteria?
   A. They are aerobic.
   B. They are in Kingdom Eubacteria.
   C. They are used in sewage treatment.
   D. They live only near deep ocean vents.

5. Which of the following foods is not processed with the help of bacteria?
   A. beef **C.** yogurt
   B. cheese **D.** pickles

Use the photo below to answer questions 6 and 7.

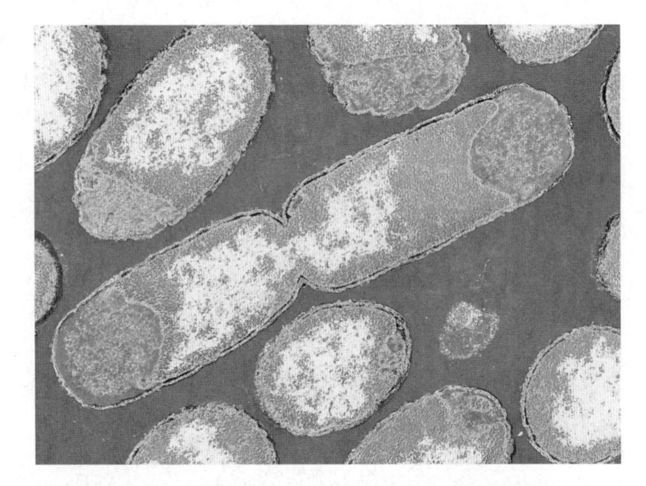

6. What process is occurring in the above photograph?
   A. mitosis **C.** fusion
   B. fission **D.** conjugation

7. The above is an example of what kind of reproduction?
   A. sexual **C.** meitotic
   B. asexual **D.** symbiotic

8. What characteristic probably was used in naming cyanobacteria?
   A. pigments **C.** cell shape
   B. slime layer **D.** cell wall

9. Each bacterium has
   A. a nucleus. **C.** ribosomes.
   B. mitochondria. **D.** a capsule.

Test-Taking Tip

Listen and Read Listen carefully to the instructions from the teacher and read the directions and each question carefully.

Part 2 Short Response/Grid In

Record your answers on the answer sheet provided by your teacher or on a sheet of paper.

10. What characteristics can be used in naming eubacteria?

11. What does an antiseptic do? Why would an antiseptic be dangerous to humans if it killed all bacteria?

Use the photo below to answer questions 12 and 13.

12. The figure above shows plant roots that have nodules, which contain nitrogen-fixing bacteria. How does this benefit the plant?

13. Explain how bacteria benefit from this relationship.

14. What happens to dead plant material that is plowed into the soil following a crop harvest? Why is this plowing beneficial to the quality of the soil?

15. What is bioremediation? Give an example of how it is used.

16. Most of the dairy products that you buy are pasteurized. What is pasteurization? How is it different from sterilization?

Part 3 Open Ended

Record your answers on a sheet of paper.

17. An antibiotic is prescribed to a patient to take for 10 days. After two days the patient feels better and stops taking the antibiotic. Several days later, the infection returns, but this time a greater amount of antibiotic was needed to cure the infection. Why? How could the patient have avoided the recurrence of the infection?

Use the photo below to answer questions 18 and 19.

18. Describe how aerobic bacteria in the wastewater treatment tank shown above clean the water. What happens to the energy that was in the waste?

19. Aerobic bacteria removed from the tanks, along with some solid waste, form sludge. After the sludge is dried, sterilized, and toxins removed, it is either burned or applied to soil. What would be the benefit of applying it to soil? Why is it important to remove toxins and sterilize the sludge?

20. What causes a bloom of cyanobacteria? Explain how it can cause fish and other organisms in a pond to die.

chapter 8

Protists and Fungi

The BIG Idea

Protists and fungi have important roles in maintaining environments.

SECTION 1
Protists

Main Idea Protists are a diverse group of one-celled eukaryotes.

SECTION 2
Fungi

Main Idea Fungi can be one-celled or many-celled eukaryotes and have unique methods of reproduction.

Fungi—Terrestrial Icebergs

A mushroom is like the tip of an iceberg; a small, visible portion of an extensive fungal network that grows under the soil. Many fungi and plant roots interact. These fungi help provide water and nutrients to these plants in exchange for carbohydrates that plants produce.

Science Journal What other ways might fungi benefit other organisms and the environment?

Start-Up Activities

Dissect a Mushroom

It is hard to tell by a mushroom's appearance whether it is safe to eat or is poisonous. Some edible mushrooms are so highly prized that people keep their location a secret for fear that others will find their treasure. Do the lab below to learn about the parts of mushrooms.

1. Obtain a mushroom from your teacher.
2. Using a magnifying lens, observe the underside of the mushroom cap. Then carefully pull off the cap and observe the gills, which are the thin, tissuelike structures. Hundreds of thousands of tiny reproductive structures called spores form on these gills.
3. Use your fingers or forceps to pull the stalk apart lengthwise. Continue this process until the pieces are as small as you can get them.
4. **Think Critically** In your Science Journal, write a description of the parts of the mushroom, and make a labeled drawing of the mushroom and its parts.

FOLDABLES™ Study Organizer

Compare Protists and Fungi Make the following Foldable to help you see how protists and fungi are similar and different.

STEP 1 **Fold** the top of a vertical piece of paper down and the bottom up to divide the paper into thirds.

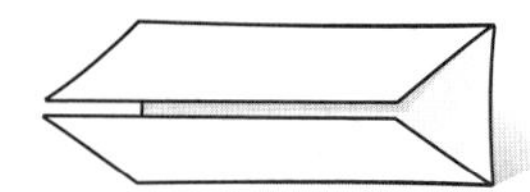

STEP 2 **Unfold and label** the three sections as shown.

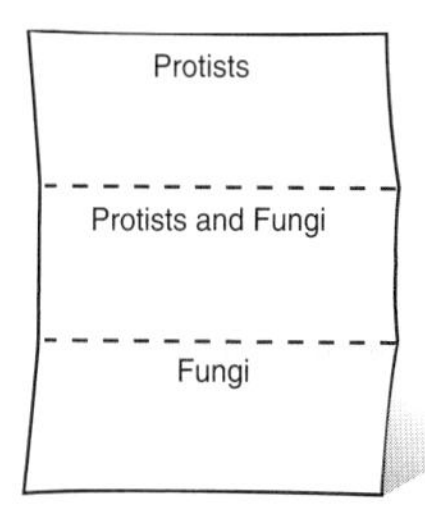

Read for Main Ideas As you read the chapter, write information about each type of organism in the appropriate section, and information that they share in the middle section.

Preview this chapter's content and activities at life.msscience.com

Get Ready to Read

Identify Cause and Effect

1 Learn It! A cause is the reason something happens. The result of what happens is called an effect. Learning to identify causes and effects helps you understand why things happen. By using graphic organizers, you can sort and analyze causes and effects as you read.

2 Practice It! Read the following paragraph. Then use the graphic organizer below to show what happens when downy mildews infect plants.

Downy mildews can have a huge effect on economies as well as social history. One of the most well-known members of this group caused the Irish potato famine during the 1840s. Potatoes were Ireland's main crop and the primary food source for its people. When the potato crop became infected with downy mildew, potatoes rotted in the fields, leaving many people with no food. A downy mildew infection of grapes in France during the 1870s nearly wiped out the entire French wine industry.

—from page 220

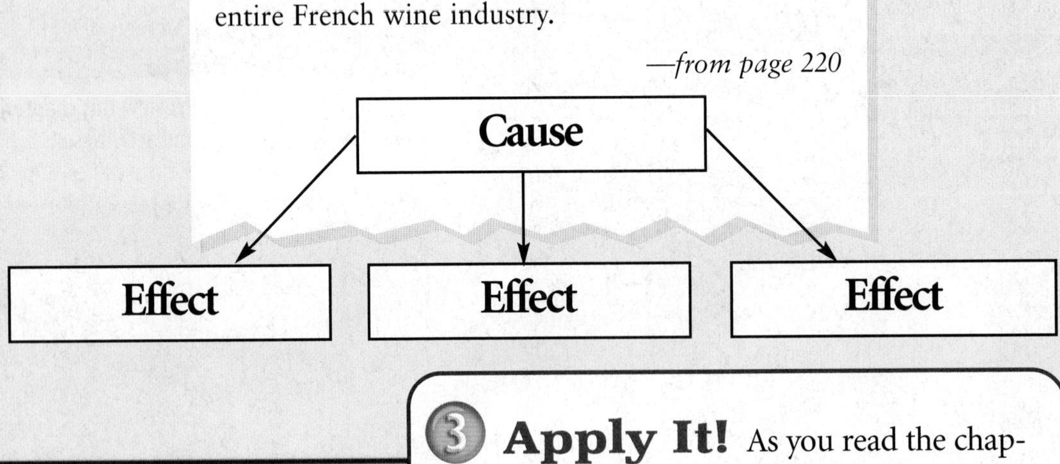

3 Apply It! As you read the chapter, be aware of causes and effects of protists or fungi on other organisms. Find five causes and their effects.

Target Your Reading

Use this to focus on the main ideas as you read the chapter.

1. **Before you read** the chapter, respond to the statements below on your worksheet or on a numbered sheet of paper.
   - Write an **A** if you **agree** with the statement.
   - Write a **D** if you **disagree** with the statement.

2. **After you read** the chapter, look back to this page to see if you've changed your mind about any of the statements.
   - If any of your answers changed, explain why.
   - Change any false statements into true statements.
   - Use your revised statements as a study guide.

Reading Tip

Graphic organizers such as the Cause-Effect organizer help you organize what you are reading so you can remember it later.

Before You Read A or D	Statement	After You Read A or D
	1 Plantlike protists cannot make their own food.	
	2 An amoeba is an animal-like protist.	
	3 Algae can be microscopic or as large as 100 m in length.	
	4 Animal-like protists do not cause diseases in humans.	
	5 Funguslike protists can damage other protists, plants, and animals.	
	6 Some fungi feed on dead organisms and others feed on living organisms.	
	7 Fungi reproduce by spores.	
	8 Lichens are fungi that help break down decaying plants.	
	9 Some fungi have an important interaction with plant roots.	
	10 Fungi are of no importance to humans.	

Science Online
Print out a worksheet of this page at life.msscience.com

section 1

Protists

as you read

What You'll Learn

- **Describe** the characteristics shared by all protists.
- **Compare and contrast** the three groups of protists.
- **List** examples of each of the three protist groups.
- **Explain** why protists are so difficult to classify.

Why It's Important

Many protists are important food sources for other organisms.

Review Vocabulary
asexual reproduction: requires only one parent to produce a new genetically identical individual

New Vocabulary
- protist
- algae
- flagellum
- protozoan
- cilia
- pseudopod

What is a protist?

Look at the organisms in **Figure 1.** As different as they appear, all of these organisms belong to one kingdom—the protist kingdom. A **protist** is a one- or many-celled organism that lives in moist or wet surroundings. All protists have eukaryotic cells—cells that have a nucleus and other internal, membrane-bound structures. Some protists are plantlike. They contain chlorophyll and make their own food. Other protists are animal-like. They do not have chlorophyll and can move. Some protists have a solid or a shell-like structure on the outside of their bodies.

Protist Reproduction Protists usually reproduce asexually by mitosis and cell division. The hereditary material in the nucleus is duplicated, the nucleus divides, and then the cytoplasm usually divides. The result is two new cells that are genetically identical. Asexual reproduction of many-celled protists occurs by regeneration. Parts of the organism can break off and grow into entirely new organisms that are genetically identical.

Most protists also can reproduce sexually. During sexual reproduction, the process of meiosis produces sex cells. Two sex cells join to form a new organism that is genetically different from the two organisms that were the sources of the sex cells. How and when sexual reproduction occurs depends on the specific type of protist.

Figure 1 The protist kingdom is made up of a variety of organisms.
Describe *the characteristics that the organisms below have in common.*

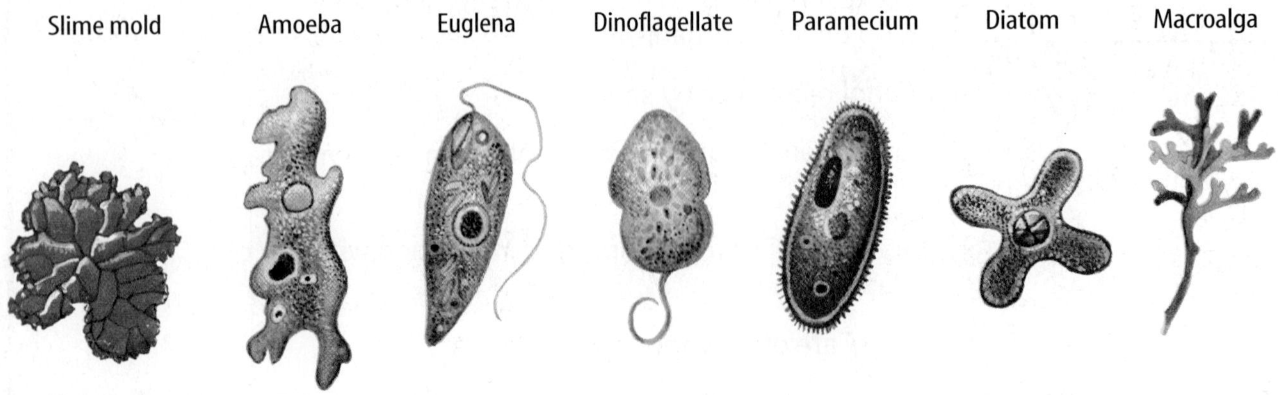

Slime mold Amoeba Euglena Dinoflagellate Paramecium Diatom Macroalga

Classification of Protists

Not all scientists agree about how to classify the organisms in this group. Protists usually are divided into three groups—plant-like, animal-like, and fungus-like—based on whether they share certain characteristics with plants, animals, or fungi. **Table 1** shows some of these characteristics. As you read this section, you will understand some of the problems of grouping protists in this way.

Table 1 Characteristics of Protist Groups

Plantlike	Animal-Like	Funguslike
Contain chlorophyll and make their own food using photosynthesis	Cannot make their own food; capture other organisms for food	Cannot make their own food; absorb food from their surroundings
Have cell walls	Do not have cell walls	Some organisms have cell walls; others do not
No specialized ways to move from place to place	Have specialized ways to move from place to place	Have specialized ways to move from place to place

Evolution of Protists Although protists that produce a hard outer covering have left many fossils, other protists lack hard parts, so few fossils of these organisms have been found. But, by studying the genetic material and structure of modern protists, scientists are beginning to understand how they are related to each other and to other organisms. Scientists hypothesize that the common ancestor of most protists was a one-celled organism with a nucleus and other cellular structures. However, evidence suggests that protists with the ability to make their own food could have had a different ancestor than protists that cannot make their own food.

Plantlike Protists

Protists in this group are called plantlike because, like plants, they contain the pigment chlorophyll in chloroplasts and can make their own food. Many of them have cell walls like plants, and some have structures that hold them in place just as the roots of a plant do, but these protists do not have roots.

Plantlike protists are known as **algae** (AL jee) (singular, *alga*). As shown in **Figure 2,** some are one cell and others have many cells. Even though all algae have chlorophyll, not all of them look green. Many have other pigments that cover up their chlorophyll.

Figure 2 Algae exist in many shapes and sizes. Microscopic algae (left photo) are found in fresh-water and salt water. You can see some types of green algae growing on rocks, washed up on the beach, or floating in the water.

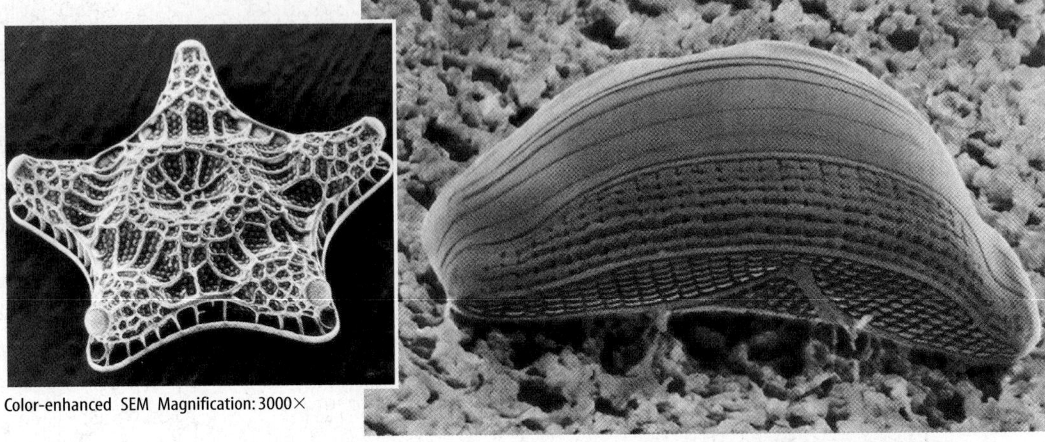
Color-enhanced SEM Magnification: 3000×
Color-enhanced SEM Magnification: 2866×

Figure 3 The cell walls of diatoms contain silica, the main element in glass. The body of a diatom is like a small box with a lid. The pattern of dots, pits, and lines on the cell wall's surface is different for each species of diatom.

Diatoms Extremely large populations of diatoms exist. Diatoms, shown in **Figure 3,** are found in freshwater and salt water. They have a golden-brown pigment that covers up the green chlorophyll. Diatoms secrete glasslike boxes around themselves. When the organisms die, these boxes sink. Over thousands of years, they can collect and form deep layers.

Dinoflagellates Another group of algae is called the dinoflagellates, which means "spinning flagellates." Dinoflagellates, as shown in **Figure 4,** have two flagella. A **flagellum** (plural, *flagella*) is a long, thin, whiplike structure used for movement. One flagellum circles the cell like a belt, and another is attached to one end like a tail. As the two flagella move, they cause the cell to spin. Because many of the species in this group produce a chemical that causes them to glow at night, they are known as fire algae. Almost all dinoflagellates live in salt water. While most contain chlorophyll, some do not and must feed on other organisms.

Euglenoids Protists that have characteristics of both plants and animals are known as the euglenoids (yoo GLEE noydz). Many of these one-celled algae have chloroplasts, but some do not. Those with chloroplasts, like *Euglena* shown in **Figure 4,** can produce their own food. However, when light is not present, *Euglena* can feed on bacteria and other protists. Although *Euglena* has no cell wall, it does have a strong, flexible layer inside the cell membrane that helps it move and change shape. Many euglenoids move by whipping their flagella. An eyespot, an adaptation that is sensitive to light, helps photosynthetic euglenoids move toward light.

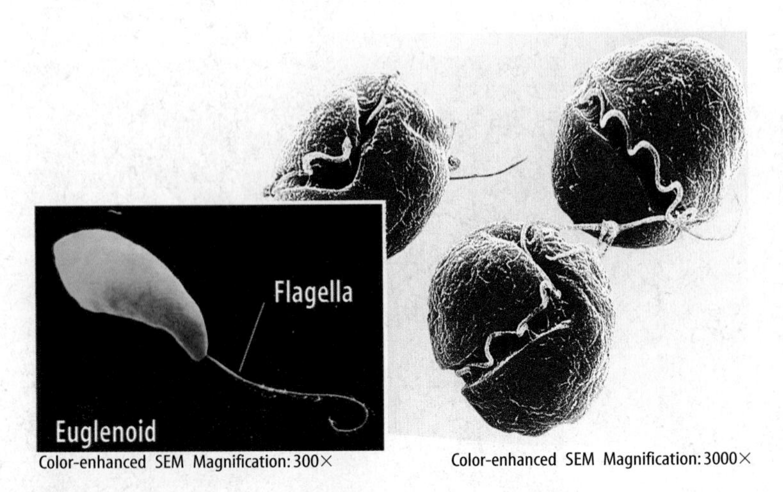

Color-enhanced SEM Magnification: 300×
Color-enhanced SEM Magnification: 3000×

Figure 4 Most dinoflagellates live in the sea. Some are free living and others are parasites. Still others, like the *Spiniferites* cyst (right photo), can produce toxins that make other organisms sick.
Determine *how euglenoids are similar to plants and animals.*

Red Algae Most red algae are many-celled and, along with the many-celled brown and green algae, sometimes are called seaweeds. Red algae contain chlorophyll, but they also produce large amounts of a red pigment. Some species of red algae can live up to 200 m deep in the ocean. They can absorb the limited amount of light at those depths to carry out the process of photosynthesis. **Figure 5** shows the depths at which different types of algae can live.

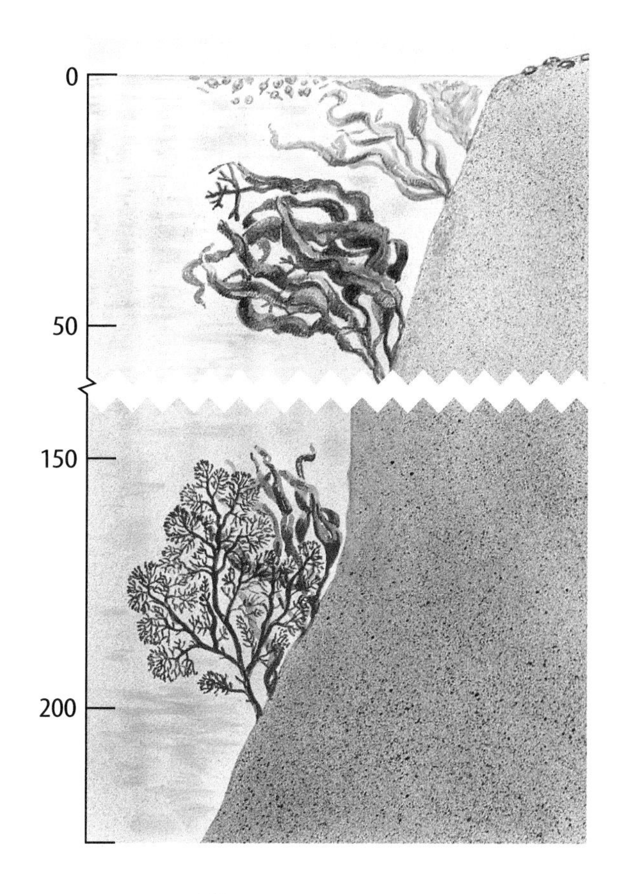

Figure 5 Green algae are found closer to the surface. Brown algae can grow from a depth of about 35 m. Red algae are found in the deepest water at 175 m to 200 m.

Green Algae There are more than 7,000 species of green algae in this diverse group of organisms. These algae, like the one shown in **Figure 6,** contain large amounts of chlorophyll. Green algae can be one-celled or many-celled. They are the most plant-like of all the algae. Because plants and green algae are similar in their structure, chlorophyll, and how they undergo photosynthesis, some scientists hypothesize that plants evolved from ancient, many-celled green algae. Although most green algae live in water, you can observe types that live in other moist environments, including on damp tree trunks and wet sidewalks.

Brown Algae As you might expect from their name, brown algae contain a brown pigment in addition to chlorophyll. They usually are found growing in cool, saltwater environments. Brown algae are many-celled and vary greatly in size. An important food source for many fish and invertebrates is a brown alga called kelp. Kelp, shown in **Figure 6,** forms a dense mat of stalks and leaflike blades where small fish and other animals live. Giant kelp is the largest organism in the protist kingdom and can grow to be 100 m in length.

Reading Check *What is kelp?*

Figure 6 Green algae (left photo) often can be seen on the surface of ponds in the summer. Giant kelp, a brown alga, can form forests like this one located off the coast of California. Extracts from kelp add to the smoothness and spreadability of products such as cheese spreads and mayonnaise.

Topic: Red Tides

Visit life.msscience.com for Web links to information about red tides and dinoflagellate blooms.

Activity Determine where red tides occur more frequently. Construct a map of the world and indicate where red tides have occurred in the last five years.

Importance of Algae

Have you thought about how important grasses are as a food source for animals that live on land? Cattle, deer, zebras, and many other animals depend on grasses as their main source of food. Algae sometimes are called the grasses of the oceans. Most animals that live in the oceans eat either algae for food or other animals that eat algae. You might think many-celled, large algae like kelp are the most important food source, but the one-celled diatoms and dinoflagellates are a more important food source for many organisms. Algae, such as *Euglena,* also are an important source of food for organisms that live in freshwater.

Algae and the Environment Algae are important in the environment because they produce oxygen as a result of photosynthesis. The oxygen produced by green algae is important for most organisms on Earth, including you.

Under certain conditions, algae can reproduce rapidly and develop into what is known as an algal bloom. Because of the large number of organisms in a bloom, the color of the water appears to change. Red tides that appear along the east and Gulf coasts of the United States are the result of dinoflagellate blooms. Toxins produced by the dinoflagellates can cause other organisms to die and can cause health problems in humans.

Algae and You People in many parts of the world eat some species of red and brown algae. You probably have eaten foods or used products made with algae. Carrageenan (kar uh JEE nuhn), a substance found in the cell walls of red algae, has gelatinlike properties that make it useful to the cosmetic and food industries. It is usually processed from the red alga Irish moss, shown in **Figure 7.** Carrageenan gives toothpastes, puddings, and salad dressings their smooth, creamy textures. Another substance, algin (AL juhn), found in the cell walls of brown algae, also has gelatinlike properties. It is used to thicken foods such as ice cream and marshmallows. Algin also is used in making rubber tires and hand lotion.

Ancient deposits of diatoms are mined and used in insulation, filters, and road paint. The cell walls of diatoms produce the sparkle that makes some road lines visible at night and the crunch you might feel in toothpaste.

Figure 7 Carrageenan, a substance extracted from Irish moss, is used for thickening dairy products such as chocolate milk.

What are some uses by humans of algae?

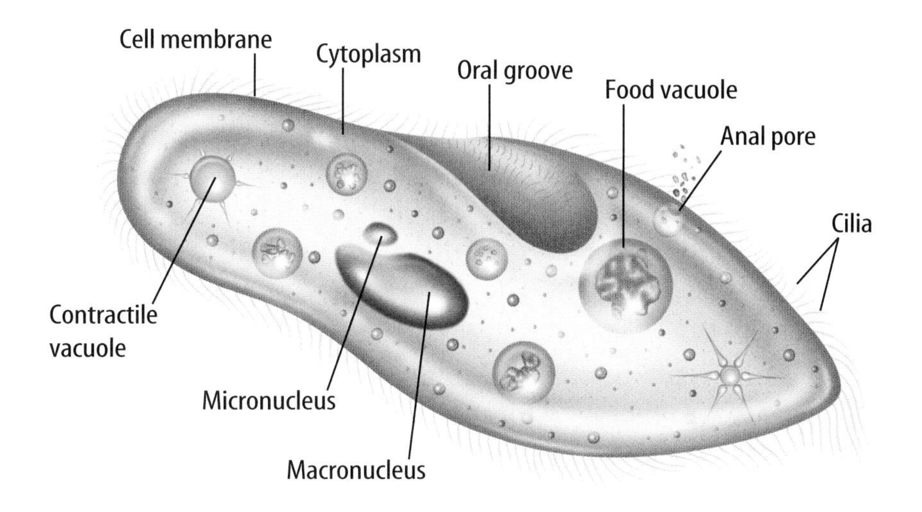

LM Magnification: 107×

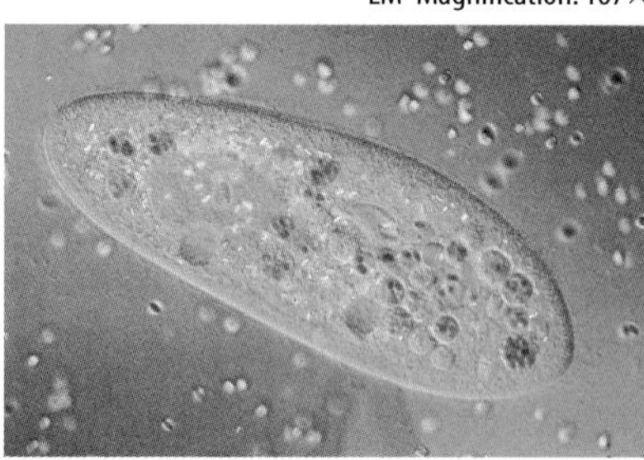

Figure 8 *Paramecium* is a typical ciliate found in many freshwater environments. These rapidly swimming protists consume bacteria.
Infer *Locate the vacuoles in the photo. What is their function?*

Animal-Like Protists

One-celled, animal-like protists are known as **protozoans.** Usually protozoans are classified by how they move. These complex organisms live in or on other living or dead organisms that are found in water or soil. Many protozoans have specialized vacuoles for digesting food and getting rid of excess water.

Ciliates As their name suggests, these protists have **cilia** (SIH lee uh)—short, threadlike structures that extend from the cell membrane. Ciliates can be covered with cilia or have cilia grouped in specific areas on the surface of the cell. The cilia beat in a coordinated way similar to rowboat oars. As a result, the organism moves swiftly in any direction. Organisms in this group include some of the most complex, one-celled protists and some of the largest, one-celled protists.

A typical ciliate is *Paramecium,* shown in **Figure 8.** *Paramecium* has two nuclei—a macronucleus and a micronucleus—another characteristic of the ciliates. The micronucleus is involved in reproduction. The macronucleus controls feeding, the exchange of oxygen and carbon dioxide, the amount of water and salts entering and leaving *Paramecium,* and other functions of *Paramecium.*

Ciliates usually feed on bacteria that are swept into the oral groove by the cilia. Once the food is inside the cell, a vacuole forms around it and the food is digested. Wastes are removed through the anal pore. Freshwater ciliates, like *Paramecium,* also have a structure called the contractile vacuole that helps get rid of excess water. When the contractile vacuole contracts, excess water is ejected from the cell.

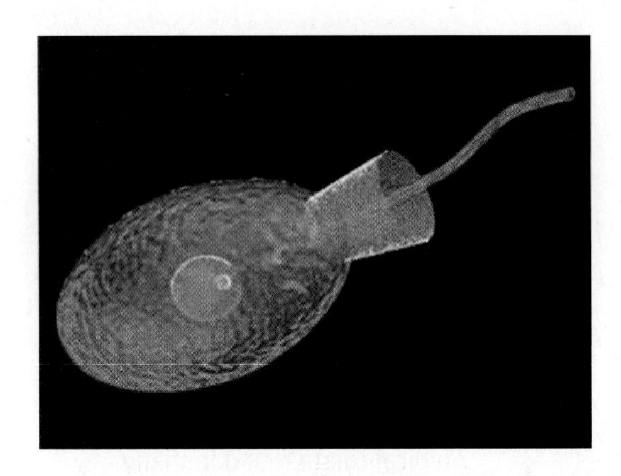

Figure 9 *Proterospongia* is a rare, freshwater protist. Some scientists hypothesize that this flagellate might share an ancestor with ancient animals.

Flagellates Protozoans called flagellates move through their watery environment by whipping their long flagella. Many species of flagellates live in freshwater, though some are parasites that harm their hosts.

Proterospongia, shown in **Figure 9,** is a member of one group of flagellates that might share an ancestor with ancient animals. These flagellates often grow in colonies of many cells that are similar in structure to cells found in animals called sponges. Like sponge cells, when *Proterospongia* cells are in colonies, they perform different functions. Moving the colony through the water and reproducing, which increases the colony's size, are two examples of jobs that the cells of *Proterospongia* carry out.

Movement with Pseudopods Some protozoans move through their environments and feed using temporary extensions of their cytoplasm called **pseudopods** (SEW duh pahdz). The word *pseudopod* means "false foot." These organisms seem to flow along as they extend their pseudopods. They are found in freshwater and saltwater environments, and certain types are parasites in animals.

The amoeba shown in **Figure 10** is a typical member of this group. To obtain food, an amoeba extends the cytoplasm of a pseudopod on either side of a food particle such as a bacterium. Then the two parts of the pseudopod flow together and the particle is trapped. A vacuole forms around the trapped food. Digestion takes place inside the vacuole.

Although some protozoans of this group, like the amoeba, have no outer covering, others secrete hard shells around themselves. The white cliffs of Dover, England are composed mostly of the remains of some of these shelled protozoans. Some shelled protozoa have holes in their shells through which the pseudopods extend.

Figure 10 In many areas of the world, a disease-causing species of amoeba lives in the water. If it enters a human body, it can cause dysentery—a condition that can lead to a severe form of diarrhea.
Infer *why an amoeba is classified as a protozoan.*

LM Magnification: 85×

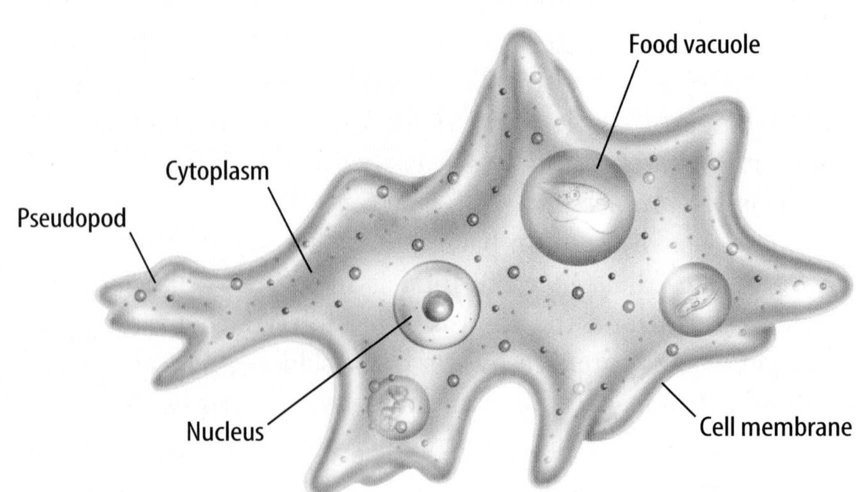

Figure 11 Asexual reproduction of the malaria parasite takes place inside a human host. Sexual reproduction takes place in the intestine of a mosquito.

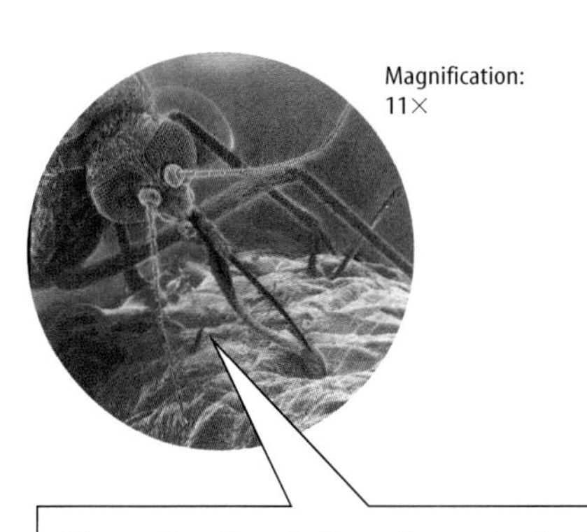

Magnification:
11×
Plasmodium lives in the salivary glands of certain female mosquitoes. The parasite can be transferred to a human's blood if an infected mosquito bites them.

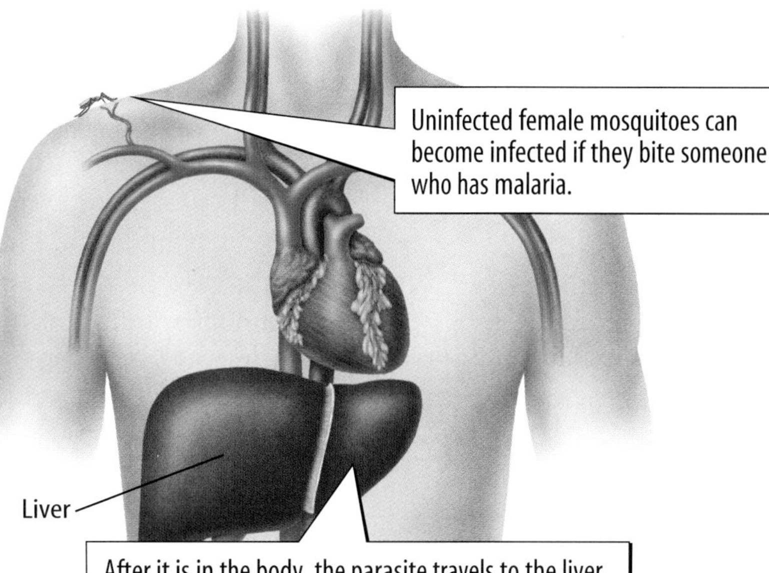

Uninfected female mosquitoes can become infected if they bite someone who has malaria.
Liver
After it is in the body, the parasite travels to the liver and then to the red blood cells where it reproduces and releases more parasites into the blood.

Other Protozoans One group of protozoans has no way of moving on their own. All of the organisms in this group are parasites of humans and other animals. These protozoans have complex life cycles that involve sexual and asexual reproduction. They often live part of their lives in one animal and part in another. The parasite that causes malaria is an example of a protozoan in this group. **Figure 11** shows the life cycle of the malaria parasite.

Importance of Protozoans

Like the algae, some protozoans are an important source of food for larger organisms. When some of the shelled protozoans die, they sink to the bottom of bodies of water and become part of the sediment. Sediment is a buildup of plant and animal remains and rock and mineral particles. The presence of these protists in sediments is used sometimes by field geologists as an indicator species. This tells them where petroleum reserves might be found beneath the surface of Earth.

Reading Check *Why are shelled protozoans important?*

One type of flagellated protozoan lives with bacteria in the digestive tract of termites. Termites feed mainly on wood. These protozoans and bacteria produce wood-digesting enzymes that help break down the wood. Without these organisms, the termites would be less able to use the chemical energy stored in wood.

African Sleeping Sickness The flagellate *Trypanosoma* is carried by the tsetse fly in Africa and causes African sleeping sickness in humans and other animals. It is transmitted to other organisms during bites from the fly. The disease affects the central nervous system. Research this disease and create a poster showing what you learn.

Observing Slime Molds

Procedure

1. Obtain live specimens of the slime mold ***Physarum polycephaalum*** from your teacher.
2. Observe the mold once each day for four days.
3. Using a **magnifying lens,** make daily drawings and observations of the mold as it grows.

Analysis

Predict the growing conditions under which the slime mold will change from the amoeboid form to the spore-producing form.

Disease in Humans The protozoans that may be most important to you are the ones that cause diseases in humans. In tropical areas, flies or other biting insects transmit many of the parasitic flagellates to humans. A flagellated parasite called *Giardia* can be found in water that is contaminated with wastes from humans or wild or domesticated animals. If you drink water directly from a stream, you could get this diarrhea-causing parasite.

Some amoebas also are parasites that cause disease. One parasitic amoeba, found in ponds and streams, can lead to a brain infection and death.

Funguslike Protists

Funguslike protists include several small groups of organisms such as slime molds, water molds, and downy mildews. Although all funguslike protists produce spores like fungi, most of them can move from place to place using pseudopods like the amoeba. All of them must take in food from an outside source.

Slime Molds As shown in **Figure 12,** slime molds are more attractive than their name suggests. Slime molds form delicate, weblike structures on the surface of their food supply. Often these structures are brightly colored. Slime molds have some protozoan characteristics. During part of their life cycle, slime molds move by means of pseudopods and behave like amoebas.

Most slime molds are found on decaying logs or dead leaves in moist, cool, shady environments. One common slime mold sometimes creeps across lawns and mulch as it feeds on bacteria and decayed plants and animals. When conditions become less favorable for slime molds, reproductive structures form on stalks and spores are produced.

Figure 12 Slime molds come in many different forms and colors ranging from brilliant yellow or orange to rich blue, violet, pink, and jet black.
Compare and contrast *how slime molds are similar to protists and fungi.*

Magnification: 3×

Magnification: 5.25×

Water Molds and Downy Mildews

Most members of this large, diverse group of funguslike protists live in water or moist places. Like fungi, they grow as a mass of threads over a plant or animal. Digestion takes place outside of these protists, then they absorb the organism's nutrients. Unlike fungi, the spores these protists produce have flagella. Their cell walls more closely resemble those of plants than those of fungi.

Some water molds are parasites of plants, and others feed on dead organisms. Most water molds appear as fuzzy, white growths on decaying matter, as shown in **Figure 13.** If you have an aquarium, you might see water molds attack a fish and cause its death. Another important type of protist is a group of plant parasites called downy mildew. Warm days and cool, moist nights are ideal growing conditions for them. They can live on aboveground parts of many plants. Downy mildews weaken plants and even can kill them.

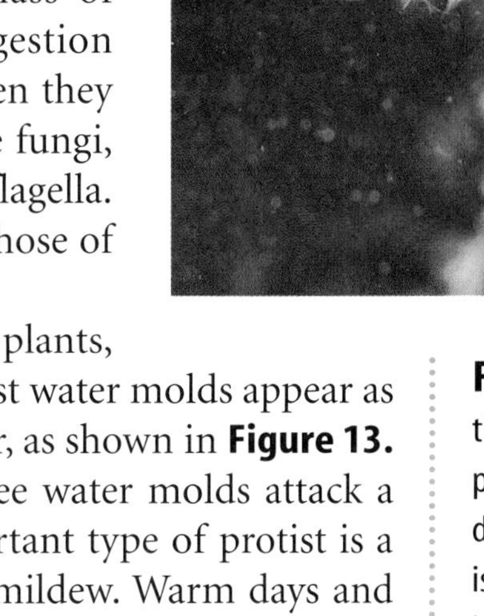

Figure 13 Water mold, the threadlike material seen in the photo, grows on a dead salamander. In this case, the water mold is acting as a decomposer. This important process will return nutrients to the water.

Reading Check *How do water molds affect organisms?*

Applying Science

Is it a fungus or a protist?

Slime molds, such as the pipe cleaner slime shown in the photograph to the right, can be found covering moist wood. They can be white or bright red, yellow, or purple. If you look at a piece of slime mold on a microscope slide, you will see that the cell nuclei move back and forth as the cytoplasm streams along. This streaming of the cytoplasm is how a slime mold creeps over the wood.

Identifying the Problem

Should slime molds be classified as protists or as fungi?

Solving the Problem

1. What characteristics do slime molds share with protists? How are slime molds similar to protozoans and algae?
2. What characteristics do slime molds share with fungi? What characteristics do slime molds have that are different from fungi?
3. What characteristics did you compare to decide what group slime molds should be classified in? What other characteristics could scientists examine to help classify slime molds?

Figure 14 Downy mildews can have a great impact on agriculture and economies when they infect potatoes, sugar beets, grapes, and melons like those above.

Importance of the Funguslike Protists

Some of the organisms in this group are important because they help break down dead organisms. However, most funguslike protists are important because of the diseases they cause in plants and animals. One species of water mold that causes lesions in fish can be a problem when the number of this species in a given area is high. Fish farms and salmon spawning in streams can be greatly affected by a water mold spreading throughout the population. Water molds cause disease in other aquatic organisms including worms and even diatoms.

Economic Effects Downy mildews can have a huge effect on economies as well as social history. One of the most well-known members of this group caused the Irish potato famine during the 1840s. Potatoes were Ireland's main crop and the primary food source for its people. When the potato crop became infected with downy mildew, potatoes rotted in the fields, leaving many people with no food. A downy mildew infection of grapes in France during the 1870s nearly wiped out the entire French wine industry. Downy mildews, as shown in **Figure 14,** continue to infect crops such as lettuce, corn, and cabbage, as well as tropical avocados and pineapples.

section 1 review

Summary

What is a protist?

- Protists can reproduce asexually or sexually.

Plantlike Protists

- Algae can make their own food.
- Euglenoids are both animal-like and plantlike.
- Seaweeds are many-celled algae.
- Green algae are an important source of oxygen and food for many organisms on Earth.
- Algae can be used to thicken food or cosmetics.

Animal-like Protists

- Protozoans are one-celled consumers. Some protozoans are parasites.
- Some protozoans form symbiotic relationships; other protozoans can cause disease.

Funguslike Protists

- Funguslike protists take in energy from other organisms.

Self Check

1. **Identify** the characteristics common to all protists.
2. **Compare and contrast** the characteristics of animal-like, plantlike, and funguslike protists.
3. **Explain** how plantlike protists are classified into different groups.
4. **Classify** What protozoan characteristics do scientists use to organize protozoans into groups?
5. **Think Critically** Why are there few fossils of certain groups of protists?

Applying Skills

6. **Make and Use a Table** Make a table of the positive and negative effects that protists might have on your life and health.
7. **Use a spreadsheet** to make a table that compares the characteristics of the three groups of protozoans. Include *example organisms, method of transportation,* and *other characteristics.*

Comparing Algae and Protozoans

Magnification: 50×

Algae and protozoans have characteristics that are similar enough to place them in the same group—the protists. However, the variety of protist forms is great. In this lab, you can observe many of the differences among protists.

Real-World Question

What are the differences between algae and protozoans?

Goals

- **Draw and label** the organisms you examine.
- **Observe** the differences between algae and protozoans.

Materials

cultures of *Paramecium, Amoeba, Euglena,* and *Spirogyra*
**prepared slides of the organisms listed above*
prepared slide of slime mold
microscope slides (4)
coverslips (4)
microscope
**stereomicroscope*
dropper
**Alternate materials*

Safefy Precautions

Procedure

1. Copy the data table in your Science Journal.
2. Make a wet mount of the *Paramecium* culture. If you need help, refer to Student Resources at the back of the book.

Protist Observations

Protist	Drawing	Observations
Paramecium	Do not write in this book.	
Amoeba		
Euglena		
Spirogyra		
Slime mold		

3. **Observe** the wet mount first under low and then under high power. Record your observations in the data table. Draw and label the organism that you observed.
4. Repeat steps 2 and 3 with the other cultures. Return all preparations to your teacher and wash your hands.
5. **Observe** the slide of slime mold under low and high power. Record your observations.

Conclude and Apply

1. **Describe** the structure used for movement by each organism that moves.
2. **List** the protists that make their own food and explain how you know that they can.
3. **Identify** the protists you observed with animal-like characteristics.

Communicating Your Data

Share the results of this lab with your classmates. **For more help, refer to the Science Skill Handbook.**

section 2

Fungi

as you read

What You'll Learn

- **Identify** the characteristics shared by all fungi.
- **Classify** fungi into groups based on their methods of reproduction.
- **Differentiate** between the imperfect fungi and all other fungi.

Why It's Important

Fungi are important sources of food and medicines, and they help recycle Earth's wastes.

Review Vocabulary

photosynthesis: a process in which chlorophyll containing organisms use energy from light and change carbon dioxide and water into simple sugars and oxygen gas

New Vocabulary

- hyphae
- saprophyte
- spore
- basidium
- ascus
- budding
- sporangium
- lichen
- mycorrhizae

What are fungi?

Do you think you can find any fungi in your house or apartment? You have fungi in your home if you have mushroom soup or fresh mushrooms. What about that package of yeast in the cupboard? Yeasts are a type of fungus used to make some breads and cheeses. You also might find fungus growing on a loaf of bread or an orange, or mildew fungus growing on your shower curtain.

Origin of Fungi Although fossils of fungi exist, most are not useful in determining how fungi are related to other organisms. Some scientists hypothesize that fungi share an ancestor with ancient, flagellated protists and slime molds. Other scientists hypothesize that their ancestor was a green or red alga.

Structure of Fungi Most species of fungi are many-celled. The body of a fungus is usually a mass of many-celled, threadlike tubes called **hyphae** (HI fee), as shown in **Figure 15.** The hyphae produce enzymes that help break down food outside of the fungus. Then, the fungal cells absorb the digested food. Because of this, most fungi are known as saprophytes. **Saprophytes** are organisms that obtain food by absorbing dead or decaying tissues of other organisms. Other fungi are parasites. They obtain their food directly from living things.

Figure 15 The hyphae of fungi are involved in the digestion of food, as well as reproduction.

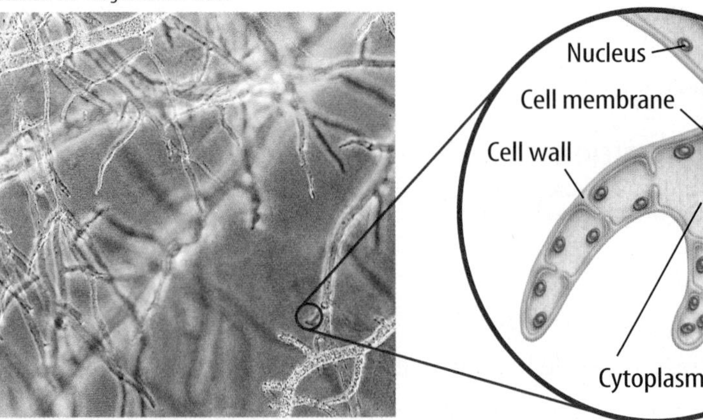

Threadlike, microscopic hyphae make up the body of a fungus.

The internal structure of hyphae.

Other Characteristics of Fungi What other characteristics do all fungi share? Because some fungi grow anchored in soil and have a cell wall around each cell, fungi once were classified as plants. But fungi don't have the specialized tissues and organs of plants, such as leaves and roots. Unlike most plants, fungi do not contain chlorophyll and cannot undergo photosynthesis.

Fungi grow best in warm, humid areas, such as tropical forests or between toes. You need a microscope to see some fungi, but in Michigan one fungus was found growing underground over an area of about 15 hectares. In the state of Washington, another type of fungus found in 1992 was growing throughout nearly 600 hectares of soil.

Topic: Unusual Fungi

Visit life.msscience.com for Web links to information about *Armillaria ostoyae* and other unusual fungi.

Activity Prepare an informational brochure about unusual fungi. Include illustrations and descriptions of at least three different kinds. Where do you find these fungi?

Reproduction Asexual and sexual reproduction in fungi usually involves the production of spores. A **spore** is a waterproof reproductive cell that can grow into a new organism. In asexual reproduction, mitosis and cell division produces spores. These spores will grow into new fungi that are genetically identical to the fungus from which the spores came.

Fungi are not identified as either male or female. Sexual reproduction can occur when the hyphae of two genetically different fungi of the same species grow close together. If the hyphae join, a reproductive structure will grow, as shown in **Figure 16.** Following meiosis in these structures, spores are produced that will grow into fungi. These fungi are genetically different from either of the two fungi whose hyphae joined during sexual reproduction. Fungi are classified into three main groups based on the type of structure formed by the joining of hyphae.

Reading Check *How are fungi classified?*

Color-enhanced LM Magnification: 30×

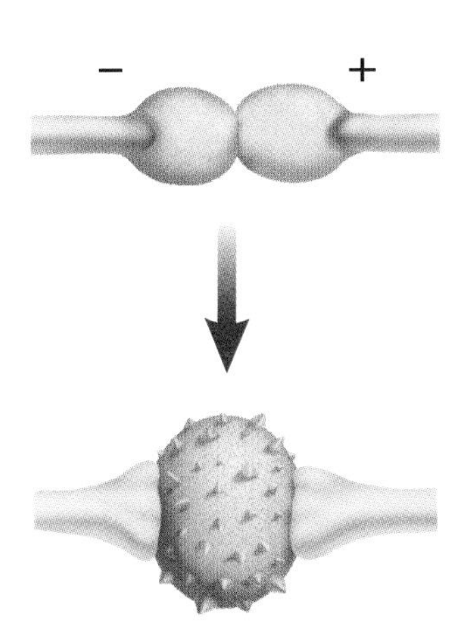

Figure 16 When two genetically different fungi of the same species meet, a reproductive structure, in this case a zygospore B, will be formed. The new fungi will be genetically different from either of the two original fungi.

Figure 17 Club fungi, like this mushroom, form a reproductive structure called a basidium. Each basidium produces four balloonlike structures called basidiospores. Spores will be released from these as the final step in sexual reproduction.

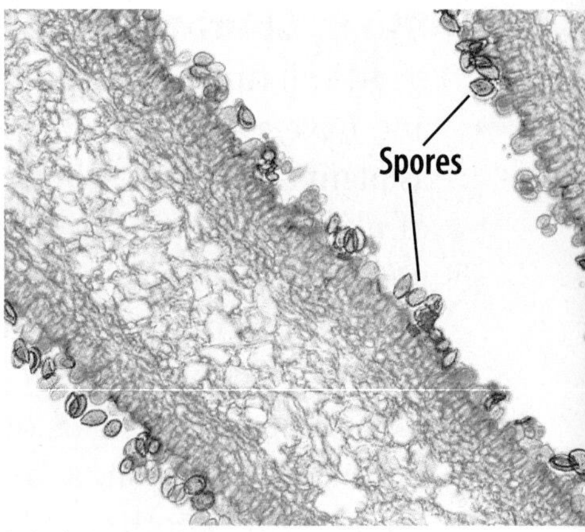

Stained LM Magnification: 18×

Club Fungi

The mushrooms shown in **Figure 17** are probably the type of fungus that you are most familiar with. The mushroom is only the reproductive structure of the fungus. Most of the fungus grows as hyphae in the soil or on the surface of its food source. These fungi commonly are known as club fungi. Their spores are produced in a club-shaped structure called a **basidium** (buh SIH dee uhm) (plural, *basidia*).

Sac Fungi

Yeasts, molds, morels, and truffles are all examples of sac fungi—a diverse group containing more than 30,000 different species. The spores of these fungi are produced in a little, saclike structure called an **ascus** (AS kus), as shown in **Figure 18.**

Although most fungi are many-celled, yeasts are one-celled organisms. Yeasts reproduce sexually by forming spores and reproduce asexually by budding, as illustrated in the right photo below. **Budding** is a form of asexual reproduction in which a new organism forms on the side of a parent organism. The two organisms are genetically identical.

Figure 18 The spores of a sac fungus are released when the tip of an ascus breaks open.

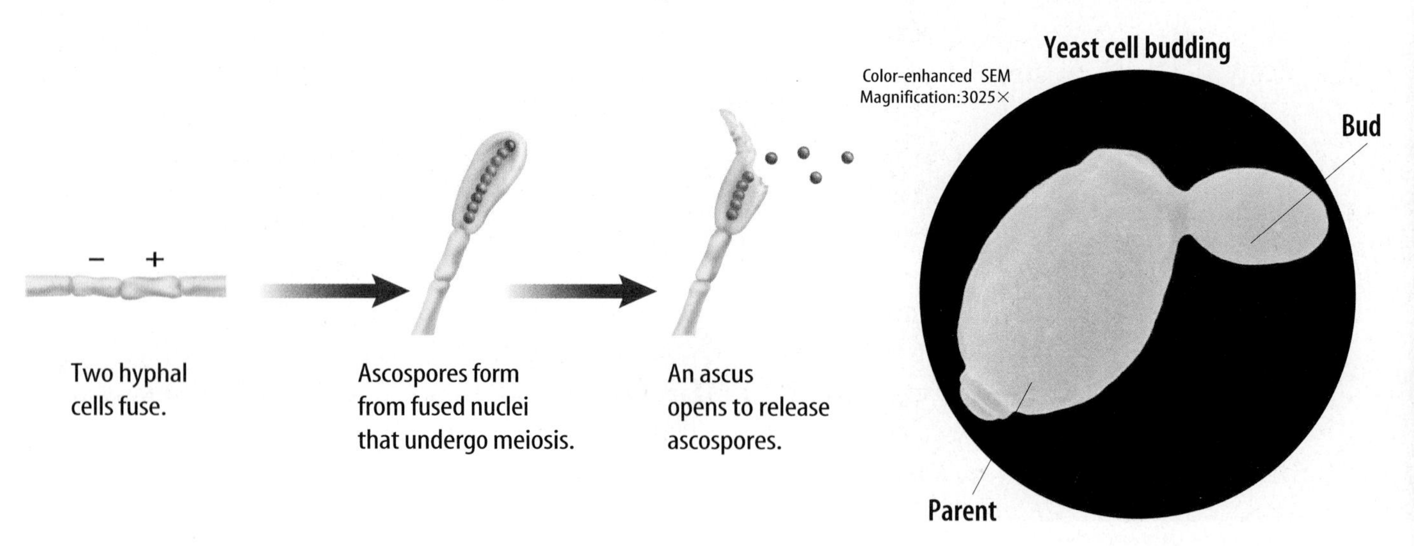

Figure 19 The black mold found growing on bread or fruit produces zygospores during sexual reproduction. Zygospores produce sporangia.

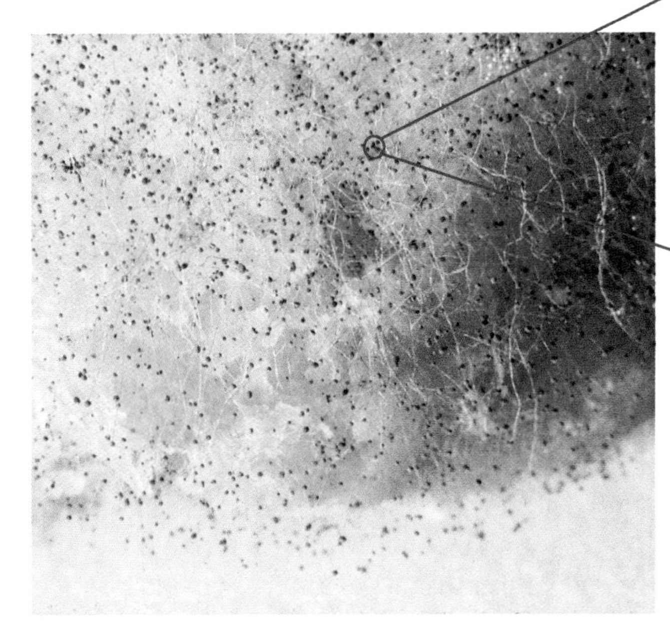

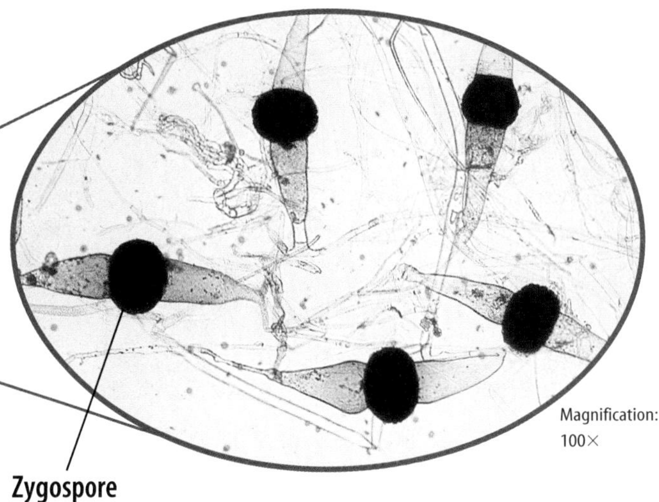

Zygote Fungi and Other Fungi

The fuzzy black mold that you sometimes find growing on a piece of fruit or an old loaf of bread, as shown in **Figure 19,** is a type of zygospore fungus. Fungi that belong to this group produce spores in a round spore case called a **sporangium** (spuh RAN jee uhm) (plural, *sporangia*) on the tips of upright hyphae. When each sporangium splits open, hundreds of spores are released into the air. Each spore will grow and reproduce if it lands in a warm, moist area that has a food supply.

Reading Check *What is a sporangium?*

Some fungi either never reproduce sexually or never have been observed reproducing sexually. Because of this, these fungi are difficult to classify. They usually are called imperfect fungi because there is no evidence that their life cycle has a sexual stage. Imperfect fungi reproduce asexually by producing spores. When the sexual stage of one of these fungi is observed, the species is classified immediately in one of the other three groups.

Penicillium is a fungus that is difficult to classify. Some scientists classify *Penicillium* as an imperfect fungi. Others think it should be classified as a sac fungus based on the type of spores it forms during asexual reproduction. Another fungus, which causes pneumonia, has been classified recently as an imperfect fungus. Like *Penicillium,* scientists do not agree about which group to place it in.

Mini LAB

Interpreting Spore Prints

Procedure

1. Obtain several **mushrooms from the grocery store** and let them age until the undersides look dark brown.
2. Remove the stems. Place the mushroom caps with the gills down on a piece of **unlined white paper.**
3. Let the mushroom caps sit undisturbed overnight and remove them from the paper the next day.

Analysis

1. **Draw** and label the results in your **Science Journal.** Describe the marks on the page and what might have made them.
2. **Estimate** the number of mushrooms that could be produced from a single mushroom cap.

Try at Home

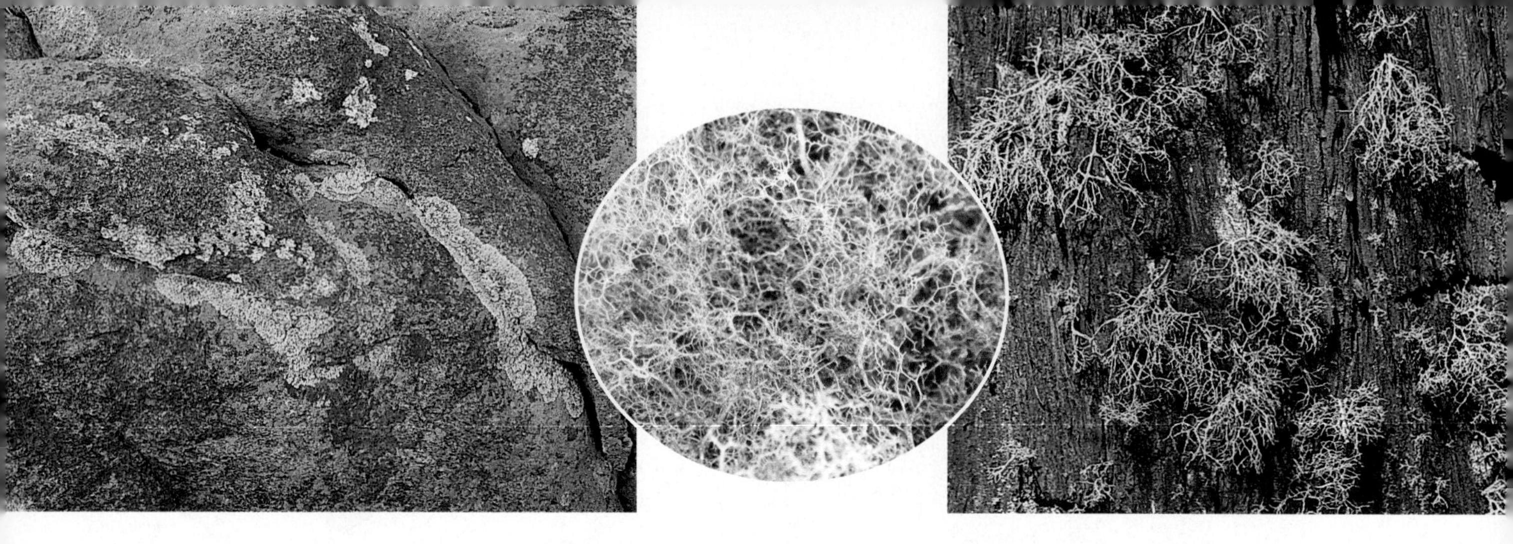

Figure 20 Lichens can look like a crust on bare rock, appear leafy, or grow upright. All three forms can grow near each other. **Determine** *one way lichens might be classified.*

Lichens

The colorful organisms in **Figure 20** are lichens. A **lichen** (LI kun) is an organism that is made of a fungus and either a green alga or a cyanobacterium. These two organisms have a relationship in which they both benefit. The alga or cyanobacterium lives among the threadlike strands of the fungus. The fungus gets food made by the green alga or cyanobacterium. The green alga or cyanobacterium gets a moist, protected place to live.

Importance of Lichens For many animals, including caribou and musk oxen, lichens are an important food source.

Lichens also are important in the weathering process of rocks. They grow on bare rock and release acids as part of their metabolism. The acids help break down the rock. As bits of rock accumulate and lichens die and decay, soil is formed. This soil supports the growth of other species.

Scientists also use lichens as indicator organisms to monitor pollution levels, as shown in **Figure 21.** Many species of lichens are sensitive to pollution. When these organisms show a decline in their health or die quickly, it alerts scientists to possible problems for larger organisms.

Figure 22 Many plants, such as these orchids, could not survive without mycorrhizae to help absorb water and important minerals from soil.

Fungi and Plants

Some fungi interact with plant roots. They form a network of hyphae and roots known as **mycorrhizae** (mi kuh RI zee). About 80 percent of plants develop mycorrhizae. The fungus helps the plant absorb more of certain nutrients from the soil better than the roots can on their own, while the plant supplies food and other nutrients to the fungi. Some plants, like the lady's slipper orchids shown in **Figure 22,** cannot grow without the development of mycorrhizae.

✔ Reading Check *Why are mycorrhizae so important to plants?*

NATIONAL GEOGRAPHIC VISUALIZING LICHENS AS AIR QUALITY INDICATORS

Figure 21

Widespread, slow growing, and with long life spans, lichens come in many varieties. Lichens absorb water and nutrients mainly from the air rather than the soil. Because certain types are extremely sensitive to toxic environments, lichens make natural, inexpensive, air-pollution detectors.

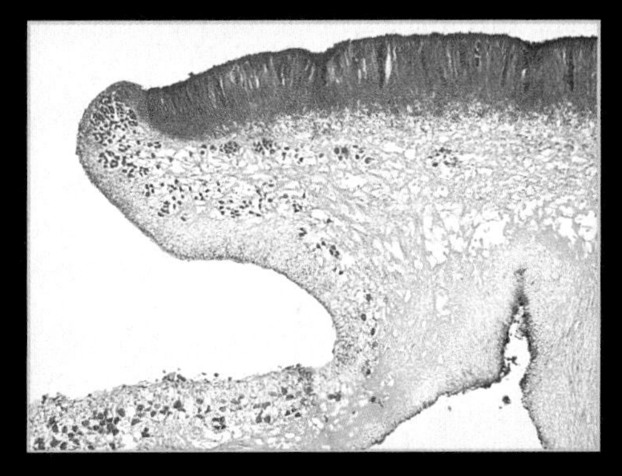

A lichen consists of a fungus and an alga or cyanobacterium living together in a partnership that benefits both organisms. In this cross section of a lichen (50x), reddish-stained bits of fungal tissue surround blue-stained algal cells.

Can you see a difference between these two red alder tree trunks? White lichens cover one trunk but not the other. Red alders are usually covered with lichens such as those seen in the photo on the left. Lichens could not survive on the tree on the right because of air pollution.

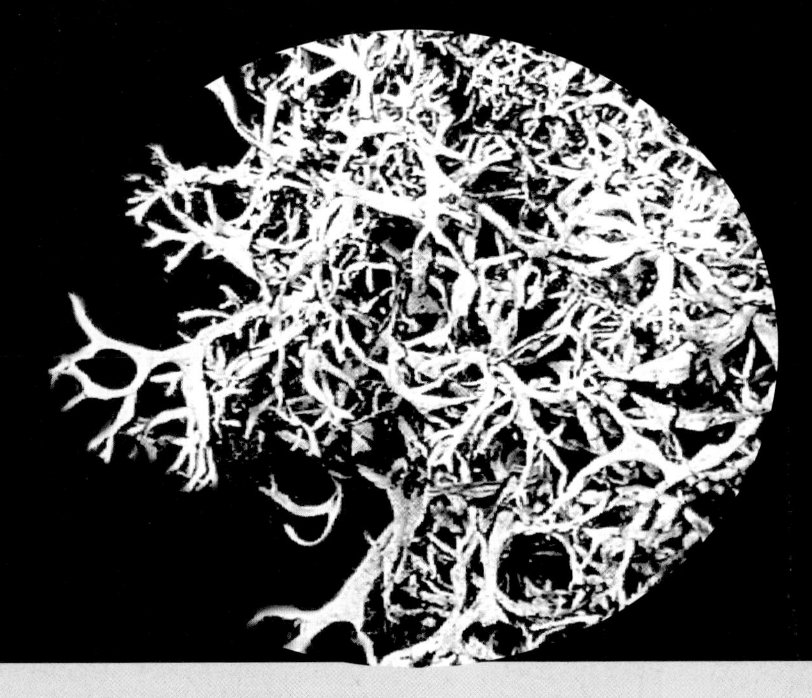

Evernia lichens, left, sicken and die when exposed to sulfur dioxide, a common pollutant emitted by coal-burning industrial plants such as the one above.

Biotechnology Living cells and materials produced by cells are used in biotechnology to develop products that benefit society. Careers in biotechnology include laboratory research and development, quality control, biostatistician, and product development. Through biotechnology, some fungi have been developed that can be used as natural pesticides to control pests like termites, tent caterpillars, aphids, and citrus mites.

Fossilized Fungus In 1999, scientists discovered a fossilized fungus in a 460 million-year-old rock. The fossil was a type of fungus that forms associations with plant roots. Scientists have known for many years that the first plants could not have survived moving from water to land alone. Early plants did not have specialized roots to absorb nutrients. Also, tubelike cells used for transporting water and nutrients to leaves were too simple.

Scientists have hypothesized that early fungi attached themselves to the roots of early plants, passing along nutrients taken from the soil. Scientists suggest that it was this relationship that allowed plants to move successfully from water onto land about 500 million years ago. Until the discovery of this fossil, no evidence had been found that this type of fungus existed at that time.

Importance of Fungi

As mentioned in the beginning of this chapter, some fungi are eaten for food. Cultivated mushrooms are an important food crop. However, wild mushrooms never should be eaten because many are poisonous. Some cheeses are produced using fungi. Yeasts are used in the baking industry. Yeasts use sugar for energy and produce alcohol and carbon dioxide as waste products. The carbon dioxide causes doughs to rise.

Agriculture Many fungi are important because they cause diseases in plants and animals. Many sac fungi are well-known by farmers because they damage or destroy plant crops. Diseases caused by sac fungi include Dutch elm disease, apple scab, and ergot disease of rye. Smuts and rust, shown in **Figure 23,** are club fungi. They cause billions of dollars worth of damage to food crops each year.

Figure 23 Rusts can infect the grains used to make many cereals including wheat (shown below), barley, rye, and oats. Not all fungi are bad for agriculture. Some are natural pesticides. This dead grasshopper (right) is infected with a fungal parasite.

Health and Medicine Fungi can cause diseases in humans and animals. Ringworm and athlete's foot are skin infections caused by species of imperfect fungi. Other fungi can cause respiratory infections.

The effects of fungi on health and medicine are not all negative. Some species of fungi naturally produce antibiotics that keep bacteria from growing on or near them. The antibiotic penicillin is produced by the imperfect fungi *Penicillium.* This fungus is grown commercially, and the antibiotic is collected to use in fighting bacterial infections. Cyclosporine, an important drug used to help fight the body's rejection of transplanted organs, also is derived from a fungus. There are many more examples of breakthroughs in medicine as a result of studying and discovering new uses of fungi. In fact, there is a worldwide effort among scientists who study fungi to investigate soil samples to find more useful drugs.

Figure 24 Fungi have an important role as decomposers in nature.
Explain *why fungi are called nature's recyclers.*

Decomposers As important as fungi are in the production of different foods and medicines, they are most important as decomposers that break down organic materials. Food scraps, clothing, and dead plants and animals are made of organic material. Often found on rotting logs, as shown in **Figure 24,** fungi break down these materials. The chemicals in these materials are returned to the soil where plants can reuse them. Fungi, along with bacteria, are nature's recyclers. They keep Earth from becoming buried under mountains of organic waste materials.

section 2 review

Summary

What are fungi?

- Fungi are consumers and reproduce both sexually and asexually.
- There are three main classifications of fungi.

Lichens

- Lichens consist of a fungus and either a green alga or a cyanobacterium.
- They help break down rocks and form soil.

Fungi and Plants

- Mycorrhizae are a network of plant roots and fungi hyphae that interact to obtain nutrients.

Importance of Fungi

- Fungi are most important as decomposers.

Self Check

1. **List** characteristics common to all fungi.
2. **Explain** how fungi are classified into different groups.
3. **Compare and contrast** imperfect fungi and other groups of fungi.
4. **List** ways lichens are important.
5. **Think Critically** If an imperfect fungus was found to produce basidia under certain environmental conditions, how would the fungus be reclassified?

Applying Math

6. **Use Proportions** Of the 100,000 fungus species, approximately 30,000 are sac fungi. What percentage of fungus species are sac fungi?

Creating a Fungus Field Guide

Goals

- **Identify** the common club fungi found in the woods or grassy areas near your home or school.
- **Create** a field guide to help future science students identify these fungi.

Possible Materials

collection jars
magnifying lens
microscopes
microscope slides and coverslips
field guide to fungi or club fungi
art supplies

Safety Precautions

WARNING: *Do not eat any of the fungi you collect. Do not touch your face during the lab.*

Real-World Question

Whether they are hiking deep into a rain forest in search of rare tropical birds, diving to coral reefs to study marine worms, or peering into microscopes to identify strains of bacteria, scientists all over the world depend on reliable field guides. Field guides are books that identify and describe certain types of organisms or the organisms living in a specific environment. Scientists find field guides for a specific area especially helpful. How can you create a field guide for the club fungi living in your area? What information would you include in a field guide of club fungi?

Make A Model

1. Decide on the locations where you will conduct your search.
2. Select the materials you will need to collect and survey club fungi.
3. Design a data table in your Science Journal to record the fungi you find.
4. Decide on the layout of your field guide. What information about the fungi you will include? What drawings you will use? How will you group the fungi?

5. **Describe** your plan to your teacher and ask your teacher how it could be improved.
6. **Present** your ideas for collecting and surveying fungi, and your layout ideas for your field guide to the class. Ask your classmates to suggest improvements in your plan.

Test Your Model

1. Search for samples of club fungi. **Record** the organisms you find in your data table. Use a fungus field guide to identify the fungi you discover. Do not pick or touch any fungi that you find unless you have permission.
2. Using your list of organisms, complete your field guide of club fungi as planned.
3. When finished, give your field guide to a classmate to identify a club fungus.

Analyze Your Data

1. **Compare** the number of fungi you found to the total number of organisms listed in the field guide you used to identify the organisms.
2. **Analyze** the problems you may have had while collecting and identifying your fungi. Suggest steps you could take to improve your collection and identification methods.
3. **Analyze** the problems you had while creating your field guide. Suggest ways your field guide could be improved.

Conclude and Apply

Infer why your field guide would be more helpful to future science students in your school than the fungus field guide you used to identify organisms.

Communicating Your Data

Compare your field guide with those assembled by your classmates. Combine all the information on local club fungi compiled by your class to create a classroom field guide to club fungi.

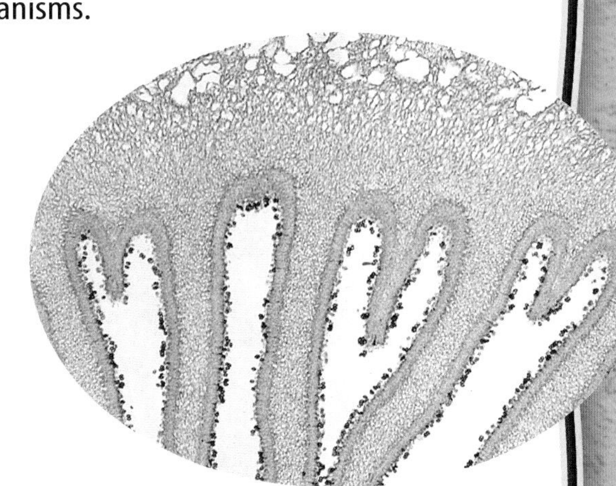

Stained LM Magnification: 80×

Chocolate SOS

Can a fungus protect cacao trees under attack?

Pods contain dozens of beans.

Losing Beans

Chocolate is made from seeds (cocoa beans) that grow in the pods on the tropical cacao tree. The monoculture (growing one type of crop) of modern fields has produced huge crops of cocoa beans, but also has enabled destructive fungi to sweep through cacao fields. A disease that attacks one plant of a species in a monoculture will rapidly spread to all plants in the monoculture. There are fewer healthy cacao trees today than there were several years ago. Since the blight began in the late 1980s, the world has lost three million tons of cocoa beans. Brazil, once the top producer and exporter of cocoa beans, harvested only 80,000 tons in 2000—the smallest harvest in 30 years.

A diseased pod from a cacao tree

Unless something stops the fungi that are destroying trees, there could be a lot less chocolate in the future. Your favorite chocolate bars could become more expensive and harder to find.

A Natural Cure

Farmers tried using traditional chemical sprays to fight the fungus, but they were ineffective because the tropical rains washed away the sprays. Now, agriculture scientists are working on a "natural solution" to the problem. They are using beneficial fungi (strains of *Trichoderma*) to fight the harmful fungi attacking the cocoa trees. When sprayed on trees, *Trichoderma* stops the spread of the harmful fungi. The test treatments in Brazil and Peru have reduced the destruction of the trees by between 30 and 50 percent.

Don't expect your favorite chocolate bars to disappear from stores anytime soon. Right now, world cocoa bean supplies still exceed demand. But if the spread of the epidemic can't be stopped, those chocolate bars could become slightly more expensive and a little harder to find.

Concept Map **Use the library and other sources to learn the steps in making chocolate—from harvesting cacao beans to packing chocolate products for sale? Draw a concept map that shows the steps. Compare your concept map with those of your classmates.**

For more information, visit life.msscience.com/time

Reviewing Main Ideas

Section 1 Protists

1. Protists are one-celled or many-celled eukaryotic organisms. They can reproduce asexually, resulting in two new cells that are genetically identical. Protists also can reproduce sexually and produce genetically different offspring.
2. The protist kingdom has members that are plantlike, animal-like, and funguslike.
3. Protists are thought to have evolved from a one-celled organism with a nucleus and other cellular structures.
4. Plantlike protists have cell walls and contain chlorophyll.
5. Animal-like protists can be separated into groups by how they move.
6. Funguslike protists have characteristics of protists and fungi.

Section 2 Fungi

1. Most species of fungi are many-celled. The body of a fungus consists of a mass of threadlike tubes.
2. Fungi are saprophytes or parasites—they feed off other things because they cannot make their own food.
3. Fungi reproduce using spores.
4. The three main groups of fungi are club fungi, sac fungi, and zygote fungi. Fungi that cannot be placed in a specific group are called imperfect fungi. Fungi are placed into one of these groups according to the structures in which they produce spores.
5. A lichen is an organism that consists of a fungus and a green alga or cyanobacterium.

Visualizing Main Ideas

Copy and complete the following concept map on a separate sheet of paper.

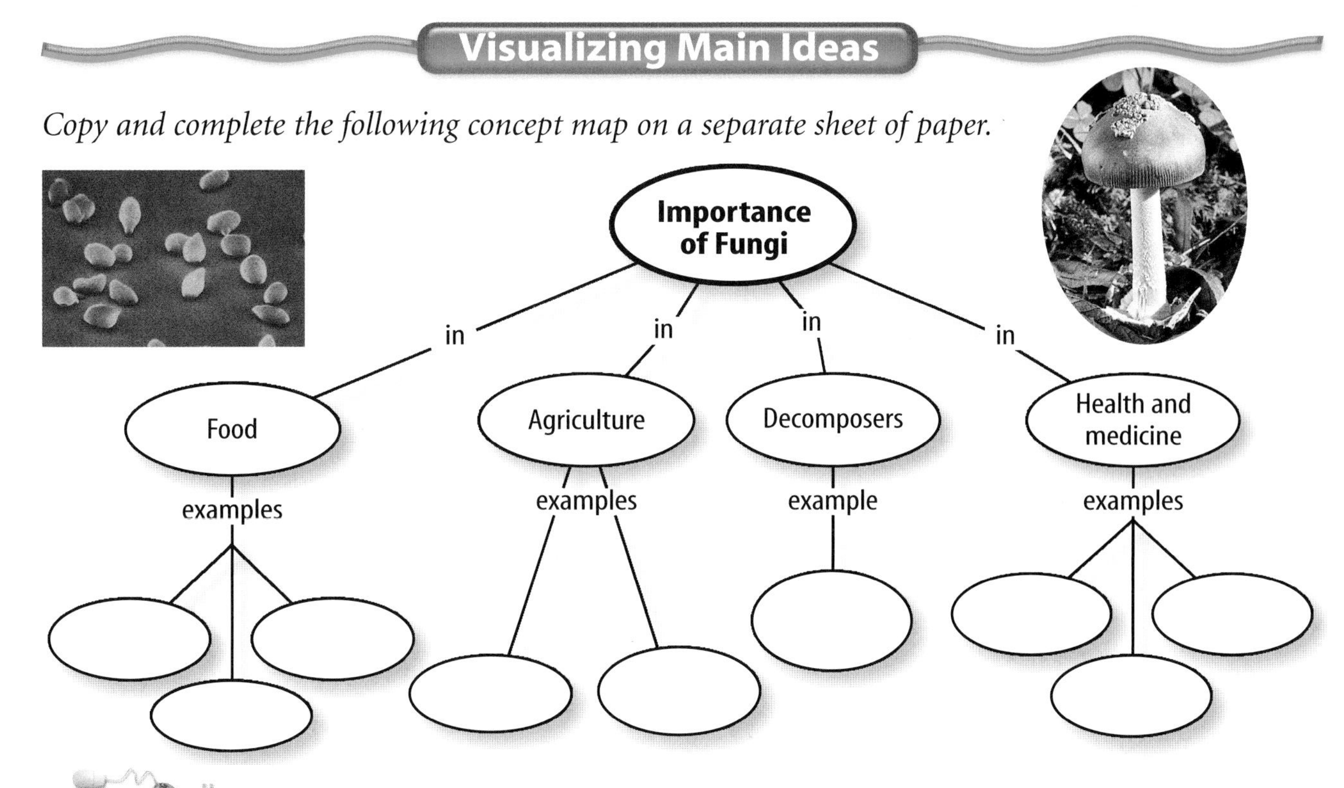

chapter 8 Review

Using Vocabulary

algae p. 211	mycorrhizae p. 226
ascus p. 224	protist p. 210
basidium p. 224	protozoan p. 215
budding p. 224	pseudopod p. 216
cilia p. 215	saprophyte p. 222
flagellum p. 212	sporangium p. 225
hyphae p. 222	spore p. 223
lichen p. 226	

Write the vocabulary word that matches each of these descriptions.

1. reproductive cell of a fungus
2. organisms that are animal-like, plantlike, or funguslike
3. threadlike structures used for movement
4. plantlike protists
5. organism made up of a fungus and an alga or a cyanobacterium
6. reproductive structure made by sac fungi
7. threadlike tubes that make up the body of a fungus
8. structure used for movement formed by oozing cytoplasm

Checking Concepts

Choose the word or phrase that best answers the question.

9. Which type of protist captures food, does not have cell walls, and can move from place to place?
   A) algae **C)** fungi
   B) protozoans **D)** lichens

10. Which of the following organisms cause red tides when found in large numbers?
   A) *Euglena* **C)** *Ulva*
   B) diatoms **D)** dinoflagellates

11. Algae are important for which of the following reasons?
   A) They are a food source for many aquatic organisms.
   B) Parts of algae are used in foods that humans eat.
   C) Algae produce oxygen as a result of the process of photosynthesis.
   D) all of the above

12. Where would you most likely find funguslike protists?
   A) on decaying logs
   B) in bright light
   C) on dry surfaces
   D) on metal surfaces

13. Where are spores produced in mushrooms?
   A) sporangia **C)** ascus
   B) basidia **D)** hyphae

14. Which of the following is used as an indicator organism?
   A) club fungus **C)** slime mold
   B) lichen **D)** imperfect fungus

Use the illustration below to answer question 15.

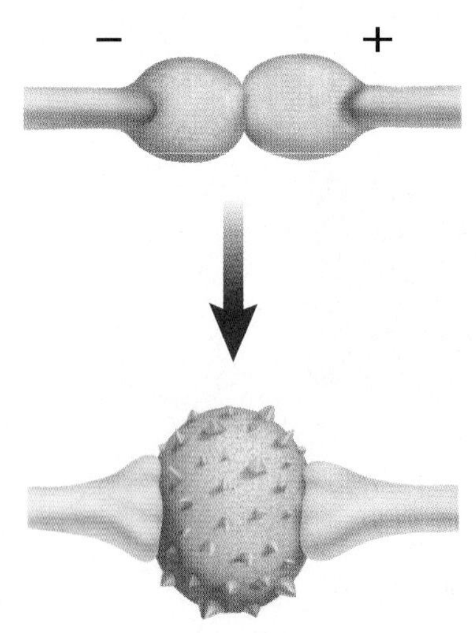

15. What is the reproductive structure, shown in the lower image above, called?
   A) hypha **C)** basidium
   B) zygospore **D)** ascus

Science Online life.msscience.com/vocabulary_puzzlemaker

Thinking Critically

16. **Identify** what kind of environment is needed to prevent fungal growth.
17. **Infer** why algae contain pigments other than just chlorophyll.
18. **Compare and contrast** the features of fungi and funguslike protists.
19. **List** some advantages plants have when they form associations with fungi.
20. **Explain** the adaptations of fungi that enable them to get food.
21. **Recognize Cause and Effect** A leaf sitting on the floor of a rain forest will decompose in six weeks. A leaf on the floor of a temperate forest will take up to a year to decompose. Explain how this is possible.
22. **Compare and Contrast** Make a chart comparing and contrasting the different ways protists and fungi can obtain food.
23. **Make and Use Tables** Copy and complete the following table that compares the different groups of fungi.

Fungi Comparisons

Fungi Group	Structure Where Sexual Spores Are Produced	Examples
Club fungi		Do not write in this book.
	Ascus	
Zygospore fungi		
	No sexual spores produced	

24. **Identify and Manipulate Variables and Controls** You find a new and unusual fungus growing in your refrigerator. Design an experiment to determine to which fungus group it should be classified.

Performance Activities

25. **Poster** Research the different types of fungi found in the area where you live. Determine to which group each fungus belongs. Create a poster to display your results and share them with your class.

Applying Math

26. **Lichen Growth** Sometimes the diameter of a lichen colony is used to estimate the age of the rock it is growing on. In some climates, it may take 100 years for a lichen colony to increase its diameter by 50 mm. Estimate how old a rock tombstone is if the largest lichen colony growing on it has a diameter of 150 mm.

Use the graph below to answer question 27.

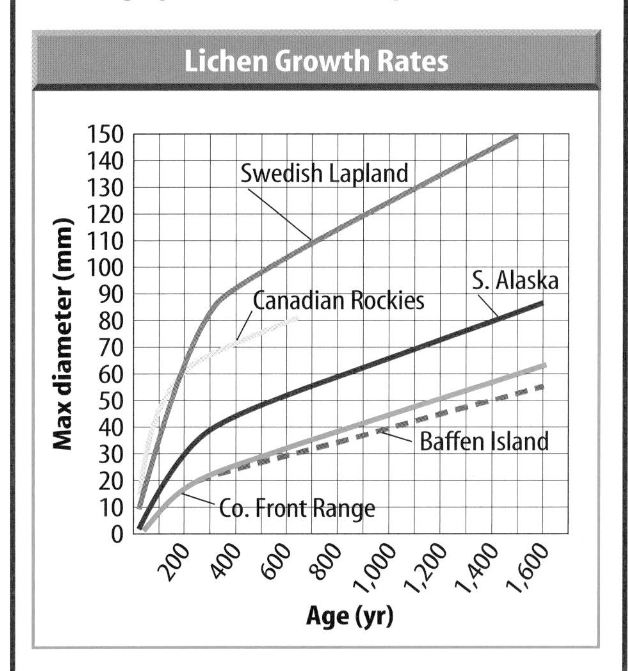

27. **Climate and Growth** The graph above illustrates lichen growth rates in different climates. According to the graph, which climate is the most favorable for lichen growth? What is the difference between the diameter of a 200-year-old colony in the Swedish Lapland compared to a 200-year-old colony on Baffen Island?

Part 1 Multiple Choice

Record your answers on the answer sheet provided by your teacher or on a sheet of paper.

Use the illustration below to answer questions 1–3.

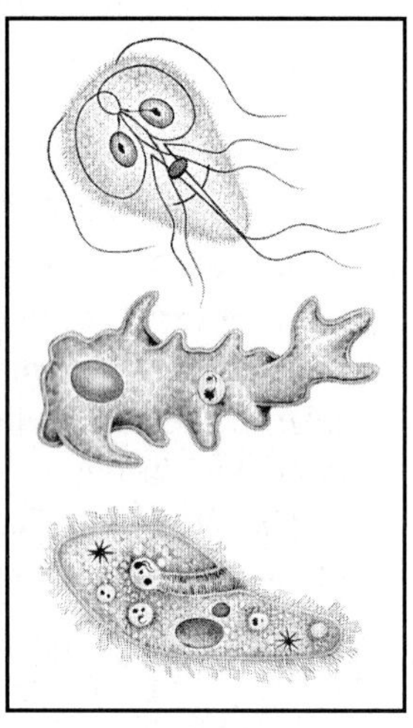

Group A Group B

1. Which of the following organisms would belong in Group B?
   A) kelp
   B) grass
   C) diatom
   D) *Paramecium*

2. Which of the following is a characteristic of Group B?
   A) makes own food
   B) has specialized ways to move
   C) absorbs food from the surroundings
   D) has cell walls

3. Which of the following is NOT a method of moving used by the protists in Group B?
   A) pseudopod
   B) cilia
   C) flagella
   D) vacuole

Test-Taking Tip

Go at Your Own Pace Stay focused during the test and don't rush, even if you notice that other students are finishing the test early.

4. Which of the following is a protozoan?
   A) ciliate
   B) diatom
   C) kelp
   D) bacteria

Use the table below to answer questions 5–7.

Diseases Caused by Protozoans		
Disease	**Source**	**Protozoan**
Giardiasis	Contaminated water	*Giardia*
Malaria	Mosquito bite	*Plasmodium*
Dysentery	Contaminated water	*Entamoeba*
Sleeping sickness	Tsetse fly bite	*Trypanosoma*
Toxoplasmosis	Contaminated soil; eating undercooked meat that contains the organism	*Toxoplasma*

5. According to the chart, which of the following disease-carrying protozoan can be transmitted to humans by a bite from another animal?
   A) *Giardia*
   B) *Plasmodium*
   C) *Entamoeba*
   D) *Toxoplasma*

6. Based on the information in the chart, which disease can be prevented by purifying water used for drinking and cooking, and by washing fruits and vegetables?
   A) malaria
   B) dysentery
   C) sleeping sickness
   D) toxoplasmosis

7. According to the chart, which protozoan disease can be prevented by cooking meat thoroughly?
   A) malaria
   B) dysentery
   C) sleeping sickness
   D) toxoplasmosis

8. Where are the spores produced in the fuzzy black mold that grows on bread?
   A) basidia
   B) sporangia
   C) ascus
   D) hyphae

Part 2 Short Response/Grid In

Record your answers on the answer sheet provided by your teacher or on a sheet of paper.

9. Brown algae can grow at an ocean depth of 35 m and red algae can grow at 200 m. Approximately how many times deeper than brown algae can red algae grow?

Use the illustration below to answer questions 10–11.

10. What type of protist is shown in the illustration above?
11. What pigment do all the organisms contain that absorbs light?
12. How are slime molds like protozoa?
13. How are saprophytes different from parasites?
14. Juan is making bread dough. "Don't forget to add fungus," Aunt Inez jokes. What did she mean?
15. One fungus found in Washington state was growing throughout 600 hectares of soil. Another fungus in Michigan was growing underground over an area of 15 hectares. How many times bigger was the fungus that was growing in Washington state?

Part 3 Open Ended

Record your answers on a sheet of paper.

16. Compare and contrast asexual and sexual reproduction.
17. Discuss the ways that algae are useful for humans.
18. Compare and contrast fungi and downy mildews.

Use the table below to answer questions 19–21.

Newly Identified Organisms		
Characteristic	**Organism A**	**Organism B**
Movement	No	Yes
One-celled or many-celled	Many-celled	One-celled
Cell walls	Yes	No
Method of reproduction	Sexual	Sexual
Makes own food	Only some cells	No
Contains chlorophyll	Only found in some cells	No
Method of obtaining food	Made by some of the organism's cells; nutrients and water absorbed from surroundings	Sweeps food into oral groove
Where found	Bare rock	Freshwater

19. Dr. Seung discovered two new organisms. Their characteristics are listed in the table above. How would you classify organism B? How do you know?
20. What specialized way of moving would you expect to see if you examined organism B? Why?
21. How would you classify organism A? How do you know?

chapter 9

Plants

The BIG Idea

The diverse plants on Earth provide humans and other organisms with food, shelter, and oxygen.

SECTION 1
An Overview of Plants

Main Idea Plants have adaptations that enable them to survive in the many environments on Earth.

SECTION 2
Seedless Plants

Main Idea Seedless plants are adapted for living in moist environments.

SECTION 3
Seed Plants

Main Idea Seed plants have adaptations that enable them to live in diverse environments.

How are all plants alike?

Plants are found nearly everywhere on Earth. A tropical rain forest like this one is crowded with lush, green plants. When you look at a plant, what do you expect to see? Do all plants have green leaves? Do all plants produce flowers and seeds?

Science Journal Write three characteristics that you think all plants have in common.

Start-Up Activities

How do you use plants?

Plants are just about everywhere—in parks and gardens, by streams, on rocks, in houses, and even on dinner plates. Do you use plants for things other than food?

1. Brainstorm with two other classmates and make a list of everything that you use in a day that comes from plants.
2. Compare your list with those of other groups in your class.
3. Search through old magazines for images of the items on your list.
4. As a class, build a bulletin board display of the magazine images.
5. **Think Critically** In your Science Journal, list things that were made from plants 100 years or more ago but today are made from plastics, steel, or some other material.

Preview this chapter's content and activities at life.msscience.com

Plants Make the following Foldable to help identify what you already know, what you want to know, and what you learned about plants.

STEP 1 **Fold** a vertical sheet of paper from side to side. Make the front edge 1.25 cm shorter than the back edge.

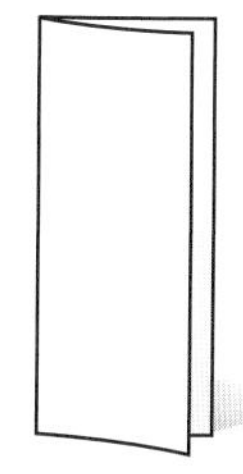

STEP 2 **Turn** lengthwise and fold into thirds.

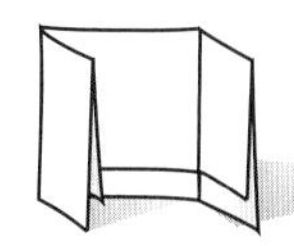

STEP 3 **Unfold and cut** only the top layer along both folds to make three tabs.

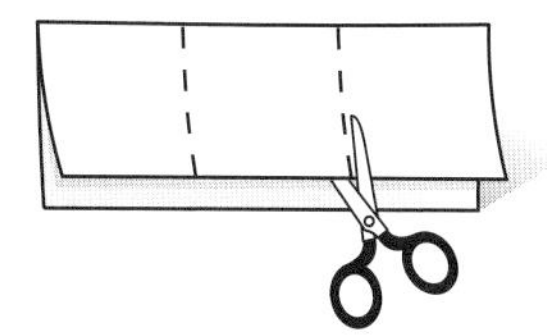

STEP 4 **Label** each tab as shown.

Identify Questions Before you read the chapter, write what you already know about plants under the left tab of your Foldable, and write questions about what you'd like to know under the center tab. After you read the chapter, list what you learned under the right tab.

Get Ready to Read

Make Connections

1 Learn It! Make connections between what you read and what you already know. Connections can be based on personal experiences (text-to-self), what you have read before (text-to-text), or events in other places (text-to-world).

As you read, ask connecting questions. Are you reminded of a personal experience? Have you read about the topic before? Did you think of a person, a place, or an event in another part of the world?

2 Practice It! Read the excerpt below and make connections to your own knowledge and experience.

What do you already know about vascular plants?

What angiosperms did you pass on your way to school?

What angiosperms are native to your state?

When people are asked to name a plant, most name an angiosperm. An **angiosperm** is a vascular plant that flowers and produces fruits with one or more seeds, such as the peaches shown in **Figure 19**. The fruit develops from a part or parts of one or more flowers. Angiosperms are familiar plants no matter where you live. They grow in parks, fields, forests, jungles, deserts, freshwater, salt water, and in the cracks of sidewalks. You might see them dangling from wires or other plants, and one species of orchid even grows underground. Angiosperms make up the plant division Anthophyta (AN thoh fi tuh). More than half of the plant species known today belong to this division.

—from page 257

3 Apply It! As you read this chapter, choose five words or phrases that make a connection to something you already know.

Make connections with memorable events, places, or people in your life. The better the connection, the more likely you will remember.

Target Your Reading

Use this to focus on the main ideas as you read the chapter.

1. **Before you read** the chapter, respond to the statements below on your worksheet or on a numbered sheet of paper.
   - Write an **A** if you **agree** with the statement.
   - Write a **D** if you **disagree** with the statement.

2. **After you read** the chapter, look back to this page to see if you've changed your mind about any of the statements.
   - If any of your answers changed, explain why.
   - Change any false statements into true statements.
   - Use your revised statements as a study guide.

Before You Read A or D	Statement	After You Read A or D
	1 All plants have roots, stems, and leaves.	
	2 A waxy covering slows the movement of water out of a plant.	
	3 Some plants have special cells in which water travels from roots to leaves.	
	4 All daisies are members of the same species.	
	5 Some mosses are adapted for desert environments.	
	6 Nonvascular plants often are the first to grow in disturbed or damaged environments.	
	7 Coal is the fossil remains of seedless plants.	
	8 Leaves, stems, and roots are organs of vascular plants.	
	9 All evergreens are conifers, such as pines and spruces.	
	10 Flowering plants are the most numerous plants on Earth.	

Science Online
Print out a worksheet of this page at life.msscience.com

section 1

An Overview of Plants

as you read

What You'll Learn

- **Identify** characteristics common to all plants.
- **Explain** which plant adaptations make it possible for plants to survive on land.
- **Compare and contrast** vascular and nonvascular plants.

Why It's Important

Plants produce food and oxygen, which are required for life by most organisms on Earth.

Review Vocabulary

species: closely related organisms that share similar characteristics and can reproduce among themselves

New Vocabulary

- cuticle
- cellulose
- vascular plant
- nonvascular plant

What is a plant?

What is the most common sight you see when you walk along nature trails in parks like the one shown in **Figure 1?** Maybe you've taken off your shoes and walked barefoot on soft, cool grass. Perhaps you've climbed a tree to see what things look like from high in its branches. In each instance, plants surrounded you.

If you named all the plants that you know, you probably would include trees, flowers, vegetables, fruits, and field crops like wheat, rice, or corn. Between 260,000 and 300,000 plant species have been discovered and identified. Scientists think many more species are still to be found, mainly in tropical rain forests. Plants are important food sources to humans and other consumers. Without plants, most life on Earth as we know it would not be possible.

Plant Characteristics Plants range in size from microscopic water ferns to giant sequoia trees that are sometimes more than 100 m in height. Most have roots or rootlike structures that hold them in the ground or onto some other object like a rock or another plant. Plants are adapted to nearly every environment on Earth. Some grow in frigid, ice-bound polar regions and others grow in hot, dry deserts. All plants need water, but some plants cannot live unless they are submerged in either freshwater or salt water.

Figure 1 All plants are many-celled and nearly all contain chlorophyll. Grasses, trees, shrubs, mosses, and ferns are all plants.

Plant Cells Like other living things, plants are made of cells. A plant cell has a cell membrane, a nucleus, and other cellular structures. In addition, plant cells have cell walls that provide structure and protection. Animal cells do not have cell walls.

Many plant cells contain the green pigment chlorophyll (KLOR uh fihl) so most plants are green. Plants need chlorophyll to make food using a process called photosynthesis. Chlorophyll is found in a cell structure called a chloroplast. Plant cells from green parts of the plant usually contain many chloroplasts.

Most plant cells have a large, membrane-bound structure called the central vacuole that takes up most of the space inside of the cell. This structure plays an important role in regulating the water content of the cell. Many substances are stored in the vacuole, including the pigments that make some flowers red, blue, or purple.

Origin and Evolution of Plants

Have plants always existed on land? The first plants that lived on land probably could survive only in damp areas. Their ancestors were probably ancient green algae that lived in the sea. Green algae are one-celled or many-celled organisms that use photosynthesis to make food. Today, plants and green algae have the same types of chlorophyll and carotenoids (kuh RAH tun oydz) in their cells. Carotenoids are red, yellow, or orange pigments that also are used for photosynthesis. These facts lead scientists to think that plants and green algae have a common ancestor.

Reading Check *How are plants and green algae alike?*

Fossil Record The fossil record for plants is not like that for animals. Most animals have bones or other hard parts that can fossilize. Plants usually decay before they become fossilized. The oldest fossil plants are about 420 million years old. **Figure 2** shows *Cooksonia,* a fossil of one of these plants. Other fossils of early plants are similar to the ancient green algae. Scientists hypothesize that some of these early plants evolved into the plants that exist today.

Cone-bearing plants, such as pines, probably evolved from a group of plants that grew about 350 million years ago. Fossils of these plants have been dated to about 300 million years ago. It is estimated that flowering plants did not exist until about 120 million years ago. However, the exact origin of flowering plants is not known.

Figure 2 This is a fossil of a plant named *Cooksonia.* These plants grew about 420 million years ago and were about 2.5 cm tall.

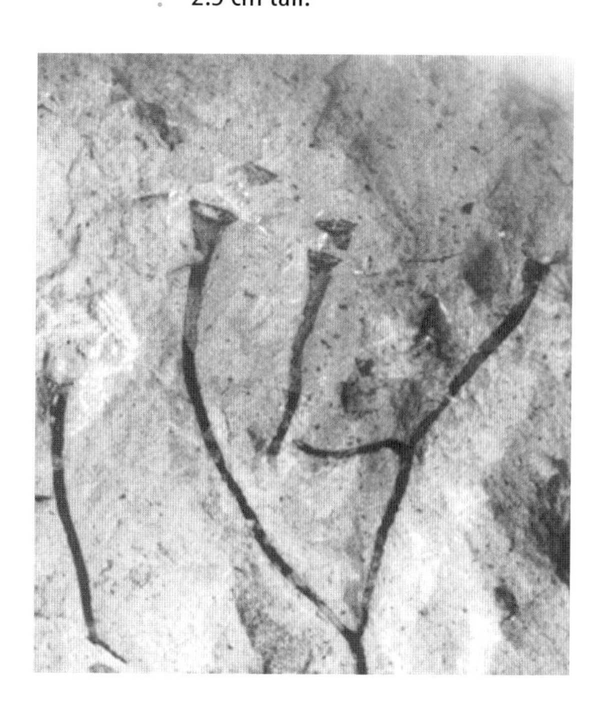

Cellulose Plant cell walls are made mostly of cellulose. Anselme Payen, a French scientist, first isolated and identified the chemical composition of cellulose in 1838, while analyzing the chemical makeup of wood. Choose a type of wood and research to learn the uses of that wood. Make a classroom display of research results.

Life on Land

Life on land has some advantages for plants. More sunlight and carbon dioxide—needed for photosynthesis—are available on land than in water. During photosynthesis, plants give off oxygen. Long ago, as more and more plants adapted to life on land, the amount of oxygen in Earth's atmosphere increased. This was the beginning for organisms that depend on oxygen.

Adaptations to Land

What is life like for green algae, shown in **Figure 3,** as they float in a shallow pool? The water in the pool surrounds and supports them as the algae make their own food through the process of photosynthesis. Because materials can enter and leave through their cell membranes and cell walls, the algae cells have everything they need to survive as long as they have water.

If the pool begins to dry up, the algae are on damp mud and are no longer supported by water. As the soil becomes drier and drier, the algae will lose water too because water moves through their cell membranes and cell walls from where there is more water to where there is less water. Without enough water in their environment, the algae will die. Plants that live on land have adaptations that allow them to conserve water, as well as other differences that make it possible for survival.

Figure 3 The alga *Spirogyra,* like all algae, must have water to survive. If the pool where it lives dries up, it will die.

LM Magnification: 22×

Protection and Support Water is important for plants. What adaptations would help a plant conserve water on land? Covering the stems, leaves, and flowers of many plants is a **cuticle** (KYEW tih kul)—a waxy, protective layer secreted by cells onto the surface of the plant. The cuticle slows the loss of water. The cuticle and other adaptations shown in **Figure 4** enable plants to survive on land.

What is the function of a plant's cuticle?

Supporting itself is another problem for a plant on land. Like all cells, plant cells have cell membranes, but they also have rigid cell walls outside the membrane. Cell walls contain **cellulose** (SEL yuh lohs), which is a chemical compound that plants can make out of sugar. Long chains of cellulose molecules form tangled fibers in plant cell walls. These fibers provide structure and support.

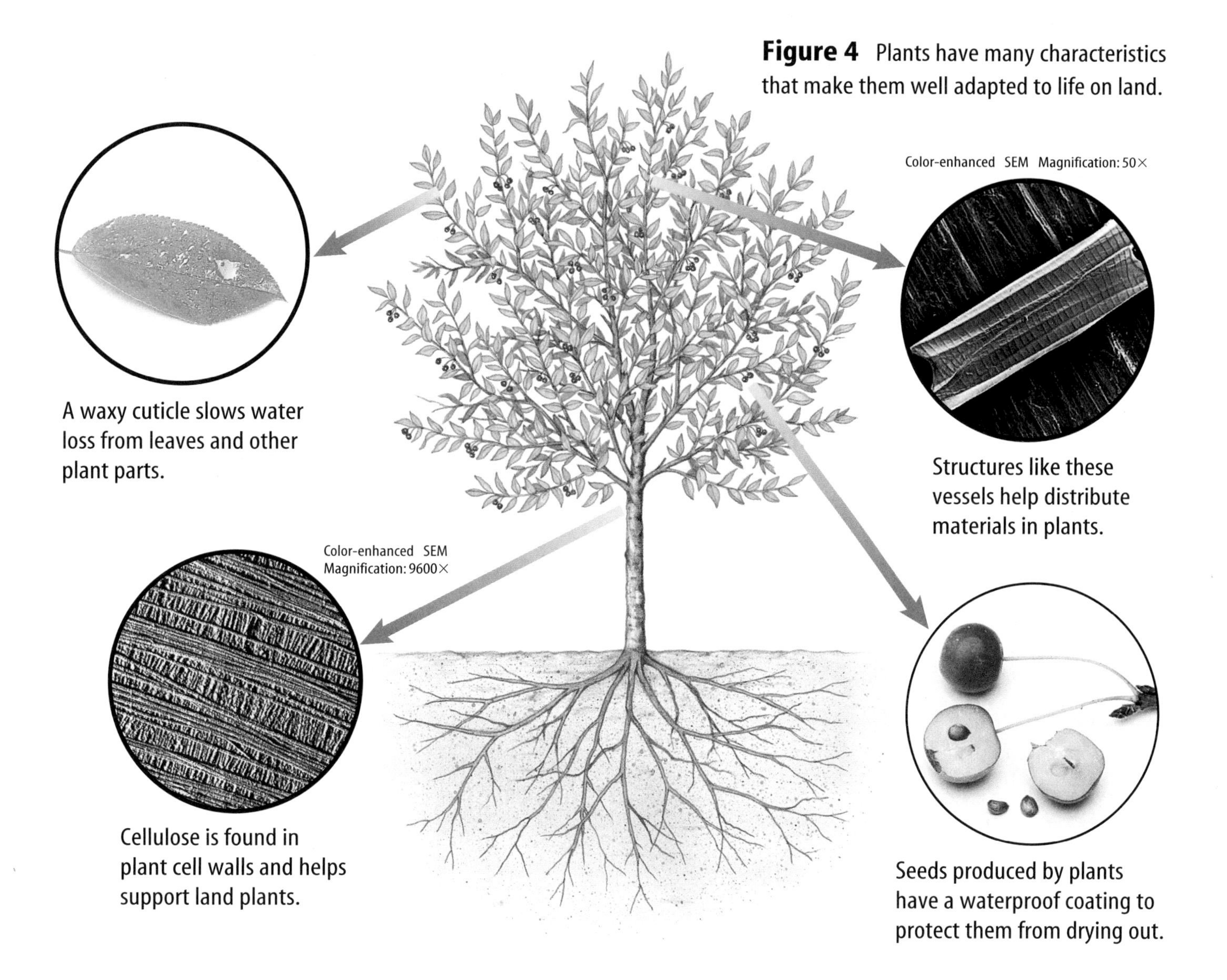

Color-enhanced SEM Magnification: 50×
Color-enhanced SEM
Magnification: 9600×
A waxy cuticle slows water loss from leaves and other plant parts.
Structures like these vessels help distribute materials in plants.
Cellulose is found in plant cell walls and helps support land plants.
Seeds produced by plants have a waterproof coating to protect them from drying out.

Figure 4 Plants have many characteristics that make them well adapted to life on land.

Other Cell Wall Substances Cells of some plants secrete other substances into the cellulose that make the cell wall even stronger. Trees, such as oaks and pines, could not grow without these strong cell walls. Wood from trees can be used for construction mostly because of strong cell walls.

Life on land means that each plant cell is not surrounded by water and dissolved nutrients that can move into the cell. Through adaptations, structures developed in many plants that distribute water, nutrients, and food to all plant cells. These structures also help provide support for the plant.

Reproduction Changes in reproduction were necessary if plants were to survive on land. The presence of water-resistant spores helped some plants reproduce successfully. Other plants adapted by producing water-resistant seeds in cones or in flowers that developed into fruits.

VISUALIZING PLANT CLASSIFICATION

Figure 5

Scientists group plants as either vascular—those with water- and food-conducting cells in their stems—or nonvascular. Vascular plants are further divided into those that produce spores and those that make seeds.

Classification of Plants

Figure 6 Although these two plants are both called daisies, they are not the same species of plant. Using their binomial names helps eliminate the confusion that might come from using their common names.

The plant kingdom is classified into major groups called divisions. A division is the same as a phylum in other kingdoms. Another way to group plants is as vascular (VAS kyuh lur) or nonvascular plants, as illustrated in **Figure 5. Vascular plants** have tubelike structures that carry water, nutrients, and other substances throughout the plant. **Nonvascular plants** do not have these tubelike structures and use other ways to move water and substances.

Naming Plants Why do biologists call a pecan tree *Carya illinoiensis* and a white oak *Quercus alba?* They are using words that accurately name the plant. In the third century B.C., most plants were grouped as trees, shrubs, or herbs and placed into smaller groups by leaf characteristics. This simple system survived until late in the eighteenth century when a Swedish botanist, Carolus Linnaeus, developed a new system. His new system used many characteristics to classify a plant. He also developed a way to name plants called binomial nomenclature (bi NOH mee ul • NOH mun klay chur). Under this system, every plant species is given a unique two-word name like the names above for the pecan tree and white oak and for the two daisies in **Figure 6.**

section 1 review

Summary

What is a plant?

- All plant cells are surrounded by a cell wall.
- Many plant cells contain chlorophyll.

Origin and Evolution of Plants

- Ancestors of land plants were probably ancient green algae.

Adaptations to Land

- A waxy cuticle helps conserve water.
- Cellulose strengthens cell walls.

Classification of Plants

- The plant kingdom is divided into two groups—nonvascular plants and vascular plants.
- Vascular tissues transport nutrients.

Self Check

1. **List** the characteristics of plants.
2. **Compare and contrast** the characteristics of vascular and nonvascular plants.
3. **Identify** three adaptations that allow plants to survive on land.
4. **Explain** why binomial nomenclature is used to name plants.
5. **Thinking Critically** If you left a board lying on the grass for a few days, what would happen to the grass underneath the board? Why?

Applying Skills

6. **Form a hypothesis** about adaptations a land plant might undergo if it lived submerged in water.

section 2

Seedless Plants

as you read

What You'll Learn

- **Distinguish** between characteristics of seedless nonvascular plants and seedless vascular plants.
- **Identify** the importance of some nonvascular and vascular plants.

Why It's Important

Seedless plants are among the first to grow in damaged or disturbed environments and help build soil for the growth of other plants.

Review Vocabulary
spore: waterproof reproductive cell

New Vocabulary
- rhizoid
- pioneer species

Seedless Nonvascular Plants

If you were asked to name the parts of a plant, you probably would list roots, stems, leaves, and flowers. You also might know that many plants grow from seeds. However, some plants, called nonvascular plants, don't grow from seeds and they do not have all of these parts. **Figure 7** shows some common types of nonvascular plants.

Nonvascular plants are usually just a few cells thick and only 2 cm to 5 cm in height. Most have stalks that look like stems and green, leaflike growths. Instead of roots, threadlike structures called **rhizoids** (RI zoydz) anchor them where they grow. Most nonvascular plants grow in places that are damp. Water is absorbed and distributed directly through their cell membranes and cell walls. Nonvascular plants also do not have flowers or cones that produce seeds. They reproduce by spores. Mosses, liverworts, and hornworts are examples of nonvascular plants.

Mosses Most nonvascular plants are classified as mosses, like the ones in **Figure 7.** They have green, leaflike growths arranged around a central stalk. Their rhizoids are made of many cells. Sometimes stalks with caps grow from moss plants. Reproductive cells called spores are produced in the caps of these stalks. Mosses often grow on tree trunks and rocks or the ground. Although they commonly are found in damp areas, some are adapted to living in deserts.

Figure 7 The seedless nonvascular plants include mosses, liverworts, and hornworts.

Figure 8 Mosses can grow in the thin layer of soil that covers these rocks.

Liverworts In the ninth century, liverworts were thought to be useful in treating diseases of the liver. The suffix *-wort* means "herb," so the word *liverwort* means "herb for the liver." Liverworts are rootless plants with flattened, leaflike bodies, as shown in **Figure 7.** They usually have one-celled rhizoids.

Hornworts Most hornworts are less than 2.5 cm in diameter and have a flattened body like liverworts, as shown in **Figure 7.** Unlike other nonvascular plants, almost all hornworts have only one chloroplast in each of their cells. Hornworts get their name from their spore-producing structures, which look like tiny horns of cattle.

INTEGRATE Environment

Nonvascular Plants and the Environment Mosses and liverworts are important in the ecology of many areas. Although they require moist conditions to grow and reproduce, many of them can withstand long, dry periods. They can grow in thin soil and in soils where other plants could not grow, as shown in **Figure 8.**

Spores of mosses and liverworts are carried by the wind. They will grow into plants if growing conditions are right. Mosses often are among the first plants to grow in new or disturbed environments, such as lava fields or after a forest fire. Organisms that are the first to grow in new or disturbed areas are called **pioneer species.** As pioneer plants grow and die, decaying material builds up. This, along with the slow breakdown of rocks, builds soil. When enough soil has formed, other organisms can move into the area.

Why are pioneer plant species important in disturbed environments?

Measuring Water Absorption by a Moss

Procedure

1. Place a few teaspoons of ***Sphagnum* moss** on a piece of **cheesecloth.** Gather the corners of the cloth and twist, then tie them securely to form a ball.
2. Weigh the ball.
3. Put 200 mL of **water** in a **container** and add the ball.
4. After 15 min, remove the ball and drain the excess water into the container.
5. Weigh the ball and measure the amount of water left in the container.
6. Wash your hands after handling the moss.

Analysis

In your **Science Journal,** calculate how much water was absorbed by the *Sphagnum* moss.

Topic: Medicinal Plants

Visit life.msscience.com for Web links to information about plants used as medicines.

Activity In your Science Journal, list four medicinal plants and their uses.

Seedless Vascular Plants

The fern in **Figure 9** is growing next to some moss plants. Ferns and mosses are alike in one way. Both reproduce by spores instead of seeds. However, ferns are different from mosses because they have vascular tissue. The vascular tissue in seedless vascular plants, like ferns, is made up of long, tubelike cells. These cells carry water, minerals, and food to cells throughout the plant. Why is vascular tissue an advantage to a plant? Nonvascular plants like the moss are usually only a few cells thick. Each cell absorbs water directly from its environment. As a result, these plants cannot grow large. Vascular plants, on the other hand, can grow bigger and thicker because the vascular tissue distributes water and nutrients to all plant cells.

Applying Science

What is the value of rain forests?

Throughout history, cultures have used plants for medicines. Some cultures used willow bark to cure headaches. Willow bark contains salicylates (suh LIH suh layts), the main ingredient in aspirin. Heart problems were treated with foxglove, which is the main source of digitalis (dih juh TAH lus), a drug prescribed for heart problems. Have all medicinal plants been identified?

Identifying the Problem

Tropical rain forests have the largest variety of organisms on Earth. Many plant species are still unknown. These forests are being destroyed rapidly. The map below shows the rate of destruction of the rain forests. Some scientists estimate that most tropical rain forests will be destroyed in 30 years.

Solving the Problem

1. What country has the most rain forest destroyed each year?
2. Where can scientists go to study rain forest plants before the plants are destroyed?
3. Predict how the destruction of rain forests might affect research on new drugs from plants.

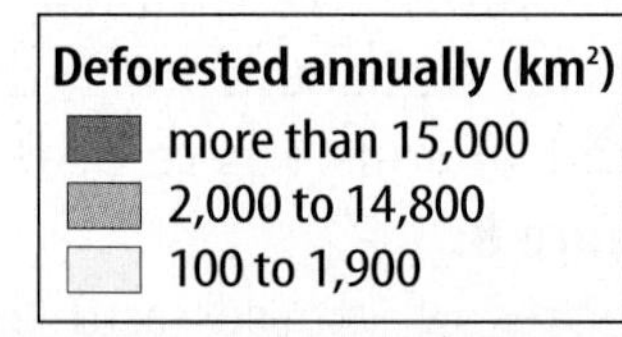

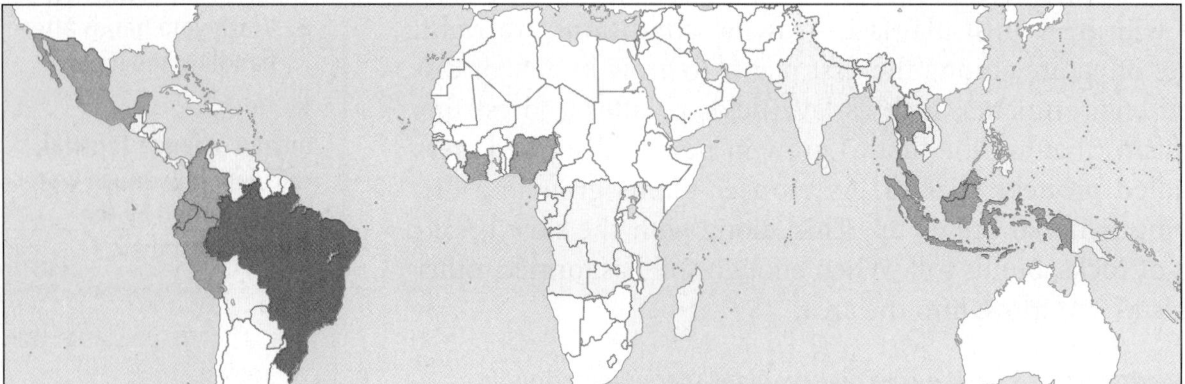

Types of Seedless Vascular Plants

Besides ferns, seedless vascular plants include ground pines, spike mosses, and horsetails. About 1,000 species of ground pines, spike mosses, and horsetails are known to exist. Ferns are more abundant, with at least 12,000 known species. Many species of seedless vascular plants are known only from fossils. They flourished during the warm, moist period 360 million to 286 million years ago. Fossil records show that some horsetails grew 15 m tall, unlike modern species, which grow only 1 m to 2 m tall.

Figure 9 The mosses and ferns pictured here are seedless plants. **Explain** *why the fern can grow taller than the moss.*

Ferns The largest group of seedless vascular plants is the ferns. They include many different forms, as shown in **Figure 10.** They have stems, leaves, and roots. Fern leaves are called fronds. Ferns produce spores in structures that usually are found on the underside of their fronds. Thousands of species of ferns now grow on Earth, but many more existed long ago. From clues left in rock layers, scientists infer that about 360 million years ago much of Earth was tropical. Steamy swamps covered large areas. The tallest plants were species of ferns. The ancient ferns grew as tall as 25 m—as tall as the tallest fern species alive today. Most modern tree ferns are about 3 m to 5 m in height and grow in tropical regions of the world.

Figure 10 Ferns come in many different shapes and sizes.

The sword fern has a typical fern shape. Spores are produced in structures on the back of the frond.

This fern grows on other plants, not in the soil.
Infer *why it's called the staghorn fern.*

Tree ferns, like this one in Hawaii, grow in tropical areas.

Figure 11 Photographers once used the dry, flammable spores of club mosses as flash powder. It burned rapidly and produced the light that was needed to take photographs.

Club Mosses Ground pines and spike mosses are groups of plants that often are called club mosses. They are related more closely to ferns than to mosses. These seedless vascular plants have needle-like leaves. Spores are produced at the end of the stems in structures that look like tiny pine cones. Ground pines, shown in **Figure 11,** are found from arctic regions to the tropics, but rarely in large numbers. In some areas, they are endangered because they have been over collected to make wreaths and other decorations.

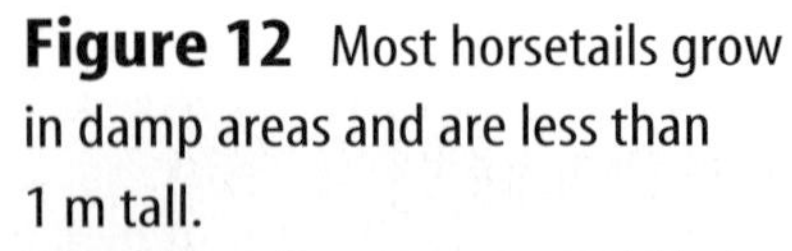

Where are spores in club mosses produced?

Spike mosses resemble ground pines. One species of spike moss, the resurrection plant, is adapted to desert conditions. When water is scarce, the plant curls up and seems dead. When water becomes available, the resurrection plant unfurls its green leaves and begins making food again. The plant can repeat this process whenever necessary.

Horsetails The stem structure of horsetails is unique among the vascular plants. The stem is jointed and has a hollow center surrounded by a ring of vascular tissue. At each joint, leaves grow out from around the stem. In **Figure 12,** you can see these joints. If you pull on a horsetail stem, it will pop apart in sections. Like the club mosses, spores from horsetails are produced in a conelike structure at the tips of some stems. The stems of the horsetails contain silica, a gritty substance found in sand. For centuries, horsetails have been used for polishing objects, sharpening tools, and scouring cooking utensils. Another common name for horsetails is scouring rush.

Figure 12 Most horsetails grow in damp areas and are less than 1 m tall.
Identify *where spores would be produced on this plant.*

Importance of Seedless Plants

When many ancient seedless plants died, they became submerged in water and mud before they decomposed. As this plant material built up, it became compacted and compressed and eventually turned into coal—a process that took millions of years.

Today, a similar process is taking place in bogs, which are poorly drained areas of land that contain decaying plants. The plants in bogs are mostly seedless plants like mosses and ferns.

Peat When bog plants die, the waterlogged soil slows the decay process. Over time, these decaying plants are compressed into a substance called peat. Peat, which forms from the remains of sphagnum moss, is mined from bogs to use as a low-cost fuel in places such as Ireland and Russia, as shown in **Figure 13.** Peat supplies about one-third of Ireland's energy requirements. Scientists hypothesize that over time, if additional layers of soil bury, compact, and compress the peat, it will become coal.

Figure 13 Peat is cut from bogs and used for a fuel in some parts of Europe.

Uses of Seedless Vascular Plants Many people keep ferns as houseplants. Ferns also are sold widely as landscape plants for shady areas. Peat and sphagnum mosses also are used for gardening. Peat is an excellent soil conditioner, and sphagnum moss often is used to line hanging baskets. Ferns also are used as weaving material for basketry.

Although most mosses are not used for food, parts of many other seedless vascular plants can be eaten. The rhizomes and young fronds of some ferns are edible. The dried stems of one type of horsetail can be ground into flour. Seedless plants have been used as folk medicines for hundreds of years. For example, ferns have been used to treat bee stings, burns, fevers, and even dandruff.

section 2 review

Summary

Seedless Nonvascular Plants

- Seedless nonvascular plants include mosses, liverworts, and hornworts.
- They are usually only a few cells thick and no more than a few centimeters tall.
- They produce spores rather than seeds.

Seedless Vascular Plants

- Seedless vascular plants include ferns, club mosses, and horsetails.
- Vascular plants grow taller and can live farther from water than nonvascular plants.

Importance of Seedless Plants

- Nonvascular plants help build new soil.
- Coal deposits formed from ancient seedless plants that were buried in water and mud before they began to decay.

Self Check

1. **Compare and contrast** the characteristics of mosses and ferns.
2. **Explain** what fossil records tell about seedless plants that lived on Earth long ago.
3. **Identify** growing conditions in which you would expect to find pioneer plants such as mosses and liverworts.
4. **Summarize** the functions of vascular tissues.
5. **Think Critically** The electricity that you use every day might be produced by burning coal. What is the connection between electricity production and seedless vascular plants?

Applying Math

6. **Use Fractions** Approximately 8,000 species of liverworts and 9,000 species of mosses exist today. Estimate what fraction of these seedless nonvascular plants are mosses.

section 3

Seed Plants

as you read

What You'll Learn

- **Identify** the characteristics of seed plants.
- **Explain** the structures and functions of roots, stems, and leaves.
- **Describe** the main characteristics and importance of gymnosperms and angiosperms.
- **Compare** similarities and differences between monocots and dicots.

Why It's Important

Humans depend on seed plants for food, clothing, and shelter.

Review Vocabulary

seed: plant embryo and food supply in a protective coating

New Vocabulary

- stomata
- guard cell
- xylem
- phloem
- cambium
- gymnosperm
- angiosperm
- monocot
- dicot

Characteristics of Seed Plants

What foods from plants have you eaten today? Apples? Potatoes? Carrots? Peanut butter and jelly sandwiches? All of these foods and more come from seed plants.

Most of the plants you are familiar with are seed plants. Most seed plants have leaves, stems, roots, and vascular tissue. They also produce seeds, which usually contain an embryo and stored food. The stored food is the source of energy for the embryo's early growth as it develops into a plant. Most of the plant species that have been identified in the world today are seed plants. The seed plants generally are classified into two major groups—gymnosperms (JIHM nuh spurmz) and angiosperms (AN jee uh spurmz).

Leaves Most seed plants have leaves. Leaves are the organs of the plant where the food-making process—photosynthesis—usually occurs. Leaves come in many shapes, sizes, and colors. Examine the structure of a typical leaf, shown in **Figure 14.**

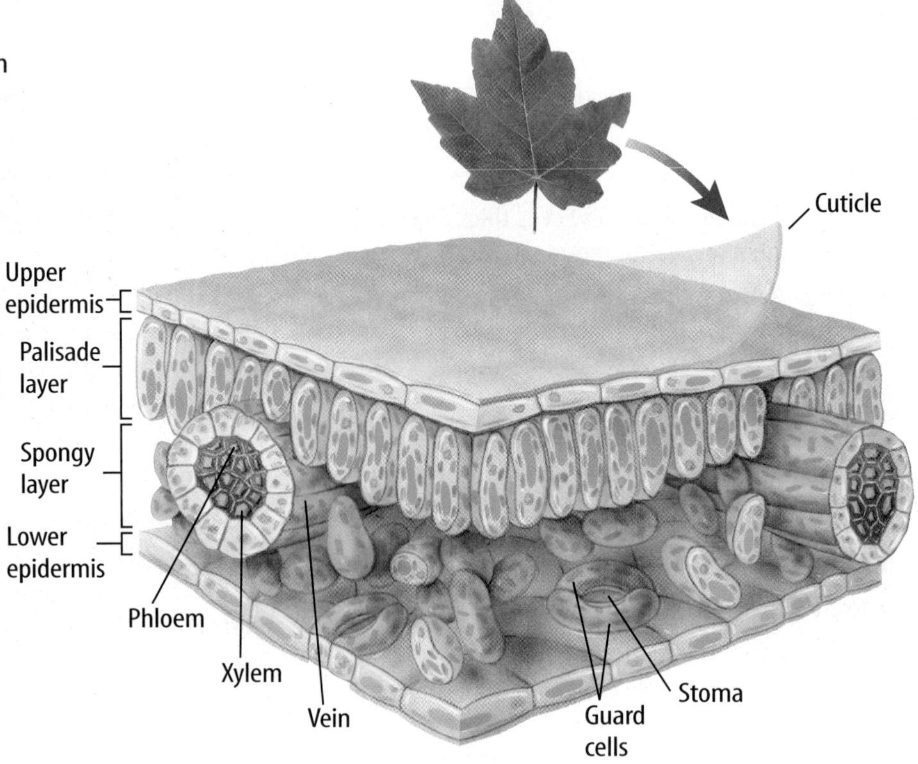

Figure 14 The structure of a typical leaf is adapted for photosynthesis. **Explain** *why cells in the palisade layer have more chloroplasts than cells in the spongy layer.*

Leaf Cell Layers A typical leaf is made of several different layers of cells. On the upper and lower surfaces of a leaf is a thin layer of cells called the epidermis, which covers and protects the leaf. A waxy cuticle coats the epidermis of some leaves. Most leaves have small openings in the epidermis called **stomata** (STOH muh tuh) (singular, *stoma*). Stomata allow carbon dioxide, water, and oxygen to enter into and exit from a leaf. Each stoma is surrounded by two **guard cells** that open and close it.

Just below the upper epidermis is the palisade layer. It consists of closely packed, long, narrow cells that usually contain many chloroplasts. Most of the food produced by plants is made in the palisade cells. Between the palisade layer and the lower epidermis is the spongy layer. It is a layer of loosely arranged cells separated by air spaces. In a leaf, veins containing vascular tissue are found in the spongy layer.

Stems The trunk of a tree is really the stem of the tree. Stems usually are located above ground and support the branches, leaves, and reproductive structures. Materials move between leaves and roots through the vascular tissue in the stem. Stems also can have other functions, as shown in **Figure 15.**

Plant stems are either herbaceous (hur BAY shus) or woody. Herbaceous stems usually are soft and green, like the stems of a tulip, while trees and shrubs have hard, rigid, woody stems. Lumber comes from woody stems.

Observing Water Moving in a Plant

Procedure

1. Into a **clear container** pour **water** to a depth of 1.5 cm. Add 25 drops of **red food coloring** to the water.
2. Put the root end of a **green onion** into the container. Do not cut the onion in any way. Wash your hands.
3. The next day, examine the outside of the onion. Peel off the onion's layers and examine them. **WARNING:** *Do not eat the onion.*

Analysis

In your **Science Journal,** infer how the location of red color inside the onion might be related to vascular tissue.

Try at Home

Figure 15 Some plants have stems with special functions.

These potatoes are stems that grow underground and store food for the plant.

The stems of this cactus store water and can carry on photosynthesis.

Some stems of this grape plant help it climb on other plants.

Roots Imagine a lone tree growing on top of a hill. What is the largest part of this plant? Maybe you guessed the trunk or the branches. Did you consider the roots, like those shown in **Figure 16?** The root systems of most plants are as large or larger than the aboveground stems and leaves.

Figure 16 The root system of a tree can be as long as the tree can be tall.
Infer *why the root system of a tree would need to be so large.*

Roots are important to plants. Water and other substances enter a plant through its roots. Roots have vascular tissue in which water and dissolved substances move from the soil through the stems to the leaves. Roots also act as anchors, preventing plants from being blown away by wind or washed away by moving water. Underground root systems support other plant parts that are aboveground—the stem, branches, and leaves of a tree. Sometimes, part of or all of the roots are aboveground, too.

Roots can store food. When you eat carrots or beets, you eat roots that contain stored food. Plants that continue growing from one year to the next use this stored food to begin new growth in the spring. Plants that grow in dry areas often have roots that store water.

Root tissues also can perform functions such as absorbing oxygen that is used in the process of cellular respiration. Because water does not contain as much oxygen as air does, plants that grow with their roots in water might not be able to absorb enough oxygen. Some swamp plants have roots that grow partially out of the water and take in oxygen from the air. In order to perform all these functions, the root systems of plants must be large.

Reading Check *What are several functions of roots in plants?*

Vascular Tissue Three tissues usually make up the vascular system in a seed plant. **Xylem** (ZI lum) tissue is made up of hollow, tubular cells that are stacked one on top of the other to form a structure called a vessel. These vessels transport water and dissolved substances from the roots throughout the plant. The thick cell walls of xylem are also important because they help support the plant.

Phloem (FLOH em) is a plant tissue also made up of tubular cells that are stacked to form structures called tubes. Tubes are different from vessels. Phloem tubes move food from where it is made to other parts of the plant where it is used or stored.

In some plants, a cambium is between xylem and phloem. **Cambium** (KAM bee um) is a tissue that produces most of the new xylem and phloem cells. The growth of this new xylem and phloem increases the thickness of stems and roots. All three tissues are illustrated in **Figure 17.**

INTEGRATE
Health

Vascular Systems Plants have vascular tissue, and you have a vascular system. Your vascular system transports oxygen, food, and wastes through blood vessels. Instead of xylem and phloem, your blood vessels include veins and arteries. In your Science Journal write a paragraph describing the difference between veins and arteries.

Figure 17 The vascular tissue of some seed plants includes xylem, phloem, and cambium. **Identify** *which of these tissues transports food throughout the plant.*

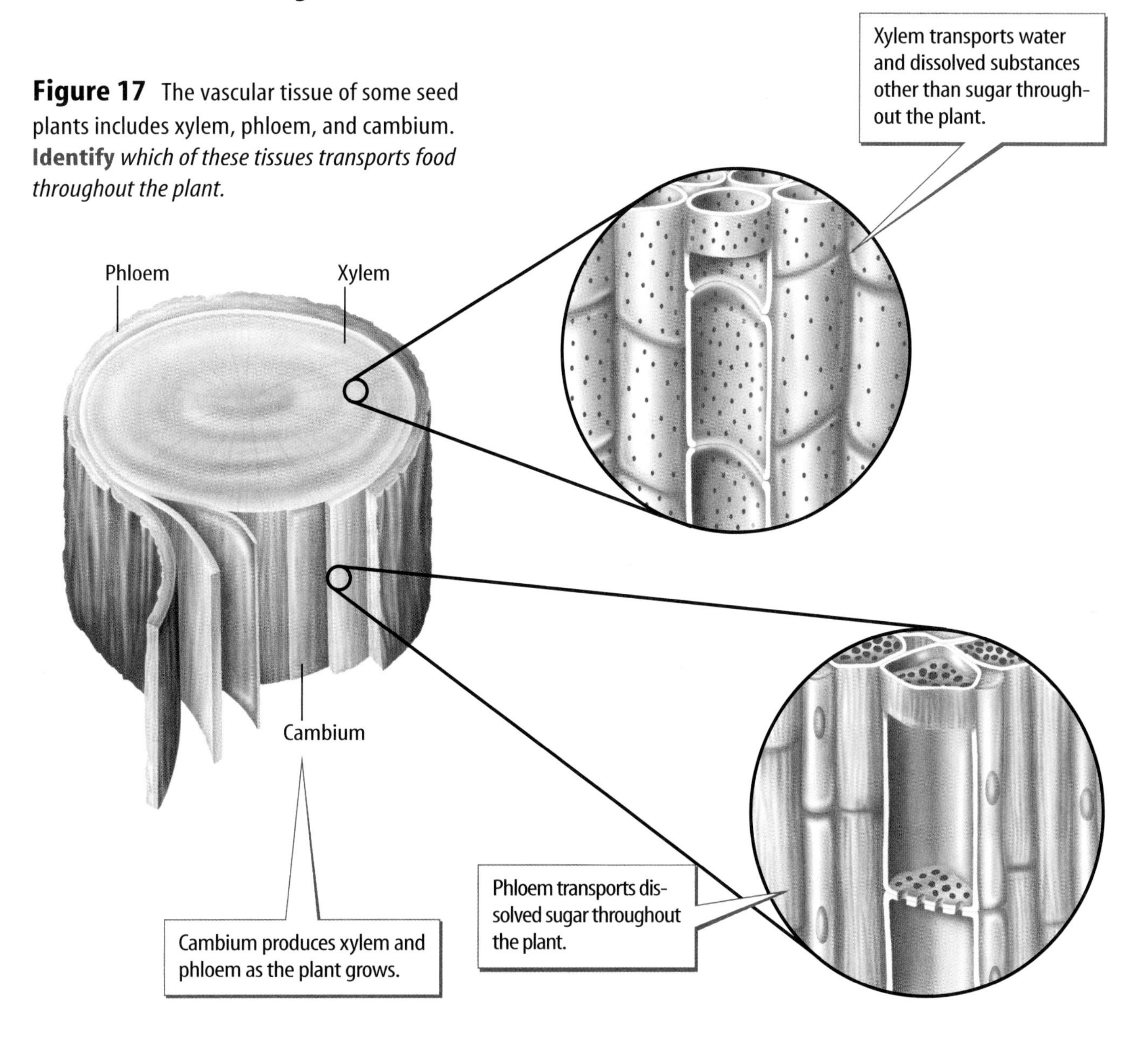
Xylem transports water and dissolved substances other than sugar throughout the plant.
Phloem
Xylem
Cambium
Cambium produces xylem and phloem as the plant grows.
Phloem transports dissolved sugar throughout the plant.

Figure 18 The gymnosperms include four divisions of plants.

About 100 species of cycads exist today. Only one genus is native to the United States.

The ginkgoes are represented by one living species. Ginkgoes lose their leaves in the fall. **Explain** *how this is different from most gymnosperms.*

Conifers are the largest, most diverse division. Most conifers are evergreen plants, such as this ponderosa pine (above).

More than half of the 70 species of gnetophytes, such as this joint fir, are in one genus.

Gymnosperms

The oldest trees alive are gymnosperms. A bristlecone pine tree in the White Mountains of eastern California is estimated to be 4,900 years old. **Gymnosperms** are vascular plants that produce seeds that are not protected by fruit. The word *gymnosperm* comes from the Greek language and means "naked seed." Another characteristic of gymnosperms is that they do not have flowers. Leaves of most gymnosperms are needlelike or scalelike. Many gymnosperms are called evergreens because some green leaves always remain on their branches.

Four divisions of plants—conifers, cycads, ginkgoes, and gnetophytes (NE tuh fites)—are classified as gymnosperms. **Figure 18** shows examples of the four divisions. You are probably most familiar with the division Coniferophyta (kuh NIH fur uh fi tuh), the conifers. Pines, firs, spruces, redwoods, and junipers belong to this division. It contains the greatest number of gymnosperm species. All conifers produce two types of cones—male and female. Both types usually are found on the same plant. Cones are the reproductive structures of conifers. Seeds develop on the female cone but not on the male cone.

What is the importance of cones to gymnosperms?

Angiosperms

When people are asked to name a plant, most name an angiosperm. An **angiosperm** is a vascular plant that flowers and produces fruits with one or more seeds, such as the peaches shown in **Figure 19.** The fruit develops from a part or parts of one or more flowers. Angiosperms are familiar plants no matter where you live. They grow in parks, fields, forests, jungles, deserts, freshwater, salt water, and in the cracks of sidewalks. You might see them dangling from wires or other plants, and one species of orchid even grows underground. Angiosperms make up the plant division Anthophyta (AN thoh fi tuh). More than half of the known plant species belong to this division.

Flowers The flowers of angiosperms vary in size, shape, and color. Duckweed, an aquatic plant, has a flower that is only 0.1 mm long. A plant in Indonesia has a flower that is nearly 1 m in diameter and can weigh 9 kg. Nearly every color can be found in some flower, although some people would not include black. Multicolored flowers are common. Some plants have flowers that are not recognized easily as flowers, such as the flowers of ash trees, shown below.

Some flower parts develop into a fruit. Most fruits contain seeds, like an apple, or have seeds on their surface, like a strawberry. If you think all fruits are juicy and sweet, there are some that are not. The fruit of the vanilla orchid, as shown to the right, contains seeds and is dry.

Angiosperms are divided into two groups—the monocots and the dicots—shortened forms of the words *monocotyledon* (mah nuh kah tuh LEE dun) and *dicotyledon* (di kah tuh LEE dun).

Figure 19 Angiosperms have a wide variety of flowers and fruits.

The fruit of the vanilla orchid is the source of vanilla flavoring.

The flowers and fruit of a peach tree are typical of many angiosperms.

Ash flowers are not large and colorful. Their fruits are small and dry.

Monocots and Dicots A cotyledon is part of a seed often used for food storage. The prefix *mono* means "one," and *di* means "two." Therefore, **monocots** have one cotyledon inside their seeds and **dicots** have two. The flowers, leaves, and stems of monocots and dicots are shown in **Figure 20.**

Many important foods come from monocots, including corn, rice, wheat, and barley. If you eat bananas, pineapple, or dates, you are eating fruit from monocots. Lilies and orchids also are monocots.

Dicots also produce familiar foods such as peanuts, green beans, peas, apples, and oranges. You might have rested in the shade of a dicot tree. Most shade trees, such as maple, oak, and elm, are dicots.

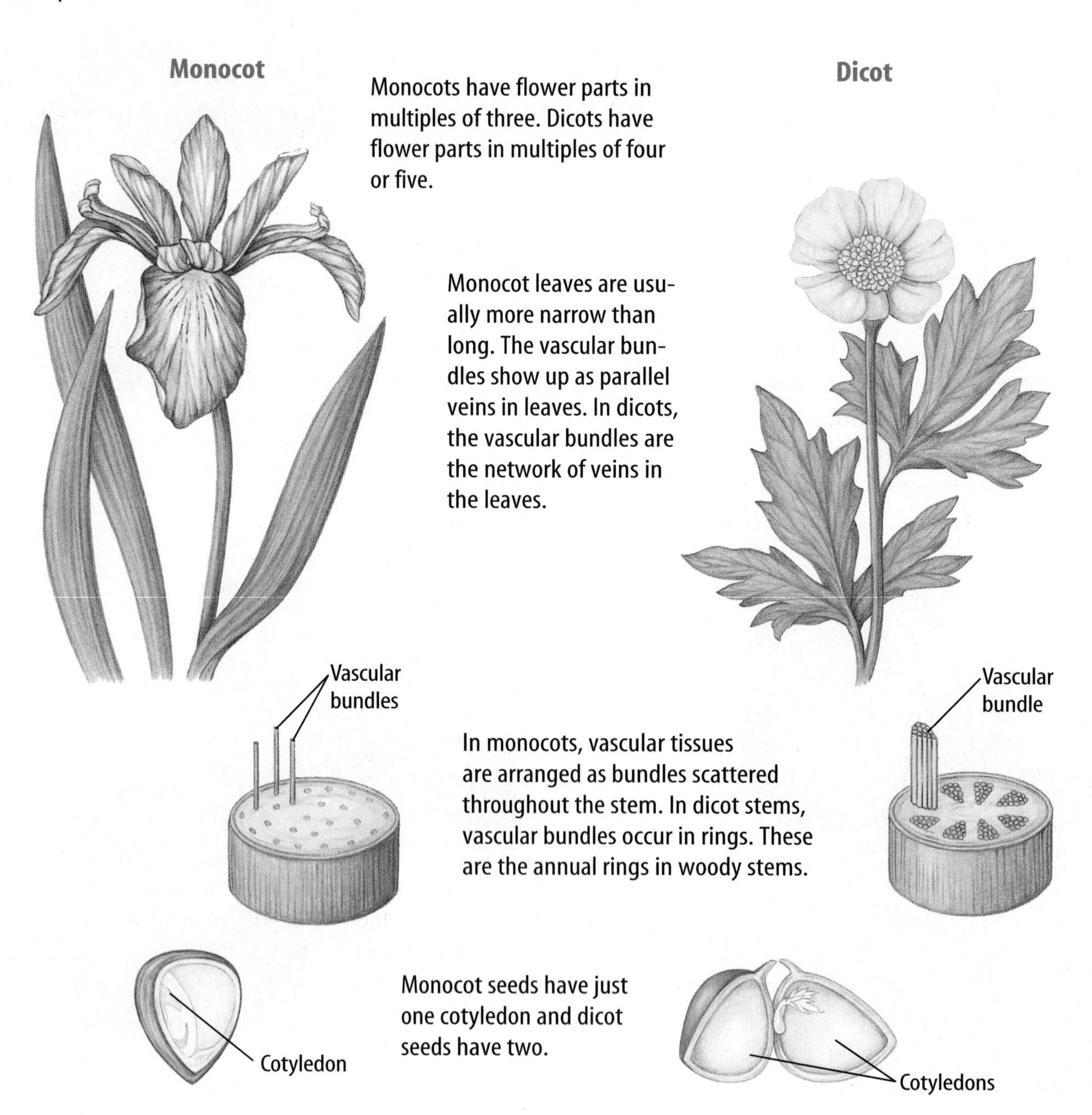

Monocot
Dicot
Monocots have flower parts in multiples of three. Dicots have flower parts in multiples of four or five.
Monocot leaves are usually more narrow than long. The vascular bundles show up as parallel veins in leaves. In dicots, the vascular bundles are the network of veins in the leaves.
Vascular bundles
Vascular bundle
In monocots, vascular tissues are arranged as bundles scattered throughout the stem. In dicot stems, vascular bundles occur in rings. These are the annual rings in woody stems.
Cotyledon
Monocot seeds have just one cotyledon and dicot seeds have two.
Cotyledons

Figure 20 By observing a monocot and a dicot, you can determine their plant characteristics.

Petunias

Parsley

Pecan tree

Figure 21 Life cycles of angiosperms include annuals, biennials, and perennials. Petunias, which are annuals, complete their life cycle in one year. Parsley plants, which are biennials, do not produce flowers and seeds the first year. Perennials, such as the pecan tree, flower and produce fruits year after year.

Life Cycles of Angiosperms Flowering plants vary greatly in appearance. Their life cycles are as varied as the kinds of plants, as shown in **Figure 21.** Some angiosperms grow from seeds to mature plants with their own seeds in less than a month. The life cycles of other plants can take as long as a century. If a plant's life cycle is completed within one year, it is called an annual. These plants must be grown from seeds each year.

Plants called biennials (bi EH nee ulz) complete their life cycles within two years. Biennials such as parsley store a large amount of food in an underground root or stem for growth in the second year. Biennials produce flowers and seeds only during the second year of growth. Angiosperms that take more than two years to grow to maturity are called perennials. Herbaceous perennials such as peonies appear to die each winter but grow and produce flowers each spring. Woody perennials such as fruit trees produce flowers and fruits on stems that survive for many years.

Importance of Seed Plants

What would a day at school be like without seed plants? One of the first things you'd notice is the lack of paper and books. Paper is made from wood pulp that comes from trees, which are seed plants. Are the desks and chairs at your school made of wood? They would need to be made of something else if no seed plants existed. Clothing that is made from cotton would not exist because cotton comes from seed plants. At lunchtime, you would have trouble finding something to eat. Bread, fruits, and potato chips all come from seed plants. Milk, hamburgers, and hot dogs all come from animals that eat seed plants. Unless you like to eat plants such as mosses and ferns, you'd go hungry. Without seed plants, your day at school would be different.

Topic: Renewable Resources
Visit life.msscience.com for Web links to information and recent news or magazine articles about the timber industry's efforts to replant trees.

Activity List in your Science Journal the species of trees that are planted and some of their uses.

Table 1 Some Products of Seed Plants

From Gymnosperms		From Angiosperms	
lumber, paper, soap, varnish, paints, waxes, perfumes, edible pine nuts, medicines		foods, sugar, chocolate, cotton cloth, linen, rubber, vegetable oils, perfumes, medicines, cinnamon, flavorings, dyes, lumber	

Products of Seed Plants Conifers are the most economically important gymnosperms. Most wood used for construction and for paper production comes from conifers. Resin, a waxy substance secreted by conifers, is used to make chemicals found in soap, paint, varnish, and some medicines.

The most economically important plants on Earth are the angiosperms. They form the basis of the diets of most animals. Angiosperms were the first plants that humans grew. They included grains, such as barley and wheat, and legumes, such as peas and lentils. Angiosperms are also the source of many of the fibers used in clothing. Besides cotton, linen fabrics come from plant fibers. **Table 1** shows just a few of the products of angiosperms and gymnosperms.

section 3 review

Summary

Characteristics of Seed Plants

- Leaves are organs in which photosynthesis takes place.
- Stems support leaves and branches and contain vascular tissues.
- Roots absorb water and nutrients from soil.

Gymnosperms

- Gymnosperms do not have flowers and produce seeds that are not protected by a fruit.

Angiosperms

- Angiosperms produce flowers that develop into a fruit with seeds.

Importance of Seed Plants

- The diets of most animals are based on angiosperms.

Self Check

1. **List** four characteristics common to all seed plants.
2. **Compare and contrast** the characteristics of gymnosperms and angiosperms.
3. **Classify** a flower with five petals as a monocot or a dicot.
4. **Explain** why the root system might be the largest part of a plant.
5. **Think Critically** The cuticle and epidermis of leaves are transparent. If they weren't, what might be the result?

Applying Skills

6. **Form a hypothesis** about what substance or substances are produced in palisade cells but not in xylem cells.

Science Online life.msscience.com/self_check_quiz

Identifying Conifers

How can you tell a pine from a spruce or a cedar from a juniper? One way is to observe their leaves. The leaves of most conifers are either needlelike—shaped like needles—or scale-like—shaped like the scales on a fish. Examine and identify some conifer branches using the key to the right.

Real-World Question

How can leaves be used to classify conifers?

Goals

- **Identify** the difference between needlelike and scalelike leaves.
- **Classify** conifers according to their leaves.

Materials

short branches of the following conifers:

pine	Douglas fir	redwood
cedar	hemlock	arborvitae
spruce	fir	juniper

**illustrations of the conifers above*

**Alternate materials*

Safety Precautions

Wash your hands after handling leaves.

Procedure

1. **Observe** the leaves or illustrations of each conifer, then use the key to identify it.
2. **Write** the number and name of each conifer you identify in your Science Journal.

Conclude and Apply

1. **Name** two traits of hemlock leaves.
2. **Compare and contrast** pine and cedar leaves.

Key to Classifying Conifer Leaves

1. All leaves are needlelike.
   a. yes, go to 2
   b. no, go to 8
2. Needles are in clusters.
   a. yes, go to 3
   b. no, go to 4
3. Clusters contain two, three, or five needles.
   a. yes, pine
   b. no, cedar
4. Needles grow on all sides of the stem.
   a. yes, go to 5
   b. no, go to 7
5. Needles grow from a woody peg.
   a. yes, spruce
   b. no, go to 6
6. Needles appear to grow from the branch.
   a. yes, Douglas fir
   b. no, hemlock
7. Most of the needles grow upward.
   a. yes, fir
   b. no, redwood
8. All the leaves are scalelike but not prickly.
   a. yes, arborvitae
   b. no, juniper

Communicating Your Data

Use the key above to identify conifers growing on your school grounds. Draw and label a map that locates these conifers. Post the map in your school. **For more help, refer to the Science Skill Handbook.**

Use the Internet

Plants as Medicine

Goals

- **Identify** two plants that can be used as a treatment for illness or as a supplement to support good health.
- **Research** the cultural and historical use of each of the two selected plants as medical treatments.
- **Review** multiple sources to understand the effectiveness of each of the two selected plants as a medical treatment.
- **Compare and contrast** the research and form a hypothesis about the medicinal effectiveness of each of the two plants.

Data Source

Science Online

Visit life.msscience.com/internet_lab for more information about plants that can be used for maintaining good health and for data collected by other students.

Real-World Question

You may have read about using peppermint to relieve an upset stomach, or taking *Echinacea* to boost your immune system and fight off illness. But did you know that pioneers brewed a cough medicine from lemon mint? In this lab, you will explore plants and their historical use in treating illness, and the benefits and risks associated with using plants as medicine. How are plants used in maintaining good health?

Echinacea

Make a Plan

1. Search for information about plants that are used as medicine and identify two plants to investigate.
2. **Research** how these plants are currently recommended for use as medicine or to promote good health. Find out how each has been used historically.
3. **Explore** how other cultures used these plants as a medicine.

Mentha

Follow Your Plan

1. Make sure your teacher approves your plan before you start.
2. **Record** data you collect about each plant in your Science Journal.

Analyze Your Data

1. **Write** a description of how different cultures have used each plant as medicine.
2. How have the plants you investigated been used as medicine historically?
3. **Record** all the uses suggested by different sources for each plant.
4. **Record** the side effects of using each plant as a treatment.

Conclude and Apply

1. After conducting your research, what do you think are the benefits and drawbacks of using these plants as alternative medicines?
2. **Describe** any conflicting information about using each of these plants as medicine.
3. Based on your analysis, would you recommend the use of each of these two plants to treat illness or promote good health? Why or why not?
4. What would you say to someone who was thinking about using any plant-based, over-the-counter, herbal supplement?

Communicating Your Data

Find this lab using the link below. Post your data for the two plants you investigated in the tables provided. **Compare** your data to those of other students. Review data that other students have entered about other plants that can be used as medicine.

life.msscience.com/internet_lab

Monarda

A LOOPY Idea Inspires a "Fastenating" Invention

A wild cocklebur plant inspired the hook-and-loop fastener.

Scientists often spend countless hours in the laboratory dreaming up useful inventions. Sometimes, however, the best ideas hit them in unexpected places at unexpected times. That's why scientists are constantly on the lookout for things that spark their curiosity.

One day in 1948, a Swiss inventor named George deMestral strolled through a field with his dog. When they returned home, deMestral discovered that the dog's fur was covered with cockleburs, parts of a prickly plant. These burs were also stuck to deMestral's jacket and pants. Curious about what made the burs so sticky, the inventor examined one under a microscope.

DeMestral noticed that the cocklebur was covered with lots of tiny hooks. By clinging to animal fur and fabric, this plant is carried to other places. While studying these burs, he got the idea to invent a new kind of fastener that could do the work of buttons, snaps, zippers, and laces—but better!

After years of experimentation, deMestral came up with a strong, durable hook-and-loop fastener made of two strips of nylon fabric. One strip has thousands of small, stiff hooks; the other strip is covered with soft, tiny loops. Today, this hook-and-loop fastening tape is used on shoes and sneakers, watchbands, hospital equipment, space suits, clothing, book bags, and more. You may have one of those hook-and-loop fasteners somewhere on you right now. They're the ones that go rrrrrrrrip when you open them.

So, if you ever get a fresh idea that clings to your mind like a hook to a loop, stick with it and experiment! Who knows? It may lead to a fabulous invention that changes the world!

This photo provides a close-up view of a hook-and-loop fastener.

List **Make a list of ten ways hook-and-loop tape is used today. Think of three new uses for it. Since you can buy strips of hook-and-loop fastening tape in most hardware and fabric stores, try out some of your favorite ideas.**

For more information, visit life.msscience.com/oops

Reviewing Main Ideas

Section 1 An Overview of Plants

1. Plants are made up of eukaryotic cells and vary greatly in size and shape.
2. Plants usually have some form of leaves, stems, and roots.
3. As plants evolved from aquatic to land environments, changes occurred in how they reproduced, supported themselves, and moved substances from one part of the plant to another.
4. The plant kingdom is classified into groups called divisions.

Section 2 Seedless Plants

1. Seedless plants include nonvascular and vascular types.
2. Most seedless nonvascular plants have no true leaves, stems, or roots. Reproduction usually is by spores.
3. Seedless vascular plants have vascular tissues that move substances throughout the plant. These plants may reproduce by spores.
4. Many ancient forms of these plants underwent a process that resulted in the formation of coal.

Section 3 Seed Plants

1. Seed plants are adapted to survive in nearly every environment on Earth.
2. Seed plants produce seeds and have vascular tissue, stems, roots, and leaves.
3. The two major groups of seed plants are gymnosperms and angiosperms. Gymnosperms generally have needlelike leaves and some type of cone. Angiosperms are plants that flower and are classified as monocots or dicots.
4. Seed plants are the most economically important plants on Earth.

Visualizing Main Ideas

Copy and complete the following concept map about the seed plants.

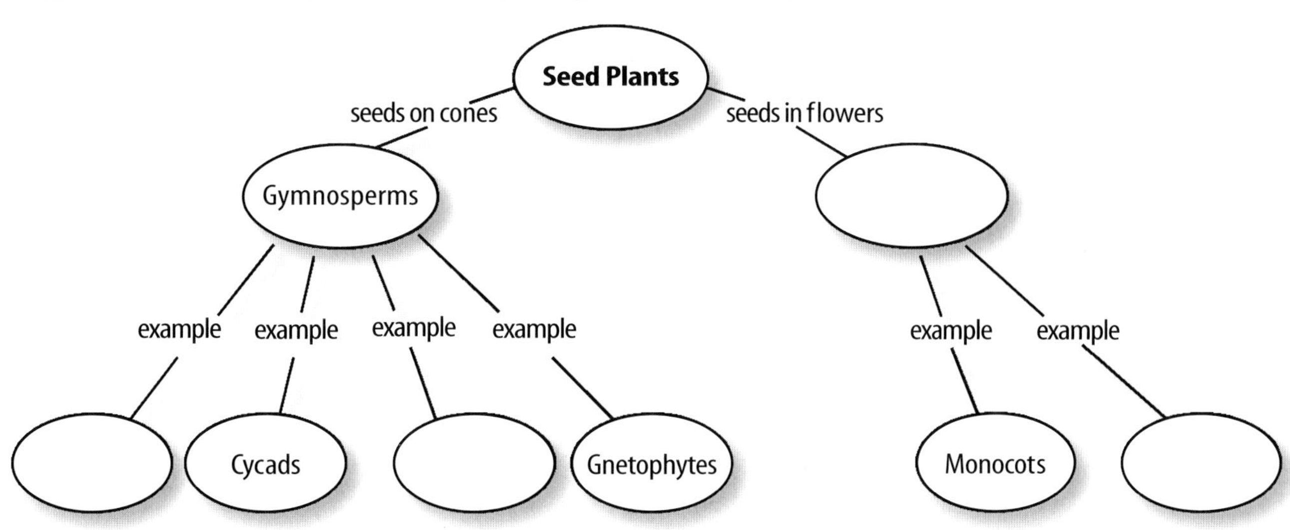

Using Vocabulary

angiosperm p. 257
cambium p. 255
cellulose p. 242
cuticle p. 242
dicot p. 258
guard cell p. 253
gymnosperm p. 256
monocot p. 258
nonvascular plant p. 245
phloem p. 255
pioneer species p. 247
rhizoid p. 246
stomata p. 253
vascular plant p. 245
xylem p. 255

Complete each analogy by providing the missing vocabulary word.

1. Angiosperm is to flower as _________ is to cone.
2. Dicot is to two seed leaves as _________ is to one seed leaf.
3. Root is to fern as _________ is to moss.
4. Phloem is to food transport as _________ is to water transport.
5. Vascular plant is to horsetail as _________ is to liverwort.
6. Cellulose is to support as _________ is to protect.
7. Fuel is to ferns as _________ is to bryophytes.
8. Cuticle is to wax as _________ is to fibers.

Checking Concepts

Choose the word or phrase that best answers the question.

9. Which of the following is a seedless vascular plant?
   A) moss **C)** horsetail
   B) liverwort **D)** pine
10. What are the small openings in the surface of a leaf surrounded by guard cells called?
    A) stomata **C)** rhizoids
    B) cuticles **D)** angiosperms
11. What are the plant structures that anchor the plant called?
    A) stems **C)** roots
    B) leaves **D)** guard cells
12. Where is most of a plant's new xylem and phloem produced?
    A) guard cell **C)** stomata
    B) cambium **D)** cuticle
13. What group has plants that are only a few cells thick?
    A) gymnosperms **C)** ferns
    B) cycads **D)** mosses
14. The oval plant parts shown to the right are found only in which plant group?

    A) nonvascular **C)** gymnosperms
    B) seedless **D)** angiosperms
15. What kinds of plants have structures that move water and other substances?
    A) vascular **C)** nonvascular
    B) protist **D)** bacterial
16. In what part of a leaf does most photosynthesis occur?
    A) epidermis **C)** stomata
    B) cuticle **D)** palisade layer
17. Which one of the following do ferns have?
    A) cones **C)** spores
    B) rhizoids **D)** seeds
18. Which of these is an advantage to life on land for plants?
    A) more direct sunlight
    B) less carbon dioxide
    C) greater space to grow
    D) less competition for food

Science Online life.msscience.com/vocabulary_puzzlemaker

Thinking Critically

19. **Predict** what might happen if a land plant's waxy cuticle was destroyed.
20. **Draw Conclusions** On a walk through the woods with a friend, you find a plant neither of you has seen before. The plant has green leaves and yellow flowers. Your friend says it is a vascular plant. How does your friend know this?
21. **Infer** Plants called succulents store large amounts of water in their leaves, stems, and roots. In what environments would you expect to find succulents growing naturally?
22. **Explain** why mosses usually are found in moist areas.
23. **Recognize Cause and Effect** How do pioneer species change environments so that other plants can grow there?
24. **Concept Map** Copy and complete this map for the seedless plants of the plant kingdom.

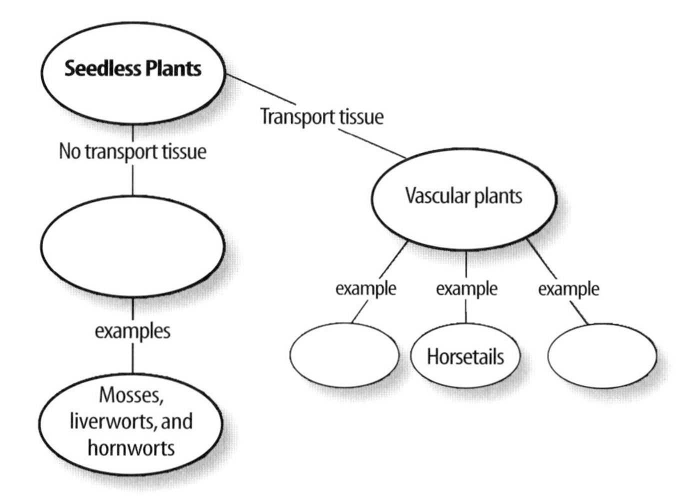

25. **Interpret Scientific Illustrations** Using **Figure 20** in this chapter, compare and contrast the number of cotyledons, bundle arrangement in the stem, veins in leaves, and number of flower parts for monocots and dicots.
26. **Sequence** Put the following events in order to show how coal is formed from plants: *living seedless plants, coal is formed, dead seedless plants decay,* and *peat is formed.*
27. **Predict** what would happen if a ring of bark and camium layer were removed from around the trunk of a tree.

Performance Activities

28. **Poem** Choose a topic in this chapter that interests you. Look it up in a reference book, in an encyclopedia, or on a CD-ROM. Write a poem to share what you learn.
29. **Display** Use dried plant material, photos, drawings, or other materials to make a poster describing the form and function of roots, stems, and leaves.

Applying Math

Use the table below to answer questions 30–32.

Number of Stomata (per mm^2)

Plant	Upper Surface	Lower Surface
Pine	50	71
Bean	40	281
Fir	0	228
Tomato	12	13

30. **Gas Exchange** What do the data in this table tell you about where gas exchange occurs in the leaf of each plant species?
31. **Compare Leaf Surfaces** Make two circle graphs—upper surface and lower surface—using the table above.
32. **Guard Cells** On average, how many guard cells are found on the lower surface of a bean leaf?

Part 1 Multiple Choice

Record your answers on the answer sheet provided by your teacher or on a sheet of paper.

1. Which of the following do plants use to photosynthesize?
   A. blood
   B. iron
   C. chlorophyll
   D. cellulose

2. Which of the following describes the function of the central vacuole in plant cells?
   A. It helps in reproduction.
   B. It helps regulate water content.
   C. It plays a key role in photosynthesis.
   D. It stores food.

Use the illustration below to answer questions 3 and 4.

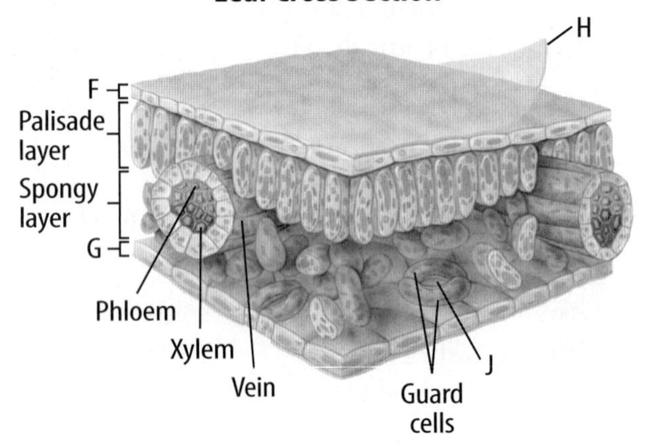

3. In the leaf cross section, what is indicated by H?
   A. upper epidermis
   B. cuticle
   C. stoma
   D. lower epidermis

4. What flows through the structure indicated by J?
   A. water only
   B. carbon dioxide and water only
   C. oxygen and carbon dioxide only
   D. water, carbon dioxide, and oxygen

5. In seed plants, vascular tissue refers to which of the following?
   A. xylem and phloem only
   B. xylem only
   C. phloem only
   D. xylem, phloem, and cambium

Use the illustration below to answer questions 6 and 7.

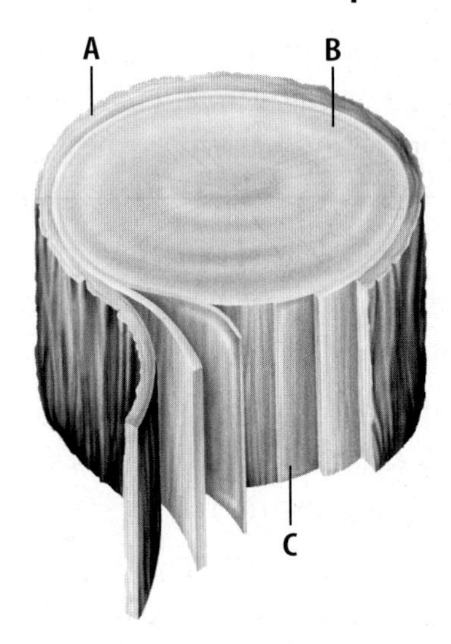

6. What is the function of the structure labeled C?
   A. It transports nutrients throughout the plant.
   B. It produces new xylem and phloem.
   C. It transports water from the roots to other parts of the plant.
   D. It absorbs water from outside the plant.

7. What type of vascular tissue is indicated by B?
   A. xylem
   B. cambium
   C. phloem
   D. cellulose

Test-Taking Tip

Eliminate Answer Choices If you don't know the answer to a multiple-choice question, eliminate as many incorrect choices as possible. Mark your best guess from the remaining answers before moving on to the next question.

Part 2 Short Response/Grid In

Record your answers on the answer sheet provided by your teacher or on a sheet of paper.

Use the two illustrations below to answer questions 8–10.

8. Identify the flowers shown above as a monocot or a dicot. Explain the differences between the flowers of monocots and dicots.

9. Give three examples of plants represented by Plant A.

10. Give three examples of plants represented by Plant B.

11. How are plants that live on land able to conserve water?

12. Explain why reproductive adaptations were necessary in order for plants to survive on land.

13. You are hiking through a dense forest area and notice some unusual plants growing on the trunk of a tall tree. The plants are no taller than about 3 cm and have delicate stalks. They do not appear to have flowers. Based on this information, what type of plants would you say you found?

14. What is a conifer? To which major group of plants does it belong?

Part 3 Open Ended

Record your answers on a sheet of paper.

Use the two diagrams below to answer questions 15–16.

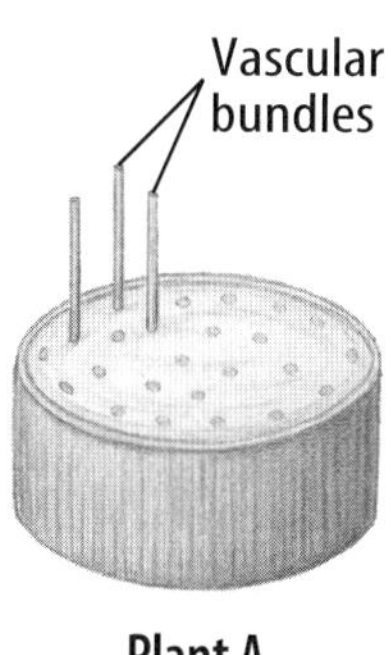

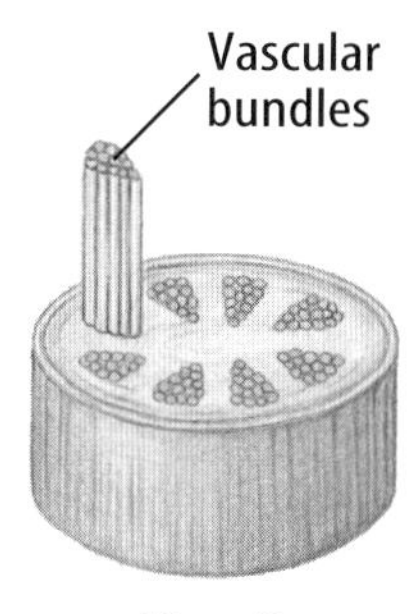

15. Two plants, A and B, have stem cross sections as shown in the diagrams above. What does the different vascular bundle arrangement tell you about each plant?

16. Describe a typical seed from each plant type.

17. Create a diagram that describes the life cycle of an annual angiosperm.

18. Discuss the importance of plants in your daily life. Give examples of plants or plant products that you use or consume regularly.

19. Compare and contrast vascular and nonvascular plants. Include examples of each type of plant.

20. Describe the group of plants known as the seedless vascular plants. How do these plants reproduce without seeds?

21. Explain what peat is and how it is formed. How is peat used today?

22. How would our knowledge of ancient plants be different if the fossil record for plants was as plentiful as it is for animals?

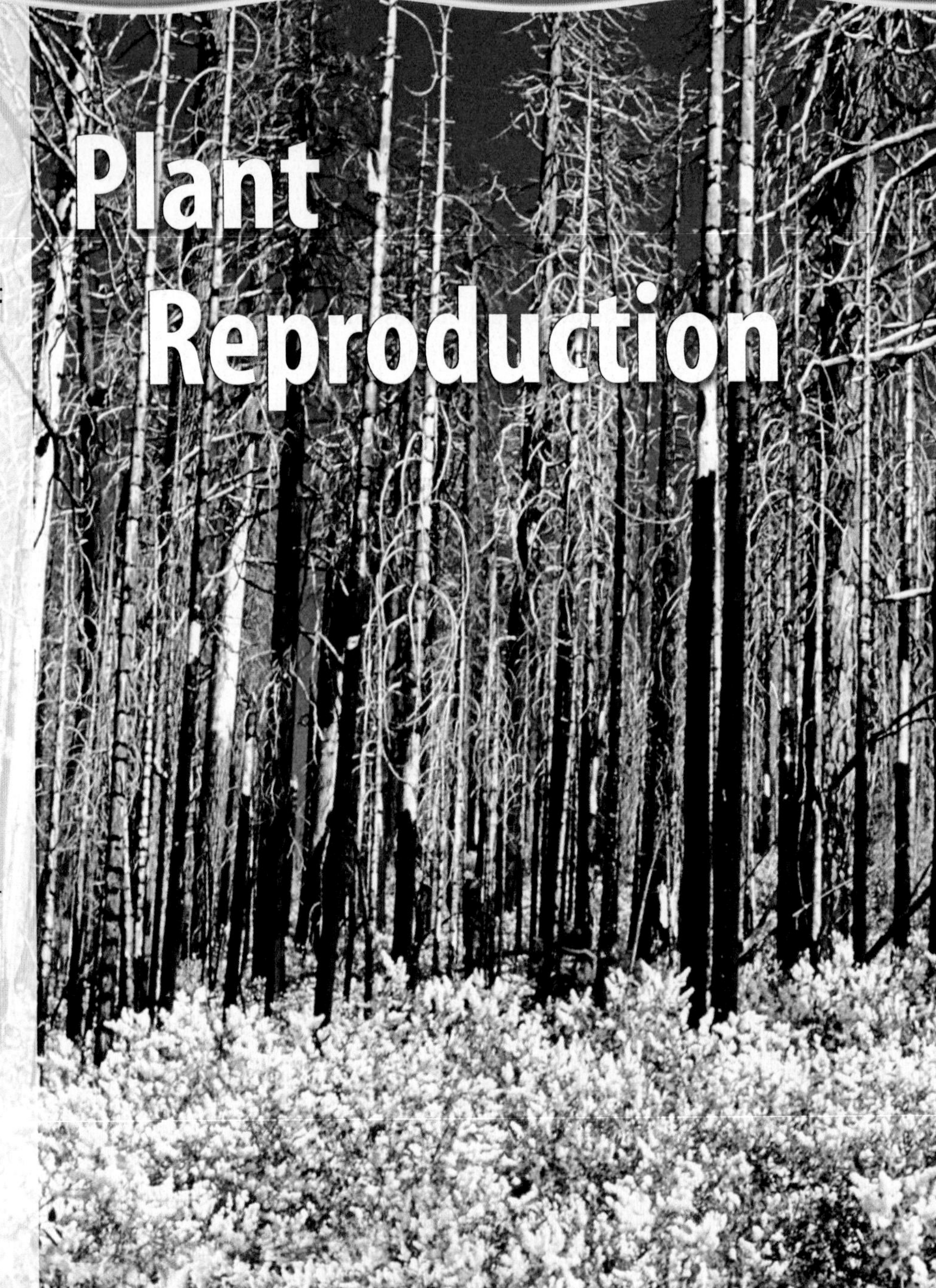

Plant Reproduction

The BIG Idea

Plants have adaptations that enable them to reproduce in specific habitats.

Section 1
Introduction to Plant Reproduction

Main Idea Reproduction in plants can be asexual or sexual. Plant life cycles include an alternation of generations.

Section 2
Seedless Reproduction

Main Idea The joining of a seedless plant's egg and sperm requires moist conditions and produces a spore.

Section 3
Seed Reproduction

Main Idea Reproduction in seed plants involves pollen grains—the sources of sperm, and ovules—the sources of the eggs. The joining of eggs and sperms can produce seeds.

A Forest from Ashes

Saplings and other plants are growing among the remains of trees destroyed by fire. Where did these new plants come from? Some may have grown from seeds, and others may have grown from roots or stems that survived underground. These plants are the result of plant reproduction.

Science Journal List three plants that reproduce by forming seeds.

Start-Up Activities

Do all fruits contain seeds?

You might know that most plants grow from seeds. Seeds are usually found in the fruits of plants. When you eat watermelon, it can contain many small seeds. Do some plants produce fruits without seeds? Do this lab to find out.

1. Obtain two grapes from your teacher. Each grape should be from a different plant.
2. Split each grape in half and examine the insides of each grape. **WARNING:** *Do not eat the grapes.*
3. **Think Critically** Were seeds found in both grapes? Hypothesize how new grape plants could be grown if no seeds are produced. In your Science Journal, list three other fruits you know of that do not contain seeds.

Preview this chapter's content and activities at life.msscience.com

FOLDABLES™ Study Organizer

Plant Reproduction Make the following Foldable to compare and contrast the sexual and asexual characteristics of a plant.

STEP 1 **Fold** one sheet of paper lengthwise.

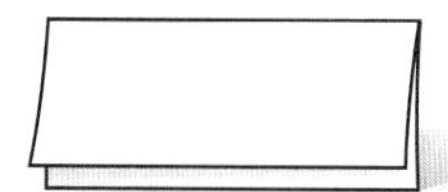

STEP 2 **Fold** into thirds.

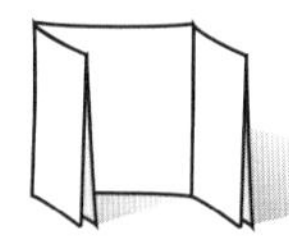

STEP 3 **Unfold and draw** overlapping ovals. **Cut** the top sheet along the folds.

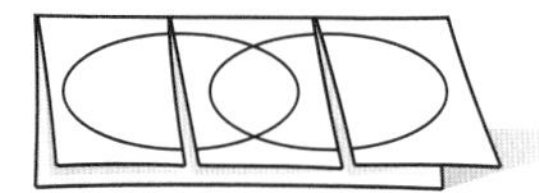

STEP 4 **Label** the ovals as shown.

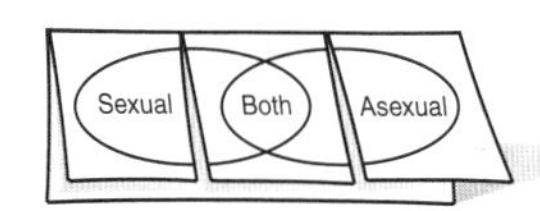

Construct a Venn Diagram As you read the chapter, list the characteristics unique to sexual reproduction under the left tab, those unique to asexual reproduction under the right tab, and those characteristics common to both under the middle tab.

Get Ready to Read

Summarize

1 Learn It! Summarizing helps you organize information, focus on main ideas, and reduce the amount of information to remember. To summarize, restate the important facts in a short sentence or paragraph. Be brief and do not include too many details.

2 Practice It! Read the text on page 290 labeled Germination. Then read the summary below and look at the important facts from that passage.

Important Facts

A series of events that results in the growth of a plant from a seed is called **germination.**

Seeds will not germinate until environmental conditions are right. Temperature, the presence or absence of light, availability of water, and amount of oxygen present can affect germination. Sometimes the seed must pass through an animal's digestive system before it will germinate.

Germination begins when seed tissues absorb water. This causes the seed to swell and the seed coat to break open.

Next, a series of chemical reactions occurs that releases energy from the stored food in the cotyledons or endosperm for growth. Eventually, a root grows from the seed, followed by a stem and leaves.

Summary

Germination is the growth of a plant from a seed. It requires certain environmental conditions. Processes occur within the seed that result in the growth of roots, a stem, and leaves.

3 Apply It! Practice summarizing as you read this chapter. Stop after each lesson and write a brief summary.

Reading Tip

Reread your summary to make sure you didn't change the author's original meaning or ideas.

Target Your Reading

Use this to focus on the main ideas as you read the chapter.

1. **Before you read** the chapter, respond to the statements below on your worksheet or on a numbered sheet of paper.
   - Write an **A** if you **agree** with the statement.
   - Write a **D** if you **disagree** with the statement.

2. **After you read** the chapter, look back to this page to see if you've changed your mind about any of the statements.
   - If any of your answers changed, explain why.
   - Change any false statements into true statements.
   - Use your revised statements as a study guide.

Before You Read A or D		Statement	After You Read A or D
	1	Both asexual and sexual plant reproduction involves sex cells.	
	2	Plants produced by asexual reproduction are genetically identical to the parent plants.	
	3	The structures of a gametophyte plant are made of haploid cells.	
	4	Insects usually move moss sperm to moss eggs.	
	5	The gametophyte stage of mosses is small and rarely observed by humans.	
	6	Generally, fern fronds have structures on them that produce spores.	
	7	An embryo, stored food, and seed coat are parts of a seed.	
	8	Gymnosperm seeds are produced inside flowers.	
	9	An angiosperm fruit that has one or more seeds grows after fertilization.	
	10	Seed germination can occur within days after seed dispersal.	

Science Online

Print out a worksheet of this page at life.msscience.com

section 1

Introduction to Plant Reproduction

as you read

What You'll Learn

- **Distinguish** between the two types of plant reproduction.
- **Describe** the two stages in a plant's life cycle.

Why It's Important

You can grow new plants without using seeds.

Review Vocabulary
fertilization: in sexual reproduction, the joining of a sperm and an egg

New Vocabulary
- spore
- gametophyte stage
- sporophyte stage

Types of Reproduction

Do people and plants have anything in common? You don't have leaves or roots, and a plant doesn't have a heart or a brain. Despite these differences, you are alike in many ways—you need water, oxygen, energy, and food to grow. Like humans, plants also can reproduce and make similar copies of themselves. Although humans have only one type of reproduction, most plants can reproduce in two different ways, as shown in **Figure 1.**

Sexual reproduction in plants and animals requires the production of sex cells—usually called sperm and eggs—in reproductive organs. The offspring produced by sexual reproduction are genetically different from either parent organism.

A second type of reproduction is called asexual reproduction. This type of reproduction does not require the production of sex cells. During asexual reproduction, one organism produces offspring that are genetically identical to it. Most plants have this type of reproduction, but humans and most other animals don't.

Figure 1 Many plants reproduce sexually with flowers that contain male and female parts. Other plants can reproduce asexually.

In crocus flowers, bees and other insects help get the sperm to the egg.

A cutting from this impatiens plant can be placed in water and will grow new roots. This new plant can then be planted in soil.

Figure 2 Asexual reproduction in plants takes many forms.

The eyes on these potatoes have begun to sprout. If a potato is cut into pieces, each piece that contains an eye can be planted and will grow into a new potato plant.

Grass plants spread by reproducing asexually.

Asexual Plant Reproduction Do you like to eat oranges and grapes that have seeds, or do you like seedless fruit? If these plants do not produce seeds, how do growers get new plants? Growers can produce new plants by asexual reproduction because many plant cells have the ability to grow into a variety of cell types. New plants can be grown from just a few cells in the laboratory. Under the right conditions, an entire plant can grow from one leaf or just a portion of the stem or root. When growers use these methods to start new plants, they must make sure that the leaf, stem, or root cuttings have plenty of water and anything else that they need to survive.

Asexual reproduction has been used to produce plants for centuries. The white potatoes shown in **Figure 2** were probably produced asexually. Many plants, such as lawn grasses also shown in **Figure 2,** can spread and cover wide areas because their stems grow underground and produce new grass plants asexually along the length of the stem.

Sexual Plant Reproduction Although plants and animals have sexual reproduction, there are differences in the way that it occurs. An important event in sexual reproduction is fertilization. Fertilization occurs when a sperm and egg combine to produce the first cell of the new organism, the zygote. How do the sperm and egg get together in plants? In some plants, water or wind help bring the sperm to the egg. For other plants, animals such as insects help bring the egg and sperm together.

How does fertilization occur in plants?

Observing Asexual Reproduction

Procedure

1. Using a pair of scissors, cut a stem with at least two pairs of leaves from a coleus or another houseplant.
2. Carefully remove the bottom pair of leaves.
3. Place the cut end of the stem into a cup that is half-filled with water for two weeks. Wash your hands.
4. Remove the new plant from the water and plant it in a small container of soil.

Analysis

1. Draw and label your results in your **Science Journal.**
2. Predict how the new plant and the plant from which it was taken are genetically related.

Figure 3 Some plants can fertilize themselves. Others require two different plants before fertilization can occur.

Flowers of pea plants contain male and female structures, and each flower can fertilize itself.

These holly flowers contain only male reproductive structures, so they can't fertilize themselves.

Compare the flowers of this female holly plant to those of the male plant.

Reproductive Organs A plant's female reproductive organs produce eggs and male reproductive organs produce sperm. Depending on the species, these reproductive organs can be on the same plant or on separate plants, as shown in **Figure 3.** If a plant has both organs, it usually can reproduce by itself. However, some plants that have both sex organs still must exchange sex cells with other plants of the same type to reproduce.

In some plant species, the male and female reproductive organs are on separate plants. For example, holly plants are either female or male. For fertilization to occur, holly plants with flowers that have different sex organs must be near each other. In that case, after the eggs in female holly flowers are fertilized, berries can form.

Another difference between you and a plant is how and when plants produce sperm and eggs. You will begin to understand this difference as you examine the life cycle of a plant.

Topic: Male and Female Plants

Visit life.msscience.com for Web links to information about male and female plants.

Activity List four plants that have male and female reproductive structures on separate plants.

Plant Life Cycles

All organisms have life cycles. Your life cycle started when a sperm and an egg came together to produce the zygote that would grow and develop into the person you are today. A plant also has a life cycle. It can start when an egg and a sperm come together, eventually producing a mature plant.

Two Stages During your life cycle, all structures in your body are formed by mitosis and cell division and are made up of diploid cells—cells with a full set of chromosomes. However, sex cells form by meiosis and are haploid—they have half a set of chromosomes.

Plants have a two-stage life cycle, as shown in **Figure 4.** The two stages are the gametophyte (guh MEE tuh fite) stage and the sporophyte (SPOHR uh fite) stage.

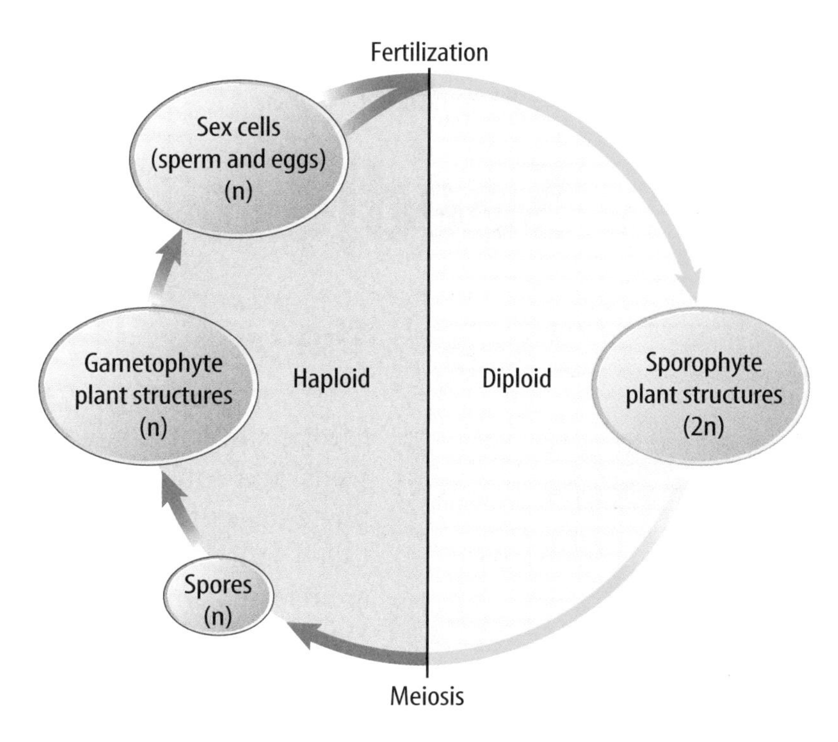

Figure 4 Plants produce diploid and haploid plant structures.
Identify *the process that begins the gametophyte stage.*

Gametophyte Stage When reproductive cells undergo meiosis and produce haploid cells called **spores,** the **gametophyte stage** begins. Spores divide by mitosis and cell division and form plant structures or an entire new plant made of haploid cells. Some of these cells undergo mitosis and cell division and form haploid sex cells.

Sporophyte Stage Fertilization—the joining of haploid sex cells—begins the **sporophyte stage.** Cells formed in this stage have the diploid number of chromosomes. Meiosis in some of these cells forms spores, and the cycle repeats.

Reading Check *What process begins the sporophyte stage?*

section 1 review

Summary

Types of Reproduction

- Asexual reproduction results in offspring that are genetically identical to the parent plant.
- Sexual reproduction requires fertilization and results in offspring that are genetically different from either parent.
- Female reproductive organs produce eggs.
- Male reproductive organs produce sperm.

Plant Life Cycles

- The gametophyte stage of the plant life cycle begins with meiosis.
- The sporophyte stage begins with fertilization.

Self Check

1. **List** three differences between the gametophyte stage and the sporophyte stage of the plant life cycle.
2. **Describe** how plants reproduce asexually.
3. **Compare and contrast** sexual reproduction in plants and animals.
4. **Think Critically** You admire a friend's houseplant. What would you do to grow an identical plant?

Applying Skills

5. **Draw Conclusions** Using a microscope, you see that the nuclei of a plant's cells contain half the usual number of chromosomes. What is the life cycle stage of this plant?

section 2

Seedless Reproduction

as you read

What You'll Learn

- **Examine** the life cycles of a moss and a fern.
- **Explain** why spores are important to seedless plants.
- **Identify** some special structures used by ferns for reproduction.

Why It's Important

Mosses help build new soil on bare rock or cooled lava, making it possible for other plants to take root.

Review Vocabulary
photosynthesis: food-making process by which plants and many other producers use light energy to produce glucose and oxygen from carbon dioxide and water

New Vocabulary
- frond
- rhizome
- sori
- prothallus

The Importance of Spores

If you want to grow ferns and moss plants, you can't go to a garden store and buy a package of seeds—they don't produce seeds. You could, however, grow them from spores. The sporophyte stage of these plants produces haploid spores in structures called spore cases. When the spore case breaks open, the spores are released and spread by wind or water. The spores, shown in **Figure 5,** can grow into plants that will produce sex cells.

Seedless plants include all nonvascular plants and some vascular plants. Nonvascular plants do not have structures that transport water and substances throughout the plant. Instead, water and substances simply move from cell to cell. Vascular plants have tubelike cells that transport water and substances throughout the plant.

Nonvascular Seedless Plants

If you walked in a damp, shaded forest, you probably would see mosses covering the ground or growing on a log. Mosses, liverworts, and hornworts are all nonvascular plants.

The sporophyte stage of most nonvascular plants is so small that it can be easily overlooked. Moss plants have a life cycle typical of how sexual reproduction occurs in this plant group.

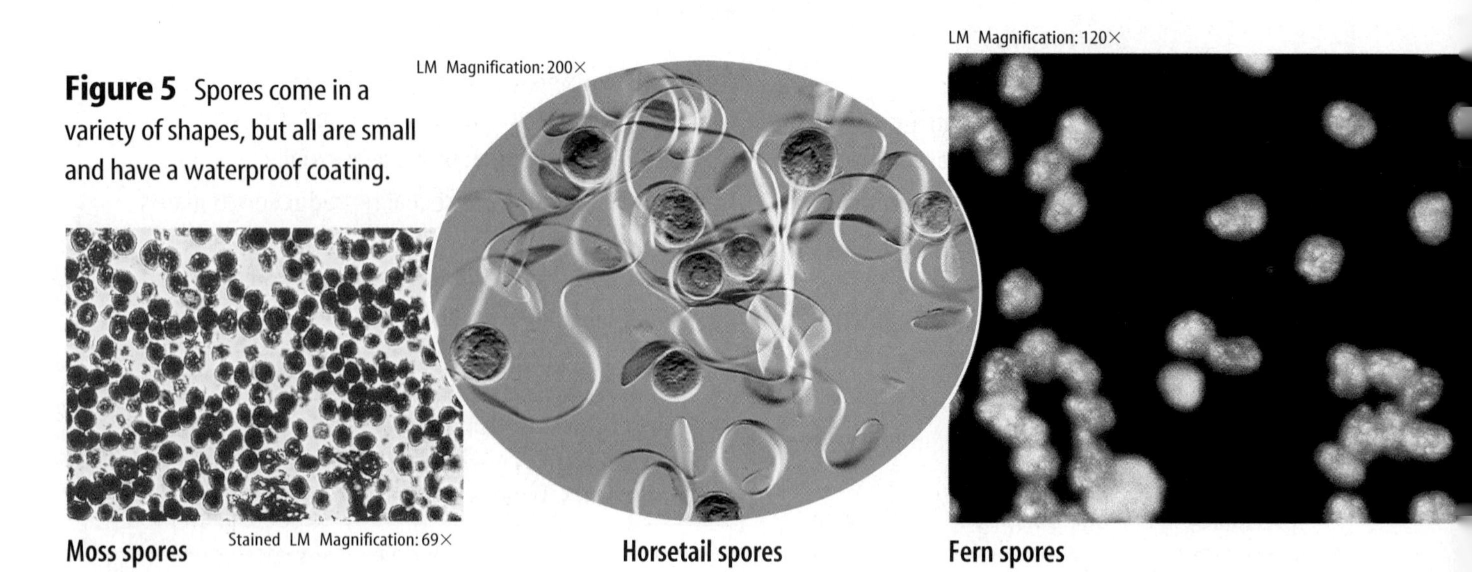

Figure 5 Spores come in a variety of shapes, but all are small and have a waterproof coating.

The Moss Life Cycle You recognize mosses as green, low-growing masses of plants. This is the gametophyte stage, which produces the sex cells. But the next time you see some moss growing, get down and look at it closely. If you see any brownish stalks growing up from the tip of the gametophyte plants, you are looking at the sporophyte stage. The sporophyte stage does not carry on photosynthesis. It depends on the gametophyte for nutrients and water. On the tip of the stalk is a tiny capsule. Inside the capsule millions of spores have been produced. When environmental conditions are just right, the capsule opens and the spores either fall to the ground or are blown away by the wind. New moss gametophytes can grow from each spore and the cycle begins again, as shown in **Figure 6.**

Figure 6 The life cycle of a moss alternates between gametophyte and sporophyte stages.
Identify *the structures that are produced by the gametophyte stage.*

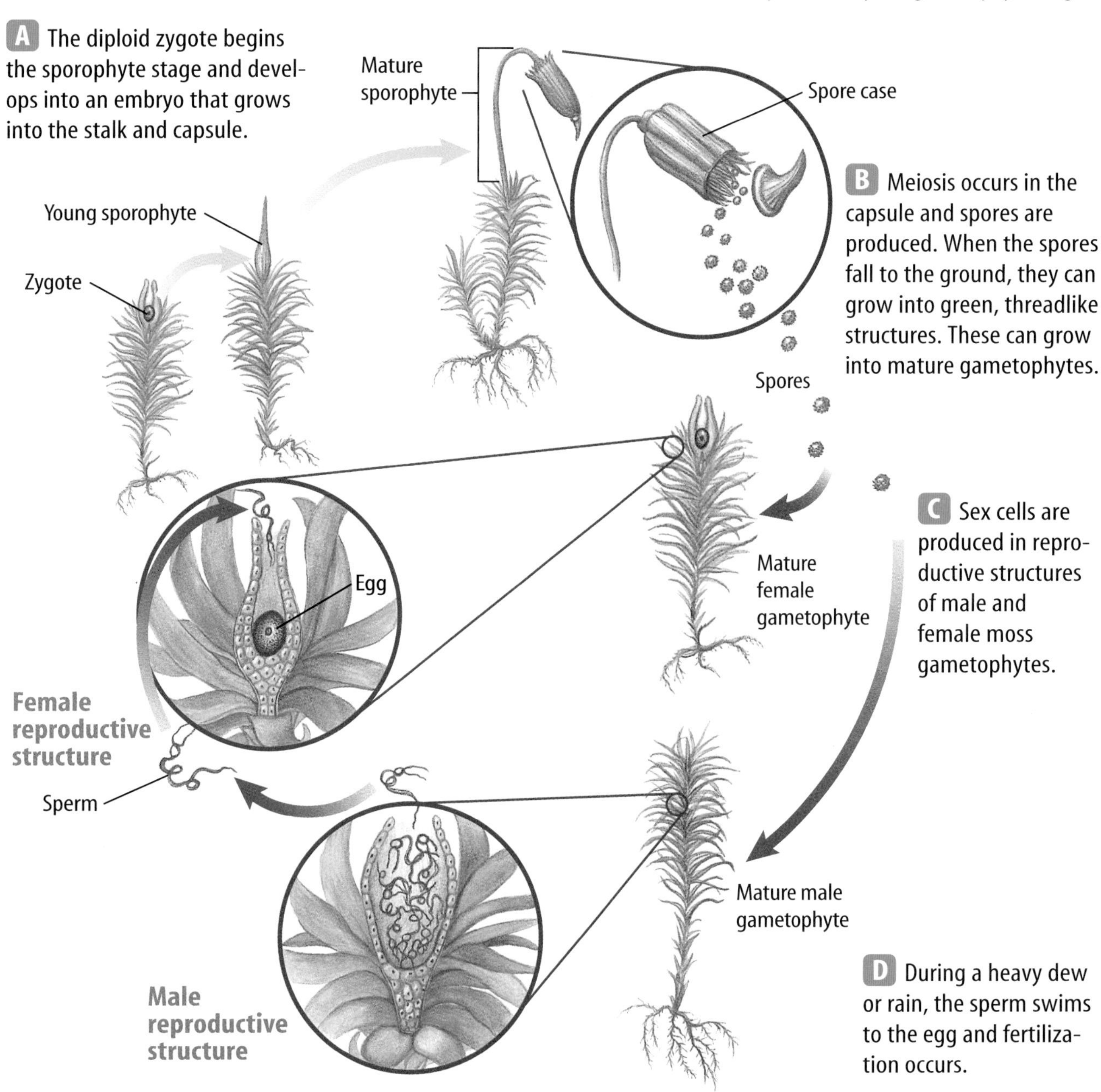

A The diploid zygote begins the sporophyte stage and develops into an embryo that grows into the stalk and capsule.
Mature sporophyte
Spore case
Young sporophyte
Zygote
B Meiosis occurs in the capsule and spores are produced. When the spores fall to the ground, they can grow into green, threadlike structures. These can grow into mature gametophytes.
Spores
Egg
Mature female gametophyte
C Sex cells are produced in reproductive structures of male and female moss gametophytes.
Female reproductive structure
Sperm
Mature male gametophyte
Male reproductive structure
D During a heavy dew or rain, the sperm swims to the egg and fertilization occurs.

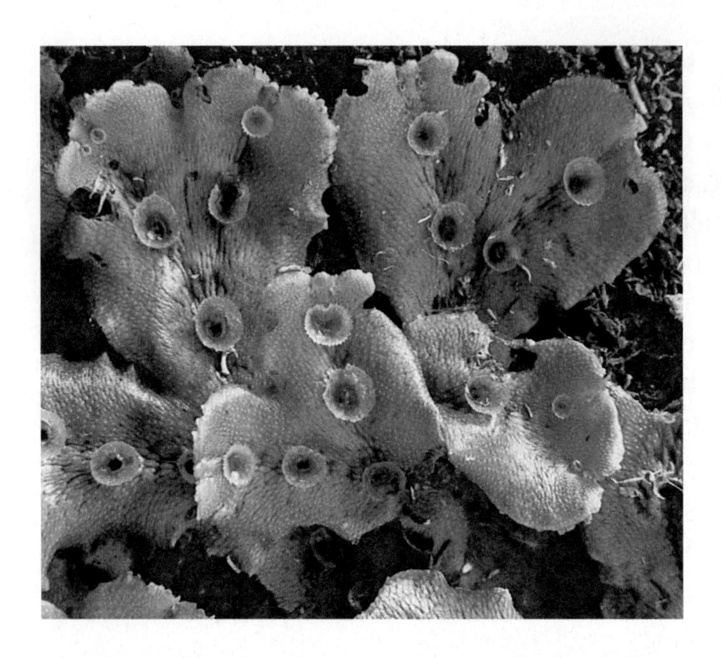

Figure 7 Small balls of cells grow in cuplike structures on the surface of the liverwort.

Nonvascular Plants and Asexual Reproduction Nonvascular plants also can reproduce asexually. For example, if a piece of a moss gametophyte plant breaks off, it can grow into a new plant. Liverworts can form small balls of cells on the surface of the gametophyte plant, as shown in **Figure 7.** These are carried away by water and grow into new gametophyte plants if they settle in a damp environment.

Vascular Seedless Plants

Millions of years ago most plants on Earth were vascular seedless plants. Today they are not as widespread.

Most vascular seedless plants are ferns. Other plants in this group include horsetails and club mosses. All of these plants have vascular tissue to transport water from their roots to the rest of the plant. Unlike the nonvascular plants, the gametophyte of vascular seedless plants is the part that is small and often overlooked.

The Fern Life Cycle The fern plants that you see in nature or as houseplants are fern sporophyte plants. Fern leaves are called **fronds.** They grow from an underground stem called a **rhizome.** Roots that anchor the plant and absorb water and nutrients also grow from the rhizome. Fern sporophytes make their own food by photosynthesis. Fern spores are produced in structures called **sori** (singular, *sorus*), usually located on the underside of the fronds. Sori can look like crusty rust-, brown-, or dark-colored bumps. Sometimes they are mistaken for a disease or for something growing on the fronds.

If a fern spore lands on damp soil or rocks, it can grow into a small, green, heart-shaped gametophyte plant called a **prothallus** (proh THA lus). A prothallus is hard to see because most of them are only about 5 mm to 6 mm in diameter. The prothallus contains chlorophyll and can make its own food. It absorbs water and nutrients from the soil. The life cycle of a fern is shown in **Figure 8.**

Catapults For thousands of years, humans have used catapults to launch objects. The spore cases of ferns act like tiny catapults as they eject their spores. In your Science Journal list tools, toys, and other objects that have used catapult technology throughout history.

Reading Check *What is the gametophyte plant of a fern called?*

Ferns may reproduce asexually, also. Fern rhizomes grow and form branches. New fronds and roots develop from each branch. The new rhizome branch can be separated from the main plant. It can grow on its own and form more fern plants.

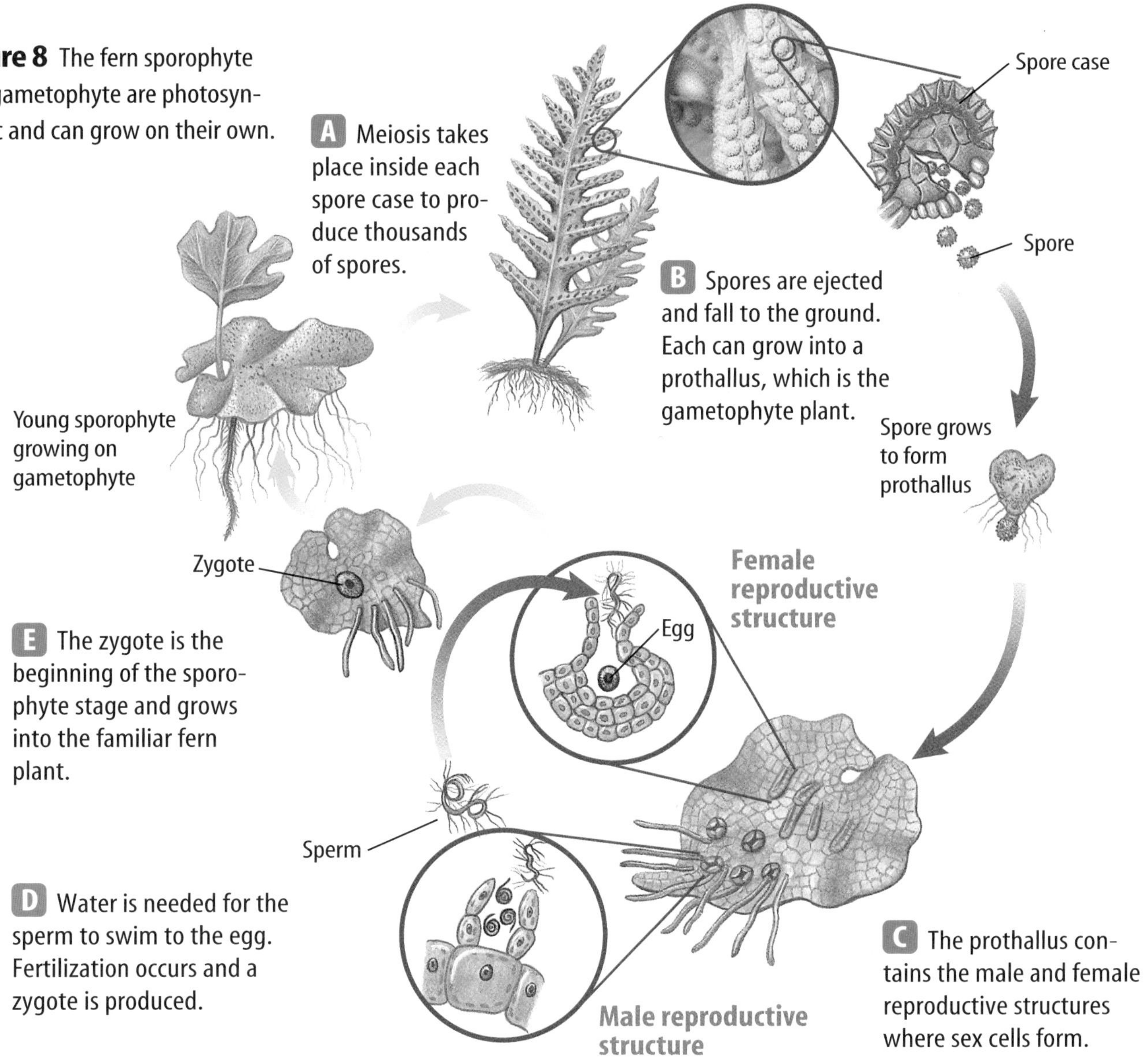

Figure 8 The fern sporophyte and gametophyte are photosynthetic and can grow on their own.

A Meiosis takes place inside each spore case to produce thousands of spores.

B Spores are ejected and fall to the ground. Each can grow into a prothallus, which is the gametophyte plant.

C The prothallus contains the male and female reproductive structures where sex cells form.

D Water is needed for the sperm to swim to the egg. Fertilization occurs and a zygote is produced.

E The zygote is the beginning of the sporophyte stage and grows into the familiar fern plant.

Spore case

Spore

Spore grows to form prothallus

Female reproductive structure

Egg

Male reproductive structure

Sperm

Zygote

Young sporophyte growing on gametophyte

section 2 review

Summary

The Importance of Spores

- Seedless plants reproduce by forming spores.
- Seedless plants include all nonvascular plants and some vascular plants.

Nonvascular Seedless Plants

- Spores are produced by the sporophyte stage and can grow into gametophyte plants.
- The sporophyte cannot photosynthesize.

Vascular Seedless Plants

- Fern sporophytes have green fronds that grow from an underground rhizome.

Self Check

1. **Describe** the life cycle of mosses.
2. **Explain** each stage in the life cycle of a fern.
3. **Compare and contrast** the gametophyte plant of a moss and the gametophyte plant of a fern.
4. **Describe** asexual reproduction in seedless plants.
5. **Think Critically** Why do some seedless plants reproduce only asexually during dry times of the year?

Applying Math

6. **Solve One-Step Equations** If moss spores are 0.1 mm in diameter, how many equal the diameter of a penny?

Comparing Seedless Plants

All seedless plants have specialized structures that produce spores. Although these sporophyte structures have a similar function, they look different. The gametophyte plants also are different from each other. Do this lab to observe the similarities and differences among three groups of seedless plants.

Real-World Question

How are the gametophyte stages and the sporophyte stages of liverworts, mosses, and ferns similar and different?

Goals

- **Describe** the sporophyte and gametophyte forms of liverworts, mosses, and ferns.
- **Identify** the spore-producing structures of liverworts, mosses, and ferns.

Materials

live mosses, liverworts, and ferns with gametophytes and sporophytes
microscope slides and coverslips (2)
magnifying lens
forceps
dropper
microscope
dissecting needle
pencil with eraser

Safety Precautions

Procedure

1. Obtain a gametophyte of each plant. With a magnifying lens, observe the rhizoids, leafy parts, and stemlike parts, if any are present.
2. Obtain a sporophyte of each plant and use a magnifying lens to observe it.
3. Locate and remove a spore structure of a moss plant. Place it in a drop of water on a slide.

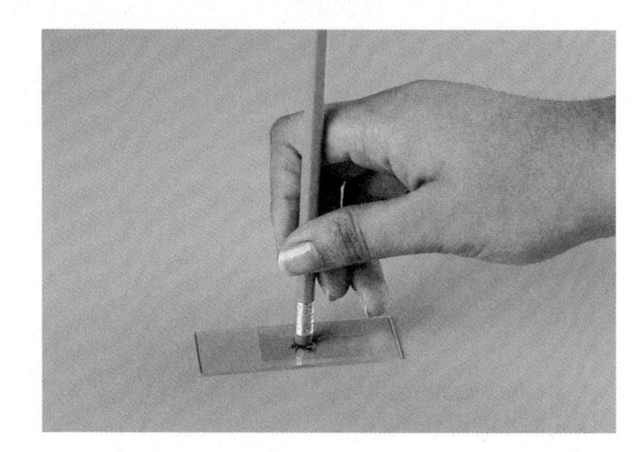

4. Place a coverslip over it. Use the eraser of a pencil to gently push on the coverslip to release the spores. **WARNING:** *Do not break the coverslip.* Observe the spores under low and high power.
5. Make labeled drawings of all observations in your Science Journal.
6. Repeat steps 3–5 using a fern.

Conclude and Apply

1. **Compare** the gametophyte's appearance to the sporophyte's appearance for each plant.
2. **List** structure(s) common to all three plants.
3. **Hypothesize** about why each plant produces a large number of spores.

Communicating Your Data

Prepare a bulletin board that shows differences between the sporophyte and gametophyte stages of liverworts, mosses, and ferns. **For more help, refer to the Science Skill Handbook.**

section 3

Seed Reproduction

The Importance of Pollen and Seeds

All the plants described so far have been seedless plants. However, the fruits and vegetables that you eat come from seed plants. Oak, maple, and other shade trees are also seed plants. All flowers are produced by seed plants. In fact, most of the plants on Earth are seed plants. How do you think they became such a successful group? Reproduction that involves pollen and seeds is part of the answer.

Pollen In seed plants, some spores develop into small structures called pollen grains. A **pollen grain,** as shown in **Figure 9,** has a water-resistant covering and contains male gametophyte parts that can produce sperm. Sperm of seed plants do not need to swim to the female part of the plant. Instead, they are carried as part of the pollen grain by gravity, wind, water currents, or animals. The transfer of pollen grains of a species to the female part of the plant of the same species is called **pollination.**

After the pollen grain reaches the female part of a plant, sperm and a pollen tube are produced. The sperm moves through the pollen tube, then fertilization can occur.

Figure 9 The waterproof covering of a pollen grain is unique and can be used to identify the plant that it came from. This pollen from a ragweed plant is a common cause of hay fever.

as you read

What You'll Learn

- **Examine** the life cycles of typical gymnosperms and angiosperms.
- **Describe** the structure and function of the flower.
- **Discuss** methods of seed dispersal in seed plants.

Why It's Important

Seeds from cones and flowers produce most plants on Earth.

Review Vocabulary

gymnosperms: vascular plants that do not flower, generally have needlelike or scalelike leaves, and produce seeds that are not protected by fruit

New Vocabulary

- pollen grain
- pollination
- ovule
- stamen
- pistil
- ovary
- germination

Figure 10 Seeds have three main parts—a seed coat, stored food, and an embryo. This pine seed also has a wing.
Infer *the function of the wing.*

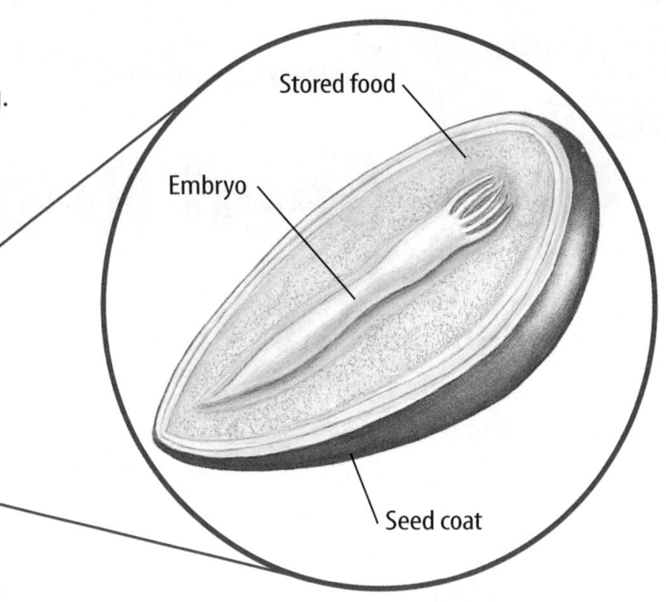

Seeds Following fertilization, the female part of a plant can develop into a seed. A seed consists of an embryo, stored food, and a protective seed coat, as shown in **Figure 10.** The embryo has structures that eventually will produce the plant's stem, leaves, and roots. In the seed, the embryo grows to a certain stage and then stops until the seed is planted. The stored food provides energy that is needed when the plant embryo begins to grow. Because the seed contains an embryo and stored food, a new plant can develop more rapidly from a seed than from a spore.

Reading Check *What are the three parts of a seed?*

Gymnosperms (JIHM nuh spurmz) and angiosperms are seed plants. One difference between the two groups is the way seeds develop. In gymnosperms, seeds usually develop in cones—in angiosperms, seeds develop in flowers and fruit.

Topic: Seed Banks
Visit life.msscience.com for Web links to information about conserving the seeds of many useful and endangered plants.

Activity List three organizations that manage seed banks, and give examples of the kinds of plants each organization works to conserve.

Gymnosperm Reproduction

If you have collected cones of pines or spruces or used them in a craft project, you probably noticed that many shapes and sizes of cones exist. You probably also noticed that some cones contain seeds. Cones are the reproductive structures of gymnosperms. Each gymnosperm species has a different cone.

Gymnosperm plants include pines, firs, spruces, cedars, cycads, and ginkgoes. The pine is a familiar gymnosperm. Production of seeds in pines is typical of most gymnosperms.

Cones A pine tree is a sporophyte plant that produces male cones and female cones as shown in **Figure 11.** Male and female gametophyte structures are produced in the cones but you'd need a magnifying lens to see these structures clearly.

A mature female cone consists of a spiral of woody scales on a short stem. At the base of each scale are two ovules. The egg is produced in the **ovule.** Pollen grains are produced in the smaller male cones. In the spring, clouds of pollen are released from the male cones. Anything near pine trees might be covered with the yellow, dustlike pollen.

Figure 11 Seed formation in pines, as in most gymnosperms, involves male and female cones.

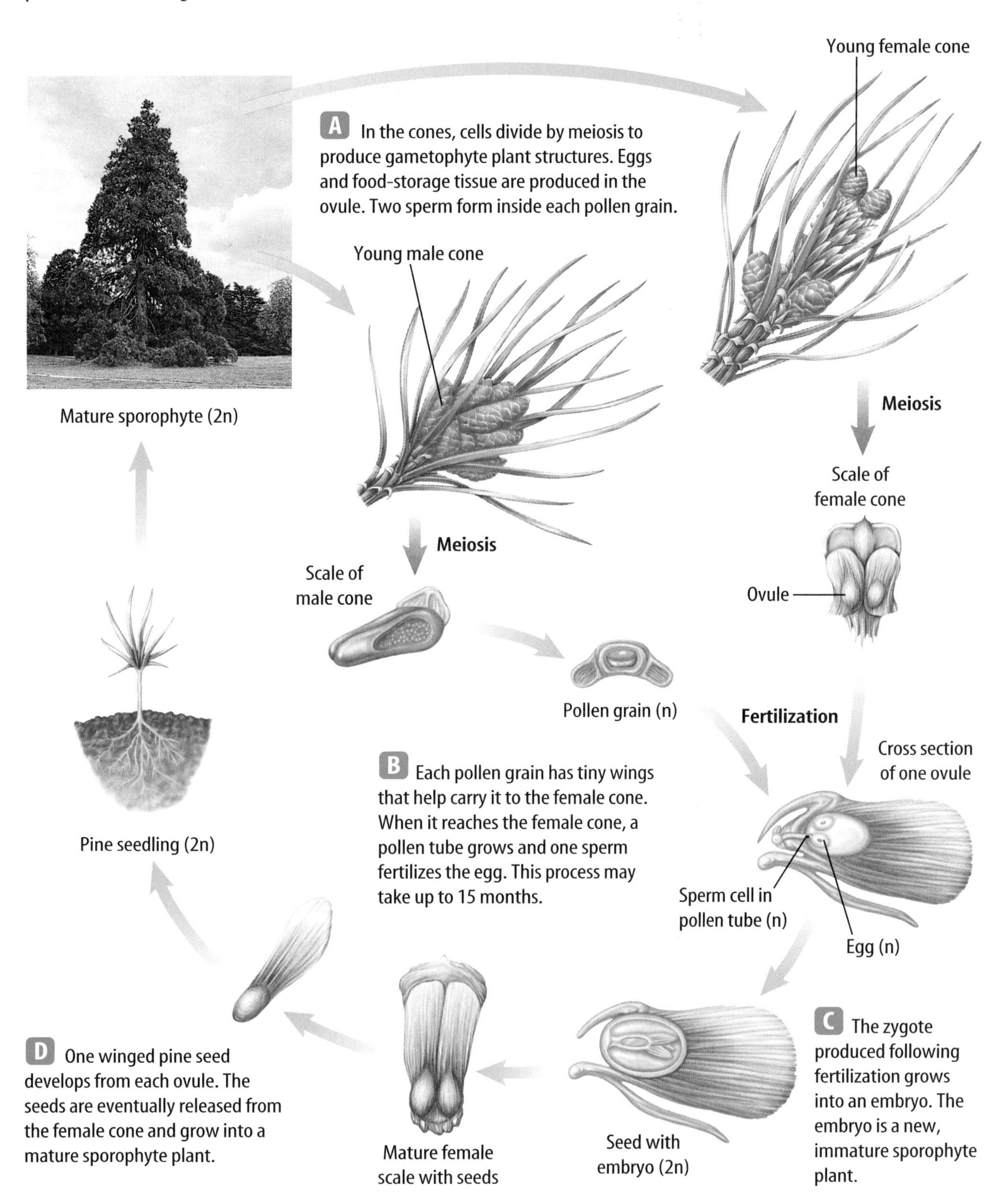

Young female cone
A In the cones, cells divide by meiosis to produce gametophyte plant structures. Eggs and food-storage tissue are produced in the ovule. Two sperm form inside each pollen grain.
Young male cone
Mature sporophyte (2n)
Meiosis
Scale of female cone
Meiosis
Scale of male cone
Ovule
Pollen grain (n)
Fertilization
Cross section of one ovule
B Each pollen grain has tiny wings that help carry it to the female cone. When it reaches the female cone, a pollen tube grows and one sperm fertilizes the egg. This process may take up to 15 months.
Pine seedling (2n)
Sperm cell in pollen tube (n)
Egg (n)
C The zygote produced following fertilization grows into an embryo. The embryo is a new, immature sporophyte plant.
D One winged pine seed develops from each ovule. The seeds are eventually released from the female cone and grow into a mature sporophyte plant.
Mature female scale with seeds
Seed with embryo (2n)

Figure 12 Seed development can take more than one year in pines. The female cone looks different at various stages of the seed-production process.

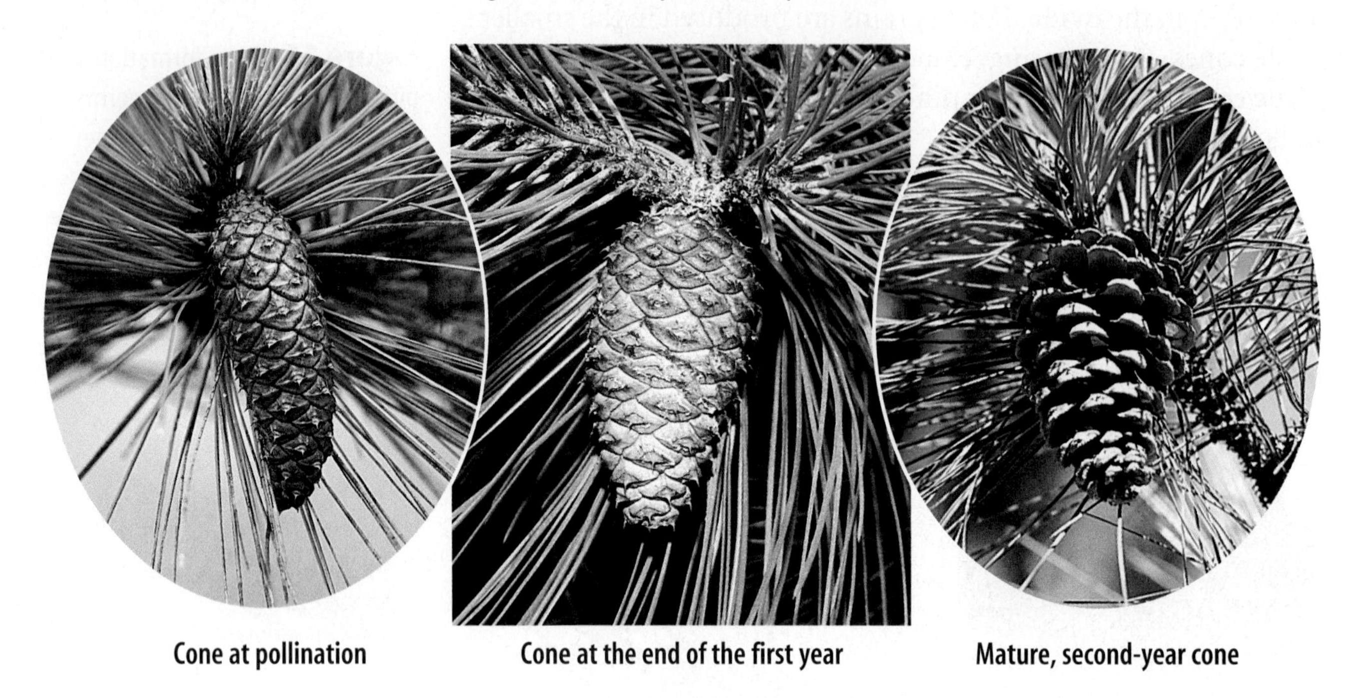

Cone at pollination

Cone at the end of the first year

Mature, second-year cone

Gymnosperm Seeds Pollen is carried from male cones to female cones by the wind. However, most of the pollen falls on other plants, the ground, and bodies of water. To be useful, the pollen has to be blown between the scales of a female cone. There it can be trapped in the sticky fluid secreted by the ovule. If the pollen grain and the female cone are the same species, fertilization and the formation of a seed can take place.

If you are near a pine tree when the female cones release their seeds, you might hear a crackling noise as the cones' scales open. It can take a long time for seeds to be released from a female pine cone. From the moment a pollen grain falls on the female cone until the seeds are released, can take two or three years, as shown in **Figure 12.** In the right environment, each seed can grow into a new pine sporophyte.

Seed Germination Some gymnosperm seeds will not germinate until the heat of a fire causes the cones to open and release the seeds. Without fires, these plants cannot reproduce. In your Science Journal, explain why some forest fires could be good for the environment.

Angiosperm Reproduction

You might not know it, but you are already familiar with angiosperms. If you had cereal for breakfast or bread in a sandwich for lunch, you ate parts of angiosperms. Flowers that you send or receive for special occasions are from angiosperms. Most of the seed plants on Earth today are angiosperms.

All angiosperms have flowers. The sporophyte plant produces the flowers. Flowers are important because they are reproductive organs. Flowers contain gametophyte structures that produce sperm or eggs for sexual reproduction.

The Flower When you think of a flower, you probably imagine something with a pleasant aroma and colorful petals. Although many such flowers do exist, some flowers are drab and have no aroma, like the flowers of the maple tree shown in **Figure 13.** Why do you think such variety among flowers exists?

Most flowers have four main parts—petals, sepals, stamen, and pistil—as shown in **Figure 14.** Generally, the colorful parts of a flower are the petals. Outside the petals are usually leaflike parts called sepals. Sepals form the outside of the flower bud. Sometimes petals and sepals are the same color.

Inside the flower are the reproductive organs of the plant. The **stamen** is the male reproductive organ. Pollen is produced in the stamen. The **pistil** is the female reproductive organ. The **ovary** is the swollen base of the pistil where ovules are found. Not all flowers have every one of the four parts. Remember the holly plants you learned about at the beginning of the chapter? What flower part would be missing on a flower from a male holly plant?

Figure 13 Maple trees produce clusters of flowers early in the spring.
Describe *how these flowers are different from those of the crocus shown in* ***Figure 1.***

Reading Check ***Where are ovules found in the flower?***

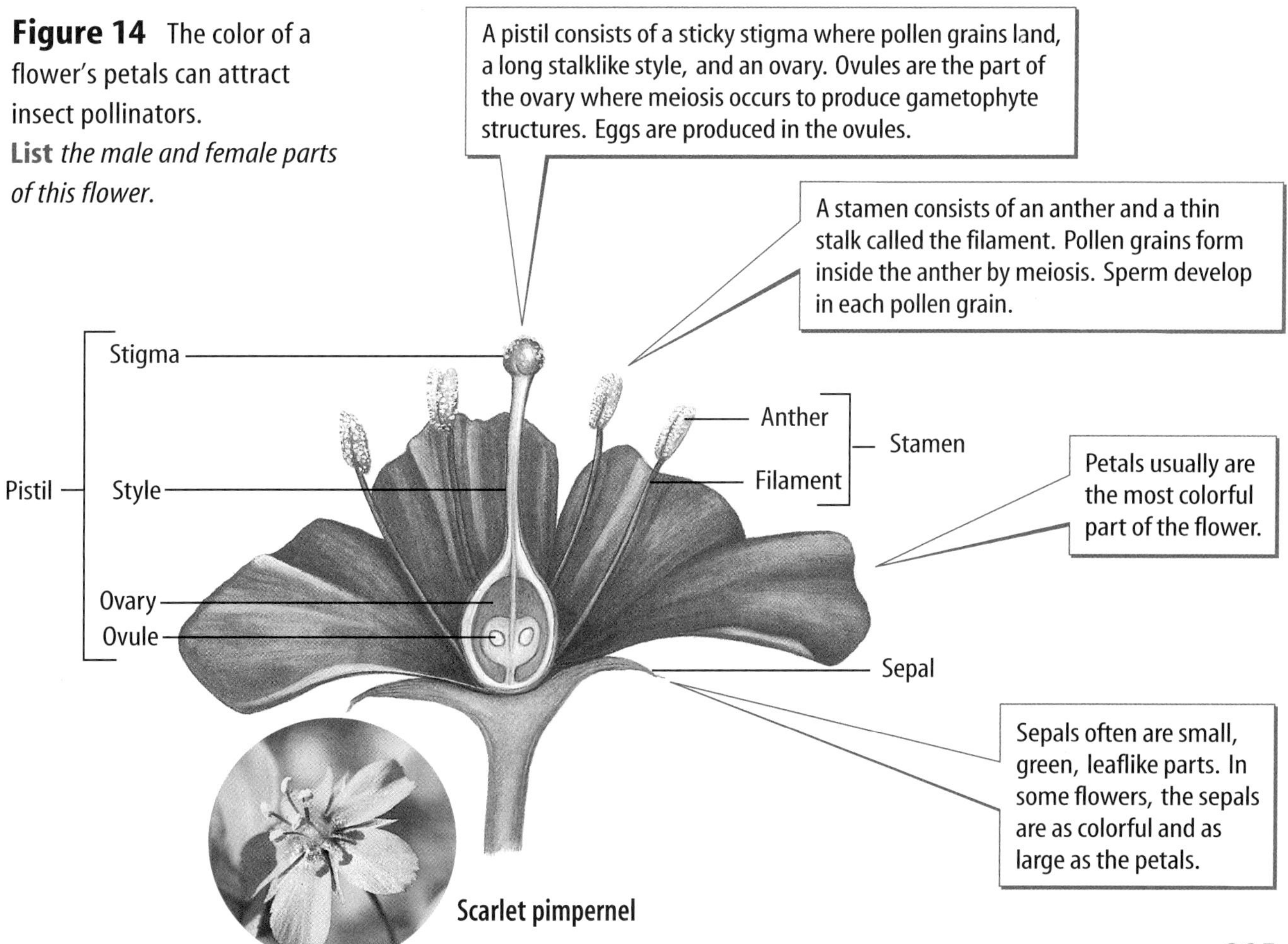

A pistil consists of a sticky stigma where pollen grains land, a long stalklike style, and an ovary. Ovules are the part of the ovary where meiosis occurs to produce gametophyte structures. Eggs are produced in the ovules.
A stamen consists of an anther and a thin stalk called the filament. Pollen grains form inside the anther by meiosis. Sperm develop in each pollen grain.
Stigma
Pistil
Style
Ovary
Ovule
Anther
Stamen
Filament
Petals usually are the most colorful part of the flower.
Sepal
Sepals often are small, green, leaflike parts. In some flowers, the sepals are as colorful and as large as the petals.
Scarlet pimpernel

Figure 14 The color of a flower's petals can attract insect pollinators.
List *the male and female parts of this flower.*

Figure 15 Looking at flowers will give you a clue about how each one is pollinated.

Honeybees are important pollinators. They are attracted to brightly colored flowers, especially blue and yellow flowers.

Flowers that are pollinated at night, like this cactus flower being pollinated by a bat, are usually white.

Flowers that are pollinated by hummingbirds usually are brightly colored, especially bright red and yellow.

Flowers that are pollinated by flies usually are dull red or brown. They often have a strong odor like rotten meat.

The flower of this wheat plant does not have a strong odor and is not brightly colored. Wind, not an animal, is the pollinator of wheat and most other grasses.

Importance of Flowers The appearance of a plant's flowers can tell you something about the life of the plant. Large flowers with brightly colored petals often attract insects and other animals, as shown in **Figure 15.** These animals might eat the flower, its nectar, or pollen. As they move about the flower, the animals get pollen on their wings, legs, or other body parts. Later, these animals spread the flower's pollen to other plants that they visit. Other flowers depend on wind, rain, or gravity to spread their pollen. Their petals can be small or absent. Flowers that open only at night, such as the cactus flower in **Figure 15,** usually are white or yellow and have strong scents to attract animal pollinators. Following pollination and fertilization, the ovules of flowers can develop into seeds.

How do animals spread pollen?

Angiosperm Seeds The development of angiosperm seeds is shown in **Figure 16.** Pollen grains reach the stigma in a variety of ways. Pollen is carried by wind, rain, or animals such as insects, birds, and mammals. A flower is pollinated when pollen grains land on the sticky stigma. A pollen tube grows from the pollen grain down through the style. The pollen tube enters the ovary and reaches an ovule. The sperm then travels down the pollen tube and fertilizes the egg in the ovule. A zygote forms and grows into the plant embryo.

Figure 16 In angiosperms, seed formation begins with the formation of sperm and eggs in the male and female flower parts.

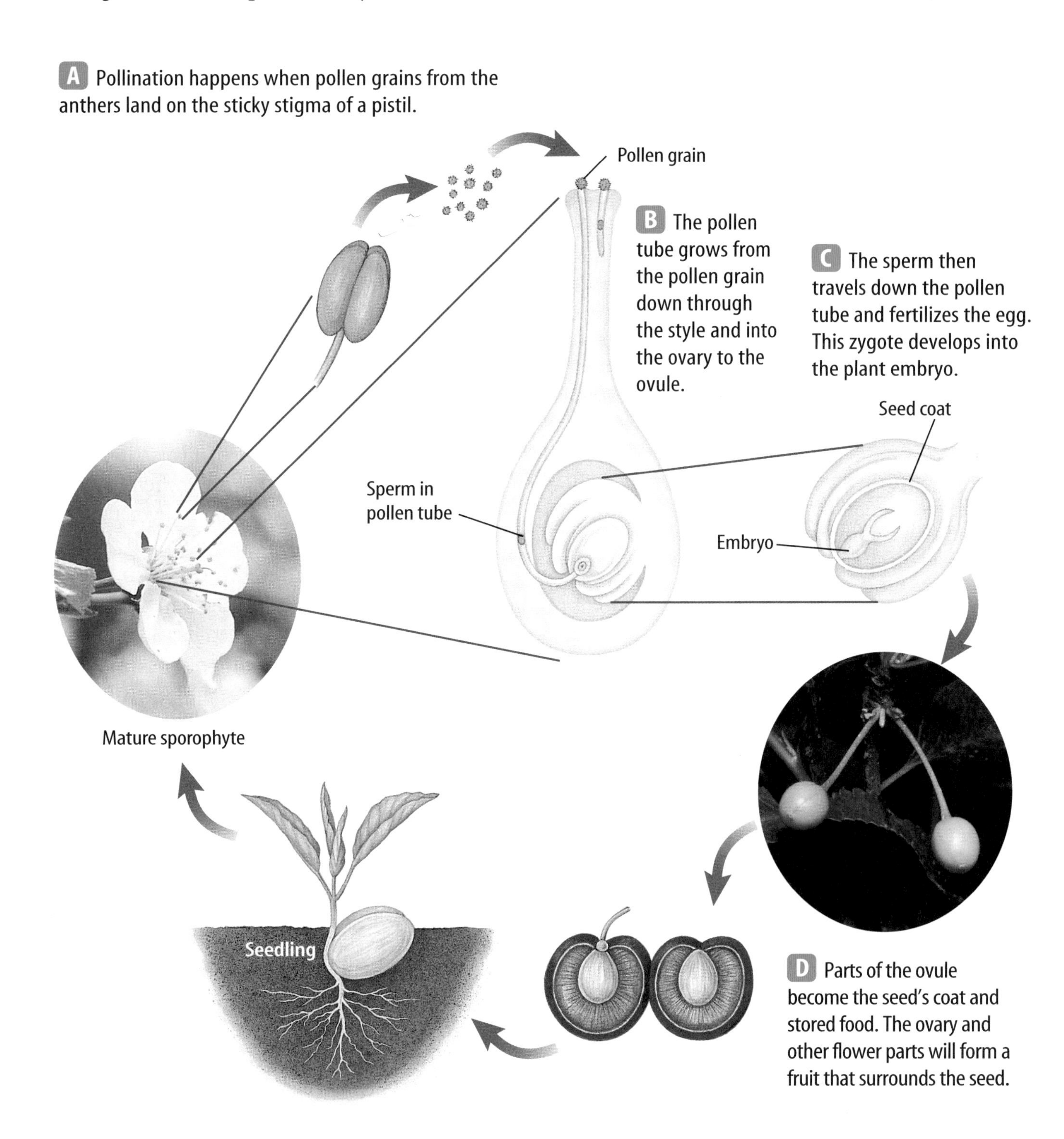

A Pollination happens when pollen grains from the anthers land on the sticky stigma of a pistil.
Pollen grain
B The pollen tube grows from the pollen grain down through the style and into the ovary to the ovule.
C The sperm then travels down the pollen tube and fertilizes the egg. This zygote develops into the plant embryo.
Seed coat
Sperm in pollen tube
Embryo
Mature sporophyte
Seedling
D Parts of the ovule become the seed's coat and stored food. The ovary and other flower parts will form a fruit that surrounds the seed.

Figure 17 Seeds of land plants are capable of surviving unfavorable environmental conditions.
1. Immature plant
2. Cotyledon(s)
3. Seed coat
4. Endosperm

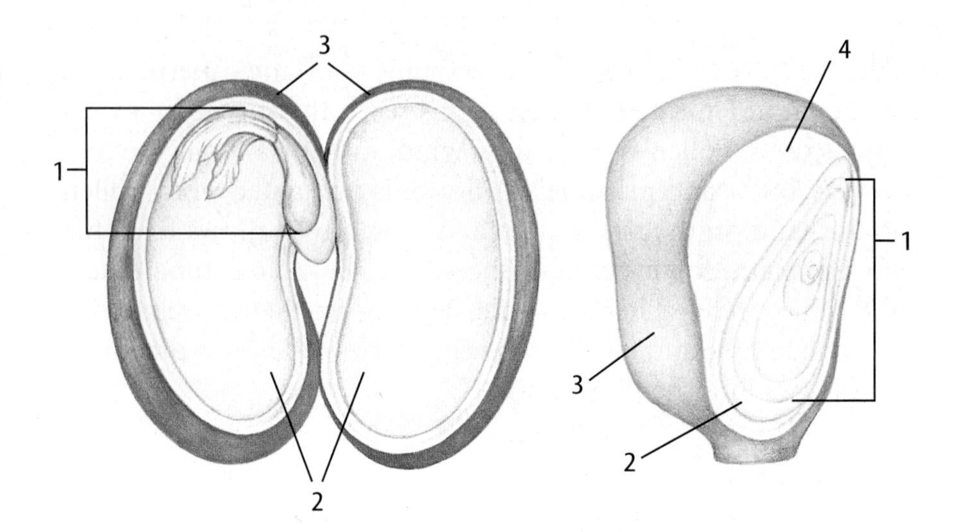

Seed Development Parts of the ovule develop into the stored food and the seed coat that surround the embryo, and a seed is formed, as shown in **Figure 17.** In the seeds of some plants, like beans and peanuts, the food is stored in structures called cotyledons. The seeds of other plants, like corn and wheat, have food stored in a tissue called endosperm.

Seed Dispersal

INTEGRATE Earth Science Sometimes, plants just seem to appear. They probably grew from a seed, but where did the seed come from? Plants have many ways of dispersing their seeds, as shown in **Figure 18.** Most seeds grow only when they are placed on or in soil. Do you know how seeds naturally get to the soil? For many seeds, gravity is the answer. They fall onto the soil from the parent plant on which they grew. However, in nature some seeds can be spread great distances from the parent plant.

Wind dispersal usually occurs because a seed has an attached structure that moves it with air currents. Some plants have very small seeds that become airborne when released by the plant.

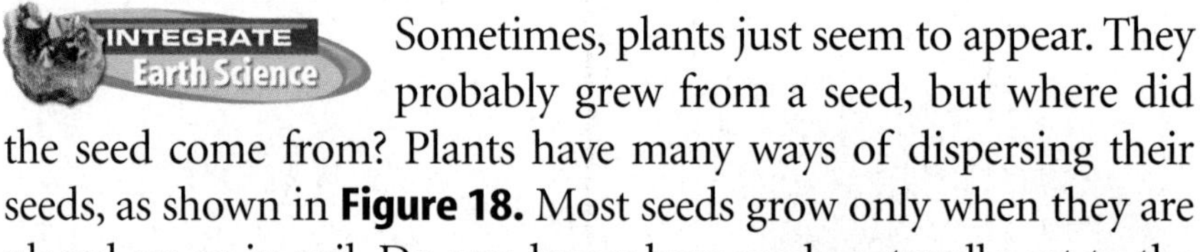

Reading Check *How can wind disperse seeds?*

Animals can disperse many seeds. Some seeds are eaten with fruits, pass through an animal's digestive system, and are dispersed as the animal moves from place to place. Seeds can be carried great distances and stored or buried by animals. Attaching to fur, feathers, and clothing is another way that seeds are dispersed by animals.

Water also disperses seeds. Raindrops can knock seeds out of a dry fruit. Some fruits and seeds float on flowing water or ocean currents. When you touch the seedpod of an impatiens flower, it explodes. The tiny seeds are ejected and spread some distance from the plant.

Mini LAB

Modeling Seed Dispersal

Procedure
1. Find a **button** you can use to represent a seed.
2. Examine the seeds pictured in **Figure 18** and invent a way that your button seed could be dispersed by wind, water, on the fur of an animal, or by humans.
3. Bring your button seed to class and demonstrate how it could be dispersed.

Analysis
1. Explain how your button seed was dispersed.
2. In your **Science Journal,** write a paragraph describing your model. Also describe other ways you could model seed dispersal.

Try at Home

NATIONAL GEOGRAPHIC VISUALIZING SEED DISPERSAL

Figure 18

Plants have many adaptations for dispersing seeds, often enlisting the aid of wind, water, or animals.

▲ Equipped with tiny hooks, burrs cling tightly to fur and feathers.

▲ Pressure builds within the seedpods of this jewelweed plant until the pod bursts, flinging seeds far and wide.

▼ Dandelion seeds are easily dislodged and sail away on a puff of wind.

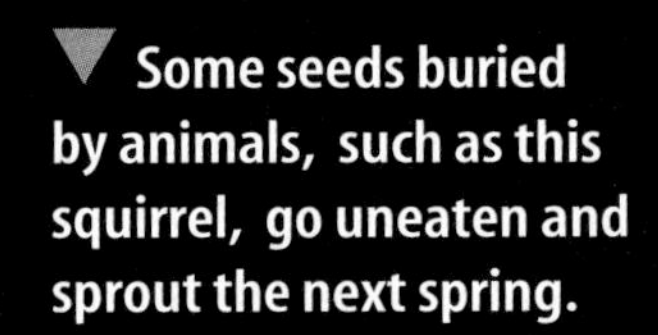

▼ Some seeds buried by animals, such as this squirrel, go uneaten and sprout the next spring.

▲ Encased in a thick, buoyant husk, a coconut may be carried hundreds of kilometers by ocean currents.

▶ Blackberry seeds eaten by this white-footed mouse will pass through its digestive tract and be deposited in a new location.

Germination A series of events that results in the growth of a plant from a seed is called **germination.** When dispersed from the plant, some seeds germinate in just a few days and other seeds take weeks or months to grow. Some seeds can stay in a resting stage for hundreds of years. In 1982, seeds of the East Indian lotus sprouted after 466 years.

Seeds will not germinate until environmental conditions are right. Temperature, the presence or absence of light, availability of water, and amount of oxygen present can affect germination. Sometimes the seed must pass through an animal's digestive system before it will germinate. Germination begins when seed tissues absorb water. This causes the seed to swell and the seed coat to break open.

Applying Math — Calculate Using Percents

HOW MANY SEEDS WILL GERMINATE? The label on a packet of carrot seeds says that it contains about 200 seeds. It also claims that 95 percent of the seeds will germinate. How many seeds should germinate if the packet is correct?

Solution

1 *This is what you know:*
- quantity = 200
- percentage = 95

2 *This is what you need to find out:*

What is 95 percent of 200?

3 *This is the procedure you need to use:*
- Set up the equation for finding percentage: $\frac{95}{100} = \frac{x}{200}$
- Solve the equation for x: $x = \frac{95 \times 200}{100}$

4 *Check your answer:*

Divide by 200 then multiply by 100. You should get the original percentage of 95.

Practice Problems

1. The label on a packet of 50 corn kernels claims that 98 percent will germinate. How many kernels will germinate if the packet is correct?
2. A seed catalog states that a packet contains 1,120 spinach seeds with a germination rate of 65 percent. How many spinach plants should the packet produce?

For more practice, visit life.msscience.com/math_practice

Figure 19 Seed germination results in a new plant.

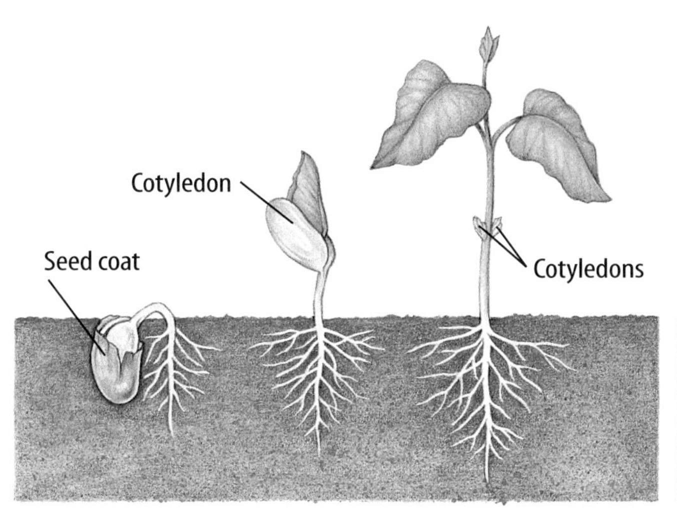

In beans, the cotyledons rise above the soil. As the stored food is used, the cotyledons shrivel and fall off.

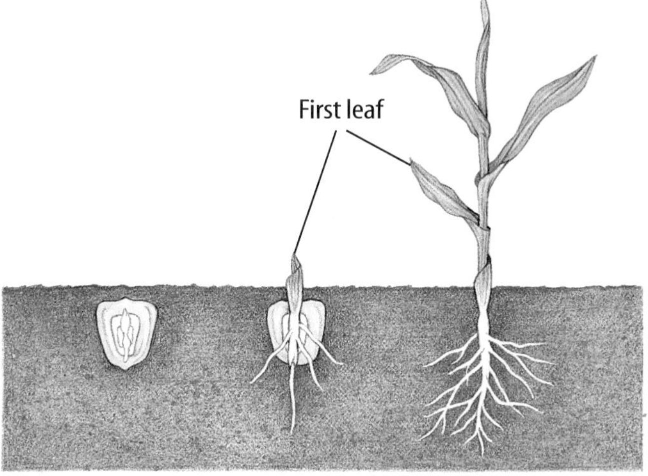

In corn, the stored food in the endosperm remains in the soil and is gradually used as the plant grows.

Next, a series of chemical reactions occurs that releases energy from the stored food in the cotyledons or endosperm for growth. Eventually, a root grows from the seed, followed by a stem and leaves as shown in **Figure 19.** After the plant emerges from the soil, photosynthesis can begin. Photosynthesis provides food as the plant continues to grow.

section 3 review

Summary

The Importance of Pollen and Seeds

- In seed plants, spores develop into pollen grains.
- Pollination is the transfer of pollen from a male plant part to a female plant part.

Gymnosperm Reproduction

- Cones are reproductive structures of gymnosperms.
- Seeds are produced in female cones.

Angiosperm Reproduction

- Flowers are reproductive structures of angiosperms.
- Female flower parts develop into seeds.

Seed Dispersal

- Seeds can be dispersed in several ways.
- Germination is the growth of a plant from a seed.

Self Check

1. **Compare and contrast** the life cycles of gymnosperms and angiosperms.
2. **Draw and label** a diagram showing all four parts of a flower.
3. **Describe** the three parts of a seed and their functions.
4. **Explain** the process of germination.
5. **Think Critically** Walnut trees produce edible seeds with a hard outer covering. Maple trees produce seeds with winglike edges. What type of seed dispersal applies to each type of tree?

Applying Skills

6. **Research information** to find out what conditions are needed for seed germination of three plants, such as corn, peas, and beans. How long does each type of seed take to germinate?

Germination Rate of Seeds

Goals

- **Design** an experiment to test the effect of an environmental factor on seed germination rate.
- **Compare** germination rates under different conditions.

Possible Materials

seeds
water
salt
potting soil
plant trays or plastic cups
**seedling warming cables*
thermometer
graduated cylinder
beakers
**Alternate materials*

Safety Precautions

WARNING: *Some kinds of seeds are poisonous. Do not place any seeds in your mouth. Be careful when using any electrical equipment to avoid shock hazards.*

Real-World Question

Many environmental factors affect the germination rate of seeds. Among these are soil temperature, air temperature, moisture content of soil, and salt content of soil. What happens to the germination rate when one of these variables is changed? How do environmental factors affect seed germination? Can you determine a way to predict the best conditions for seed germination?

Form a Hypothesis

Based on your knowledge of seed germination, state a hypothesis to explain how environmental factors affect germination rates.

Matt Meadows

Test Your Hypothesis

Make a Plan

1. As a group, agree upon and write your hypothesis and decide how you will test it. Identify which results will confirm the hypothesis.
2. **List** the steps you need to take to test your hypothesis. Be specific, and describe exactly what you will do at each step. List your materials.
3. **Prepare** a data table in your Science Journal to record your observations.
4. Reread your entire experiment to make sure that all of the steps are in a logical order.
5. **Identify** all constants, variables, and controls of the experiment.

Follow Your Plan

1. Make sure your teacher approves your plan and your data table before you proceed.
2. Use the same type and amount of soil in each tray.
3. While the experiment is going on, record your observations accurately and complete the data table in your Science Journal.

Analyze Your Data

1. **Compare** the germination rate in the two groups of seeds.
2. **Compare** your results with those of other groups.
3. Did changing the variable affect germination rates? Explain.
4. Make a bar graph of your experimental results.

Conclude and Apply

1. **Interpret** your graph to estimate the conditions that give the best germination rate.
2. **Describe** the conditions that affect germination rate.

Communicating Your Data

Write a short article for a local newspaper telling about this experiment. Give some ideas about when and how to plant seeds in the garden and the conditions needed for germination.

TIME SCIENCE AND Society

SCIENCE ISSUES THAT AFFECT YOU!

Genetic Engineering

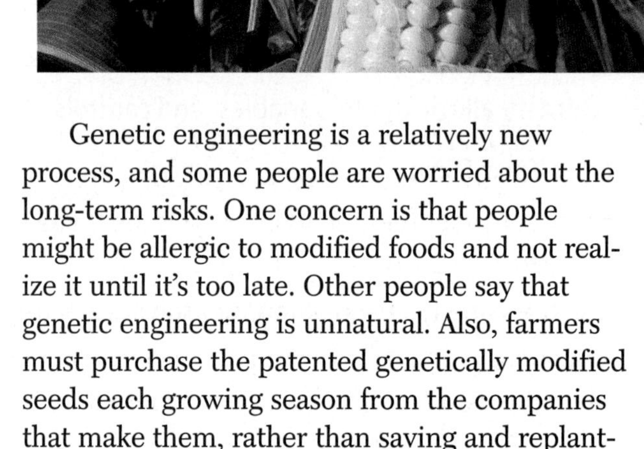

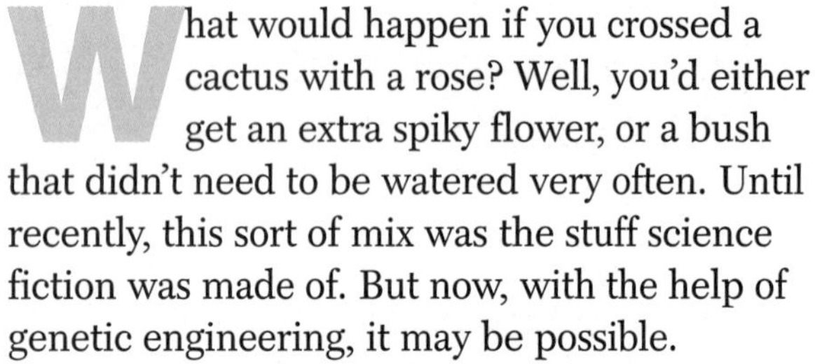

Genetically modified "super" corn can resist heat, cold, drought, and insects.

What would happen if you crossed a cactus with a rose? Well, you'd either get an extra spiky flower, or a bush that didn't need to be watered very often. Until recently, this sort of mix was the stuff science fiction was made of. But now, with the help of genetic engineering, it may be possible.

Genetic engineering is a way of taking genes—sections of DNA that produce certain traits, like the color of a flower or the shape of a nose—from one species and giving them to another.

In 1983, the first plant was genetically modified, or changed. Since then, many crops in the U.S. have been modified in this way, including soybeans, potatoes, tomatoes, and corn.

One purpose of genetic engineering is to transfer an organism's traits. For example, scientists have changed lawn grass by adding to it the gene from another grass species. This gene makes lawn grass grow slowly, so it doesn't have to be mowed very often. Genetic engineering can also make plants that grow bigger and faster, repel insects, or resist herbicides. These changes could allow farmers to produce more crops with fewer chemicals. Scientists predict that genetic engineering soon will produce crops that are more nutritious and that can resist cold, heat, or even drought.

Genetic engineering is a relatively new process, and some people are worried about the long-term risks. One concern is that people might be allergic to modified foods and not realize it until it's too late. Other people say that genetic engineering is unnatural. Also, farmers must purchase the patented genetically modified seeds each growing season from the companies that make them, rather than saving and replanting the seeds from their current crops.

People in favor of genetic engineering reply that there are always risks with new technology, but proper precautions are being taken. Each new plant is tested and then approved by U.S. governmental agencies. And they say that most "natural" crops aren't really natural. They are really hybrid plants bred by agriculturists, and they couldn't survive on their own.

As genetic engineering continues, so does the debate.

Debate **Research the pros and cons of genetic engineering at the link shown to the right. Decide whether you are for or against genetic engineering. Debate your decision with a classmate.**

Science Online
For more information, visit
life.msscience.com/time

Reviewing Main Ideas

Section 1 Introduction to Plant Reproduction

1. Plants reproduce sexually and asexually. Sexual reproduction involves the formation of sex cells and fertilization.
2. Asexual reproduction does not involve sex cells and produces plants genetically identical to the parent plant.
3. Plant life cycles include a gametophyte and a sporophyte stage. The gametophyte stage begins with meiosis. The sporophyte stage begins when the egg is fertilized by a sperm.
4. In some plant life cycles, the sporophyte and gametophyte stages are separate and not dependent on each other. In other plant life cycles, they are part of the same organism.

Section 2 Seedless Reproduction

1. For liverworts and mosses, the gametophyte stage is the familiar plant form. The sporophyte stage produces spores.
2. In ferns, the sporophyte stage is the familiar plant form.
3. Seedless plants, like mosses and ferns, use sexual reproduction to produce spores.

Section 3 Seed Reproduction

1. In seed plants the male reproductive organs produce pollen grains that eventually contain sperm. Eggs are produced in the ovules of the female reproductive organs.
2. The male and female reproductive organs of gymnosperms are called cones. Wind usually moves pollen from the male cone to the female cone for pollination.
3. The reproductive organs of angiosperms are in a flower. The male reproductive organ is the stamen, and the female reproductive organ is the pistil. Gravity, wind, rain, and animals can pollinate a flower.
4. Seeds of gymnosperms and angiosperms are dispersed in many ways. Wind, water, and animals spread seeds. Some plants can eject their seeds.
5. Germination is the growth of a plant from a seed.

Visualizing Main Ideas

Copy and complete the following table that compares reproduction in different plant groups.

Color-enhanced SEM Magnification: 100×

Plant Reproduction

Plant Group	Seeds?	Pollen?	Cones?	Flowers?
Mosses				
Ferns		Do not write in this book.		
Gymnosperms				
Angiosperms				

Using Vocabulary

frond p. 278
gametophyte stage p. 275
germination p. 290
ovary p. 285
ovule p. 283
pistil p. 285
pollen grain p. 281
pollination p. 281
prothallus p. 278
rhizome p. 278
sori p. 278
spore p. 275
sporophyte stage p. 275
stamen p. 285

Fill in the blank with the correct vocabulary word or words.

1. A(n) _________ is the leaf of a fern.
2. In seed plants, the _________ contains the egg.
3. The plant structures in the _________ are made up of haploid cells.
4. The green, leafy moss plant is part of the _________ in the moss life cycle.
5. Two parts of a sporophyte fern are a frond and _________.
6. The female reproductive organ of the flower is the _________.
7. The _________ is the swollen base of the pistil.

Checking Concepts

Choose the word or phrase that best answers the question.

8. How are colorful flowers usually pollinated?
   A) insects **C)** clothing
   B) wind **D)** gravity
9. What type of reproduction produces plants that are genetically identical?
   A) asexual **C)** spore
   B) sexual **D)** flower
10. Which of the following terms describes the cells in the gametophyte stage?
   A) haploid **C)** diploid
   B) prokaryote **D)** missing a nucleus
11. What structures do ferns form when they reproduce sexually?
   A) spores **C)** seeds
   B) anthers **D)** flowers
12. What contains food for the plant embryo?
   A) endosperm **C)** stigma
   B) pollen grain **D)** root

Use the photo below to answer question 13.

13. What disperses the seeds shown above?
   A) rain **C)** wind
   B) animals **D)** insects
14. What is the series of events that results in a plant growing from a seed?
   A) pollination **C)** germination
   B) prothallus **D)** fertilization
15. In seedless plants, meiosis produces what kind of plant structure?
   A) prothallus **C)** flowers
   B) seeds **D)** spores
16. Ovules and pollen grains take part in what process?
   A) germination
   B) asexual reproduction
   C) seed dispersal
   D) sexual reproduction
17. What part of the flower receives the pollen grain from the anther?
   A) sepal **C)** stamen
   B) petal **D)** stigma

Science Online life.msscience.com/vocabulary_puzzlemaker

Thinking Critically

18. **Explain** why male cones produce so many pollen grains.

19. **Predict** whether a seed without an embryo could germinate. Explain your answer.

20. **Discuss** the importance of water in the sexual reproduction of nonvascular plants and ferns.

21. **Infer** why the sporophyte stage in mosses is dependent on the gametophyte stage.

22. **List** the features of flowers that ensure pollination.

23. **Compare and contrast** the fern sporophyte and gametophyte stages.

24. **Interpret Scientific Illustrations** Using **Figure 16,** sequence these events.
    - pollen is trapped on the stigma
    - pollen tube reaches the ovule
    - fertilization
    - pollen released from the anther
    - pollen tube forms through the style
    - a seed forms

25. **Concept Map** Copy and complete this concept map of a typical plant life cycle.

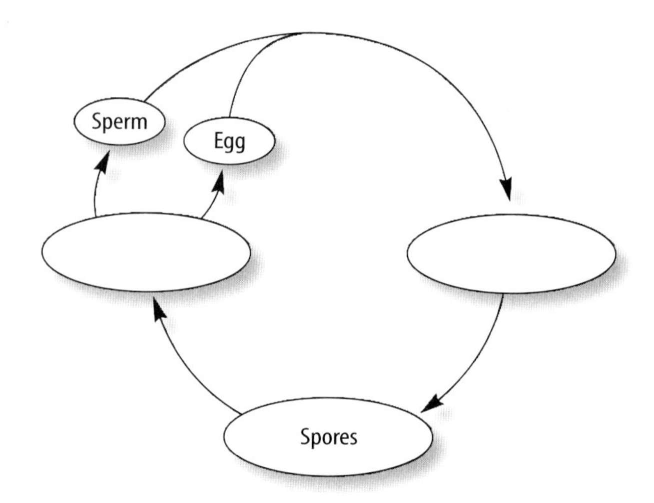

26. **Predict** Observe pictures of flowers or actual flowers and predict how they are pollinated. Explain your prediction.

Performance Activities

27. **Display** Collect several different types of seeds and use them to make a mosaic picture of a flower.

28. **Technical Writing** Write a newspaper story to tell people about the importance of gravity, water, wind, insects, and other animals in plant life cycles.

Applying Math

29. **Germination Rates** A seed producer tests a new batch of corn seeds before putting them on the market. The producer plants a sample of 150 seeds, and 110 of the seeds germinate. What is the germination rate for this batch of corn seeds?

30. **Seed Production** Each blossom on an apple tree, if fertilized, can become a fruit. Suppose an apple tree bears 1,200 blossoms in the spring. If 95 percent are pollinated, how many apples could the tree produce? If each apple contains five seeds, how many seeds would the tree produce?

Use the table below to answer question 31.

Onion Seed Data

Temperature (°C)	10	15	20	25	30	35
Days to germinate	13	7	5	4	4	13

31. **Onion Seeds** Make a bar graph for the following data table about onion seeds. Put days on the horizontal axis and temperature on the vertical axis.

Part 1 Multiple Choice

Record your answers on the answer sheet provided by your teacher or on a sheet of paper.

1. Which statement applies to asexual reproduction?
A. Sperm and egg are required.
B. Offspring are genetically different from the parents.
C. Most animals reproduce in this way.
D. Offspring are genetically identical to the parent.

2. Which term describes the uniting of a sperm and egg to form a zygote?
A. fertilization **C.** pollination
B. meiosis **D.** germination

Use the picture below to answer questions 3 and 4.

3. What is the primary method by which these horsetail spores are dispersed?
A. water **C.** wind
B. insects **D.** grazing animals

Test-Taking Tip

Come Back To It Never skip a question. If you are unsure of an answer, mark your best guess on another sheet of paper and mark the question in your test booklet to remind you to come back to it at the end of the test.

4. The horsetail plant that produced these spores uses tubelike cells to transport water and other substances from one part of the plant to another. What type of plant is a horsetail?
A. vascular **C.** nonvascular
B. seed **D.** pollinated

5. Which of the following is a characteristic of angiosperms?
A. production of cones
B. seeds not protected by fruit
C. growth from a rhizome
D. production of flowers

Use the illustration below to answer questions 6 and 7.

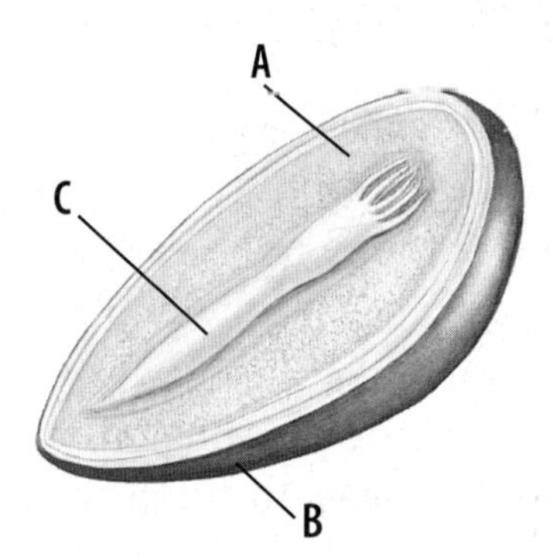

6. Structure B represents which part of this seed?
A. stored food **C.** seed coat
B. embryo **D.** ovary

7. Which part(s) of this seed will grow into stems, roots, and leaves?
A. A **C.** C
B. B **D.** A and B

8. What causes seed germination to begin?
A. warm temperature
B. exposure to water
C. at least 9 hours of daylight in a 24-hour period
D. soil rich in organic material

Part 2 | Short Response/Grid In

Record your answers on the answer sheet provided by your teacher or on a sheet of paper.

9. Sperm and eggs are found in different parts of plants. Explain why it is important for these cells to unite, and describe some factors in an environment that help unite them.

10. Make a sketch of a fern plant. Label the fronds, rhizome, roots, and sori.

Use the illustration below to answer questions 11 and 12.

11. What type of seed plant produces the structure shown here? Describe how it is involved in the reproduction of this plant.

12. Why are the scales open?

13. Describe the importance of flowers in angiosperms. What factors can differ from one flower to another?

14. Explain the role played by animals that eat fruits in the dispersal and germination of seeds.

15. Describe the characteristics of certain plants, such as grasses, that enable them to be distributed widely in an environment.

Part 3 | Open Ended

Record your answers on a sheet of paper.

16. You have a holly plant in your yard which, despite having ample water, sunlight, and fertilizer, has never produced berries. The flowers produced by this plant have only female structures. What could you do to help this plant produce berries?

17. Why is it important that spores produced during the gametophyte stage of a plant's life cycle be haploid cells?

18. Describe some of the factors that have contributed to the success of seed plants.

Use the illustration below to answer questions 19 and 20.

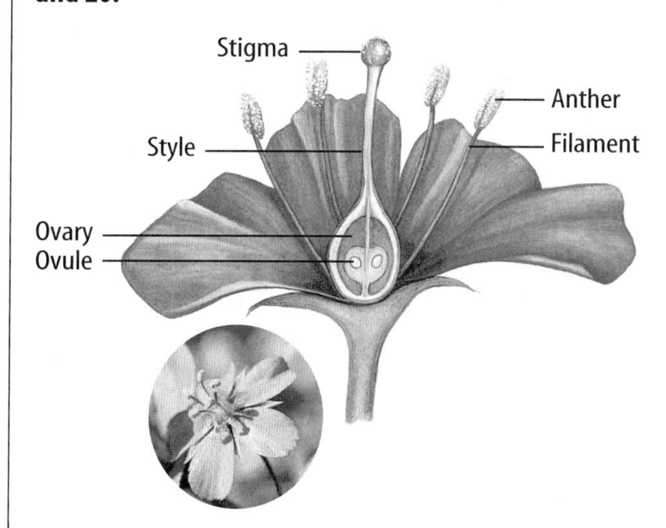

19. Describe the role of each structure labeled in this picture in the production of eggs or sperm.

20. Describe the process of pollination of this plant by insects.

21. Explain how a flower's appearance can indicate its method of pollination. Give three examples of flowers and the method of pollination for each.

Plant Processes

The BIG Idea

Plant processes are necessary for maintaining life on Earth.

SECTION 1
Photosynthesis and Cellular Respiration

Main Idea Photosynthesis and cellular respiration help cycle carbon dioxide and oxygen in the environment.

SECTION 2
Plant Responses

Main Idea Plant responses can result from internal and/or external stimuli.

How did it get so big?

From crabgrass to giant sequoias, many plants start as small seeds. Some trees may grow to be more than 20 m tall. One tree can produce enough lumber to build a house. Where does all that wood come from? Did you know that plants are essential to the survival of all animals on Earth?

Science Journal Describe what would happen to life on Earth if all the green plants disappeared.

Start-Up Activities

Do plants lose water?

Plants, like all other living organisms, are made of cells, reproduce, and need water to live. What would happen if you forgot to water a houseplant? From your own experiences, you probably know that the houseplant would wilt. Do the following lab to discover one way plants lose water.

1. Obtain a self-sealing plastic bag, some aluminum foil, and a small potted plant from your teacher.
2. Using the foil, carefully cover the soil around the plant in the pot. Place the potted plant in the plastic bag.
3. Seal the bag and place it in a sunny window. Wash your hands.
4. Observe the plant at the same time every day for a few days.
5. **Think Critically** Write a paragraph that describes what happened in the bag. If enough water is lost by a plant and not replaced, predict what will happen to the plant.

Photosynthesis and Respiration Make the following Foldable to help you distinguish between photosynthesis and respiration.

STEP 1 **Fold** a vertical sheet of paper in half from top to bottom.

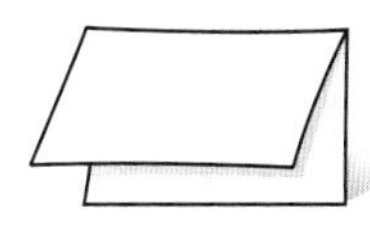

STEP 2 **Fold** in half from side to side with the fold at the top.

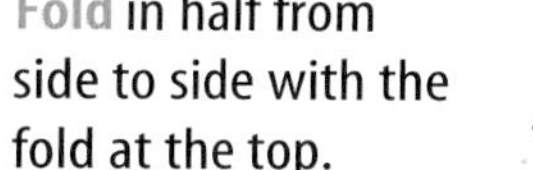

STEP 3 **Unfold** the paper once. **Cut** only the fold of the top flap to make two tabs.

STEP 4 **Turn** the paper vertically and **label** the front tabs as shown.

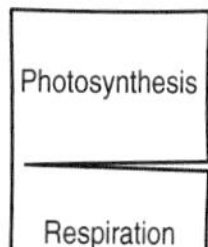

Compare and Contrast As you read the chapter, write the characteristics of respiration and photosynthesis under the appropriate tab.

Preview this chapter's content and activities at life.msscience.com

Get Ready to Read

Compare and Contrast

1 Learn It! Good readers compare and contrast information as they read. This means they look for similarities and differences to help them to remember important ideas. Look for signal words in the text to let you know when the author is comparing or contrasting.

Compare and Contrast Signal Words	
Compare	**Contrast**
as	but
like	or
likewise	unlike
similarly	however
at the same time	although
in a similar way	on the other hand

2 Practice It! Read the excerpt below and notice how the author uses contrast signal words to describe the differences between plant hormones.

> **Like** gibberellins, cytokinins (si tuh KI nunz) also cause rapid growth. Cytokinins promote growth by causing faster cell divisions. **Like** ethylene, the effect of cytokinins on the plant also is controlled by auxins. Interestingly, cytokinins can be sprayed on stored vegetables to keep them fresh longer.

3 Apply It! Compare and contrast plant hormones on page 315.

Reading Tip

As you read, use other skills, such as summarizing and connecting, to help you understand comparisons and contrasts.

Target Your Reading

Use this to focus on the main ideas as you read the chapter.

1. **Before you read** the chapter, respond to the statements below on your worksheet or on a numbered sheet of paper.
   - Write an **A** if you **agree** with the statement.
   - Write a **D** if you **disagree** with the statement.

2. **After you read** the chapter, look back to this page to see if you've changed your mind about any of the statements.
   - If any of your answers changed, explain why.
   - Change any false statements into true statements.
   - Use your revised statements as a study guide.

Before You Read A or D	Statement	After You Read A or D
	1 Photosynthesis generally occurs in a plant's leaves.	
	2 All plant cells can carry out photosynthesis.	
	3 Carbon dioxide and oxygen enter a leaf through stomata.	
	4 A product of photosynthesis is oxygen.	
	5 Plants transform light energy into chemical energy during photosynthesis.	
	6 Cellular respiration transforms stored chemical energy into usable energy for plant cells.	
	7 Plants respond quickly to stimuli in their environments.	
	8 Phototropism is a plant response to light.	
	9 A large quantity of a plant hormone must be present to have any effect on plant growth.	
	10 The number of hours of light influences when a plant flowers.	

Science Online
Print out a worksheet of this page at life.msscience.com

section 1

Photosynthesis and Cellular Respiration

as you read

What You'll Learn

- **Explain** how plants take in and give off gases.
- **Compare and contrast** photosynthesis and cellular respiration.
- **Discuss** why photosynthesis and cellular respiration are important.

Why It's Important

Understanding photosynthesis and respiration in plants will help you understand how life exists on Earth.

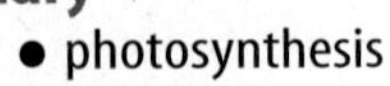

Review Vocabulary

cellulose: chemical compound made of sugar; forms tangled fibers in plant cell walls and provides structure and support

New Vocabulary

- stomata
- chlorophyll
- photosynthesis
- cellular respiration

Taking in Raw Materials

Sitting in the cool shade under a tree, you eat lunch. Food is one of the raw materials you need to grow. Oxygen is another. It enters your lungs and eventually reaches every cell in your body. Your cells use oxygen to help release energy from the food that you eat. The process that uses oxygen to release energy from food produces carbon dioxide and water as wastes. These wastes move in your blood to your lungs, where they are removed as gases when you exhale. You look up at the tree and wonder, "Does a tree need to eat? Does it use oxygen? How does it get rid of wastes?"

Movement of Materials in Plants Trees and other plants don't take in foods the way you do. Plants make their own foods using the raw materials water, carbon dioxide, and inorganic chemicals in the soil. Just like you, plants also produce waste products.

Most of the water used by plants is taken in through roots, as shown in **Figure 1.** Water moves into root cells and then up through the plant to where it is used. When you pull up a plant, its roots are damaged and some are lost. If you replant it, the plant will need extra water until new roots grow to replace those that were lost.

Leaves, instead of lungs, are where most gas exchange occurs in plants. Most of the water taken in through the roots exits through the leaves of a plant. Carbon dioxide, oxygen, and water vapor exit and enter the plant through openings in the leaf. The leaf's structure helps explain how it functions in gas exchange.

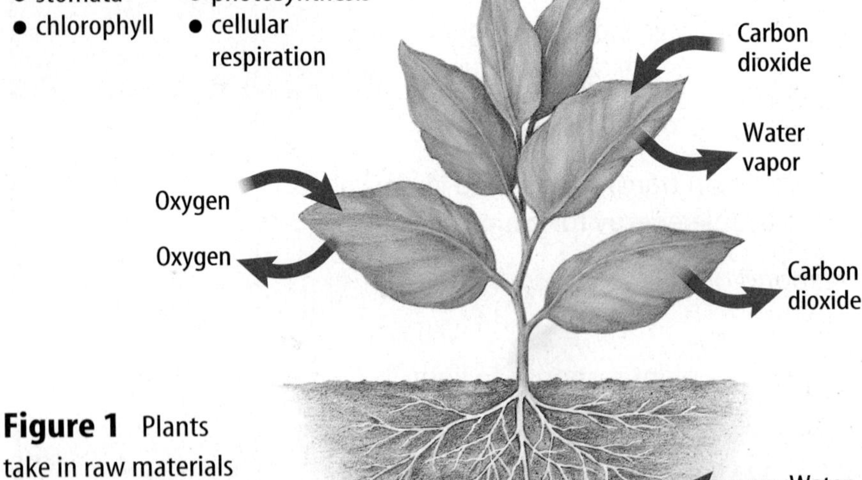

Figure 1 Plants take in raw materials through their roots and leaves and get rid of wastes through their leaves.

Figure 2 A leaf's structure determines its function. Food is made in the inner layers. Most stomata are found on the lower epidermis.
Identify *the layer that contains most of the cells with chloroplasts.*

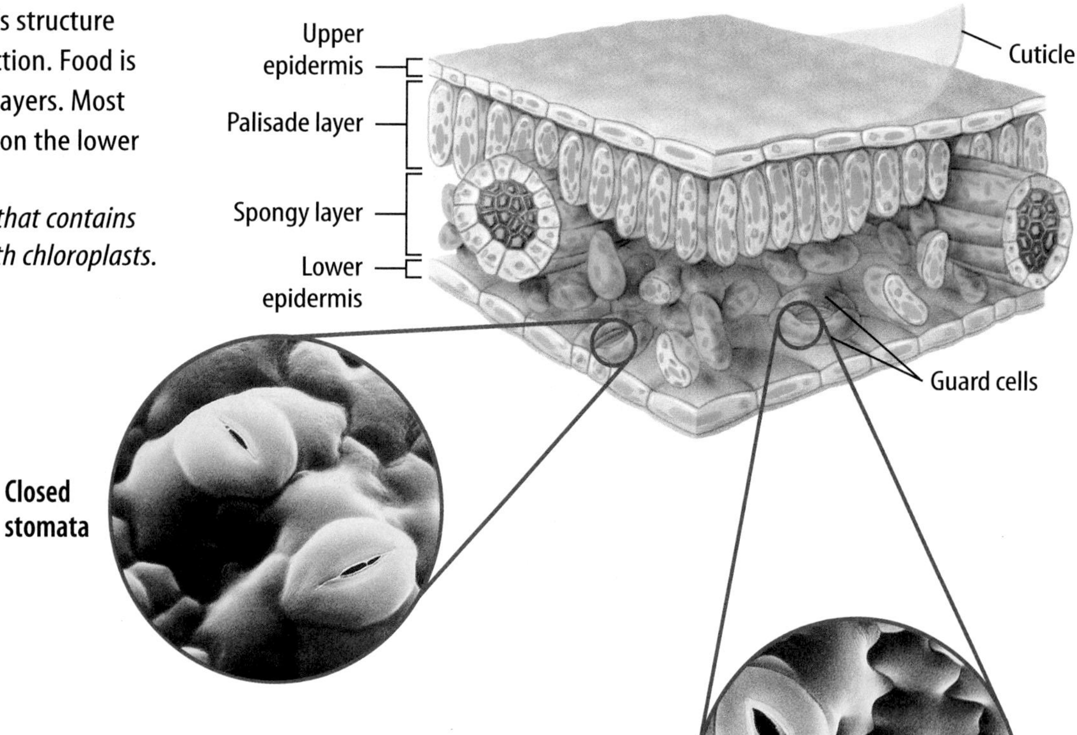

Leaf Structure and Function A leaf is made up of many different layers, as shown in **Figure 2.** The outer cell layer of the leaf is the epidermis. A waxy cuticle that helps keep the leaf from drying out covers the epidermis. Because the epidermis is nearly transparent, sunlight—which is used during food production—reaches the cells inside the leaf. If you examine the epidermis under a microscope, you will see that it contains many small openings. These openings, called **stomata** (stoh MAH tuh) (singular, *stoma*), act as doorways for raw materials such as carbon dioxide, water vapor, and waste gases to enter and exit the leaf. Sto-mata also are found on the stems of many plants. More than 90 percent of the water plants take in through their roots is lost through the stomata. In one day, a growing tomato plant can lose up to 1 L of water.

Two cells called guard cells surround each stoma and control its size. As water moves into the guard cells, they swell and bend apart, opening a stoma. When guard cells lose water, they deflate and close the stoma. **Figure 2** shows closed and open stomata.

Stomata usually are open during the day, when most plants need to take in raw materials to make food. They usually are closed at night when food making slows down. Stomata also close when a plant is losing too much water. This adaptation conserves water, because less water vapor escapes from the leaf.

Inside the leaf are two layers of cells, the spongy layer and the palisade layer. Carbon dioxide and water vapor, which are needed in the food-making process, fill the spaces of the spongy layer. Most of the plant's food is made in the palisade layer.

Nutritionist Vitamins are substances needed for good health. Nutritionists promote healthy eating habits. Research to learn about other roles that nutritionists fulfill. Create a pamphlet to promote the career of nutritionist.

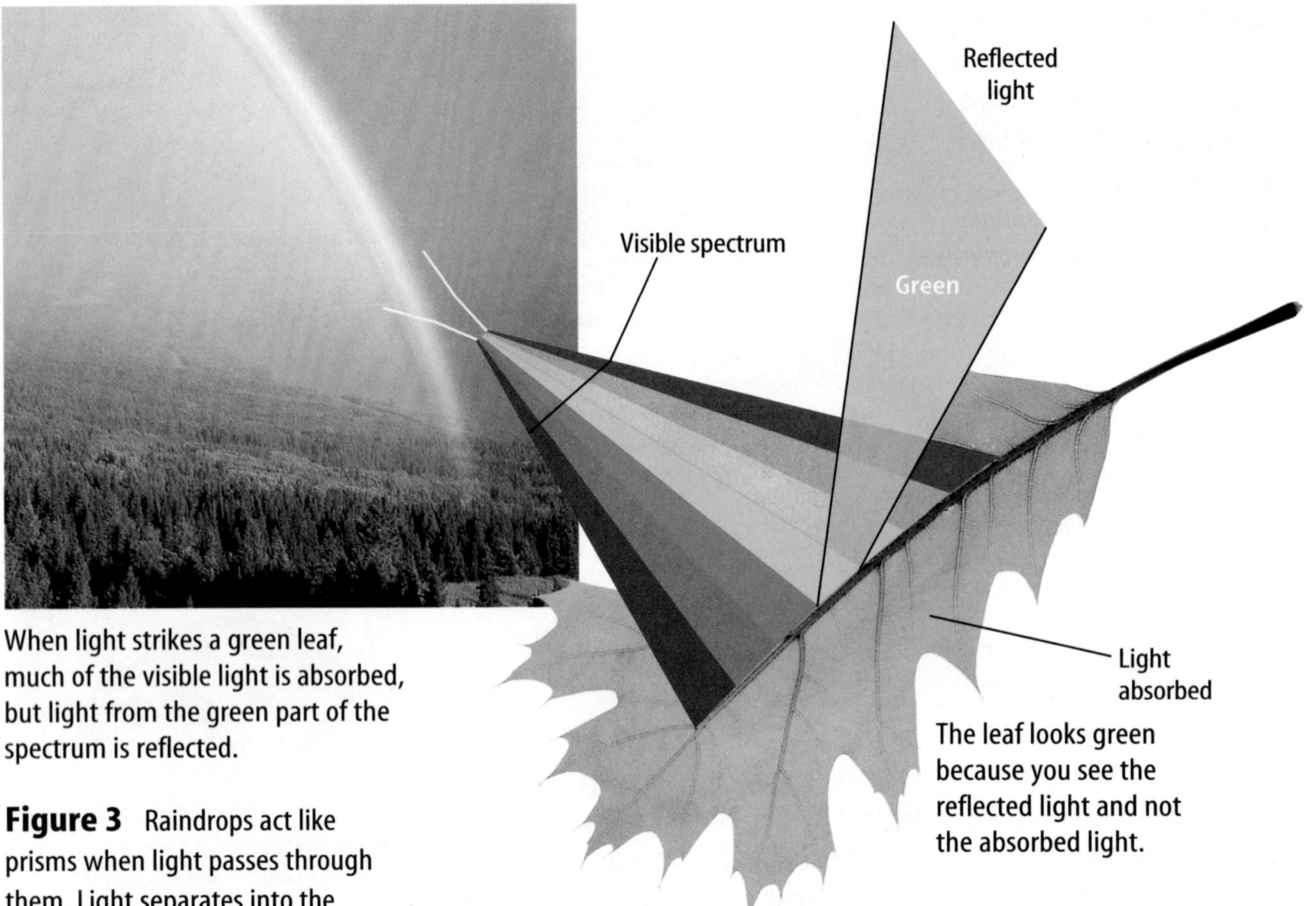
Reflected
light
Visible spectrum
Green
Light
absorbed
The leaf looks green
because you see the
reflected light and not
the absorbed light.

When light strikes a green leaf, much of the visible light is absorbed, but light from the green part of the spectrum is reflected.

Figure 3 Raindrops act like prisms when light passes through them. Light separates into the colors of the visible spectrum. You see a rainbow when this happens.

Chloroplasts and Plant Pigments If you look closely at the leaf in **Figure 2,** you'll see that some of the cells contain small, green structures called chloroplasts. Most leaves look green because some of their cells contain so many chloroplasts. Chloroplasts are green because they contain a green pigment called **chlorophyll** (KLOR uh fihl).

Reading Check *Why are chloroplasts green?*

As shown in **Figure 3,** light from the Sun contains all colors of the visible spectrum. A pigment is a substance that reflects a particular part of the visible spectrum and absorbs the rest. When you see a green leaf, you are seeing green light energy reflected from chlorophyll. Most of the other colors of the spectrum, especially red and blue, are absorbed by chlorophyll. In the spring and summer, most leaves have so much chlorophyll that it hides all other pigments. In fall, the chlorophyll in some leaves breaks down and the leaves change color as other pigments become visible. Pigments, especially chlorophyll, are important to plants because the light energy that they absorb is used while making food. For plants, this food-making process—photosynthesis—happens in the chloroplasts.

The Food-Making Process

Photosynthesis (foh toh SIHN thuh suhs) is the process during which a plant's chlorophyll traps light energy and sugars are produced. In plants, photosynthesis occurs only in cells with chloroplasts. For example, photosynthesis occurs only in a carrot plant's lacy green leaves, shown in **Figure 4.** Because a carrot's root cells lack chlorophyll and normally do not receive light, they can't perform photosynthesis. But excess sugar produced in the leaves is stored in the familiar orange root that you and many animals eat.

Besides light, plants also need the raw materials carbon dioxide and water for photosynthesis. The overall chemical equation for photosynthesis is shown below. What happens to each of the raw materials in the process?

$$\underset{\text{carbon dioxide}}{6CO_2} + \underset{\text{water}}{6H_2O} + \text{light energy} \xrightarrow{\text{chlorophyll}} \underset{\text{glucose}}{C_6H_{12}O_6} + \underset{\text{oxygen}}{6O_2}$$

Light-Dependent Reactions Some of the chemical reactions that take place during photosynthesis require light, but others do not. Those that need light can be called the light-dependent reactions of photosynthesis. During light-dependent reactions, chlorophyll and other pigments trap light energy that eventually will be stored in sugar molecules. Light energy causes water molecules, which were taken up by the roots, to split into oxygen and hydrogen. The oxygen leaves the plant through the stomata. This is the oxygen that you breathe. Hydrogen produced when water is split is used in photosynthesis reactions that occur when there is no light.

Inferring What Plants Need to Produce Chlorophyll

Procedure

1. Cut two pieces of **black construction paper** large enough so that each one completely covers one leaf on a **plant**.
2. Cut a square out of the center of each piece of paper.
3. Sandwich the leaf between the two paper pieces and **tape** the pieces together along their edges.
4. Place the plant in a sunny area. Wash your hands.
5. After seven days, carefully remove the paper and observe the leaf.

Analysis

In your **Science Journal,** describe how the color of the areas covered by paper compare to the areas not covered. Infer why this happened.

Try at Home

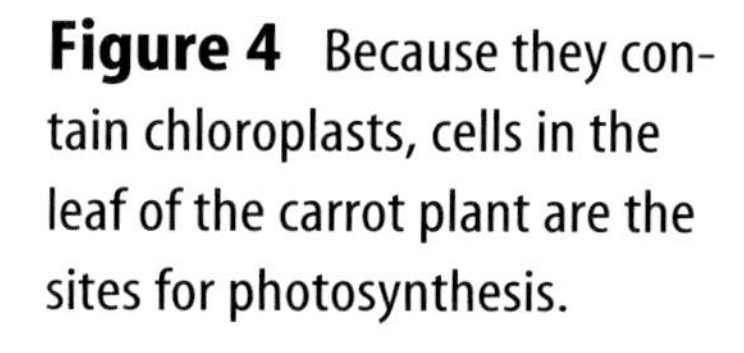

Figure 4 Because they contain chloroplasts, cells in the leaf of the carrot plant are the sites for photosynthesis.

Topic: Plant Sugars

Visit life.msscience.com for Web links to information about sugars and related molecules produced by plants.

Activity List three sugar-containing molecules that plants produce.

Light-Independent Reactions Reactions that don't need light are called the light-independent reactions of photosynthesis. Carbon dioxide, the raw material from the air, is used in these reactions. The light energy trapped during the light-dependent reactions is used to combine carbon dioxide and hydrogen to make sugars. One important sugar that is made is glucose. The chemical bonds that hold glucose and other sugars together are stored energy. **Figure 5** compares what happens during each stage of photosynthesis.

What happens to the oxygen and glucose that were made during photosynthesis? Most of the oxygen from photosynthesis is a waste product and is released through stomata. Glucose is the main form of food for plant cells. A plant usually produces more glucose than it can use. Excess glucose is stored in plants as other sugars and starches. When you eat carrots, as well as beets, potatoes, or onions, you are eating the stored product of photosynthesis.

Glucose also is the basis of a plant's structure. You don't grow larger by breathing in and using carbon dioxide. However, that's exactly what plants do as they take in carbon dioxide gas and convert it into glucose. Cellulose, an important part of plant cell walls, is made from glucose. Leaves, stems, and roots are made of cellulose and other substances produced using glucose. The products of photosynthesis are used for plant growth.

Figure 5 Photosynthesis includes two sets of reactions, the light-dependent reactions and the light-independent reactions.
Describe *what happens to the glucose produced during photosynthesis.*

Standard plant cell

Chloroplast

Light

H_2O

O_2

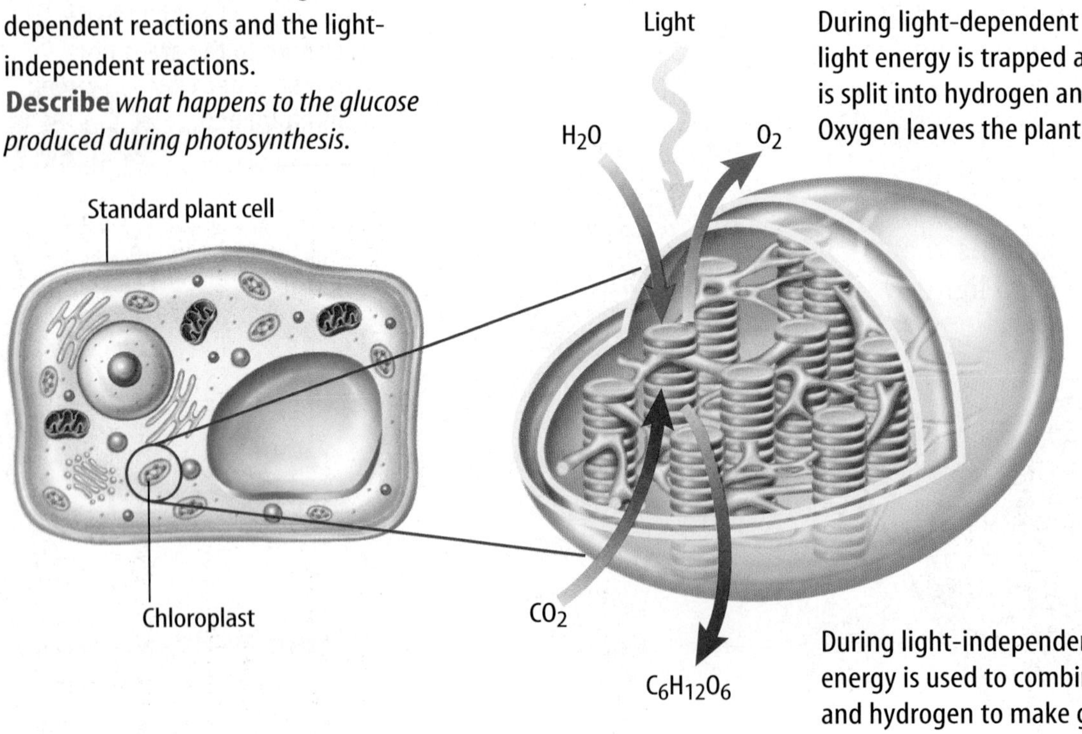

During light-dependent reactions, light energy is trapped and water is split into hydrogen and oxygen. Oxygen leaves the plant.

CO_2

$C_6H_{12}O_6$

During light-independent reactions, energy is used to combine carbon dioxide and hydrogen to make glucose and other sugars.

Figure 6 Tropical rain forests contain large numbers of photosynthetic plants.
Infer *why tropical forests are considered an important source of oxygen.*

Importance of Photosynthesis Why is photosynthesis important to living things? First, photosynthesis produces food. Organisms that carry on photosynthesis provide food directly or indirectly for nearly all the other organisms on Earth. Second, photosynthetic organisms, like the plants in **Figure 6,** use carbon dioxide and release oxygen. This removes carbon dioxide from the atmosphere and adds oxygen to it. Most organisms, including humans, need oxygen to stay alive. As much as 90 percent of the oxygen entering the atmosphere today is a result of photosynthesis.

The Breakdown of Food

Look at the photograph in **Figure 7.** Do the fox and the plants in the photograph have anything in common? They don't look alike, but the fox and the plants are made of cells that break down food and release energy in a process called **cellular respiration.** How does this happen?

Cellular respiration is a series of chemical reactions that breaks down food molecules and releases energy. Cellular respiration occurs in cells of most organisms and uses oxygen to break down food chemically and release energy. In plants and many organisms that have one or more cells, a nucleus, and other organelles, cellular respiration occurs in the mitochondria (singular, *mitochondrion*). The overall chemical equation for cellular respiration is shown below.

$$\underset{\text{glucose}}{C_6H_{12}O_6} + \underset{\text{oxygen}}{6O_2} \longrightarrow \underset{\text{carbon dioxide}}{6CO_2} + \underset{\text{water}}{6H_2O} + \text{energy}$$

Figure 7 You know that animals such as this red fox carry on cellular respiration, but so do all the plants that surround the fox.

Figure 8 Cellular respiration takes place in the mitochondria of plant cells. **Describe** *what happens to a molecule before it enters a mitochondrion.*

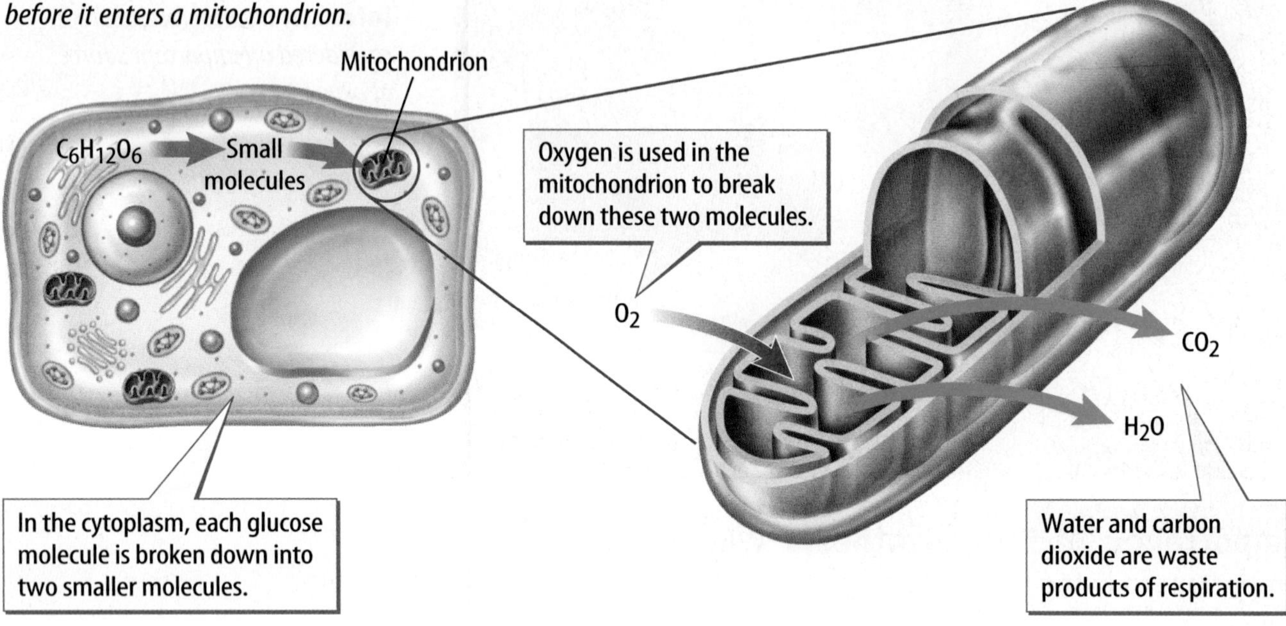

Cellular Respiration Before cellular respiration begins, glucose molecules are broken down into two smaller molecules. This happens in the cytoplasm. The smaller molecules then enter a mitochondrion, where cellular respiration takes place. Oxygen is used in the chemical reactions that break down the small molecules into water and carbon dioxide. The reactions also release energy. Every cell in the organism needs this energy. **Figure 8** shows cellular respiration in a plant cell.

Figure 9 Plants use the energy released during cellular respiration to carry out many functions.

Importance of Respiration Although food contains energy, it is not in a form that can be used by cells. Cellular respiration changes food energy into a form all cells can use. This energy drives the life processes of almost all organisms on Earth.

Reading Check *What organisms use cellular respiration?*

Plants use energy released by cellular respiration to transport sugars and to open and close stomata. Some of the energy is used to produce substances needed for photosynthesis, such as chlorophyll. **Figure 9** shows some uses of energy in plants.

The waste product carbon dioxide is also important. Cellular respiration returns carbon dioxide to the atmosphere, where it can be used again by plants and some other organisms for photosynthesis.

Table 1 Comparing Photosynthesis and Cellular Respiration

	Energy	Raw Materials	End Products	Where
Photosynthesis	stored	water and carbon dioxide	glucose and oxygen	cells with chlorophyll
Cellular respiration	released	glucose and oxygen	water and carbon dioxide	cells with mitochondria

Comparison of Photosynthesis and Cellular Respiration

Look back in the section to find the equations for photosynthesis and cellular respiration. You can see that cellular respiration is almost the reverse of photosynthesis. Photosynthesis combines carbon dioxide and water by using light energy. The end products are glucose (food) and oxygen. During photosynthesis, energy is stored in food. Photosynthesis occurs only in cells that contain chlorophyll, such as those in the leaves of plants. Cellular respiration combines oxygen and food and releases the energy in the chemical bonds of the food. The end products of cellular respiration are energy, carbon dioxide, and water. All plant cells contain mitochondria. Any cell with mitochondria can use the process of cellular respiration. **Table 1** compares photosynthesis and cellular respiration.

Summary

Taking in Raw Materials

- Leaves take in carbon dioxide that is used in photosynthesis.
- Oxygen and carbon dioxide are waste products of photosynthesis and cellular respiration.

The Food-Making Process

- Photosynthesis takes place in chloroplasts.
- Photosynthesis is a series of chemical reactions that transforms energy from light into energy stored in the chemical bonds of sugar molecules.

The Breakdown of Food

- Cellular respiration uses oxygen to release energy from food.
- Cellular respiration takes place in mitochondria.

Self Check

1. **Describe** how gases enter and exit a leaf.
2. **Explain** why photosynthesis and cellular respiration are important.
3. **Identify** what must happen to glucose molecules before cellular respiration begins.
4. **Compare and contrast** the number of organisms that respire and the number that photosynthesize.
5. **Think Critically** Humidity is water vapor in the air. Infer how plants contribute to humidity.

Applying Math

6. **Solve One-Step Equations** How many CO_2 molecules result from the breakdown of a glucose molecule ($C_6H_{12}O_6$)? Refer to the equation in this section.

Stomata in Leaves

Stomata open and close, which allows gases into and out of a leaf. These openings are usually invisible without the use of a microscope. Do this lab to see some stomata.

Real-World Question

Where are stomata in lettuce leaves?

Goals

- **Describe** guard cells and stomata.
- **Infer** the conditions that make stomata open and close.

Materials

lettuce in dish of water
coverslip
microscope
microscope slide
salt solution
forceps

Safety Precautions

WARNING: *Never eat or taste any materials used in the laboratory.*

Procedure

1. Copy the Stomata Data table into your Science Journal.
2. From a head of lettuce, tear off a piece of an outer, crisp, green leaf.
3. Bend the piece of leaf in half and carefully use a pair of forceps to peel off some of the epidermis, the transparent tissue that covers a leaf. Prepare a wet mount of this tissue.
4. Examine your prepared slide under low and high power on the microscope.
5. Count the total number of stomata in your field of view and then count the number of open stomata. Enter these numbers in the data table.
6. Make a second slide of the lettuce leaf epidermis. This time place a few drops of salt solution on the leaf instead of water.
7. Wait a few minutes. Repeat steps 4 and 5.
8. **Calculate** the percent of open stomata using the following equation:

$$\frac{\text{number of open stomata}}{\text{total number of stomata}} \times 100 = \text{percent open}$$

Stomata Data (Sample data only)

	Wet Mount	Salt-Solution Mount
Total Number of Stomata		
Number of Open Stomata	Do not write in this book.	
Percent Open		

Conclude and Apply

1. **Determine** which slide preparation had a greater percentage of open stomata.
2. **Infer** why fewer stomata were open in the salt-solution mount.
3. What can you infer about the function of stomata in a leaf?

Communicating Your Data

Collect data from your classmates and compare it to your data. Discuss any differences you find and why they occurred. **For more help, refer to the Science Skill Handbook.**

section 2

Plant Responses

What are plant responses?

It's dark. You're alone in a room watching a horror film on television. Suddenly, the telephone near you rings. You jump, and your heart begins to beat faster. You've just responded to a stimulus. A stimulus is anything in the environment that causes a response in an organism. The response often involves movement either toward the stimulus or away from the stimulus. A stimulus may come from outside (external) or inside (internal) the organism. The ringing telephone is an example of an external stimulus. It caused you to jump, which is a response. Your beating heart is a response to an internal stimulus. Internal stimuli are usually chemicals produced by organisms. Many of these chemicals are hormones. Hormones are substances made in one part of an organism for use somewhere else in the organism.

All living organisms, including plants, respond to stimuli. Many different chemicals are known to act as hormones in plants. These internal stimuli have a variety of effects on plant growth and function. Plants respond to external stimuli such as touch, light, and gravity. Some responses, such as the response of the Venus's-flytrap plant in **Figure 10,** are rapid. Other plant responses are slower because they involve changes in growth.

as you read

What You'll Learn

- **Identify** the relationship between a stimulus and a tropism in plants.
- **Compare and contrast** long-day and short-day plants.
- **Explain** how plant hormones and responses are related.

Why It's Important

You will be able to grow healthier plants if you understand how they respond to certain stimuli.

Review Vocabulary
behavior: the way in which an organism interacts with other organisms and its environment

New Vocabulary
- tropism
- auxin
- photoperiodism
- long-day plant
- short-day plant
- day-neutral plant

Figure 10 A Venus's-flytrap has three small trigger hairs on the surface of its toothed leaves. When two hairs are touched at the same time, the plant responds by closing its trap in less than 1 second.

Figure 11 Tropisms are responses to external stimuli.
Identify *the part of a plant that shows negative gravitropism.*

The pea plant's tendrils respond to touch by coiling around things.

This plant is growing toward the light, an example of positive phototropism.

This plant was turned on its side. With the roots visible, you can see that they are showing positive gravitropism.

Tropisms

Some responses of a plant to an external stimuli are called tropisms. A **tropism** (TROH pih zum) can be seen as movement caused by a change in growth and can be positive or negative. For example, plants might grow toward a stimulus—a positive tropism—or away from a stimulus—a negative tropism.

Touch One stimulus that can result in a change in a plant's growth is touch. When the pea plant, as shown in **Figure 11,** touches a solid object, it responds by growing faster on one side of its stem than on the other side. As a result the stem bends and twists around any object it touches.

Light Did you ever see a plant leaning toward a window? Light is an important stimulus to plants. When a plant responds to light, the cells on the side of the plant opposite the light get longer than the cells facing the light. Because of this uneven growth, the plant bends toward the light. This response causes the leaves to turn in such a way that they can absorb more light. When a plant grows toward light it is called a positive response to light, or positive phototropism, shown in **Figure 11.**

Gravity Plants respond to gravity. The downward growth of plant roots, as shown in **Figure 11,** is a positive response to gravity. A stem growing upward is a negative response to gravity. Plants also may respond to electricity, temperature, and darkness.

Gravity and Plants
Gravity is a stimulus that affects how plants grow. Can plants grow without gravity? In space the force of gravity is low. Write a paragraph in your Science Journal that describes your idea for an experiment aboard a space shuttle to test how low gravity affects plant growth.

Plant Hormones

Hormones control the changes in growth that result from tropisms and affect other plant growth. Plants often need only millionths of a gram of a hormone to stimulate a response.

Ethylene Many plants produce the hormone ethylene (EH thuh leen) gas and release it into the air around them. Ethylene is produced in cells of ripening fruit, which stimulates the ripening process. Commercially, fruits such as oranges and bananas are picked when they are unripe and the green fruits are exposed to ethylene during shipping so they will ripen. Another plant response to ethylene causes a layer of cells to form between a leaf and the stem. The cell layer causes the leaf to fall from the stem.

Applying Math — Calculate Averages

GROWTH HORMONES Gibberellins are plant hormones that increase growth rate. The graphs on the right show data from an experiment to determine how gibberellins affect the growth of bean seedlings. What is the average height of control bean seedlings after 14 days?

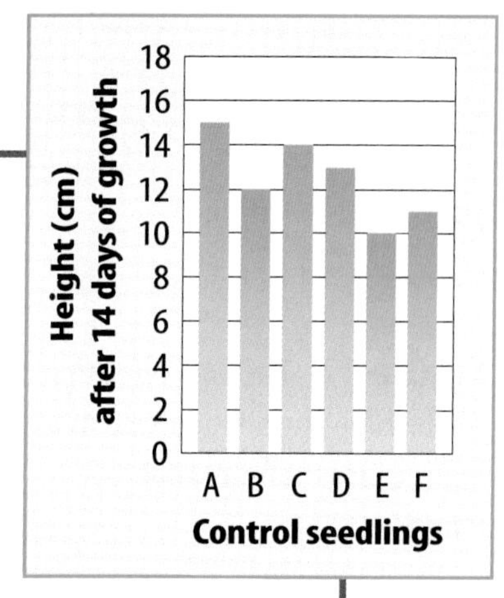

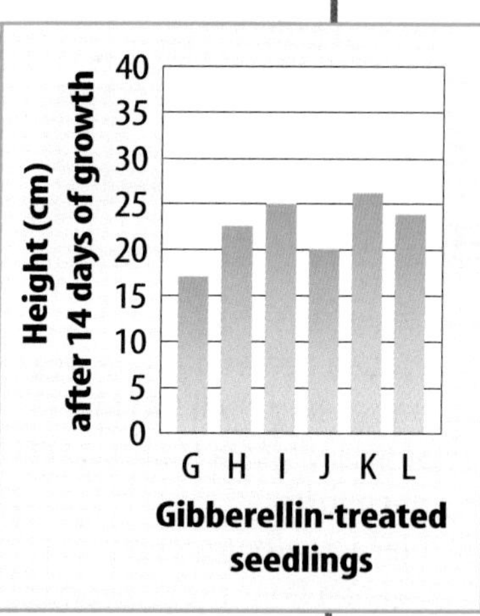

Solution

1. *This is what you know:*
   - height of control seedlings after 14 days
   - number of control seedlings
2. *This is what you need to find out:* What is the average height of control seedlings after 14 days?
3. *This is the procedure you need to use:*
   - Find the total of the seedling heights. $15 + 12 + 14 + 13 + 10 + 11 = 75$ cm
   - Divide the height total by the number of control seedlings to find the average height. $75 \text{ cm}/6 = 12.5$ cm
4. *Check your answer:* Multiply 12.5 cm by 6 and you should get 75 cm.

Practice Problems

1. Calculate the average height of seedlings treated with gibberellin.
2. In an experiment, the heights of gibberellin-treated rose stems were 20, 26, 23, 24, 23, 25, and 26 cm. The average height of the controls was 23 cm. Did gibberellin have an effect?

For more practice, visit life.msscience.com/math_practice

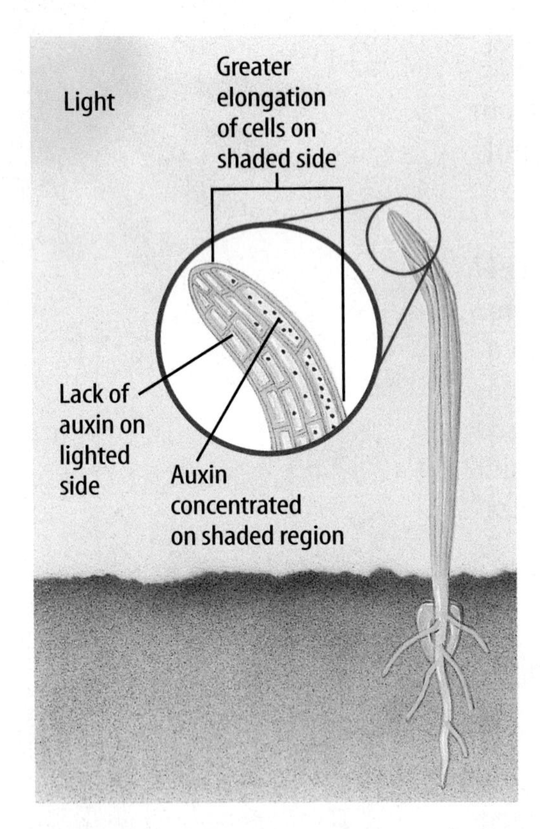

Figure 12 The concentration of auxin on the shaded side of a plant causes cells on that side to lengthen.

Auxin Scientists identified the plant hormone, **auxin** (AWK sun) more than 100 years ago. Auxin is a type of plant hormone that causes plant stems and leaves to exhibit positive response to light. When light shines on a plant from one side, the auxin moves to the shaded side of the stem where it causes a change in growth, as shown in **Figure 12.** Auxins also control the production of other plant hormones, including ethylene.

Reading Check *How are auxins and positive response to light related?*

Development of many parts of the plant, including flowers, roots, and fruit, is stimulated by auxins. Because auxins are so important in plant development, synthetic auxins have been developed for use in agriculture. Some of these synthetic auxins are used in orchards so that all plants produce flowers and fruit at the same time. Other synthetic auxins damage plants when they are applied in high doses and are used as weed killers.

Observing Ripening

Procedure

1. Place a **green banana** in a **paper bag.** Roll the top shut and place it on a table or counter.
2. Place another green banana near the paper bag.
3. After two days check the bananas to see how they have ripened. **WARNING:** *Do not eat the materials used in the lab.*

Analysis

Which banana ripened more quickly? Why?

Try at Home

Gibberellins and Cytokinins Two other groups of plant hormones that also cause changes in plant growth are gibberellins (jih buh REH lunz) and cytokinins (si tuh KI nunz). Gibberellins are chemical substances that were isolated first from a fungus. The fungus caused a disease in rice plants called "foolish seedling" disease. The fungus infects the stems of plants and causes them to grow too tall. Gibberellins can be mixed with water and sprayed on plants and seeds to stimulate plant stems to grow and seeds to germinate.

Like gibberellins, cytokinins also cause rapid growth. Cytokinins promote growth by causing faster mitosis and cell divisions. Like ethylene, the effect of cytokinins on the plant also is controlled by auxin. Interestingly, cytokinins can be sprayed on stored vegetables to keep them fresh longer.

Abscisic Acid Because hormones that cause growth in plants were known to exist, biologists suspected that substances that have the reverse effect also must exist. Abscisic (ab SIH zihk) acid is one such substance. Many plants grow in areas that have cold winters. Normally, if seeds germinate or buds develop on plants during the winter, they will die. Abscisic acid is the substance that keeps seeds from sprouting and buds from developing during the winter. This plant hormone also causes stomata to close and helps plants respond to water loss on hot summer days. **Figure 13** summarizes how plant hormones affect plants and how hormones are used.

NATIONAL GEOGRAPHIC

VISUALIZING PLANT HORMONES

Figure 13

Chemical compounds called plant hormones help determine how a plant grows. There are five main types of hormones. They coordinate a plant's growth and development, as well as its responses to environmental stimuli, such as light, gravity, and changing seasons. Most changes in plant growth are a result of plant hormones working together, but exactly how hormones cause these changes is not completely understood.

▲ ETHYLENE By controlling the exposure of these tomatoes to ethylene, a hormone that stimulates fruit ripening, farmers are able to harvest unripe fruit and make it ripen just before it arrives at the supermarket.

◀ GIBBERELLINS The larger mustard plant in the photo at left was sprayed with gibberellins, plant hormones that stimulate stem elongation and fruit development.

◀ CYTOKININS Lateral buds do not usually develop into branches. However, if a plant's main stem is cut, as in this bean plant, naturally occurring cytokinins will stimulate the growth of lateral branches, causing the plant to grow "bushy."

▼ AUXINS Powerful growth hormones called auxins regulate responses to light and gravity, stem elongation, and root growth. The root growth on the plant cuttings, center and right, is the result of auxin treatment.

▶ ABA (ABSCISIC ACID) In plants such as the American basswood, right, abscisic acid causes buds to remain dormant for the winter. When spring arrives, ABA stops working and the buds sprout.

Topic: Plant Pigments

Visit life.msscience.com for Web links to information about how plant pigments are involved in photoperiodism.

Activity Explain how day length affects a hormone that promotes flowering.

Photoperiods

Sunflowers bloom in the summer, and cherry trees flower in the spring. Some plant species produce flowers at specific times during the year. A plant's response to the number of hours of daylight and darkness it receives daily is **photoperiodism** (foh toh PIHR ee uh dih zum).

Earth revolves around the Sun once each year. As Earth moves in its orbit, it also rotates. One rotation takes about 24 h. Because Earth is tilted about 23.5° from a line perpendicular to its orbit, the hours of daylight and darkness vary with the seasons. As you probably have noticed, the Sun sets later in summer than in winter. These changes in the number of hours of daylight and darkness affect plant growth.

Darkness and Flowers Many plants require a certain number of hours of darkness to begin the flowering process. Generally, plants that require less than 10 h to 12 h of darkness to flower are called **long-day plants.** Some long-day plants are spinach, lettuce, and beets. Plants that need 12 or more hours of darkness to flower are called **short-day plants.** Some short-day plants are poinsettias, strawberries, and ragweed. **Figure 14** shows what happens when a short-day plant receives fewer hours of darkness than it needs to flower.

Reading Check *What is needed to begin the flowering process?*

Day-Neutral Plants Plants like dandelions and roses are **day-neutral plants.** They have no specific photoperiod, and the flowering process can begin within a range of hours of darkness.

In nature, photoperiodism affects where flowering plants can grow and produce flowers and fruit. Even if a particular environment has the proper temperature and other growing conditions for a plant, it will not flower and produce fruit without the correct photoperiod. **Table 2** shows how day length affects flowering in all three types of plants.

Sometimes the photoperiod of a plant has a narrow range. For example, some soybeans will flower with 9.5 h of darkness but will not flower with 10 h of darkness. Farmers must choose the variety of soybeans with a photoperiod that matches the hours of darkness in the part of the country where they plant their crop.

Figure 14 When short-day plants receive less darkness than required to produce flowers, they produce larger leaves instead.

Table 2 Photoperiodism

	Long-Day Plants	Short-Day Plants	Day-Neutral Plants
Early Summer Noon 6 AM 6 PM Midnight			
Late Fall Noon 6 AM 6 PM Midnight			
	An iris is a long-day plant that is stimulated by short nights to flower in the early summer.	Goldenrod is a short-day plant that is stimulated by long nights to flower in the fall.	Roses are day-neutral plants and have no specific photoperiod.

Today, greenhouse growers are able to provide any number of hours of artificial daylight or darkness. This means that you can buy short-day flowering plants during the summer and long-day flowering plants during the winter.

section 2 review

Summary

What are plant responses?

- Plants respond to both internal and external stimuli.

Tropisms

- Tropisms are plant responses to external stimuli, including touch, light, and gravity.

Plant Hormones

- Hormones control changes in plant growth, including changes that result from tropisms.

Photoperiods

- Long-day plants flower in late spring or summer.
- Short-day plants flower in late fall or winter.

Self Check

1. **List** one example of an internal stimulus and one example of an external stimulus in plants.
2. **Compare and contrast** photoperiodism and phototropism.
3. **Identify** the term that describes the photoperiod of red raspberries that produce fruit in late spring and in the fall.
4. **Distinguish** between abscisic acid and gibberellins.
5. **Think Critically** Describe the relationship between hormones and tropisms.

Applying Skills

6. **Compare and contrast** the responses of roots and stems to gravity.

Tropism in Plants

Goals
- **Describe** how roots and stems respond to gravity.
- **Observe** how changing the stimulus changes the growth of plants.

Materials
paper towel
30-cm × 30-cm sheet of aluminum foil
water
mustard seeds
marking pen
1-L clear-glass or plastic jar

Safety Precautions

WARNING: *Some kinds of seeds are poisonous. Do not put any seed in your mouth.*

Real-World Question

Grapevines can climb on trees, fences, or other nearby structures. This growth is a response to the stimulus of touch. Tropisms are specific plant responses to stimuli outside of the plant. One part of a plant can respond positively while another part of the same plant can respond negatively to the same stimulus. Gravitropism is a response to gravity. Why might it be important for some plant parts to have a positive response to gravity while other plant parts have a negative response? Do stems and roots respond to gravity in the same way? You can design an experiment to test how some plant parts respond to the stimulus of gravity.

Procedure

1. Copy the data table on the right in your Science Journal.
2. Moisten the paper towel with water so that it's damp but not dripping. Fold it in half twice.
3. Place the folded paper towel in the center of the foil and sprinkle mustard seeds in a line across the center of the towel.
4. Fold the foil around the towel and seal each end by folding the foil over. Make sure the paper towel is completely covered by the foil.
5. Use a marking pen to draw an arrow on the foil, and place the foil package in the jar with the arrow pointing upward.
6. After five days, carefully open the package and record your observations in the data table. (Note: *If no stems or roots are growing yet, reseal the package and place it back in the jar, making sure that the arrow points upward. Reopen the package in two days.*)
7. Reseal the foil package, being careful not to disturb the seedlings. Place it in the jar so that the arrow points downward instead of upward.
8. After five more days, reopen the package and observe any new growth of the seedlings' roots and stems. Record your observations in your data table.

Response to Gravity

Position of Arrow on Foil Package	Observations of Seedling Roots	Observations of Seedling Stems
Arrow up		
Arrow down	Do not write in this book.	

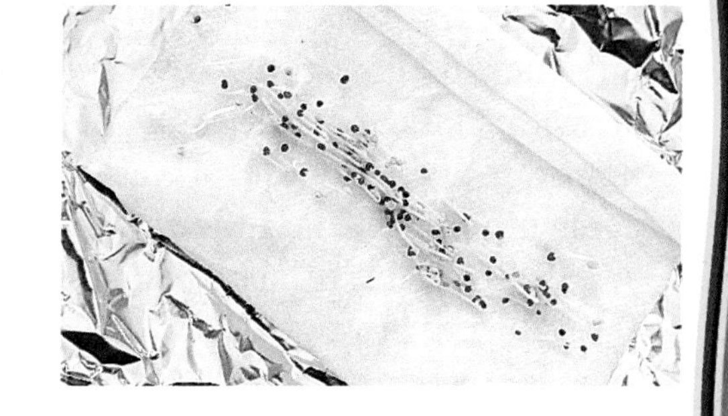

Analyze Your Data

1. **Classify** the responses you observed as positive or negative tropisms.
2. **Explain** why the plants' growth changed when you placed them upside down.

Conclude and Apply

1. **Infer** why it was important that no light reach the seedlings during your experiment.
2. **Describe** some other ways you could have changed the position of the foil package to test the seedlings' response.

Communicating Your Data

Compare drawings you make of the growth of the seedlings before and after you turned the package. **Compare** your drawings with those of other students in your class. **For more help, refer to the Science Skill Handbook.**

"Sunkissed: An Indian Legend"

as told by Alberto and Patricia De La Fuente

A long time ago, deep down in the very heart of the old Mexican forests, so far away from the sea that not even the largest birds ever had time to fly that far, there was a small, beautiful valley. A long chain of snow-covered mountains stood between the valley and the sea. . . . Each day the mountains were the first ones to tell everybody that Tonatiuh, the King of Light, was coming to the valley. . . .

"Good morning, Tonatiuh!" cried a little meadow. . . .

The wild flowers always started their fresh new day with a kiss of golden sunlight from Tonatiuh, but it was necessary to first wash their sleepy baby faces with the dew that Metztli, the Moon, sprinkled for them out of her bucket onto the nearby leaves during the night. . . .

. . . All night long, then, Metztli Moon would walk her night-field making sure that by sun-up all flowers had the magic dew that made them feel beautiful all day long.

However, much as flowers love to be beautiful as long as possible, they want to be happy too. So every morning Tonatiuh himself would give each one a single golden kiss of such power that it was possible to be happy all day long after it. As you can see, then, a flower needs to feel beautiful in the first place, but if she does not feel beautiful, she will not be ready for her morning sun-kiss. If she cannot wash her little face with the magic dew, the whole day is lost.

Understanding Literature

Legends and Oral Traditions A legend is a traditional story often told orally and believed to be based on actual people and events. Legends are believed to be true even if they cannot be proved. "Sunkissed: An Indian Legend" is a legend about a little flower that is changed forever by the Sun. This legend also is an example of an oral tradition. Oral traditions are stories or skills that are handed down by word of mouth. What in this story indicates that it is a legend?

Respond to the Reading

1. What does this passage tell you about the relationship between the Sun and plants?
2. What does this passage tell you about the relationship between water and the growth of flowers?
3. **Linking Science and Writing** Create an idea for a fictional story that explains why the sky becomes so colorful during a sunset. Then retell your story to your classmates.

The passage from "Sunkissed: An Indian Legend" does not teach us the details about photosynthesis or respiration. However, it does show how sunshine and water are important to plant life. The difference between the legend and the information contained in your textbook is this—photosynthesis and respiration can be proved scientifically, and the legend, although fun to read, cannot.

Reviewing Main Ideas

Section 1 Photosynthesis and Cellular Respiration

1. Carbon dioxide and water vapor enter and leave a plant through openings in the epidermis called stomata. Guard cells cause a stoma to open and close.
2. Photosynthesis takes place in the chloroplasts of plant cells. Light energy is used to produce glucose and oxygen from carbon dioxide and water.
3. Photosynthesis provides the food for most organisms on Earth.
4. Many organisms use cellular respiration to release the energy stored in food molecules. Cellular respiration uses oxygen and occurs in the mitochondria of cells. Carbon dioxide and water are waste products.
5. The energy released by cellular respiration is used for the life processes of most living organisms, including plants.
6. Photosynthesis and cellular respiration are almost reverse processes. The end products of photosynthesis are the raw materials needed for cellular respiration. The end products of cellular respiration are the raw materials needed for photosynthesis.

Section 2 Plant Responses

1. Plants respond positively and negatively to stimuli. The response may be a movement, a change in growth, or the beginning of some process such as flowering.
2. A stimulus from outside the plant is called a tropism. Outside stimuli include light, gravity, and touch.
3. Plant hormones cause responses in plants. Some hormones cause plants to exhibit tropisms. Other hormones cause changes in plant growth rates.
4. The number of hours of darkness each day can affect flowering times of plants.

Visualizing Main Ideas

Copy and complete the following concept map on photosynthesis and cellular respiration.

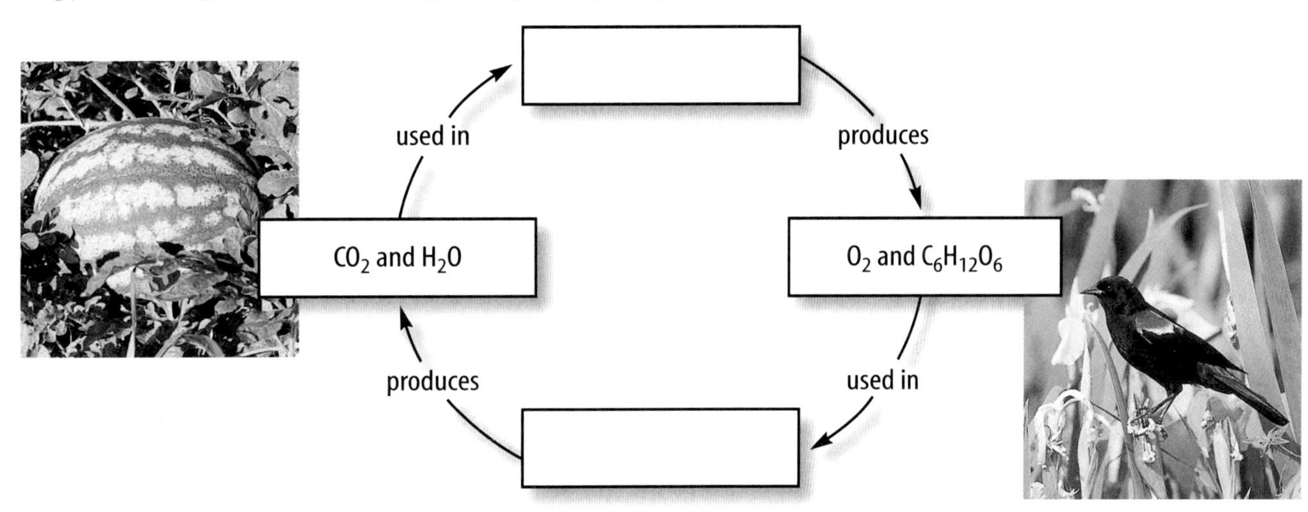

Using Vocabulary

auxin p. 314
cellular respiration p. 307
chlorophyll p. 304
day-neutral plant p. 316
long-day plant p. 316
photoperiodism p. 316
photosynthesis p. 305
short-day plant p. 316
stomata p. 303
tropism p. 312

Fill in the blanks with the correct vocabulary word(s) from the list above.

1. ________ is a hormone that causes plant stems and leaves to exhibit positive phototropism.
2. ________ is a light-dependent process conducted by green plants but not by animals.
3. ________ is required for photosynthesis.
4. A poinsettia, often seen flowering during December holidays, is a(n) ________.
5. In most living things, energy is released from food by ________.
6. Spinach requires only ten hours of darkness to flower, which makes it a(n) ________.
7. A(n) ________ can cause a plant to bend toward light.
8. Plants usually take in carbon dioxide through ________.
9. ________ controls a plant's response to the number of hours of darkness.
10. Plants that flower without a specific photo period are ________.

Checking Concepts

Choose the word or phrase that best answers the question.

11. What raw material needed by plants enters through open stomata?
    A) sugar
    B) chlorophyll
    C) carbon dioxide
    D) cellulose

12. What is a function of stomata?
    A) photosynthesis
    B) to guard the interior cells
    C) to allow sugar to escape
    D) to permit the release of oxygen

13. What plant process produces water, carbon dioxide, and energy?
    A) cell division
    B) photosynthesis
    C) growth
    D) cellular respiration

14. What are the products of photosynthesis?
    A) glucose and oxygen
    B) carbon dioxide and water
    C) chlorophyll and glucose
    D) carbon dioxide and oxygen

15. What are plant substances that affect plant growth called?
    A) tropisms
    B) glucose
    C) germination
    D) hormones

16. Leaves change colors because what substance breaks down?
    A) hormone
    B) carotenoid
    C) chlorophyll
    D) cytoplasm

17. Which is a product of cellular respiration?
    A) CO_2
    B) O_2
    C) C_2H_4
    D) H_2

Use the photo below to answer question 18.

18. What stimulus is this plant responding to?
    A) light
    B) gravity
    C) touch
    D) water

Science Online life.msscience.com/vocabulary_puzzlemaker

Thinking Critically

19. **Predict** You buy pears at the store that are not completely ripe. What could you do to help them ripen more rapidly?
20. **Name** each tropism and state whether it is positive or negative.
    - **a.** Stem grows up.
    - **b.** Roots grow down.
    - **c.** Plant grows toward light.
    - **d.** A vine grows around a pole.
21. **Infer** Scientists who study sedimentary rocks and fossils suggest that oxygen was not in Earth's atmosphere until plantlike, one-celled organisms appeared. Why?
22. **Explain** why apple trees bloom in the spring but not in the summer.
23. **Discuss** why day-neutral and long-day plants grow best in countries near the equator.
24. **Form a hypothesis** about when guard cells open and close in desert plants.
25. **Concept Map** Copy and complete the following concept map about photoperiodism using the following information: flower year-round—*corn, dandelion, tomato*; flower in the spring, fall, or winter—*chrysanthemum, rice, poinsettia*; flower in summer—*spinach, lettuce, petunia.*

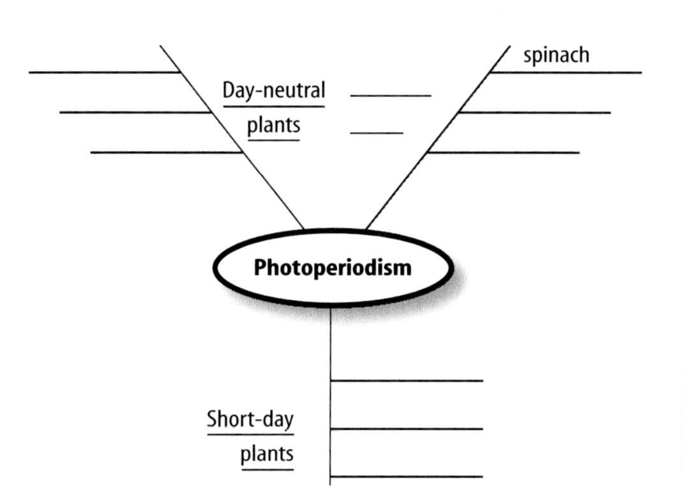

26. **Compare and contrast** the action of auxin and the action of ethylene on a plant.

Performance Activities

27. **Coloring Book** Create a coloring book of day-neutral plants, long-day plants, and short-day plants. Use pictures from magazines and seed catalogs to get your ideas. Label the drawings with the plant's name and how it responds to darkness. Let a younger student color the flowers in your book.

Applying Math

28. **Stomata** A houseplant leaf has 1,573 stomata. During daylight hours, when the plant is well watered, about 90 percent of the stomata were open. During daylight hours when the soil was dry, about 25 percent of the stomata remained open. How many stomata were open (a) when the soil was wet and (b) when it was dry?

Use the graph below to answer question 29.

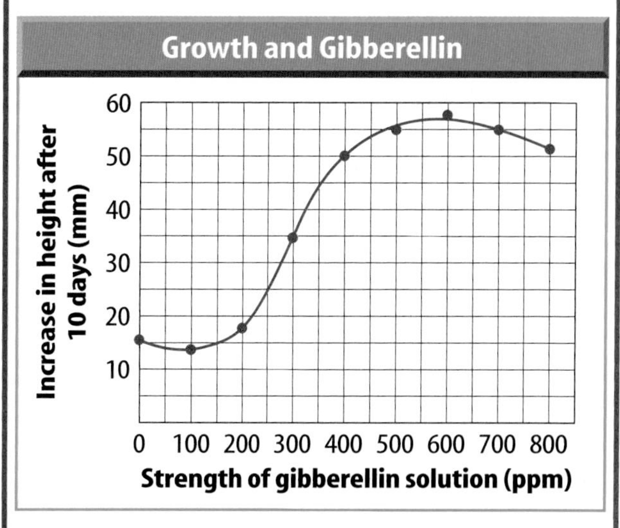

29. **Gibberellins** The graph above shows the results of applying different amounts of gibberellin to the roots of bean plants. What effect did a 100-ppm solution of gibberellin have on bean plant growth? Which gibberellin solution resulted in the tallest plants?

chapter 11 Standardized Test Practice

Part 1 Multiple Choice

Record your answers on the answer sheet provided by your teacher or on a sheet of paper.

1. Which statement correctly describes the leaf epidermis?
   A. This is an inner cell layer of the leaf.
   B. This layer is nearly transparent.
   C. Food is made in this layer.
   D. Sunlight cannot penetrate this layer.

2. What happens when a plant is losing too much water?
   A. stomata close
   B. guard cells swell
   C. stomata open
   D. respiration increases

3. Which statement is TRUE?
   A. Changes in number of hours of darkness have no effect on plant growth.
   B. Plants that need less than 10 to 12 hours of darkness to flower are called short-day plants.
   C. Plants that need 12 or more hours of darkness to flower are called short-day plants.
   D. Very few plants rely on a specific length of darkness to flower.

Use the illustration below to answer question 4.

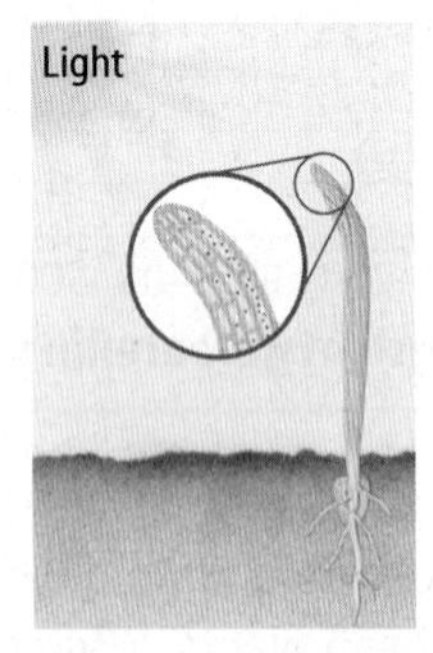

4. What controls the plant growth response shown above?
   A. auxin **C.** abscisic acid
   B. gravity **D.** length of darkness

Use the illustration below to answer questions 5 and 6.

5. What type of response is displayed by this plant?
   A. negative phototropism
   B. positive gravitropism
   C. positive phototropism
   D. negative gravitropism

6. What plant hormone is responsible for the response shown here?
   A. abscisic acid **C.** a gibberellin
   B. auxin **D.** a cytokinin

7. In which plant cell structure does cellular respiration take place?
   A. nucleus **C.** vacuole
   B. mitochondrion **D.** cell wall

8. Which is NOT produced through cellular respiration?
   A. glucose **C.** water
   B. energy **D.** carbon dioxide

9. Which plant hormone prevents the development of buds during the winter?
   A. abscisic acid **C.** gibberellin
   B. auxin **D.** cytokinin

10. What chemical absorbs light energy which plants use in photosynthesis?
    A. oxygen **C.** chlorophyll
    B. hydrogen **D.** glucose

Part 2 Short Response/Grid In

Record your answers on the answer sheet provided by your teacher or on a sheet of paper.

Use the illustration below to answer questions 11 and 12.

$$6CO_2 + 6H_2O + \text{light energy} \xrightarrow{\text{chlorophyll}} C_6H_{12}O_6 + 6O_2$$

11. Identify this process. How would this process change if the amount of available water was limited?

12. Based on this equation, what is the main food source for plant cells? How do animals use this food source?

13. Why is cellular respiration necessary for plants? Describe some plant processes which require energy.

14. What advantage do growers gain by picking and shipping unripe fruit? What role does ethylene play in this commercial process?

15. Identify specific stimuli to which plants respond in the natural environment.

16. Many people who save poinsettia plants from Christmas cannot get them to flower the following Christmas. Why?

17. What effect have commercial greenhouses had on the availability of long-day and short-day plants year-round?

18. Where are stomata found on the leaf? What function do these structures perform?

19. Describe the relationship between chlorophyll and the color of leaves in spring and summer.

Part 3 Open Ended

Record your answers on a sheet of paper.

20. Cellulose is an important component of plants. Describe its relationship to glucose. Identify cell and plant structures which contain significant amounts of cellulose.

21. Organisms which make their own food generate most of the oxygen in Earth's atmosphere. Trace the path of this element from a component of water in the soil to a gas in the air.

Use the illustration below to answer questions 22 and 23.

22. Explain how the tropism shown by this plant could help a gardener incorporate a larger number of plants into a small vegetable garden plot.

23. What advantages might thigmotropism, the response shown in this picture, provide for some plants?

24. The destruction of large areas of rain forest concerns scientists on many levels. Describe the relationship between environmental conditions for plant growth in rainforest regions, their relative rate of photosynthesis, and the amount of oxygen this process adds to the atmosphere.

Test-Taking Tip

Pace Yourself If you are taking a timed test, keep track of time during the test. If you find that you're spending too much time on a multiple-choice question, mark your best guess and move on.

unit 3 Animal Diversity

How Are Animals & Airplanes Connected?

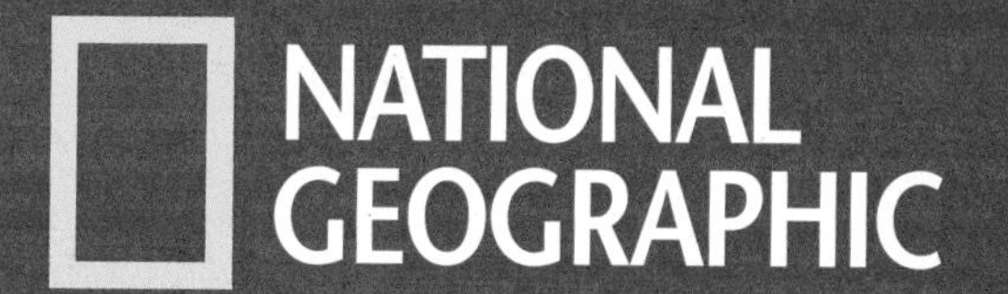

For thousands of years, people dreamed of flying like birds. Detailed sketches of flying machines were made about 500 years ago. Many of these machines featured mechanical wings that were intended to flap like the wings of a bird. But human muscles are not powerful enough to make such wings flap. Later, inventors studied birds such as eagles, which often glide through the air on outstretched wings. Successful gliders were built in the 1800s, but the gliders had no source of power to get them off the ground—and they were hard to control. Around 1900, two inventors studied bird flight more carefully and discovered that birds steer by changing the shape and position of their wings. The inventors built an engine-powered flying machine equipped with wires that could cause small changes in the shape and position of the wings. Though hardly as graceful as a soaring bird, the first powered, controlled flight took place in 1903, in the airplane seen here.

unit projects

Visit life.msscience.com/unit_project to find project ideas and resources. Projects include:

- **History** Research Charles Darwin and his system for classifying animals. Write a time-travel interview to express your new knowledge.
- **Technology** Explore a biology-related career, and then write a want ad for a new job position.
- **Model** Study an animal, its characteristics, and habitat. Design a lunch bag with your new knowledge. A snack related to your animal may be placed inside to share with your classmates.

WebQuest Investigate *Origins of Birds* to learn about the theory that birds descended from theropod dinosaurs. Compare and contrast bird characteristics with other animals.

chapter 12

Introduction to Animals

The BIG Idea

The animal kingdom includes diverse groups of organisms.

SECTION 1
Is it an animal?
Main Idea All animals have similar characteristics but different adaptations.

SECTION 2
Sponges and Cnidarians
Main Idea Sponges and cnidarians have adaptations for living in aquatic environments.

SECTION 3
Flatworms and Roundworms
Main Idea Flatworms and roundworms have adaptations for different functions and environments.

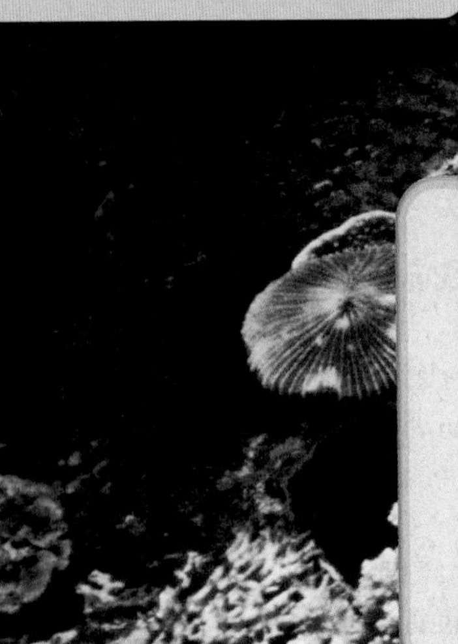

Plant or Animal?

There are many animals on Earth, and not all look like a cat or a dog. A coral is an animal, and a coral reef is made of millions of these animals. By studying how animals are classified today, scientists can identify the relationships that exist among different animal groups.

Science Journal List all of the animals that you can identify in this picture.

Start-Up Activities

Animal Symmetry

The words *left* and *right* have meaning to you because your body has a left and a right side. But what is left or right to a jellyfish or sea star? How an animal's body parts are arranged is called symmetry. In the following lab, you will compare three types of symmetry found in animals.

1. On a piece of paper, draw three shapes—a circle, a triangle with two equal sides, and a free-form shape—then cut them out.
2. See how many different ways you can fold each shape through the center to make similar halves with each fold.
3. **Think Critically** Record which shapes can be folded into equal halves and which shapes cannot. Can any of the shapes be folded into equal halves more than one way? Which shape would be similar to a human? A sea star? A sponge?

FOLDABLES™ Study Organizer

Animal Classification Make the following Foldable to help you classify the main characteristics of different animals.

STEP 1 **Fold** a piece of paper in half from top to bottom and then fold it in half again to divide it into fourths.

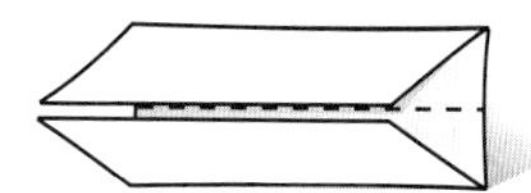

STEP 2 **Turn** the paper vertically, **unfold and label** the four columns as shown.

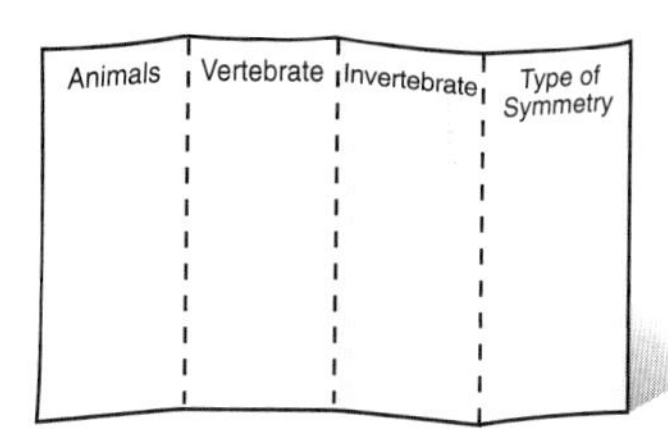

Read for Main Ideas As you read this chapter, list the characteristics of different animals in the appropriate column.

Preview this chapter's content and activities at life.msscience.com

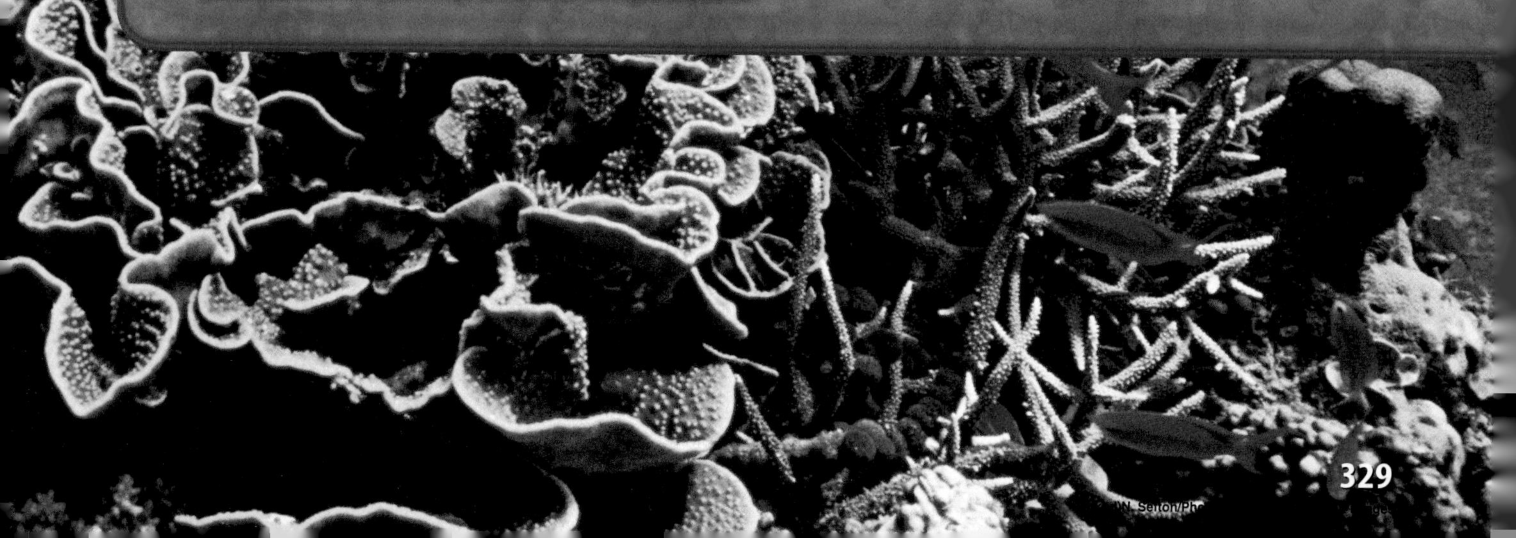

Get Ready to Read

Make Inferences

1 Learn It! When you make inferences, you draw conclusions that are not directly stated in the text. This means you "read between the lines." You interpret clues and draw upon prior knowledge. Authors rely on a reader's ability to infer because all the details are not always given.

2 Practice It! Read the excerpt below and pay attention to highlighted words as you make inferences. Use this Think-Through chart to help you make inferences.

After an animal is classified as an invertebrate or vertebrate and its symmetry is determined, **other characteristics** are identified that place it in **one of the groups of animals** with which it has the most characteristics in common. Sometimes, a newly discovered animal is different from any existing group and **a new classification group is formed** for that animal.

—*from page 335*

Text	Question	Inferences
Other characteristics	What other characteristics?	presence of certain organs, type of reproduction
One of the groups of animals	What are the groups of animals?	mammals, fish, arthropods, reptiles, amphibians
a new classification group is formed	What new classification group?	Is it a new species or new phylum?

3 Apply It! As you read this chapter, practice your skill at making inferences by making connections and asking questions.

Sometimes you make inferences by using other reading skills, such as questioning and predicting.

Target Your Reading

Use this to focus on the main ideas as you read the chapter.

1. **Before you read** the chapter, respond to the statements below on your worksheet or on a numbered sheet of paper.
   - Write an **A** if you **agree** with the statement.
   - Write a **D** if you **disagree** with the statement.

2. **After you read** the chapter, look back to this page to see if you've changed your mind about any of the statements.
   - If any of your answers changed, explain why.
   - Change any false statements into true statements.
   - Use your revised statements as a study guide.

Before You Read A or D		Statement	After You Read A or D
	1	Animals are organisms with prokaryotic cells.	
	2	An animal with bilateral symmetry has two similar sides.	
	3	Invertebrate animals include the greatest number of species.	
	4	Sponges only can be found in warm, shallow, saltwater environments.	
	5	Sponges lack tissues, organs, and organ systems.	
	6	Cnidarians capture their prey using stinging cells.	
	7	A cnidarian has two body forms during its life cycle.	
	8	A planarian, a flatworm, can reproduce sexually and asexually.	
	9	Flatworms do not harm humans.	
	10	Most roundworms are parasites.	

Science Online
Print out a worksheet of this page at life.msscience.com

section 1

Is it an animal?

as you read

What You'll Learn

- **Identify** the characteristics common to most animals.
- **Determine** how animals meet their needs.
- **Distinguish** between invertebrates and vertebrates.

Why It's Important

Animals provide food, medicines, and companionship in your daily life.

Review Vocabulary
adapation: any variation that makes an organism better suited to its environment

New Vocabulary
- herbivore
- carnivore
- omnivore
- vertebrate
- invertebrate
- radial symmetry
- bilateral symmetry

Animal Characteristics

From microscopic worms to giant whales, the animal kingdom includes an amazing variety of living things, but all of them have certain characteristics in common. What makes the animals in **Figure 1** different from plants? Is it because animals eat other living things? Is this enough information to identify them as animals? What characteristics do animals have?

1. Animals are made of many cells. Different kinds of cells carry out different functions such as sensing the environment, getting rid of wastes, and reproducing.
2. Animal cells have a nucleus and specialized structures inside the cells called organelles.
3. Animals depend on other living things in the environment for food. Some eat plants, some eat other animals, and some eat plants and animals.
4. Animals digest their food. The proteins, carbohydrates, and fats in foods are broken down into simpler molecules that can move into the animal's cells.
5. Many animals move from place to place. They can escape from their enemies and find food, mates, and places to live. Animals that move slowly or not at all have adaptations that make it possible for them to take care of these needs in other ways.
6. All animals are capable of reproducing sexually. Some animals also can reproduce asexually.

Figure 1 These organisms look like plants, but they're one of the many plantlike animals that can be found growing on shipwrecks and other underwater surfaces.
Infer *how these animals obtain food.*

Figure 2 Animals eat a variety of foods.

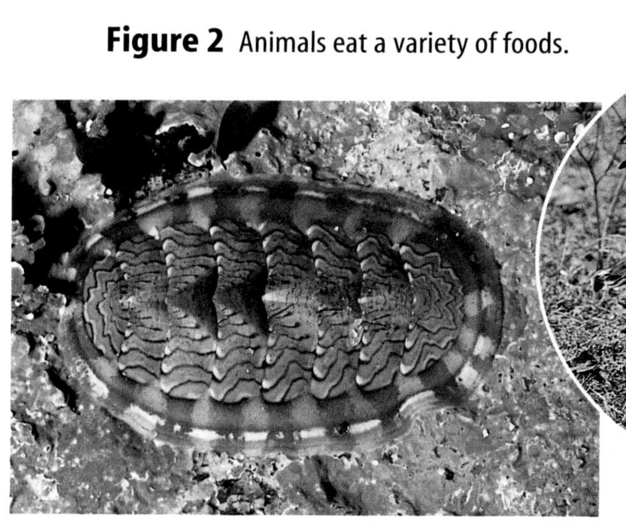

Chitons eat algae from rocks.

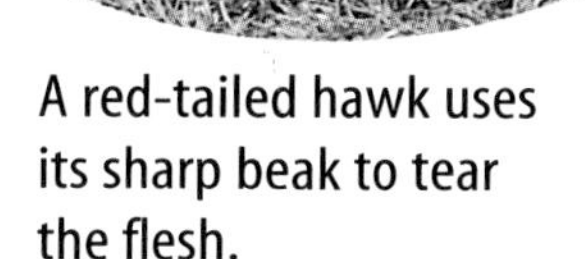

A red-tailed hawk uses its sharp beak to tear the flesh.

Cardinal fish eat small invertebrates and some plant material.

How Animals Meet Their Needs

Any structure, process, or behavior that helps an organism survive in its environment is an adaptation. Adaptations are inherited from previous generations. In a changing environment, adaptations determine which individuals are more likely to survive and reproduce.

Adaptations for Obtaining Energy One of the most basic needs of animals is the need for food. All animals have adaptations that allow them to obtain, eat, and digest different foods. The chiton, shown in **Figure 2,** deer, some fish, and many insects are examples of herbivores. **Herbivores** eat only plants or parts of plants. In general, herbivores eat more often and in greater amounts than other animals because plants don't supply as much energy as other types of food.

Reading Check *Why are butterflies considered to be herbivores?*

Animals that eat only other animals, like the red-tailed hawk in **Figure 2,** are **carnivores.** Most carnivores capture and kill other animals for food. But some carnivores, called scavengers, eat only the remains of other animals. Animal flesh supplies more energy than plants do, so carnivores don't need to eat as much or as often as herbivores.

Animals that eat plants and animals or animal flesh are called **omnivores.** Bears, raccoons, robins, humans, and the cardinal fish in **Figure 2** are examples of omnivores.

Many beetles and other animals such as millipedes feed on tiny bits of decaying matter called detritus (dih TRI tus). They are called detritivores (dih TRI tih vorz).

Carnivore Lore
Carnivores have always been written about as having great power and strength. Find a poem or short story about a carnivore and interpret what the author is trying to convey about the animal.

Figure 3 The pill bug's outer covering protects it and reduces moisture loss from its body.

Physical Adaptations Some prey species have physical features that enable them to avoid predators. Outer coverings protect some animals. Pill bugs, as seen in **Figure 3,** have protective plates. Porcupines have sharp quills that prevent most predators from eating them. Turtles and many animals that live in water have hard shells that protect them from predators.

Size is also a type of defense. Large animals are usually safer than small animals. Few predators will attack animals such as moose or bison simply because they are so large.

Mimicry is an adaptation in which one animal closely resembles another animal in appearance or behavior. If predators cannot distinguish between the two, they usually will not eat either animal. The venomous coral snake and the nonvenomous scarlet king snake, shown in **Figure 4,** look alike. In some cases, this is a disadvantage for scarlet king snakes because people mistake them for coral snakes and kill them.

✔ Reading Check ***How might mimicry be an advantage and a disadvantage for an animal?***

Many animals, like the flounder in **Figure 5,** blend into their surrounding environment, enabling them to hide from their predators. English peppered moths are brown and speckled like the lichens (LI kunz) on trees, making it difficult for their predators to see them. Many freshwater fish, like the trout also in **Figure 5,** have light bellies and dark, speckled backs that blend in with the gravelly bottoms of their habitats when they are viewed from above. Any marking or coloring that helps an animal hide from other animals is called camouflage. Some animals, like the cuttlefish in **Figure 5,** have the ability to change their color depending on their surroundings.

Modeling Animal Camouflage

Procedure

1. Pretend that a room in your home is the world of some fictitious animal. From **materials you can find around your home,** build a fictitious animal that would be camouflaged in this world.
2. Put your animal into its world and ask someone to find it.

Analysis

1. In how many places was your animal camouflaged?
2. What changes would increase its chances of surviving in its world?

Try at Home

Coral snake

Scarlet king snake

Figure 4 Mimicry helps some animals survive.
Describe *the difference between the two snakes.*

Figure 5 Many types of animals blend with their surroundings.

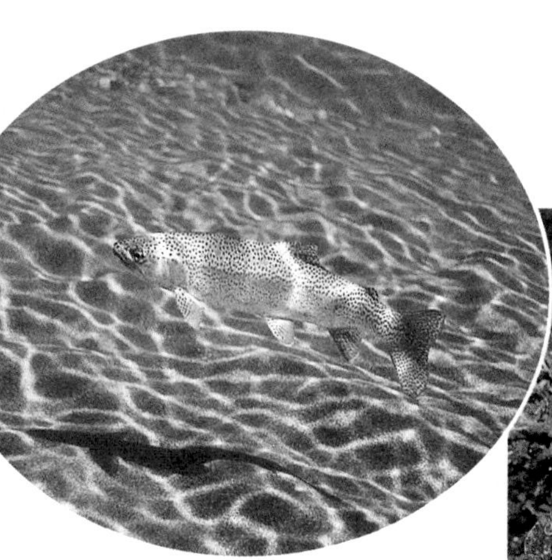

Bottom fish like this flounder, blend with the ocean floor.

A trout blends with the bottom of a stream.

Cuttlefish can be especially difficult to find because they can change color to blend with their surroundings.

Predator Adaptations Camouflage is an adaptation for many predators so they can sneak up on their prey. Tigers have stripes that hide them in tall grasses. Killer whales are black on their upper surface and white underneath. When seen from above, the whale blends into the darkness of the deep ocean. The killer whale's white underside appears to be nearly the same color as the bright sky overhead when viewed from below. Adaptations such as these enable predators to hunt prey more successfully.

Behavioral Adaptations In addition to physical adaptations, animals have behavioral adaptations that enable them to capture prey or to avoid predators. Chemicals are used by some animals to escape predators. Skunks spray attacking animals with a bad-smelling liquid. Some ants and beetles also use this method of defense. When squid and octopuses are threatened, they release a cloud of ink so they can escape, as shown in **Figure 6.**

Some animals are able to run faster than most of their predators. The Thomson's gazelle can run at speeds up to 80 km/h. A lion can run only about 36 km/h, so speed is a factor in the Thomson's gazelle's survival.

Traveling in groups is a behavior that is demonstrated by predators and prey. Herring swim in groups called schools that resemble an organism too large for a predator fish to attack. On the other hand, when wolves travel in packs, they can successfully hunt large prey that one predator alone could not capture.

Figure 6 An octopus's cloud of ink confuses a predator long enough for the octopus to escape.

Figure 7 Animals can be classified into two large groups. These groups can be broken down further based on different animal characteristics.

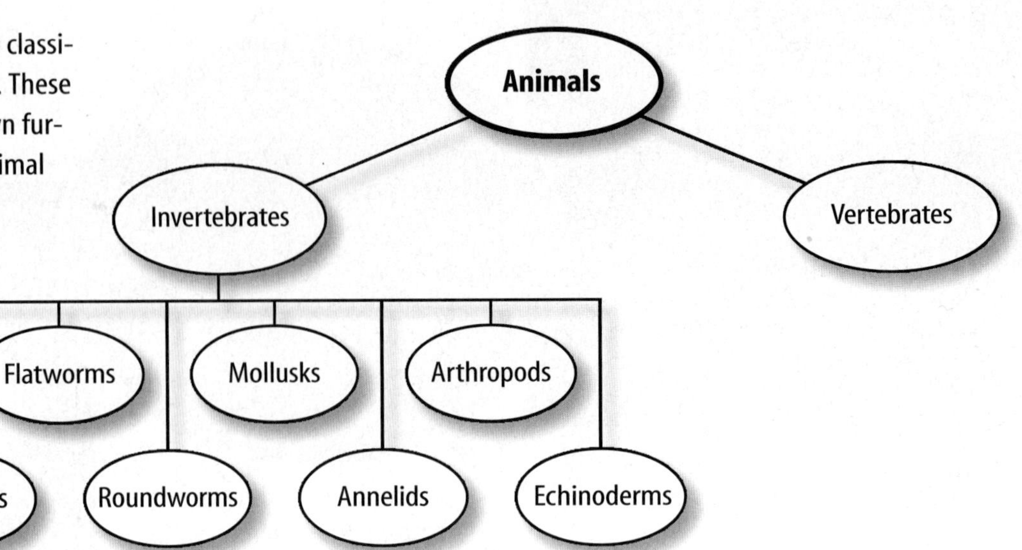

Animal Classification

Scientists have identified and named more than 1.8 million species of animals. It is estimated that there are another 3 million to 30 million more to identify and name. Animals can be classified into two major groups, as shown in **Figure 7.** All animals have common characteristics, but those in one group have more, similar characteristics because all the members of a group probably descended from a common ancestor. When a scientist finds a new animal, how does he or she begin to classify it?

Check for a Backbone To classify an animal, a scientist first looks to see whether or not the animal has a backbone. Animals with backbones are called **vertebrates.** Their backbones are made up of a stack of structures called vertebrae that support the animal. The backbone also protects and covers the spinal cord—a bundle of nerves that is connected to the brain. The spinal cord carries messages to all other parts of the body. It also carries messages from other parts of the body to the brain. Examples of vertebrates include fish, frogs, snakes, birds, and humans.

An animal without a backbone is classified as an **invertebrate.** About 97 percent of all animal species are invertebrates. Sponges, jellyfish, worms, insects, and clams are examples of invertebrates. Many invertebrates are well protected by their outer coverings. Some have shells, some have a skeleton on the outside of their body, and others have a spiny outer covering.

Topic: Animal Classification

Visit life.msscience.com for Web links to information about how the classification of an animal can change as new information is learned.

Activity Name a recent reclassification of an animal and one reason it was reclassified.

Symmetry After determining whether or not a backbone is present, a scientist might look at an animal's symmetry (SIH muh tree). Symmetry is how the body parts of an animal are arranged. Organisms that have no definite shape are called asymmetrical. Most sponges are asymmetrical animals.

Figure 8 Symmetry is a characteristic of all animals. Sea urchins can sense things from all directions.

Most animals have bilateral symmetry like this crayfish. **Name** *the type of symmetry you have.*

Animals that have body parts arranged in a circle around a center point, the way spokes of a bicycle wheel are arranged, have **radial symmetry.** Hydras, jellyfish, sea urchins, like the one in **Figure 8,** and some sponges have radial symmetry.

Most animals have bilateral symmetry. In Latin, the word *bilateral* means "two sides." An animal with **bilateral symmetry,** like the crayfish shown in **Figure 8,** can be divided into right and left halves that are nearly mirror images of each other.

After an animal is classified as an invertebrate or a vertebrate and its symmetry is determined, other characteristics are identified that place it in one of the groups of animals with which it has the most characteristics in common. Sometimes a newly discovered animal is different from any existing group, and a new classification group is formed for that animal.

section 1 review

Summary

Animal Characteristics

- Animals are made of many eukaryotic cells.
- Animals obtain and digest food, reproduce and most move from place to place.

How Animals Meet Their Needs

- Animals have many different physical, predatory, and behavioral adaptations.
- Animals can be herbivores, carnivores, omnivores, or detritivores depending on what they eat.

Animal Classification

- Scientists classify animals in two large groups: vertebrates and invertebrates.
- An animal's symmetry plays a role in its classification.

Self Check

1. **Explain** different adaptations for obtaining food.
2. **Compare and contrast** invertebrates and vertebrates.
3. **List** the three types of symmetry. Give an example for each type.
4. **Think Critically** Radial symmetry is found among species that live in water. Why might radial symmetry be an uncommon adaptation of animals that live on land?

Applying Skills

5. **Concept Map** Make an events-chain concept map showing the steps used to classify a new animal.
6. **Communicate** Choose an animal you are familiar with. Describe the adaptations it has for getting food and avoiding predators.

section 2

Sponges and Cnidarians

as you read

What You'll Learn

- **Describe** the characteristics of sponges and cnidarians.
- **Explain** how sponges and cnidarians obtain food and oxygen.
- **Determine** the importance of living coral reefs.

Why It's Important

Sponges and cnidarians are important to medical research because they are sources of chemicals that fight disease.

Review Vocabulary
flagella: long, thin whiplike structures that grow from a cell

New Vocabulary
- sessile
- hermaphrodite
- polyp
- medusa
- tentacle
- stinging cell

Sponges

In their watery environments, sponges play many roles. They interact with many other animals such as worms, shrimp, snails, and sea stars. These animals live on, in, and under sponges. Sponges also are important as a food source for some snails, sea stars, and fish. Certain sponges contain photosynthetic bacteria and protists that provide oxygen and remove wastes for the sponge.

Only about 17 species of sponges are commercially important. Humans have long used the dried and cleaned bodies of some sponges for bathing and cleaning. Most sponges you see today are synthetic sponges or vegetable loofah sponges, but natural sea sponges like those in **Figure 9** still are available.

Today scientists are finding other uses for sponges. Chemicals made by sponges are being tested and used to make drugs that fight disease-causing bacteria, fungi, and viruses. These chemicals also might be used to treat certain forms of arthritis.

Origin of Sponges Fossil evidence shows that sponges appeared on Earth about 600 million years ago. Because sponges have little in common with other animals, many scientists have concluded that sponges probably evolved separately from all other animals. Sponges living today have many of the same characteristics as their fossilized ancestors.

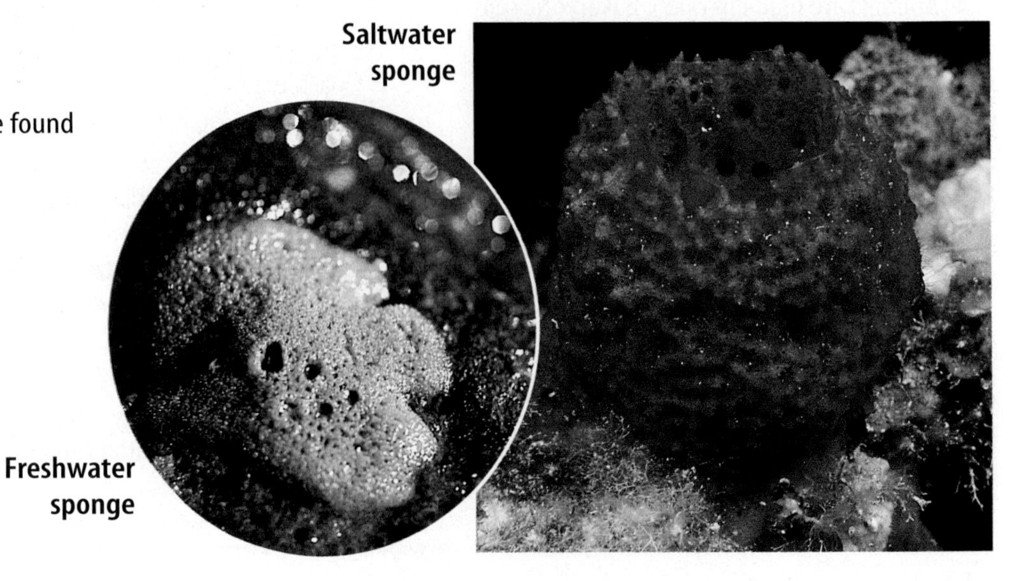

Figure 9 Sponges can be found in a variety of habitats.

Characteristics of Sponges

Most of the 5,000 species of sponges are found in warm, shallow salt water near coastlines, although some are found at ocean depths of 8,500 m or more. A few species, like the one in **Figure 9,** live in freshwater rivers, lakes, and streams. The colors, shapes, and sizes of sponges vary. Saltwater sponges are brilliant red, orange, yellow, or blue, while freshwater sponges are usually a dull brown or green. Some sponges have radial symmetry, but most are asymmetrical. Sponges can be smaller than a marble or larger than a compact car.

Adult sponges live attached to one place unless they are washed away by strong waves or currents. Organisms that remain attached to one place during their lifetimes are called **sessile** (SE sile). They often are found with other sponges in permanent groups called colonies. Early scientists classified sponges as plants because they didn't move. As microscopes were improved, scientists observed that sponges couldn't make their own food, so sponges were reclassified as animals.

Body Structure A sponge's body, like the one in **Figure 10,** is a hollow tube that is closed at the bottom and open at the top. The sponge has many small openings in its body. These openings are called pores.

Sponges have less complex body organization than other groups of animals. They have no tissues, organs, or organ systems. The body wall has two cell layers made up of several different types of cells. Those that line the inside of the sponge are called collar cells. The beating motion of the collar cells' flagella moves water through the sponge.

Many sponge bodies contain sharp, pointed structures called spicules (SPIH kyewlz). The soft-bodied, natural sponges that some people use for bathing or washing their cars have skeletons of a fibrous material called spongin. Other sponges contain spicules and spongin. Spicules and spongin provide support for a sponge and protection from predators.

Figure 10 Specialized cells, called collar cells, have flagella that move water through the pores in a sponge. Other cells filter microscopic food from the water as it passes through.

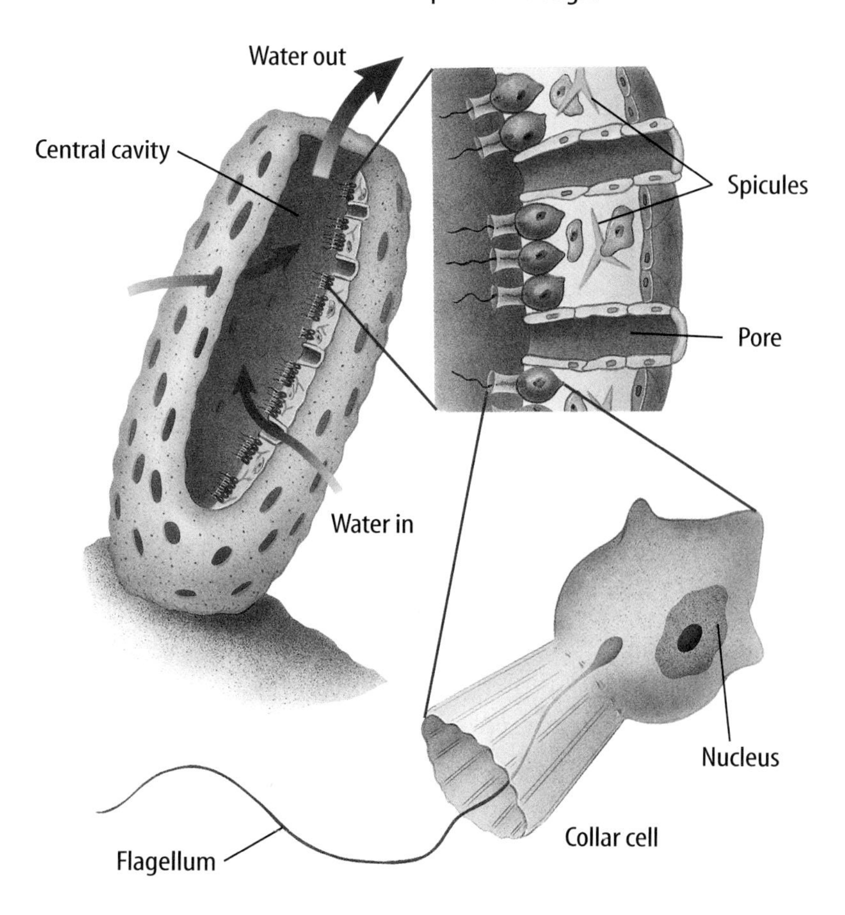

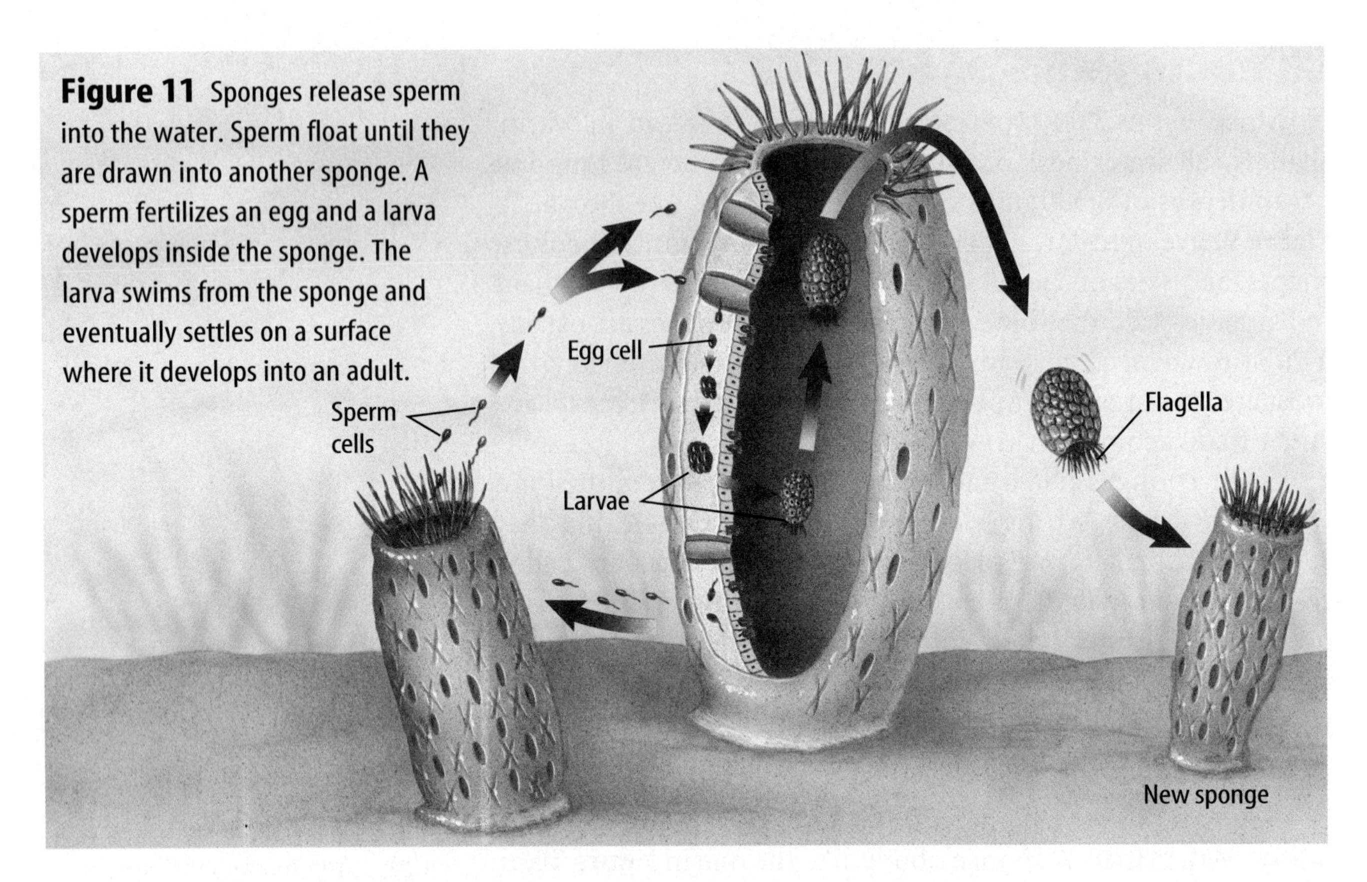

Figure 11 Sponges release sperm into the water. Sperm float until they are drawn into another sponge. A sperm fertilizes an egg and a larva develops inside the sponge. The larva swims from the sponge and eventually settles on a surface where it develops into an adult.

Obtaining Food and Oxygen Sponges filter microscopic food particles such as bacteria, algae, protists, and other materials from the water as it is pulled in through their pores. Oxygen also is removed from the water. The filtered water carries away wastes through an opening in the top of the sponge.

Reading Check *How do sponges get oxygen?*

Reproduction Sponges can reproduce sexually, as shown in **Figure 11.** Some species of sponges have separate sexes, but most sponge species are **hermaphrodites** (hur MA fruh dites)—animals that produce sperm and eggs in the same body. However, a sponge's sperm cannot fertilize its own eggs. After an egg is released, it might be fertilized and then develop into a larva (plural, *larvae*). The larva usually looks different from the adult form. Sponge larvae have cilia that allow them to swim. After a short time, the larvae settle down on objects where they will remain and grow into adult sponges.

Asexual reproduction occurs by budding or regeneration. A bud forms on a sponge, then drops from the parent sponge to grow on its own. New sponges also can grow by regeneration from small pieces of a sponge. Regeneration occurs when an organism grows new body parts to replace lost or damaged ones. Sponge growers cut sponges into pieces, attach weights to them, and put them back into the ocean to regenerate.

Spicule Composition
Spicules of glass sponges are composed of silica. Other sponges have spicules of calcium carbonate. Relate the composition of spicules to the composition of the water in which the sponge lives. Write your answer in your Science Journal.

Cnidarians

Another group of invertebrates includes colorful corals, flowerlike sea anemones, tiny hydras, delicate jellyfish, and the iridescent Portuguese man-of-war, shown in **Figure 12.** These animals are classified as cnidarians (ni DAR ee uhnz).

Cnidarian Environments Most cnidarians live in salt water, although many types of hydras live in freshwater. Sea anemones and most jellyfish, also called jellies, live as individual organisms. Hydras and corals tend to form colonies.

Two Body Forms Cnidarians have two different body forms. The **polyp** (PAH lup) form, shown in **Figure 13** on the left, is shaped like a vase and usually is sessile. Sea anemones, corals, and hydras are cnidarians that live most of their lives as polyps. The **medusa** (mih DEW suh) form, shown in **Figure 13** on the right, is bell-shaped and free-swimming. A jelly spends most of its life as a medusa floating on ocean currents. Some species of jellies have tentacles that grow to 30 m and trail behind the animal.

Reading Check *What are some possible benefits of having a medusa and a polyp form?*

Figure 12 The Portuguese man-of-war also is called the bluebottle. This animal is not one organism. It is four kinds of cnidarians that depend on one another for survival.

Figure 13 Cnidarians have medusa and polyp body forms.

Adult sea anemones are polyps that grow attached to the ocean bottom, a rock, coral, or any surface. They depend on the movement of water to bring them food.

Jellies can perform upward movements but must float to move downward.

Topic: Cnidarian Ecology

Visit life.msscience.com for Web links to information about cnidarian ecology.

Activity Describe a cnidarian that is endangered, and the reasons why it is endangered.

Body Structure All cnidarians have one body opening and radial symmetry. They have more complex bodies than sponges do. They have two cell layers that are arranged into tissues and a digestive cavity where food is broken down. In the two-cell-layer body plan of cnidarians, no cell is ever far from the water. In each cell, oxygen from the water is exchanged for carbon dioxide and other cell wastes.

Cnidarians have a system of nerve cells called a nerve net. The nerve net carries impulses and connects all parts of the organism. This makes cnidarians capable of some simple responses and movements. Hydras can somersault away from a threatening situation.

Armlike structures called **tentacles** (TEN tih kulz) surround the mouths of most cnidarians. Certain fish, shrimp, and other small animals live unharmed among the tentacles of large sea anemones, as shown in **Figure 14A.** The tentacles have stinging cells. A **stinging cell,** as shown in **Figure 14B,** has a capsule with a coiled, threadlike structure that helps the cnidarian capture food. Animals that live among an anemone's tentacles are not affected by the stinging cells. The animals are thought to help clean the sea anemone and protect it from certain predators.

Obtaining Food Cnidarians are predators. Some can stun their prey with nerve toxins produced by stinging cells. The threadlike structure in the stinging cell is sticky or barbed. When a cnidarian is touched or senses certain chemicals in its environment, the threadlike structures discharge and capture the prey. The tentacles bring the prey to the mouth, and the cnidarian ingests the food. Because cnidarians have only one body opening, undigested food goes back out through the mouth.

Figure 14 Tentacles surround the mouth of a sea anemone.

A Clown fish are protected from the sea anemone's sting by a special mucous covering. The anemone eats scraps that the fish drop, and the fish are protected from predators by the anemone's sting.

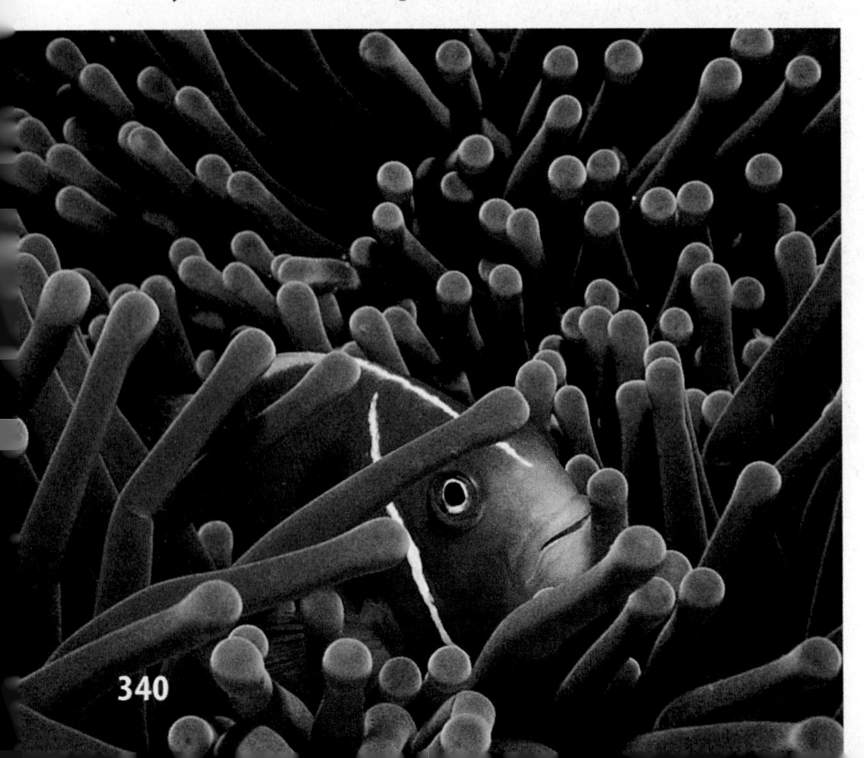

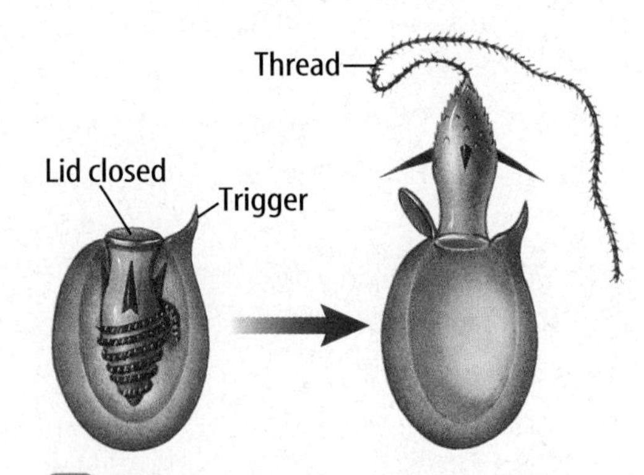

B A sea anemone's stinging cells have triggerlike structures. When prey brushes against the trigger, the thread is released into the prey. A toxin in the stinging cell stuns the prey.
Identify *the type of adaptation this is: physical, behavior, or predatory. Explain your answer.*

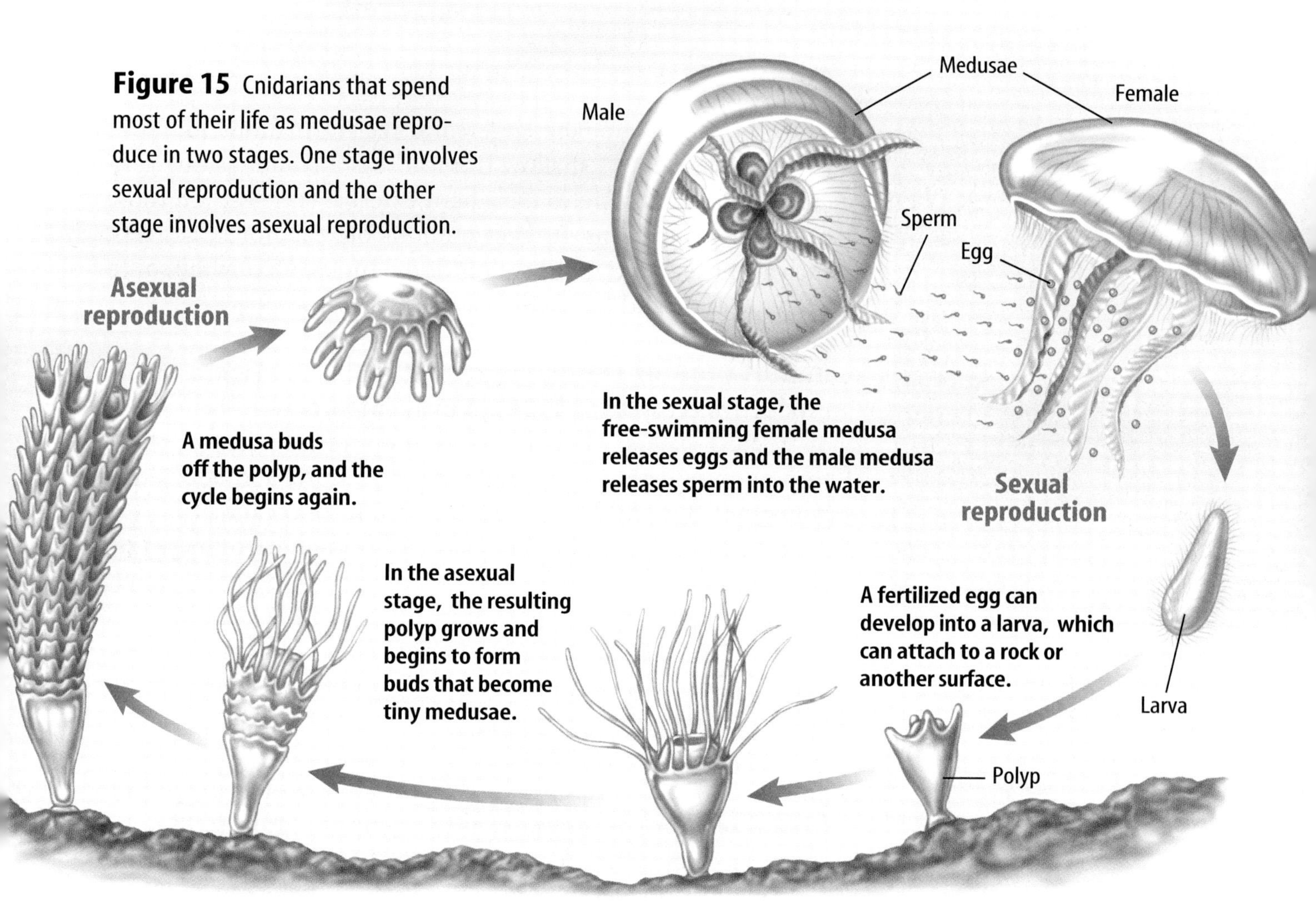

Figure 15 Cnidarians that spend most of their life as medusae reproduce in two stages. One stage involves sexual reproduction and the other stage involves asexual reproduction.

Reproduction Cnidarians reproduce asexually and sexually, as shown in **Figure 15.** Polyp forms reproduce asexually by producing buds that eventually fall off the cnidarian and develop into new polyps. Polyps also reproduce sexually by producing eggs or sperm. Sperm are released into the water and fertilize the eggs, which also are released into the water.

Medusa (plural, *medusae*) forms of cnidarians have two stages of reproduction—a sexual stage and an asexual stage. Free-swimming medusae produce eggs or sperm and release them into the water. The eggs are fertilized by sperm from another medusa of the same species and develop into larvae. The larvae eventually settle down and grow into polyps. When young medusae bud off the polyp, the cycle begins again.

Origin of Cnidarians

The first cnidarians might have been on Earth more than 600 million years ago. Scientists hypothesize that the medusa body was the first form of cnidarian. Polyps could have formed from larvae of medusae that became permanently attached to a surface. Most of the cnidarian fossils are corals.

Figure 16 Coral reefs are colonies made up of many individual corals.
Infer *the benefit of living in a colony for the corals.*

Corals

The large coral reef formations found in shallow tropical seas are built as one generation of corals secretes their hard external skeletons on those of earlier generations. It takes millions of years for large reefs, such as those found in the waters of the Indian Ocean, the south Pacific Ocean, and the Caribbean Sea, to form.

Importance of Corals Coral reefs, shown in **Figure 16,** are productive ecosystems and extremely important in the ecology of tropical waters. They have a diversity of life comparable to tropical rain forests. Some of the most beautiful and fascinating animals of the world live in the formations of coral reefs.

Beaches and shorelines are protected from much of the action of waves by coral reefs. When coral reefs are destroyed or severely damaged, large amounts of shoreline can be washed away.

If you go scuba diving or snorkeling, you might explore a coral reef. Coral reefs are home for organisms that provide valuable shells and pearls. Fossil reefs can give geologists clues about the location of oil deposits.

Like sponges, corals produce chemicals to protect themselves from diseases or to prevent other organisms from settling on them. Medical researchers are learning that some of these chemicals might provide humans with drugs to fight cancer. Some coral is even used as a permanent replacement for missing sections of bone in humans.

section 2 review

Summary

Sponges

- Most sponges live in salt water, are sessile, and vary in size, color, and shape.
- A sponge has no tissues, organs, or organ systems.
- Sponges filter food from the water, and reproduce sexually and asexually.

Cnidarians

- Cnidarians live mostly in salt water and have two body forms: polyp and medusa.
- Cnidarians have nerve cells, tissues, and a digestive cavity.
- Corals are cnidarians that make up a diverse ecosystem called a coral reef.

Self Check

1. **Compare and contrast** how sponges and cnidarians get their food.
2. **Describe** the two body forms of cnidarians and tell how each reproduces.
3. **Infer** why most fossils of cnidarians are coral fossils. Would you expect to find a fossil sponge? Explain.
4. **Think Critically** What effect might the destruction of a large coral reef have on other ocean life?

Applying Math

5. **Solve One-Step Equations** A sponge 1 cm in diameter and 10 cm tall can move 22.5 L of water through its body each day. What volume of water will it pump through its body in 1 h? In 1 min?

Science Online life.msscience.com/self_check_quiz

Observing a Cnidarian

The hydra has a body cavity that is a simple, hollow sac. It is one of the few freshwater cnidarians.

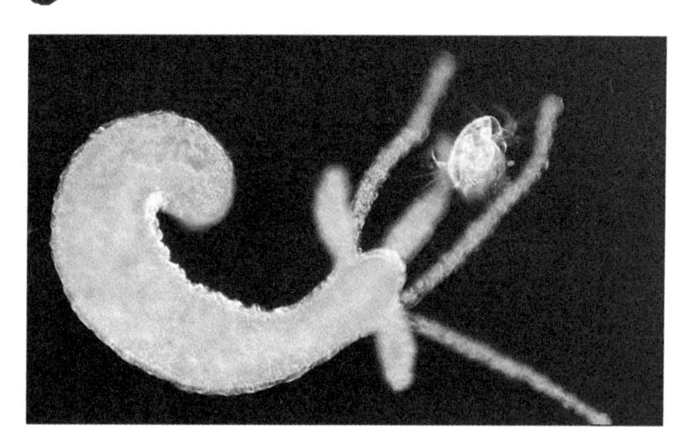

Real-World Question

How does a hydra react to food and other stimuli?

Goals

- **Predict** how a hydra will respond to various stimuli.
- **Observe** how a hydra responds to stimuli.

Materials

dropper
hydra culture
small dish
toothpick
Daphnia or brine shrimp
stereomicroscope

Safety Precautions

Procedure

1. Copy the data table and use it to record your observations.

Hydra Observations

Features	Observations
Color	Do not write in this book.
Number of tentacles	
Reaction to touch	
Reaction to food	

2. Use a dropper to place a hydra and some of the water in which it is living into a dish.
3. Place the dish on the stage of a stereomicroscope. Bring the hydra into focus. Record the hydra's color.
4. **Identify** and count the number of tentacles. Locate the mouth.
5. Study the basal disk by which the hydra attaches itself to a surface.
6. **Predict** what will happen if the hydra is touched with a toothpick. Carefully touch the tentacles with a toothpick. Describe the reaction in the data table.
7. Drop a *Daphnia* or a small amount of brine shrimp into the dish. Observe how the hydra takes in food. Record your observations.
8. Return the hydra to the culture.

Conclude and Apply

1. **Analyze** what happened when the hydra was touched. What happened to other areas of the animal?
2. **Describe** the advantages tentacles provide for hydra.

Communicating Your Data

Compare your results with those of other students. Discuss whether all of the hydras studied had the same responses, and how the responses aid hydras in survival.

section 3

Flatworms and Roundworms

as you read

What You'll Learn

- **List** the characteristics of flatworms and roundworms.
- **Distinguish** between free-living and parasitic organisms.
- **Identify** disease-causing flatworms and roundworms.

Why It's Important

Many species of flatworms and roundworms cause disease in plants and animals.

Review Vocabulary
cilia: short, threadlike structures that aid in locomotion

New Vocabulary
- free-living organism
- anus

What is a worm?

Worms are invertebrates with soft bodies and bilateral symmetry. They have three tissue layers, as shown in **Figure 17,** which are organized into organs and organ systems.

Flatworms

As their name implies, flatworms have flattened bodies. Members of this group include planarians, flukes, and tapeworms. Some flatworms are free-living, but most are parasites, which means that they depend on another organism for food and a place to live. Unlike a parasite, a **free-living organism** doesn't depend on another organism for food or a place to live.

Planarians An example of a free-living flatworm is the planarian, as shown in **Figure 18.** It has a triangle-shaped head with two eyespots. Its one body opening—a mouth—is on the underside of the body. A muscular tube called the pharynx connects the mouth and the digestive tract. A planarian feeds on small organisms and dead bodies of larger organisms. Most planarians live under rocks, on plant material, or in freshwater. They vary in length from 3 mm to 30 cm. Their bodies are covered with fine, hairlike structures called cilia. As the cilia move, the worm is moved along in a slimy mucous track that is secreted from the underside of the planarian.

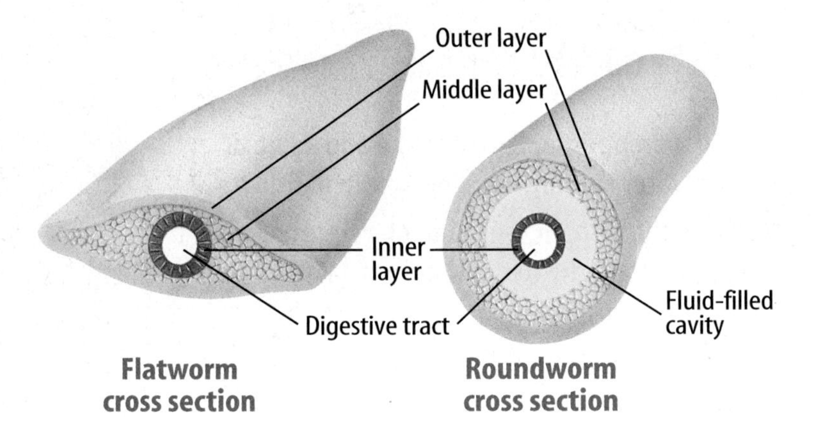

Figure 17 Worms have cells that are arranged into three specialized tissue layers and organs.

Figure 18 The planarian is a common freshwater flatworm.

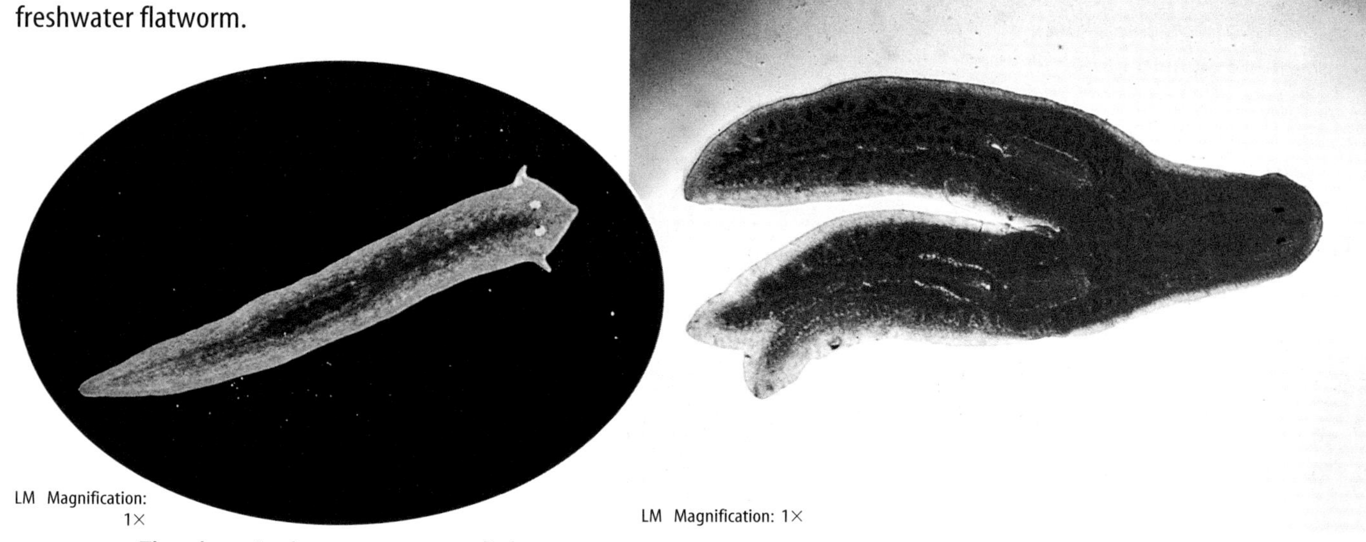

The planarian's eyespots sense light.

Planarians can reproduce asexually by splitting, then regenerating the other half.

Planarians reproduce asexually by dividing in two, as shown in **Figure 18.** A planarian can be cut in two, and each piece will grow into a new worm. They also have the ability to regenerate. Planarians reproduce sexually by producing eggs and sperm. Most are hermaphrodites and exchange sperm with one another. They lay fertilized eggs that hatch in a few weeks.

Flukes All flukes are parasites with complex life cycles that require more than one host. Most flukes reproduce sexually. The male worm deposits sperm in the female worm. She lays the fertilized eggs inside the host. The eggs leave the host in its urine or feces. If the eggs end up in water, they usually infect snails. After they leave the snail, the young worms can burrow into the skin of a new host, such as a human, while he or she is standing or swimming in the water.

Of the many diseases caused by flukes, the most widespread one affecting humans is schistosomiasis (shis tuh soh MI uh sus). It is caused by blood flukes—flatworms that live in the blood, as shown in **Figure 19.** More than 200 million people, mostly in developing countries, are infected with blood flukes. It is estimated that 1 million people die each year because of them. Other types of flukes can infect the lungs, liver, eyes, and other organs of their host.

What is the most common disease that is caused by flukes?

Figure 19 Female blood flukes deposit their eggs in the blood of their host. The eggs travel through the host and eventually end up in the host's digestive system.

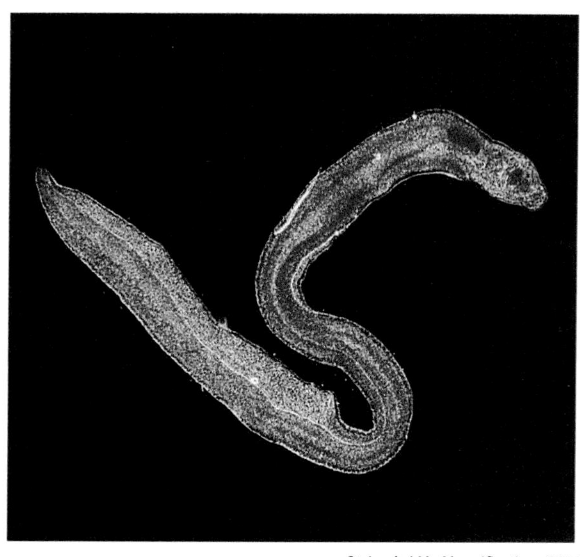

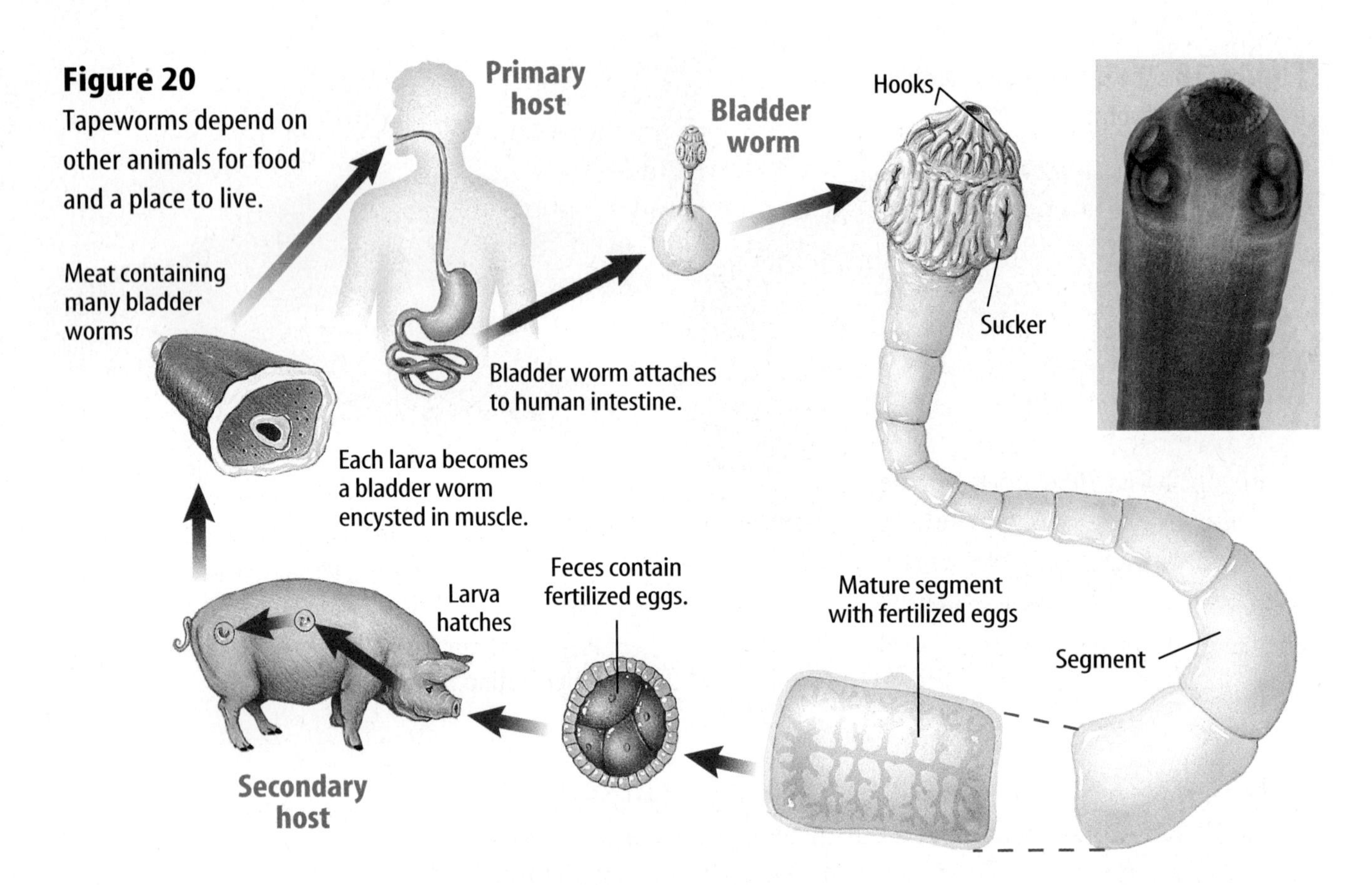

Figure 20
Tapeworms depend on other animals for food and a place to live.

Tapeworms Another type of flatworm is the tapeworm. These worms are parasites. The adult form uses hooks and suckers to attach itself to the intestine of a host organism, as illustrated in **Figure 20.** Dogs, cats, humans, and other animals are hosts for tapeworms. A tapeworm doesn't have a mouth or a digestive system. Instead, the tapeworm absorbs food that is digested by the host from its host's intestine.

A tapeworm grows by producing new body segments immediately behind its head. Its ribbonlike body can grow to be 12 m long. Each body segment has both male and female reproductive organs. The eggs are fertilized by sperm in the same segment. After a segment is filled with fertilized eggs, it breaks off and passes out of the host's body with the host's wastes. If another host eats a fertilized egg, the egg hatches and develops into an immature tapeworm called a bladder worm.

Observing Planarian Movement

Procedure

1. Use a **dropper** to transfer a **planarian** to a **watch glass.**
2. Add enough **water** so the planarian can move freely.
3. Place the glass under a **stereomicroscope** and observe the planarian.

Analysis

1. Describe how a planarian moves in the water.
2. What body parts appear to be used in movement?
3. Explain why a planarian is a free-living flatworm.

Origin of Flatworms

Because of the limited fossil evidence, the evolution of flatworms is uncertain. Evidence suggests that they were the first group of animals to evolve bilateral symmetry with senses and nerves in the head region. They also were probably the first group of animals to have a third tissue layer that develops into organs and systems. Some scientists hypothesize that flatworms and cnidarians might have had a common ancestor.

Roundworms

If you own a dog, you've probably had to get medicine from your veterinarian to protect it from heartworms—a type of roundworm. Roundworms also are called nematodes and more nematodes live on Earth than any other type of many-celled organism. It is estimated that more than a half million species of roundworms exist. They are found in soil, animals, plants, freshwater, and salt water. Some are parasitic, but most are free-living.

Roundworms are slender and tapered at both ends like the one in **Figure 21.** The body is a tube within a tube, with fluid in between. Most nematode species have male and female worms and reproduce sexually. Nematodes have two body openings, a mouth, and an anus. The **anus** is an opening at the end of the digestive tract through which wastes leave the body.

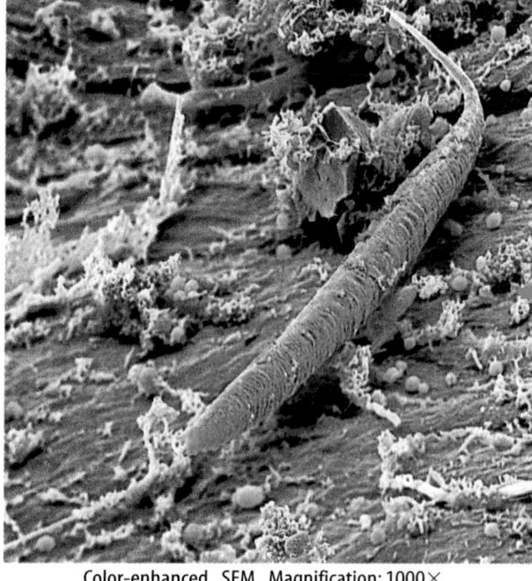

Color-enhanced SEM Magnification: 1000×

Figure 21 Some roundworms infect humans and other animals. Others infect plants, and some are free-living in the soil.

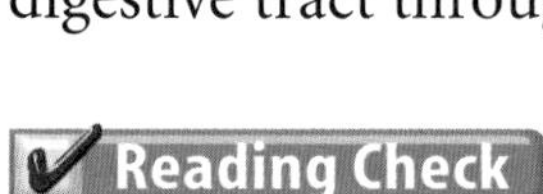

What characteristics of roundworms might contribute to the success of the group?

Applying Math Use Percentages

SPECIES COUNTS In a forest ecosystem, about four percent of the 400 different animal species are roundworm species. How many roundworm species are in this ecosystem?

Solution

1. *This is what you know:*
   - total animal species = 400
   - roundworms species = 4% of total animal species
2. *This is what you must find out:* How many roundworm species are in the ecosystem?
3. *This is the procedure you need to use:*
   - Change 4% to a decimal. $\frac{4}{100} = 0.04$
   - Use following equation: (roundworm-species percent as a decimal) × (total animal species) = number of roundworm species
   - Substitute in known values: $0.04 \times 400 = 16$ roundworm species
4. *Check your answer:* Divide 16 by 0.04 and you should get 400.

Practice Problems

1. Flatworms make up 1.5 percent of all animal species in the forest ecosystem. How many flatworms species probably are present?
2. If there are 16 bird species present, what percent of the animal species are the bird species?

For more practice, visit life.msscience.com/math_practice

NATIONAL GEOGRAPHIC

VISUALIZING PARASITIC WORMS

Figure 22

Many diseases are caused by parasitic roundworms and flatworms that take up residence in the human body. Some of these diseases result in diarrhea, weight loss, and fatigue; others, if left untreated, can be fatal. Micrographs of several species of roundworms and flatworms and their magnifications are shown here.

▶ 78× BLOOD FLUKE These parasites live as larvae in lakes and rivers and penetrate the skin of people wading in the water. After maturing in the liver, the flukes settle in veins in the intestine and bladder, causing schistosomiasis (shis tuh soh MI uh sus), which damages the liver and spleen.

▲ 6× LIVER FLUKE Humans and other mammals ingest the larvae of these parasites by eating contaminated plant material. Immature flukes penetrate the intestinal wall and pass via the liver into the bile ducts. There they mature into adults that feed on blood and tissue.

▼ 125× PINWORMS Typically inhabiting the large intestine, the female pinworm lays her eggs near the host's anus, causing discomfort. The micrograph below shows pinworm eggs on a piece of clear tape.

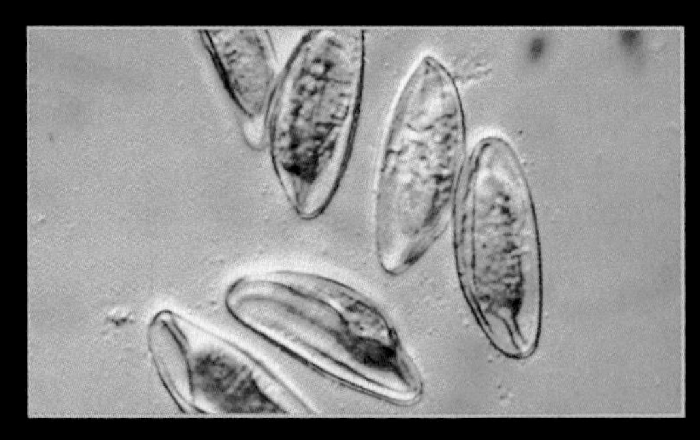

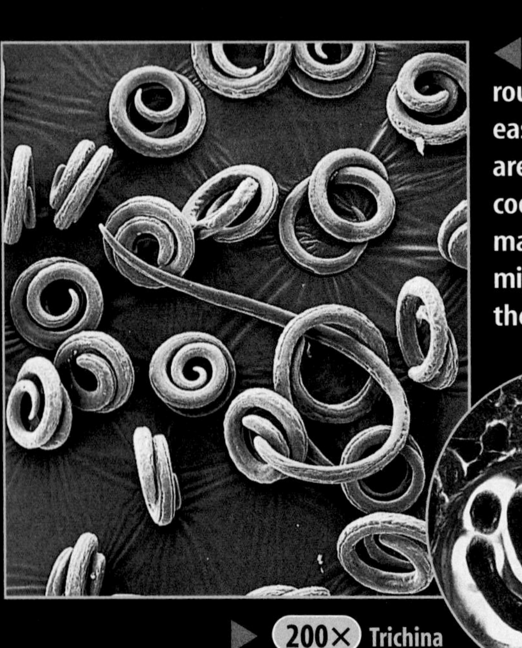

◀ 170× ROUNDWORMS The roundworms that cause the disease trichinosis (trih kuh NOH sus) are eaten as larvae in undercooked infected meat. They mature in the intestine, then migrate to muscle tissue, where they form painful cysts.

▶ 200× Trichina larvae in muscle tissue

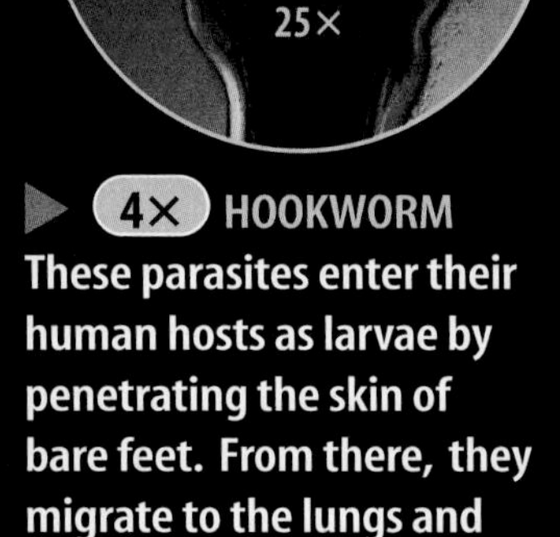

▶ 4× HOOKWORM These parasites enter their human hosts as larvae by penetrating the skin of bare feet. From there, they migrate to the lungs and eventually to the intestine, where they mature.

Origin of Roundworms More than 550 million years ago, roundworms appeared early in animal evolution. They were the first group of animals to have a digestive system with a mouth and an anus. Scientists hypothesize that roundworms are more closely related to arthropods than to vertebrates. However, it is still unclear how roundworms fit into the evolution of animals.

Importance of Roundworms Some roundworms, shown in **Figure 22,** cause diseases in humans. Others are parasites of plants or of other animals, such as the fish shown in **Figure 23.** Some nematodes cause damage to fiber, agricultural products, and food. It is estimated that the worldwide annual amount of nematode damage is in the millions of dollars.

Not all roundworms are a problem for humans, however. In fact, many species are beneficial. Some species of roundworms feed on termites, fleas, ants, beetles, and many other types of insects that cause damage to crops and human property. Some species of beneficial nematodes kill other pests. Research is being done with nematodes that kill deer ticks that cause Lyme disease.

Roundworms also are important because they are essential to the health of soil. They provide nutrients to the soil as they break down organic material. They also help in cycling nutrients such as nitrogen.

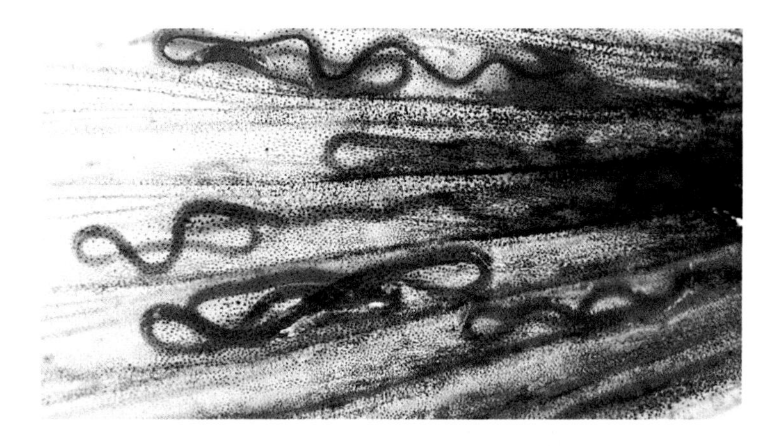

Figure 23 This fish's fin is infected with parasitic roundworms. These roundworms damage the fin, which makes it difficult for the fish to swim and escape from predators.

section 3 review

Summary

Common Characteristics

- Both flatworms and roundworms are invertebrates with soft bodies, bilateral symmetry, and three tissue layers that are organized into organs and organ systems.

Flatworms

- Flatworms have flattened bodies, and can be free-living or parasitic. They generally have one body opening.

Roundworms

- Also called nematodes, roundworms have a tube within a tube body plan. They have two openings: a mouth and an anus.

Self Check

1. **Compare and contrast** the body plan of a flatworm to the body plan of a roundworm.
2. **Distinguish** between a free-living flatworm and a parasitic flatworm.
3. **Explain** how tapeworms get energy.
4. **Identify** three roundworms that cause diseases in humans. How can humans prevent infection from each?
5. **Think Critically** Why is a flatworm considered to be more complex than a hydra?

Applying Skills

6. **Concept Map** Make an events-chain concept map for tapeworm reproduction.

Design Your Own

Comparing Free-Living and Parasitic Flatworms

Goals

- **Compare and contrast** the body parts and functions of free-living and parasitic flatworms.
- **Observe** how flatworms are adapted to their environments.

Possible Materials

petri dish with a planarian
compound microscope
prepared slide of a tapeworm
stereomicroscope
light source, such as a lamp
small paintbrush
small piece of liver
dropper
water

Safety Precautions

Real-World Question

How are the body parts of flatworms adapted to the environment in which they live? Are the adaptations of free-living flatworms and parasitic flatworms the same?

Form a Hypothesis

Form a hypothesis about what adaptations you think free-living and parasitic worms might have. What would be the benefits of these adaptations?

Using Scientific Methods

Test Your Hypothesis

Make a Plan

1. As a group, make a list of possible ways you might design a procedure to compare and contrast types of flatworms. Your teacher will provide you with information on handling live flatworms.
2. Choose one of the methods you described in step 1. List the steps you will need to take to follow the procedure. Be sure to describe exactly what you will do at each step of the activity.
3. **List** the materials that you will need to complete your experiment.
4. If you need a data table, design one in your Science Journal so it is ready to use when your group begins to collect data.

Follow Your Plan

1. Make sure your teacher approves your plan before you start.
2. Carry out the experiment according to the approved plan.
3. While the experiment is going on, record any observations that you make and complete the data table in your Science Journal.

Analyze Your Data

1. **Explain** how parasitic and free-living flatworms are similar.
2. **Describe** the differences between parasitic and free-living worms.

Conclude and Apply

1. **Identify** which body systems are more developed in free-living flatworms.
2. **Identify** which body system is more complex in parasitic flatworms.
3. **Infer** which adaptations allow some flatworms to live as free-living organisms.

Communicating Your Data

Compare and discuss your experiment design and conclusions with other students. **For more help, refer to the Science Skill Handbook.**

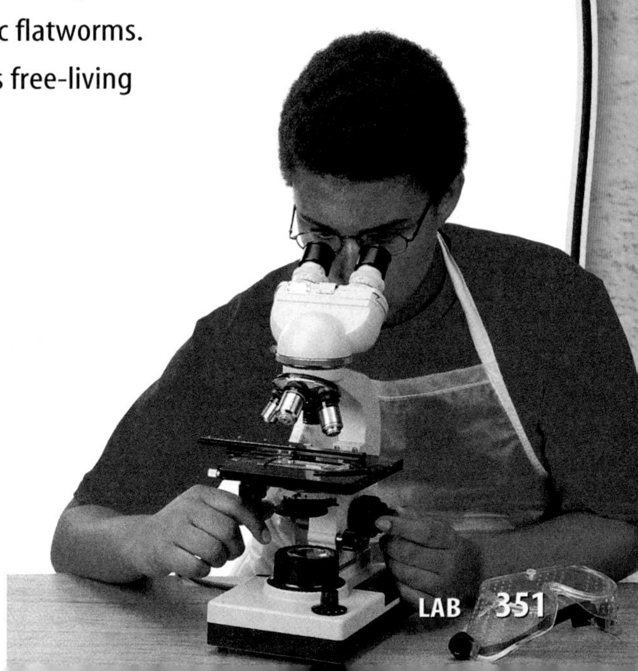

TIME SCIENCE AND HISTORY

SCIENCE CAN CHANGE THE COURSE OF HISTORY!

A natural sponge

SPONGES

A common household item contains a lot of history

Sponges and baths. They go together like a hammer and nails. But sponges weren't always used just to scrub people and countertops. Some Greek artists dipped sponges into paint to dab on their artwork and crafts. Greek and Roman soldiers padded their helmets with soft sponges similar to modern padded bicycle helmets to soften enemies' blows. The Roman soldiers also used sponges like a canteen to soak up water from a nearby stream and squeeze it into their mouths. Sponges have appeared in artwork from prehistoric times and the Middle Ages, and are mentioned in Shakespeare's play Hamlet.

Natural sponges have been gathered over time from the Mediterranean, Caribbean Sea, and off the coast of Florida. Divers used to carry up the sponges from deep water, but today sponges are harvested in shallower water. Synthetic sponges, made of rubber or cellulose, are used more today than natural ones. Natural sponges absorb more water and last longer, but synthetic sponges are less expensive. Natural sponges may also cure diseases. Medical researchers hypothesize that an enzyme produced by sponges might help cure cancer. Who says natural sponges are washed up?

Brainstorm Work with your classmates to come up with as many sayings and phrases as you can using the word *sponge.* Use some of them in a story about sponges. Share your stories with the class.

For more information, visit life.msscience.com/time

Reviewing Main Ideas

Section 1 Is it an animal?

1. Animals are many-celled organisms that must find and digest their food.
2. Herbivores eat plants, carnivores eat animals or animal flesh, omnivores eat plants and animals, and detritivores feed on decaying plants and animals.
3. Animals have many ways to escape from predators such as speed, mimicry, protective outer coverings, and camouflage.
4. Invertebrates are animals without backbones. Animals that have backbones are called vertebrates.
5. When body parts are arranged the same way on both sides of the body, it is called bilateral symmetry. If body parts are arranged in a circle around a central point, it is known as radial symmetry. Animals without a specific central point are asymmetrical.

Section 2 Sponges and Cnidarians

1. Adult sponges are sessile and obtain food by filtering water through their pores. Sponges can reproduce sexually and asexually.
2. Cnidarians are hollow-bodied animals with radial symmetry. Most have tentacles with stinging cells to obtain food.
3. Coral reefs have been deposited by reef-building corals over millions of years.

Section 3 Flatworms and Roundworms

1. Flatworms have bilateral symmetry. Free-living and parasitic forms exist.
2. Roundworms have a tube-within-a-tube body plan and bilateral symmetry.
3. Flatworm and roundworm species can cause disease in humans.

Visualizing Main Ideas

Copy and complete the following concept map.

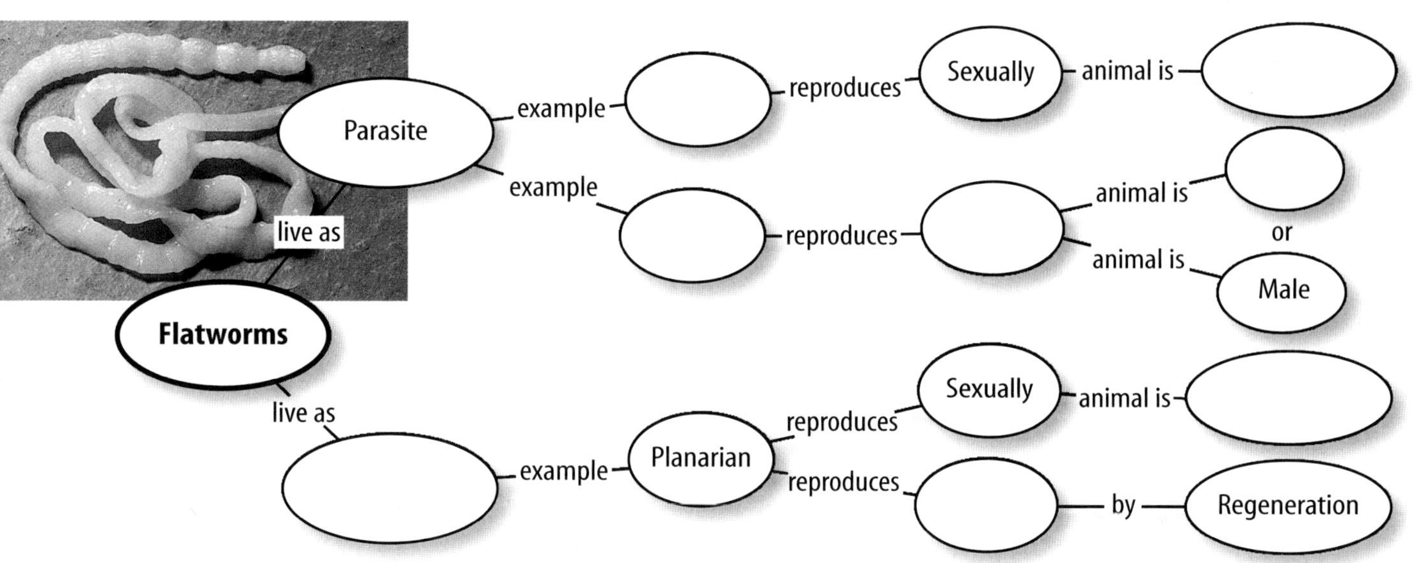

chapter 12 Review

Using Vocabulary

anus p. 347	medusa p. 339
bilateral symmetry p. 335	omnivore p. 331
carnivore p. 331	polyp p. 339
free-living organism p. 344	radial symmetry p. 335
herbivore p. 331	sessile p. 337
hermaphrodite p. 338	stinging cell p. 340
invertebrate p. 334	tentacle p. 340
	vertebrate p. 334

Find the correct vocabulary word(s).

1. animal without backbones
2. body parts arranged around a central point
3. animal that eat only other animals
4. animal that eat just plants
5. animal that produce sperm and eggs in one body
6. animal with backbones
7. body parts arranged similarly on both sides of the body
8. cnidarian body that is vase shaped
9. attached to one place
10. cnidarian body that is bell shaped

Checking Concepts

Choose the word or phrase that best answers the question.

11. Which of the following animals is sessile?
 A) jellyfish **C)** planarian
 B) roundworm **D)** sponge

12. What characteristic do all animals have?
 A) digest their food
 B) radial symmetry
 C) free-living
 D) polyp and medusa forms

13. Which term best describes a hydra?
 A) carnivore **C)** herbivore
 B) filter feeder **D)** parasite

14. Which animal has a mouth and an anus?
 A) roundworm **C)** planarian
 B) jellyfish **D)** tapeworm

15. What characteristic do scientists use to classify sponges?
 A) material that makes up their skeletons
 B) method of obtaining food
 C) reproduction
 D) symmetry

16. Which animal is a cnidarian?
 A) fluke **C)** jellyfish
 B) heartworm **D)** sponge

Use the photo below to answer question 17.

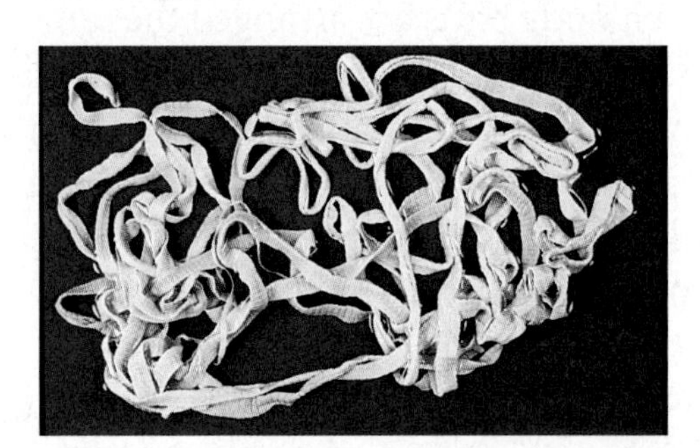

17. The photo above shows which hermaphroditic invertebrate organism?
 A) fluke **C)** tapeworm
 B) coral **D)** roundworm

18. How do sponges reproduce asexually?
 A) budding **C)** medusae
 B) polyps **D)** eggs and sperm

19. What is the young organism that the fertilized egg of a sponge develops into?
 A) bud **C)** medusa
 B) larva **D)** polyp

20. Which group do roundworms belong to?
 A) cnidarians **C)** planarians
 B) nematodes **D)** sponges

Science Online life.msscience.com/vocabulary_puzzlemaker

Thinking Critically

21. **Compare and contrast** the body organization of a sponge to that of a flatworm.
22. **Infer** the advantages of being able to reproduce sexually and asexually for animals like sponges, cnidarians, and flatworms.
23. **List** the types of food that sponges, hydras, and planarians eat. Explain why each organism eats a different size of particle.
24. **Compare and contrast** the medusa and polyp body forms of cnidarians.
25. **Infer** why scientists think the medusa stage was the first stage of the cnidarians.
26. **Form a hypothesis** about why cooking pork at high temperatures prevents harmful roundworms from developing, if they are present in the uncooked meat.
27. **Predict** what you can about the life of an organism that has no mouth or digestive system but has suckers and hooks on its head.
28. **Interpret Scientific Illustrations** Look at the photograph below. This animal escapes from predators by mimicry. Where in nature might you find the animal in this photo?

Performance Activities

29. **Report** Research tapeworms and other parasitic worms that live in humans. Find out how they are able to live in the intestines without being digested by the human host. Report your findings to the class.
30. **Video Presentation** Create a video presentation using computer software or slides to illustrate the variety of sponges and cnidarians found on a coral reef.

Applying Math

31. **Reef Ecology** Coral reefs are considered the "rain forests of the ocean" due to the number of different species that depend on them. If scientists estimate that out of 4,000 species, 1,000 are from the coral reef ecosystem, what percentage of life is dependent on the reef?

Use the table below to answer questions 32 and 33.

Reef Area Data

Country/Geographical Location	Reef Area [km^2]
Indonesia	51,000
Australia	49,000
Philippines	25,100
France	14,300
Papua, New Guinea	13,800
Fiji	10,000
Maledives	8,900
Saudi Arabia	6,700
Marshall Islands	6,100
India	5,800
United States	3,800
Other	89,800

32. **Reef Disappearance** Coral reefs are disappearing for many reasons, such as increased temperatures, physical damage, and pollution. In 2003, scientists predict that at the current rate of disappearance, in 2100 coral reefs will be gone. Use the table above to calculate the current rate of coral reef disappearance.
33. **Reef Locations** What percentage of coral reefs are off of the Australian coast?

Standardized Test Practice

Part 1 Multiple Choice

Record your answers on the answer sheet provided by your teacher or on a sheet of paper.

1. An animal that kills and then only eats other animals is an
   A) omnivore. **C)** carnivore.
   B) herbivore. **D)** scavenger.

2. An animal that does not have a backbone is called
   A) a vertebrate. **C)** a hermaphrodite.
   B) an invertebrate. **D)** a medusa.

Use the illustration below to answer questions 3 and 4.

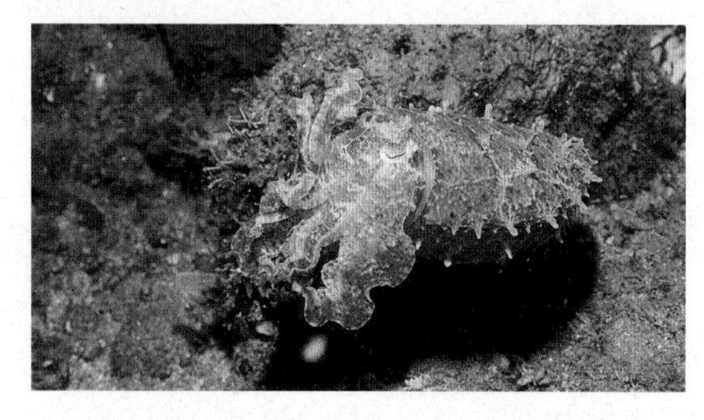

3. This animal escapes from predators by using
   A) behavioral adaptation.
   B) predator adaptation.
   C) mimicry.
   D) physical adaptation.

4. Markings that help an animal hide from its predators are called
   A) camouflage. **C)** behaviour.
   B) sessile. **D)** mimicry.

5. Which of the following is not a cnidarian?
   A) coral **C)** sea anemone
   B) hydra **D)** sponge

Test-Taking Tip

Study Advice Do not "cram" the night before a test. It can hamper your memory and make you tired.

6. Animals that have body parts arranged around a center point
   A) exhibit radial symmetry.
   B) exhibit bilateral symmetry.
   C) exhibit asymmetry.
   D) exhibit no symmetry.

Use the photo below to answer questions 7 and 8.

7. This organism would typically be found in which environment?
   A) lake **C)** ocean
   B) river **D)** pond

8. It would likely spend most of its life
   A) floating on water currents.
   B) attached to rock or coral.
   C) grouped in a colony.
   D) dependent on three other cnidarians.

9. Worms have which type of symmetry?
   A) asymmetry
   B) bilateral symmetry
   C) radial symmetry
   D) no symmetry

10. Which animal's body has the least complex body organization?
    A) cnidarians **C)** worms
    B) nematodes **D)** sponges

Part 2 Short Response/Grid In

Record your answers on the answer sheet provided by your teacher or on a sheet of paper.

11. Name three adaptations, and give an example for each.

Use the photo below to answer questions 12 and 13.

12. What types of classification and symmetry does this animal have and why?
13. Compare and contrast this with the type of classification and symmetry you have as a human.
14. How do sponges get food and oxygen?
15. Explain how sponges reproduce. Do they have more than one method of reproduction?
16. Explain the two forms of cnidarians.
17. Describe structure of the stinging cells unique to cnidarians. What are the purposes of these cells?
18. Explain the primary difference between a roundworm and a flatworm.
19. Roundworms and flatworms are the simplest organism to have what feature?

Part 3 Open Ended

Record your answers on a sheet of paper.

20. What are the characteristics that animals have in common and causes them to be included in their own kingdom?
21. Animals need energy. They get this from food. Explain the differences between herbivores, carnivores, omnivores and detritivores. Be sure to include examples of all categories.
22. Compare and contrast mimicry and camouflage. Give an example of both mimicry and camouflage.

Use the photo below to answer question 23.

23. How is this animal useful to humans?
24. Why is coral so important to us?
25. Explain how two types of animals may interact in a host and parasitic relationship. Include humans in this discussion.
26. Compare and contrast flatworms and roundworms. In your opinion, which are more developed? Defend your answer by providing examples.

chapter 13

Mollusks, Worms, Arthropods, Echinoderms

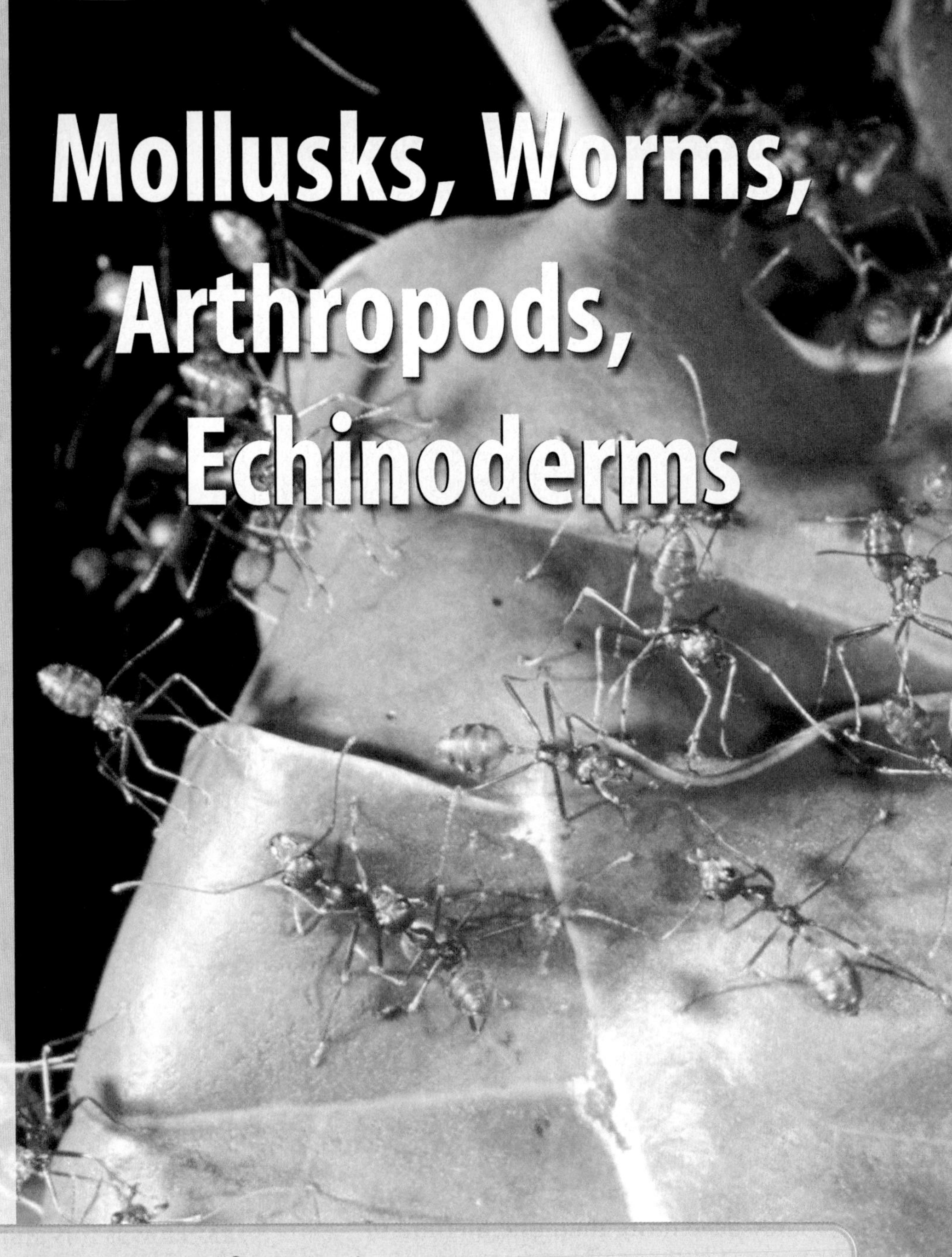

The BIG Idea

Mollusks, worms, arthropods, and echinoderms are all groups of invertebrates.

SECTION 1
Mollusks
Main Idea Mollusks are characterized by their soft bodies.

SECTION 2
Segmented Worms
Main Idea Segmented worms are tube-like and live on land or in water.

SECTION 3
Arthropods
Main Idea Arthropods have segmented bodies and are the largest group of animals on Earth.

SECTION 4
Echinoderms
Main Idea Echinoderms have a hard endoskeleton covered with bumpy skin.

An Army of Ants!

These green weaver worker ants are working together to defend their nest. These ants, and more than a million other species, are members of the largest and most diverse group of animals, the arthropods. In this chapter, you will be studying these animals, as well as mollusks, worms, and echinoderms.

Science Journal Write three animals from each animal group that you will be studying: mollusks, worms, arthropods, and echinoderms.

Start-Up Activities

Mollusk Protection

If you've ever walked along a beach, especially after a storm, you've probably seen many seashells. They come in different colors, shapes, and sizes. If you look closely, you will see that some shells have many rings or bands. In the following lab, find out what the bands tell you about the shell and the organism that made it.

1. Use a magnifying lens to examine a clam's shell.
2. Count the number of rings or bands on the shell. Count as number one the large, top point called the crown.
3. Compare the distances between the bands of the shell.
4. **Think Critically** Do other students' shells have the same number of bands? Are all of the bands on your shell the same width? What do you think the bands represent, and why are some wider than others? Record your answers in your Science Journal.

Preview this chapter's content and activities at life.msscience.com

Invertebrates Make the following Foldable to help you organize the main characteristics of the four groups of complex invertebrates.

STEP 1 **Draw** a mark at the midpoint of a sheet of paper along the side edge. Then **fold** the top and bottom edges in to touch the midpoint.

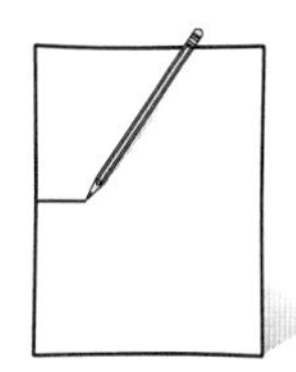

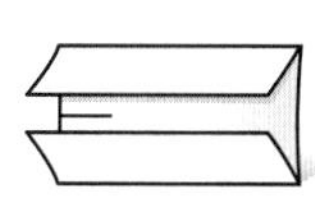

STEP 2 **Fold** in half from side to side.

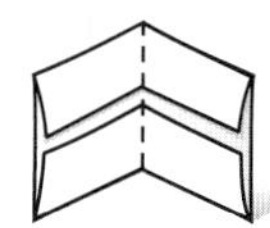

STEP 3 **Turn** the paper vertically. **Open and cut** along the inside fold lines to form four tabs.

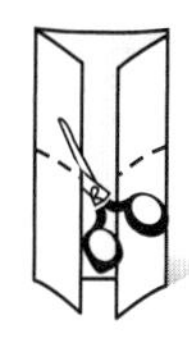

STEP 4 **Label** the tabs *Mollusks, Worms, Arthropods,* and *Echinoderms.*

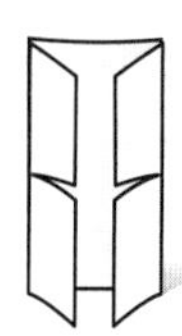

Classify As you read the chapter, list the characteristics of the four groups of invertebrates under the appropriate tab.

Get Ready to Read

Take Notes

1 Learn It! The best way for you to remember information is to write it down, or take notes. Good note-taking is useful for studying and research. When you are taking notes, it is helpful to

- phrase the information in your own words;
- restate ideas in short, memorable phrases;
- stay focused on main ideas and only the most important supporting details.

2 Practice It! Make note-taking easier by using a chart to help you organize information clearly. Write the main ideas in the left column. Then write at least three supporting details in the right column. Read the text from Section 2 of this chapter under the heading *Earthworm Body Systems*, page 365. Then take notes using a chart, such as the one below.

Main Idea	Supporting Details
	1. 2. 3. 4. 5.
	1. 2. 3. 4. 5.

3 Apply It! As you read this chapter, make a chart of the main ideas. Next to each main idea, list at least two supporting details.

Read one or two paragraphs first and take notes after you read. You are likely to take down too much information if you take notes as you read.

Target Your Reading

Use this to focus on the main ideas as you read the chapter.

1. **Before you read** the chapter, respond to the statements below on your worksheet or on a numbered sheet of paper.
   - Write an **A** if you **agree** with the statement.
   - Write a **D** if you **disagree** with the statement.

2. **After you read** the chapter, look back to this page to see if you've changed your mind about any of the statements.
   - If any of your answers changed, explain why.
   - Change any false statements into true statements.
   - Use your revised statements as a study guide.

Before You Read A or D		Statement	After You Read A or D
	1	All mollusks have shells.	
	2	Every mollusk lives in water.	
	3	Earthworms eat soil.	
	4	Segmented worms can be found on land, in freshwater, and in oceans.	
	5	Insects have difficulty surviving because they are small and must compete with each other for food and space.	
	6	Spiders are arthropods.	
	7	Insects breathe through holes in their abdomen.	
	8	All echinoderms have radial symmetry.	
	9	Every echinoderm can repair its body through regeneration.	

Science Online
Print out a worksheet of this page at life.msscience.com

section 1

Mollusks

as you read

What You'll Learn

- **Identify** the characteristics of mollusks.
- **Describe** gastropods, bivalves, and cephalopods.
- **Explain** the environmental importance of mollusks.

Why It's Important

Mollusks are a food source for many animals. They also filter impurities from the water.

Review Vocabulary

visceral mass: contains the stomach and other organs

New Vocabulary

- mantle
- gill
- open circulatory system
- radula
- closed circulatory system

Characteristics of Mollusks

Mollusks (MAH lusks) are soft-bodied invertebrates with bilateral symmetry and usually one or two shells. Their organs are in a fluid-filled cavity. The word *mollusk* comes from the Latin word meaning "soft." Most mollusks live in water, but some live on land. Snails, clams, and squid are examples of mollusks. More than 110,000 species of mollusks have been identified.

Body Plan All mollusks, like the one in **Figure 1,** have a thin layer of tissue called a mantle. The **mantle** covers the body organs, which are located in the visceral (VIH suh rul) mass. Between the soft body and the mantle is a space called the mantle cavity. It contains **gills**—the organs in which carbon dioxide from the mollusk is exchanged for oxygen in the water.

The mantle also secretes the shell or protects the body if the mollusk does not have a shell. The shell is made up of several layers. The inside layer is the smoothest. It is usually the thickest layer because it's added to throughout the life of the mollusk. The inside layer also protects the soft body.

The circulatory system of most mollusks is an open system. In an **open circulatory system,** the heart moves blood out into the open spaces around the body organs. The blood, which contains nutrients and oxygen, completely surrounds and nourishes the body organs.

Most mollusks have a well-developed head with a mouth and some sensory organs. Some mollusks, such as squid, have tentacles. On the underside of a mollusk is the muscular foot, which is used for movement.

Figure 1 The general mollusk body plan is shown by this snail. Most mollusks have a head, foot, and visceral mass.

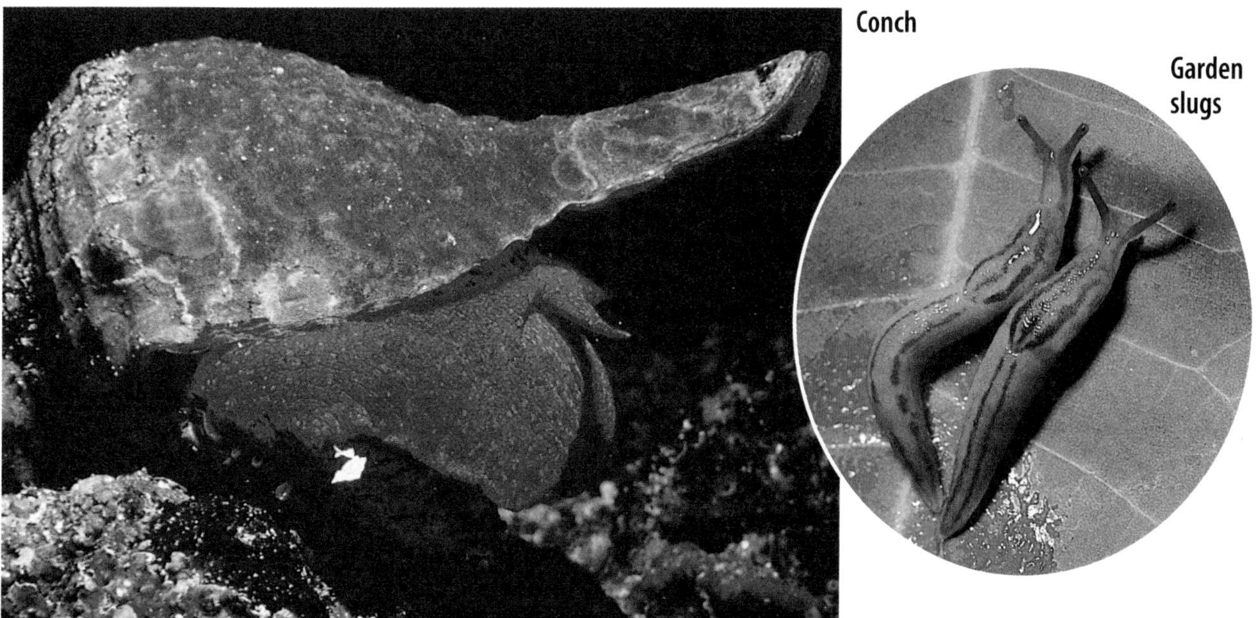

Figure 2 Conchs, sometimes called marine snails, have a single shell covering their internal organs. Garden slugs are mollusks without a shell.
Identify *the mollusk group that both conchs and garden slugs belong to.*

Classification of Mollusks

The first thing scientists look at when they classify mollusks is whether or not the animal has a shell. Mollusks that have shells are then classified by the kind of shell and kind of foot that they have. The three most common groups of mollusks are gastropods, bivalves, and cephalopods.

Gastropods The largest group of mollusks, the gastropods, includes snails, conchs like the one in **Figure 2,** abalones, whelks, sea slugs, and garden slugs, also shown in **Figure 2.** Conchs are sometimes called univalves. Except for slugs, which have no shell, gastropods have a single shell. Many have a pair of tentacles with eyes at the tips. Gastropods use a **radula** (RA juh luh)—a tonguelike organ with rows of teeth—to obtain food. The radula works like a file to scrape and tear food materials. That's why snails are helpful to have in an aquarium—they scrape the algae off the walls and keep the tank clean.

Reading Check *How do gastropods get food?*

Slugs and many snails are adapted to life on land. They move by rhythmic contractions of the muscular foot. Glands in the foot secrete a layer of mucus on which they slide. Slugs and snails are most active at night or on cloudy days when they can avoid the hot Sun. Slugs do not have shells but are protected by a layer of mucus instead, so they must live in moist places. Slugs and land snails damage plants as they eat leaves and stems.

Figure 3 Scallops force water between their valves to move away from sea stars and other predators. They can move up to 1 m with each muscular contraction.

Bivalves Mollusks that have a hinged, two-part shell joined by strong muscles are called bivalves. Clams, oysters, and scallops are bivalve mollusks and are a familiar source of seafood. These animals pull their shells closed by contracting powerful muscles near the hinge. To open their shells, they relax these muscles.

Bivalves are well adapted for living in water. For protection, clams burrow deep into the sand by contracting and relaxing their muscular foot. Mussels and oysters attach themselves with a strong thread or cement to a solid surface. This keeps waves and currents from washing them away. Scallops, shown in **Figure 3,** escape predators by rapidly opening and closing their shells. As water is forced out, the scallop moves rapidly in the opposite direction.

Cephalopods The most specialized and complex mollusks are the cephalopods (SE fuh luh pawdz). Squid, octopuses, cuttlefish, and chambered nautiluses belong to this group. The word *cephalopod* means "head-footed" and describes the body structure of these invertebrates. Cephalopods, like the cuttlefish in **Figure 4,** have a large, well-developed head. Their foot is divided into many tentacles with strong suction cups or hooks for capturing prey. All cephalopods are predators. They feed on fish, crustaceans, worms, and other mollusks.

Squid and octopuses have a well-developed nervous system and large eyes similar to human eyes. Unlike other mollusks, cephalopods have closed circulatory systems. In a **closed circulatory system,** blood containing food and oxygen moves through the body in a series of closed vessels, just as your blood moves through your blood vessels.

Reading Check *What makes a cephalopod different from other mollusks?*

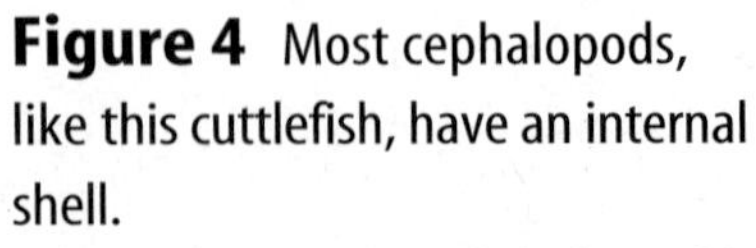

Figure 4 Most cephalopods, like this cuttlefish, have an internal shell.
Infer *why an internal shell would be a helpful adaptation.*

Figure 5 Squid and other cephalopods use jet propulsion to move quickly away from predators.

INTEGRATE Physics

Cephalopod Propulsion All cephalopods live in oceans and are adapted for swimming. Squid and other cephalopods have a water-filled cavity between an outer muscular covering and its internal organs. When the cephalopod tightens its muscular covering, water is forced out through an opening near the head, as shown in **Figure 5.** The jet of water propels the cephalopod backwards, and it moves away quickly. According to Newton's third law of motion, when one object exerts a force on a second object, the second object exerts a force on the first that is equal and opposite in direction. The movement of cephalopods is an example of this law. Muscles exert force on water under the mantle. Water being forced out exerts a force that results in movement backwards.

A squid can propel itself at speeds of more than 6 m/s using this jet propulsion and can briefly outdistance all but whales, dolphins, and the fastest fish. A squid even can jump out of the water and reach heights of almost 5 m above the ocean's surface. It then can travel through the air as far as 15 m. However, squid can maintain their top speed for just a few pulses. Octopuses also can swim by jet propulsion, but they usually use their tentacles to creep more slowly over the ocean floor.

Origin of Mollusks Some species of mollusks, such as the chambered nautilus, have changed little from their ancestors. Mollusk fossils date back more than 500 million years. Many species of mollusks became extinct about 65 million years ago. Today's mollusks are descendants of ancient mollusks.

Mollusk Extinction By about 65 million years ago, many mollusks had become extinct. What were the major physical events of the time that could have contributed to changing the environment? Write your answers in your Science Journal.

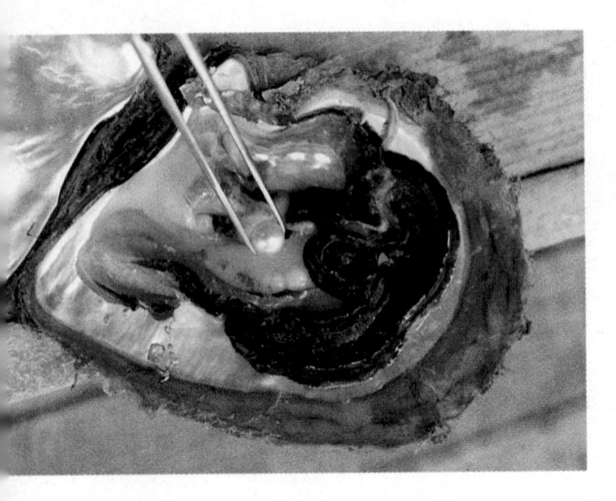

Figure 6 A pearl starts as an irritant—a grain of sand or a parasite—to an oyster. The oyster coats the irritant with a material that forms smooth, hard layers. It can take years for a pearl to form. Culturing pearls is a commercial industry in some countries.

Value of Mollusks

Mollusks have many uses. They are food for fish, sea stars, birds, and humans. Many people make their living raising or collecting mollusks to sell for food. Other invertebrates, such as hermit crabs, use empty mollusk shells as shelter. Many mollusk shells are used for jewelry and decoration. Pearls are produced by several species of mollusks, but most are made by mollusks called pearl oysters, shown in **Figure 6.** Mollusk shells also provide information about the conditions in an ecosystem, including the source and distribution of water pollutants. The internal shell of a cuttlefish is the cuttlebone, which is used in birdcages to provide birds with calcium. Squid and octopuses are able to learn tasks, so scientists are studying their nervous systems to understand how learning takes place and how memory works.

Even though mollusks are beneficial in many ways, they also can cause problems for humans. Land slugs and snails damage plants. Certain species of snails are hosts of parasites that infect humans. Shipworms, a type of bivalve, make holes in submerged wood of docks and boats, causing millions of dollars in damage each year. Because clams, oysters, and other mollusks are filter feeders, bacteria, viruses, and toxic protists from the water can become trapped in the animals. Eating these infected mollusks can result in sickness or even death.

section 1 review

Summary

Mollusks

- The body plans of mollusks include a mantle, visceral mass, head, and foot.

Classification of Mollusks

- Gastropods typically have one shell, a foot, and eat using a radula.
- Bivalves have a hinged two-part shell, a muscular foot, and eat by filtering their food from the water.
- Cephalopods have a head, a foot which has been modified into tentacles, and a well-developed nervous system.

Value of Mollusks

- Mollusks are food for many animals, have commercial uses, and are used for research.

Self Check

1. **Explain** how a squid and other cephalopods can move so rapidly.
2. **Identify** some positive and negative ways that mollusks affect humans.
3. **Think Critically** Why is it unlikely that you would find garden slugs or land snails in a desert?

Applying Skills

4. **Interpret Scientific Illustrations** Observe the images of gastropods and bivalves in this section, and infer how bivalves are not adapted to life on land, but gastropods are.
5. **Use a Computer** Make a data table that compares and contrasts the following for gastropods, bivalves, and cephalopods: *methods for obtaining food, movement, circulation,* and *habitat.*

Science Online life.msscience.com/self_check_quiz

section 2

Segmented Worms

Segmented Worm Characteristics

The worms you see crawling across sidewalks after a rain and those used for fishing are called annelids (A nuh ludz). The word *annelid* means "little rings" and describes the bodies of these worms. They have tube-shaped bodies that are divided into many segments.

Have you ever watched a robin try to pull an earthworm out of the ground or tried it yourself? Why don't they slip out of the soil easily? On the outside of each body segment are bristlelike structures called **setae** (SEE tee). Segmented worms use their setae to hold on to the soil and to move. Segmented worms also have bilateral symmetry, a body cavity that holds the organs, and two body openings—a mouth and an anus. Annelids can be found in freshwater, salt water, and moist soil. Earthworms, like the one in **Figure 7,** marine worms, and leeches are examples of annelids.

Reading Check *What is the function of setae?*

as you read

What You'll Learn

- **Identify** the characteristics of segmented worms.
- **Describe** the structures of an earthworm and how it takes in and digests food.
- **Explain** the importance of segmented worms.

Why It's Important

Earthworms condition and aerate the soil, which helps increase crop yields.

Review Vocabulary
aerate: to supply with air

New Vocabulary
- setae
- crop
- gizzard

Earthworm Body Systems

The most well-known annelids are earthworms. They have a definite anterior, or front end, and a posterior, or back end. Earthworms have more than 100 body segments. The segments can be seen on the outside and the inside of the body cavity. Each body segment, except for the first and last segments, has four pairs of setae. Earthworms move by using their setae and two sets of muscles in the body wall. One set of muscles runs the length of the body, and the other set circles the body. When an earthworm contracts its long muscles, it causes some of the segments to bunch up and the setae to stick out. This anchors the worm to the soil. When the circular muscles contract, the setae are pulled in and the worm can move forward.

Figure 7 One species of earthworm that lives in Australia can grow to be 3.3 m long.

Digestion and Excretion

As an earthworm burrows through the soil, it takes soil into its mouth. Earthworms get energy from the bits of leaves and other organic matter found in the soil. The soil ingested by an earthworm moves to the **crop,** which is a sac used for storage. Behind the crop is a muscular structure called the **gizzard,** which grinds the soil and the bits of organic matter. This ground material passes to the intestine, where the organic matter is broken down and the nutrients are absorbed by the blood. Wastes leave the worm through the anus. When earthworms take in soil, they provide spaces for air and water to flow through it and mix the soil. Their wastes pile up at the openings to their burrows. These piles are called castings. Castings, like those in **Figure 8,** help fertilize the soil.

Figure 8 Earthworm castings—also called vermicompost—are used as an organic fertilizer in gardens.
Infer *why earthworms are healthy to have in a garden or compost pile.*

Circulation and Respiration

Earthworms have a closed circulatory system, as shown in **Figure 9.** Two blood vessels along the top of the body and one along the bottom of the body meet in the front end of the earthworm. There, they connect to heartlike structures called aortic arches, which pump blood through the body. Smaller vessels go into each body segment.

Earthworms don't have gills or lungs. Oxygen and carbon dioxide are exchanged through their skin, which is covered with a thin film of watery mucus. It's important never to touch earthworms with dry hands or remove their thin mucous layer, because they could suffocate. But as you can tell after a rainstorm, earthworms don't survive in puddles of water either.

Figure 9 An earthworm's circulatory system includes five aortic arches that pump blood throughout its body.

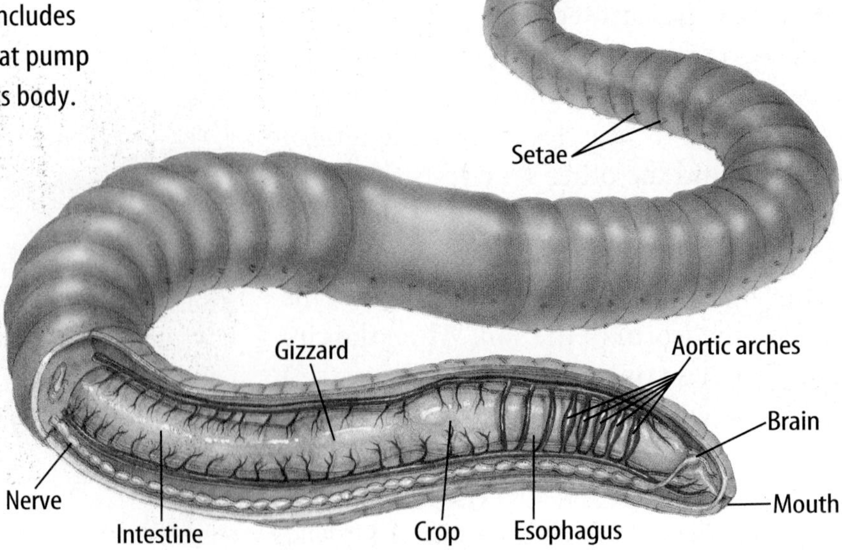

Nerve Response and Reproduction Earthworms have a small brain in their front segment. Nerves in each segment join to form a main nerve cord that connects to the brain. Earthworms respond to light, temperature, and moisture.

Earthworms are hermaphrodites (hur MA fruh dites)—meaning they produce sperm and eggs in the same body. Even though each worm has male and female reproductive structures, an individual worm can't fertilize its own eggs. Instead, it has to receive sperm from another earthworm in order to reproduce.

Marine Worms

More than 8,000 species of marine worms, or polychaetes, (PAH lee keets) exist, which is more than any other kind of annelid. Marine worms float, burrow, build structures, or walk along the ocean floor. Some polychaetes even produce their own light. Others, like the ice worms in **Figure 10,** are able to live 540 m deep. Polychaetes, like earthworms, have segments with setae. However, the setae occur in bundles on these worms. The word polychaete means "many bristles."

Sessile, bottom-dwelling polychaetes, such as the Christmas tree worms shown in **Figure 11,** have specialized tentacles that are used for exchanging oxygen and carbon dioxide and gathering food. Some marine worms build tubes around their bodies. When these worms are startled, they retreat into their tubes. Free-swimming polychaetes, such as the bristleworm shown in **Figure 11,** have a head with eyes; a tail; and parapodia (per uh POH dee uh). Parapodia are paired, fleshy outgrowths on each segment, which aid in feeding and locomotion.

Figure 10 Ice worms, a type of marine polychaete, were discovered first in 1997 living 540 m deep in the Gulf of Mexico.

Figure 11 These Christmas tree worms filter microorganisms from the water to eat. This bristleworm swims backwards and forwards, so it has eyes at both ends of its body.

Leeches

A favorite topic for scary movies is leeches. If you've ever had to remove a leech from your body after swimming in a freshwater pond, lake, or river, you know it isn't fun. Leeches are segmented worms, but their bodies are not as round or as long as earthworms are, and they don't have setae. They feed on the blood of other animals. A sucker at each end of a leech's body is used to attach itself to an animal. If a leech attaches to you, you probably won't feel it. Leeches produce many chemicals, including an anesthetic (a nus THEH tihk) that numbs the wound so you don't feel its bite. After the leech has attached itself, it cuts into the animal and sucks out two to ten times its own weight in blood. Even though leeches prefer to eat blood, they can survive by eating aquatic insects and other organisms instead.

Topic: Beneficial Leeches
Visit life.msscience.com for Web links to information about the uses of chemicals from leech saliva.

Activity Describe a possible use for leech saliva, and design a 30-second commercial on how you might sell it.

Why is producing an anesthetic an advantage to a leech?

Leeches and Medicine

Sometimes, leeches are used after surgery to keep blood flowing to the repaired area, as shown in **Figure 12.** For example, the tiny blood vessels in the ear quickly can become blocked with blood clots after surgery. To keep blood flowing in such places, physicians might attach leeches to the surgical site. As the leeches feed on the blood, chemicals in their saliva prevent the blood from coagulating. Besides the anti-clotting chemical, leech saliva also contains a chemical that dilates blood vessels, which improves the blood flow and allows the wound to heal more quickly. These chemicals are being studied to treat patients with heart or circulatory diseases, strokes, arthritis, or glaucoma.

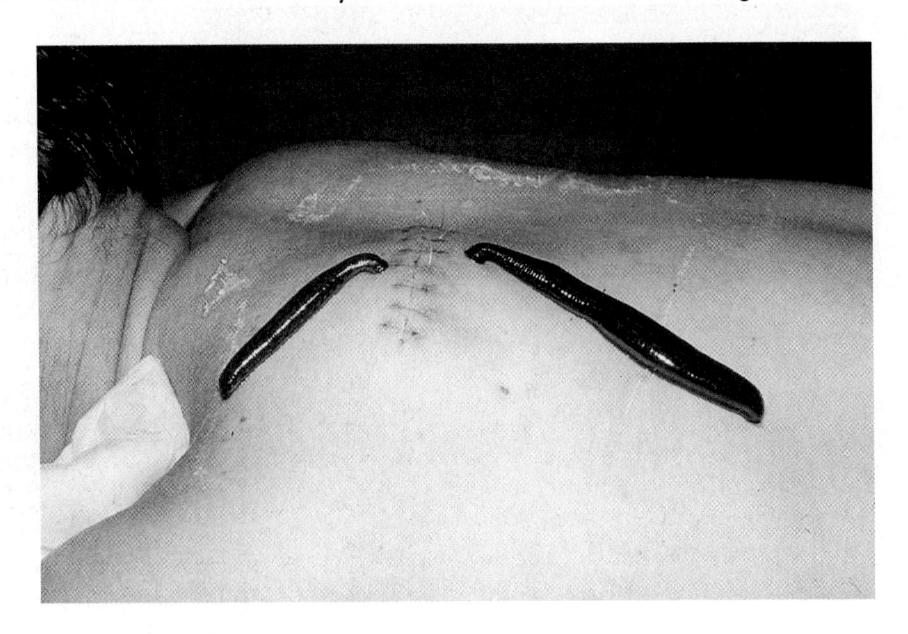

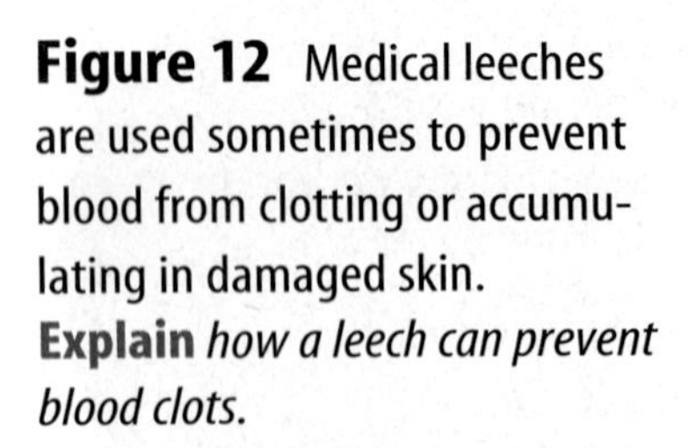

Figure 12 Medical leeches are used sometimes to prevent blood from clotting or accumulating in damaged skin.
Explain *how a leech can prevent blood clots.*

Value of Segmented Worms

Different kinds of segmented worms are helpful to other animals in a variety of ways. Earthworms help aerate the soil by constantly burrowing through it. By grinding and partially digesting the large amount of plant material in soil, earthworms speed up the return of nitrogen and other nutrients to the soil for use by plants.

Researchers are developing drugs based on the chemicals that come from leeches because leech saliva prevents blood clots. Marine worms and their larvae are food for many fish, invertebrates, and mammals.

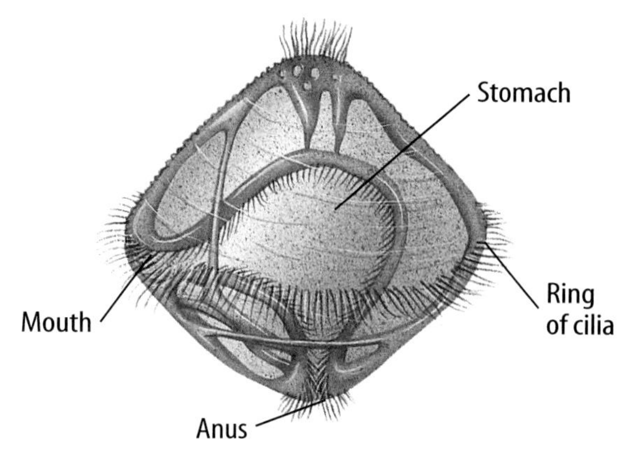

Mollusk larva

Origin of Segmented Worms

Some scientists hypothesize that segmented worms evolved in the sea. The fossil record for segmented worms is limited because of their soft bodies. The tubes of marine worms are the most common fossils of the segmented worms. Some of these fossils date back about 620 million years.

Similarities between mollusks and segmented worms suggest that they could have a common ancestor. These groups were the first animals to have a body cavity with space for body organs to develop and function. Mollusks and segmented worms have a one-way digestive system with a separate mouth and anus. Their larvae, shown in **Figure 13,** are similar and are the best evidence that they have a common ancestor.

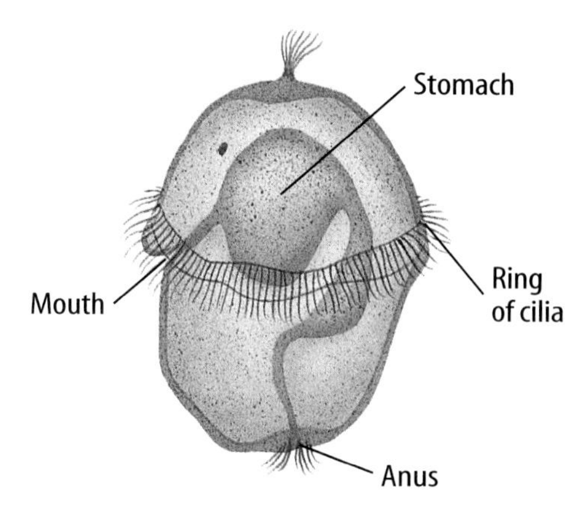

Annelid larva

Figure 13 Some mollusk larvae have many structures that are similar to those of some annelid larvae.

section 2 review

Summary

Segmented Worm Characteristics

- Segmented worms have tube-shaped bodies divided into many segments, bilateral symmetry, a body cavity with organs, and two body openings.
- Earthworms have a definite anterior and posterior end, a closed circulatory system, a small brain, and eat organic matter in the soil. Most segments have four pairs of setae.
- Polychaetes are marine worms with many setae occurring in bundles.
- Leeches are segmented worms that feed on the blood of other animals. They have no setae, but a sucker at each end of the body.

Self Check

1. **Define** setae and state their function.
2. **Describe** how an earthworm takes in and digests its food.
3. **Compare and contrast** how earthworms and marine worms exchange oxygen and carbon dioxide.
4. **Think Critically** What advantages do marine worms with tubes have over free-swimming polychaetes?

Applying Math

5. **Use Proportions** Suppose you find six earthworms in 10 cm^3 of soil. Based on this sample, calculate the number of earthworms you would find in 10 m^3 of soil.

section 3

Arthropods

as you read

What You'll Learn

- **Determine** the characteristics that are used to classify arthropods.
- **Explain** how the structure of the exoskeleton relates to its function.
- **Distinguish** between complete and incomplete metamorphosis.

Why It's Important

Arthropods, such as those that carry diseases and eat crops, affect your life every day.

Review Vocabulary

venom: toxic fluid injected by an animal

New Vocabulary

- appendage
- exoskeleton
- molting
- spiracle
- metamorphosis

Characteristics of Arthropods

There are more than a million different species of arthropods, (AR thruh pahdz) making them the largest group of animals. The word *arthropoda* means "jointed foot." The jointed **appendages** of arthropods can include legs, antennae, claws, and pincers. Arthropod appendages are adapted for moving about, capturing prey, feeding, mating, and sensing their environment. Arthropods also have bilateral symmetry, segmented bodies, an exoskeleton, a body cavity, a digestive system with two openings, and a nervous system. Most arthropod species have separate sexes and reproduce sexually. Arthropods are adapted to living in almost every environment. They vary in size from microscopic dust mites to the large, Japanese spider crab, shown in **Figure 14.**

Segmented Bodies The bodies of arthropods are divided into segments similar to those of segmented worms. Some arthropods have many segments, but others have segments that are fused together to form body regions, such as those of insects, spiders, and crabs.

Exoskeletons All arthropods have a hard, outer covering called an **exoskeleton.** It covers, supports, and protects the internal body and provides places for muscles to attach. In many land-dwelling arthropods, such as insects, the exoskeleton has a waxy layer that reduces water loss from the animal.

An exoskeleton cannot grow as the animal grows. From time to time, the exoskeleton is shed and replaced by a new one in a process called **molting.** While the animals are molting, they are not well protected from predators because the new exoskeleton is soft. Before the new exoskeleton hardens, the animal swallows air or water to increase its exoskeleton's size. This way the new exoskeleton allows room for growth.

Figure 14 The Japanese spider crab has legs that can span more than 3 m.

Insects

More species of insects exist than all other animal groups combined. More than 700,000 species of insects have been classified, and scientists identify more each year. Insects have three body regions—a head, a thorax, and an abdomen, as shown in **Figure 15.** However, it is almost impossible on some insects to see where one region stops and the next one begins.

Figure 15 One of the largest types of ants is the carpenter ant. Like all insects, it has a head, thorax, and abdomen.

Head An insect's head has a pair of antennae, eyes, and a mouth. The antennae are used for touch and smell. The eyes are simple or compound. Simple eyes detect light and darkness. Compound eyes, like those in **Figure 16,** contain many lenses and can detect colors and movement. The mouthparts of insects vary, depending on what the insect eats.

Thorax Three pairs of legs and one or two pairs of wings, if present, are attached to the thorax. Some insects, such as silverfish and fleas, don't have wings, and other insects have wings only for part of their lives. Insects are the only invertebrate animals that can fly. Flying allows insects to find places to live, food sources, and mates. Flight also helps them escape from their predators.

Reading Check *How does flight benefit insects?*

Abdomen The abdomen has neither wings nor legs but it is where the reproductive structures are found. Females lay thousands of eggs, but only a fraction of the eggs develop into adults. Think about how overproduction of eggs might ensure that each insect species will continue.

Insects have an open circulatory system that carries digested food to cells and removes wastes. However, insect blood does not carry oxygen because it does not have hemoglobin. Instead, insects have openings called **spiracles** (SPIHR ih kulz) on the abdomen and thorax through which air enters and waste gases leave the insect's body.

Stained LM Magnification: 400×

Figure 16 Each compound eye is made up of small lenses that fit together. Each lens sees a part of the picture to make up the whole scene. Insects can't focus their eyes. Their eyes are always open and can detect movements.

Observing Metamorphosis

Procedure

1. Place a 2-cm piece of ripe **banana** in a **jar** and leave it open.
2. Check the jar every day for two weeks. When you see fruit flies, cover the mouth of the jar with **cheesecloth.**
3. Identify, describe, and draw all the stages of metamorphosis that you observe.

Analysis

1. What type of metamorphosis do fruit flies undergo?
2. In which stages are the flies the most active?

From Egg to Adult Many insects go through changes in body form as they grow. This series of changes is called **metamorphosis** (me tuh MOR fuh sihs). Grasshoppers, silverfish, lice, and crickets undergo incomplete metamorphosis, shown in **Figure 17.** The stages of incomplete metamorphosis are egg, nymph, and adult. The nymph form molts several times before becoming an adult. Many insects—butterflies, beetles, ants, bees, moths, and flies—undergo complete metamorphosis, also shown in **Figure 17.** The stages of complete metamorphosis are egg, larva, pupa, and adult. Caterpillar is the common name for the larva of a moth or butterfly. Other insect larvae are called grubs, maggots, or mealworms. Only larval forms molt.

Reading Check *When do grasshoppers molt?*

Figure 17 The two types of metamorphosis are shown here.

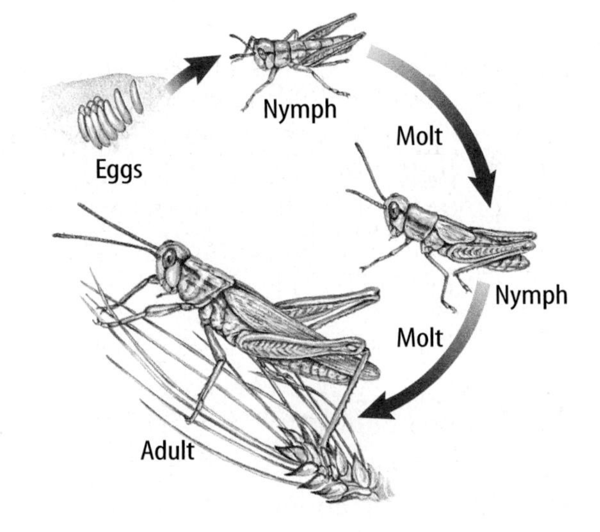

In incomplete metamorphosis, nymphs are smaller versions of their parents.

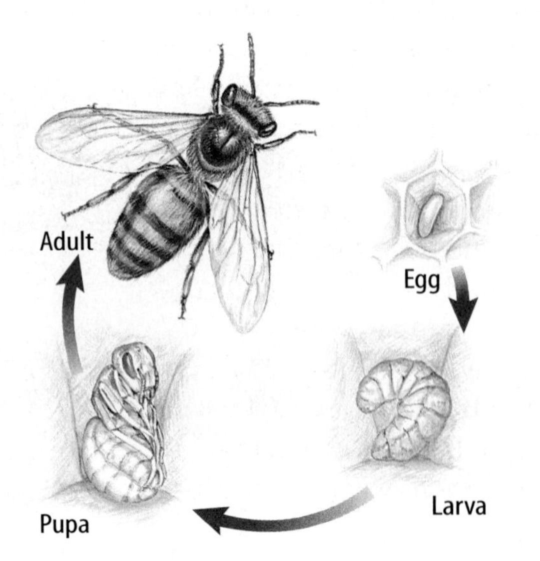

Many insects go through complete metamorphosis.

Figure 18 Feeding adaptations of insects include different mouthparts.

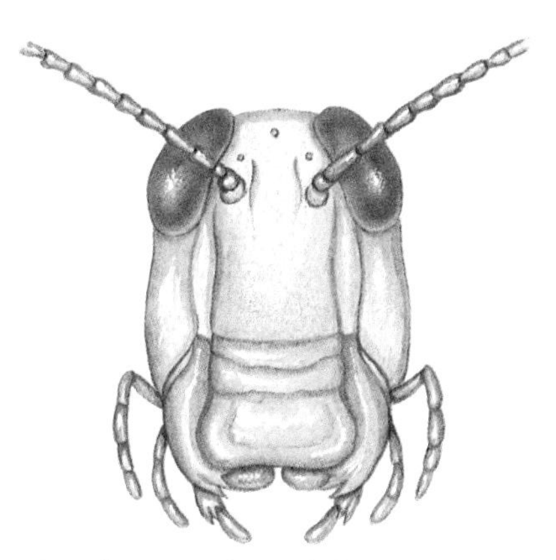

Grasshoppers have left and right mouthparts called mandibles that enable them to chew through tough plant tissues.

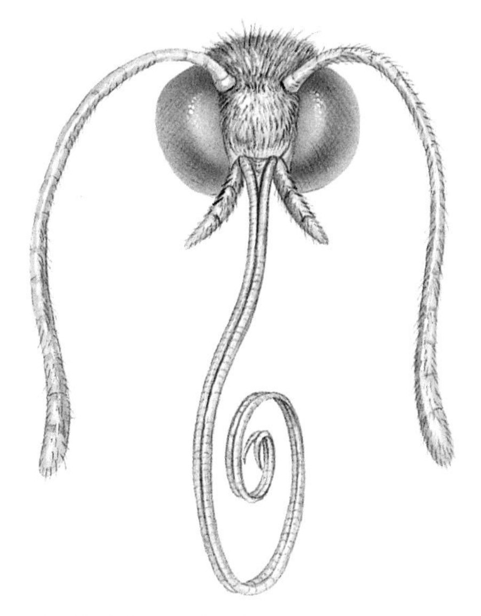

Butterflies and other nectar eaters have a long siphon that enables them to drink nectar from flowers.

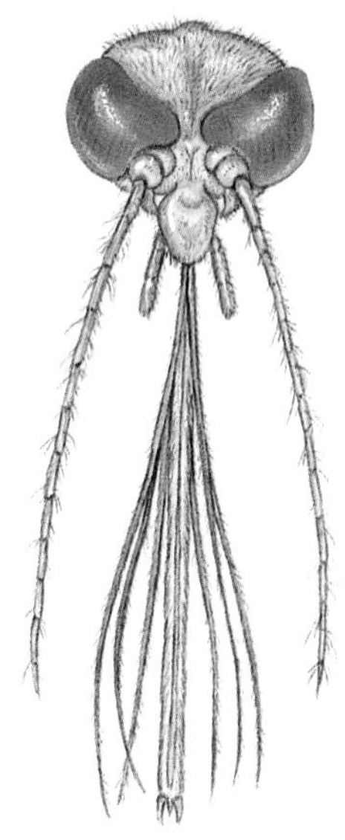

Mosquitoes have mouths that are adapted for piercing skin and sucking blood.

Obtaining Food Insects feed on plants, the blood of animals, nectar, decaying materials, wood in houses, and clothes. Mouthparts of insects, such as those in **Figure 18,** are as diverse as the insects themselves. Grasshoppers and ants have large mandibles (MAN duh bulz) for chewing plant tissue. Butterflies and honeybees are equipped with siphons for lapping up nectar in flowers. Aphids and cicadas pierce plant tissues and suck out plant fluids. Praying mantises eat other animals. External parasites, such as mosquitoes, fleas, and lice, drink the blood and body fluids of other animals. Silverfish eat things that contain starch and some moth larvae eat wool clothing.

Insect Success Because of their tough, flexible, waterproof exoskeletons; their ability to fly; rapid reproductive cycles; and small sizes, insects are extremely successful. Most insects have short life spans, so genetic traits can change more quickly in insect populations than in organisms that take longer to reproduce. Because insects generally are small, they can live in a wide range of environments and avoid their enemies. Many species of insects can live in the same area and not compete with one another for food, because many are so specialized in what they eat.

Protective coloration, or camouflage, allows insects to blend in with their surroundings. Many moths resting on trees look like tree bark or bird droppings. Walking sticks and some caterpillars resemble twigs. When a leaf butterfly folds its wings it looks like a dead leaf.

Disease Carriers Some insects may carry certain diseases to humans. Some species of mosquitoes can carry malaria or yellow fever, and can cause problems around the world. Research to learn about one disease that is carried by an insect, where it is a problem, and the steps that are being taken for prevention and treatment. Make a bulletin board of all the information that you and your classmates gather.

Arachnids

Spiders, scorpions, mites, and ticks are examples of arachnids (uh RAK nudz). They have two body regions—a head-chest region called the cephalothorax (se fuh luh THOR aks) and an abdomen. Arachnids have four pairs of legs but no antennae. Many arachnids are adapted to kill prey with venom glands, stingers, or fangs. Others are parasites.

Scorpions Arachnids that have a sharp, venom-filled stinger at the end of their abdomen are called scorpions. The venom from the stinger paralyzes the prey. Unlike other arachnids, scorpions have a pair of well-developed appendages—pincers—with which they grab their prey. The sting of a scorpion is painful and can be fatal to humans.

Applying Math — Use Percentages

SILK ELASTICITY A strand of spider's silk can be stretched from 65 cm to 85 cm before it loses its elasticity—the ability to snap back to its original length. Calculate the percent of elasticity of spider's silk.

Solution

1. *This is what you know:*
   - original length of silk strand = 65 cm
   - stretched length of silk strand = 85 cm
2. *This is what you need to find out:* percent of elasticity
3. *This is the procedure you need to use:*
   - Find the difference between the stretched and original length. 85 cm − 65 cm = 20 cm
   - $\frac{\text{difference in length}}{\text{original length}} \times 100 = \%$ of elasticity
   - $\frac{20 \text{ cm}}{65 \text{ cm}} \times 100 = 30.7\ \%$ of elasticity
4. *Check your answer:* Multiply 30.7% by 65 cm and you should get 20 cm.

Practice Problems

1. A 40-cm strand of nylon can be stretched to a length of 46.5 cm before losing its elasticity. Calculate the percent of elasticity for nylon and compare it to that of spider's silk.
2. Knowing the elasticity of spider's silk, what was the original length of a silk strand when the difference between the two strands is 44 cm?

Science Online — For more practice, visit life.msscience.com/math_practice

Spiders Because spiders can't chew their food, they release enzymes into their prey that help digest it. The spider then sucks the predigested liquid into its mouth.

Oxygen and carbon dioxide are exchanged in book lungs, illustrated in **Figure 19.** Openings on the abdomen allow these gases to move into and out of the book lungs.

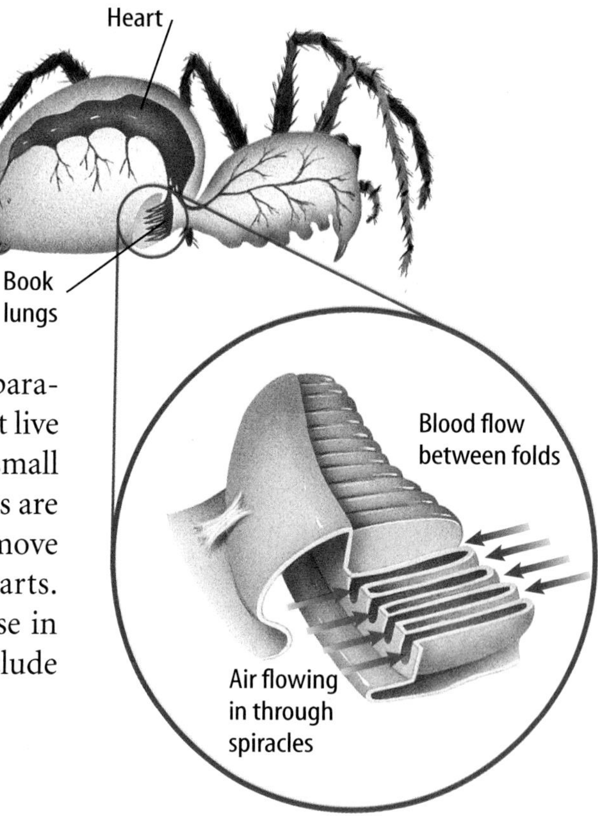

Figure 19 Air circulates between the moist folds of the book lungs bringing oxygen to the blood.

Mites and Ticks Most mites are animal or plant parasites. However, some are not parasites, like the mites that live in the follicles of human eyelashes. Most mites are so small that they look like tiny specs to the unaided eye. All ticks are animal parasites. Ticks attach to their host's skin and remove blood from their hosts through specialized mouthparts. Ticks often carry bacteria and viruses that cause disease in humans and other animals. Diseases carried by ticks include Lyme disease and Rocky Mountain spotted fever.

Centipedes and Millipedes

Two groups of arthropods—centipedes and millipedes—have long bodies with many segments and many legs, antennae, and simple eyes. They can be found in damp environments, including in woodpiles, under vegetation, and in basements. Centipedes and millipedes reproduce sexually. They make nests for their eggs and stay with them until the eggs hatch.

Compare the centipede and millipede in **Figure 20.** How many pairs of legs does the centipede have per segment? How many pairs of legs does the millipede have per segment? Centipedes hunt for their prey, which includes snails, slugs, and worms. They have a pair of venomous claws that they use to inject venom into their prey. Their pinches are painful to humans but usually aren't fatal. Millipedes feed on plants and decaying material and often are found under the damp plant material.

Figure 20 Centipedes are predators—they capture and eat other animals. Millipedes eat plants or decaying plant material.

VISUALIZING ARTHROPOD DIVERSITY

Figure 21

Some 600 million years ago, the first arthropods lived in Earth's ancient seas. Today, they inhabit nearly every environment on Earth. Arthropods are the most abundant and diverse group of animals on Earth. They range in size from nearly microscopic mites to spindly, giant Japanese spider crabs with legs spanning more than 3 m.

▲ LOBSTER Like crabs, lobsters are crustaceans that belong to the group called Decapoda, which means "ten legs." It's the lobster's tail, however, that interests most seafood lovers.

◀ GRASS SPIDER Grass spiders spin fine, nearly invisible webs just above the ground.

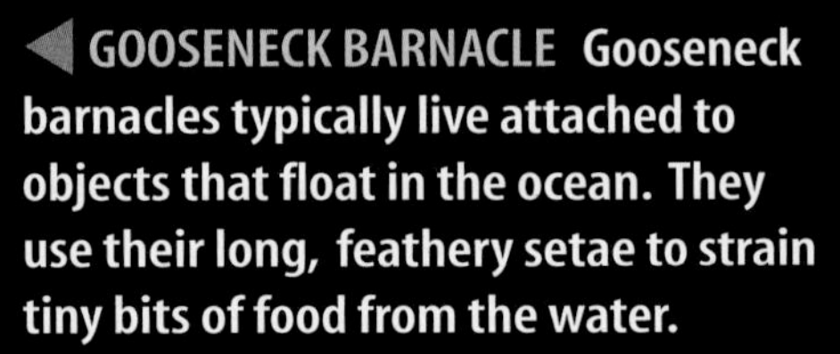

◀ GOOSENECK BARNACLE Gooseneck barnacles typically live attached to objects that float in the ocean. They use their long, feathery setae to strain tiny bits of food from the water.

▼ MONARCH BUTTERFLY Monarchs are a common sight in much of the United States during the summer. In fall, they migrate south to warmer climates.

◀ HISSING COCKROACH Most cockroaches are considered to be pests by humans, but hissing cockroaches, such as this one, are sometimes kept as pets.

▶ HORSESHOE CRAB Contrary to their name, horseshoe crabs are not crustaceans. They are more closely related to spiders than to crabs.

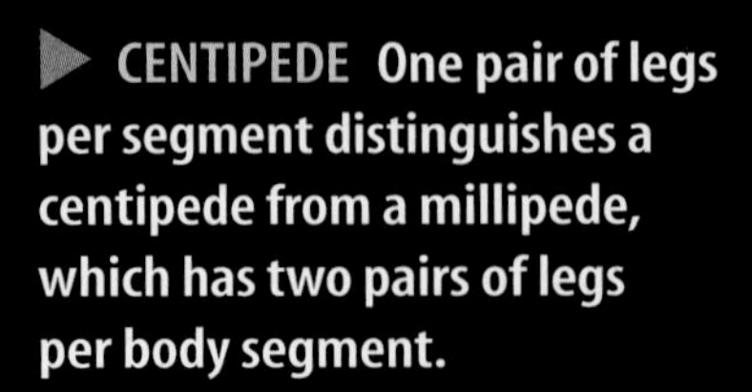

▶ CENTIPEDE One pair of legs per segment distinguishes a centipede from a millipede, which has two pairs of legs per body segment.

Crustaceans

Crabs, crayfish, shrimp, barnacles, pill bugs, and water fleas are crustaceans. Crustaceans and other arthropods are shown in **Figure 21.** Crustaceans have one or two pairs of antennae and mandibles, which are used for crushing food. Most crustaceans live in water, but some, like the pill bugs shown in **Figure 22,** live in moist environments on land. Pill bugs are common in gardens and around house foundations. They are harmless to humans.

Crustaceans, like the blue crab shown in **Figure 22,** have five pairs of legs. The first pair of legs are claws that catch and hold food. The other four pairs are walking legs. They also have five pairs of appendages on the abdomen called swimmerets. They help the crustacean move and are used in reproduction. In addition, the swimmerets force water over the feathery gills where the oxygen and carbon dioxide are exchanged. If a crustacean loses an appendage, it will grow back, or regenerate.

Figure 22 The segments in some crustaceans, such as this crab, aren't obvious because they are covered by a shieldlike structure. Pill bugs—also called roly polys—are crustaceans that live on land.
Compare and contrast *pill bugs to centipedes and millipedes.*

Value of Arthropods

Arthropods play several roles in the environment. They are a source of food for many animals, including humans. Some humans consider shrimp, crab, crayfish, and lobster as food delicacies. In Africa and Asia, many people eat insect larvae and insects such as grasshoppers, termites, and ants, which are excellent sources of protein.

Agriculture would be impossible without bees, butterflies, moths, and flies that pollinate crops. Bees manufacture honey, and silkworms produce silk. Many insects and spiders are predators of harmful animal species, such as stableflies. Useful chemicals are obtained from some arthropods. For example, bee venom is used to treat rheumatic arthritis.

Not all arthropods are useful to humans. Almost every cultivated crop has some insect pest that feeds on it. Many arthropods—mosquitoes, tsetse flies, fleas, and ticks—carry human and other animal diseases. In addition, weevils, cockroaches, carpenter ants, clothes moths, termites, and carpet beetles destroy food, clothing, and property.

Insects are an important part of the ecological communities in which humans live. Removing all of the insects would cause more harm than good.

Controlling Insects One common way to control problem insects is by insecticides. However, many insecticides also kill helpful insects. Another problem is that many toxic substances that kill insects remain in the environment and accumulate in the bodies of animals that eat them. As other animals eat the contaminated animals, the insecticides can find their way into human food. Humans also are harmed by these toxins.

Different types of bacteria, fungi, and viruses are being used to control some insect pests. Natural predators and parasites of insect pests have been somewhat successful in controlling them. Other biological controls include using sterile males or naturally occurring chemicals that interfere with the reproduction or behavior of insect pests.

Figure 23 More than 15,000 species of trilobites have been classified. They are one of the most recognized types of fossils.

Origin of Arthropods Because of their hard body parts, arthropod fossils, like the one in **Figure 23,** are among the oldest and best-preserved fossils of many-celled animals. Some are more than 500 million years old. Because earthworms and leeches have individual body segments, scientists hypothesize that arthropods probably evolved from an ancestor of segmented worms. Over time, groups of body segments fused and became adapted for locomotion, feeding, and sensing the environment. The hard exoskeleton and walking legs allowed arthropods to be among the first successful land animals.

section 3 review

Summary

Characteristics of Arthropods

- All arthropods have jointed appendages, bilateral symmetry, a body cavity, a digestive system, a nervous system, segmented bodies, and an exoskeleton.

Arthropod Types

- Insects have three body segments—head, thorax, and abdomen—a pair of antennae, and three pairs of legs. They go through complete or incomplete metamorphosis.
- Arachnids have two body segments—a cephalothorax and an abdomen—four pairs of legs, and no antennae.
- Centipedes and millipedes have long bodies with many segments and legs.
- Crustaceans have five pairs of legs and five pairs of appendages called swimmerets.

Self Check

1. **Infer** the advantages and disadvantages of an exoskeleton.
2. **Compare and contrast** the stages of complete and incomplete metamorphosis.
3. **List** four ways arthropods obtain food.
4. **Evaluate** the impact of arthropods.
5. **Concept Map** Make an events-chain concept map of complete metamorphosis and one of incomplete metamorphosis.
6. **Think Critically** Choose an insect you are familiar with and explain how it is adapted to its environment.

Applying Math

7. **Make a Graph** Of the major arthropod groups, 88% are insects, 7% are arachnids, 3% are crustaceans, 1% are centipedes and millipedes, and all others make up 1%. Show this data in a circle graph.

Science Online life.msscience.com/self_check_quiz

Observing a Crayfish

A crayfish has a segmented body and a fused head and thorax. It has a snout and eyes on movable eyestalks. Most crayfish have pincers.

Real-World Question

How does a crayfish use its appendages?

Goals

- **Observe** a crayfish.
- **Determine** the function of pincers.

Materials

crayfish in a small aquarium
uncooked ground beef
stirrer

Safety Precautions

Procedure

1. Copy the data table and use it to record all of your observations during this lab.

Crayfish Observations

Body Region	Number of Appendages	Function
Head		
Thorax	Do not write in this book.	
Abdomen		

2. Your teacher will provide you with a crayfish in an aquarium. Leave the crayfish in the aquarium while you do the lab. Draw your crayfish.

3. Gently touch the crayfish with the stirrer. How does the body feel?
4. **Observe** how the crayfish moves in the water.
5. **Observe** the compound eyes. On which body region are they located?
6. Drop a small piece of ground beef into the aquarium. Observe the crayfish's reaction. Wash your hands.
7. Return the aquarium to its proper place.

Conclude and Apply

1. **Infer** how the location of the eyes is an advantage for the crayfish.
2. **Explain** how the structure of the pincers aids in getting food.
3. **Infer** how the exoskeleton provides protection.

Compare your observations with those of other students in your class. **For more help, refer to the** Science Skill Handbook.

section 4

Echinoderms

as you read

What You'll Learn

- **List** the characteristics of echinoderms.
- **Explain** how sea stars obtain and digest food.
- **Discuss** the importance of echinoderms.

Why It's Important

Echinoderms are a group of animals that affect oceans and coastal areas.

Review Vocabulary
epidermis: outer, thinnest layer of skin

New Vocabulary
- water-vascular system
- tube feet

Echinoderm Characteristics

Echinoderms (ih KI nuh durm) are found in oceans all over the world. The term *echinoderm* is from the Greek words *echinos* meaning "spiny" and *derma* meaning "skin." Echinoderms have a hard endoskeleton covered by a thin, bumpy, or spiny epidermis. They are radially symmetrical, which allows them to sense food, predators, and other things in their environment from all directions.

All echinoderms have a mouth, stomach, and intestines. They feed on a variety of plants and animals. For example, sea stars feed on worms and mollusks, and sea urchins feed on algae. Others feed on dead and decaying matter called detritus (de TRI tus) found on the ocean floor.

Echinoderms have no head or brain, but they do have a nerve ring that surrounds the mouth. They also have cells that respond to light and touch.

Water-Vascular System A characteristic unique to echinoderms is their water-vascular system. It allows them to move, exchange carbon dioxide and oxygen, capture food, and release wastes. The **water-vascular system,** as shown in **Figure 24,** is a network of water-filled canals with thousands of tube feet connected to it. **Tube feet** are hollow, thin-walled tubes that each end in a suction cup. As the pressure in the tube feet changes, the animal is able to move along by pushing out and pulling in its tube feet.

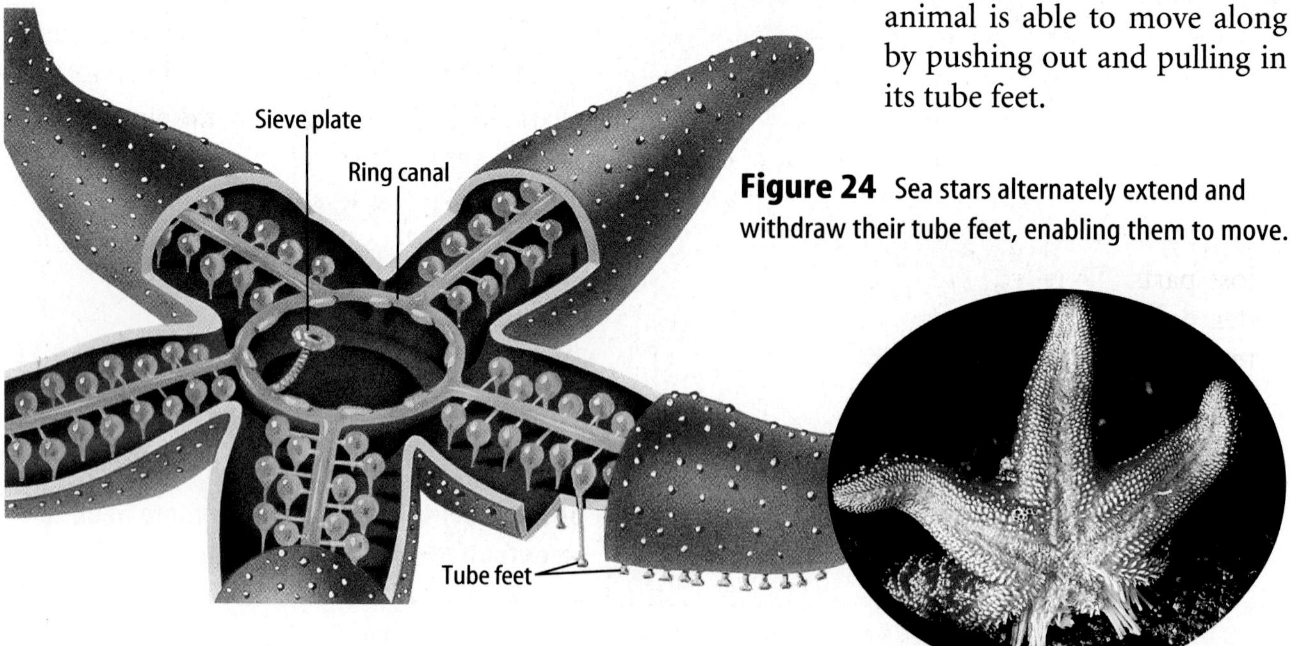

Figure 24 Sea stars alternately extend and withdraw their tube feet, enabling them to move.

Types of Echinoderms

Approximately 6,000 species of echinoderms are living today. Of those, more than one-third are sea stars. The other groups include brittle stars, sea urchins, sand dollars, and sea cucumbers.

Sea Stars Echinoderms with at least five arms arranged around a central point are called sea stars. The arms are lined with thousands of tube feet. Sea stars use their tube feet to open the shells of their prey. When the shell is open slightly, the sea star pushes its stomach through its mouth and into its prey. The sea star's stomach surrounds the soft body of its prey and secretes enzymes that help digest it. When the meal is over, the sea star pulls its stomach back into its own body.

What is unusual about the way that sea stars eat their prey?

Sea stars reproduce sexually when females release eggs and males release sperm into the water. Females can produce millions of eggs in one season.

Sea stars also can repair themselves by regeneration. If a sea star loses an arm, it can grow a new one. If enough of the center disk is left attached to a severed arm, a whole new sea star can grow from that arm.

Modeling the Strength of Tube Feet

Procedure

1. Hold your arm straight out, palm up.
2. Place a **heavy book** on your hand.
3. Have your partner time how long you can hold your arm up with the book on it.

Analysis

1. Describe how your arm feels after a few minutes.
2. If the book models the sea star and your arm models the clam, infer how a sea star successfully overcomes a clam to obtain food.

Try at Home

Brittle Stars Like the one in **Figure 25,** brittle stars have fragile, slender, branched arms that break off easily. This adaptation helps a brittle star survive attacks by predators. While the predator is eating a broken arm, the brittle star escapes. Brittle stars quickly regenerate lost parts. They live hidden under rocks or in litter on the ocean floor. Brittle stars use their flexible arms for movement instead of their tube feet. Their tube feet are used to move particles of food into their mouth.

Figure 25 A brittle star's arms are so flexible that they wave back and forth in the ocean currents. They are called brittle stars because their arms break off easily if they are grabbed by a predator.

Figure 26 Like all echinoderms, sand dollars and sea urchins are radially symmetrical.

Sand dollars live on ocean floors where they can burrow into the sand.

Sea urchins use tube feet and their spines to move around on the bottom of the ocean.

Sea Urchins and Sand Dollars Another group of echinoderms includes sea urchins, sea biscuits, and sand dollars. They are disk- or globe-shaped animals covered with spines. They do not have arms, but sand dollars have a five-pointed pattern on their surface. **Figure 26** shows living sand dollars, covered with stiff, hairlike spines, and sea urchins with long, pointed spines that protect them from predators. Some sea urchins have sacs near the end of the spines that contain toxic fluid that is injected into predators. The spines also help in movement and burrowing. Sea urchins have five toothlike structures around their mouth.

Science Online

Topic: Humans and Echinoderms

Visit life.msscience.com for Web links to information about how echinoderms are used by humans.

Activity Choose one or two uses and write an essay on why echinoderms are important to you.

Sea Cucumbers The animal shown in **Figure 27** is a sea cucumber. Sea cucumbers are soft-bodied echinoderms that have a leathery covering. They have tentacles around their mouth and rows of tube feet on their upper and lower surfaces. When threatened, sea cucumbers may expel their internal organs. These organs regenerate in a few weeks. Some sea cucumbers eat detritus, and others eat plankton.

Reading Check *What makes sea cucumbers different from other echinoderms?*

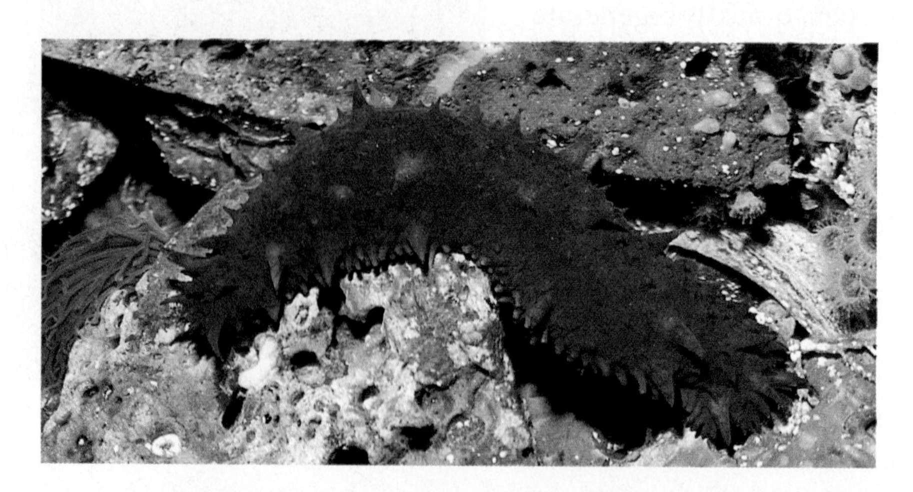

Figure 27 Sea cucumbers have short tube feet, which they use to move around.
Describe *the characteristics of sea cucumbers.*

Value of Echinoderms

Echinoderms are important to the marine environment because they feed on dead organisms and help recycle materials. Sea urchins control the growth of algae in coastal areas. Sea urchin eggs and sea cucumbers are used for food in some places. Many echinoderms are used in research and some might be possible sources of medicines. Sea stars are important predators that control populations of other animals. However, because sea stars feed on oysters and clams, they also destroy millions of dollars' worth of mollusks each year.

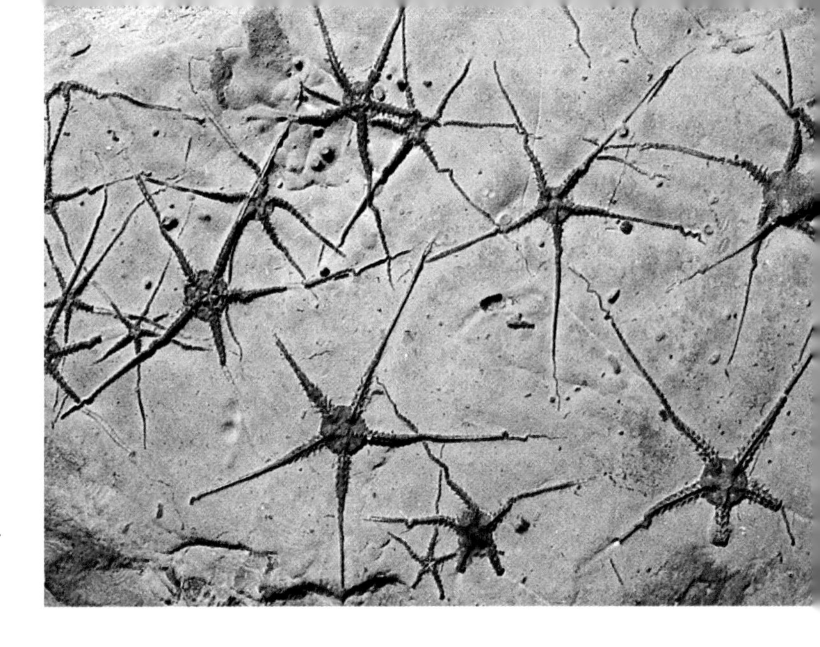

Figure 28 *Ophiopinna elegans* was a brittle star that lived about 165 million years ago.
Explain *the origins of echinoderms.*

Origin of Echinoderms Like the example in **Figure 28,** a good fossil record exists for echinoderms. Echinoderms date back more than 400 million years. The earliest echinoderms might have had bilateral symmetry as adults and may have been attached to the ocean floor by stalks. Many larval forms of modern echinoderms are bilaterally symmetrical.

Scientists hypothesize that echinoderms more closely resemble animals with backbones than any other group of invertebrates. This is because echinoderms have complex body systems and an embryo that develops the same way that the embryos of animals with backbones develop.

section 4 review

Summary

Echinoderm Characteristics

- Echinoderms have a hard endoskeleton and are covered by thin, spiny skin.
- They are radially symmetrical. They have no brain or head, but have a nerve ring, and respond to light and touch.
- They have a specialized water-vascular system, which helps them move, exchange gases, capture food, and release wastes.

Types of Echinoderms

- The largest group of echinoderms is sea stars.
- Other groups include brittle stars, sea urchins and sand dollars, and sea cucumbers.

Self Check

1. **Explain** how echinoderms move and get their food.
2. **Infer** how sea urchins are beneficial.
3. **List** the methods of defense that echinoderms have to protect themselves from predators.
4. **Think Critically** Why would the ability to regenerate lost body parts be an important adaptation for sea stars, brittle stars, and other echinoderms?

Applying Skills

5. **Form a Hypothesis** Why do you think echinoderms live on the ocean floor?
6. **Communicate** Choose an echinoderm and write about it. Describe its appearance, how it gets food, where it lives, and other interesting facts.

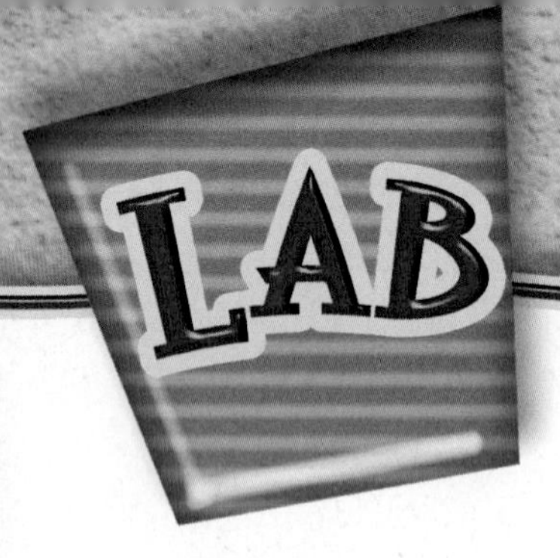

What do worms eat?

Goals
- **Construct** five earthworm habitats.
- **Test** different foods to determine which ones earthworms eat.

Materials
orange peels
apple peels
banana skin
kiwi fruit skin
watermelon rind
**skins of five different fruits*
widemouthed jars (5)
potting soil
water
humus
**peat moss*
earthworms
black construction paper (5 sheets)
masking tape
marker
rubber bands (5)
**Alternate materials*

Safety Precautions

WARNING: *Do not handle earthworms with dry hands. Do not eat any materials used in the lab.*

Real-World Question

Earthworms are valuable because they improve the soil in which they live. There can be 50,000 earthworms living in one acre. Their tunnels increase air movement through the soil and improve water drainage. As they eat the decaying material in soil, their wastes can enrich the soil. Other than decaying material, what else do earthworms eat? Do they have favorite foods?

Procedure

1. Pour equal amounts of soil into each of the jars. Do not pack the soil. Leave several centimeters of space at the top of each jar.
2. Sprinkle equal amounts of water into each jar to moisten the soil. Avoid pouring too much water into the jars.
3. Pour humus into each of your jars to a depth of 2 cm. The humus should be loose.
4. Add watermelon rinds to the first jar, orange peels to the second, apple peels to the third, kiwi fruit skins to the fourth, and a banana peel to the fifth jar. Each jar should have 2 cm of fruit skins on top of the layer of humus.

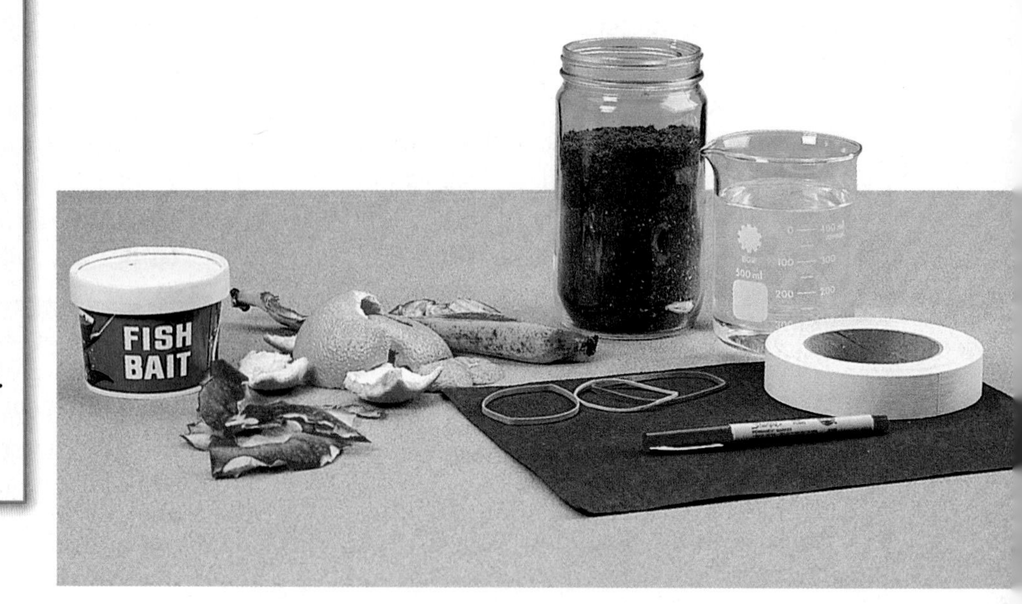

5. Add five earthworms to each jar.
6. Wrap a sheet of black construction paper around each jar and secure it with a rubber band.
7. Using the masking tape and marker, label each jar with the type of fruit it contains.
8. Copy the data table below in your Science Journal.
9. Place all of your jars in the same cool, dark place. Observe your jars every other day for a week and record your observations in your data table.

Fruit Wastes

Date	Watermelon rind	Orange peels	Apple peels	Kiwi skins	Banana peels
	Do not write in this book.				

Analyze Your Data

1. **Record** the changes in your data table.
2. **Compare** the amount of skins left in each jar.
3. **Record** which fruit skin had the greatest change. The least?

Conclude and Apply

1. **Infer** the type of food favored by earthworms.
2. **Infer** why some of the fruit skins were not eaten by the earthworms.
3. **Identify** a food source in each jar other than the fruit skins.
4. **Predict** what would happen in the jars over the next month if you continued the experiment.

Communicating Your Data

Use the results of your experiment and information from your reading to help you write a recipe for an appetizing dinner that worms would enjoy. Based on the results of your experiment, add other fruit skins or foods to your menu you think worms might like.

Science and Language Arts

from "The Creatures on My Mind"

by Ursula K. Le Guin

When I stayed for a week in New Orleans... I had an apartment with a balcony... But when I first stepped out on it, the first thing I saw was a huge beetle. It lay on its back directly under the light fixture. I thought it was dead, then saw its legs twitch and twitch again. Big insects horrify me. As a child I feared moths and spiders, but adolescence cured me, as if those fears evaporated in the stew of hormones. But I never got enough hormones to make me easy with the large, hard-shelled insects: wood roaches, June bugs, mantises, cicadas. This beetle was a couple of inches long; its abdomen was ribbed, its legs long and jointed; it was dull reddish brown; it was dying. I felt a little sick seeing it lie there twitching, enough to keep me from sitting out on the balcony that first day... And if I had any courage or common sense, I kept telling myself, I'd... put it out of its misery. We don't know what a beetle may or may not suffer...

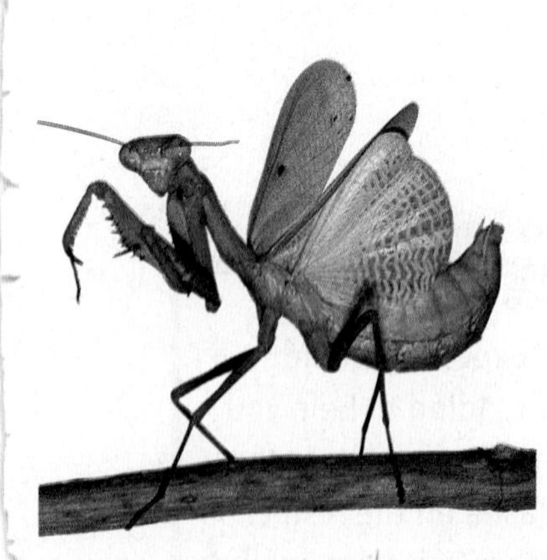

Understanding Literature

Personal Experience Narrative In this passage, the author uses her personal experience to consider her connection to other living things. In this piece, the author recounts a minor event in her life when she happens upon a dying beetle. The experience allows the author to pose some important questions about another species and to think about how beetles might feel when they die. How do you think the beetle is feeling?

Respond to the Reading

1. How do you suppose the beetle injured itself?
2. From the author's description, in what stage of development is the beetle?
3. **Linking Science and Writing** Write about a personal experience that caused you to think about an important question or topic in your life.

The author names several arthropod species in the passage, including insects and an arachnid. Beetles, June bugs, mantises, cicadas, and moths are all insects. The spider is an arachnid. Of the arthropods the author names, can you tell which ones go through a complete metamorphosis?

Reviewing Main Ideas

Section 1 Mollusks

1. Mollusks are soft-bodied invertebrates that usually are covered by a hard shell. They move using a muscular foot.
2. Mollusks with one shell are gastropods. Bivalves have two shells. Cephalopods have an internal shell and a foot that is divided into tentacles.

Section 2 Segmented Worms

1. Segmented worms have tube-shaped bodies divided into sections, a body cavity that holds the internal organs, and bristlelike structures called setae to help them move.
2. An earthworm's digestive system has a mouth, crop, gizzard, intestine, and anus. Polychaetes are marine worms. Leeches are parasites that attach to animals and feed on their blood.

Section 3 Arthropods

1. More than a million species of arthropods exist, which is more than any other group of animals. Most arthropods are insects.
2. Arthropods are grouped by number of body segments and appendages. Exoskeletons cover, protect, and support arthropod bodies.
3. Young arthropods develop either by complete metamorphosis or incomplete metamorphosis.

Section 4 Echinoderms

1. Echinoderms have a hard, spiny exoskeleton covered by a thin epidermis.
2. Most echinoderms have a water-vascular system that enables them to move, exchange carbon dioxide and oxygen, capture food, and give off wastes.

Visualizing Main Ideas

Copy and complete the following concept map about insects.

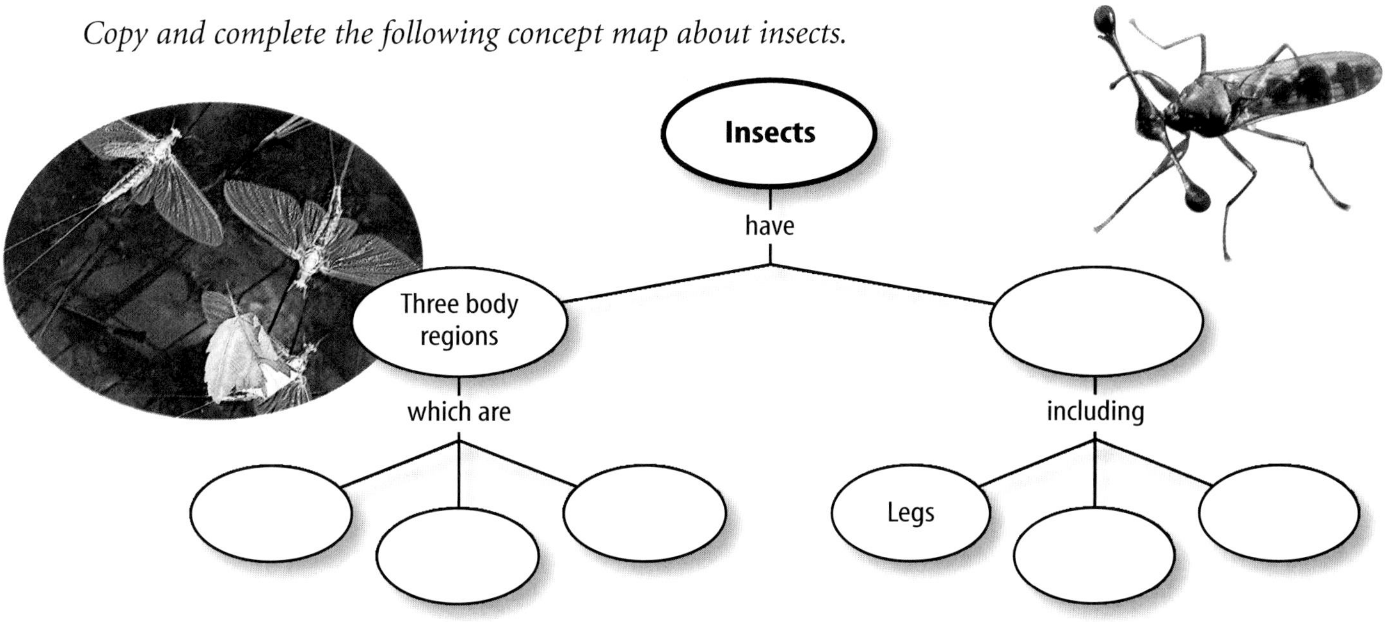

Insects
have
Three body regions
which are
including
Legs

chapter 13 Review

Using Vocabulary

appendage p. 370	molting p. 370
closed circulatory system p. 362	open circulatory system p. 360
crop p. 366	radula p. 361
exoskeleton p. 370	setae p. 365
gill p. 360	spiracle p. 371
gizzard p. 366	tube feet p. 380
mantle p. 360	water-vascular system p. 380
metamorphosis p. 372	

Fill in the blanks with the correct vocabulary word or words.

1. Mollusk shells are secreted by the _________.
2. As earthworms move through soil using their _________, they take in soil, which is stored in the _________.
3. The _________ covers and protects arthropod bodies.
4. Insects exchange oxygen and carbon dioxide through _________.
5. _________ act like suction cups and help sea stars move and feed.
6. Snails use a(n) _________ to get food.
7. The blood of mollusks moves in a(n) _________.

Checking Concepts

Choose the word or phrase that best answers the question.

8. What structure covers organs of mollusks?
   A) gills
   B) food
   C) mantle
   D) visceral mass
9. What structures do echinoderms use to move and to open shells of mollusks?
   A) mantle
   B) calcium plates
   C) spines
   D) tube feet
10. Which organism has a closed circulatory system?
   A) earthworm
   B) octopus
   C) slug
   D) snail
11. What evidence suggests that arthropods might have evolved from annelids?
   A) Arthropods and annelids have gills.
   B) Both groups have species that live in salt water.
   C) Segmentation is present in both groups.
   D) All segmented worms have setae.

Use the photo below to answer questions 12 and 13.

12. Which of the following correctly describes the arthropod pictured above?
   A) three body regions, six legs
   B) two body regions, eight legs
   C) many body segments, ten legs
   D) many body segments, one pair of legs per segment
13. What type of arthropod is this animal?
   A) annelid
   B) arachnid
   C) insect
   D) mollusk
14. Which is an example of an annelid?
   A) earthworm
   B) octopus
   C) slug
   D) snail
15. Which sequence shows incomplete metamorphosis?
   A) egg—larvae—adult
   B) egg—nymph—adult
   C) larva—pupa—adult
   D) nymph—pupa—adult

Science Online life.msscience.com/vocabulary_puzzlemaker

Thinking Critically

Use the photo below to answer question 16.

16. **Describe** how this animal obtains food.

17. **Compare** the ability of clams, oysters, scallops, and squid to protect themselves.

18. **Compare and contrast** an earthworm gizzard to teeth in other animals.

19. **Explain** the evidence that mollusks and annelids may share a common ancestor.

20. **Infer** how taking in extra water or air after molting, but before the new exoskeleton hardens, helps an arthropod.

21. **Classify** the following animals into arthropod groups: *spider, pill bug, crayfish, grasshopper, crab, silverfish, cricket, wasp, scorpion, shrimp, barnacle, tick,* and *butterfly.*

22. **Compare and Contrast** Copy and complete this Venn diagram to compare and contrast arthropods to annelids.

Annelids — Arthropods

Closed circulatory system | Bilateral symmetry | Open circulatory system

23. **Recognize Cause and Effect** If all the earthworms were removed from a hectare of soil, what would happen to the soil? Why?

24. **Research Information** The suffix *-ptera* means "wings." Research the meaning of the prefix listed below and give an example of a member of each insect group.

Diptera	Homoptera
Orthoptera	Hemiptera
Coleoptera	

Performance Activities

25. **Construct** Choose an arthropod that develops through complete metamorphosis and construct a three-dimensional model for each of the four stages.

Applying Math

Use the table below to answer questions 26 and 27.

Described Species

Type of Organism	Number of Described Species
Anthropods	1,065,000
Land plants	270,000
Fungi	72,000
Mollusks	70,000
Nematodes	25,000
Birds	10,000
Mammals	5,000
Bacteria	4,000
Other	145,000

26. **Arthropods** Using the table above, what percentage of organisms are arthropods? Mollusks?

27. **Species Distribution** Make a bar graph that shows the number of described species listed in the table above.

chapter 13 Standardized Test Practice

Part 1 Multiple Choice

Record your answers on the answer sheet provided by your teacher or on a sheet of paper.

1. Which of the following is not a mollusk?

A. clam
B. snail
C. crab
D. squid

Use the illustration below to answer questions 2 and 3.

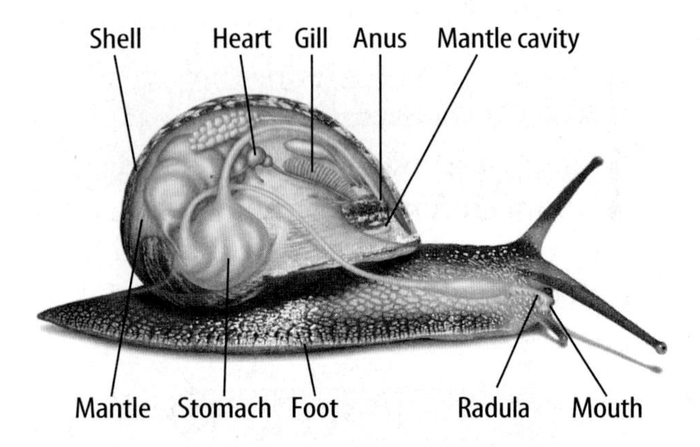

2. This mollusk uses which of the following to exchange carbon dioxide with oxygen from the water?

A. radula
B. gill
C. mantle
D. shell

3. Which structure covers the body organs of this mollusk?

A. radula
B. gill
C. mantle
D. shell

4. Which is the largest group of mollusks?

A. cephalopods
B. bivalves
C. monovalves
D. gastropods

5. Which openings allow air to enter an insect's body?

A. spiracles
B. gills
C. thorax
D. setae

Use the photo below to answer questions 6 and 7.

6. This organism is an example of what type of mollusk?

A. gastropod
B. bivalve
C. cephalopod
D. monovalve

7. How do these animals move?

A. a muscular foot
B. tentacles
C. contraction and relaxation
D. jet propulsion

8. What does the word annelid mean?

A. segmented
B. bristled
C. little rings
D. worms

9. What are bristlelike structures on the outside of each body segment of annelids called?

A. crops
B. gizzards
C. radula
D. setae

10. What is the largest group of animals?

A. arthropods
B. cephalopods
C. gastropods
D. annelids

11. What is it called when an arthropod loses its exoskeleton and replaces it with a new one?

A. shedding
B. molting
C. manging
D. exfoliating

Part 2 Short Response/Grid In

Record your answers on the answer sheet provided by your teacher or on a sheet of paper.

12. Describe how a sea star captures and consumes its prey.

13. Explain how sea stars repair or replace lost or damaged body parts.

14. Describe how gastropods, such as snails and garden slugs, eat.

Use the photo below to answer questions 15 and 16.

15. Describe this animal's vascular system. How is it used?

16. This animal has a unique method of movement. What is it and how does it work?

17. Describe the type of reproductive system found in earthworms.

18. What is an open circulatory system? Give three examples of animals that have an open circulatory system.

19. How are pearls formed in clams, oysters, and some other gastropods?

Part 3 Open Ended

Record your answers on a sheet of paper.

Use the photo below to answer questions 20 and 21.

20. Name and describe the phylum that this sea star belongs to.

21. This animal has a vascular system that is unique. Describe it.

22. What structures allow an earthworm to move? Describe its locomotion.

23. There are more species of insects than all other animal groups combined. In all environments, they have to compete with one another for survival. How do so many insects survive?

24. Insect bodies are divided into three segments. What are these three segments and what appendages and organs are in/on each part?

Test-Taking Tip

Show Your Work For constructed-response questions, show all of your work and any calculations on your answer sheet.

Question 22 Write out all of the adaptations that insects have for survival and determine which are the most beneficial to the success of the group.

chapter 14

Fish, Amphibians, and Reptiles

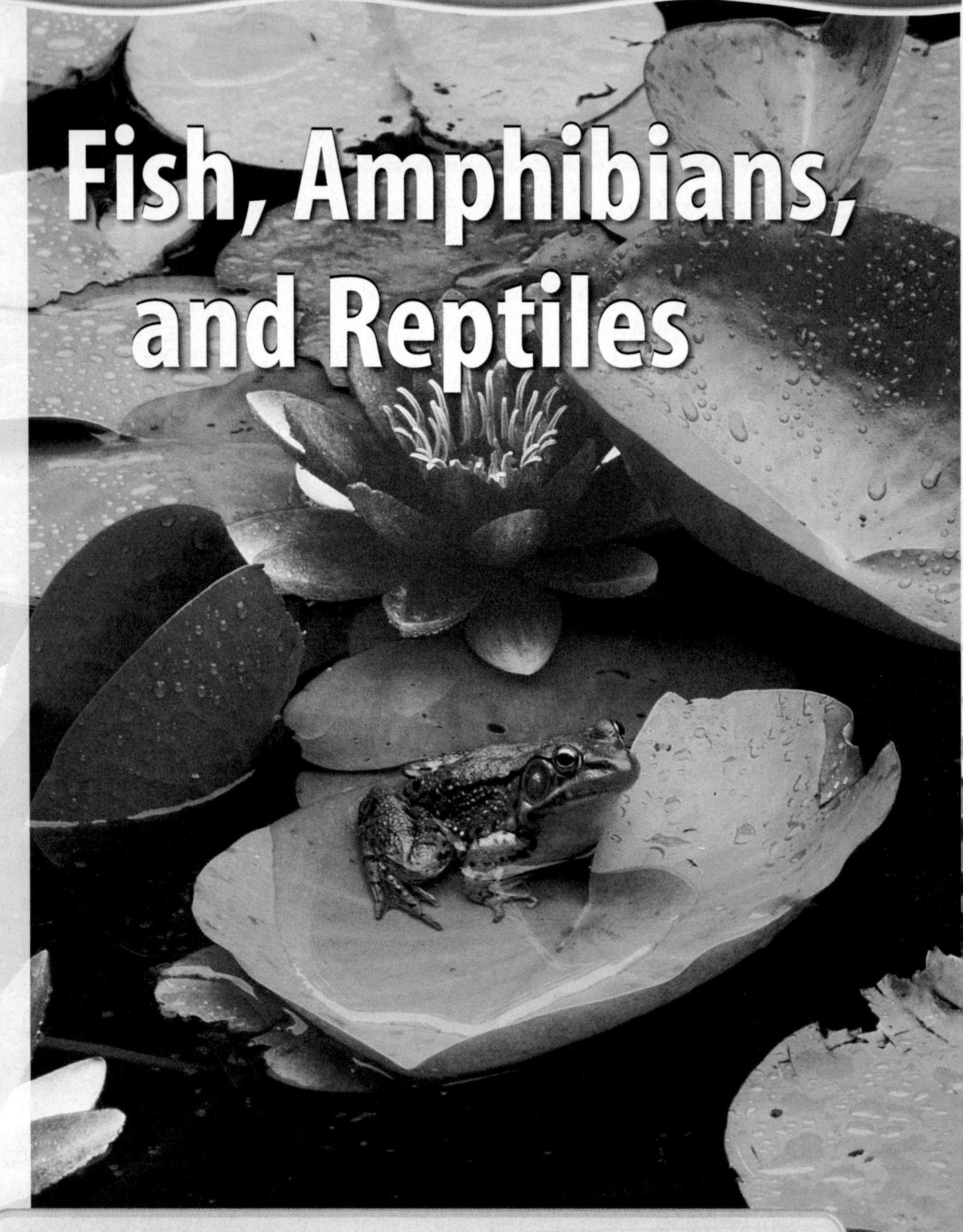

The BIG Idea

Fish, amphibians, and reptiles are three groups of vertebrates.

SECTION 1
Chordates and Vertebrates

Main Idea Pharyngeal pouches, postanal tails, nerve cords, and notochords are characteristics of both chordates and vertebrates.

SECTION 2
Fish

Main Idea There are three types of fish: jawless, jawless cartilaginous (kar tuh LA juh nus), and bony.

SECTION 3
Amphibians

Main Idea Most amphibians live on land and reproduce in water.

SECTION 4
Reptiles

Main Idea Reptiles have thick, waterproof skin with scales.

Can I find one?

If you want to find a frog or salamander—two types of amphibians—visit a nearby pond or stream. By studying fish, amphibians, and reptiles, scientists can learn about a variety of vertebrate characteristics, including how these animals reproduce, develop, and are classified.

Science Journal List two unique characteristics for each animal group you will be studying.

Start-Up Activities

Snake Hearing

How much do you know about reptiles? For example, do snakes have eyelids? Why do snakes flick their tongues in and out? How can some snakes swallow animals that are larger than their own heads? Snakes don't have ears, so how do they hear? In this lab, you will discover the answer to one of these questions.

1. Hold a tuning fork by the stem and tap it on a hard piece of rubber, such as the sole of a shoe.
2. Hold it next to your ear. What, if anything, do you hear?
3. Tap the tuning fork again. Press the base of the stem firmly against your chin. In your Science Journal, describe what happens.
4. **Think Critically** Using the results from step 3, infer how a snake detects vibrations. In your Science Journal, predict how different animals can use vibrations to hear.

Preview this chapter's content and activities at
life.msscience.com

FOLDABLES™ Study Organizer

Fish, Amphibians, and Reptiles Make the following Foldable to help you organize information about the animals you will be studying.

STEP 1 **Fold** one piece of paper lengthwise into thirds.

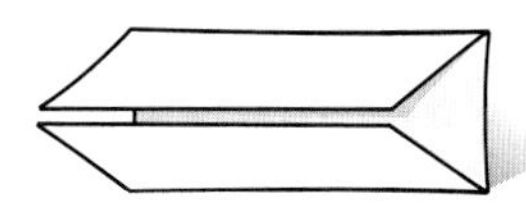

STEP 2 **Fold** the paper widthwise into fourths.

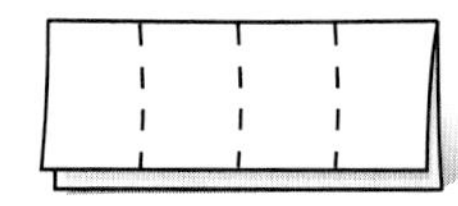

STEP 3 **Unfold,** lay the paper lengthwise, and draw lines along the folds.

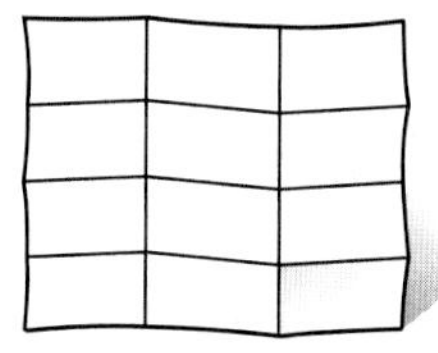

STEP 4 **Label** your table as shown.

Fish	Amphibians	Reptiles

Make a Table As you read this chapter, complete the table describing characteristics of each type of animal.

Get Ready to Read

Questions and Answers

1 Learn It! Knowing how to find answers to questions will help you on reviews and tests. Some answers can be found in the textbook, while other answers require you to go beyond the textbook. These answers might be based on knowledge you already have or things you have experienced.

2 Practice It! Read the excerpt below. Answer the following questions and then discuss them with a partner.

> The three-chambered heart in amphibians is an important change from the circulatory system of fish. In the three-chambered heart, one chamber receives oxygen-filled blood from the lungs and skin, and another chamber receives carbon dioxide-filled blood from the body tissues. Blood moves from both of these chambers to the third chamber, which pumps oxygen-filled blood to body tissues and carbon dioxide-filled blood back to the lungs. Limited mixing of these two bloods occurs.
>
> —*from page 408*

- How does the heart of an amphibian differ from the heart of a fish?
- From where do the first two chambers of the amphibian heart receive blood?
- What does the third chamber do?

3 Apply It! Look at some questions in the text. Which questions can be answered directly from the text? Which require you to go beyond the text?

As you read, keep track of questions you answer in the chapter. This will help you remember what you read.

Target Your Reading

Use this to focus on the main ideas as you read the chapter.

1. **Before you read** the chapter, respond to the statements below on your worksheet or on a numbered sheet of paper.
   - Write an **A** if you **agree** with the statement.
   - Write a **D** if you **disagree** with the statement.

2. **After you read** the chapter, look back to this page to see if you've changed your mind about any of the statements.
   - If any of your answers changed, explain why.
   - Change any false statements into true statements.
   - Use your revised statements as a study guide.

Before You Read A or D		Statement	After You Read A or D
	1	All vertebrates can maintain a nearly constant internal body temperature.	
	2	All vertebrates are covered with skin.	
	3	Counting the rings on the scales of some species of fish can help determine the fish's age.	
	4	Fish have a heart and lungs similar to humans but smaller.	
	5	Fish have an organ that keeps them from sinking or floating to the surface in water.	
	6	Amphibians breathe through their skin.	
	7	All amphibians live on land but reproduce in water.	
	8	Like amphibians, reptiles also can breathe through their skin.	
	9	Reptiles can maintain a constant internal body temperature.	

Science Online
Print out a worksheet of this page at life.msscience.com

section 1

Chordates and Vertebrates

as you read

What You'll Learn

- **List** the characteristics of all chordates.
- **Identify** characteristics shared by vertebrates.
- **Differentiate** between ectotherms and endotherms.

Why It's Important

Humans are vertebrates. Other vertebrates play important roles in your life because they provide food, companionship, and labor.

Review Vocabulary

motor responses: responses that involve muscular movement

New Vocabulary

- chordate
- notochord
- postanal tail
- nerve cord
- pharyngeal pouch
- endoskeleton
- cartilage
- vertebrae
- ectotherm
- endotherm

Chordate Characteristics

During a walk along the seashore at low tide, you often can see jellylike masses of animals clinging to rocks. Some of these animals may be sea squirts, as shown in **Figure 1,** which is one of the many types of animals known as chordates (KOR dayts). **Chordates** are animals that have four characteristics present at some stage of their development—a notochord, postanal tail, nerve cord, and pharyngeal pouches.

Notochord All chordates have an internal **notochord** that supports the animal and extends along the upper part of its body, as shown in **Figure 2.** The notochord is flexible but firm because it is made up of fluid-filled cells that are enclosed in a stiff covering. The notochord also extends into the **postanal tail**—a muscular structure at the end of the developing chordate. Some chordates, such as fish, amphibians, reptiles, birds, and mammals, develop backbones that partly or entirely replace the notochord. They are called vertebrates. In some chordates, such as the sea squirt, other tunicates, and the lancelets, the notochord is kept into adulthood.

Reading Check *What happens to the notochord as a bat develops?*

Figure 1 Sea squirts get their name because when they're taken out of the ocean, they squirt water out of their body.
Determine *what you have in common with a sea squirt.*

Nerve Cord Above the notochord and along the length of a developing chordate's body is a tubelike structure called the **nerve cord,** also shown in **Figure 2.** As most chordates develop, the front end of the nerve cord enlarges to form the brain and the remainder becomes the spinal cord. These two structures become the central nervous system that develops into complex systems for sensory and motor responses.

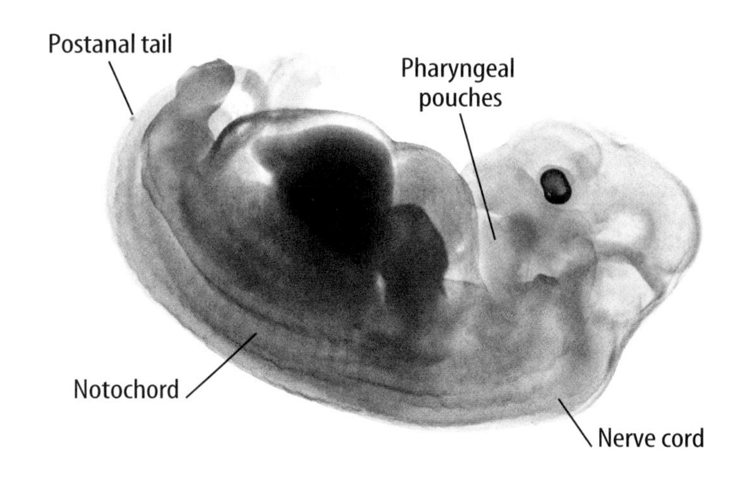

Figure 2 At some time during its development, a chordate has a notochord, postanal tail, nerve cord, and pharyngeal pouches.

Pharyngeal Pouches All developing chordates have **pharyngeal pouches.** They are found in the region between the mouth and the digestive tube as pairs of openings to the outside. Many chordates have several pairs of pharyngeal pouches. Ancient invertebrate chordates used them for filter feeding. This is still their purpose in some living chordates such as lancelets. In fish, they have developed into internal gills where oxygen and carbon dioxide are exchanged. In humans, pharyngeal pouches are present only during embryonic development. However, one pair becomes the tubes that go from the ears to the throat.

Vertebrate Characteristics

Besides the characteristics common to all chordates, vertebrates have distinct characteristics. These traits set vertebrates apart from other chordates.

Structure All vertebrates have an internal framework called an **endoskeleton.** It is made up of bone and/or flexible tissue called **cartilage.** Your ears and the tip of your nose are made of cartilage. The endoskeleton provides a place for muscle attachment and supports and protects the organs. Part of the endoskeleton is a flexible, supportive column called the backbone, as shown in **Figure 3.** It is a stack of **vertebrae** alternating with cartilage. The backbone surrounds and protects the spinal nerve cord. Vertebrates also have a head with a skull that encloses and protects the brain.

Most of a vertebrate's internal organs are found in a central body cavity. A protective skin covers a vertebrate. Hair, feathers, scales, or horns sometimes grow from the skin.

Figure 3 Vertebrae are separated by soft disks of cartilage.

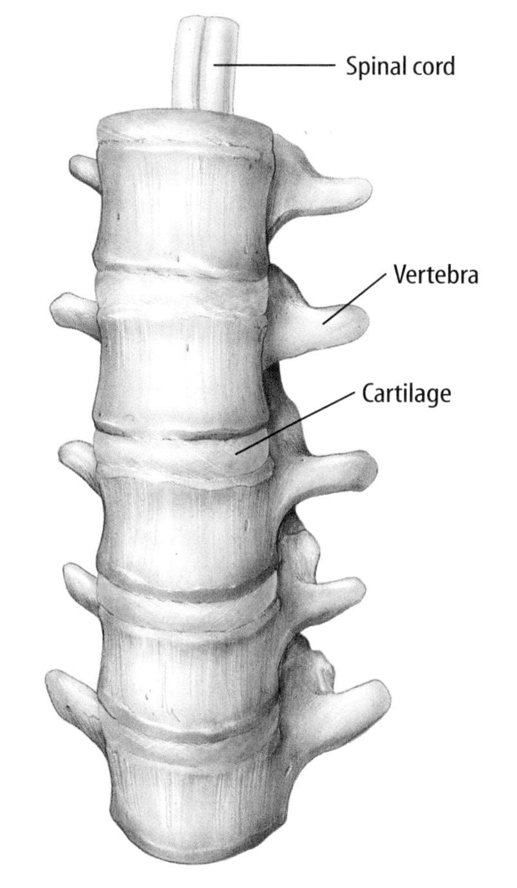

Vertebrae column

Table 1 Types of Vertebrates		
Group	Estimated Number of Species	Examples
Jawless fish	60	lamprey, hagfish
Jawed cartilaginous fish	500 to 900	shark, ray, skate
Bony fish	20,000	salmon, bass, guppy, sea horse, lungfish
Amphibians	4,000	frog, toad, salamander
Reptiles	7,970	turtle, lizard, snake, crocodile, alligator
Birds	8,700	stork, eagle, sparrow, turkey, duck, ostrich
Mammals	4,600	human, whale, bat, mouse, lion, cow, otter

Vertebrate Groups Seven main groups of vertebrates are found on Earth today, as shown in **Table 1.** Vertebrates are either ectotherms or endotherms. Fish, amphibians, and reptiles are ectotherms, also known as cold-blooded animals. An **ectotherm** has an internal body temperature that changes with the temperature of its surroundings. Birds and mammals are endotherms, which sometimes are called warm-blooded animals. An **endotherm** has a nearly constant internal body temperature.

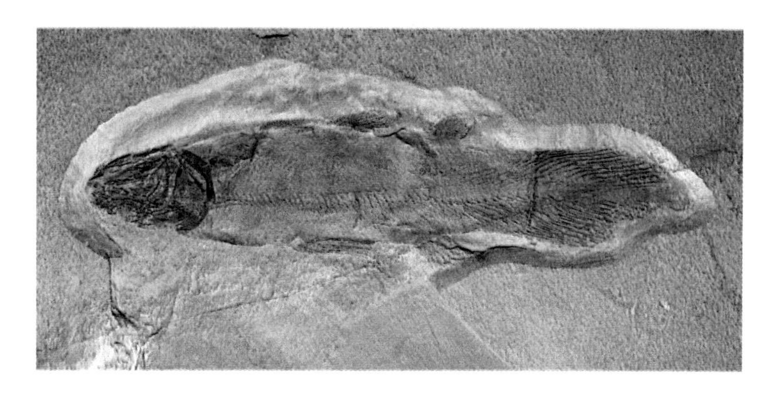

Figure 4 Placoderms were the first fish with jaws. These predatory fish were covered with heavy armor.

Vertebrate Origins Some vertebrate fossils, like the one in **Figure 4,** are of water-dwelling, armored animals that lived about 420 million years ago (mya). Lobe-finned fish appeared in the fossil record about 395 mya. The oldest known amphibian fossils date from about 370 mya. Reptile fossils have been found in deposits about 350 million years old. One well-known group of reptiles—the dinosaurs—first appeared about 230 mya.

In 1861, a fossil imprint of an animal with scales, jaws with teeth, claws on its front limbs, and feathers was found. The 150-million-year-old fossil was an ancestor of birds, and was named *Archaeopteryx* (ar kee AHP tuh rihks).

Mammal-like reptiles appeared about 235 mya. However, true mammals appeared about 190 mya, and modern mammals originated about 38 million years ago.

section 1 review

Summary

Chordate Characteristics

- Chordates have four common characteristics at some point in their development: a notochord, postanal tail, nerve cord, and pharyngeal pouches.

Vertebrate Characteristics

- All vertebrates have an endoskeleton, a backbone, a head with a skull to protect the brain, internal organs in a central body cavity, and a protective skin.
- Vertebrates can be ectothermic or endothermic.
- There are seven main groups of vertebrates.

Self Check

1. **Explain** the difference between a vertebra and a notochord.
2. **Compare and contrast** some of the physical differences between ectotherms and endotherms.
3. **Think Critically** If the outside temperature decreases by 20°C, what will happen to a reptile's body temperature?

Applying Skills

4. **Concept Map** Construct a concept map using these terms: *chordates, bony fish, amphibians, cartilaginous fish, reptiles, birds, mammals, lancelets, tunicates, invertebrate chordates, jawless fish,* and *vertebrates.*

Endotherms and Ectotherms

Birds and mammals are endotherms. Fish, amphibians and reptiles are ectotherms.

Real-World Question

How can you determine whether an animal you have never seen before is an endotherm or an ectotherm? What tests might you conduct to find the answer?

Goals

- **Construct** an imaginary animal.
- **Determine** whether your animal is an endotherm or an ectotherm.

Materials

fiberfill
cotton balls
old socks
tissue
cloth
thermometer
Alternate materials

Safety Precautions

Procedure

1. Design an animal that has a thermometer inside. Construct the animal using cloth and some kind of stuffing material. Make sure that you will be able to remove and reinsert the thermometer.
2. Draw a picture of your animal and record data about its size and shape.
3. Copy the data table in your Science Journal.
4. Place your animal in three locations that have different temperatures. Record the locations in the data table.
5. In each location, record the time and the temperature of your animal at the beginning and after 10 min.

Animal Temperature

Location	Beginning Time/ Temperature	Ending Time/ Temperature
	Do not write in this book.	

Conclude and Apply

1. **Describe** your results. Did the animal's temperature vary depending upon the location?
2. Based on your results, is your animal an endotherm or an ectotherm? Explain.
3. **Compare** your results to those of others in your class. Were the results the same for animals of different sizes? Did the shape of the animal, such as one being flatter and another more cylindrical, matter?
4. Based on your results and information in the chapter, do you think your animal is most likely a bird, a mammal, a reptile, an amphibian, or a fish? Explain.

Communicating Your Data

Compare your conclusions with those of other students in your class. **For more help, refer to the Science Skill Handbook.**

section 2

Fish

Fish Characteristics

Did you know that more differences appear among fish than among any other vertebrate group? In fact, there are more species of fish than species of other vertebrate groups. All fish are ectotherms. They are adapted for living in nearly every type of water environment on Earth—freshwater and salt water. Some fish, such as salmon, spend part of their life in freshwater and part of it in salt water. Fish are found at varying depths, from shallow pools to deep oceans.

A streamlined shape, a muscular tail, and fins allow most fish to move rapidly through the water. **Fins** are fanlike structures attached to the endoskeleton. They are used for steering, balancing, and moving. Paired fins on the sides allow fish to move right, left, backward, and forward. Fins on the top and bottom of the body give the fish stability. Most fish secrete a slimy mucus that also helps them move through the water.

Most fish have scales. **Scales** are hard, thin plates that cover the skin and protect the body, similar to shingles on the roof of a house. Most fish scales are made of bone. **Figure 5** illustrates how they can be tooth shaped, diamond shaped, cone shaped, or round. The shape of the scales can be used to help classify fish. The age of some species can be estimated by counting the annual growth rings of the scales.

as you read

What You'll Learn

- **List** the characteristics of the three classes of fish.
- **Explain** how fish obtain food and oxygen and reproduce.
- **Describe** the importance and origin of fish.

Why It's Important

Fish are an important food source for humans as well as many other animals.

Review Vocabulary
streamline: formed to reduce resistance to motion through a fluid or air

New Vocabulary
- fin
- scale

Figure 5 Four types of fish scales are shown here.

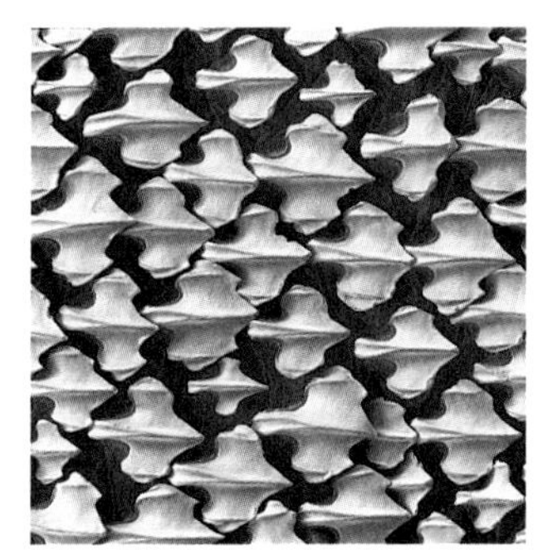

Sharks are covered with placoid scales such as these. Shark teeth are modified forms of these scales.

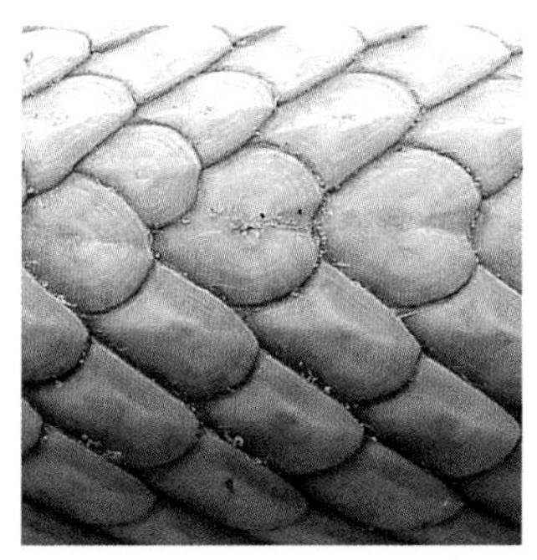

Lobe-finned fish and gars are covered by ganoid scales. These scales don't overlap like other fish scales.

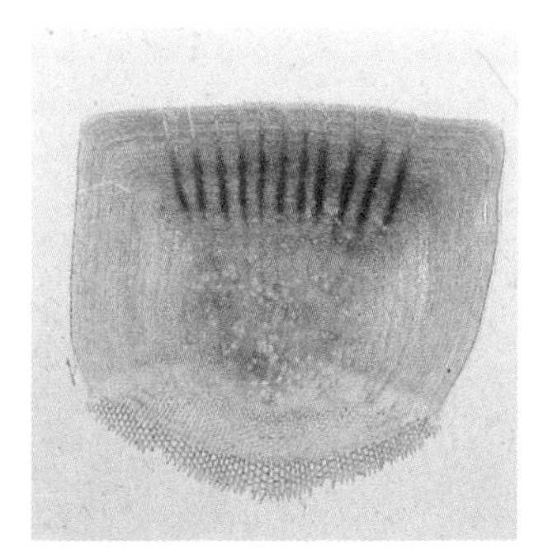

Ctenoid (TEN oyd) scales have a rough edge, which is thought to reduce drag as the fish swims through the water.

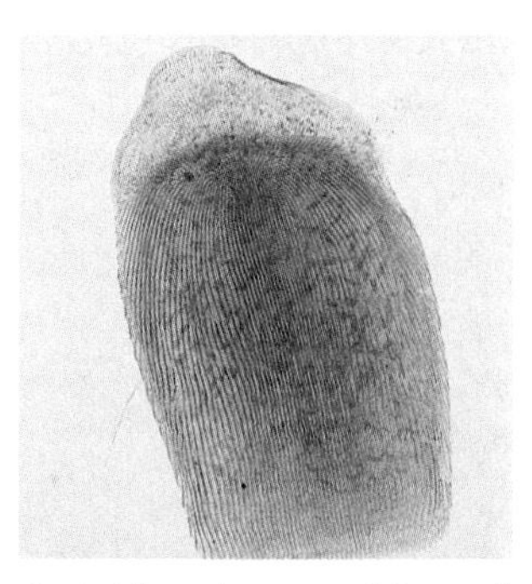

Cycloid scales are thin and overlap, giving the fish flexibility. These scales grow as the fish grows.

Figure 6 The sensory organs in the lateral line of a fish send messages to the fish's brain.

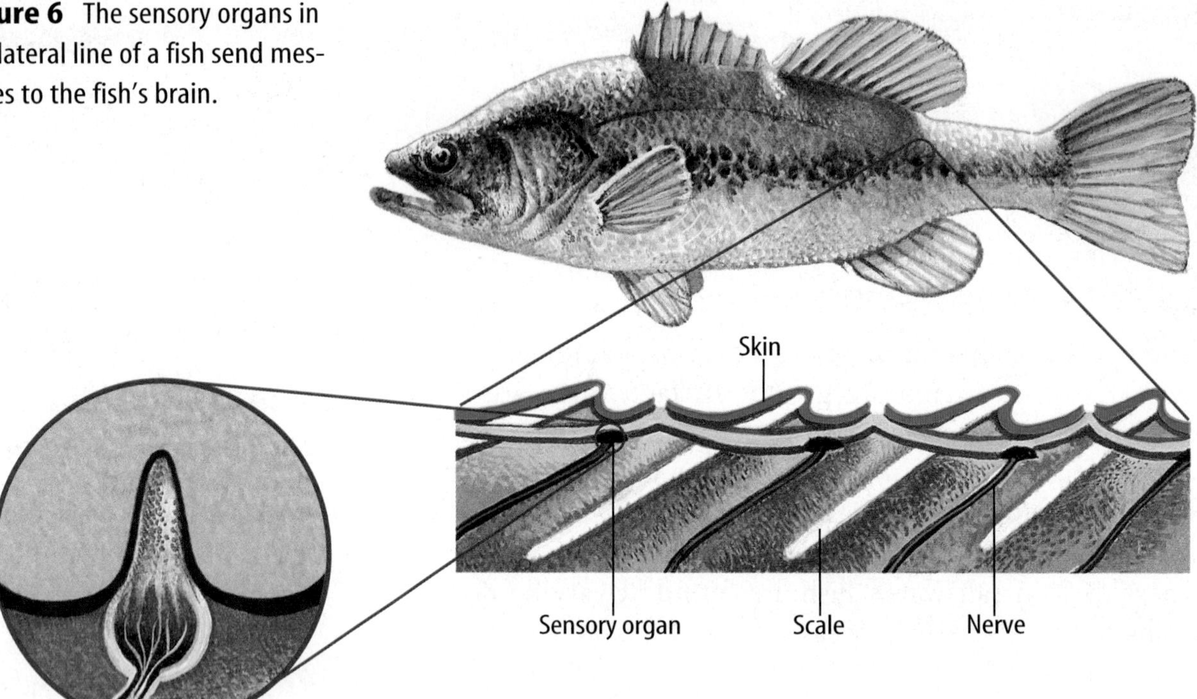

Body Systems All fish have highly developed sensory systems. Most fish have a lateral line system, as shown in **Figure 6.** A lateral line system is made up of a shallow, canal-like structure that extends along the length of the fish's body and is filled with sensory organs. The lateral line enables a fish to sense its environment and to detect movement. Some fish, such as sharks, also have a strong sense of smell. Sharks can detect blood in the water from several kilometers away.

Fish have a two-chambered heart in which oxygen-filled blood mixes with carbon dioxide-filled blood. A fish's blood isn't carrying as much oxygen as blood that is pumped through a three- or four-chambered heart.

Gas Exchange Most fish have organs called gills for the exchange of carbon dioxide and oxygen. Gills are located on both sides of the fish's head and are made up of feathery gill filaments that contain many tiny blood vessels. When a fish takes water into its mouth, the water passes over the gills, where oxygen from the water is exchanged with carbon dioxide in the blood. The water then passes out through slits on each side of the fish. Many fish, such as the halibut in **Figure 7,** are able to take in water while lying on the ocean floor.

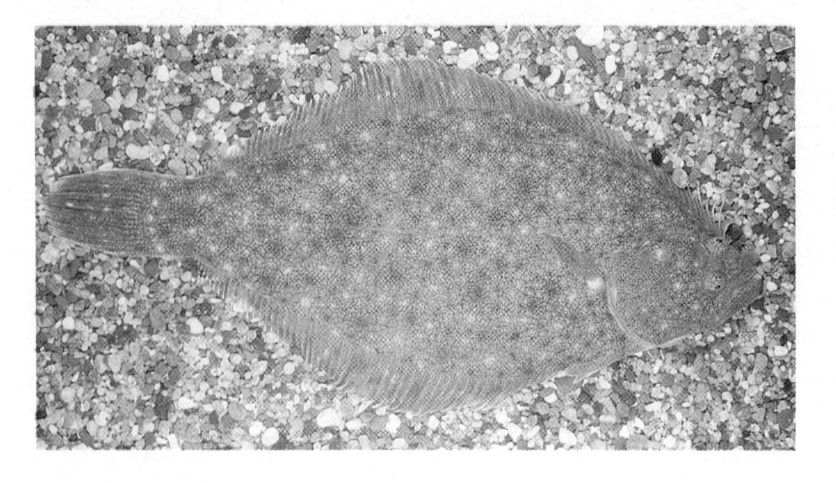

Figure 7 Even though a halibut's eyes are on one side of the fish, gills are on both sides.
Describe *how a fish breathes.*

Figure 8 Fish obtain food in different ways.

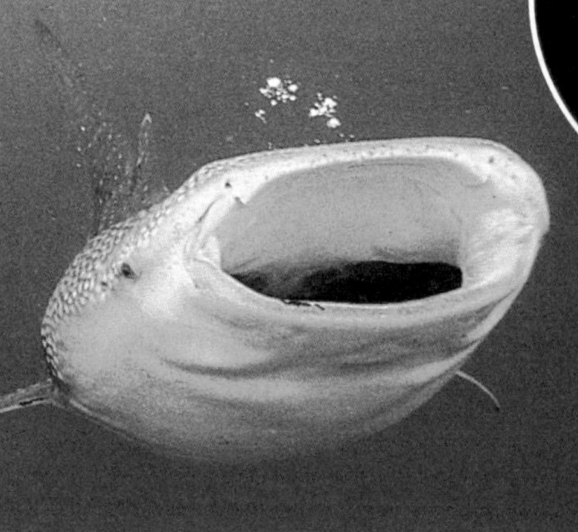

A whale shark's mouth can open to 1.4 m wide.

Sawfish are rare. They use their toothed snouts to root out bottom fish to eat.

Parrot fish use their hard beaks to bite off pieces of coral.

Electric eels produce a powerful electric shock that stuns their prey.

Feeding Adaptations Some of the adaptations that fish have for obtaining food are shown in **Figure 8.** Some of the largest sharks are filter feeders that take in small animals as they swim. The archerfish shoots down insects by spitting drops of water at them. Even though some fish have strong teeth, most do not chew their food. They use their teeth to capture their prey or to tear off chunks of food.

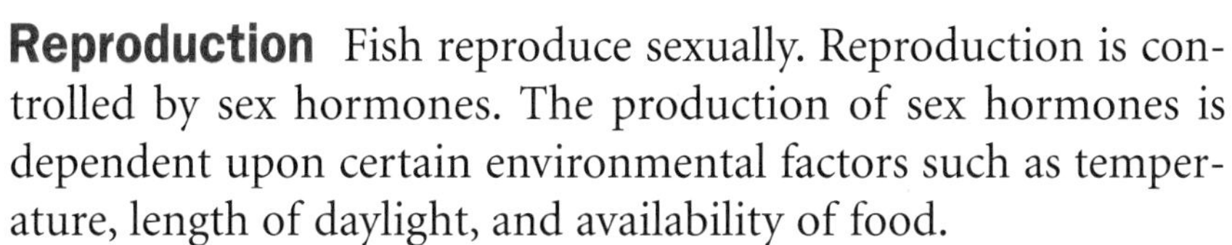

Reproduction Fish reproduce sexually. Reproduction is controlled by sex hormones. The production of sex hormones is dependent upon certain environmental factors such as temperature, length of daylight, and availability of food.

Female fish release large numbers of eggs into the water. Males then swim over the eggs and release sperm. This behavior is called spawning. The joining of the egg and sperm cells outside the female's body is called external fertilization. Certain species of sharks and rays have internal fertilization and lay fertilized eggs. Some fish, such as guppies and other sharks, have internal fertilization but the eggs develop and hatch inside the female's body. After they hatch, they leave her body.

Some species do not take care of their young. They release hundreds or even millions of eggs, which increases the chances that a few offspring will survive to become adults. Fish that care for their young lay fewer eggs. Some fish, including some catfish, hold their eggs and young in their mouths. Male sea horses keep the fertilized eggs in a pouch until they hatch.

Types of Fish

Fish vary in size, shape, color, living environments, and other factors. Despite their diversity, fish are grouped into only three categories—jawless fish, jawed cartilaginous (kar tuh LA juh nuss) fish, and bony fish.

Fish Fats Many fish contain oil with omega-3 fatty acids, which seems to reverse the effects of too much cholesterol. A diet rich in fish that contain this oil might prevent the formation of fatty deposits in the arteries of humans. In your Science Journal, develop a menu for a meal that includes fish.

Jawless Fish

Lampreys, along with the hagfish in **Figure 9,** are jawless fish. Jawless fish have round, toothed mouths and long, tubelike bodies covered with scaleless, slimy skin. Most lampreys are parasites. They attach to other fish with their suckerlike mouth. They then feed by removing blood and other body fluids from the host fish. Hagfish feed on dead or dying fish and other aquatic animals.

Jawless fish have flexible endoskeletons made of cartilage. Hagfish live only in salt water, but some species of lamprey live in salt water and other species live in freshwater.

Jawed Cartilaginous Fish

Sharks, skates, and rays are jawed cartilaginous fish. These fish have endoskeletons made of cartilage like jawless fish. Unlike jawless fish, these fish have movable jaws that usually have well-developed teeth. Their bodies are covered with tiny scales that make their skins feel like fine sandpaper.

Sharks are top predators in many ocean food chains. They are efficient at finding and killing their food, which includes other fish, mammals, and some reptiles. Because of overfishing and the fact that shark reproduction is slow, shark populations are decreasing at an alarming rate.

Reading Check *Why are shark populations decreasing?*

Figure 9 Hagfish have cartilaginous skeletons. They feed on marine worms, mollusks, and crustaceans, in addition to dead and dying fish.
Infer *how hagfish eat.*

Figure 10 Bony fish come in many sizes, shapes, and colors. However, all bony fish have the same basic body structure.

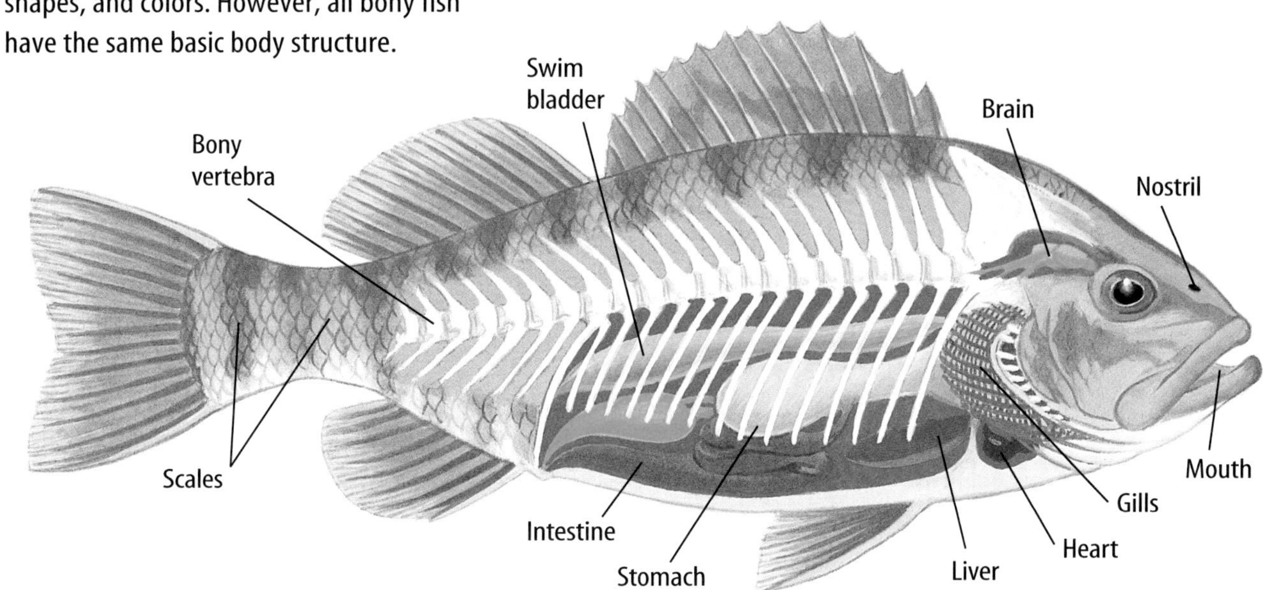

Bony Fish

About 95 percent of all species of fish are bony fish. They have skeletons made of bone. The body structure of a typical bony fish is shown in **Figure 10.** A bony flap covers and protects the gills. It closes as water moves into the mouth and over the gills. When it opens, water exits from the gills.

Swim Bladder An important adaptation in most bony fish is the swim bladder. It is an air sac that allows the fish to adjust its density in response to the density of the surrounding water. The density of matter is found by dividing its mass by its volume. If the density of the object is greater than that of the liquid it is in, the object will sink. If the density of the object is equal to the density of the liquid, the object will neither sink nor float to the surface. If the density of the object is less than the density of the liquid, the object will float on the liquid's surface.

The transfer of gases—mostly oxygen in deepwater fish and nitrogen in shallow-water fish—between the swim bladder and the blood causes the swim bladder to inflate and deflate. As the swim bladder fills with gases, the fish's density decreases and it rises in the water. When the swim bladder deflates, the fish's density increases and it sinks. Glands regulate the gas content in the swim bladder, enabling the fish to remain at a specific depth with little effort. Deepwater fish often have oil in their swim bladders rather than gases. Some bottom-dwelling fish and active fish that frequently change depth have no swim bladders.

Modeling How Fish Adjust to Different Depths

Procedure

1. Fill a **balloon** with air.
2. Place it in a **bowl of water.**
3. Fill **another balloon** partially with water, then blow air into it until it is the same size as the air-filled balloon.
4. Place the second balloon in the bowl of water.

Analysis

1. Infer what structure these balloons model.
2. Compare where in the water (on the surface, or below the surface) two fish would be if they had swim bladders similar to the two balloons.

Try at Home

Lobe-Finned Fish One of the three types of bony fish is the lobe-finned fish, as shown in **Figure 11.** Lobe-finned fish have fins that are lobelike and fleshy. These organisms were thought to have been extinct for more than 70 million years. But in 1938, some South African fishers caught a lobe-finned fish in a net. Several living lobe-finned fish have been studied since. Lobe-finned fish are important because scientists hypothesize that fish similar to these were the ancestors of the first land vertebrates—the amphibians.

Figure 11 Coelacanths (SEE luh kanthz) have been found living in the Indian Ocean north of Madagascar.

Applying Math — Solve a One-Step Equation

DENSITY OF A FISH A freshwater fish has a mass of 645 g and a volume of 700 cm^3. What is the fish's density, and will it sink or float in freshwater?

Solution

1. *This is what you know:*
   - density of freshwater = 1g/cm^3
   - mass of fish = 645 g
   - volume of fish = 700 cm^3
2. *This is the equation you need to use:* $\frac{\text{mass of object (g)}}{\text{volume of object (cm}^3\text{)}} = \text{density of object (g/cm}^3\text{)}$
3. *Substitute the known values:* $\frac{645 \text{ g}}{700 \text{ cm}^3} = 0.921 \text{ g/cm}^3$
4. *Check your answer:* Multiply 0.921 g/cm^3 by 700 cm^3. You should get 645 g. The fish will float in freshwater. Its density is less than that of freshwater.

Practice Problems

1. Calculate the density of a saltwater fish that has a mass of 215 g and a volume of 180 cm^3. Will this fish float or sink in salt water? The density of ocean salt water is about 1.025 g/cm^3.
2. A fish with a mass of 440 g and a volume of 430 cm^3 floats in its water. Is it a freshwater fish or a saltwater fish?

For more practice, visit life.msscience.com/math_practice

Figure 12 Australian lungfish are one of the six species of lungfish.
Identify *the unique adaptation of a lungfish.*

Lungfish A lungfish, as shown in **Figure 12,** has one lung and gills. This adaptation enables them to live in shallow waters that have little oxygen. The lung enables the lungfish to breathe air when the water evaporates. Drought conditions stimulate lungfish to burrow into the mud and cover themselves with mucus until water returns. Lungfish have been found along the coasts of South America and Australia.

Ray-Finned Fish Most bony fish have fins made of long, thin bones covered with skin. Ray-finned fish, like those in **Figure 13,** have a lot of variation in their body plans. Most predatory fish have long, flexible bodies, which enable them to pursue prey quickly. Many bottom fish have flattened bodies and mouths adapted for eating off the bottom. Fish with unusual shapes, like the sea horse and anglerfish, also can be found. Yellow perch, tuna, salmon, swordfish, and eels are ray-finned fish.

Figure 13 Bony fish have a diversity of body plans.

Most bony fish are ray-finned fish, like this rainbow trout.

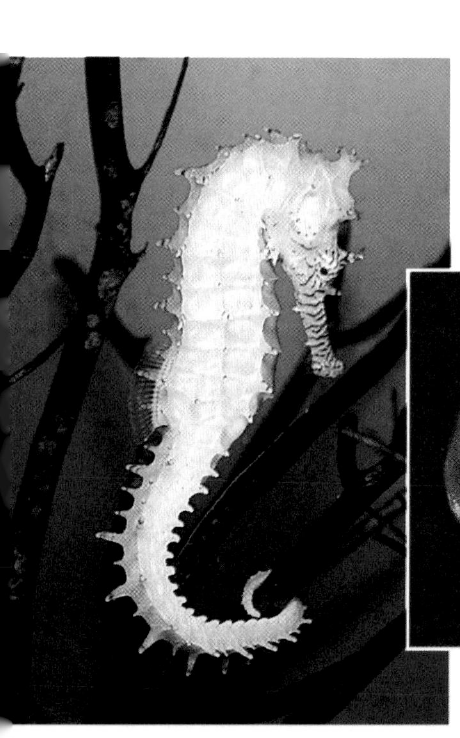

Sea horses use their tails to anchor themselves to sea grass. This prevents the ocean currents from washing them away.

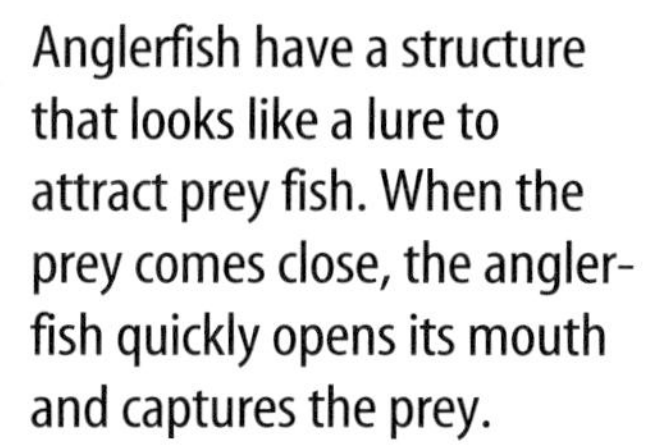

Anglerfish have a structure that looks like a lure to attract prey fish. When the prey comes close, the anglerfish quickly opens its mouth and captures the prey.

Figure 14 Lancelets are small, eel-like animals. They spend most of their time buried in the sand and mud at the bottom of the ocean.

Importance and Origin of Fish

Fish play a part in your life in many ways. They provide food for many animals, including humans. Fish farming and commercial fishing also are important to the U.S. economy. Fishing is a method of obtaining food as well as a form of recreation enjoyed by many people. Many fish eat large amounts of insect larvae, such as mosquitoes, which keeps insect populations in check. Some, such as grass carp, are used to keep the plant growth from clogging waterways. Captive fish are kept in aquariums for humans to admire their bright colors and exotic forms.

Reading Check *How are fish helpful to humans?*

Most scientists agree that fish evolved from small, soft-bodied, filter-feeding organisms similar to present-day lancelets, shown in **Figure 14.** The earliest fossils of fish are those of jawless fish that lived about 450 million years ago. Fossils of these early fish usually are found where ancient streams emptied into the sea. This makes it difficult to tell whether these fish ancestors evolved in freshwater or in salt water.

Today's bony fish are probably descended from the first jawed fish called the acanthodians (a kan THOH dee unz). They appeared in the fossil record about 410 mya. Another group of ancient fish—the placoderms—appeared about 400 mya. For about 50 million years, placoderms dominated most water ecosystems then disappeared. Modern sharks and rays probably descended from the placoderms.

section 2 review

Summary

Fish Characteristics

- All fish have a streamlined shape, a muscular tail, fins, scales, well-developed sensory systems, and gills.
- All fish reproduce sexually and feed in many different ways.

Types of Fish

- There are three categories of fish: jawless fish, jawed cartilaginous fish, and bony fish.
- There are three types of bony fish: lobe-finned fish, lungfish, and ray-finned fish.

Self Check

1. **List** examples for each of the three classes of fish.
2. **Explain** how jawless fish and cartilaginous jawed fish take in food.
3. **Describe** the many ways that fish are important to humans.
4. **Think Critically** Female fish lay thousands of eggs. Why aren't lakes and oceans overcrowded with fish?

Applying Skills

5. **Concept Map** Make an events-chain concept map to show what must take place for the fish to rise from the bottom to the surface of the lake.

Science Online life.msscience.com/self_check_quiz

section 3

Amphibians

Amphibian Characteristics

The word *amphibian* comes from the Greek word *amphibios*, which means "double life." They are well named, because amphibians spend part of their lives in water and part on land. Frogs, toads, and the salamander shown in **Figure 15** are examples of amphibians. What characteristics do these animals have that allow them to live on land and in water?

Amphibians are ectotherms. Their body temperature changes when the temperature of their surroundings changes. In cold weather, amphibians become inactive and bury themselves in mud or leaves until the temperature warms. This period of inactivity during cold weather is called **hibernation.** Amphibians that live in hot, dry environments become inactive and hide in the ground when temperatures become too hot. Inactivity during the hot, dry months is called **estivation.**

Reading Check *How are hibernation and estivation similar?*

Respiration Amphibians have moist skin that is smooth, thin, and without scales. They have many capillaries directly beneath the skin and in the lining of the mouth. This makes it possible for oxygen and carbon dioxide to be exchanged through the skin and the mouth lining. Amphibians also have small, simple, saclike lungs in the chest cavity for the exchange of oxygen and carbon dioxide. Some salamanders have no lungs and breathe only through their skin.

as you read

What You'll Learn

- **Describe** the adaptations amphibians have for living in water and living on land.
- **List** the kinds of amphibians and the characteristics of each.
- **Explain** how amphibians reproduce and develop.

Why It's Important

Because amphibians are sensitive to changes in the environment, they can be used as biological indicators.

Review Vocabulary
habitat: place where an organism lives and that provides the types of food, shelter, moisture, and temperature needed for survival

New Vocabulary
- hibernation
- estivation

Figure 15 Salamanders often are mistaken for lizards because of their shape. However, like all amphibians, they have a moist, scaleless skin that requires them to live in a damp habitat.

Figure 16 Red-eyed tree frogs are found in forests of Central and South America. They eat a variety of foods, including insects and even other frogs.

Circulation The three-chambered heart in amphibians is an important change from the circulatory system of fish. In the three-chambered heart, one chamber receives oxygen-filled blood from the lungs and skin, and another chamber receives carbon dioxide-filled blood from the body tissues. Blood moves from both of these chambers to the third chamber, which pumps oxygen-filled blood to body tissues and carbon dioxide-filled blood back to the lungs. Limited mixing of these two bloods occurs.

Reproduction Even though amphibians are adapted for life on land, they depend on water for reproduction. Because their eggs do not have a protective, waterproof shell, they can dry out easily, so amphibians must have water to reproduce.

Amphibian eggs are fertilized externally by the male. As the eggs come out of the female's body, the male releases sperm over them. In most species the female lays eggs in a pond or other body of water. However, many species have developed special reproductive adaptations, enabling them to reproduce away from bodies of water. Red-eyed tree frogs, like the ones in **Figure 16,** lay eggs in a thick gelatin on the underside of leaves that hang over water. After the tadpoles hatch, they fall into the water below, where they continue developing. The Sonoran Desert toad waits for small puddles to form in the desert during the rainy season. It takes tadpoles only two to 12 days to hatch in these temporary puddles.

Figure 17 Amphibians go through metamorphosis as they develop.

After hatching, most young amphibians, like these tadpoles, do not look like adult forms.

Amphibian eggs are laid in a jellylike material to keep them moist.

Development Most amphibians go through a developmental process called metamorphosis (me tuh MOR fuh sus). Fertilized eggs hatch into tadpoles, the stage that lives in water. Tadpoles have fins, gills, and a two-chambered heart similar to fish. As tadpoles grow into adults, they develop legs, lungs, and a three-chambered heart. **Figure 17** shows this life cycle.

The tadpole of some amphibian species, such as salamanders, are not much different from the adult stage. Young salamanders look like adult salamanders, but they have external gills and usually a tail fin.

Topic: Biological Indicators
Visit life.msscience.com for Web links to information about amphibians as biological indicators.

Activity What factors make amphibians good biological indicators?

Frogs and Toads

Adult frogs and toads have short, broad bodies with four legs but no neck or tail. The strong hind legs are used for swimming and jumping. Bulging eyes and nostrils on top of the head let frogs and toads see and breathe while the rest of their body is submerged in water. On spring nights, males make their presence known with loud, distinctive croaking sounds. On each side of the head, just behind the eyes, are round tympanic membranes. These membranes vibrate somewhat like an eardrum in response to sounds and are used by frogs and toads to hear.

Most frog and toad tongues are attached at the front of their mouths. When they see prey, their tongue flips out and contacts the prey. The prey gets stuck in the sticky saliva on the tongue and the tongue flips back into the mouth. Toads and frogs eat a variety of insects, worms, and spiders, and one tropical species eats berries.

Amphibians go through metamorphosis, which means they change form from larval stage to adult.

Most adult amphibians are able to move about and live on land.

Describing Frog Adaptations

Procedure

1. Carefully observe a **frog** in a **jar.** Notice the position of its legs as it sits. Record all of your observations in your **Science Journal.**
2. Observe its mouth, eyes, nostrils and ears.
3. Observe the color of its back and belly.
4. Return the frog to your teacher.

Analysis

1. Describe the adaptations the frog has for living in water.
2. What adaptations does it have for living on land?

Salamanders

Most species of salamanders and newts live in North America. These amphibians often are mistaken for lizards because of their long, slender bodies. The short legs of salamanders and newts appear to stick straight out from the sides of their bodies.

Land-living species of salamanders and newts usually are found near water. These amphibians hide under leaf litter and rocks during the day to avoid the drying heat of the Sun. At night, they use their well-developed senses of smell and vision to find and feed on worms, crustaceans, and insects.

Many species of salamanders breed on land, where fertilization is internal. Aquatic species of salamanders and newts release and fertilize their eggs in the water.

Importance of Amphibians

Most adult amphibians are insect predators and are helpful in keeping some insect populations in check. They also are a source of food for other animals, including other amphibians. Some people consider frog legs a delicacy.

Poison frogs, like the one in **Figure 18,** produce a poison that can kill large animals. They also are known as poison dart frogs or poison arrow frogs. The toxin is secreted through their skin and can affect muscles and nerves of animals that come in contact with it. Native people of the Emberá Chocó in Colombia, South America, cover blowgun darts that they use for hunting with the poison of one species of these frogs. Researchers are studying the action of these toxins to learn more about how the nervous system works. Researchers also are using amphibians in regeneration studies in hopes of developing new ways of treating humans who have lost limbs or were born without limbs.

Figure 18 Poison frogs are brightly colored to show potential predators that they are poisonous. Toxins from poison frogs have been used in hunting for centuries.

Biological Indicators Because they live on land and reproduce in water, amphibians are affected directly by changes in the environment, including pesticides and other pollution. Amphibians also absorb gases through their skin, making them susceptible to air pollutants. Amphibians, like the one in **Figure 19,** are considered to be biological indicators. Biological indicators are species whose overall health reflects the health of a particular ecosystem.

What is a biological indicator?

Figure 19 Beginning in 1995, deformed frogs such as this were found. Concerned scientists hypothesize that an increase in the number of deformed frogs could be a warning of environmental problems for other organisms.

Origin of Amphibians The fossil record shows that ancestors of modern fish were the first vertebrates on Earth. For about 150 million years, they were the only vertebrates. Then as the climate changed and competition for food and space increased, some lobe-finned fish might have traveled across land searching for water as their ponds dried up. Fossil evidence shows that from these lobe-finned fish evolved aquatic animals with four limbs. Amphibians probably evolved from these aquatic animals about 350 mya.

Because competition on land from other animals was minimal, evolution favored the development of amphibians. Insects, spiders, and other invertebrates were an abundant source of food on land. Land was almost free of predators, so amphibians were able to reproduce in large numbers, and many new species evolved. For 100 million years or more, amphibians were the dominant land animals.

section 3 review

Summary

Amphibian Characteristics

- Amphibians have two phases of life—one in water and one on land.
- All amphibians have a three-chambered heart, reproduce in the water by laying eggs, and go through metamorphosis.

Types and Importance of Amphibians

- Frogs and toads have short, broad bodies, while salamanders have long, slender bodies.
- Amphibians are used for food, research, and are important as biological indicators.

Self Check

1. **List** the adaptations amphibians have for living in water and for living on land.
2. **Explain** how tadpole and frog hearts differ.
3. **Describe** two different environments where amphibians lay eggs.
4. **Think Critically** Why do you suppose frogs and toads seem to appear suddenly after a rain?

Applying Skills

5. **Concept Map** Make an events-chain concept map of frog metamorphosis. Describe each stage in your Science Journal.

section 4

Reptiles

as you read

What You'll Learn

- **List** the characteristics of reptiles.
- **Determine** how reptile adaptations enable them to live on land.
- **Explain** the importance of the amniotic egg.

Why It's Important

Reptiles provide information about how body systems work during extreme weather conditions

Review Vocabulary
bask: to warm by continued exposure to heat

New Vocabulary
- amniotic egg

Reptile Characteristics

Reptiles are ectotherms with a thick, dry, waterproof skin. Their skin is covered with scales that help reduce water loss and protect them from injury. Even though reptiles are ectotherms, they are able to modify their internal body temperatures by their behavior. When the weather is cold, they bask in the Sun, which warms them. When the weather is warm and the Sun gets too hot, they move into the shade to cool down.

Reading Check *How are reptiles able to modify their body temperature?*

Some reptiles, such as turtles, crocodiles, and lizards, like the skink in **Figure 20,** move on four legs. Claws are used to dig, climb, and run. Reptiles, such as snakes and some lizards, move without legs.

Body Systems Scales on reptiles prevent the exchange of oxygen and carbon dioxide through the skin. Reptiles breathe with lungs. Even turtles and sea snakes that live in water must come to the surface to breathe.

The circulatory system of reptiles is more highly developed than that of amphibians. Most reptiles have a three-chambered heart with a partial wall inside the main chamber. This means that less mixing of oxygen-filled blood and carbon dioxide-filled blood occurs than in amphibians. This type of circulatory system provides more oxygen to all parts of the body. Crocodilians have a four-chambered heart that completely separates the oxygen-filled blood and the carbon dioxide-filled blood and keeps them from mixing.

Figure 20 Skinks, like this northern blue-tongue skink, are one of the largest lizard families with around 800 species.

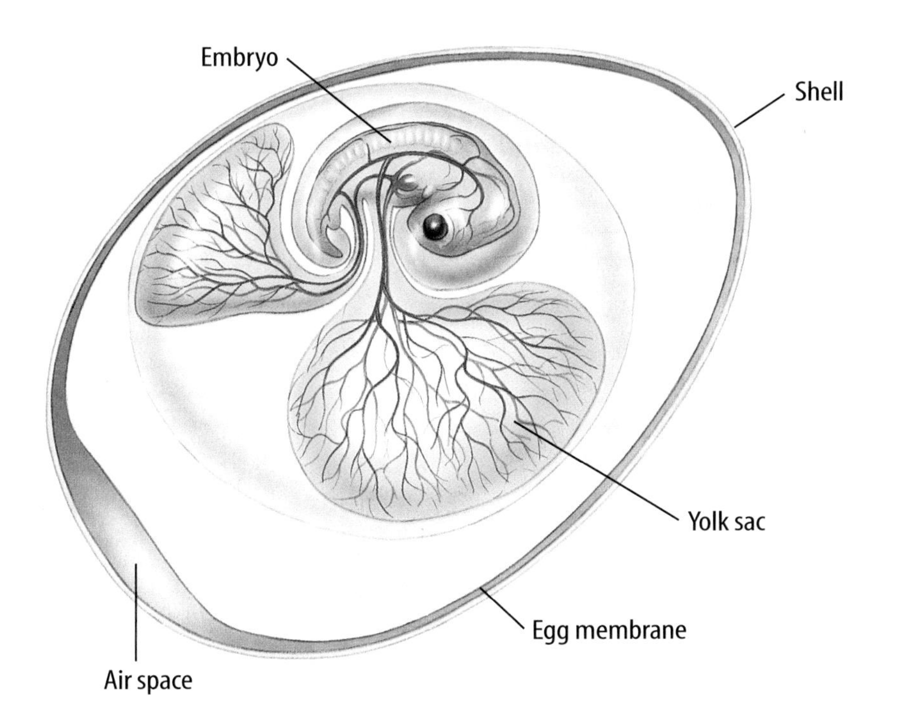

Figure 21 The development of amniotic eggs enabled reptiles to reproduce on land.
Infer *how an amniotic egg helps reptiles be a more successful group.*

Amniotic Egg One of the most important adaptations of reptiles for living on land is the way they reproduce. Unlike the eggs of most fish and amphibians, eggs of reptiles are fertilized internally—inside the body of the female. After fertilization, the females of many reptiles lay eggs that are covered by tough, leathery shells. The shell prevents the eggs from drying out. This adaptation enables reptiles to lay their eggs on land.

The **amniotic egg** provides a complete environment for the embryo's development. **Figure 21** shows the structures in a reptilian egg. This type of egg contains membranes that protect and cushion the embryo and help it get rid of wastes. It also contains a large food supply—the yolk—for the embryo. Minute holes in the shell, called pores, allow oxygen and carbon dioxide to be exchanged. By the time it hatches, a young reptile looks like a small adult.

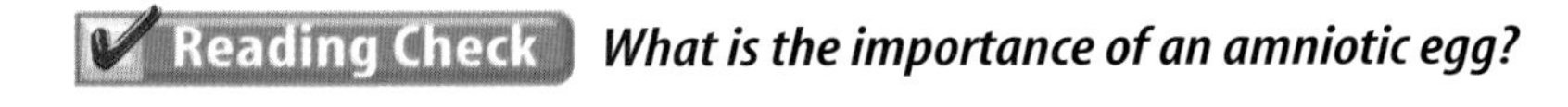

What is the importance of an amniotic egg?

Types of Modern Reptiles

Reptiles live on every continent except Antarctica and in all the oceans except those in the polar regions. They vary greatly in size, shape, and color. Reticulated pythons, 10 m in length, can swallow small deer whole. Some sea turtles weigh more than 350 kg and can swim faster than humans can run. Three-horned lizards have movable eye sockets and tongues as long as their bodies. The three living groups of reptiles are lizards and snakes, turtles, and crocodilians.

Visit life.msscience.com for Web links to recent news about the nesting sites of turtles.

Activity Name two conservation organizations that are giving the turtles a helping hand, and how they are doing it.

Lizards and Snakes Some animals in the largest group of reptiles—the lizards and snakes like those shown in **Figure 22**—have a type of jaw not found in other reptiles, like the turtle also shown in **Figure 22.** The jaw has a special joint that unhinges and increases the size of their mouths. This enables them to swallow their prey whole. Lizards have movable eyelids, external ears, and legs with clawed toes on each foot. They feed on plants, other reptiles, insects, spiders, worms, and mammals.

Snakes have developed ways of moving without legs. They have poor hearing and most have poor eyesight. Recall how you could feel the vibrations of the tuning fork in the Launch Lab. Snakes do not hear sound waves in the air. They "hear" vibrations in the ground that are picked up by the lower jawbone and conducted to the bones of the snake's inner ear. From there, the vibrations are transferred to the snake's brain, where the sounds are interpreted.

Snakes are meat eaters. Some snakes wrap around and constrict their prey. Others inject their prey with venom. Many snakes feed on small mammals, and as a result, help control those populations.

Most snakes lay eggs after they are fertilized internally. In some species, eggs develop and hatch inside the female's body then leave her body shortly thereafter.

Figure 22 Examples of reptiles are shown below.

When frilled lizards are threatened, they flare out their collar. This behavior helps keep predators away.

Diamondback terrapins are one of the few species of turtles that live in brackish—slightly salty—water.

Rosy boas are one of only two species of boas found in the United States.

Turtles The only reptiles that have a two-part shell made of hard, bony plates are turtles. The vertebrae and ribs are fused to the inside of the top part of the shell. The muscles are attached to the lower and upper part of the inside of the shell. Most turtles can withdraw their heads and legs into the shell for protection against predators.

Reading Check *What is the purpose of a turtle's shell?*

Turtles have no teeth but they do have powerful jaws with a beaklike structure used to crush food. They feed on insects, worms, fish, and plants. Turtles live in water and on land. Those that live on land are called tortoises.

Like most reptiles, turtles provide little or no care for their young. Turtles dig out a nest, deposit their eggs, cover the nest, and leave. Turtles never see their own hatchlings. Young turtles, like those in **Figure 23,** emerge from the eggs fully formed and live on their own.

Figure 23 Most turtles are eaten shortly after they hatch. Only a few sea turtles actually make it into the ocean.

Crocodilians Found in or near water in warm climates, crocodilians, such as crocodiles, gavials, and alligators, are similar in appearance. They are lizardlike in shape, and their backs have large, deep scales. Crocodilians can be distinguished from each other by the shape of their heads. Crocodiles have a narrow head with a triangular-shaped snout. Alligators have a broad head with a rounded snout. Gavials, as shown in **Figure 24,** have a very slender snout with a bulbous growth on the end. Crocodiles are aggressive and can attack animals as large as cattle. Alligators are less aggressive than crocodiles, and feed on fish, turtles, and waterbirds. Gavials primarily feed on fish. Crocodilians are among the world's largest living reptiles.

Crocodilians are some of the few reptiles that care for their young. The female guards the nest of eggs and when the eggs hatch, the male and female protect the young. A few crocodilian females have been photographed opening their nests in response to noises made by hatchlings. After the young hatch, a female carries them in her huge mouth to the safety of the water. She continues to keep watch over the young until they can protect themselves.

Figure 24 Indian gavials are one of the rarest crocodilian species on Earth. Adults are well adapted for capturing fish.

NATIONAL GEOGRAPHIC

VISUALIZING EXTINCT REPTILES

Figure 25

If you're like most people, the phrase "prehistoric reptiles" probably brings dinosaurs to mind. But not all ancient reptiles were dinosaurs. The first dinosaurs didn't appear until about 115 million years after the first reptiles. Paleontologists have unearthed the fossils of a variety of reptilian creatures that swam through the seas and waterways of ancient Earth. Several examples of these extinct aquatic reptiles are shown here.

▲ MOSASAUR (MOH zuh sawr) Marine-dwelling mosasaurs had snakelike bodies, large skulls, and long snouts. They also had jointed jawbones, an adaptation for grasping and swallowing large prey.

▲ ICHTHYOSAUR (IHK thee uh sawr) Ichthyosaurs resembled a cross between a dolphin and a shark, with large eyes, four paddlelike limbs, and a fishlike tail that moved from side to side. These extinct reptiles were fearsome predators with long jaws armed with numerous sharp teeth.

◀ ELASMOSAURUS (uh laz muh SAWR us) Predatory ***Elasmosaurus*** had a long neck—with as many as 76 vertebrae—topped by a small head.

▲ CHAMPOSAUR (CHAM puh sawr) This ancient reptile looked something like a modern crocodile, with a long snout studded with razor-sharp teeth. Champosaurs lived in freshwater lakes and streams and preyed on fish and turtles.

▲ PLESIOSAUR (PLEE zee uh sawr) These marine reptiles had stout bodies, paddlelike limbs, and long necks. Plesiousaurs might have fed by swinging their heads from side to side through schools of fish.

The Importance of Reptiles

Reptiles are important predators in many environments. In farming areas, snakes eat rats and mice that destroy grains. Small lizards eat insects, and large lizards eat small animals that are considered pests.

Humans in many parts of the world eat reptiles and their eggs or foods that include reptiles, such as turtle soup. The number of reptile species is declining in areas where swamps and other lands are being developed for homes and recreation areas. Coastal nesting sites of sea turtles are being destroyed by development or are becoming unusable because of pollution. For years, many small turtles were collected in the wild and then sold as pets. People now understand that such practices disturb turtle populations. Today most species of turtles and their habitats are protected by law.

Origin of Reptiles Reptiles first appeared in the fossil record about 345 mya. The earliest reptiles did not depend upon water for reproduction. As a result, they began to dominate the land about 200 mya. Some reptiles even returned to the water to live, although they continued to lay their eggs on land. Dinosaurs—descendants of the early reptiles—ruled Earth during this era, then died out about 65 mya. Some of today's reptiles, such as the crocodilians, have changed little from their ancestors, some of which are illustrated in **Figure 25.**

A Changing Environment Dinosaurs, reptiles that ruled Earth for 160 million years, died out about 65 million years ago. In your Science Journal, describe what changes in the environment could have caused the extinction of the dinosaurs.

section 4 review

Summary

Reptile Characteristics

- Reptiles are ectotherms with a thick, dry, waterproof skin that is covered with scales.
- Most have a three-chambered heart with a partial wall in the main chamber.
- Reptile young develop in an amniotic egg.

Types and Importance of Reptiles

- Lizards and snakes are the largest group of reptiles. Most lizards have legs, while snakes do not.
- Turtles have a two-part bony shell.
- Crocodilians are large reptiles and one of the few reptiles that care for their young.
- Reptiles are important predators. Some reptiles are food sources.

Self Check

1. **Describe** reptilian adaptations for living on land.
2. **Explain** how turtles differ from other reptiles.
3. **Infer** why early reptiles, including dinosaurs, were so successful as a group.
4. **Draw** the structure of an amniotic egg.
5. **Think Critically** Venomous coral snakes and some nonvenomous snakes have bright red, yellow, and black colors. How is this an advantage and a disadvantage to the nonvenomous snake?

Applying Math

6. **Solve One-Step Equations** *Brachiosaurus*, a dinosaur, was about 12 m tall and 22 m long. The average elephant is 3 m tall and 6 m long. How much taller and longer is the *Brachiosaurus* compared to an elephant?

Design Your Own

Water Temperature and the Respiration Rate of Fish

Goals

- **Design** and carry out an experiment to measure the effect of water temperature on the rate of respiration of fish.
- **Observe** the breathing rate of fish.

Possible Materials

goldfish
aquarium water
small fishnet
600-mL beakers
container of ice water
stirring rod
thermometer
aquarium

Safety Precautions

Protect your clothing. Use the fishnet to transfer fish into beakers.

Real-World Question

Imagine that last summer was hot with few storms. One day after many sunny, windless days, you noticed that a lot of dead fish were floating on the surface of your neighbor's pond. What might have caused these fish to die? How does water temperature affect the respiration rate of fish?

Form a Hypothesis

Fish obtain oxygen from the water. State a hypothesis about how water temperature affects the respiration rate of fish.

Test Your Hypothesis

Make a Plan

1. As a group, agree upon and write out a plan. You might make a plan that relates the amount of oxygen dissolved in water at different water temperatures and how this affects fish.

2. As a group, list the steps that you need to take to follow your plan. Be specific and describe exactly what you will do at each step. List your materials.
3. How will you measure the breathing rate of fish?
4. **Explain** how you will change the water temperature in the beakers. Fish respond better to a gradual change in temperature than an abrupt change. How will you measure the response of fish to changes in water temperature?
5. What data will your group collect? Prepare a data table in your Science Journal to record the data you collect. How many times will you run your experiment?
6. Read over your entire experiment to make sure the steps are in logical order. Identify any constants, variables, and controls.

Follow Your Plan

1. Make sure your teacher approves your setup and your plan before you start.
2. Carry out the experiment according to the approved plan.
3. While the experiment is going on, write down any observations that you make and complete the data table in your Science Journal.

Analyze Your Data

1. **Compare** your results with the results of other groups in your class. Were the results similar?
2. **Infer** what you were measuring when you counted mouth or gill cover openings.
3. **Describe** how a decrease in water temperature affects respiration rate and behavior of the fish.
4. **Explain** how your results could be used to determine the kind of environment in which a fish can live.

Conclude and Apply

1. **Explain** how fish can live in water that is totally covered by ice.
2. **Predict** what would happen to a fish if the water were to become very warm.

Construct a graph of your data on poster board and share your results with your classmates.

TIME SCIENCE AND Society

SCIENCE ISSUES THAT AFFECT YOU!

Venom

Bumble bee

hissssssss

Gila monster

Pit viper

Venom as Medicine

Hiss, rattle… Run! Just the sound of a snake sends most people on a sprint to escape what could be a painful bite. Why? The bites could contain venom, a toxic substance injected into prey or an enemy. Venom can harm—or even kill—the victim. Some venomous creatures use it to stun, kill, and digest their prey, while others use it as a means of protection.

Venom is produced by a gland in the body. Some fish use their sharp, bony spines to inject venom. Venomous snakes, such as pit vipers, have fangs. Venom passes through these hollow teeth into a victim's body. The Gila monster, the largest lizard in the United States, has enlarged, grooved teeth in its lower jaw through which its venom travels. It is one of only two species of venomous lizards.

Doctors and scientists have discovered a shocking surprise within this sometimes deadly liquid. Oddly enough, the very same toxin that harms and weakens people can heal, too. In fact, doctors use the deadliest venom—that of some pit viper species—to treat certain types of heart attacks. Cobra venom has been used to soothe the effects of cancer, and other snake venoms reduce the spasms of epilepsy and asthma.

Some venoms also contain substances that help clot blood. Hemophiliacs—people whose blood will not clot naturally—rely on the medical benefits that venom-based medicines supply. Venoms also are used in biological research. For instance, venoms that affect the nervous system help doctors and researchers learn more about how nerves function.

It's still smart to steer clear of the rattle or the stinger—but it's good to know that the venom in them might someday help as many as it can hurt.

Research Besides venom, what other defenses do animals use to protect themselves or to subdue their prey? Explore how some animals that are native to your region use their built-in defenses.

For more information, visit life.msscience.com/time

Reviewing Main Ideas

Section 1 Chordates and Vertebrates

1. Chordates include lancelets, tunicates, and vertebrates. Chordates have a notochord, a nerve cord, pharyngeal pouches, and a postanal tail.
2. All vertebrates have an endoskeleton that includes a backbone and a skull that protects the brain.
3. An endotherm is an animal that has a nearly constant internal body temperature. An ectotherm has a body temperature that changes with the temperature of its environment.

Section 2 Fish

1. Fish are vertebrates that have a streamlined body, fins, gills for gas exchange, and a highly developed sensory system.
2. Fish are divided into three groups—jawless fish, jawed cartilaginous fish, and bony fish.
3. Bony fish, with scales and a swim bladder, have the greatest number of known fish species.

Section 3 Amphibians

1. The first vertebrates to live on land were the amphibians.
2. Amphibians have adaptations that allow them to live on land and in the water. The adaptations include moist skin, mucous glands, and lungs. Most amphibians are dependent on water to reproduce.
3. Most amphibians go through a metamorphosis from egg, to larva, to adult. During metamorphosis, legs develop, lungs replace gills, and the tail is lost.

Section 4 Reptiles

1. Reptiles are land animals with thick, dry, scaly skin. They lay amniotic eggs with leathery shells.
2. Turtles with tough shells, meat-eating crocodilians, and snakes and lizards make up the reptile groups.
3. Early reptiles were successful because of their adaptations to living on land.

Visualizing Main Ideas

Copy and complete the concept map below that describes chordates.

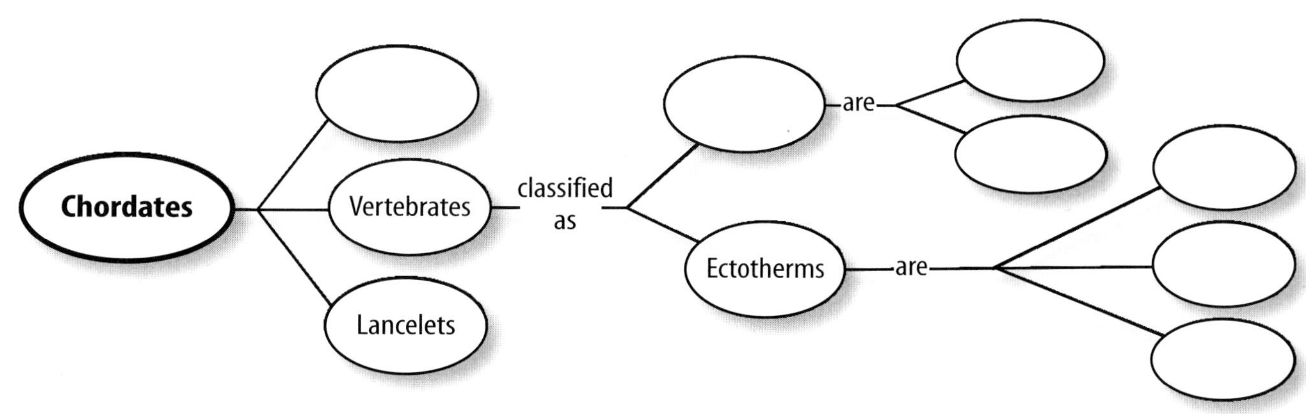

chapter 14 Review

Using Vocabulary

amniotic egg p. 413	hibernation p. 407
cartilage p. 395	nerve cord p. 395
chordate p. 394	notochord p. 394
ectotherm p. 397	pharyngeal pouch p. 395
endoskeleton p. 395	postanal tail p. 394
endotherm p. 397	scale p. 399
estivation p. 407	vertebrae p. 395
fin p. 399	

Fill in the blanks with the correct vocabulary word or words.

1. All chordates have a notochord, pharyngeal pouches, postanal tail, and a(n) ________.
2. The inactivity of amphibians during hot, dry weather is ________.
3. All animals with a constant internal temperature are ________.
4. Reptiles are ________ with scaly skin.
5. Jawless fish have skeletons made of a tough, flexible tissue called ________.
6. Reptiles lay ________.
7. The structure that becomes the backbone in vertebrates is the ________.

Checking Concepts

Choose the word or phrase that best answers the question.

8. Which animals have fins, scales, and gills?
   A) amphibians **C)** reptiles
   B) crocodiles **D)** fish
9. Which is an example of a cartilaginous fish?
   A) hagfish **C)** perch
   B) tuna **D)** goldfish
10. What fish group has the greatest number of species?
   A) bony **C)** jawed cartilaginous
   B) jawless **D)** amphibians
11. Which of these fish have gills and lungs?
   A) shark **C)** lungfish
   B) ray **D)** perch
12. Biological indicators include which group of ectothermic vertebrates?
   A) amphibians
   B) cartilaginous fish
   C) bony fish
   D) reptiles

Use the photo below to answer question 13.

13. Which kinds of reptiles are included with the animal above?
   A) snakes **C)** turtles
   B) crocodiles **D)** alligators
14. What term best describes eggs of reptiles?
   A) amniotic **C)** jellylike
   B) brown **D)** hard-shelled
15. Vertebrates that have lungs and moist skin belong to which group?
   A) amphibians **C)** reptiles
   B) fish **D)** lizards
16. How can crocodiles be distinguished from alligators?
   A) care of the young
   B) scales on the back
   C) shape of the head
   D) habitats in which they live

Science Online life.msscience.com/vocabulary_puzzlemaker

Thinking Critically

17. **Infer** Populations of frogs and toads are decreasing in some areas. What effects could this decrease have on other animal populations?

18. **Explain** why some amphibians are considered to be biological indicators.

19. **Compare and contrast** the ways tunicates and lancelets are similar to humans.

20. **Describe** the physical features common to all vertebrates.

21. **Compare and contrast** endotherms and ectotherms.

22. **Explain** how the development of the amniotic egg led to the success of early reptiles.

23. **Communicate** In your Science Journal, sequence the order in which these structures appeared in evolutionary history, then explain what type of organism had this adaptation and the advantage it provided: skin has mucous glands; skin has scales; dry, scaly skin.

24. **Compare and Contrast** Copy and complete this chart that compares the features of some vertebrate groups.

Vertebrate Groups

Feature	Fish	Amphibians	Reptiles
Heart			
Respiratory organ(s)	Do not write in this book.		
Reproduction requires water			

25. **Explain** how a fish uses its swim bladder.

26. **Identify and Manipulate Variables and Controls** Design an experiment to find out the effect of water temperature on frog egg development.

27. **Classify** To what animal group does an animal with a two-chambered heart belong?

28. **Identify** why it is necessary for a frog to live in a moist environment.

Performance Activities

29. **Conduct a Survey** Many people are wary of reptiles. Write questions about reptiles to find out how people feel about these animals. Give the survey to your classmates, then graph the results and share them with your class.

30. **Display** Cut out pictures of fish from magazines and mount them on poster board. Letter the names of each fish on 3-in × 5-in cards. Have your classmates try to match the names of the fish with their pictures. To make this activity more challenging, use only the scientific names of each fish.

Applying Math

Use the table below to answer questions 31 and 32.

Fish Species

Kinds of Fish	Number of Species
Jawless	45
Jawed cartilaginous	500
Bony	20,000

31. **Fish Species** Make a circle graph of the species of fish in the table above.

32. **Fish Percentages** What percent of fish species is bony fish? Jawed cartilaginous? Jawless?

chapter 14 Standardized Test Practice

Part 1 Multiple Choice

Record your answers on the answer sheet provided by your teacher or on a sheet of paper.

1. What are fins attached to?
   A. ectoskeleton **C.** endoskeleton
   B. notochord **D.** spine

2. How many chambers does a fish heart contain, and does it carry more or less oxygen than other types of hearts?
   A. two, less **C.** three, less
   B. four, less **D.** four, more

Use the photo below to answer question 3.

3. What type of fish is shown in this picture?
   A. bony
   B. jawed cartilaginous
   C. large-mouth bass
   D. jawless

4. How do amphibians exchange carbon dioxide and oxygen?
   A. lungs only **C.** gills only
   B. lungs and skin **D.** lungs and gills

5. How do frogs and toads hear?
   A. eardrum
   B. tympanic membrane
   C. skin
   D. tongue

6. Fish and amphibians do not have this type of egg so they must reproduce near water.
   A. external **C.** porous
   B. membranous **D.** amniotic

Use the photos below to answer questions 7 and 8.

7. What is the developmental process shown in this diagram?
   A. metamorphosis
   B. respiration
   C. ectotherm
   D. asexual reproduction

8. Where does this transition take place?
   A. land to air **C.** water to land
   B. air to land **D.** land to water

9. What is one way to distinguish a crocodile from an alligator?
   A. the shape of the snout
   B. number of eggs in nest
   C. size of teeth
   D. placement of nostrils

10. What are turtles missing that all other reptiles have?
    A. hair
    B. three-chambered heart
    C. teeth
    D. shelled eggs

Test-Taking Tip

Marking on Tests Be sure to ask if it is okay to mark in the test booklet when taking the test, but make sure you mark all of the answers on your answer sheet.

Question 6 Cross out answers you know are wrong or circle answers you know are correct. This will help you narrow your choices.

Part 2 | Short Response/Grid In

Record your answers on the answer sheet provided by your teacher or on a sheet of paper.

Use the illustration below to answer question 11.

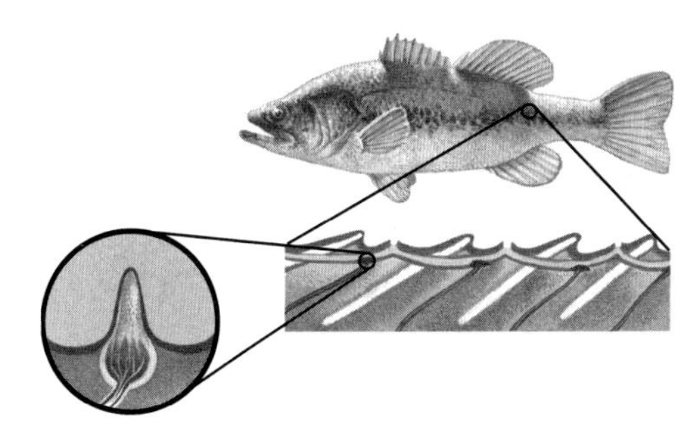

11. Describe the body system of the fish shown in this diagram. Why is it important to the fish?
12. What organs do fish have for the exchange of carbon dioxide and oxygen? How does the exchange take place?
13. How does a fish's swim bladder regulate its depth in water?
14. What is the difference between hibernation and estivation?
15. As an amphibian goes through metamorphosis, how do their heart and lungs change?
16. What is one possible reason for the decline in the number of reptiles in swamps and coastal areas?
17. What is the relationship between the number of young produced and the amount of care given by the parents in fish, amphibians, and reptiles?
18. How do snakes hear?
19. Are reptiles endothermic or ectothermic? Can reptiles modify their body temperature?
20. During chordate development, what structures originate from the nerve chord?

Part 3 | Open Ended

Record your answers on a sheet of paper.

21. Describe the composition and function of fish scales.
22. How did amphibians evolve and why were they the dominant land animals for a period of time?
23. Compare and contrast the circulatory systems of fish, amphibians, and reptiles. Which system provides the most oxygenated blood to the organs?

Use the illustration below to answer question 24.

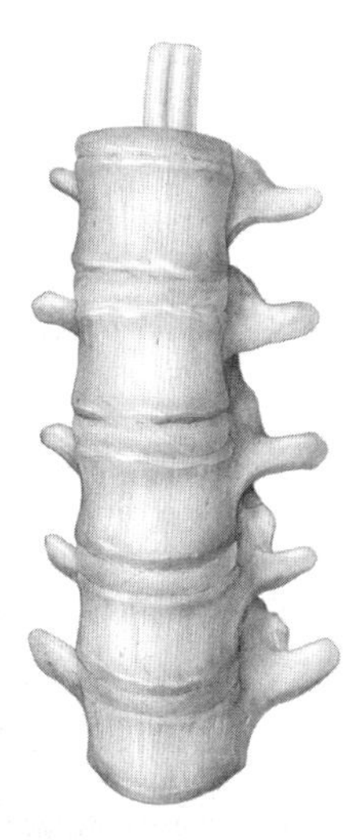

24. Explain the composition and significance of the structure in this diagram.
25. Discuss the structure of a turtle's shell and what other body parts are attached to it.
26. What are two classifications of how organisms regulate body temperature? How does regulation of body temperature help to determine the climate in which an organism is found?
27. What are pharyngeal pouches and what animal group has them at some point during development? How has their function changed over time?

chapter 15

Birds and Mammals

The BIG Idea

Birds and mammals are both diverse groups of animals.

SECTION 1
Birds

Main Idea Birds have feathers, lay eggs, and live in many different ecosystems.

SECTION 2
Mammals

Main Idea Mammals share some characteristics, such as having hair, producing milk, and caring for their young.

More Alike than Not!

Birds and mammals have adaptations that allow them to live on every continent and in every ocean. Some of these animals have adapted to withstand the coldest or hottest conditions. These adaptations help to make these animal groups successful.

Science Journal List similar characteristics of a mammal and a bird. What characteristics are different?

Start-Up Activities

Bird Gizzards

You may have observed a variety of animals in your neighborhood. Maybe you have watched birds at a bird feeder. Birds don't chew their food because they don't have teeth. Instead, many birds swallow small pebbles, bits of eggshells, and other hard materials that go into the gizzard—a muscular digestive organ. Inside the gizzard, they help grind up the seeds. The lab below models the action of a gizzard.

1. Place some cracked corn, sunflower seeds, nuts or other seeds, and some gravel in an old sock.
2. Roll the sock on a hard surface and tightly squeeze it.
3. Describe the appearance of the seeds after rolling.
4. **Think Critically** Describe in your Science Journal how a bird's gizzard helps digest the bird's food.

Preview this chapter's content and activities at life.msscience.com

FOLDABLES™ Study Organizer

Birds and Mammals Make the following Foldable to help you organize information about the behaviors of birds and mammals.

STEP 1 **Fold** one piece of paper widthwise into thirds.

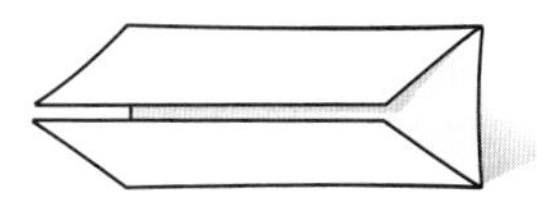

STEP 2 **Fold** down 2.5 cm from the top. (Hint: From the tip of your index finger to your middle knuckle is about 2.5 cm.)

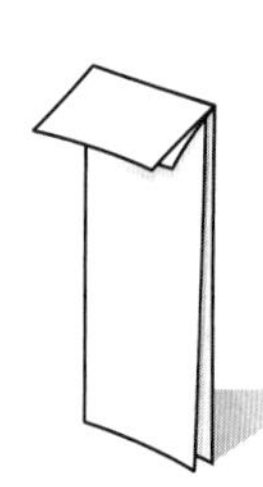

STEP 3 **Fold** the rest into fifths.

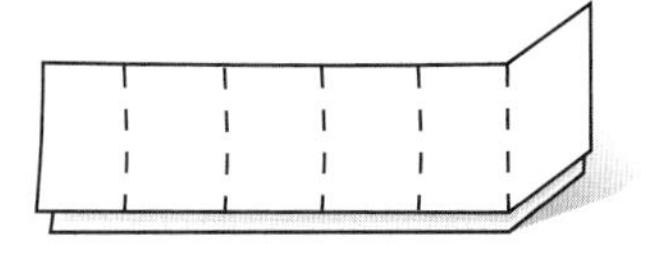

STEP 4 **Unfold,** lay the paper lengthwise, and draw lines along the folds. **Label** your table as shown.

	Birds	Mammals
Habitats		
Diet		
Movement		
Body Systems		
Young		

Make a Table As you read the chapter, complete the table describing the behaviors of birds and mammals.

Get Ready to Read

Identify the Main Idea

1 Learn It! Main ideas are the most important ideas in a paragraph, a section, or a chapter. Supporting details are facts or examples that explain the main idea. Understanding the main idea allows you to grasp the whole picture.

2 Picture It! Read the following paragraph. Draw a graphic organizer like the one below to show the main idea and supporting details.

> All adult mammals have hair on their bodies. It may be thick fur that covers all or part of the animal's body, or just a few hairs around the mouth. Fur traps air and helps keep the animal warm. Whiskers located near the mouth help many mammals sense their environments. Whales have almost no hair. They rely on a thick layer of fat under the skin, called blubber, to keep them warm. Porcupine quills and hedgehog spines are modified hairs that offer protection from predators.
>
> —*from page 438*

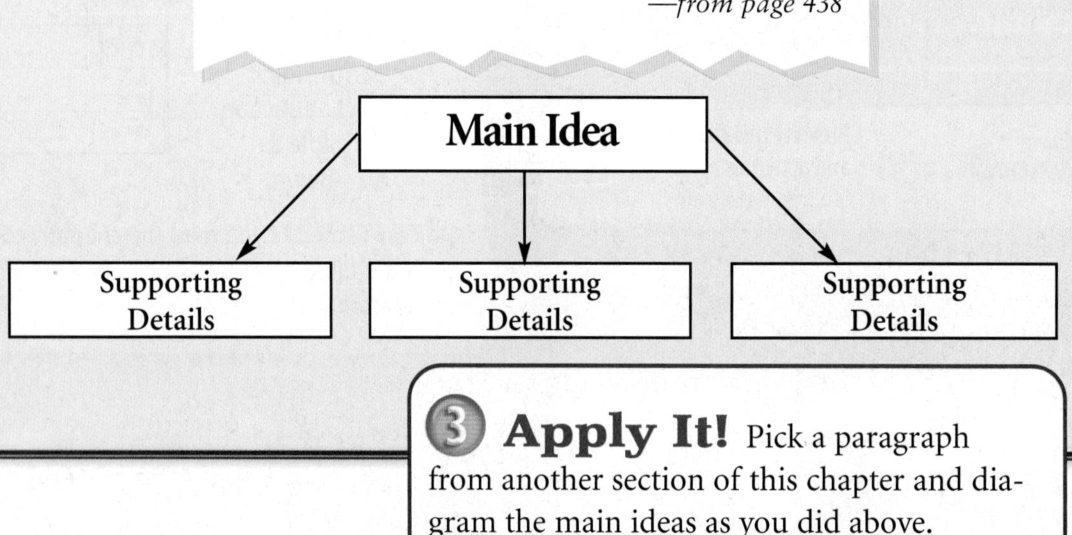

3 Apply It! Pick a paragraph from another section of this chapter and diagram the main ideas as you did above.

The main idea is often the first sentence in a paragraph but not always.

Target Your Reading

Use this to focus on the main ideas as you read the chapter.

1. **Before you read** the chapter, respond to the statements below on your worksheet or on a numbered sheet of paper.
   - Write an **A** if you **agree** with the statement.
   - Write a **D** if you **disagree** with the statement.

2. **After you read** the chapter, look back to this page to see if you've changed your mind about any of the statements.
   - If any of your answers changed, explain why.
   - Change any false statements into true statements.
   - Use your revised statements as a study guide.

Before You Read A or D		Statement	After You Read A or D
	1	All birds can fly.	
	2	Many birds have hollow bones.	
	3	Birds take in oxygen both while inhaling and exhaling.	
	4	Birds swallow small stones to help them digest their food.	
	5	All mammals have glands in their skin, though they are not important.	
	6	All mammals have the same number and type of teeth.	
	7	All mammals give birth to live young.	
	8	Mammals have well-developed brains.	

Science Online
Print out a worksheet of this page at life.msscience.com

section 1

Birds

as you read

What You'll Learn

- **Identify** the characteristics of birds.
- **Identify** the adaptations birds have for flight.
- **Explain** how birds reproduce and develop.

Why It's Important

Most birds demonstrate structural and behavioral adaptations for flight.

Review Vocabulary

thrust: for an object moving through air, the horizontal force that pushes or pulls the object forward

New Vocabulary

- contour feather
- down feather
- endotherm
- preening

Bird Characteristics

Birds are versatile animals. Geese have been observed flying at an altitude of 9,000 m, and penguins have been seen underwater at a depth of 543 m. An ostrich might weigh 155,000 g, while a hummingbird might weigh only 2 g. Some birds can live in the tropics and others can live in polar regions. Their diets vary and include meat, fish, insects, fruit, seeds, and nectar. Birds have feathers and scales and they lay eggs. Which of these characteristics is unique to birds?

Bird Eggs Birds lay amniotic (am nee AH tihk) eggs with hard shells, as shown in **Figure 1.** This type of egg provides a moist, protective environment for the developing embryo. The hard shell is made of calcium carbonate, the same chemical that makes up seashells, limestone, and marble. The egg is fertilized internally before the shell forms around it. The female bird lays one or more eggs usually in some type of nest, also shown in **Figure 1.** A group of eggs is called a clutch. One or both parents may keep the eggs warm, or incubate them, until they hatch. The length of time for incubation varies from species to species. The young are cared for by one or both parents.

Figure 1 This robin's round nest is built of grasses and mud in a tree.

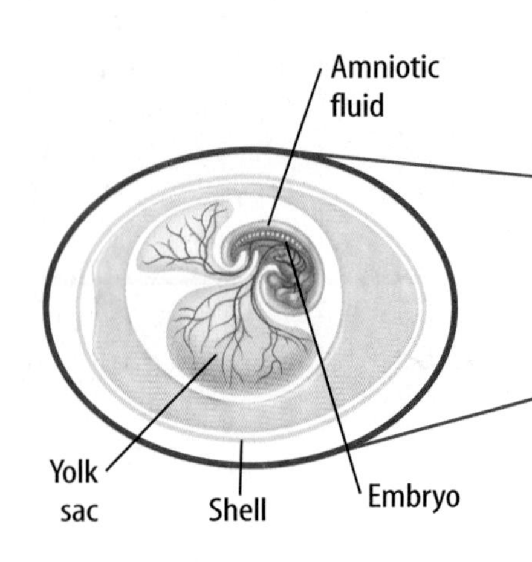

Sternum The sternum has a structure called a keel, which is where flight muscles attach.
Tail A bird does not have a bony tail.
Leg bone
Hollow leg bone

Figure 2 The hollow bones of birds are an adaptation for flight.
Infer *what advantages thin cross braces provide.*

Flight Adaptations People have always been fascinated by the ability of birds to fly. Flight in birds is made possible by their almost hollow but strong skeleton, wings, feathers, strong flight muscles, and an efficient respiratory system. Well-developed senses, especially eyesight, and tremendous amounts of energy also are needed for flight.

Hollow Bones One adaptation that birds have for flight is a unique internal skeleton, as shown in **Figure 2.** Many bones of some birds are joined together. This provides more strength and more stability for flight. Most bones of birds that fly are almost hollow. These bones have thin cross braces inside that also strengthen the bones. The hollow spaces inside of the bones are filled with air.

Reading Check *What features strengthen a bird's bones?*

A large sternum, or breastbone, supports the powerful chest muscles needed for flight. The last bones of the spine support the tail feathers, which play an important part in steering and balancing during flight and landing.

INTEGRATE
Astronomy

Star Navigation Many theories have been proposed about how birds navigate at night. Some scientists hypothesize that star positions help night-flying birds find their way. Research the location of the North Star. In your Science Journal, infer how the North Star might help birds fly at night.

Modeling Feather Function

Procedure

1. Wrap **polyester fiber** or **cotton** around the bulb of an **alcohol thermometer.** Place it into a **plastic bag.** Record its temperature in your **Science Journal.**
2. Place a second **alcohol thermometer** into a **plastic bag** and record its temperature.
3. Simultaneously submerge the thermometers into a **container** of **cold water,** keeping the top of each bag above the water's surface.
4. After 2 min, record the temperature of each thermometer.

Analysis

1. Which thermometer had the greater change in temperature?
2. Infer the type of feather that the fiber or cotton models.

Feathers Birds are the only animals that have feathers. Their bodies are covered with two main types of feathers—contour feathers and down feathers. Strong, lightweight **contour feathers** give a bird its coloring and smooth shape. These are also the feathers that a bird uses when flying. The contour feathers on the wings and tail help the bird steer and keep it from spinning out of control.

Have you ever wondered how ducks can swim in a pond on a freezing cold day and keep warm? Soft, fluffy **down feathers** provide an insulating layer next to the skin of adult birds and cover the bodies of young birds. Birds are **endotherms,** meaning they maintain a constant body temperature. Feathers help birds maintain their body temperature, and grow in much the same way as your hair grows. Each feather grows from a microscopic pit in the skin called a follicle (FAH lih kul). When a feather falls out, a new one grows in its place. As shown in **Figure 3,** the shaft of a feather has many branches called barbs. Each barb has many branches called barbules that give the feather strength.

Reading Check *Why are some young birds covered with down feathers?*

A bird has an oil gland located just above the base of its tail. Using its bill or beak, a bird rubs oil from the gland over its feathers in a process called **preening.** The oil conditions the feathers and helps them last longer.

Figure 3 Down feathers help keep birds warm. Contour feathers are the feathers used for flight, and the feathers that cover the body.

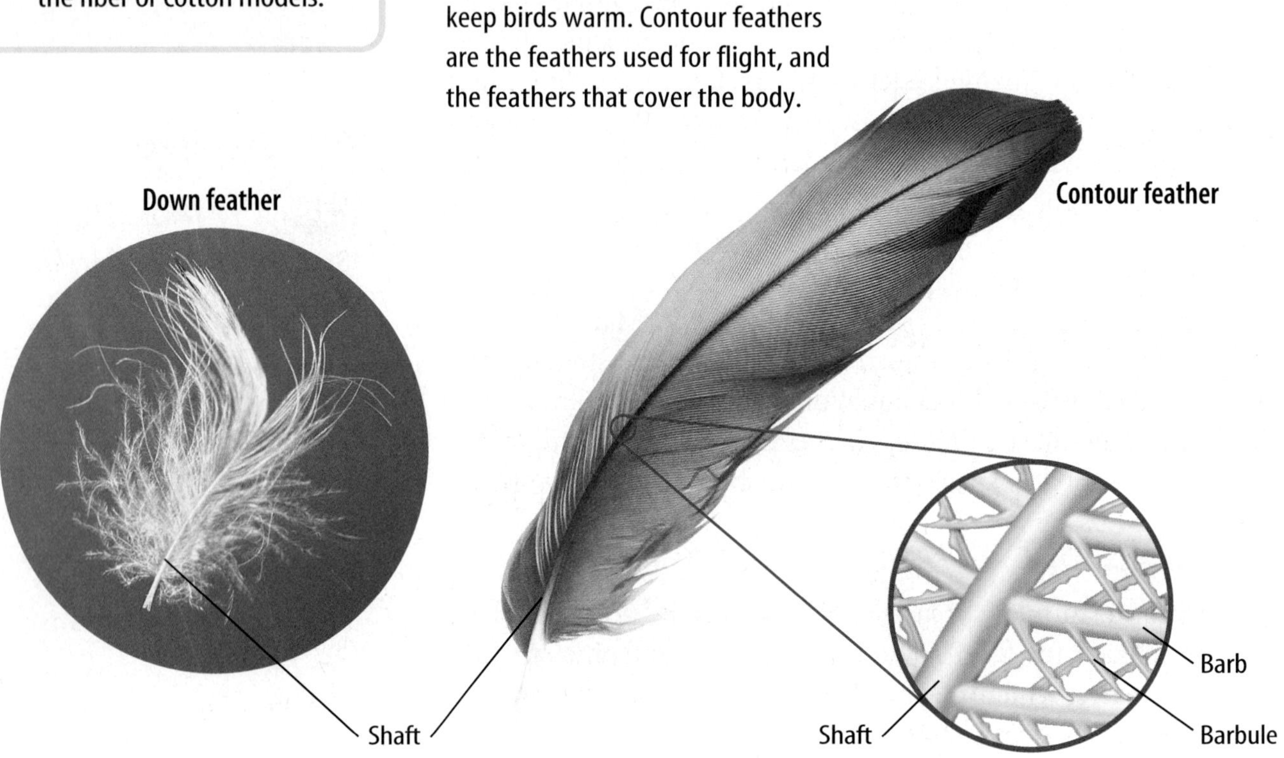

Figure 4 Wings provide an upward force called lift for birds and airplanes.
Describe *how birds are able to fly.*

Bald eagles are able to soar for long periods of time because their wings have a large surface area to provide lift.

This glider gets lift from its wings the same way a bald eagle gets lift.

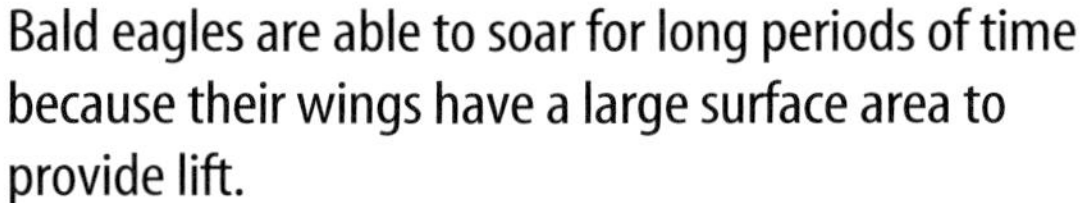

Wings Although not all birds fly, most wings are adapted for flight. Wings are attached to powerful chest muscles. By flapping its wings, a bird attains thrust to go forward and lift to stay in the air. Its wings move up and down, as well as back and forth.

The shape of a bird's wings helps it fly. The wings are curved on top and flat or slightly curved on the bottom. Humans copied this shape to make airplane wings, as shown in **Figure 4.** When a bird flies, air moves more slowly across the bottom than across the top of its wings. Slow-moving air has greater pressure than fast-moving air, resulting in an upward push called lift. The amount of lift depends on the total surface area of the wing, the speed at which air moves over the wing, and the angle of the wing to the moving air. Once birds with large wings, such as vultures, reach high altitudes, they can soar and glide for a long time without having to beat their wings.

Wings also serve important functions for birds that don't fly. Penguins are birds that use their wings to swim underwater. Ostriches use their wings in courtship and to maintain their balance while running or walking.

Bird Pests Some birds have become pests in urban areas. Research to learn what birds are considered pests in urban areas, what effect they have on the urban environment, and what measures are taken to reduce the problems they create. Build a bulletin board showing your results.

Body Systems

Whether they fly, swim, or run, most birds are extremely active. Their body systems are adapted for these activities.

Topic: Homing Pigeons

Visit life.msscience.com for Web links to information about homing pigeons.

Activity Name two past uses for homing pigeons, and write an advertisement for a new use in the future.

Digestive System Because flying uses large amounts of energy, birds need large amounts of high energy foods, such as nuts, seeds, nectar, insects, and meat. Food is broken down quickly in the digestive system to supply this energy. In some birds, digestion can take less than an hour—for humans digestion can take more than a day.

From a bird's mouth, unchewed food passes into a digestive organ called the crop. The crop stores the food until it absorbs enough moisture to move on. The food enters the stomach where it is partially digested before it moves into the muscular gizzard. In the gizzard, food is ground and crushed by small stones and grit that the bird has swallowed. Digestion is completed in the intestine, and then the food's nutrients move into the bloodstream.

Respiratory System Body heat is generated when energy in food is combined with oxygen. A bird's respiratory system efficiently obtains oxygen, which is needed to power flight muscles and to convert food into energy. Birds have two lungs. Each lung is connected to balloonlike air sacs that reach into different parts of the body, including some of the bones. Most of the air inhaled by a bird passes into the air sacs behind the lungs. When a bird exhales, air with oxygen passes from these air sacs into the lungs. Air flows in only one direction through a bird's lungs. Unlike other vertebrates, birds receive air with oxygen when they inhale and when they exhale. This provides a constant supply of oxygen for the flight muscles.

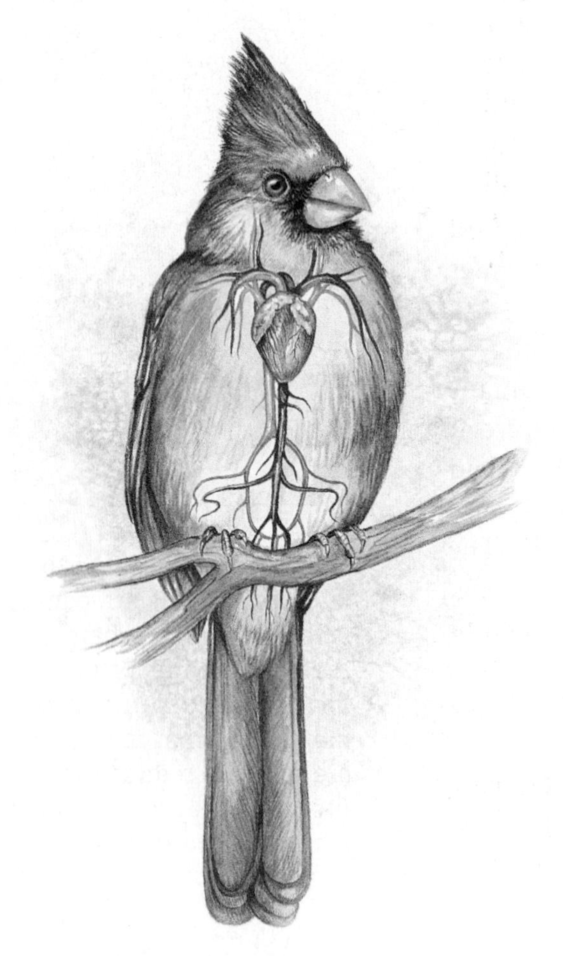

Figure 5 A bird's blood is circulated quickly so enough oxygen-filled blood is carried to the bird's muscles.

Circulatory System A bird's circulatory system consists of a heart, arteries, capillaries, and veins, as shown in **Figure 5.** Their four-chambered heart is large compared to their body. On average, a sparrow's heart is 1.68 percent of its body weight. The average human heart is only 0.42 percent of the human's body weight. Oxygen-filled blood is kept separate from carbon dioxide-filled blood as both move through a bird's heart and blood vessels. A bird's heart beats rapidly—an active hummingbird's heart can beat more than 1,000 times per minute.

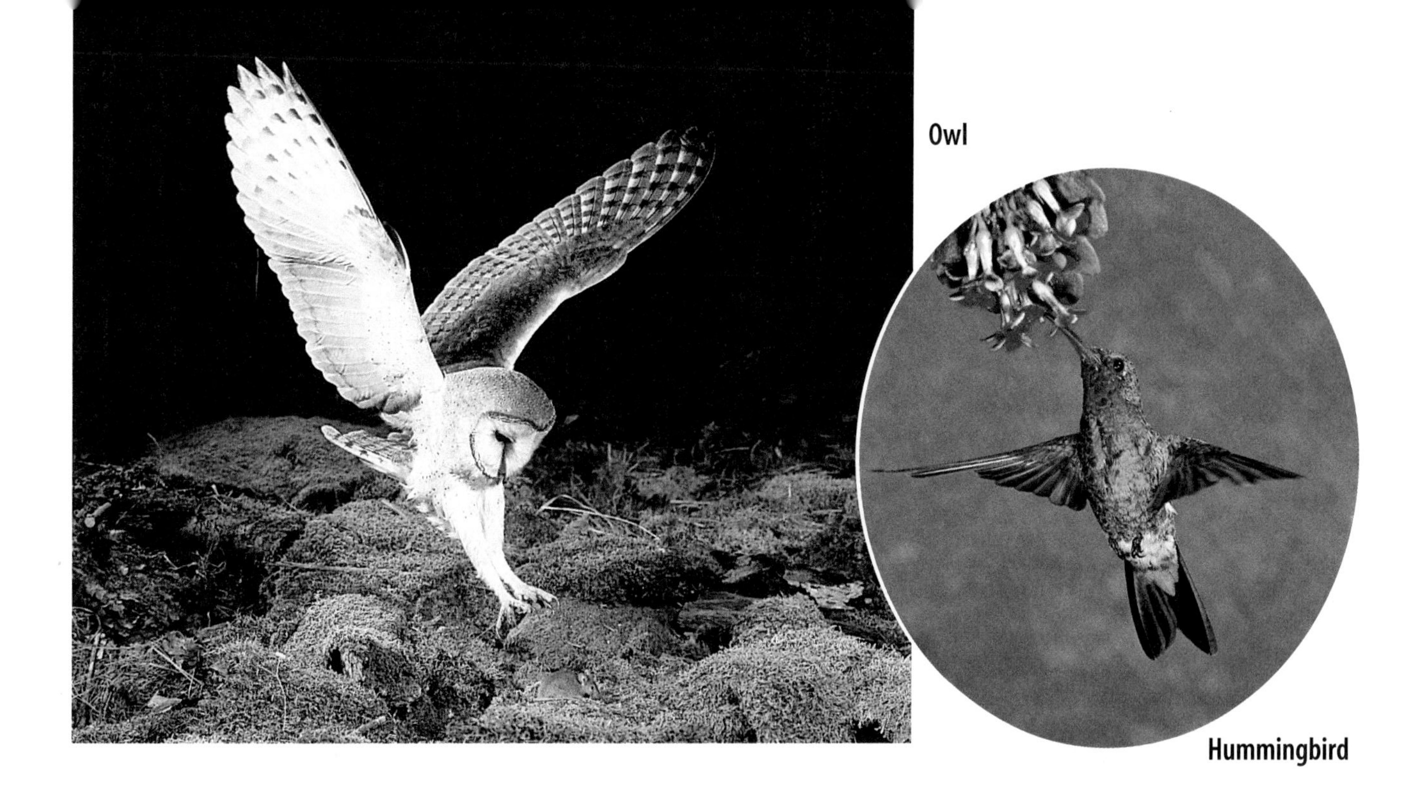

Figure 6 In nature, some birds, like the owl on the left, help control pests. Others, like the hummingbird above, pollinate flowers. **Identify** *other important uses of birds.*

The Importance of Birds

Birds play important roles in nature. Some are sources of food and raw materials, and others are kept as pets. Some birds, like the owl in **Figure 6,** help control pests, such as destructive rodents. Barn swallows and other birds help keep insect populations in check by eating them. Some birds, like the hummingbird in **Figure 6,** are pollinators for many flowers. As they feed on the flower's nectar, pollen collects on their feathers and is deposited on the next flower they visit. Other birds eat fruits, then their seeds are dispersed in the birds' droppings. Seed-eating birds help control weeds. Birds can be considered pests when their populations grow too large. In cities where large numbers of birds roost, their droppings can damage buildings. Some droppings also can contain microorganisms that can cause diseases in humans.

Uses of Birds Humans have hunted birds for food and fancy feathers for centuries. Eventually, wild birds such as chickens and turkeys were domesticated and their meat and eggs became a valuable part of human diets. Feathers are used in mattresses and pillows because of their softness and ability to be fluffed over and over. Down feathers are good insulators. Even bird droppings, called guano (GWAH noh), are collected from seabird colonies and used as fertilizer.

Parakeets, parrots, and canaries often are kept as pets because many sing or can be taught to imitate sounds and human voices. Most birds sold as pets are bred in captivity, but some wild birds still are collected illegally, which threatens many species.

NATIONAL GEOGRAPHIC VISUALIZING BIRDS

Figure 7

There are almost 9,000 living species of birds. Birds are subdivided into smaller groups based on characteristics such as beak size and shape, foot structure, and diet. Birds belonging to several groups are shown here.

INSECT EATERS This nuthatch has a pointed beak that can pry up bark or bore into wood to find insects.

BIRDS OF PREY This osprey has large claws that grasp and a sharp beak that tears flesh.

WATERBIRDS Wood ducks have webbed feet that propel them through the water.

FLIGHTLESS BIRDS The ostrich evolved in places where there were once few mammal predators. Though they cannot fly, some flightless birds are fast runners.

WADING BIRDS The great blue heron's long legs allow it to walk in shallow water.

SEED EATERS This cardinal's thick, strong beak can crack seeds.

Origin of Birds Birds, like those in **Figure 7,** have some characteristics of reptiles, including scales on their feet and legs. Scientists learn about the origins of most living things by studying their fossils; however, few fossils of birds have been found. Some scientists hypothesize that birds developed from reptiles millions of years ago.

Archaeopteryx (ar kee AHP tuh rihks)—a birdlike fossil—is about 150 million years old. Although it is not known that *Archaeopteryx* was a direct ancestor of modern birds, evidence shows that it had feathers and wings similar to modern birds. However, it had solid bones, teeth, a long bony tail, and clawed front toes, like some reptiles.

In 1991 in Texas, scientists discovered a fossil that had hollow bones and a well-developed sternum with a keel. *Protoavis* (proh toh AY vihs) lived about 225 million years ago. No fossil feathers were found with *Protoavis.* Scientists do not know if it was an ancestor of modern birds or a type of ground-living dinosaur. **Figure 8** shows an artist's idea of what *Archaeopteryx* and *Protoavis* may have looked like.

Figure 8 The first *Archaeopteryx* fossil was found more than 100 years ago. *Archaeopteryx,* to the left, is considered a link between reptiles and birds. *Protoavis,* below, may be an ancestor of birds.

section 1 review

Summary

Bird Characteristics

- Birds are feathered vertebrates. Females lay hard-shelled amniotic eggs, which hatch after an incubation period.
- Most birds are capable of flight, due to a hollow strong skeleton, wings, feathers, strong flight muscles, and efficient body systems.

Body Systems

- Birds have no teeth, but grind food in their gizzard.
- Birds take in oxygen while inhaling and exhaling, which provides more oxygen for flight.

Importance of Birds

- Birds pollinate plants, are a food source, keep pest populations low, and are used commercially.

Self Check

1. **Describe** the type of feather that helps birds maintain their body temperature.
2. **Sequence** Make a network-tree concept map about birds using the following terms: *birds, beaks, hollow bones, wings, eggs, adaptations for flight, feathers,* and *air sacs.*
3. **Think Critically** Hypothesize why most birds eat nuts, berries, insects, nectar, or meat, but not grass and leaves.

Applying Skills

4. **Venn Diagram** Draw a Venn diagram to compare and contrast the characteristics of birds that fly and birds that do not fly.
5. **Communicate** Many expressions mention birds, such as "proud as a peacock" and "wise as an owl." Make a list of several of these expressions and then decide which are accurate.

Science Online life.msscience.com/self_check_quiz

section 2

Mammals

as you read

What You'll Learn

- **Identify** the characteristics of mammals and explain how they have enabled mammals to adapt to different environments.
- **Distinguish** among monotremes, marsupials, and placentals.
- **Explain** why many species of mammals are becoming threatened or endangered.

Why It's Important

Mammals, including humans, have many characteristics in common.

Review Vocabulary
gland: a cell or group of cells that releases fluid

New Vocabulary
- mammal
- mammary gland
- omnivore
- carnivore
- herbivore
- monotreme
- marsupial
- placental
- gestation period
- placenta
- umbilical cord

Characteristics of Mammals

You probably can name dozens of mammals, but can you list a few of their common characteristics? **Mammals** are endothermic vertebrates that have hair and produce milk to feed their young, as shown in **Figure 9.** Like birds, mammals care for their young. Mammals can be found almost everywhere on Earth. Each mammal species is adapted to its unique way of life.

Skin and Glands Skin covers and protects the bodies of all mammals. A mammal's skin is an organ that produces hair and in some species, horns, claws, nails, or hooves. The skin also contains different kinds of glands. One type of gland found in all mammals is the mammary gland. Female mammals have **mammary glands** that produce milk for feeding their young. Oil glands produce oil that lubricates and conditions the hair and skin. Sweat glands in some species remove wastes and help keep them cool. Many mammals have scent glands that secrete substances that can mark their territory, attract a mate, or be a form of defense.

Figure 9 Mammals, such as this moose, care for their young after they are born.
Explain *how mammals feed their young.*

Figure 10 Mammals have teeth that are shaped specifically for the food they eat.

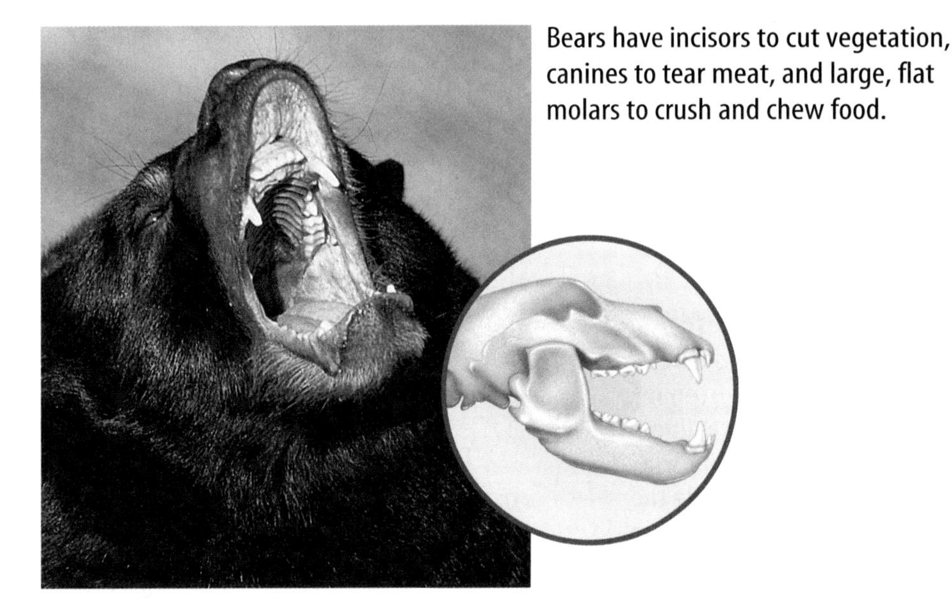

Bears have incisors to cut vegetation, canines to tear meat, and large, flat molars to crush and chew food.

A tiger easily can tear away the flesh of an animal because of large, sharp canine teeth and strong jaw muscles.

A horse's back teeth, called molars, are large. **Infer** *how a horse chews.*

Teeth Notice that each mammal in **Figure 10** has different kinds of teeth. Almost all mammals have specialized teeth. Scientists can determine a mammal's diet by examining its teeth. Front teeth, called incisors, bite and cut. Sometimes the teeth next to the incisors, called canine teeth, are well developed to grip and tear. Premolars and molars at the back of the mouth shred, grind, and crush. Animals, like the bear in **Figure 10,** and humans, have all four kinds of teeth. They eat plants and other animals, so they are called **omnivores.** A **carnivore,** like the tiger in **Figure 10,** has large canine teeth and eats only the flesh of other animals. **Herbivores,** such as the horse in **Figure 10,** eat only plants. Their large premolars and molars grind the tough fibers in plants.

Inferring How Blubber Insulates

Procedure

1. Fill a **self-sealing plastic bag** about one-third full with **vegetable shortening.**
2. Turn another self-sealing plastic bag inside out. Carefully place it inside the bag with the shortening so that you are able to seal one bag to the other. This is a blubber mitten.
3. Put your hand in the blubber mitten and place it in **ice water** for 5 s. Remove the blubber mitten when finished.
4. Put your bare hand in the same bowl of ice water for 5 s.

Analysis

1. Which hand seemed colder?
2. Infer how a layer of blubber provides protection against cold water.

Try at Home

Hair All adult mammals have hair on their bodies. It may be thick fur that covers all or part of the animal's body, or just a few hairs around the mouth. Fur traps air and helps keep the animal warm. Whiskers located near the mouth help many mammals sense their environments. Whales have almost no hair. They rely on a thick layer of fat under their skin, called blubber, to keep them warm. Porcupine quills and hedgehog spines are modified hairs that offer protection from predators.

Body Systems

The body systems of mammals are adapted to their activities and enable them to survive in many environments.

Mammals have four-chambered hearts that pump oxygen-filled blood directly throughout the body in blood vessels. Mammals have lungs made of millions of microscopic sacs. These sacs increase the lungs' surface area, allowing a greater exchange of carbon dioxide and oxygen.

A mammal's nervous system consists of a brain, spinal cord, and nerves. In mammals, the part of the brain involved in learning, problem solving, and remembering is larger than in other animals. Another large part of the mammal brain controls its muscle coordination.

The digestive systems of mammals vary according to the kinds of food they eat. Herbivores, like the one shown in **Figure 11,** have long digestive tracts compared to carnivores because plants take longer to digest than meat.

Figure 11 Herbivores, like this elk, have four-chambered stomachs and long intestinal tracts that contain microorganisms, which help break down the plant material.
Explain *why herbivores need a longer digestive system than carnivores.*

Reproduction and Caring for Young All mammals reproduce sexually. Most mammals give birth to live young after a period of development inside the female reproductive organ called the uterus. Many mammals are nearly helpless, and sometimes even blind, when they are born. They can't care for themselves for the first several days or even years. If you've seen newborn kittens or human babies, you know they just eat, sleep, grow, and develop. However, the young of some mammals, such as antelope, deer, and elephants, are well developed at birth and are able to travel with their constantly moving parents. These young mammals usually can stand by the time they are a few minutes old. Marine mammals, such as the whales, shown in **Figure 12,** can swim as soon as they are born.

Figure 12 When a whale is born, the female whale must quickly push the newborn whale to the water's surface to breathe. Otherwise, the newborn whale will drown.

Is a house cat or a deer more developed at birth?

During the time that young mammals are dependent on their female parent's milk, they learn many of the skills needed for their survival. Defensive skills are learned while playing with other young of their own kind. Other skills are learned by imitating adults. In many mammal species only females raise the young. Males of some species, such as wolves and humans, help provide shelter, food, and protection for their young.

Applying Science

Does a mammal's heart rate determine how long it will live?

Some mammals live long lives, but other mammals live for only a few years. Do you think that a mammal's life span might be related to how fast its heart beats? Use your ability to interpret a data table to answer this question.

Identifying the Problem

The table on the right lists the average heart rates and life spans of several different mammals. Heart rate is recorded as the number of heartbeats per minute, and life span is recorded as the average maximum years. When you examine the data, look for a relationship between the two variables.

Mammal Heart Rates and Life Spans

Mammal	Heart Rate (beats/min)	Life Span (years)
Mouse	400	2
Large dog	80	15
Bear	40	15–20
Elephant	25	75

Solving the Problem

1. Infer how heart rate and life span are related in mammals.
2. Humans have heart rates of about 70 beats per minute. Some humans may live for more than 100 years. Is this consistent with the data in the table? Explain.

Types of Mammals

Mammals are classified into three groups based on how their young develop. The three mammal groups are monotremes (MAH nuh treemz), marsupials (mar SEW pee ulz), and placentals (pluh SEN tulz).

Monotremes The duck-billed platypus, shown in **Figure 13,** is a monotreme. **Monotremes** are mammals that lay eggs with leathery shells. The female incubates the eggs for about ten days. After the young hatch, they nurse by licking the female's skin and hair where milk oozes from the mammary glands. Monotreme mammary glands do not have nipples.

Figure 13 A duck-billed platypus is a mammal, yet it lays eggs. **Explain** *why the duck-billed platypus is classified as a mammal.*

Marsupials Many of the mammals that are classified as marsupials live in Australia, New Guinea, or South America. Only one type of marsupial, the opossum, lives in North America. **Marsupials** give birth to immature young that usually crawl into an external pouch on the female's abdomen. However, not all marsupials have pouches. Whether an immature marsupial is in a pouch or not, it instinctively crawls to a nipple. It stays attached to the nipple and feeds until it is developed. In pouched marsupials, the developed young return to the pouch for feeding and protection. Examples of marsupials are kangaroos and opossums, as shown in **Figure 14,** wallabies, koalas, bandicoots, and Tasmanian devils.

Figure 14 Opossums are the only marsupials found in North America. A joey, or young kangaroo, returns to its mother's pouch when danger is near.

Placentals In **placentals,** embryos completely develop inside the female's uterus. The time during which the embryo develops in the uterus is called the **gestation period.** Gestation periods range from 16 days in hamsters to 650 days in elephants. Placentals are named for the **placenta,** an organ that develops from tissues of the embryo and tissues that line the inside of the uterus. The placenta absorbs oxygen and food from the mother's blood. An umbilical cord connects the embryo to the placenta, as shown in **Figure 15.** Several blood vessels make up the umbilical cord. Blood in the **umbilical cord** transports food and oxygen from the placenta to the embryo and removes waste products from the embryo. The female parent's blood doesn't mix with the embryo's blood. Examples of placentals are shown in **Table 1** on the following two pages.

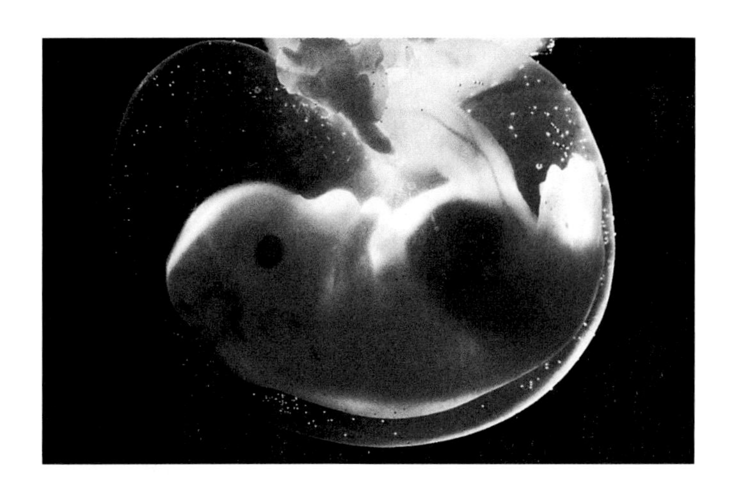

Figure 15 An unborn mammal receives nutrients and oxygen through the umbilical cord. **Compare and contrast** *placental, marsupial, and monotreme development.*

How does an embryo receive the things it needs to grow?

Some placental groups include unusual animals such as the manatee shown in **Figure 16.** Dugongs and manatees are aquatic mammals. They have no back legs, and their front legs are modified into flippers. Another group includes small, rabbit-like animals called hyraxes that have hooves and molars for grinding vegetation. The aardvark is the only member of its group. Aardvarks have tubelike teeth and dig termites for food. Many Southeast Asian islands are home to members of a group that includes gliding lemurs. Pangolins, another group of placentals, look like anteaters covered with scales.

Figure 16 A manatee swims slowly below the surface of the water.

Topic: Manatee Habitats
Visit life.msscience.com for Web links to recent news or magazine articles about manatees and their habitats.

Activity Create a pamphlet about the manatees' habitat, their threats, and what people can do to help them.

Table 1 Placentals

Order	Examples		Major Characteristics
Rodentia (roh DEN chuh)	beavers, mice, rats, squirrels		one pair of chisel-like front teeth adapted for gnawing; incisors grow throughout life; herbivores
Chiroptera (ki RAHP tuh ruh)	bats		front limbs adapted for flying; active at night; different species feed on fruit, insects, fish, or other bats
Insectivora (ihn sek TIH vuh ruh)	moles, shrews, hedgehogs		small; feed on insects, earthworms, and other small animals; most have long skulls and narrow snouts; high metabolic rate
Carnivora (kar NIH vuh ruh)	cats, dogs, bears, foxes, raccoons		long, sharp canine teeth for tearing flesh; most are predators, some are omnivores
Primates (PRI maytz)	apes, monkeys, humans		arms with grasping hands and opposable thumbs; eyes are forward facing; large brains; omnivores
Artiodactyla (ar tee oh DAHK tih luh)	deer, moose, pigs, camels, giraffes, cows		hooves with an even number of toes; most are herbivores with large, flat molars; complex stomachs and intestines

Order	Examples		Major Characteristics
Cetacea (sih TAY shuh)	whales, dolphins, porpoises		one or two blowholes on top of the head for breathing; forelimbs are modified into flippers; teeth or baleen
Lagomorpha (la guh MOR fuh)	rabbits, hares, pikas		some with long hind legs adapted for jumping and running; one pair of large, upper incisors; one pair of small, peglike incisors
Pinnipedia (pih nih PEE dee uh)	sea lions, seals, walruses		marine carnivores; limbs modified for swimming
Edentata (ee dehn TAH tuh)	anteaters, sloths, armadillos		eat insects and other small animals; most are toothless or have tiny, peglike teeth
Perissodactyla (puh ris oh DAHK tih luh)	horses, zebras, tapirs, rhinoceroses		hooves with an odd number of toes; skeletons adapted for running; herbivores with large, grinding molars
Proboscidea (proh boh SIH dee uh)	elephants		a long nose called a trunk; herbivores; upper incisor teeth grow to form tusks; thick, leathery skin

Importance of Mammals

Mammals, like other organisms, are important in maintaining a balance in the environment. Carnivores, such as lions, help control populations of other animals. Bats help pollinate flowers and control insects. Other mammals pick up plant seeds in their fur and distribute them. However, mammals and other animals are in trouble today. As millions of wildlife habitats are destroyed for shopping centers, recreational areas, housing, and roads, many mammals are left without food, shelter, and space to survive. Because humans have the ability to reason, they have a responsibility to learn that their survival is related closely to the survival of all mammals. What can you do to protect the mammals in your community?

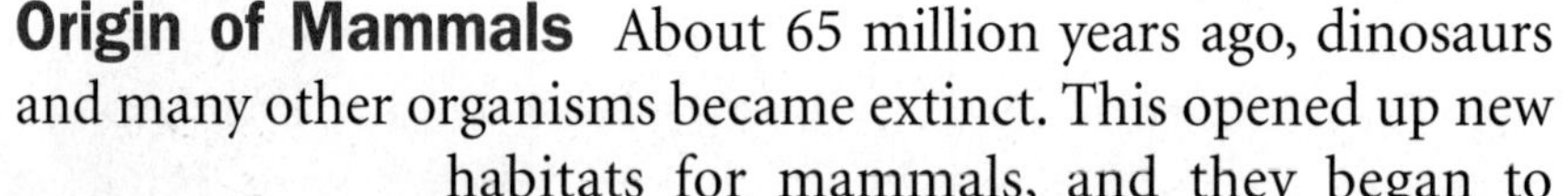

Origin of Mammals About 65 million years ago, dinosaurs and many other organisms became extinct. This opened up new habitats for mammals, and they began to branch out into many different species. Some of these species gave rise to modern mammals. Today, more than 4,000 species of mammals have evolved from animals similar to the one in **Figure 17,** which lived about 200 million years ago.

Figure 17 The *Dvinia* was an ancestor of ancient mammals.

section 2 review

Summary

Characteristics of Mammals

- Mammals have mammary glands, hair covering all or part of the body, and teeth specialized to the foods they eat.
- A mammal's body systems are well-adapted to the environment it lives in.

Types of Mammals

- There are three types of mammals: monotremes, which lay eggs; marsupials, which give birth to immature young that are nursed until developed, usually in a pouch; and placentals, which completely develop inside the female.

Importance of Mammals

- Mammals help maintain a balance in the environment. They are a food source, pollinators, and used commercially.

Self Check

1. **Describe** five characteristics of mammals and explain how they allow mammals to survive in different environments.
2. **Compare and contrast** birds and mammals.
3. **Describe** the differences between herbivores, carnivores, and omnivores.
4. **Classify** the following animals into the three mammal groups: whales, koalas, horses, elephants, opossums, kangaroos, rabbits, bats, bears, platypuses, and monkeys.
5. **Think Critically** How have humans contributed to the decrease in many wildlife populations?

Applying Math

6. **Solve One-Step Equations** The tallest land mammal is the giraffe at 5.6 m. Calculate your height in meters, and determine how many of you it would take to be as tall as the giraffe.

Mammal Footprints

Have you ever seen an animal footprint in the snow or soft soil? In this lab, you will observe pictures of mammal footprints and identify the mammal that made the footprint.

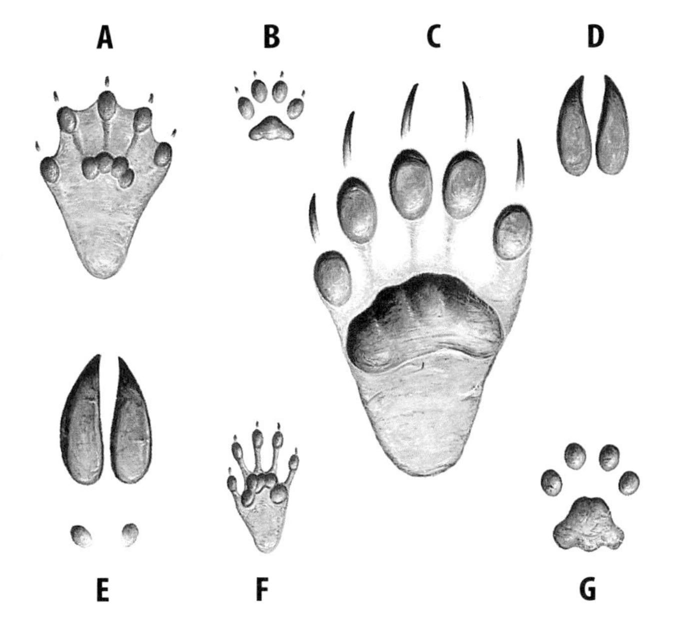

Real-World Question

How do mammal footprints differ?

Goals

- **Identify** mammal footprints.
- **Predict** where mammals live based on their footprints.

Materials

diagram of footprints

Procedure

1. Copy the following data table in your Science Journal.

Identifying Mammal Footprints

Animal	Letter of Footprint	Traits of Footprint
Bear		
Beaver		
Cougar	Do not write in this book.	
Coyote		
Deer		
Moose		
Raccoon		

2. Compare and contrast the different mammal footprints in the above diagram.
3. Based on your observations, match each footprint to an animal listed in the first column of the data table.
4. **Write** your answers in the column labeled *Letter of Footprint*. Complete the data table.

Conclude and Apply

1. Which mammals have hoofed feet?
2. Which mammals have clawed toes?
3. Which mammals have webbed feet?
4. **Explain** how the different feet are adapted to the areas in which these different mammals live.
5. What are the differences between track B and track E? How does that help you identify the track?

Communicating Your Data

Compare your conclusions with those of other students in your class. **For more help, refer to the Science Skill Handbook.**

Bird Counts

Goals
- **Research** how to build a bird feeder and attract birds to it.
- **Observe** the types of birds that visit your feeder.
- **Identify** the types of birds that you observe at your bird feeder.
- **Graph** your results and then communicate them to other students.

Data Source

Visit life.msscience.com/internet_lab for Web links to more information about how to build a bird feeder, hints on bird-watching, and data from other students who do this lab.

Safety Precautions

Real-World Question
What is the most common bird in your neighborhood? Think about the types of birds that you observe around your neighborhood. What types of food do they eat? Do all birds come to a bird feeder? Form a hypothesis about the type of bird that you think you will see most often at your bird feeder.

Cardinal

Make a Plan
1. **Research** general information about how to attract and identify birds. Determine where you will make your observations.
2. **Search** reference sources to find out how to build a bird feeder. Do all birds eat the same types of seeds?
3. **Select** the type of feeder you will build and the seed you will use based on your research.
4. What variables can you control in this lab? Do seasonal changes, length of time, or weather conditions affect your observations?
5. What will you do to identify the birds that you do not know?

Follow Your Plan
1. Make sure your teacher approves your plan before you start.
2. **Record** your data in your Science Journal each time you make an observation of the birds at your bird feeder.

American Goldfinch

Analyze Your Data

1. **Write** a description of where you placed your feeder and when you made your bird observations.
2. **Record** the total number of birds you observed each day.
3. **Record** the total number of each type of bird species you observed each day.
4. **Graph** your data using a line graph, a circle graph, or a bar graph.

Black-capped Chickadee

Conclude and Apply

1. **Interpret Data** What type of bird was most common to your feeder?
2. **Explain** if all of your classmates' data agree with yours. Why or why not?
3. **Review** your classmates' data and determine if the location of bird observations affected the number of birds observed.
4. **Infer** if the time of day affected the number of birds observed. Explain.
5. **Infer** Many birds eat great numbers of insects. What might humans do to maintain a healthy environment for birds?

Birds at a feeder

Communicating Your Data

Find this lab using the link below. Post your data in the table provided. **Compare** your data to those of other students. Combine your data with those of other students and plot the combined data on a map to recognize patterns in bird populations.

life.msscience.com/internet_lab

SCIENCE Stats

Eggciting Facts

Did you know...

...The ostrich lays the biggest egg of all birds now living. Its egg is 15 cm to 20 cm long and 10 cm to 15 cm wide. The volume of the ostrich egg is about equal to 24 chicken eggs. It can have a mass from approximately 1 kg to a little more than 2 kg. The shell of an ostrich egg is 1.5 mm thick and can support the weight of an adult human.

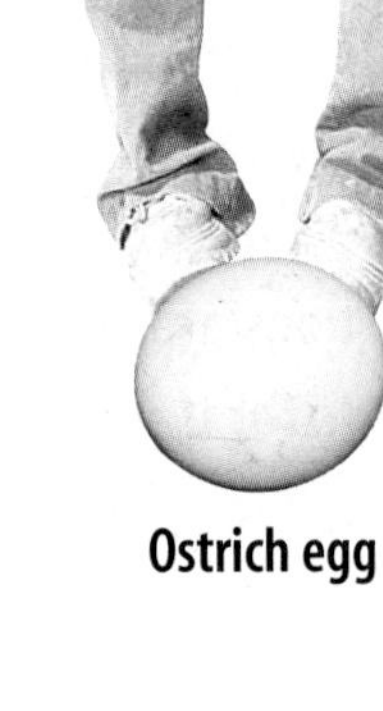

Ostrich egg

...The bird that lays the smallest egg is the hummingbird. Hummingbird eggs are typically 1.3 cm long and 0.8 cm wide. The smallest hummingbird egg on record was less than 1 cm long and weighed 0.36 g.

Hummingbird egg and nest

...The elephant bird, extinct within the last 1,000 years, laid an egg that was seven times larger than an ostrich egg. These eggs weighed about 12 kg. They were 30 cm long and could hold up to 8.5 L of liquid. It could hold the equivalent of 12,000 hummingbird eggs.

Applying Math How many elephant bird eggs would it take to equal a dog weighing 48 kg?

Graph It

Go to life.msscience.com/science_stats and research the egg length of an American robin, a house sparrow, a bald eagle, and a Canada goose. Make a bar graph of this information.

Reviewing Main Ideas

Section 1 Birds

1. Birds are endothermic animals that are covered with feathers and lay eggs.
2. Adaptations that enable most birds to fly include wings; feathers; a strong, lightweight skeleton; and efficient body systems.
3. Birds lay eggs enclosed in hard shells. All birds' eggs are incubated until they hatch.

Section 2 Mammals

1. Mammals are endothermic animals with fur or hair.
2. Mammary glands of female mammals can produce milk.
3. Mammals have teeth that are specialized for eating certain foods. Herbivores eat plants, carnivores eat meat, and omnivores eat plants and meat.
4. There are three groups of mammals. Monotremes lay eggs. Most marsupials have pouches for the development of their young. Placental offspring develop within a uterus and are nourished through a placenta.
5. Mammals are important in maintaining balance in the environment.

Visualizing Main Ideas

Copy and complete the following concept map on mammals.

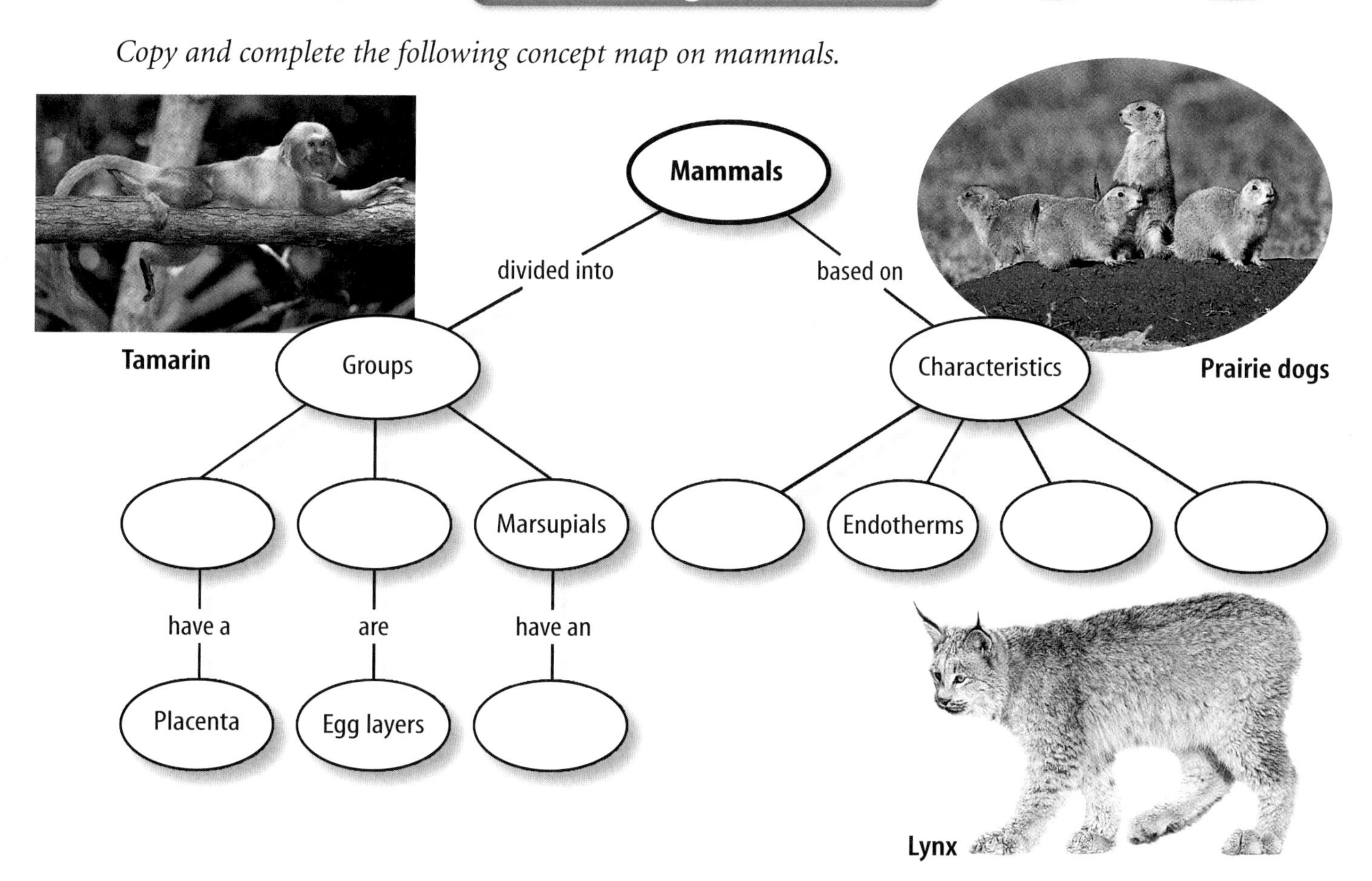

chapter 15 Review

Using Vocabulary

carnivore p. 437
contour feather p. 430
down feather p. 430
endotherm p. 430
gestation period p. 441
herbivore p. 437
mammal p. 436
mammary gland p. 436
marsupial p. 440
monotreme p. 440
omnivore p. 437
placenta p. 441
placental p. 441
preening p. 430
umbilical cord p. 441

Explain the difference between the vocabulary words in each of the following sets.

1. omnivore—carnivore—herbivore
2. contour feather—down feather
3. monotreme—marsupial
4. placenta—umbilical cord
5. endotherm—mammal
6. placental—monotreme
7. mammary gland—mammal
8. mammal—omnivore
9. endotherm—down feather
10. preening—down feather

Checking Concepts

Choose the word or phrase that best answers the question.

11. Which of the following birds has feet adapted for moving on water?
    A) duck **C)** owl
    B) oriole **D)** rhea
12. Birds do NOT use their wings for which of the following activities?
    A) flying **C)** balancing
    B) swimming **D)** eating
13. Which of these mammals lay eggs?
    A) carnivores **C)** monotremes
    B) marsupials **D)** placentals
14. Birds use which of the following organs to crush and grind their food?
    A) crop **C)** gizzard
    B) stomach **D)** small intestine
15. Which of the following mammals is classified as a marsupial?
    A) cat **C)** kangaroo
    B) human **D)** camel

Use the photo below to answer question 16.

16. What are mammals with pouches, like the koala pictured above, called?
    A) marsupials **C)** placentals
    B) monotremes **D)** chiropterans
17. Which of the following have mammary glands without nipples?
    A) marsupials **C)** monotremes
    B) placentals **D)** omnivores
18. Teeth that are used for tearing food are called what?
    A) canines **C)** molars
    B) incisors **D)** premolars
19. Bird eggs do NOT have which of the following structures?
    A) hard shells **C)** placentas
    B) yolks **D)** membranes
20. Which of the following animals eat only plant materials?
    A) carnivores **C)** omnivores
    B) herbivores **D)** endotherms

Science Online life.msscience.com/vocabulary_puzzlemaker

Thinking Critically

21. **Compare and contrast** bird and mammal reproduction.

22. **Classify** You are a paleontologist studying fossils. One fossil appears to have hollow bones, a keeled breastbone, and a short, bony tail. How would you classify it?

23. **Explain** which type of bird, a duck or an ostrich, would have lighter bones.

24. **List** the features of birds that allow them to be fully adapted to life on land.

25. **Concept Map** Copy and complete this concept map about birds.

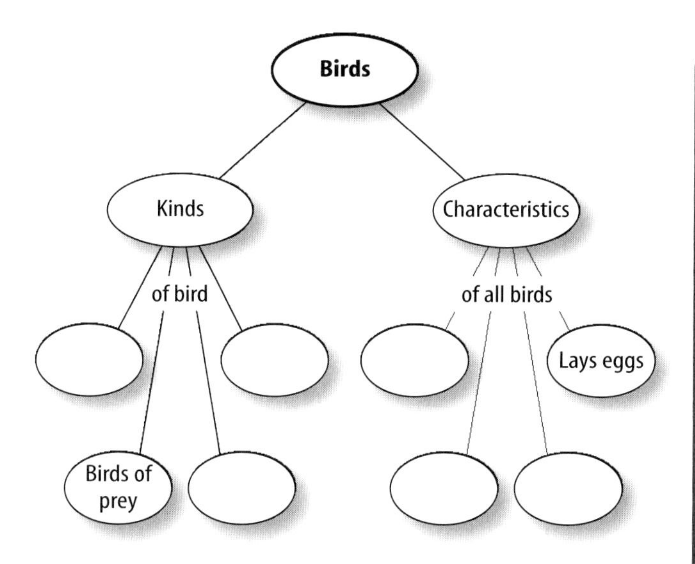

26. **Describe** A mammal's teeth are similar in size and include all four types of teeth. What kind of mammal has teeth like this?

27. **Classify** You discover three new species of placentals, with the following traits. Using **Table 1** in this chapter, place each placental into the correct order.
Placental 1 swims and eats meat.
Placental 2 flies and eats fruit.
Placental 3 runs on four legs and hunts.

28. **Classify** Group the following mammals as herbivore, carnivore, or omnivore:
bear, tiger, opossum, raccoon, mouse, rabbit, seal, and ape.

29. **Compare and contrast** the teeth of herbivores, carnivores, and omnivores. How are their types of teeth adapted to their diets?

Performance Activities

30. **Song with Lyrics** Create a song about bird adaptations for flight by changing the words to a song that you know. Include in your song as many adaptations as possible.

Applying Math

Use the graph below to answer questions 31 and 32.

Record of Canada Geese

Year	Number of Geese
1996	550
1997	600
1998	575
1999	750
2000	825

31. **Number of Geese** This table is a record of the approximate number of Canada geese that wintered at a midwestern wetland area over a five-year time period. Construct a line graph from these data.

32. **Population Increase** What percent increase occurred in the Canada goose population between 1996 and 2000? What percent increase occurred each year?

chapter 15 Standardized Test Practice

Part 1 Multiple Choice

Record your answers on the answer sheet provided by your teacher or on a sheet of paper.

Use the illustration below to answer questions 1–2.

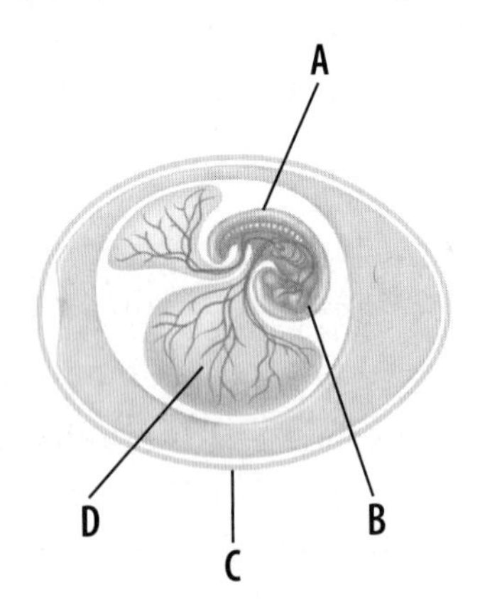

1. Which letter represents amniotic fluid?
   A. A **C.** C
   B. B **D.** D

2. Which letter represents a developing embryo?
   A. A **C.** C
   B. B **D.** D

3. Which of the following features is an adaptation that allows birds to fly?
   A. a gizzard
   B. bones that are almost hollow
   C. a crop
   D. a four-chambered heart

4. Which of the following is a monotreme?
   A. a penguin **C.** a kangaroo
   B. an eagle **D.** a platypus

5. What is a characteristic that sets mammals apart from birds?
   A. They help pollinate flowers.
   B. They have a four-chambered heart.
   C. They have special glands that produce milk for feeding their young.
   D. They have a special skeleton that is lightweight but strong.

Use the illustration below to answer questions 6–7.

6. What part of the mammal's body is indicated by 1–4 in the diagram?
   A. small intestine **C.** stomach
   B. large intestine **D.** gizzard

7. Which animals have this type of digestive system?
   A. carnivorous birds
   B. carnivorous mammals
   C. herbivorous mammals
   D. birds that eat only nuts and seeds

8. What is the significance of *Archaeopteryx?*
   A. It was the first birdlike fossil found.
   B. It represents the direct ancestor of birds.
   C. It was probably a ground-living dinosaur with wings.
   D. It is the oldest birdlike fossil.

9. What is the unique characteristic of a marsupial?
   A. They all live in Australia.
   B. Their young develop in a pouch.
   C. They lay eggs.
   D. They provide milk for their young.

Part 2 Short Response/Grid In

Record your answers on the answer sheet provided by your teacher or on a sheet of paper.

10. What are two examples of body features that enable birds to fly?
11. What does it mean if an animal is an endotherm?

Use the photos below to answer questions 12 and 13.

12. What is the purpose of feather B?
13. What is the purpose of feather A?
14. In a bird's digestive system, what purpose do the crop and gizzard serve?
15. What adaptation in birds provides a constant supply of oxygen for the flight muscles?
16. Give two examples of special structures produced by the skin of mammals.
17. Give the names of the three groups of mammals based on how their young develop. Give an example of each one.

Test-Taking Tip

Essay Organization For essay questions, spend a few minutes listing and organizing the main points that you plan to discuss. Make sure to do all of this work on your scratch paper, not on the answer sheet.

Question 18 List the characteristics that you want to discuss in one column, and the advantage in flight in another column.

Part 3 Open Ended

Record your answers on a sheet of paper.

18. Describe the physical characteristics of birds' bones that make flight possible.
19. Compare the barbs of a contour feather with those of a down feather.
20. Describe the function of wings in flightless birds such as penguins and ostriches.
21. Explain how feathers help a bird fly.

Use the illustration below to answer questions 22 and 23.

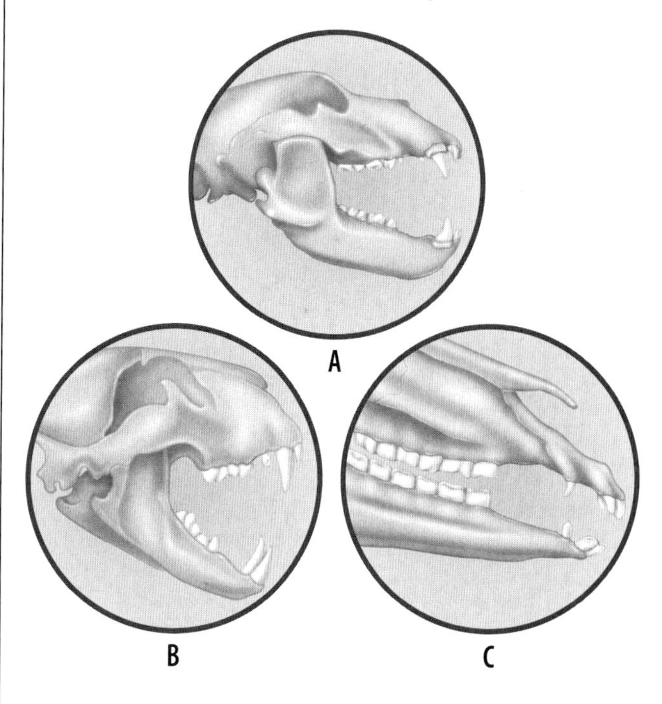

22. How can you tell that diagram C does NOT represent a carnivore? What can you tell about the diet of the animal that would have the type of teeth shown in diagram C?
23. What can you say about the diets of the animals represented by the teeth shown in diagrams A and B?

chapter

Animal Behavior

The BIG Idea

All the different behaviors of an animal are important for its survival.

SECTION 1

Types of Behavior

Main Idea Some behaviors are inherited, while others are learned.

SECTION 2

Behavioral Interactions

Main Idea Social behaviors are important for the survival of a species and each of its individuals.

Why do animals fight?

Animals often defend territories from other members of the same species. Fighting is usually a last resort to protect a territory that contains food, shelter, and potential mates.

Science Journal What other behaviors might an animal use to signal that a territory is occupied?

Start-Up Activities

How do animals communicate?

One way humans communicate is by speaking. Other animals communicate without the use of sound. For example, a gull chick pecks at its parent's beak to get food. Try the lab below to see if you can communicate without speaking.

1. Form groups of students. One at a time, have each student choose an object and describe that object using gestures.
2. The other students observe and try to identify the object that is being described.
3. **Think Critically** In your Science Journal, describe how you and the other students were able to communicate without speaking to one another.

Preview this chapter's content and activities at life.msscience.com

FOLDABLES™ Study Organizer

Behavior As you study behaviors, make the following Foldable to help you find the similarities and differences between the behaviors of two animals.

STEP 1 **Fold** a vertical sheet of paper in half from top to bottom.

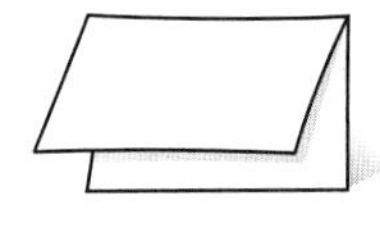

STEP 2 **Fold** in half from side to side with the fold at the top.

STEP 3 **Unfold** the paper once. **Cut** only the fold of the top flap to make two tabs.

STEP 4 **Turn** the paper vertically and **label** the front tabs as shown.

Observed Behaviors of Animal 1
Observed Behaviors of Animal 2

Read and Write Before you read the chapter, choose two animals to compare. As you read the chapter, list the behaviors you learn about Animal 1 and Animal 2 under the appropriate tab.

Get Ready to Read

New Vocabulary

1 Learn It! What should you do if you find a word you don't know or understand? Here are some suggested strategies:

1. Use context clues (from the sentence or the paragraph) to help you define it.
2. Look for prefixes, suffixes, or root words that you already know.
3. Write it down and ask for help with the meaning.
4. Guess at its meaning.
5. Look it up in the glossary or a dictionary.

2 Practice It! Look at the word submission in the following passage. See how context clues can help you understand its meaning.

To avoid being attacked and injured by an individual of its own species, an animal shows submission. Postures that make an animal appear smaller often are used to communicate surrender. In some animal groups, one individual is usually dominant. Members of the group show submissive behavior toward the dominant individual. This stops further aggressive behavior by the dominant animal. Young animals also display submissive behaviors toward parents or dominant animals.

—from page 464

Context Clue
Submission is shown with a posture that makes an animal appear smaller.

Context Clue
Submission is used to communicate surrender.

Context Clue
Submission is shown toward a dominant individual.

3 Apply It! Make a vocabulary bookmark with a strip of paper. As you read, keep track of words you do not know or want to learn more about.

Target Your Reading

Use this to focus on the main ideas as you read the chapter.

1. **Before you read** the chapter, respond to the statements below on your worksheet or on a numbered sheet of paper.
   - Write an **A** if you **agree** with the statement.
   - Write a **D** if you **disagree** with the statement.

2. **After you read** the chapter, look back to this page to see if you've changed your mind about any of the statements.
   - If any of your answers changed, explain why.
   - Change any false statements into true statements.
   - Use your revised statements as a study guide.

Reading Tip

Read a paragraph containing a vocabulary word from beginning to end. Then, go back to determine the meaning of the word.

Before You Read A or D		Statement	After You Read A or D
	1	Domesticated dogs recognize that people are not part of their pack.	
	2	All behaviors must be learned.	
	3	Behaviors are learned from both experience and practice.	
	4	A reflex is a complex, learned behavior.	
	5	Living in groups can create added safety for animals.	
	6	Many animals are able to communicate through the use of chemicals.	
	7	Sleeping during the day is an innate behavior for bats.	
	8	Bees dance to show other bees where food is located.	

Science Online
Print out a worksheet of this page at life.msscience.com

section 1

Types of Behavior

as you read

What You'll Learn

- **Identify** the differences between innate and learned behavior.
- **Explain** how reflexes and instincts help organisms survive.
- **Identify** examples of imprinting and conditioning.

Why It's Important

Innate behavior helps you survive on your own.

Review Vocabulary
salivate: to secrete saliva in anticipation of food

New Vocabulary
- behavior
- innate behavior
- reflex
- instinct
- imprinting
- conditioning
- insight

Behavior

When you come home from school, does your dog run to meet you? Your dog barks and wags its tail as you scratch behind its ears. Sitting at your feet, it watches every move you make. Why do dogs do these things? In nature, dogs are pack animals that generally follow a leader. They have been living with people for about 12,000 years. Domesticated dogs treat people as part of their own pack, as shown in **Figure 1.**

Animals are different from one another in their behavior. They are born with certain behaviors, and they learn others. **Behavior** is the way an organism interacts with other organisms and its environment. Anything in the environment that causes a reaction is called a stimulus. A stimulus can be external, such as a rival male entering another male's territory; or internal, such as hunger or thirst. You are the stimulus that causes your dog to bark and wag its tail. Your dog's reaction to you is a response.

Figure 1 Dogs are pack animals by nature. A pack of wild dogs must work together to survive. This domesticated dog (right) has accepted a human as its leader.

Cliff swallows build nests out of mud.

Hummingbirds build delicate cup-shaped nests on branches of trees.

Figure 2 Bird nests come in different sizes and shapes. This male weaverbird is knotting the ends of leaves together to secure the nest.

Innate Behavior

A behavior that an organism is born with is called an **innate behavior.** These types of behaviors are inherited. They don't have to be learned.

Innate behavior patterns occur the first time an animal responds to a particular internal or external stimulus. For birds like the swallows and the hummingbird in **Figure 2** building a nest is innate behavior. When it's time for the female weaverbird to lay eggs, the male weaverbird builds an elaborate nest, also shown in **Figure 2.** Although a young male's first attempt may be messy, the nest is constructed correctly.

The behavior of animals that have short life spans is mostly innate behavior. Most insects do not learn from their parents. In many cases, the parents have died or moved on by the time the young hatch. Yet every insect reacts innately to its environment. A moth will fly toward a light, and a cockroach will run away from it. They don't learn this behavior. Innate behavior allows animals to respond instantly. This quick response often means the difference between life and death.

Reflexes The simplest innate behaviors are reflex actions. A **reflex** is an automatic response that does not involve a message from the brain. Sneezing, shivering, yawning, jerking your hand away from a hot surface, and blinking your eyes when something is thrown toward you are all reflex actions.

In humans a reflex message passes almost instantly from a sense organ along the nerve to the spinal cord and back to the muscles. The message does not go to the brain. You are aware of the reaction only after it has happened. Your body reacts on its own. A reflex is not the result of conscious thinking.

Reflex A tap on a tendon in your knee causes your leg to straighten. This is known as the knee-jerk reflex. Abnormalities in this reflex tell doctors of a possible problem in the central nervous system. Research other types of reflexes and write a report about them in your Science Journal.

Instincts An **instinct** is a complex pattern of innate behavior. Spinning a web like the one in **Figure 3** is complicated, yet spiders spin webs correctly on the first try. Unlike reflexes, instinctive behaviors can take weeks to complete. Instinctive behavior begins when the animal recognizes a stimulus and continues until all parts of the behavior have been performed.

Reading Check *What is the difference between a reflex and an instinct?*

Figure 3 Spiders, like this orb weaver spider, know how to spin webs as soon as they hatch.

Learned Behavior

All animals have innate and learned behaviors. Learned behavior develops during an animal's lifetime. Animals with more complex brains exhibit more behaviors that are the result of learning. However, the behavior of insects, spiders, and other arthropods is mostly instinctive behavior. Fish, reptiles, amphibians, birds, and mammals all learn. Learning is the result of experience or practice.

Learning is important for animals because it allows them to respond to changing situations. In changing environments, animals that have the ability to learn a new behavior are more likely to survive. This is especially important for animals with long life spans. The longer an animal lives, the more likely it is that the environment in which it lives will change.

Learning also can modify instincts. For example, grouse and quail chicks, shown in **Figure 4,** leave their nests the day they hatch. They can run and find food, but they can't fly. When something moves above them, they instantly crouch and keep perfectly still until the danger has passed. They will crouch without moving even if the falling object is only a leaf. Older birds have learned that leaves will not harm them, but they freeze when a hawk moves overhead.

Figure 4 As they grow older, these quail chicks will learn which organisms to avoid.
Describe *why it is important for young quail to react the same toward all organisms.*

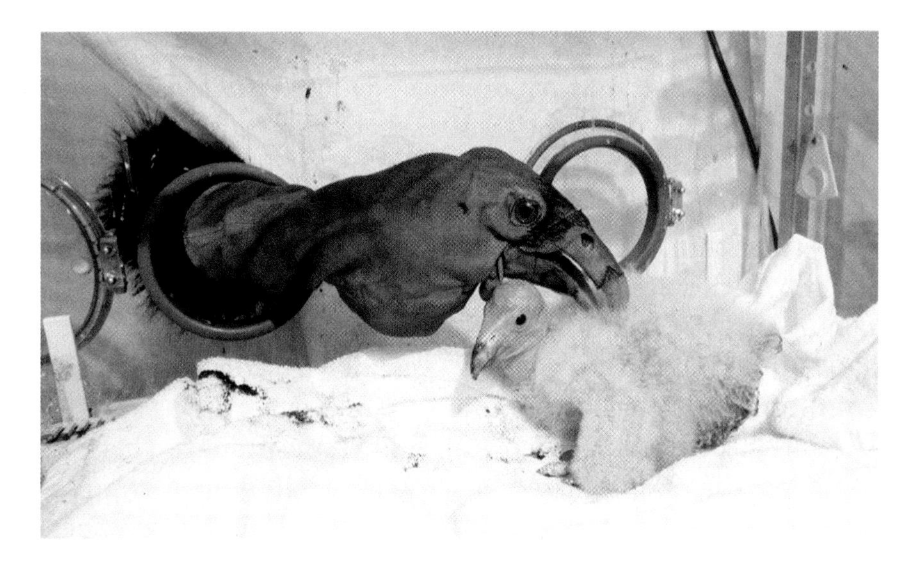

Figure 5 When feeding chicks in captivity, puppets of adult condors are used so the chicks don't learn to associate humans with food.

Topic: Captive Breeding

Visit life.msscience.com for Web links to information about captive breeding.

Activity Identify and describe techniques used to raise captive species and introduce them into the wild.

Imprinting Learned behavior includes imprinting, trial and error, conditioning, and insight. Have you ever seen young ducks following their mother? This is an important behavior because the adult bird has had more experience in finding food, escaping predators, and getting along in the world. **Imprinting** occurs when an animal forms a social attachment, like the condor in **Figure 5,** to another organism within a specific time period after birth or hatching.

Konrad Lorenz, an Austrian naturalist, developed the concept of imprinting. Working with geese, he discovered that a gosling follows the first moving object it sees after hatching. The moving object, whatever it is, is imprinted as its parent. This behavior works well when the first moving object a gosling sees is an adult female goose. But goslings hatched in an incubator might see a human first and become imprinted on that human. Animals that become imprinted toward animals of another species have difficulty recognizing members of their own species.

Trial and Error Can you remember when you learned to ride a bicycle? You probably fell many times before you learned how to balance on the bicycle. After a while you could ride without having to think about it. You have many skills that you learned through trial and error, such as feeding yourself and tying your shoes, as shown in **Figure 6.**

Behavior that is modified by experience is called trial-and-error learning. Many animals learn by trial and error. When baby chicks first try to feed themselves, they peck at many stones before they get any food. As a result of trial and error, they learn to peck only at food particles.

Figure 6 Were you able to tie your shoes on the first attempt? **List** *other things you do every day that require learning.*

Observing Conditioning

Procedure

1. Obtain several **photos of different foods and landscapes** from your teacher.
2. Show each picture to a classmate for 20 s.
3. Record how each photo made your partner feel.

Analysis

1. How did your partner feel after looking at the photos of food?
2. What effect did the landscape pictures have on your partner?
3. Infer how advertising might condition consumers to buy specific food products.

Conditioning Do you have an aquarium in your school or home? If you put your hand above the tank, the fish probably will swim to the top of the tank, expecting to be fed. They have learned that a hand shape above them means food. What would happen if you tapped on the glass right before you fed them? Soon the fish probably will swim to the top of the tank if you just tap on the glass. Because they are used to being fed after you tap on the glass, they associate the tap with food.

Animals often learn new behaviors by conditioning. In **conditioning,** behavior is modified so that a response to one stimulus becomes associated with a different stimulus. There are two types of conditioning. One type introduces a new stimulus before the usual stimulus. Russian scientist Ivan P. Pavlov performed experiments using this type of conditioning. He knew that the sight and smell of food made hungry dogs secrete saliva. Pavlov added another stimulus. He rang a bell before he fed the dogs. The dogs began to connect the sound of the bell with food. Then Pavlov rang the bell without giving the dogs food. They salivated when the bell was rung even though he did not give them food. The dogs, like the one in **Figure 7,** were conditioned to respond to the bell.

In the second type of conditioning, the new stimulus is given after the affected behavior. Getting an allowance for doing chores is an example of this type of conditioning. You do your chores because you want to receive your allowance. You have been conditioned to perform an activity that you may not have done if you had not been offered a reward.

Reading Check *How does conditioning modify behavior?*

Figure 7 In Pavlov's experiment, a dog was conditioned to salivate when a bell was rung. It associated the bell with food.

Insight How does learned behavior help an animal deal with a new situation? Suppose you have a new math problem to solve. Do you begin by acting as though you've never seen it before, or do you use what you have learned previously in math to solve the problem? If you use what you have learned, then you have used a kind of learned behavior called insight. **Insight** is a form of reasoning that allows animals to use past experiences to solve new problems. In experiments with chimpanzees, as shown in **Figure 8,** bananas were placed out of the chimpanzees' reach. Instead of giving up, they piled up boxes found in the room, climbed them, and reached the bananas. At some time in their lives, the chimpanzees must have solved a similar problem. The chimpanzees demonstrated insight during the experiment. Much of adult human learning is based on insight. When you were a baby, you learned by trial and error. As you grow older, you will rely more on insight.

Figure 8 This illustration shows how chimpanzees may use insight to solve problems.

section 1 review

Summary

Behavior

- Animals are born with certain behaviors, while other behaviors are learned.
- A stimulus is anything in the environment that causes a reaction.

Innate and Learned Behaviors

- Innate behaviors are those behaviors an organism inherits, such as reflexes and instincts.
- Learned behavior allows animals to respond to changing situations.
- Imprinting, trial and error, conditioning, and insight are examples of learned behavior.

Self Check

1. **Compare and contrast** a reflex and an instinct.
2. **Compare and contrast** imprinting and conditioning.
3. **Think Critically** Use what you know about conditioning to explain how the term *mouthwatering food* might have come about.

Applying Skills

4. **Use a Spreadsheet** Make a spreadsheet of the behaviors in this section. Sort the behaviors according to whether they are innate or learned behaviors. Then identify the type of innate or learned behavior.

section 2

Behavioral Interactions

as you read

What You'll Learn

- **Explain** why behavioral adaptations are important.
- **Describe** how courtship behavior increases reproductive success.
- **Explain** the importance of social behavior and cyclic behavior.

Why It's Important

Organisms must be able to communicate with each other to survive.

Review Vocabulary

nectar: a sweet liquid produced in a plant's flower that is the main raw material of honey

New Vocabulary

- social behavior
- society
- aggression
- courtship behavior
- pheromone
- cyclic behavior
- hibernation
- migration

Instinctive Behavior Patterns

Complex interactions of innate behaviors between organisms result in many types of animal behavior. For example, courtship and mating within most animal groups are instinctive ritual behaviors that help animals recognize possible mates. Animals also protect themselves and their food sources by defending their territories. Instinctive behavior, just like natural hair color, is inherited.

Social Behavior

Animals often live in groups. One reason, shown in **Figure 9,** is that large numbers provide safety. A lion is less likely to attack a herd of zebras than a lone zebra. Sometimes animals in large groups help keep each other warm. Also, migrating animal groups are less likely to get lost than animals that travel alone.

Interactions among organisms of the same species are examples of **social behavior.** Social behaviors include courtship and mating, caring for the young, claiming territories, protecting each other, and getting food. These inherited behaviors provide advantages that promote survival of the species.

Reading Check *Why is social behavior important?*

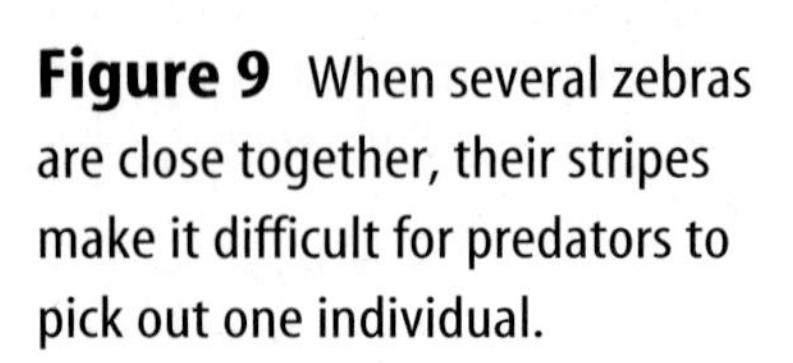

Figure 9 When several zebras are close together, their stripes make it difficult for predators to pick out one individual.

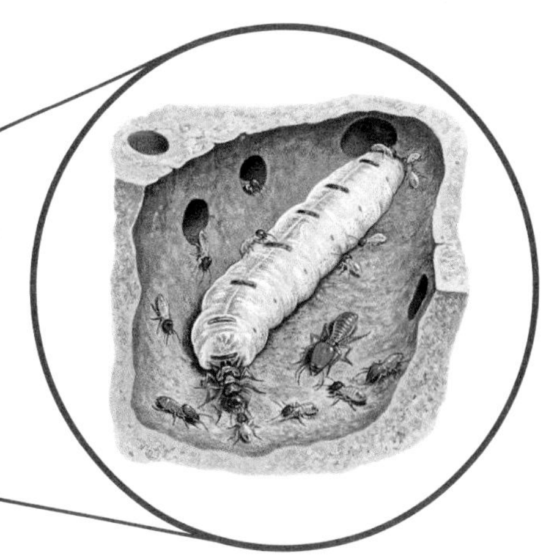

Figure 10 Termites built this large mound in Australia. The mound has a network of tunnels and chambers for the queen termite to deposit eggs into.

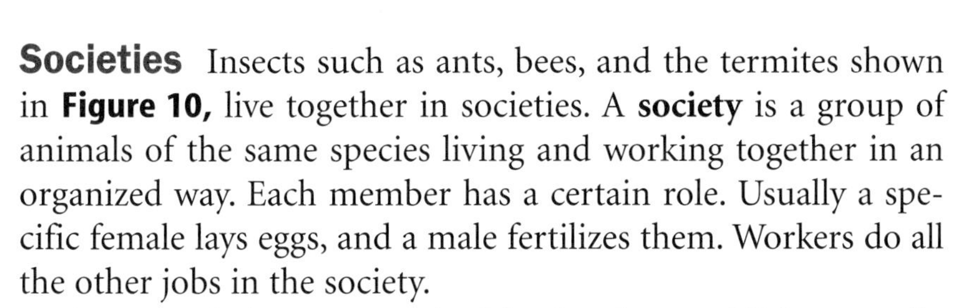

Societies Insects such as ants, bees, and the termites shown in **Figure 10,** live together in societies. A **society** is a group of animals of the same species living and working together in an organized way. Each member has a certain role. Usually a specific female lays eggs, and a male fertilizes them. Workers do all the other jobs in the society.

Some societies are organized by dominance. Wolves usually live together in packs. A wolf pack has a dominant female. The top female controls the mating of the other females. If plenty of food is available, she mates and then allows the others to do so. If food is scarce, she allows less mating. During such times, she is usually the only one to mate.

Territorial Behavior

Many animals set up territories for feeding, mating, and raising young. A territory is an area that an animal defends from other members of the same species. Ownership of a territory occurs in different ways. Songbirds sing, sea lions bellow, and squirrels chatter to claim territories. Other animals leave scent marks. Some animals, like the tiger in **Figure 11,** patrol an area and attack other animals of the same species who enter their territory. Why do animals defend their territories? Territories contain food, shelter, and potential mates. If an animal has a territory, it will be able to mate and produce offspring. Defending territories is an instinctive behavior. It improves the survival rate of an animal's offspring.

Figure 11 A tiger's territory may cover several miles. It will confront any other tiger who enters it. **Explain** *what may be happening in this photo.*

Aggression Have you ever watched as one dog approached another dog that was eating a bone? What happened to the appearance of the dog with the bone? Did its hair on its back stick up? Did it curl its lips and make growling noises? This behavior is called aggression. **Aggression** is a forceful behavior used to dominate or control another animal. Fighting and threatening are aggressive behaviors animals use to defend their territories, protect their young, or to get food.

Many animals demonstrate aggression. Some birds let their wings droop below their tail feathers. It may take another bird's perch and thrust its head forward in a pecking motion as a sign of aggression. Cats lay their ears flat, arch their backs, and hiss.

Figure 12 Young wolves roll over and make themselves as small as possible to show their submission to adult wolves.

Submission Animals of the same species seldom fight to the death. Teeth, beaks, claws, and horns are used for killing prey or for defending against members of a different species.

To avoid being attacked and injured by an individual of its own species, an animal shows submission. Postures that make an animal appear smaller often are used to communicate surrender. In some animal groups, one individual is usually dominant. Members of the group show submissive behavior toward the dominant individual. This stops further aggressive behavior by the dominant animal. Young animals also display submissive behaviors toward parents or dominant animals, as shown in **Figure 12.**

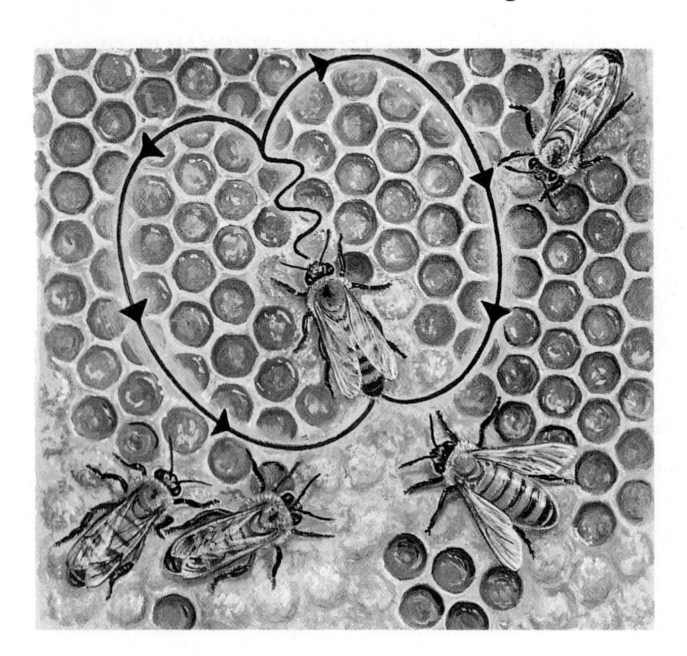

Figure 13 During the waggle dance, if the food source is far from the hive, the dance takes the form of a figure eight. The angle of the waggle is equal to the angle from the hive between the Sun and nectar source.

Communication

In all social behavior, communication is important. Communication is an action by a sender that influences the behavior of a receiver. How do you communicate with the people around you? You may talk, make noises, or gesture like you did in this chapter's Launch Lab. Honeybees perform a dance, as shown in **Figure 13,** to communicate to other bees in the hive the location of a food source. Animals in a group communicate with sounds, scents, and actions. Alarm calls, chemicals, speech, courtship behavior, and aggression are forms of communication.

Figure 14 This male Emperor of Germany bird of paradise attracts mates by posturing and fanning its tail.
List *other behaviors animals use to attract mates.*

Courtship Behavior A male bird of paradise, shown in **Figure 14,** spreads its tail feathers and struts. A male sage grouse fans its tail, fluffs its feathers, and blows up its two red air sacs. These are examples of behavior that animals perform before mating. This type of behavior is called **courtship behavior.** Courtship behaviors allow male and female members of a species to recognize each other. These behaviors also stimulate males and females so they are ready to mate at the same time. This helps ensure reproductive success.

In most species the males are more colorful and perform courtship displays to attract a mate. Some courtship behaviors allow males and females to find each other across distances.

INTEGRATE Chemistry

Chemical Communication Ants are sometimes seen moving single file toward a piece of food. Male dogs frequently urinate on objects and plants. Both behaviors are based on chemical communication. The ants have laid down chemical trails that others of their species can follow. The dog is letting other dogs know he has been there. In these behaviors, the animals are using chemicals called pheromones (FER uh mohnz) to communicate. A chemical that is produced by one animal to influence the behavior of another animal of the same species is called a **pheromone.** They are powerful chemicals needed only in small amounts. They remain in the environment so that the sender and the receiver can communicate without being in the same place at the same time. They can advertise the presence of an animal to predators, as well as to the intended receiver of the message.

Males and females use pheromones to establish territories, warn of danger, and attract mates. Certain ants, mice, and snails release alarm pheromones when injured or threatened.

Demonstrating Chemical Communication

Procedure

1. Obtain a **sample of perfume or air freshener.**
2. Spray it into the air to leave a scent trail as you move around the house or apartment to a hiding place.
3. Have someone try to discover where you are by following the scent of the substance.

Analysis

1. What was the difference between the first and last room you were in?
2. Would this be an efficient way for humans to communicate? Explain.

Try at Home

Figure 15 Many animals use sound to communicate.

Frogs often croak loud enough to be heard far away.

Pileated woodpecker calls often can be heard above everything else in the forest.

Howler monkeys got their name because of the sounds they make.

Sound Communication Male crickets rub one forewing against the other forewing. This produces chirping sounds that attract females. Each cricket species produces several calls that are different from other cricket species. These calls are used by researchers to identify different species. Male mosquitoes have hairs on their antennae that sense buzzing sounds produced by females of their same species. The tiny hairs vibrate only to the frequency emitted by a female of the same species.

Vertebrates use a number of different forms of sound communication. Rabbits thump the ground, gorillas pound their chests, beavers slap the water with their flat tails, and frogs, like the one in **Figure 15,** croak. Do you think that sound communication in noisy environments is useful? Seabirds that live where waves pound the shore rather than in some quieter place must rely on visual signals, not sound, for communication.

Morse Code Samuel B. Morse created a code in 1838 using numbers to represent letters. His early work led to Morse code. Naval ships today still use Morse code to communicate with each other using huge flashlights mounted on the ships' decks. In your Science Journal, write what reasons you believe that Morse code is still used by the Navy.

Light Communication Certain kinds of flies, marine organisms, and beetles have a special form of communication called bioluminescence. Bioluminescence, shown in **Figure 16,** is the ability of certain living things to give off light. This light is produced through a series of chemical reactions in the organism's body. Probably the most familiar bioluminescent organisms in North America are fireflies. These insects are not flies, but beetles. The flash of light that is produced on the underside of the last abdominal segments is used to locate a prospective mate. Each species has its own characteristic flashing. Males fly close to the ground and emit flashes of light. Females must flash an answer at exactly the correct time to attract males.

NATIONAL GEOGRAPHIC

VISUALIZING BIOLUMINESCENCE

Figure 16

Many marine organisms use bioluminescence as a form of communication. This visible light is produced by a chemical reaction and often confuses predators or attracts mates. Each organism on this page is shown in its normal and bioluminescent state.

▼ KRILL The blue dots shown below this krill are all that are visible when krill bioluminesce. The krill may use bioluminescence to confuse predators.

▲ JELLYFISH This jellyfish lights up like a neon sign when it is threatened.

◀ BLACK DRAGONFISH The black dragonfish lives in the deep ocean where light doesn't penetrate. It has light organs under its eyes that it uses like a flashlight to search for prey.

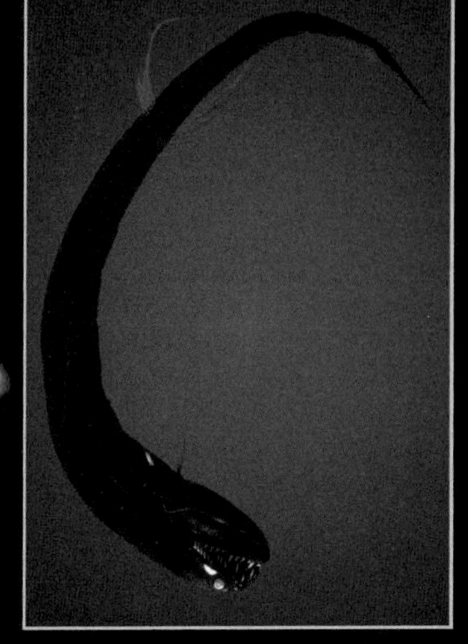

▲ DEEP-SEA SEA STAR The sea star uses light to warn predators of its unpleasant taste.

Uses of Bioluminescence Many bioluminescent animals are found deep in oceans where sunlight does not reach. The ability to produce light may serve several functions. One species of fish dangles a special luminescent organ in front of its mouth. This lures prey close enough to be caught and eaten. Deep-sea shrimp secrete clouds of a luminescent substance when disturbed. This helps them escape their predators. Patterns of luminescence on an animal's body may serve as marks of recognition similar to the color patterns of animals that live in sunlit areas.

Topic: Owl Behavior
Visit life.msscience.com for Web links to information about owl behavior.

Activity List five different types of owl behavior and describe how each behavior helps the owl survive.

Cyclic Behavior

Why do most songbirds rest at night while some species of owls rest during the day? Some animals like the owl in **Figure 17** show regularly repeated behaviors such as sleeping in the day and feeding at night.

A **cyclic behavior** is innate behavior that occurs in a repeating pattern. It often is repeated in response to changes in the environment. Behavior that is based on a 24-hour cycle is called a circadian rhythm. Most animals come close to this 24-hour cycle of sleeping and wakefulness. Experiments show that even if animals can't tell whether it is night or day, they continue to behave in a 24-hour cycle.

Animals that are active during the day are diurnal (dy UR nul). Animals that are active at night are nocturnal. Owls are nocturnal. They have round heads, big eyes, and flat faces. Their flat faces reflect sound and help them navigate at night. Owls also have soft feathers that make them almost silent while flying.

Reading Check *What is a diurnal behavior?*

Figure 17 Barn owls usually sleep during the day and hunt at night.
Identify *the type of behavior the owl is exhibiting.*

Hibernation Some cyclic behaviors also occur over long periods of time. **Hibernation** is a cyclic response to cold temperatures and limited food supplies. During hibernation, an animal's body temperature drops to near that of its surroundings, and its breathing rate is greatly reduced. Animals in hibernation, such as the bats in **Figure 18,** survive on stored body fat. The animal remains inactive until the weather becomes warm in the spring. Some mammals and many amphibians and reptiles hibernate.

Animals that live in desertlike environments also go into a state of reduced activity. This period of inactivity is called estivation. Desert animals sometimes estivate due to extreme heat, lack of food, or periods of drought.

Figure 18 Many bats find a frost-free place like this abandoned coal mine to hibernate for the winter when food supplies are low.

Applying Science

How can you determine which animals hibernate?

Many animals hibernate in the winter. During this period of inactivity, they survive on stored body fat. While they are hibernating, they undergo several physical changes. Heart rate slows down and body temperature decreases. The degree to which the body temperature decreases varies among animals. Scientists disagree about whether some animals truly hibernate or if they just reduce their activity and go into a light sleep. Usually, a true hibernator's body temperature will decrease significantly while it is hibernating.

Identifying the Problem

The table on the right shows the difference between the normal body temperature and the hibernating body temperature of several animals. What similarities do you notice?

Average Body Temperatures of Hibernating Animals

Animal	Normal Body Temperature (°C)	Hibernating Body Temperature (°C)
Woodchuck	37	3
Squirrel	32	4
Grizzly bear	32–37	27–32
Whippoorwill	40	18
Hoary marmot	37	10

Solving the Problem

1. Which animals would you classify as true hibernators and which would you classify as light sleepers? Explain.
2. Some animals such as snakes and frogs also hibernate. Why would it be difficult to record their normal body temperature?

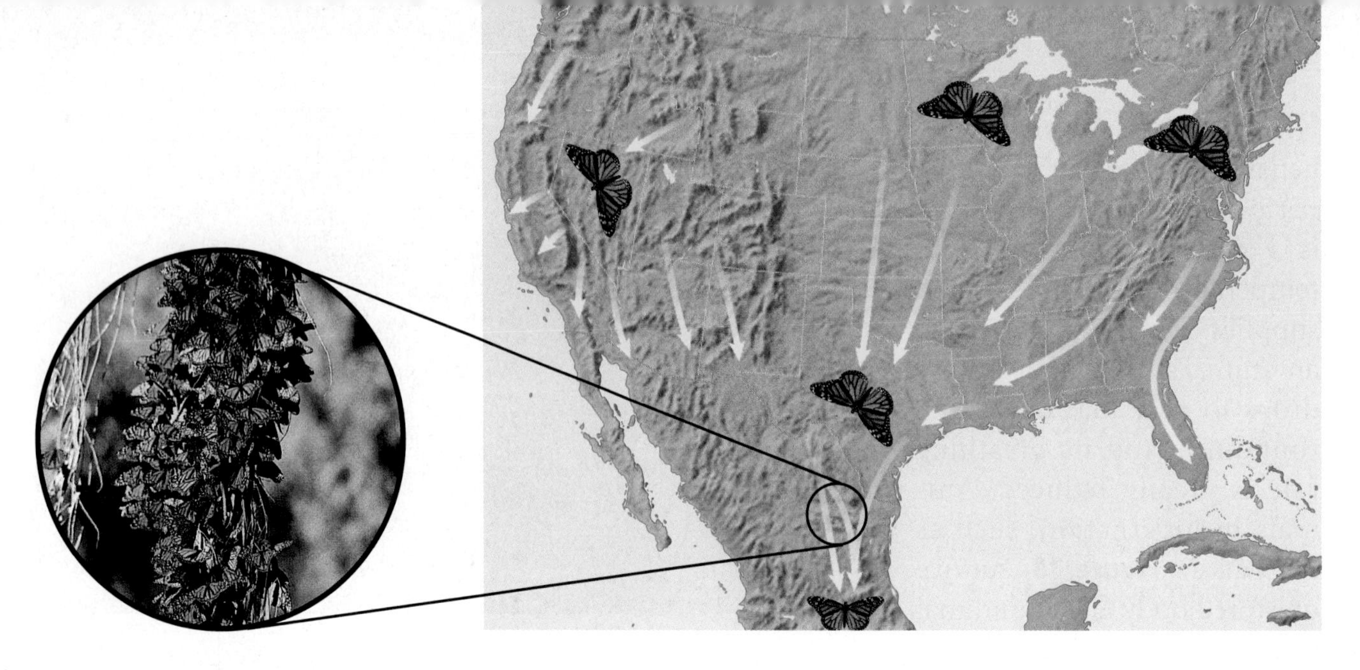

Figure 19 Many monarch butterflies travel from the United States to Mexico for the winter.

Migration Instead of hibernating, many animals move to new locations when the seasons change. This instinctive seasonal movement of animals is called **migration.** Most animals migrate to find food or to reproduce in environments that are more favorable for the survival of offspring. Many bird species fly for hours or days without stopping. The blackpoll warbler flies more than 4,000 km, nearly 90 hours nonstop from North America to its winter home in South America. Monarch butterflies, shown in **Figure 19,** can migrate as far as 2,900 km. Gray whales swim from arctic waters to the waters off the coast of northern Mexico. After the young are born, they make the return trip.

section 2 review

Summary

Instinctive Behavior Patterns

- Instinctive behavior patterns are inherited.
- Courtship and mating are instinctive for most animal groups.

Social and Territorial Behaviors

- Interactions among organisms of a group are examples of social behavior.
- Many animals protect a territory for feeding, mating, and raising young.

Communication and Cyclic Behavior

- Species can communicate with each other using behavior, chemicals, sound, or bioluminescence.
- Cyclic behaviors occur in response to environmental changes.

Self Check

1. **Describe** some examples of courtship behavior and how this behavior helps organisms survive.
2. **Identify** and **explain** two reasons that animals migrate.
3. **Compare and contrast** hibernation and migration.
4. **Think Critically** Suppose a species of frog lives close to a loud waterfall. It often waves a bright blue foot in the air. What might the frog be doing?

Applying Math

5. **Solve One-Step Equations** Some cicadas emerge from the ground every 17 years. The population of one type of caterpillar peaks every five years. If the peak cycle of the caterpillars and the emergence of cicadas coincided in 1990, in what year will they coincide again?

Science Online life.msscience.com/self_check_quiz

Observing Earthworm Behavior

Earthworms can be seen at night wriggling across wet sidewalks and driveways. Why don't you see many earthworms during the day?

Real-World Question

How do earthworms respond to light?

Goals

- **Predict** how earthworms will behave in the presence of light.

Materials

scissors
shoe box with lid
flashlight
tape
paper
moist paper towels
earthworms
timer

Safety Precautions

Procedure

1. Cut a round hole, smaller than the end of the flashlight, near one end of the lid.
2. Tape a sheet of paper to the lid so it hangs just above the bottom of the box and about 10 cm away from the end with the hole in it.
3. Place the moist paper towels in the bottom of the box.
4. Place the earthworms in the end of the box that has the hole in it.
5. Hold the flashlight over the hole and turn it on.
6. Leave the box undisturbed for 30 minutes, then open the lid and observe the worms.
7. **Record** the results of your experiment in your Science Journal.

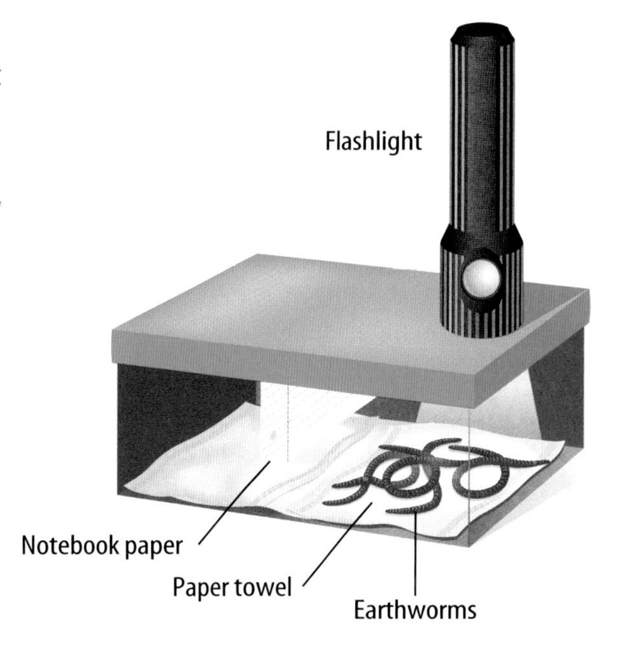

Conclude and Apply

1. **Identify** which direction the earthworms moved when the light was turned on.
2. **Infer** Based on your observations, what can you infer about earthworms?
3. **Explain** what type of behavior the earthworms exhibited.
4. **Predict** where you would need to go to find earthworms during the day.

Communicating Your Data

Write a story that describes a day in the life of an earthworm. List activities, dangers, and problems an earthworm might face. Include a description of its habitat. **For more help, refer to the** Science Skill Handbook.

Animal Habitats

Goals

- **Research** the natural habitat and basic needs of one animal.
- **Design** and model an appropriate zoo, animal park, or aquarium environment for this animal. Working cooperatively with your classmates, design an entire zoo or animal park.

Possible Materials

poster board
markers or colored pencils
materials that can be used to make a scale model

Real-World Question

Zoos, animal parks, and aquariums are safe places for wild animals. Years ago, captive animals were kept in small cages or behind glass windows. Almost no attempt was made to provide natural habitats for the animals. People who came to see the animals could not observe the animal's normal behavior. Now, most captive animals are kept in exhibit areas that closely resemble their natural habitats. These areas provide suitable environments for the animals so that they can interact with members of their same species and have healthier, longer lives. What types of environments are best suited for raising animals in captivity? How can the habitats provided at an animal park affect the behavior of animals?

Make a Model

1. Choose an animal to research. Find out where this animal is found in nature. What does it eat? What are its natural predators? Does it exhibit unique territorial, courtship, or other types of behavior? How is this animal adapted to its natural environment?

2. **Design** a model of a proposed habitat in which this animal can live successfully. Don't forget to include all of the things, such as shelter, food, and water, that your animal will need to survive. Will there be any other organisms in the habitat?
3. **Research** how zoos, animal parks, or aquariums provide habitats for animals. Information may be obtained by contacting scientists who work at zoos, animal parks, and aquariums.
4. **Present** your design to your class in the form of a poster, slide show, or video. Compare your proposed habitat with that of the animal's natural environment. Make sure you include a picture of your animal in its natural environment.

Test Your Model

1. Using all of the information you have gathered, create a model exhibit area for your animal.
2. Indicate what other plants and animals may be present in the exhibit area.

Analyze Your Data

1. **Decide** whether all of the animals studied in this lab can coexist in the same zoo or wildlife preserve.
2. **Analyze** problems that might exist in your design. Suggest some ways you might want to improve your design.

Conclude and Apply

1. **Interpret Data** Using the information provided by the rest of your classmates, design an entire zoo or aquarium that could include the majority of animals studied.
2. **Predict** which animals could be grouped together in exhibit areas.
3. **Determine** how large your zoo or wildlife preserve needs to be. Which animals require a large habitat?

Communicating Your Data

Give an oral presentation to another class on the importance of providing natural habitats for captive animals. **For more help, refer to the Science Skill Handbook.**

Oops! Accidents in SCIENCE

SOMETIMES GREAT DISCOVERIES HAPPEN BY ACCIDENT!

Going to the Dogs

A simple and surprising stroll showed that dogs really are humans' best friends

You've probably seen visually impaired people walking with their trusted "seeing-eye" dogs. Over 85 years ago, a doctor and his patient discovered this canine ability entirely by accident!

Near the end of World War I in Germany, Dr. Gerhard Stalling and his dog strolled with a patient—a German soldier who had been blinded—around hospital grounds.

A dog safely guides its owner across a street.

While they were walking, the doctor was called away. A few moments later, the doctor returned but the dog and the soldier were gone! Searching the paths frantically, Dr. Stalling made an astonishing discovery. His pet had led the soldier safely around the hospital grounds. Inspired by what his dog could do, Dr. Stalling set up the first school in the world dedicated to training dogs as guides.

German shepherds make excellent guide dogs.

German shepherds, golden retrievers, and Labrador retrievers seem to make the best guide dogs. They learn hand gestures and simple commands to lead visually impaired people safely across streets and around obstacles. This is what scientists call "learned behavior." Animals gain learned behavior through experience. But, a guide dog doesn't just learn to respond to special commands; it also must learn when *not* to obey. If its human owner urges the dog to cross the street and the dog sees that a car is approaching, the dog refuses because it has learned to disobey the command. This trait, called "intelligent disobedience," ensures the safety of the owner and the dog—a sure sign that dogs are still humans' best friends.

Write Lead a blindfolded partner around the classroom. Help your partner avoid obstacles. Then trade places. Write in your Science Journal about your experience leading and being led.

Science Online

For more information, visit life.msscience.com/oops

Reviewing Main Ideas

Section 1 Types of Behavior

1. Behavior that an animal has when it's born is innate behavior. Other animal behaviors are learned through experience.
2. Reflexes are simple innate behaviors. An instinct is a complex pattern of innate behavior.
3. Learned behavior includes imprinting, in which an animal forms a social attachment immediately after birth.
4. Behavior modified by experience is learning by trial and error.
5. Conditioning occurs when the response to one stimulus becomes associated with another. Insight is the ability to use past experiences to solve new problems.

Section 2 Behavioral Interactions

1. Behavioral adaptations such as defense of territory, courtship behavior, and social behavior help species of animals survive and reproduce.
2. Courtship behaviors allow males and females to recognize each other and prepare to mate.
3. Interactions among members of the same species are social behaviors.
4. Communication among organisms occurs in several forms, including chemical, sound, and light.
5. Cyclic behaviors are behaviors that occur in repeating patterns. Animals that are active during the day are diurnal. Animals that are active at night are nocturnal.

Visualizing Main Ideas

Copy and complete the following concept map on types of behavior.

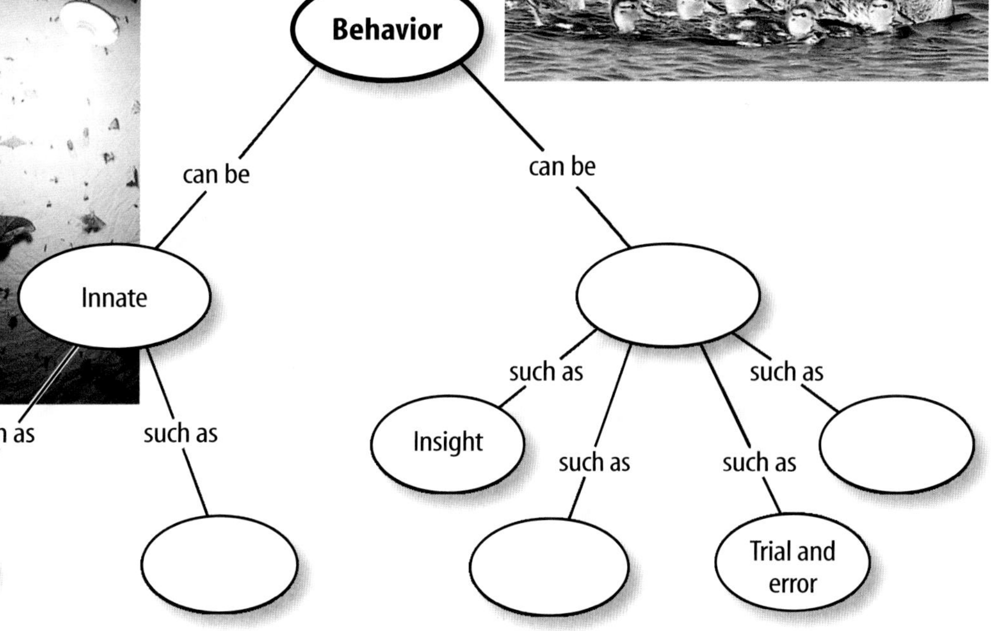

Behavior
can be
can be
Innate
such as
such as
such as
Insight
such as
such as
such as
Trial and error

Using Vocabulary

aggression p. 464	insight p. 461
behavior p. 456	instinct p. 458
conditioning p. 460	migration p. 470
courtship behavior p. 465	pheromone p. 465
cyclic behavior p. 468	reflex p. 457
hibernation p. 469	social behavior p. 462
imprinting p. 459	society p. 463
innate behavior p. 457	

Explain the differences between the pairs of vocabulary words given below. Then explain how the words are related.

1. conditioning—imprinting
2. innate behavior—social behavior
3. insight—instinct
4. social behavior—society
5. instinct—reflex
6. hibernation—migration
7. courtship behavior—pheromone
8. cyclic behavior—migration
9. aggression—social behavior
10. behavior—reflex

Checking Concepts

Choose the word or phrase that best answers the question.

11. What is an instinct an example of?
    A) innate behavior
    B) learned behavior
    C) imprinting
    D) conditioning

12. What is an area that an animal defends from other members of the same species called?
    A) society **C)** migration
    B) territory **D)** aggression

13. Which animals depend least on instinct and most on learning?
    A) birds **C)** mammals
    B) fish **D)** amphibians

14. What is a spider spinning a web an example of?
    A) conditioning **C)** learned behavior
    B) imprinting **D)** an instinct

15. What is a forceful act used to dominate or control another called?
    A) courtship **C)** aggression
    B) reflex **D)** hibernation

16. What is an organized group of animals doing specific jobs called?
    A) community **C)** society
    B) territory **D)** circadian rhythm

17. What is the response of inactivity and slowed metabolism that occurs during cold conditions?
    A) hibernation **C)** migration
    B) imprinting **D)** circadian rhythm

18. Which of the following is a reflex?
    A) writing **C)** sneezing
    B) talking **D)** riding a bicycle

Use the photo below to answer question 19.

19. The photo above is an example of what type of communication?
    A) light communication
    B) sound communication
    C) chemical communication
    D) cyclic behavior

Science Online life.msscience.com/vocabulary_puzzlemaker

Thinking Critically

20. **Explain** the type of behavior involved when the bell rings at the end of class.
21. **Describe** the advantages and disadvantages of migration as a means of survival.
22. **Explain** how a habit, such as tying your shoes, is different from a reflex.
23. **Explain** how behavior increases an animal's chance for survival using one example.
24. **Infer** Hens lay more eggs in the spring when the number of daylight hours increases. How can farmers use this knowledge of behavior to their advantage?
25. **Record Observations** Make observations of a dog, cat, or bird for a week. Record what you see. How did the animal communicate with other animals and with you?
26. **Classify** Make a list of 25 things that you do regularly. Classify each as an innate or learned behavior. Which behaviors do you have more of?
27. **Concept Map** Copy and complete the following concept map about communication. Use these words: *sound, chirping, bioluminescence,* and *buzzing.*

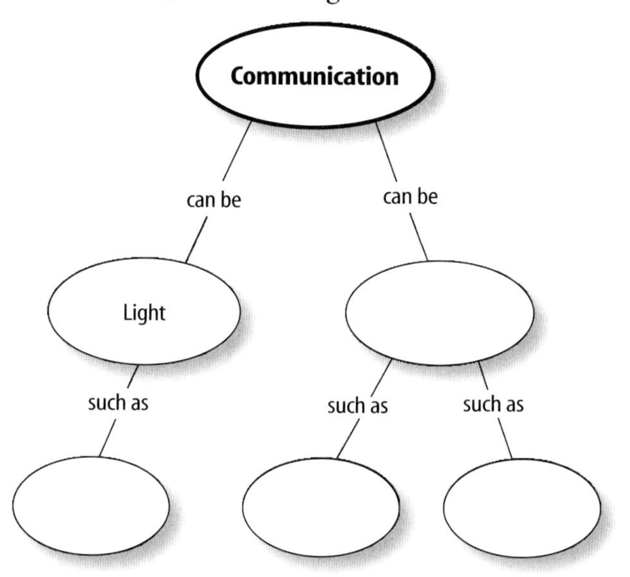

Performance Activities

28. **Poster** Draw a map showing the migration route of monarch butterflies, gray whales, or blackpoll warblers.

Applying Math

Use the graphs below to answer question 29.

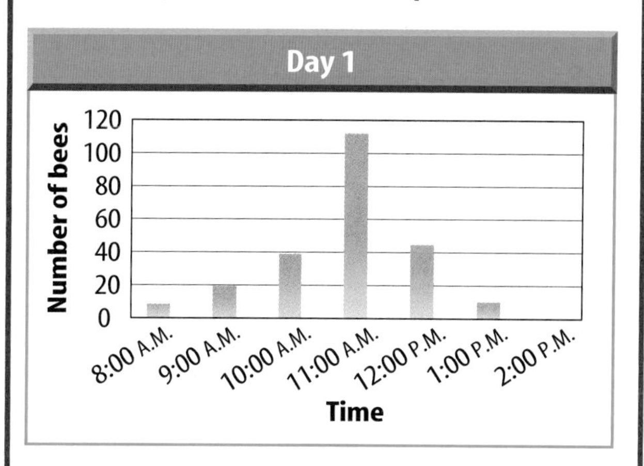

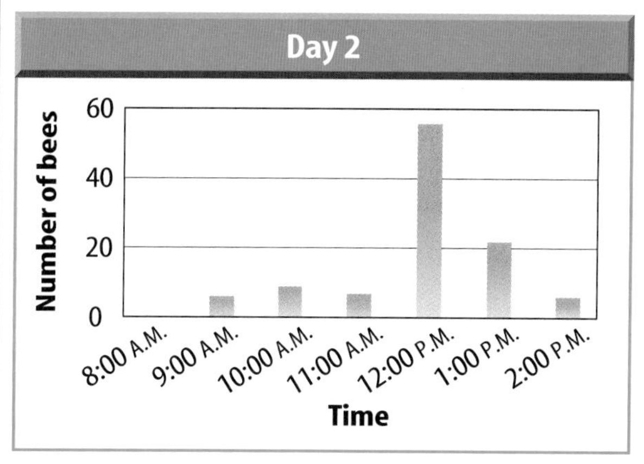

29. **Bee Foraging** Bees were trained to forage from 1:00 P.M. to 2:30 P.M. in New York and then were flown to California. The graphs above show the number of bees looking for food during the first two days in California. What was the difference in peak activity from day 1 to day 2? Was there a difference in the proportion of bees active during peak hours ?
30. **Bird Flight** A blackpoll warbler flies 4,000 km nonstop from North America to South America in about 90 hours. What is its rate of speed?

chapter 16 Standardized Test Practice

Part 1 Multiple Choice

Record your answers on the answer sheet provided by your teacher or on a sheet of paper.

1. Which of the following is true about innate behaviors?
   A. They are learned behaviors.
   B. They are observed in only some animals.
   C. They are the result of conscious thought.
   D. They include reflexes.

2. A spider spinning its web is an example of a(n)
   A. reflex.
   B. instinct.
   C. imprinting.
   D. conditioning.

Use the illustration below to answer questions 3 and 4.

3. The illustration above describes what kind of learned behavior?
   A. conditioning
   B. trial and error
   C. imprinting
   D. insight

4. Which of the following best describes this learned behavior?
   A. The dog learns to salivate when presented with food.
   B. The dog learns to eat only if the bell is rung.
   C. The dog is conditioned to stop salivating when a bell is rung.
   D. The dog is conditioned to salivate when a bell is rung.

5. Which of the following is an example of territorial behavior?
   A. A honeybee performs a waggle dance when it returns to the hive.
   B. A peacock fans his tail while approaching a peahen.
   C. A mountain goat charges and attacks an unfamiliar mountain goat.
   D. A group of bats remain in hibernation for the winter.

Use the photo below to answer questions 6 and 7.

6. The male wolf lying on its back is displaying what kind of behavior to the other male wolf?
   A. aggressive behavior
   B. submissive behavior
   C. cyclic behavior
   D. courtship behavior

7. Which of the following statements best describes the behavior of the wolf that is standing?
   A. The wolf is displaying its dominance over the wolf on the ground.
   B. The wolf is displaying courtship behavior to the other wolf.
   C. The wolf is using bioluminescence to communicate with the other wolf.
   D. The wolf is watching the other wolf perform the waggle dance.

Part 2 Short Response/Grid In

Record your answers on the answer sheet provided by your teacher or on a sheet of paper.

8. Give an example of an innate behavior in a hummingbird.

9. Which is simpler and more automatic, instincts or reflexes?

Use the illustration below to answer questions 10 and 11.

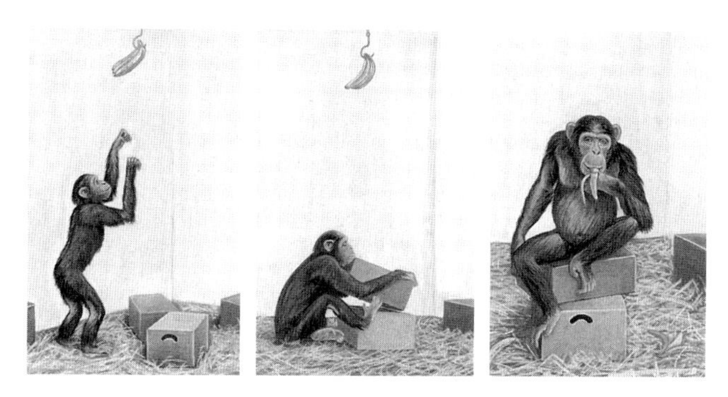

10. What type of learning is shown above?

11. What is required in order for an animal to use this type of learning to solve a problem?

12. Could a young child solve a problem using insight? Why or why not?

13. Give three examples of social behaviors.

14. Why might an animal be submissive to another animal?

Test-Taking Tip

Compare and Contrast Make sure each part of the question is answered when listing discussion points. For example, if the question asks you to compare and contrast, make sure you list both similarities and differences.

Part 3 Open Ended

Record your answers on a sheet of paper.

15. Compare and contrast the innate behaviors of animals with short life spans and animals with long life spans.

16. Give three examples of ways bioluminescence is used for communication.

17. Explain the difference between a diurnal animal and a nocturnal animal. Give an example of each.

18. Compare and contrast hibernation and estivation.

Use the photo below to answer questions 19 and 20.

19. Explain the type of behavior that is shown above.

20. How is this behavior related to why zoos feed newborn condors with hand puppets that look like adult condors?

21. A male antelope approaches a female antelope during the breeding season. Is the male antelope responding to an external stimulus, an internal stimulus, or both? Explain.

Human Body Systems

How are Chickens & Rice Connected?

Back in the 1800s, a mysterious disease called beriberi affected people in certain parts of Asia. One day, a doctor in Indonesia noticed some chickens staggering around, a symptom often seen in people with beriberi. It turned out that the chickens had been eating white rice—the same kind of rice that was being eaten by human beriberi sufferers. White rice has had the outer layers, including the bran, removed. When the sick chickens were fed rice that still had its bran, they quickly recovered. It turned out that the same treatment worked for people with beriberi! Research eventually showed that rice bran contains a vitamin, B_1, which is essential for good health. Today, white rice usually is "vitamin-enriched" to replace B_1 and other nutrients lost in processing.

unit projects

Visit **life.msscience.com/unit_project** to find project ideas and resources. Projects include:

- **History** Contribute to a class "remedy journal" with interesting, out-dated medical treatments, and how techniques have improved.
- **Technology** Investigate rare and interesting medical conditions, including their history, characteristics, and treatments. Present a colorful poster with photos and information for class display.
- **Model** Research and create a menu that includes vitamin-rich foods. Prepare a sample and a recipe card for a class food fair.

Understand the *History of Disease Prevention,* and how science has progressed through history. Become acquainted with famous scientists and learn how healthy lifestyles prevent disease.

chapter 17

Structure and Movement

The BIG Idea

Our bones, muscles, and skin give our bodies structure and enable us to move.

SECTION 1
The Skeletal System

Main Idea Bones support our bodies, protect internal organs, and store minerals.

SECTION 2
Muscular System

Main Idea Muscles provide motion for internal organs, perform many tasks, and enable us to move from place to place.

SECTION 3
The Skin

Main Idea The skin protects us, senses stimuli, forms vitamin D, helps regulate body temperature, and excretes wastes.

How are you like a building?

Internal and external structures support both buildings and the human body. Bones support us instead of steel or wood. The covering of a building protects the inside from the outside environment. Your skin protects your body's internal environment.

Science Journal Imagine that your body did not have a support system. Describe how you might perform your daily activities.

Start-Up Activities

Effect of Muscles on Movement

The expression "Many hands make light work" is also true when it comes to muscles in your body. In fact, hundreds of muscles and bones work together to bring about smooth, easy movement. Muscle interactions enable you to pick up a penny or lift a 10-kg weight.

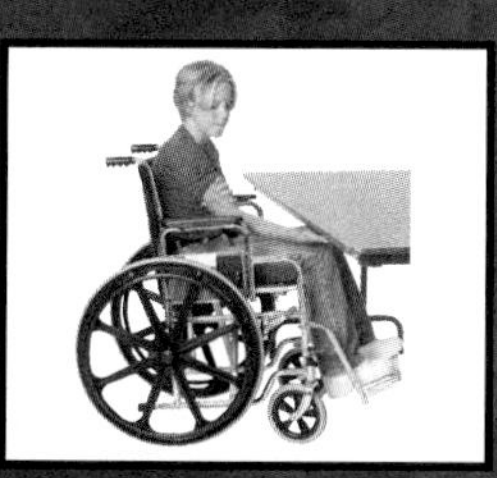

1. Sit on a chair at an empty table and place the palm of one hand under the edge of the table.
2. Push your hand up against the table. Do not push too hard.
3. Use your other hand to feel the muscles located on both sides of your upper arm, as shown in the photo.
4. Next, place your palm on the top of the table and push down. Again, feel the muscles in your upper arm.
5. **Think Critically** Describe in your Science Journal how the different muscles in your upper arm were working during each movement.

FOLDABLES™ Study Organizer

Structure and Movement Without skin, muscle and bone each of us would be a formless mass. Make the following Foldable to help you understand the function of skin, muscle and bone in structure and movement.

STEP 1 **Fold** a sheet of paper in half length-wise. Make the back edge about 5 cm longer than the front edge.

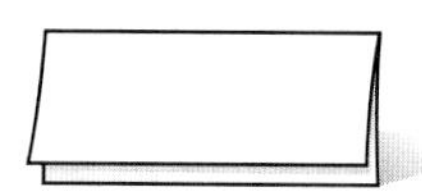

STEP 2 **Turn** the paper so the fold is on the bottom. Then, **fold** it into thirds.

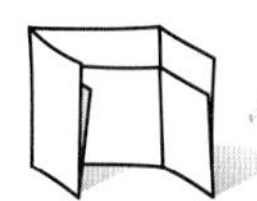

STEP 3 **Unfold and cut** only the top layer along both folds to make three tabs. **Label** the Foldable as shown.

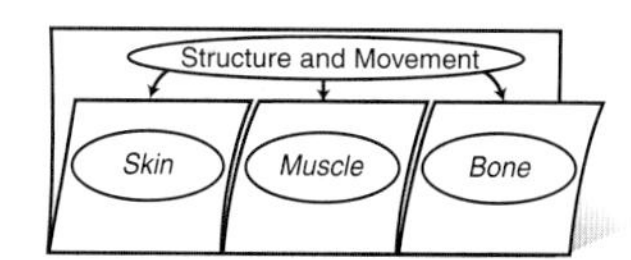

Read and Write As you read this chapter, write the functions that skin, muscle, and bone each have in structure and movement.

Preview this chapter's content and activities at life.msscience.com

Get Ready to Read

Monitor

1 Learn It! An important strategy to help you improve your reading is monitoring, or finding your reading strengths and weaknesses. As you read, monitor yourself to make sure the text makes sense. Discover different monitoring techniques you can use at different times, depending on the type of test and situation.

2 Practice It! The paragraph below appears in Section 2. Read the passage and answer the questions that follow. Discuss your answers with other students to see how they monitor their reading.

> How do muscles allow you to move your body? You move because pairs of skeletal muscles work together. When one muscle of a pair contracts, the other muscle relaxes, or returns to its original length, as shown in **Figure 10.** Muscles always pull. They never push. When the muscles on the back of your upper leg contract, they pull your lower leg back and up. When you straighten your leg, the back muscles lengthen and relax, and the muscles on the front of your upper leg contract. Compare how the muscles of your legs work with how the muscles of your arms work.
>
> *—from page 494*

- What questions do you still have after reading?
- Do you understand all of the words in the passage?
- Did you have to stop reading often? Is the reading level appropriate for you?

3 Apply It! Identify one paragraph that is difficult to understand. Discuss it with a partner to improve your understanding.

Target Your Reading

Use this to focus on the main ideas as you read the chapter.

Reading Tip

Monitor your reading by slowing down or speeding up depending on your understanding of the text.

1. **Before you read** the chapter, respond to the statements below on your worksheet or on a numbered sheet of paper.
   - Write an **A** if you **agree** with the statement.
   - Write a **D** if you **disagree** with the statement.

2. **After you read** the chapter, look back to this page to see if you've changed your mind about any of the statements.
   - If any of your answers changed, explain why.
   - Change any false statements into true statements.
   - Use your revised statements as a study guide.

Before You Read A or D	Statement	After You Read A or D
	1 Bones are hard, nonliving structures.	
	2 Red blood cells form in the centers of some bones.	
	3 Bones rub against each other at joints.	
	4 Your arm muscles are the same as your heart muscles.	
	5 Movement occurs because muscles relax and contract.	
	6 Muscles increase in size mostly because the number of muscle cells increases.	
	7 The skin is the largest organ of the human body.	
	8 The different skin colors result from different pigments in skin.	
	9 A bruise forms when the skin breaks.	
	10 Human skin grafts can be grown from donor skin tissue.	

Science Online
Print out a worksheet of this page at life.msscience.com

section 1

The Skeletal System

as you read

What You'll Learn

- **Identify** five functions of the skeletal system.
- **Compare and contrast** movable and immovable joints.

Why It's Important

You'll begin to understand how your bones and joints allow you to move.

Review Vocabulary
skeleton: a framework of living bones that supports your body

New Vocabulary
- skeletal system
- periosteum
- cartilage
- joint
- ligament

Living Bones

Often in a horror movie, a mad scientist works frantically in his lab while a complete human skeleton hangs silently in the corner. When looking at a skeleton, you might think that bones are dead structures made of rocklike material. Although these bones are no longer living, the bones in your body are very much alive. Each is a living organ made of several different tissues. Like all the other living tissues in your body, bone tissue is made of cells that take in nutrients and use energy. Bone cells have the same needs as other body cells.

Functions of Your Skeletal System All the bones in your body make up your **skeletal system,** as shown in **Figure 1.** It is the framework of your body and has five major functions.

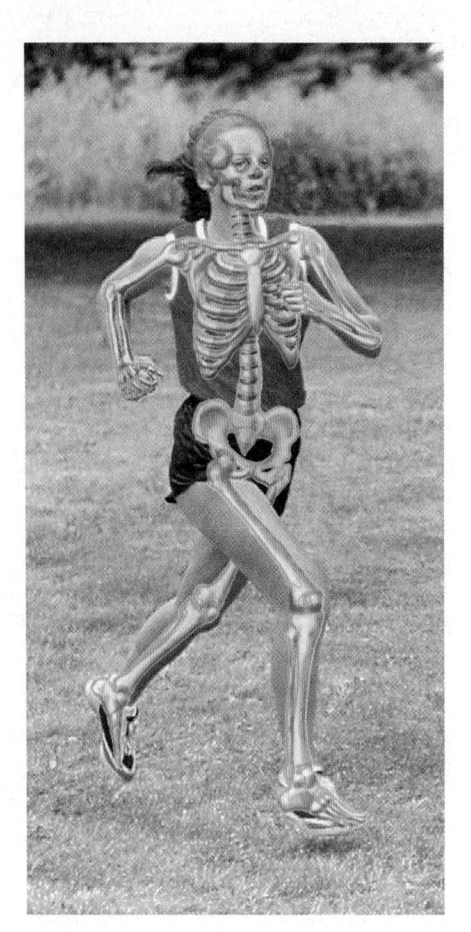

Figure 1 The 206 bones of the human body are connected, forming a framework called the skeleton.

1. The skeleton gives shape and support to your body.
2. Bones protect your internal organs. For example, ribs surround the heart and lungs, and the skull encloses the brain.
3. Major muscles are attached to bone and help them move.
4. Blood cells are formed in the center of many bones in soft tissue called red marrow.
5. Major quantities of calcium and phosphorous compounds are stored in the skeleton for later use. Calcium and phosphorus make bones hard.

Bone Structure

Several characteristics of bones are noticeable. The most obvious are the differences in their sizes and shapes. The shapes of bones are inherited. However, a bone's shape can change when the attached muscles are used.

Looking at bone through a magnifying glass will show you that it isn't smooth. Bones have bumps, edges, round ends, rough spots, and many pits and holes. Muscles and ligaments attach to some of the bumps and pits. In your body blood vessels and nerves enter and leave through the holes. Internal characteristics, how a bone looks from the inside, and external characteristics, how the same bone looks from the outside, are shown in **Figure 2.**

A living bone's surface is covered with a tough, tight-fitting membrane called the **periosteum** (per ee AH stee um). Small blood vessels in the periosteum carry nutrients into the bone. Cells involved in the growth and repair of bone also are found in the periosteum. Under the periosteum are two different types of bone tissue—compact bone and spongy bone.

Compact Bone Directly under the periosteum is a hard, strong layer called compact bone. Compact bone gives bones strength. It has a framework containing deposits of calcium phosphate. These deposits make the bone hard. Bone cells and blood vessels also are found in this layer. This framework is living tissue and even though it's hard, it keeps bone from being too rigid, brittle, or easily broken.

Figure 2 Bone is made of layers of living tissue. Compact bone is arranged in circular structures called Haversian systems—tiny, connected channels through which blood vessels and nerve fibers pass.

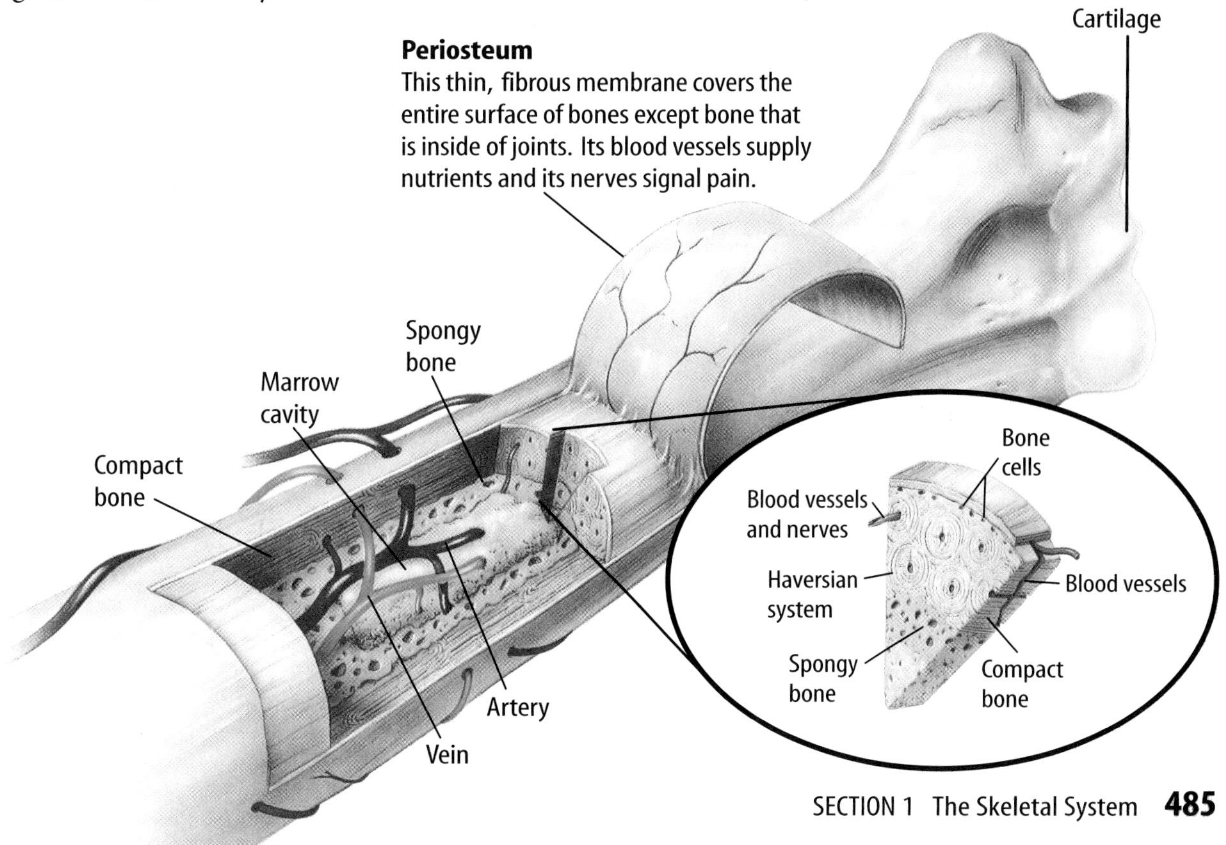

Topic: Bone Fractures

Visit life.msscience.com for Web links to information about new techniques for treating bone fractures.

Activity Describe one of these new techniques in your Science Journal.

Spongy Bone Spongy bone is located toward the ends of long bones such as those in your thigh and upper arm. Spongy bone has many small, open spaces that make bones lightweight. If all your bones were completely solid, you'd have greater mass. In the centers of long bones are large openings called cavities. These cavities and the spaces in spongy bone are filled with a substance called marrow. Some marrow is yellow and is composed of fat cells. Red marrow produces red blood cells at a rate of two million to three million cells per second.

Cartilage The ends of bones are covered with a smooth, slippery, thick layer of tissue called **cartilage.** Cartilage does not contain blood vessels or minerals. Nutrients are delivered to cartilage by nearby blood vessels. Cartilage is flexible and important in joints because it acts as a shock absorber. It also makes movement easier by reducing friction that would be caused by bones rubbing together. Cartilage can be damaged because of disease, injury, or years of use. People with damaged cartilage experience pain when they move.

What is cartilage?

Figure 3 Cartilage is replaced slowly by bone as solid tissue grows outward. Over time, the bone reshapes to include blood vessels, nerves, and marrow.
Describe *the type of bone cell that builds up bone.*

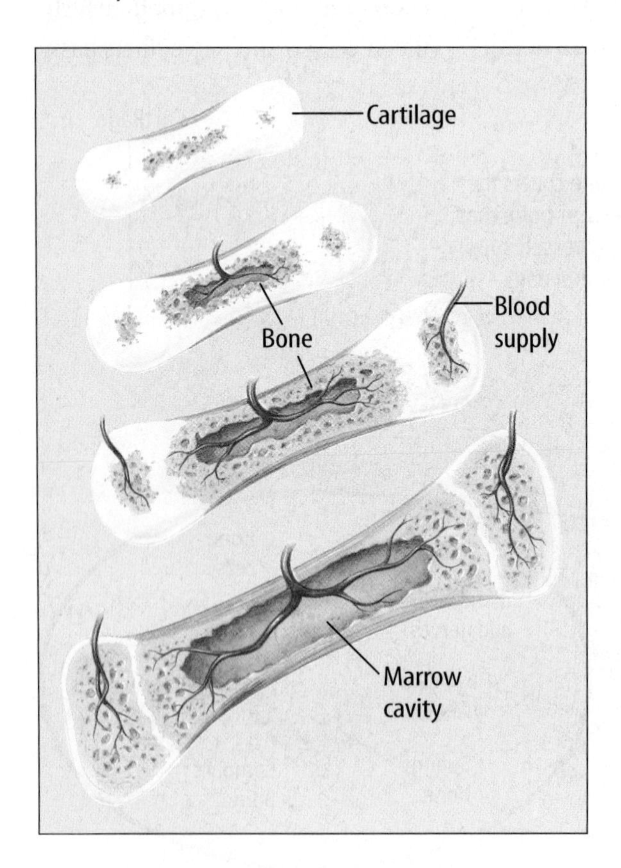

Bone Formation

Although your bones have some hard features, they have not always been this way. Months before your birth, your skeleton was made of cartilage. Gradually the cartilage broke down and was replaced by bone, as illustrated in **Figure 3.** Bone-forming cells called osteoblasts (AHS tee oh blasts) deposit the minerals calcium and phosphorus in bones, making the bone tissue hard. At birth, your skeleton was made up of more than 300 bones. As you developed, some bones fused, or grew together, so that now you have only 206 bones.

Healthy bone tissue is always being formed and re-formed. Osteoblasts build up bone. Another type of bone cell, called an osteoclast, breaks down bone tissue in other areas of the bone. This is a normal process in a healthy person. When osteoclasts break bone down, they release the elements calcium and phosphorus into the bloodstream. This process maintains calcium and phosphorus in your blood at about the levels they need to be. These elements are necessary for the working of your body, including the movement of your muscles.

Joints

What will you do during your lunch break today? You may sit at a table, pick up a sandwich, bite off a piece of a carrot and chew it, and walk to class. All of these motions are possible because your skeleton has joints.

Any place where two or more bones come together is a **joint.** The bones making up healthy joints are kept far enough apart by a thin layer of cartilage so that they do not rub against each other as they move. The bones are held in place at these joints by tough bands of tissue called **ligaments.** Many joints, such as your knee, are held together by several ligaments. Muscles move bones by moving joints.

Applying Math — Calculate Volume

VOLUME OF BONES The Haversian systems found in the cross section of your bones are arranged in long cylinders. This cylindrical shape allows your bones to withstand great pressure. Estimate the volume of a bone that is 36 cm long and is 7 cm in diameter.

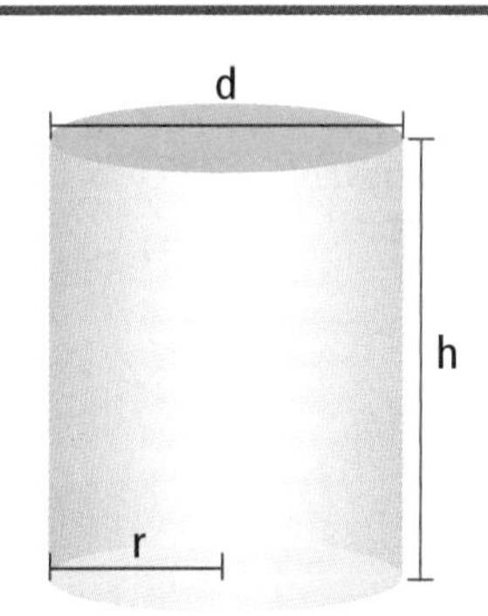

Solution

1 *This is what you know:* The bone has a shape of a cylinder whose height, h, measures 36 cm and whose diameter is 7.0 cm.

2 *This is what you need to find out:* What is the volume of the cylinder?

3 *This is the procedure you need to use:*

- Volume = $\pi \times (\text{radius})^2 \times$ height, or $V = \pi \times r^2 \times h$
- A radius is one-half the diameter $\left(\frac{1}{2} \times 7 \text{ cm}\right)$, so $r = 3.5$ cm, $h = 36$ cm, and $\pi = 3.14$.
- Substitute in known values and solve. $3.14 \times (3.5 \text{ cm})^2 \times 36 \text{ cm} = 1{,}384.74 \text{ cm}^3$
- The volume of the bone is approximately $1{,}384.74 \text{ cm}^3$.

4 *Check your answer:* Divide your answer by 3.14 and then divide that number by $(3.5)^2$. This number should be the height of the bone.

Practice Problems

1. Estimate the volume of a bone that has a height of 12 cm and a diameter of 2.4 cm.
2. If a bone has a volume of 314 cm^3 and a diameter of 4 cm, what is its height?

For more practice, visit life.msscience.com/math_practice

Immovable Joints Refer to **Figure 4** as you learn about different types of joints. Joints are broadly classified as immovable or movable. An immovable joint allows little or no movement. The joints of the bones in your skull and pelvis are classified as immovable joints.

Movable Joints All movements, including somersaulting and working the controls of a video game, require movable joints. A movable joint allows the body to make a wide range of motions. There are several types of movable joints—pivot, ball and socket, hinge, and gliding. In a pivot joint, one bone rotates in a ring of another bone that does not move. Turning your head is an example of a pivot movement.

A ball-and-socket joint consists of a bone with a rounded end that fits into a cuplike cavity on another bone. A ball-and-socket joint provides a wider range of motion than a pivot joint does. That's why your legs and arms can swing in almost any direction.

A third type of joint is a hinge joint. This joint has a back-and-forth movement like hinges on a door. Elbows, knees, and fingers have hinge joints. Hinge joints have a smaller range of motion than the ball-and-socket joint. They are not dislocated as easily, or pulled apart, as a ball-and-socket joint can be.

A fourth type of joint is a gliding joint in which one part of a bone slides over another bone. Gliding joints also move in a back-and-forth motion and are found in your wrists and ankles and between vertebrae. Gliding joints are used the most in your body. You can't write a word, use a joystick, or take a step without using a gliding joint.

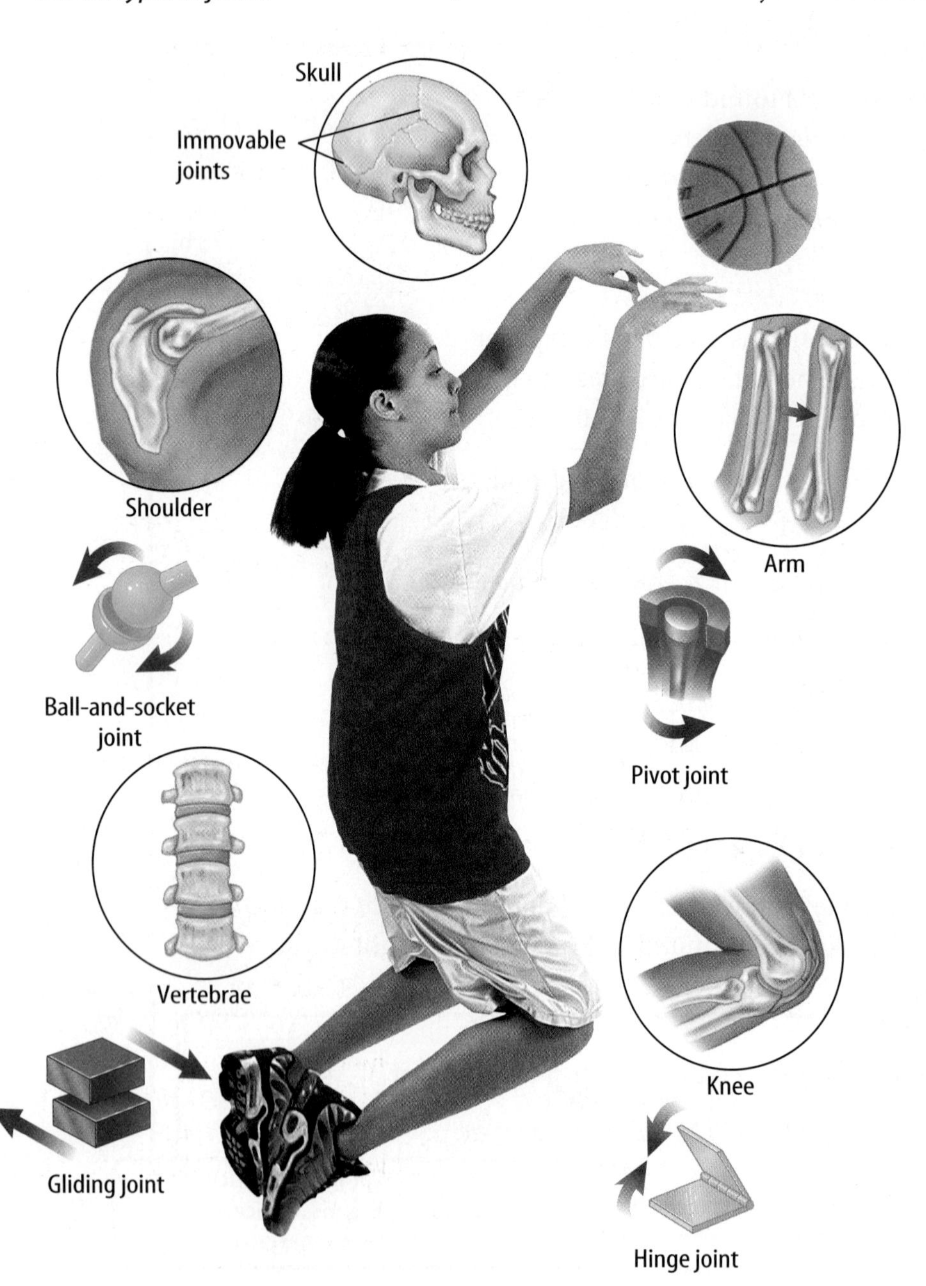

Figure 4 When a basketball player shoots a ball, several types of joints are in action.
Describe *other activities that use several types of joints.*

Moving Smoothly When you rub two pieces of chalk together, their surfaces begin to wear away, and they get reshaped. Without the protection of the cartilage at the end of your bones, they also would wear away at the joints. Cartilage helps make joint movement easier. It reduces friction and allows bones to slide more easily over each other. Shown in **Figure 5,** pads of cartilage, called disks, are located between the vertebrae in your back. They act as a cushion and prevent injury to your spinal cord. A fluid that comes from nearby blood vessels also lubricates the joint.

Reading Check *Why is cartilage important?*

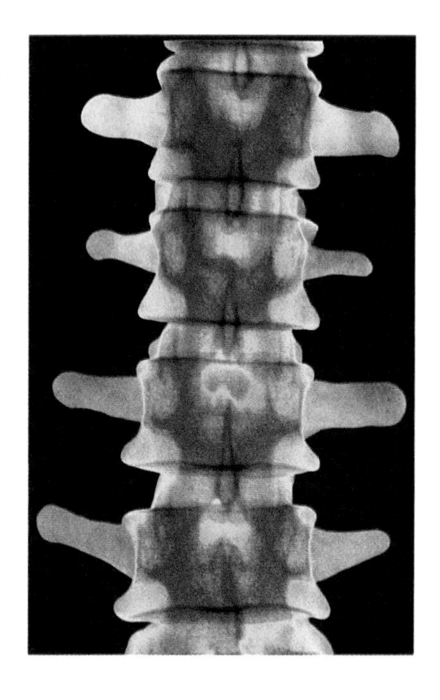

Figure 5 A colored X ray of the human backbone shows disks of cartilage between the vertebrae.

Common Joint Problems Arthritis is the most common joint problem. The term *arthritis* describes more than 100 different diseases that can damage the joints. About one out of every seven people in the United States suffers from arthritis. All forms of arthritis begin with the same symptoms: pain, stiffness, and swelling of the joints.

Two types of arthritis are osteoarthritis and rheumatoid arthritis. Osteoarthritis results when cartilage breaks down because of years of use. Rheumatoid arthritis is an ongoing condition in which the body's immune system tries to destroy its own tissues.

section 1 review

Summary

Living Bones

- The skeletal system is the framework of your body and has five major functions.

Bone Structure

- A tough membrane called the periosteum covers a bone and supplies nutrients to it.
- Compact bone is hard bone located directly under the periosteum.
- Spongy bone is lightweight and located toward the ends of long bones.
- Cartilage covers the ends of bones and acts as a shock absorber.

Bone Formation

- Osteoblasts are bone-forming cells and osteoclasts are cells that break down bone.
- A joint is any place where two or more bones come together.
- Ligaments are tough bands of tissue that hold bones together at joints.

Self Check

1. **List** the five major functions of the skeletal system.
2. **Name** and give an example of both a movable joint and an immovable joint.
3. **Explain** the functions of cartilage in your skeletal system.
4. **Describe** ligaments.
5. **Think Critically** A thick band of bone forms around a broken bone as it heals. In time, the thickened band disappears. Explain how this extra bone can disappear over time.

Applying Skills

6. **Make and Use Tables** Use a table to classify the bones of the human body as follows: *long, short, flat,* or *irregular.*
7. **Use graphics software** to make a circle graph that shows how an adult's bones are distributed: *29 skull bones, 26 vertebrae, 25 ribs, four shoulder bones, 60 arm and hand bones, two hip bones,* and *60 leg and feet bones.*

section 2

The Muscular System

as you read

What You'll Learn

- **Identify** the major function of the muscular system.
- **Compare and contrast** the three types of muscles.
- **Explain** how muscle action results in the movement of body parts.

Why It's Important

The muscular system is responsible for how you move and the production of thermal energy in your body. Muscles also give your body its shape.

Review Vocabulary

bone: dense, calcified tissue of the skeleton, that is moved by muscles

New Vocabulary

- muscle
- voluntary muscle
- involuntary muscle
- skeletal muscle
- tendon
- cardiac muscle
- smooth muscle

Movement of the Human Body

The golfer looks down the fairway and then at the golf ball. With intense concentration and muscle coordination, the golfer swings the club along a graceful arc and connects with the ball. The ball sails through the air, landing inches away from the flag. The crowd applauds. A few minutes later, the golfer makes the final putt and wins the tournament. The champion has learned how to use controlled muscle movement to bring success.

Muscles help make all of your daily movements possible. **Figure 6** shows which muscles connect some of the bones in your body. A **muscle** is an organ that can relax, contract, and provide the force to move your body parts. In the process, energy is used and work is done. Imagine how much energy the more than 600 muscles in your body use each day. No matter how still you might try to be, some muscles in your body are always moving. You're breathing, your heart is beating, and your digestive system is working.

Figure 6 Your muscles come in many shapes and sizes. Even simple movements require the coordinated use of several muscles. The muscles shown here are only those located directly under the skin. Beneath these muscles are middle and deep layers of muscles.

Figure 7 Facial expressions generally are controlled by voluntary muscles. It takes only 13 muscles to smile, but 43 muscles to frown.

Muscle Control Your hand, arm, and leg muscles are voluntary. So are the muscles of your face, shown in **Figure 7.** You can choose to move them or not move them. Muscles that you are able to control are called **voluntary muscles.** In contrast, **involuntary muscles** are muscles you can't control consciously. They go on working all day long, all your life. Blood gets pumped through blood vessels, and food is moved through your digestive system by the action of involuntary muscles.

What is a body activity that is controlled by involuntary muscles?

Your Body's Simple Machines—Levers

Your skeletal system and muscular system work together when you move, in the same way that the parts of a bicycle work together when it moves. A machine, such as a bicycle, is any device that makes work easier. A simple machine does work with only one movement, like a hammer. The hammer is a type of simple machine called a lever, which is a rod or plank that pivots or turns about a point. This point is called a fulcrum. The action of muscles, bones, and joints working together is like a lever. In your body, bones are rods, joints are fulcrums, and contraction and relaxation of muscles provide the force to move body parts. Levers are classified into three types—first-class, second-class, and third-class. Examples of the three types of levers that are found in the human body are shown in **Figure 8.**

Topic: Joint Replacement
Visit life.msscience.com for Web links to recent news or magazine articles about replacing diseased joints.

Activity Make a list in your Science Journal of the most commonly replaced joints.

NATIONAL GEOGRAPHIC

VISUALIZING HUMAN BODY LEVERS

Figure 8

All three types of levers—first-class, second-class, and third-class—are found in the human body. In the photo below, a tennis player prepares to serve a ball. As shown in the accompanying diagrams, the tennis player's stance demonstrates the operation of all three classes of levers in the human body.

▲ Fulcrum

Input force

Output force

FIRST-CLASS LEVER
The fulcrum lies between the input force and the output force. This happens when the tennis player uses his neck muscles to tilt his head back.

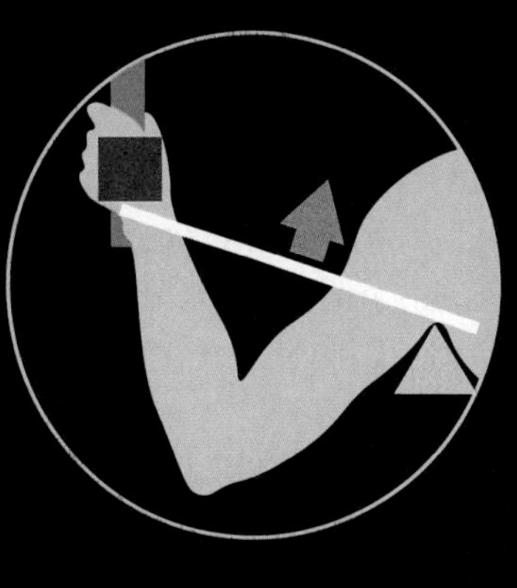

THIRD-CLASS LEVER
The input force is between the fulcrum and the output force. This happens when the tennis player flexes the muscles in his arm and shoulder.

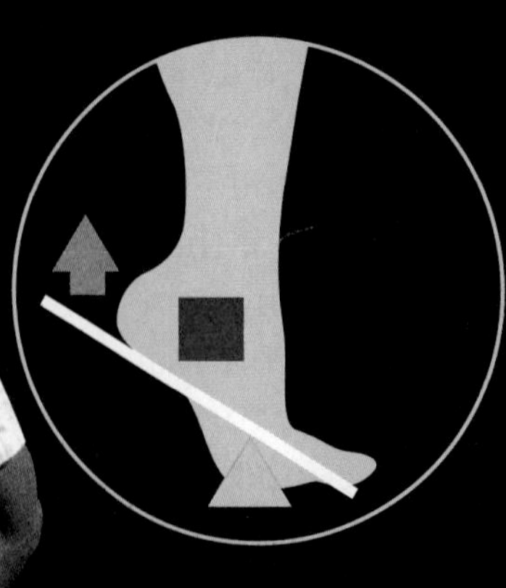

SECOND-CLASS LEVER
The output force lies between the fulcrum and the input force. This happens when the tennis player's calf muscles lift the weight of his body up on his toes.

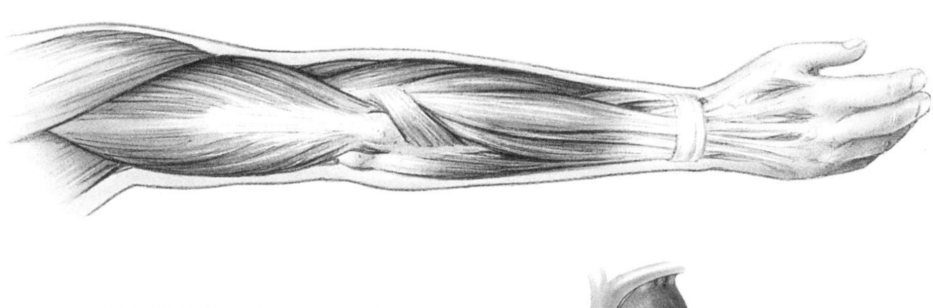

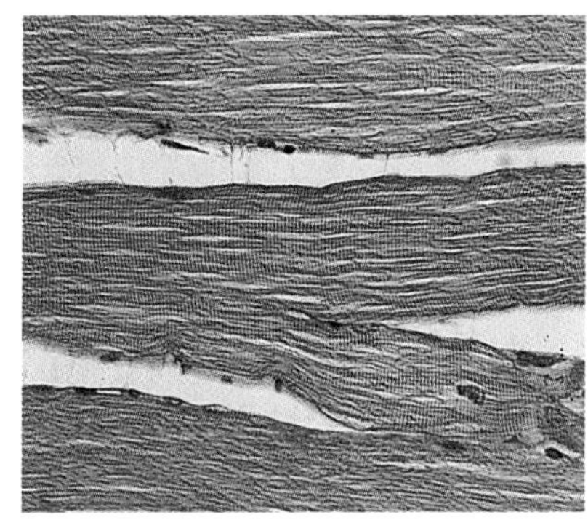

A Skeletal muscles move bones. The muscle tissue is striated, and attached to bone.

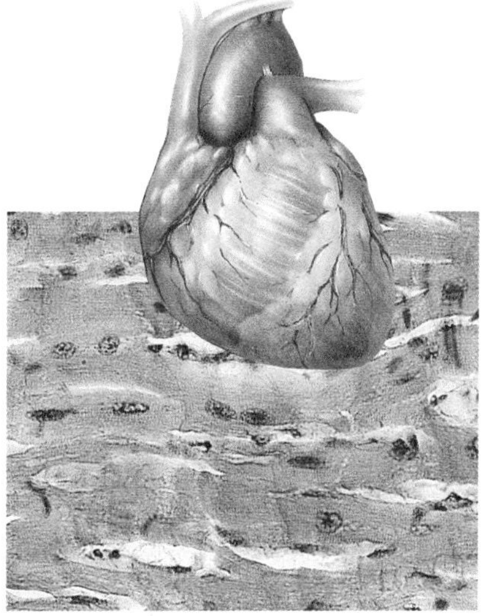

B Cardiac muscle is found only in the heart. The muscle tissue has striations.

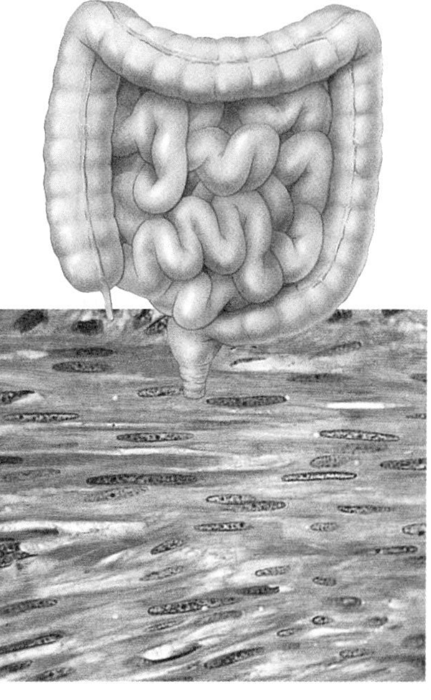

C Smooth muscle is found in many of your internal organs, such as the digestive tract. This muscle tissue is nonstriated.

Figure 9 The three types of muscle tissue are skeletal muscle, cardiac muscle, and smooth muscle.

Classification of Muscle Tissue

All the muscle tissue in your body is not the same. The three types of muscles are skeletal, smooth, and cardiac. The muscles that move bones are **skeletal muscles.** They are more common than other muscle types and are attached to bones by thick bands of tissue called **tendons.** When viewed under a microscope, skeletal muscle cells are striated (STRI ay tud), and appear striped. You can see the striations in **Figure 9A.** Skeletal muscles are voluntary muscles. You choose when to walk or when not to walk. Skeletal muscles tend to contract quickly and tire more easily than involuntary muscles do.

The remaining two types of muscles are shown in **Figures 9B** and **9C. Cardiac muscle** is found only in the heart. Like skeletal muscle, cardiac muscle is striated. This type of muscle contracts about 70 times per minute every day of your life. **Smooth muscles** are found in your intestines, bladder, blood vessels, and other internal organs. They are nonstriated, involuntary muscles that slowly contract and relax. Internal organs are made of one or more layers of smooth muscles.

Figure 10 **A** When the flexor (hamstring) muscles of your thigh contract, the lower leg is brought toward the thigh. **B** When the extensor (quadriceps) muscles contract, the lower leg is straightened.
Describe *the class of lever shown to the right.*

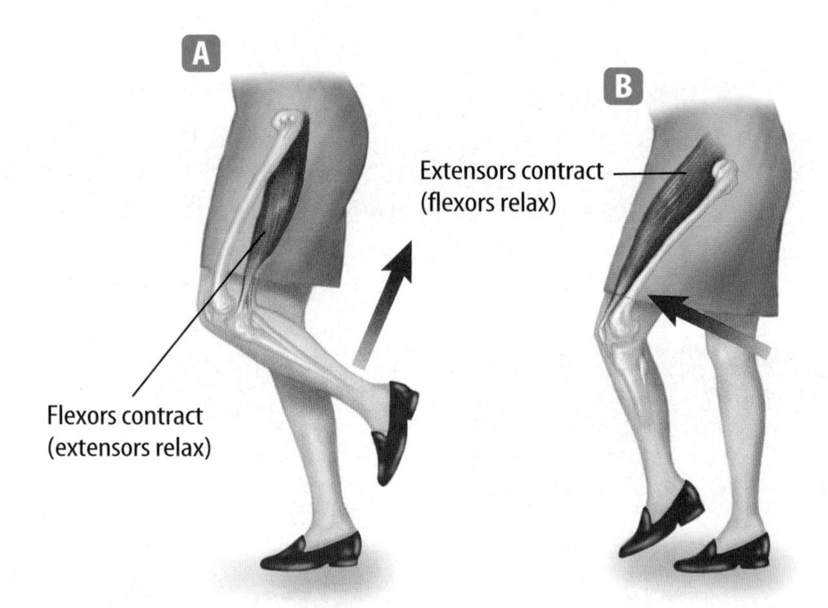

Working Muscles

How do muscles allow you to move your body? You move because pairs of skeletal muscles work together. When one muscle of a pair contracts, the other muscle relaxes, or returns to its original length, as shown in **Figure 10.** Muscles always pull. They never push. When the muscles on the back of your upper leg contract, they shorten and pull your lower leg back and up. When you straighten your leg, the back muscles lengthen and relax, and the muscles on the front of your upper leg contract. Compare how the muscles of your legs work with how the muscles of your arms work.

Changes in Muscles Over a period of time, muscles can become larger or smaller, depending on whether or not they are used. Skeletal muscles that do a lot of work, such as those in your writing hand, become strong and large. For example, many soccer and basketball players have noticeably larger, defined leg muscles. Muscles that are given regular exercise respond quickly to stimuli. Some of this change in muscle size is because of an increase in the number of muscle cells. However, most of this change in muscle size is because individual muscle cells become larger.

In contrast, if you participate only in nonactive pastimes such as watching television or playing computer games, your muscles will become soft and flabby and will lack strength. Muscles that aren't exercised become smaller in size. When someone is paralyzed, his or her muscles become smaller due to lack of use.

How do muscles increase their size?

Comparing Muscle Activity

Procedure

1. Hold a **book** in your outstretched hand over a dining or kitchen **table.**
2. Lift the book from this position to a height of 30 cm from the table 20 times.

Analysis

1. Compare your arm muscle activity to the continuous muscle activity of the heart.
2. Infer whether heart muscles become tired.

How Muscles Move Your muscles need energy to contract and relax. Your blood carries energy-rich molecules to your muscle cells where the chemical energy stored in these molecules is released. As the muscle contracts, this released energy changes to mechanical energy (movement) and thermal energy (warmth), as shown in **Figure 11.** When the supply of energy-rich molecules in a muscle is used up, the muscle becomes tired and needs to rest. During this resting period, your blood supplies more energy-rich molecules to your muscle cells. The thermal energy of muscle contractions helps keep your body temperature constant.

Figure 11 Chemical energy is needed for muscle activity. During activity, chemical energy supplied by food is changed into mechanical energy (movement) and thermal energy (warmth).

section 2 review

Summary

Movement of the Human Body

- Muscles are organs that relax, contract, and provide force to move your body parts.

Classification of Muscle Tissue

- Skeletal muscles are striated muscles that move bones.
- Cardiac muscles are striated muscles which are found only in the heart.
- Smooth muscles are found in your internal organs and are nonstriated muscles.

Working Muscles

- Muscles always pull and when one muscle of a pair contracts, the other muscle relaxes.
- Chemical energy is needed for muscle activity.

Self Check

1. **Describe** the function of muscles.
2. **Compare and contrast** the three types of muscle tissue.
3. **Name** the type of muscle tissue found in your heart.
4. **Describe** how a muscle attaches to a bone.
5. **Think Critically** What happens to your upper-arm muscles when you bend your arm at the elbow?

Applying Skills

6. **Concept Map** Using a concept map, sequence the activities that take place when you bend your leg at the knee.
7. **Communicate** Write a paragraph in your Science Journal about the three forms of energy involved in a muscle contraction.

section 3

The Skin

as you read

What You'll Learn

- **Distinguish** between the epidermis and dermis of the skin.
- **Identify** the skin's functions.
- **Explain** how skin protects the body from disease and how it heals itself.

Why It's Important

Skin plays a vital role in protecting your body.

Review Vocabulary

vitamin: an inorganic nutrient needed by the body in small quantities for growth, disease prevention, and/or regulation of body functions

New Vocabulary

- epidermis
- dermis
- melanin

Your Largest Organ

What is the largest organ in your body? When you think of an organ, you might imagine your heart, stomach, lungs, or brain. However, your skin is the largest organ of your body. Much of the information you receive about your environment comes through your skin. You can think of your skin as your largest sense organ.

Skin Structures

Skin is made up of three layers of tissue—the epidermis, the dermis, and a fatty layer—as shown in **Figure 12.** Each layer of skin is made of different cell types. The **epidermis** is the outer, thinnest layer of your skin. The epidermis's outermost cells are dead and water repellent. Thousands of epidermal cells rub off every time you take a shower, shake hands, blow your nose, or scratch your elbow. New cells are produced constantly at the base of the epidermis. These new cells move up and eventually replace those that are rubbed off.

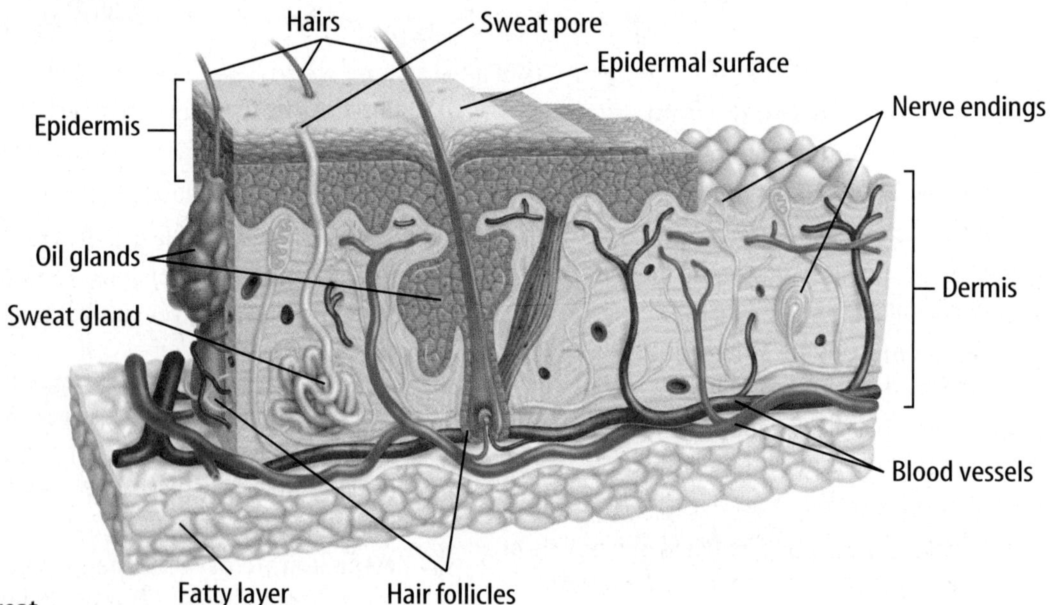

Figure 12 Hair, sweat glands, and oil glands are part of your body's largest organ, the skin.

Melanin Cells in the epidermis produce the chemical melanin (MEL uh nun). **Melanin** is a pigment that protects your skin from the ultraviolet (UV) rays in sunlight and gives it color. The different amounts of melanin produced by cells result in differences in skin color, as shown in **Figure 13.** When your skin is exposed to UV rays, melanin production increases and your skin becomes darker. Lighter skin tones have less protection. Such skin burns more easily and may be more susceptible to skin cancer.

Other Skin Layers The **dermis** is the layer of cells directly below the epidermis. This layer is thicker than the epidermis and contains many blood vessels, nerves, muscles, oil and sweat glands, and other structures. Below the dermis is a fatty region that insulates the body. This is where much of the fat is deposited when a person gains weight.

Skin Functions

Your skin is not only the largest organ of your body, it also carries out several major functions, including protection, sensory response, formation of vitamin D, regulation of body temperature, and ridding the body of wastes. The most important function of the skin is protection. The skin forms a protective covering over the body that prevents physical and chemical injury. Some bacteria and other disease-causing organisms cannot pass through the skin as long as it is unbroken. Glands in the skin secrete fluids that can damage or destroy some bacteria. The skin also slows down water loss from body tissues.

Specialized nerve cells in the skin detect and relay information to the brain, making the skin a sensory organ, too. Because of these cells, you are able to sense the softness of a cat, the sharpness of a pin, or the heat of a frying pan.

Mountain Climber
Research the effects of ultraviolet radiation on skin. Mountain climbers risk becoming severely sunburned even in freezing temperatures due to increased ultraviolet (UV) radiation. Research other careers that increase your risk of sunburn. Record your answers in your Science Journal.

Figure 13 Melanin gives skin and eyes their color. The more melanin that is present, the darker the color is. This pigment provides protection from damage caused by harmful UV rays.

Figure 14 Normal human body temperature is about 37°C. Temperature varies throughout the day. The highest body temperature is reached at about 11 A.M. and the lowest at around 4 A.M. At 43°C (109.5°F) internal bleeding results, causing death.

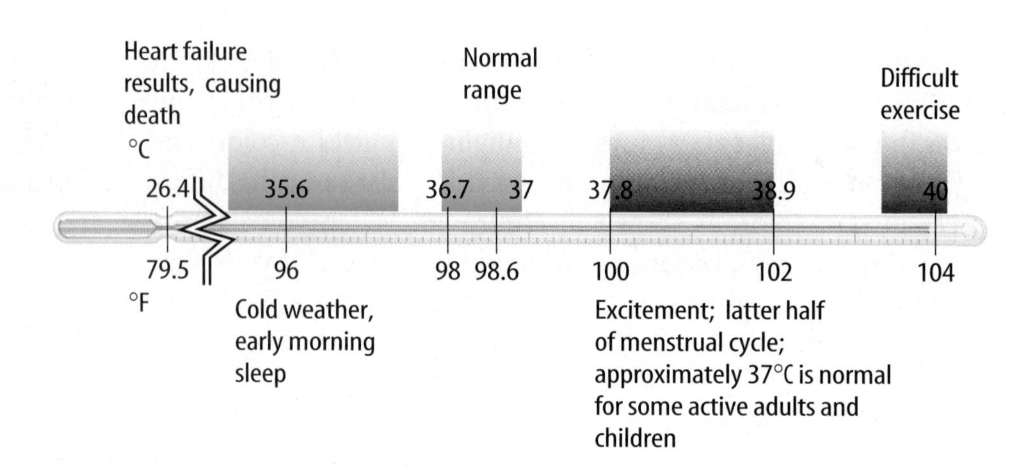

Vitamin D Formation Another important function of skin is the formation of vitamin D. Small amounts of this vitamin are produced in the presence of ultraviolet light from a fatlike molecule in your epidermis. Vitamin D is essential for good health because it helps your body absorb calcium into your blood from food in your digestive tract.

Recognizing Why You Sweat

Procedure

1. Examine the epidermis and the pores of your skin using a **magnifying lens.**
2. Place a **clear-plastic sandwich bag** on your hand. Use tape to seal the bag around your wrist. **WARNING:** *Do not wrap the tape too tightly.*
3. Quietly study your **text** for 10 min, then look at your hand. Remove the bag.
4. Describe what happened to your hand while it was inside the bag.

Analysis

1. Identify what formed inside the bag. Where did this substance come from?
2. Why does this substance form even when you are not active?

Thermal Energy and Waste Exchange Humans can withstand a limited range of body temperatures, as shown in **Figure 14.** Your skin plays an important role in regulating your body temperature. Blood vessels in the skin can help release or hold thermal energy. If the blood vessels expand, or dilate, blood flow increases and thermal energy is released. In contrast, less thermal energy is released when the blood vessels constrict. Think of yourself after running—are you flushed red or pale and shivering?

The adult human dermis has about 3 million sweat glands. These glands help regulate the body's temperature and excrete wastes. When the blood vessels dilate, pores open in the skin that lead to the sweat glands. Perspiration, or sweat, moves out onto the skin. Thermal energy transfers from the body to the sweat on the skin. Eventually, this sweat evaporates, removing the thermal energy and cooling the skin. This system eliminates excess thermal energy produced by muscle contractions.

Reading Check *What are two functions of sweat glands?*

As your cells use nutrients for energy, they produce wastes. Such wastes, if not removed from your body, can act as poisons. In addition to helping regulate your body's temperature, sweat glands release water, salt, and other waste products. If too much water and salt are released by sweating during periods of extreme heat or physical exertion, you might feel light-headed or may even faint.

Skin Injuries and Repair

Your skin often is bruised, scratched, burned, ripped, and exposed to harsh conditions like cold and dry air. In response, the skin produces new cells in its epidermis and repairs tears in the dermis. When the skin is injured, disease-causing organisms can enter the body rapidly. An infection often results.

Bruises Bruises are common, everyday events. Playing sports or working around your house often results in minor injuries. What is a bruise and how does your body repair it?

When you have a bruise, your skin is not broken but the tiny blood vessels underneath the skin have burst. Red blood cells from these broken blood vessels leak into the surrounding tissue. These blood cells then break down, releasing a chemical called hemoglobin. The hemoglobin gradually breaks down into its components, called pigments. The color of these pigments causes the bruised area to turn shades of blue, red, and purple, as shown in **Figure 15.** Swelling also may occur. As the injury heals, the bruise eventually turns yellow as the pigment in the red blood cells is broken down even more and reenters the bloodstream. After all of the pigment is absorbed into the bloodstream, the bruise disappears and the skin looks normal again.

Reading Check *What is the source of the yellow color of a bruise that is healing?*

Acidic Skin Oil and sweat glands in your skin cause the skin to be acidic. With a pH between 3 and 5, the growth of potential disease-causing micro-organisms on your skin is reduced. What does pH mean? What common substances around your home have a pH value similar to that of your skin? Research to find these answers and then record them in your Science Journal.

Cuts Any tear in the skin is called a cut. Blood flows out of the cut until a clot forms over it. A scab then forms, preventing bacteria from entering the body. Cells in the surrounding blood vessels fight infection while the skin cells beneath the scab grow to fill the gap in the skin. In time, the scab falls off, leaving the new skin behind. If the cut is large enough, a scar may develop because of the large amounts of thick tissue fibers that form.

The body generally can repair bruises and small cuts. What happens when severe burns, some diseases, and surgeries result in injury to large areas of skin? Sometimes, not enough skin cells are left that can divide to replace this lost layer. If not treated, this can lead to rapid water loss from skin and muscle tissues, leading to infection and possible death. Skin grafts can prevent such problems. What are skin grafts?

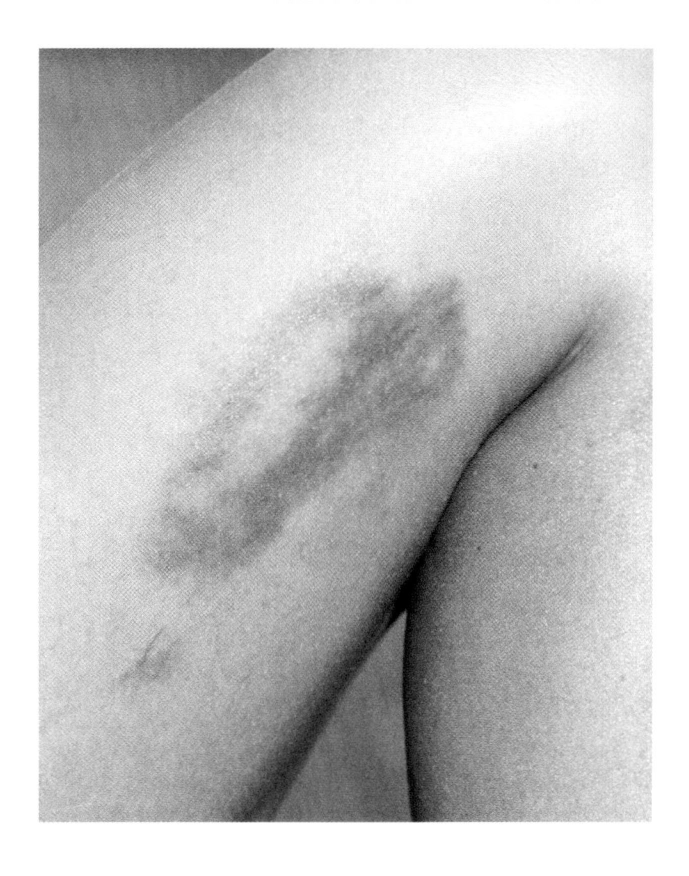

Figure 15 Bruising occurs when capillaries and other tiny blood vessels beneath the skin burst.

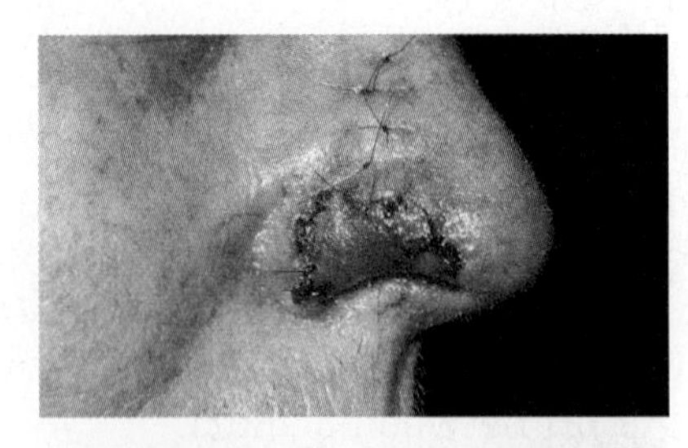

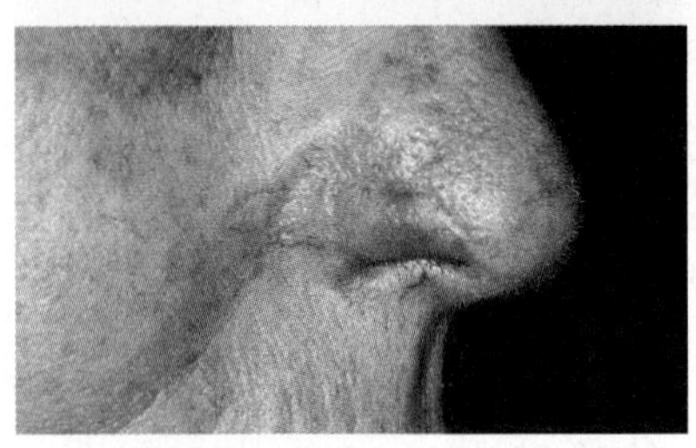

Figure 16 A cancerous growth was removed from the nose of a 69-year-old woman. A piece of skin removed from her scalp was grafted onto her nose to replace the lost skin (top). The skin graft is healing after only one month (bottom).

Skin Grafts Pieces of skin that are cut from one part of a person's body and then moved to the injured or burned area where there is no skin are called skin grafts. This skin graft is kept alive by nearby blood vessels and soon becomes part of the surrounding skin. Successful skin grafts, shown in **Figure 16,** must be taken from the victim's own body or possibly an identical twin. Skin transplants from other sources are rejected in about three weeks.

What can be done for severe burn victims who have little healthy skin left? Since the 1880s, doctors have used the skin from dead humans, called cadavers, to treat such burns temporarily. However, the body usually rejects this skin, so it must be replaced continually until the burn heals.

A recent advancement in skin repair uses temporary grafts from cadavers to prevent immediate infections, while scientists grow large sheets of epidermis from small pieces of the burn victim's healthy skin. After 19 to 21 days, the cadaver skin patch is removed and the new epidermis is applied. With new technologies, severe cases of skin loss or damage that cannot be repaired may no longer be fatal.

section 3 review

Summary

Skin Structures

- The epidermis is the thinnest, outermost layer of skin.
- The dermis is the thick layer below the epidermis. It contains blood vessels, nerves, muscles, oil, and sweat glands.
- Melanin is a pigment that protects your skin and gives it color.

Skin Functions

- Your skin provides protection, and eliminates body wastes.

Skin Injuries and Repair

- A bruise is caused by tiny broken blood vessels underneath the skin.
- When you cut your skin, blood flows out of the cut until a clot forms, causing a scab to protect against bacteria.
- Skin grafts can be made from a cadaver or a victim's healthy skin to repair the epidermis.

Self Check

1. **Compare and contrast** the epidermis and dermis.
2. **List** five of the major functions of the body's largest organ, skin.
3. **Explain** how skin helps prevent disease in the body.
4. **Describe** one way in which doctors are able to repair severe skin damage from burns, injuries, or surgeries.
5. **Think Critically** Why is a person who has been severely burned in danger of dying from loss of water?

Applying Math

6. **Solve One-Step Equations** The skin of eyelids is about 0.5 mm thick. On the soles of your feet, skin is up to 0.4 cm thick. How many times thicker is the skin on the soles of your feet compared to your eyelids?
7. **Calculate** The outermost layers of your skin are replaced every 27 days. How many times per year are your outermost layers of skin replaced?

Measuring Skin Surface

Skin covers the entire surface of your body and is your body's largest organ. Skin cells make up a layer of skin about 2 mm thick. These cells are continually lost and re-formed. Skin cells are shed daily at a rate of an average of 50,000 cells per minute. In one year, humans lose about 2 kg of skin and hair. How big is this organ? Find the surface area of human skin.

Real-World Question

How much skin covers your body?

Goal

- **Estimate** the surface area of skin that covers the body of a middle-school student.

Materials

10 large sheets of newspaper
scissors
tape
meterstick or ruler

Safety Precautions

Procedure

1. Form groups of three or four, either all female or all male. Select one person from your group to measure the surface area of his or her skin.
2. **Estimate** how much skin covers the average student in your classroom. In your Science Journal, record your estimation.
3. Wrap newspaper snugly around each part of your classmate's body. Overlap sheets of paper and use tape to secure them. Cover entire hands and feet. Small body parts, such as fingers and toes, do not need to be wrapped individually.
   WARNING: *Do not cover face. May cause suffocation.*
4. After your classmate is completely covered with paper, carefully cut the newspaper off his or her body. **WARNING:** *Do not cut any clothing or skin.*
5. Lay all of the overlapping sheets of newspaper on the floor. Using scissors and more tape, cut and piece the paper suit together to form a rectangle.
6. Using a meterstick, measure the length and width of the resulting rectangle. Multiply these two measurements for an estimate of the surface area of your classmate's skin.

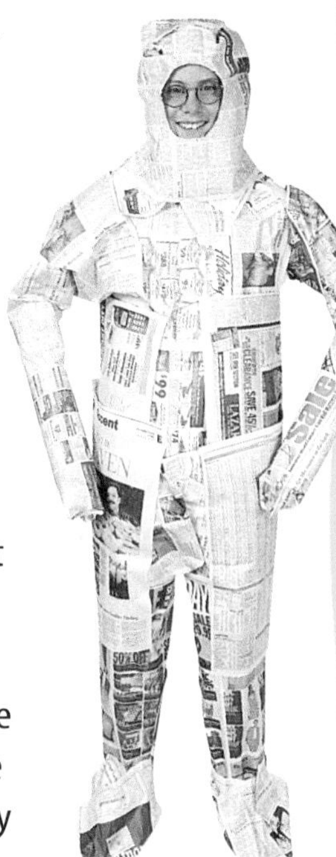

Conclude and Apply

1. Was your estimation correct? Explain.
2. How accurate are your measurements of your classmate's skin surface area? How could your measurements be improved?
3. **Calculate** the skin's volume using 2 mm as the average skin thickness and your calculated surface area from this lab.

Communicating Your Data

Make a table of all data. Find the average area for male groups and then for female groups. Discuss the differences. **For more help, refer to the Math Skill Handbook.**

Use the Internet

Similar Skeletons

Goals

- **Identify** a skeletal structure in the human body.
- **Write** a list of mammals with which you are familiar.
- **Compare** the identified human skeletal structure to a skeletal structure in each of the mammals.
- **Determine** if the mammal skeletal structure that you selected is similar to the human skeletal structure you identified.
- **Describe** how the mammal skeletal structure is similar to or different from the skeletal structure in a human.

Data Source

Science Online

Visit life.msscience.com/internet_lab for Web links to more information about skeletal structures, and for data collected by other students.

Real-World Question

Humans and other mammals share many similar characteristics, including similar skeletal structures. Think about all the different types of mammals you have seen or read about. Tigers, dogs, and household cats are meat-eating mammals. Whales and dolphins live in water. Primates, which include gorillas, chimpanzees, and humans, can walk on two legs. Mammals live in different environments, eat different types of food, and even look different, but they all have hair, possess the ability to maintain fairly constant body temperatures, and have similar skeletal structures. Which skeletal structures are similar among humans and other mammals? How many bones do you have in your hand? What types of bones are they? Do other mammals have similar skeletal structures? Form a hypothesis about the skeletal structures that humans and other mammals have in common.

Make a Plan

1. Choose a specific part of the human skeletal structure to study, such as your hand, foot, skull, leg, or arm.

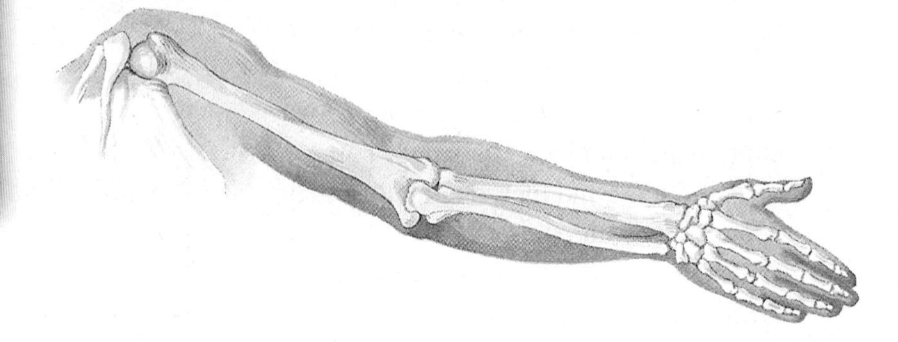

2. **List** four to six different mammals.
3. Do these mammals possess skeletal structures similar to the human skeleton? Remember, the mammals' skeletons can be similar to that of the human, but the structures can have different functions.
4. **Compare and contrast** the mammal and human skeletal structures. Are the types of bone similar? Is the number of bones the same? Where are these structures located?

Follow Your Plan

1. Make sure your teacher approves your plan before you start.
2. Visit the link below to post your data.

Analyze Your Data

1. **Describe** how each mammal's skeletal structure is similar to or different from the human skeletal structure you chose.
2. **Record** your data in the data table provided on the Web site.

Conclude and Apply

1. Visit the link below and compare your data to that of other students. Do other students agree with your conclusions?
2. Do the structures studied have similar functions in the human and the mammals you researched?

Communicating Your Data

Find this lab using the link below. Post your data in the table provided. Compare your data with that posted by other students.

life.msscience.com/internet_lab

Oops! Accidents in SCIENCE

SOMETIMES GREAT DISCOVERIES HAPPEN BY ACCIDENT!

First Aid Dolls

A fashion doll is doing her part for medical science! It turns out that the plastic joints that make it possible for one type of doll's legs to bend make good joints in prosthetic (artificial) fingers for humans.

Jane Bahor (photo at right) works at Duke University Medical Center in Durham, North Carolina. She makes lifelike body parts for people who have lost legs, arms, or fingers. A few years ago, she met a patient named Jennifer Jordan, an engineering student who'd lost a finger. The artificial finger that Bahor made looked real, but it couldn't bend. She and Jordan began to discuss the problem.

"If only the finger could bend like a doll's legs bend," said Bahor. "It would be so much more useful to you!"

Jordan's eyes lit up. "That's it!" Jordan said. The engineer went home and borrowed one of her sister's dolls. Returning with it to Bahor's office, she and Bahor did "surgery." They operated on the fashion doll's legs and removed the knee joints from their vinyl casings.

"It turns out that the doll's knee joints flexed the same way that human finger joints do," says Bahor. "We could see that using these joints would allow patients more use and flexibility with their 'new' fingers."

Holding On

The new, fake, flexible fingers can bend in the same way that a doll's legs bend. A person can use his or her other hand to bend and straighten the joint. When the joint bends, it makes a sound similar to a cracking knuckle.

Being able to bend prosthetic fingers allows wearers to hold a pen, pick up a cup, or grab a steering wheel. These are tasks that were impossible before the plastic knee joints were implanted in the artificial fingers. "We've even figured out how to insert three joints in each finger, so that now its wearer can almost make a fist," adds Bahor. Just like the doll's legs, the prosthetic fingers stay bent until the wearer straightens them.

Bahor removes a knee joint from a doll. The joint will soon be in a human's prosthetic finger!

Invent **Choose a "problem" you can solve. Use what Bahor calls "commonly found materials" to solve the problem. Then make a model or a drawing of the problem-solving device.**

For more information, visit life.msscience.com/oops

Reviewing Main Ideas

Section 1 The Skeletal System

1. Bones are living structures that protect, support, make blood, store minerals, and provide for muscle attachment.
2. The skull and pelvic joints in adults do not move and are classified as immovable.
3. Movable joints move freely, and include pivot, hinge, ball-and-socket, and gliding joints.

Section 2 The Muscular System

1. Skeletal muscle is voluntary and moves bones. Smooth muscle is involuntary and controls movement of internal organs. Cardiac muscle is involuntary and located only in the heart.
2. Muscles contract—they pull, not push, to move body parts.
3. Skeletal muscles work in pairs—when one contracts, the other relaxes.

Section 3 The Skin

1. The epidermis has dead cells on its surface. Melanin is produced in the epidermis. Cells at the base of the epidermis produce new skin cells. The dermis is the inner layer where nerves, sweat and oil glands, and blood vessels are located.
2. The functions of skin include protection, reduction of water loss, production of vitamin D, and maintenance of body temperature.
3. Glands in the epidermis produce substances that destroy bacteria.
4. Severe damage to skin, including injuries and burns, can lead to infection and death if it is not treated.

Visualizing Main Ideas

Copy and complete the following concept map on body movement.

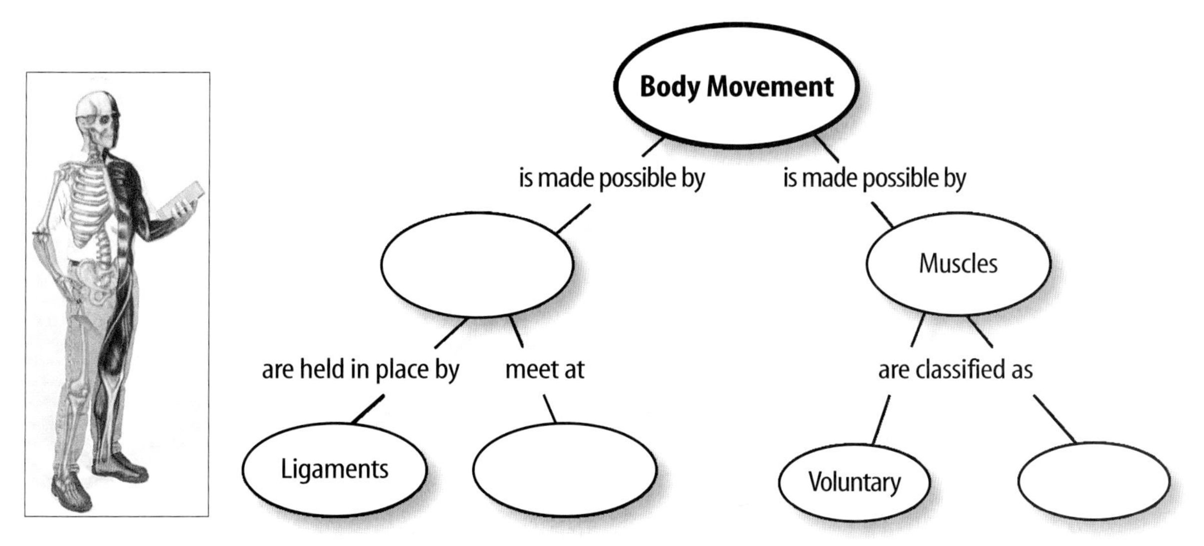

Using Vocabulary

Match the definitions with the correct vocabulary word.

1. tough outer covering of bone
2. internal framework of the body
3. outer layer of skin
4. thick band of tissue that attaches muscle to a bone
5. muscle found only in the heart
6. a tough band of tissue that holds two bones together
7. organ that can relax and contract to aid in the movement of the body
8. a muscle that you control

Checking Concepts

Choose the word or phrase that best answers the question.

9. Which of the following is the most solid form of bone?
   A) compact **C)** spongy
   B) periosteum **D)** marrow

10. Where are blood cells made?
   A) compact bone **C)** cartilage
   B) periosteum **D)** marrow

11. Where are minerals stored?
   A) bone **C)** muscle
   B) skin **D)** blood

12. What are the ends of bones covered with?
   A) cartilage **C)** ligaments
   B) tendons **D)** muscle

13. Where are immovable joints found in the human body?
   A) at the elbow **C)** in the wrist
   B) at the neck **D)** in the skull

14. What kind of joints are the knees, toes, and fingers?
   A) pivot **C)** gliding
   B) hinge **D)** ball and socket

15. Which vitamin is made in the skin?
   A) A **C)** D
   B) B **D)** K

16. Where are dead skin cells found?
   A) dermis **C)** epidermis
   B) marrow **D)** periosteum

17. Which of the following is found in bone?
   A) iron **C)** vitamin D
   B) calcium **D)** vitamin K

18. Which of the following structures helps retain fluids in the body?
   A) bone **C)** skin
   B) muscle **D)** a joint

Use the illustration below to answer question 19.

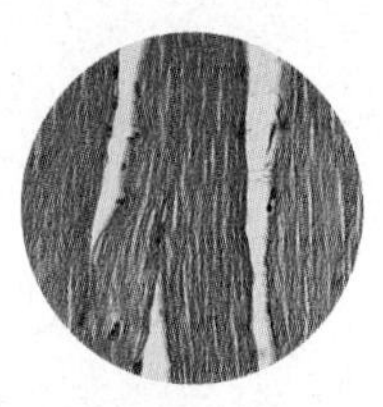

19. Where would this type of muscle tissue be found in your body?
   A) heart **C)** stomach
   B) esophagus **D)** leg

Science Online life.msscience.com/vocabulary_puzzlemaker

Thinking Critically

20. **Explain** why skin might not be able to produce enough vitamin D.
21. **List** what factors a doctor might consider before choosing a method of skin repair for a severe burn victim.
22. **Explain** what a lack of calcium would do to bones.
23. **Concept Map** Copy and complete the following concept map that describes the types and functions of bone cells.

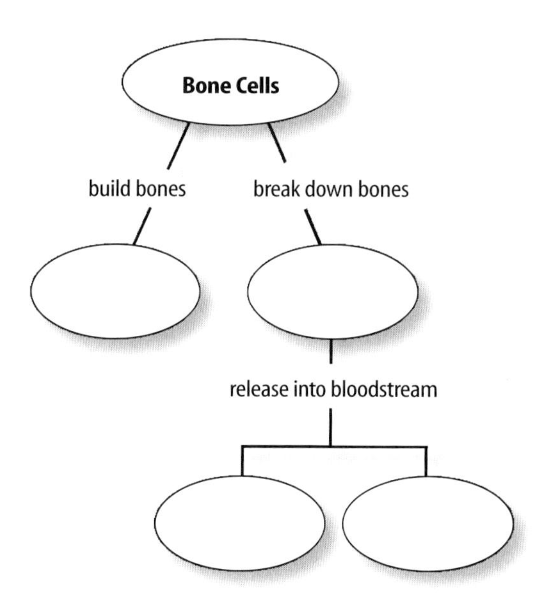

24. **Name** the function of your lower lip's skin that changes when a dentist gives you novocaine before filling a bottom tooth. Why?
25. **Draw Conclusions** The joints in the skull of a newborn baby are flexible, but those of a teenager have fused together and are immovable. Conclude why the infant's skull joints are flexible.
26. **Predict** what would happen if a person's sweat glands didn't produce sweat.
27. **Compare and contrast** the functions of ligaments and tendons.
28. **Form a Hypothesis** Your body has about 3 million sweat glands. Make a hypothesis about where these sweat glands are on your body. Are they distributed evenly throughout your body?

Performance Activities

29. **Display** Research the differences among first-, second-, and third-degree burns. A local hospital's burn unit or a fire department are possible sources of information about burns. Display pictures of each type of burn and descriptions of treatments on a three-sided, free-standing poster.

Applying Math

30. **Bone Volume** Estimate the volume of a hand bone that is 7 cm long and is 1.5 cm in diameter.

Use the graph below to answer question 31.

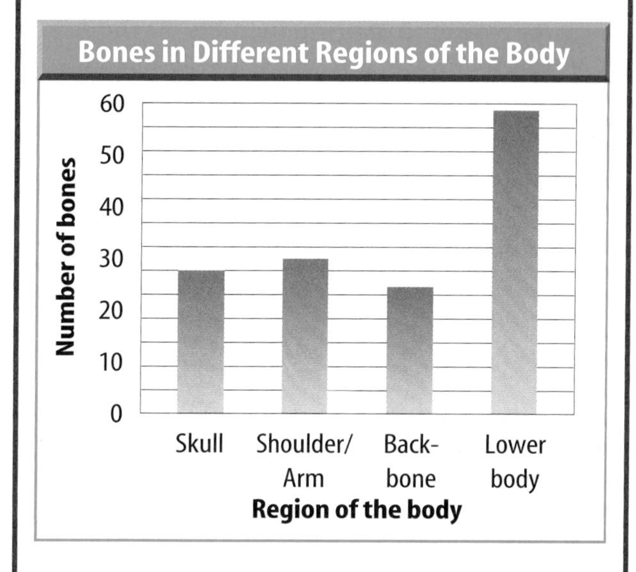

31. **Bone Quantity** The total number of bones in the human body is 206. Approximately what percentage of bones is located in the backbone?
    A) 2%
    B) 12%
    C) 50%
    D) 75%

Standardized Test Practice

Part 1 Multiple Choice

Record your answers on the answer sheet provided by your teacher or on a sheet of paper.

1. Which type of muscle tends to contract quickly and tire more easily?

A. cardiac muscle **C.** skeletal muscle
B. bladder **D.** smooth muscle

Use the table below to answer questions 2 and 3.

Number of Bicycle Deaths per Year		
Year	**Male**	**Female**
1996	654	107
1997	712	99
1998	658	99
1999	656	94
2000	605	76

Data from Insurance Institute for Highway Safety

2. If 99% of the people who die in bicycle accidents were not wearing helmets, to the nearest whole number, how many people who died in 1998 were wearing bicycle helmets?

A. 7 **C.** 8
B. 6 **D.** 9

3. Which year had the greatest total number of bicycle deaths?

A. 1996 **C.** 1998
B. 1997 **D.** 1999

4. Which of the following is NOT released by sweat glands?

A. water **C.** waste products
B. salt **D.** oil

Test-Taking Tip

Ease Nervousness Stay calm during the test. If you feel yourself getting nervous, close your eyes and take five slow, deep breaths.

Use the illustration below to answer questions 5 and 6.

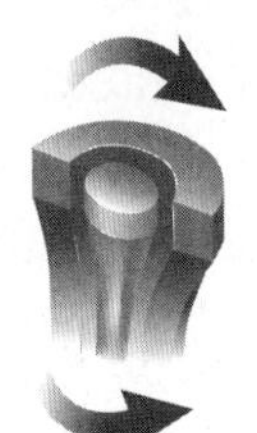

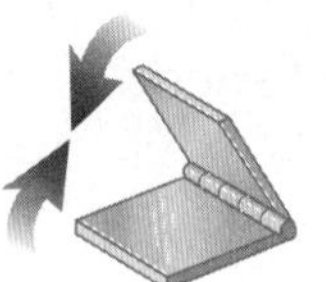

5. Which type of joint do your elbows have?

A. hinge **C.** ball-and-socket
B. gliding **D.** pivot

6. Which type of joint allows your legs and arms to swing in almost any direction?

A. hinge **C.** ball-and-socket
B. gliding **D.** pivot

7. What is the name of the pigment that gives your skin color?

A. hemoglobin **C.** melanin
B. keratin **D.** calcium

8. What does the periosteum do?

A. connects bones together
B. covers the surface of bones
C. produces energy
D. makes vitamin D

9. Which type of muscle is found in the intestines?

A. skeletal muscle
B. smooth muscle
C. cardiac muscle
D. tendon

Part 2 Short Response/Grid In

Record your answers on the answer sheet provided by your teacher or on a sheet of paper.

10. At birth, your skeleton had approximately 300 bones. As you developed, some bones fused together. Now you have 206 bones. How many fewer bones do you have now?

11. One in seven people in the United States suffers from arthritis. Calculate the percentage of people that suffer from arthritis.

12. Explain the difference between voluntary and involuntary muscles.

Use the illustration below to answer questions 13 and 14.

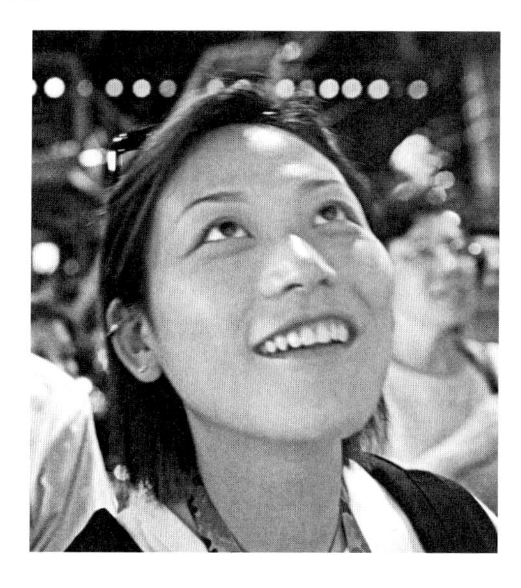

13. What type of lever is shown in the photo?

14. Where is the fulcrum?

15. How do muscles help maintain body temperature?

16. Explain what happens when your skin is exposed to ultraviolet rays.

Part 3 Open Ended

Record your answers on a sheet of paper.

17. Compare and contrast compact and spongy bone.

18. Explain how bone cells help maintain homeostasis.

19. Describe the changes that occur in muscles that do a lot of work. Compare these muscles to the muscles of a person who only does inactive pastimes.

Use the illustration below to answer questions 20 and 21.

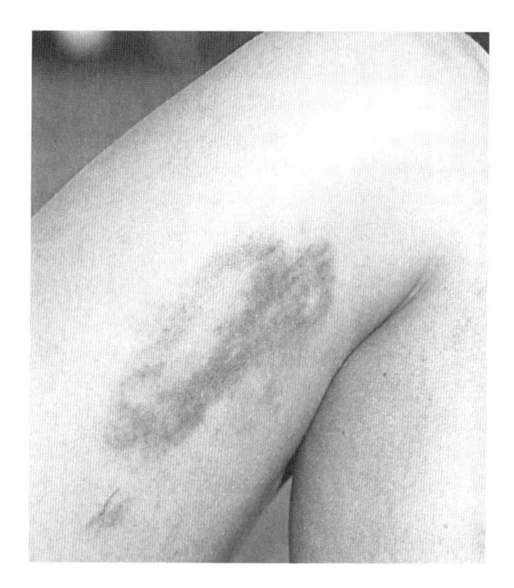

20. Identify the injury in the photograph. Describe the sequence of events from the time of injury until the injury disappears.

21. Contrast the injury in the photograph with a cut. Explain why a cut needs to be cleaned but the injury in the photograph does not.

22. What might happen to your body temperature if blood vessels in the skin did not contain smooth muscle?

chapter 18

Nutrients and Digestion

The BIG Idea

Our bodies can use nutrients in foods because of the structures and functions of the digestive system.

SECTION 1
Nutrition

Main Idea A balanced diet provides nutrients and energy for a healthy lifestyle.

SECTION 2
The Digestive System

Main Idea The digestive organs process and absorb nutrients and then eliminate wastes.

Intestinal Landscape

This photo may look like a pile of potatoes, but it is a close-up of your small intestine. The wall of the small intestine has many fingerlike projections that soak up substances from digested food. The small intestine is just one of many organs that make up your digestive system.

Science Journal Make a list of all the organs you think are part of your digestive system.

Start-Up Activities

Model the Digestive Tract

Imagine taking a bite of your favorite food. When you eat, your body breaks down food to release energy. How long does it take?

Organs of the Digestive System

Organ	Length	Time
Mouth	8 cm	5 s to 30 s
Pharynx and esophagus	25 cm	10 s
Stomach	16 cm	2 h to 4 h
Small intestine	4.75 m	3 h
Large intestine	1.25 m	2 days

1. Make a label for each of the digestive organs listed here. Include the organ's name, length, and the time it takes for food to pass through it.
2. Working with a partner, place a piece of masking tape that is 6.5 m long on the classroom floor.
3. Beginning at one end of the tape, and in the same order as they are listed in the table, mark the length for each organ. Place each label next to its section.
4. **Think Critically** In your Science Journal, suggest reasons why food spends a different amount of time in each organ.

FOLDABLES™ Study Organizer

Nutrients in Foods Make the following Foldable to help you organize foods based on the nutrients that they contain.

STEP 1 **Fold** the top of a vertical piece of paper down and the bottom up to divide the paper into thirds. Then, **fold** the paper in half from top to bottom.

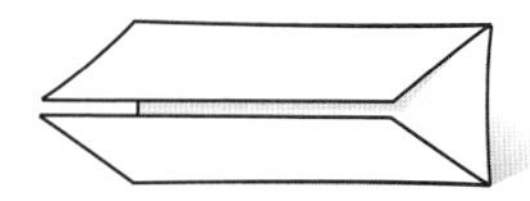

STEP 2 **Turn** the paper horizontally, **unfold and label** the six columns as follows: *Proteins, Carbohydrates, Lipids, Water, Vitamins,* and *Minerals.*

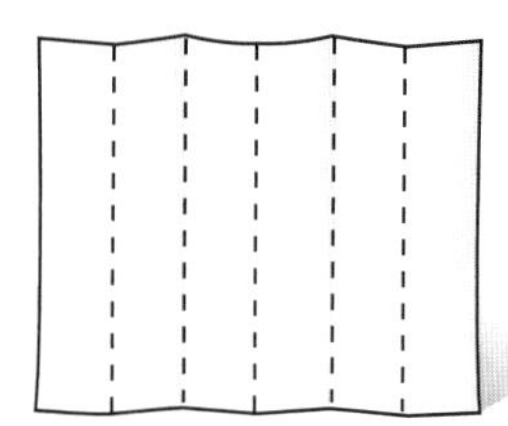

Read for Main Ideas As you read the chapter, list foods you eat that provide each of these nutrients in the proper columns.

Preview this chapter's content and activities at life.msscience.com

Get Ready to Read

Visualize

① Learn It! Visualize by forming mental images of the text as you read. Imagine how the text descriptions look, sound, feel, smell, or taste. Look for any pictures or diagrams on the page that may help you add to your understanding.

② Practice It! Read the following paragraph. As you read, use the underlined details to form a picture in your mind.

> The stomach, shown in **Figure 14,** is a muscular bag. When empty, it is somewhat sausage shaped with folds on the inside. As food enters from the esophagus, the stomach expands and the folds smooth out. Mechanical and chemical digestions take place in the stomach. Mechanically, food is mixed in the stomach by peristalsis. Chemically, food also is mixed with enzymes and strong digestive solutions, such as hydrochloric acid solution, to help break it down.
>
> *—from page 527*

Based on the description above, try to visualize the stomach. Now look at **Figure 14** on page 527.

- How closely do these images match your mental picture?
- Reread the passage and look at the picture again. Did your ideas change?
- Compare your image with what others in your class visualized.

③ Apply It! Read the chapter and list three subjects you were able to visualize. Make a rough sketch showing what you visualized.

Reading Tip
Forming your own mental images will help you remember what you read.

Target Your Reading

Use this to focus on the main ideas as you read the chapter.

1. **Before you read** the chapter, respond to the statements below on your worksheet or on a numbered sheet of paper.
   - Write an **A** if you **agree** with the statement.
   - Write a **D** if you **disagree** with the statement.

2. **After you read** the chapter, look back to this page to see if you've changed your mind about any of the statements.
   - If any of your answers changed, explain why.
   - Change any false statements into true statements.
   - Use your revised statements as a study guide.

Before You Read A or D	Statement	After You Read A or D
	1 Foods with many Calories have few nutrients.	
	2 Proteins primarily form and maintain bones.	
	3 Carbohydrates usually are the main sources of energy for your body.	
	4 You can live longer without water than without food.	
	5 Most Americans do not eat enough fruits and vegetables.	
	6 Enzymes digest foods.	
	7 Digestion of some food begins and ends in the mouth.	
	8 Water is absorbed into your blood mostly in your small intestine.	
	9 Chewing is a type of mechanical digestion.	
	10 Bacteria that live in your large intestine produce vitamin D.	

Science Online
Print out a worksheet of this page at life.msscience.com

section 1

Nutrition

as you read

What You'll Learn

- **Distinguish** among the six classes of nutrients.
- **Identify** the importance of each type of nutrient.
- **Explain** the relationship between diet and health.

Why It's Important

You can make healthful food choices if you know what nutrients your body uses daily.

Review Vocabulary
molecule: the smallest particle of a substance that retains the properties of the substance and is composed of one or more atoms

New Vocabulary

- nutrient
- protein
- amino acid
- carbohydrate
- fat
- vitamin
- mineral
- food group

Why do you eat?

You're listening to a favorite song on the radio, maybe even singing along. Then all of a sudden, the music stops. You examine the radio to see what happened. The batteries died. You hunt for more batteries and quickly put in the new ones. In the same way that the radio needs batteries to work, you need food to carry out your daily activities—but not just any food. When you are hungry, you probably choose food based on taste and the amount of time you have to eat it. However, as much as you don't want to admit it, the nutritional value of the food you choose is more important than the taste. A chocolate-iced donut might be tasty and quick to eat, yet it provides few of the nutrients your body needs. **Nutrients** (NEW tree unts) are substances in foods that provide energy and materials for cell development, growth, and repair.

Energy Needs Your body needs energy for every activity that it performs. Muscle activities such as the beating of your heart, blinking your eyes, and lifting your backpack require energy. How much energy you need depends on several factors, such as body mass, age, and activity level. This energy comes from the foods you eat. The amount of energy available in food is measured in Calories. A Calorie (Cal) is the amount of heat necessary to raise the temperature of 1 kg of water 1°C. As shown in **Figure 1,** different foods contain different numbers of Calories. A raw carrot may have 30 Cal. This means that when you eat a carrot, your body has 30 Cal of energy available to use. A slice of cheese pizza might have 170 Cal, and one hamburger might have 260 Cal. The number of Calories varies due to the kinds of nutrients a food provides.

Figure 1 Foods vary in the number of Calories they contain. A hamburger has the same number of Calories as 8.5 average-sized carrots.

Classes of Nutrients

Six kinds of nutrients are available in food—proteins, carbohydrates, fats, vitamins, minerals, and water. Proteins, carbohydrates, vitamins, and fats all contain carbon and are called organic nutrients. In contrast, inorganic nutrients, such as water and minerals, do not contain carbon. Foods containing carbohydrates, fats, and proteins need to be digested or broken down before your body can use them. Water, vitamins, and minerals don't require digestion and are absorbed directly into your bloodstream.

Figure 2 Meats, poultry, eggs, fish, peas, beans, and nuts are all rich in protein.

Proteins Your body uses proteins for replacement and repair of body cells and for growth. **Proteins** are large molecules that contain carbon, hydrogen, oxygen, nitrogen and sometimes sulfur. A molecule of protein is made up of a large number of smaller units, or building blocks, called **amino acids.** In **Figure 2** you can see some sources of proteins. Different foods contain different amounts of protein, as shown in **Figure 3.**

Your body needs only 20 amino acids in various combinations to make the thousands of proteins used in your cells. Most of these amino acids can be made in your body's cells, but eight of them cannot. These eight are called essential amino acids. They have to be supplied by the foods you eat. Complete proteins provide all of the essential amino acids. Eggs, milk, cheese, and meat contain complete proteins. Incomplete proteins are missing one or more of the essential amino acids. If you are a vegetarian, you can get all of the essential amino acids by eating a wide variety of protein-rich vegetables, fruits, and grains.

Figure 3 The amount of protein in a food is not the same as the number of Calories in the food. A taco has nearly the same amount of protein as a slice of pizza, but it usually has about 100 fewer Calories.

Topic: Dietary Fiber

Visit life.msscience.com for Web links to recent news or magazine articles about the importance of fiber in your diet.

Activity In your Science Journal, classify your favorite foods into two groups—*Good source of fiber* and *Little or no fiber.*

Carbohydrates Study the nutrition label on several boxes of cereal. You'll notice that the number of grams of carbohydrates found in a typical serving of cereal is higher than the amounts of the other nutrients. **Carbohydrates** (kar boh HI drayts) usually are the main sources of energy for your body. Each carbohydrate molecule is made of carbon, hydrogen, and oxygen atoms. Energy holds the atoms together. When carbohydrates are broken down in the presence of oxygen in your cells, this energy is released for use by your body.

Three types of carbohydrates are sugar, starch, and fiber, as shown in **Figure 4.** Sugars are called *simple carbohydrates.* You're probably most familiar with table sugar. However, fruits, honey, and milk also contain forms of sugar. Your cells break down glucose, a simple sugar. The other two types of carbohydrates—starch and fiber—are called *complex carbohydrates.* Starch is found in potatoes and foods made from grains such as pasta. Starches are made up of many simple sugars in long chains. Fiber, such as cellulose, is found in the cell walls of plant cells. Foods like whole-grain breads and cereals, beans, peas, and other vegetables and fruits are good sources of fiber. Because different types of fiber are found in foods, you should eat a variety of fiber-rich plant foods. You cannot digest fiber, but it is needed to keep your digestive system running smoothly.

Nutritious snacks can help your body get the nutrients it needs, especially when you are growing rapidly and are physically active. Choose snacks that provide nutrients such as complex carbohydrates, proteins, and vitamins, as well as fiber. Foods high in sugar and fat can have lots of Calories that supply energy, but they provide only some of the nutrients your body needs.

Figure 4 These foods contain carbohydrates that provide energy for all the things that you do.
List *the carbohydrates that you've eaten today.*

Figure 5 Fat is stored in certain cells in your body. The cytoplasm and nucleus are pushed to the edge of the cell by the fat deposits.

Nucleus
Fat deposit
Cytoplasm
Some foods you might choose for lunch or snacks that are high in fat are outlined in red.

Fats The term fat has developed a negative meaning for some people. However, **fats,** also called lipids, are necessary because they provide energy and help your body absorb vitamins. Fat tissue cushions your internal organs. A major part of every cell membrane is made up of fat. A gram of fat can release more than twice as much energy as a gram of carbohydrate can. During the digestion process, fat is broken down into smaller molecules called fatty acids and glycerol (GLIH suh rawl). Because fat is a good storage unit for energy, excess energy from the foods you eat is converted to fat and stored for later use, as shown in **Figure 5.**

Reading Check *Why is fat a good storage unit for energy?*

Fats are classified as unsaturated or saturated based on their chemical structure. Unsaturated fats are usually liquid at room temperature. Vegetable oils as well as fats found in seeds are unsaturated fats. Saturated fats are found in meats, animal products, and some plants and are usually solid at room temperature. Although fish contains saturated fat, it also has some unsaturated fats that your body needs. Saturated fats have been associated with high levels of blood cholesterol. Your body makes cholesterol in your liver. Cholesterol is part of the cell membrane in all of your cells. However, a diet high in cholesterol may result in deposits forming on the inside walls of blood vessels. These deposits can block the blood supply to organs and increase blood pressure. This can lead to heart disease and strokes.

Mini LAB

Comparing the Fat Content of Foods

Procedure

1. Collect three pieces of each of the following foods: **potato chips; pretzels; peanuts;** and **small cubes of fruits, cheese, vegetables, and meat.**
2. Place the food items on a piece of **brown grocery bag.** Label the paper with the name of each food. Do not taste the foods.
3. Allow foods to sit for 30 min.
4. Remove the items, properly dispose of them, and observe the paper.

Analysis

1. Which items left a translucent (greasy) mark? Which left a wet mark?
2. How are the foods that left a greasy mark on the paper alike?
3. Use this test to determine which other foods contain fats. A greasy mark means the food contains fat. A wet mark means the food contains a lot of water.

Vitamins Organic nutrients needed in small quantities for growth, regulating body functions, and preventing some diseases are called **vitamins.** For instance, your bone cells need vitamin D to use calcium, and your blood needs vitamin K in order to clot.

Most foods supply some vitamins, but no food has them all. Some people feel that taking extra vitamins is helpful, while others feel that eating a well-balanced diet usually gives your body all the vitamins it needs.

Vitamins are classified into two groups, as shown in **Figure 6.** Some vitamins dissolve easily in water and are called water-soluble vitamins. They are not stored by your body so you have to take them daily. Other vitamins dissolve only in fat and are called fat-soluble vitamins. These vitamins are stored by your body. Although you eat or drink most vitamins, some are made by your body. Vitamin D is made when your skin is exposed to sunlight. Some vitamin K and two of the B vitamins are made with the help of bacteria that live in your large intestine.

Applying Science

Is it unhealthy to snack between meals?

Most children eat three meals each day accompanied by snacks in between. Grabbing a bite to eat to satisfy you until your next meal is a common occurrence in today's society, and 20 percent of our energy and nutrient needs comes from snacking. While it would be best to select snacks consisting of fruits and vegetables, most children prefer to eat a bag of chips or a candy bar. Although these quick snacks are highly convenient, many times they are high in fat, as well.

Identifying the Problem

The table on the right lists several snack foods that are popular among adolescents. They are listed alphabetically, and the grams of fat per individual serving is shown. As you examine the chart, can you conclude which snacks would be a healthier choice based on their fat content?

Fat in Snack Foods

One Serving	Fat (g)
Candy bar	12
Frozen pizza	30
Ice cream	8
Potato chips	10
Pretzels	1

Solving the Problem

1. Looking at the data, what can you conclude about the snack foods you eat? What other snack foods do you eat that are not listed on the chart? How do you think they compare in nutritional value? Which snack foods are healthiest?
2. Pizza appears to be the unhealthiest choice on the chart because of the amount of the fat it contains. Why do you think pizza contains so much fat? List at least three ways to make pizza a healthier snack food.

NATIONAL GEOGRAPHIC VISUALIZING VITAMINS

Figure 6

Vitamins come in two groups—water soluble, which should be replaced daily, and fat soluble, which can be stored in the body. The sources and benefits of both groups are shown below.

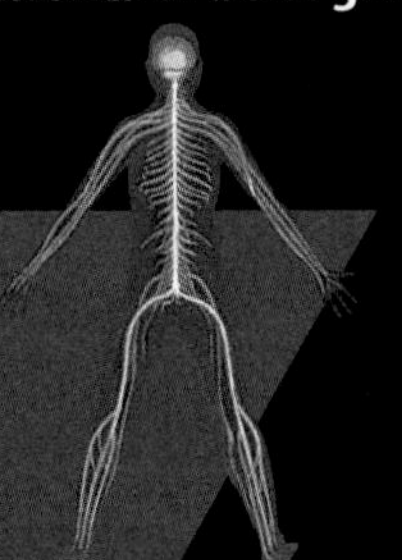

WATER SOLUBLE

Need to be replenished every day because they are excreted by the body

B (B_6, B_{12}, riboflavin, niacin, thiamine, etc.)

Aids in growth, healthy nervous system, use of carbohydrates, and red blood cell production

C

Aids in growth, healthy bones and teeth, wound recovery

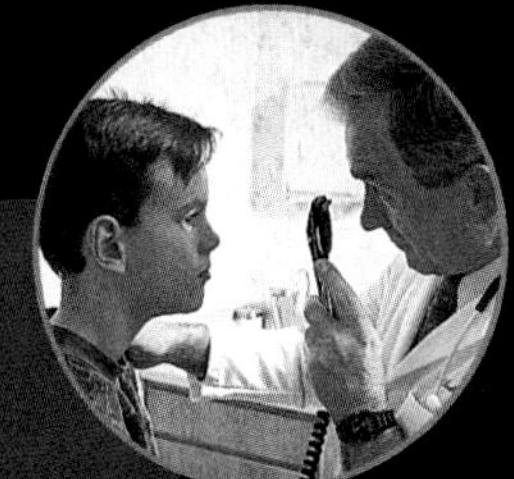

FAT SOLUBLE

Stored in the body in fatty tissue

A

Aids in growth, eyesight, healthy skin

D

Aids in absorption of calcium and phosphorus by bones and teeth

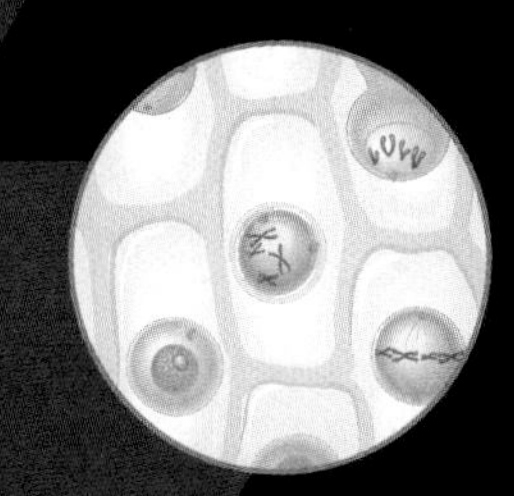

E

Aids in formation of cell membranes

K

Aids in blood clotting and wound recovery

Minerals Inorganic nutrients—nutrients that lack carbon and regulate many chemical reactions in your body—are called **minerals.** Your body uses about 14 minerals. Minerals build cells, take part in chemical reactions in cells, send nerve impulses throughout your body, and carry oxygen to body cells. In **Figure 7,** you can see how minerals can get from the soil into your body. Of the 14 minerals, calcium and phosphorus are used in the largest amounts for a variety of body functions. One of these functions is the formation and maintenance of bone. Some minerals, called trace minerals, are required only in small amounts. Copper and iodine usually are listed as trace minerals. Several minerals, what they do, and some food sources for them are listed in **Table 1.**

Reading Check *Why is copper considered a trace mineral?*

Figure 7 The roots of the wheat take in phosphorus from the soil. Then the mature wheat is harvested and used in bread and cereal. Your body gets phosphorus when you eat the cereal.

Table 1 Minerals

Mineral	Health Effect	Food Sources
Calcium	strong bones and teeth, blood clotting, muscle and nerve activity	dairy products, eggs, green leafy vegetables, soy
Phosphorus	strong bones and teeth, muscle contraction, stores energy	cheese, meat, cereal
Potassium	balance of water in cells, nerve impulse conduction, muscle contraction	bananas, potatoes, nuts, meat, oranges
Sodium	fluid balances in tissues, nerve impulse conduction	meat, milk, cheese, salt, beets, carrots, nearly all foods
Iron	oxygen is transported in hemoglobin by red blood cells	red meat, raisins, beans, spinach, eggs
Iodine (trace)	thyroid activity, metabolic stimulation	seafood, iodized salt

Figure 8 About two-thirds of your body water is located within your body cells. Water helps maintain the cells' shapes and sizes. The water that is lost through perspiration and respiration must be replaced.

Water Loss

Method of Loss	Amount (mL/day)
Exhaled air	350
Feces	150
Skin (mostly as sweat)	500
Urine	1,500

Water Have you ever gone on a bike ride on a hot summer day without a bottle of water? You probably were thirsty and maybe you even stopped to get some water. Water is important for your body. Next to oxygen, water is the most important factor for survival. Different organisms need different amounts of water to survive. You could live for a few weeks without food but for only a few days without water because your cells need water to carry out their work. Most of the nutrients you have studied in this chapter can't be used by your body unless they are carried in a solution. This means that they have to be dissolved in water. In cells, chemical reactions take place in solutions.

The human body is about 60 percent water by weight. About two thirds of your body water is located in your body cells. Water also is found around cells and in body fluids such as blood. As shown in **Figure 8,** your body loses water as perspiration. When you exhale, water leaves your body as water vapor. Water also is lost every day when your body gets rid of wastes. To replace water lost each day, you need to drink about 2 L of liquids. However, drinking liquids isn't the only way to supply cells with water. Most foods have more water than you realize. An apple is about 80 percent water, and many meats are 90 percent water.

Salt Mines The mineral halite is processed to make table salt. In the United States, most salt comes from underground mines. Research to find the locations of these mines, then label them on a map.

Why do you get thirsty? Your body is made up of systems that operate together. When your body needs to replace lost water, messages are sent to your brain that result in a feeling of thirst. Drinking water satisfies your thirst and usually restores the body's homeostasis (hoh mee oh STAY sus). Homeostasis is the regulation of the body's internal environment, such as temperature and amount of water. When homeostasis is restored, the signal to the brain stops and you no longer feel thirsty.

Food Groups

Because no naturally occurring food has every nutrient, you need to eat a variety of foods. Nutritionists at the U.S. Department of Agriculture developed dietary guidelines, as listed in **Table 2,** to help people select foods that supply all the nutrients needed for energy and growth.

Foods that contain the same type of nutrient belong to a **food group.** There are five food groups—bread and cereal, vegetable, fruit, milk, and meat. The recommended daily amount for each food group will supply your body with the nutrients it needs for good health. Using the dietary guideline to make choices when you eat will help you maintain good health.

Table 2 Dietary Guidelines for Americans 2005 from the USDA

Food Group	Recommendations
Fruits	Eat a variety of fruits—whether fresh, frozen, canned, or dried—rather than fruit juice for most of your fruit choices. For a 2,000-Calorie diet, you will need two cups of fruit each day (for example, a small banana, a large orange, and 1/4 cup of dried apricots or peaches).
 Vegetables	Eat more dark green vegetables, such as broccoli, kale, and other dark leafy greens; orange vegetables, such as carrots, sweet potatoes, pumpkin, and winter squash; and beans and peas, such as pinto beans, kidney beans, black beans, garbanzo beans, split peas, and lentils.
Calcium- rich foods	Get three cups of low-fat or fat-free milk products every day. If you don't or can't consume milk products, choose lactose-free milk products and/or calcium fortified foods and beverages.
Grains	Eat at least three ounces of whole-grain cereals, breads, crackers, rice, or pasta every day. Look to see that grains such as wheat, rice, oats, or corn are referred to as "whole" in the list of ingredients. In general, at least half the grains should come from whole grains with the remainder from enriched or whole-grain products.
Proteins	Choose lean meats and poultry. Bake it, broil it, or grill it. And vary your protein choices–with more fish, beans, peas, nuts, and seeds.

Other Recommendations You should eat 2 cups of fruit and 2.5 cups of vegetables per day for a 2,000-Calorie intake, with higher or lower amounts depending on the Calorie level. Choose a variety of fruits and vegetables each day. In particular, select from dark green, orange, legumes, starchy vegetables, and other vegetables several times a week. Eat three or more one-ounce servings of whole-grain products per day. A one-ounce serving is about one slice of bread, one cup of breakfast cereal, or one-half cup of cooked rice or pasta. Consume three cups per day of fat-free or low-fat milk or an equivalent amount of low-fat yogurt and/or low-fat cheese (1.5 ounces of cheese equals one cup of milk). Remember to limit fats, salt, and sugars. Select foods low in saturated fats and trans fats. Choose and prepare foods and beverages with little salt and/or added sweeteners.

Children and adolescents should consume whole-grain products often; at least half the grains should be whole grains. Children two to eight years should consume two cups per day of fat-free or low-fat milk or equivalent milk products. Children nine years of age and older should consume three cups per day of fat-free or low-fat milk or equivalent milk products.

Food Labels The nutritional facts found on all packaged foods make it easier to make healthful food choices. These labels, as shown in **Figure 9,** can help you plan meals that supply the daily recommended amounts of nutrients and meet special dietary requirements (for example, a low-fat diet).

Figure 9 The information on a food label can help you decide what to eat.

Nutrition Facts
Serving Size 1 Meal

Amount Per Serving	
Calories 330	Calories from Fat 60
	% Daily Value*
Total Fat 7g	**10%**
Saturated Fat 3.5g	**17%**
Polyunsaturated Fat 1g	
Monounsaturated Fat 2.5g	
Cholesterol 35mg	**12%**
Sodium 460mg	**19%**
Total Carbohydrate 52g	**18%**
Dietary Fiber 6g	**24%**
Sugars 17g	
Protein 15g	

Vitamin A 15% • Vitamin C 70%
Calcium 4% • Iron 10%

* Percent Daily Values are based on a 2,000 calorie diet. Your daily values may be higher or lower depending on your calorie needs.

	Calories	2,000	2,500
Total Fat	Less than	65g	80g
Sat Fat	Less than	20g	25g
Cholesterol	Less than	300mg	300mg
Sodium	Less than	2,400mg	2,400mg
Total Carbohydrate		300g	375g
Dietary Fiber		25g	30g

section 1 review

Summary

Why do you eat?

- Food provides the energy for your body.

Classes of Nutrients

- The nutrients in food fall into six classes.
- Organic nutrients—proteins, vitamins, fats, and carbohydrates—contain carbon.
- Inorganic nutrients—water and minerals—do not contain carbon.

Food Groups

- Foods are divided into groups based on the type of nutrient in the foods.
- The five food groups are bread and cereal, vegetable, fruit, milk, and meat.

Self Check

1. **List** the six classes of nutrients. Give one example of a food source for each class.
2. **Describe** a major function of each class of nutrient.
3. **Discuss** how food choices can positively and negatively affect your health.
4. **Explain** the importance of water in the body.
5. **Think Critically** What foods from each food group would provide a balanced breakfast? Explain.

Applying Skills

6. **Interpret Data** Nutritional information can be found on the labels of most foods. Interpret the labels found on three different types of food products.

Identifying Vitamin C Content

Vitamin C is found in many fruits and vegetables. Oranges have a high vitamin C content. Try this lab to test which orange juice has the highest vitamin C content.

Real-World Question

Which orange juice contains the most vitamin C?

Goals

- **Observe** the vitamin C content of different orange juices.

Materials

test tubes (4)
paper cups
test-tube rack
masking tape
wooden stirrers (13)
graduated cylinder
graduated container
2% tincture of iodine
dropper
cornstarch
triple-beam balance
weighing paper
water (50 mL)
glass-marking pencil

dropper bottles (4) containing:
(1) freshly squeezed orange juice
(2) orange juice made from frozen concentrate
(3) canned orange juice
(4) dairy carton orange juice

Alternate materials

Safety Precautions

WARNING: *Do not taste any of the juices. Iodine is poisonous, can stain skin and clothing, and is an irritant that can cause damage if it comes in contact with your eyes. Notify your teacher if a spill occurs.*

Sample Data

Drops of Iodine Needed to Change Color

Juice	Trial 1	Trial 2	Trial 3	Average
1 Fresh juice				
2 Frozen juice				
3 Canned juice	Do not write in this book.			
4 Carton juice				

Procedure

1. Copy the data table shown above.
2. Label four test tubes as shown in the table above and place them in the test-tube rack.
3. **Measure** and pour 5 mL of juice from each bottle into its labeled test tube.
4. **Measure** 0.3 g of cornstarch, then put it in a container. Slowly mix in 50 mL of water until the cornstarch completely dissolves.
5. Add 5 mL of the cornstarch solution to each of the four test tubes. Stir well.
6. Add iodine to test tube 1, one drop at a time. Stir after each drop. Record the number of drops needed to change the juice to purple. The more vitamin C, the more drops needed.
7. Repeat step 6 with test tubes 2, 3, and 4.
8. Empty and clean the test tubes. Repeat steps 3 through 7 two more times, then average your results.

Conclude and Apply

1. **Compare and contrast** the amount of vitamin C in the orange juices tested.
2. **Infer** why the amount of vitamin C varied.

section 2

The Digestive System

Functions of the Digestive System

You are walking through a park on a cool, autumn afternoon. Birds are searching in the grass for insects. A squirrel is eating an acorn. Why are the animals so busy? Like you, they need food to supply their bodies with energy. Food is processed in your body in four stages—ingestion, digestion, absorption, and elimination. Whether it is a piece of fruit or an entire meal, all the food you eat is treated to the same processes in your body. As soon as food enters your mouth, or is ingested as shown in **Figure 10,** breakdown begins. **Digestion** is the process that breaks down food into small molecules so that they can be absorbed and moved into the blood. From the blood, food molecules are transported across the cell membrane to be used by the cell. Unused molecules pass out of your body as wastes.

Digestion is mechanical and chemical. **Mechanical digestion** takes place when food is chewed, mixed, and churned. **Chemical digestion** occurs when chemical reactions occur that break down large molecules of food into smaller ones.

as you read

What You'll Learn

- **Distinguish** the differences between mechanical digestion and chemical digestion.
- **Identify** the organs of the digestive system and what takes place in each.
- **Explain** how homeostasis is maintained in digestion.

Why It's Important

The processes of the digestive system make the food you eat available to your cells.

Review Vocabulary
bacteria: one-celled organism without membrane-bound organelles

New Vocabulary
- digestion
- mechanical digestion
- chemical digestion
- enzyme
- peristalsis
- chyme
- villi

Figure 10 Humans have to chew solid foods before swallowing them, but snakes have adaptations that allow them to swallow their food whole.

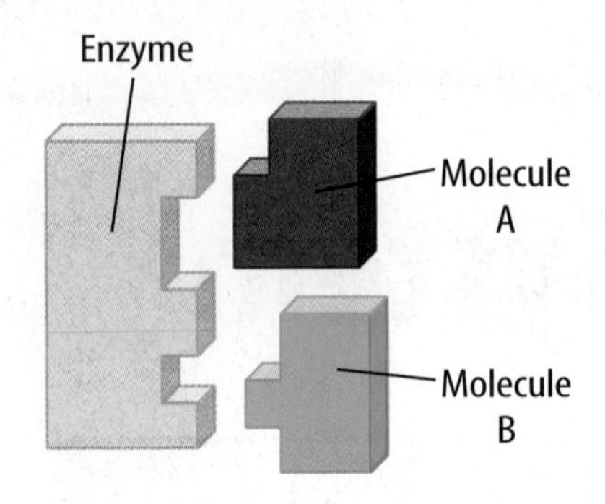

The surface shape of an enzyme fits the shape of specific molecules that take part in the reaction.

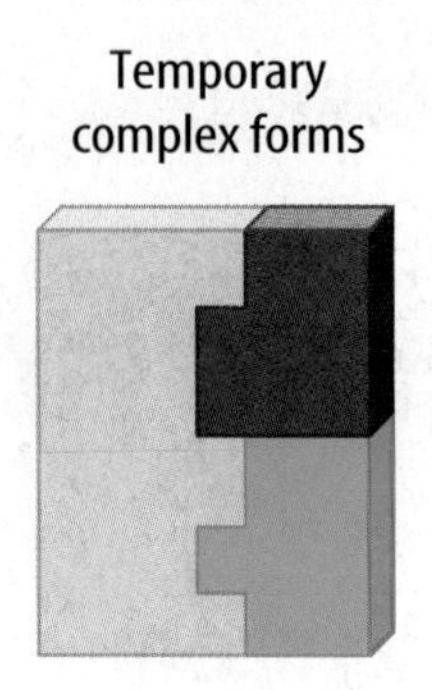

The enzyme and the molecules join and the reaction occurs between the two molecules.

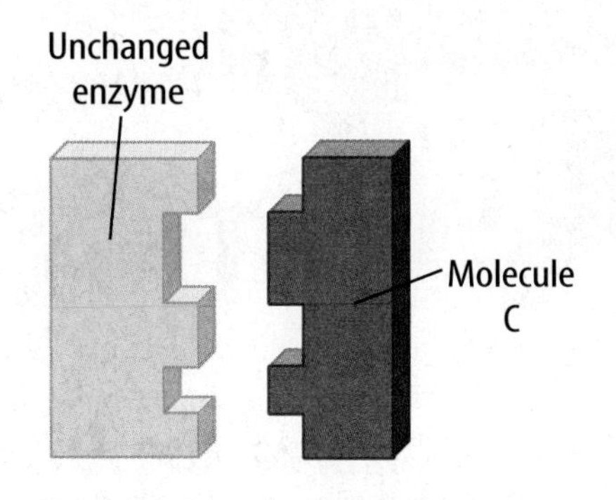

Following the reaction, the enzyme and the new molecule separate. The enzyme is not changed by the reaction. The resulting new molecule has a new chemical structure.

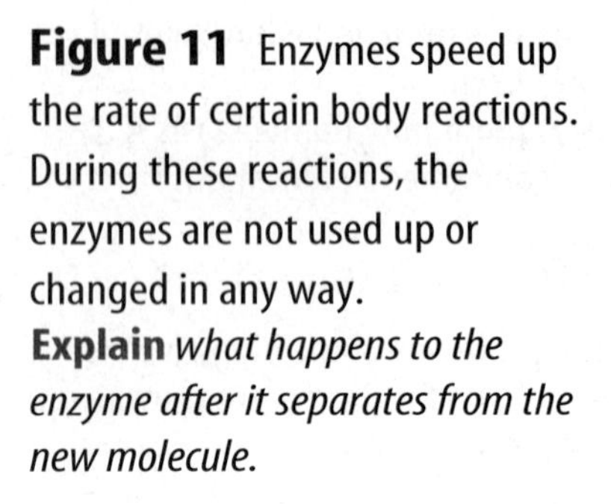

Figure 11 Enzymes speed up the rate of certain body reactions. During these reactions, the enzymes are not used up or changed in any way.
Explain *what happens to the enzyme after it separates from the new molecule.*

Enzymes

Chemical digestion is possible only because of enzymes (EN zimez). An **enzyme** is a type of protein that speeds up the rate of a chemical reaction in your body. One way enzymes speed up reactions is by reducing the amount of energy necessary for a chemical reaction to begin. If enzymes weren't there to help, the rate of chemical reactions would slow down. Some might not even happen at all. As shown in **Figure 11,** enzymes work without being changed or used up.

Enzymes in Digestion Many enzymes help you digest carbohydrates, proteins, and fats. Amylase (AM uh lays) is an enzyme produced by glands near the mouth. This enzyme helps speed up the breakdown of complex carbohydrates, such as starch, into simpler carbohydrates—sugars. In your stomach, the enzyme pepsin aids the chemical reactions that break down complex proteins into less complex proteins. In your small intestine, a number of other enzymes continue to speed up the breakdown of proteins into amino acids.

The pancreas, an organ on the back outside of the stomach, releases several enzymes through a tube into the small intestine. Some of these enzymes continue to aid the process of starch breakdown that started in the mouth. The resulting sugars are turned into glucose and are used by your body's cells. Different enzymes from the pancreas are involved in the breakdown of fats into fatty acids. Others help in the reactions that break down proteins.

What is the role of enzymes in the chemical digestion of food?

Other Enzyme Actions Enzyme-aided reactions are not limited to the digestive process. Enzymes also help speed up chemical reactions responsible for building your body. They are involved in the energy production activities of your muscle and nerve cells. Enzymes also aid in the blood-clotting process. Without enzymes, the chemical reactions of your body would not happen. In fact, you would not exist.

Organs of the Digestive System

Your digestive system has two parts—the digestive tract and the accessory organs. The major organs of your digestive tract—mouth, esophagus (ih SAH fuh guhs), stomach, small intestine, large intestine, rectum, and anus—are shown in **Figure 12.** Food passes through all of these organs. The tongue, teeth, salivary glands, liver, gallbladder, and pancreas, also shown in **Figure 12,** are the accessory organs. Although food doesn't pass through them, they are important in mechanical and chemical digestion. Your liver, gallbladder, and pancreas produce or store enzymes and chemicals that help break down food as it passes through the digestive tract.

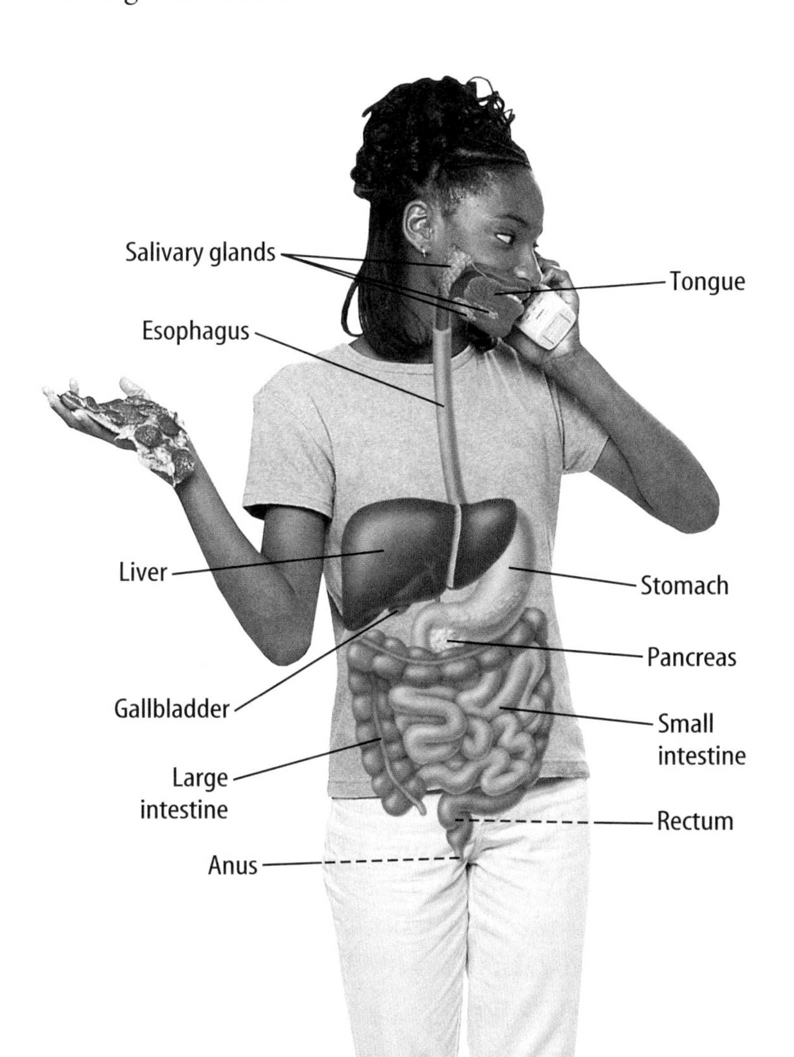

Figure 12 The human digestive system can be described as a tube divided into several specialized sections. If stretched out, an adult's digestive system is 6 m to 9 m long.

Figure 13 About 1.5 L of saliva are produced each day by salivary glands in your mouth.
Describe *what happens in your mouth when you think about a food you like.*

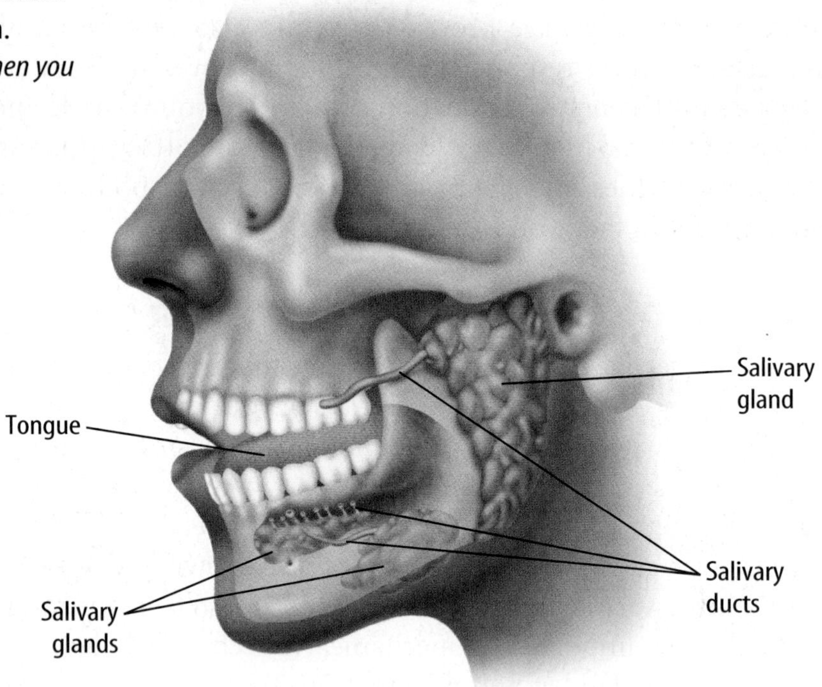

The Mouth Mechanical and chemical digestion begin in your mouth. Mechanical digestion happens when you chew your food with your teeth and mix it with your tongue. Chemical digestion begins with the addition of a watery substance called saliva (suh LI vuh). As you chew, your tongue moves food around and mixes it with saliva. Saliva is produced by three sets of glands near your mouth, as shown in **Figure 13.** Although saliva is mostly water, it also contains mucus and an enzyme that aids in the breakdown of starch into sugar. Food mixed with saliva becomes a soft mass and is moved to the back of your mouth by your tongue. It is swallowed and passes into your esophagus. Now ingestion is complete, but the process of digestion continues.

Topic: The Stomach
Visit life.msscience.com for Web links to information about the role of the stomach during digestion.

Activity Research to find out how antacids provide relief from stomach discomfort.

The Esophagus Food moving into the esophagus passes over the epiglottis (ep uh GLAH tus). This structure automatically covers the opening to the windpipe to prevent food from entering it, otherwise you would choke. Your esophagus is a muscular tube about 25 cm long. It takes about 4 s to 10 s for food to move down the esophagus to the stomach. No digestion takes place in the esophagus. Mucous glands in the wall of the esophagus keep the food moist. Smooth muscles in the wall move food downward with a squeezing action. These waves of muscle contractions, called **peristalsis** (per uh STAHL sus), move food through the entire digestive tract.

The Stomach The stomach, shown in **Figure 14,** is a muscular bag. When empty, it is somewhat sausage shaped with folds on the inside. As food enters from the esophagus, the stomach expands and the folds smooth out. Mechanical and chemical digestion take place in the stomach. Mechanically, food is mixed in the stomach by peristalsis. Chemically, food is mixed with enzymes and strong digestive solutions, such as hydrochloric acid solution, to help break it down.

Specialized cells in the walls of the stomach release about 2 L of hydrochloric acid solution each day. The acidic solution works with the enzyme pepsin to digest protein. The acidic solution has another important purpose—it destroys bacteria that are present in the food. The stomach also produces mucus, which makes food more slippery and protects the stomach from the strong, digestive solutions. Food moves through your stomach in 2 hours to 4 hours and is changed into a thin, watery liquid called **chyme** (KIME). Little by little, chyme moves out of your stomach and into your small intestine.

Reading Check ***Why isn't your stomach digested by the acidic digestive solution?***

Figure 14 A band of muscle is at the entrance of the stomach to control the entry of food from the esophagus. Muscles at the end of the stomach control the flow of the partially digested food into the first part of the small intestine.

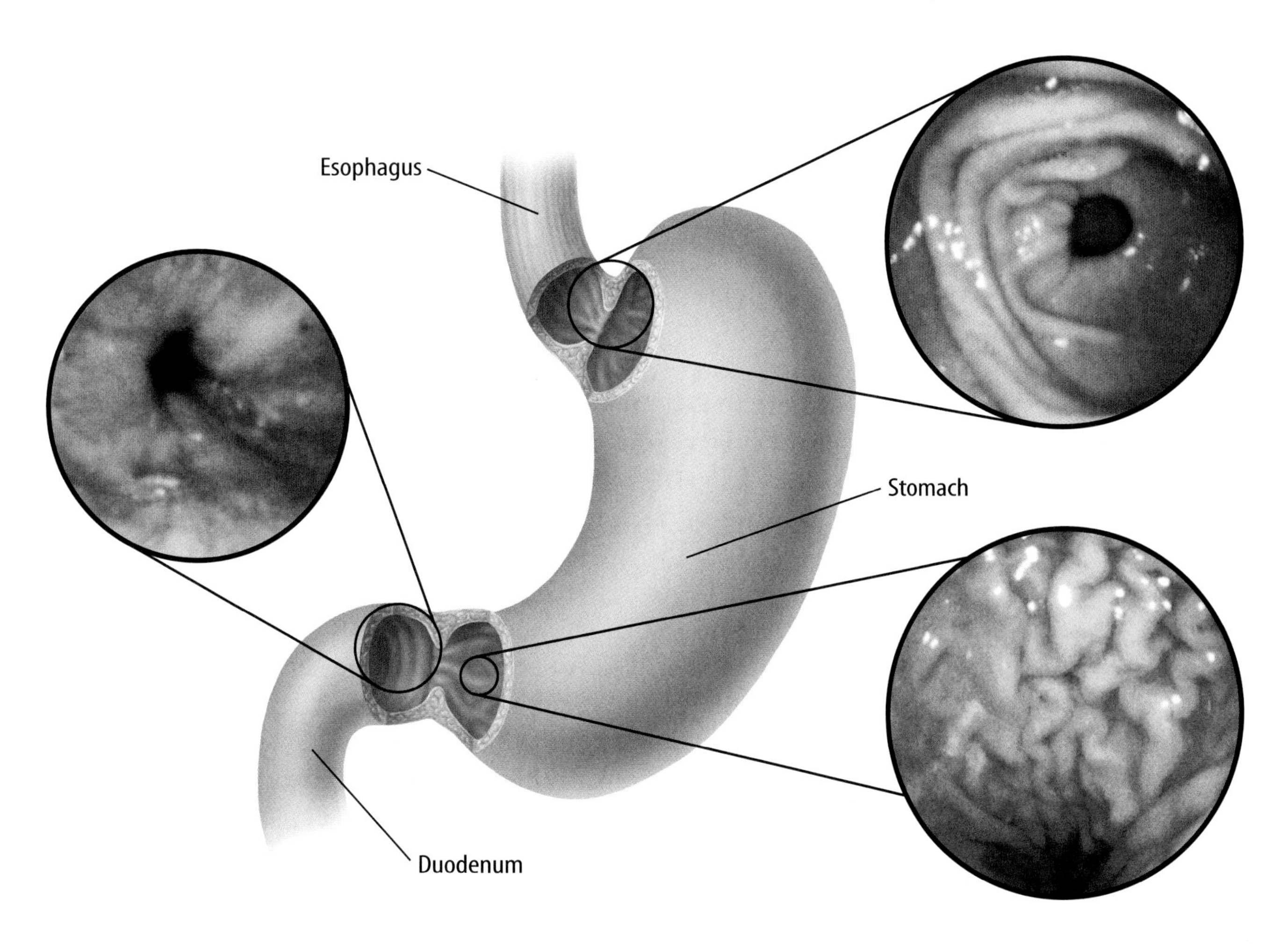

Figure 15 Hundreds of thousands of densely packed villi give the impression of a velvet cloth surface. If the surface area of your villi could be stretched out, it would cover an area the size of a tennis court.
Infer *what would happen to a person's weight if the number of villi were drastically reduced. Why?*

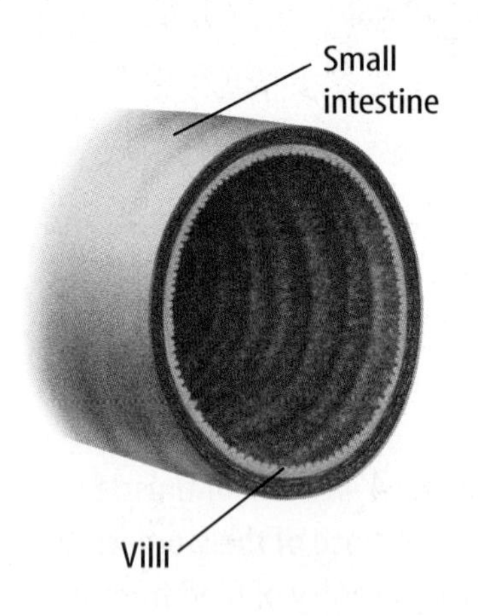

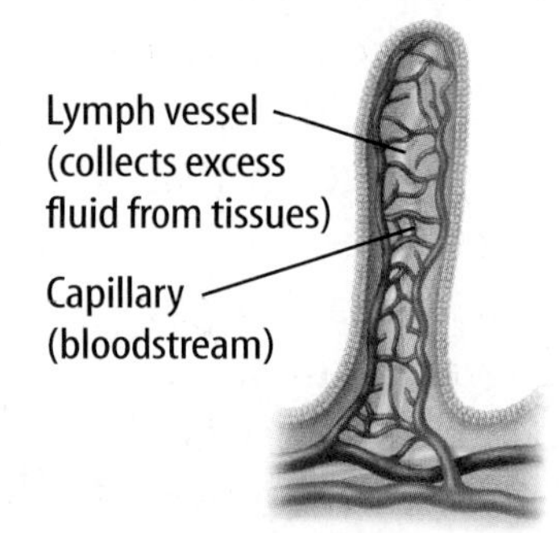

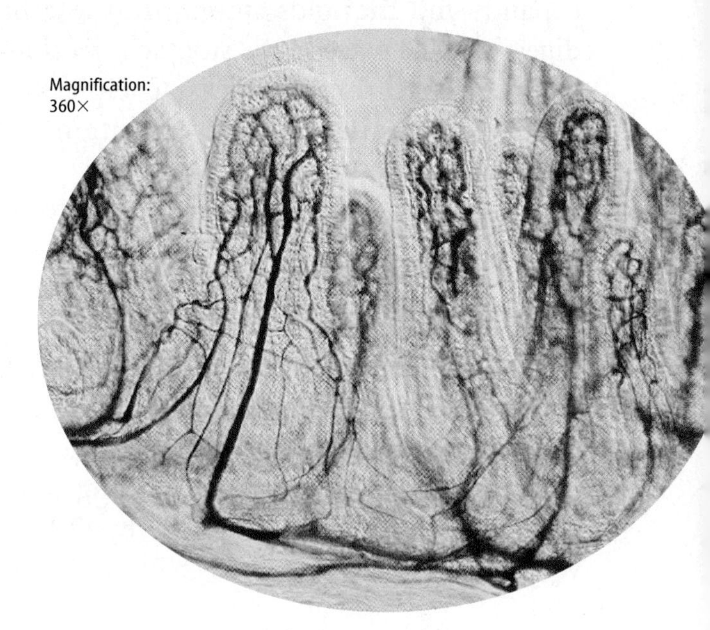

Mini LAB

Modeling Absorption in the Small Intestine

Procedure

1. Place one piece of smooth **cotton cloth** (about 25 cm × 25 cm) and a similar-sized piece of **cotton terry cloth** into a **bowl of water.**
2. Soak each for 30 s.
3. Remove the cloths and drain for 1 minute.
4. Wring out each cloth into different containers. Measure the amount of water collected in each.

Analysis

1. Which cloth absorbed the most water?
2. How does the surface of the terry cloth compare to the internal surface of the small intestine?

Try at Home

The Small Intestine Your small intestine is small in diameter, but it measures 4 m to 7 m in length. As chyme leaves your stomach, it enters the first part of your small intestine, called the duodenum (doo AH duh num). Most digestion takes place in your duodenum. Here, a greenish fluid from the liver, called bile, is added. The acid from the stomach makes large fat particles float to the top of the liquid. Bile breaks up the large fat particles, similar to the way detergent breaks up grease.

Chemical digestion of carbohydrates, proteins, and fats occurs when a digestive solution from the pancreas is mixed in. This solution contains bicarbonate ions and enzymes. The bicarbonate ions help neutralize the stomach acid that is mixed with chyme. Your pancreas also makes insulin, a hormone that allows glucose to pass from the bloodstream into your cells.

Absorption of food takes place in the small intestine. Look at the wall of the small intestine in **Figure 15.** The wall has many ridges and folds that are covered with fingerlike projections called **villi** (VIH li). Villi increase the surface area of the small intestine so that nutrients in the chyme have more places to be absorbed. Peristalsis continues to move and mix the chyme. The villi move and are bathed in the soupy liquid. Nutrients move into blood vessels within the villi. From here, blood transports the nutrients to all cells of your body. Peristalsis continues to force the remaining undigested and unabsorbed materials slowly into the large intestine.

The Large Intestine When the chyme enters the large intestine, it is still a thin, watery mixture. The main job of the large intestine is to absorb water from the undigested mass. This keeps large amounts of water in your body and helps maintain homeostasis. Peristalsis usually slows down in the large intestine. The chyme might stay there for as long as three days. After the excess water is absorbed, the remaining undigested materials become more solid. Muscles in the rectum, which is the last section of the large intestine, and the anus control the release of semisolid wastes from the body in the form of feces (FEE seez).

Bacteria The species of bacteria that live in your large intestine are adapted to their habitat. What do you think would happen to the bacteria if their environment were to change? How would this affect your large intestine? Discuss your ideas with a classmate and write your answers in your Science Journal.

Bacteria Are Important

Many types of bacteria live in your body. Bacteria live in many of the organs of your digestive tract including your mouth and large intestine. Some of these bacteria live in a relationship that is beneficial to the bacteria and to your body. The bacteria in your large intestine feed on undigested material like cellulose. In turn, bacteria make vitamins you need—vitamin K and two B vitamins. Vitamin K is needed for blood clotting. The two B vitamins, niacin and thiamine, are important for your nervous system and for other body functions. Bacterial action also converts bile pigments into new compounds. The breakdown of intestinal materials by bacteria produces gas.

section 2 review

Summary

Functions of the Digestive System

- To process food, your body must ingest, digest, absorb, and eliminate it.

Enzymes

- Enzymes aid the chemical reactions of digestion.

Organs of the Digestive System

- The mouth, esophagus, stomach, small intestine, large intestine, rectum, and anus are the major organs of this system.
- The digestive system also has accessory organs that are important to mechanical and chemical digestion.

Bacteria Are Important

- Bacteria help break down intestinal materials and make vitamins.

Self Check

1. **Compare and contrast** mechanical digestion and chemical digestion.
2. **Explain** how activities in the large intestine help maintain homeostasis.
3. **Describe** the function of each of the organs in the digestive tract.
4. **Explain** how the accessory organs aid digestion.
5. **Think Critically** A cracker contains starch. Explain why a cracker begins to taste sweet after it is in your mouth for five minutes without being chewed.

Applying Skills

6. **Recognize Cause and Effect** What would happen to some of the nutrients in chyme if the pancreas did not secrete its solution into the small intestine?

Particle Size and Absorption

Goals
- **Compare** the dissolving rates of different sized particles.
- **Predict** the dissolving rate of sugar particles larger than sugar cubes.
- **Predict** the dissolving rate of sugar particles smaller than particles of ground sugar.
- Using the lab results, infer why the body must break down and dissolve food particles.

Materials
250-mL beakers or jars (3)
thermometers (3)
sugar granules
mortar and pestle
triple-beam balance
stirring rod
sugar cubes
weighing paper
warm water
stopwatch

Safety Precautions

WARNING: *Do not taste, eat, or drink any materials used in the lab.*

Real-World Question
Before food reaches the small intestine, it is digested mechanically in the mouth and the stomach. The food mass is reduced to small particles. You can chew an apple into small pieces, but you would feed applesauce to a small child who didn't have teeth. What is the advantage of reducing the size of the food material? How does reducing the size of food particles aid the process of digestion?

Procedure
1. Copy the data table below into your Science Journal.

Dissolving Time of Sugar Particles

Size of Sugar Particles	Mass	Time Until Dissolved
Sugar cube	Do not write in this book.	
Sugar granules		
Ground sugar particles		

2. Place a sugar cube into your mortar and grind up the cube with the pestle until the sugar becomes powder.
3. Using the triple-beam balance and weighing paper, measure the mass of the powdered sugar from your mortar. Using separate sheets of weighing paper, measure the mass of a sugar cube and the mass of a sample of the granular sugar. The masses of the powdered sugar, sugar cube, and granular sugar should be approximately equal to each other. Record the three masses in your data table.
4. Place warm water into the three beakers. Use the thermometers to be certain the water in each beaker is the same temperature.

5. Place the sugar cube in a beaker, the powdered sugar in a second beaker, and the granular sugar in the third beaker. Place all the sugar samples in the beakers at the same time and start the stopwatch when you put the sugar samples in the beaker.
6. Stir each sample equally.
7. **Measure** the time it takes each sugar sample to dissolve and record the times in your data table.

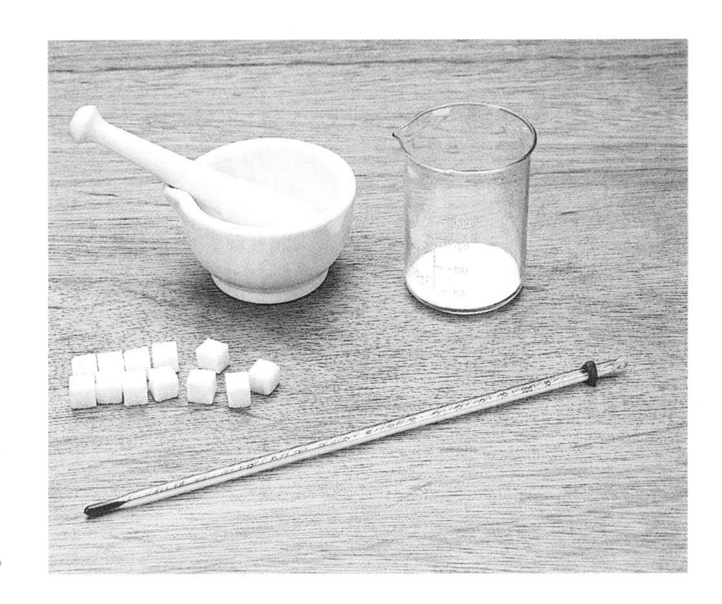

Analyze Your Data

1. **Identify** the experiment's constants and variables.
2. **Compare** the rate at which the sugar samples dissolved. What type of sugar dissolved most rapidly? Which was the slowest to dissolve?

Conclude and Apply

1. **Predict** how long it would take sugar particles larger than the sugar cubes to dissolve. Predict how long it would take sugar particles smaller than the powdered sugar to dissolve.
2. **Infer and explain** the reason why small particles dissolve more rapidly than large particles.
3. **Infer** why you should thoroughly chew your food.
4. **Explain** how reducing the size of food particles aids the process of digestion.

Communicating Your Data

Write a news column for a health magazine explaining to health-conscious people what they can do to digest their food better.

Eating Well

Does the same diet work for everyone?

Growing up in India in the first half of the twentieth century, R. Rajalakshmi (RAH jah lok shmee) saw many people around her who did not get enough food. Breakfast for a poor child might have been a cup of tea. Lunch might have consisted of a slice of bread. For dinner, a child might have eaten a serving of rice with a small piece of fish. This type of diet, low in calories and nutrients, produced children who were often sick and died young.

Good Diet, Wrong Place

R. Rajalakshmi studied biochemistry and nutrition at universities in India and in Canada. In the 1960s, she was asked to help manage a program to improve nutrition in her country. At that time, North American and European nutritionists suggested foods that were common and worked well for people who lived in these nations.

Thanks to R. Rajalakshmi and other nutritionists, many children in India are eating well and staying healthy.

For example, they told poor Indian women to eat more meat and eggs and drink more orange juice. But Rajalakshmi knew this advice was useless in a country such as India. People there didn't eat such foods. They weren't easy to find. And for the poor, such foods were too expensive.

The Proper Diet for India

Rajalakshmi knew that for the program to work, it had to fit Indian culture. So she decided to restructure the nutrition program. She first found out what healthy middle class people in India ate. She took note of the nutrients available in those foods. Then she looked for cheap, easy-to-find foods that would provide the same nutrients.

Rajalakshmi created a balanced diet of cheap, locally grown fruits, vegetables, and grains. Legumes (plants related to peas and peanuts), vegetables, and an Indian food called dhokla (DOH kluh) were basics. Dhokla is made of grains, legumes, and leafy vegetables. The grains and legumes provided protein, and the vegetables added vitamins and minerals.

Rajalakshmi's ideas were thought unusual in the 1960s. For example, she insisted that a diet without meat could provide all major nutrients. Now we know she was right. But it took persistence to get others to accept her diet about 40 years ago. Because of Rajalakshmi's program, Indian children almost doubled their food intake. And many children who would have been hungry and ill grew healthy and strong.

Report **Choose a continent and research what foods are native to that area. Share your findings with your classmates and compile a list of the foods and where they originated. Using the class list, mark the origins of the different foods on a world map.**

For more information, visit life.msscience.com/time

Reviewing Main Ideas

Section 1 Nutrition

1. Proteins, carbohydrates, fats, vitamins, minerals, and water are the six nutrients found in foods.
2. Carbohydrates provide energy, proteins are needed for growth and repair, and fats store energy and cushion organs. Vitamins and minerals regulate functions. Water makes up about 60 percent of your body's mass and is used for a variety of homeostatic functions.
3. Health is affected by the combination of foods that make up a diet.

Section 2 The Digestive System

1. Mechanical digestion breaks down food through chewing and churning.
2. Enzymes and other chemicals aid chemical digestion.
3. Digestion breaks down food into substances that cells can absorb and use. Carbohydrates break down into simple sugars; proteins into amino acids; and fats into fatty acids and glycerol.
4. Food is ingested in the mouth. Digestion occurs in the mouth, stomach, and small intestine, with absorption occurring in the small and large intestines. Wastes are excreted through the anus.
5. The accessory digestive organs move and cut up food and supply digestive enzymes and other chemicals, such as bile, needed for digestion.
6. The large intestine absorbs water, which helps the body maintain homeostasis.

Visualizing Main Ideas

Copy and complete the following table indicating good sources of vitamins and minerals.

Vitamin and Mineral Sources

Food Type	Source of Vitamin	Source of Mineral
Milk	D	
Spinach		iron
Meat		calcium, potassium
Eggs	E	
Carrots		sodium

Using Vocabulary

- amino acid p. 513
- carbohydrate p. 514
- chemical digestion p. 523
- chyme p. 527
- digestion p. 523
- enzyme p. 524
- fat p. 515
- food group p. 520
- mechanical digestion p. 523
- mineral p. 518
- nutrient p. 512
- peristalsis p. 526
- protein p. 513
- villi p. 528
- vitamin p. 516

Fill in the blanks with the correct vocabulary word or words.

1. ________ is the muscular contractions of the esophagus.
2. The ________ increase the surface area of the small intestine.
3. The building blocks of proteins are ________.
4. The liquid product of digestion is called ________.
5. ________ is the breakdown of food.
6. Your body's main source of energy is ________.
7. ________ are inorganic nutrients.
8. Pears and apples belong to the same ________.
9. ________ is when food is chewed and mixed.
10. A(n) ________ is a nutrient needed in small quantities for growth and for regulating body functions.

Checking Concepts

Choose the word or phrase that best answers the question.

11. In which organ is water absorbed?
    - **A)** liver
    - **B)** esophagus
    - **C)** small intestine
    - **D)** large intestine

12. What beneficial substances are produced by bacteria in the large intestine?
    - **A)** fats
    - **B)** minerals
    - **C)** vitamins
    - **D)** proteins

13. Which organ makes bile?
    - **A)** gallbladder
    - **B)** liver
    - **C)** stomach
    - **D)** small intestine

14. Where in humans does most chemical digestion occur?
    - **A)** duodenum
    - **B)** stomach
    - **C)** liver
    - **D)** large intestine

15. Which of these organs is an accessory organ?
    - **A)** mouth
    - **B)** stomach
    - **C)** small intestine
    - **D)** liver

16. Which vitamin is found most abundantly in citrus fruits?
    - **A)** A
    - **B)** B
    - **C)** C
    - **D)** K

17. Where is hydrochloric acid solution added to the food mass?
    - **A)** mouth
    - **B)** stomach
    - **C)** small intestine
    - **D)** large intestine

18. Which of the following is in the same food group as yogurt and cheese?

A) 

C) 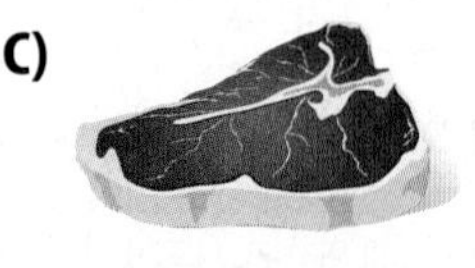

B) 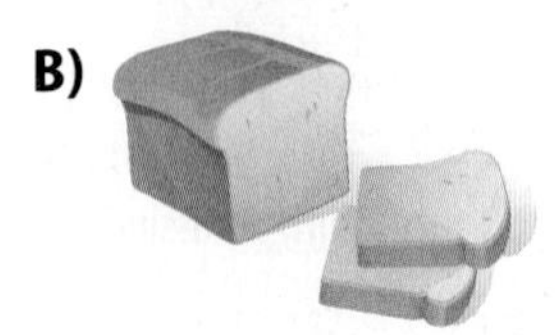

D) 

19. Which organ produces enzymes that help in digestion of proteins, fats, and carbohydrates?
    - **A)** mouth
    - **B)** pancreas
    - **C)** large intestine
    - **D)** gallbladder

Science Online life.msscience.com/vocabulary_puzzlemaker

Thinking Critically

Use the figure below to answer question 20.

Nutrition Facts
Serving Size 1 Meal

Amount Per Serving
Calories 330 Calories from Fat 60

	% Daily Value*
Total Fat 7g	**10%**
Saturated Fat 3.5g	**17%**
Polyunsaturated Fat 1g	
Monounsaturated Fat 2.5g	
Cholesterol 35mg	**12%**
Sodium 460mg	**19%**
Total Carbohydrate 52g	**18%**
Dietary Fiber 6g	**24%**
Sugars 17g	
Protein 15g	

Vitamin A 15% • Vitamin C 70%
Calcium 4% • Iron 10%

* Percent Daily Values are based on a 2,000 calorie diet. Your daily values may be higher or lower depending on your calorie needs.

	Calories	2,000	2,500
Total Fat	Less than	65g	80g
Sat Fat	Less than	20g	25g
Cholesterol	Less than	300mg	300mg
Sodium	Less than	2,400mg	2,400mg
Total Carbohydrate		300g	375g
Dietary Fiber		25g	30g

20. **Explain** how the information on the food label above can help you make healthful food choices.

21. **Infer** Food does not enter your body until it is absorbed into the blood. Explain why.

22. **Discuss** the meaning of the familiar statement "You are what you eat." Base your answer on your knowledge of food groups and nutrients.

23. **Explain** Bile's action is similar to that of soap. Use this information to explain how bile works on fats.

24. **Compare and contrast** the three types of carbohydrates—sugar, starch, and fiber.

Performance Activities

25. **Project** Research the ingredients used in antacid medications. Identify the compounds used to neutralize the excess stomach acid. Note the time, and then place an antacid tablet in a glass of vinegar—an acid. Using pH paper, check when the acid is neutralized. Record the time it took for the antacid to neutralize the vinegar. Repeat this procedure with different antacids. Compare your results.

Applying Math

26. **Villi Surface Area** The surface area of the villi in your small intestine is comparable to the area of a tennis court. A tennis court measures 11.0 m by 23.8 m. What is the area of a tennis court—and the surface area of the small intestine's villi—in square meters?

Use the table below to answer question 27.

Recommended Dietary Allowances

Nutrient	Percent U.S. RDA
Protein	2
Vitamin A	20
Vitamin C	25
Vitamin D	15
Calcium (Ca)	less than 2
Iron (Fe)	25
Zinc (Zn)	15
Total fat	5
Saturated fat	3
Cholesterol	0
Sodium	3

27. **Nutrients** A product nutrient label is shown above. Make a bar graph of this information.

chapter 18 Standardized Test Practice

Part 1 Multiple Choice

Record your answers on the answer sheet provided by your teacher or on a sheet of paper.

1. How many amino acids are required by your body?

A. 5
B. 12
C. 20
D. 50

Use the illustration below to answer questions 2 and 3.

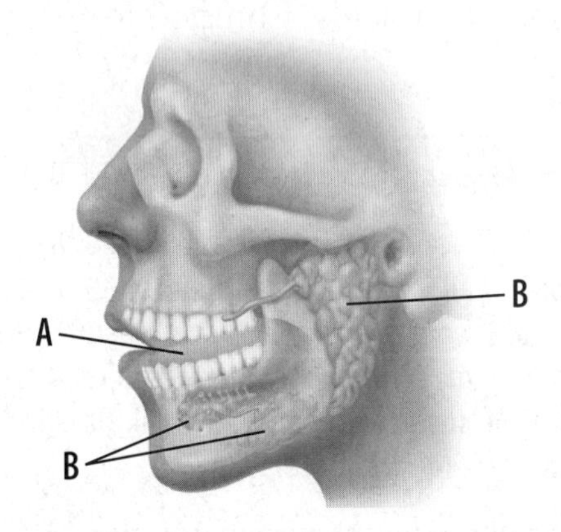

2. How does the organ labeled "A" help break down food?

A. produces enzymes
B. produces saliva
C. moves food around
D. produces mucus

3. Which of the following is produced by the organs labeled "B"?

A. saliva
B. bile
C. hydrochloric acid
D. chyme

Test-Taking Tip

Using Tables Concentrate on what the question is asking from a table, not all the information in the table.

Question 4 Look in the column titled *DV (Daily Values)* to find what the question asks, then follow the row to the *Item* column to find the answer.

Use the table below to answer questions 4 and 5.

Nutrition Facts of Vanilla Ice Cream		
Item	**Amount**	**DV (Daily Values)**
Serving Size	112 g	0
Calories	208	0
Total Fat	19 g	29%
Saturated Fat	11 g	55%
Cholesterol	0.125 g	42%
Sodium	0.90 g	4%
Total Carbohydrates	22 g	7%
Fiber	0 g	0%
Sugars	22 g	n/a
Protein	5 g	n/a
Calcium	0.117 g	15%
Iron	n/a	0%

4. According to the table above, which mineral has the greatest DV?

A. sodium
B. cholesterol
C. iron
D. calcium

5. If you had two servings of this vanilla ice cream, how many grams of saturated fat and Daily Value (DV) percentage would you eat?

A. 11 g, 110%
B. 22 g, 110%
C. 21 g, 55%
D. 5.5 g, 110%

6. Which of the following is the correct sequence of the organs of the digestive tract?

A. mouth, stomach, esophagus, small intestine, large intestine
B. esophagus, mouth, stomach, small intestine, large intestine
C. mouth, esophagus, small intestine, stomach, large intestine
D. mouth, esophagus, stomach, small intestine, large intestine

Part 2 | Short Response/Grid In

Record your answers on the answer sheet provided by your teacher or on a sheet of paper.

7. Explain the difference between organic and inorganic nutrients. Name a class of nutrients for each.

Use the photos below to answer questions 8 and 9.

8. During the activity shown above, which of the two teens is losing more body water? Why?
9. Based on the activity shown above, which teen may need more food energy (Calories)? Why?
10. Name three food sources that contain complete proteins.
11. How does bile help in digestion?
12. What is meant by an "essential amino acid"?
13. How do bacteria that live in the large intestine help your body?
14. Explain the importance of fats in the body.
15. Enzymes play an important role in the digestive process. But enzyme-aided reactions are also involved in other body systems. Give an example of how enzymes are used by the body in a way that does not involve the digestive system.
16. A taco has 180 Calories (Cal) and an ice cream sundae has 540 Cal. How many tacos could you eat to equal the number of Calories in the ice cream sundae?

Part 3 | Open Ended

Record your answers on a sheet of paper.

17. Explain what might happen to a child who is deficient in vitamin D. What foods should be eaten to prevent a deficiency in vitamin D?
18. Certain bacteria that do not normally live in the body can make toxins that affect intestinal absorption. Explain what might happen if these bacteria were present in the small and large intestines.

Use the photo below to answer question 19.

19. Identify the food group shown above. Explain why it is important for children and adolescents to eat adequate amounts of food from this group.
20. Identify the foods that should be consumed on a limited basis.
21. Antibiotics may be given to help a person fight off a bacterial infection. If a person is taking antibiotics, what might happen to the normal bacteria living in the large intestine? How would this affect the body?
22. Sometimes the esophagus can be affected by a disease in which peristalsis is not normal and the band of muscle at the entrance to the stomach does not work properly. What do you think would happen to food that the person swallowed?

Circulation

The BIG Idea

Your circulatory system moves needed materials to all cells and, along with your lymphatic system, removes wastes.

SECTION 1
The Circulatory System
Main Idea Your circulatory system provides each cell in your body with a continuous supply of oxygen and nutrients and takes away carbon dioxides and other wastes.

SECTION 2
Blood
Main Idea Blood is needed by all human body systems.

SECTION 3
The Lymphatic System
Main Idea The lymphatic system maintains a fluid balance in your body and fights infections.

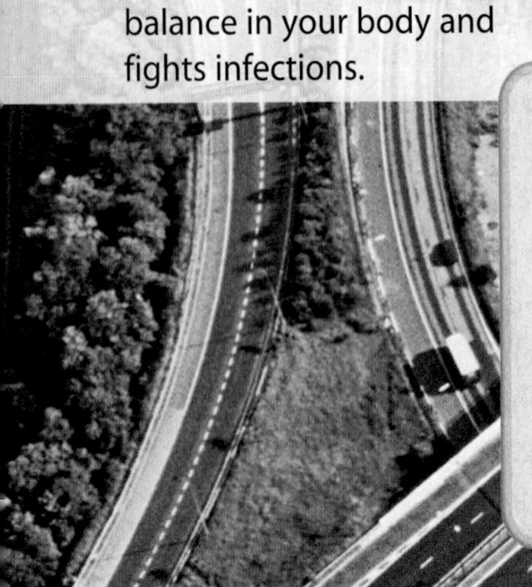

What does a highway have to do with circulation?

Think of this interchange as a simplified way to visualize how your blood travels through your body. Your complex circulatory system also plays an important role in protecting you from disease.

Science Journal Infer how the circulatory system provides your body with the nutrients it needs to stay healthy?

Start-Up Activities

Comparing Circulatory and Road Systems

If you look at an aerial view of a road system, as shown in the photograph, you see roads leading in many directions. These roads provide a way to carry people and goods from one place to another. Your circulatory system is like a road system. Just as roads are used to transport goods to homes and factories, your blood vessels transport substances throughout your body.

1. Look at a map of your city, county, or state.
2. Identify roads that are interstates, as well as state and county routes, using the map key.
3. Plan a route to a destination that your teacher describes. Then plan a different return trip.
4. Draw a diagram in your Science Journal showing your routes to and from the destination.
5. **Think Critically** If the destination represents your heart, what do the routes represent? Draw a comparison between a blocked road on your map and a clogged artery in your body.

Circulation Your body is supplied with nutrients by blood circulating through your blood vessels. Make the following Foldable to help you organize information about circulation.

STEP 1 **Fold** a sheet of paper in half lengthwise. Make the back edge about 5 cm longer than the front edge.

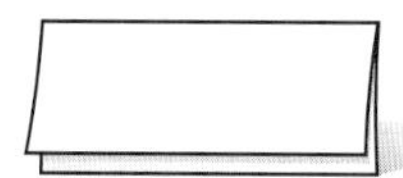

STEP 2 **Turn** the paper so the fold is on the bottom. Then, **fold** it into thirds.

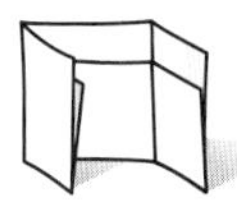

STEP 3 **Unfold and cut** only the top layer along both folds to make three tabs. **Label** the top of the page *Circulation,* and label the three tabs *Pulmonary, Coronary,* and *Systemic.*

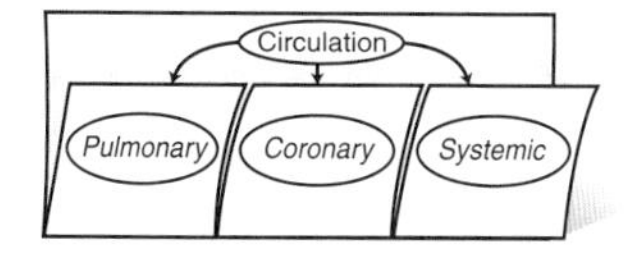

Read and Write As you read the chapter, write about each section under its tab.

Preview this chapter's content and activities at life.msscience.com

Get Ready to Read

Questioning

1 Learn It! Asking questions helps you to understand what you read. As you read, think about the questions you'd like answered. Often you can find the answer in the next paragraph or lesson. Learn to ask good questions by asking who, what, when, where, why, and how.

2 Practice It! Read the following passage from Section 2.

> People can inherit one of four types of blood: A, B, AB, or O, as shown in **Table 1.** Types A, B, and AB have chemical identification tags called antigens (AN tih junz) on their red blood cells. Type O red blood cells have no antigens.
>
> —*from page 553*

Here are some questions you might ask about this paragraph:

- How are antigens different on A, B, and AB blood?
- Where are the antigens in blood?
- Why does the presence of an antigen affect the blood?

3 Apply It! As you read the chapter, look for answers to Reading Check questions.

Test yourself. Create questions and then read to find answers to your own questions.

Target Your Reading

Use this to focus on the main ideas as you read the chapter.

1. **Before you read** the chapter, respond to the statements below on your worksheet or on a numbered sheet of paper.
   - Write an **A** if you **agree** with the statement.
   - Write a **D** if you **disagree** with the statement.

2. **After you read** the chapter, look back to this page to see if you've changed your mind about any of the statements.
   - If any of your answers changed, explain why.
   - Change any false statements into true statements.
   - Use your revised statements as a study guide.

Before You Read A or D	Statement	After You Read A or D
	1 Oxygen-poor blood flows through veins when it leaves the heart.	
	2 The heart has four compartments.	
	3 Blood flows to and from the lungs before circulating throughout the body.	
	4 Capillaries connect veins and arteries.	
	5 A heart-healthy lifestyle includes regular check-ups, a healthful diet, and regular exercise.	
	6 Red blood cells are the same in all humans.	
	7 A function of blood is to help fight infections.	
	8 Anemia only affects the circulatory system.	
	9 Lymph nodes function as filters for your body.	
	10 Blood contains a liquid called plasma that is mostly white blood cells.	

Science Online
Print out a worksheet of this page at life.msscience.com

section 1

The Circulatory System

as you read

What You'll Learn

- **Compare and contrast** arteries, veins, and capillaries.
- **Explain** how blood moves through the heart.
- **Identify** the functions of the pulmonary and systemic circulation systems.

Why It's Important

Your body's cells depend on the blood vessels to bring nutrients and remove wastes.

Review Vocabulary

heart: organ that circulates blood through your body continuously

New Vocabulary

- atrium
- ventricle
- coronary circulation
- pulmonary circulation
- systemic circulation
- artery
- vein
- capillary

How Materials Move Through the Body

It's time to get ready for school, but your younger sister is taking a long time in the shower. "Don't use up all the water," you shout. Water is carried throughout your house in pipes that are part of the plumbing system. The plumbing system supplies water for all your needs and carries away wastes. Just as you expect water to flow when you turn on the faucet, your body needs a continuous supply of oxygen and nutrients and a way to remove wastes. In a similar way materials are moved throughout your body by your cardiovascular (kar dee oh VAS kyuh lur) system. It includes your heart, kilometers of blood vessels, and blood.

Blood vessels carry blood to every part of your body, as shown in **Figure 1.** Blood moves oxygen and nutrients to cells and carries carbon dioxide and other wastes away from the cells. Sometimes blood carries substances made in one part of the body to another part of the body where these substances are needed. Movement of materials into and out of your cells occurs by diffusion (dih FYEW zhun) and active transport. Diffusion occurs when a material moves from an area where there is more of it to an area where there is less of it. Active transport is the opposite of diffusion. Active transport requires an input of energy from the cell, but diffusion does not.

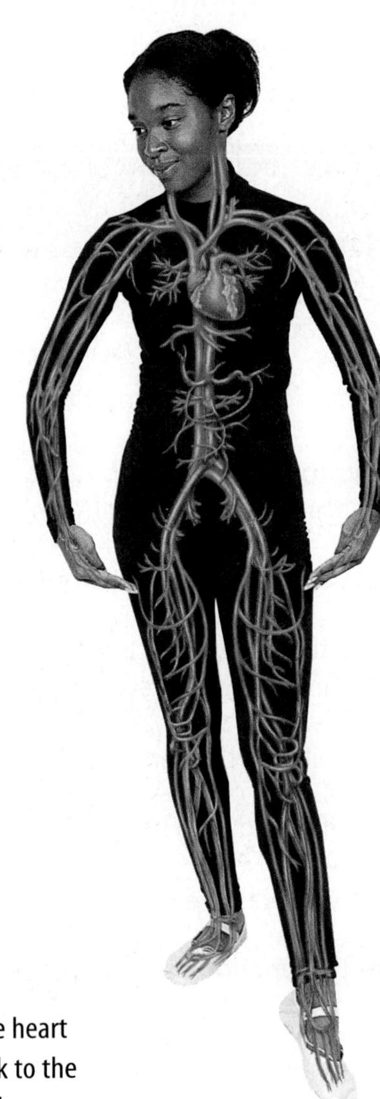

Figure 1 The blood is pumped by the heart to all the cells of the body and then back to the heart through a network of blood vessels.

The Heart

Your heart is an organ made of cardiac muscle tissue. It is located behind your breastbone, called the sternum, and between your lungs. Your heart has four compartments called chambers. The two upper chambers are called the right and left **atriums** (AY tree umz). The two lower chambers are called the right and left **ventricles** (VEN trih kulz). During one heartbeat, both atriums contract at the same time. Then, both ventricles contract at the same time. A one-way valve separates each atrium from the ventricle below it. The blood flows only in one direction from an atrium to a ventricle, then from a ventricle into a blood vessel. A wall prevents blood from flowing between the two atriums or the two ventricles. This wall keeps blood rich in oxygen separate from blood low in oxygen. If oxygen-rich blood and oxygen-poor blood were to mix, your body's cells would not get all the oxygen they need.

Scientists have divided the circulatory system into three sections—coronary circulation, pulmonary (PUL muh ner ee) circulation, and systemic circulation. The beating of your heart controls blood flow through each section.

Coronary Circulation Your heart has its own blood vessels that supply it with nutrients and oxygen and remove wastes. **Coronary** (KOR uh ner ee) **circulation,** as shown in **Figure 2,** is the flow of blood to and from the tissues of the heart. When the coronary circulation is blocked, oxygen and nutrients cannot reach all the cells of the heart. This can result in a heart attack.

Inferring How Hard the Heart Works

Procedure

1. Make a fist and observe its size, which is approximately the size of your heart.
2. Place your fist in a **bowl of water.** Then clench and unclench your fist to cause water to squirt out between your thumb and forefinger.
3. Continue the squeezing action for 3 min. Determine the number of squeezes per minute.

Analysis

1. State how many times you squeezed your fist in 1 min. A resting heart beats approximately 70 times per minute.
2. What can you do when the muscles of your hand and arm get tired? Explain why cardiac muscle does not get tired.

Try at Home

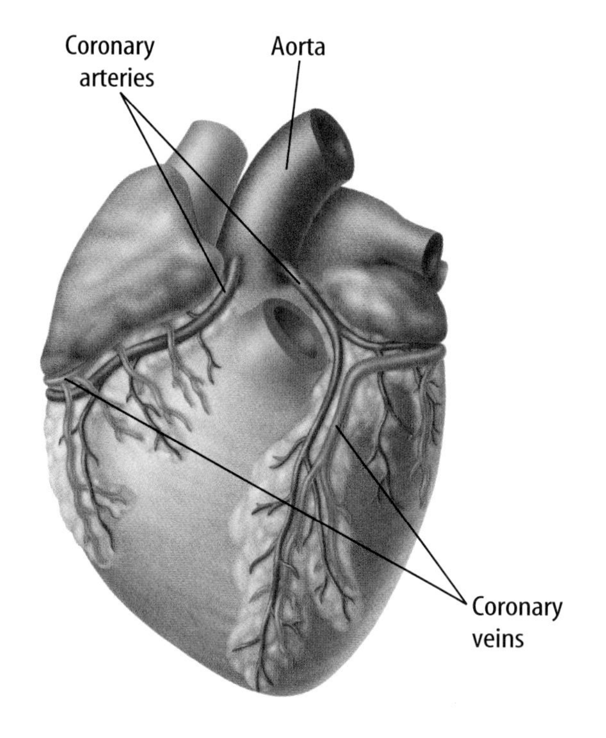

Figure 2 Like the rest of the body, the heart receives the oxygen and nutrients that it needs from the blood. The blood also carries away wastes from the heart's cells. On the diagram, you can see the coronary arteries, which nourish the heart.

Blood, high in carbon dioxide and low in oxygen, returns from the body to the heart. It enters the right atrium through the superior and inferior vena cavae.

Oxygen-rich blood travels from the lungs through the pulmonary veins and into the left atrium. The pulmonary veins are the only veins that carry oxygen-rich blood.

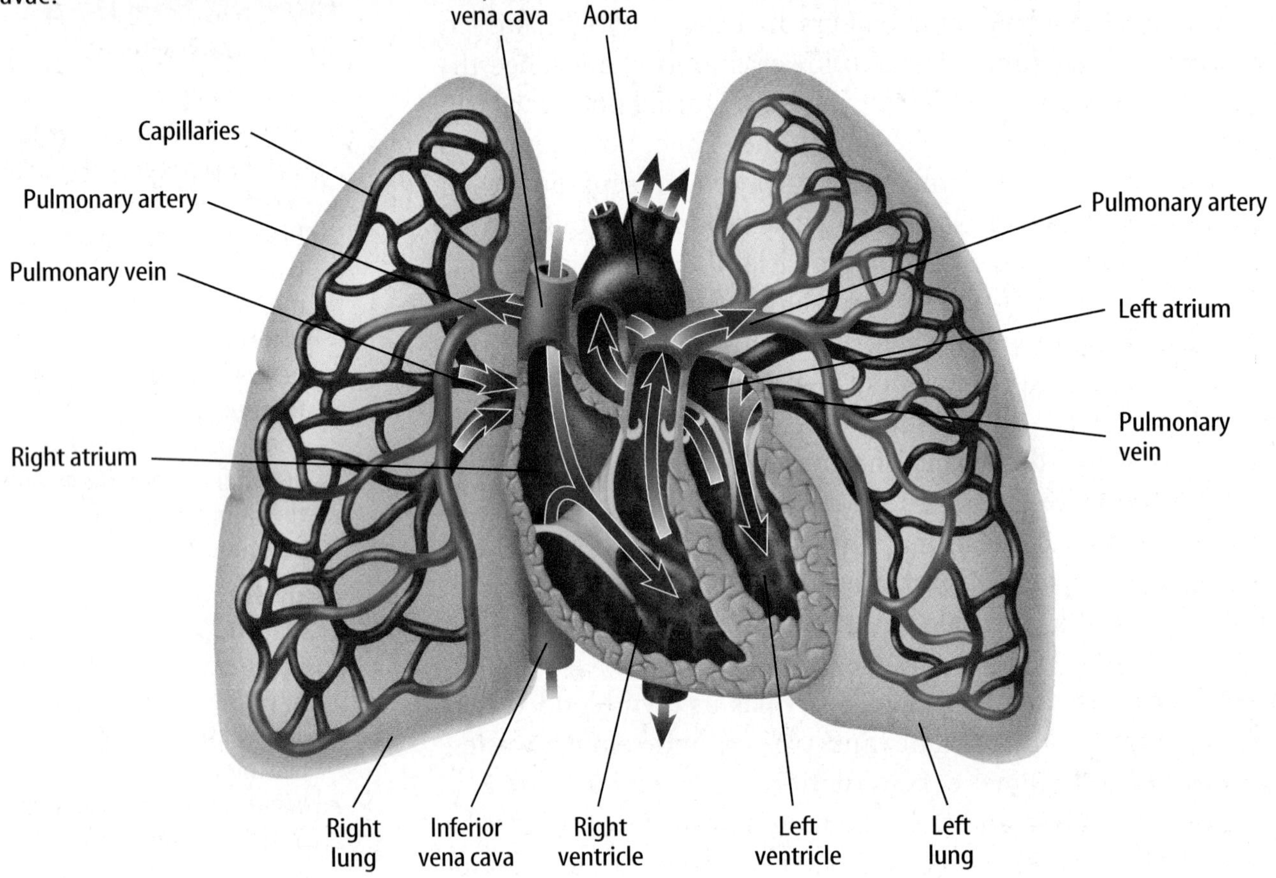

The right atrium contracts, forcing the blood into the right ventricle. When the right ventricle contracts, the blood leaves the heart and goes through the pulmonary arteries to the lungs. The pulmonary arteries are the only arteries that carry blood that is high in carbon dioxide.

The left atrium contracts and forces the blood into the left ventricle. The left ventricle contracts, forcing the blood out of the heart and into the aorta.

Figure 3 Pulmonary circulation moves blood between the heart and lungs.

Pulmonary Circulation The flow of blood through the heart to the lungs and back to the heart is **pulmonary circulation.** Use **Figure 3** to trace the path blood takes through this part of the circulatory system. The blood returning from the body through the right side of the heart and to the lungs contains cellular wastes. The wastes include molecules of carbon dioxide and other substances. In the lungs, gaseous wastes diffuse out of the blood, and oxygen diffuses into the blood. Then the blood returns to the left side of the heart. In the final step of pulmonary circulation, the oxygen-rich blood is pumped from the left ventricle into the aorta (ay OR tuh), the largest artery in your body. Next, the oxygen-rich blood flows to all parts of your body.

Systemic Circulation Oxygen-rich blood moves to all of your organs and body tissues, except the heart and lungs, by **systemic circulation,** and oxygen-poor blood returns to the heart. Systemic circulation is the largest of the three sections of your circulatory system. **Figure 4** shows the major arteries (AR tuh reez) and veins (VAYNZ) of the systemic circulation system. Oxygen-rich blood flows from your heart in the arteries of this system. Then nutrients and oxygen are delivered by blood to your body cells and exchanged for carbon dioxide and wastes. Finally, the blood returns to your heart in the veins of the systemic circulation system.

Reading Check *What are the functions of the systemic circulation system in your body?*

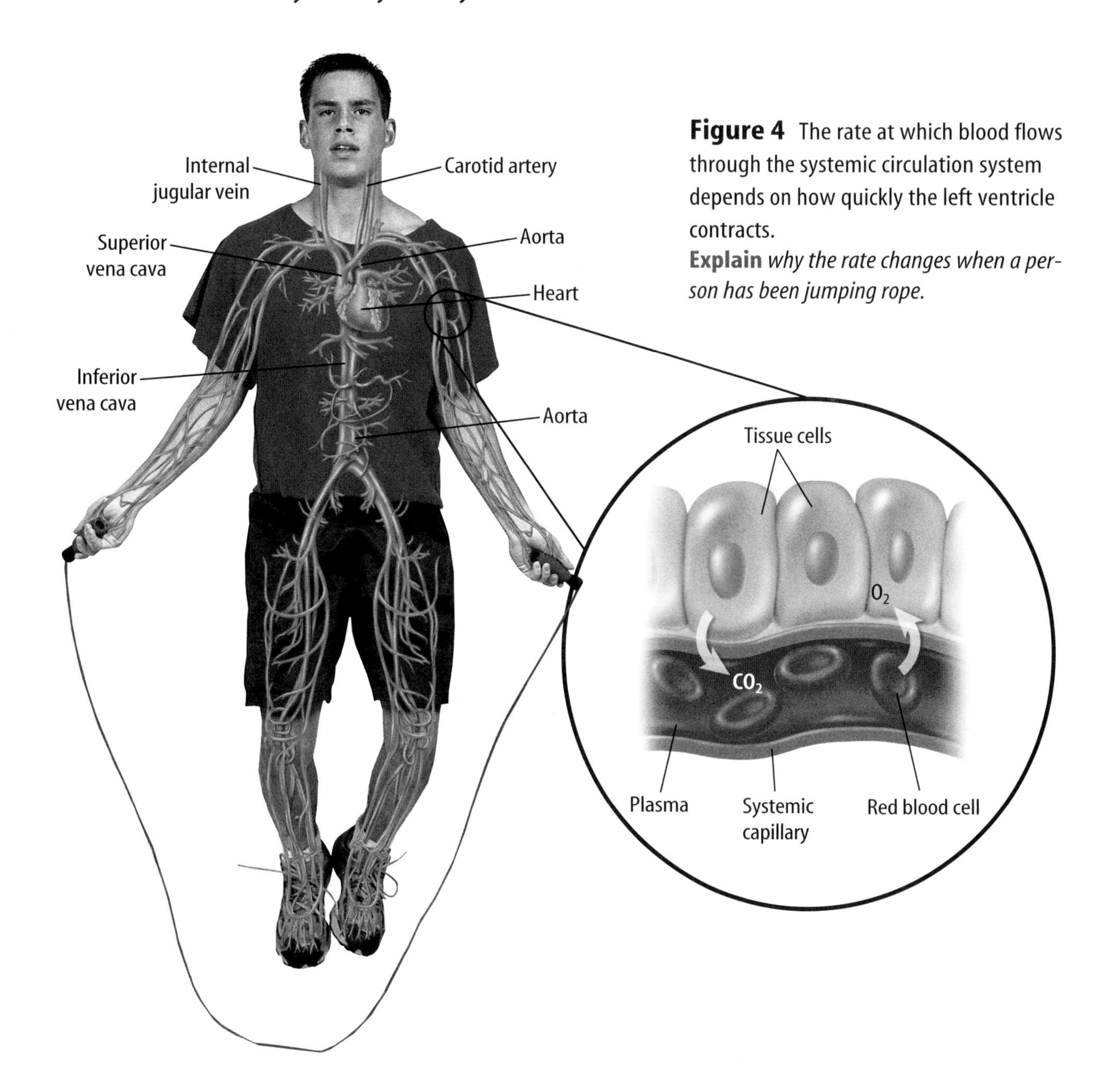

Figure 4 The rate at which blood flows through the systemic circulation system depends on how quickly the left ventricle contracts.
Explain *why the rate changes when a person has been jumping rope.*

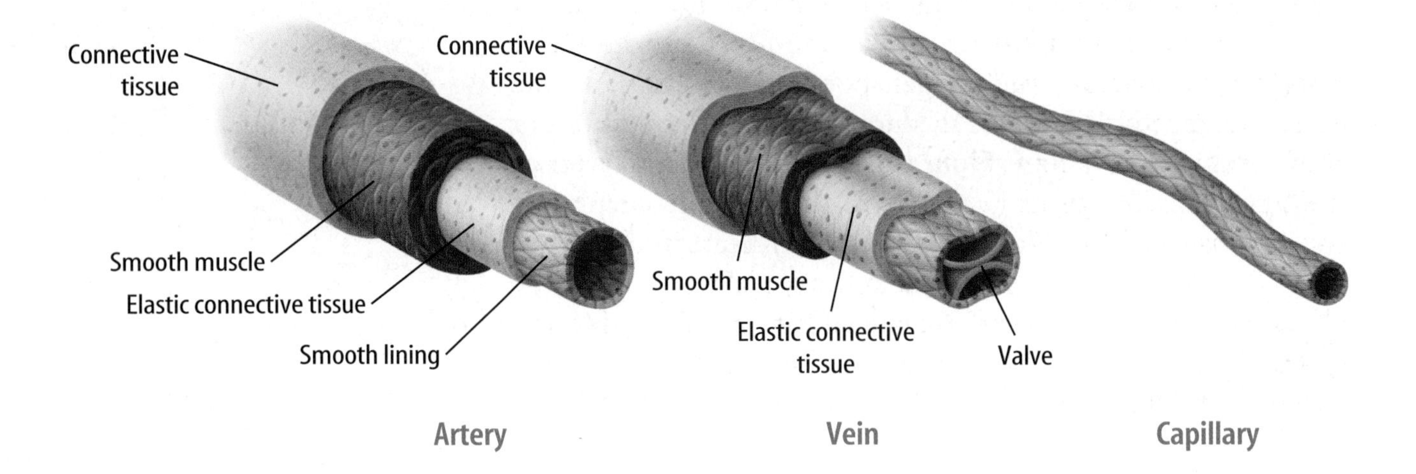

Figure 5 The structures of arteries, veins, and capillaries are different. Valves in veins prevent blood from flowing backward. Capillaries are much smaller. Capillary walls are only one cell thick.

Blood Vessels

In the middle 1600s, scientists proved that blood moves in one direction in a blood vessel, like traffic on a one-way street. They discovered that blood moves by the pumping of the heart and flows from arteries to veins. But, they couldn't explain how blood gets from arteries to veins. Using a new invention of that time, the microscope, scientists discovered capillaries (KAP uh ler eez), the connection between arteries and veins.

Arteries As blood is pumped out of the heart, it travels through arteries, capillaries, and then veins. **Arteries** are blood vessels that carry blood away from the heart. Arteries, shown in **Figure 5,** have thick, elastic walls made of connective tissue and smooth muscle tissue. Each ventricle of the heart is connected to an artery. The right ventricle is connected to the pulmonary artery, and the left ventricle is attached to the aorta. Every time your heart contracts, blood is moved from your heart into arteries.

Veins The blood vessels that carry blood back to the heart are called **veins,** as shown in **Figure 5.** Veins have one-way valves that keep blood moving toward the heart. If blood flows backward, the pressure of the blood against the valves causes them to close. The flow of blood in veins also is helped by your skeletal muscles. When skeletal muscles contract, the veins in these muscles are squeezed and help blood move toward the heart. Two major veins return blood from your body to your heart. The superior vena cava returns blood from your head and neck. Blood from your abdomen and lower body returns through the inferior vena cava.

What are the similarities and differences between arteries and veins?

Capillaries Arteries and veins are connected by microscopic blood vessels called **capillaries,** as shown in **Figure 5.** The walls of capillaries are only one cell thick. You can see capillaries when you have a bloodshot eye. They are the tiny red lines you see in the white area of your eye. Nutrients and oxygen diffuse into body cells through the thin capillary walls. Waste materials and carbon dioxide diffuse from body cells into the capillaries.

Blood Pressure Some molecules of nutrients are forced through capillary walls by the force of blood pressure. What is the cause of the pressure? Discuss your answer with a classmate. Then write your answer in your Science Journal.

Blood Pressure

INTEGRATE Physics

If you fill a balloon with water and then push on it, the pressure moves through the water in all directions, as shown in **Figure 6.** Your circulatory system is like the water balloon. When your heart pumps blood through the circulatory system, the pressure of the push moves through the blood. The force of the blood on the walls of the blood vessels is called blood pressure. This pressure is highest in arteries and lowest in veins. When you take your pulse, you can feel the waves of pressure. This rise and fall of pressure occurs with each heartbeat. Normal resting pulse rates are 60 to 100 heartbeats per minute for adults, and 80 to 100 beats per minute for children.

Measuring Blood Pressure Blood pressure is measured in large arteries and is expressed by two numbers, such as 120 over 80. The first number is a measure of the pressure caused when the ventricles contract and blood is pushed out of the heart. This is called the systolic (sihs TAH lihk) pressure. Then, blood pressure drops as the ventricles relax. The second number is a measure of the diastolic (di uh STAH lihk) pressure that occurs as the ventricles fill with blood just before they contract again.

Controlling Blood Pressure Your body tries to keep blood pressure normal. Special nerve cells in the walls of some arteries sense changes in blood pressure. When pressure is higher or lower than normal, messages are sent to your brain by these nerve cells. Then messages are sent by your brain to raise or lower blood pressure—by speeding up or slowing the heart rate for example. This helps keep blood pressure constant within your arteries. When blood pressure is constant, enough blood reaches all organs and tissues in your body and delivers needed nutrients to every cell.

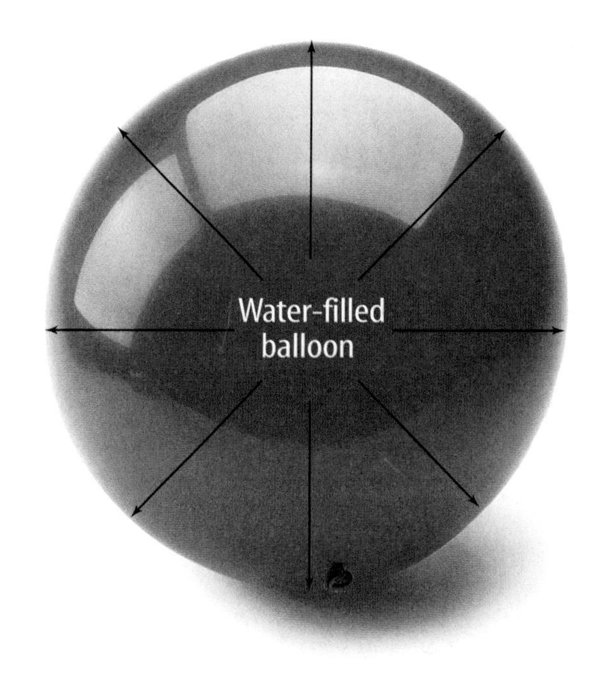

Figure 6 When pressure is exerted on a fluid in a closed container, the pressure is transmitted through the liquid in all directions. Your circulatory system is like a closed container.

NATIONAL GEOGRAPHIC

VISUALIZING ATHEROSCLEROSIS

Figure 7

Healthy blood vessels have smooth, unobstructed interiors like the one at the right. Atherosclerosis is a disease in which fatty substances build up in the walls of arteries, such as the coronary arteries that supply the heart muscle with oxygen-rich blood. As illustrated below, these fatty deposits can gradually restrict—and ultimately block—the life-giving river of blood that flows through an artery.

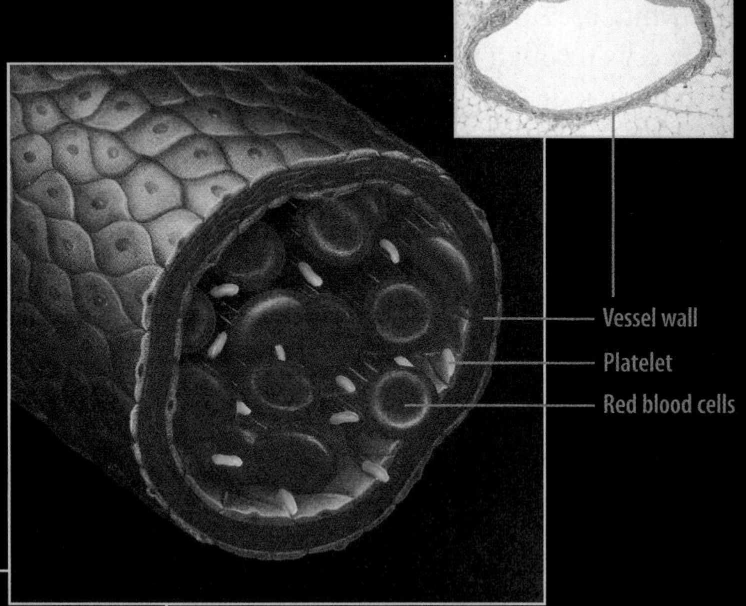

▲ HEALTHY ARTERY The illustration and photo above show a normal functioning artery.

Vessel wall

Plaque

◀ PARTIALLY CLOGGED ARTERY The illustration and inset photo at left show fatty deposits, called plaques, that have formed along the artery's inner wall. As the diagram illustrates, plaques narrow the pathway through the artery, restricting and slowing blood flow. As blood supply to the heart muscle cells dwindles, they become starved for oxygen and nutrients.

▶ NEARLY BLOCKED ARTERY In the illustration and photo at right, fatty deposits have continued to build. The pathway through the coronary artery has gradually narrowed until blood flow is very slow and nearly blocked. Under these conditions, the heart muscle cells supplied by the artery are greatly weakened. If blood flow stops entirely, a heart attack will result.

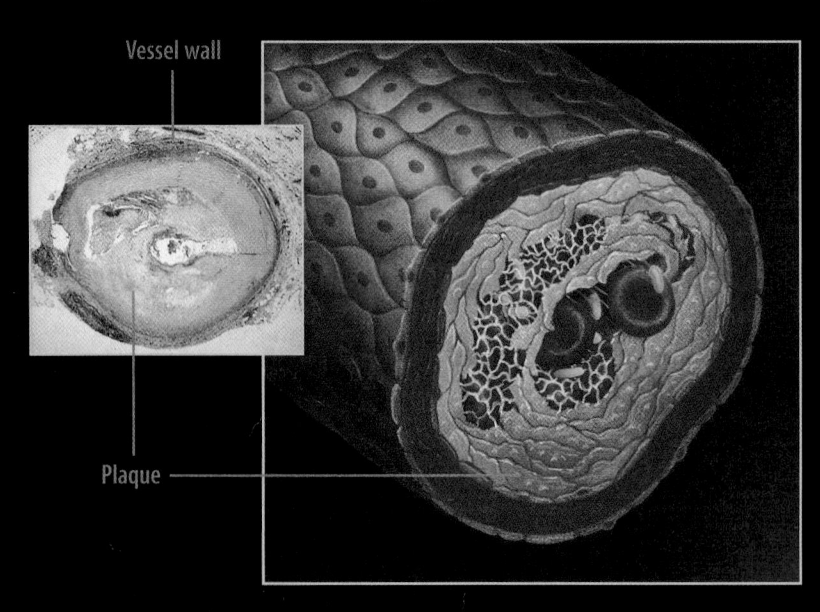

Cardiovascular Disease

Any disease that affects the cardiovascular system—the heart, blood vessels, and blood—can seriously affect the health of your entire body. People often think of cancer and automobile accidents as the leading causes of death in the United States. However, heart diseases are the leading cause of deaths when you factor in all age groups.

Atherosclerosis One leading cause of heart disease is called atherosclerosis (ah thuh roh skluh ROH sus). In this condition, shown in **Figure 7,** fatty deposits build up on arterial walls. Eating foods high in cholesterol and saturated fats can cause these deposits to form. Atherosclerosis can occur in any artery in the body, but deposits in coronary arteries are especially serious. If a coronary artery is blocked, a heart attack can occur. Open heart surgery may then be needed to correct the problem.

Hypertension Another condition of the cardiovascular system is called hypertension (HI pur TEN chun), or high blood pressure. **Figure 8** shows the instruments used to measure blood pressure. When blood pressure is higher than normal most of the time, extra strain is placed on the heart. The heart must work harder to keep blood flowing. One cause of hypertension is atherosclerosis. A clogged artery can increase pressure within the vessel. The walls become stiff and hard, like a metal pipe. The artery walls no longer contract and dilate easily because they have lost their elasticity.

Heart Failure Heart failure results when the heart cannot pump blood efficiently. It might be caused when heart muscle tissue is weakened by disease or when heart valves do not work properly. When the heart does not pump blood properly, fluids collect in the arms, legs, and lungs. People with heart failure usually are short of breath and tired.

Reading Check *What is heart failure?*

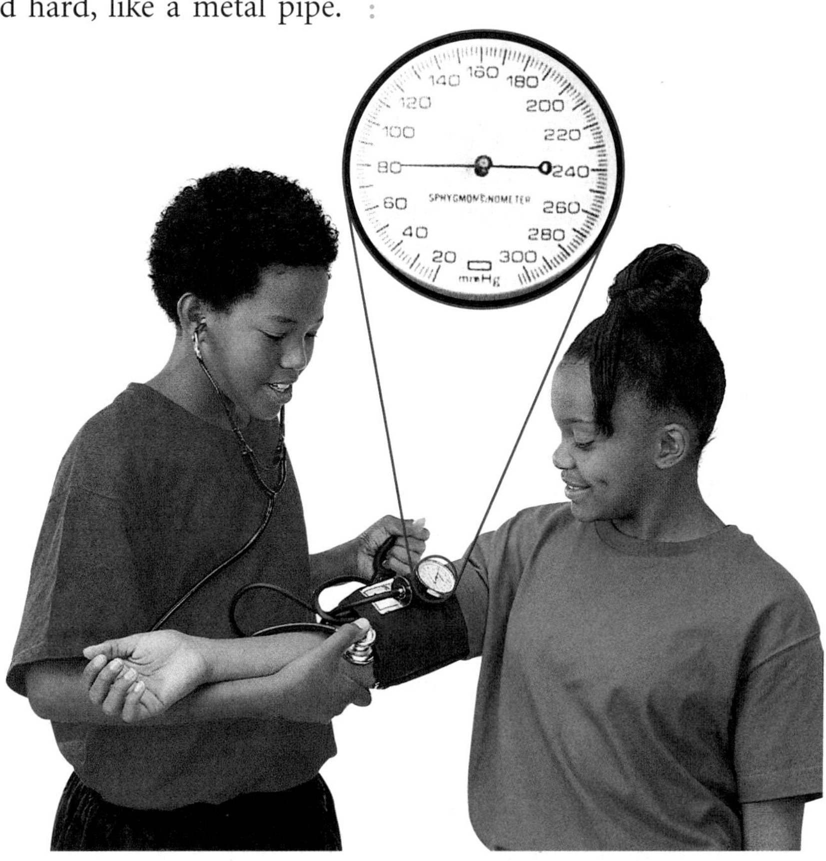

Figure 8 Blood pressure is measured in large arteries using a blood pressure cuff and stethoscope.

Topic: Cardiovascular Disease

Visit life.msscience.com for Web links to recent news or magazine articles about cardiovascular disease.

Activity In your Science Journal, list five steps you can take to lead a healthy life style.

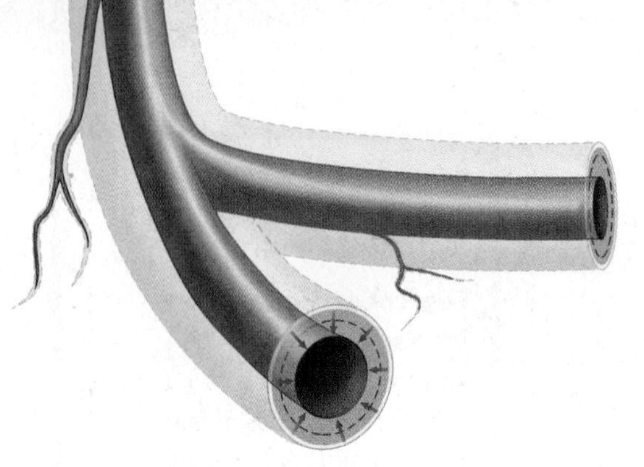

Figure 9 Nicotine, present in tobacco, contracts blood vessels and causes the body to release hormones that raise blood pressure. **Name** *another substance that raises blood pressure.*

Preventing Cardiovascular Disease Having a healthy lifestyle is important for the health of your cardiovascular system. The choices you make to maintain good health may reduce your risk of future serious illness. Regular checkups, a healthful diet, and exercise are part of a heart-healthy lifestyle.

Many diseases, including cardiovascular disease, can be prevented by following a good diet. Choose foods that are low in salt, sugar, cholesterol, and saturated fats. Being overweight is associated with heart disease and high blood pressure. Large amounts of body fat force the heart to pump faster.

Learning to relax and having a regular program of exercise can help prevent tension and relieve stress. Exercise also strengthens the heart and lungs, helps in controlling cholesterol, tones muscles, and helps lower blood pressure.

Another way to prevent cardiovascular disease is to not smoke. Smoking causes blood vessels to contract, as shown in **Figure 9,** and makes the heart beat faster and harder. Smoking also increases carbon monoxide levels in the blood. Not smoking helps prevent heart disease and a number of respiratory system problems, too.

section 1 review

Summary

Cardiovascular System

- Coronary circulation is the flow of blood to and from the tissues of the heart.
- Pulmonary circulation is the flow of blood through the heart, to the lungs, and back to the heart.
- Oxygen-rich blood is moved to all tissues and organs of the body, except the heart and lungs, by systemic circulation.

Blood Vessels

- Arteries carry blood away from the heart.
- Veins carry blood back to the heart.
- Arteries and veins are connected by capillaries.

Blood Pressure

- The force of the blood on the walls of the blood vessels is called blood pressure.

Cardiovascular Disease

- Atherosclerosis occurs when fatty deposits build up on arterial walls.
- High blood pressure is called hypertension.

Self Check

1. **Compare and contrast** the structure of the three types of blood vessels.
2. **Explain** the pathway of blood through the heart.
3. **Contrast** pulmonary and systemic circulation. Identify which vessels carry oxygen-rich blood.
4. **Explain** how exercise can help prevent heart disease.
5. **Think Critically** What waste product builds up in blood and cells when the heart is unable to pump blood efficiently?

Applying Skills

6. **Concept Map** Make an events-chain concept map to show pulmonary circulation beginning at the right atrium and ending at the aorta.
7. **Use a Database** Research diseases of the circulatory system. Make a database showing what part of the circulatory system is affected by each disease. Categories should include the organs and vessels of the circulatory system.

Science Online life.msscience.com/self_check_quiz

The Heart as a Pump

The heart is a pumping organ. Blood is forced through the arteries as heart muscles contract and then relax. This creates a series of waves in blood as it flows through the arteries. These waves are called the pulse. Try this lab to learn how physical activity affects your pulse.

Real-World Question

What does the pulse rate tell you about the work of the heart?

Goals

- **Observe** pulse rate.
- **Compare** pulse rate at rest to rate after jogging.

Materials

watch or clock with a second hand
*stopwatch
*Alternate materials

Pulse Rate

Pulse Rate	Partner's	Yours
At rest	Do not write in this book.	
After jogging		

Procedure

1. Make a table like the one shown. Use it to record your data.
2. Sit down to take your pulse. Your partner will serve as the recorder.
3. Find your pulse by placing your middle and index fingers over the radial artery in your wrist as shown in the photo.
   WARNING: *Do not press too hard.*
4. **Count** each beat of the radial pulse silently for 15 s. Multiply the number of beats by four to find your pulse rate per minute. Have your partner record the number in the data table.
5. Now jog in place for 1 min and take your pulse again. Count the beats for 15 s.
6. **Calculate** this new pulse rate and have your partner record it in the data table.
7. Reverse roles with your partner and repeat steps 2 through 6.
8. **Collect** and record the new data.

Conclude and Apply

1. **Describe** why the pulse rate changes.
2. **Infer** what causes the pulse rate to change.
3. **Explain** why the heart is a pumping organ.

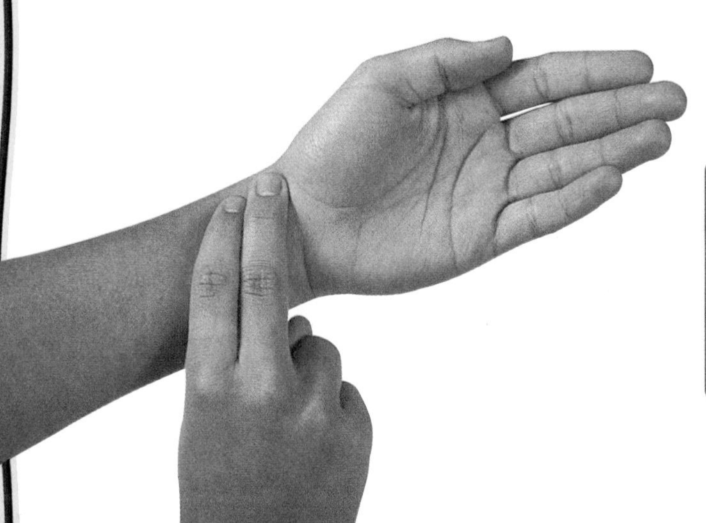

Communicating Your Data

Record the class average for pulse rate at rest and after jogging. Compare the class averages to your data. **For more help, refer to the Science Skill Handbook.**

section 2

Blood

as you read

What You'll Learn

- **Identify** the parts and functions of blood.
- **Explain** why blood types are checked before a transfusion.
- **Give examples** of diseases of blood.

Why It's Important

Blood plays a part in every major activity of your body.

Review Vocabulary

blood vessels: Structures that include arteries, veins, and capillaries, which transport blood

New Vocabulary

- plasma
- hemoglobin
- platelet

Functions of Blood

You take a last, deep, calming breath before plunging into a dark, vessel-like tube. The water transports you down the slide much like the way blood carries substances to all parts of your body. Blood has four important functions.

1. Blood carries oxygen from your lungs to all your body cells. Carbon dioxide diffuses from your body cells into your blood. Your blood carries carbon dioxide to your lungs to be exhaled.
2. Blood carries waste products from your cells to your kidneys to be removed.
3. Blood transports nutrients and other substances to your body cells.
4. Cells and molecules in blood fight infections and help heal wounds.

Anything that disrupts or changes these functions affects all the tissues of your body. Can you understand why blood is sometimes called the tissue of life?

Parts of Blood

As shown in **Figure 10,** blood is a tissue made of plasma (PLAZ muh), platelets (PLAYT luts), and red and white blood cells. Blood makes up about eight percent of your body's total mass. If you weigh 45 kg, you have about 3.6 kg of blood moving through your body. The amount of blood in an adult would fill five 1-L bottles.

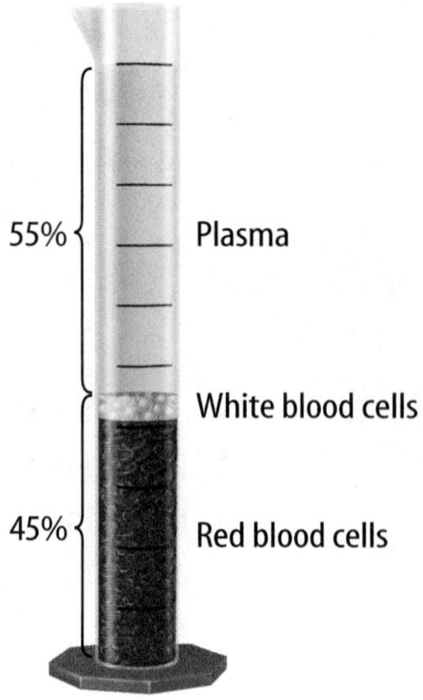

55% Plasma
White blood cells
45% Red blood cells

Figure 10 The blood in this graduated cylinder has separated into its parts. Each part plays a key role in body functions.

Plasma The liquid part of blood is mostly water and is called **plasma.** It makes up more than half the volume of blood. Nutrients, minerals, and oxygen are dissolved in plasma and carried to cells. Wastes from cells are also carried in plasma.

Blood Cells A cubic millimeter of blood has about five million red blood cells. These disk-shaped blood cells, shown in **Figure 11,** are different from other cells in your body because they have no nuclei. They contain **hemoglobin** (HEE muh gloh bun), which is a molecule that carries oxygen and carbon dioxide, and are made of an iron compound that gives blood its red color. Hemoglobin carries oxygen from your lungs to your body cells. Then it carries some of the carbon dioxide from your body cells back to your lungs. The rest of the carbon dioxide is carried in the cytoplasm of red blood cells and in plasma. Red blood cells have a life span of about 120 days. They are made at a rate of 2 million to 3 million per second in the center of long bones like the femur in your thigh. Red blood cells wear out and are destroyed at about the same rate.

In contrast to red blood cells, a cubic millimeter of blood has about 5,000 to 10,000 white blood cells. White blood cells fight bacteria, viruses, and other invaders of your body. Your body reacts to invaders by increasing the number of white blood cells. These cells leave the blood through capillary walls and go into the tissues that have been invaded. Here, they destroy bacteria and viruses and absorb dead cells. The life span of white blood cells varies from a few days to many months.

Circulating with the red and white blood cells are platelets. **Platelets** are irregularly shaped cell fragments that help clot blood. A cubic millimeter of blood can contain as many as 400,000 platelets. Platelets have a life span of five to nine days.

Topic: White Blood Cells
Visit life.msscience.com for Web links to information about types of human white blood cells and their functions.

Activity Write a brief summary describing how white blood cells destroy bacteria and viruses in your Science Journal.

Figure 11 Red blood cells supply your body with oxygen, and white blood cells and platelets have protective roles.

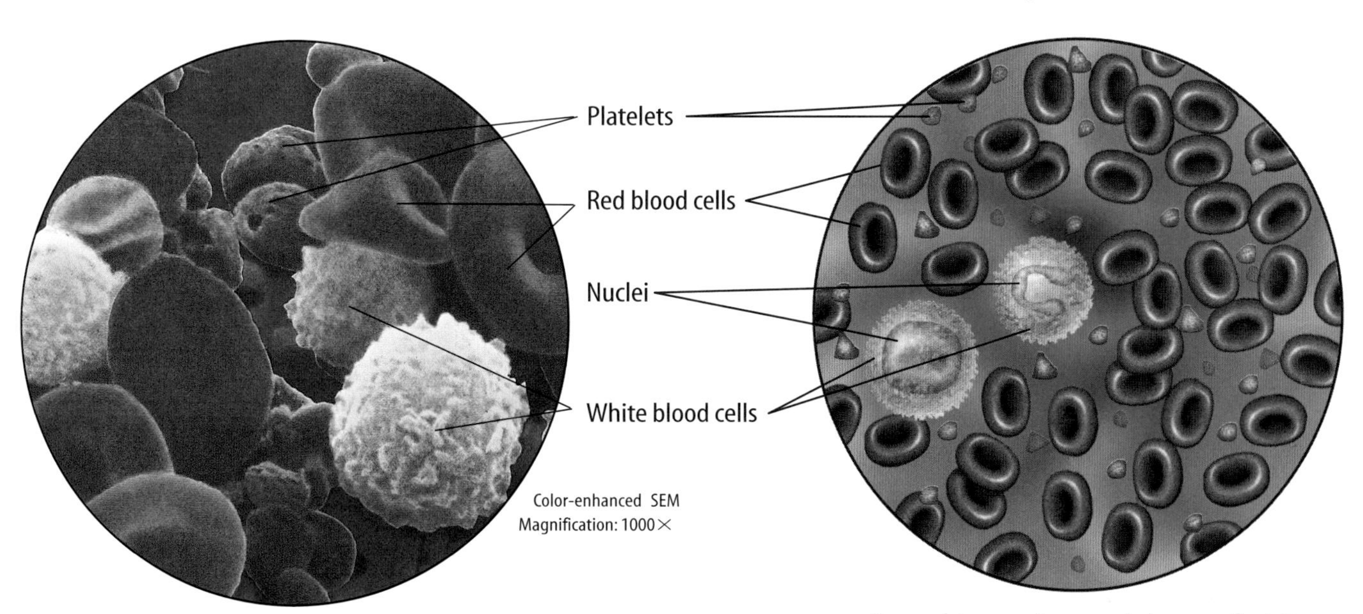

Platelets help stop bleeding. Platelets not only plug holes in small vessels, they also release chemicals that help form filaments of fibrin.

Several types, sizes, and shapes of white blood cells exist. These cells destroy bacteria, viruses, and foreign substances.

Figure 12 When the skin is damaged, a sticky blood clot seals the leaking blood vessel. Eventually, a scab forms to protect the wound from further damage and allow it to heal.

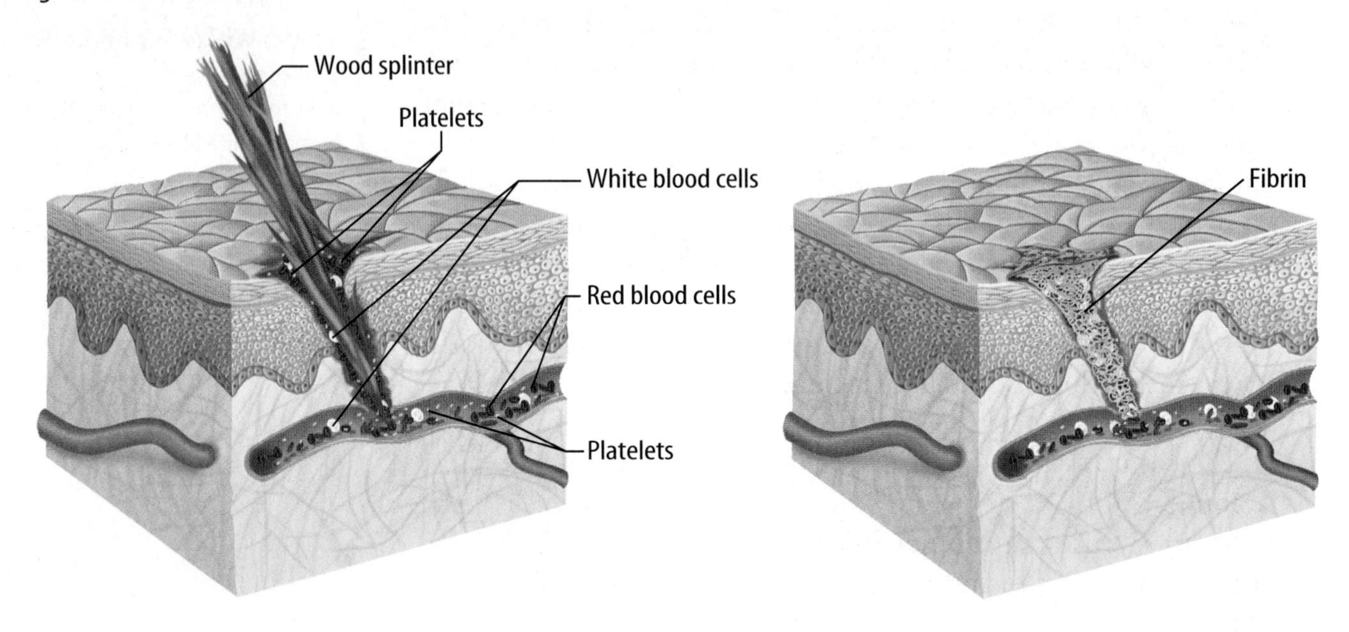

Modeling Scab Formation

Procedure

1. Place a 5-cm × 5-cm square of **gauze** on a piece of **aluminum foil.**
2. Place several drops of a **liquid bandage solution** onto the gauze and let it dry. Keep the liquid bandage away from eyes and mouth.
3. Use a **dropper** to place one drop of **water** onto the area of the liquid bandage. Place another drop of water in another area of the gauze.

Analysis

1. Compare the drops of water in both areas.
2. Describe how the treated area of the gauze is like a scab.

Blood Clotting

You're running with your dog in a park, when all of a sudden you trip and fall down. Your knee starts to bleed, but the bleeding stops quickly. Already the wounded area has begun to heal. Bleeding stops because platelets and clotting factors in your blood make a blood clot that plugs the wounded blood vessels. A blood clot also acts somewhat like a bandage. When you cut yourself, platelets stick to the wound and release chemicals. Then substances called clotting factors carry out a series of chemical reactions. These reactions cause threadlike fibers called fibrin (FI brun) to form a sticky net, as shown in **Figure 12.** This net traps escaping blood cells and plasma and forms a clot. The clot helps stop more blood from escaping. After the clot is in place and becomes hard, skin cells begin the repair process under the scab. Eventually, the scab is lifted off. Bacteria that might get into the wound during the healing process are destroyed by white blood cells.

Reading Check *What blood components help form blood clots?*

Most people will not bleed to death from a minor wound, such as a cut or scrape. However, some people have a genetic condition called hemophilia (hee muh FIH lee uh). Their plasma lacks one of the clotting factors that begins the clotting process. A minor injury can be a life threatening problem for a person with hemophilia.

Table 1 Blood Types		
Blood Type	Antigen	Antibody
A	A	Anti-B
B	B	Anti-A
AB	A, B	None
O	None	Anti-B, Anti-A

Blood Types

Blood clots stop blood loss quickly in a minor wound, but a person with a serious wound might lose a lot of blood and need a blood transfusion. During a blood transfusion, a person receives donated blood or parts of blood. The medical provider must be sure that the right type of blood is given. If the wrong type is given, the red blood cells will clump together. Then, clots form in the blood vessels and the person could die.

The ABO Identification System People can inherit one of four types of blood: A, B, AB, or O, as shown in **Table 1.** Types A, B, and AB have chemical identification tags called antigens (AN tih junz) on their red blood cells. Type O red blood cells have no antigens.

Each blood type also has specific antibodies in its plasma. Antibodies are proteins that destroy or neutralize substances that do not belong in or are not part of your body. Because of these antibodies, certain blood types cannot be mixed. This limits blood transfusion possibilities as shown in **Table 2.** If type A blood is mixed with type B blood, the type A antibodies determine that type B blood does not belong there. The type A antibodies cause the type B red blood cells to clump. In the same way, type B antibodies cause type A blood to clump. Type AB blood has no antibodies, so people with this blood type can receive blood from A, B, AB, and O types. Type O blood has both A and B antibodies.

Reading Check *Why are people with type O blood called universal donors?*

Table 2 Blood Transfusion Options

Type	Can Receive	Can Donate To
A	O, A	A, AB
B	O, B	B, AB
AB	all	AB
O	O	all

The Rh Factor Another chemical identification tag in blood is the Rh factor. The Rh factor also is inherited. If the Rh factor is on red blood cells, the person has Rh-positive (Rh+) blood. If it is not present, the person's blood is called Rh-negative (Rh−). If an Rh− person receives a blood transfusion from an Rh+ person, he or she will produce antibodies against the Rh factor. These antibodies can cause Rh+ cells to clump. Clots then form in the blood vessels and the person could die.

When an Rh− mother is pregnant with an Rh+ baby, the mother might make antibodies to the child's Rh factor. Close to the time of birth, Rh antibodies from the mother can pass from her blood into the baby's blood. These antibodies can destroy the baby's red blood cells. If this happens, the baby must receive a blood transfusion before or right after birth. At 28 weeks of pregnancy and immediately after the birth, an Rh− mother can receive an injection that blocks the production of antibodies to the Rh+ factor. These injections prevent this life-threatening situation. To prevent deadly results, blood groups and Rh factor are checked before transfusions and during pregnancies.

Blood Transfusions The first blood transfusions took place in the 1600s and were from animal to animal, and then from animal to human. In 1818, James Blundell, a British obstetrician, performed the first successful transfusion of human blood to a patient for the treatment of hemorrhage.

Reading Check *Why is it important to check Rh factor?*

Applying Science

Will there be enough blood donors?

Successful human blood transfusions began during World War II. This practice is much safer today due to extensive testing of the donated blood prior to transfusion. Health care professionals have determined that each blood type can receive certain other blood types as illustrated in **Table 2.**

Blood Type Distribution

	Rh+ (%)	Rh− (%)
0	37	7
A	36	6
B	9	1
AB	3	1

Identifying the Problem

The table on the right lists the average distribution of blood types in the United States. The data are recorded as percents, or a sample of 100 people. By examining these data and the data in **Table 2,** can you determine safe donors for each blood type? Recall that people with Rh− blood cannot receive a transfusion from an Rh+ donor.

Solving the Problem

1. If a Type B, Rh+ person needs a blood transfusion, how many possible donors are there?
2. Frequently, the supply of donated blood runs low. Which blood type and Rh factor would be most affected in such a shortage? Explain your answer.

Diseases of Blood

Because blood circulates to all parts of your body and performs so many important functions, any disease of the blood is a cause for concern. One common disease of the blood is anemia (uh NEE mee uh). In this disease of red blood cells, body tissues can't get enough oxygen and are unable to carry on their usual activities. Anemia has many causes. Sometimes, anemia is caused by the loss of large amounts of blood. A diet lacking iron or certain vitamins also might cause anemia. Anemia also can be the result of another disease or a side effect of treatment for a disease. One type of anemia results from sickle-cell disease, an inherited condition. A person with this disease has abnormally shaped red blood cells, as shown in **Figure 13,** that cannot function properly.

Leukemia (lew KEE mee uh) is a disease in which one or more types of white blood cells are made in excessive numbers. These cells are immature and do not fight infections well. They fill the bone marrow and crowd out the normal cells. Then not enough red blood cells, normal white blood cells, and platelets can be made. Types of leukemia affect children or adults. Medicines, blood transfusions, and bone marrow transplants are used to treat this disease. If the treatments are not successful, the person eventually will die from related complications.

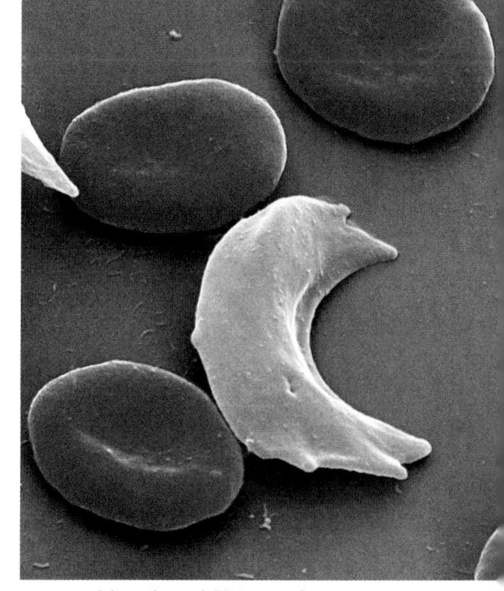

Color-enhanced TEM Magnification: 7400×

Figure 13 Persons with sickle-cell disease have misshapened red blood cells. The sickle-shaped cells clog the capillaries of a person with this disease. Oxygen cannot reach tissues served by the capillaries, and wastes cannot be removed. **Describe** *how this damages the affected tissues.*

section 2 review

Summary

Parts of Blood

- Plasma is made mostly of water, with nutrients, minerals, and oxygen dissolved in it.
- Red blood cells contain hemoglobin, which carries oxygen and carbon dioxide.
- White blood cells control infections and viruses.
- Blood clotting factors and platelets help blood to clot.

Blood Types

- People can inherit one of four types of blood and an Rh factor.
- Type A, B, and AB blood all have antigens. Type O blood has no antigens.

Diseases of Blood

- Anemia is a disease of red blood cells.
- Leukemia is a disease that produces immature white blood cells that don't fight infections.

Self Check

1. **List** the four functions of blood in the body.
2. **Infer** why blood type and Rh factor are checked before a transfusion.
3. **Interpret Data** Look at the data in **Table 2** about blood group interactions. To which group(s) can blood type AB donate blood, and which blood type(s) can AB receive blood from?
4. **Think Critically** Think about the main job of your red blood cells. If red blood cells couldn't deliver oxygen to your cells, what would be the condition of your body tissues?

Applying Math

5. **Use Percentages** Find the total number of red blood cells, white blood cells, and platelets in 1 mm^3 of blood. Calculate what percentage of the total each type is.

section 3

The Lymphatic System

as you read

What You'll Learn

- **Describe** functions of the lymphatic system.
- **Identify** where lymph comes from.
- **Explain** how lymph organs help fight infections.

Why It's Important

The lymphatic system helps protect you from infections and diseases.

Review Vocabulary
smooth muscles: muscles found in your internal organs and digestive track

New Vocabulary
- lymph
- lymphatic system
- lymphocyte
- lymph node

Functions of the Lymphatic System

You're thirsty so you turn on the water faucet and fill a glass with water. The excess water runs down the drain. In a similiar way, your body's excess tissue fluid is removed by the lymphatic (lihm FA tihk) system. The nutrient, water, and oxygen molecules in blood diffuse through capillary walls to nearby cells. Water and other substances become part of the tissue fluid that is found between cells. This fluid is collected and returned to the blood by the lymphatic system.

After tissue fluid diffuses into the lymphatic capillaries it is called **lymph** (LIHMF). Your **lymphatic system,** as shown in **Figure 14,** carries lymph through a network of lymph capillaries and larger lymph vessels. Then, the lymph drains into large veins near the heart. No heartlike structure pumps the lymph through the lymphatic system. The movement of lymph depends on the contraction of smooth muscles in lymph vessels and skeletal muscles. Lymphatic vessels, like veins, have valves that keep lymph from flowing backward. If the lymphatic system is not working properly, severe swelling occurs because the tissue fluid cannot get back to the blood.

In addition to water and dissolved substances, lymph also contains **lymphocytes** (LIHM fuh sites), a type of white blood cell. Lymphocytes help your body defend itself against disease-causing organisms.

What are the differences and similarities between lymph and blood?

Lymphatic Organs

Before lymph enters the blood, it passes through lymph nodes, which are bean-shaped organs of varying sizes found throughout the body. **Lymph nodes** filter out microorganisms and foreign materials that have been taken up by lymphocytes. When your body fights an infection, lymphocytes fill the lymph nodes. The lymph nodes become warm, reddened, and tender to the touch. After the invaders are destroyed, the redness, warmth, and tenderness in the lymph nodes goes away.

Besides lymph nodes, the tonsils, the thymus, and the spleen are important lymphatic organs. Tonsils are in the back of your throat and protect you from harmful microorganisms that enter through your mouth and nose. Your thymus is a soft mass of tissue located behind the sternum. It makes lymphocytes that travel to other lymph organs. The spleen is the largest lymphatic organ. It is located behind the upper-left part of the stomach and filters the blood by removing worn out and damaged red blood cells. Cells in the spleen take up and destroy bacteria and other substances that invade your body.

A Disease of the Lymphatic System

HIV is a deadly virus. When HIV enters a person's body, it attacks and destroys lymphocytes called helper T cells that help make antibodies to fight infections. This affects a person's immunity to some diseases. Usually, the person dies from these diseases, not from the HIV infection.

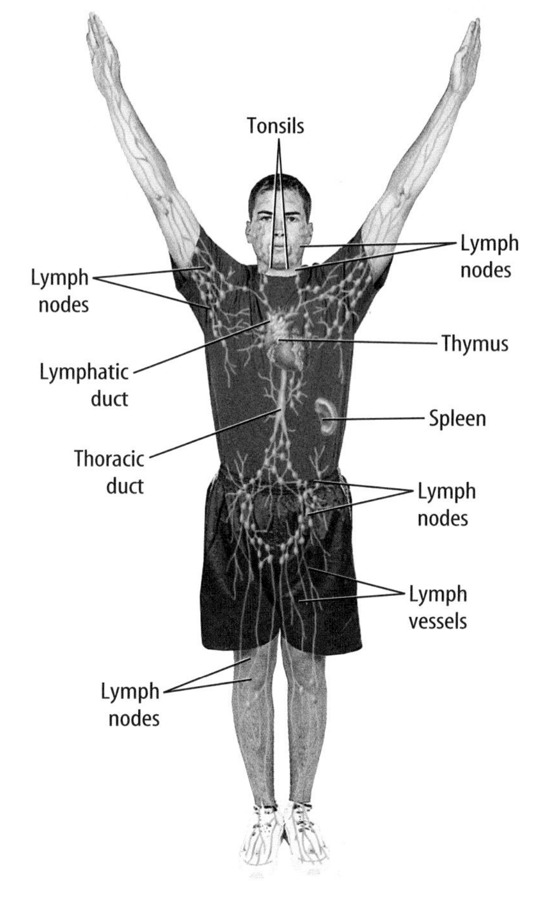

Figure 14 The lymphatic system is connected by a network of vessels.
Describe *how muscles help move lymph.*

section 3 review

Summary

Functions of the Lymphatic System

- Fluid is collected and returned from the body tissues to the blood by the lymphatic system.
- After fluid from tissues diffuses into the lymphatic capillaries it is called lymph.
- Lymphocytes are a type of white blood cell that helps your body defend itself against disease.

Lymphatic Organs

- Lymph nodes filter out microorganisms and foreign materials taken up by lymphocytes.
- The tonsils, thymus, and spleen also protect your body from harmful microorganisms that enter through your mouth and nose.

A Disease of the Lymphatic System

- HIV destroys helper T cells that help make antibodies to fight infections.

Self Check

1. **Describe** where lymph comes from and how it gets into the lymphatic capillaries.
2. **Explain** how lymphatic organs fight infection.
3. **Sequence** the events that occur when HIV enters the body.
4. **Think Critically** When the amount of fluid in the spaces between cells increases, so does the pressure in these spaces. What do you infer will happen?

Applying Skills

5. **Concept Map** The circulatory system and the lymphatic system work together in several ways. Make a concept map comparing the two systems.
6. **Communicate** An infectious microorganism enters your body. In your Science Journal, describe how the lymphatic system protects the body against the microorganism.

Blood Type Reactions

Goals

- **Design** an experiment that simulates the reactions between different blood types.
- **Identify** which blood types can donate to which other blood types.

Possible Materials

simulated blood (10 mL low-fat milk and 10 mL water plus red food coloring)
lemon juice as antigen A (for blood types B and O)
water as antigen A (for blood types A and AB)
droppers
small paper cups
marking pen
10-mL graduated cylinder

Safety Precautions

WARNING: *Do not taste, eat, or drink any materials used in the lab.*

Real-World Question

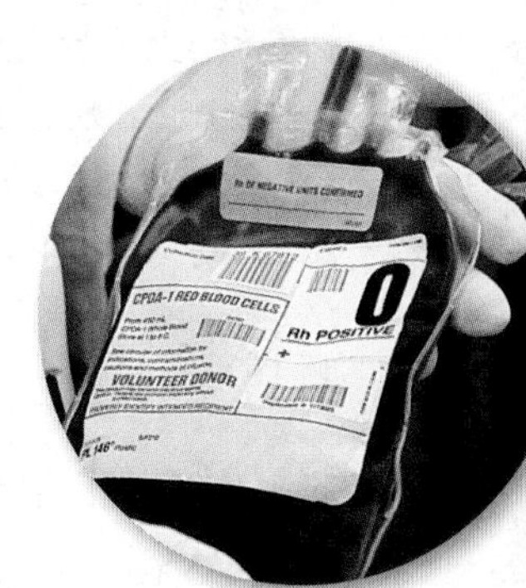

Human blood can be classified into four main blood types—A, B, AB, and O. These types are determined by the presence or absence of antigens on the red blood cells. After blood is collected into a transfusion bag, it is tested to determine the blood type. The type is labeled clearly on the bag. Blood is refrigerated to keep it fresh and available for transfusion. What happens when two different blood types are mixed?

Form a Hypothesis

Based on your reading and observations, state a hypothesis about how different blood types will react to each other.

Test Your Hypothesis

Make a Plan

1. As a group, agree upon the hypothesis and decide how you will test it. Identify the results that will confirm the hypothesis.
2. **List** the steps you must take and the materials you will need to test your hypothesis. Be specific. Describe exactly what you will do in each step.
3. **Prepare** a data table like the one at the right in your Science Journal to record your observations.
4. Reread the entire experiment to make sure all steps are in logical order.
5. **Identify** constants and variables. Blood type O will be the control.

Blood Type Reactions

Blood Type	Clumping (Yes or No)
A	Do not write in this book.
B	
AB	
O	

Follow Your Plan

1. While doing the experiment, record your observations and complete the data table in your Science Journal.

Analyze Your Data

1. **Compare** the reactions of each blood type (A, B, AB, and O) when antigen A was added to the blood.
2. **Observe** where clumping took place.
3. **Compare** your results with those of other groups.
4. What was the control factor in this experiment?
5. What were your variables?

Conclude and Apply

1. Did the results support your hypothesis? Explain.
2. **Predict** what might happen to a person if other antigens are not matched properly.
3. What would happen in an investigation with antigen B added to each blood type?

Communicating Your Data

Write a brief report on how blood is tested to determine blood type. **Describe** why this is important to know before receiving a blood transfusion. **For more help, refer to the Science Skill Handbook.**

Have a Heart

Dr. Daniel Hale Williams was a pioneer in open-heart surgery.

People didn't always know where blood came from or how it moved through the body

"Ouch!" You prick your finger, and when blood starts to flow out of the cut, you put on a bandage. But if you were a scientist living long ago, you might have also asked yourself some questions: How did your blood get to the tip of your finger? And why and how does it flow through (and sometimes out of!) your body?

As early as the 1500s, a Spanish scientist named Miguel Serveto (mee GEL • ser VE toh) asked that question. His studies led him to the theory that blood circulated throughout the human body, but he didn't know how or why.

About 100 years later, William Harvey, an English doctor, explored Serveto's idea. Harvey studied animals to develop a theory about how the heart and the circulatory system work.

Harvey hypothesized, from his observations of animals, that blood was pumped from the heart throughout the body, and that it returned to the heart and recirculated. He published his ideas in 1628 in his famous book, *On the Motion of the Heart and Blood in Animals.* His theories were correct, but many of Harvey's patients left him. His patients thought his ideas were ridiculous. His theories were correct, and over time, Harvey's book became the basis for all modern research on heart and blood vessels.

Medical Pioneer

More than two centuries later, another pioneer stepped forward and used Harvey's ideas to change the science frontier again. His name was Dr. Daniel Hale Williams. In 1893, Williams used what he knew about heart and blood circulation to become a new medical pioneer. He performed the first open-heart surgery by removing a knife from the heart of a stabbing victim. He stitched the wound to the fluid-filled sac surrounding the heart, and the patient lived several more years. In 1970, the U.S. recognized Williams by issuing a stamp in his honor.

Report **Identify a pioneer in science or medicine who has changed our lives for the better. Find out how this person started in the field, and how they came to make an important discovery. Give a presentation to the class.**

For more information, visit life.msscience.com/time

Reviewing Main Ideas

Section 1 The Circulatory System

1. Arteries carry blood away from the heart. Capillaries allow the exchange of nutrients, oxygen, and wastes in cells. Veins return blood to the heart.
2. Carbon-dioxide-rich blood enters the right atrium, moves to the right ventricle, and then goes to the lungs through the pulmonary artery. Oxygen-rich blood returns to the left atrium, moves to the left ventricle, and then leaves through the aorta.
3. Pulmonary circulation is the path of blood between the heart and lungs. Circulation through the rest of the body is called systemic circulation. Coronary circulation is the flow of blood to tissues of the heart.

Section 2 Blood

1. Plasma carries nutrients, blood cells, and other substances.
2. Red blood cells carry oxygen and carbon dioxide, platelets form clots, and white blood cells fight infection.
3. A, B, AB, and O blood types are determined by the presence or absence of antigens on red blood cells.
4. Anemia is a disease of red blood cells, in which not enough oxygen is carried to the body's cells.
5. Leukemia is a disease where one or more types of white blood cells are present in excessive numbers. These cells are immature and do not fight infection well.

Section 3 The Lymphatic System

1. Lymph structures filter blood, produce white blood cells that destroy bacteria and viruses, and destroy worn out blood cells.
2. HIV attacks helper T cells, which are a type of lymphocyte. The person is unable to fight infections well.

Visualizing Main Ideas

Copy and complete this concept map on the functions of the parts of the blood.

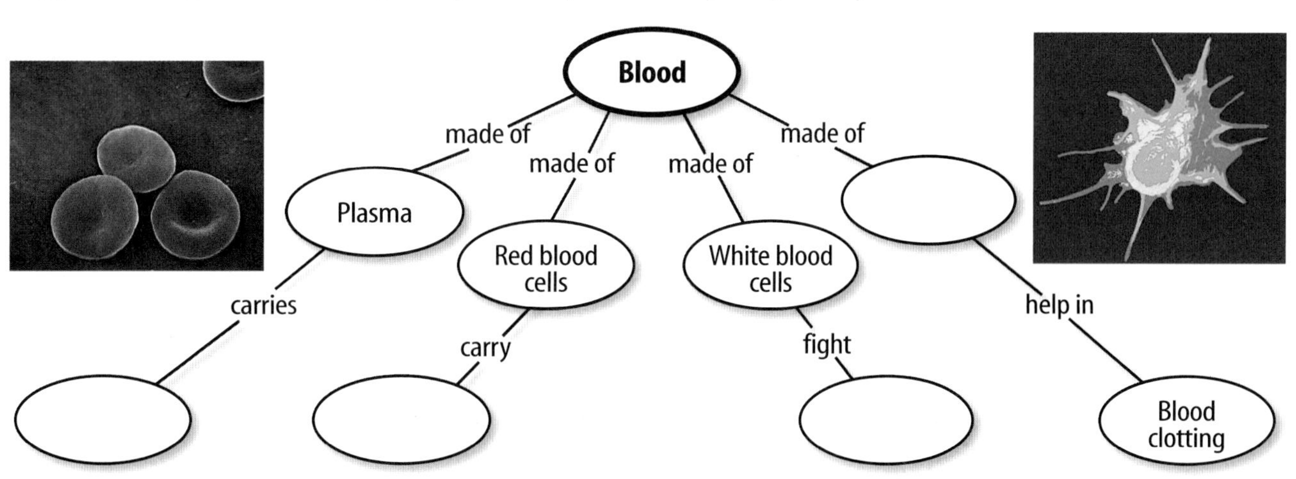

chapter 19 Review

Using Vocabulary

artery p. 544
atrium p. 541
capillary p. 545
coronary circulation p. 541
hemoglobin p. 551
lymph p. 556
lymph node p. 556
lymphatic system p. 556
lymphocyte p. 556
plasma p. 550
platelet p. 551
pulmonary circulation p. 542
systemic circulation p. 543
vein p. 544
ventricle p. 541

Fill in the blanks with the correct vocabulary word(s).

1. The ________ carries blood to the heart.
2. The ________ transports tissue fluid through a network of vessels.
3. ________ is the chemical in red blood cells.
4. ________ are cell fragments.
5. The smallest blood vessels are called the ________.
6. The flow of blood to and from the lungs is called ________.
7. ________ helps protect your body against infections.
8. The largest section of the circulatory system is the ________.
9. ________ are blood vessels that carry blood away from the heart.
10. The two lower chambers of the heart are called the right and left ________.

Checking Concepts

Choose the word or phrase that best answers the question.

11. Where does the exchange of food, oxygen, and wastes occur?
    A) arteries **C)** veins
    B) capillaries **D)** lymph vessels
12. What is circulation to all body organs called?
    A) coronary **C)** systemic
    B) pulmonary **D)** organic
13. Where is blood under greatest pressure?
    A) arteries **C)** veins
    B) capillaries **D)** lymph vessels
14. Which cells fight off infection?
    A) red blood **C)** white blood
    B) bone **D)** nerve
15. Of the following, which carries oxygen in blood?
    A) red blood cells **C)** white blood cells
    B) platelets **D)** lymph
16. What is required to clot blood?
    A) plasma **C)** platelets
    B) oxygen **D)** carbon dioxide
17. What kind of antigen does type O blood have?
    A) A **C)** A and B
    B) B **D)** no antigen

Use the figure below to answer question 18.

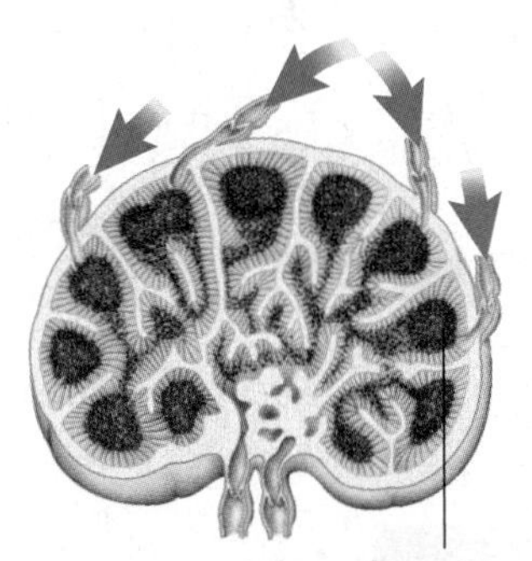

18. What is the bean-shaped organ above that filters out microorganisms and foreign materials taken up by lymphocytes?
    A) kidney **C)** lung
    B) lymph **D)** lymph node
19. What is the largest filtering lymph organ?
    A) spleen **C)** tonsil
    B) thymus **D)** node

Science Online life.msscience.com/vocabulary_puzzlemaker

Thinking Critically

20. **Identify** the following as having oxygen-rich or carbon dioxide-filled blood: *aorta, coronary arteries, coronary veins, inferior vena cava, left atrium, left ventricle, right atrium, right ventricle,* and *superior vena cava.*

21. **Compare and contrast** the three types of blood vessels.

22. **Compare and contrast** the life spans of the red blood cells, white blood cells, and platelets.

23. **Describe** the sequence of blood clotting from the wound to forming a scab.

24. **Compare and contrast** the functions of arteries, veins, and capillaries.

25. **Concept Map** Copy and complete the events-chain concept map showing how lymph moves in your body.

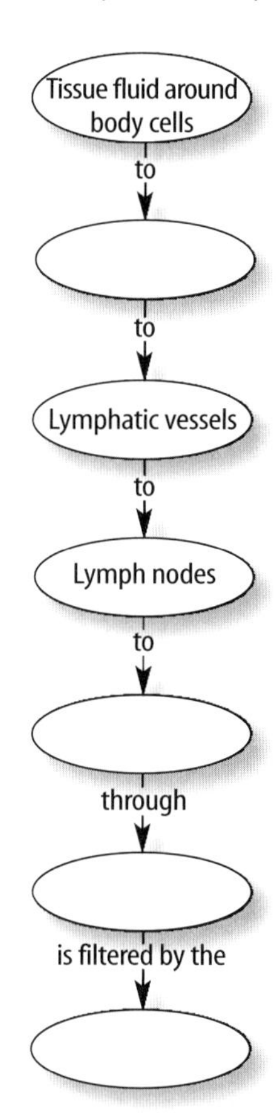

26. **Explain** how the lymphatic system works with the cardiovascular system.

27. **Infer** why cancer of the blood cells or lymph nodes is hard to control.

28. **Explain** why a pulse is usually taken at the neck or wrist, when arteries are distributed throughout the body.

Performance Activities

29. **Poster** Prepare a poster illustrating heart transplants. Include an explanation of why the patient is given drugs that suppress the immune system and describe the patient's life after the operation.

30. **Scientific Illustrations** Prepare a drawing of the human heart and label its parts.

Applying Math

Use the table below to answer question 31.

Gender and Heart Rate

Sex	Pulse/Minute
Male 1	72
Male 2	64
Male 3	65
Female 1	67
Female 2	84
Female 3	74

31. **Heart Rates** Using the table above, find the average heart rate of the three males and the three females. Compare the two averages.

32. **Blood Mass** Calculate how many kilograms of blood is moving through your body, if blood makes up about eight percent of your body's total mass and you weigh 38 kg.

Standardized Test Practice

Part 1 Multiple Choice

Record your answers on the answer sheet provided by your teacher or on a sheet of paper.

1. Which of the following is a function of blood?
   A. carry saliva to the mouth
   B. excrete salts from the body
   C. transport nutrients and other substances to cells
   D. remove lymph from around cells

Use the table below to answer questions 2 and 3.

Results from Ashley's Activities			
Activity	**Pulse Rate (beats/min)**	**Body Temperature**	**Degree of Sweating**
1	80	98.6°F	None
2	90	98.8°F	Minimal
3	100	98.9°F	Little
4	120	99.1°F	Moderate
5	150	99.5°F	Considerable

2. Which of the following activities caused Ashley's pulse to be less than 100 beats per minute?
   A. Activity 2 **C.** Activity 4
   B. Activity 3 **D.** Activity 5

3. A reasonable hypothesis based on these data, is that during Activity 2, Ashley was probably
   A. sprinting **C.** sitting down
   B. marching **D.** walking slowly

4. Which of the following activities contributes to cardiovascular disease?
   A. smoking **C.** sleeping
   B. jogging **D.** balanced diet

5. Where does blood low in oxygen enter first?
   A. right atrium **C.** left ventricle
   B. left atrium **D.** right ventricle

6. Which of the following is an artery?
   A. left ventricle **C.** superior vena cava
   B. aorta **D.** inferior vena cava

7. Which of the following is NOT a part of the lymphatic system?
   A. lymph nodes **C.** heartlike structure
   B. valves **D.** lymph capillaries

Use the table below to answer questions 8 and 9.

Blood Cell Counts (per 1 mm^3)			
Patient	**Red Blood Cells**	**White Blood Cells**	**Platelets**
Normal	3.58–4.99 million	3,400–9,600	162,000–380,000
Mrs. Stein	3 million	8,000	400,000
Mr. Chavez	5 million	7,500	50,000

8. What problem might Mrs. Stein have?
   A. low oxygen levels in tissues
   B. inability to fight disease
   C. poor blood clotting
   D. irregular heart beat

9. If Mr. Chavez cut himself, what might happen?
   A. minimal bleeding
   B. prolonged bleeding
   C. infection
   D. quick healing

10. Which lymphatic organ protects your body from harmful microorganisms that enter through your mouth?
    A. spleen **C.** node
    B. thymus **D.** tonsils

Test-Taking Tip

Don't Stray During the test, keep your eyes on your own paper. If you need to rest them, close them or look up at the ceiling.

Part 2 Short Response/Grid In

Record your answers on the answer sheet provided by your teacher or on a sheet of paper.

11. If red blood cells are made at the rate of 2 million per second in the center of long bones, how many red blood cells are made in one hour?

12. If a cubic milliliter of blood has 10,000 white blood cells and 400,000 platelets, how many times more platelets than white blood cells are present in a cubic milliliter of blood?

13. What would happen if type A blood was given to a person with type O blood?

Use the illustration below to answer questions 14 and 15.

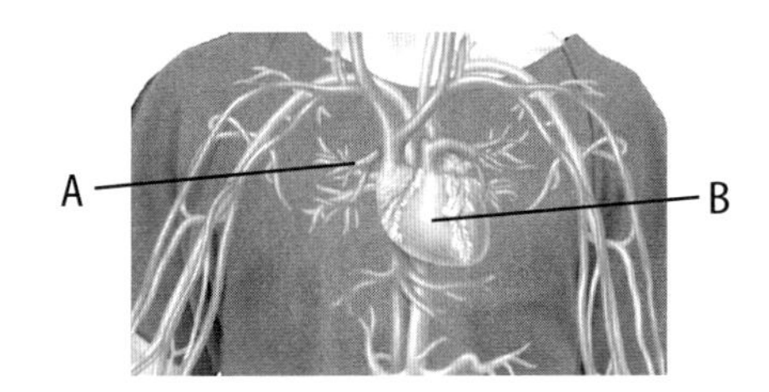

14. What might happen if there was a blood clot blocking vessel "A"?

15. What might happen if there was a blood clot blocking vessel "B"?

16. Why don't capillaries have thick, elastic walls?

17. Why would a cut be dangerous for a person with hemophilia?

18. Why would a person with leukemia have low numbers of red blood cells, normal white blood cells, and platelets in the blood?

Part 3 Open Ended

Record your answers on a sheet of paper.

Use the illustration below to answer questions 19 and 20.

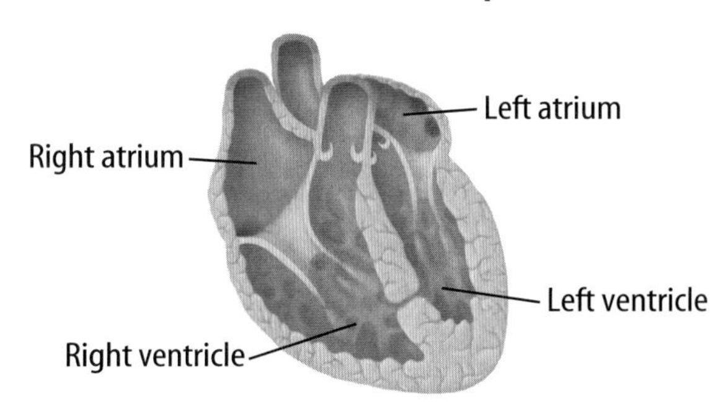

19. What is wrong with this heart? How do you know?

20. The left ventricle pumps blood under higher pressure than the right ventricle does. In which direction would you predict blood would flow through the hole in the heart? Compare the circulation in this heart with that of a normal heart.

21. What are some ways to prevent cardiovascular disease?

22. Compare and contrast diffusion and active transport.

23. Describe the role of the brain in blood pressure homeostasis. Why is this important?

24. Thrombocytopenia is a condition in which the number of platelets in the blood is decreased. Hemophilia is a genetic condition where blood plasma lacks one of the clotting factors. Compare how a small cut would affect a person with thrombocytopenia and someone with hemophilia.

chapter 20

Respiration and Excretion

The BIG Idea

The respiratory and excretory systems exchange some of the body's needed substances and all of its wastes.

SECTION 1
The Respiratory System

Main Idea Organs of the respiratory system supply your body with oxygen and remove carbon dioxide and other gaseous wastes.

SECTION 2
The Excretory System

Main Idea The excretory system removes your body's liquid, gaseous, and solid wastes.

Why do you sweat?

How do you feel when you've just finished running a mile, sliding into home base, or scoring a soccer goal? Maybe you felt that your lungs would burst. You need a constant supply of oxygen to keep your body cells functioning, and your body is adapted to meet that need.

Science Journal How do you think your body adapts to meet your needs while you are playing sports?

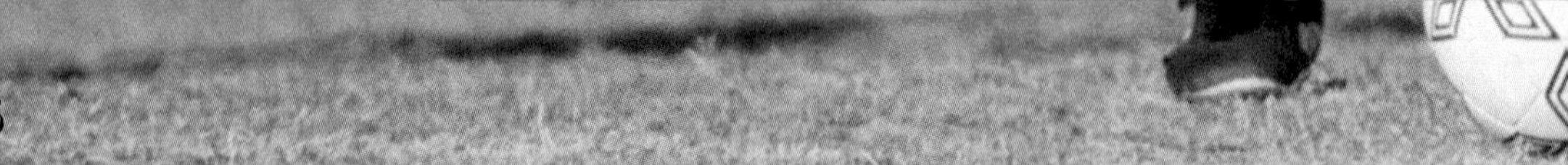

Start-Up Activities

Effect of Activity on Breathing

Your body can store food and water, but it cannot store much oxygen. Breathing brings oxygen into your body. In the following lab, find out about one factor that can change your breathing rate.

1. Put your hand on the side of your rib cage. Take a deep breath. Notice how your rib cage moves out and upward when you inhale.
2. Count the number of breaths you take for 15 s. Multiply this number by four to calculate your normal breathing rate for 1 min.
3. Repeat step 2 two more times, then calculate your average breathing rate.
4. Do a physical activity described by your teacher for 1 min and repeat step 2 to determine your breathing rate now.
5. Time how long it takes for your breathing rate to return to normal.
6. **Think Critically** Explain how breathing rate appears to be related to physical activity.

Preview this chapter's content and activities at life.msscience.com

Respiration and Excretion Make the following Foldable to help you identify what you already know, what you want to know, and what you learned about respiration.

STEP 1 **Fold** a vertical sheet of paper from side to side. Make the front edge about 1.25 cm shorter than the back edge.

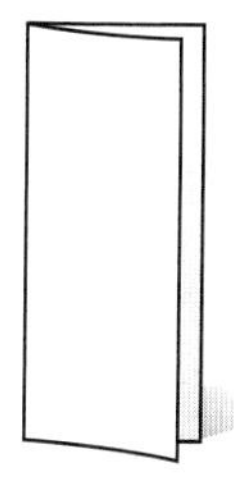

STEP 2 **Turn** lengthwise and **fold** into thirds.

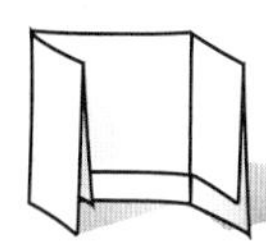

STEP 3 **Unfold and cut** only the top layer along both folds to make three tabs.

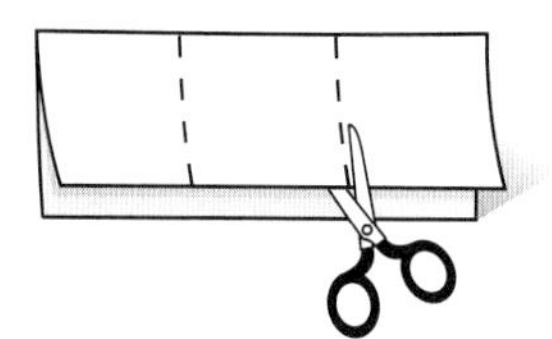

STEP 4 **Label** each tab.

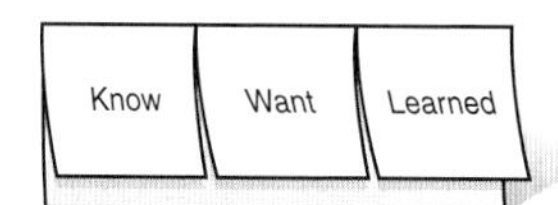

Read and Write Before you read the chapter, write what you already know about respiration under the left tab of your Foldable, and write questions about what you'd like to know under the center tab. After you read the chapter, list what you learned under the right tab.

Get Ready to Read

Make Predictions

1 Learn It! A prediction is an educated guess based on what you already know. One way to predict while reading is to guess what you believe the author will tell you next. As you are reading, each new topic should make sense because it is related to the previous paragraph or passage.

2 Practice It! Read the excerpt below from Section 2. Based on what you have read, make predictions about what you will read in the rest of the section. After you read Section 2, go back to your predictions to see if they were correct.

Predict how normal blood pressure is maintained.

Predict what happens when the brain detects too little water in blood.

Predict how urine forms.

To stay in good health, the fluid levels within the body must be balanced and normal blood pressure must be maintained. An area in the brain, the hypothalamus (hi poh THAL uh mus), constantly monitors the amount of water in the blood. When the brain detects too much water in the blood, the hypothalamus releases a lesser amount of a specific hormone. This signals the kidneys to return less water to the blood and increase the amount of wastewater, called **urine** that is excreted.

—from page 578

3 Apply It! Before you read, skim the questions in the Chapter Review. Choose three questions and predict the answers.

As you read, check the predictions you made to see if they were correct.

Target Your Reading

Use this to focus on the main ideas as you read the chapter.

1. **Before you read** the chapter, respond to the statements below on your worksheet or on a numbered sheet of paper.
   - Write an **A** if you **agree** with the statement.
   - Write a **D** if you **disagree** with the statement.

2. **After you read** the chapter, look back to this page to see if you've changed your mind about any of the statements.
   - If any of your answers changed, explain why.
   - Change any false statements into true statements.
   - Use your revised statements as a study guide.

Before You Read A or D	Statement	After You Read A or D
	1 The exchange of oxygen and carbon dioxide happens by diffusion.	
	2 Breathing is the same as respiration.	
	3 Vocal sounds are made by the trachea.	
	4 Air enters and leaves your body when you diaphragm contracts and relaxes.	
	5 Respiratory problems have no effect on other body systems.	
	6 Your skin is part of your excretory system.	
	7 Kidneys filter wastes from your blood.	
	8 It takes about two hours for all of your blood to move through your kidneys.	
	9 Most urinary infections begin in the bladder.	
	10 Your circulatory system does not connect with your excretory system.	

Science Online

Print out a worksheet of this page at life.msscience.com

section 1

The Respiratory System

as you read

What You'll Learn

- **Describe** the functions of the respiratory system.
- **Explain** how oxygen and carbon dioxide are exchanged in the lungs and in tissues.
- **Identify** the pathway of air in and out of the lungs.
- **Explain** the effects of smoking on the respiratory system.

Why It's Important

Your body's cells depend on your respiratory system to supply oxygen and remove carbon dioxide.

Review Vocabulary

lungs: saclike respiratory organs that function with the heart to remove carbon dioxide from blood and provide it with oxygen

New Vocabulary

- pharynx
- larynx
- trachea
- bronchi
- alveoli
- diaphragm
- emphysema
- asthma

Functions of the Respiratory System

Can you imagine an astronaut walking on the Moon without a space suit or a diver exploring the ocean without scuba gear? Of course not. You couldn't survive in either location under those conditions because you need to breathe air. Earth is surrounded by a layer of gases called the atmosphere (AT muh sfihr). You breathe atmospheric gases that are closest to Earth. As shown in **Figure 1,** oxygen is one of those gases.

For thousands of years people have known that air, food, and water are needed for life. However, the gas in the air that is necessary for life was not identified as oxygen until the late 1700s. At that time, a French scientist experimented and discovered that an animal breathed in oxygen and breathed out carbon dioxide. He measured the amount of oxygen that the animal used and the amount of carbon dioxide produced by its bodily processes. After his work with animals, the French scientist used this knowledge to study the way that humans use oxygen. He measured the amount of oxygen that a person uses when resting and when exercising. These measurements were compared, and he discovered that more oxygen is used by the body during exercise.

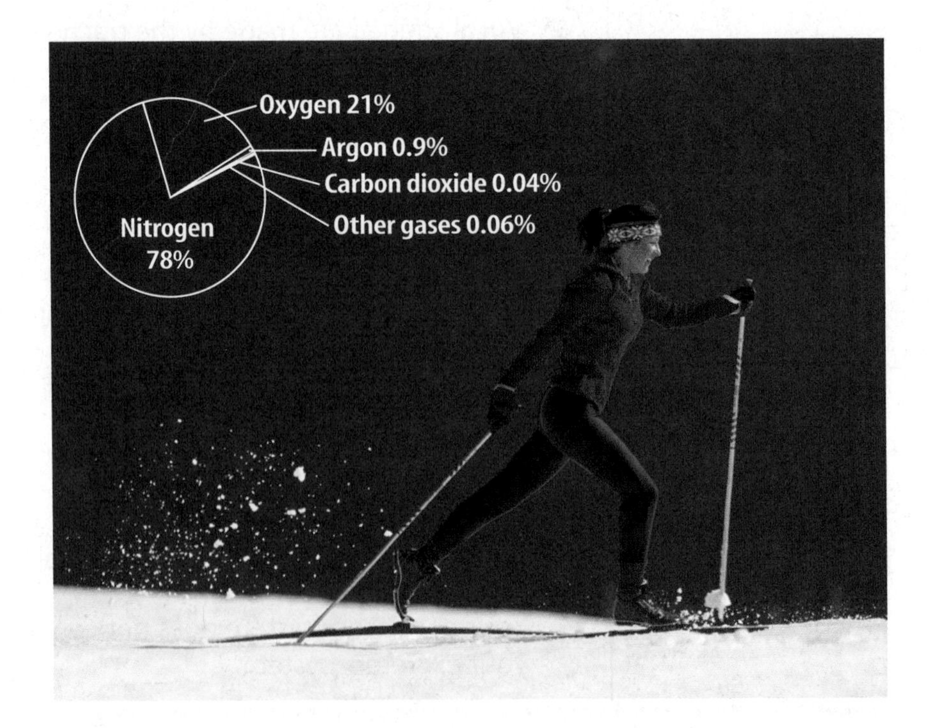

Figure 1 Air, which is needed by most organisms, is only 21 percent oxygen.

Figure 2 Several processes are involved in how the body obtains, transports, and uses oxygen.

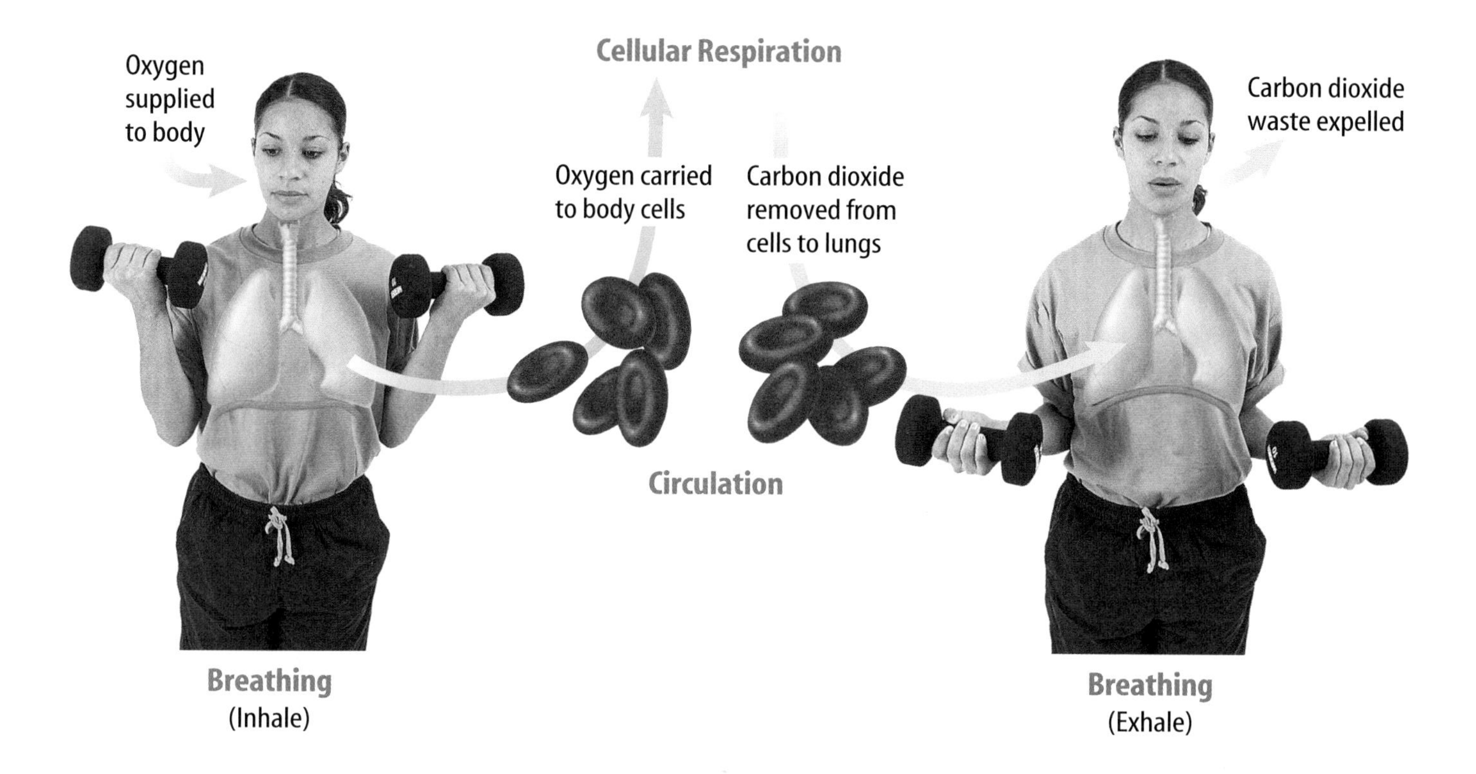

Breathing and Cellular Respiration People often confuse the terms *breathing* and *respiration.* Breathing is the movement of the chest that brings air into the lungs and removes waste gases. The air entering the lungs contains oxygen. It passes from the lungs into the circulatory system because there is less oxygen in the blood than in cells of the lungs. Blood carries oxygen to individual cells. At the same time, the digestive system supplies glucose from digested food to the same cells. The oxygen delivered to the cells is used to release energy from glucose. This chemical reaction, shown in the equation in **Figure 2,** is called cellular respiration. Without oxygen, this reaction would not take place. Carbon dioxide and water molecules are waste products of cellular respiration. They are carried back to the lungs in the blood. Exhaling, or breathing out, eliminates waste carbon dioxide and some water molecules.

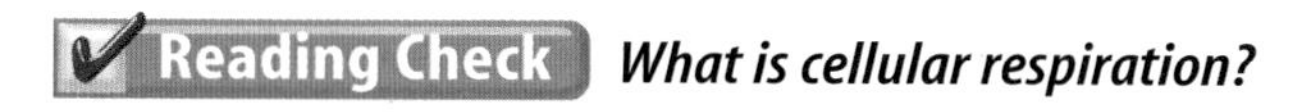

What is cellular respiration?

Water Vapor The amount of water vapor in the atmosphere varies from almost none over deserts to nearly four percent in tropical rain forest areas. This means that every 100 molecules that make up air include only four molecules of water. In your Science Journal, infer how breathing dry air can stress your respiratory system.

Organs of the Respiratory System

The respiratory system, shown in **Figure 3,** is made up of structures and organs that help move oxygen into the body and waste gases out of the body. Air enters your body through two openings in your nose called nostrils or through the mouth. Fine hairs inside the nostrils trap dust from the air. Air then passes through the nasal cavity and is warmed by the body's thermal energy and moistened. Glands that produce sticky mucus line the nasal cavity. The mucus traps dust, pollen, and other materials that were not trapped by nasal hairs. This process helps filter and clean the air you breathe. Tiny, hairlike structures, called cilia (SIH lee uh), sweep mucus and trapped material to the back of the throat where it can be swallowed.

Pharynx Warmed, moist air then enters a tubelike passageway used by food, liquid, and air called the **pharynx** (FER ingks). At the lower end of the pharynx is a flap of tissue called the epiglottis (eh puh GLAH tus). When you swallow, your epiglottis folds down to prevent food or liquid from entering your airway. The food enters your esophagus instead. If you began to choke, what do you think has happened?

Figure 3 Air can enter the body through the nostrils and the mouth.
Explain *the advantages of having air enter through the nostrils.*

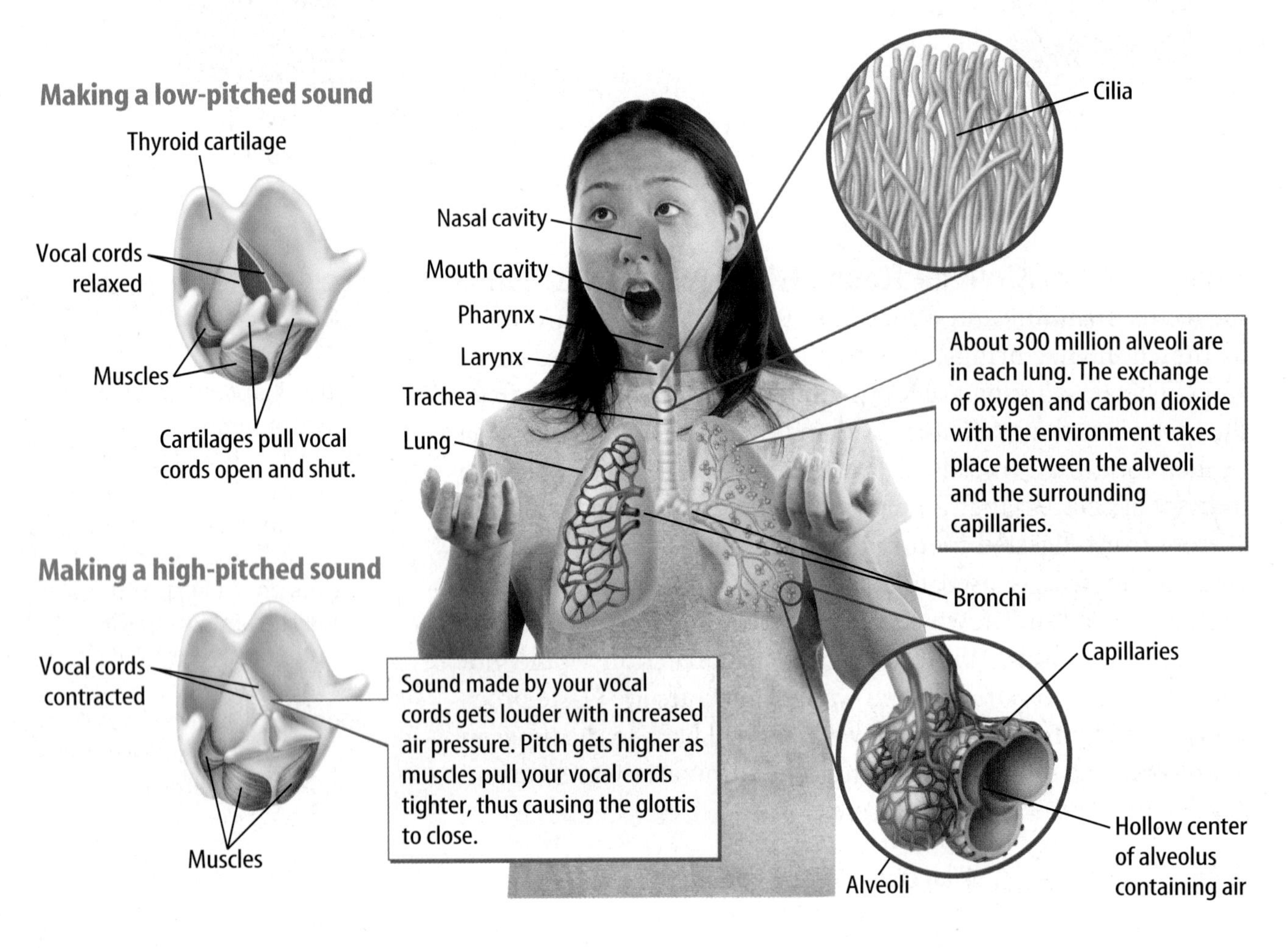

Larynx and Trachea Next, the air moves into your larynx (LER ingks). The **larynx** is the airway to which two pairs of horizontal folds of tissue, called vocal cords, are attached as shown in **Figure 3.** Forcing air between the cords causes them to vibrate and produce sounds. When you speak, muscles tighten or loosen your vocal cords, resulting in different sounds. Your brain coordinates the movement of the muscles in your throat, tongue, cheeks, and lips when you talk, sing, or just make noise. Your teeth also are involved in forming letter sounds and words.

From the larynx, air moves into the **trachea** (TRAY kee uh), which is a tube about 12 cm in length. Strong, C-shaped rings of cartilage prevent the trachea from collapsing. The trachea is lined with mucous membranes and cilia, as shown in **Figure 3,** that trap dust, bacteria, and pollen. Why must the trachea stay open all the time?

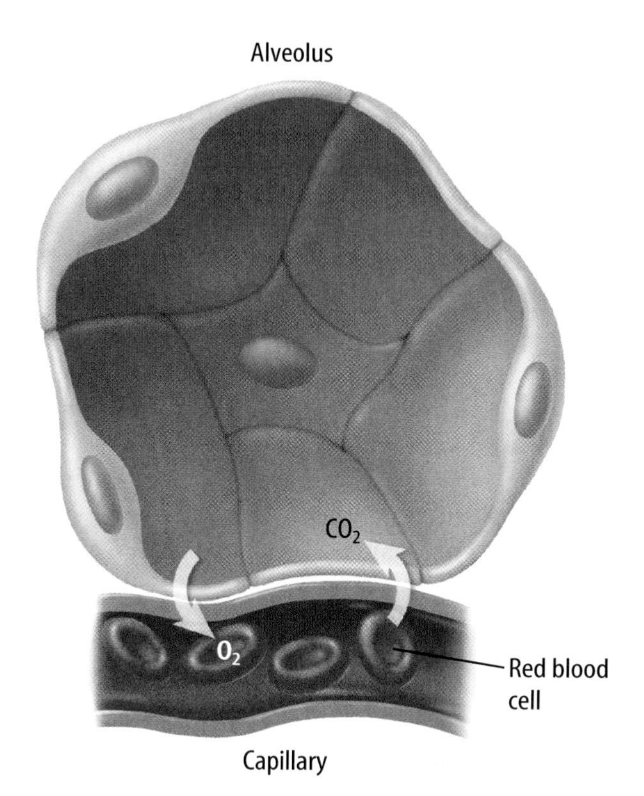

Figure 4 The thin capillary walls allow gases to be exchanged easily between the alveoli and the capillaries.

Bronchi and the Lungs Air is carried into your lungs by two short tubes called **bronchi** (BRAHN ki) (singular, *bronchus*) at the lower end of the trachea. Within the lungs, the bronchi branch into smaller and smaller tubes. The smallest tubes are called bronchioles (BRAHN kee ohlz). At the end of each bronchiole are clusters of tiny, thin-walled sacs called **alveoli** (al VEE uh li). Air passes into the bronchi, then into the bronchioles, and finally into the alveoli. Lungs are masses of alveoli arranged in grapelike clusters. The capillaries surround the alveoli like a net, as shown in **Figure 3.**

The exchange of oxygen and carbon dioxide takes place between the alveoli and capillaries. This easily happens because the walls of the alveoli (singular, *alveolus*) and the walls of the capillaries are each only one cell thick, as shown in **Figure 4.** Oxygen moves through the cell membranes of the alveoli and then through the cell membranes of the capillaries into the blood. There the oxygen is picked up by hemoglobin (HEE muh gloh bun), a molecule in red blood cells, and carried to all body cells. At the same time, carbon dioxide and other cellular wastes leave the body cells. The wastes move through the cell membranes of the capillaries. Then they are carried by the blood. In the lungs, waste gases move through the cell membranes of the capillaries and through the cell membranes of the alveoli. Then waste gases leave the body during exhalation.

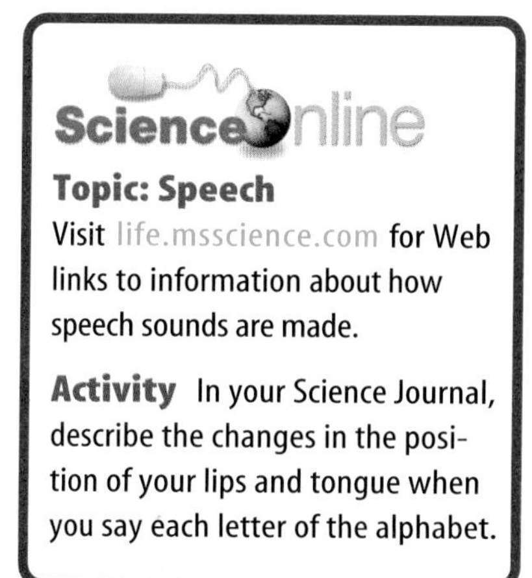

Topic: Speech
Visit life.msscience.com for Web links to information about how speech sounds are made.

Activity In your Science Journal, describe the changes in the position of your lips and tongue when you say each letter of the alphabet.

Comparing Surface Area

Procedure

1. Stand a **bathroom-tissue cardboard tube** in an **empty bowl.**
2. Drop **marbles** into the tube, filling it to the top.
3. Empty the tube and count the number of marbles.
4. Repeat steps 2 and 3 two more times. Calculate the average number of marbles needed to fill the tube.
5. The tube's inside surface area is approximately 161.29 cm^2. Each marble has a surface area of approximately 8.06 cm^2. Calculate the surface area of the average number of marbles.

Analysis

1. Compare the inside surface area of the tube with the surface area of the average number of marbles needed to fill the tube.
2. If the tube represents a bronchus, what do the marbles represent?
3. Using this model, explain what makes gas exchange in the lungs efficient.

Try at Home

Why do you breathe?

Signals from your brain tell the muscles in your chest and abdomen to contract and relax. You don't have to think about breathing to breathe, just like your heart beats without you telling it to beat. Your brain can change your breathing rate depending on the amount of carbon dioxide present in your blood. As carbon dioxide increases, your breathing rate increases. When there is less carbon dioxide in your blood, your breathing rate decreases. You do have some control over your breathing—you can hold your breath if you want to. Eventually, though, your brain will respond to the buildup of carbon dioxide in your blood. The brain's response will tell your chest and abdomen muscles to work automatically, and you will breathe whether you want to or not.

Inhaling and Exhaling Breathing is partly the result of changes in air pressure. Under normal conditions, a gas moves from an area of high pressure to an area of low pressure. When you squeeze an empty, soft-plastic bottle, air is pushed out. This happens because air pressure outside the top of the bottle is less than the pressure you create inside the bottle when you squeeze it. As you release your grip on the bottle, the air pressure inside the bottle becomes less than it is outside the bottle. Air rushes back in, and the bottle returns to its original shape.

Your lungs work in a similar way to the squeezed bottle. Your **diaphragm** (DI uh fram) is a muscle beneath your lungs that contracts and relaxes to help move gases into and out of your lungs. **Figure 5** illustrates breathing.

Reading Check *How does your diaphragm help you breathe?*

When a person is choking, a rescuer can use abdominal thrusts, as shown in **Figure 6,** to save the life of the choking victim.

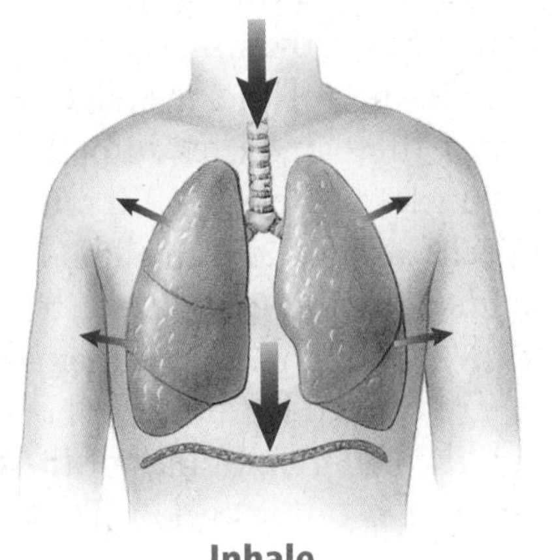

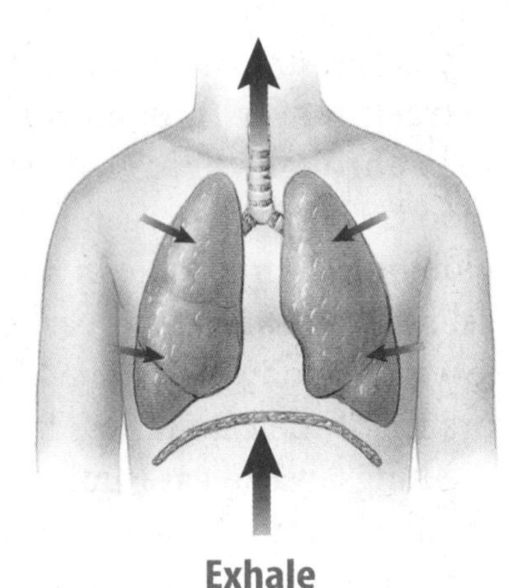

Figure 5 Your lungs inhale and exhale about 500 mL of air with an average breath. This increases to 2,000 mL of air per breath when you do strenuous activity.

VISUALIZING ABDOMINAL THRUSTS

Figure 6

When food or other objects become lodged in the trachea, airflow between the lungs and the mouth and nasal cavity is blocked. Death can occur in minutes. However, prompt action by someone can save the life of a choking victim. The rescuer uses abdominal thrusts to force the victim's diaphragm up. This decreases the volume of the chest cavity and forces air up in the trachea. The result is a rush of air that dislodges and expels the food or other object. The victim can breathe again. This technique is shown at right and should only be performed in emergency situations.

Food is lodged in the victim's trachea.

The rescuer places her fist against the victim's stomach.

The rescuer's second hand adds force to the fist.

An upward thrust dislodges the food from the victim's trachea.

A The rescuer stands behind the choking victim and wraps her arms around the victim's upper abdomen. She places a fist (thumb side in) against the victim's stomach. The fist should be below the ribs and above the navel.

B With a violent, sharp movement, the rescuer thrusts her fist up into the area below the ribs. This action should be repeated as many times as necessary.

Table 1 Smokers' Risk of Death from Disease

Disease	Smokers' Risk Compared to Nonsmokers' Risk
Lung cancer	23 times higher for males, 11 times higher for females
Chronic bronchitis and emphysema	5 times higher
Heart disease	2 times higher

Diseases and Disorders of the Respiratory System

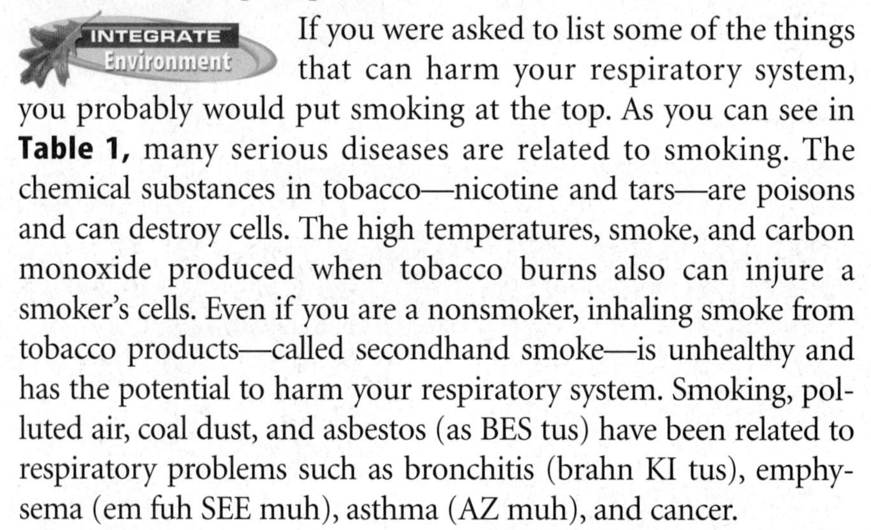

INTEGRATE Environment If you were asked to list some of the things that can harm your respiratory system, you probably would put smoking at the top. As you can see in **Table 1,** many serious diseases are related to smoking. The chemical substances in tobacco—nicotine and tars—are poisons and can destroy cells. The high temperatures, smoke, and carbon monoxide produced when tobacco burns also can injure a smoker's cells. Even if you are a nonsmoker, inhaling smoke from tobacco products—called secondhand smoke—is unhealthy and has the potential to harm your respiratory system. Smoking, polluted air, coal dust, and asbestos (as BES tus) have been related to respiratory problems such as bronchitis (brahn KI tus), emphysema (em fuh SEE muh), asthma (AZ muh), and cancer.

Topic: Second-Hand Smoke
Visit life.msscience.com for Web links to information about the health concerns of second-hand smoke.

Activity Make a poster to teach younger students about the dangers of second-hand smoke.

Respiratory Infections Bacteria, viruses, and other microorganisms can cause infections that affect any of the organs of the respiratory system. The common cold usually affects the upper part of the respiratory system—from the nose to the pharynx. The cold virus also can cause irritation and swelling in the larynx, trachea, and bronchi. The cilia that line the trachea and bronchi can be damaged. However, cilia usually heal rapidly. A virus that causes influenza, or flu, can affect many of the body's systems. The virus multiplies in the cells lining the alveoli and damages them. Pneumonia is an infection in the alveoli that can be caused by bacteria, viruses, or other microorganisms. Before antibiotics were available to treat these infections, many people died from pneumonia.

What parts of the respiratory system are affected by the cold virus?

Chronic Bronchitis When bronchial tubes are irritated and swell, and too much mucus is produced, a disease called bronchitis develops. Sometimes, bacterial infections occur in the bronchial tubes because the mucus there provides nearly ideal conditions for bacteria to grow. Antibiotics are effective treatments for this type of bronchitis.

Many cases of bronchitis clear up within a few weeks, but the disease sometimes lasts for a long time. When this happens, it is called chronic (KRAH nihk) bronchitis. A person who has chronic bronchitis must cough often to try to clear the excess mucus from the airway. However, the more a person coughs, the more the cilia and bronchial tubes can be harmed. When cilia are damaged, they cannot move mucus, bacteria, and dirt particles out of the lungs effectively. Then harmful substances, such as sticky tar from burning tobacco, build up in the airways. Sometimes, scar tissue forms and the respiratory system cannot function properly.

Emphysema A disease in which the alveoli in the lungs enlarge is called **emphysema** (em fuh SEE muh). When cells in the alveoli are reddened and swollen, an enzyme is released that causes the walls of the alveoli to break down. As a result, alveoli can't push air out of the lungs, so less oxygen moves into the bloodstream from the alveoli. When blood becomes low in oxygen and high in carbon dioxide, shortness of breath occurs. Some people with emphysema require extra oxygen as shown in **Figure 7.** Because the heart works harder to supply oxygen to body cells, people who have emphysema often develop heart problems, as well.

Figure 7 Lung diseases can have major effects on breathing.

A normal, healthy lung can exchange oxygen and carbon dioxide effectively.

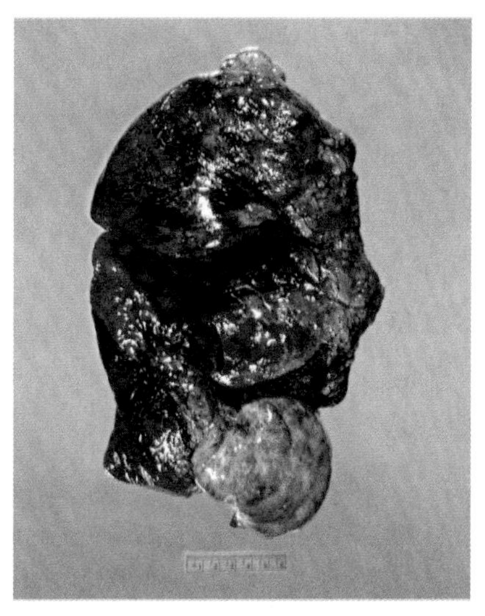

A diseased lung carries less oxygen to body cells.

Emphysema may take 20 to 30 years to develop.

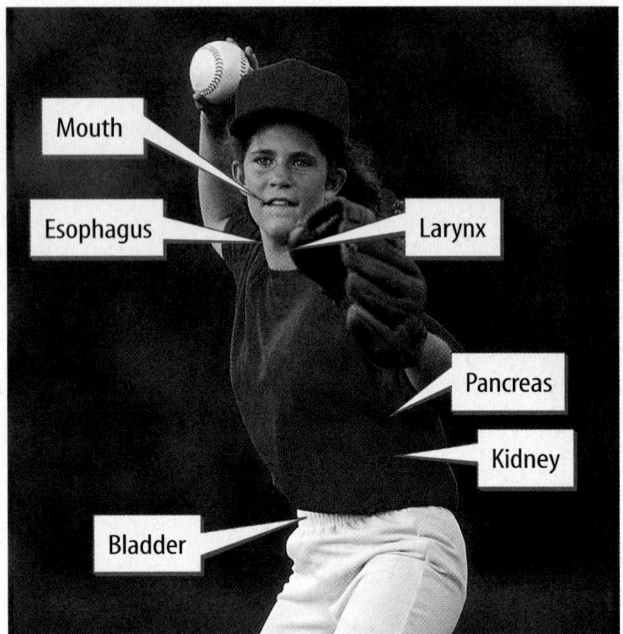

Figure 8 More than 85 percent of all lung cancer is related to smoking. Smoking also can play a part in the development of cancer in other body organs indicated above.

Lung Cancer The leading cause of cancer-related deaths in men and women in the United States is lung cancer. Inhaling the tar in cigarette smoke is the greatest contributing factor to lung cancer. Tar and other ingredients found in smoke act as carcinogens (kar SIH nuh junz) in the body. Carcinogens are substances that can cause an uncontrolled growth of cells. In the lungs, this is called lung cancer. As represented in **Figure 8,** smoking also has been linked to the development of cancers of the esophagus, mouth, larynx, pancreas, kidney, and bladder.

Reading Check ***What happens to the lungs of people who begin smoking?***

Asthma Shortness of breath, wheezing, or coughing can occur in a lung disorder called **asthma.** When a person has an asthma attack, the bronchial tubes contract quickly. Inhaling medicine that relaxes the bronchial tubes is the usual treatment for an asthma attack. Asthma is often an allergic reaction. An allergic reaction occurs when the body overreacts to a foreign substance. An asthma attack can result from breathing certain substances such as cigarette smoke or certain plant pollen, eating certain foods, or stress in a person's life.

section 1 review

Summary

Functions of the Respiratory System

- Breathing brings air into the lungs and removes waste gases.
- Cellular respiration converts oxygen and glucose to carbon dioxide, water, and energy.

Organs of the Respiratory System

- Air is carried into the lungs by bronchi.
- Bronchioles are smaller branches of bronchi, and at the ends of these are alveoli.

Diseases and Disorders of the Respiratory System

- Emphysema is a disease that causes the alveoli to enlarge.
- Lung cancer occurs when carcinogens cause an uncontrolled growth of cells.

Self Check

1. **Describe** the main function of the respiratory system.
2. **Explain** how oxygen, carbon dioxide, and other waste gases are exchanged in the lungs and body tissues.
3. **Identify** how air moves into and out of the lungs.
4. **Think Critically** How is the work of the digestive and circulatory systems related to the respiratory system?

Applying Skills

5. **Research Information** Nicotine in tobacco is a poison. Using library references, find out how nicotine affects the body.
6. **Communicate** Use references to find out about lung disease common among coal miners, stonecutters, and sandblasters. Find out what safety measures are required now for these trades. In your Science Journal, write a paragraph about these safety measures.

Science Online life.msscience.com/self_check_quiz

section 2

The Excretory System

Functions of the Excretory System

It's your turn to take out the trash. You carry the bag outside and put it in the trash can. The next day, you bring out another bag of trash, but the trash can is full. When trash isn't collected, it piles up. Just as trash needs to be removed from your home to keep it livable, your body must eliminate wastes to remain healthy. Undigested material is eliminated by your large intestine. Waste gases are eliminated through the combined efforts of your circulatory and respiratory systems. Some salts are eliminated when you sweat. These systems function together as parts of your excretory system. If wastes aren't eliminated, toxic substances build up and damage organs. If not corrected, serious illness or death occurs.

The Urinary System

The **urinary system** rids the blood of wastes produced by the cells. **Figure 9** shows how the urinary system functions as a part of the excretory system. The urinary system also controls blood volume by removing excess water produced by body cells during respiration.

as you read

What You'll Learn

- **Distinguish** between the excretory and urinary systems.
- **Describe** how the kidneys work.
- **Explain** what happens when urinary organs don't work.

Why It's Important

The urinary system helps clean your blood of cellular wastes.

Review Vocabulary

blood: tissue that transports oxygen, nutrients, and waste materials throughout your body

New Vocabulary

- urinary system
- urine
- kidney
- nephron
- ureter
- bladder
- urethra

Figure 9 The excretory system includes other body systems.

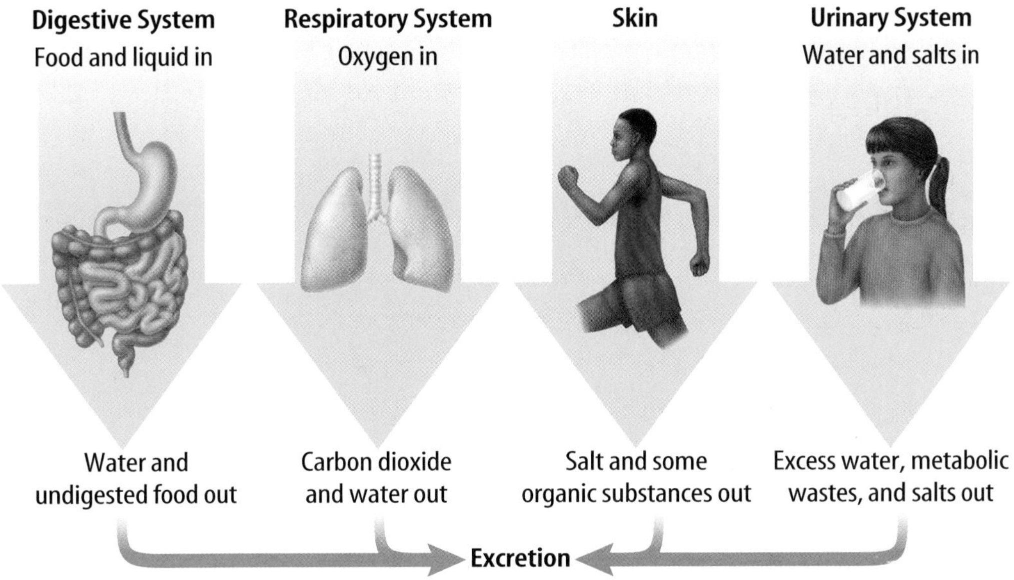

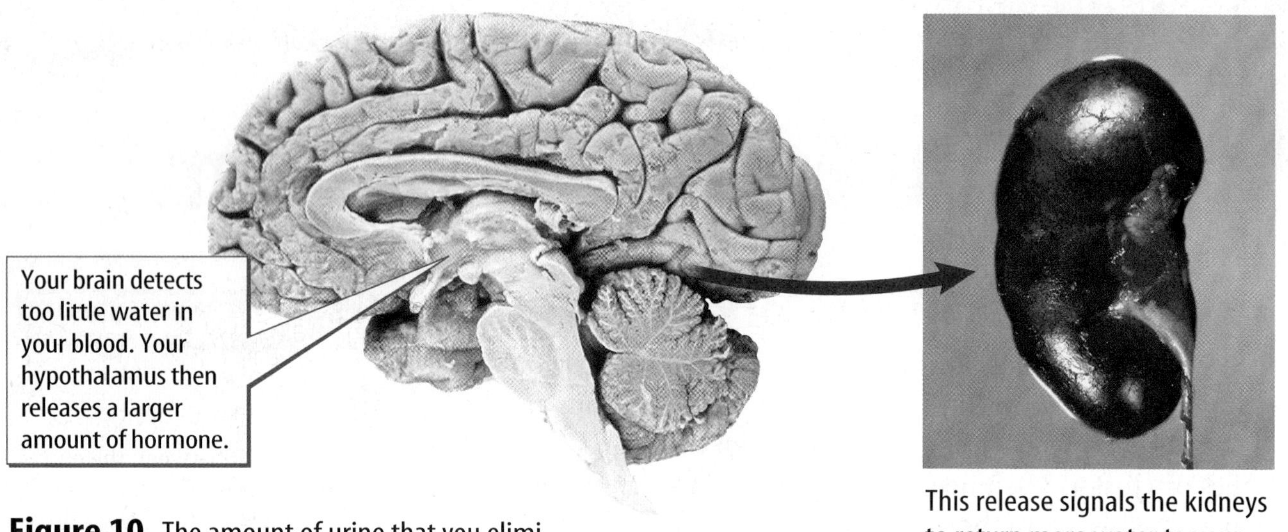

Your brain detects too little water in your blood. Your hypothalamus then releases a larger amount of hormone.
This release signals the kidneys to return more water to your blood and decrease the amount of urine excreted.

Figure 10 The amount of urine that you eliminate each day is determined by the level of a hormone that is produced by your hypothalamus.

Regulating Fluid Levels To stay in good health, the fluid levels within the body must be balanced and normal blood pressure must be maintained. An area in the brain, the hypothalamus (hi poh THA luh mus), constantly monitors the amount of water in the blood. When the brain detects too much water in the blood, the hypothalamus releases a lesser amount of a specific hormone. This signals the kidneys to return less water to the blood and increase the amount of wastewater, called **urine,** that is excreted. **Figure 10** indicates how the body reacts when too little water is in the blood.

Reading Check *How does the urinary system control the volume of water in the blood?*

A specific amount of water in the blood is also important for the movement of gases and excretion of solid wastes from the body. The urinary system also balances the amounts of certain salts and water that must be present for all cell activities to take place.

Organs of the Urinary System Excretory organs is another name for the organs of the urinary system. The main organs of the urinary system are two bean-shaped **kidneys.** Kidneys are located on the back wall of the abdomen at about waist level. The kidneys filter blood that contains wastes collected from cells. In approximately 5 min, all of the blood in your body passes through the kidneys. The red-brown color of the kidneys is due to their enormous blood supply. In **Figure 11,** you can see that blood enters the kidneys through a large artery and leaves through a large vein.

Filtration in the Kidney The kidney, as shown in **Figure 11A,** is a two-stage filtration system. It is made up of about 1 million tiny filtering units called **nephrons** (NEF rahnz), which are shown in **Figure 11B.** Each nephron has a cuplike structure and a tubelike structure called a duct. Blood moves from a renal artery to capillaries in the cuplike structure. The first filtration occurs when water, sugar, salt, and wastes from the blood pass into the cuplike structure. Left behind in the blood are red blood cells and proteins. Next, liquid in the cuplike structure is squeezed into a narrow tubule. Capillaries that surround the tubule perform the second filtration. Most of the water, sugar, and salt are reabsorbed and returned to the blood. These collection capillaries merge to form small veins, which merge to form a renal vein in each kidney. Purified blood is returned to the main circulatory system. The liquid left behind flows into collecting tubules in each kidney. This wastewater, or urine, contains excess water, salts, and other wastes that are not reabsorbed by the body. An average-sized person produces about 1 L of urine per day.

Mini LAB

Modeling Kidney Function

Procedure

1. Mix a small amount of **soil** and **fine gravel** with **water** in a **clean cup.**
2. Place the **funnel** into a **second cup.**
3. Place a small piece of **wire screen** in the funnel.
4. Carefully pour the mud-water-gravel mixture into the funnel. Let it drain.
5. Remove the screen and replace it with a piece of **filter paper.**
6. Place the funnel in **another clean cup.**
7. Repeat step 4.

Analysis

1. What part of the blood does the gravel represent?
2. How does this experiment model the function of a person's kidneys?

Figure 11 The urinary system removes wastes from the blood and includes the kidneys, the bladder, and the connecting tubes.

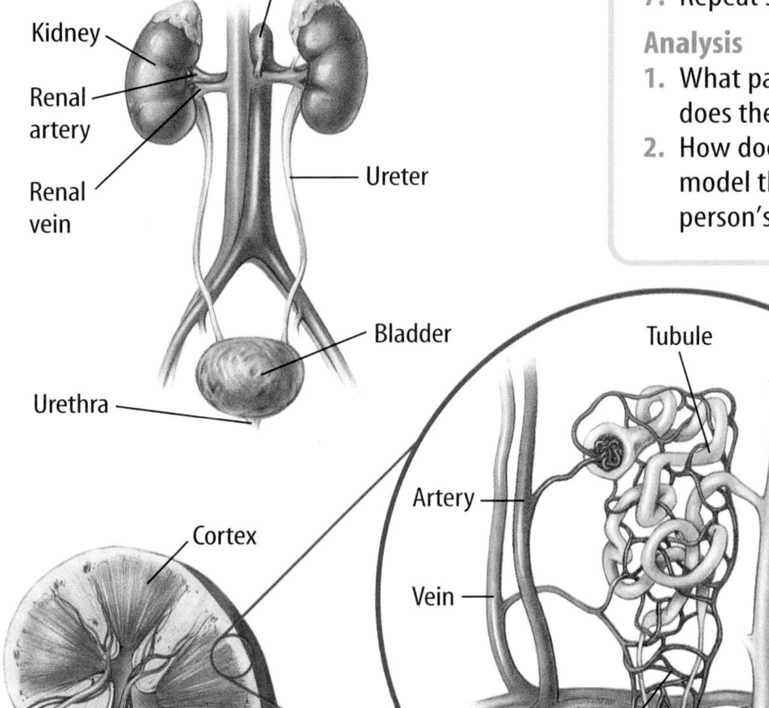

A Kidneys are made up of many nephrons.

B A single nephron is shown in detail. **Describe** *the main function of the nephron.*

Urine Collection and Release The urine in each collecting tubule drains into a funnel-shaped area of each kidney that leads to the ureter (YOO ruh tur). **Ureters** are tubes that lead from each kidney to the bladder. The **bladder** is an elastic, muscular organ that holds urine until it leaves the body. The elastic walls of the bladder can stretch to hold up to 0.5 L of urine. When empty, the bladder looks wrinkled and the cells lining the bladder are thick. When full, the bladder looks like an inflated balloon and the cells lining the bladder are stretched and thin. A tube called the **urethra** (yoo REE thruh) carries urine from the bladder to the outside of the body.

Applying Science

How does your body gain and lose water?

Your body depends on water. Without water, your cells could not carry out their activities and body systems could not function. Water is so important to your body that your brain and other body systems are involved in balancing water gain and water loss.

Identifying the Problem

Table A shows the major sources by which your body gains water. Oxidation of nutrients occurs when energy is released from nutrients by your body's cells. Water is a waste product of these reactions. **Table B** lists the major sources by which your body loses water. The data show you how daily gain and loss of water are related.

Table A

Major Sources by Which Body Water is Gained		
Source	Amount (mL)	Percent
Oxidation of nutrients	250	10
Foods	750	30
Liquids	1,500	60
Total	**2,500**	**100**

Table B

Major Sources by Which Body Water is Lost		
Source	Amount (mL)	Percent
Urine	1,500	60
Skin	500	20
Lungs	350	14
Feces	150	6
Total	**2,500**	**100**

Solving the Problem

1. What is the greatest source of water gained by your body?
2. Explain how the percentages of water gained and lost would change in a person who was working in extremely warm temperatures. In this case, what organ of the body would be the greatest contributor to water loss?

Other Organs of Excretion

Large amounts of liquid wastes are lost every day by your body in other ways, as shown in **Figure 12.** The liver also filters the blood to remove wastes. Certain wastes are converted to other substances. For example, excess amino acids are changed to a chemical called urea (yoo REE uh) that is excreted in urine. Hemoglobin from broken-down red blood cells becomes part of bile, which is the digestive fluid from the liver.

Figure 12 On average, the volume of water lost daily by exhaling is a little more than the volume of a soft-drink can. The volume of water lost by your skin each day is about the volume of a 591 mL soft-drink bottle.

Urinary Diseases and Disorders

What happens when someone's kidneys don't work properly or stop working? Waste products that are not removed build up and act as poisons in body cells. Water that normally is removed from body tissues accumulates and causes swelling of the ankles and feet. Sometimes these fluids also build up around the heart, causing it to work harder to move blood to the lungs.

Without excretion, an imbalance of salts occurs. The body responds by trying to restore this balance. If the balance isn't restored, the kidneys and other organs can be damaged. Kidney failure occurs when the kidneys don't work as they should. This is always a serious problem because the kidneys' job is so important to the rest of the body.

Infections caused by microorganisms can affect the urinary system. Usually, the infection begins in the bladder. However, it can spread and involve the kidneys. Most of the time, these infections can be cured with antibiotics.

Because the ureters and urethra are narrow tubes, they can be blocked easily in some disorders. A blockage of one of these tubes can cause serious problems because urine cannot flow out of the body properly. If the blockage is not corrected, the kidneys can be damaged.

Why is a blocked ureter or urethra a serious problem?

Detecting Urinary Diseases Urine can be tested for any signs of a urinary tract disease. A change in the urine's color can suggest kidney or liver problems. High levels of glucose can be a sign of diabetes. Increased amounts of a protein called albumin (al BYOO mun) indicate kidney disease or heart failure. When the kidneys are damaged, albumin can get into the urine, just as a leaky water pipe allows water to drip.

Desalination Nearly 80 percent of Earth's surface is covered by water. Ninety-seven percent of this water is salt water. Humans cannot drink salt water. Desalination is a process that removes salt from salt water making it safe for human consumption. Research to learn which countries use desalination as a source of drinking water. Mark the countries' locations on a world map.

Figure 13 A dialysis machine can replace or help with some of the activities of the kidneys in a person with kidney failure. Like the kidney, the dialysis machine removes wastes from the blood.

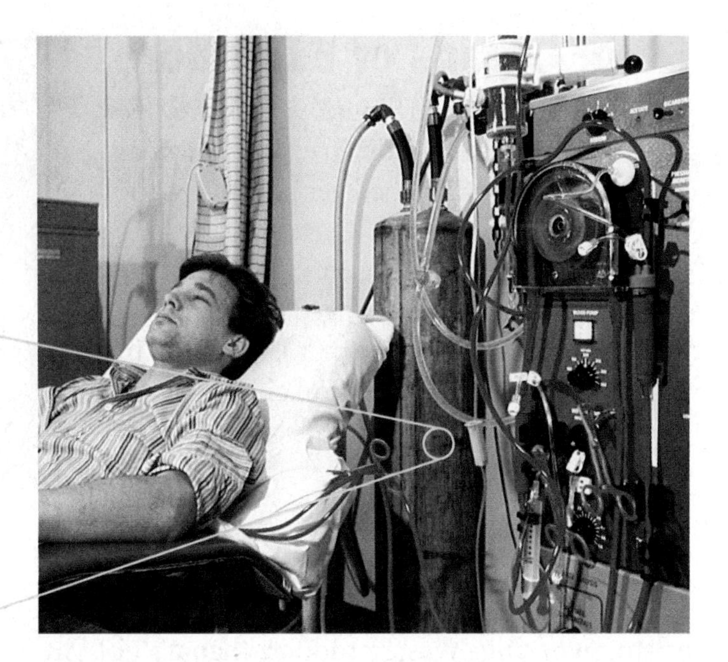

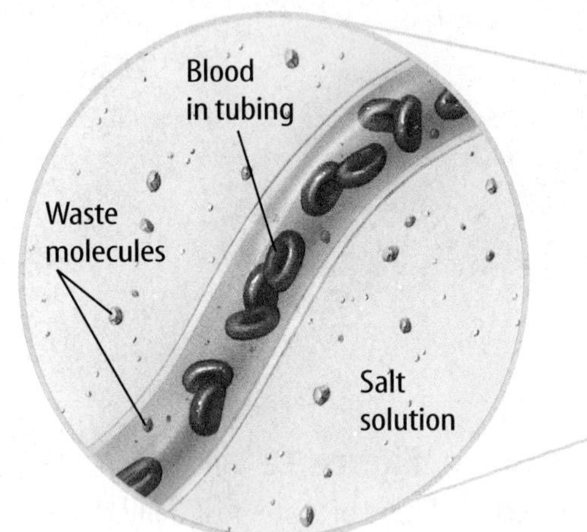

Dialysis A person who has only one kidney still can live normally. The remaining kidney increases in size and works harder to make up for the loss of the other kidney. However, if both kidneys fail, the person will need to have his or her blood filtered by an artificial kidney machine in a process called dialysis (di AH luh sus), as shown in **Figure 13.**

section 2 review

Summary

The Urinary System

- The urinary system rids the blood of wastes produced by your cells.
- The hypothalamus monitors and regulates the amount of water in the blood.
- Nephrons are tiny filtering units in the kidneys that remove water, sugar, salt, and wastes from blood.
- Urine from the kidneys drains into the ureter, then into the bladder, and is carried outside the body by the urethra.

Urinary Diseases and Disorders

- Waste products that are not removed build up and act as poisons in your cells.
- If both kidneys fail, your blood will need to be filtered using a process called dialysis.
- Urine can be tested for kidney and liver problems, heart failure, and diabetes.

Self Check

1. **Explain** how the kidneys remove wastes and keep fluids and salts in balance.
2. **Describe** what happens when the urinary system does not function properly.
3. **Compare** the excretory system and urinary system.
4. **Concept Map** Using a network-tree concept map, compare the excretory functions of the kidneys and the lungs.
5. **Think Critically** Explain why reabsorption of certain materials in the kidneys is important to your health.

Applying Math

6. **Solve One-Step Equations** In approximately 5 min, all 5 L of blood in the body pass through the kidneys. Calculate the average rate of flow through the kidneys in liters per minute.

Science Online life.msscience.com/self_check_quiz

Kidney Structure

As your body uses nutrients, wastes are created. One role of the kidneys is to filter waste products out of the bloodstream and excrete this waste outside the body. How can these small structures filter all the blood in the body in 5 min?

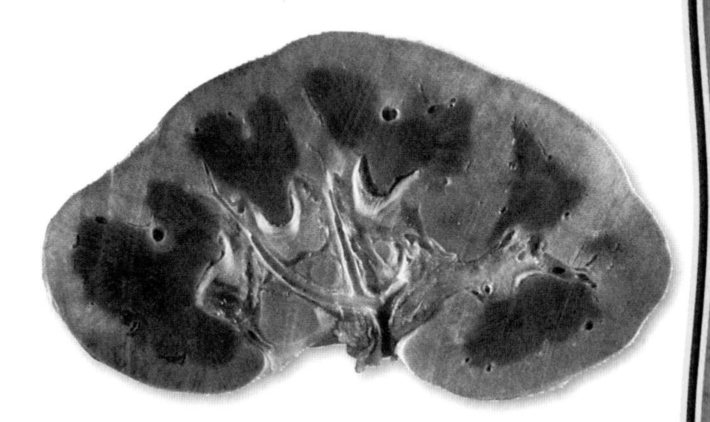

Real-World Question

How does the structure of the kidney relate to the function of a kidney?

Goals

- **Observe** the external and internal structures of a kidney.

Materials

large animal kidney
**model of a kidney*
scalpel
magnifying lens
disposable gloves
dissecting tray
**Alternate materials*

Safety Precautions

WARNING: *Use extreme care when using sharp instruments. Wear disposable gloves. Wash your hands with soap after completing this lab.*

Procedure

1. **Examine** the outside of the kidney supplied by your teacher.
2. If the kidney still is encased in fat, peel off the fat carefully.
3. Using a scalpel, carefully cut the tissue in half lengthwise around the outline of the kidney. This cut should result in a section similar to the illustration on this page.
4. **Observe** the internal features of the kidney using a magnifying lens, or view these features in a model.
5. **Compare** the specimen or model with the kidney in the illustration.
6. **Draw** the kidney in your Science Journal and label its structures.

Conclude and Apply

1. What part makes up the cortex of the kidney? Why is this part red?
2. **Describe** the main function of nephrons.
3. The medulla of the kidney is made up of a network of tubules that come together to form the ureter. What is the function of this network of tubules?
4. How can the kidney be compared to a portable water-purifying system?

Communicating Your Data

Compare your conclusions with those of other students in your class. **For more help, refer to the Science Skill Handbook.**

Simulating the Abdominal Thrust Maneuver

Goals

- **Construct** a model of the trachea with a piece of food stuck in it.
- **Demonstrate** what happens when the abdominal thrust maneuver is performed on someone.
- **Predict** another way that air could get into the lungs if the food could not be dislodged with an abdominal thrust maneuver.

Possible Materials

paper towel roll or other tube
paper (wadded into a ball)
clay
bicycle pump
sports bottle
scissors

Safety Precautions

Always be careful when you use scissors.

Real-World Question

Have you ever taken a class in CPR or learned about how to help a choking victim? Using the abdominal thrust maneuver, or Heimlich maneuver, is one way to remove food or another object that is blocking someone's airway. What happens internally when the maneuver is used? What can you use to make a model of the trachea? How can you simulate what happens during an abdominal thrust maneuver using your model?

Make a Model

1. **List** the materials that you will need to construct your model. What will represent the trachea and a piece of food or other object blocking the airway?
2. How can you use your model to simulate the effects of an abdominal thrust maneuver?
3. Suggest a way to get air into the lungs if the food could not be dislodged. How would you simulate this method in your model?

4. **Compare** your plans for the model and the abdominal thrust maneuver simulation with those of other students in your class. Discuss why each of you chose the plans and materials that you did.
5. Make sure your teacher approves your plan and materials for your model before you start.

Test the Model

1. **Construct** your model of a trachea with an object stuck in it. Make sure that air cannot get through the trachea if you try blowing softly through it.
2. Simulate what happens when an abdominal thrust maneuver is used. Record your observations. Was the object dislodged? How hard was it to dislodge the object?
3. Replace the object in the trachea. Use your model to simulate how you could get air into the lungs if an abdominal thrust maneuver did not remove the object. Is it easy to blow air through your model now?
4. Model a crushed trachea. Is it easy to blow air through the trachea in this case?

Analyze Your Data

1. **Describe** how easy it was to get air through the trachea in each step in the Make the Model section above. Include any other observations that you made as you worked with your model.
2. Think about what you did to get air into the trachea when the object could not be dislodged with an abdominal thrust maneuver. How could this be done to a person? Do you know what this procedure is called?

Conclude and Apply

Explain why the trachea has cartilage around it to protect it. What might happen if it did not?

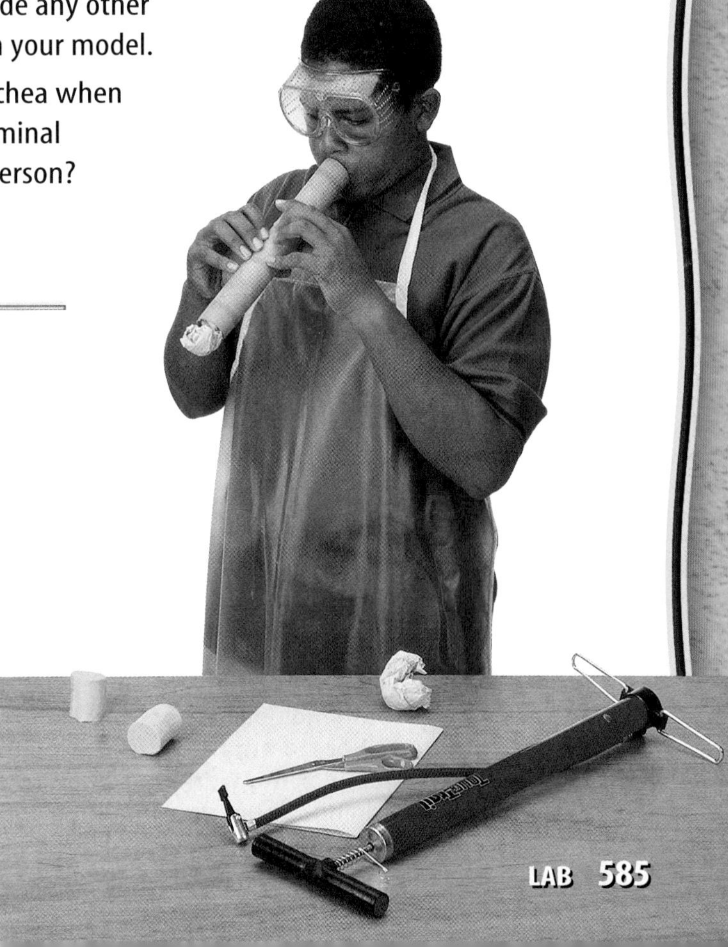

Communicating Your Data

Explain to your family or friends what you have learned about how the abdominal thrust maneuver can help choking victims.

TIME SCIENCE AND HISTORY

SCIENCE CAN CHANGE THE COURSE OF HISTORY!

Guts and determination helped one pioneering doctor to save the lives of thousands

Overcoming the odds is a challenge that many people face. Dr. Samuel Lee Kountz, Jr. had the odds stacked against him. Thanks to his determination he beat them.

Dr. Kountz was interested in kidney transplants, a process that was still brand new in the 1950s. For many patients, a kidney transplant added months or a year to one's life. But then a patient's body would reject the kidney, and the patient would die. Dr. Kountz was determined to see that kidney transplants saved lives and kept patients healthy for years.

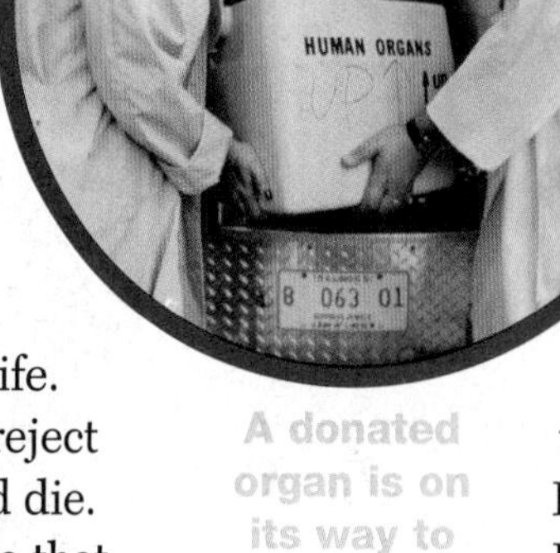

A donated organ is on its way to save a life.

Fixing the Problem

Kountz discovered the root of the problem—why and how a patient's body rejected the transplanted kidney. He discovered that the patient's cells attacked and destroyed the small blood vessels of the transplanted kidney. So the new kidney would die from lack of blood-supplied oxygen. From this, doctors knew when to give patients the right kinds of drugs, so that their bodies could overcome the rejection process.

In 1959, Kountz performed the first successful kidney transplant. He went on to develop a procedure to keep body organs healthy for up to 60 hours after being taken from a donor. He also set up a system of organ donor cards through the National Kidney Foundation. And in his career, Dr. Kountz transplanted more than 1,000 kidneys himself—and paved the way for thousands more.

Research **What kinds of medical breakthroughs has the last century brought? Locate an article that explains either a recent advance in medicine or the work that doctors and medical researchers are doing. Share your findings with your class.**

For more information, visit life.msscience.com/time

Reviewing Main Ideas

Section 1 The Respiratory System

1. The respiratory system brings oxygen into the body and removes carbon dioxide.
2. Inhaled air passes through the nasal cavity, pharynx, larynx, trachea, bronchi, and into the alveoli of the lungs.
3. Breathing brings air into the lungs and removes waste gases.
4. The chemical reaction in the cells that needs oxygen to release energy from glucose is called cellular respiration.
5. The exchange of oxygen and carbon dioxide between aveoli and capillaries, and between capillaries and body cells, happens by the process of diffusion.
6. Smoking causes many problems throughout the respiratory system, including chronic bronchitis, emphysema, and lung cancer.

Section 2 The Excretory System

1. The kidneys are the major organs of the urinary system. They filter wastes from all of the blood in the body.
2. The first stage of kidney filtration occurs when water, sugar, salt, and wastes from the blood pass into the cuplike part of the nephron. The capillaries surrounding the tubule part of the nephron perform the second filtration, returning most of the water, sugar, and salt to the blood.
3. The urinary system is part of the excretory system. The skin, lungs, liver, and large intestine are also excretory organs.
4. Urine can be tested for signs of urinary tract disease and other diseases.
5. A person who has only one kidney still can live normally. When kidneys fail to work, an artificial kidney can be used to filter the blood in a process called dialysis.

Visualizing Main Ideas

Copy and complete the following table on the respiratory and excretory systems.

Human Body Systems

	Respiratory System	Excretory System
Major Organs		
Wastes Eliminated	Do not write in this book.	
Disorders		

chapter 20 Review

Using Vocabulary

alveoli p. 571
asthma p. 576
bladder p. 580
bronchi p. 571
diaphragm p. 572
emphysema p. 575
kidney p. 578
larynx p. 571
nephron p. 579
pharynx p. 570
trachea p. 571
ureter p. 580
urethra p. 580
urinary system p. 577
urine p. 578

For each set of vocabulary words below, explain the relationship that exists.

1. alveoli—bronchi
2. bladder—urine
3. larynx—pharynx
4. ureter—urethra
5. alveoli—emphysema
6. nephron—kidney
7. urethra—bladder
8. asthma—bronchi
9. kidney—urine
10. diaphragm—alveoli

Checking Concepts

Choose the word or phrase that best answers the question.

11. When you inhale, which of the following contracts and moves down?
    A) bronchioles **C)** nephrons
    B) diaphragm **D)** kidneys

12. Air is moistened, filtered, and warmed in which of the following structures?
    A) larynx **C)** nasal cavity
    B) pharynx **D)** trachea

13. Exchange of gases occurs between capillaries and which of the following structures?
    A) alveoli **C)** bronchioles
    B) bronchi **D)** trachea

14. Which of the following is a lung disorder that can occur as an allergic reaction?
    A) asthma **C)** atherosclerosis
    B) cancer **D)** emphysema

15. When you exhale, which way does the rib cage move?
    A) up **C)** out
    B) down **D)** stays the same

16. Which of the following conditions does smoking worsen?
    A) arthritis **C)** excretion
    B) respiration **D)** emphysema

17. In the illustration to the right, what is the name of the organ labeled A?
    A) kidneys
    B) bladder
    C) ureter
    D) urethra

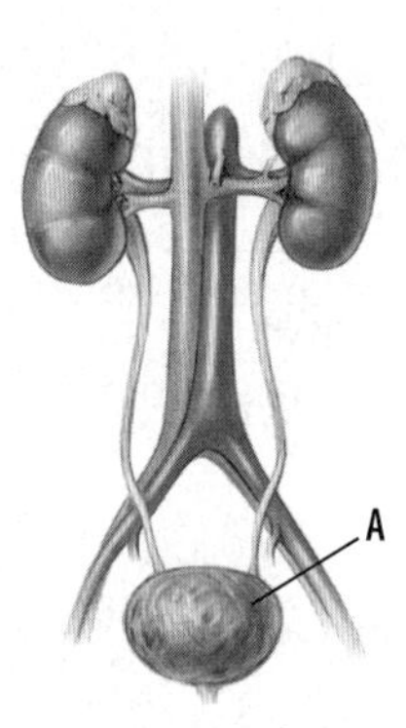

18. What are the filtering units of the kidneys?
    A) nephrons **C)** neurons
    B) ureters **D)** alveoli

19. Approximately 1 L of water is lost per day through which of the following?
    A) sweat **C)** urine
    B) lungs **D)** large intestine

20. Which of the following substances is not reabsorbed by blood after it passes through the kidneys?
    A) salt **C)** wastes
    B) sugar **D)** water

Science Online life.msscience.com/vocabulary_puzzlemaker

Thinking Critically

21. **Explain** why certain foods, such as peanuts, can cause choking in small children.

22. **Infer** why it is an advantage to have lungs with many smaller air sacs instead of having just two large sacs, like balloons.

23. **Explain** the damage to cilia, alveoli, and lungs from smoking.

24. **Describe** what happens to the blood if the kidneys stop working.

25. **Explain** why it is often painful when small, solid particles called kidney stones, pass into the ureter.

Use the table below to answer question 26.

Materials Filtered by the Kidneys

Substance Filtered in Urine	Amount Moving Through Kidney	Amount Excreted
Water	125 L	1 L
Salt	350 g	10 g
Urea	1 g	1 g
Glucose	50 g	0 g

26. **Interpret Data** Study the data above. How much of each substance is reabsorbed into the blood in the kidneys? What substance is excreted completely in the urine?

27. **Recognize Cause and Effect** Discuss how lack of oxygen is related to lack of energy.

28. **Form a hypothesis** about the number of breaths a person might take per minute in each of these situations: sleeping, exercising, and standing on top of Mount Everest. Give a reason for each hypothesis.

Performance Activities

29. **Questionnaire and Interview** Prepare a questionnaire that can be used to interview a health specialist who works with lung cancer patients.

Applying Math

30. **Lung Capacity** Make a circle graph of total lung capacity using the following data:
    - volume of air in a normal inhalation or exhalation = 500 mL
    - volume of additional air that can be inhaled forcefully after a normal inhalation = 3,000 mL
    - volume of additional air that can be exhaled forcefully after a normal expiration = 1,100 mL
    - volume of air still left in the lungs after all the air that can be exhaled has been forcefully exhaled = 1,200 mL

Use the table below to answer question 31.

Death Rates in Industry

Industry	Number of Deaths (1999)	Current Smokers (2000)
Construction	3336	37.4%
Eating and drinking places	907	39.7%
Engineering and science	55	18.7%
Mining	327	32.6%
Railroads	385	24.8%
Trucking service	1004	33.2%

31. **Lung Cancer Deaths** The table above shows the number of lung cancer deaths and the percentage of smokers for specified industries. How many times higher are the death rates for the construction industry than for the eating-and-drinking-places industry?

Part 1 Multiple Choice

Record your answers on the answer sheet provided by your teacher or on a sheet of paper.

1. Which of the following diseases is caused by smoking?
   A. lung cancer **C.** dialysis
   B. diabetes **D.** bladder infection

Use the table below to answer questions 2 and 3.

Major Sources by Which Body Water is Lost		
Source	Amount per day (mL)	Percent
Urine	1,500	60
Skin	500	20
Lungs	350	14
Feces	150	6
Total	2,500	100

2. If the amount of body water lost in the urine increased by 500 mL, what percent of the total body water lost would now be lost in the urine?
   A. 60% **C.** 67%
   B. 75% **D.** 66%

3. If a person had diarrhea, which source of body water loss would increase?
   A. urine **C.** skin
   B. lungs **D.** feces

4. The movement of the chest that brings air into the lungs and removes waste gases is called
   A. oxidation. **C.** respiration.
   B. breathing. **D.** expiration.

5. What traps dust, pollen, and other materials in your nose?
   A. glands
   B. vocal cords
   C. nasal hairs and mucus
   D. epiglottis

Use the illustration below to answer question 6.

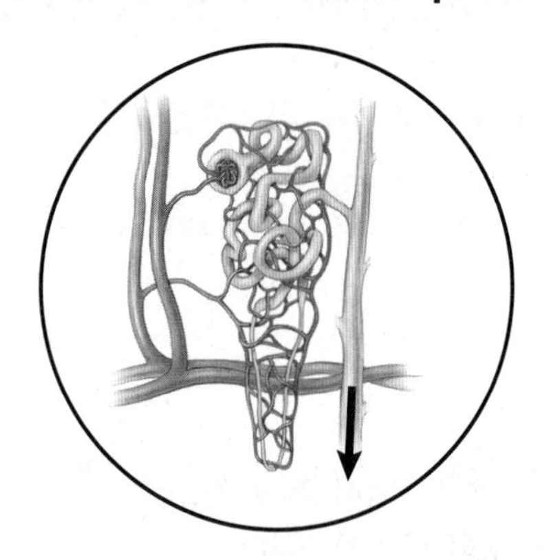

6. What is the structure shown above and to what body system does it belong?
   A. capillary—circulatory
   B. alveolus—respiratory
   C. nephron—urinary
   D. ureter—excretory

7. What is the correct order of steps in the abdominal thrust maneuver?
   A. Rescuer stands behind victim and wraps arms around victim's upper abdomen; rescuer places fist against victim's stomach; rescuer thrusts fist up into area below ribs; rescuer repeats action as many times as necessary.
   B. Rescuer places fist against victim's stomach; rescuer thrusts fist up into area below ribs; rescuer stands behind victim and wraps arms around victim's upper abdomen; rescuer repeats action as many times as necessary.
   C. Rescuer places fist against victim's stomach; rescuer thrusts fist up into area below ribs; rescuer repeats action as many times as necessary.
   D. Rescuer stands in front of victim; rescuer places fist against victim's stomach; rescuer thrusts fist up into area below ribs; rescuer repeats action as needed.

Part 2 Short Response/Grid In

Record your answers on the answer sheet provided by your teacher or on a sheet of paper.

Use the paragraph and table below to answer questions 8–11.

For one week, research scientists collected and accurately measured the amount of body water lost and gained per day for four different patients. The following table lists results from their investigation.

Body Water Gained (+) and Lost (−)				
Person	**Day 1 (L)**	**Day 2 (L)**	**Day 3 (L)**	**Day 4 (L)**
Mr. Stoler	+0.15	+0.15	−0.35	+0.12
Mr. Jemma	−0.01	0.00	−0.20	−0.01
Mr. Lowe	0.00	+0.20	−0.28	+0.01
Mr. Cheng	−0.50	−0.50	−0.55	−0.32

8. What was Mr. Cheng's average daily body water loss for the 4 days shown in the table?
9. Which patient had the greatest amount of body water gained on days 1 and 2?
10. According to the data in the table, on which day was the temperature in each patient's hospital room probably the hottest?
11. Which patient had the highest total gain in body water over the 4-day period?
12. What chemical substances in tobacco can destroy cells?
13. What effect can plant pollen have on the respiratory system?
14. Why do alveoli have thin walls?
15. How is energy released from glucose? What also is produced?

Part 3 Open Ended

Record your answers on a sheet of paper.

16. Explain the role of cilia in the respiratory system. Give an example of a disease in which cilia are damaged. What effects does this damage have on the respiratory system?

Use the table below to answer questions 17–19.

Urine Test Results				
Test Items	**Normal Results**	**Mrs. Beebe**	**Mrs. Chavez**	**Mrs. Jelton**
Glucose	Absent	High	Absent	Absent
Albumin	Absent	Absent	Absent	Absent
Urine volume per 24 hours	1 L	1 L	1 L	0.5 L

17. Mrs. Jelton's urine tests were done when outside temperatures had been higher than 35°C for several days. When Mrs. Jelton came to Dr. Marks' office after the urine test, he asked her about the amount of liquid that she had been drinking. Infer why Dr. Marks asked this question.
18. Assuming that Mrs. Jelton is healthy, form a hypothesis that would explain what had happened.
19. Dr. Marks called another patient to come in for more testing. Who was it? How do you know?

Test-Taking Tip

Understand Symbols Be sure you understand all symbols on a table or graph before attempting to answer any questions about the table or graph.

Questions 21–23. Notice that the unit of volume is in liters (L).

chapter 21

Control and Coordination

The BIG Idea

Organs of the nervous system control and coordinate all body functions.

SECTION 1
The Nervous System
Main Idea The nervous system functions by responding to internal and external stimuli.

SECTION 2
The Senses
Main Idea Your body's senses enable you to enjoy your environment, help maintain homeostasis, and protect you from harm.

Could you stop the puck?

One second, the puck is across the ice rink. In the next second the goalie is trying to stop a goal. A goalie needs to be able to respond quickly, without even thinking about it. In this chapter, you will learn how your body senses and responds to stimuli in the world around you.

Science Journal Which senses do you think are at work when you respond to a glass crashing on a tile floor?

Start-Up Activities

How quick are your responses?

If the weather is cool, you might put on a jacket. If you see friends, you might call out to them. You also might pick up a crying baby. Every second of the day you react to different sights, sounds, and smells in your environment. You control some of these reactions, but others take place in your body without thought. Some reactions protect you from harm.

1. Wearing safety goggles, sit on a chair 1 m away from a partner.
2. Ask your partner to toss a wadded-up piece of paper at your face without warning you.
3. Switch positions and repeat the activity.
4. **Think Critically** Describe in your Science Journal how you reacted to the ball of paper being thrown at you. Explain how your anticipation of being hit altered your body's response.

Preview this chapter's content and activities at life.msscience.com

Senses Your body is constantly responding to stimuli around you. Make the following Foldable to help you understand your five senses.

STEP 1 **Collect** three sheets of paper and layer them about 1.25 cm apart vertically. Keep the edges level.

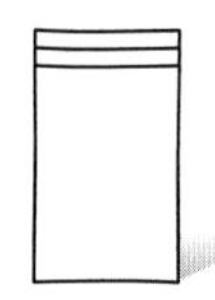

STEP 2 **Fold** up the bottom edges of the paper to form five equal tabs.

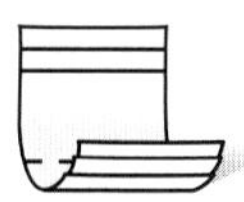

STEP 3 **Crease** the fold, and then staple along the fold. **Label** the tabs *Five Senses, Vision, Hearing, Smell, Taste*, and *Touch*.

Read and Write As you read the chapter, write what you learn about each of your senses under the appropriate tab.

Get Ready to Read

Identify Cause and Effect

① Learn It! A cause is the reason something happens. The result of what happens is called an effect. Learning to identify causes and effects helps you understand why things happen. By using graphic organizers, you can sort and analyze causes and effects as you read.

② Practice It! Read the following paragraph. Then use the graphic organizer below to show how the inner ear helps to maintain balance.

> The cristae ampullaris react to rotating body movements. Fluid in the semicircular canals swirls while the body rotates. This causes the gel-like fluid around the hair cells to move and a stimulus is sent to the brain. In a similar way, when the head tips, the gel-like fluid surrounding the hair cells in the maculae is pulled down by gravity. The hair cells are then stimulated and the brain interprets that the head has tilted.
>
> —*from page 608*

Cause

Effect | Effect | Effect

③ Apply It! As you read the chapter, be aware of causes and effects of stimuli and the nervous system. Find five causes and their effects.

Graphic organizers such as the Cause-Effect organizer help you organize what you are reading so you can remember it later.

Target Your Reading

Use this to focus on the main ideas as you read the chapter.

1. **Before you read** the chapter, respond to the statements below on your worksheet or on a numbered sheet of paper.
   - Write an **A** if you **agree** with the statement.
   - Write a **D** if you **disagree** with the statement.

2. **After you read** the chapter, look back to this page to see if you've changed your mind about any of the statements.
   - If any of your answers changed, explain why.
   - Change any false statements into true statements.
   - Use your revised statements as a study guide.

Before You Read A or D	Statement	After You Read A or D
	1 The nervous system includes nerve cells, the brain, and the spinal cord.	
	2 A neuron only moves messages from the brain to the body.	
	3 Motor neurons receive messages from muscles.	
	4 The peripheral nervous system connects the body to the central nervous system.	
	5 Damage to the left side of your brain affects the function of the right side of your body.	
	6 Optic nerves connect the ears and brain.	
	7 Farsightedness is the condition when distant objects are more in focus than near objects.	
	8 You can identify most foods using only your sense of taste.	
	9 Internal organs have sensory receptors.	

Science Online
Print out a worksheet of this page at life.msscience.com

section 1

The Nervous System

as you read

What You'll Learn

- **Describe** the basic structure of a neuron and how an impulse moves across a synapse.
- **Compare** the central and peripheral nervous systems.
- **Explain** how drugs affect the body.

Why It's Important

Your body is able to react to your environment because of your nervous system.

Review Vocabulary
response: a reaction to a specific stimulus

New Vocabulary
- homeostasis
- neuron
- dendrite
- axon
- synapse
- central nervous system
- peripheral nervous system
- cerebrum
- cerebellum
- brain stem
- reflex

How the Nervous System Works

After doing the dishes and finishing your homework, you settle down in your favorite chair and pick up that mystery novel you've been trying to finish. Only three pages to go . . . Who did it? Why did she do it? Crash! You scream. What made that unearthly noise? You turn around to find that your dog's wagging tail has just swept the lamp off the table. Suddenly, you're aware that your heart is racing and your hands are shaking. After a few minutes though, your breathing returns to normal and your heartbeat is back to its regular rate. What's going on?

Responding to Stimuli The scene described above is an example of how your body responds to changes in its environment. Any internal or external change that brings about a response is called a stimulus (STIHM yuh lus). Each day, you're bombarded by thousands of stimuli, as shown in **Figure 1.** Noise, light, the smell of food, and the temperature of the air are all stimuli from outside your body. Chemical substances such as hormones are examples of stimuli from inside your body. Your body adjusts to changing stimuli with the help of your nervous system.

Figure 1 Stimuli are everywhere and all the time, even when you're with your friends.
List *the types of stimuli present at this party.*

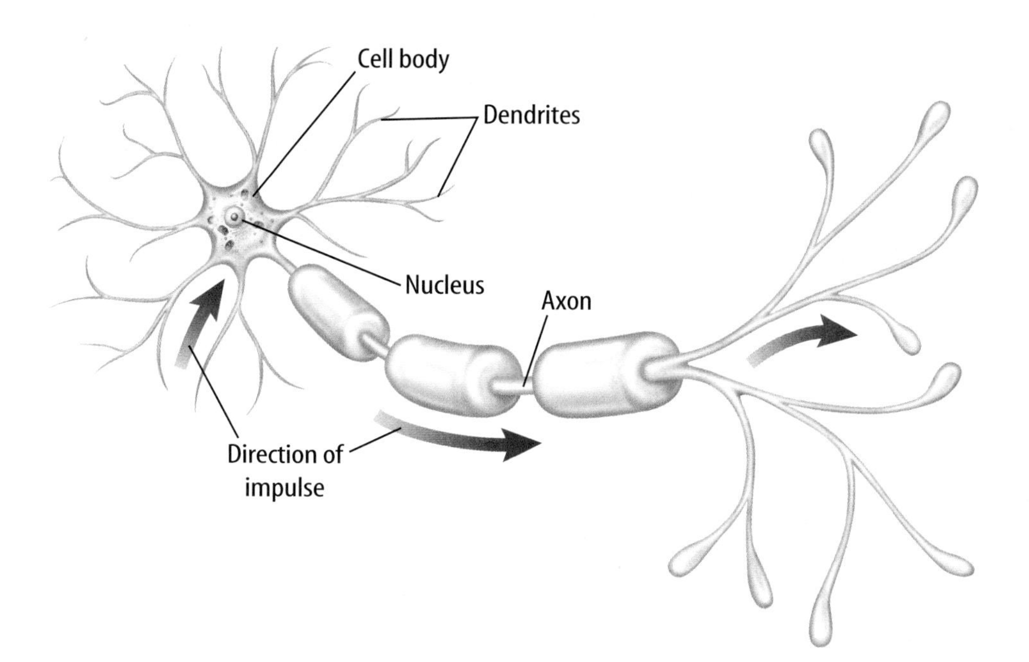

Figure 2 A neuron is made up of a cell body, dendrites, and axons.
Explain *how the branching of the dendrites allows for more impulses to be picked up by the neuron.*

Homeostasis It's amazing how your body handles all these stimuli. Control systems maintain steady internal conditions. The regulation of steady, life-maintaining conditions inside an organism, despite changes in its environment, is called **homeostasis.** Examples of homeostasis are the regulation of your breathing, heartbeat, and digestion. Your nervous system is one of several control systems used by your body to maintain homeostasis.

Nerve Cells

The basic functioning units of the nervous system are nerve cells, or **neurons** (NOOR ahnz). As shown in **Figure 2,** a neuron is made up of a cell body and branches called dendrites and axons. Any message carried by a neuron is called an impulse. **Dendrites** receive impulses from other neurons and send them to the cell body. **Axons** (AK sahns) carry impulses away from the cell body. Notice the branching at the end of the axon. This allows the impulses to move to many other muscles, neurons, or glands.

Types of Nerve Cells Your body has sensory receptors that produce electrical impulses and respond to stimuli, such as changes in temperature, sound, pressure, and taste. Three types of neurons—sensory neurons, motor neurons, and interneurons—transport impulses. Sensory neurons receive information and send impulses to the brain or spinal cord, where interneurons relay these impulses to motor neurons. Motor neurons then conduct impulses from the brain or spinal cord to muscles or glands throughout your body.

Multiple Sclerosis In 1868, Jean Martin Charcot, a neurology professor in Paris, was the first to scientifically describe, document, and name the disease multiple sclerosis. It was named because of the many scars found widely dispersed throughout the central nervous system. Research to find out the symptoms of multiple sclerosis.

NATIONAL GEOGRAPHIC

VISUALIZING NERVE IMPULSE PATHWAYS

Figure 3

Millions of nerve impulses are moving throughout your body as you read this page. In response to stimuli, many impulses follow a specific pathway—from sensory neuron to interneuron to motor neuron—to bring about a response. Like a relay team, these three types of neurons work together. The illustration on this page shows how the sound of a breaking window might startle you and cause you to drop a glass of water.

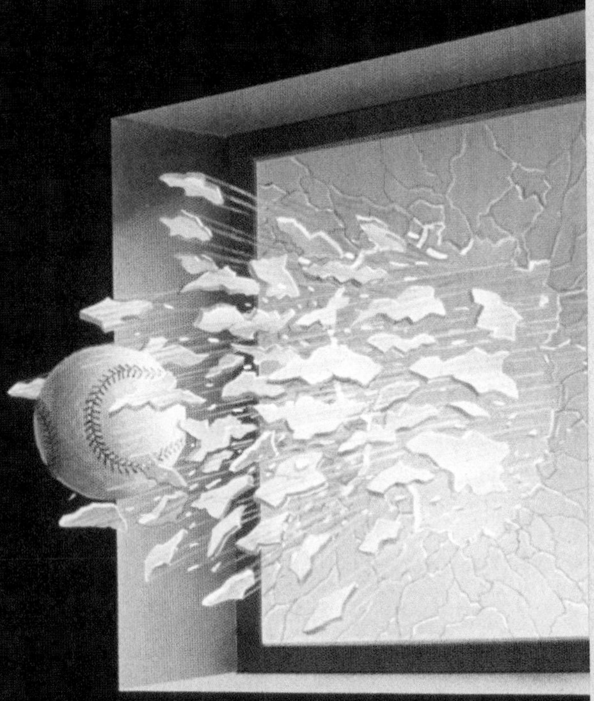

SENSORY NEURONS When you hear a loud noise, receptors in your ears—the specialized endings of sensory neurons—are stimulated. These sensory neurons produce nerve impulses that travel to your brain.

INTERNEURONS Interneurons in your brain receive the impulses from sensory neurons and pass them along to motor neurons.

MOTOR NEURONS Impulses travel down the axons of motor neurons to muscles—in this case, your biceps—which contract to jerk your arms in response to the loud noise.

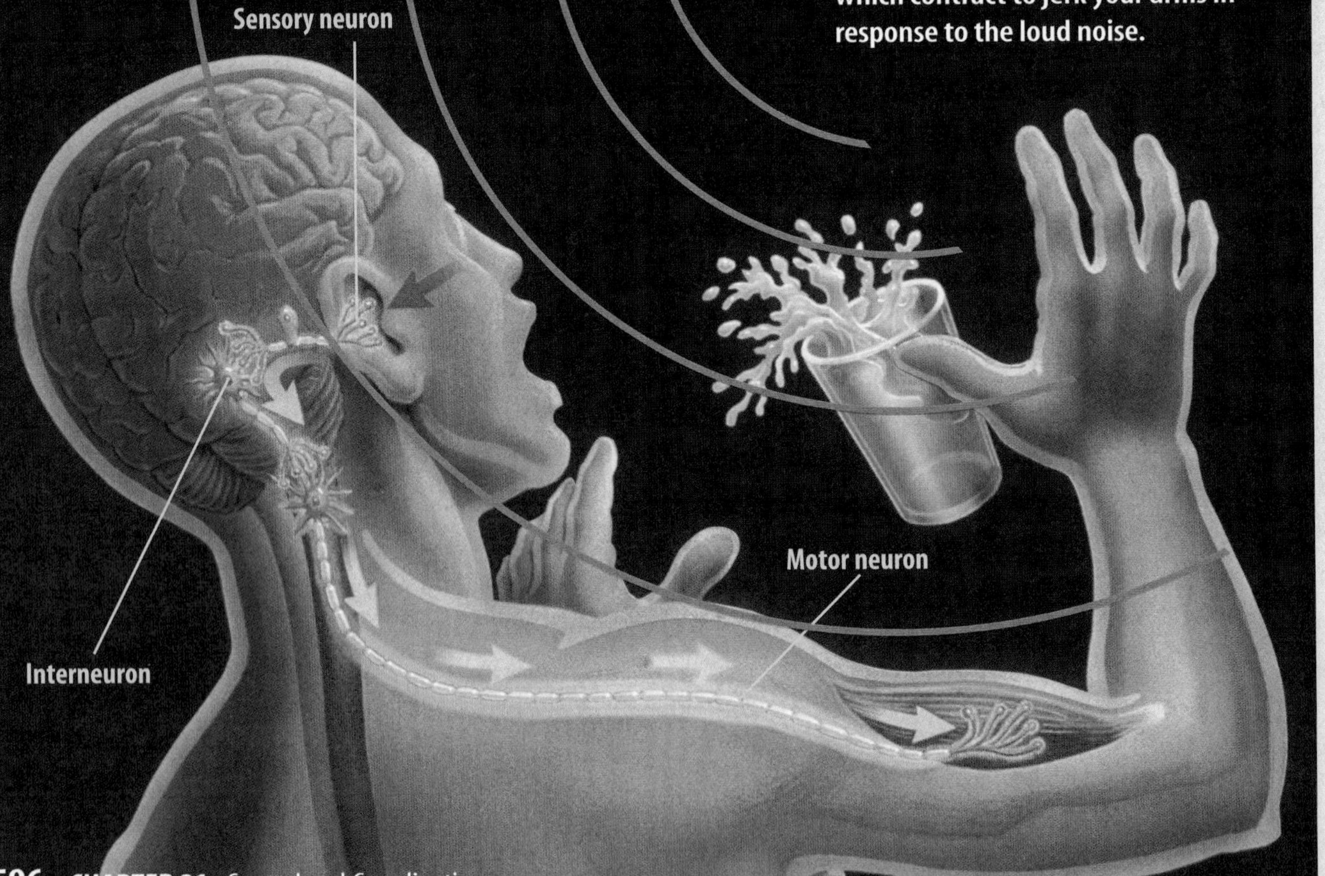

Figure 4 An impulse moves in only one direction across a synapse—from an axon to the dendrites or cell body of another neuron.

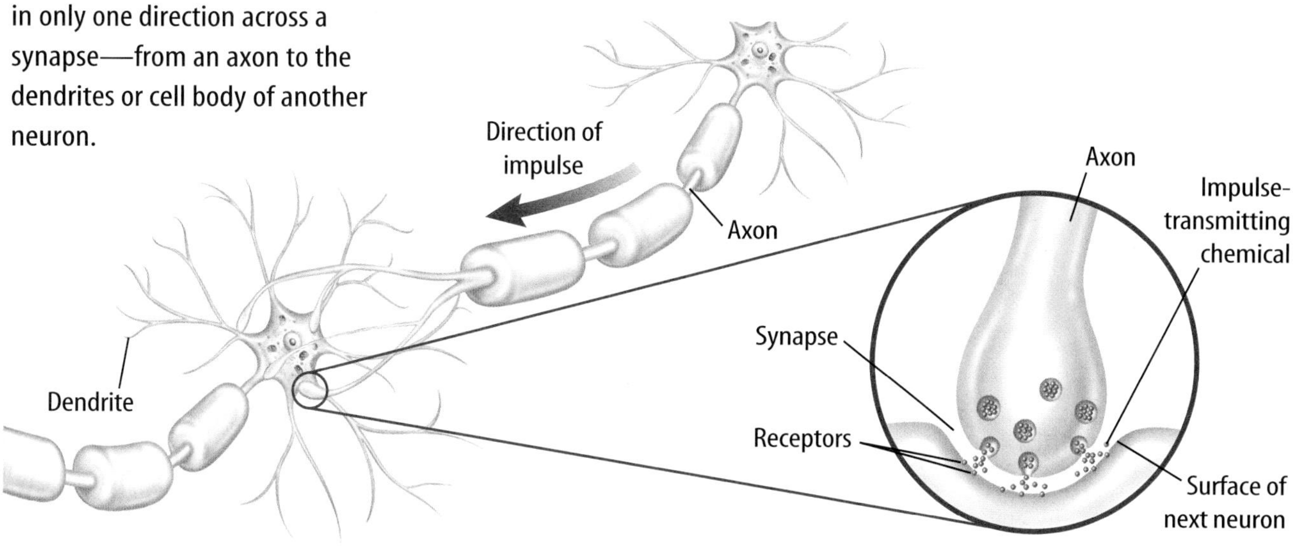

Synapses In a relay race, the first runner sprints down the track with a baton in his or her hand. As the runner rounds the track, he or she hands the baton off to the next runner. The two runners never physically touch each other. The transfer of the baton signals the second runner to continue the race.

As shown in **Figure 3,** your nervous system works in a similar way. Like the runners in a relay race, neurons don't touch each other. How does an impulse move from one neuron to another? To move from one neuron to the next, an impulse crosses a small space called a **synapse** (SIH naps). In **Figure 4,** note that when an impulse reaches the end of an axon, the axon releases a chemical. This chemical flows across the synapse and stimulates the impulse in the dendrite of the next neuron. Your neurons are adapted in such a way that impulses move in only one direction. An impulse moves from neuron to neuron just like a baton moves from runner to runner in a relay race. The baton represents the chemical at the synapse.

The Central Nervous System

Figure 5 shows the organs of the central nervous system (CNS) and the peripheral (puh RIH fuh rul) nervous system (PNS). The **central nervous system** is made up of the brain and spinal cord. The **peripheral nervous system** is made up of all the nerves outside the CNS. These nerves include those in your head, called cranial nerves, and the nerves that come from your spinal cord, called spinal nerves. The peripheral nervous system connects the brain and spinal cord to other body parts. Sensory neurons send impulses to the brain or spinal cord.

Figure 5 The brain and spinal cord (yellow) form the central nervous system (CNS). All other nerves (green) are part of the peripheral nervous system (PNS).

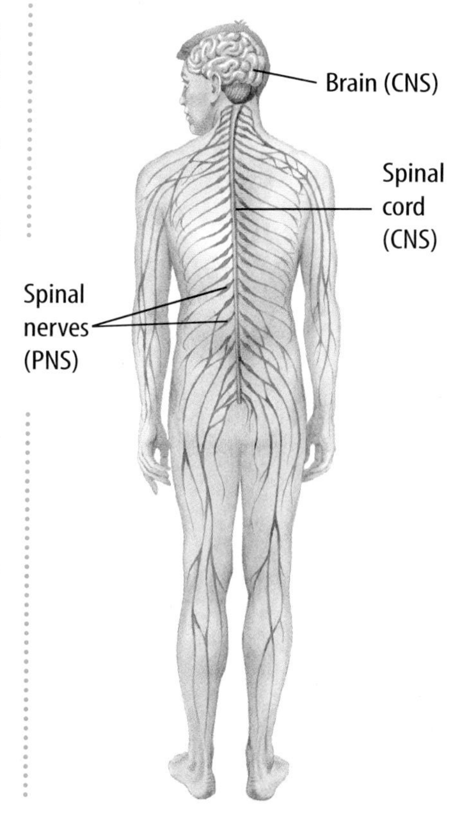

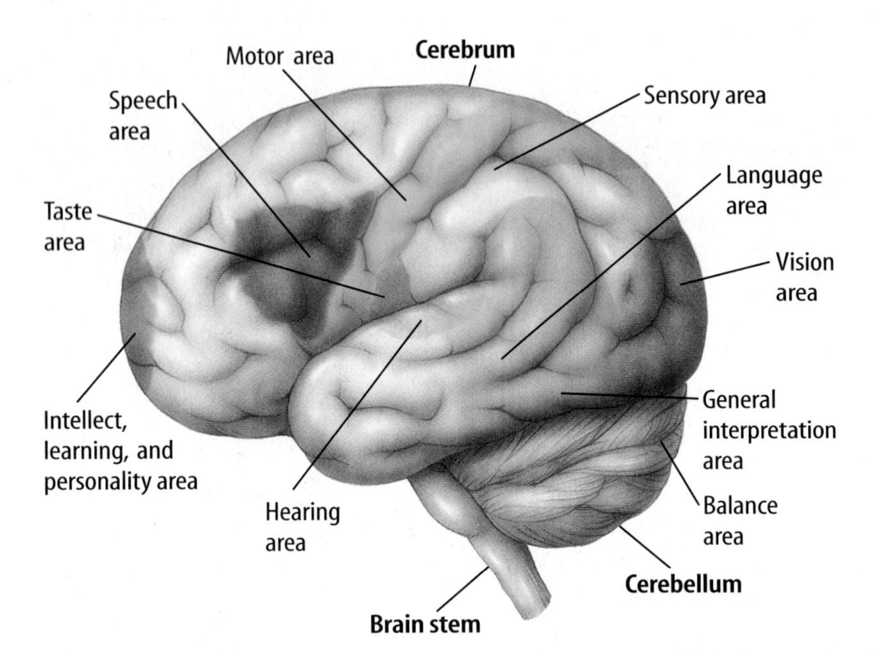

The Brain The brain coordinates all of your body activities. If someone tickles your feet, why does your whole body seem to react? The brain is made up of approximately 100 billion neurons, which is nearly ten percent of all the neurons in the human body. Surrounding and protecting the brain are a bony skull, three membranes, and a layer of fluid. As shown in **Figure 6,** the brain is divided into three major parts—the brain stem, the cerebellum (ser uh BE lum), and the cerebrum (suh REE brum).

Figure 6 Different areas of the brain control specific body activities.
Describe *the three major parts of the brain, and their functions.*

Cerebrum Thinking takes place in the **cerebrum,** which is the largest part of the brain. This also is where impulses from the senses are interpreted, memory is stored, and movements are controlled. The outer layer of the cerebrum, called the cortex, is marked by many ridges and grooves. These structures increase the surface area of the cortex, allowing more complex thoughts to be processed. **Figure 6** shows some of the motor and sensory tasks that the cortex controls.

What major activity takes place within the cerebrum?

Cerebellum Stimuli from the eyes and ears and from muscles and tendons, which are the tissues that connect muscles to bones, are interpreted in the **cerebellum.** With this information, the cerebellum is able to coordinate voluntary muscle movements, maintain muscle tone, and help maintain balance. A complex activity, such as riding a bike, requires a lot of coordination and control of your muscles. The cerebellum coordinates your muscle movements so that you can maintain your balance.

Brain Stem At the base of the brain is the **brain stem.** It extends from the cerebrum and connects the brain to the spinal cord. The brain stem is made up of the midbrain, the pons, and the medulla (muh DUH luh). The midbrain and pons act as pathways connecting various parts of the brain with each other. The medulla controls involuntary actions such as heartbeat, breathing, and blood pressure. The medulla also is involved in such actions as coughing, sneezing, swallowing, and vomiting.

Impulses Acetylcholine (uh see tul KOH leen) is a chemical produced by neurons, which carries an impulse across a synapse to the next neuron. After the impulse is started, the acetylcholine breaks down rapidly. In your Science Journal, hypothesize why the breakdown of acetylcholine is important.

The Spinal Cord Your spinal cord, illustrated in **Figure 7,** is an extension of the brain stem. It is made up of bundles of neurons that carry impulses from all parts of the body to the brain and from the brain to all parts of your body. The adult spinal cord is about the width of an adult thumb and is about 43 cm long.

Topic: Nervous System
Visit life.msscience.com for Web links to information about the nervous system.

Activity In your Science Journal, make a brochure outlining recent medical advances.

The Peripheral Nervous System

Your brain and spinal cord are connected to the rest of your body by the peripheral nervous system. The PNS is made up of 12 pairs of nerves from your brain called cranial nerves and 31 pairs from your spinal cord called spinal nerves. Spinal nerves are made up of bundles of sensory and motor neurons bound together by connective tissue. For this reason, a single spinal nerve can have impulses going to and from the brain at the same time. Some nerves contain only sensory neurons, and some contain only motor neurons, but most nerves contain both types of neurons.

Somatic and Autonomic Systems The peripheral nervous system has two major divisions. The somatic system controls voluntary actions. It is made up of the cranial and spinal nerves that go from the central nervous system to your skeletal muscles. The autonomic system controls involuntary actions—those not under conscious control—such as your heart rate, breathing, digestion, and glandular functions. These two divisions, along with the central nervous system, make up your body's nervous system.

Figure 7 A column of vertebrae, or bones, protects the spinal cord. The spinal cord is made up of bundles of neurons that carry impulses to and from all parts of the body, similar to a telephone cable.

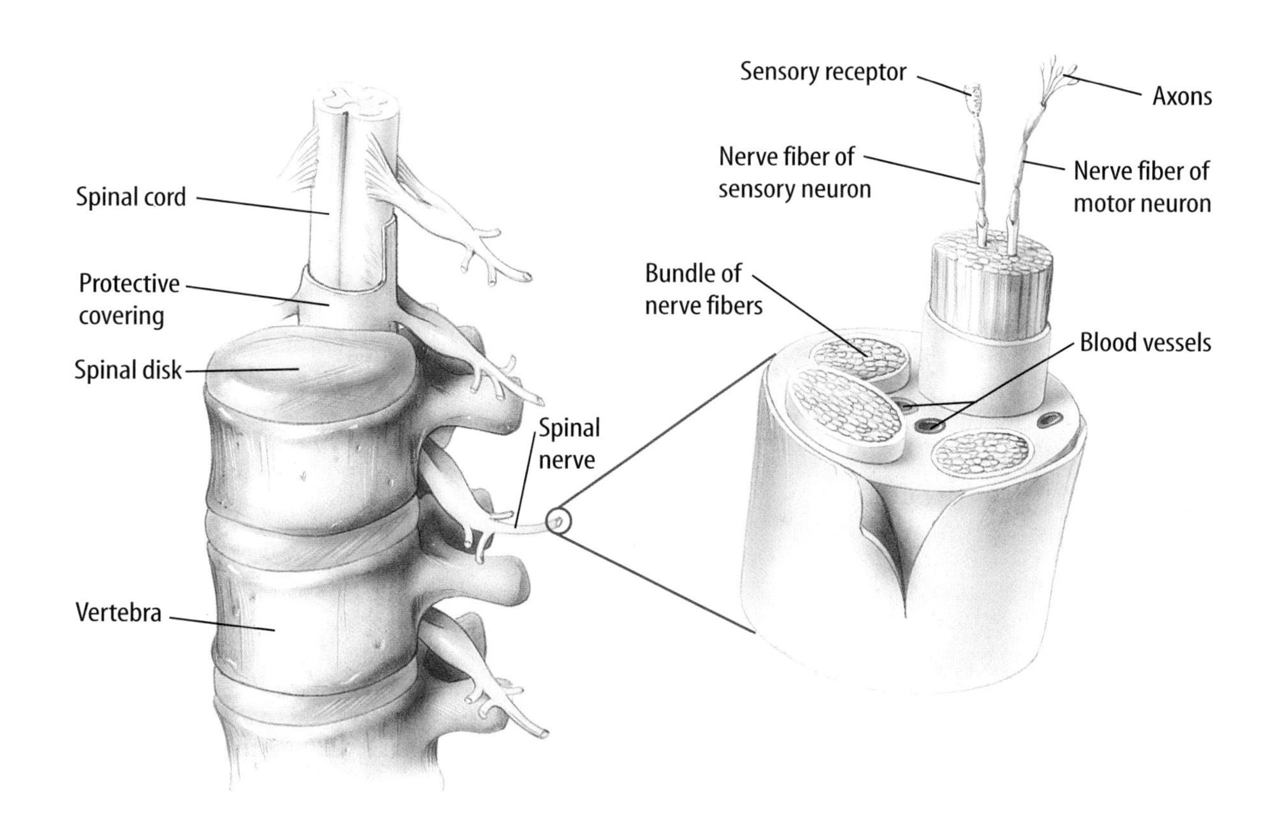

Safety and the Nervous System

Every mental process and physical action of the body is associated with the structures of the central and peripheral nervous systems. Therefore, any injury to the brain or the spinal cord can be serious. A severe blow to the head can bruise the brain and cause temporary or permanent loss of mental and physical abilities. For example, the back of the brain controls vision. An injury in this region could result in the loss of vision.

Although the spinal cord is surrounded by the vertebrae of your spine, spinal cord injuries do occur. They can be just as dangerous as a brain injury. Injury to the spine can bring about damage to nerve pathways and result in paralysis (puh RA luh suhs), which is the loss of muscle movement. As shown in **Figure 8,** a neck injury that damages certain nerves could prevent a person from breathing. Major causes of head and spinal injuries include automobile, motorcycle, and bicycle accidents, as well as sports injuries. Just like wearing safety belts in automobiles, it is important to wear the appropriate safety gear for a sport or when riding a bicycle or motorized vehicle.

Figure 8 Head and spinal cord damage can result in paralysis, depending on where the injury occurs.
Explain *why it is important to wear safety equipment and safety belts.*

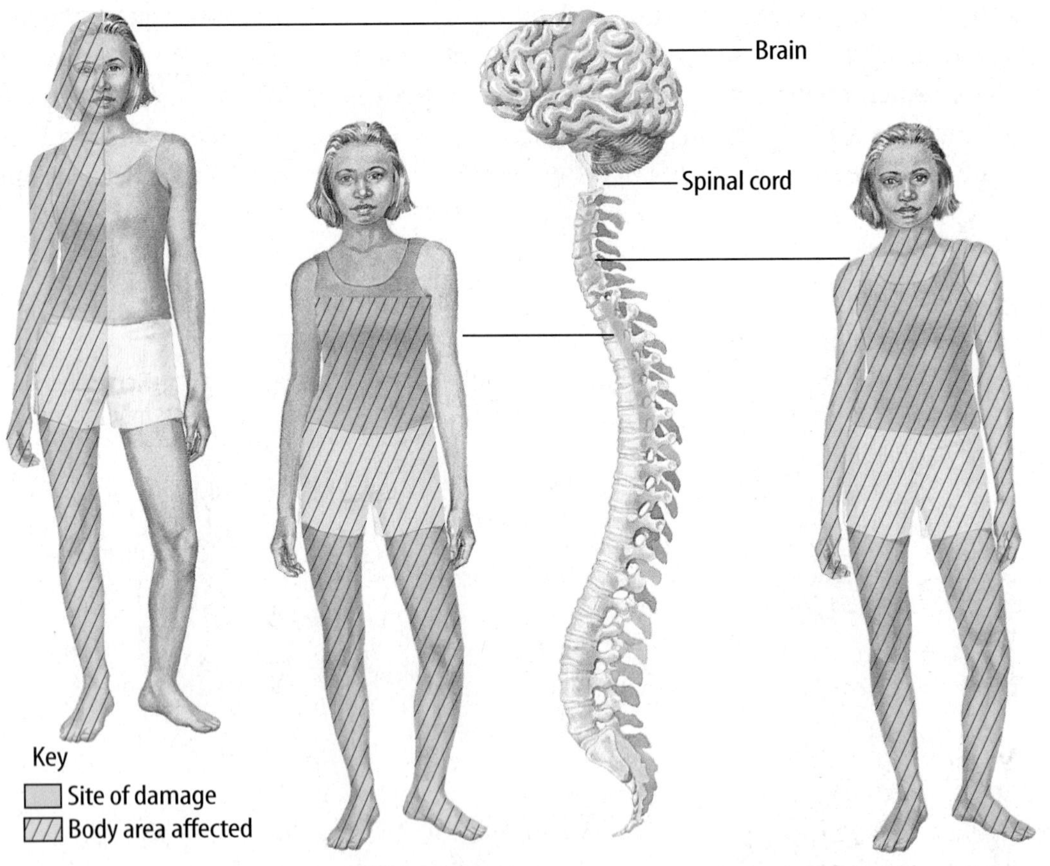

Damage to one side of the brain can result in the paralysis of the opposite side of the body.

Damage to the middle or lower spinal cord can result in the legs and possibly part of the torso being paralyzed.

Damage to the spinal cord in the lower neck area can cause the body to be paralyzed from the neck down.

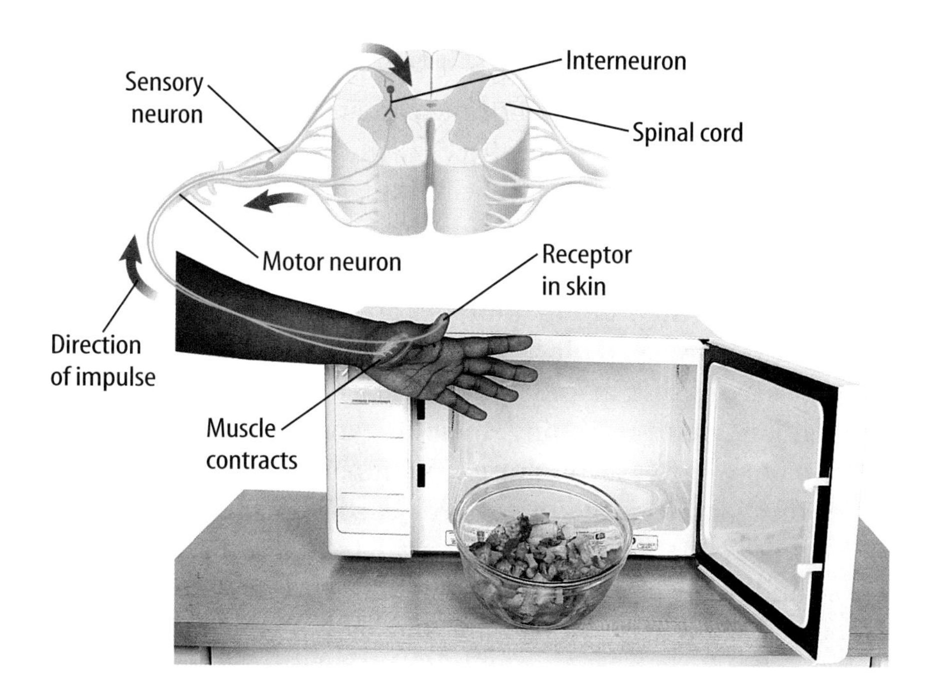

Figure 9 Your response in a reflex is controlled in your spinal cord, not your brain.

Reflexes You experience a reflex if you accidentally touch something sharp, something extremely hot or cold, or when you cough or vomit. A **reflex** is an involuntary, automatic response to a stimulus. You can't control reflexes because they occur before you know what has happened. A reflex involves a simple nerve pathway called a reflex arc, as illustrated in **Figure 9.**

Imagine that while walking on a sandy beach, a pain suddenly shoots through your foot as you step on the sharp edge of a broken shell. Sensory receptors in your foot respond to this sharp object, and an impulse is sent to the spinal cord. As you just learned, the impulse passes to an interneuron in the spinal cord that immediately relays the impulse to motor neurons. Motor neurons transmit the impulse to muscles in your leg. Instantly, without thinking, you lift up your leg in response to the sharp-edged shell. This is a withdrawal reflex.

A reflex allows the body to respond without having to think about what action to take. Reflex responses are controlled in your spinal cord, not in your brain. Your brain acts after the reflex to help you figure out what to do to make the pain stop.

Topic: Reflexes and Paralysis

Visit life.msscience.com for Web links to information about reflexes and paralysis.

Activity Make a small poster that illustrates what you learn.

Reading Check *Why are reflexes important?*

Do you remember reading at the beginning of this chapter about being frightened after a lamp was broken? What would have happened if your breathing and heart rate didn't calm down within a few minutes? Your body systems can't be kept in a state of continual excitement. The organs of your nervous system control and coordinate body responses. This helps maintain homeostasis within your body.

Figure 10 Caffeine, a substance found in colas, coffee, chocolate, and some teas, can cause excitability and sleeplessness.

Drugs and the Nervous System

Many drugs, such as alcohol and caffeine, directly affect your nervous system. When swallowed, alcohol passes directly through the walls of the stomach and small intestine into the circulatory system. After it is inside the circulatory system, it can travel throughout your body. Upon reaching neurons, alcohol moves through their cell membranes and disrupts their normal cell functions. As a result, this drug slows the activities of the central nervous system and is classified as a depressant. Muscle control, judgment, reasoning, memory, and concentration also are impaired. Heavy alcohol use destroys brain and liver cells.

A stimulant is a drug that speeds up the activity of the central nervous system. Caffeine is a stimulant found in coffee, tea, cocoa, and many soft drinks, as shown in **Figure 10.** Too much caffeine can increase heart rate and aggravates restlessness, tremors, and insomnia in some people. It also can stimulate the kidneys to produce more urine.

Think again about a scare from a loud noise. The organs of your nervous system control and coordinate responses to maintain homeostasis within your body. This task might be more difficult when your body must cope with the effects of drugs.

section 1 review

Summary

How the Nervous System Works

- The nervous system responds to stimuli to maintain homeostasis.
- To move from one neuron to another, an impulse crosses a synapse.

The Central Nervous System

- The brain controls all body activities.
- Spinal neurons carry impulses from all parts of the body to the brain.

The Peripheral Nervous System

- The somatic system controls voluntary actions and the autonomic system controls involuntary actions.

Safety and the Nervous System

- The spinal cord controls reflex responses.

Drugs and the Nervous System

- Many drugs affect your nervous system.

Self Check

1. **Draw and label** the parts of a neuron.
2. **Compare and contrast** the central and peripheral nervous systems.
3. **Explain** why you have trouble falling asleep after drinking several cups of hot cocoa.
4. **Explain** the advantage of having reflexes controlled by the spinal cord.
5. **Think Critically** Explain why many medications caution the consumer not to operate heavy machinery.

Applying Skills

6. **Concept Map** Prepare an events-chain concept map of the different kinds of neurons that pass an impulse from a stimulus to a response.
7. **Use a Word Processor** Create a flowchart showing the reflex pathway of a nerve impulse when you step on a sharp object. Label the body parts involved in each step.

life.msscience.com/self_check_quiz

IMPROVING REACTION TIME

Your reflexes allow you to react quickly without thinking. Sometimes you can improve how quickly you react. Complete this lab to see if you can decrease your reaction time.

Real-World Question

How can reaction time be improved?

Goals

- **Observe** reflexes.
- **Identify** stimuli and responses.

Materials

metric ruler

Procedure

1. Make a data table in your Science Journal to record where the ruler is caught during this lab. Possible column heads are *Trial, Right Hand,* and *Left Hand.*
2. Have a partner hold the ruler as shown.
3. Hold the thumb and index finger of your right hand apart at the bottom of the ruler. Do not touch the ruler.
4. Your partner must let go of the ruler without warning you.
5. Catch the ruler between your thumb and finger by quickly bringing them together.
6. Repeat this lab several times and record in a data table where the ruler was caught.
7. Repeat this lab with your left hand.

Conclude and Apply

1. **Identify** the stimulus, response, and variable in this lab.
2. Use the table on the right to determine your reaction time.
3. **Calculate** the average reaction times for both your right and left hand.
4. **Compare** the response of your writing hand and your other hand for this lab.
5. Draw a conclusion about how practice relates to stimulus-response time.

Reaction Time

Where Caught (cm)	Reaction Time(s)
5	0.10
10	0.14
15	0.17
20	0.20
25	0.23
30	0.25

Communicating Your Data

Compare your conclusions with those of other students in your class. **For more help, refer to the Science Skill Handbook.**

section 2

The Senses

as you read

What You'll Learn

- **List** the sensory receptors in each sense organ.
- **Explain** what type of stimulus each sense organ responds to and how.
- **Explain** why healthy senses are needed.

Why It's Important

Your senses make you aware of your environment, enable you to enjoy your world, and help keep you safe.

Review Vocabulary

sense organ: specialized organ that, when stimulated, initiates a process of sensory perception

New Vocabulary

- retina
- cochlea
- olfactory cell
- taste bud

The Body's Alert System

"Danger . . . danger . . . code-red alert! An unidentified vessel has entered the spaceship's energy force field. All crew members are to be on alert!" Like spaceships in science fiction movies, your body has an alert system, too—your sense organs. You might see a bird, hear a dog bark, or smell popcorn. You can enjoy the taste of salt on a pretzel, the touch of a fuzzy peach, or feel warmth of a cozy fire. Light rays, sound waves, thermal energy, chemicals, or pressure that comes into your personal territory will stimulate your sense organs. Sense organs are adapted for intercepting these different stimuli. They are then converted into impulses by the nervous system.

Vision

Think about the different kinds of objects you might look at every day. It's amazing that with one glance you might see the words on this page, the color illustrations, and your classmate sitting next to you. The eye, shown in **Figure 11,** is the vision sense organ. Your eyes have unique adaptations that usually enable you to see shapes of objects, shadows, and color.

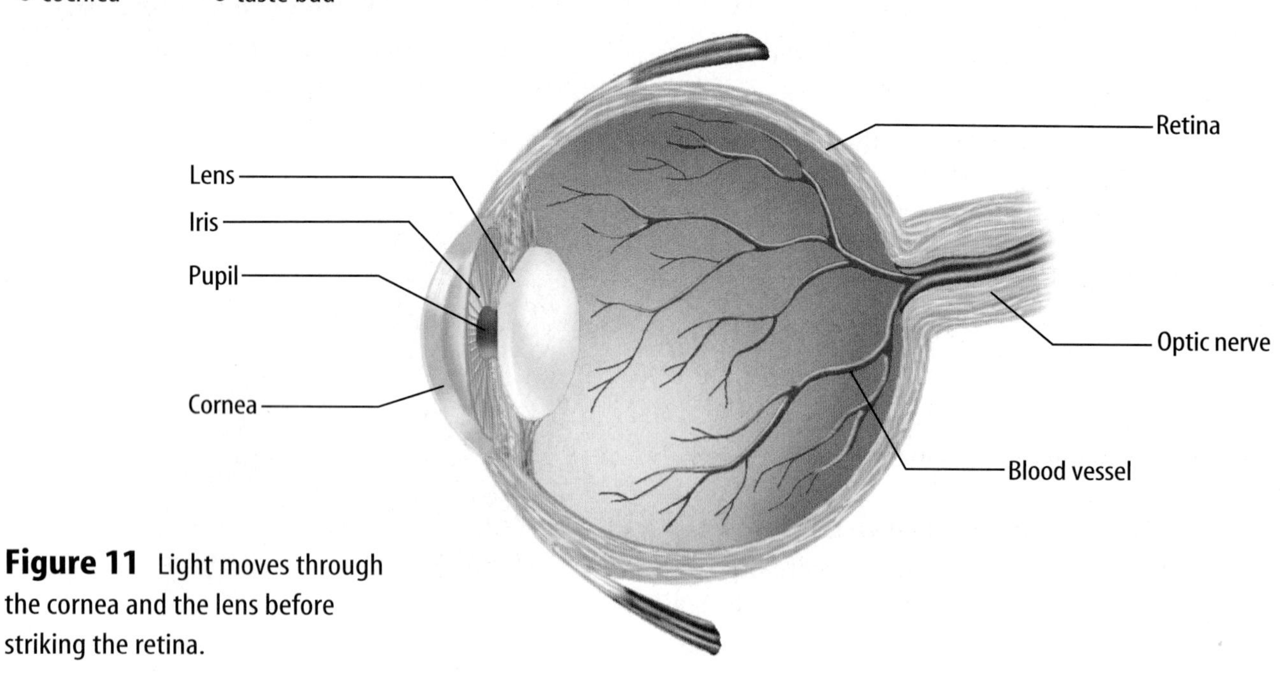

Figure 11 Light moves through the cornea and the lens before striking the retina.

How do you see? Light travels in a straight line unless something causes it to refract or change direction. Your eyes are equipped with structures that refract light. Two of these structures are the cornea and the lens. As light enters the eye, it passes through the cornea—the transparent section at the front of the eye—and is refracted. Then light passes through a lens and is refracted again. The lens directs the light onto the retina (RET nuh). The **retina** is a tissue at the back of the eye that is sensitive to light energy. Two types of cells called rods and cones are found in the retina. Cones respond to bright light and color. Rods respond to dim light. They are used to help you detect shape and movement. Light energy stimulates impulses in these cells.

The impulses pass to the optic nerve. This nerve carries the impulses to the vision area of the cortex, located on your brain's cerebrum. The image transmitted from the retina to the brain is upside down and reversed. The brain interprets the image correctly, and you see what you are looking at. The brain also interprets the images received by both eyes. It blends them into one image that gives you a sense of distance. This allows you to tell how close or how far away something is.

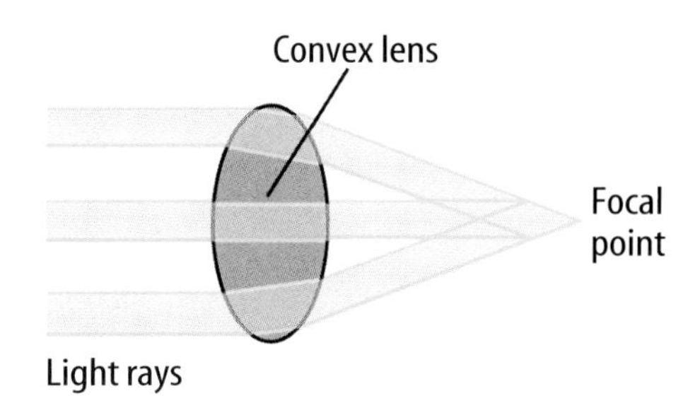

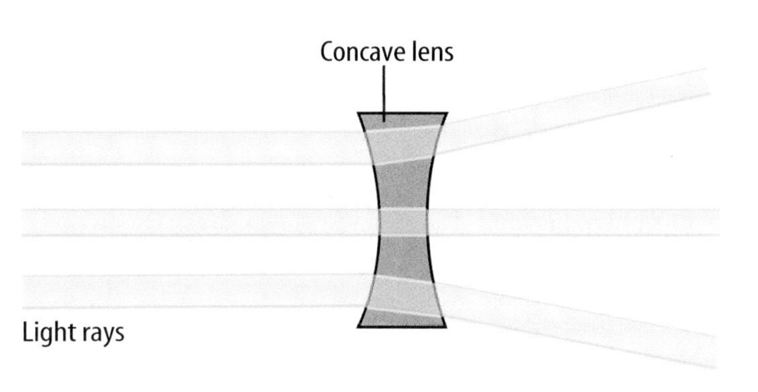

Figure 12 Light passing through a convex lens is refracted toward the center and passes through a focal point. Light that passes through a concave lens is refracted outward.
Name *the type of lens found in a microscope.*

✔ Reading Check *What difficulties would a person who had vision only in one eye encounter?*

Lenses

Light is refracted when it passes through a lens. The way it refracts depends on the type of lens it passes through. A lens that is thicker in the middle and thinner on the edges is called a convex lens. As shown in **Figure 12,** the lens in your eye refracts light so that it passes through a point, called a focal point. Convex lenses can be used to magnify objects. The light passes through a convex lens and enters the eye in such a way that your brain interprets the image as enlarged.

A lens that is thicker at its edges than in its middle is called a concave lens. Follow the light rays in **Figure 12** as they pass through a concave lens. You'll see that this kind of lens causes the parallel light to spread out.

Figure 13 Glasses and contact lenses sharpen your vision.

A nearsighted person cannot see distant objects because the image is focused in front of the retina.

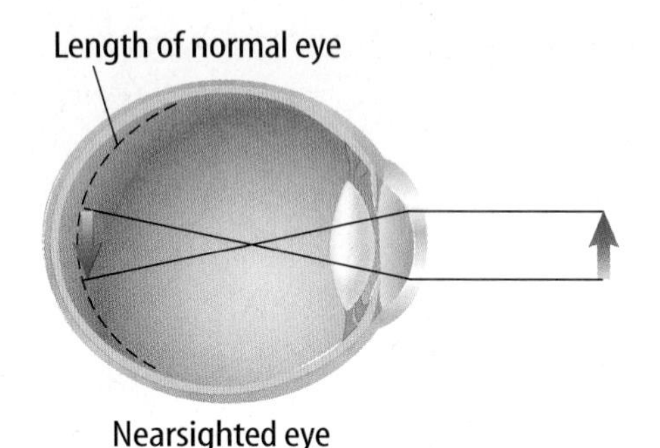

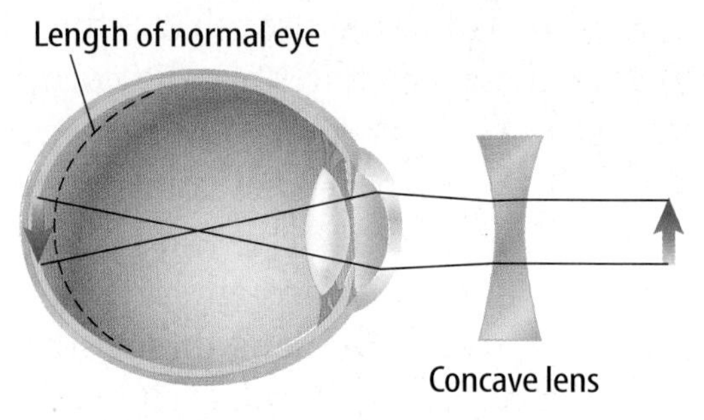

A concave lens corrects nearsightedness.

A farsighted person cannot see close objects because the image is focused behind the retina.

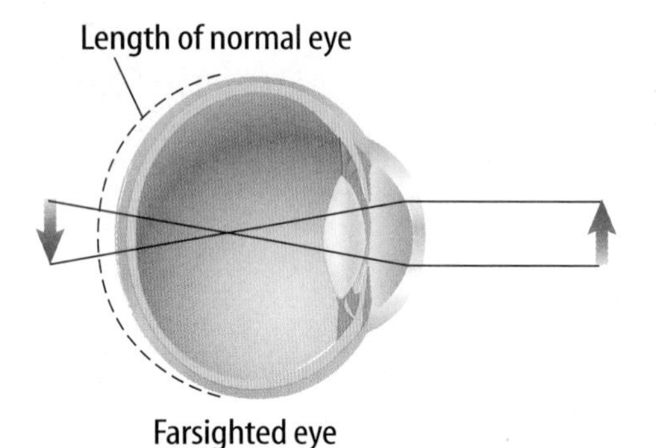

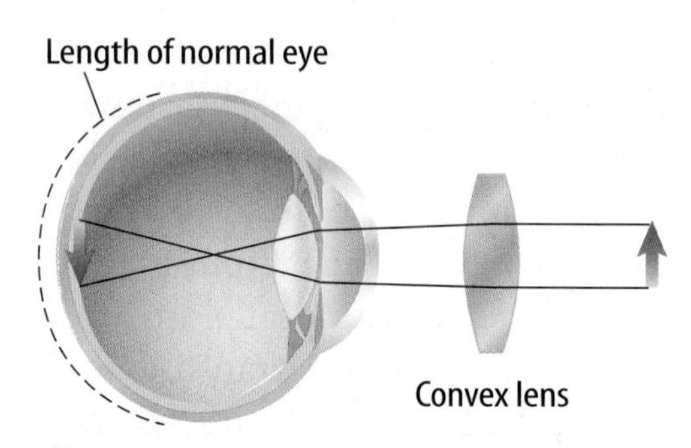

A convex lens corrects farsightedness.

Correcting Vision Problems Do you wear contact lenses or eyeglasses to correct your vision? Are you nearsighted or farsighted? In an eye with normal vision, light rays are focused onto the retina by the coordinated actions of the eye muscles, the cornea, and the lens. The image formed on the retina is interpreted by the brain as being sharp and clear. However, if the eyeball is too long from front to back, as illustrated in **Figure 13,** light from objects is focused in front of the retina. This happens because the shape of the eyeball and lens cannot be changed enough by the eye muscles to focus a sharp image onto the retina. The image that reaches the retina is blurred. This condition is called nearsightedness—near objects are seen more clearly than distant objects. To correct nearsightedness, concave lenses are used to help focus images sharply on the retina.

Similarly, vision correction is needed when the eyeball is too short from front to back. In this case, light from objects is focused behind the retina despite the coordinated actions of the eye muscles, cornea, and lens. This condition is called farsightedness, also as illustrated in **Figure 13,** because distant objects are clearer than near objects. Convex lenses correct farsightedness.

Telescopes Refracting telescopes have two convex lenses for viewing objects in space. The larger lens collects light and forms an inverted, or upside-down, image of the object. The second lens magnifies the inverted image. In your Science Journal, hypothesize why telescopes used to view things on Earth have three lenses, not two.

Hearing

Whether it's the roar of a rocket launch, the cheers at a football game, or the distant song of a robin in a tree, sound waves are necessary for hearing sound. Sound energy is to hearing as light energy is to vision. When an object vibrates, sound waves are produced. These waves can travel through solids, liquids, and gases as illustrated in **Figure 14.** When the waves reach your ear, they usually stimulate nerve cells deep within your ear. Impulses are sent to the brain. When the sound impulse reaches the hearing area of the cortex, it responds and you hear a sound.

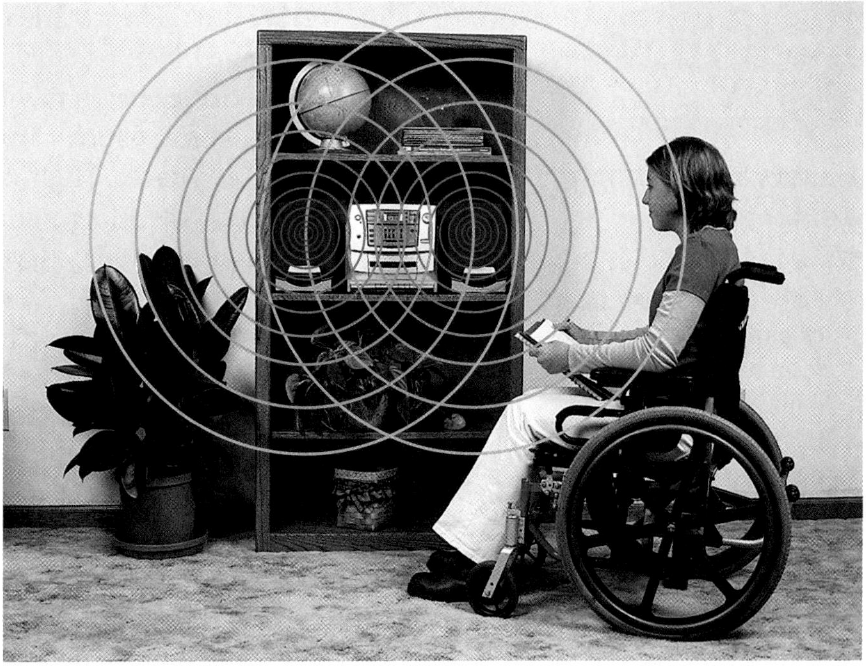

Figure 14 Objects produce sound waves that can be heard by your ears.

The Outer Ear and Middle Ear **Figure 15** shows that your ear is divided into three sections—the outer ear, middle ear, and inner ear. Your outer ear intercepts sound waves and funnels them down the ear canal to the middle ear. The sound waves cause the eardrum to vibrate much like the membrane on a musical drum vibrates when you tap it. These vibrations then move through three tiny bones called the hammer, anvil, and stirrup. The stirrup bone rests against a second membrane on an opening to the inner ear.

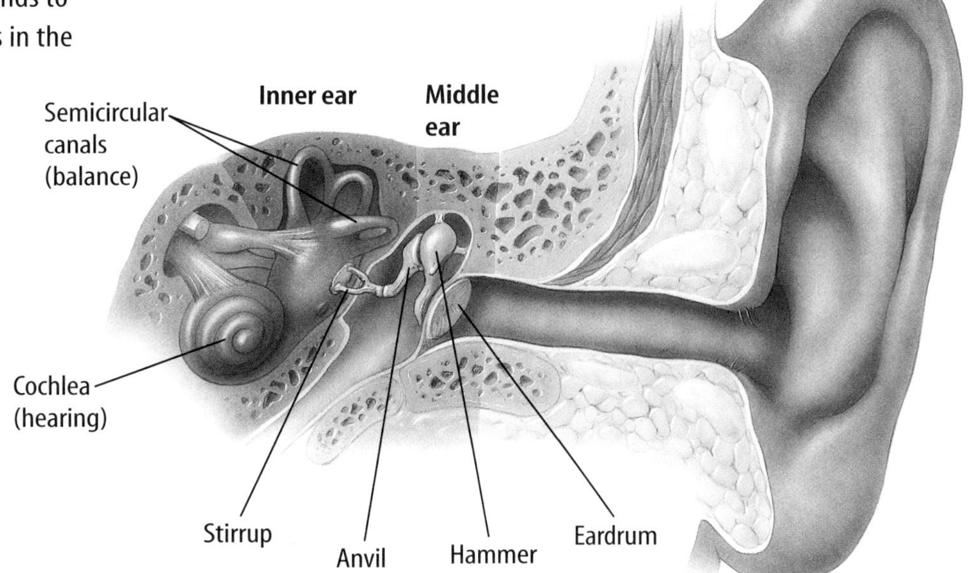

Figure 15 Your ear responds to sound waves and to changes in the position of your head.

Observing Balance Control

Procedure

1. Place **two narrow strips of paper** on the wall to form two parallel vertical lines 20–25 cm apart. Have a person stand between them for 3 min, without leaning on the wall.
2. Observe how well balance is maintained.
3. Have the person close his or her eyes, then stand within the lines for 3 min.

Analysis

1. When was balance more difficult to maintain? Why?
2. What other factors might cause a person to lose his or her sense of balance?

Try at Home

The Inner Ear The **cochlea** (KOH klee uh) is a fluid-filled structure shaped like a snail's shell. When the stirrup vibrates, fluids in the cochlea begin to vibrate. These vibrations bend hair cells in the cochlea, which causes electrical impulses to be sent to the brain by a nerve. High-pitched sounds make the endings move differently than lower sounds do. Depending on how the nerve endings are stimulated, you hear a different type of sound.

Balance Structures in your inner ear also control your body's balance. Structures called the cristae ampullaris (KRIHS tee • am pyew LEER ihs) and the maculae (MA kyah lee), illustrated in **Figure 16,** sense different types of body movement.

Both structures contain tiny hair cells. As your body moves, gel-like fluid surrounding the hair cells moves and stimulates the nerve cells at the base of the hair cells. This produces nerve impulses that are sent to the brain, which interprets the body movements. The brain, in turn, sends impulses to skeletal muscles, resulting in other body movements that maintain balance.

The cristae ampullaris react to rotating body movements. Fluid in the semicircular canals swirls when the body rotates. This causes the gel-like fluid around the hair cells to move and a stimulus is sent to the brain. In a similar way, when the head tips, the gel-like fluid surrounding the hair cells in the maculae is pulled down by gravity. The hair cells are then stimulated and the brain interprets that the head has tilted.

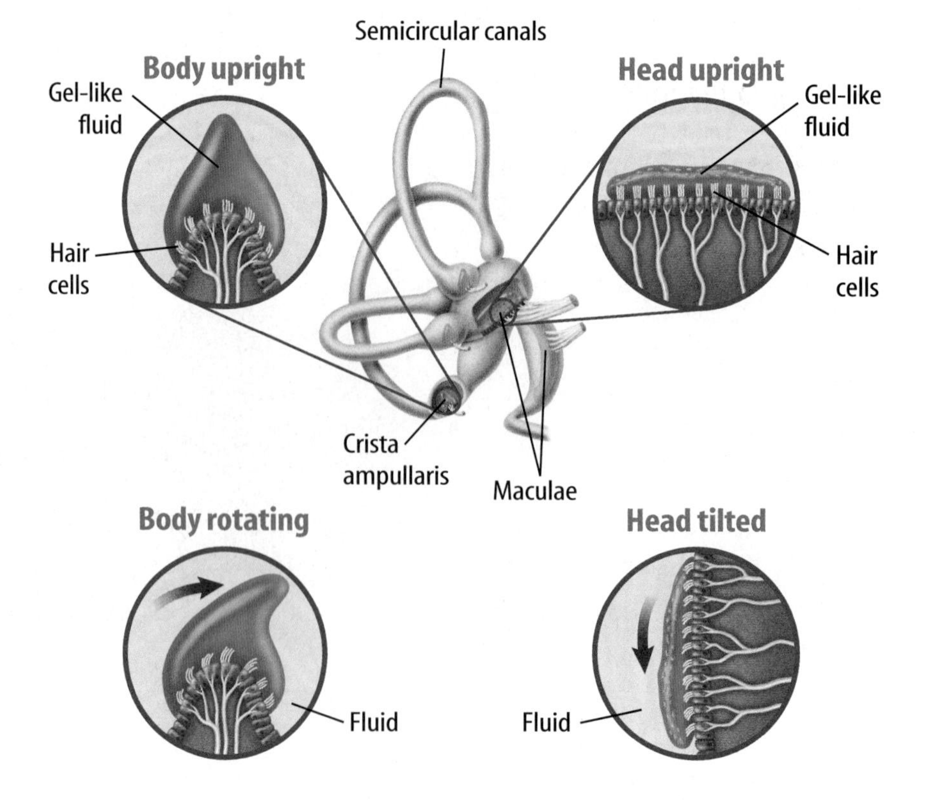

Figure 16 In your inner ear, the cristae ampullaris react to rotating movements of your body, and the maculae check the position of your head with respect to the ground.
Explain *why spinning around makes you dizzy.*

Smell

Some sharks can sense as few as ten drops of tuna liquid in an average-sized swimming pool. Even though your ability to detect odors is not as good as a shark's, your sense of smell is still important. Smell can determine which foods you eat. Strong memories or feelings also can be responses to something you smell.

You smell food because it gives off molecules into the air. These molecules stimulate sensitive nerve cells, called **olfactory** (ohl FAK tree) **cells,** in your nasal passages. Olfactory cells are kept moist by mucus. When molecules in the air dissolve in this moisture, the cells become stimulated. If enough molecules are present, an impulse starts in these cells, then travels to the brain where the stimulus is interpreted. If the stimulus is recognized from a previous experience, you can identify the odor. If you don't recognize a particular odor, it is remembered and may be identified the next time you encounter it.

Topic: Sense of Smell
Visit life.msscience.com for Web links to information about the sense of smell in humans compared to that of other mammals.

Activity In your Science Journal, summarize your research.

Applying Math — Solve a One-Step Equation

SPEED OF SOUND You see the flash of fireworks and then four seconds later, you hear the boom because light waves travel faster than sound waves. Light travels so fast that you see it almost instantaneously. Sound, on the other hand, travels at 340 m/s. How far away are you from the source of the fireworks?

Solution

1 *This is what you know:*
- time: $t = 4$ s; speed of sound: $v = 340$ m/s

2 *This is what you need to find out:*
How far are you away from the fireworks?

3 *This is the procedure you need to use:*
- Use the equation: $d = vt$
- Substitute known values and solve:
  $d = (340 \text{ m/s})(4 \text{ s})$
  $d = 1360 \text{ m}$

4 *Check your answer:*
Divide your answer by time. You should get speed.

Practice Problems

1. A hiker standing at one end of a lake hears his echo 2.5 s after he shouts. It was reflected by a cliff at the end of the lake. How long is the lake?
2. If you see a flash of lightning during a thunderstorm and it takes 5 s to hear the thunder, how far away is the lightning?

Science Online — For more practice, visit life.msscience.com/math_practice

Comparing Sense of Smell

Procedure

1. To test your classmates' abilities to recognize different odors, blindfold them one at a time, then pass near their noses small **samples of different foods, colognes, or household products.** **WARNING:** *Do not eat or drink anything in the lab. Do not use any products that give off noxious fumes.*
2. Ask each student to identify the different samples.
3. Record each student's response in a data table according to his or her gender.

Analysis

1. Compare the numbers of correctly identified odors for males and females.
2. What can you conclude about the differences between males and females in their abilities to recognize odors?

Taste

Sometimes you taste a new food with the tip of your tongue and find that it tastes sweet. Then when you chew it, you are surprised to find that it tastes bitter. **Taste buds** on your tongue, like the one in **Figure 17,** are the major sensory receptors for taste. About 10,000 taste buds are found all over your tongue, enabling you to tell one taste from another.

Tasting Food Taste buds respond to chemical stimuli. Most taste buds respond to several taste sensations. However, certain areas of the tongue are more receptive to one taste than another. The five taste sensations on the tongue are sweet, salty, sour, bitter, and the taste of MSG (monosodium glutamate). When you think of hot french fries, your mouth begins to water. This response is helpful because in order to taste something, it has to be dissolved in water. Saliva begins this process. This solution of saliva and food washes over the taste buds, and impulses are sent to your brain. The brain interprets the impulses, and you identify the tastes.

Reading Check *What needs to happen to food before you are able to taste it?*

Smell and Taste Smell and taste are related. The sense of smell is needed to identify some foods such as chocolate. When saliva in your mouth mixes with the chocolate, odors travel up the nasal passage in the back of your throat. The olfactory cells are stimulated, and the taste and smell of chocolate are sensed. So when you have a stuffy nose and some foods seem tasteless, it may be because the food's molecules are blocked from contacting the olfactory cells in your nasal passages.

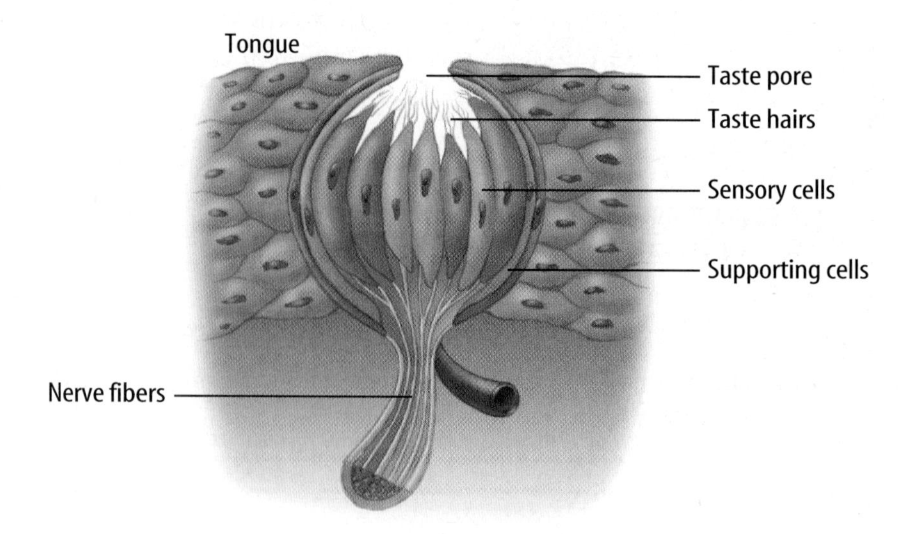

Figure 17 Taste buds are made up of a group of sensory cells with tiny taste hairs projecting from them. When food is taken into the mouth, it is dissolved in saliva. This mixture then stimulates receptor sites on the taste hairs, and an impulse is sent to the brain.

Other Sensory Receptors in the Body

As you are reading at school, you suddenly experience a bad pain in your lower right abdomen. The pain is not going away and you yell for help. Several hours later, you are resting in a hospital bed. The doctor has removed the source of your problem—your appendix. If not removed, a burst appendix can spread poison throughout your body.

Your internal organs have several kinds of sensory receptors. These receptors respond to touch, pressure, pain, and temperature. They pick up changes in touch, pressure, and temperature and transmit impulses to the brain or spinal cord. In turn, your body responds to this new information.

Sensory receptors also are located throughout your skin. As shown in **Figure 18,** your fingertips have many different types of receptors for touch. As a result, you can tell whether an object is rough or smooth, hot or cold, and hard or soft. Your lips are sensitive to temperature and prevent you from drinking something so hot that it would burn you. Pressure-sensitive skin cells warn you of danger and enable you to move to avoid injury.

The body responds to protect itself from harm. All of your body's senses work together to maintain homeostasis. Your senses help you enjoy or avoid things around you. You constantly react to your environment because of information received by your senses.

Figure 18 Many of the sensations picked up by receptors in the skin are stimulated by mechanical energy. Pressure, motion, and touch are examples.

section 2 review

Summary

Vision

- Light causes impulses that pass to the optic nerve. Your brain interprets the image.

Lenses

- Convex lenses and concave lenses are used to correct vision.

Hearing

- Sound waves stimulate nerve cells in the inner ear.
- Structures in the inner ear sense body movements.

Smell

- Molecules in the air stimulate nasal nerve cells, which allow you to smell.

Taste

- Taste buds are sensory receptors.

Self Check

1. **List** the types of stimuli your ears respond to.
2. **Describe** the sensory receptors for the eyes and nose.
3. **Explain** why it is important to have sensory receptors for pain and pressure in your internal organs.
4. **Outline** the role of saliva in tasting.
5. **Think Critically** Unlike many other organs, the brain is insensitive to pain. What is the advantage of this?

Applying Skills

6. **Make and Use Tables** Organize the information on senses in a table that names the sense organs and which stimuli they respond to.
7. **Communicate** Write a paragraph in your Science Journal that describes what each of the following objects would feel like: ice cube, snake, silk blouse, sandpaper, jelly, and smooth rock.

Design Your Own

Skin Sensitivity

Goals

- **Observe** the sensitivity to touch on specific areas of the body.
- **Design** an experiment that tests the effects of a variable, such as how close the contact points are, to determine which body areas can distinguish which stimuli are closest to one another.

Possible Materials

3-in × 5-in index card
toothpicks
tape
glue
metric ruler
**Alternate materials*

Safety Precautions

WARNING: *Do not apply heavy pressure when touching the toothpicks to the skin of your classmates.*

Real-World Question

Your body responds to touch, pressure, temperature, and other stimuli. Not all parts of your body are equally sensitive to stimuli. Some areas are more sensitive than others are. For example, your lips are sensitive to temperature. This protects you from burning your mouth and tongue. Now think about touch. How sensitive is the skin on various parts of your body to touch? Which areas can distinguish the smallest amount of distance between stimuli? What areas of the body are most sensitive to touch?

Form a Hypothesis

Based on your experiences, state a hypothesis about which of the following five areas of the body—fingertip, forearm, back of the neck, palm, and back of the hand—you believe to be most sensitive. Rank the areas from 5 (most sensitive) to 1 (least sensitive).

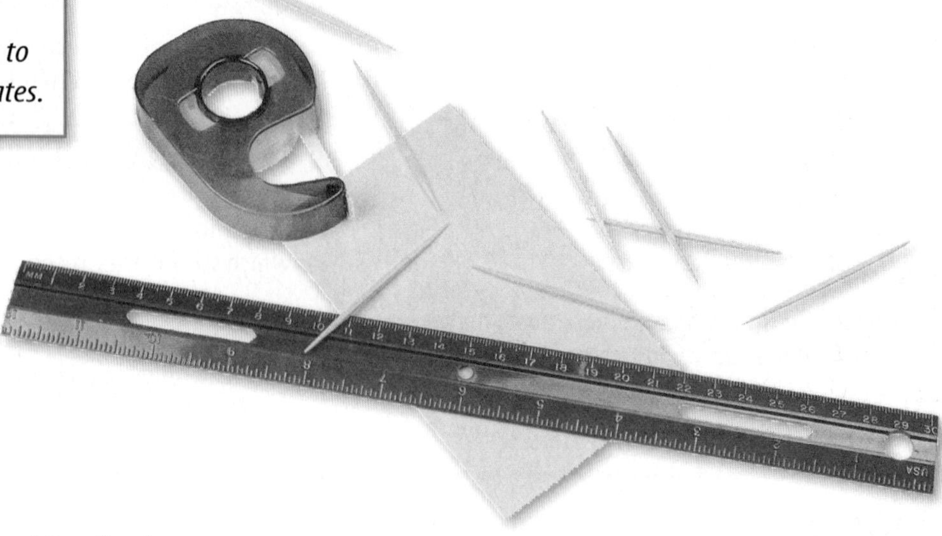

Test Your Hypothesis

Make a Plan

1. As a group, agree upon and write the hypothesis statement.
2. As a group, list the steps you need to test your hypothesis. Describe exactly what you will do at each step. Consider the following as you list the steps. How will you know that sight is not a factor? How will you use the card shown on the right to determine sensitivity to touch? How will you determine that one or both points are sensed?
3. **Design** a data table in your Science Journal to record your observations.
4. Reread your entire experiment to make sure that all steps are in the correct order.
5. **Identify** constants, variables, and controls of the experiment.

Follow Your Plan

1. Make sure your teacher approves your plan before you start.
2. Carry out the experiment as planned.
3. While the experiment is going on, write down any observations that you make and complete the data table in your Science Journal.

Analyze Your Data

1. **Identify** which part of the body is least sensitive and which part is most sensitive.
2. **Identify** which part of the body tested can distinguish between the closest stimuli.
3. **Compare** your results with those of other groups.
4. Rank body parts tested from most to least sensitive. Did your results from this investigation support your hypothesis? Explain.

Conclude and Apply

1. Based on the results of your investigation, what can you infer about the distribution of touch receptors on the skin?
2. What other parts of your body would you predict to be less sensitive? Explain your predictions.

Communicating Your Data

Write a report to share with your class about body parts of animals that are sensitive to touch. **For more help, refer to the Science Skill Handbook.**

Science and Language Arts

Sula

by Toni Morrison

In the following passage from Sula, *a novel by Toni Morrison, the author describes Nel's response to the arrival of her old friend Sula.*

Nel alone noticed the peculiar quality of the May that followed the leaving of the birds. It had a sheen, a glimmering as of green, rain-soaked Saturday nights (lit by the excitement of newly installed street lights); of lemon-yellow afternoons bright with iced drinks and splashes of daffodils. It showed in the damp faces of her children and the river-smoothness of their voices. Even her own body was not immune to the magic. She would sit on the floor to sew as she had done as a girl, fold her legs up under her or do a little dance that fitted some tune in her head. There were easy sun-washed days and purple dusks . . . .

Although it was she alone who saw this magic, she did not wonder at it. She knew it was all due to Sula's return to the Bottom.

Understanding Literature

Diction and Tone An author's choice of words, or diction, can help convey a certain tone in the writing. In the passage, Toni Morrison's word choices help the reader understand that the character Nel is enjoying the month of May. Find two more examples in which diction conveys a pleasant tone.

Respond to the Reading

1. Describe, in your own words, how Nel feels about the return of her friend Sula.
2. What parts of the passage help you determine Nel's feelings?
3. **Linking Science and Writing** Write a paragraph describing the month of January that clearly shows a person's dislike for the month.

In the passage, Nel has a physical reaction to her environment. She is moved to "do a little dance" in response to the sights and sounds of May. This action is an example of a voluntary response to stimuli from outside the body. Movement of the body is a coordinated effort of the skeletal, muscular, and nervous system. Nel can dance because motor neurons conduct impulses from the brain to her muscles.

Reviewing Main Ideas

Section 1 The Nervous System

1. Your body constantly is receiving a variety of stimuli from inside and outside the body. The nervous system responds to these stimuli to maintain homeostasis.
2. A neuron is the basic unit of structure and function of the nervous system.
3. A stimulus is detected by sensory neurons. Electrical impulses are carried to the interneurons and transmitted to the motor neurons. The result is the movement of a body part.
4. A response that is made automatically is a reflex.
5. The central nervous system contains the brain and spinal cord. The peripheral nervous system is made up of cranial and spinal nerves.
6. Many drugs, such as alcohol and caffeine, have a direct effect on your nervous system.

Section 2 The Senses

1. Your senses respond to stimuli. The eyes respond to light energy, and the ears respond to sound waves.
2. Olfactory cells of the nose and taste buds of the tongue are stimulated by chemicals.
3. Sensory receptors in your internal organs and skin respond to touch, pressure, pain, and temperature.
4. Your senses enable you to enjoy or avoid things around you. You are able to react to the changing conditions of your environment.

Visualizing Main Ideas

Copy and complete the following concept map about the nervous system.

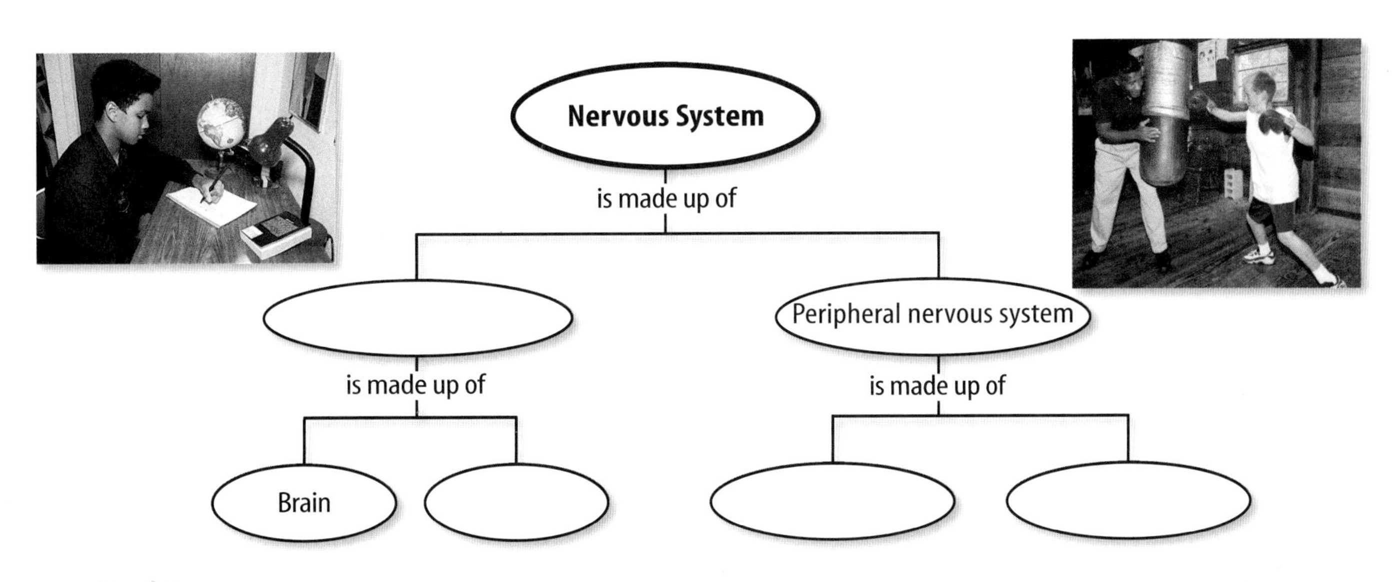

chapter 21 Review

Using Vocabulary

axon p. 595	neuron p. 595
brain stem p. 598	olfactory cell p. 609
central nervous system p. 597	peripheral nervous system p. 597
cerebellum p. 598	reflex p. 601
cerebrum p. 598	retina p. 605
cochlea p. 608	synapse p. 597
dendrite p. 595	taste bud p. 610
homeostasis p. 595	

Explain the difference between the vocabulary words in each of the following sets.

1. axon—dendrite
2. central nervous system—peripheral nervous system
3. cerebellum—cerebrum
4. reflex—synapse
5. brain stem—neuron
6. olfactory cell—taste bud
7. dendrite—synapse
8. cerebrum—central nervous system
9. retina—cochlea
10. synapse—neuron

Checking Concepts

Choose the word or phrase that best answers the question.

11. How do impulses cross synapses between neurons?
    A) by osmosis
    B) through interneurons
    C) through a cell body
    D) by a chemical

12. Which is in the inner ear?
    A) anvil **C)** eardrum
    B) hammer **D)** cochlea

13. What are neurons called that detect stimuli in the skin and eyes?
    A) interneurons **C)** motor neurons
    B) synapses **D)** sensory neurons

14. Which stimulus does the skin not sense?
    A) pain **C)** temperature
    B) pressure **D)** taste

15. What part of the brain controls voluntary muscles?
    A) cerebellum **C)** cerebrum
    B) brain stem **D)** pons

16. What part of the brain has an outer layer called the cortex?
    A) pons **C)** cerebrum
    B) brain stem **D)** spinal cord

17. What does the somatic system of the PNS control?
    A) skeletal muscles **C)** glands
    B) heart **D)** salivary glands

18. What part of the eye is light finally focused on?
    A) lens **C)** pupil
    B) retina **D)** cornea

19. What is the largest part of the brain?
    A) cerebellum **C)** cerebrum
    B) brain stem **D)** pons

Use the illustration below to answer question 20.

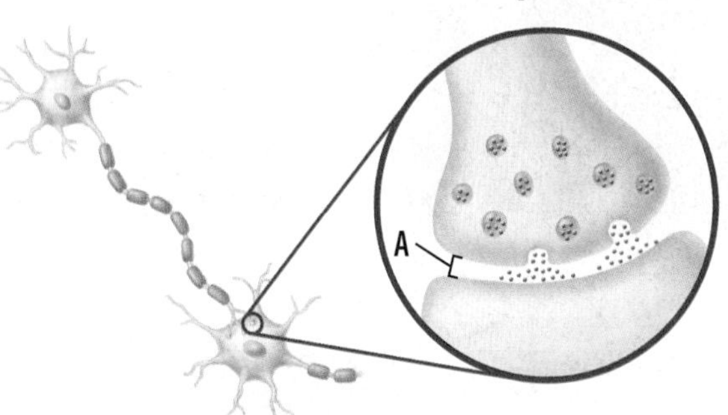

20. What is the name given to A?
    A) axon **C)** synapse
    B) dendrite **D)** nucleus

Science Online life.msscience.com/vocabulary_puzzlemaker

Thinking Critically

21. **Describe** why it is helpful to have impulses move only in one direction in a neuron.
22. **Describe** how smell and taste are related.
23. **Form a Hypothesis** If a fly were to land on your face and another one on your back, which might you feel first? How could you test your choice?
24. **Concept Map** Copy and complete this events-chain concept map to show the correct sequence of the structures through which light passes in the eye.

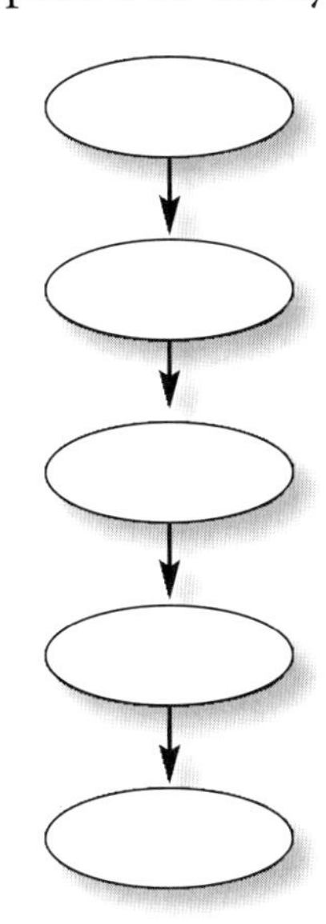

25. **Classify** Group the types of neurons as to their location and direction of impulse.
26. **Compare and contrast** the structures and functions of the cerebrum, cerebellum, and brain stem. Include in your discussion the following functions: balance, involuntary muscle movements, muscle tone, memory, voluntary muscles, thinking, and senses.
27. **Draw Conclusions** If an impulse traveled down one neuron but failed to move on to the next neuron, what might you conclude about the first neuron?

Performance Activities

28. **Illustrate** In an emergency room, the doctor notices that a patient has uncoordinated body movements and has difficulty maintaining his balance. Draw and label the part of the brain that may have been injured.

Applying Math

29. **Sound Waves** How deep is a cave if you shout into the cave and it takes 1.5 seconds to hear your echo? Remember that the speed of sound is 340 m/s.

Use the paragraph and table below to answer question 30.

A police officer brought the following table into a school to educate students about the dangers of drinking and driving.

Approximate Blood Alcohol Percentage for Men

Drinks	Body Weight in Kilograms							
	45.4	54.4	63.5	72.6	81.6	90.7	99.8	108.9
1	0.04	0.03	0.03	0.02	0.02	0.02	0.02	0.02
2	0.08	0.06	0.05	0.05	0.04	0.04	0.03	0.03
3	0.11	0.09	0.08	0.07	0.06	0.06	0.05	0.05
4	0.15	0.12	0.11	0.09	0.08	0.08	0.07	0.06
5	0.19	0.16	0.13	0.12	0.11	0.09	0.09	0.08

Subtract 0.01% for each 40 minutes of drinking. One drink is 40 mL of 80-proof liquor, 355 mL of beer, or 148 mL of table wine.

30. **Blood Alcohol** A 72-kg man has been tested for blood alcohol content. His blood alcohol percentage is 0.07. Based upon the information in the table above, about how much has he had to drink?

A) 628 mL of 80-proof liquor
B) 1,064 mL of beer
C) 295 mL of table wine
D) four drinks

Standardized Test Practice

Part 1 Multiple Choice

Record your answers on the answer sheet provided by your teacher or on a sheet of paper.

1. An internal or external change that brings about a response is called a
   A. reflex. **C.** receptor.
   B. stimulus. **D.** heartbeat.

Use the illustration below to answer questions 2–4.

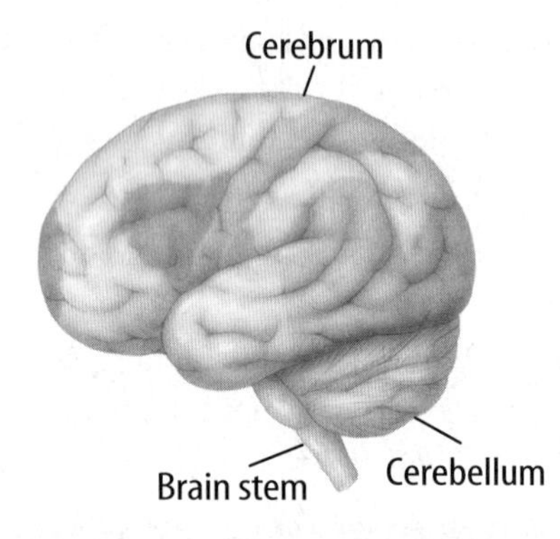

2. Which part of the brain helps in maintaining balance?
   A. cerebrum **C.** brain stem
   B. cerebellum **D.** no brain region

3. Which part of the brain controls involuntary actions?
   A. cerebrum **C.** brain stem
   B. cerebellum **D.** no brain region

4. In which part of the brain is memory stored?
   A. cerebrum **C.** brain stem
   B. cerebellum **D.** no brain region

5. Which structure helps control the body's balance?
   A. retina **C.** cochlea
   B. eardrum **D.** cristae ampullaris

Test-Taking Tip

Focus On Your Test During the test, keep your eyes on your own paper. If you need to rest them, close them or look up at the ceiling.

Use the table below to answer questions 6 and 7.

Approximate Blood Alcohol Percentage for Men

	Body Weight in Kilograms							
Drinks	45.4	54.4	63.5	72.6	81.6	90.7	99.8	108.9
1	0.04	0.03	0.03	0.02	0.02	0.02	0.02	0.02
2	0.08	0.06	0.05	0.05	0.04	0.04	0.03	0.03
3	0.11	0.09	0.08	0.07	0.06	0.06	0.05	0.05
4	0.15	0.12	0.11	0.09	0.08	0.08	0.07	0.06
5	0.19	0.16	0.13	0.12	0.11	0.09	0.09	0.08

Subtract 0.01% for each 40 minutes of drinking. One drink is 40 mL of 80-proof liquor, 355 mL of beer, or 148 mL of table wine.

6. In Michigan, it is illegal for drivers under 21 years of age to drink alcohol. Underage drivers can be arrested for drinking and driving if their blood alcohol percentage is more than 0.02 percent. According to this information, how many drinks would it take a 72-kg man to exceed this limit?
   A. three **C.** one
   B. two **D.** zero

7. In some states, the legal blood alcohol percentage limit for driving while under the influence of alcohol is 0.08 percent. According to this information, how many drinks would a 54-kg man have to consume to exceed this limit?
   A. four **C.** two
   B. three **D.** one

8. What neurons conduct impulses from the brain to glands?
   A. dendrites **C.** sensory neurons
   B. interneurons **D.** motor neurons

9. What carries impulses from the eye to the vision area of the brain?
   A. visual cortex **C.** sensory neurons
   B. optic nerve **D.** dendrites

Part 2 | Short Response/Grid In

Record your answers on the answer sheet provided by your teacher or on a sheet of paper.

10. Meningitis is an infection of the fluid that surrounds the spinal cord and the brain. Meningitis can be caused by bacteria or viruses. In bacterial meningitis, the amount of fluid in the brain and spinal cord may increase. What do you think might happen if the amount of fluid increases?

Use the table below to answer questions 11–13.

Number of Bicycle Deaths per Year		
Year	**Male**	**Female**
1996	654	107
1997	712	99
1998	658	99
1999	656	94
2000	605	76

Data from Insurance Institute for Highway Safety

11. Head injuries are the most serious injuries that are found in people who died in bicycle accidents. Ninety percent of the deaths were in people who were not wearing bicycle helmets. Using the data in the table, approximately how many of the people (male and female) who died in bicycle accidents in 1998 were wearing bicycle helmets?

12. In 2000, what percentage of the people who died were women?

13. Which of the years from 1996 to 2000 had the greatest total number of bicycle deaths?

14. Jeremy used a magnifying lens to study an earthworm. What kind of lens did he use? What part of this type of lens is the thickest?

15. Explain why alcohol is classified as a depressant.

Part 3 | Open Ended

Record your answers on a sheet of paper.

16. Compare and contrast the somatic and autonomic systems.

Use the illustration below to answer question 17.

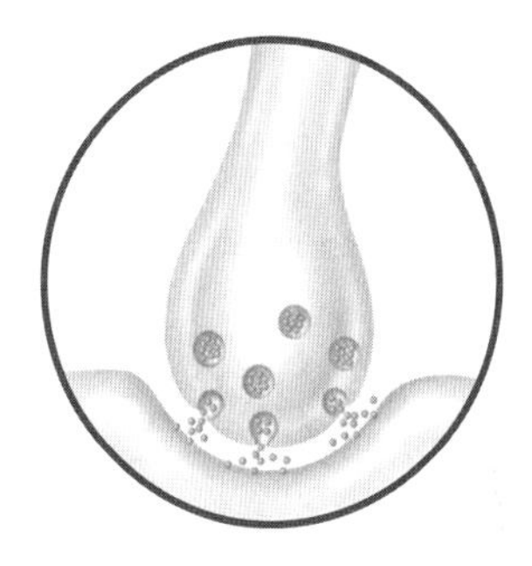

17. Identify the structure in the illustration. Explain what might happen if chemical was not released here.

18. Inez and Maria went to the ice cream parlor. They both ordered strawberry sundaes. Marie thought that the sundae was made with fresh strawberries, because it tasted so great. Inez thought that her sundae did not have much flavor. What could be the reason that Inez's sundae was tasteless? Explain why.

Use the illustration below to answer question 19.

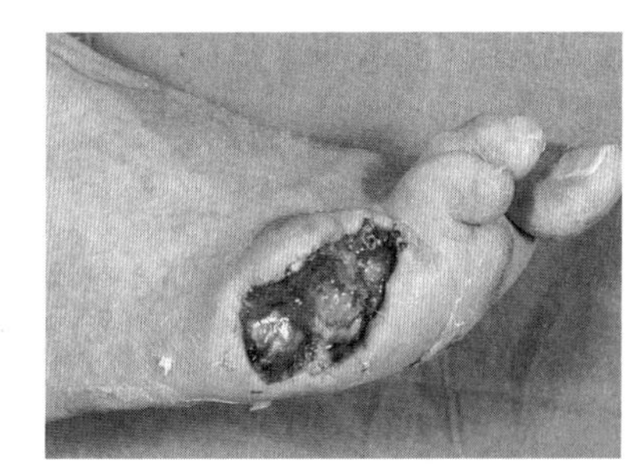

19. The person with this foot sore has diabetes. People with diabetes often lose sensation in their feet. Explain why a sore like the one in the photograph might develop if skin sensory receptors were not working properly.

chapter 22

Regulation and Reproduction

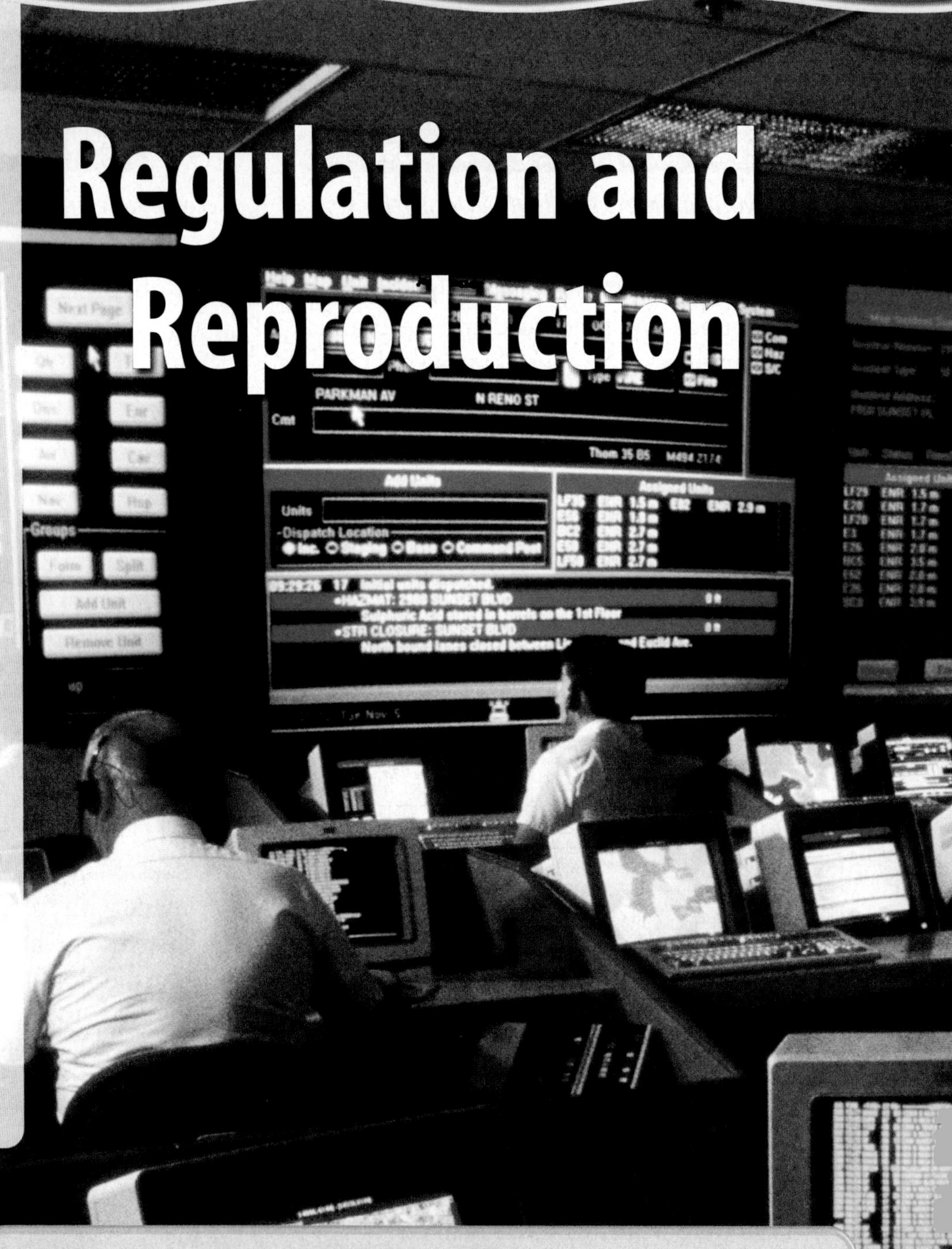

The BIG Idea

Human reproduction and growth and development involve the interactions of all body systems.

SECTION 1
The Endocrine System
Main Idea Hormones from endocrine glands affect many body functions including reproduction.

SECTION 2
The Reproductive System
Main Idea Males and females have different reproductive structures and functions.

SECTION 3
Human Life Stages
Main Idea Before birth and until death, a human changes continuously.

Where's the emergency?

This fire station control room has panels of blinking buttons and monitors. Dispatchers can access and relay emergency information quickly using this complex monitoring system. In a similar way, your body's endocrine system monitors and controls the actions of many of your body's functions.

Science Journal Write a paragraph describing how an emergency call might be handled at a fire station.

Start-Up Activities

Model a Chemical Message

Your body has systems that work together to coordinate your body's activities. One of these systems sends chemical messages through your blood to certain tissues, which, in turn, respond. Do the lab below to see how a chemical signal can be sent.

1. Cut a 10-cm-tall *Y* shape from filter paper and place it on a plastic, ceramic, or glass plate.
2. Sprinkle baking soda on one arm of the *Y* and salt on the other arm.
3. Using a dropper, place five or six drops of vinegar halfway up the leg of the *Y*.
4. **Think Critically** Describe in your Science Journal how the chemical moves along the paper and the reaction(s) it causes.

Preview this chapter's content and activities at
life.msscience.com

Stages of Life Make the following Foldable to help you predict the stages of life.

STEP 1 **Fold** a vertical sheet of paper in half from top to bottom. Then fold it in half again top to bottom two more times. Unfold all the folds.

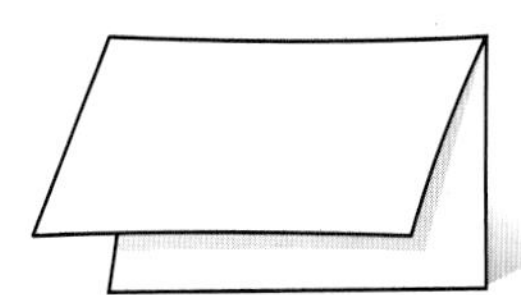

STEP 2 **Refold** the paper into a fan, using the folds as a guide. Unfold all the folds again.

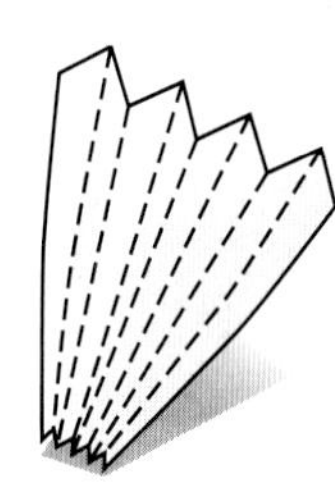

STEP 3 **Label** as shown.

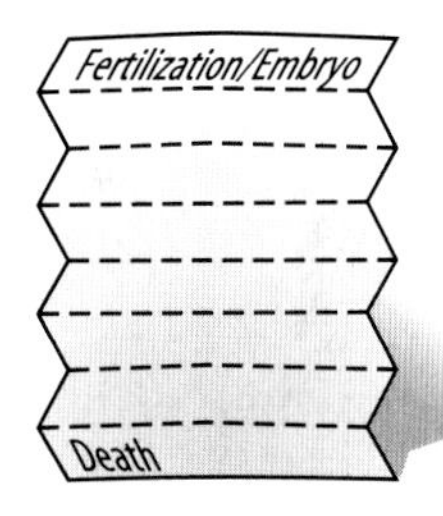

Read and Write Before you read the chapter, list as many stages of life as you can on your Foldable. Add to your list as you read the chapter.

Get Ready to Read

Make Connections

1 Learn It! Make connections between what you read and what you already know. Connections can be based on personal experiences (text-to-self), what you have read before (text-to-text), or events in other places (text-to-world).

As you read, ask connecting questions. Are you reminded of a personal experience? Have you read about the topic before? Did you think of a person, a place, or an event in another part of the world?

2 Practice It! Read the excerpt below and make connections to your own knowledge and experience.

Text-to-self: **How do hormones affect growth?**

Text-to-self: **How much have you grown in the last year? Has your growth spurt begun?**

Text-to-self: **Which of your friends have begun their growth spurts?**

> Adolescence usually is when the final growth spurt occurs. Because the time when hormones begin working varies among individuals and between males and females, growth rates differ. Girls often begin their final growth phase at about age 11 and end around age 16. Boys usually start their growth spurt at age 13 and end around 18 years of age.
>
> *—from page 640*

3 Apply It! As you read this chapter, choose five words or phrases that make a connection to something you already know.

Reading Tip

Make connections with memorable events, places, or people in your life. The better the connection, the more likely you will remember.

Target Your Reading

Use this to focus on the main ideas as you read the chapter.

1. **Before you read** the chapter, respond to the statements below on your worksheet or on a numbered sheet of paper.
   - Write an **A** if you **agree** with the statement.
   - Write a **D** if you **disagree** with the statement.

2. **After you read** the chapter, look back to this page to see if you've changed your mind about any of the statements.
   - If any of your answers changed, explain why.
   - Change any false statements into true statements.
   - Use your revised statements as a study guide.

Before You Read A or D	Statement	After You Read A or D
	1 One hormone can affect several types of tissues.	
	2 Chemical messages travel among endocrine glands and coordinate their functions.	
	3 The endocrine system regulates the function of the reproductive system.	
	4 Sperm form in the prostate gland.	
	5 The head of a sperm contains genetic material.	
	6 Eggs form in females before birth.	
	7 Fertilization of an egg by a sperm occurs in the uterus.	
	8 The monthly reproductive cycle of a female is menopause.	
	9 The umbilical cord connects the fetus to the mother.	
	10 Adulthood is the stage of development when a person stops growing.	

Science Online
Print out a worksheet of this page at life.msscience.com

section 1

The Endocrine System

as you read

What You'll Learn

- **Define** how hormones function.
- **Identify** different endocrine glands and the effects of the hormones they produce.
- **Describe** how a feedback system works in your body.

Why It's Important

The endocrine system uses chemicals to control many systems in your body.

Review Vocabulary

tissue: groups of cells that work together to perform a specific function

New Vocabulary

- hormone

Functions of the Endocrine System

You go through the dark hallways of a haunted house. You can't see a thing. Your heart is pounding. Suddenly, a monster steps out in front of you. You scream and jump backwards. Your body is prepared to defend itself or get away. Preparing the body for fight or flight in times of emergency, as shown in **Figure 1,** is one of the functions of the body's control systems.

Control Systems All of your body's systems work together, but the endocrine (EN duh krun) and the nervous systems are your body's control systems. The endocrine system sends chemical messages in your blood that affect specific tissues called target tissues. The nervous system sends rapid impulses to and from your brain, then throughout your body. Your body does not respond as quickly to chemical messages as it does to impulses.

Endocrine Glands

Tissues found throughout your body called endocrine glands produce the chemical messages called **hormones** (HOR mohnz). Hormones can speed up or slow down certain cellular processes. Some glands in your body release their products through small tubes called ducts. Endocrine glands are ductless and each endocrine gland releases its hormone directly into the blood. Then, the blood transports the hormone to the target tissue. A target tissue usually is located in the body far from the location of the endocrine gland that produced the hormone to which it responds.

Reading Check *What is the function of hormones?*

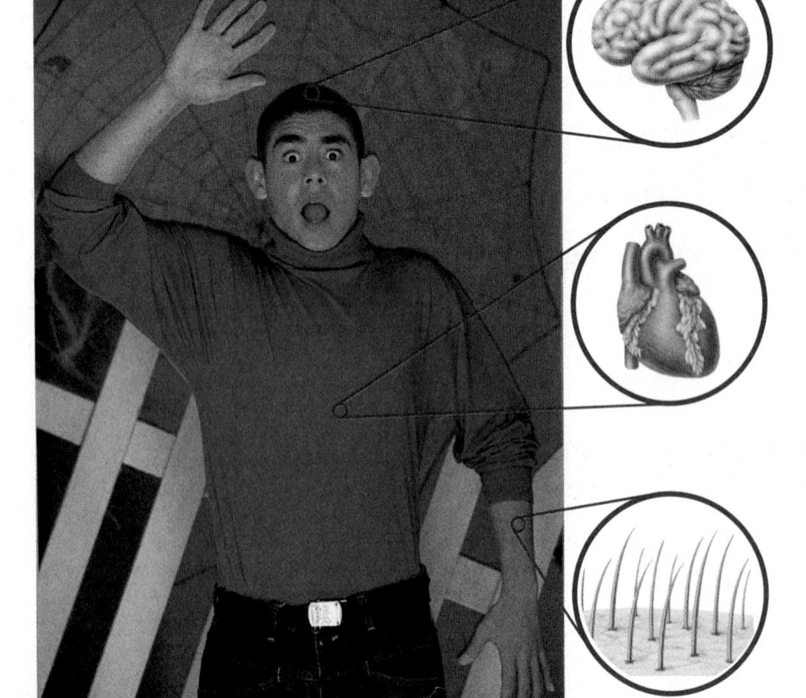

Figure 1 Your endocrine system enables many parts of your body to respond immediately in a fearful situation.

Gland Functions Endocrine glands have many functions in the body. The functions include the regulation of its internal environment, adaptation to stressful situations, promotion of growth and development, and the coordination of circulation, digestion, and the absorption of food. **Figure 2** on the next two pages shows some of the body's endocrine glands.

Applying Math Use Percentages

GLUCOSE LEVELS Calculate how much higher the blood sugar (glucose) level of a diabetic is before breakfast when compared to a nondiabetic before breakfast. Express this number as a percentage of the nondiabetic sugar level before breakfast.

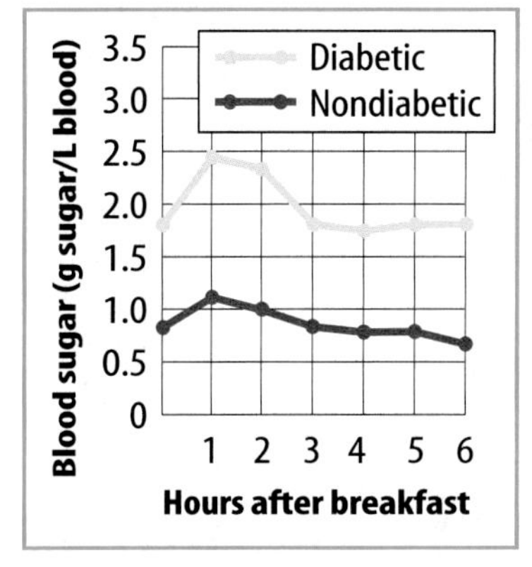

Solution

1 *This is what you know:*

- nondiabetic blood sugar at 0 h = 0.85 g sugar/L blood
- diabetic blood sugar at 0 h = 1.8 g sugar/L blood

2 *This is what you need to find out:*

How much higher is the glucose level of a diabetic person than that of a nondiabetic person before breakfast?

3 *This is the procedure you need to use:*

- Find the difference in glucose levels: 1.8 g/L − 0.85 g/L = 0.95 g/L
- Use this equation:

$$\frac{\text{difference between values}}{\text{nondiabetic value}} \times 100\% \ 5 \text{ percent difference}$$

- Substitute in the known values:

$$\frac{0.95}{0.85} \times 100\% = 112\%$$

- Before breakfast, a diabetic's blood sugar is about 112 percent higher than that of a nondiabetic.

4 *Check your answer:*

Change 112% to a decimal then multiply it by 0.85. You should get 0.95.

Practice Problems

1. Express as a percentage how much higher the blood sugar value is for a diabetic person compared to a nondiabetic person 1 h after breakfast.
2. Express as a percentage how much higher the blood sugar value is for a diabetic person compared to a nondiabetic person 3 h and 6 h after breakfast.

For more practice, visit life.msscience.com/math_practice

NATIONAL GEOGRAPHIC VISUALIZING THE ENDOCRINE SYSTEM

Figure 2

Your endocrine system is involved in regulating and coordinating many body functions, from growth and development to reproduction. This complex system consists of many diverse glands and organs, including the nine shown here. Endocrine glands produce chemical messenger molecules, called hormones, that circulate in the bloodstream. Hormones exert their influence only on the specific target cells to which they bind.

PINEAL GLAND Shaped like a tiny pinecone, the pineal gland lies deep in the brain. It produces melatonin, a hormone that may function as a sort of body clock by regulating wake/sleep patterns.

PITUITARY GLAND A pea-size structure attached to the hypothalamus of the brain, the pituitary gland produces hormones that affect a wide range of body activities, from growth to reproduction.

THYMUS The thymus is located in the upper chest, just behind the sternum. Hormones produced by this organ stimulate the production of certain infection-fighting cells.

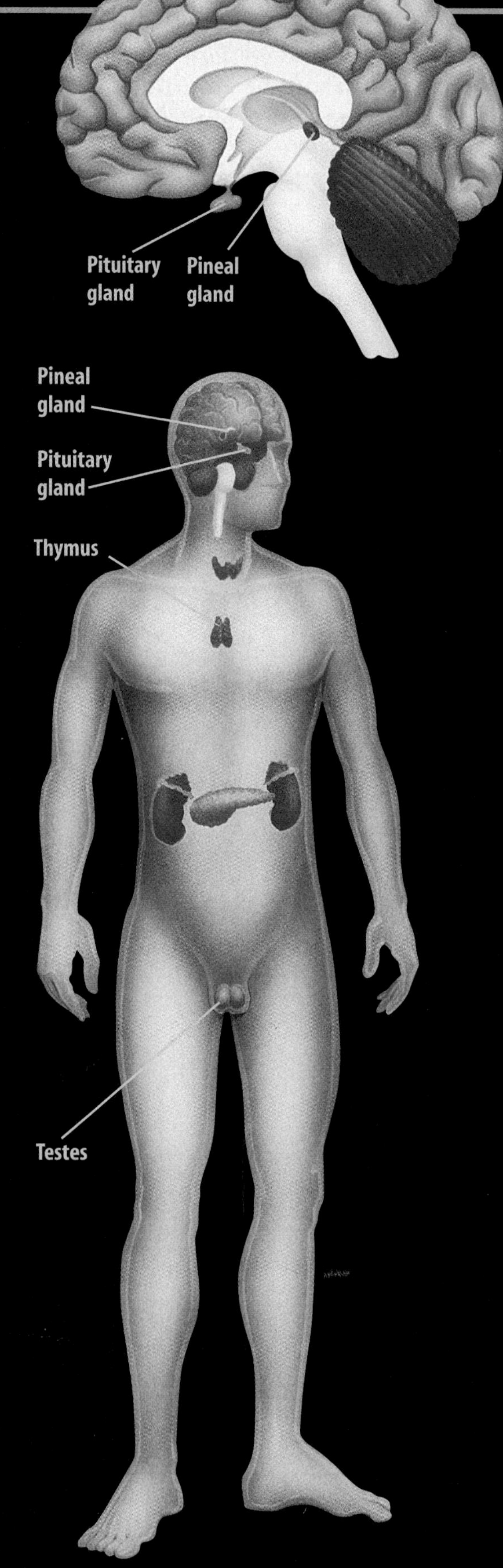

TESTES These paired male reproductive organs primarily produce testosterone, a hormone that controls the development and maintenance of male sexual traits. Testosterone also plays an important role in the production of sperm.

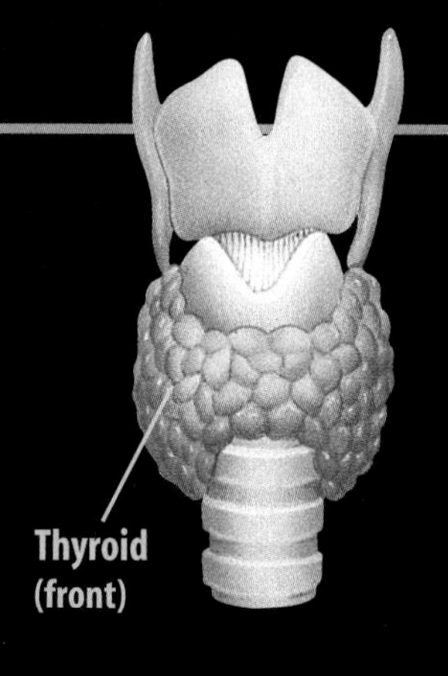

THYROID GLAND Located below the larynx, the bi-lobed thyroid gland is richly supplied with blood vessels. It produces hormones that regulate metabolic rate, control the uptake of calcium by bones, and promote normal nervous system development.

PARATHYROID GLANDS Attached to the back surface of the thyroid are tiny parathyroids, which help regulate calcium levels in the body. Calcium is important for bone growth and maintenance, as well as for muscle contraction and nerve impulse transmission.

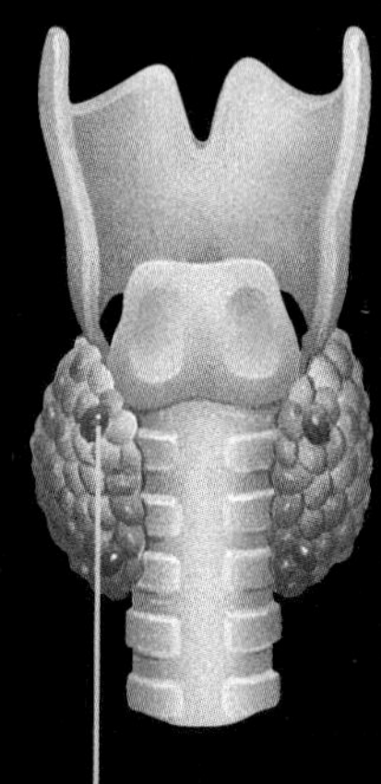

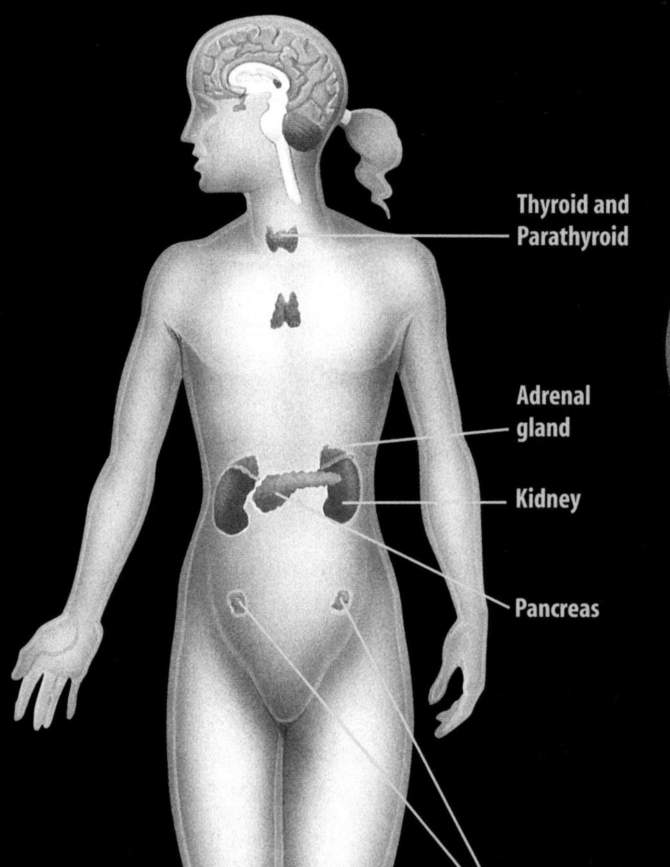

ADRENAL GLANDS On top of each of your kidneys is an adrenal gland. This complex endocrine gland produces a variety of hormones. Some play a critical role in helping your body adapt to physical and emotional stress. Others help stabilize blood sugar levels.

PANCREAS Scattered throughout the pancreas are millions of tiny clusters of endocrine tissue called the islets of Langerhans. Cells that make up the islets produce hormones that help control sugar levels in the bloodstream.

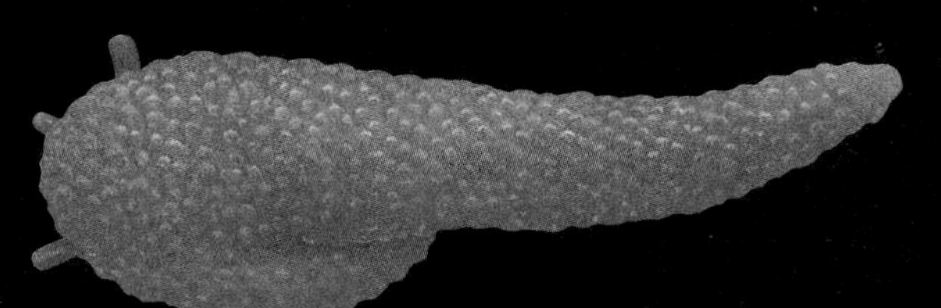

OVARIES Found deep in the pelvic cavity, ovaries produce female sex hormones known as estrogen and progesterone. These hormones regulate the female reproductive cycle and are responsible for producing and maintaining female sex characteristics.

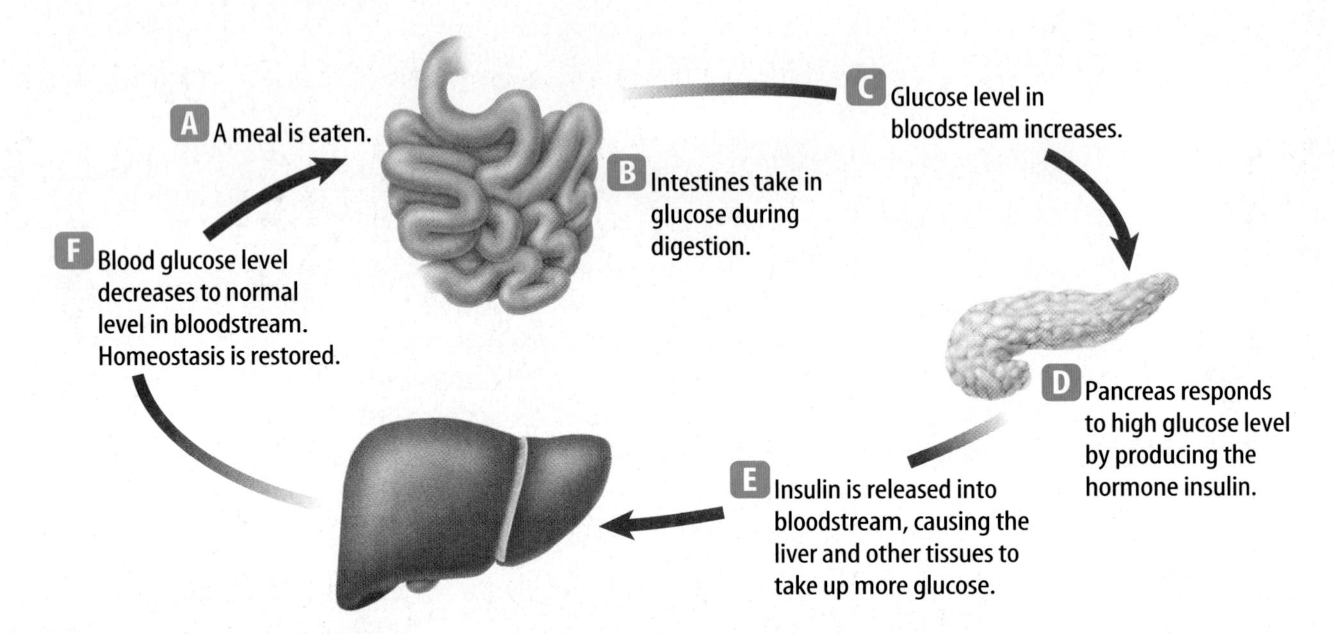

Figure 3 Many internal body conditions, such as hormone level, blood sugar level, and body temperature, are controlled by negative-feedback systems.

A Negative-Feedback System

To control the amount of hormones that are in your body, the endocrine system sends chemical messages back and forth within itself. This is called a negative-feedback system. It works much the way a thermostat works. When the temperature in a room drops below a set level, the thermostat signals the furnace to turn on. Once the furnace has raised the temperature in the room to the set level, the thermostat signals the furnace to shut off. It will continue to stay off until the thermostat signals that the temperature has dropped again. **Figure 3** shows how a negative-feedback system controls the level of glucose in your bloodstream.

section 1 review

Summary

Functions of the Endocrine System

- The nervous system and the endocrine system are the control systems of your body.
- The endocrine system uses hormones to deliver messages to the body.

Endocrine Glands

- Endocrine glands release hormones directly into the bloodstream.

A Negative-Feedback System

- The endocrine system uses a negative-feedback system to control the amount of hormones in your body.

Self Check

1. **Explain** the function of hormones.
2. **Choose** one endocrine gland. How does it work?
3. **Describe** a negative-feedback system.
4. **Think Critically** Glucose is required for cellular respiration, the process that releases energy within cells. How would a lack of insulin affect this process?

Applying Skills

5. **Predict** why the circulatory system is a good mechanism for delivering hormones throughout the body.
6. **Research** recent treatments for growth disorders involving the pituitary gland. Write a brief paragraph of your results in your Science Journal.

section 2

The Reproductive System

Reproduction and the Endocrine System

Reproduction is the process that continues life on Earth. Most human body systems, such as the digestive system and the nervous system, are the same in males and females, but this is not true for the reproductive system. Males and females each have structures specialized for their roles in reproduction. Although structurally different, both the male and female reproductive systems are adapted to allow for a series of events that can lead to the birth of a baby.

Hormones are the key to how the human reproductive system functions, as shown in **Figure 4.** Sex hormones are necessary for the development of sexual characteristics, such as breast development in females and facial hair growth in males. Hormones from the pituitary gland also begin the production of eggs in females and sperm in males. Eggs and sperm transfer hereditary information from one generation to the next.

as you read

What You'll Learn

- **Identify** the function of the reproductive system.
- **Compare and contrast** the major structures of the male and female reproductive systems.
- **Sequence** the stages of the menstrual cycle.

Why It's Important

Human reproductive systems help ensure that human life continues on Earth.

Review Vocabulary
cilia: short, hairlike structures that extend from a cell

New Vocabulary
- testes
- sperm
- semen
- ovary
- ovulation
- uterus
- vagina
- menstrual cycle
- menstruation

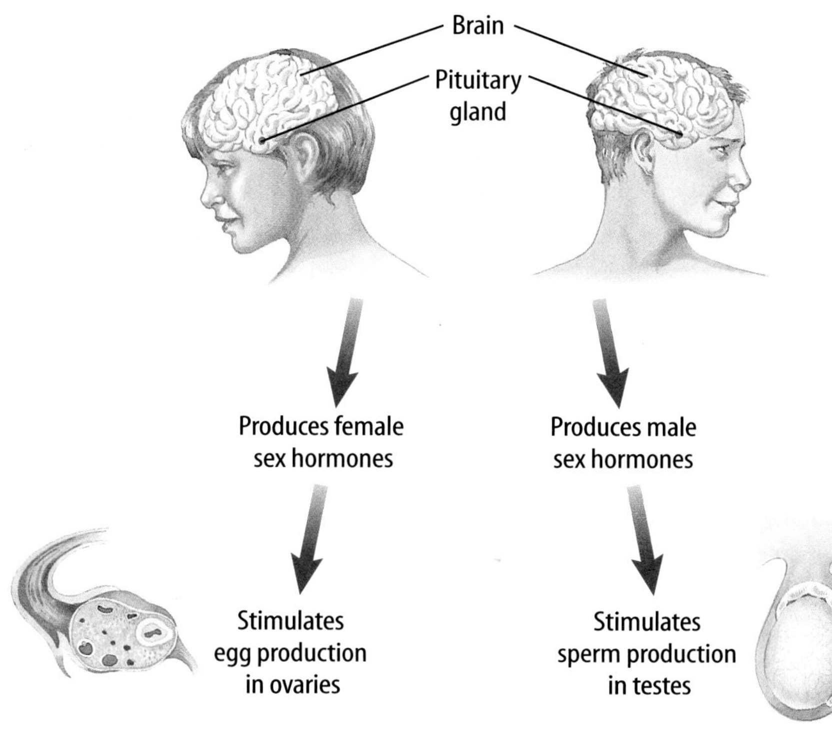

Figure 4 The pituitary gland produces hormones that control the male and female reproductive systems.

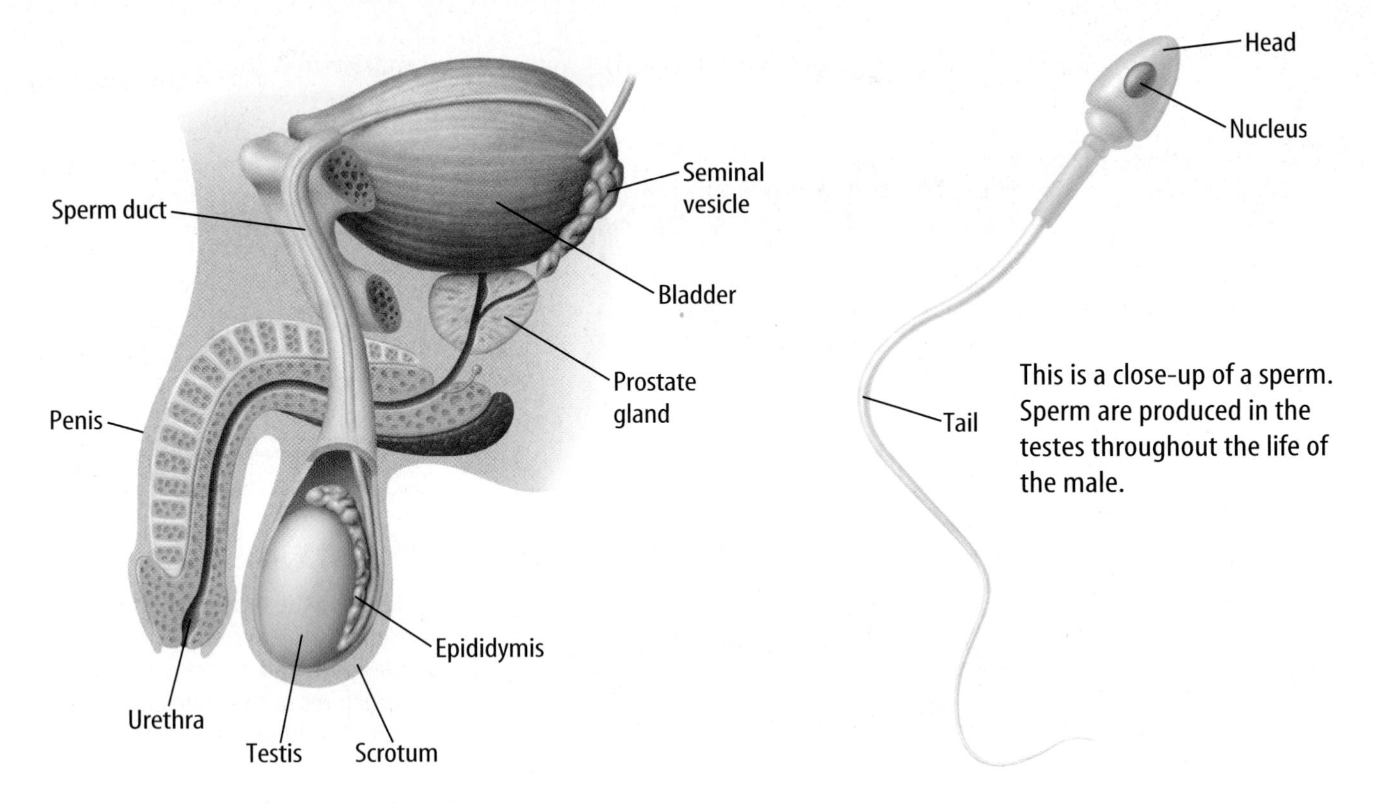

Figure 5 The structures of the male reproductive system are shown from the side of the body.

The Male Reproductive System

The male reproductive system is made up of external and internal organs. The external organs of the male reproductive system are the penis and scrotum, shown in **Figure 5.** The scrotum contains two organs called testes (TES teez). As males mature sexually, the **testes** begin to produce testosterone, the male hormone, and **sperm,** which are male reproductive cells.

Sperm Each sperm cell has a head and tail. The head contains hereditary information, and the tail moves the sperm. Because the scrotum is located outside the body cavity, the testes, where sperm are produced, are kept at a lower temperature than the rest of the body. Sperm are produced in greater numbers at lower temperatures.

Many organs help in the production, transportation, and storage of sperm. After sperm are produced, they travel from the testes through sperm ducts that circle the bladder. Behind the bladder, a gland called the seminal vesicle provides sperm with a fluid. This fluid supplies the sperm with an energy source and helps them move. This mixture of sperm and fluid is called **semen** (SEE mun). Semen leaves the body through the urethra, which is the same tube that carries urine from the body. However, semen and urine never mix. A muscle at the back of the bladder contracts to prevent urine from entering the urethra as sperm leave the body.

The Female Reproductive System

Unlike male reproductive organs, most of the reproductive organs of the female are inside the body. The **ovaries**—the female sex organs—are located in the lower part of the body cavity. Each of the two ovaries is about the size and shape of an almond. **Figure 6** shows the different organs of the female reproductive system.

Topic: Ovarian Cysts
Visit life.msscience.com for Web links to information about ovarian cysts.

Activity Make a small pamphlet explaining what cysts are and how they can be treated.

The Egg When a female is born, she already has all of the cells in her ovaries that eventually will develop into eggs—the female reproductive cells. At puberty, eggs start to develop in her ovaries because of specific sex hormones.

About once a month, an egg is released from an ovary in a hormone-controlled process called **ovulation** (ahv yuh LAY shun). The two ovaries release eggs on alternating months. One month, an egg is released from an ovary. The next month, the other ovary releases an egg, and so on. After the egg is released, it enters the oviduct. If a sperm fertilizes the egg, it usually happens in an oviduct. Short, hairlike structures called cilia help sweep the egg through the oviduct toward the uterus (YEW tuh rus).

Reading Check *When are eggs released by the ovaries?*

The **uterus** is a hollow, pear-shaped, muscular organ with thick walls in which a fertilized egg develops. The lower end of the uterus, the cervix, narrows and is connected to the outside of the body by a muscular tube called the **vagina** (vuh JI nuh). The vagina also is called the birth canal because during birth, a baby travels through this tube from the uterus to the outside of the mother's body.

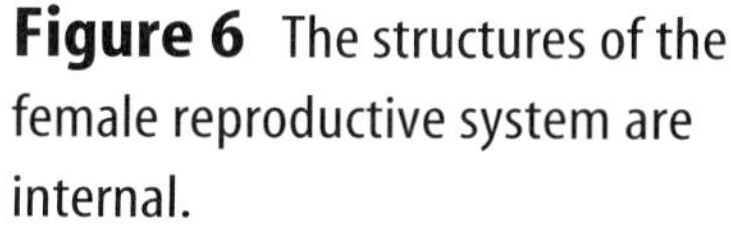

Figure 6 The structures of the female reproductive system are internal.
Name *where eggs develop in the female reproductive system.*

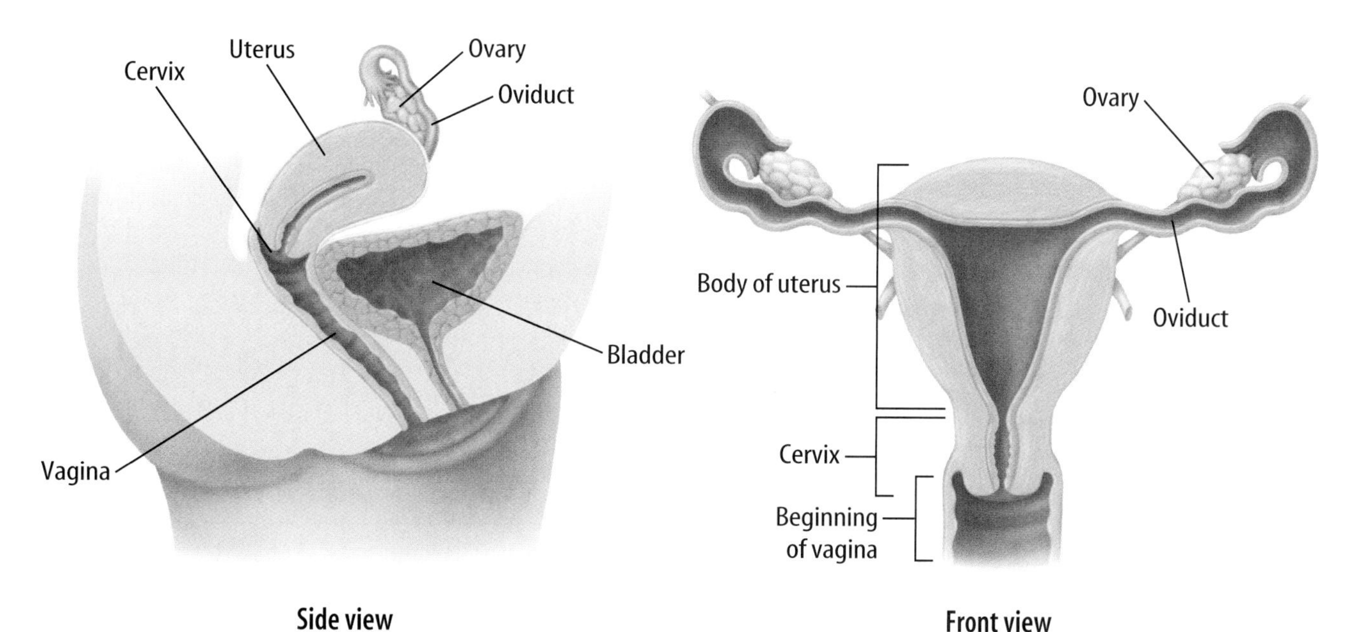

Graphing Hormone Levels

Procedure

Make a line graph of this table.

Hormone Changes	
Day	Level of Hormone
1	12
5	14
9	15
13	70
17	13
21	12
25	8

Analysis

1. On what day is the highest level of hormone present?
2. What event takes place around the time of the highest hormone level?

The Menstrual Cycle

How is the female body prepared for having a baby? The **menstrual cycle** is the monthly cycle of changes in the female reproductive system. Before and after an egg is released from an ovary, the uterus undergoes changes. The menstrual cycle of a human female averages 28 days. However, the cycle can vary in some individuals from 20 to 40 days. Changes include the maturing of an egg, the production of female sex hormones, the preparation of the uterus to receive a fertilized egg, and menstrual flow.

Reading Check *What is the menstrual cycle?*

Endocrine Control Hormones control the entire menstrual cycle. The pituitary gland responds to chemical messages from the hypothalamus by releasing several hormones. These hormones start the development of eggs in the ovary. They also start the production of other hormones in the ovary, including estrogen (ES truh jun) and progesterone (proh JES tuh rohn). The interaction of all these hormones results in the physical processes of the menstrual cycle.

Phase One As shown in **Figure 7,** the first day of phase 1 starts when menstrual flow begins. Menstrual flow consists of blood and tissue cells released from the thickened lining of the uterus. This flow usually continues for four to six days and is called **menstruation** (men STRAY shun).

Figure 7 The three phases of the menstrual cycle make up the monthly changes in the female reproductive system.
Explain *why the uterine lining thickens.*

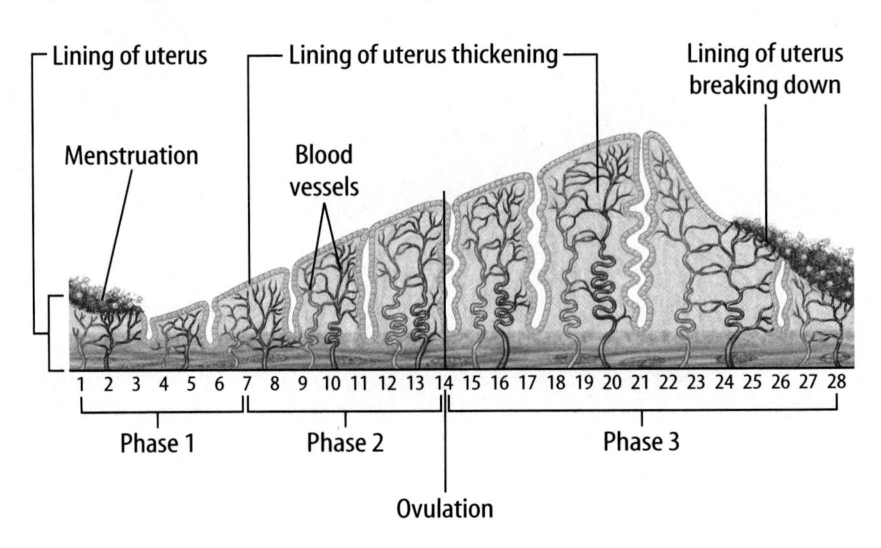

Phase Two Hormones cause the lining of the uterus to thicken in phase 2. Hormones also control the development of an egg in the ovary. Ovulation occurs about 14 days before menstruation begins. Once the egg is released, it must be fertilized within 24 h or it usually begins to break down. Because sperm can survive in a female's body for up to three days, fertilization can occur soon after ovulation.

Phase Three Hormones produced by the ovaries continue to cause an increase in the thickness of the uterine lining during phase 3. If a fertilized egg does arrive, the uterus is ready to support and nourish the developing embryo. If the egg is not fertilized, the lining of the uterus breaks down as the hormone levels decrease. Menstruation begins and the cycle repeats itself.

Menopause For most females, the first menstrual period happens between ages nine years and 13 years and continues until 45 years of age to 60 years of age. Then, a gradual reduction of menstruation takes place as hormone production by the ovaries begins to shut down. Menopause occurs when both ovulation and menstrual periods end. It can take several years for the completion of menopause. As **Figure 8** indicates, menopause does not inhibit a woman's ability to enjoy an active life.

Figure 8 This older woman enjoys exercising with her granddaughter.

section 2 review

Summary

Reproduction and the Endocrine System

- Reproduction is the process that continues life.
- The human reproductive system needs hormones to function.

The Male Reproductive System

- Sperm are produced in the testes and leave the male through the penis.

The Female Reproductive System

- Eggs are produced in the ovaries and, if fertilized, can develop in the uterus.

The Menstrual Cycle

- A female's menstrual cycle occurs approximately every 28 days.
- If an egg is not fertilized, the lining of the uterus breaks down and is shed in a process called menstruation.

Self Check

1. **Identify** the major function of male and female reproductive systems in humans.
2. **Explain** the movement of sperm through the male reproductive system.
3. **Compare and contrast** the major organs and structures of the male and female reproductive systems.
4. **Sequence** the stages of the menstrual cycle in a human female using diagrams and captions.
5. **Think Critically** Adolescent females often require additional amounts of iron in their diet. Explain.

Applying Math

6. **Order of Operations** Usually, one egg is released each month during a female's reproductive years. If menstruation begins at 12 years of age and ends at 50 years of age, calculate the number of eggs her body can release during her reproductive years.

Interpreting Diagrams

Starting in adolescence, hormones cause the development of eggs in the ovary and changes in the uterus. These changes prepare the uterus to accept a fertilized egg that can attach itself in the wall of the uterus. What happens to an unfertilized egg?

Real-World Question

What changes occur to the uterus during a female's monthly menstrual cycle?

Goals

- **Observe** the stages of the menstrual cycle in the diagram.
- **Relate** the process of ovulation to the cycle.

Materials

paper
pencil

Procedure

1. The diagrams below illustrate the menstrual cycle.
2. Copy and complete the data table using information in this chapter and diagrams below.
3. On approximately what day in a 28-day cycle is the egg released from the ovary?

Menstruation Cycle

Days	Condition of Uterus	What Happens
1–6		
7–12	Do not write in this book.	
13–14		
15–18		

Conclude and Apply

1. **Infer** how many days the average menstrual cycle lasts.
2. **State** on what days the lining of the uterus builds up.
3. **Infer** why this process is called a cycle.
4. **Calculate** how many days before menstruation ovulation usually occurs.

Communicating Your Data

Compare your data table with those of other students in your class. **For more help, refer to the Science Skill Handbook.**

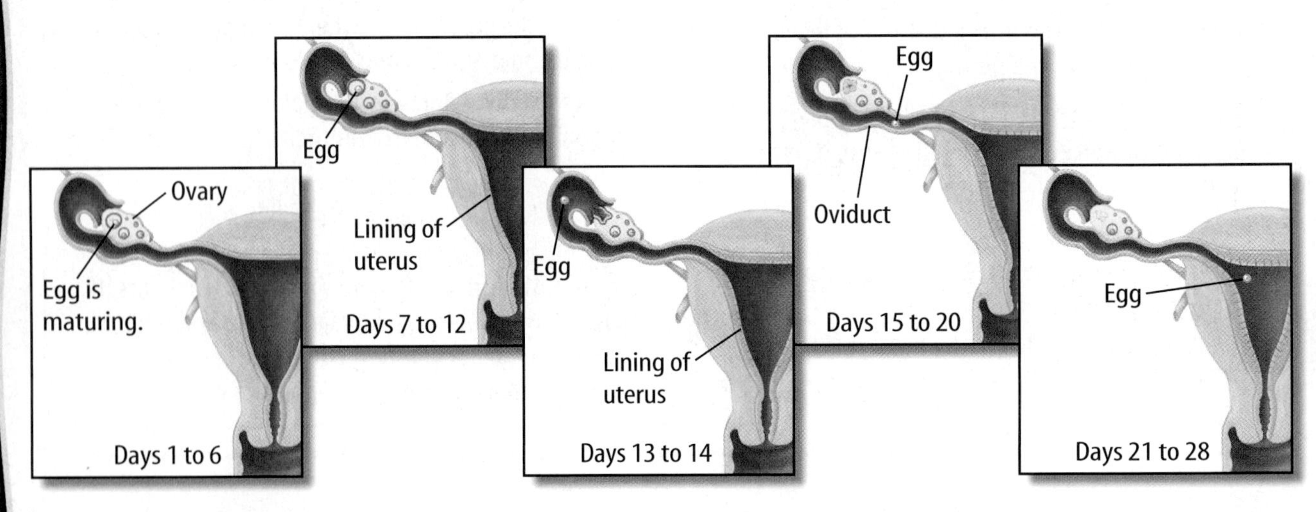

section 3

Human Life Stages

The Function of the Reproductive System

Before the invention of powerful microscopes, some people imagined an egg or a sperm to be a tiny person that grew inside a female. In the latter part of the 1700s, experiments using amphibians showed that contact between an egg and sperm is necessary for the development of life. With the development of the cell theory in the 1800s, scientists recognized that a human develops from an egg that has been fertilized by a sperm. The uniting of a sperm and an egg is known as fertilization. Fertilization, as shown in **Figure 9,** usually takes place in the oviduct.

as you read

What You'll Learn

- **Describe** the fertilization of a human egg.
- **List** the major events in the development of an embryo and fetus.
- **Describe** the developmental stages of infancy, childhood, adolescence, and adulthood.

Why It's Important

Fertilization begins the entire process of human growth and development.

Review Vocabulary

nutrient: substance in food that provides energy and materials for cell development, growth, and repair

New Vocabulary

- pregnancy
- embryo
- amniotic sac
- fetus
- fetal stress

Fertilization

INTEGRATE Chemistry

Although 200 million to 300 million sperm can be deposited in the vagina, only several thousand reach an egg in the oviduct. As they enter the female, the sperm come into contact with chemical secretions in the vagina. It appears that this contact causes a change in the membrane of the sperm. The sperm then become capable of fertilizing the egg. The one sperm that makes successful contact with the egg releases an enzyme from the saclike structure on its head. Enzymes help speed up chemical reactions that have a direct effect on the protective membranes on the egg's surface. The structure of the egg's membrane is disrupted, and the sperm head can enter the egg.

Zygote Formation Once a sperm has entered the egg, changes in the electric charge of the egg's membrane prevent other sperm from entering the egg. At this point, the nucleus of the successful sperm joins with the nucleus of the egg. This joining of nuclei creates a fertilized cell called the zygote. It begins to undergo mitosis and cell division.

Figure 9 After the sperm releases enzymes that disrupt the egg's membrane, it penetrates the egg.

Color-enhanced SEM Magnification: 340×

Figure 10 The development of fraternal and identical twins is different.

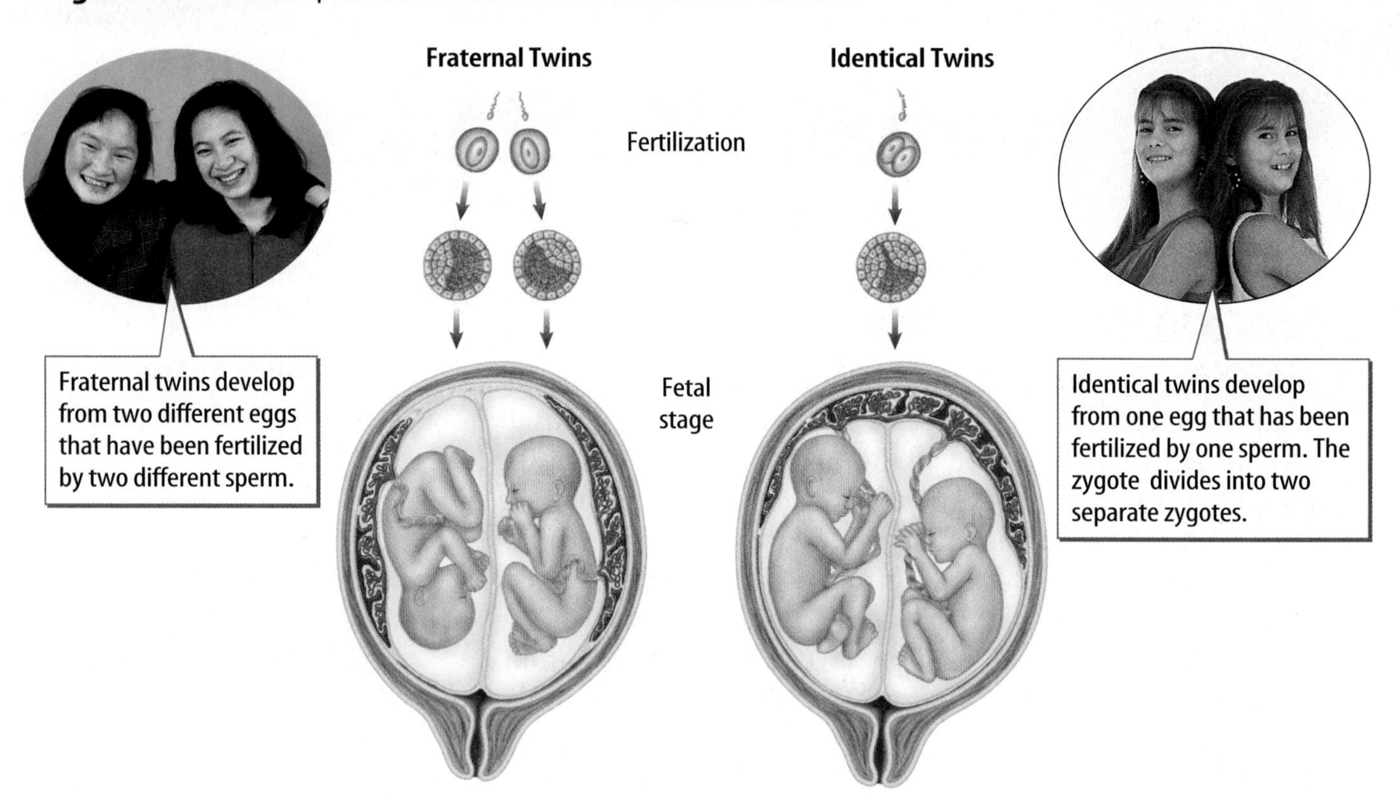

Fraternal Twins
Identical Twins
Fertilization
Fetal stage
Fraternal twins develop from two different eggs that have been fertilized by two different sperm.
Identical twins develop from one egg that has been fertilized by one sperm. The zygote divides into two separate zygotes.

Multiple Births

Sometimes two eggs leave the ovary at the same time. If both eggs are fertilized and both develop, fraternal twins are born. Fraternal twins, as shown in **Figure 10,** can be two girls, two boys, or a boy and a girl. Because fraternal twins come from two eggs, they only resemble each other.

Because identical twin zygotes develop from the same egg and sperm, as explained in **Figure 10,** they have the same hereditary information. These identical zygotes develop into identical twins, which are either two girls or two boys. Multiple births also can occur when three or more eggs are produced at one time or when the zygote separates into three or more parts.

Midwives Some women choose to deliver their babies at home rather than at a hospital. An at-home birth can be attended by a certified nurse-midwife. Research to find the educational and skill requirements of a nurse-midwife.

Development Before Birth

After fertilization, the zygote moves along the oviduct to the uterus. During this time, the zygote is dividing and forming into a ball of cells. After about seven days, the zygote attaches to the wall of the uterus, which has been thickening in preparation to receive a zygote, as shown in **Figure 11.** If attached to the wall of the uterus, the zygote will develop into a baby in about nine months. This period of development from fertilized egg to birth is known as **pregnancy.**

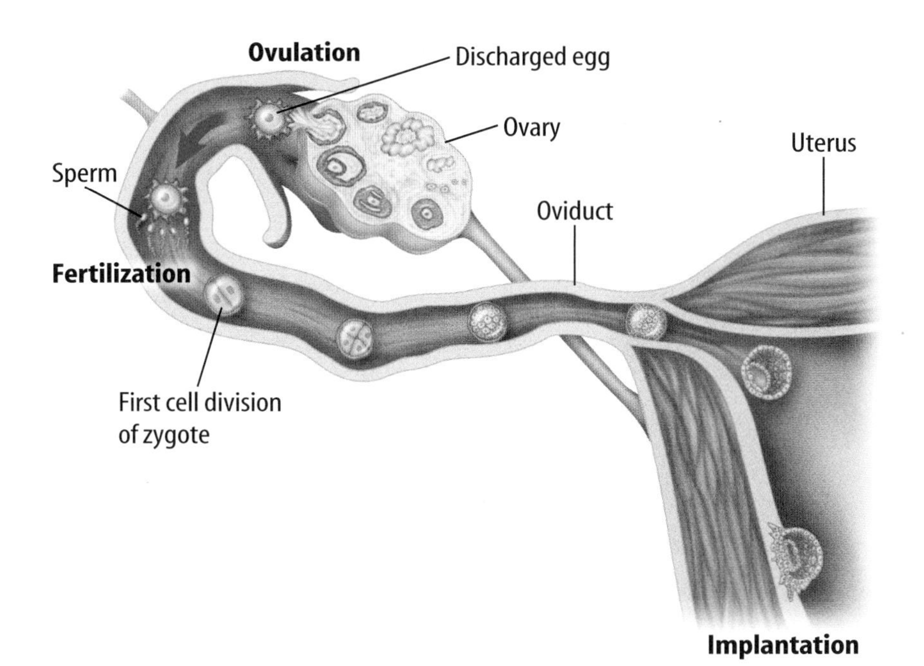

Figure 11 After a few days of rapid mitosis and cell division, the zygote, now a ball of cells, reaches the lining of the uterus, where it attaches itself to the lining for development.

The Embryo After the zygote attaches to the wall of the uterus, it is known as an **embryo,** illustrated in **Figure 12.** It receives nutrients from fluids in the uterus until the placenta (plu SEN tuh) develops from tissues of the uterus and the embryo. An umbilical cord develops that connects the embryo to the placenta. In the placenta, materials diffuse between the mother's blood and the embryo's blood, but their bloods do not mix. Blood vessels in the umbilical cord carry nutrients and oxygen from the mother's blood through the placenta to the embryo. Other substances in the mother's blood can move into the embryo, including drugs, toxins, and disease organisms. Wastes from the embryo are carried in other blood vessels in the umbilical cord through the placenta to the mother's blood.

Reading Check ***Why must a pregnant woman avoid alcohol, tobacco, and harmful drugs?***

Pregnancy in humans lasts about 38 to 39 weeks. During the third week, a thin membrane called the **amniotic** (am nee AH tihk) **sac** begins to form around the embryo. The amniotic sac is filled with a clear liquid called amniotic fluid, which acts as a cushion for the embryo and stores nutrients and wastes.

During the first two months of development, the embryo's major organs form and the heart structure begins to beat. At five weeks, the embryo has a head with eyes, nose, and mouth features. During the sixth and seventh weeks, fingers and toes develop.

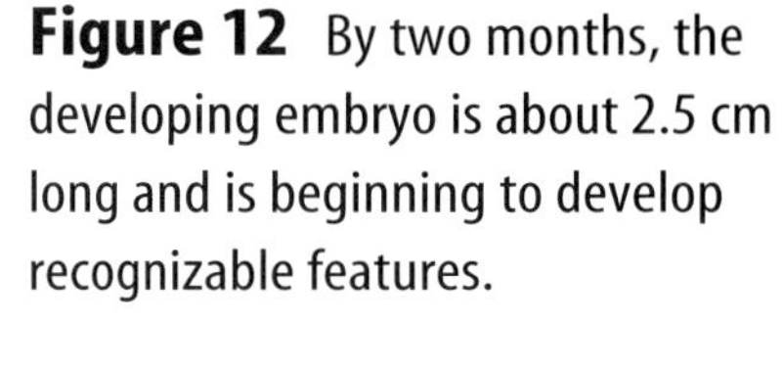
Figure 12 By two months, the developing embryo is about 2.5 cm long and is beginning to develop recognizable features.

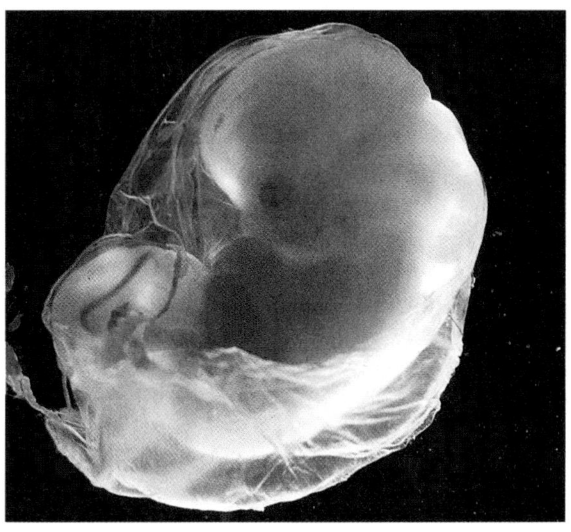

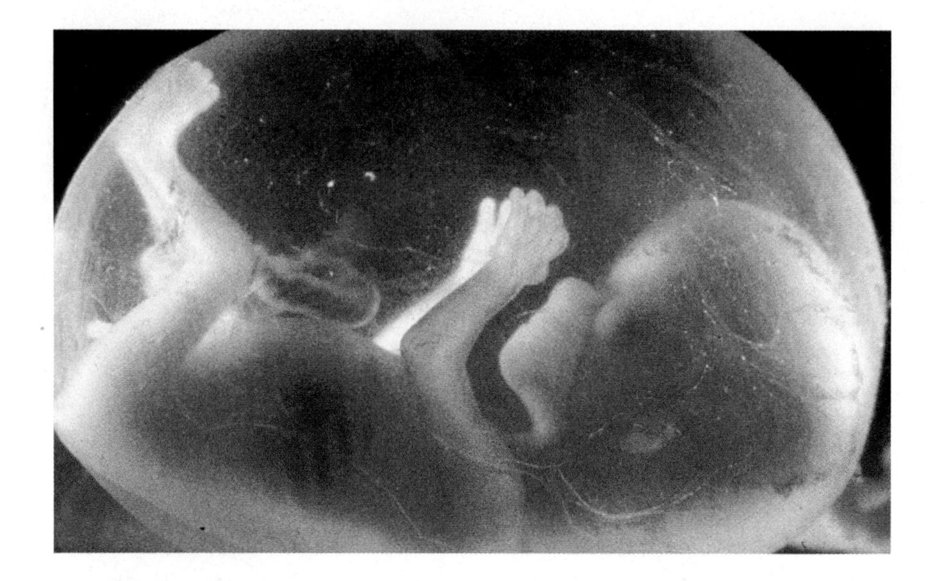

Figure 13 A fetus at about 16 weeks is approximately 15 cm long and weighs 140 g.
Describe *the changes that take place in a fetus by the end of the seventh month.*

The Fetus After the first two months of pregnancy, the developing embryo is called a **fetus,** shown in **Figure 13.** At this time, body organs are present. Around the third month, the fetus is 8 cm to 9 cm long. The mother may feel the fetus move. The fetus can even suck its thumb. By the fourth month, an ultrasound test can determine the sex of the fetus. The fetus is 30 cm to 38 cm in length by the end of the seventh month of pregnancy. Fatty tissue builds up under the skin, and the fetus looks less wrinkled. By the ninth month, the fetus usually has shifted to a head-down position within the uterus, a position beneficial for delivery. The head usually is in contact with the opening of the uterus to the vagina. The fetus is about 50 cm in length and weighs from 2.5 kg to 3.5 kg.

Interpreting Fetal Development

Procedure
Make a bar graph of the following data.

Fetal Development

End of Month	Length (cm)
3	8
4	15
5	25
6	30
7	35
8	40
9	51

Analysis
1. During which month does the greatest increase in length occur?
2. On average, how many centimeters does the baby grow per month?

Try at Home

The Birthing Process

The process of childbirth, as shown in **Figure 14,** begins with labor, the muscular contractions of the uterus. As the contractions increase in strength and number, the amniotic sac usually breaks and releases its fluid. Over a period of hours, the contractions cause the opening of the uterus to widen. More powerful and more frequent contractions push the baby out through the vagina into its new environment.

Delivery Often a mother is given assistance by a doctor during the delivery of the baby. As the baby emerges from the birth canal, a check is made to determine if the umbilical cord is wrapped around the baby's neck or any body part. When the head is free, any fluid in the baby's nose and mouth is removed by suction. After the head and shoulders appear, contractions force the baby out completely. Up to an hour after delivery, contractions occur that push the placenta out of the mother's body.

Cesarean Section Sometimes a baby must be delivered before labor begins or before it is completed. At other times, a baby cannot be delivered through the birth canal because the mother's pelvis might be too small or the baby might be in the wrong birthing position. In cases like these, surgery called a cesarean (suh SEER ee uhn) section is performed. An incision is made through the mother's abdominal wall, then through the wall of the uterus. The baby is delivered through this opening.

Reading Check *What is a cesarean section?*

Topic: Cesarean Sections
Visit life.msscience.com for Web links to information about cesarean section delivery.

Activity Make a chart listing the advantages and disadvantages of a cesarean section delivery.

After Birth When the baby is born, it is attached to the umbilical cord. The person assisting with the birth clamps the cord in two places and cuts it between the clamps. The baby does not feel any pain from this procedure. The baby might cry, which is the result of air being forced into its lungs. The scar that forms where the cord was attached is called the navel.

Figure 14 Childbirth begins with labor. The opening to the uterus widens, and the baby passes through.

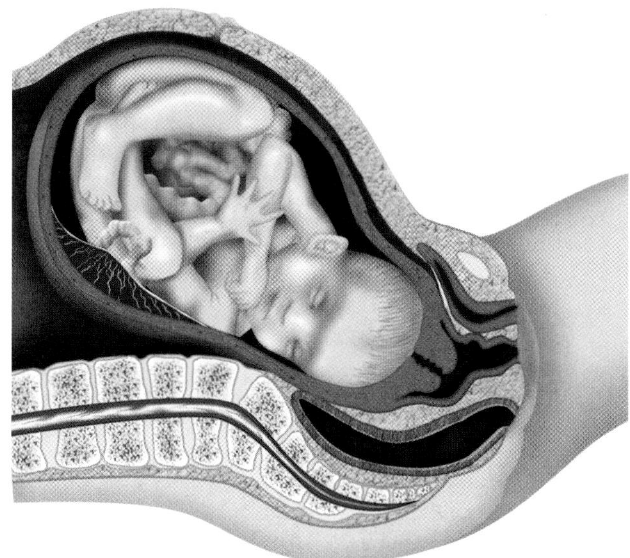

The fetus moves into the opening of the birth canal, and the uterus begins to widen.

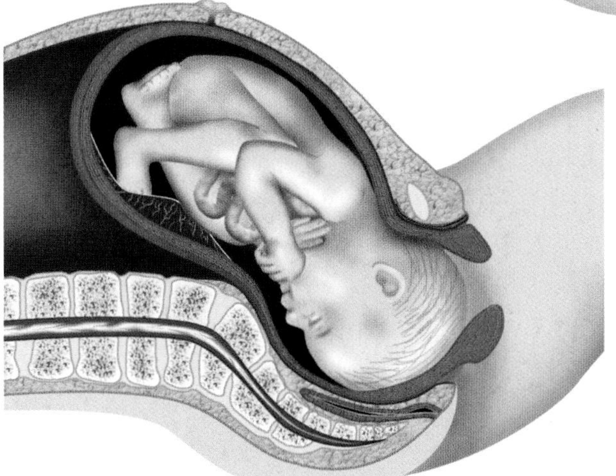

The base of the uterus is completely dilated.

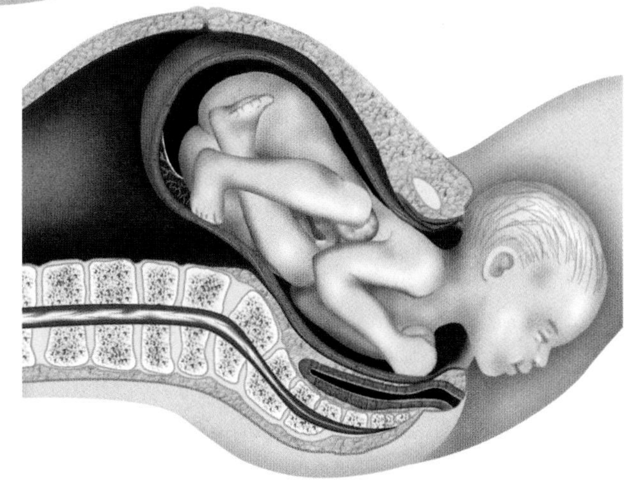

The fetus is pushed out through the birth canal.

Stages After Birth

Defined stages of development occur after birth, based on the major developments that take place during those specific years. Infancy lasts from birth to around 18 months of age. Childhood extends from the end of infancy to sexual maturity, or puberty. The years of adolescence vary, but they usually are considered to be the teen years. Adulthood covers the years of age from the early 20s until life ends, with older adulthood considered to be over 60. The age spans of these different stages are not set, and scientists differ in their opinions regarding them.

Infancy What type of environment must the infant adjust to after birth? The experiences the fetus goes through during birth cause **fetal stress.** The fetus has emerged from an environment that was dark, watery, a constant temperature, and nearly soundless. In addition, the fetus might have been forced through the constricted birth canal. However, in a short period of time, the infant's body becomes adapted to its new world.

The first four weeks after birth are known as the neonatal (nee oh NAY tul) period. The term *neonatal* means "newborn." During this time, the baby's body begins to function normally. Unlike the newborn of some other animals, human babies, such as the one shown in **Figure 15,** depend on other humans for their survival. In contrast, many other animals, such as the young horse also shown in **Figure 15,** begin walking a few hours after they are born.

Figure 15 Human babies are more dependent upon their caregivers than many other mammals are.

Infants and toddlers are completely dependent upon caregivers for all their needs.

Other young mammals are more self-sufficient. This colt is able to stand within an hour after birth.

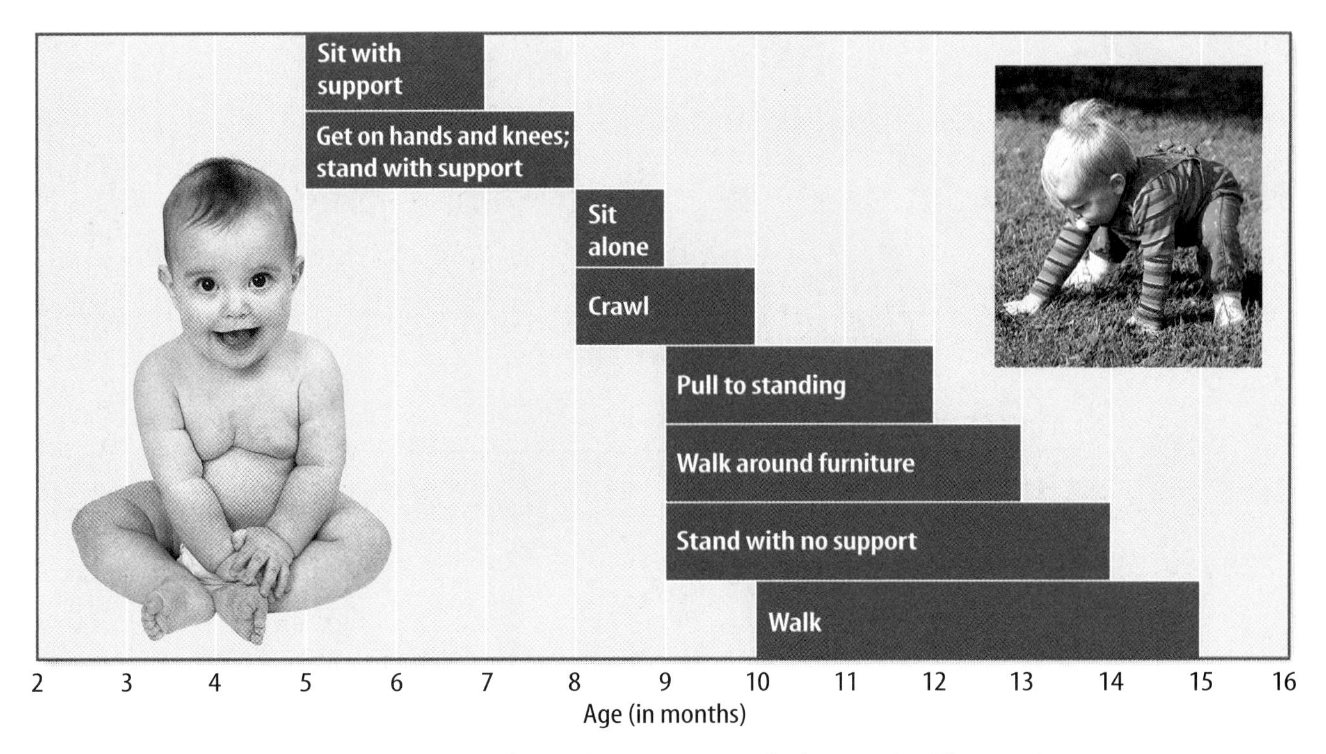

Figure 16 Infants show rapid development in their nervous and muscular systems through 18 months of age.

During these first 18 months, infants show increased physical coordination, mental development, and rapid growth. Many infants will triple their weight in the first year. **Figure 16** shows the extremely rapid development of the nervous and muscular systems during this stage, which enables infants to start interacting with the world around them.

Childhood After infancy is childhood, which lasts until about puberty, or sexual maturity. Sexual maturity occurs around 12 years of age. Overall, growth during early childhood is rather rapid, although the physical growth rate for height and weight is not as rapid as it is in infancy. Between two and three years of age, the child learns to control his or her bladder and bowels. At age two to three, most children can speak in simple sentences. Around age four, the child is able to get dressed and undressed with some help. By age five, many children can read a limited number of words. By age six, children usually have lost their chubby baby appearance, as seen in **Figure 17.** However, muscular coordination and mental abilities continue to develop. Throughout this stage, children develop their abilities to speak, read, write, and reason. These ages of development are only guidelines because each child develops at a different rate.

Figure 17 Children, like these kindergartners, grow and develop at different rates.

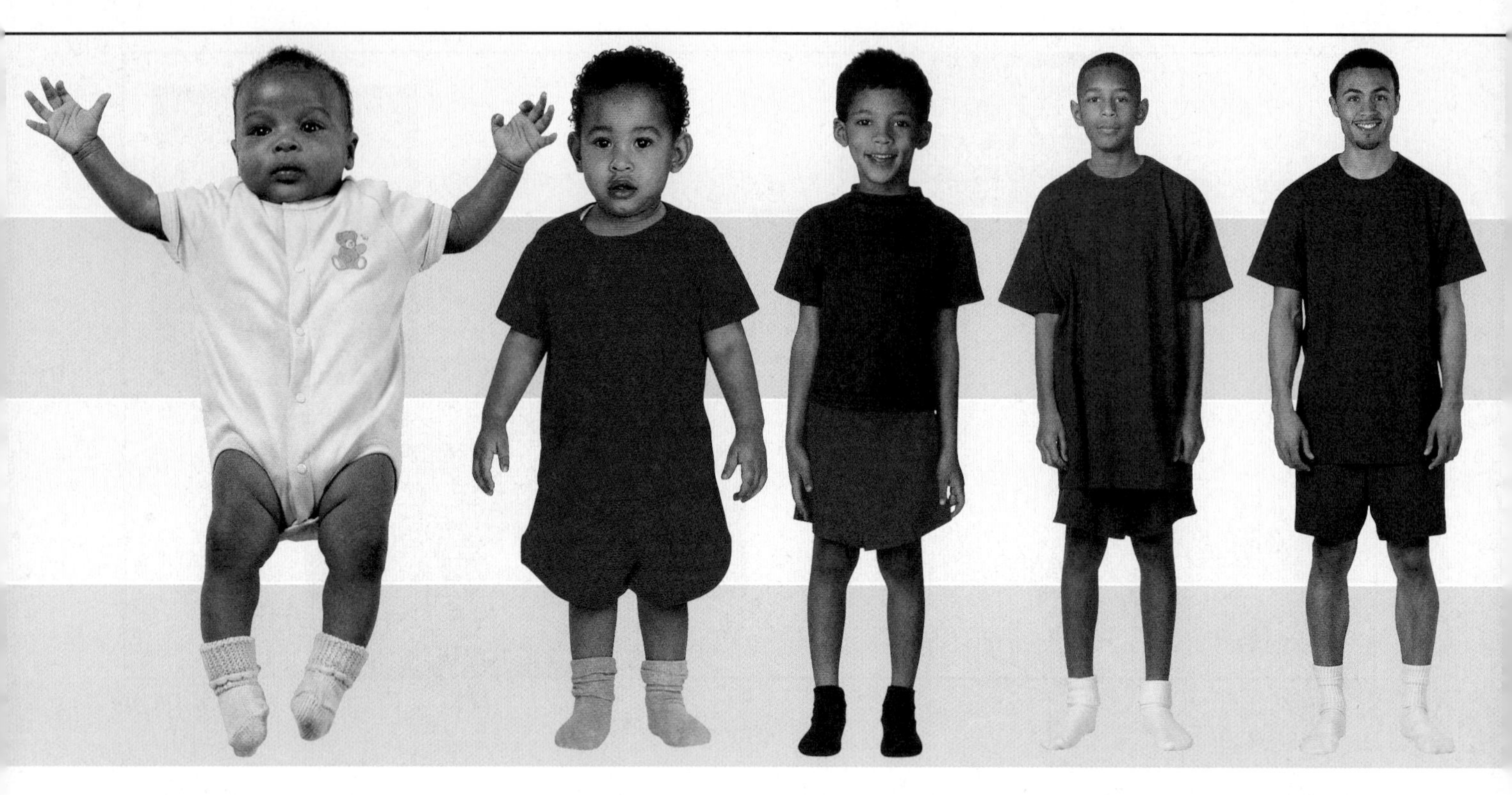

Figure 18 The proportions of body parts change over time as the body develops.
Describe *how the head changes proportion.*

Adolescent Growth During adolescence, body parts do not all grow at the same rate. Legs grow longer before the upper body lengthens. This changes the body's center of gravity, the point at which the body maintains its balance. This is one cause of teenager clumsiness. In your Science Journal, write a paragraph about how this might affect playing sports.

Adolescence Adolescence usually begins around age 12 or 13. A part of adolescence is puberty—the time of development when a person becomes physically able to reproduce. For girls, puberty occurs between ages nine and 13. For boys, puberty occurs between ages 13 and 16. During puberty, hormones produced by the pituitary gland cause changes in the body. These hormones produce reproductive cells and sex hormones. Secondary sex characteristics also develop. In females, the breasts develop, pubic and underarm hair appears, and fatty tissue is added to the buttocks and thighs. In males, the hormones cause a deepened voice, an increase in muscle size, and the growth of facial, pubic, and underarm hair.

Adolescence usually is when the final growth spurt occurs. Because the time when hormones begin working varies among individuals and between males and females, growth rates differ. Girls often begin their final growth phase at about age 11 and end around age 16. Boys usually start their growth spurt at age 13 and end around 18 years of age.

Adulthood The final stage of development, adulthood, begins with the end of adolescence and continues through old age. This is when the growth of the muscular and skeletal system stops. **Figure 18** shows how body proportions change as you age.

People from age 45 to age 60 are sometimes considered middle-aged adults. During these years, physical strength begins to decline. Circulatory and respiratory systems become less efficient. Bones become more brittle, and the skin wrinkles.

Older Adulthood People over the age of 60 may experience an overall decline in their physical body systems. The cells that make up these systems no longer function as well as they did at a younger age. Connective tissues lose their elasticity, causing muscles and joints to be less flexible. Bones become thinner and more brittle. Hearing and vision are less sensitive. The lungs and heart work less efficiently. However, exercise and eating well over a lifetime can help extend the health of one's body systems. Many healthy older adults enjoy full lives and embrace challenges, as shown in **Figure 19.**

Figure 19 Astronaut and Senator John Glenn traveled into space twice. In 1962, at age 40, he was the first U.S. citizen to orbit Earth. He was part of the space shuttle crew in 1998 at age 77. Senator Glenn has helped change people's views of what many older adults are capable of doing.

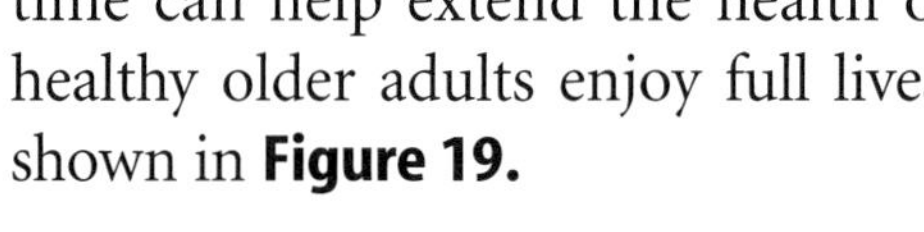

What physical changes occur during late adulthood?

Human Life Spans Seventy-seven years is the average life span—from birth to death—of humans in the United States, although an increasing number of people live much longer. However, body systems break down with age, resulting in eventual death. Death can occur earlier than old age for many reasons, including diseases, accidents, and bad health choices.

section 3 review

Summary

Fertilization

- Fertilization is the uniting of a sperm and an egg.

Development Before Birth

- Pregnancy begins when an egg is fertilized and lasts until birth.

The Birthing Process

- Birth begins with labor. Contractions force the baby out of the mother's body.

Stages After Birth

- Infancy (birth to 18 months) and childhood (until age 12) are periods of physical and mental growth.
- A person becomes physically able to reproduce during adolescence. Adulthood is the final stage of development.

Self Check

1. **Describe** what happens when an egg is fertilized in a female.
2. **Explain** what happens to an embryo during the first two months of pregnancy.
3. **Describe** the major events that occur during childbirth.
4. **Name** the stage of development that you are in. What physical changes have occurred or will occur during this stage of human development?
5. **Think Critically** Why is it hard to compare the growth and development of different adolescents?

Applying Skills

6. **Use a Spreadsheet** Using your text and other resources, make a spreadsheet for the stages of human development from a zygote to a fetus. Title one column *Zygote,* another *Embryo*, and a third *Fetus.* Complete the spreadsheet.

Changing Body Proportions

Goals
- **Measure** specific body proportions of adolescents.
- **Infer** how body proportions differ between adolescent males and females.

Materials
tape measure
erasable pencil
graph paper

Real-World Question

The ancient Greeks believed that the perfect body was completely balanced. Arms and legs should not be too long or short. A person's head should not be too large or small. The extra-large muscles of a body builder would have been ugly to the Greeks. How do you think they viewed the bodies of infants and children? Infants and young children have much different body proportions than adults, and teenagers often go through growth spurts that quickly change their body proportions. How do the body proportions differ between adolescent males and females?

Procedure

1. Copy the data table in your Science Journal and record the gender of each person that you measure.
2. Measure each person's head circumference by starting in the middle of the forehead and wrapping the tape measure once around the head. Record these measurements.

3. Measure each person's arm length from the top of the shoulder to the tip of the middle finger while the arm is held straight out to the side of the body. Record these measurements.
4. Ask each person to remove his or her shoes and stand next to a wall. Mark their height with an erasable pencil and measure their height from the floor to the mark. Record these measurements in the data table.

Age and Body Measurements

Gender of Person	Head Circumference (cm)	Arm Length (cm)	Height (cm)
	Do not write in this book.		

5. **Combine** your data with that of your classmates. Find the averages of head circumference, arm length, and height. Then, find these averages for males and females.
6. Make a bar graph of your calculations in step 5. Plot the measurements on the *y*-axis and plot all of the averages along the *x*-axis.
7. **Calculate** the proportion of average head circumference to average height for everyone in your class by dividing the average head circumference by the average height. Repeat this calculation for males and females.
8. **Calculate** the proportion of average arm length to average height for everyone in your class by dividing the average arm length by the average height. Repeat this calculation for males and females.

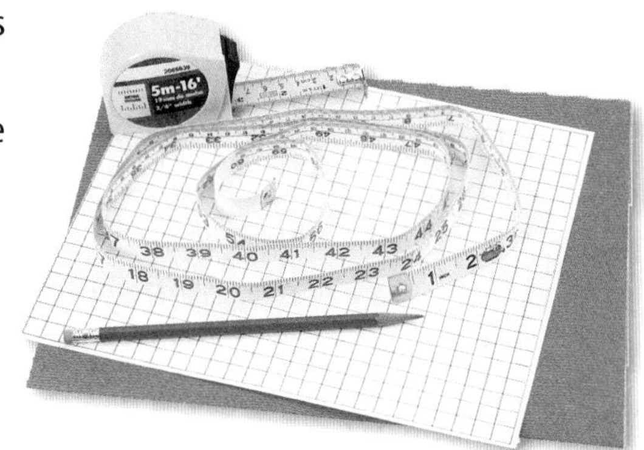

Analyze Your Data

Analyze whether adolescent males or females have larger head circumferences or longer arms. Which group has the larger proportion of head circumference or arm length to height?

Conclude and Apply

Explain if this lab supports the information in this chapter about the differences between growth rates of adolescent males and females.

Communicating Your Data

Construct data tables on poster board showing your results and those of your classmates. Discuss with your classmates why these results might be different.

SCIENCE Stats

Facts About Infants

Did you know...

...Humans and chimpanzees share about 99 percent of their genes. Although humans look different than chimps, reproduction is similar and gestation is the same—about nine months. Youngsters of both species lose their baby teeth at about six years of age.

Mammal Facts

Mammal	Average Gestation	Average Birth Weight	Average Adult Weight	Average Life Span (years)
African elephant	22 months	136 kg	4,989.5 kg	35
Blue whale	12 months	1,800 kg	135,000 kg	60
Human	**9 months**	**3.3 kg**	**59–76 kg**	**77***
Brown bear	7 months	0.23–0.5 kg	350 kg	22.5
Cat	2 months	99 g	2.7–7 kg	13.5
Kangaroo	1 month	0.75 –1.0 g	45 kg	5
Golden hamster	2.5 weeks	0.3 g	112 g	2

*In the United States

Applying Math Assume that a female of each mammal listed in the table above is pregnant once during her life. Which mammal is pregnant for the greatest proportion of her life?

Echidna

...Of about 4,000 species of mammals, only three lay eggs: the platypus, the short-beaked echidna (ih KIHD nuh), and the long-beaked echidna.

Find Out About It

Visit life.msscience.com/science_stats to research which species of vertebrate animals has the longest life span and which has the shortest. Present your findings in a table that also shows the life span of humans.

Reviewing Main Ideas

Section 1 The Endocrine System

1. Endocrine glands secrete hormones directly into the bloodstream. They affect specific tissues in the body.
2. A change in the body causes an endocrine gland to function. Hormone production slows or stops when homeostasis is reached.

Section 2 The Reproductive System

1. Reproductive systems allow new organisms to be formed.
2. The testes produce sperm, which leave the male body through the penis.
3. The female ovaries produce eggs. If fertilized, an egg develops into a fetus within the uterus.
4. An unfertilized egg and the built-up lining of the uterus are shed in menstruation.

Section 3 Human Life Stages

1. After fertilization, the zygote becomes an embryo, then a fetus. Twins occur when two eggs are fertilized or when a zygote divides after fertilization.
2. Birth begins with labor. The amniotic sac breaks. Then, usually after several hours, contractions force the baby out of the mother's body.
3. Infancy, from birth to 18 months of age, is a period of rapid growth of mental and physical skills. Childhood lasts until age 12 and involves further physical and mental development.
4. Adolescence is when a person becomes physically able to reproduce. In adulthood, physical development is complete and body systems become less efficient. Death occurs at the end of life.

Visualizing Main Ideas

Copy and complete the following table on life stages.

Human Development

Stage of Life	Age Range	Physical Development
Infant		sits, stands, speaks words
		walks, speaks, writes, reads
Adolescent		
		end of muscular and skeletal growth

Using Vocabulary

amniotic sac p. 635
embryo p. 635
fetal stress p. 638
fetus p. 636
hormone p. 622
menstrual cycle p. 630
menstruation p. 630
ovary p. 629
ovulation p. 629
pregnancy p. 634
semen p. 628
sperm p. 628
testes p. 628
uterus p. 629
vagina p. 629

Fill in the blank with the correct vocabulary word or words.

1. _________ is a mixture of sperm and fluid.
2. The time of the development until the birth of a baby is known as _________.
3. During the first two months of pregnancy, the unborn child is known as a(n) _________.
4. The _________ is a hollow, pear-shaped muscular organ.
5. The _________ is the membrane that protects the unborn child.
6. The _________ is the organ that produces eggs.

Checking Concepts

Choose the word or phrase that best answers the question.

7. Where is the egg usually fertilized?
   A) oviduct **C)** vagina
   B) uterus **D)** ovary
8. What are the chemicals produced by the endocrine system?
   A) enzymes **C)** hormones
   B) target tissues **D)** saliva
9. Which gland produces melatonin?
   A) adrenal **C)** pancreas
   B) thyroid **D)** pineal
10. Where does the embryo develop?
   A) oviduct **C)** uterus
   B) ovary **D)** vagina

Use the figure below to answer question 11.

Prevalence of Diabetes per 100 Adults, United States, 2001

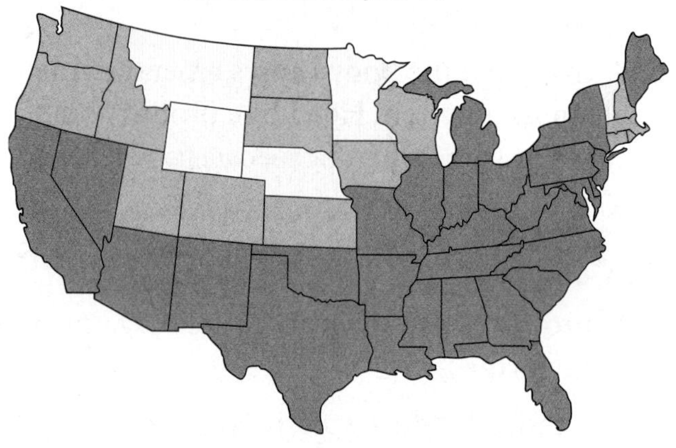

KEY: < 4% 4–4.9% 5–5.9% 6+%

11. Using the figure above, which state has the lowest incidence of diabetes?
   A) Wyoming **C)** Michigan
   B) Florida **D)** Washington
12. What is the monthly process that releases an egg called?
   A) fertilization **C)** menstruation
   B) ovulation **D)** puberty
13. What is the union of an egg and a sperm?
   A) fertilization **C)** menstruation
   B) ovulation **D)** puberty
14. During what stage of development does the amniotic sac form?
   A) zygote **C)** fetus
   B) embryo **D)** newborn
15. When does puberty occur?
   A) childhood **C)** adolescence
   B) adulthood **D)** infancy
16. During which period does growth stop?
   A) childhood **C)** adolescence
   B) adulthood **D)** infancy

Science Online life.msscience.com/vocabulary_puzzlemaker

Thinking Critically

17. **List** the effects that adrenal gland hormones can have on your body as you prepare to run a race.

18. **Explain** the similar functions of the ovaries and testes.

Use the diagram below to answer question 19.

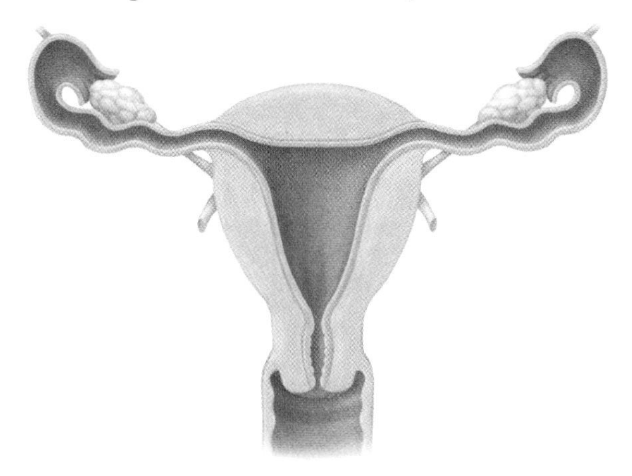

19. **Identify** the structure in the above diagram in which each process occurs: ovulation, fertilization, and implantation.

20. **Compare and contrast** your endocrine system with the thermostat in your home.

21. **Explain** if quadruplets—four babies born at one birth—are always identical or always fraternal, or if they can be either.

22. **Predict** During the ninth month of pregnancy, the fetus develops a white, greasy coating. Predict what the function of this coating might be.

23. **Form a hypothesis** about the effect of raising identical twins apart from each other.

24. **Classify** each of the following structures as female or male and internal or external: ovary, penis, scrotum, testes, uterus, and vagina.

Performance Activities

25. **Letter** Find newspaper or magazine articles on the effects of smoking on the health of the developing embryo and newborn. Write a letter to the editor about why a mother's smoking is damaging to her unborn baby's health.

Applying Math

26. **Blood Sugar Levels** Carol is diabetic and has a fasting blood sugar level of 180 mg/dL. Luisa does not have diabetes and has a fasting blood sugar level of 90 mg/dL. Express as a percentage how much higher the fasting blood sugar level is for Carol as compared to that for Luisa.

Use the graph below to answer questions 27 and 28.

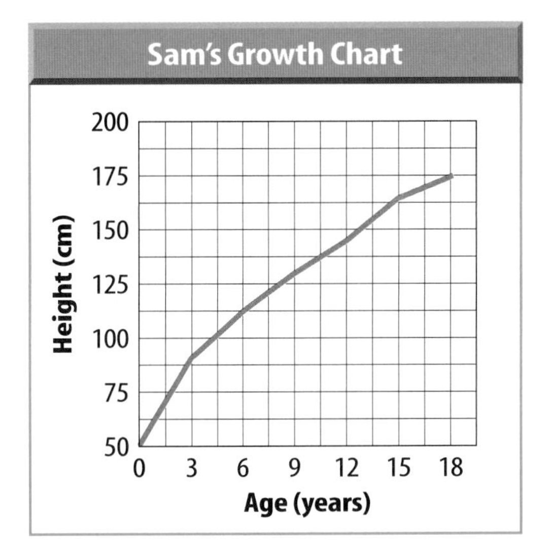

27. **Early Childhood Growth** The graph above charts Sam's growth from birth to 18 years of age. According to the graph, how much taller was Sam at 12 years of age than he was at 3 years of age?

28. **Adolescent Growth** According to the graph, how much did Sam grow between 12 and 18 years of age?

Standardized Test Practice

Part 1 Multiple Choice

Record your answers on the answer sheet provided by your teacher or on a sheet of paper.

1. When do eggs start to develop in the ovaries?
A. before birth
B. at puberty
C. during childhood
D. during infancy

Use the graph below to answers questions 2 and 3.

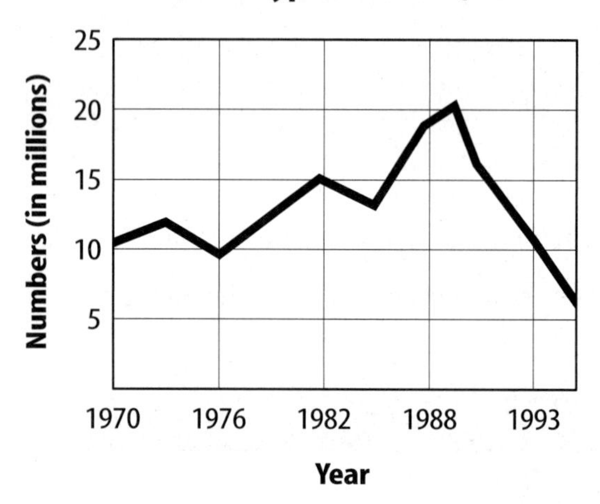

2. According to the information in the graph, in which year was the syphilis rate the lowest?
A. 1976
B. 1982
C. 1988
D. 1993

3. According to the information in the graph, during which years was there a decrease in the syphilis rate?
A. 1970–1972
B. 1976–1982
C. 1988–1990
D. 1990–1993

4. Which of the following glands is found in the neck?
A. pineal
B. adrenal
C. thyroid
D. pancreas

Test-Taking Tip

Bar Graphs On a bar graph, line up each bar with its corresponding value by laying your pencil between the two points.

5. What is the mixture of sperm and fluid called?
A. semen
B. testes
C. seminal vesicle
D. epididymis

Use the table below to answer questions 6–8.

Results of Folic Acid on Development of Neural Tube Defect		
Group	**Babies with Neural Tube Defect**	**Babies without Neural Tube Defect**
Group I—Folic Acid	6	497
Group II—No Folic Acid	21	581

(From CDC)

6. Researchers have found that the B-vitamin folic acid can prevent neural tube defects. In a study done in Europe in 1991, one group of pregnant women was given extra folic acid, and the other group did not receive extra folic acid. What percentage of babies were born with a neural tube defect in Group II?
A. 1.0%
B. 2.5%
C. 3.0%
D. 4.0%

7. What percentage of babies were born with a neural tube defect in Group I?
A. 1.0%
B. 2.5%
C. 3.0%
D. 4.0%

8. Which of the following statements is true regarding the data in this table?
A. Folic acid had no effect on the percentage of babies with a neural tube defect.
B. Extra folic acid decreased the percentage of babies with a neural tube defect.
C. Extra folic acid increased the percentage of babies with a neural tube defect.
D. Group I and Group II had the same percentage of babies born with a neural tube defect.

Part 2 Short Response/Grid In

Record your answers on the answer sheet provided by your teacher or on a sheet of paper.

9. How are endocrine glands different from salivary glands?

10. What does parathyroid hormone do for the body?

11. What is the function of the cilia in the oviduct?

Use the illustration below to answer questions 12 and 13.

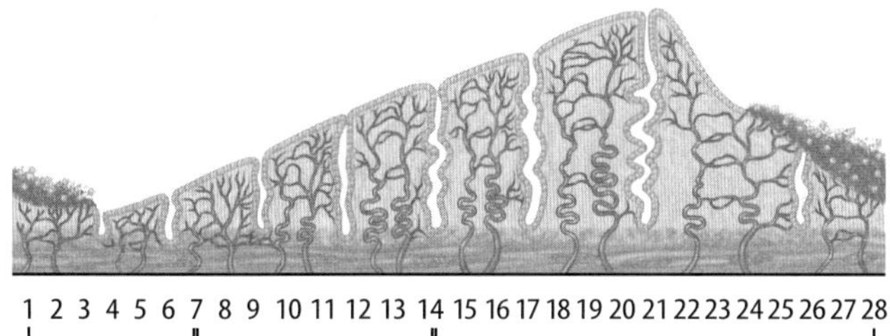

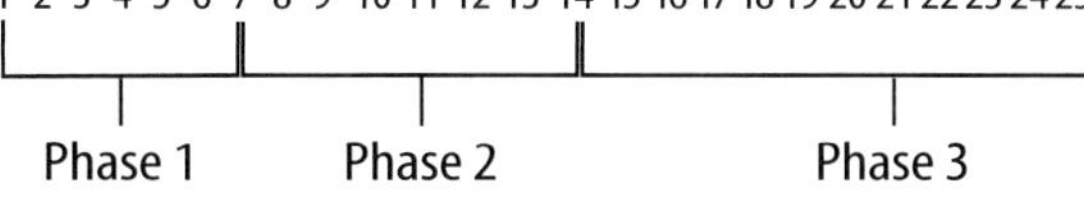

12. According to the illustration, what percentage of the menstrual cycle is phase 3?

13. According to the illustration, what percentage of the menstrual cycle is phase 2?

14. According to the illustration, on which day does ovulation occur?

15. During which stage of development before birth does amniotic fluid develop? What is the purpose of amniotic fluid?

16. During which stage of development after birth is physical growth and development the most rapid?

17. Rubella, also know as German measles, is caused by a virus. If a pregnant woman is infected with rubella, the virus can affect the formation of major organs, such as the heart, in the fetus. During which stage of development before birth would a rubella infection be most dangerous?

Part 3 Open Ended

Record your answers on a sheet of paper.

18. Predict how each of the following factors may affect sperm production: hot environment, illness with fever, testes located inside the body cavity, and injury to the testes. Explain your answer.

19. Sexually transmitted diseases can cause infection of the female reproductive organs, including the oviduct. Infection of the oviduct can result in scarring. What might happen to an egg that enters a scarred oviduct?

Use the table below to answer question 20.

Pre-eclampsia Risk in Pregnancy

Risk Factors	Risk Ratio
First pregnancy	3:1
Over 40 years of age	3:1
Family history	5:1
Chronic hypertension	10:1
Chronic renal disease	20:1
Antiphospholipid syndrome	10:1
Diabetes mellitus	2:1
Twin birth	4:1
Angiotensinogen gene T235	
Homozygous	20:1
Heterozygous	4:1

20. Pre-eclampsia is a condition that can develop in a woman after 20 weeks of pregnancy. It involves the development of hypertension or high blood pressure, an abnormal amount of protein in urine, and swelling. Infer why a woman with chronic hypertension has a higher risk of developing pre-eclampsia than a woman without hypertension.

chapter 23

Immunity and Disease

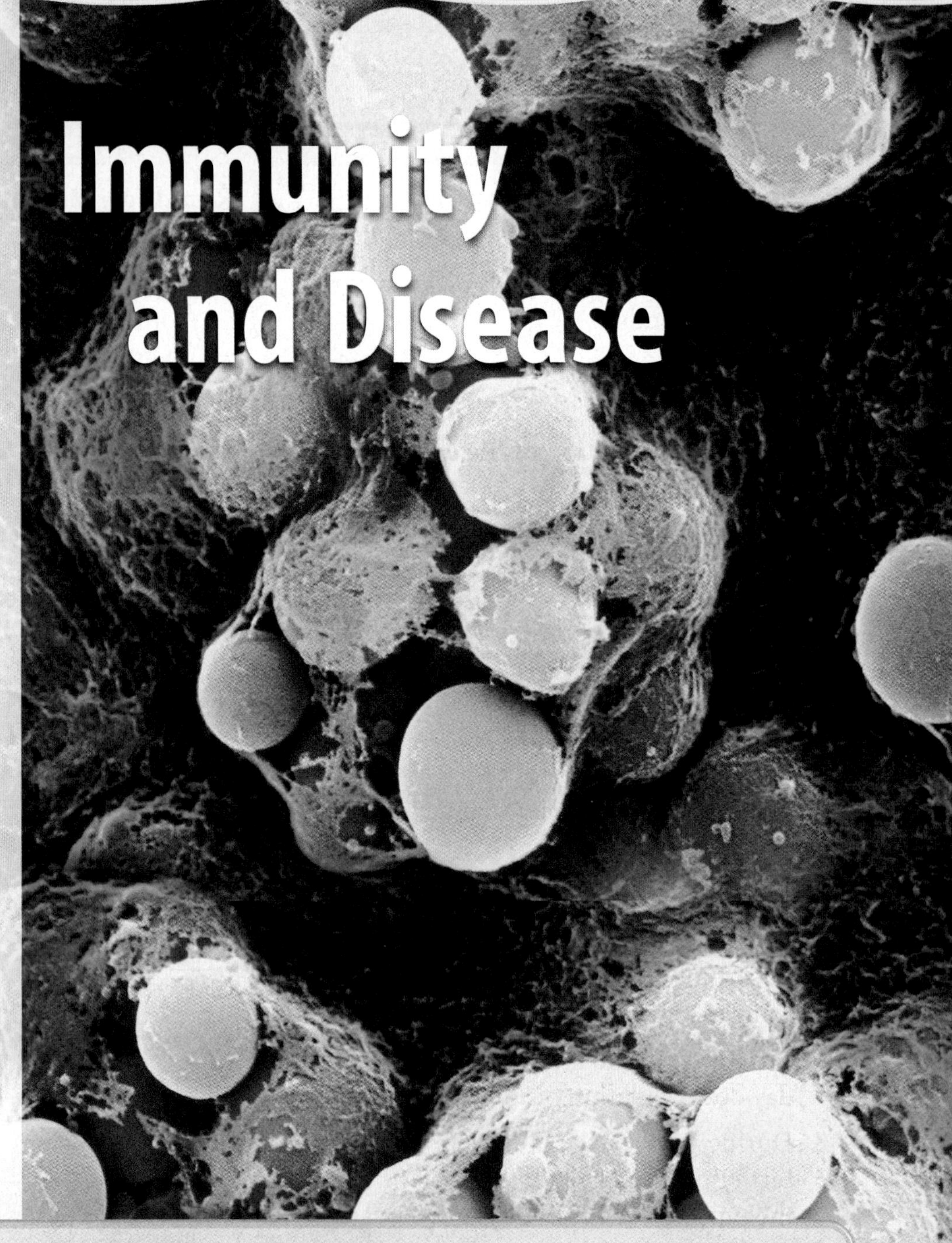

The BIG Idea

Your immune system provides defenses against infectious and noninfectious diseases.

SECTION 1
The Immune System

Main Idea Any organ, tissue, or cell that prevents pathogens from entering or surviving in the body is part of the immune system.

SECTION 2
Infectious Diseases

Main Idea A disease that is caused by a pathogen and moves from an organism or the environment to another organism is an infectious disease.

SECTION 3
Noninfectious Diseases

Main Idea A noninfectious disease occurs when cells, tissues, or organs do not function normally.

Attacked by Bacteria

You may not know it, but there's a war being fought in your body. Every second of your life your body is fighting harmful attacks. Sometimes your white blood cells are strong enough to fight alone. But, sometimes your body needs help from the laboratory —vaccines or medicines.

Science Journal Write a paragraph describing a battle between your white cells and a foreign invader.

Start-Up Activities

How do diseases spread?

Knowing how diseases are spread will help you understand how your body fights disease. You can discover one way diseases are spread by doing the following lab.

1. Wash your hands before and after this lab. Don't touch your face until the lab is completed and your hands are washed.
2. Work with a partner. Place a drop of peppermint food flavoring on a cotton ball. Pretend that the flavoring is a mass of cold viruses.
3. Use the cotton ball to rub an X over the palm of your right hand. Let it dry.
4. Shake hands with your partner.
5. Have your partner shake hands with another student. Then each student should smell their hands.
6. **Think Critically** In your Science Journal, note how many persons your "virus" infected. Write a paragraph describing some ways the spread of diseases could be stopped.

Classifying Diseases Make the following Foldable to classify human diseases as either infectious or noninfectious.

STEP 1 **Fold** a sheet of paper in half lengthwise.

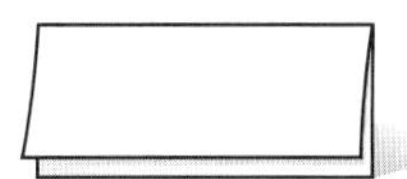

STEP 2 **Fold** paper down 2.5 cm from the top. (Hint: From the tip of your index finger to your middle knuckle is about 2.5 cm.)

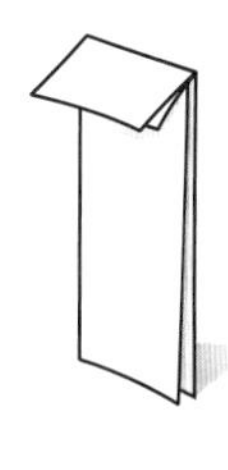

STEP 3 **Open and draw** lines along the 2.5-cm fold. **Label** as shown.

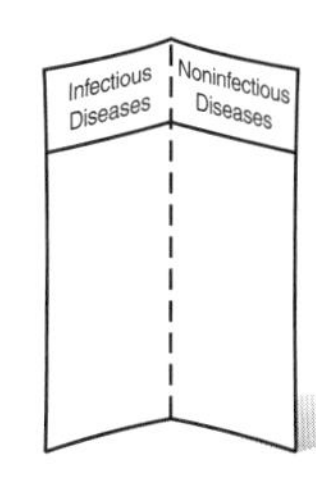

Read and Write As you read the chapter, classify human diseases as infectious or noninfectious by listing them on the proper fold.

Preview this chapter's content and activities at life.msscience.com

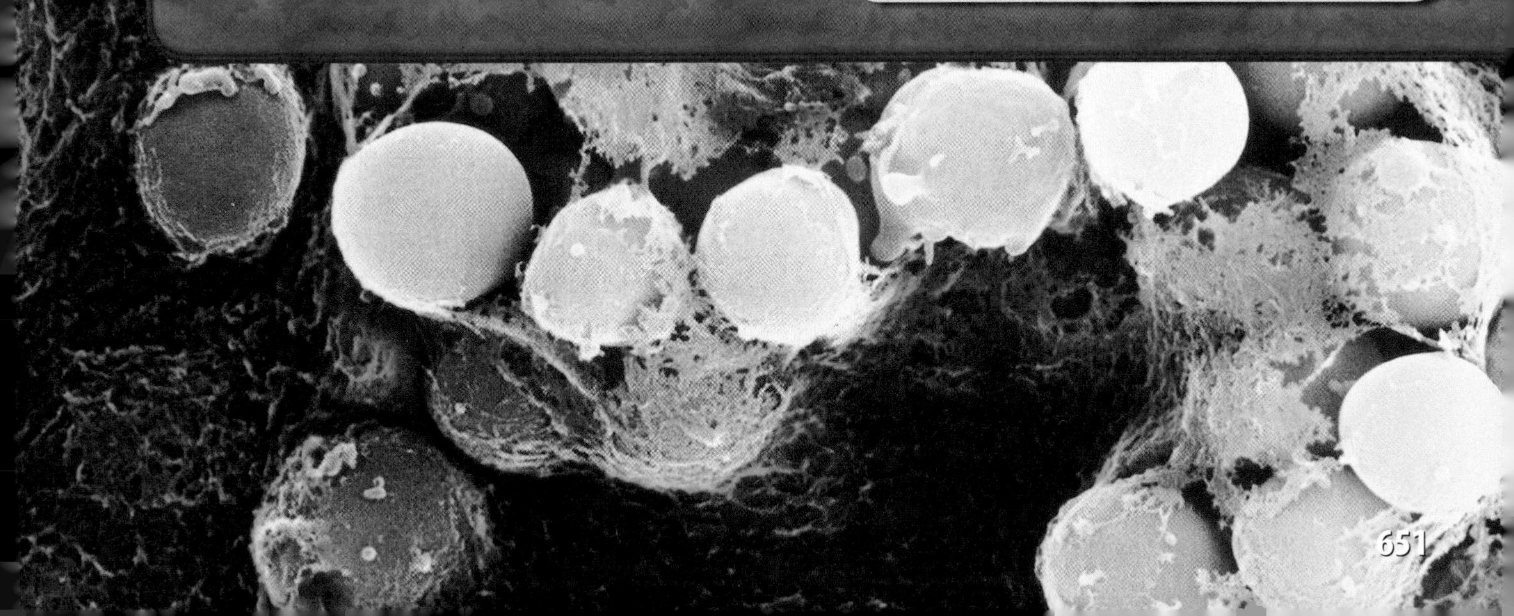

Get Ready to Read

Summarize

1 Learn It! Summarizing helps you organize information, focus on main ideas, and reduce the amount of information to remember. To summarize, restate the important facts in a short sentence or paragraph. Be brief and do not include too many details.

2 Practice It! Read the text on page 669 labeled Cancer. Then read the summary below and look at the important facts from that passage.

Summary

Cancer has been known since ancient times. It is a complicated disease. One characteristic of cancer cells is uncontrolled growth. This growth can result in a tumor anywhere in the body. Cancerous cells from a tumor can invade other tissues in the body.

Important Facts

Cancer has been a disease of humans since ancient times. Egyptian mummies... medieval manuscripts report details about the disease.

Cancer is the name given to a group of closely related diseases that result from uncontrolled cell growth. It is a complicated disease, and no one fully understands how cancers form.

Certain regulatory molecules in the body control the beginning and ending of cell division. If this control is lost, a mass of cells called a tumor (TEW mur) results from this abnormal growth. Tumors may occur anywhere in your body.

Cancerous cells may leave a tumor and spread uncontrollably through the blood and lymph vessels, then invade other tissues.

3 Apply It! Practice summarizing as you read this chapter. Stop after each section and write a brief summary.

Reading Tip

Reread your summary to make sure you didn't change the author's original meaning or ideas.

Target Your Reading

Use this to focus on the main ideas as you read the chapter.

1. **Before you read** the chapter, respond to the statements below on your worksheet or on a numbered sheet of paper.
   - Write an **A** if you **agree** with the statement.
   - Write a **D** if you **disagree** with the statement.

2. **After you read** the chapter, look back to this page to see if you've changed your mind about any of the statements.
   - If any of your answers changed, explain why.
   - Change any false statements into true statements.
   - Use your revised statements as a study guide.

Before You Read A or D	Statement	After You Read A or D
	1 A pathogen is an organism or a virus that causes disease.	
	2 Active immunity occurs when you get a vaccination.	
	3 Your skin can protect you from disease.	
	4 Antigens form in response to antibodies.	
	5 White blood cells patrol your body and destroy pathogens.	
	6 Infectious diseases can move from one organism to another organism by a third organism.	
	7 All sexually transmitted diseases are infectious.	
	8 Symptoms appear immediately when HIV infects your body.	
	9 Overexposure to toxins can cause chronic diseases.	
	10 Cilia and mucus do not help fight diseases.	

Science Online

Print out a worksheet of this page at life.msscience.com

section 1

The Immune System

as you read

What You'll Learn

- **Describe** the natural defenses your body has against disease.
- **Explain** the difference between an antigen and an antibody.
- **Compare and contrast** active and passive immunity.

Why It's Important

Your body's defenses fight the pathogens that you are exposed to every day.

Review Vocabulary
enzyme: a type of protein that speeds up chemical reactions in the body

New Vocabulary
- immune system
- antigen
- antibody
- active immunity
- passive immunity
- vaccination

Lines of Defense

The Sun has just begun to peek over the horizon, casting an orange glow on the land. A skunk ambles down a dirt path. Behind the skunk, you and your dog come over a hill for your morning exercise. Suddenly, the skunk stops and raises its tail high in the air. Your dog creeps forward. "No!" you shout. The dog ignores your command. Without further warning, the skunk sprays your dog. Yelping pitifully and carrying an awful stench, your dog takes off. The skunk used its scent to protect itself. Its first-line defense was to warn your dog with its posture. Its second-line defense was its spray. Just as the skunk protects itself from predators, your body also protects itself from harm.

Your body has many ways to defend itself. Its first-line defenses work against harmful substances and all types of disease-causing organisms, called pathogens (PA thuh junz). Your second-line defenses are specific and work against specific pathogens. This complex group of defenses is called your **immune system.** Tonsils, shown in **Figure 1,** are one of the immune system organs that protect your body.

Reading Check *What types of defenses does your body have?*

First-Line Defenses Your skin and respiratory, digestive, and circulatory systems are first-line defenses against pathogens. As shown in **Figure 2,** the skin is a barrier that prevents many pathogens from entering your body. Although most pathogens can't get through unbroken skin, they can get into your body easily through a cut or through your mouth and the membranes in your nose and eyes. The conditions on the skin can affect pathogens. Perspiration contains substances that can slow the growth of some pathogens. At times, secretions from the skin's oil glands and perspiration are acidic. Some pathogens cannot grow in this acidic environment.

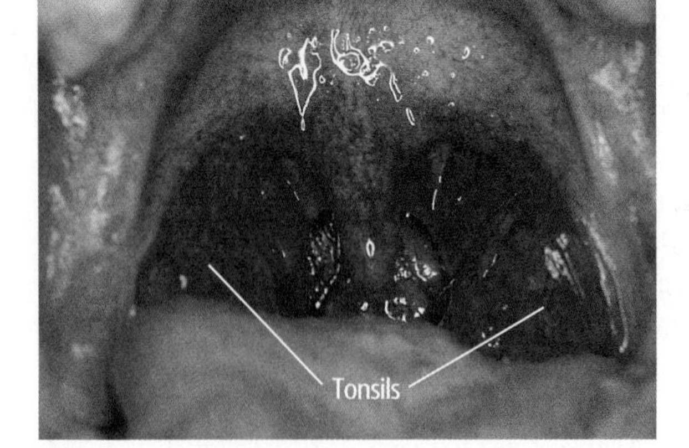

Figure 1 Tonsils help prevent infection in your respiratory and digestive tract.

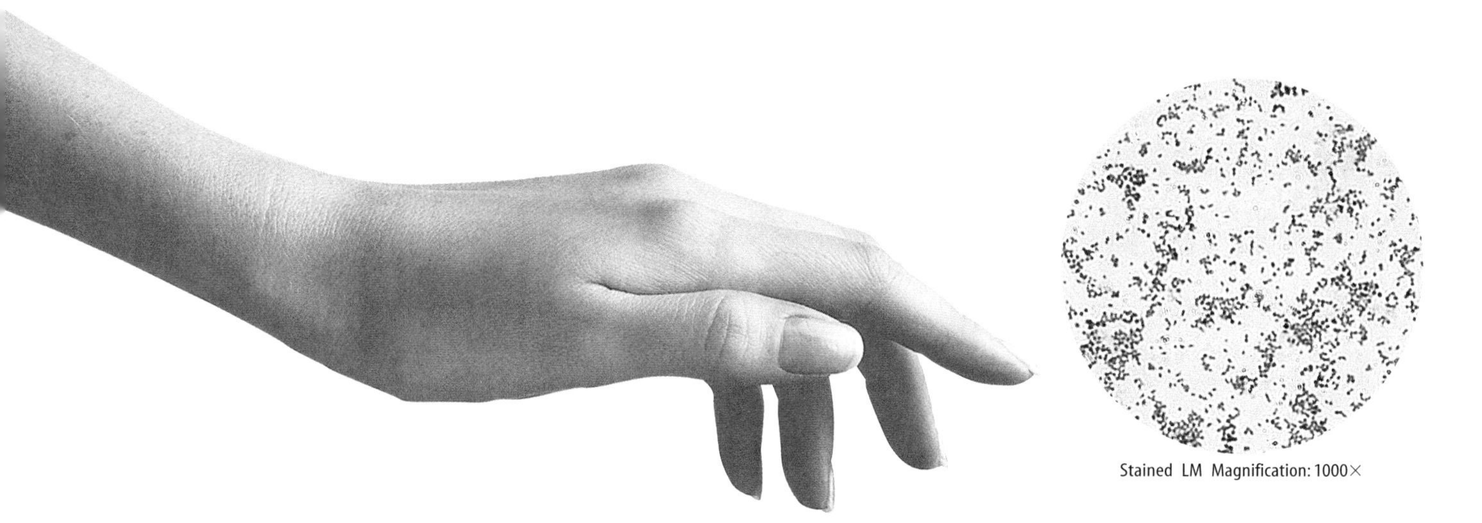

Stained LM Magnification: 1000×

Figure 2 Most pathogens, like the staphylococci bacteria shown above, cannot pass through unbroken skin.
Infer *what happens if staphylococci bacteria enter your body through your skin.*

Internal First-Line Defenses Your respiratory system traps pathogens with hairlike structures, called cilia (SIH lee uh), and mucus. Mucus contains an enzyme that weakens the cell walls of some pathogens. When you cough or sneeze, you get rid of some of these trapped pathogens.

Your digestive system has several defenses against pathogens—saliva, enzymes, hydrochloric acid, and mucus. Saliva in your mouth contains substances that kill bacteria. Also, enzymes (EN zimez) in your stomach, pancreas, and liver help destroy pathogens. Hydrochloric acid in your stomach helps digest your food. It also kills some bacteria and stops the activity of some viruses that enter your body on the food that you eat. The mucus found on the walls of your digestive tract contains a chemical that coats bacteria and prevents them from binding to the inner lining of your digestive organs.

Your circulatory system contains white blood cells, like the one in **Figure 3,** that surround and digest foreign organisms and chemicals. These white blood cells constantly patrol your body, sweeping up and digesting bacteria that invade. They slip between cells of tiny blood vessels called capillaries. If the white blood cells cannot destroy the bacteria fast enough, you might develop a fever. Many pathogens are sensitive to temperature. A slight increase in body temperature slows their growth and activity but speeds up your body's defenses.

Inflammation When tissue is damaged by injury or infected by pathogens, it becomes inflamed. Signs of inflammation include redness, temperature increase, swelling, and pain. Chemical substances released by damaged cells cause capillary walls to expand, allowing more blood to flow into the area. Other chemicals released by damaged tissue attract certain white blood cells that surround and take in pathogenic bacteria. If pathogens get past these first-line defenses, your body uses another line of defense called specific immunity.

Figure 3 A white blood cell leaves a capillary. It will search out and destroy harmful microorganisms in your body tissues.

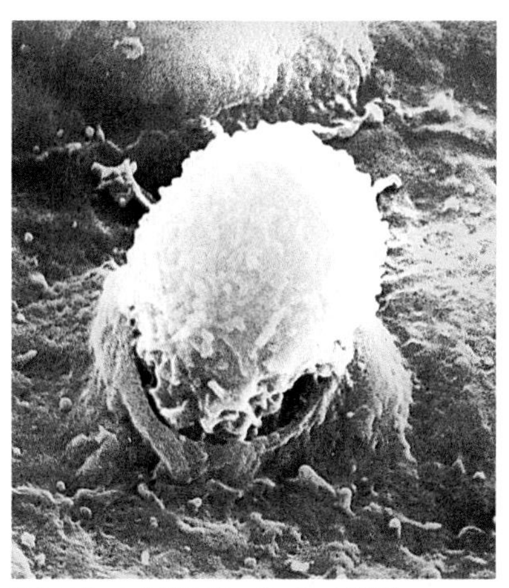

Color-enhanced SEM Magnification: 3450×

Topic: Disease Theory
Visit life.msscience.com for Web links to information about one of the historical theories of disease—the four body humors.

Activity Make a picture book describing the humoral theory of disease.

Specific Immunity When your body fights disease, it is battling complex molecules that don't belong there. Molecules that are foreign to your body are called **antigens** (AN tih junz). Antigens can be separate molecules or they can be found on the surface of a pathogen. For example, the protein in the cell membrane of a bacterium can be an antigen. When your immune system recognizes molecules as being foreign to your body, as in **Figure 4,** special lymphocytes called T cells respond. Lymphocytes are a type of white blood cell. One type of T cells, called killer T cells, releases enzymes that help destroy invading foreign matter. Another type of T cells, called helper T cells, turns on the immune system. They stimulate other lymphocytes, known as B cells, to form antibodies.

An **antibody** is a protein made in response to a specific antigen. The antibody attaches to the antigen and makes it useless. This can happen in several ways. The pathogen might not be able to stay attached to a cell. It might be changed in such a way that a killer T cell can capture it more easily or the pathogen can be destroyed.

Reading Check *What is an antibody?*

Another type of lymphocyte, called memory B cells, also has antibodies for the specific pathogen. Memory B cells remain in the blood ready to defend against an invasion by that same pathogen another time.

Figure 4 The response of your immune system to disease-causing organisms can be divided into four steps—recognition, mobilization, disposal, and immunity.
Explain *the function of B cells.*

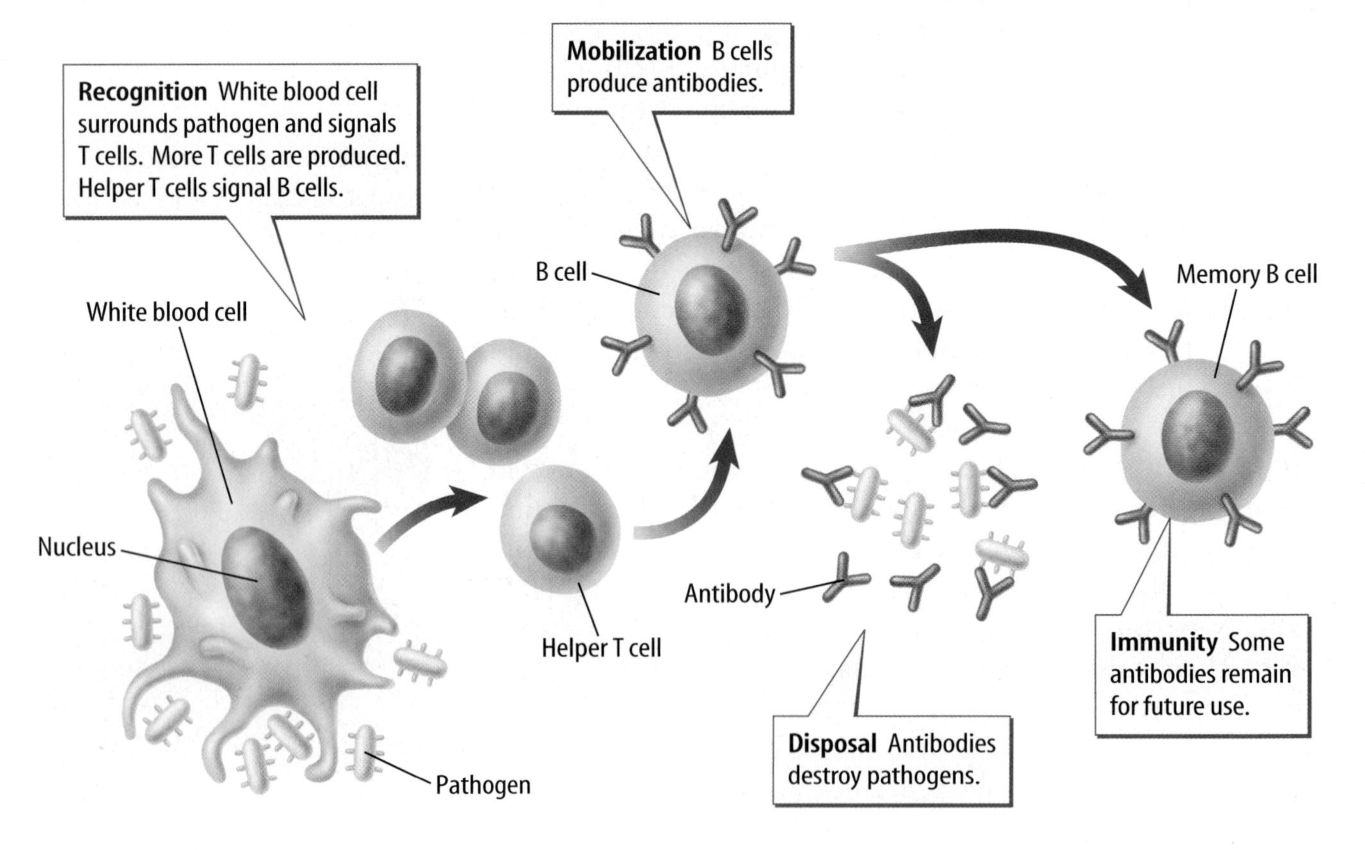

Active Immunity Antibodies help your body build defenses in two ways—actively and passively. In **active immunity** your body makes its own antibodies in response to an antigen. **Passive immunity** results when antibodies that have been produced in another animal are introduced into your body.

When a pathogen invades your body and quickly multiplies, you get sick. Your body immediately starts to make antibodies to attack the pathogen. After enough antibodies form, you usually get better. Some antibodies stay on duty in your blood, and more are produced rapidly if the pathogen enters your body again. Because of this defense system you usually get certain diseases such as chicken pox only once. Why can you catch a cold over and over? There are many different cold viruses that give you similar symptoms. As you grow older and are exposed to many more types of pathogens, you will build immunity to each one.

Vaccination A vaccine is a form of an antigen that gives you immunity against a disease. A vaccine only can prevent a disease, not cure it. The process of giving a vaccine by injection or by mouth is called **vaccination.** If a specific vaccine is injected into your body, your body forms antibodies against that pathogen. If you later encounter the same pathogen, your bloodstream already has antibodies that are needed to fight and destroy it. Vaccines have helped reduce cases of childhood diseases, as shown in **Table 1.**

Determining Reproduction Rates

Procedure

1. Place **one penny** on a table. Imagine that the penny is a bacterium that can divide every 10 min.
2. Place **two pennies** below the first penny to form a triangle. These represent the two new bacteria after the first bacterium divides.
3. Repeat three more divisions, placing two pennies under each penny as described above.
4. Calculate how many bacteria you would have after 5 h of reproduction. Graph your data.

Analysis

1. How many bacteria are present after 5 h?
2. Why is it important to take antibiotics promptly if you have an infection?

Try at Home

Table 1 Annual Cases of Disease Before and After Vaccine Availability in the U.S.

Disease	Before	After
Measles	503,282	89
Diptheria	175,885	1
Tetanus	1,314	34
Mumps	152,209	606
Rubella	47,745	345
Pertussis (whooping cough)	147,271	6,279

Data from the National Immunization Program, CDC

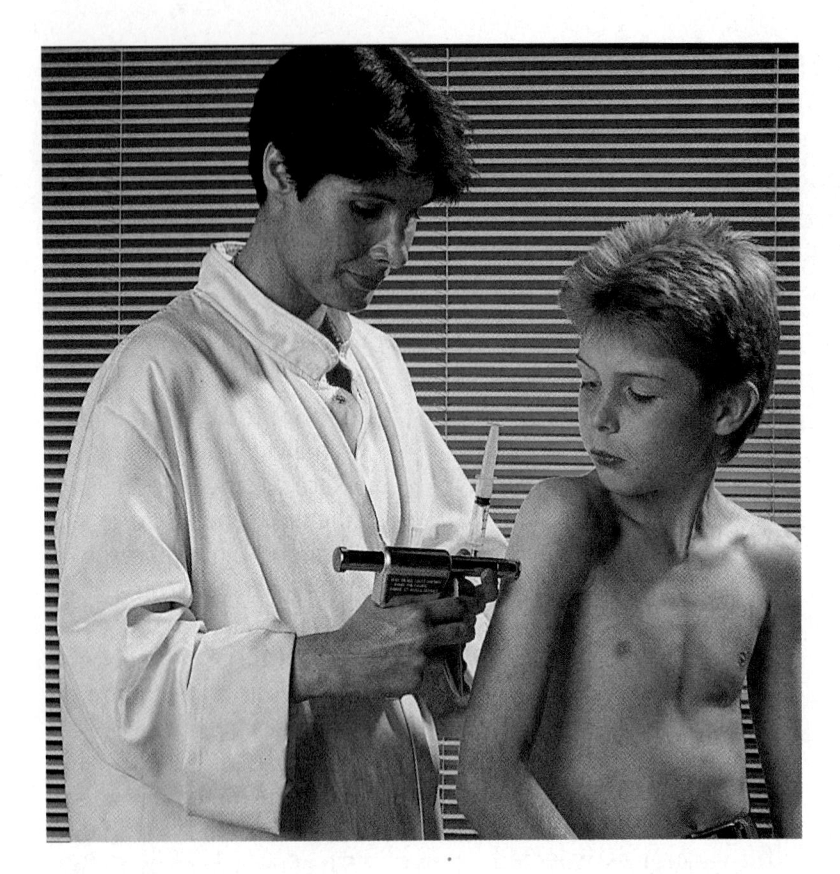

Figure 5 The Td vaccine, which protects against tetanus and diphtheria, usually is injected into the arm.

Passive Immunity Passive immunity does not last as long as active immunity does. For example, you were born with all the antibodies that your mother had in her blood. However, these antibodies stayed with you for only a few months. Because newborn babies lose their passive immunity in a few months, they need to be vaccinated to develop their own immunity.

Tetanus Tetanus is a disease caused by a common soil bacterium. The bacterium produces a chemical that paralyzes muscles. Puncture wounds, deep cuts, and other wounds can be infected by this bacterium. Several times in early childhood you received active vaccines, as shown in **Figure 5,** that stimulated antibody production to tetanus toxin. You should continue to get vaccines or boosters every ten years to maintain protection. Booster shots for diphtheria, which is a dangerous infectious respiratory disease, are given in the same vaccine with tetanus.

section 1 review

Summary

Lines of Defense

- Your body's immune system protects you from harmful substances called pathogens.
- First-line defenses work against harmful substances and all types of pathogens.
- Second-line defenses work against specific pathogens.
- Antibodies help protect your body against specific foreign molecules called antigens.
- Your body gets antibodies through active immunity and passive immunity.
- Vaccines help you develop active immunity against a disease.
- You need to receive booster shots for some vaccines to maintain protection.

Self Check

1. **Describe** how harmful bacteria cause infections in your body.
2. **List** the natural defenses your body has against disease.
3. **Explain** how your immune system reacts when it detects an antigen.
4. **Compare and contrast** active and passive immunity.
5. **Think Critically** Several diseases have symptoms similar to those of measles. Why doesn't the measles vaccine protect you from all of these diseases?

Applying Skills

6. **Make Models** Create models of the different types of T cells, antigens, and B cells from clay, construction paper, or other art materials. Use them to explain how T cells function in the immune system.

Science Online life.msscience.com/self_check_quiz

Infectious Diseases

Disease in History

For centuries, people have feared outbreaks of disease. The plague, smallpox, and influenza have killed millions of people worldwide. Today, the causes of these diseases are known, and treatments can prevent or cure them. But even today, there are diseases such as the Ebola virus in Africa that cannot be cured. Outbreaks of new diseases, such as severe acute respiratory syndrome (SARS), shown in **Table 2,** also occur.

Microorganisms With the invention of the microscope in the latter part of the seventeenth century, bacteria, yeast, and mold spores were seen for the first time. However, it took almost 200 years more to discover the relationship between some of them and disease. Scientists gradually learned that microorganisms were responsible for fermentation and decay. If decay-causing microorganisms could cause changes in other organisms, it was hypothesized that microorganisms could cause diseases and carry them from one person to another. Scientists did not make a connection between viruses and disease transmission until the late 1800s and early 1900s.

Table 2 Probable Cases of SARS (November 1, 2002 to July 7, 2003)

Country	Number of Cases	Number of Deaths
Canada	251	38
China	7,756	730
Singapore	206	32
United States	73	0
Vietnam	63	5
Other countries	90	7

Data from the World Health Organization

as you read

***What* You'll Learn**

- **Describe** the work of Pasteur, Koch, and Lister in the discovery and prevention of disease.
- **Identify** diseases caused by viruses and bacteria.
- **List** sexually transmitted diseases, their causes, and treatments.
- **Explain** how HIV affects the immune system.

***Why* It's Important**

You can help prevent certain illnesses if you know what causes disease and how disease spreads.

Review Vocabulary

protist: a one- or many-celled organism that lives in moist or wet surroundings

New Vocabulary

- pasteurization
- virus
- infectious disease
- biological vector
- sexually transmitted disease (STD)

Disease Organisms The French chemist Louis Pasteur learned that microorganisms cause disease in humans. Many scientists of his time did not believe that microorganisms could harm larger organisms, such as humans. However, Pasteur discovered that microorganisms could spoil wine and milk. He then realized that microorganisms could attack the human body in the same way. Pasteur invented **pasteurization** (pas chuh ruh ZAY shun), which is the process of heating a liquid to a specific temperature that kills most bacteria.

Today, it is known that many diseases are caused by bacteria, certain viruses, protists (PROH tihsts), or fungi. Bacteria cause tetanus, tuberculosis, strep throat, and bacterial pneumonia. Malaria and sleeping sickness are caused by protists. Fungi are the pathogens for athlete's foot and ringworm. Viruses are the cause of many common diseases—colds, influenza, AIDS, measles, mumps, smallpox, and SARS.

Many harmful bacteria that infect your body can reproduce rapidly. The conditions in your body, such as temperature and available nutrients, help the bacteria grow and multiply. Bacteria can slow down the normal growth and metabolic activities of body cells and tissues. Some bacteria even produce toxins that kill cells on contact.

A **virus** is a minute piece of genetic material surrounded by a protein coating that infects and multiplies in host cells. The host cells die when the viruses break out of them. These new viruses infect other cells, leading to the destruction of tissues or the interruption of vital body activities.

Disease Immunity Edward Jenner demonstrated that a vaccine could be produced to prevent smallpox. However, it wasn't until Louis Pasteur applied his germ theory to the process that the mechanism of vaccinations was understood. Pasteur demonstrated that germs cause diseases and that vaccines, which contained small amounts of disease organisms, could cause the body to build immunity to that disease without causing it. Research Jenner and write a summary in your Science Journal about his discovery of the smallpox vaccine.

✔ Reading Check *What is the relationship between a virus and a host cell?*

Pathogenic protists, such as the organisms that cause malaria, can destroy tissues and blood cells or interfere with normal body functions. In a similar manner, fungus infections can cause athlete's foot, nonhealing wounds, chronic lung disease, or inflammation of the membranes of the brain.

Koch's Rules Many diseases caused by pathogens can be treated with medicines. In many cases, these organisms need to be identified before specific treatment can begin. Today, a method developed in the nineteenth century still is used to identify organisms.

Pasteur may have shown that bacteria cause disease, but he didn't know how to tell which specific organism causes which disease. It was a young German doctor, Robert Koch, who first developed a way to isolate and grow one type of bacterium at a time, as shown in **Figure 6.**

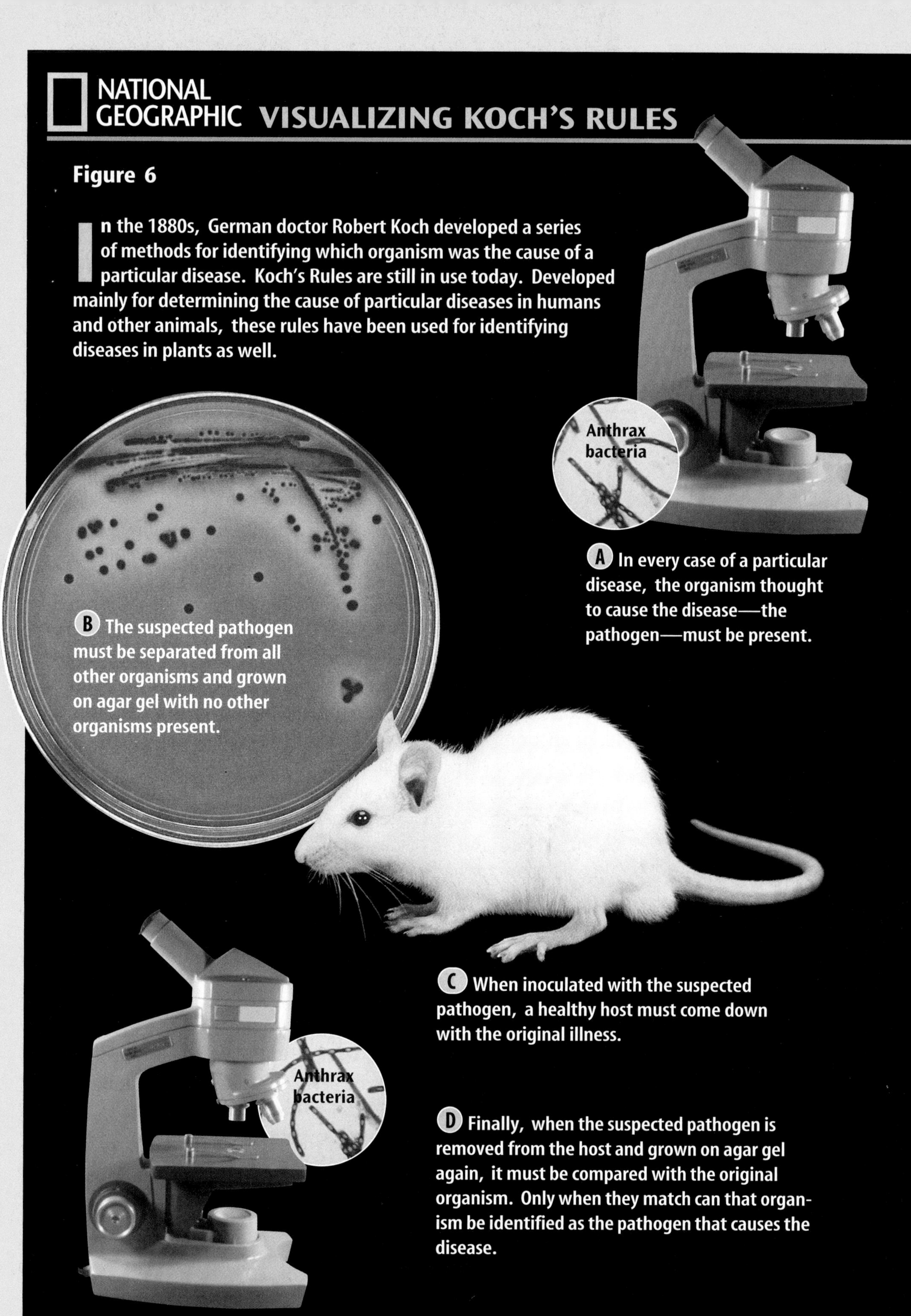

VISUALIZING KOCH'S RULES

Figure 6

In the 1880s, German doctor Robert Koch developed a series of methods for identifying which organism was the cause of a particular disease. Koch's Rules are still in use today. Developed mainly for determining the cause of particular diseases in humans and other animals, these rules have been used for identifying diseases in plants as well.

A In every case of a particular disease, the organism thought to cause the disease—the pathogen—must be present.

B The suspected pathogen must be separated from all other organisms and grown on agar gel with no other organisms present.

C When inoculated with the suspected pathogen, a healthy host must come down with the original illness.

D Finally, when the suspected pathogen is removed from the host and grown on agar gel again, it must be compared with the original organism. Only when they match can that organism be identified as the pathogen that causes the disease.

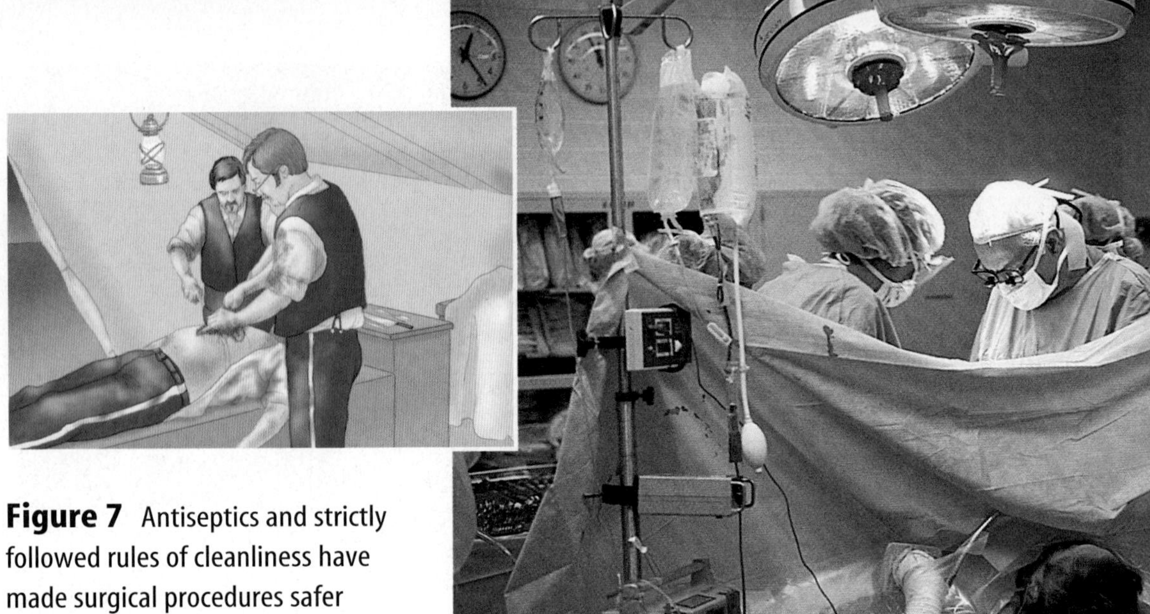

Figure 7 Antiseptics and strictly followed rules of cleanliness have made surgical procedures safer than they once were.
Describe *the differences you see in the two operating scenes shown.*

Keeping Clean Washing your hands before or after certain activities should be part of your daily routine. Restaurant employees are required to wash their hands immediately after using the rest room. Medical professionals wash their hands before examining each patient. However, hand washing was not always a routine, even for doctors. Into the late 1800s, doctors such as those in **Figure 7** regularly operated in their street clothes and with bare, unwashed hands. A bloody apron and well-used tools were considered signs of prestige for a surgeon. More patients died from the infections that they contracted during or after the surgery than from the surgery itself.

Joseph Lister, an English surgeon, recognized the relationship between the infection rate and cleanliness. Lister dramatically reduced the number of deaths among his patients by washing their skin and his hands with carbolic (kar BAH lihk) acid, which is a liquid that kills pathogens. Lister also used carbolic acid to clean his instruments and soak bandages, and he even sprayed the air with it. The odor was strong and it irritated the skin, but more and more people began to survive surgical procedures.

Modern Operating Procedures Today antiseptics and antiseptic soaps are used to kill pathogens on skin. Every person on the surgical team washes his or her hands thoroughly and wears sterile gloves and a covering gown. The patient's skin is cleaned around the area of the body to be operated on and then covered with sterile cloths. Tools that are used to operate on the patient and all operating room equipment also are sterilized. Even the air is filtered.

What are three ways that pathogens are reduced in today's operating room?

Mini LAB

Observing Antiseptic Action

Procedure

1. Place a few grains of **dried yeast** onto a **glass plate** or a saucer.
2. Add two drops of **hydrogen peroxide** to the yeast. Observe.
3. Clean up and wash hands before removing goggles.

Analysis

1. How does the action of hydrogen peroxide mechanically clean a wound?
2. Explain why hydrogen peroxide is classified as an antiseptic.

How Diseases Are Spread

You walk into your kitchen before school. Your younger sister sits at the table eating a bowl of cereal. She has a fever, a runny nose, and a cough. She coughs loudly. "Hey, cover your mouth! I don't want to catch your cold," you tell her. A disease that is caused by a virus, bacterium, protist, or fungus and is spread from an infected organism or the environment to another organism is called an **infectious disease.** Infectious diseases are spread by direct contact with the infected organism, through water and air, on food, by contact with contaminated objects, and by disease-carrying organisms called **biological vectors.** Examples of vectors that have been sources of disease are rats, birds, cats, dogs, mosquitoes, fleas, and flies, as shown in **Figure 8.**

People also can be carriers of disease. When you have influenza and sneeze, you expel thousands of virus particles into the air. Colds and many other diseases are spread through contact. Each time you turn a doorknob, press the button on a water fountain, or use a telephone, your skin comes in contact with bacteria and viruses, which is why regular handwashing is recommended. The Centers for Disease Control and Prevention (CDC) in Atlanta, Georgia, monitors the spread of diseases throughout the United States. The CDC also tracks worldwide epidemics and watches for diseases brought into the United States.

Figure 8 When flies land on food, they can transport pathogens from one location to another.

Applying Science

Has the annual percentage of deaths from major diseases changed?

Each year, many people die from diseases. Medical science has found numerous ways to treat and cure disease. Have new medicines, improved surgery techniques, and healthier lifestyles helped decrease the number of deaths from disease? By using your ability to interpret data tables, you can find out.

Identifying the Problem

The table to the right shows the percentage of total deaths due to six major diseases for a 50-year period. Study the data. Can you see any trends in the percentage of deaths?

Percentage of Deaths Due to Major Disease

Disease	Year			
	1950	1980	1990	2000
Heart	37.1	38.3	33.5	29.6
Cancer	14.6	20.9	23.5	23.0
Stroke	10.8	8.6	6.7	7.0
Diabetes	1.7	1.8	2.2	2.9
Pneumonia and flu	3.3	2.7	3.7	2.7

Solving the Problem

1. Has the percentage increased for any disease that is listed?
2. What factors could have contributed to this increase?

Sexually Transmitted Diseases

Infectious diseases that are passed from person to person during sexual contact are called **sexually transmitted diseases (STDs).** STDs are caused by bacteria or viruses.

Bacterial STDs Gonorrhea (gah nuh REE uh), chlamydia (kluh MIH dee uh), and syphilis (SIH fuh lus) are STDs caused by bacteria. The bacteria that cause gonorrhea and syphilis are shown in **Figure 9.** A person may have gonorrhea or chlamydia for some time before symptoms appear. When symptoms do appear, they can include painful urination, genital discharge, and genital sores. Antibiotics are used to treat these diseases. Some of the bacteria that cause gonorrhea may be resistant to the antibiotics usually used to treat the infection. However, the disease usually can be treated with other antibiotics. If they are untreated, gonorrhea and chlamydia can leave a person sterile because the reproductive organs can be damaged permanently.

Syphilis has three stages. In stage 1, a sore that lasts 10 to 14 days appears on the mouth or genitals. Stage 2 may involve a rash, fever, and swollen lymph glands. Within weeks to a year, these symptoms usually disappear. The person with syphilis often believes that the disease has gone away, but it hasn't. If he or she does not seek treatment, the disease advances to stage 3, when syphilis may infect the cardiovascular and nervous systems. In all stages, syphilis is treatable with antibiotics. However, the damage to body organs in stage 3 cannot be reversed and death can result.

Figure 9 Bacteria that cause gonorrhea and syphilis can be destroyed with antibiotics. **Explain** *why a person might not get treatment for a syphilis infection.*

Gonorrhea bacteria

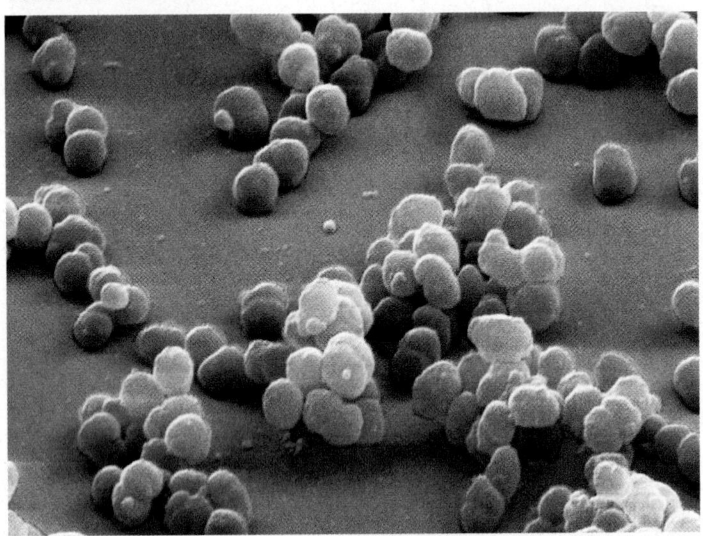

Color-enhanced TEM Magnification: 12000×

Syphilis bacteria

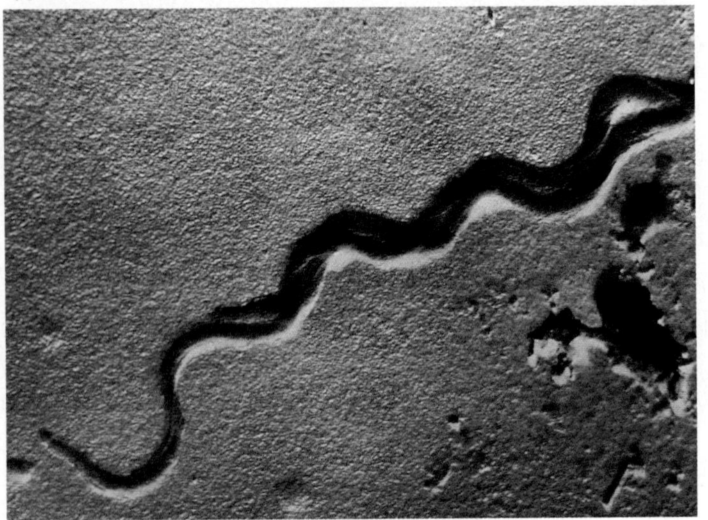

Color-enhanced SEM Magnification: 45000×

Viral STDs Genital herpes, a lifelong viral disease, causes painful blisters on the sex organs. This type of herpes can be transmitted during sexual contact or from an infected mother to her child during birth. The herpes virus hides in the body for long periods of time and then reappears suddenly. Herpes has no cure, and no vaccine can prevent it. However, the symptoms of herpes can be treated with antiviral medicines.

HIV and Your Immune System

Human immunodeficiency virus (HIV) can exist in blood and body fluids. This virus can hide in body cells, sometimes for years. You can become infected with HIV by having sex with an HIV-infected person or by reusing an HIV-contaminated hypodermic needle for an injection. However, a freshly unwrapped sterile needle cannot transmit infection. The risk of getting HIV through blood transfusion is small because all donated blood is tested for the presence of HIV. A pregnant female with HIV can infect her child when the virus passes through the placenta. The child also may become infected from contacts with blood during the birth process or when nursing after birth.

What are ways that a person can become infected with HIV?

HIV cannot multiply outside the body, and it does not survive long in the environment. The virus cannot be transmitted by touching an infected person, by handling objects used by the person unless they are contaminated with body fluids, or from contact with a toilet seat.

AIDS An HIV infection can lead to Acquired Immune Deficiency Syndrome (AIDS), which is a disease that attacks the body's immune system. HIV, as shown in **Figure 10,** is different from other viruses. It attacks the helper T cells in the immune system. The virus enters the T cell and multiplies. When the infected cell bursts open, it releases more HIV. These infect other T cells. Soon, so many T cells are destroyed that not enough B cells are stimulated to produce antibodies. The body no longer has an effective way to fight invading antigens. The immune system then is unable to fight HIV or any other pathogen. For this reason, when people with AIDS die it is from other diseases such as tuberculosis (too bur kyuh LOH sus), pneumonia, or cancer.

Through 2004, more than 944,000 cases of AIDS were documented in the United States. At this time the disease has no known cure. However, several medications help treat AIDS in some patients. One group of medicines interferes with the way that the virus multiplies in the host cell and is effective if it is used in the early stages of the disease. Another group of medicines that is being tested blocks the entrance of HIV into the host cell. These medicines prevent the pathogen from binding to the cell's surface.

Topic: AIDS

Visit life.msscience.com for Web links to information about the number of AIDS cases worldwide.

Activity Make a graph showing the number of AIDS cases in seven countries.

Figure 10 A person can be infected with HIV and not show any symptoms of the infection for several years.
Infer *why this characteristic makes the spread of AIDS more likely.*

Color-enhanced TEM Magnification: 40000×

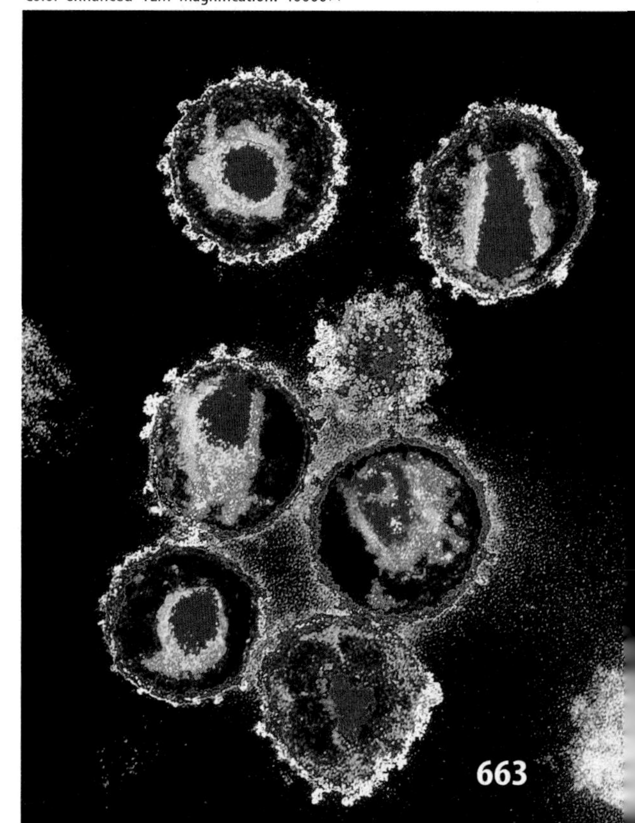

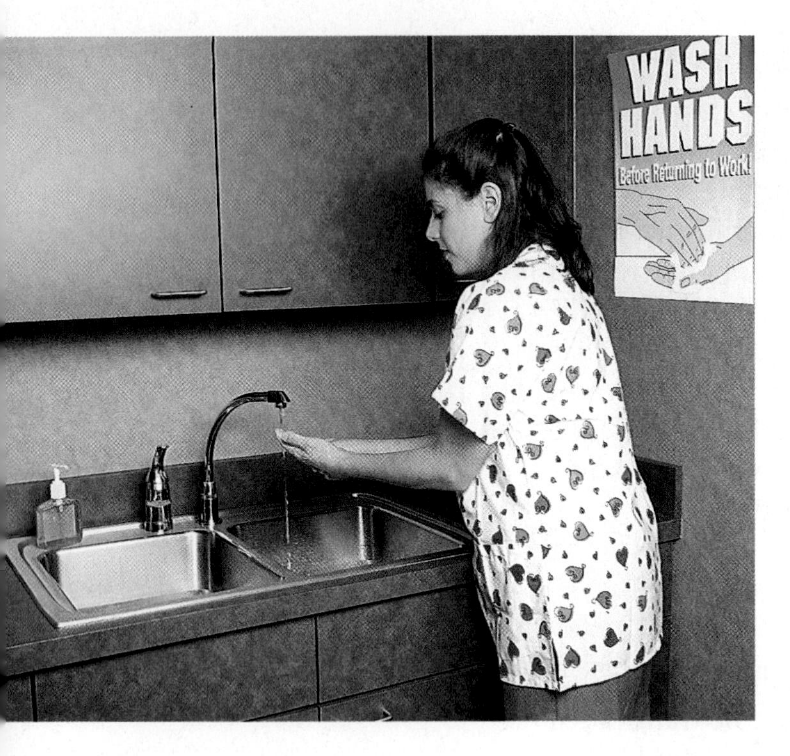

Figure 11 Proper hand washing includes using warm water and soap. The soapy lather must be rubbed over the hands, wrists, fingers, and thumbs for 15–20 s. Thoroughly rinse and dry with a clean towel.

Fighting Disease

Washing a small wound with soap and water is the first step in preventing an infection. Cleaning the wound with an antiseptic and covering it with a bandage are other steps. Is it necessary to wash your body to help prevent diseases? Yes! In addition to reducing body odor, washing your body removes and destroys some surface microorganisms. In medical facilities, hand washing, shown in **Figure 11,** is important to reduce the spread of pathogens. It is also important for everyone to wash his or her hands to reduce the spread of disease.

In your mouth, microorganisms are responsible for mouth odor and tooth decay. Using dental floss and routine tooth brushing keep these organisms under control.

Exercise and good nutrition help the circulatory and respiratory systems work more effectively. Good health habits, including getting enough rest and eating well-balanced meals, can make you less susceptible to the actions of disease organisms such as those that cause colds and flu. Keeping up with recommended immunizations and having annual health checkups also can help you stay healthy.

section 2 review

Summary

Disease in History

- Pasteur, Koch, and Lister played key roles in disease discovery and prevention.

How Diseases Are Spread

- Diseases are spread by air, water, food, animals, and through contact with pathogens.

Sexually Transmitted Diseases

- STDs, such as gonorrhea and herpes, are caused by either bacteria or viruses.

HIV and Your Immune System

- HIV can lead to AIDS, a disease of the immune system.

Fighting Disease

- Cleanliness, exercise, and good health habits can help prevent disease.

Self Check

1. **Explain** how the discoveries of Pasteur, Koch, and Lister help in the battle against the spread of disease.
2. **Identify** three infectious diseases caused by a virus and three caused by a bacterium.
3. **Define** sexually transmitted diseases. How are they contracted and treated?
4. **Describe** the way HIV affects the immune system and how it is different from other viruses.
5. **Think Critically** In what ways does Koch's procedure demonstrate the use of scientific methods?

Applying Skills

6. **Recognize Cause and Effect** How is poor cleanliness related to the spread of disease? Write your answer in your Science Journal.

Science Online life.msscience.com/self_check_quiz

MICROORGANISMS AND DISEASE

Microorganisms are everywhere. Washing your hands and disinfecting items you use helps remove some of these organisms.

Real-World Question

How do microorganisms cause infection?

Goals

- **Observe** the transmission of microorganisms.
- **Relate** microorganisms to infections.

Materials

fresh apples (6)
rotting apple
rubbing alcohol (5 mL)
self-sealing plastic bags (6)
labels and pencil
gloves
paper towels
sandpaper
cotton ball
soap and water
newspaper

Safety Precautions

WARNING: *Do not eat the apples.* When you complete the experiment, give all bags to your teacher for disposal.

Procedure

1. **Label** the plastic bags *1* through *6*. Put on gloves. Place a fresh apple in bag *1*.
2. Rub the rotting apple over the other five apples. This is your source of microorganisms. **WARNING:** *Do not touch your face.*
3. Put one apple in bag *2*.
4. Hold one apple 1.5 m above a newspaper on the floor and drop it. Put it in bag *3*.
5. Rub one apple with sandpaper. Place this apple in bag *4*.
6. Wash one apple with soap and water. Dry well and put it in bag *5*.
7. Use a cotton ball to spread alcohol over the last apple. Let it air dry. Place it in bag *6*.
8. Seal all bags and put them in a dark place.
9. Copy the data table below. On days 3 and 7, compare all apples without removing them from the bags. **Record** your observations.

Apple Observations

Condition	Day 3	Day 7
1. Fresh		
2. Untreated		
3. Dropped		
4. Rubbed with sandpaper	Do not write in this book.	
5. Washed with soap and water		
6. Covered with alcohol		

Conclude and Apply

1. **Infer** how this experiment relates to infections on your skin.
2. **Explain** why it is important to clean a wound.

Communicating Your Data

Prepare a poster illustrating the advantages of washing hands to avoid the spread of disease. Get permission to put the poster near a school rest room.

section 3

Noninfectious Diseases

as you read

What You'll Learn

- **Define** noninfectious diseases and list causes of them.
- **Describe** the basic characteristics of cancer.
- **Explain** what happens during an allergic reaction.
- **Explain** how chemicals in the environment can be harmful to humans.

Why It's Important

Knowing the causes of noninfectious diseases can help you understand their prevention and treatment.

Review Vocabulary

gene: a section of DNA on a chromosome that carries instructions for making a specific protein

New Vocabulary

- noninfectious disease
- allergy
- allergen
- chemotherapy

Chronic Disease

It's a beautiful, late-summer day. Flowers are blooming everywhere. You and your cousin hurry to get to the ballpark before the first pitch of the game. "Achoo!" Your cousin sneezes. Her eyes are watery and red. "Oh no! I sure don't want to catch that cold," you mutter. "I don't have a cold," she responds, "it's my allergies." Not all diseases are caused by pathogens. Diseases and disorders such as diabetes, allergies, asthma, cancer, and heart disease are **noninfectious diseases.** They are not spread from one person to another. Many are chronic (KRAH nihk). This means that they can last for a long time. Although some chronic diseases can be cured, others cannot.

Some infectious diseases can be chronic too. For example, deer ticks carry a bacterium that causes Lyme disease. This bacterium can affect the nervous system, heart, and joints for weeks to years. It can become chronic if not treated. Antibiotics will kill the bacteria, but some damage cannot be reversed.

Allergies

If you've had an itchy rash after eating a certain food, you probably have an allergy to that food. An **allergy** is an overly strong reaction of the immune system to a foreign substance. Many people have allergic reactions, such as the one shown in **Figure 12,** to cosmetics, shellfish, strawberries, peanuts, and insect stings. Most allergic reactions are minor. However, severe allergic reactions can occur, causing shock and even death if they aren't treated promptly.

Figure 12 Allergic reactions are caused by many things.

Hives are one kind of allergic reaction.

Some common substances stimulate allergic responses in people.

Allergens Substances that cause an allergic response are called **allergens.** Some chemicals, certain foods, pollen, molds, some antibiotics, and dust are allergens for some people. Some foods cause hives or stomach cramps and diarrhea. Pollen can cause a stuffy nose, breathing difficulties, watery eyes, and a tired feeling in some people. Dust can contain cat and dog dander and dust mites, as shown in **Figure 13.** Asthma (AZ muh) is a lung disorder that is associated with reactions to allergens. A person with asthma can have shortness of breath, wheezing, and coughing when he or she comes into contact with something they are allergic to.

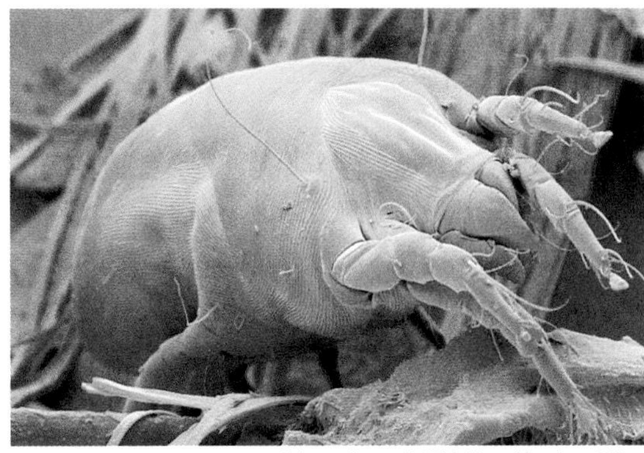

Color-enhanced SEM Magnification: 245×

Figure 13 Dust mites are smaller than a period at the end of a sentence. They can live in pillows, mattresses, carpets, furniture, and other places.

When you come in contact with an allergen, your immune system usually forms antibodies. Your body reacts by releasing chemicals called histamines (HIHS tuh meenz) that promote red, swollen tissues. Antihistamines are medications that can be used to treat allergic reactions and asthma. Some severe allergies are treated with repeated injections of small doses of the allergen. This allows your body to become less sensitive to the allergen.

What does your body release in response to an allergen?

Diabetes

A chronic disease associated with the levels of insulin produced by the pancreas is diabetes. Insulin is a hormone that enables glucose to pass from the bloodstream into your cells. Doctors recognize two types of diabetes—Type 1 and Type 2. Type 1 diabetes is the result of too little or no insulin production. In Type 2 diabetes, your body cannot properly process the insulin. Symptoms of diabetes include fatigue, excessive thirst, frequent urination, and tingling sensations in the hands and feet.

If glucose levels in the blood remain high for a long time, health problems can develop. These problems can include blurred vision, kidney failure, heart attack, stroke, loss of feeling in the feet, and the loss of consciousness (diabetic coma). Patients with Type 1 diabetes, as shown in **Figure 14,** must monitor their intake of sugars and usually require daily injections of insulin to control their glucose levels. Careful monitoring of diet and weight usually are enough to control Type 2 diabetes. Since 1980, there has been an increase in the number of people with diabetes. Although the cause of diabetes is unknown, scientists have discovered that Type 2 diabetes is more common in people who are overweight and that it might be inherited.

Figure 14 Type 1 diabetes requires daily monitoring by either checking the amount of glucose in blood or the amount excreted in urine.

Chemicals and Disease

INTEGRATE Chemistry Chemicals are everywhere—in your body, the foods you eat, cosmetics, cleaning products, pesticides, fertilizers, and building materials. Of the thousands of chemical substances used by consumers, less than two percent are harmful. Those chemicals that are harmful to living things are called toxins, as shown in **Figure 15.** Toxins can cause birth defects, cell mutations, cancers, tissue damage, chronic diseases, and death.

The Effects The amount of a chemical that is taken into your body and how long your body is in contact with it determine how it affects you. For example, low levels of a toxin might cause cardiac or respiratory problems. However, higher levels of the same toxin might cause death. Some chemicals, such as the asbestos shown in **Figure 15,** can be inhaled over a long period of time. Eventually, the asbestos can cause chronic diseases of the lungs. Lead-based paints, if ingested, can accumulate in your body and eventually cause damage to the central nervous system. Another toxin, ethyl (EH thul) alcohol, is found in beer, wine, and liquor. It can cause birth defects in the children of mothers who drink alcohol during pregnancy.

Manufacturing, mining, transportation, and farming produce chemical wastes. These chemical substances interfere with the ability of soil, water, and air to support life. Pollution, caused by harmful chemicals, sometimes produces chronic diseases in humans. For example, long-term exposure to carbon monoxide, sulfur oxides, and nitrogen oxides in the air might cause a number of diseases, including bronchitis, emphysema (em fuh ZEE muh), and lung cancer.

Figure 15 Toxins can be in the environment.

Chemical spills can be dangerous and might end up in groundwater.

These scientists are testing the contents of barrels found in a dump.

Asbestos, if inhaled into the lungs over a long period of time, can cause chronic diseases of the lungs. Protective clothing must be worn when removing asbestos.

Table 3 Characteristics of Cancer Cells
Cell growth is uncontrolled.
These cells do not function as part of your body.
The cells take up space and interfere with normal bodily functions.
The cells travel throughout your body.
The cells produce tumors and abnormal growths anywhere in your body.

Cancer

Cancer has been a disease of humans since ancient times. Egyptian mummies show evidence of bone cancer. Ancient Greek scientists described several different kinds of cancers. Even medieval manuscripts report details about the disease.

Cancer is the name given to a group of closely related diseases that result from uncontrolled cell growth. It is a complicated disease, and no one fully understands how cancers form. Characteristics of cancer cells are shown in **Table 3.** Certain regulatory molecules in the body control the beginning and ending of cell division. If this control is lost, a mass of cells called a tumor (TEW mur) results from this abnormal growth. Tumors can occur anywhere in your body. Cancerous cells can leave a tumor, spread throughout the body via blood and lymph vessels, and then invade other tissues.

Reading Check *How do cancers spread?*

Types of Cancers Cancers can develop in any body tissue or organ. Leukemia (lew KEE mee uh) is a cancer of white blood cells. The cancerous white blood cells are immature and are no longer effective in fighting disease. The cancer cells multiply in the bone marrow and crowd out red blood cells, normal white blood cells, and platelets. Cancer of the lungs often starts in the bronchi and then spreads into the lungs. The surface area for air exchange in the lungs is reduced and breathing becomes difficult. Colorectal cancer, or cancer of the large intestine, is one of the leading causes of death among men and women. Changes in bowel movements and blood in the feces may be indications of the disease. In breast cancer, tumors grow in the breast. The second most common cancer in males is cancer of the prostate gland, which is an organ that surrounds the urethra.

Dioxin Danger Dioxin is a dangerous chemical found in small amounts in certain herbicides. It can cause miscarriages, cancers, and liver disorders. Research to find out about the dioxin contamination in Times Beach, Missouri. Write a brief report in your Science Journal.

Figure 16 Tobacco products have been linked directly to lung cancer. Some chemicals around the home are carcinogenic. **Explain** *why labels should not be removed from cleaning products.*

Causes In the latter part of the eighteenth century, a British physician recognized the association of soot to cancer in chimney sweeps. Since that time, scientists have learned more about causes of cancer. Research done in the 1940s and 1950s related genes to cancer.

Although not all the causes of cancer are known, many causes have been identified. Smoking has been linked to lung cancer. Lung cancer is the leading cause of cancer deaths for adults in the United States. Expo-sure to certain chemicals also can increase your chances of developing cancer. These substances, called carcinogens (kar SIH nuh junz), include asbestos, various solvents, heavy metals, alcohol, and home and garden chemicals, as shown in **Figure 16.**

Exposure to X rays, nuclear radiation, and ultraviolet radiation of the Sun also increases your risk of getting cancer. Exposure to ultraviolet radiation might lead to skin cancer. Certain foods that are cured, or smoked, including barbecued meats, can give rise to cancers. Some food additives and certain viruses are suspected of causing cancers. Some people have a genetic predisposition for cancer, meaning that they have genes that make them more susceptible to the disease. This does not mean that they definitely will have cancer, but if it is triggered by certain factors they have a greater chance of developing cancer.

Treatment Surgery to remove cancerous tissue, radiation with X rays to kill cancer cells, and chemotherapy are some treatments for cancer. **Chemotherapy** (kee moh THER uh pee) is the use of chemicals to destroy cancer cells. However, early detection of cancer is the key to any successful treatment.

Research in the science of immune processes, called immunology, has led to some new approaches for treating cancer. For example, specialized antibodies produced in the laboratory are being tested as anticancer agents. These antibodies are used as carriers to deliver medicines and radioactive substances directly to cancer cells. In another test, killer T cells are removed from a cancer patient and treated with chemicals that stimulate T cell production. The treated cells are then reinjected into the patient. Trial tests have shown some success in destroying certain types of cancer cells with this technique.

Prevention Knowing some causes of cancer might help you prevent it. The first step is to know the early warning signs, shown in **Table 4.** Medical attention and treatments such as chemotherapy or surgery in the early stages of some cancers can cure or keep them inactive.

A second step in cancer prevention concerns lifestyle choices. Choosing not to use tobacco and alcohol products can help prevent mouth and lung cancers and the other associated respiratory and circulatory system diseases. Selecting a healthy diet without many foods that are high in fats, salt, and sugar also might reduce your chances of developing cancer. Using sunscreen lotions and limiting the amount of time that you expose your skin to direct sunlight are good preventive measures against skin cancer. Before using harmful home or garden chemicals, carefully read the entire label and precisely follow precautions and directions for use.

Inhaling certain air pollutants such as carbon monoxide, sulfur dioxide, and asbestos fibers is dangerous to your health. To keep the air cleaner, the U.S. Government has regulations such as the Clean Air Act. These laws are intended to reduce the amount of these substances that are released into the air.

Table 4 Early Warning Signs of Cancer
Changes in bowel or bladder habits
A sore that does not heal
Unusual bleeding or discharge
Thickening or lump in the breast or elsewhere
Indigestion or difficulty swallowing
Obvious change in a wart or mole
Nagging cough or hoarseness

Provided by the National Cancer Institute

section 3 review

Summary

Chronic Disease

- Chronic diseases last for a long time.
- Allergies are strong reactions to foreign substances.
- Diabetes is a chronic disease associated with your body's insulin levels.

Chemicals and Disease

- Harmful chemicals can cause birth defects, cancers, chronic diseases, and death.

Cancer

- Cancer results from uncontrolled cell growth.
- Early detection and healthy lifestyle choices can help in the treatment or prevention of some cancers.

Self Check

1. **Infer** why diabetes is classified as a noninfectious disease.
2. **Describe** how toxins in the environment can be harmful to your body.
3. **Explain** how cancer cells affect body organ functions.
4. **Identify** some ways your body can respond to allergens.
5. **Think Critically** Joel has an ear infection. The doctor prescribes an antibiotic. After taking the antibiotic, Joel breaks out in a rash. What is happening to him?

Applying Math

6. **Make and Use Tables** Make a table that relates several causes of cancer and their effects on your body.

Defensive Saliva

Goals

- **Design** an experiment to test the reaction of a bicarbonate to acids and bases.
- **Test** the reaction of a bicarbonate to acids and bases.

Possible Materials

head of red cabbage
cooking pot
coffee filter
drinking glasses
clear household ammonia
baking soda
water
spoon
white vinegar
lemon juice
orange juice

Safety Precautions

WARNING: *Never eat or drink anything used in an investigation.*

Real-World Question

What happens when you think about a juicy cheeseburger or smell freshly baked bread? Your mouth starts making saliva. Saliva is the first line of defense for fighting harmful bacteria, acids, and bases entering your body. Saliva contains salts, including bicarbonates. An example of a bicarbonate found in your kitchen is baking soda. Bicarbonates help to maintain normal pH levels in your mouth. When surfaces in your mouth have normal pH levels, the growth of bacteria is slowed and the effects of acids and bases are reduced. In this activity, you will design your own experiment to show the importance of saliva bicarbonates. How do the bicarbonates in saliva work to protect your mouth from harmful bacteria, acids, and bases?

Form a Hypothesis

Based on your reading in the text, form a hypothesis to explain how the bicarbonates in saliva react to acids and bases.

Test Your Hypothesis

Make a Plan

1. **List** the materials you will need for your experiment. Red cabbage juice can be used as an indicator to test for acids and bases. Vinegar and citrus juices are acids, ammonia is a base, and baking soda (bicarbonate of soda) is a bicarbonate.
2. **Describe** how you will prepare the red cabbage juice and how you will use it to test for the presence of acids and bases.
3. **Describe** how you will test the effect of bicarbonate on acids and bases.
4. **List** the steps you will take to set up and complete your experiment. Describe exactly what you will do in each step.
5. Prepare a data table in your Science Journal to record your observations.
6. Examine the steps of your experiment to make certain they are in logical order.

Follow Your Plan

1. Ask your teacher to examine the steps of your experiment and data table before you start.
2. Conduct your experiment according to the approved plan.
3. **Record** your observations in your data table.

Analyze Your Data

1. **Compare** the color change of the acids and bases in the cabbage juice.
2. **Describe** how well the bicarbonate neutralized the acids and bases.
3. **Identify** any problems you had while setting up and conducting your experiment.

Conclude and Apply

1. **Conclude** whether or not your results support your hypothesis.
2. **Explain** why your saliva contains a bicarbonate based on your experiment.
3. **Predict** how quickly bacteria would grow in your glass containing acid compared to another glass containing acid and the bicarbonate.
4. **Describe** how saliva protects your mouth from bacteria.
5. **Predict** what would happen if your saliva were made of only water.

Using what you learned in this experiment, create a poster about the importance of good dental hygiene. Invite a dental hygienist to speak to your class.

SCIENCE Stats

Battling Bacteria

Did you know...

... The term *antibiotic* was first coined by an American microbiologist. The scientist received a Nobel prize in 1952 for the discovery of streptomycin (strep toh MY suhn), an antibiotic used against tuberculosis.

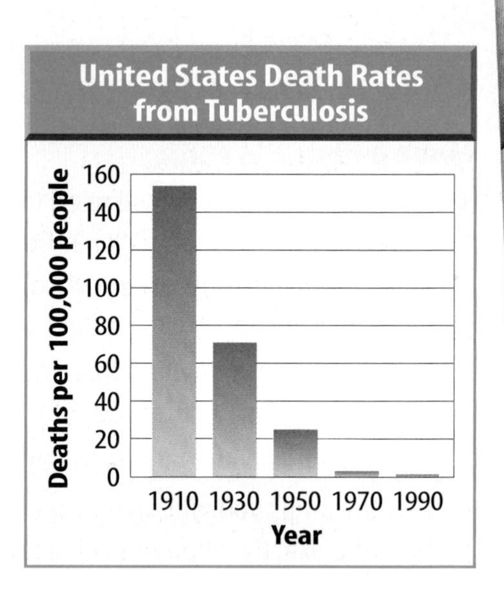

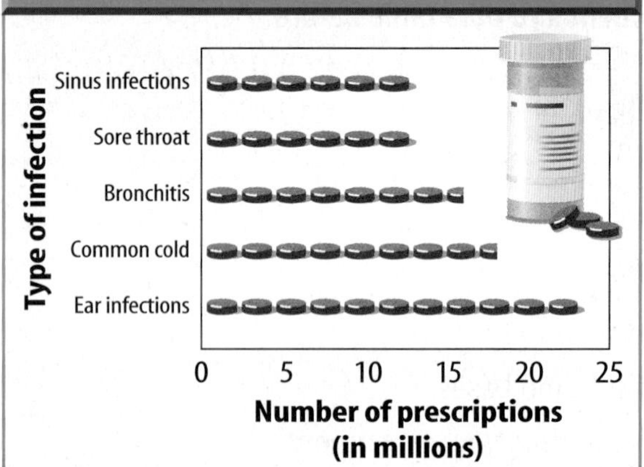

...In recent decades many bacteria have become resistant to antibiotics. For example, one group of bacteria that cause illnesses of the stomach and intestines—*Shigella* (shih GEL uh)—became harder to control. In 1985, less than one third of *Shigella* were resistant to the antibiotic ampicillin (am puh SI luhn). By 1991, however, more than two thirds of *Shigella* were resistant to the drug.

Applying Math It is believed that 30 percent of the antibiotics prescribed for ear infections are unnecessary. Using the graph, calculate the number of unnecessary prescriptions.

...People have long used natural remedies to treat infections. These remedies include garlic, *Echinacea* (purple coneflower), and an antibiotic called squalamine, found in sharks' stomachs.

Find Out About It

Visit life.msscience.com/science_stats to research the production of four antibiotics. Create a graph comparing the number of kilograms of each antibiotic produced in one year.

Reviewing Main Ideas

Section 1 The Immune System

1. Your body is protected against most pathogens by the immune system.
2. Active immunity is long lasting, but passive immunity is not.
3. Antigens are foreign molecules in your body. Your body makes an antibody that attaches to an antigen, making it harmless.

Section 2 Infectious Diseases

1. Pasteur and Koch discovered that microorganisms cause diseases. Lister learned that cleanliness helps control microorganisms.
2. Pathogens can be spread by air, water, food, and animal contact. Bacteria, viruses, fungi, and protists can cause infectious diseases.
3. Sexually transmitted diseases can be passed between persons during sexual contact.
4. HIV damages your body's immune system.

Section 3 Noninfectious Diseases

1. Causes of noninfectious diseases, such as diabetes and cancer, include genetics, chemicals, poor diet, and uncontrolled cell growth.
2. An allergy is a reaction of the immune system to a foreign substance.
3. Cancer results from uncontrolled cell growth, causing cells to multiply, spread through the body, and invade normal tissue.
4. Cancer is treated with surgery, chemotherapy, and radiation. Early detection can help cure or slow some cancers.

Visualizing Main Ideas

Copy and complete the following concept map on infectious diseases.

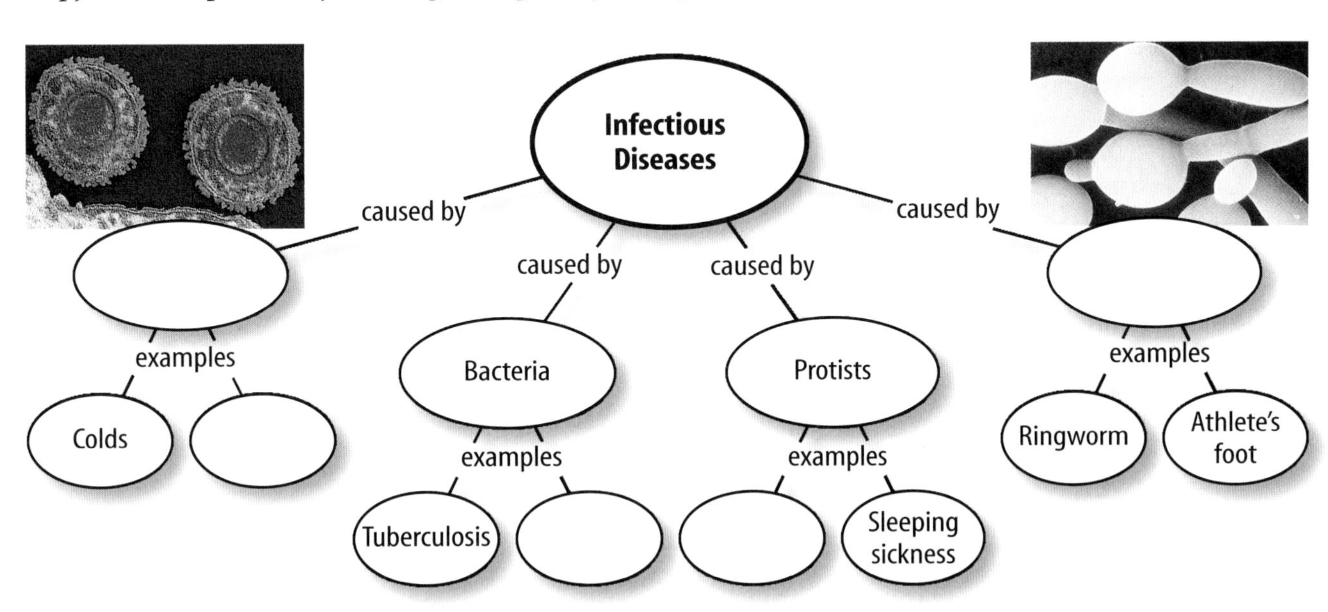

Using Vocabulary

active immunity p. 655	infectious disease p. 661
allergen p. 667	noninfectious disease p. 666
allergy p. 666	passive immunity p. 655
antibody p. 654	pasteurization p. 658
antigen p. 654	sexually transmitted disease (STD) p. 662
biological vector p. 661	vaccination p. 655
chemotherapy p. 670	virus p. 658
immune system p. 652	

Fill in the blanks with the correct vocabulary words.

1. A(n) _________ can cause infectious diseases.
2. A disease-carrying organism is called a(n) _________.
3. Measles is an example of _________.
4. Injection of weakened viruses is called _________.
5. _________ occurs when your body makes its own antibodies.
6. A(n) _________ stimulates histamine release.
7. Heating a liquid to kill harmful bacteria is called _________.
8. Diabetes is an example of a(n) _________ disease.

Checking Concepts

Choose the word or phrase that best answers the question.

9. Which has not been found to be a biological vector?

A) B) C) D)

10. How can infectious diseases be caused?
    - **A)** heredity
    - **B)** allergies
    - **C)** chemicals
    - **D)** organisms
11. How do scientists know that a pathogen causes a specific disease?
    - **A)** It is present in all cases of the disease.
    - **B)** It does not infect other animals.
    - **C)** It causes other diseases.
    - **D)** It is treated with heat.
12. What is formed in the blood to fight invading antigens?
    - **A)** hormones
    - **B)** allergens
    - **C)** pathogens
    - **D)** antibodies
13. Which is one of your body's general defenses against some pathogens?
    - **A)** stomach enzymes
    - **B)** HIV
    - **C)** some vaccines
    - **D)** hormones
14. Which is known as an infectious disease?
    - **A)** allergies
    - **B)** asthma
    - **C)** syphilis
    - **D)** diabetes
15. Which disease is caused by a virus that attacks white blood cells?
    - **A)** AIDS
    - **B)** measles
    - **C)** flu
    - **D)** polio
16. Which is a characteristic of cancer cells?
    - **A)** controlled cell growth
    - **B)** help your body stay healthy
    - **C)** interfere with normal body functions
    - **D)** do not multiply or spread
17. Which is caused by a virus?
    - **A)** AIDS
    - **B)** gonorrhea
    - **C)** ringworm
    - **D)** syphilis
18. How can cancer cells be destroyed?
    - **A)** chemotherapy
    - **B)** antigens
    - **C)** vaccines
    - **D)** viruses

Science Online life.msscience.com/vocabulary_puzzlemaker

Thinking Critically

19. **Explain** if it is better to vaccinate people or to wait until they build up their own immunity.

20. **Infer** what advantage a breast-fed baby might have compared to a formula-fed baby.

21. **Describe** how your body protects itself from antigens.

22. **Explain** how helper T cells and B cells work to eliminate antigens.

23. **Compare and contrast** antibodies, antigens, and antibiotics.

Use the graph below to answer question 24.

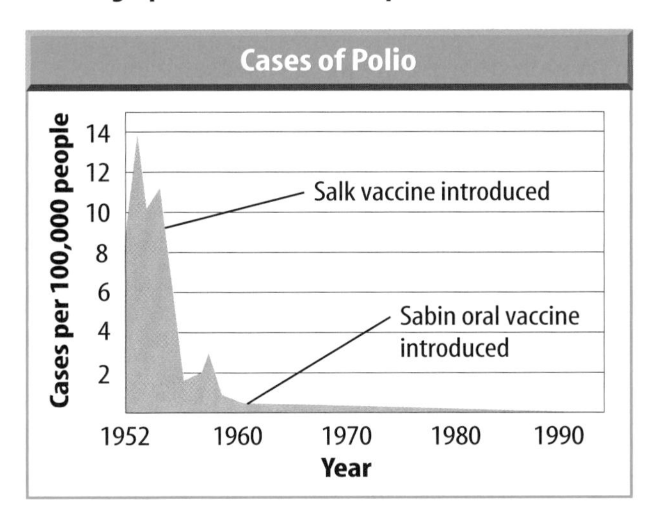

24. **Interpret Data** Using the graph above, explain the rate of polio cases between 1952 and 1965. What conclusions can you draw about the effectiveness of the polio vaccines?

25. **Concept Map** Make a network-tree concept map that compares the various defenses your body has against diseases. Compare general defenses, active immunity, and passive immunity.

Performance Activities

26. **Poster** Design and construct a poster to illustrate how a person with the flu could spread the disease to family members, classmates, and others.

Applying Math

27. **Antibiotic Tablets** You have an earache and your doctor prescribes an antibiotic to treat the infection. The antibiotic can be taken as a tablet at dosages of 400 mg or 1,000 mg. How many 400 mg tablets are needed to equal one 1,000 mg tablet?

Use the graph below to answer questions 28 and 29.

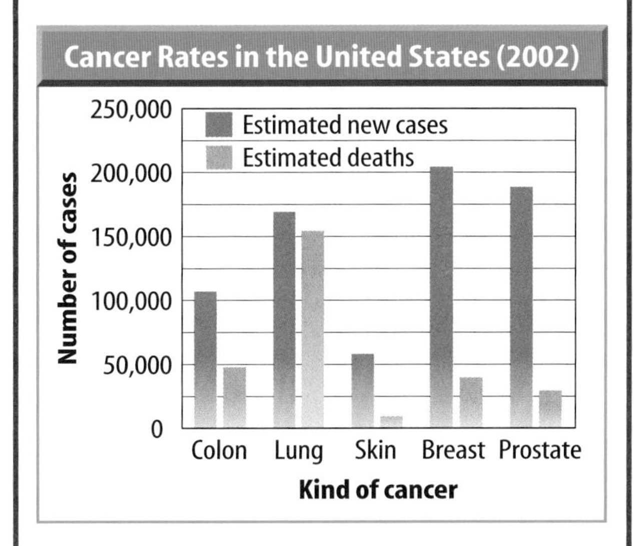

28. **Cancer Cases** The graph above shows the estimated number of new cases and estimated number of deaths for various cancers in the year 2002. Which cancer occurs most frequently? Most infrequently? Estimate the difference between new cases of colon cancer and new cases of skin cancer.

29. **Cancer Deaths** Estimate the difference between deaths from lung cancer and deaths from prostate cancer.

Part 1 Multiple Choice

Record your answers on the answer sheet provided by your teacher or on a sheet of paper.

Use the graph below to answer questions 1 and 2.

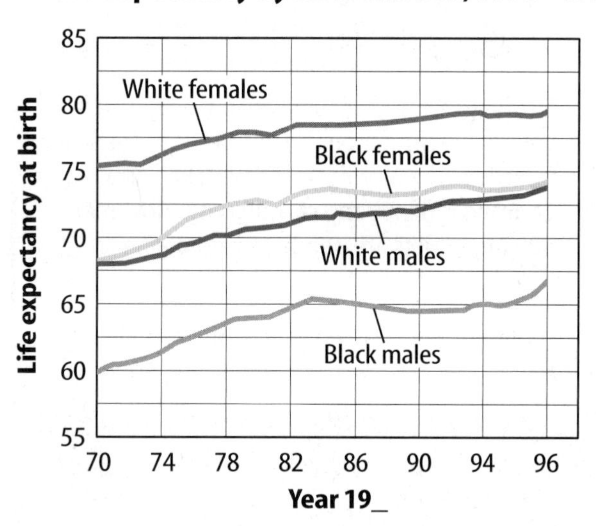

1. According to the information in the graph, which group had the lowest life expectancy in both 1975 and 1994?
   A. white males
   B. black females
   C. white females
   D. black males

2. A reasonable hypothesis based on the information in the graph is that life expectancy
   A. decreased for black males between 1970 and 1984.
   B. is longer for females than for males.
   C. decreased for white males between 1970 and 1980.
   D. is longer for males than for females.

3. Which of the following is NOT a sign of inflammation?
   A. redness **C.** bleeding
   B. pain **D.** swelling

Use the table below to answer questions 4–6.

Causes of Disease Before and After Vaccine Availability in the U.S.

Disease	Average Number of Cases per Year Before Vaccine Available	Cases in 1998 After Vaccine Available
Measles	503,282	89
Diphtheria	175,885	1
Tetanus	1,314	34
Mumps	152,209	606
Rubella	47,745	345
Pertussis (whooping cough)	147,271	6,279

Data from the National Immunization Program, CDC

4. Which of the following diseases had the highest number of cases before vaccine?
   A. diphtheria **C.** rubella
   B. mumps **D.** pertussis

5. Which of the following diseases had the highest number of cases after vaccine?
   A. measles **C.** mumps
   B. tetanus **D.** rubella

6. Which of the diseases in the table are caused by bacteria?
   A. measles, rubella, mumps
   B. measles, tetanus, mumps
   C. mumps, pertussis, rubella
   D. tetanus, pertussis, diphtheria

Test-Taking Tip

Missing Information Questions will often ask about missing information. Notice what is missing as well as what is given.

Question 6 Base your answer on choices that can be found in the text, such as *measles* and *tetanus.*

Part 2 Short Response/Grid In

Record your answers on the answer sheet provided by your teacher or on a sheet of paper.

7. What are some health practices that can help fight infectious disease?
8. How does mucus help defend your body?
9. Why are the body's second-line defenses called specific immunity?

Use the table below to answer questions 10–12.

Teen Opinions on Smoking			
All numbers are percentages	**Agree**	**Disagree**	**No opinion or don't know**
Seeing someone smoke turns me off	67	22	10
I'd rather date people who don't smoke	86	8	6
It's safe to smoke for only a year or two	7	92	1
Smoking can help you when you're bored	7	92	1
Smoking helps reduce stress	21	78	3
Smoking helps keep your weight down	18	80	2
Chewing tobacco and snuff cause cancer	95	2	3
I strongly dislike being around smokers	65	22	13

Data from CDC

10. According to the table, which statement had the highest percentage of teen agreement?
11. According to the table, which pairs of statements had the same percentages of teen disagreement?
12. According to the information in the table, do teens generally have positive or negative opinions about smoking? Explain.

Part 3 Open Ended

Record your answers on a sheet of paper.

13. Which is longer lasting—active immunity or passive immunity? Why?
14. Dr. Cavazos has isolated a bacterium that she thinks causes a recently discovered disease. How can she prove it? What steps should she follow?
15. Compare and contrast infectious and noninfectious diseases.
16. Would a vaccination against measles be helpful if a person already had the disease a year ago? Explain.
17. Compare and contrast Type 1 and Type 2 diabetes.

Use the illustration below to answer questions 18 and 19.

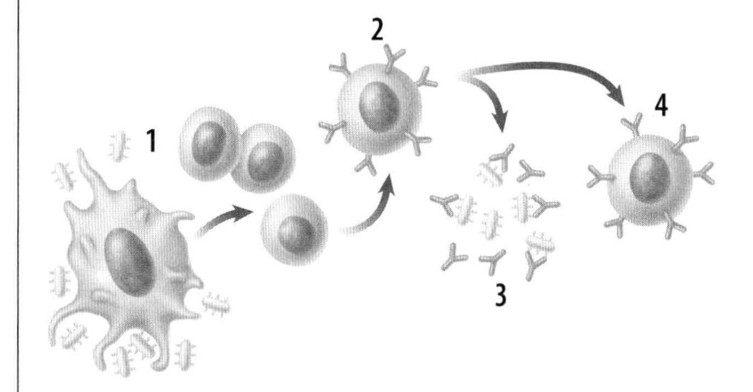

18. Explain the four steps of the immune system response.
19. Sometimes a person is born without the cells labeled *2* in the illustration above. If this person was given a vaccination for tetanus, what results would be expected? Explain.

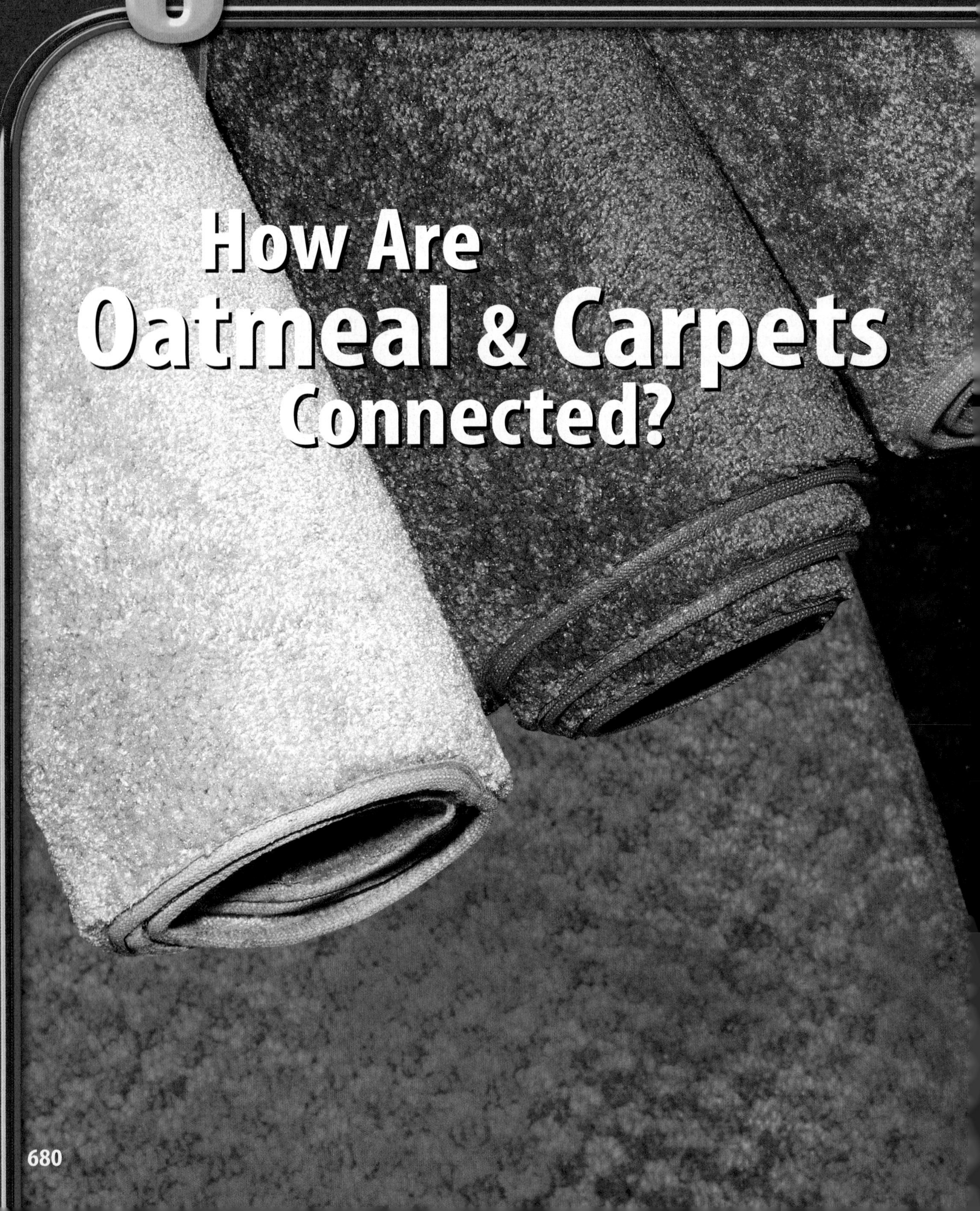

unit 5 Ecology

How Are Oatmeal & Carpets Connected?

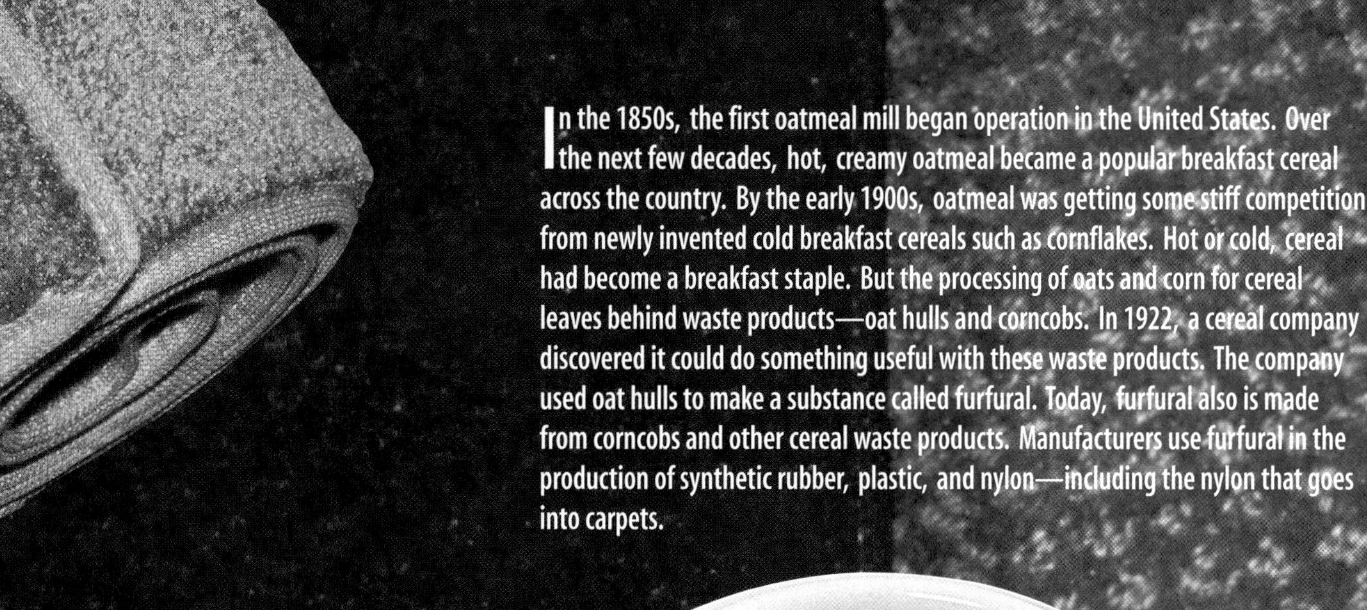

In the 1850s, the first oatmeal mill began operation in the United States. Over the next few decades, hot, creamy oatmeal became a popular breakfast cereal across the country. By the early 1900s, oatmeal was getting some stiff competition from newly invented cold breakfast cereals such as cornflakes. Hot or cold, cereal had become a breakfast staple. But the processing of oats and corn for cereal leaves behind waste products—oat hulls and corncobs. In 1922, a cereal company discovered it could do something useful with these waste products. The company used oat hulls to make a substance called furfural. Today, furfural also is made from corncobs and other cereal waste products. Manufacturers use furfural in the production of synthetic rubber, plastic, and nylon—including the nylon that goes into carpets.

unit projects

Visit life.msscience.com/unit_project to find project ideas and resources. Projects include:

- **Career** You are an environmental scientist as you design your own ecosystem-interaction web to demonstrate relationships from birth to death of your specific organism.
- **Technology** Chart your research results on the manufacturing of different materials. Compare cost, energy use, resources, and environmental concerns.
- **Model** Design your own two-week personal conservation project. Decide how you can make a difference as you reduce, reuse, and recycle.

Investigate the *Barrier Islands* ecosystem, then form an opinion as to whether developers should build on these environmentally fragile islands.

chapter 24

Interactions of Life

The BIG Idea

Living organisms interact with their environment and with one another in many ways.

SECTION 1
Living Earth

Main Idea All living and nonliving things on Earth are organized into levels, such as communities and ecosystems.

SECTION 2
Populations

Main Idea A population's size is affected by many things, including competition.

SECTION 3
Interactions Within Communities

Main Idea Every organism has a role in its environment.

Are these birds in danger?

The birds are a help to the rhinoceros. They feed on ticks and other parasites plucked from the rhino's hide. When the birds sense danger, they fly off, giving the rhino an early warning. Earth's living organisms supply one another with food, shelter, and other requirements for life.

Science Journal Describe how a familiar bird, insect, or other animal depends on other organisms.

Start-Up Activities

How do lawn organisms survive?

You probably have taken thousands of footsteps on grassy lawns or playing fields. If you look closely at the grass, you'll see that each blade is attached to roots in the soil. How do grass plants obtain everything they need to live and grow? What other kinds of organisms live in the grass? The following lab will give you a chance to take a closer look at the life in a lawn.

1. Examine a section of sod from a lawn.
2. How do the roots of the grass plants hold the soil?
3. Do you see signs of other living things besides grass?
4. **Think Critically** In your Science Journal, answer the above questions and describe any organisms that are present in your section of sod. Explain how these organisms might affect the growth of grass plants. Draw a picture of your section of sod.

Preview this chapter's content and activities at life.msscience.com

Ecology Make the following Foldable to help organize information about one of your favorite wild animals and its role in an ecosystem.

STEP 1 **Fold** a vertical sheet of paper from side to side. Make the front edge 1.25 cm shorter than the back edge.

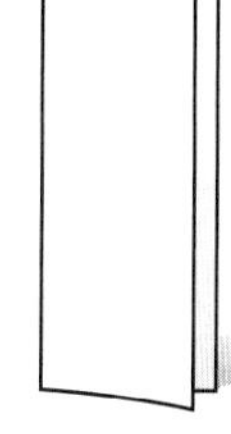

STEP 2 **Turn** lengthwise and **fold** into thirds.

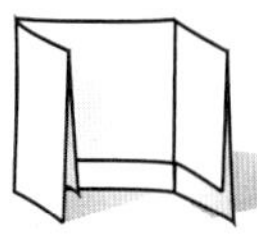

STEP 3 **Unfold and cut** only the top layer along both folds to make three tabs. **Label** each tab.

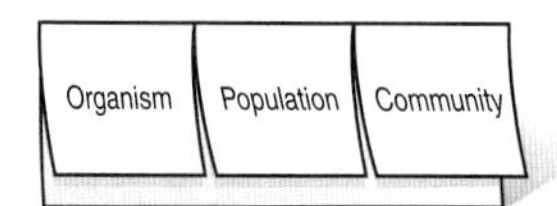

Identify Questions Before you read the chapter, write what you already know about your favorite animal under the left tab of your Foldable. As you read the chapter, write how the animal is part of a population and a community under the appropriate tabs.

Get Ready to Read

Compare and Contrast

1 Learn It! Good readers compare and contrast information as they read. This means they look for similarities and differences to help them to remember important ideas. Look for signal words in the text to let you know when the author is comparing or contrasting.

Compare and Contrast Signal Words	
Compare	**Contrast**
as	but
like	or
likewise	unlike
similarly	however
at the same time	although
in a similar way	on the other hand

2 Practice It! Read the excerpt below and notice how the author uses contrast signal words to describe the differences between the biotic potentials of species.

> The highest rate of reproduction under ideal conditions is a population's biotic potential. The **larger** the number of offspring that are produced by parent organisms, the **higher** the biotic potential of the species will be. Compare an avocado tree to a tangerine tree.
>
> *—from page 692*

3 Apply It! Compare and contrast the different types of symbiotic relationships on page 698.

Reading Tip

As you read, use other skills, such as summarizing and connecting, to help you understand comparisons and contrasts.

Target Your Reading

Use this to focus on the main ideas as you read the chapter.

1. **Before you read** the chapter, respond to the statements below on your worksheet or on a numbered sheet of paper.
   - Write an **A** if you **agree** with the statement.
   - Write a **D** if you **disagree** with the statement.

2. **After you read** the chapter, look back to this page to see if you've changed your mind about any of the statements.
   - If any of your answers changed, explain why.
   - Change any false statements into true statements.
   - Use your revised statements as a study guide.

Before You Read A or D		Statement	After You Read A or D
	1	A community is all the populations of species that live in an ecosystem.	
	2	All deserts are hot and dry environments.	
	3	An ecosystem is made up of only the living things in an area.	
	4	Organisms living in the wild always have enough food and living space.	
	5	The greatest competition in nature is among organisms of the same species.	
	6	Both nonliving and living parts of an ecosystem can limit the number of individuals in a population.	
	7	Living organisms do not need a constant supply of energy.	
	8	All consumers are predators.	
	9	Relationships between organisms of different species cannot benefit both organisms.	

Science Online
Print out a worksheet of this page at life.msscience.com

section 1

Living Earth

as you read

What You'll Learn

- **Identify** places where life is found on Earth.
- **Define** ecology.
- **Observe** how the environment influences life.

Why It's Important

All living things on Earth depend on each other for survival.

Review Vocabulary

adaptation: any variation that makes an organism better suited to its environment

New Vocabulary

- biosphere
- ecosystem
- ecology
- population
- community
- habitat

The Biosphere

What makes Earth different from other planets in the solar system? One difference is Earth's abundance of living organisms. The part of Earth that supports life is the **biosphere** (BI uh sfihr). The biosphere includes the top portion of Earth's crust, all the waters that cover Earth's surface, and the atmosphere that surrounds Earth.

Reading Check *What three things make up the biosphere?*

As **Figure 1** shows, the biosphere is made up of different environments that are home to different kinds of organisms. For example, desert environments receive little rain. Cactus plants, coyotes, and lizards are included in the life of the desert. Tropical rain forest environments receive plenty of rain and warm weather. Parrots, monkeys, and tens of thousands of other organisms live in the rain forest. Coral reefs form in warm, shallow ocean waters. Arctic regions near the north pole are covered with ice and snow. Polar bears, seals, and walruses live in the arctic.

Figure 1 Earth's biosphere consists of many environments, including ocean waters, polar regions, and deserts.

INTEGRATE Astronomy

Life on Earth In our solar system, Earth is the third planet from the Sun. The amount of energy that reaches Earth from the Sun helps make the temperature just right for life. Mercury, the planet closest to the Sun, is too hot during the day and too cold at night to make life possible there. Venus, the second planet from the Sun, has a thick, carbon dioxide atmosphere and high temperatures. It is unlikely that life could survive there. Mars, the fourth planet, is much colder than Earth because it is farther from the Sun and has a thinner atmosphere. It might support microscopic life, but none has been found. The planets beyond Mars probably do not receive enough heat and light from the Sun to have the right conditions for life.

Ecosystems

On a visit to Yellowstone National Park in Wyoming, you might see a prairie scene like the one shown in **Figure 2.** Bison graze on prairie grass. Cowbirds follow the bison, catching grasshoppers that jump away from the bisons' hooves. This scene is part of an ecosystem. An **ecosystem** consists of all the organisms living in an area, as well as the nonliving parts of that environment. Bison, grass, birds, and insects are living organisms of this prairie ecosystem. Water, temperature, sunlight, soil, and air are nonliving features of this prairie ecosystem. **Ecology** is the study of interactions that occur among organisms and their environments. Ecologists are scientists who study these interactions.

Reading Check *What is an ecosystem?*

Figure 2 Ecosystems are made up of living organisms and the nonliving factors of their environment. In this prairie ecosystem, cowbirds eat insects and bison graze on grass.
List *other kinds of organisms that might live in this ecosystem.*

Science Online

Topic: Human Population Data

Visit life.msscience.com for Web links to information about the estimated human population size for the world today.

Activity Create a graph that shows how the human population has changed between the year 2000 and this year.

Populations

Suppose you meet an ecologist who studies how a herd of bison moves from place to place and how the female bison in the herd care for their young. This ecologist is studying the members of a population. A **population** is made up of all organisms of the same species that live in an area at the same time. For example, all the bison in a prairie ecosystem are one population. All the cowbirds in this ecosystem make up a different population. The grasshoppers make up yet another population.

Ecologists often study how populations interact. For example, an ecologist might try to answer questions about several prairie species. How does grazing by bison affect the growth of prairie grass? How does grazing influence the insects that live in the grass and the birds that eat those insects? This ecologist is studying a community. A **community** is all the populations of all species living in an ecosystem. The prairie community is made of populations of bison, grasshoppers, cowbirds, and all other species in the prairie ecosystem. An arctic community might include populations of fish, seals that eat fish, and polar bears that hunt and eat seals. **Figure 3** shows how organisms, populations, communities, and ecosystems are related.

Figure 3 The living world is arranged in several levels of organization.

Figure 4 The trees of the forest provide a habitat for woodpeckers and other birds. This salamander's habitat is the moist forest floor.

Habitats

Each organism in an ecosystem needs a place to live. The place in which an organism lives is called its **habitat.** The animals shown in **Figure 4** live in a forest ecosystem. Trees are the woodpecker's habitat. These birds use their strong beaks to pry insects from tree bark or break open acorns and nuts. Woodpeckers usually nest in holes in dead trees. The salamander's habitat is the forest floor, beneath fallen leaves and twigs. Salamanders avoid sunlight and seek damp, dark places. This animal eats small worms, insects, and slugs. An organism's habitat provides the kinds of food and shelter, the temperature, and the amount of moisture the organism needs to survive.

section 1 review

Summary

The Biosphere

- The biosphere is the portion of Earth that supports life.

Ecosystems

- An ecosystem is made up of the living organisms and nonliving parts of an area.

Populations

- A population is made up of all members of a species that live in the same ecosystem.
- A community consists of all the populations in an ecosystem.

Habitats

- A habitat is where an organism lives.

Self Check

1. **List** three parts of the Earth included in the biosphere.
2. **Define** the term *ecology*.
3. **Compare and contrast** the terms *habitat* and *biosphere*.
4. **Identify** the major difference between a community and a population, and give one example of each.
5. **Think Critically** Does the amount of rain that falls in an area determine which kinds of organisms can live there? Why or why not?

Applying Skills

6. **Form a hypothesis** about how a population of dandelion plants might be affected by a population of rabbits.

section 2

Populations

as you read

What You'll Learn

- **Identify** methods for estimating population sizes.
- **Explain** how competition limits population growth.
- **List** factors that influence changes in population size.

Why It's Important

Competition caused by population growth reduces the amount of food, living space, and other resources available to organisms, including humans.

Review Vocabulary
natural selection: hypothesis that states organisms with traits best suited to their environment are more likely to survive and reproduce

New Vocabulary
- limiting factor
- carrying capacity

Competition

Wild crickets feed on plant material at night. They hide under leaves or in dark damp places during the day. In some pet shops, crickets are raised in cages and fed to pet reptiles. Crickets require plenty of food, water, and hiding places. As a population of caged crickets grows, extra food and more hiding places are needed. To avoid crowding, some crickets might have to be moved to other cages.

Food and Space Organisms living in the wild do not always have enough food or living space. The Gila woodpecker, shown in **Figure 5,** lives in the Sonoran Desert of Arizona and Mexico. This woodpecker makes its nest by drilling a hole in a saguaro (suh GWAR oh) cactus. Woodpeckers must compete with each other for nesting spots. Competition occurs when two or more organisms seek the same resource at the same time.

Growth Limits Competition limits population size. If available nesting spaces are limited, some woodpeckers will not be able to raise young. Gila woodpeckers eat cactus fruit, berries, and insects. If food becomes scarce, some woodpeckers might not survive to reproduce. Competition for food, living space, or other resources can limit population growth.

In nature, the most intense competition is usually among individuals of the same species, because they need the same kinds of food and shelter. Competition also takes place among different species. For example, after a Gila woodpecker has abandoned its nest, owls, flycatchers, snakes, and lizards might compete for the shelter of the empty hole.

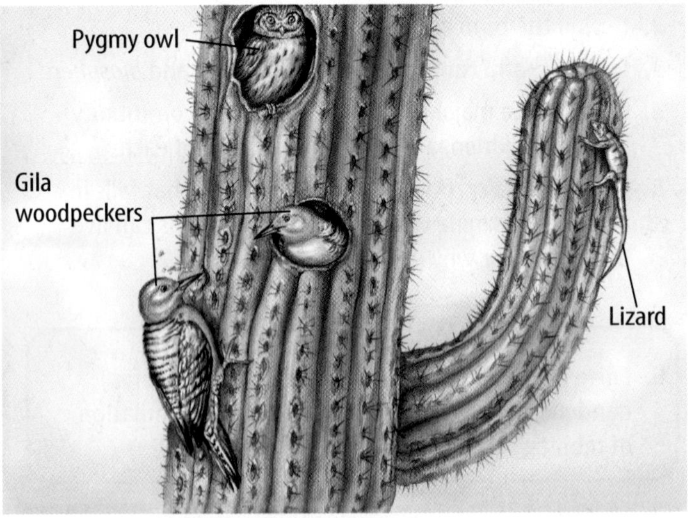

Figure 5 Gila woodpeckers make nesting holes in the saguaro cactus. Many animals compete for the shelter these holes provide.

Population Size

Ecologists often need to measure the size of a population. This information can indicate whether or not a population is healthy and growing. Population counts can help identify populations that could be in danger of disappearing.

Some populations are easy to measure. If you were raising crickets, you could measure the size of your cricket population simply by counting all the crickets in the container. What if you wanted to compare the cricket populations in two different containers? You would calculate the number of crickets per square meter (m^2) of your container. The number of individuals of one species per a specific area is called population density. **Figure 6** shows Earth's human population density.

Reading Check *What is population density?*

Measuring Populations Counting crickets can be tricky. They look alike, move a lot, and hide. The same cricket could be counted more than once, and others could be completely missed. Ecologists have similar problems when measuring wildlife populations. One of the methods they use is called trap-mark-release. Suppose you want to count wild rabbits. Rabbits live underground and come out at dawn and dusk to eat. Ecologists set traps that capture rabbits without injuring them. Each captured rabbit is marked and released. Later, another sample of rabbits is captured. Some of these rabbits will have marks, but many will not. By comparing the number of marked and unmarked rabbits in the second sample, ecologists can estimate the population size.

Observing Seedling Competition

Procedure

1. Fill **two plant pots** with **moist potting soil.**
2. Plant **radish seeds** in one pot, following the spacing instructions on the seed packet. Label this pot *Recommended Spacing*.
3. Plant **radish seeds** in the second pot, spaced half the recommended distance apart. Label this pot *Densely Populated*. Wash your hands.
4. Keep the soil moist. When the seeds sprout, move them to a well-lit area.
5. Measure and record in your **Science Journal** the height of the seedlings every two days for two weeks.

Analysis

1. Which plants grew faster?
2. Which plants looked healthiest after two weeks?
3. How did competition influence the plants?

Try at Home

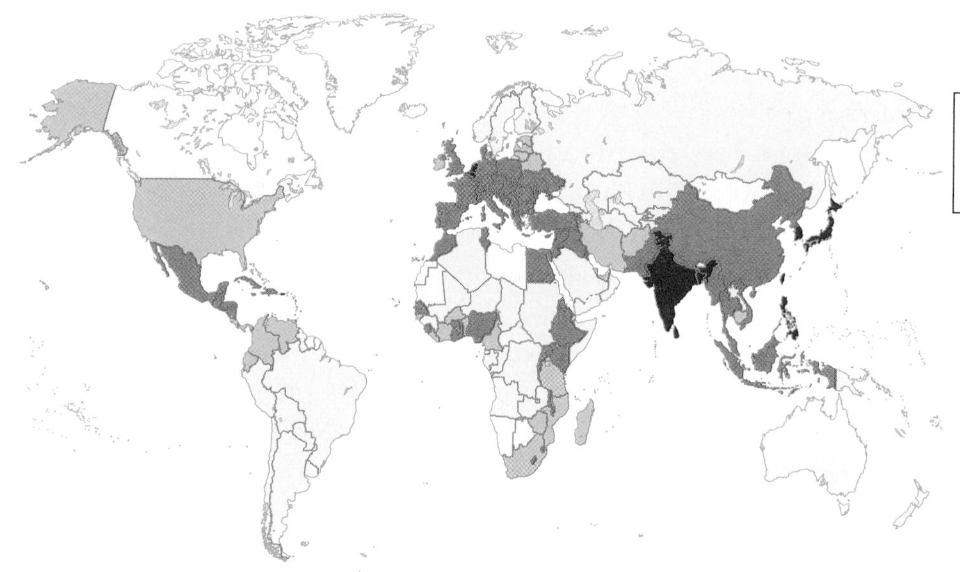

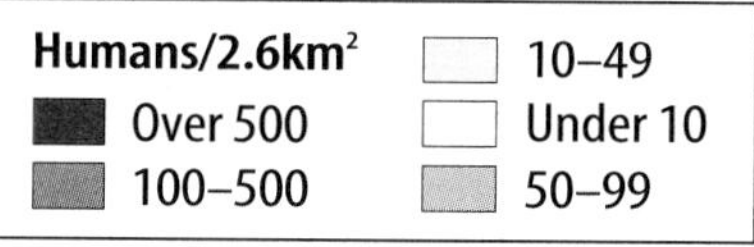

Figure 6 This map shows human population density.
Interpret Illustrations *Which countries have the highest population density?*

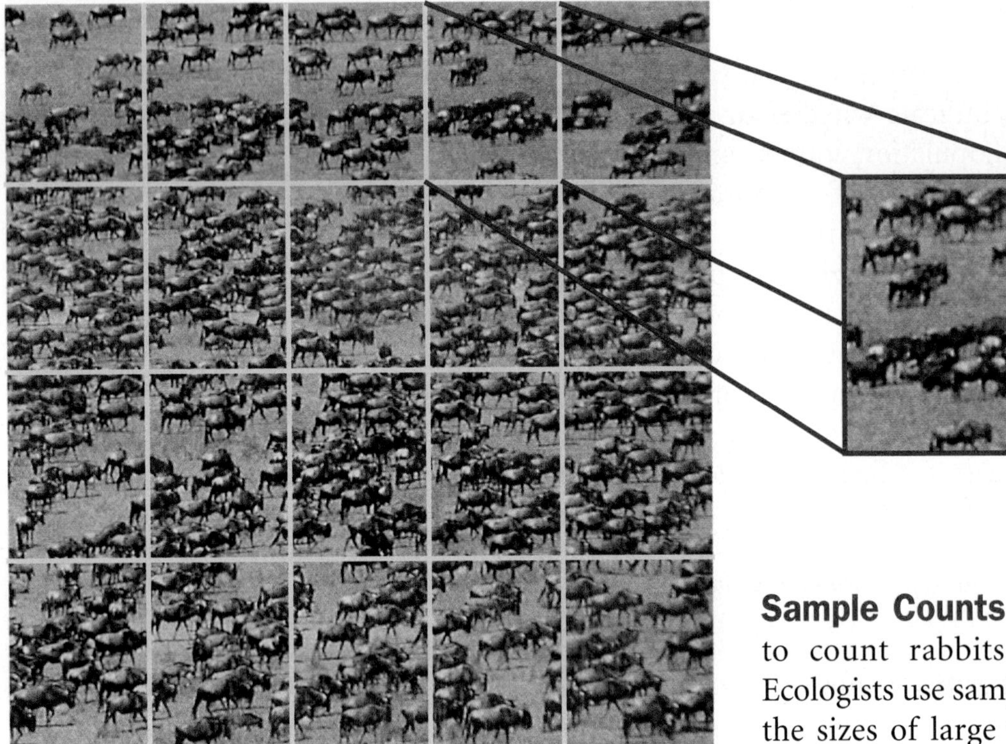

Figure 7 Ecologists can estimate population size by making a sample count. Wildebeests graze on the grassy plains of Africa. **Draw Conclusions** *How could you use the enlarged square to estimate the number of wildebeests in the entire photograph?*

Sample Counts What if you wanted to count rabbits over a large area? Ecologists use sample counts to estimate the sizes of large populations. To estimate the number of rabbits in an area of 100 acres, for example, you could count the rabbits in one acre and multiply by 100 to estimate the population size. **Figure 7** shows another approach to sample counting.

Limiting Factors One grass plant can produce hundreds of seeds. Imagine those seeds drifting onto a vacant field. Many of the seeds sprout and grow into grass plants that produce hundreds more seeds. Soon the field is covered with grass. Can this grass population keep growing forever? Suppose the seeds of wildflowers or trees drift onto the field. If those seeds sprout, trees and flowers would compete with grasses for sunlight, soil, and water. Even if the grasses did not have to compete with other plants, they might eventually use up all the space in the field. When no more living space is available, the population cannot grow.

In any ecosystem, the availability of food, water, living space, mates, nesting sites, and other resources is often limited. A **limiting factor** is anything that restricts the number of individuals in a population. Limiting factors include living and nonliving features of the ecosystem.

A limiting factor can affect more than one population in a community. Suppose a lack of rain limits plant growth in a meadow. Fewer plants produce fewer seeds. For seed-eating mice, this reduction in the food supply could become a limiting factor. A smaller mouse population could, in turn, become a limiting factor for the hawks and owls that feed on mice.

Carrying Capacity A population of robins lives in a grove of trees in a park. Over several years, the number of robins increases and nesting space becomes scarce. Nesting space is a limiting factor that prevents the robin population from getting any larger. This ecosystem has reached its carrying capacity for robins. **Carrying capacity** is the largest number of individuals of one species that an ecosystem can support over time. If a population begins to exceed the environment's carrying capacity, some individuals will not have enough resources. They could die or be forced to move elsewhere, like the deer shown in **Figure 8.**

Figure 8 These deer might have moved into a residential area because a nearby forest's carrying capacity for deer has been reached.

How are limiting factors related to carrying capacity?

Applying Science

Do you have too many crickets?

You've decided to raise crickets to sell to pet stores. A friend says you should not allow the cricket population density to go over 210 crickets/m^2. Use what you've learned in this section to measure the population density in your cricket tanks.

Identifying the Problem

The table on the right lists the areas and populations of your three cricket tanks. How can you determine if too many crickets are in one tank? If a tank contains too many crickets, what could you do? Explain why too many crickets in a tank might be a problem.

Cricket Population

Tank	Area (m^2)	Number of Crickets
1	0.80	200
2	0.80	150
3	1.5	315

Solving the Problem

1. Do any of the tanks contain too many crickets? Could you make the population density of the three tanks equal by moving crickets from one tank to another? If so, which tank would you move crickets into?
2. Wild crickets living in a field have a population density of 2.4 crickets/m^2. If the field's area is 250 m^2, what is the approximate size of the cricket population? Why would the population density of crickets in a field be lower than the population density of crickets in a tank?

Topic: Birthrates and Death Rates
Visit life.msscience.com for Web links to information about birthrates and death rates for the human population.

Activity Find out whether the human population worldwide is increasing because of rising birthrates or declining death rates.

Biotic Potential What would happen if no limiting factors restricted the growth of a population? Think about a population that has an unlimited supply of food, water, and living space. The climate is favorable. Population growth is not limited by diseases, predators, or competition with other species. Under ideal conditions like these, the population would continue to grow.

The highest rate of reproduction under ideal conditions is a population's biotic potential. The larger the number of offspring that are produced by parent organisms, the higher the biotic potential of the species will be. Compare an avocado tree to a tangerine tree. Assume that each tree produces the same number of fruits. Each avocado fruit contains one large seed. Each tangerine fruit contains a dozen seeds or more. Because the tangerine tree produces more seeds per fruit, it has a higher biotic potential than the avocado tree.

Changes in Populations

Birthrates and death rates also influence the size of a population and its rate of growth. A population gets larger when the number of individuals born is greater than the number of individuals that die. When the number of deaths is greater than the number of births, populations get smaller. Take the squirrels living in New York City's Central Park as an example. In one year, if 900 squirrels are born and 800 die, the population increases by 100. If 400 squirrels are born and 500 die, the population decreases by 100.

The same is true for human populations. **Table 1** shows birthrates, death rates, and population changes for several countries around the world. In countries with faster population growth, birthrates are much higher than death rates. In countries with slower population growth, birthrates are only slightly higher than death rates. In Germany, where the population is getting smaller, the birthrate is lower than the death rate.

Table 1 Population Growth

	Birthrate*	Death Rate*	Population Increase (percent)
Rapid-Growth Countries			
Jordan	38.8	5.5	3.3
Uganda	50.8	21.8	2.9
Zimbabwe	34.3	9.4	5.2
Slow-Growth Countries			
Germany	9.4	10.8	−1.5
Sweden	10.8	10.6	0.1
United States	14.8	8.8	0.6

*Number per 1,000 people

Figure 9 Mangrove seeds sprout while they are still attached to the parent tree. Some sprouted seeds drop into the mud below the parent tree and continue to grow. Others drop into the water and can be carried away by tides and ocean currents. When they wash ashore, they might start a new population of mangroves or add to an existing mangrove population.

Moving Around Most animals can move easily from place to place, and these movements can affect population size. For example, a male mountain sheep might wander many miles in search of a mate. After he finds a mate, their offspring might establish a completely new population of mountain sheep far from the male's original population.

Many bird species move from one place to another during their annual migrations. During the summer, populations of Baltimore orioles are found throughout eastern North America. During the winter, these populations disappear because the birds migrate to Central America. They spend the winter there, where the climate is mild and food supplies are plentiful. When summer approaches, the orioles migrate back to North America.

Even plants and microscopic organisms can move from place to place, carried by wind, water, or animals. The tiny spores of mushrooms, mosses, and ferns float through the air. The seeds of dandelions, maple trees, and other plants have feathery or winglike growths that allow them to be carried by wind. Spine-covered seeds hitch rides by clinging to animal fur or people's clothing. Many kinds of seeds can be transported by river and ocean currents. Mangrove trees growing along Florida's Gulf Coast, shown in **Figure 9,** provide an example of how water moves seeds.

Comparing Biotic Potential

Procedure

1. Remove all the seeds from a **whole fruit.** Do not put fruit or seeds in your mouth.
2. Count the total number of seeds in the fruit. Wash your hands, then record these data in your **Science Journal.**
3. Compare your seed totals with those of classmates who examined other types of fruit.

Analysis

1. Which type of fruit had the most seeds? Which had the fewest seeds?
2. What is an advantage of producing many seeds? Can you think of a possible disadvantage?
3. To estimate the total number of seeds produced by a tomato plant, what would you need to know?

NATIONAL GEOGRAPHIC

VISUALIZING POPULATION GROWTH

Figure 10

When a species enters an ecosystem that has abundant food, water, and other resources, its population can flourish. Beginning with a few organisms, the population increases until the number of organisms and available resources are in balance. At that point, population growth slows or stops. A graph of these changes over time produces an S-curve, as shown here for coyotes.

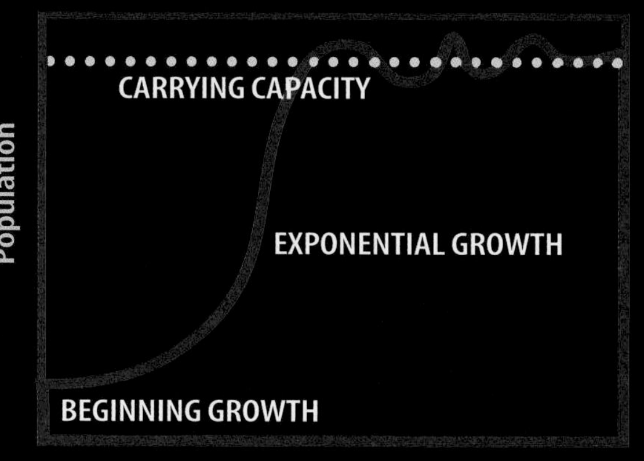

BEGINNING GROWTH During the first few years, population growth is slow, because there are few adults to produce young. As the population grows, so does the number of breeding adults.

EXPONENTIAL GROWTH As the number of adults in the population grows, so does the number of births. The coyote population undergoes exponential growth, quickly increasing in size.

CARRYING CAPACITY As resources become less plentiful, the birthrate declines and the death rate may rise. Population growth slows. The coyote population has reached the environmental carrying capacity—the maximum number of coyotes that the environment can sustain.

Exponential Growth When a species moves into a new area with plenty of food, living space, and other resources, the population grows quickly, in a pattern called exponential growth. Exponential growth means that the larger a population gets, the faster it grows. Over time, the population will reach the ecosystem's carrying capacity for that species. **Figure 10** shows each stage in this pattern of population growth.

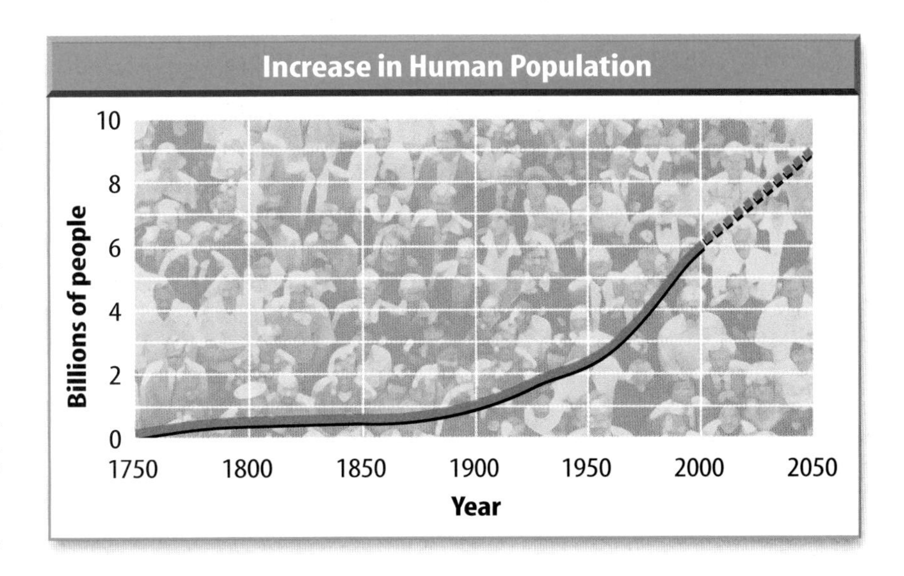

Figure 11 The size of the human population is increasing by about 1.6 percent per year.
Identify *the factors that affect human population growth.*

As a population approaches its ecosystem's carrying capacity, competition for living space and other resources increases. As you can see in **Figure 11,** Earth's human population shows exponential growth. By the year 2050, the population could reach 9 billion. You probably have read about or experienced some of the competition associated with human population growth, such as freeway traffic jams, crowded subways and buses, or housing shortages. As population density increases, people are forced to live closer to one another. Infectious diseases can spread easily when people are crowded together.

section 2 review

Summary

Competition

- When more than one organism needs the same resource, competition occurs.
- Competition limits population size.

Population Size

- Population density is the number of individuals per unit area.
- Limiting factors are resources that restrict population size.
- An ecosystem's carrying capacity is the largest population it can support.
- Biotic potential is the highest possible rate of growth for a population.

Changes in Populations

- Birthrates, death rates, and movement from place to place affect population size.

Self Check

1. **Describe** three ways in which ecologists can estimate the size of a population.
2. **Explain** how birthrates and death rates influence the size of a population.
3. **Explain** how carrying capacity influences the number of organisms in an ecosystem.
4. **Think Critically** Why are food and water the limiting factors that usually have the greatest effect on population size?

Applying Skills

5. **Make and use a table** on changes in the size of a deer population in Arizona. Use the following data. In 1910 there were 6 deer; in 1915, 36 deer; in 1920, 143 deer; in 1925, 86 deer; and in 1935, 26 deer. Explain what might have caused these changes.

section 3

Interactions Within Communities

as you read

What You'll Learn

- **Describe** how organisms obtain energy for life.
- **Explain** how organisms interact.
- **Recognize** that every organism occupies a niche.

Why It's Important

Obtaining food, shelter, and other needs is crucial to the survival of all living organisms, including you.

Review Vocabulary
social behavior: interactions among members of the same species

New Vocabulary
- producer
- consumer
- symbiosis
- mutualism
- commensalism
- parasitism
- niche

Obtaining Energy

Just as a car engine needs a constant supply of gasoline, living organisms need a constant supply of energy. The energy that fuels most life on Earth comes from the Sun. Some organisms use the Sun's energy to create energy-rich molecules through the process of photosynthesis. The energy-rich molecules, usually sugars, serve as food. They are made up of different combinations of carbon, hydrogen, and oxygen atoms. Energy is stored in the chemical bonds that hold the atoms of these molecules together. When the molecules break apart—for example, during digestion—the energy in the chemical bonds is released to fuel life processes.

Producers Organisms that use an outside energy source like the Sun to make energy-rich molecules are called **producers.** Most producers contain chlorophyll (KLOR uh fihl), a chemical that is required for photosynthesis. As shown in **Figure 12,** green plants are producers. Some producers do not contain chlorophyll and do not use energy from the Sun. Instead, they make energy-rich molecules through a process called chemosynthesis (kee moh SIHN thuh sus). These organisms can be found near volcanic vents on the ocean floor. Inorganic molecules in the water provide the energy source for chemosynthesis.

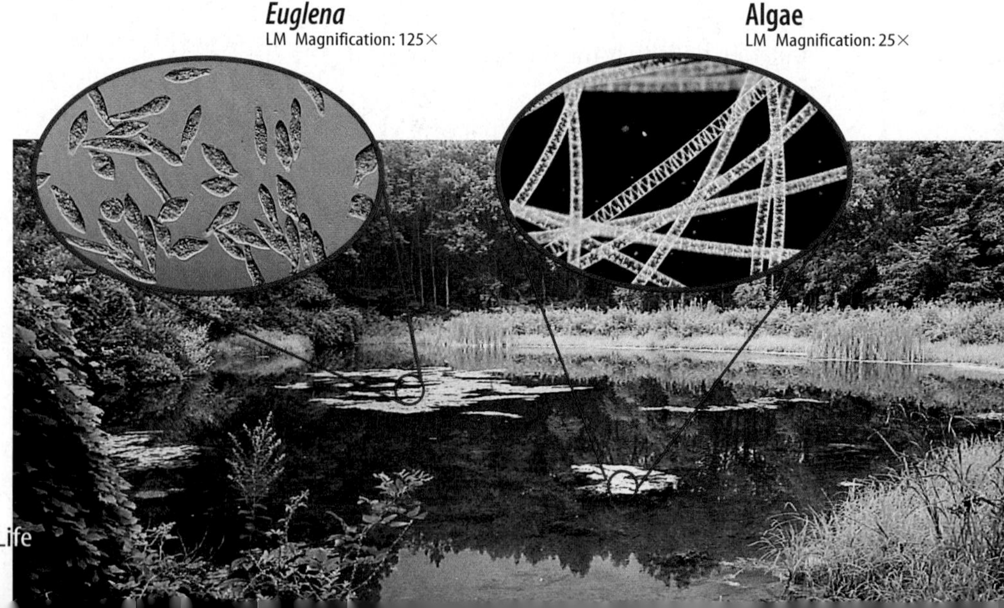

Figure 12 Green plants, including the grasses that surround this pond, are producers. The pond water also contains producers, including microscopic organisms like *Euglena* and algae.

Figure 13 Four categories of consumers are shown.
Identify *the consumer category that would apply to a bear. What about a mushroom?*

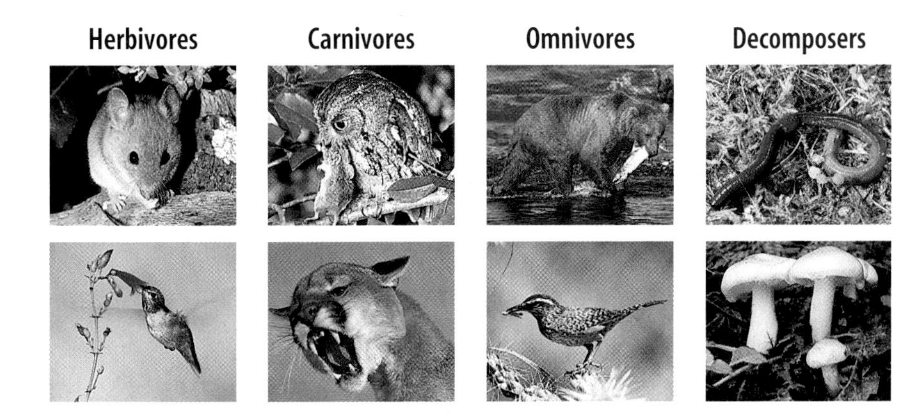

Consumers Organisms that cannot make their own energy-rich molecules are called **consumers.** Consumers obtain energy by eating other organisms. **Figure 13** shows the four general categories of consumers. Herbivores are the vegetarians of the world. They include rabbits, deer, and other plant eaters. Carnivores are animals that eat other animals. Frogs and spiders are carnivores that eat insects. Omnivores, including pigs and humans, eat mostly plants and animals. Decomposers, including fungi, bacteria, and earthworms, consume wastes and dead organisms. Decomposers help recycle once-living matter by breaking it down into simple, energy-rich substances. These substances might serve as food for decomposers, be absorbed by plant roots, or be consumed by other organisms.

Glucose The nutrient molecule produced during photosynthesis is glucose. Look up the chemical structure of glucose and draw it in your Science Journal.

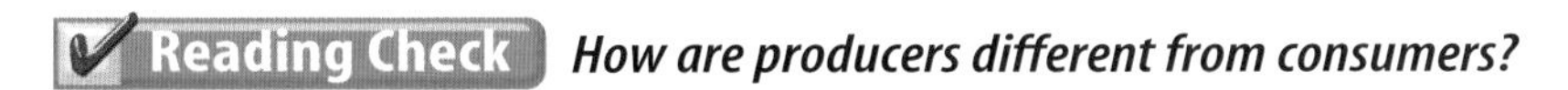

How are producers different from consumers?

Food Chains Ecology includes the study of how organisms depend on each other for food. A food chain is a simple model of the feeding relationships in an ecosystem. For example, shrubs are food for deer, and deer are food for mountain lions, as illustrated in **Figure 14.** What food chain would include you?

Figure 14 Food chains illustrate how consumers obtain energy from other organisms in an ecosystem.

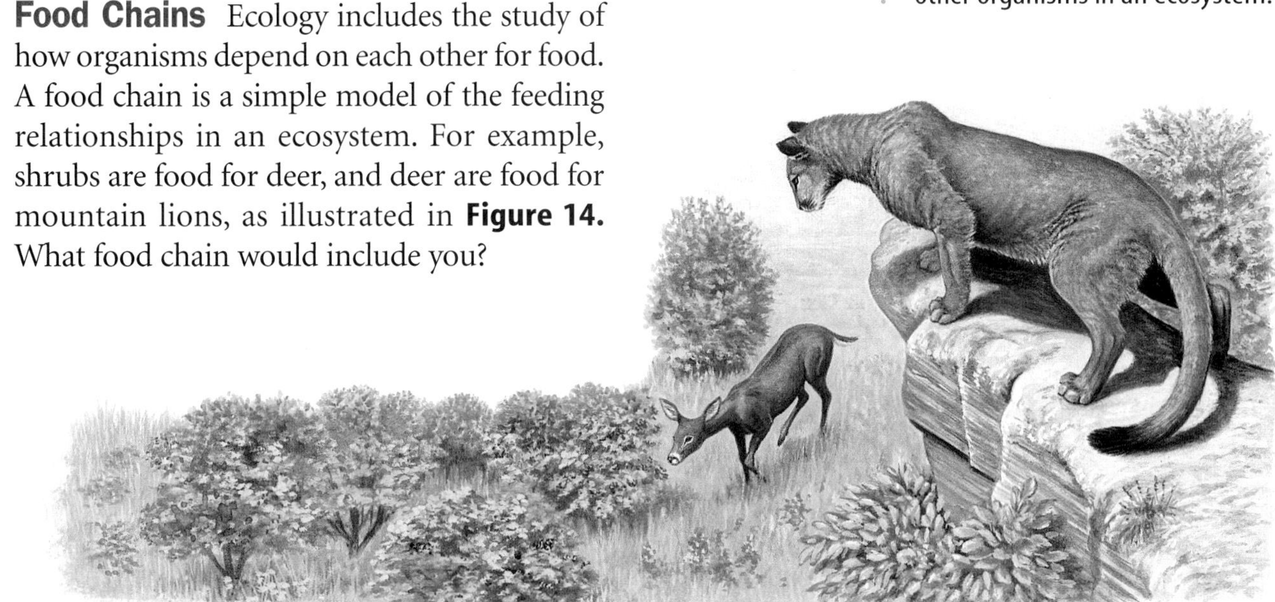

Symbiotic Relationships

Figure 15 Many examples of symbiotic relationships exist in nature.

Not all relationships among organisms involve food. Many organisms live together and share resources in other ways. Any close relationship between species is called **symbiosis.**

Lichens are a result of mutualism.

Mutualism You may have noticed crusty lichens growing on fences, trees, or rocks. Lichens, like those shown in **Figure 15,** are made up of an alga or a cyanobacterium that lives within the tissues of a fungus. Through photosynthesis, the cyanobacterium or alga supplies energy to itself and the fungus. The fungus provides a protected space in which the cyanobacterium or alga can live. Both organisms benefit from this association. A symbiotic relationship in which both species benefit is called **mutualism** (MYEW chuh wuh lih zum).

Clown fish and sea anemones have a commensal relationship.

Commensalism If you've ever visited a marine aquarium, you might have seen the ocean organisms shown in **Figure 15.** The creature with gently waving, tubelike tentacles is a sea anemone. The tentacles contain a mild poison. Anemones use their tentacles to capture shrimp, fish, and other small animals to eat. The striped clown fish can swim among the tentacles without being harmed. The anemone's tentacles protect the clown fish from predators. In this relationship, the clown fish benefits but the sea anemone is not helped or hurt. A symbiotic relationship in which one organism benefits and the other is not affected is called **commensalism** (kuh MEN suh lih zum).

LM Magnification: 128×

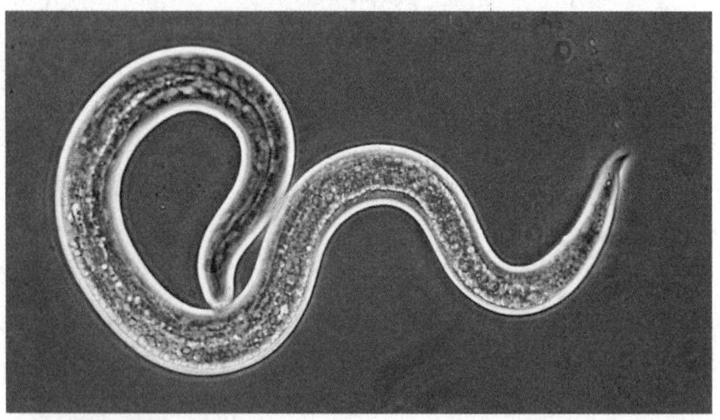

Some roundworms are parasites that rob nutrients from their hosts.

Parasitism Pet cats or dogs sometimes have to be treated for worms. Roundworms, like the one shown in **Figure 15,** are common in puppies. This roundworm attaches itself to the inside of the puppy's intestine and feeds on nutrients in the puppy's blood. The puppy may have abdominal pain, bloating, and diarrhea. If the infection is severe, the puppy might die. A symbiotic relationship in which one organism benefits but the other is harmed is called **parasitism** (PER uh suh tih zum).

Niches

One habitat might contain hundreds or even thousands of species. Look at the rotting log habitat shown in **Figure 16.** A rotting log in a forest can be home to many species of insects, including termites that eat decaying wood and ants that feed on the termites. Other species that live on or under the rotting log include millipedes, centipedes, spiders, and worms. You might think that competition for resources would make it impossible for so many species to live in the same habitat. However, each species has different requirements for its survival. As a result, each species has its own niche (NICH). An organism's **niche** is its role in its environment—how it obtains food and shelter, finds a mate, cares for its young, and avoids danger.

Reading Check *Why does each species have its own niche?*

Special adaptations that improve survival are often part of an organism's niche. Milkweed plants contain a poison that prevents many insects from feeding on them. Monarch butterfly caterpillars have an adaptation that allows them to eat milkweed. Monarchs can take advantage of a food resource that other species cannot use. Milkweed poison also helps protect monarchs from predators. When the caterpillars eat milkweed, they become slightly poisonous. Birds avoid eating monarchs because they learn that the caterpillars and adult butterflies have an awful taste and can make them sick.

Plant Poisons The poison in milkweed is similar to the drug digitalis. Small amounts of digitalis are used to treat heart ailments in humans, but it is poisonous in large doses. Research the history of digitalis as a medicine. In your Science Journal, list diseases for which it was used but is no longer used.

Figure 16 Different adaptations enable each species living in this rotting log to have its own niche. Termites eat wood. They make tunnels inside the log. Millipedes feed on plant matter and find shelter beneath the log. Wolf spiders capture insects living in and around the log.

Figure 17 The alligator is a predator. The turtle is its prey.

Predator and Prey When you think of survival in the wild, you might imagine an antelope running away from a lion. An organism's niche includes how it avoids being eaten and how it finds or captures its food. Predators, like the one shown in **Figure 17,** are consumers that capture and eat other consumers. The prey is the organism that is captured by the predator. The presence of predators usually increases the number of different species that can live in an ecosystem. Predators limit the size of prey populations. As a result, food and other resources are less likely to become scarce, and competition between species is reduced.

Cooperation Individual organisms often cooperate in ways that improve survival. For example, a white-tailed deer that detects the presence of wolves or coyotes will alert the other deer in the herd. Many insects, such as ants and honeybees, live in social groups. Different individuals perform different tasks required for the survival of the entire nest. Soldier ants protect workers that go out of the nest to gather food. Worker ants feed and care for ant larvae that hatch from eggs laid by the queen. These cooperative actions improve survival and are a part of the specie's niche.

section 3 review

Summary

Obtaining Energy

- All life requires a constant supply of energy.
- Most producers make food by photosynthesis using light energy.
- Consumers cannot make food. They obtain energy by eating producers or other consumers.
- A food chain models the feeding relationships between species.

Symbiotic Relationships

- Symbiosis is any close relationship between species.
- Mutualism, commensalism, and parasitism are types of symbiosis.
- An organism's niche describes the ways in which the organism obtains food, avoids danger, and finds shelter.

Self Check

1. **Explain** why all consumers depend on producers for food.
2. **Describe** a mutualistic relationship between two imaginary organisms. Name the organisms and explain how each benefits.
3. **Compare and contrast** the terms *habitat* and *niche.*
4. **Think Critically** A parasite can obtain food only from a host organism. Explain why most parasites weaken, but do not kill, their hosts.

Applying Skills

5. **Design an experiment** to classify the symbiotic relationship that exists between two hypothetical organisms. Animal A definitely benefits from its relationship with Plant B, but it is not clear whether Plant B benefits, is harmed, or is unaffected.

Feeding Habits of Planaria

You probably have watched minnows darting about in a stream. It is not as easy to observe organisms that live at the bottom of a stream, beneath rocks, logs, and dead leaves. Countless stream organisms, including insect larvae, worms, and microscopic organisms, live out of your view. One such organism is a type of flatworm called a planarian. In this lab, you will find out about the eating habits of planarians.

Real-World Question

What food items do planarians prefer to eat?

Goals

- **Observe** the food preference of planarians.
- **Infer** what planarians eat in the wild.

Materials

small bowl
planarians (several)
lettuce leaf
raw liver or meat
guppies (several)
pond or stream water
magnifying lens

Safety Precautions

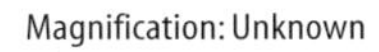

Magnification: Unknown

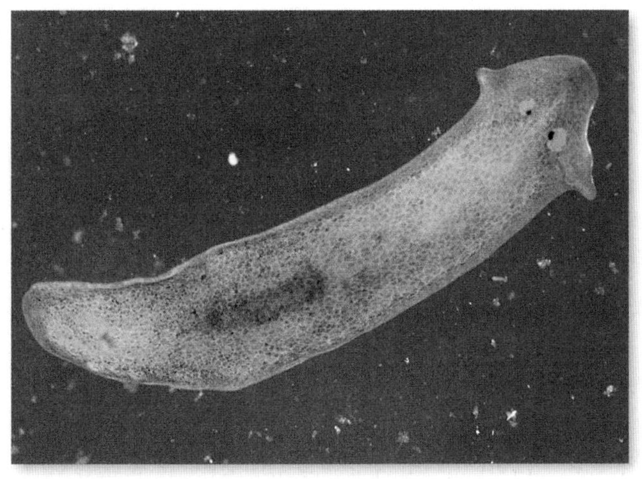

Procedure

1. Fill the bowl with stream water.
2. Place a lettuce leaf, piece of raw liver, and several guppies in the bowl. Add the planarians. Wash your hands.
3. **Observe** what happens inside the bowl for at least 20 minutes. Do not disturb the bowl or its contents. Use a magnifying lens to look at the planarians.
4. **Record** all of your observations in your Science Journal.

Conclude and Apply

1. **Name** the food the planarians preferred.
2. **Infer** what planarians might eat when in their natural environment.
3. **Describe,** based on your observations during this lab, a planarian's niche in a stream ecosystem.
4. **Predict** where in a stream you might find planarians. Use references to find out whether your prediction is correct.

Communicating Your Data

Share your results with other students in your class. Plan an adult-supervised trip with several classmates to a local stream to search for planarians in their native habitat. **For more help, refer to the Science Skill Handbook.**

POPULATION GROWTH IN FRUIT FLIES

Goals

- **Identify** the environmental factors needed by a population of fruit flies.
- **Design** an experiment to investigate how a change in one environmental factor affects in any way the size of a fruit fly population.
- **Observe** and **measure** changes in population size.

Possible Materials

fruit flies
standard fruit fly culture kit
food items (banana, orange peel, or other fruit)
water
heating or cooling source
culture containers
cloth, plastic, or other tops for culture containers
magnifying lens

Safety Precautions

Real-World Question

Populations can grow at an exponential rate only if the environment provides the right amount of food, shelter, air, moisture, heat, living space, and other factors. You probably have seen fruit flies hovering near ripe bananas or other fruit. Fruit flies are fast-growing organisms often raised in science laboratories. The flies are kept in culture tubes and fed a diet of specially prepared food flakes. Can you improve on this standard growing method to achieve faster population growth? Will a change in one environmental factor affect the growth of a fruit fly population?

Form a Hypothesis

Based on your reading about fruit flies, state a hypothesis about how changing one environmental factor will affect the rate of growth of a fruit fly population.

Test Your Hypothesis

Make a Plan

1. As a group, decide on one environmental factor to investigate. Agree on a hypothesis about how a change in this factor will affect population growth. Decide how you will test your hypothesis, and identify the experimental results that would support your hypothesis.
2. **List** the steps you will need to take to test your hypothesis. Describe exactly what you will do. List your materials.
3. **Determine** the method you will use to measure changes in the size of your fruit fly populations.

4. Prepare a data table in your Science Journal to record weekly measurements of your fruit fly populations.
5. Read the entire experiment and make sure all of the steps are in a logical order.
6. **Research** the standard method used to raise fruit flies in the laboratory. Use this method as the control in your experiment.
7. **Identify** all constants, variables, and controls in your experiment.

Follow Your Plan

1. Make sure your teacher approves your plan before you start.
2. Carry out your experiment.
3. **Measure** the growth of your fruit fly populations weekly and record the data in your data table.

Analyze Your Data

1. **Identify** the constants and the variables in your experiment.
2. **Compare** changes in the size of your control population with changes in your experimental population. Which population grew faster?
3. **Make and Use Graphs** Using the information in your data table, make a line graph that shows how the sizes of your two fruit fly populations changed over time. Use a different colored pencil for each population's line on the graph.

Conclude and Apply

1. **Explain** whether or not the results support your hypothesis.
2. **Compare** the growth of your control and experimental populations. Did either population reach exponential growth?
   How do you know?

Communicating Your Data

Compare the results of your experiment with those of other students in your class. **For more help, refer to the** Science Skill Handbook.

The Census measures a human population

Counting people is important to the United States and to many other countries around the world. It helps governments determine the distribution of people in the various regions of a nation. To obtain this information, the government takes a census—a count of how many people are living in their country on a particular day at a particular time, and in a particular place. A census is a snapshot of a country's population.

Counting on the Count

When the United States government was formed, its founders set up the House of Representatives based on population. Areas with more people had more government representatives, and areas with fewer people had fewer representatives. In 1787, the requirement for a census became part of the U.S. Constitution. A census must be taken every ten years so the proper number of representatives for each state can be calculated.

The Short Form

Before 1970, United States census data was collected by field workers. They went door to door to count the number of people living in each household. Since then, the census has been done mostly by mail. Census data are important in deciding how to distribute government services and funding.

The 2000 Snapshot

One of the findings of the 2000 Census is that the U.S. population is becoming more equally spread out across age groups. Census officials estimate that by 2020 the population of children, middle-aged people, and senior citizens will be about equal. It's predicted also that there will be more people who are over 100 years old than ever before. Federal, state, and local governments will be using the results of the 2000 Census for years to come as they plan our future.

Census **Develop a school census. What questions will you ask? (Don't ask questions that are too personal.) Who will ask them? How will you make sure you counted everyone? Using the results, can you make any predictions about your school's future or its current students?**

For more information, visit life.msscience.com/time

Reviewing Main Ideas

Section 1 Living Earth

1. Ecology is the study of interactions that take place in the biosphere.
2. A population is made up of all organisms of one species living in an area at the same time.
3. A community is made up of all the populations living in one ecosystem.
4. Living and nonliving factors affect an organism's ability to survive in its habitat.

Section 2 Populations

1. Population size can be estimated by counting a sample of a total population.
2. Competition for limiting factors can restrict the size of a population.
3. Population growth is affected by birthrate, death rate, and the movement of individuals into or out of a community.
4. Exponential population growth can occur in environments that provide a species with plenty of food, shelter, and other resources.

Section 3 Interactions Within Communities

1. All life requires energy.
2. Most producers use light to make food in the form of energy-rich molecules. Consumers obtain energy by eating other organisms.
3. Mutualism, commensalism, and parasitism are the three kinds of symbiosis.
4. Every species has its own niche, which includes adaptations for survival.

Visualizing Main Ideas

Copy and complete the following concept map on communities.

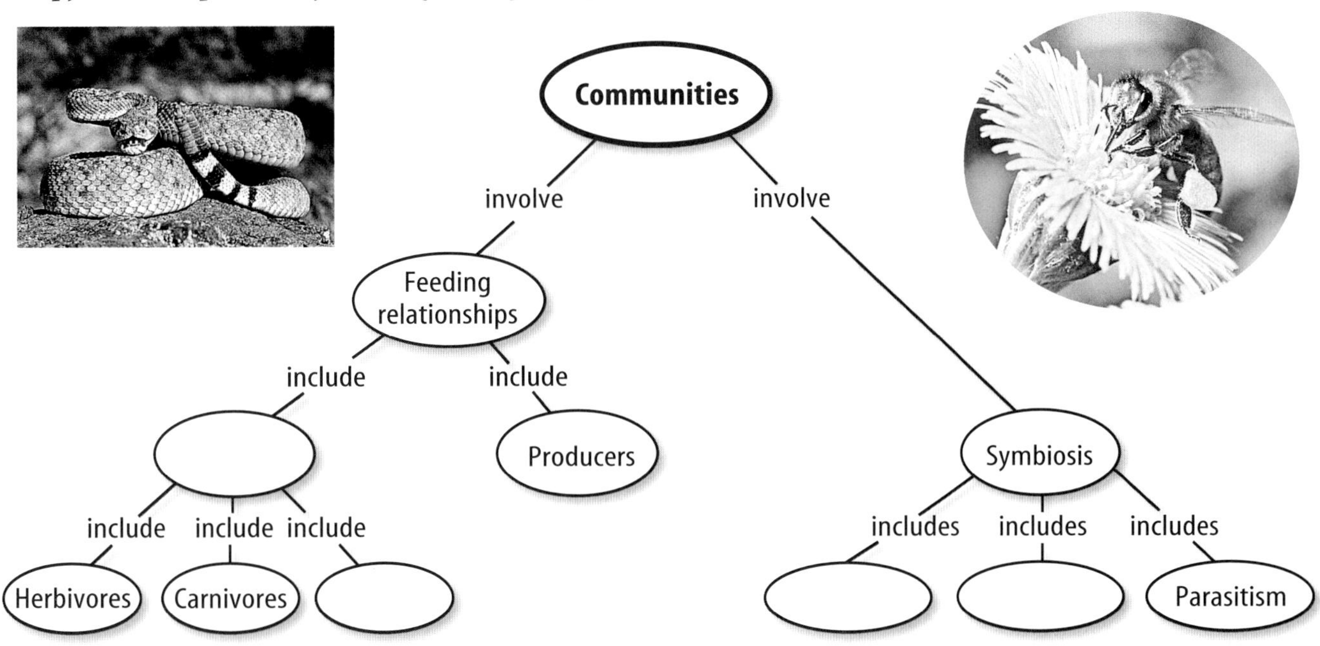

Using Vocabulary

biosphere p. 684	limiting factor p. 690
carrying capacity p. 691	mutualism p. 698
commensalism p. 698	niche p. 699
community p. 686	parasitism p. 698
consumer p. 697	population p. 686
ecology p. 685	producer p. 696
ecosystem p. 685	symbiosis p. 698
habitat p. 687	

Explain the difference between the vocabulary words in each of the following sets.

1. niche—habitat
2. mutualism—commensalism
3. limiting factor—carrying capacity
4. biosphere—ecosystem
5. producer—consumer
6. population—ecosystem
7. community—population
8. parasitism—symbiosis
9. ecosystem—ecology
10. parasitism—commensalism

Checking Concepts

Choose the word or phrase that best answers the question.

11. Which of the following is a living factor in the environment?
    A) animals **C)** sunlight
    B) air **D)** soil

12. What is made up of all the populations in an area?
    A) niches **C)** community
    B) habitats **D)** ecosystem

13. What does the number of individuals in a population that occupies an area of a specific size describe?
    A) clumping **C)** spacing
    B) size **D)** density

14. Which of the following animals is an example of an herbivore?
    A) wolf **C)** tree
    B) moss **D)** rabbit

15. What term best describes a symbiotic relationship in which one species is helped and the other is harmed?
    A) mutualism **C)** commensalism
    B) parasitism **D)** consumerism

16. Which of the following conditions tends to increase the size of a population?
    A) births exceed deaths
    B) population size exceeds the carrying capacity
    C) movements out of an area exceed movements into the area
    D) severe drought

17. Which of the following is most likely to be a limiting factor in a population of fish living in the shallow water of a large lake?
    A) sunlight **C)** food
    B) water **D)** soil

18. In which of the following categories does the pictured organism belong?

    A) herbivore
    B) carnivore
    C) producer
    D) consumer

19. Which pair of words is incorrect?
    A) black bear—carnivore
    B) grasshopper—herbivore
    C) pig—omnivore
    D) lion—carnivore

Science Online life.msscience.com/vocabulary_puzzlemaker

Thinking Critically

20. **Infer** why a parasite has a harmful effect on the organism it infects.
21. **Explain** what factors affect carrying capacity.
22. **Describe** your own habitat and niche.
23. **Make and Use Tables** Copy and complete the following table.

Types of Symbiosis

Organism A	Organism B	Relationship
Gains	Doesn't gain or lose	
Gains		Mutualism
Gains	Loses	

24. **Explain** how several different niches can exist in the same habitat.
25. **Make a model** of a food chain using the following organisms: grass, snake, mouse, and hawk.
26. **Predict** Dandelion seeds can float great distances on the wind with the help of white, featherlike attachments. Predict how a dandelion seed's ability to be carried on the wind helps reduce competition among dandelion plants.
27. **Classify** the following relationships as parasitism, commensalism, or mutualism: a shark and a remora fish that cleans and eats parasites from the shark's gills; head lice and a human; a spiny sea urchin and a tiny fish that hides from predators by floating among the sea urchin's spines.
28. **Compare and contrast** the diets of omnivores and herbivores. Give examples of each.
29. **List** three ways exponential growth in the human population affects people's lives.

Performance Activities

30. **Poster** Use photographs from old magazines to create a poster that shows at least three different food chains. Illustrate energy pathways from organism to organism and from organisms to the environment. Display your poster for your classmates.

Applying Math

31. **Measuring Populations** An ecologist wants to know the size of a population of wild daisy plants growing in a meadow that measures 1,000 m^2. The ecologist counts 30 daisy plants in a sample area of 100 m^2. What is the estimated population of daisies in the entire meadow?

Use the table below to answer question 32.

Lynx and Hare Pelts Sold to the Hudson's Bay Company*

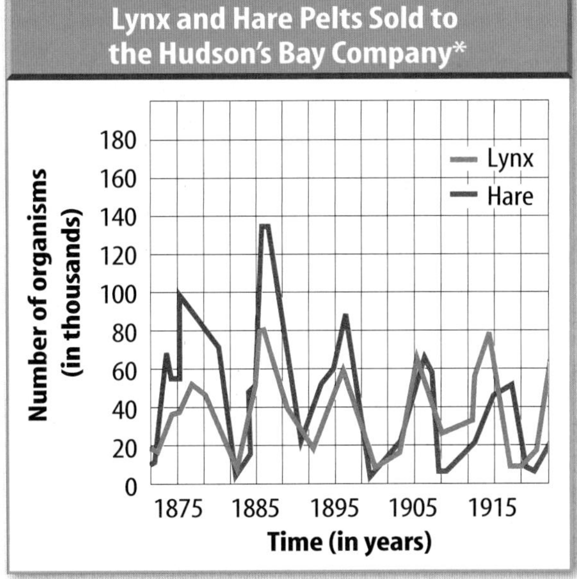

* Data from 1875 through 1904 reflects actual pelts counted. Data from 1905 through 1915 is based on answers to questionnaire.

32. **Changes in Populations** The graph above shows changes over time in the sizes of lynx and rabbit populations in an ecosystem. What does the graph tell you about the relationship between these two species? Explain how they influence each other's population size.

chapter 24 Standardized Test Practice

Part 1 Multiple Choice

Record your answers on the answer sheet provided by your teacher or on a sheet of paper.

1. Which of the following terms is defined in part by nonliving factors?
   A. population
   B. community
   C. ecosystem
   D. niche

2. Which of the follow terms would include all places where organisms live on Earth?
   A. ecosystem
   B. habitat
   C. biosphere
   D. community

3. Which of the following is not a method of measuring populations?
   A. total count
   B. trap-release
   C. sample count
   D. trap-mark-release

Use the photo below to answer questions 4 and 5.

4. Dead plants at the bottom of this pond are consumed by
   A. omnivores.
   B. herbivores.
   C. carnivores.
   D. decomposers.

5. If the pond shrinks in size, what effect will this have on the population density of the pond's minnow species?
   A. It will increase.
   B. It will decrease.
   C. It will stay the same.
   D. No effect; it is not a limiting factor.

6. Which of the following includes organisms that can directly convert energy from the Sun into food?
   A. producers
   B. decomposers
   C. omnivores
   D. consumers

7. You have a symbiotic relationship with bacteria in your digestive system. These bacteria break down food you ingest, and you get vital nutrients from them. Which type of symbiosis is this?
   A. mutualism
   B. barbarism
   C. commensalism
   D. parasitism

Use the photo below to answer questions 8 and 9.

8. An eastern screech owl might compete with which organism most intensely for resources?
   A. mouse
   B. hawk
   C. mountain lion
   D. wren

9. Which of the following organisms might compete with the mouse for seeds?
   A. hawk
   B. lion
   C. fox
   D. sparrow

10. Which of the following is an example of a community?
    A. all the white-tailed deer in a forest
    B. all the trees, soil, and water in a forest
    C. all the plants and animals in a wetland
    D. all the cattails in a wetland

Part 2 Short Response/Grid In

Record your answers on the answer sheet provided by your teacher or on a sheet of paper.

Use the graph below to answer question 11.

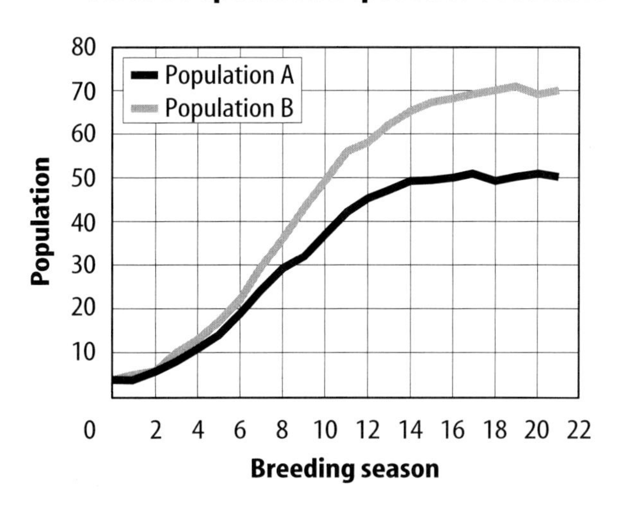

11. The graph depicts the growth of two white-footed mice populations, one exposed to hawks (population A) and one without hawks (population B). Are hawks a limiting factor for either mouse population? If not, then what other factor could be a limiting factor for that population?

12. Diagram the flow of energy through an ecosystem. Include the sources of energy, producers, consumers, and decomposers in the ecosystem.

Test-Taking Tip

Understand the Question Be sure you understand the question before you read the answer choices. Make special note of words like NOT or EXCEPT. Read and consider choices before you mark your answer sheet.

Question 11 Make sure you understand which mouse population is subject to predation by hawks and which mouse population do hawks not affect.

Part 3 Open Ended

Record your answers on a sheet of paper.

13. The colors and patterns of the viceroy butterfly are similar to the monarch butterfly, however, the viceroy caterpillars don't feed on milkweed. How does the viceroy butterfly benefit from this adaptation of its appearance? Under what circumstance would this adaptation not benefit the viceroy? Why?

Use the illustration below to answer question 14.

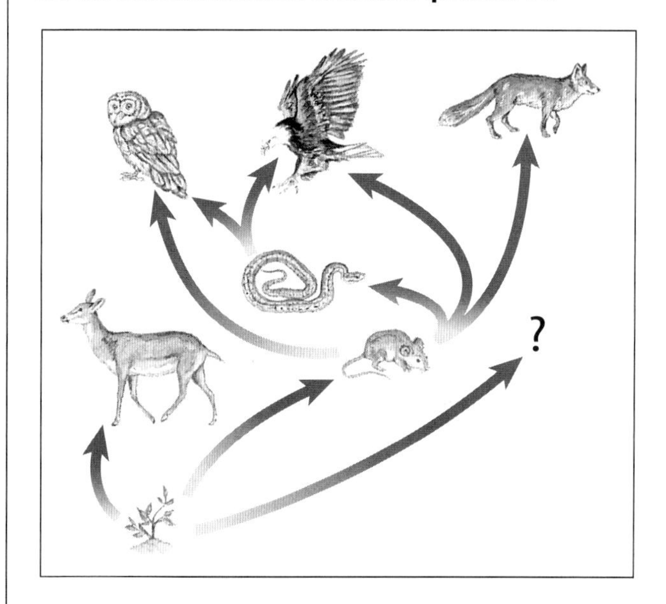

14. The illustration depicts a food web for a particular ecosystem. If the "?" is another mouse species population that is introduced into the ecosystem, explain what impact this would have on the species populations in the ecosystem.

15. Identify and explain possible limiting factors that would control the size of an ant colony.

16. How would you measure the size of a population of gray squirrels in a woodland? Explain which method you would choose and why.

chapter

The Nonliving Environment

The BIG Idea

Environments have both living and nonliving elements.

SECTION 1
Abiotic Factors
Main Idea Both living and nonliving parts of an environment are needed for organisms to survive.

SECTION 2
Cycles in Nature
Main Idea Many nonliving elements on Earth, such as water and oxygen, are recycled over and over.

SECTION 3
Energy Flow
Main Idea All living things use energy.

Sun, Surf, and Sand

Living things on this coast directly or indirectly depend on nonliving things, such as sunlight, water, and rocks, for energy and raw materials needed for their life processes. In this chapter, you will read how these and other nonliving things affect life on Earth.

Science Journal List all the nonliving things that you can see in this picture in order of importance. Explain your reasoning for the order you chose.

Start-Up Activities

Earth Has Many Ecosystems

Do you live in a dry, sandy region covered with cactus plants or desert scrub? Is your home in the mountains? Does snow fall during the winter? In this chapter, you'll learn why the nonliving factors in each ecosystem are different. The following lab will get you started.

1. Locate your city or town on a globe or world map. Find your latitude. Latitude shows your distance from the equator and is expressed in degrees, minutes, and seconds.
2. Locate another city with the same latitude as your city but on a different continent.
3. Locate a third city with latitude close to the equator.
4. Using references, compare average annual precipitation and average high and low temperatures for all three cities.
5. **Think Critically** Hypothesize how latitude affects average temperatures and rainfall.

Preview this chapter's content and activities at life.msscience.com

Nonliving Factors Make the following Foldable to help you understand the cause and effect relationships within the nonliving environment.

STEP 1 **Fold** two vertical sheets of paper in half from top to bottom. **Cut** the papers in half along the folds.

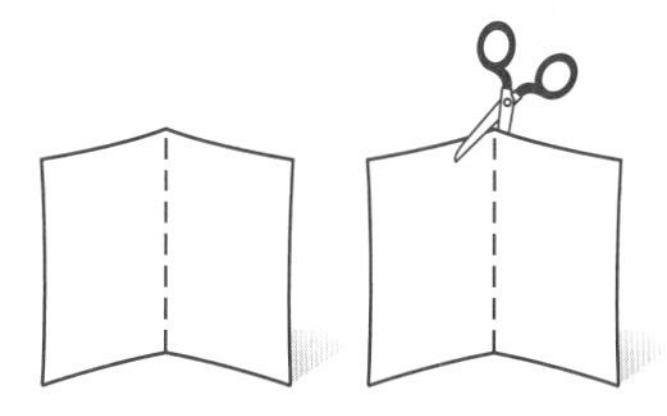

STEP 2 Discard one piece and **fold** the three vertical pieces in half from top to bottom.

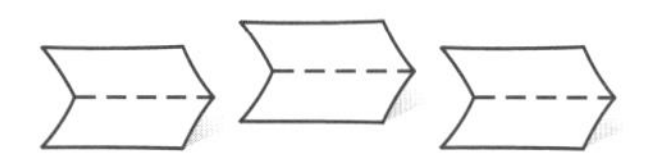

STEP 3 **Turn** the papers horizontally. **Tape** the short ends of the pieces together (overlapping the edges slightly).

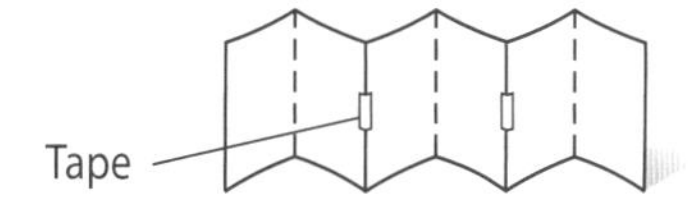

STEP 4 On one side, **label** the folds: *Nonliving, Water, Soil, Wind, Temperature,* and *Elevation.* **Draw** a picture of a familiar ecosystem on the other side.

Sequence As you read the chapter, write on the folds how each nonliving factor affects the environment that you draw.

Get Ready to Read

Make Inferences

1 Learn It! When you make inferences, you draw conclusions that are not directly stated in the text. This means you "read between the lines." You interpret clues and draw upon prior knowledge. Authors rely on a reader's ability to infer because all the details are not always given.

2 Practice It! Read the excerpt below and pay attention to highlighted words as you make inferences. Use this Think-Through chart to help you make inferences.

Water vapor that has been released into the atmosphere eventually comes into contact with colder air. The temperature of the water vapor drops. Over time, the water vapor cools enough to change back into liquid water. The process of changing from a gas to a liquid is called condensation. Water vapor condenses on particles of dust in the air, forming tiny droplets. At first, the droplets clump together to form clouds.

—from page 721

Text	Question	Inferences
Released into the atmosphere	How is the water vapor released back into the atmosphere?	Evaporation? Animals?
Comes into contact with colder air	Where does it come into contact with cold air?	High in the atmosphere? Cool air from cool weather?
Condenses on particles of dust	Where have the dust particles come from?	Pollution? Wind?

3 Apply It! As you read this chapter, practice your skill at making inferences by making connections and asking questions.

Reading Tip

Sometimes you make inferences by using other reading skills, such as questioning and predicting.

Target Your Reading

Use this to focus on the main ideas as you read the chapter.

1. **Before you read** the chapter, respond to the statements below on your worksheet or on a numbered sheet of paper.
   - Write an **A** if you **agree** with the statement.
   - Write a **D** if you **disagree** with the statement.

2. **After you read** the chapter, look back to this page to see if you've changed your mind about any of the statements.
   - If any of your answers changed, explain why.
   - Change any false statements into true statements.
   - Use your revised statements as a study guide.

Before You Read A or D	Statement	After You Read A or D
	1 The nonliving part on an environment often determines what living organisms are found there.	
	2 Most living organisms are mainly made up of water.	
	3 Heat from the Sun is responsible for wind.	
	4 Animals do not release water vapor.	
	5 The air we breathe mostly contains nitrogen.	
	6 Photosynthesis uses oxygen to produce energy.	
	7 Energy can be both converted to other forms and recycled.	
	8 Matter can be converted to other forms, but cannot be recycled.	
	9 The majority of energy is found at the bottom on an energy pyramid.	
	10 Water is a living part of the environment.	

Science Online
Print out a worksheet of this page at life.msscience.com

section 1

Abiotic Factors

as you read

What You'll Learn

- **Identify** common abiotic factors in most ecosystems.
- **List** the components of air that are needed for life.
- **Explain** how climate influences life in an ecosystem.

Why It's Important

Knowing how organisms depend on the nonliving world can help humans maintain a healthy environment.

Review Vocabulary
environment: everything, such as climate, soil, and living things, that surrounds and affects an organism

New Vocabulary
- biotic
- abiotic
- atmosphere
- soil
- climate

Environmental Factors

Living organisms depend on one another for food and shelter. The leaves of plants provide food and a home for grasshoppers, caterpillars, and other insects. Many birds depend on insects for food. Dead plants and animals decay and become part of the soil. The features of the environment that are alive, or were once alive, are called **biotic** (bi AH tihk) factors. The term *biotic* means "living."

Biotic factors are not the only things in an environment that are important to life. Most plants cannot grow without sunlight, air, water, and soil. Animals cannot survive without air, water, or the warmth that sunlight provides. The nonliving, physical features of the environment are called **abiotic** (ay bi AH tihk) factors. The prefix *a* means "not." The term *abiotic* means "not living." Abiotic factors include air, water, soil, sunlight, temperature, and climate. The abiotic factors in an environment often determine which kinds of organisms can live there. For example, water is an important abiotic factor in the environment, as shown in **Figure 1.**

Figure 1 Abiotic factors—air, water, soil, sunlight, temperature, and climate—influence all life on Earth.

Air

Air is invisible and plentiful, so it is easily overlooked as an abiotic factor of the environment. The air that surrounds Earth is called the **atmosphere.** Air contains 78 percent nitrogen, 21 percent oxygen, 0.94 percent argon, 0.03 percent carbon dioxide, and trace amounts of other gases. Some of these gases provide substances that support life.

Carbon dioxide (CO_2) is required for photosynthesis. Photosynthesis—a series of chemical reactions—uses CO_2, water, and energy from sunlight to produce sugar molecules. Organisms, like plants, that can use photosynthesis are called producers because they produce their own food. During photosynthesis, oxygen is released into the atmosphere.

When a candle burns, oxygen from the air chemically combines with the molecules of candle wax. Chemical energy stored in the wax is converted and released as heat and light energy. In a similar way, cells use oxygen to release the chemical energy stored in sugar molecules. This process is called respiration. Through respiration, cells obtain the energy needed for all life processes. Air-breathing animals aren't the only organisms that need oxygen. Plants, some bacteria, algae, fish, and other organisms need oxygen for respiration.

Water

Water is essential to life on Earth. It is a major ingredient of the fluid inside the cells of all organisms. In fact, most organisms are 50 percent to 95 percent water. Respiration, digestion, photosynthesis, and many other important life processes can take place only in the presence of water. As **Figure 2** shows, environments that have plenty of water usually support a greater diversity of and a larger number of organisms than environments that have little water.

Figure 2 Water is an important abiotic factor in deserts and rain forests.

Life in deserts is limited to species that can survive for long periods without water.

Thousands of species can live in lush rain forests where rain falls almost every day.

Determining Soil Makeup

Procedure

1. Collect 2 cups of **soil.** Remove large pieces of debris and break up clods.
2. Put the soil in a **quart jar** or **similar container that has a lid.**
3. Fill the container with **water** and add 1 teaspoon of **dishwashing liquid.**
4. Put the lid on tightly and shake the container.
5. After 1 min, measure and record the depth of sand that settled on the bottom.
6. After 2 h, measure and record the depth of silt that settles on top of the sand.
7. After 24 h, measure and record the depth of the layer between the silt and the floating organic matter.

Analysis

1. Clay particles are so small that they can remain suspended in water. Where is the clay in your sample?
2. Is sand, silt, or clay the greatest part of your soil sample?

Try at Home

Soil

Soil is a mixture of mineral and rock particles, the remains of dead organisms, water, and air. It is the topmost layer of Earth's crust, and it supports plant growth. Soil is formed, in part, of rock that has been broken down into tiny particles.

Soil is considered an abiotic factor because most of it is made up of nonliving rock and mineral particles. However, soil also contains living organisms and the decaying remains of dead organisms. Soil life includes bacteria, fungi, insects, and worms. The decaying matter found in soil is called humus. Soils contain different combinations of sand, clay, and humus. The type of soil present in a region has an important influence on the kinds of plant life that grow there.

Sunlight

All life requires energy, and sunlight is the energy source for almost all life on Earth. During photosynthesis, producers convert light energy into chemical energy that is stored in sugar molecules. Consumers are organisms that cannot make their own food. Energy is passed to consumers when they eat producers or other consumers. As shown in **Figure 3,** photosynthesis cannot take place if light is never available.

Shady forest

Bottom of deep ocean

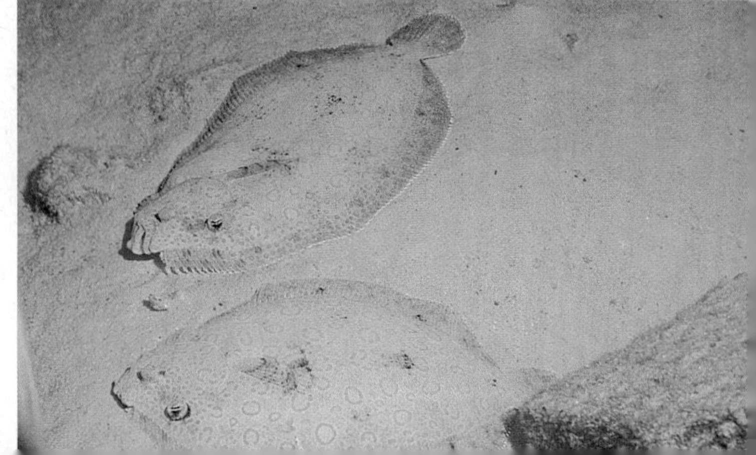

Figure 3 Photosynthesis requires light. Little sunlight reaches the shady forest floor, so plant growth beneath trees is limited. Sunlight does not reach into deep lake or ocean waters. Photosynthesis can take place only in shallow water or near the water's surface.
Infer *how fish that live at the bottom of the deep ocean obtain energy.*

Figure 4 Temperature is an abiotic factor that can affect an organism's survival.

The penguin has a thick layer of fat to hold in heat and keep the bird from freezing. These emperor penguins huddle together for added warmth.

The Arabian camel stores fat only in its hump. This way, the camel loses heat from other parts of its body, which helps it stay cool in the hot desert.

Temperature

Sunlight supplies life on Earth with light energy for photosynthesis and heat energy for warmth. Most organisms can survive only if their body temperatures stay within the range of 0°C to 50°C. Water freezes at 0°C. The penguins in **Figure 4** are adapted for survival in the freezing Antarctic. Camels can survive the hot temperatures of the Arabian Desert because their bodies are adapted for staying cool. The temperature of a region depends in part on the amount of sunlight it receives. The amount of sunlight depends on the land's latitude and elevation.

What does sunlight provide for life on Earth?

Latitude In this chapter's Launch Lab, you discovered that temperature is affected by latitude. You found that cities located at latitudes farther from the equator tend to have colder temperatures than cities at latitudes nearer to the equator. As **Figure 5** shows, polar regions receive less of the Sun's energy than equatorial regions. Near the equator, sunlight strikes Earth directly. Near the poles, sunlight strikes Earth at an angle, which spreads the energy over a larger area.

Figure 5 Because Earth is curved, latitudes farther from the equator are colder than latitudes near the equator.

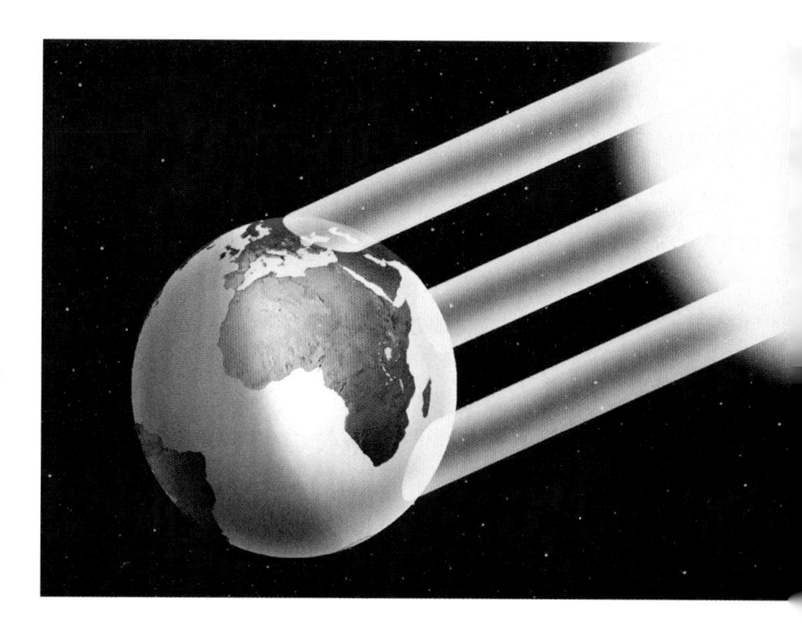

Figure 6 The stunted growth of these trees is a result of abiotic factors.

Elevation If you have climbed or driven up a mountain, you probably noticed that the temperature got cooler as you went higher. A region's elevation, or distance above sea level, affects its temperature. Earth's atmosphere acts as insulation that traps the Sun's heat. At higher elevations, the atmosphere is thinner than it is at lower elevations. Air becomes warmer when sunlight heats molecules in the air. Because there are fewer molecules at higher elevations, air temperatures there tend to be cooler.

At higher elevations, trees are shorter and the ground is rocky, as shown in **Figure 6.** Above the timberline—the elevation beyond which trees do not grow—plant life is limited to low-growing plants. The tops of some mountains are so cold that no plants can survive. Some mountain peaks are covered with snow year-round.

Applying Math — Solve for an Unknown

TEMPERATURE CHANGES You climb a mountain and record the temperature every 1,000 m of elevation. The temperature is 30°C at 304.8 m, 25°C at 609.6 m, 20°C at 914.4 m, 15°C at 1,219.2 m, and 5°C at 1,828.8 m. Make a graph of the data. Use your graph to predict the temperature at an altitude of 2,133.6 m.

Solution

1. *This is what you know:* The data can be written as ordered pairs (elevation, temperature). The ordered pairs for these data are (304.8, 30), (609.6, 25), (914.4, 20), (1,219.2, 15), (1,828.8, 5).
2. *This is what you want to find:* Predict the temperature at an elevation of 2,133.6 m.
3. *This is what you need to do:* Graph the data by plotting elevation on the x-axis and temperature on the y-axis.
4. *Predict the temperature at 2,133.6 m:* Extend the graph line to predict the temperature at 2,133.6 m.

Practice Problems

1. Temperatures on another mountain are 33°C at sea level, 31°C at 125 m, 29°C at 250 m, and 26°C at 425 m. Graph the data and predict the temperature at 550 m.
2. Predict what the temperature would be at 375 m.

For more practice, visit life.msscience.com/math_practice

Climate

In Fairbanks, Alaska, winter temperatures may be as low as −52°C, and more than a meter of snow might fall in one month. In Key West, Florida, snow never falls and winter temperatures rarely dip below 5°C. These two cities have different climates. **Climate** refers to an area's average weather conditions over time, including temperature, rainfall or other precipitation, and wind.

For the majority of living things, temperature and precipitation are the two most important components of climate. The average temperature and rainfall in an area influence the type of life found there. Suppose a region has an average temperature of 25°C and receives an average of less than 25 cm of rain every year. It is likely to be the home of cactus plants and other desert life. A region with similar temperatures that receives more than 300 cm of rain every year is probably a tropical rain forest.

Wind Heat energy from the Sun not only determines temperature, but also is responsible for the wind. The air is made up of molecules of gas. As the temperature increases, the molecules spread farther apart. As a result, warm air is lighter than cold air. Colder air sinks below warmer air and pushes it upward, as shown in **Figure 7.** These motions create air currents that are called wind.

Farmer Changes in weather have a strong influence in crop production. Farmers sometimes adapt by changing planting and harvesting dates, selecting a different crop, or changing water use. In your Science Journal, describe another profession affected by climate.

Topic: Weather Data
Visit life.msscience.com for Web links to information about recent weather data for your area.

Activity In your Science Journal, describe how these weather conditions affect plants or animals that live in your area.

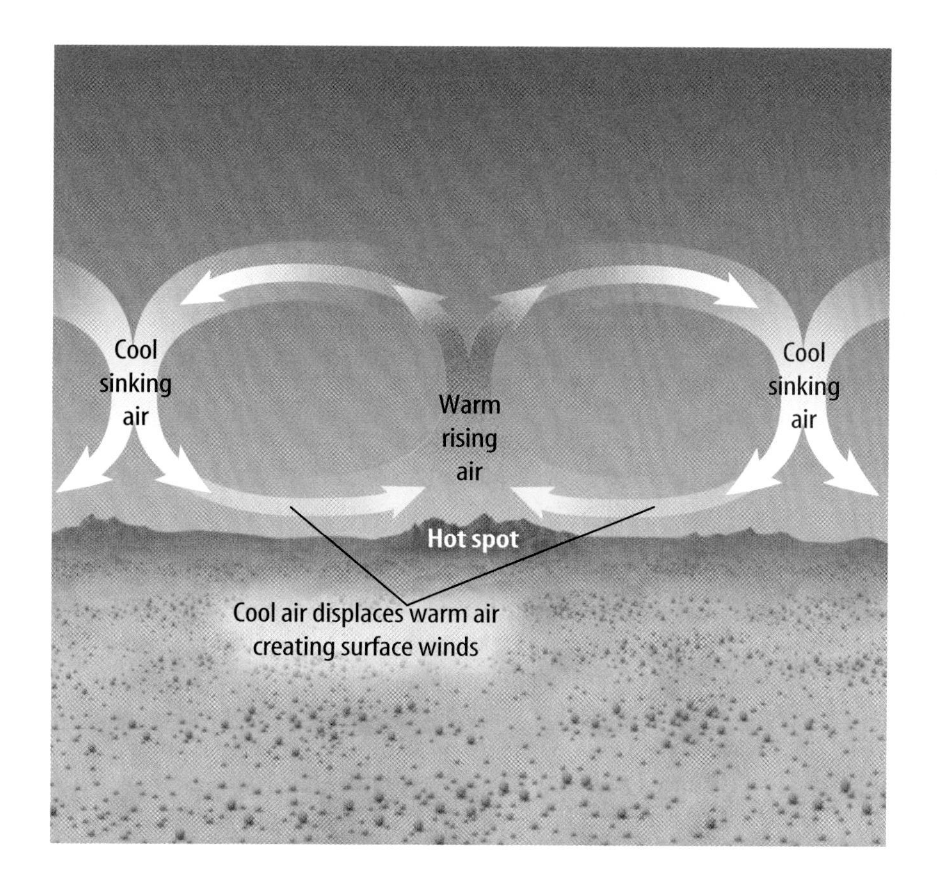

Figure 7 Winds are created when sunlight heats some portions of Earth's surface more than others. In areas that receive more heat, the air becomes warmer. Cold air sinks beneath the warm air, forcing the warm air upward.

Figure 8 In Washington State, the western side of the Cascade Mountains receives an average of 101 cm of rain each year. The eastern side of the Cascades is in a rain shadow that receives only about 25 cm of rain per year.

The Rain Shadow Effect The presence of mountains can affect rainfall patterns. As **Figure 8** shows, wind blowing toward one side of a mountain is forced upward by the mountain's shape. As the air nears the top of the mountain, it cools. When air cools, the moisture it contains falls as rain or snow. By the time the cool air crosses over the top of the mountain, it has lost most of its moisture. The other side of the mountain range receives much less precipitation. It is not uncommon to find lush forests on one side of a mountain range and desert on the other side.

section 1 review

Summary

Environmental Factors

- Organisms depend on one another as well as sunlight, air, water, and soil.

Air, Water, and Soil

- Some of the gases in air provide substances to support life.
- Water is a major component of the cells in all organisms.
- Soil supports plant growth.

Sunlight, Temperature, and Climate

- Light energy supports almost all life on Earth.
- Most organisms require temperature between 0°C and 50°C to survive.
- For most organisms, temperature and precipitation are the two most important components of climate.

Self Check

1. **Compare and contrast** biotic factors and abiotic factors in ecosystems.
2. **Explain** why soil is considered an abiotic factor and a biotic factor.
3. **Think Critically** On day 1, you hike in shade under tall trees. On day 2, the trees are shorter and farther apart. On day 3, you see small plants but no trees. On day 4, you see snow. What abiotic factors might contribute to these changes?

Applying Math

4. **Use an Electronic Spreadsheet** Obtain two months of temperature and precipitation data for two cities in your state. Enter the data in a spreadsheet and calculate average daily temperature and precipitation. Use your calculations to compare the two climates.

life.msscience.com/self_check_quiz

Humus Farm

Besides abiotic factors, such as rock particles and minerals, soil also contains biotic factors, including bacteria, molds, fungi, worms, insects, and decayed organisms. Crumbly, dark brown soil contains a high percentage of humus that is formed primarily from the decayed remains of plants, animals, and animal droppings. In this lab, you will cultivate your own humus.

Real-World Question

How does humus form?

Goals

- **Observe** the formation of humus.
- **Observe** biotic factors in the soil.
- **Infer** how humus forms naturally.

Materials

widemouthed jar
soil
grass clippings or green leaves
water
marker
metric ruler
graduated cylinder

Safety Precautions

Wash your hands thoroughly after handling soil, grass clippings, or leaves.

Humus Formation

Date	Observations
	Do not write in this book.

Procedure

1. Copy the data table below into your Science Journal.
2. Place 4 cm of soil in the jar. Pour 30 mL of water into the jar to moisten the soil.
3. Place 2 cm of grass clippings or green leaves on top of the soil in the jar.
4. Use a marker to mark the height of the grass clippings or green leaves in the jar.
5. Put the jar in a sunny place. Every other day, add 30 mL of water to it. In your Science Journal, write a prediction of what you think will happen in your jar.
6. **Observe** your jar every other day for four weeks. Record your observations in your data table.

Conclude and Apply

1. **Describe** what happened during your investigation.
2. **Infer** how molds and bacteria help the process of humus formation.
3. **Infer** how humus forms on forest floors or in grasslands.

Communicating Your Data

Compare your humus farm with those of your classmates. With several classmates, write a recipe for creating the richest humus. Ask your teacher to post your recipe in the classroom. **For more help, refer to the Science Skill Handbook.**

section 2

Cycles in Nature

as you read

What You'll Learn

- **Explain** the importance of Earth's water cycle.
- **Diagram** the carbon cycle.
- **Recognize** the role of nitrogen in life on Earth.

Why It's Important

The recycling of matter on Earth demonstrates natural processes.

Review Vocabulary
biosphere: the part of the world in which life can exist

New Vocabulary

- evaporation
- condensation
- water cycle
- nitrogen fixation
- nitrogen cycle
- carbon cycle

The Cycles of Matter

Imagine an aquarium containing water, fish, snails, plants, algae, and bacteria. The tank is sealed so that only light can enter. Food, water, and air cannot be added. Will the organisms in this environment survive? Through photosynthesis, plants and algae produce their own food. They also supply oxygen to the tank. Fish and snails take in oxygen and eat plants and algae. Wastes from fish and snails fertilize plants and algae. Organisms that die are decomposed by the bacteria. The organisms in this closed environment can survive because the materials are recycled. A constant supply of light energy is the only requirement. Earth's biosphere also contains a fixed amount of water, carbon, nitrogen, oxygen, and other materials required for life. These materials cycle through the environment and are reused by different organisms.

The Water Cycle

If you leave a glass of water on a sunny windowsill, the water will evaporate. **Evaporation** takes place when liquid water changes into water vapor, which is a gas, and enters the atmosphere, shown in **Figure 9.** Water evaporates from the surfaces of lakes, streams, puddles, and oceans. Water vapor enters the atmosphere from plant leaves in a process known as transpiration (trans puh RAY shun). Animals release water vapor into the air when they exhale. Water also returns to the environment from animal wastes.

Figure 9 Water vapor is a gas that is present in the atmosphere.

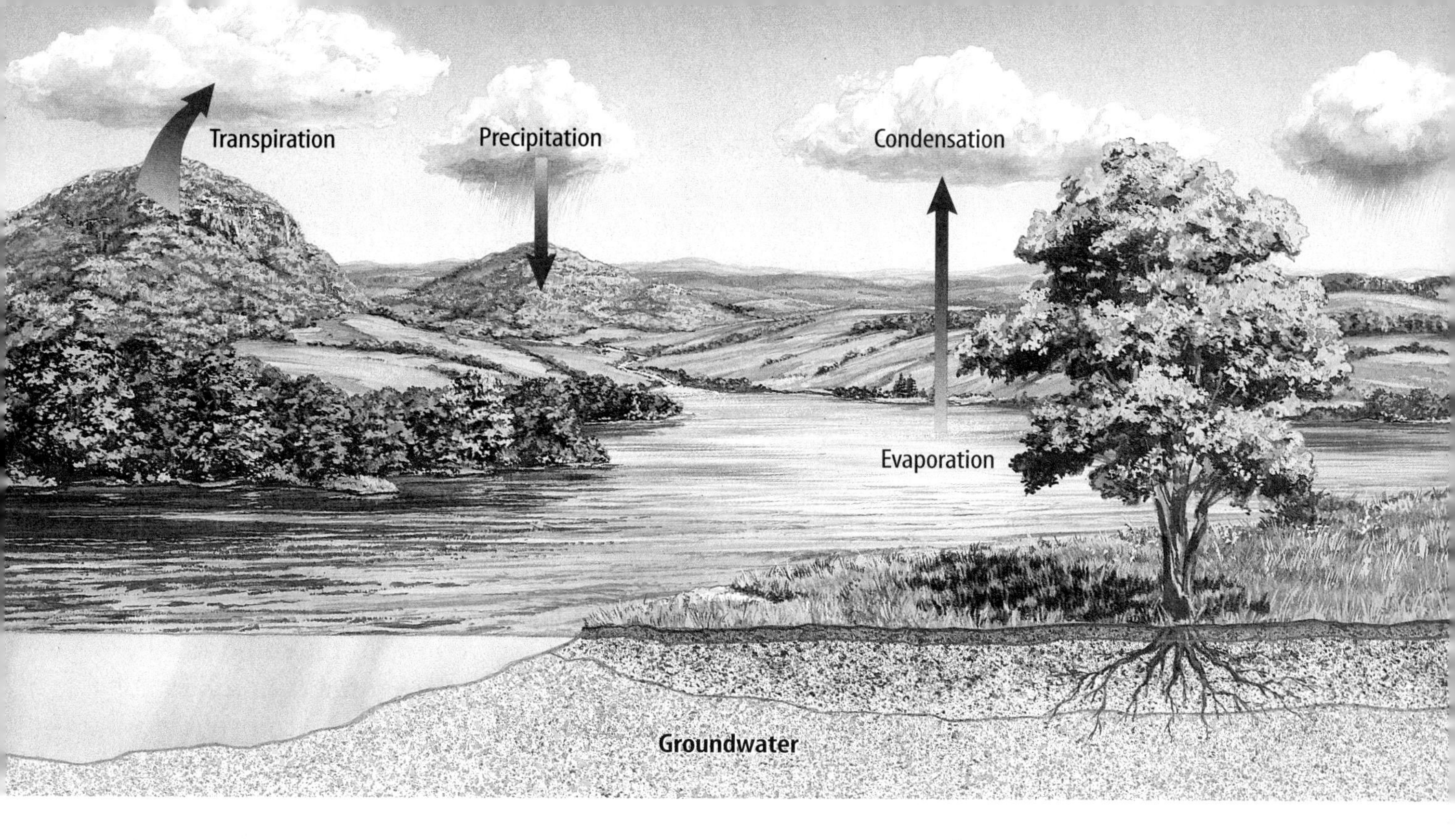

Condensation Water vapor that has been released into the atmosphere eventually comes into contact with colder air. The temperature of the water vapor drops. Over time, the water vapor cools enough to change back into liquid water. The process of changing from a gas to a liquid is called **condensation.** Water vapor condenses on particles of dust in the air, forming tiny droplets. At first, the droplets clump together to form clouds. When they become large and heavy enough, they fall to the ground as rain or other precipitation. As the diagram in **Figure 10** shows, the **water cycle** is a model that describes how water moves from the surface of Earth to the atmosphere and back to the surface again.

Figure 10 The water cycle involves evaporation, condensation, and precipitation. Water molecules can follow several pathways through the water cycle.
Identify *as many water cycle pathways as you can from this diagram.*

Water Use Data about the amount of water people take from reservoirs, rivers, and lakes for use in households, businesses, agriculture, and power production is shown in **Table 1.** These actions can reduce the amount of water that evaporates into the atmosphere. They also can influence how much water returns to the atmosphere by limiting the amount of water available to plants and animals.

Table 1 U.S. Estimated Water Use in 1995

Water Use	Millions of Gallons per Day	Percent of Total
Homes and Businesses	41,600	12.2
Industry and Mining	28,000	8.2
Farms and Ranches	139,200	40.9
Electricity Production	131,800	38.7

The Nitrogen Cycle

The element nitrogen is important to all living things. Nitrogen is a necessary ingredient of proteins. Proteins are required for the life processes that take place in the cells of all organisms. Nitrogen is also an essential part of the DNA of all organisms. Although nitrogen is the most plentiful gas in the atmosphere, most organisms cannot use nitrogen directly from the air. Plants need nitrogen that has been combined with other elements to form nitrogen compounds. Through a process called **nitrogen fixation,** some types of soil bacteria can form the nitrogen compounds that plants need. Plants absorb these nitrogen compounds through their roots. Animals obtain the nitrogen they need by eating plants or other animals. When dead organisms decay, the nitrogen in their bodies returns to the soil or to the atmosphere. This transfer of nitrogen from the atmosphere to the soil, to living organisms, and back to the atmosphere is called the **nitrogen cycle,** shown in **Figure 11.**

Reading Check *What is nitrogen fixation?*

Figure 11 During the nitrogen cycle, nitrogen gas from the atmosphere is converted to a soil compound that plants can use. **State** *one source of recycled nitrogen.*

Nitrogen gas is changed into usable compounds by lightning or by nitrogen-fixing bacteria that live on the roots of certain plants.
Plants use nitrogen compounds to build cells.
Animals eat plants. Animal wastes return some nitrogen compounds to the soil.
Animals and plants die and decompose, releasing nitrogen compounds back into the soil.

Figure 12 The swollen nodules on the roots of soybean plants contain colonies of nitrogen-fixing bacteria that help restore nitrogen to the soil.The bacteria depend on the plant for food, while the plant depends on the bacteria to form the nitrogen compounds the plant needs.

Soil Nitrogen Human activities can affect the part of the nitrogen cycle that takes place in the soil. If a farmer grows a crop, such as corn or wheat, most of the plant material is taken away when the crop is harvested. The plants are not left in the field to decay and return their nitrogen compounds to the soil. If these nitrogen compounds are not replaced, the soil could become infertile. You might have noticed that adding fertilizer to soil can make plants grow greener, bushier, or taller. Most fertilizers contain the kinds of nitrogen compounds that plants need for growth. Fertilizers can be used to replace soil nitrogen in crop fields, lawns, and gardens. Compost and animal manure also contain nitrogen compounds that plants can use. They also can be added to soil to improve fertility.

Another method farmers use to replace soil nitrogen is to grow nitrogen-fixing crops. Most nitrogen-fixing bacteria live on or in the roots of certain plants. Some plants, such as peas, clover, and beans, including the soybeans shown in **Figure 12,** have roots with swollen nodules that contain nitrogen-fixing bacteria. These bacteria supply nitrogen compounds to the soybean plants and add nitrogen compounds to the soil.

Mini LAB

Comparing Fertilizers

Procedure

1. Examine the three numbers (e.g., 5-10-5) on the **labels of three brands of houseplant fertilizer.** The numbers indicate the percentages of nitrogen, phosphorus, and potassium, respectively, that the product contains.
2. Compare the prices of the three brands of fertilizer.
3. Compare the amount of each brand needed to fertilize a typical houseplant.

Analysis

1. **Identify** the brand with the highest percentage of nitrogen.
2. **Calculate** which brand is the most expensive source of nitrogen. The least expensive.

NATIONAL GEOGRAPHIC

VISUALIZING THE CARBON CYCLE

Figure 13

Carbon—in the form of different kinds of carbon-containing molecules—moves through an endless cycle. The diagram below shows several stages of the carbon cycle. It begins when plants and algae remove carbon from the environment during photosynthesis. This carbon returns to the atmosphere via several carbon-cycle pathways.

Ⓐ Air contains carbon in the form of carbon dioxide gas. Plants and algae use carbon dioxide to make sugars, which are energy-rich, carbon-containing compounds.

Ⓑ Organisms break down sugar molecules made by plants and algae to obtain energy for life and growth. Carbon dioxide is released as a waste.

Ⓒ Burning fossil fuels and wood releases carbon dioxide into the atmosphere.

Ⓓ When organisms die, their carbon-containing molecules become part of the soil. The molecules are broken down by fungi, bacteria, and other decomposers. During this decay process, carbon dioxide is released into the air.

Ⓔ Under certain conditions, the remains of some dead organisms may gradually be changed into fossil fuels such as coal, gas, and oil. These carbon compounds are energy rich.

The Carbon Cycle

Carbon atoms are found in the molecules that make up living organisms. Carbon is an important part of soil humus, which is formed when dead organisms decay, and it is found in the atmosphere as carbon dioxide gas (CO_2). The **carbon cycle** describes how carbon molecules move between the living and nonliving world, as shown in **Figure 13.**

The carbon cycle begins when producers remove CO_2 from the air during photosynthesis. They use CO_2, water, and sunlight to produce energy-rich sugar molecules. Energy is released from these molecules during respiration—the chemical process that provides energy for cells. Respiration uses oxygen and releases CO_2. Photosynthesis uses CO_2 and releases oxygen. These two processes help recycle carbon on Earth.

Reading Check *How does carbon dioxide enter the atmosphere?*

Human activities also release CO_2 into the atmosphere. Fossil fuels such as gasoline, coal, and heating oil are the remains of organisms that lived millions of years ago. These fuels are made of energy-rich, carbon-based molecules. When people burn these fuels, CO_2 is released into the atmosphere as a waste product. People also use wood for construction and for fuel. Trees that are harvested for these purposes no longer remove CO_2 from the atmosphere during photosynthesis. The amount of CO_2 in the atmosphere is increasing. Extra CO_2 could trap more heat from the Sun and cause average temperatures on Earth to rise.

Topic: Life Processes
Visit life.msscience.com for Web links to information about chemical equations that describe photosynthesis and respiration.

Activity Use these equations to explain how respiration is the reverse of photosynthesis.

section 2 review

Summary

The Cycles of Matter

- Earth's biosphere contains a fixed amount of water, carbon, nitrogen, oxygen, and other materials that cycle through the environment.

The Water Cycle

- Water cycles through the environment using several pathways.

The Nitrogen Cycle

- Some types of bacteria can form nitrogen compounds that plants and animals can use.

The Carbon Cycle

- Producers remove CO_2 from the air during photosynthesis and produce O_2.
- Consumers remove O_2 and produce CO_2.

Self Check

1. **Describe** the water cycle.
2. **Infer** how burning fossil fuels might affect the makeup of gases in the atmosphere.
3. **Explain** why plants, animals, and other organisms need nitrogen.
4. **Think Critically** Most chemical fertilizers contain nitrogen, phosphorous, and potassium. If they do not contain carbon, how do plants obtain carbon?

Applying Skills

5. **Identify and Manipulate Variables and Controls** Describe an experiment that would determine whether extra carbon dioxide enhances the growth of tomato plants.

section 3

Energy Flow

as you read

What You'll Learn

- **Explain** how organisms produce energy-rich compounds.
- **Describe** how energy flows through ecosystems.
- **Recognize** how much energy is available at different levels in a food chain.

Why It's Important

All living things, including people, need a constant supply of energy.

Review Vocabulary

energy: the capacity for doing work

New Vocabulary

- chemosynthesis
- food web
- energy pyramid

Converting Energy

All living things are made of matter, and all living things need energy. Matter and energy move through the natural world in different ways. Matter can be recycled over and over again. The recycling of matter requires energy. Energy is not recycled, but it is converted from one form to another. The conversion of energy is important to all life on Earth.

Photosynthesis During photosynthesis, producers convert light energy into the chemical energy in sugar molecules. Some of these sugar molecules are broken down as energy. Others are used to build complex carbohydrate molecules that become part of the producer's body. Fats and proteins also contain stored energy.

Chemosynthesis Not all producers rely on light for energy. During the 1970s, scientists exploring the ocean floor were amazed to find communities teeming with life. These communities were at a depth of almost 3.2 km and living in total darkness. They were found near powerful hydrothermal vents like the one shown in **Figure 14.**

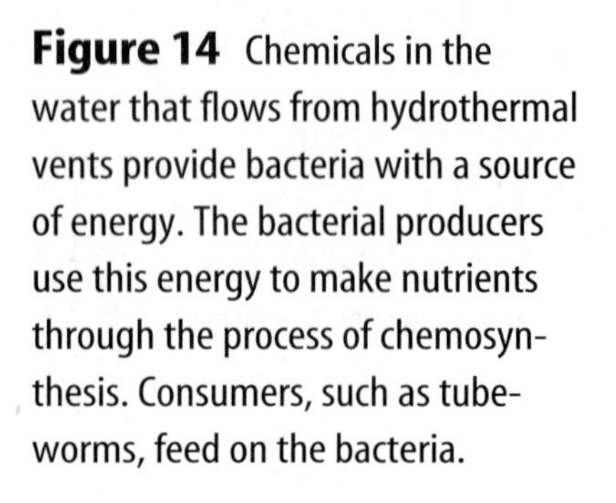

Figure 14 Chemicals in the water that flows from hydrothermal vents provide bacteria with a source of energy. The bacterial producers use this energy to make nutrients through the process of chemosynthesis. Consumers, such as tubeworms, feed on the bacteria.

Hydrothermal Vents A hydrothermal vent is a deep crack in the ocean floor through which the heat of molten magma can escape. The water from hydrothermal vents is extremely hot from contact with molten rock that lies deep in Earth's crust.

Because no sunlight reaches these deep ocean regions, plants or algae cannot grow there. How do the organisms living in this community obtain energy? Scientists learned that the hot water contains nutrients such as sulfur molecules that bacteria use to produce their own food. The production of energy-rich nutrient molecules from chemicals is called **chemosynthesis** (kee moh SIHN thuh sus). Consumers living in the hydrothermal vent communities rely on chemosynthetic bacteria for nutrients and energy. Chemosynthesis and photosynthesis allow producers to make their own energy-rich molecules.

Hydrothermal Vents The first hydrothermal vent community discovered was found along the Galápagos rift zone. A rift zone forms where two plates of Earth's crust are spreading apart. In your Science Journal, describe the energy source that heats the water in the hydrothermal vents of the Galápagos rift zone.

Reading Check *What is chemosynthesis?*

Energy Transfer

Energy can be converted from one form to another. It also can be transferred from one organism to another. Consumers cannot make their own food. Instead, they obtain energy by eating producers or other consumers. The energy stored in the molecules of one organism is transferred to another organism. That organism can oxidize food to release energy that it can use for maintenance and growth or is transformed into heat. At the same time, the matter that makes up those molecules is transferred from one organism to another.

Food Chains A food chain is a way of showing how matter and energy pass from one organism to another. Producers—plants, algae, and other organisms that are capable of photosynthesis or chemosynthesis—are always the first step in a food chain. Animals that consume producers such as herbivores are the second step. Carnivores and omnivores—animals that eat other consumers—are the third and higher steps of food chains. One example of a food chain is shown in **Figure 15.**

Figure 15 In this food chain, grasses are producers, marmots are herbivores that eat the grasses, and grizzly bears are consumers that eat marmots. The arrows show the direction in which matter and energy flow.
Infer *what might happen if grizzly bears disappeared from this ecosystem.*

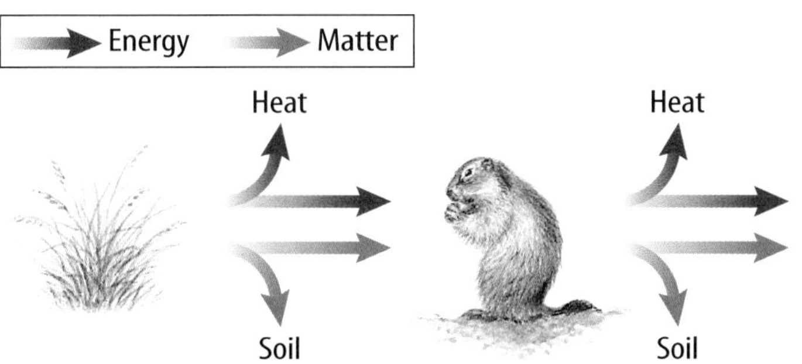

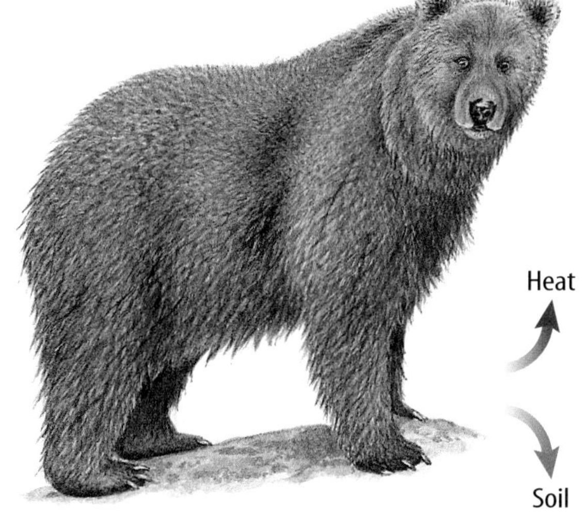

Food Webs A forest community includes many feeding relationships. These relationships can be too complex to show with a food chain. For example, grizzly bears eat many different organisms, including berries, insects, chipmunks, and fish. Berries are eaten by bears, birds, insects, and other animals. A bear carcass might be eaten by wolves, birds, or insects. A **food web** is a model that shows all the possible feeding relationships among the organisms in a community. A food web is made up of many different food chains, as shown in **Figure 16.**

Energy Pyramids

Food chains usually have at least three links, but rarely more than five. This limit exists because the amount of available energy is reduced as you move from one level to the next in a food chain. Imagine a grass plant that absorbs energy from the Sun. The plant uses some of this energy to grow and produce seeds. Some of the energy is stored in the seeds.

Figure 16 Compared to a food chain, a food web provides a more complete model of the feeding relationships in a community.

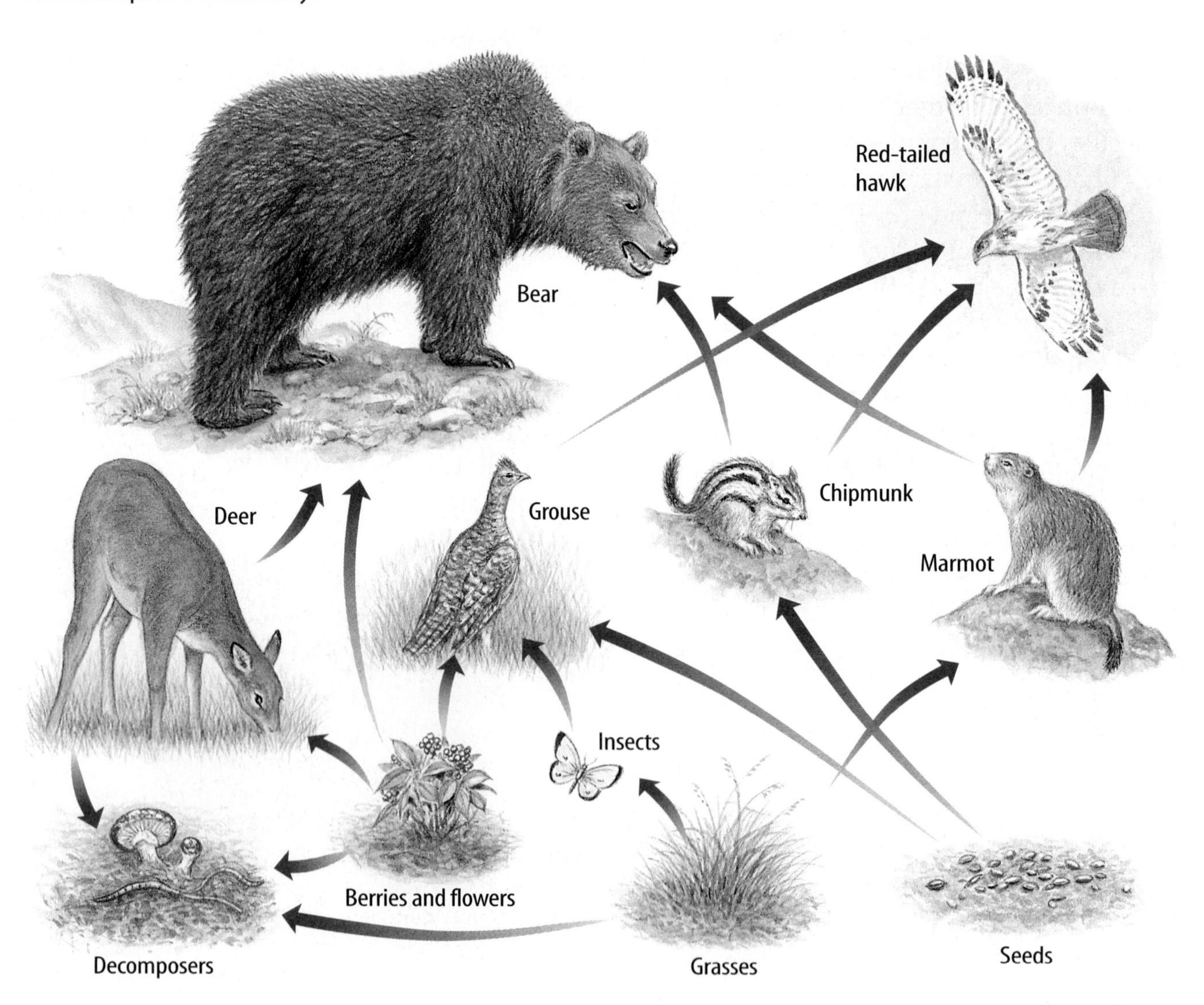

Available Energy When a mouse eats grass seeds, energy stored in the seeds is transferred to the mouse. However, most of the energy the plant absorbed from the Sun was used for the plant's growth. The mouse uses energy from the seed for its own life processes, including respiration, digestion, and growth. Some of this energy was given off as heat. A hawk that eats the mouse obtains even less energy. The amount of available energy is reduced from one feeding level of a food chain to another.

An **energy pyramid,** like the one in **Figure 17,** shows the amount of energy available at each feeding level in an ecosystem. The bottom of the pyramid, which represents all of the producers, is the first feeding level. It is the largest level because it contains the most energy and the largest number of organisms. As you move up the pyramid, the transfer of energy is less efficient and each level becomes smaller. Only about ten percent of the energy available at each feeding level of an energy pyramid is transferred to the next higher level.

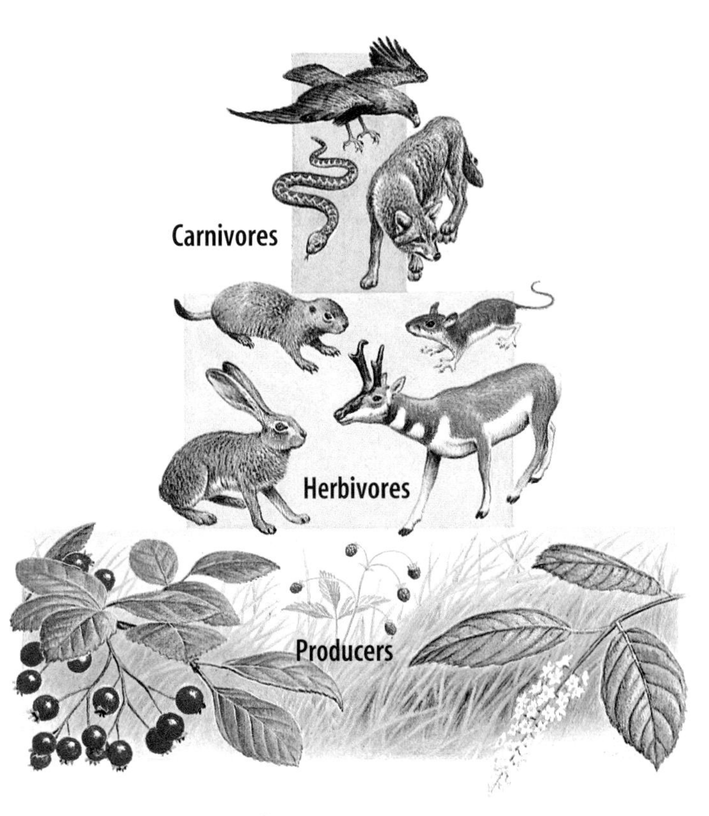

Figure 17 An energy pyramid shows that each feeding level has less energy than the one below it. **Describe** *what would happen if the hawks and snakes outnumbered the rabbits and mice in this ecosystem.*

Why does the first feeding level of an energy pyramid contain the most energy?

section 3 review

Summary

Converting Energy

- Most producers convert light energy into chemical energy.
- Some producers can produce their own food using energy in chemicals such as sulfur.

Energy Transfer

- Producers convert energy into forms that other organisms can use.
- Food chains show how matter and energy pass from one organism to another.

Energy Pyramids

- Energy pyramids show the amount of energy available at each feeding level.
- The amount of available energy decreases from the base to the top of the energy pyramid.

Self Check

1. **Compare and contrast** a food web and an energy pyramid.
2. **Explain** why there is a limit to the number of links in a food chain.
3. **Think Critically** Use your knowledge of food chains and the energy pyramid to explain why the number of mice in a grassland ecosystem is greater than the number of hawks.

Applying Math

4. **Solve One-Step Equations** A forest has 24,055,000 kilocalories (kcals) of producers, 2,515,000 kcals of herbivores, and 235,000 kcals of carnivores. How much energy is lost between producers and herbivores? Between herbivores and carnivores?

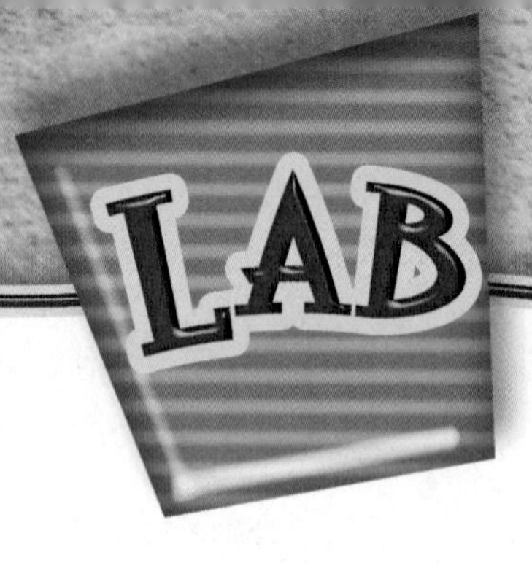

Where does the mass of a plant come from?

Goals
- **Measure** the mass of soil before and after radish plants have been grown in it.
- **Measure** the mass of radish plants grown in the soil.
- **Analyze** the data to determine whether the mass gained by the plants equals the mass lost by the soil.

Materials
8-oz plastic or paper cup
potting soil to fill cup
scale or balance
radish seeds (4)
water
paper towels

Safety Precautions

Real-World Question

An enormous oak tree starts out as a tiny acorn. The acorn sprouts in dark, moist soil. Roots grow down through the soil. Its stem and leaves grow up toward the light and air. Year after year, the tree grows taller, its trunk grows thicker, and its roots grow deeper. It becomes a towering oak that produces thousands of acorns of its own. An oak tree has much more mass than an acorn. Where does this mass come from? The soil? The air? In this activity, you'll find out by conducting an experiment with radish plants. Does all of the matter in a radish plant come from the soil?

Procedure

1. Copy the data table into your Science Journal.
2. Fill the cup with dry soil.
3. Find the mass of the cup of soil and record this value in your data table.
4. Moisten the soil in the cup. Plant four radish seeds 2 cm deep in the soil. Space the seeds an equal distance apart. Wash your hands.
5. Add water to keep the soil barely moist as the seeds sprout and grow.
6. When the plants have developed four to six true leaves, usually after two to three weeks, carefully remove the plants from the soil. Gently brush the soil off the roots. Make sure all the soil remains in the cup.
7. Spread the plants out on a paper towel. Place the plants and the cup of soil in a warm area to dry out.
8. When the plants are dry, measure their mass and record this value in your data table. Write this number with a plus sign in the Gain or Loss column.
9. When the soil is dry, find the mass of the cup of soil. Record this value in your data table. Subtract the End mass from the Start mass and record this number with a minus sign in the Gain or Loss column.

Mass of Soil and Radish Plants

	Start	End	Gain (+) or Loss (−)
Mass of dry soil and cup			
Mass of dried radish plants	0 g		

Do not write in this book.

Analyze Your Data

1. **Calculate** how much mass was gained or lost by the soil. By the radish plants.
2. Did the mass of the plants come completely from the soil? How do you know?

Conclude and Apply

1. In the early 1600s, a Belgian scientist named J. B. van Helmont conducted this experiment with a willow tree. What is the advantage of using radishes instead of a tree?
2. **Predict** where all of the mass gained by the plants came from.

Compare your conclusions with those of other students in your class. **For more help, refer to the** Science Skill Handbook.

SCIENCE Stats

Extreme Climates

Did you know...

... The greatest snowfall in one year occurred at Mount Baker in Washington State. Approximately 2,896 cm of snow fell during the 1998–99, 12-month snowfall season. That's enough snow to bury an eight-story building.

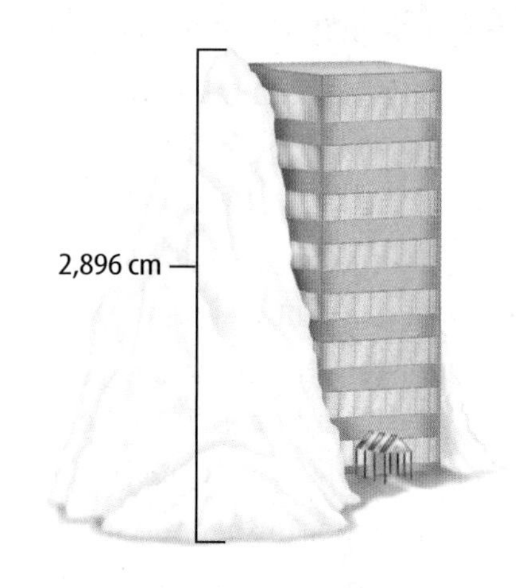

Applying Math What was the average monthly snowfall at Mount Baker during the 1998–99 snowfall season?

... The hottest climate in the United States is found in Death Valley, California. In July 1913, Death Valley reached approximately 57°C. As a comparison, a comfortable room temperature is about 20°C.

... The record low temperature of a frigid −89°C was set in Antarctica in 1983. As a comparison, the temperature of a home freezer is about −15°C.

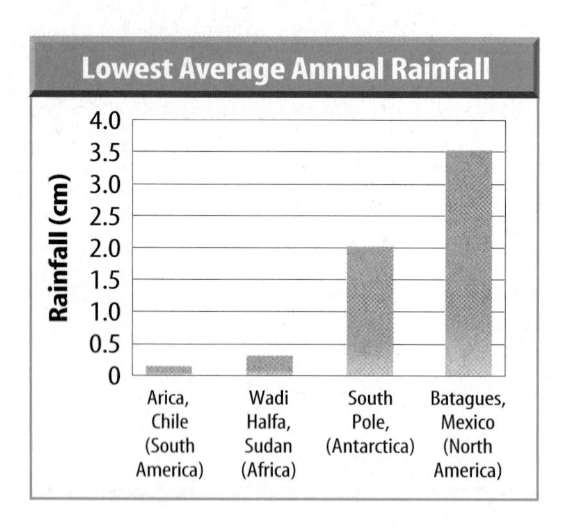

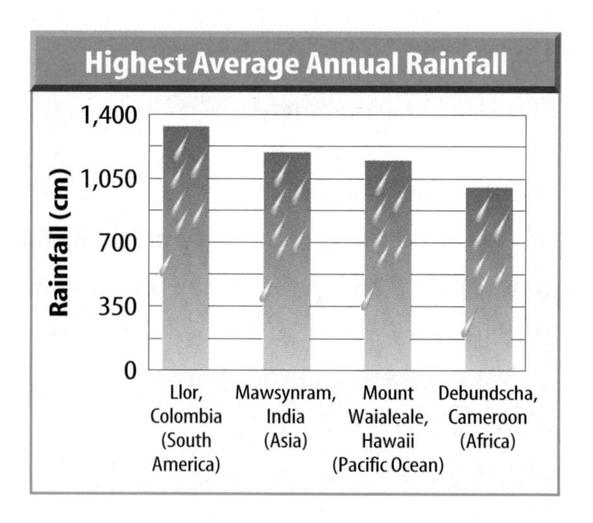

Graph It

Visit life.msscience.com/science_stats to find the average monthly rainfall in a tropical rain forest. Make a line graph to show how the amount of precipitation changes during the 12 months of the year.

Reviewing Main Ideas

Section 1 Abiotic Factors

1. Abiotic factors include air, water, soil, sunlight, temperature, and climate.
2. The availability of water and light influences where life exists on Earth.
3. Soil and climate have an important influence on the types of organisms that can survive in different environments.
4. High latitudes and elevations generally have lower average temperatures.

Section 2 Cycles in Nature

1. Matter is limited on Earth and is recycled through the environment.
2. The water cycle involves evaporation, condensation, and precipitation.
3. The carbon cycle involves photosynthesis and respiration.
4. Nitrogen in the form of soil compounds enters plants, which then are consumed by other organisms.

Section 3 Energy Flow

1. Producers make energy-rich molecules through photosynthesis or chemosynthesis.
2. When organisms feed on other organisms, they obtain matter and energy.
3. Matter can be recycled, but energy cannot.
4. Food webs are models of the complex feeding relationships in communities.
5. Available energy decreases as you go to higher feeding levels in an energy pyramid.

Visualizing Main Ideas

This diagram represents photosynthesis in a leaf. Match each letter with one of the following terms: light, carbon dioxide, *or* oxygen.

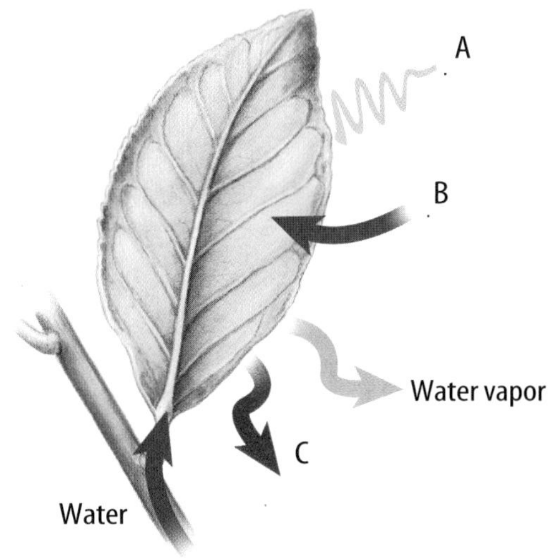

Using Vocabulary

abiotic p. 712	energy pyramid p. 729
atmosphere p. 713	evaporation p. 720
biotic p. 712	food web p. 728
carbon cycle p. 725	nitrogen cycle p. 722
chemosynthesis p. 727	nitrogen fixation p. 722
climate p. 717	soil p. 714
condensation p. 721	water cycle p. 721

Which vocabulary word best corresponds to each of the following events?

1. A liquid changes to a gas.
2. Some types of bacteria form nitrogen compounds in the soil.
3. Decaying plants add nitrogen to the soil.
4. Chemical energy is used to make energy-rich molecules.
5. Decaying plants add carbon to the soil.
6. A gas changes to a liquid.
7. Water flows downhill into a stream. The stream flows into a lake, and water evaporates from the lake.
8. Burning coal and exhaust from automobiles release carbon into the air.

Checking Concepts

Choose the word or phrase that best answers the question.

9. Which of the following is an abiotic factor?
   A) penguins **C)** soil bacteria
   B) rain **D)** redwood trees

Use the equation below to answer question 10.

$$CO_2 + H_2O \xrightarrow[\text{energy}]{\text{light}} \text{sugar} + O_2$$

10. Which of the following processes is shown in the equation above?
    A) condensation **C)** burning
    B) photosynthesis **D)** respiration

11. Which of the following applies to latitudes farther from the equator?
    A) higher elevations
    B) higher temperatures
    C) higher precipitation levels
    D) lower temperatures

12. Water vapor forming droplets that form clouds directly involves which process?
    A) condensation **C)** evaporation
    B) respiration **D)** transpiration

13. Which one of the following components of air is least necessary for life on Earth?
    A) argon **C)** carbon dioxide
    B) nitrogen **D)** oxygen

14. Which group makes up the largest level of an energy pyramid?
    A) herbivores **C)** decomposers
    B) producers **D)** carnivores

15. Earth receives a constant supply of which of the following items?
    A) light energy **C)** nitrogen
    B) carbon **D)** water

16. Which of these is an energy source for chemosynthesis?
    A) sunlight **C)** sulfur molecules
    B) moonlight **D)** carnivores

Use the illustration below to answer question 17.

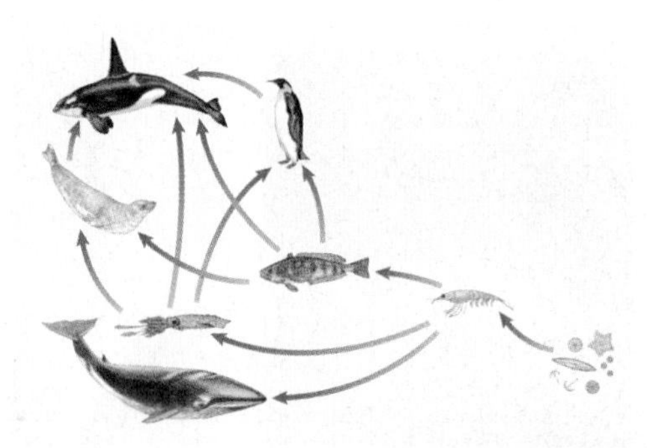

17. What is the illustration above an example of?
    A) food chain **C)** energy pyramid
    B) food web **D)** carbon cycle

Thinking Critically

18. **Draw a Conclusion** A country has many starving people. Should they grow vegetables and corn to eat, or should they grow corn to feed cattle so they can eat beef? Explain.

19. **Explain** why a food web is a better model of energy flow than a food chain.

20. **Infer** Do bacteria need nitrogen? Why or why not?

21. **Describe** why it is often easier to walk through an old, mature forest of tall trees than through a young forest of small trees.

22. **Explain** why giant sequoia trees grow on the west side of California's Inyo Mountains and Death Valley, a desert, is on the east side of the mountains.

23. **Concept Map** Copy and complete this food web using the following information: *caterpillars and rabbits eat grasses, raccoons eat rabbits and mice, mice eat grass seeds,* and *birds eat caterpillars.*

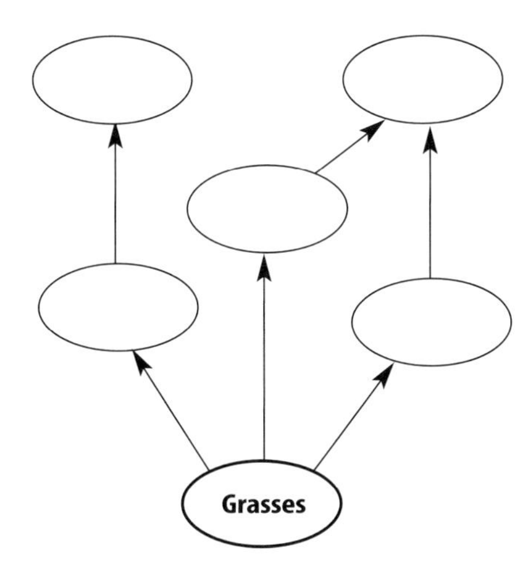

24. **Form a Hypothesis** For each hectare of land, ecologists found 10,000 kcals of producers, 10,000 kcals of herbivores, and 2,000 kcals of carnivores. Suggest a reason why producer and herbivore levels are equal.

25. **Recognize Cause and Effect** A lake in Kenya has been taken over by a floating weed. How could you determine if nitrogen fertilizer runoff from farms is causing the problem?

Performance Activities

26. **Poster** Use magazine photographs to make a visual representation of the water cycle.

Applying Math

27. **Energy Budget** Raymond Lindeman, from the University of Minnesota, was the first person to calculate the total energy budget of an entire community at Cedar Bog Lake in MN. He found the total amount of energy produced by producers was 1,114 kilocalories per meter squared per year. About 20% of the 1,114 kilocalories were used up during respiration. How many kilocalories were used during respiration?

28. **Kilocalorie Use** Of the 600 kilocalories of producers available to a caterpillar, the caterpillar consumes about 150 kilocalories. About 25% of the 150 kilocalories is used to maintain its life processes and is lost as heat, while 16% cannot be digested. How many kilocalories are lost as heat? What percentage of the 600 kilocalories is available to the next feeding level?

Use the table below to answer question 29.

Mighty Migrators

Species	Distance (km)
Desert locust	4,800
Caribou	800
Green turtle	1,900
Arctic tern	35,000
Gray whale	19,000

29. **Make and Use Graphs** Climate can cause populations to move from place to place. Make a bar graph of migration distances shown above.

Part 1 Multiple Choice

Record your answers on the answer sheet provided by your teacher or on a sheet of paper.

1. The abiotic factor that provides energy for nearly all life on Earth is
   A. air. **C.** water.
   B. sunlight. **D.** soil.

2. Which of the following is characteristic of places at high elevations?
   A. fertile soil
   B. fewer molecules in the air
   C. tall trees
   D. warm temperatures

Use the diagram below to answer questions 3 and 4.

3. The air at point C is
   A. dry and warm.
   B. dry and cool.
   C. moist and warm.
   D. moist and cool.

4. The air at point A is
   A. dry and warm.
   B. dry and cool.
   C. moist and warm.
   D. moist and cool.

5. What process do plants use to return water vapor to the atmosphere?
   A. transpiration **C.** respiration
   B. evaporation **D.** condensation

6. Clouds form as a result of what process?
   A. evaporation **C.** respiration
   B. transpiration **D.** condensation

Use the illustration of the nitrogen cycle below to answer questions 7 and 8.

7. Which of the following items shown in the diagram contribute to the nitrogen cycle by releasing AND absorbing nitrogen?
   A. the decaying organism only
   B. the trees only
   C. the trees and the grazing cows
   D. the lightning and the decaying organism

8. Which of the following items shown in the diagram contribute to the nitrogen cycle by ONLY releasing nitrogen?
   A. the decaying organism only
   B. the trees only
   C. the trees and the grazing cows
   D. the lightning and the decaying organism

9. Where is most of the energy found in an energy pyramid?
   A. at the top level
   B. in the middle levels
   C. at the bottom level
   D. all levels are the same

10. What organisms remove carbon dioxide gas from the air during photosynthesis?
    A. consumers **C.** herbivores
    B. producers **D.** omnivores

Part 2 Short Response/Grid In

Record your answers on the answer sheet provided by your teacher or on a sheet of paper.

11. Give two examples of abiotic factors and describe how each one is important to biotic factors.

Use the table below to answer questions 12 and 13.

U.S. Estimated Water Use in 1995		
Water Use	**Millions of Gallons per Day**	**Percent of Total**
Homes and Businesses	41,600	12.2
Industry and Mining	28,000	8.2
Farms and Ranches	139,200	40.9
Electricity Production	131,800	38.7

12. According to the table above, what accounted for the highest water use in the U.S. in 1995?

13. What percentage of the total amount of water use results from electricity production and homes and business combined?

14. Where are nitrogen-fixing bacteria found?

15. Describe two ways that carbon is released into the atmosphere.

16. How are organisms near hydrothermal vents deep in the ocean able to survive?

17. Use a diagram to represent the transfer of energy among these organisms: a weasel, a rabbit, grasses, and a coyote.

Test-Taking Tip

Answer All Questions Never leave any answer blank.

Part 3 Open Ended

Record your answers on a sheet of paper.

18. Explain how a decrease in the amount of sunlight would affect producers that use photosynthesis, and producers that use chemosynthesis.

19. Describe how wind and wind currents are produced.

20. Use the water cycle to explain why beads of water form on the outside of a glass of iced water on a hot day.

21. Draw a flowchart that shows how soy beans, deer, and nitrogen-fixing bacteria help cycle nitrogen from the atmosphere, to the soil, to living organisms, and back to the atmosphere.

Use the diagram below to answer questions 22 and 23.

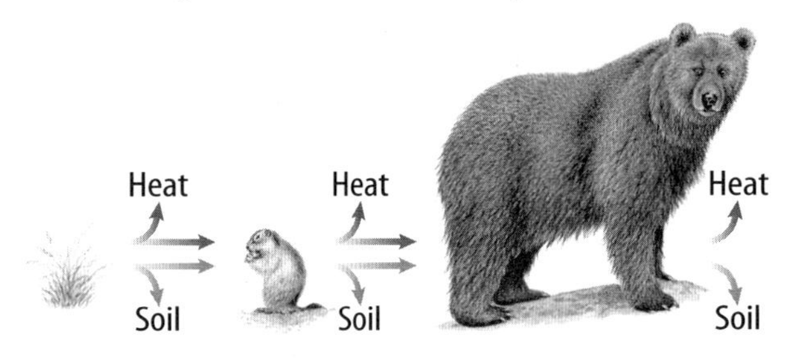

22. What term is used for the diagram above? Explain how the diagram represents energy transfer.

23. Explain how the grass and bear populations would be affected if the marmot population suddenly declined.

24. Compare and contrast an energy pyramid and a food web.

25. What happens to the energy in organisms at the top of an energy pyramid when they die?

Ecosystems

The BIG Idea

Earth has many diverse ecosystems on land and in water.

SECTION 1
How Ecosystems Change
Main Idea Ecosystems gradually change over time.

SECTION 2
Biomes
Main Idea Land on Earth is divided into large geographic areas that have similar climates and ecosystems.

SECTION 3
Acquatic Ecosystems
Main Idea Both Earth's salt water and freshwater are divided into a variety of ecosystems.

The Benefits of Wildfires

Ecosystems are places where organisms, including humans, interact with each other and with their physical environment. In some ecosystems, wildfires are an essential part of the physical environment. Organisms in these ecosystems are well adapted to the changes that fire brings, and can benefit from wildfires.

Science Journal What traits might plants on this burning Montana hillside have that enable them to survive?

Start-Up Activities

What environment do houseplants need?

The plants growing in your classroom or home may not look like the same types of plants that you find growing outside. Many indoor plants don't grow well outside in most North American climates. Do the lab below to determine what type of environment most houseplants thrive in.

1. Examine a healthy houseplant in your classroom or home.
2. Describe the environmental conditions found in your classroom or home. For example, is the air humid or dry? Is the room warm or cool? Does the temperature stay about the same, or change during the day?
3. Using observations from step 1 and descriptions from step 2, hypothesize about the natural environment of the plants in your classroom or home.
4. **Think Critically** In your Science Journal, record the observations that led to your hypothesis. How would you design an experiment to test your hypothesis?

Primary and Secondary Succession Make the following Foldable to help you illustrate the main ideas about succession.

STEP 1 **Fold** a vertical sheet of paper in half from top to bottom.

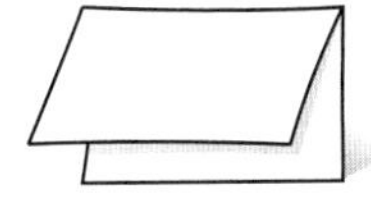

STEP 2 **Fold** in half from side to side with the fold at the top.

STEP 3 **Unfold** the paper once. **Cut** only the fold of the top flap to make two tabs.

STEP 4 **Turn** the paper vertically and **label** on the front tabs as shown.

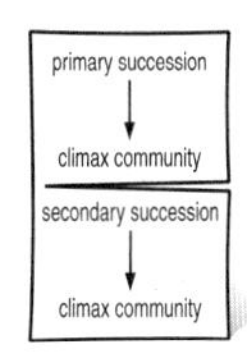

Illustrate and Label As you read the chapter, define terms and collect information under the appropriate tabs.

Preview this chapter's content and activities at life.msscience.com

Get Ready to Read

Take Notes

① Learn It! The best way for you to remember information is to write it down, or take notes. Good note-taking is useful for studying and research. When you are taking notes, it is helpful to

- phrase the information in your own words;
- restate ideas in short, memorable phrases;
- stay focused on main ideas and only the most important supporting details.

② Practice It! Make note-taking easier by using a chart to help you organize information clearly. Write the main ideas in the left column. Then write at least three supporting details in the right column. Read the text from Section 3 of this chapter under the heading *Water Pollution*, page 755. Then take notes using a chart, such as the one below.

Main Idea	Supporting Details
	1. 2. 3. 4. 5.
	1. 2. 3. 4. 5.

③ Apply It! As you read this chapter, make a chart of the main ideas. Next to each main idea, list at least two supporting details.

Read one or two paragraphs first and take notes after you read. You are likely to take down too much information if you take notes as you read.

Target Your Reading

Use this to focus on the main ideas as you read the chapter.

1. **Before you read** the chapter, respond to the statements below on your worksheet or on a numbered sheet of paper.
   - Write an **A** if you **agree** with the statement.
   - Write a **D** if you **disagree** with the statement.

2. **After you read** the chapter, look back to this page to see if you've changed your mind about any of the statements.
   - If any of your answers changed, explain why.
   - Change any false statements into true statements.
   - Use your revised statements as a study guide.

Before You Read A or D	Statement		After You Read A or D
	1	Gradual changes to the types of species in an ecosystem always follow the same pattern.	
	2	A variety of organisms can live on bare rock.	
	3	Plant communities are always changing and unstable.	
	4	Deserts in different parts of the world do not have any similarities with one another.	
	5	Rainforests are all located near the equator.	
	6	Most of the soil in the tundra is frozen year round.	
	7	Aquatic ecosystems are either freshwater or saltwater.	
	8	Coral reefs are durable ecosystems that adapt quickly to stress.	
	9	The amount of sunlight available in ocean and lake waters affects the number of organisms found there.	

Science Online
Print out a worksheet of this page at life.msscience.com

section 1

How Ecosystems Change

as you read

What You'll Learn

- **Explain** how ecosystems change over time.
- **Describe** how new communities begin in areas without life.
- **Compare** pioneer species and climax communities.

Why It's Important

Understanding ecosystems and your role in them can help you manage your impact on them and predict the changes that may happen in the future.

Review Vocabulary

ecosystem: community of living organisms interacting with each other and their physical environment

New Vocabulary

- succession
- pioneer species
- climax community

Ecological Succession

What would happen if the lawn at your home were never cut? The grass would get longer, as in **Figure 1,** and soon it would look like a meadow. Later, larger plants would grow from seeds brought to the area by animals or wind. Then, trees might sprout. In fact, in 20 years or less you wouldn't be able to tell that the land was once a mowed lawn. An ecologist can tell you what type of ecosystem your lawn would become. If it would become a forest, they can tell you how long it would take and predict the type of trees that would grow there. **Succession** refers to the normal, gradual changes that occur in the types of species that live in an area. Succession occurs differently in different places around the world.

Primary Succession As lava flows from the mouth of a volcano, it is so hot that it destroys everything in its path. When it cools, lava forms new land composed of rock. It is hard to imagine that this land eventually could become a forest or grassland someday.

The process of succession that begins in a place previously without plants is called primary succession. It starts with the arrival of living things such as lichens (LI kunz). These living things, called **pioneer species,** are the first to inhabit an area. They survive drought, extreme heat and cold, and other harsh conditions and often start the soil-building process.

Figure 1 Open areas that are not maintained will become overgrown with grasses and shrubs as succession proceeds.

Figure 2 Lichens, like these in Colorado, are fragile and take many years to grow. They often cling to bare rock where many other organisms can't survive. **Describe** *how lichens form soil.*

New Soil During primary succession, shown in **Figure 2,** soil begins to form as lichens and the forces of weather and erosion help break down rocks into smaller pieces. When lichens die, they decay, adding small amounts of organic matter to the rock. Plants such as mosses and ferns can grow in this new soil. Eventually, these plants die, adding more organic material. The soil layer thickens, and grasses, wildflowers, and other plants begin to take over. When these plants die, they add more nutrients to the soil. This buildup is enough to support the growth of shrubs and trees. All the while, insects, small birds, and mammals have begun to move in. What was once bare rock now supports a variety of life.

Secondary Succession What happens when a fire, such as the one in **Figure 3,** disturbs a forest or when a building is torn down in a city? After a forest fire, not much seems to be left except dead trees and ash-covered soil. After the rubble of a building is removed, all that remains is bare soil. However, these places do not remain lifeless for long. The soil already contains the seeds of weeds, grasses, and trees. More seeds are carried to the area by wind and birds. Other wildlife may move in. Succession that begins in a place that already has soil and was once the home of living organisms is called secondary succession. Because soil already is present, secondary succession occurs faster and has different pioneer species than primary succession does.

Which type of succession usually starts without soil?

Topic: Eutrophication

Visit life.msscience.com for Web links to information about eutrophication (yoo truh fih KAY shun)—secondary succession in an aquatic ecosystem.

Activity Using the information that you find, illustrate or describe in your Science Journal this process for a small freshwater lake.

NATIONAL GEOGRAPHIC VISUALIZING SECONDARY SUCCESSION

Figure 3

In the summer of 1988, wind-driven flames like those shown in the background photo swept through Yellowstone National Park, scorching nearly a million acres. The Yellowstone fire was one of the largest forest fires in United States history. The images on this page show secondary succession—the process of ecological regeneration—triggered by the fire.

▶ After the fire, burned timber and blackened soil seemed to be all that remained. However, the fire didn't destroy the seeds that were protected under the soil.

◀ Within weeks, grasses and other plants were beginning to grow in the burned areas. Ecological succession was underway.

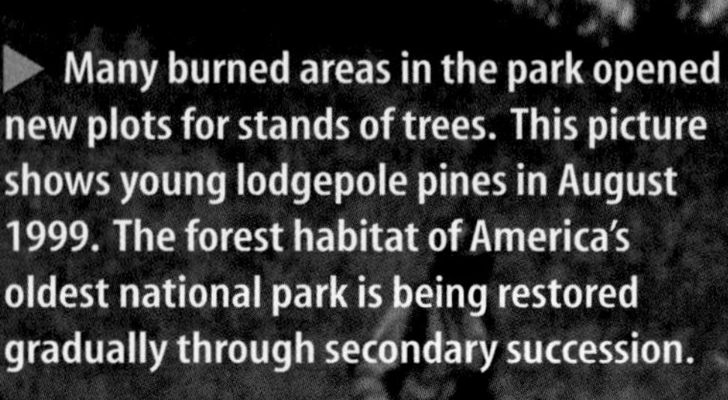

▶ Many burned areas in the park opened new plots for stands of trees. This picture shows young lodgepole pines in August 1999. The forest habitat of America's oldest national park is being restored gradually through secondary succession.

Figure 4 This beech-maple forest is an example of a climax community.

Climax Communities A community of plants that is relatively stable and undisturbed and has reached a stage of succession is called a **climax community.** The beech-maple forest shown in **Figure 4** is an example of a community that has reached the end of succession. New trees grow when larger, older trees die. The individual trees change, but the species remain stable. There are fewer changes of species in a climax community over time, as long as the community isn't disturbed by wildfire, avalanche, or human activities.

Primary succession begins in areas with no previous vegetation. It can take hundreds or even thousands of years to develop into a climax community. Secondary succession is usually a shorter process, but it still can take a century or more.

section 1 review

Summary

Ecological Succession

- Succession is the natural, gradual changes over time of species in a community.
- Primary succession occurs in areas that previously were without soil or plants.
- Secondary succession occurs in areas where soil has been disturbed.
- Climax communities have reached an end stage of succession and are stable.
- Climax communities have less diversity than communities in mid-succession.

Self Check

1. **Compare** primary and secondary succession.
2. **Describe** adaptations of pioneer species.
3. **Infer** the kind of succession that will take place on an abandoned, unpaved country road.
4. **Think Critically** Show the sequence of events in primary succession. Include the term *climax community.*

Applying Math

5. **Solve One-Step Equations** A tombstone etched with 1802 as the date of death has a lichen on it that is 6 cm in diameter. If the lichen began growing in 1802, calculate its average yearly rate of growth.

section 2

Biomes

as you read

What You'll Learn

- **Explain** how climate influences land environments.
- **Identify** seven biomes of Earth.
- **Describe** the adaptations of organisms found in each biome.

Why It's Important

Resources that you need to survive are found in a variety of biomes.

Review Vocabulary
climate: the average weather conditions of an area over many years

New Vocabulary
- biome
- tundra
- taiga
- temperate deciduous forest
- temperate rain forest
- tropical rain forest
- desert
- grassland

Factors That Affect Biomes

Does a desert in Arizona have anything in common with a desert in Africa? Both have heat, little rain, poor soil, water-conserving plants with thorns, and lizards. Even widely separated regions of the world can have similar biomes because they have similar climates. Climate is the average weather pattern in an area over a long period of time. The two most important climatic factors that affect life in an area are temperature and precipitation.

Major Biomes

Large geographic areas that have similar climates and ecosystems are called **biomes** (BI ohmz). Seven common types of land biomes are mapped in **Figure 5.** Areas with similar climates produce similar climax communities. Tropical rain forests are climax communities found near the equator, where temperatures are warm and rainfall is plentiful. Coniferous forests grow where winter temperatures are cold and rainfall is moderate.

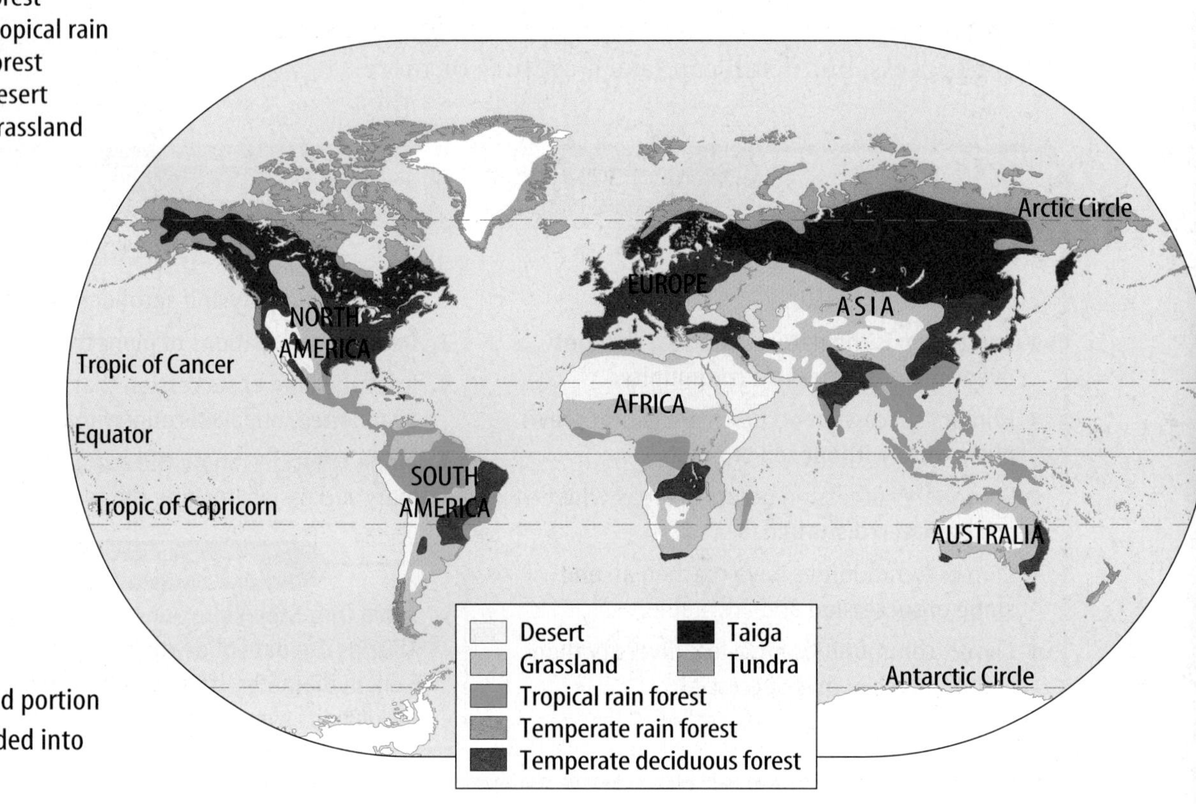

Figure 5 The land portion of Earth can be divided into seven biomes.

Tundra At latitudes just south of the north pole or at high elevations, a biome can be found that receives little precipitation but is covered with ice most of the year. The **tundra** is a cold, dry, treeless region, sometimes called a cold desert. Precipitation averages less than 25 cm per year. Winters in the Arctic can be six to nine months long. For some of these months, the Sun never appears above the horizon and it is dark 24 hours a day. The average daily temperature is about –12°C. For a few days during the short, cold summer, the Sun is always visible. Only the top portion of soil thaws in the summer. Below the thawed surface is a layer of permanently frozen soil called permafrost, shown in **Figure 6.** Alpine tundra, found above the treeline on high mountains, have similar climates. Tundra soil has few nutrients because the cold temperatures slow the process of decomposition.

Figure 6 This permafrost in Alaska is covered by soil that freezes in the winter and thaws in the summer.
Infer *what types of problems this might cause for people living in this area.*

Tundra Life Tundra plants are adapted to drought and cold. They include mosses, grasses, and small shrubs, as seen in **Figure 7.** Many lichens grow on the tundra. During the summer, mosquitoes, blackflies, and other biting insects fill the air. Migratory birds such as ducks, geese, shorebirds, and songbirds nest on the Arctic tundra during the summer. Other inhabitants include hawks, snowy owls, and willow grouse. Mice, voles, lemmings, arctic hares, caribou, reindeer, and musk oxen also are found there.

People are concerned about overgrazing by animals on the tundra. Fences, roads, and pipelines have disrupted the migratory routes of some animals and forced them to stay in a limited area. Because the growing season is so short, plants and other vegetation can take decades to recover from damage.

Figure 7 Lichens, mosses, grasses, and small shrubs thrive on the tundra. Ptarmigan also live on the tundra. In winter, their feathers turn white. Extra feathers on their feet keep them warm and prevent them from sinking into the snow.

Tundra

Ptarmigan

Taiga South of the tundra—between latitudes 50°N and 60°N and stretching across North America, northern Europe, and Asia—is the world's largest biome. The **taiga** (TI guh), shown in **Figure 8,** is a cold, forest region dominated by cone-bearing evergreen trees. Although the winter is long and cold, the taiga is warmer and wetter than the tundra. Precipitation is mostly snow and averages 35 cm to 100 cm each year.

Most soils of the taiga thaw completely during the summer, making it possible for trees to grow. However, permafrost is present in the extreme northern regions of the taiga. The forests of the taiga might be so dense that little sunlight penetrates the trees to reach the forest floor. However, some lichens and mosses do grow on the forest floor. Moose, lynx, shrews, bears, and foxes are some of the animals that live in the taiga.

Figure 8 The taiga is dominated by cone-bearing trees. The lynx, a mammal adapted to life in the taiga, has broad, heavily furred feet that act like snowshoes to prevent it from sinking in the snow.
Infer *why "snowshoe feet" are important for a lynx.*

Temperate Deciduous Forest Temperate regions usually have four distinct seasons each year. Annual precipitation ranges from about 75 cm to 150 cm and is distributed throughout the year. Temperatures range from below freezing during the winter to 30°C or more during the warmest days of summer.

Figure 9 White-tailed deer are one of many species that you can find in a deciduous forest. In autumn, the leaves on deciduous trees change color and fall to the ground.

Temperate Forest Life Many evergreen trees grow in the temperate regions of the world. However, most of the temperate forests in Europe and North America are dominated by climax communities of deciduous trees, which lose their leaves every autumn. These forests, like the one in **Figure 9,** are called **temperate deciduous forests.** In the United States, most of them are located east of the Mississippi River.

When European settlers first came to America, they cut trees to create farmland and to supply wood. As forests were cut, organisms lost their habitats. When agriculture shifted from the eastern to the midwestern and western states, secondary succession began, and trees eventually returned to some areas. Now, nearly as many trees grow in the New England states as did before the American Revolutionary War. Many trees are located in smaller patches. Yet, the recovery of large forests such as those in the Adirondack Mountains in New York State shows the result of secondary succession.

Figure 10 In the Olympic rain forest in Washington State, mosses and lichens blanket the ground and hang from the trees. Wet areas are perfect habitats for amphibians like the Pacific giant salamander above.

Temperate Rain Forest New Zealand, southern Chile, and the Pacific Northwest of the United States are some of the places where **temperate rain forests,** shown in **Figure 10,** are found. The average temperature of a temperate rain forest ranges from 9°C to 12°C. Precipitation ranges from 200 cm to 400 cm per year.

Trees with needlelike leaves dominate these forests, including the Douglas fir, western red cedar, and spruce. Many grow to great heights. Animals of the temperate rain forest include the black bear, cougar, bobcat, northern spotted owl, and marbled murrelet. Many species of amphibians also inhabit the temperate rain forest, including salamanders.

The logging industry in the Northwest provides jobs for many people. However, it also removes large parts of the temperate rain forest and destroys the habitat of many organisms. Many logging companies now are required to replant trees to replace the ones they cut down. Also, some rain forest areas are protected as national parks and forests.

Figure 11 Tropical rain forests are lush environments that contain such a large variety of species that many have not been discovered.

Modeling Rain Forest Leaves

Procedure

1. Draw an oval leaf about 10 cm long on a piece of **poster board.** Cut it out.
2. Draw a second leaf the same size but make one end pointed. This is called a drip tip. Cut this leaf out.
3. Hold your hands palm-side up over a **sink** and have someone lay a leaf on each one. Point the drip tip away from you. Tilt your hands down but do not allow the leaves to fall off.
4. Have someone gently spray water on the leaves and observe what happens.

Analysis

1. From which leaf does water drain faster?
2. Infer why it is an advantage for a leaf to get rid of water quickly in a rain forest.

Try at Home

Tropical Rain Forest Warm temperatures, wet weather, and lush plant growth are found in **tropical rain forests.** These forests are warm because they are near the equator. The average temperature, about 25°C, doesn't vary much between night and day. Most tropical rain forests receive at least 200 cm of rain annually. Some receive as much as 600 cm of rain each year.

Tropical rain forests, like the one in **Figure 11,** are home to an astonishing variety of organisms. They are one of the most biologically diverse places in the world. For example, one tree in a South American rain forest might contain more species of ants than exist in all of the British Isles.

Tropical Rain Forest Life Different animals and plants live in different parts of the rain forest. Scientists divide the rain forest into zones based on the types of plants and animals that live there, just as a library separates books about different topics onto separate shelves. The zones include: forest floor, understory, canopy, and emergents, as shown in **Figure 12.** These zones often blend together, but their existence provide different habitats for many diverse organisms to live in the tropical rain forest.

Reading Check *What are the four zones of a tropical rain forest?*

Although tropical rain forests support a huge variety of organisms, the soil of the rain forest contains few nutrients. Over the years, nutrients have been washed out of the soil by rain. On the forest floor, decomposers immediately break down organic matter, making nutrients available to the plants again.

Human Impact Farmers that live in tropical areas clear the land to farm and to sell the valuable wood. After a few years, the crops use up the nutrients in the soil, and the farmers must clear more land. As a result, tropical rain forest habitats are being destroyed. Through education, people are realizing the value and potential value of preserving the species of the rain forest. In some areas, logging is prohibited. In other areas, farmers are taught new methods of farming so they do not have to clear rain forest lands continually.

Emergents These giant trees are much higher than the average canopy tree. Birds, such as the macaw, and insects are found here.
Canopy The canopy includes the upper parts of the trees. It's full of life—insects, birds, reptiles, and mammals.
Understory This dark, cool environment is under the canopy leaves but above the ground. Many insects, reptiles, and amphibians live in the understory.
Forest Floor The forest floor is home to many insects and the largest mammals in the rain forest generally live here.

Figure 12 Tropical rain forests contain abundant and diverse organisms.

Desertification When vegetation is removed from soil in areas that receive little rain, the dry, unprotected surface can be blown away. If the soil remains bare, a desert might form. This process is called desertification. Look on a biome map and hypothesize about which areas of the United States are most likely to become deserts.

Desert The driest biome on Earth is the **desert.** Deserts receive less than 25 cm of rain each year and support little plant life. Some desert areas receive no rain for years. When rain does come, it quickly drains away. Any water that remains on the ground evaporates rapidly.

Most deserts, like the one in **Figure 13,** are covered with a thin, sandy, or gravelly soil that contains little organic matter. Due to the lack of water, desert plants are spaced far apart and much of the ground is bare. Barren, windblown sand dunes are characteristics of the driest deserts.

Reading Check *Why is much of a desert bare ground?*

Desert Life Desert plants are adapted for survival in the extreme dryness and hot and cold temperatures of this biome. Most desert plants are able to store water. Cactus plants are probably the most familiar desert plants of the western hemisphere. Desert animals also have adaptations that help them survive the extreme conditions. Some, like the kangaroo rat, never need to drink water. They get all the moisture they need from the breakdown of food during digestion. Most animals are active only during the night, late afternoon, or early morning when temperatures are less extreme. Few large animals are found in the desert.

In order to provide water for desert cities, rivers and streams have been diverted. When this happens, wildlife tends to move closer to cities in their search for food and water. Education about desert environments has led to an awareness of the impact of human activities. As a result, large areas of desert have been set aside as national parks and wilderness areas to protect desert habitats.

Figure 13 Desert plants, like these in the Sonoran Desert, are adapted for survival in the extreme conditions of the desert biome. The giant hairy scorpion found in some deserts has a venomous sting.

Grasslands Temperate and tropical regions that receive between 25 cm and 75 cm of precipitation each year and are dominated by climax communities of grasses are called **grasslands.** Most grasslands have a dry season, when little or no rain falls. This lack of moisture prevents the development of forests. Grasslands are found in many places around the world, and they have a variety of names. The prairies and plains of North America, the steppes of Asia, the savannas of Africa shown in **Figure 14,** and the pampas of South America are types of grasslands.

Figure 14 Animals such as zebras and wildebeests are adapted to life on the savannas in Africa.

Grasslands Life The most noticeable animals in grassland biomes are usually mammals that graze on the stems, leaves, and seeds of grass plants. Kangaroos graze in the grasslands of Australia. In Africa, communities of animals such as wildebeests, impalas, and zebras thrive in the savannas.

Grasslands are perfect for growing many crops such as wheat, rye, oats, barley, and corn. Grasslands also are used to raise cattle and sheep. However, overgrazing can result in the death of grasses and the loss of valuable topsoil from erosion. Most farmers and ranchers take precautions to prevent the loss of valuable habitats and soil.

section 2 review

Summary

Major Biomes

- Tundra, sometimes called a cold desert, can be divided into two types: arctic and alpine.
- Taiga is the world's largest biome. It is a cold forest region with long winters.
- Temperate regions have either a deciduous forest biome or a rain forest biome.
- Tropical rain forests are one of the most biologically diverse biomes.
- Humans have a huge impact on tropical rain forests.
- The driest biome is the desert. Desert organisms are adapted for extreme dryness and temperatures.
- Grasslands provide food for wildlife, livestock, and humans.

Self Check

1. **Determine** which two biomes are the driest.
2. **Compare and contrast** tundra organisms and desert organisms.
3. **Identify** the biggest climatic difference between a temperate rain forest and a tropical rain forest.
4. **Explain** why the soil of tropical rain forests make poor farmland.
5. **Think Critically** If you climb a mountain in Arizona, you might reach an area where the trees resemble the taiga trees in northern Canada. Why would a taiga forest exist in Arizona?

Applying Skills

6. **Record Observations** Animals have adaptations that help them survive in their environments. Make a list of animals that live in your area, and record the physical or behavioral adaptations that help them survive.

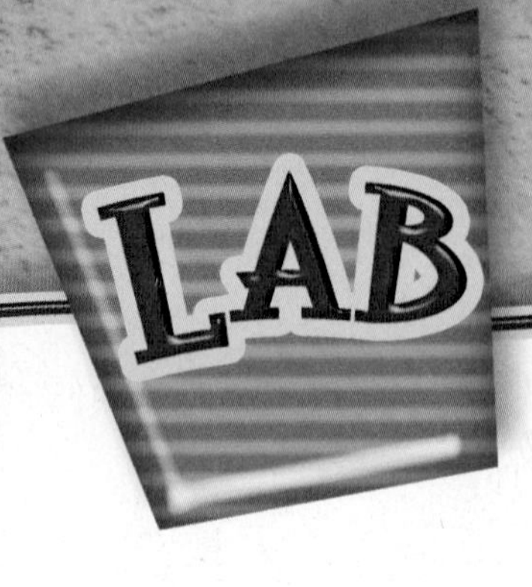

Studying a Land Ecosystem

An ecological study includes observation and analysis of organisms and the physical features of the environment.

Real-World Question

How do you study an ecosystem?

Goals

- **Observe** biotic factors and abiotic factors of an ecosystem.
- **Analyze** the relationships among organisms and their environments.

Materials

graph paper
binoculars
thermometer
pencil
magnifying lens
field guides
notebook
compass
tape measure

Safety Precautions

Procedure

1. Choose a portion of an ecosystem to study. You might choose a decaying log, a pond, a garden, or even a crack in the sidewalk.
2. Determine the boundaries of your study area.
3. Using a tape measure and graph paper, make a map of your area. Determine north.
4. **Record** your observations in a table similar to the one shown on this page.
5. **Observe** the organisms in your study area. Use field guides to identify them. Use a magnifying lens to study small organisms and binoculars to study animals you can't get near. Look for evidence (such as tracks or feathers) of organisms you do not see.
6. Measure and record the air temperature in your study area.
7. Visit your study area many times and at different times of day for one week. At each visit, make the same measurements and record all observations. Note how the living and nonliving parts of the ecosystem interact.

Environmental Observations		
Date		
Time of day		
Temperature		
Organisms observed	Do not write in this book.	
Comments		

Conclude and Apply

1. **Predict** what might happen if one or more abiotic factors were changed suddenly.
2. **Infer** what might happen if one or more populations of plants or animals were removed from the area.
3. **Form a hypothesis** to explain how a new population of organisms might affect your ecosystem.

Communicating Your Data

Make a classroom display of all data recorded. **For more help, refer to the Science Skill Handbook.**

section 3

Aquatic Ecosystems

Freshwater Ecosystems

In a land environment, temperature and precipitation are the most important factors that determine which species can survive. In aquatic environments, water temperature, the amount of sunlight present, and the amounts of dissolved oxygen and salt in the water are important. Earth's freshwater ecosystems include flowing water such as rivers and streams and standing water such as lakes, ponds, and wetlands.

Rivers and Streams Flowing freshwater environments vary from small, gurgling brooks to large, slow-moving rivers. Currents can quickly wash loose particles downstream, leaving a rocky or gravelly bottom. As the water tumbles and splashes, as shown in **Figure 15,** air from the atmosphere mixes in. Naturally fast-flowing streams usually have clearer water and higher oxygen content than slow-flowing streams.

Most nutrients that support life in flowing-water ecosystems are washed into the water from land. In areas where the water movement slows, such as in the pools of streams or in large rivers, debris settles to the bottom. These environments tend to have higher nutrient levels and more plant growth. They contain organisms that are not as well adapted to swiftly flowing water, such as freshwater mussels, minnows, and leeches.

as you read

What You'll Learn

- **Compare** flowing freshwater and standing freshwater ecosystems.
- **Identify** and describe important saltwater ecosystems.
- **Identify** problems that affect aquatic ecosystems.

Why It's Important

All of the life processes in your body depend on water.

Review Vocabulary

aquatic: growing or living in water

New Vocabulary

- wetland
- intertidal zone
- coral reef
- estuary

Figure 15 Streams like this one are high in oxygen because of the swift, tumbling water.
Determine *where most nutrients in streams come from.*

Modeling Freshwater Environments

Procedure

1. Obtain a sample of **pond sediment or debris, plants, water, and organisms** from your teacher.
2. Cover the bottom of a **clear-plastic container** with about 2 cm of the debris.
3. Add one or two plants to the container.
4. Carefully pour pond water into the container until it is about two-thirds full.
5. Use a **net** to add several organisms to the water. Seal the container.
6. Using a **magnifying lens,** observe as many organisms as possible. Record your observations. Return your sample to its original habitat.

Analysis

Write a short paragraph describing the organisms in your sample. How did the organisms interact with each other?

Human Impact People use rivers and streams for many activities. Once regarded as a free place to dump sewage and other pollutants, many people now recognize the damage this causes. Treating sewage and restricting pollutants have led to an improvement in the water quality in some rivers.

Lakes and Ponds When a low place in the land fills with rainwater, snowmelt, or water from an overflowing stream, a lake or pond might form. Pond or lake water hardly moves. It contains more plants than flowing-water environments contain.

Lakes, such as the one shown in **Figure 16,** are larger and deeper than ponds. They have more open water because most plant growth is limited to shallow areas along the shoreline. In fact, organisms found in the warm, sunlit waters of the shorelines often are similar to those found in ponds. If you were to dive to the bottom, you would discover few, if any, plants or algae growing. Colder temperatures and lower light levels limit the types of organisms that can live in deep lake waters. Floating in the warm, sunlit waters near the surface of freshwater lakes and ponds are microscopic algae, plants, and other organisms known as plankton.

A pond is a small, shallow body of water. Because ponds are shallow, they are filled with animal and plant life. Sunlight usually penetrates to the bottom. The warm, sunlit water promotes the growth of plants and algae. In fact, many ponds are filled almost completely with plant material, so the only clear, open water is at the center. Because of the lush growth in pond environments, they tend to be high in nutrients.

Figure 16 Ponds contain more vegetation than lakes contain. The population of organisms in the shallow water of lakes is high. Fewer types of organisms live in the deeper water.

Figure 17 Life in the Florida Everglades was threatened due to pollution, drought, and draining of the water. Conservation efforts are being made in an attempt to preserve this ecosystem.

Water Pollution Human activities can harm freshwater environments. Fertilizer-filled runoff from farms and lawns, as well as sewage dumped into the water, can lead to excessive growth of algae and plants in lakes and ponds. The growth and decay of these organisms reduces the oxygen level in the water, which makes it difficult for some organisms to survive. To prevent problems, sewage is treated before it is released. People also are being educated about problems associated with polluting lakes and ponds. Fines and penalties are issued to people caught polluting waterways. These controls have led to the recovery of many freshwater ecosystems.

Wetlands As the name suggests, **wetlands,** shown in **Figure 17,** are regions that are wet for all or most of a year. They are found in regions that lie between landmasses and water. Other names for wetlands include swamps, bogs, and fens. Some people refer to wetlands as biological supermarkets. They are fertile ecosystems, but only plants that are adapted to waterlogged soil survive there. Wetland animals include beavers, muskrats, alligators, and the endangered bog turtle. Many migratory bird populations use wetlands as breeding grounds.

Reading Check *Where are wetlands found?*

Wetlands once were considered to be useless, disease-ridden places. Many were drained and destroyed to make roads, farmland, shopping centers, and housing developments. Only recently have people begun to understand the importance of wetlands. Products that come from wetlands, including fish, shellfish, cranberries, and plants, are valuable resources. Now many developers are restoring wetlands, and in most states access to land through wetlands is prohibited.

Environmental Author Rachel Carson (1907–1964) was a scientist that turned her knowledge and love of the environment into articles and books. After 15 years as an editor for the U.S. Fish and Wildlife Service, she resigned and devoted her time to writing. She probably is known best for her book *Silent Spring,* in which she warned about the long-term effects of the misuse of pesticides. In your Science Journal, compile a list of other authors who write about environmental issues.

Saltwater Ecosystems

About 95 percent of the water on the surface of Earth contains high concentrations of various salts. The amount of dissolved salts in water is called salinity. The average ocean salinity is about 35 g of salts per 1,000 g of water. Saltwater ecosystems include oceans, seas, a few inland lakes such as the Great Salt Lake in Utah, coastal inlets, and estuaries.

Applying Math — Convert Units

TEMPERATURE Organisms that live around hydrothermal vents in the ocean deal with temperatures that range from 1.7°C to 371°C. You have probably seen temperatures measured in degrees Celsius (°C) and degrees Fahrenheit (°F). Which one are you familiar with? If you know the temperature in one system, you can convert it to the other.

You have a Fahrenheit thermometer and measure the water temperature of a pond at 59°F. What is that temperature in degrees Celsius?

Solution

1. *This is what you know:* water temperature in degrees Fahrenheit = 59°F
2. *This is what you need to find out:* The water temperature in degrees Celsius.
3. *This is the procedure you need to use:*
   - Solve the equation for degrees Celsius:
     $(°C \times 1.8) + 32 = °F$
     $°C = (°F - 32)/1.8$
   - Substitute the known value:
     $°C = (59°F - 32)/1.8 = 15°C$
   - Water temperature that is 59°F is 15°C.
4. *Check your answer:* Substitute the Celsius temperature back into the original equation. You should get 59.

Practice Problems

1. The thermometer outside your classroom reads 78°F. What is the temperature in degrees Celsius?
2. If lake water was 12°C in October and 23°C in May, what is the difference in degrees Fahrenheit?

For more practice, visit life.msscience.com/math_practice

Open Oceans Life abounds in the open ocean. Scientists divide the ocean into different life zones, based on the depth to which sunlight penetrates the water. The lighted zone of the ocean is the upper 200 m or so. It is the home of the plankton that make up the foundation of the food chain in the open ocean. Below about 200 m is the dark zone of the ocean. Animals living in this region feed on material that floats down from the lighted zone, or they feed on each other. A few organisms are able to produce their own food.

Topic: Coral Reefs
Visit life.msscience.com for Web links to information about coral reef ecosystems.

Activity Construct a diorama of a coral reef. Include as many different kinds of organisms as you can for a coral reef ecosystem.

Coral Reefs One of the most diverse ecosystems in the world is the coral reef. **Coral reefs** are formed over long periods of time from the calcium carbonate skeletons secreted by animals called corals. When corals die, their skeletons remain. Over time, the skeletal deposits form reefs such as the Great Barrier Reef off the coast of Australia, shown in **Figure 18.**

Reefs do not adapt well to long-term stress. Runoff from fields, sewage, and increased sedimentation from cleared land harm reef ecosystems. Organizations like the Environmental Protection Agency have developed management plans to protect the diversity of coral reefs. These plans treat a coral reef as a system that includes all the areas that surround the reef. Keeping the areas around reefs healthy will result in a healthy environment for the coral reef ecosystem.

Figure 18 The lighter areas around this island are part of the Great Barrier Reef. It comprises about 3,000 reefs and about 900 islands. Reefs contain colorful fish and a large variety of other organisms.

Figure 19 As the tide recedes, small pools of seawater are left behind. These pools contain a variety of organisms such as sea stars and periwinkles.

Sea star

Periwinkles

INTEGRATE Earth Science

Seashores All of Earth's landmasses are bordered by ocean water. The shallow waters along the world's coastlines contain a variety of saltwater ecosystems, all of which are influenced by the tides and by the action of waves. The gravitational pull of the Moon, and to a lesser extent, the Sun, on Earth causes the tides to rise and fall each day. The height of the tides varies according to the phases of the Moon, the season, and the slope of the shoreline. The **intertidal zone** is the portion of the shoreline that is covered with water at high tide and exposed to the air during low tide. Organisms that live in the intertidal zone, such as those in **Figure 19,** must be adapted to dramatic changes in temperature, moisture, and salinity and must be able to withstand the force of wave action.

Estuaries Almost every river on Earth eventually flows into an ocean. The area where a river meets an ocean and contains a mixture of freshwater and salt water is called an **estuary** (ES chuh wer ee). Other names for estuaries include bays, lagoons, harbors, inlets, and sounds. They are located near coastlines and border the land. Salinity in estuaries changes with the amount of freshwater brought in by rivers and streams, and with the amount of salt water pushed inland by the ocean tides.

Estuaries, shown in **Figure 20,** are extremely fertile, productive environments because freshwater streams bring in tons of nutrients washed from inland soils. Therefore, nutrient levels in estuaries are higher than in freshwater ecosystems or other saltwater ecosystems.

Figure 20 The Chesapeake Bay is an estuary rich in resources. Fish and shrimp are harvested by commercial fishing boats.
Describe *what other resources can be found in estuaries.*

Estuary Life Organisms found in estuaries include many species of algae, salt-tolerant grasses, shrimp, crabs, clams, oysters, snails, worms, and fish. Estuaries also serve as important nurseries for many species of ocean fish. Estuaries provide much of the seafood consumed by humans.

Why are estuaries more fertile than other aquatic ecosystems?

section 3 review

Summary

Freshwater Ecosystems

- Temperature, light, salt, and dissolved oxygen are important factors.
- Rivers, streams, lakes, ponds, and wetlands are freshwater ecosystems.
- Human activities, such as too much lawn fertilizer, can pollute aquatic ecosystems.

Saltwater Ecosystems

- About 95 percent of Earth's water contains dissolved salts.
- Saltwater ecosystems include open oceans, coral reefs, seashores, and estuaries.
- Organisms that live on seashores have adaptations that enable them to survive dramatic changes in temperature, moisture, and salinity.
- Estuaries serve as nursery areas for many species of ocean fish.

Self Check

1. **Identify** the similarities and differences between a lake and a stream.
2. **Compare and contrast** the dark zone of the ocean with the forest floor of a tropical rain forest. What living or nonliving factors affect these areas?
3. **Explain** why fewer plants are at the bottom of deep lakes.
4. **Infer** what adaptations are necessary for organisms that live in the intertidal zone.
5. **Think Critically** Would you expect a fast moving mountain stream or the Mississippi River to have more dissolved oxygen? Explain.

Applying Skills

6. **Communicate** Wetlands trap and slowly release rain, snow, and groundwater. Describe in your Science Journal what might happen to a town located on a floodplain if nearby wetlands are destroyed.

Use the Internet

Exploring Wetlands

Goals

- **Identify** wetland regions in the United States.
- **Describe** the significance of the wetland ecosystem.
- **Identify** plant and animal species native to a wetland region.
- **Identify** strategies for supporting the preservation of wetlands.

Data Source

Science Online

Visit life.msscience.com/ for more information about wetland environments and for data collected by other students.

Real-World Question

Wetlands, such as the one shown below, are an important part of the environment. These fertile ecosystems support unique plants and animals that can survive only in wetland conditions. The more you understand the importance of wetlands, the more you can do to preserve and protect them. Why are wetlands an important part of the ecosystem?

Make a Plan

1. **Determine** where some major wetlands are located in the United States.
2. **Identify** one wetland area to study in depth. Where is it located? Is it classified as a marsh, bog, or something else?
3. **Explain** the role this ecosystem plays in the overall ecology of the area.
4. **Research information** about the plants and animals that live in the wetland environment you are researching.
5. **Investigate** what laws protect the wetland you are studying.

Follow Your Plan

1. Make sure your teacher approves your plan before you start.
2. Perform the investigation.
3. Post your data at the link shown below.

Analyze Your Data

1. **Describe** the wetland area you have researched. What region of the United States is it located in? What other ecological factors are found in that region?
2. **Outline** the laws protecting the wetland you are investigating. How long have the laws been in place?
3. **List** the plants and animals native to the wetland area you are researching. Are those plants and animals found in other parts of the region or the United States? What adaptations do the plants and animals have that help them survive in a wetland environment?

Conclude and Apply

1. **Infer** Are all wetlands the same?
2. **Determine** what the ecological significance of the wetland area that you studied for that region of the country is.
3. **Draw Conclusions** Why should wetland environments be protected?
4. **Summarize** what people can do to support the continued preservation of wetland environments in the United States.

Find this lab using the link below. **Post** your data in the table provided. **Review** other students' data to learn about other wetland environments in the United States.

life.msscience.com/internet_lab

TIME SCIENCE AND Society

SCIENCE ISSUES THAT AFFECT YOU!

Creating Wetlands to Purify Wastewater

Pebbles were added to the Corrales wetlands to help with drainage.

Water irises thrived in the wetlands, less than a year after planting.

Students enjoy pure water from the Corrales wetlands after it is filtered.

When you wash your hands or flush the toilet, do you think about where the wastewater goes? In most places, it eventually ends up being processed in a traditional sewage-treatment facility. But some places are experimenting with a new method that processes wastewater by creating wetlands. Wetlands are home to filtering plants, such as cattails, and sewage-eating bacteria.

In 1996, school officials at the Corrales Elementary School in Albuquerque, New Mexico, faced a big problem. The old wastewater-treatment system had failed. Replacing it was going to cost a lot of money. Instead of constructing a new sewage-treatment plant, school officials decided to create a natural wetlands system. The wetlands system could do the job less expensively, while protecting the environment.

Today, this wetlands efficiently converts polluted water into cleaner water that's good for the environment. U.S. government officials are monitoring this alternative sewage-treatment system to see if it is successful. So far, so good!

Wetlands filter water through the actions of the plants and microorganisms that live there. When plants absorb water into their roots, some also take up pollutants. The plants convert the pollutants to forms that are not dangerous. At the same time, bacteria and other microorganisms are filtering water as they feed. Water moves slowly through wetlands, so the organisms have plenty of time to do their work. Wetlands built by people to filter small amounts of pollutants are called "constructed wetlands". In many places, constructed wetlands are better at cleaning wastewater than sewers or septic systems.

Visit and Observe **Visit a wetlands and create a field journal of your observations. Draw the plants and animals you see. Use a field guide to help identify the wildlife. If you don't live near a wetlands, use resources to research wetlands environments.**

For more information, visit life.msscience.com/time

Reviewing Main Ideas

Section 1 How Ecosystems Change

1. Ecological succession is the gradual change from one plant community to another.
2. Primary succession begins in a place where no plants were before.
3. Secondary succession begins in a place that has soil and was once the home of living organisms.
4. A climax community has reached a stable stage of ecological succession.

Section 2 Biomes

1. Temperature and precipitation help determine the climate of a region.
2. Large geographic areas with similar climax communities are called biomes.
3. Earth's land biomes include tundra, taiga, temperate deciduous forest, temperate rain forest, tropical rain forest, grassland, and desert.

Section 3 Aquatic Ecosystems

1. Freshwater ecosystems include streams, rivers, lakes, ponds, and wetlands.
2. Wetlands are areas that are covered with water most of the year. They are found in regions that lie between land-masses and water.
3. Saltwater ecosystems include estuaries, seashores, coral reefs, a few inland lakes, and the deep ocean.
4. Estuaries are fertile transitional zones between freshwater and saltwater environments.

Visualizing Main Ideas

Copy and complete this concept map about land biomes.

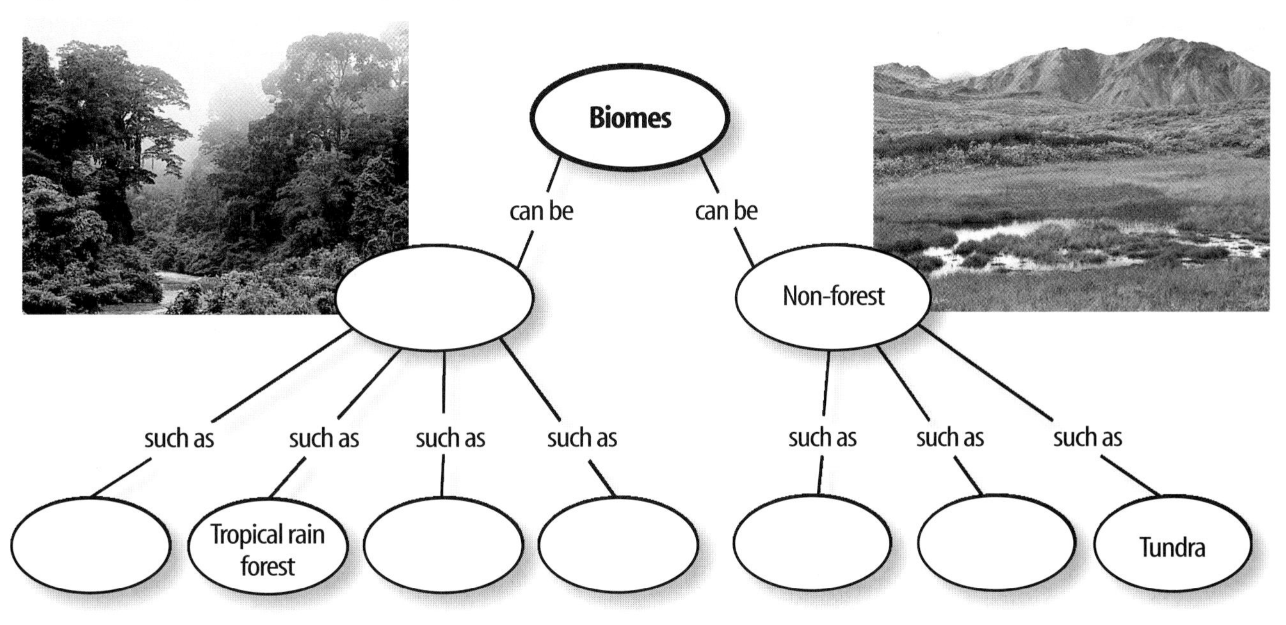

chapter 26 Review

Using Vocabulary

biome p. 744	succession p. 740
climax community p. 743	taiga p. 746
coral reef p. 757	temperate deciduous forest p. 747
desert p. 750	temperate rain forest p. 747
estuary p. 758	tropical rain forest p. 748
grassland p. 751	tundra p. 745
intertidal zone p. 758	wetland p. 755
pioneer species p. 740	

Fill in the blanks with the correct vocabulary word or words.

1. ________ refers to the normal changes in the types of species that live in communities.
2. A(n) ________ is a group of organisms found in a stable stage of succession.
3. Deciduous trees are dominant in the ________.
4. The average temperature in ________ is between 9°C and 12°C.
5. ________ are the most biologically diverse biomes in the world.
6. A(n) ________ is an area where freshwater meets the ocean.

Checking Concepts

Choose the word or phrase that best answers the question.

7. What are tundra and desert examples of?
   A) ecosystems **C)** habitats
   B) biomes **D)** communities
8. What is a hot, dry biome called?
   A) desert **C)** coral reef
   B) tundra **D)** grassland
9. Where would organisms that are adapted to live in slightly salty water be found?
   A) lake **C)** open ocean
   B) estuary **D)** intertidal zone
10. Which biome contains mostly frozen soil called permafrost?
    A) taiga
    B) temperate rain forest
    C) tundra
    D) temperate deciduous forest
11. A new island is formed from a volcanic eruption. Which species probably would be the first to grow and survive?
    A) palm trees **C)** grasses
    B) lichens **D)** ferns
12. What would the changes in communities that take place on a recently formed volcanic island best be described as?
    A) primary succession
    B) secondary succession
    C) tertiary succession
    D) magma
13. What is the stable end stage of succession?
    A) pioneer species
    B) climax community
    C) limiting factor
    D) permafrost

Use the illustration below to answer question 14.

Observed Fire Danger Class—June, 2003

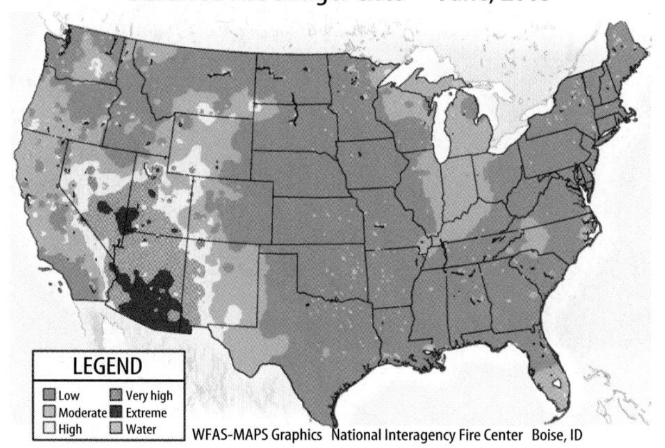

14. Which area of the U.S. had the highest observed fire danger on June 20, 2003?
    A) northeast **C)** northwest
    B) southeast **D)** southwest

Thinking Critically

15. **Explain** In most cases, would a soil sample from a temperate deciduous forest be more or less nutrient-rich than a soil sample from a tropical rain forest?
16. **Explain** why some plant seeds need fire in order to germinate. How does this give these plants an advantage in secondary succession?
17. **Determine** A grassy meadow borders a beech-maple forest. Is one of these ecosystems undergoing succession? Why?
18. **Infer** why tundra plants are usually small.
19. **Make and Use a Table** Copy and complete the following table about aquatic ecosystems. Include these terms: *intertidal zone, lake, pond, coral reef, open ocean, river, estuary,* and *stream.*

Aquatic Ecosystems

Saltwater	Freshwater
Do not write	in this book.

20. **Recognize Cause and Effect** Wildfires like the one in Yellowstone National Park in 1988, cause many changes to the land. Determine the effect of a fire on an area that has reached its climax community.

Performance Activities

21. **Oral Presentation** Research a biome not in this chapter. Find out about its climate and location, and which organisms live there. Present this information to your class.

Applying Math

Use the table below to answer question 22.

Rainfall Amounts

Biome	Average Precipitation/Year (cm)
Taiga	50
Temperate rain forest	200
Tropical rain forest	400
Desert	25
Temperate deciduous forest	150
Tundra	25

22. **Biome Precipitation** How many times more precipitation does the tropical rain forest biome receive than the taiga or desert?

Use the graph below to answer question 23.

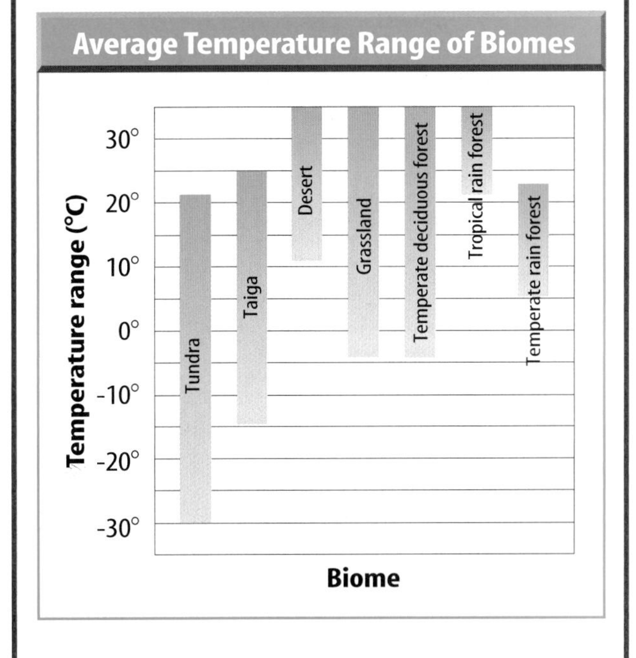

23. **Biome Temperatures** According to the graph, which biome has the greatest and which biome has the least variation in temperature throughout the year? Estimate the difference between the two.

Part 1 Multiple Choice

Record your answers on the answer sheet provided by your teacher or on a sheet of paper.

1. What two factors are most responsible for limiting life in a particular area?
   A. sunlight and temperature
   B. precipitation and temperature
   C. precipitation and sunlight
   D. soil conditions and precipitation

2. Which of the following forms during primary succession?
   A. trees
   B. soil
   C. wildlife
   D. grasses

Use the illustrations below to answer questions 3 and 4.

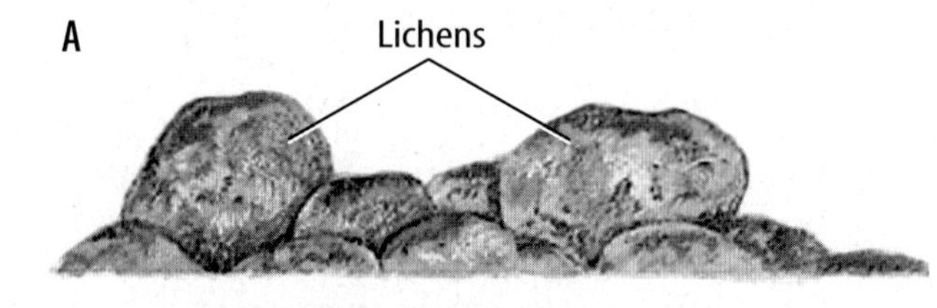

3. Which of the following statements best describes what is represented by A?
   A. Primary succession is occurring.
   B. Secondary succession is occurring.
   C. A forest fire has probably occurred.
   D. The climax stage has been reached.

4. Which of the following statements best describes what is represented by B?
   A. The climax stage has been reached.
   B. Pioneer species are forming soil.
   C. Bare rock covers most of the area.
   D. Secondary succession is occurring.

Use the map below to answer questions 5 and 6.

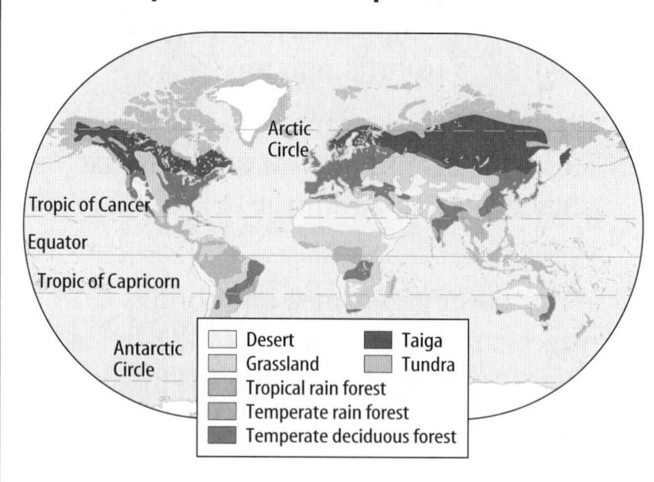

5. What biome is located in the latitudes just south of the north pole?
   A. taiga
   B. temperate deciduous rain forest
   C. tundra
   D. temperate rain forest

6. The tropical rainforest biome is found primarily near the
   A. Arctic Circle.
   B. Tropic of Cancer.
   C. equator.
   D. Tropic of Capricorn.

7. Which of the following is composed of a mix of salt water and freshwater?
   A. an intertidal zone
   B. an estuary
   C. a seashore
   D. a coral reef

Test-Taking Tip

Come Prepared Bring at least two sharpened No. 2 pencils and a good eraser to the test. Check to make sure that your eraser completely removes all pencil marks.

Part 2 Short Response/Grid In

Record your answers on the answer sheet provided by your teacher or on a sheet of paper.

8. Name two products that come from wetlands.

9. Which takes longer, primary succession or secondary succession? Why?

Use the photos below to answer questions 10 and 11.

10. In what biome would you most likely find A? How is this animal adapted to survive in its biome?

11. In what biome would you most likely find B? How is this animal adapted to survive in its biome?

12. Which biome receives the most rainfall per year? Which receives the least rainfall?

13. Why are forests unlikely to develop in grasslands?

14. What are two kinds of wetlands? What kinds of animals are found in wetlands?

15. What organisms inhabit the upper zone of the open ocean and why are they so important?

Part 3 Open Ended

Record your answers on a sheet of paper.

16. Explain how lichens contribute to the process of soil formation.

17. Compare and contrast a freshwater lake ecosystem with a freshwater pond ecosystem.

18. What special adaptations must all of the organisms that live in the intertidal zone have?

19. What are the differences between the temperate rain forest biome and the tropical rain forest biome?

Use the illustration below to answer questions 20 and 21.

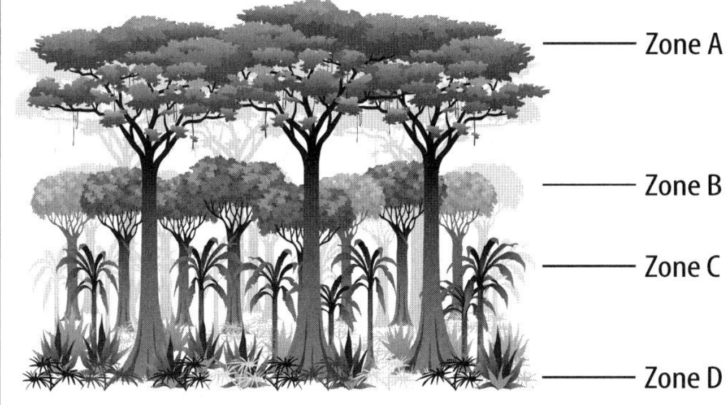

20. Identify and describe zone C in the diagram. What kinds of wildlife are found there?

21. Identify zone D and zone A. Describe the environment in each zone. Why might an organism that lives in zone A not be able to survive in zone D?

22. Discuss the effects of human impact on freshwater environments like lakes and ponds.

chapter 27

Conserving Resources

The BIG Idea

Many of Earth's resources are limited.

SECTION 1
Resources
Main Idea Earth has natural resources that can be replenished as well as natural resources that cannot be replenished.

SECTION 2
Pollution
Main Idea Air, water, and soil pollution have many causes, including hazardous waste and the burning of fossil fuels.

SECTION 3
The Three Rs of Conservation
Main Idea Natural resources can be conserved by following the three Rs of conservation: reduce, reuse, and recycle.

Resources Fuel Our Lives

Resources, such as clean water and air, are commonly taken for granted. We depend on water and air to survive. Fossil fuels are another type of resource, and we depend on them for energy. However, fossil fuels can pollute our air and water.

Science Journal List some other resources that we depend on and describe how we use them.

Start-Up Activities

What happens when topsoil is left unprotected?

Plants grow in the top, nutrient-rich layer, called topsoil. Plants help keep topsoil in place by protecting it from wind and rain. Try the following experiment to find out what happens when topsoil is left unprotected.

1. Use a mixture of moist sand and potting soil to create a miniature landscape in a plastic basin or aluminum-foil baking pan. Form hills and valleys in your landscape.
2. Use clumps of moss to cover areas of your landscape. Leave some sloping portions without plant cover.
3. Simulate a rainstorm over your landscape by spraying water on it from a spray bottle or by pouring a slow stream of water on it from a beaker.
4. **Think Critically** In your Science Journal, record your observations and describe what happened to the land that was not protected by plant cover.

Preview this chapter's content and activities at life.msscience.com

Resources Make the following Foldable to help you organize information and diagram ideas about renewable and nonrenewable resources.

STEP 1 **Fold** a sheet of paper in half lengthwise. Make the back edge about 5 cm longer than the front edge.

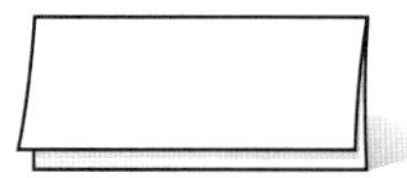

STEP 2 **Turn** the paper so the fold is on the bottom. Then **fold** in half.

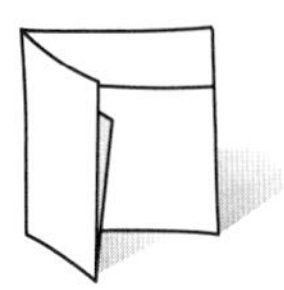

STEP 3 **Unfold and cut** only the top layer along the fold to make two tabs. **Label** the Foldable as shown.

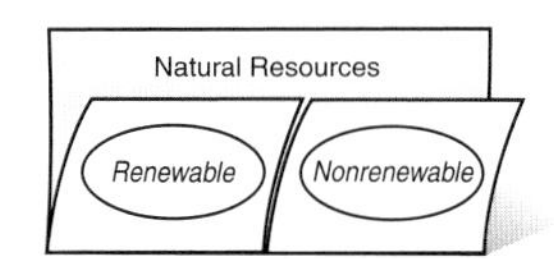

Make a Concept Map Before you read the chapter, list examples of each type of natural resource you already know on the back of the appropriate tabs. As you read the chapter, add to your lists.

Get Ready to Read

Questions and Answers

1 Learn It! Knowing how to find answers to questions will help you on reviews and tests. Some answers can be found in the textbook, while other answers require you to go beyond the textbook. These answers might be based on knowledge you already have or things you have experienced.

2 Practice It! Read the excerpt below. Answer the following questions and then discuss them with a partner.

> Even though renewable resources are recycled or replaced, they are sometimes in short supply. Rain and melted snow replace the water in streams, lakes, and reservoirs. Sometimes there may not be enough rain or snowmelt to meet all the needs of people, plants, and animals. In some parts of the world, especially desert regions, water and other resources usually are scarce. Other resources can be used instead.
>
> *—from page 771*

- How is water in streams, lakes, and reservoirs replenished?
- What happens if there is not enough rain or snowmelt to replace water that has been used?
- How can you conserve natural resources, such as water?

3 Apply It! Look at some questions in the text. Which questions can be answered directly from the text? Which require you to go beyond the text?

As you read, keep track of questions you answer in the chapter. This will help you remember what you read.

Target Your Reading

Use this to focus on the main ideas as you read the chapter.

1. **Before you read** the chapter, respond to the statements below on your worksheet or on a numbered sheet of paper.
   - Write an **A** if you **agree** with the statement.
   - Write a **D** if you **disagree** with the statement.

2. **After you read** the chapter, look back to this page to see if you've changed your mind about any of the statements.
   - If any of your answers changed, explain why.
   - Change any false statements into true statements.
   - Use your revised statements as a study guide.

Before You Read A or D		Statement	After You Read A or D
	1	All Earth's resources are able to be replenished.	
	2	Fossil fuels must be burned to release the energy they hold.	
	3	The only unlimited source of energy for Earth is the Sun.	
	4	Air pollution can be cleaned when it reacts with sunlight.	
	5	Underground water supplies are safe from pollution.	
	6	Some household items, such as batteries and paint, are hazardous and may pollute the environment if disposed incorrectly.	
	7	Conservation can help prevent shortages of natural resources.	
	8	Certain plastics can be recycled into items such as carpeting and clothing.	
	9	Aluminum is the only metal that can be recycled.	

Science Online
Print out a worksheet of this page at life.msscience.com

section 1

Resources

as you read

What You'll Learn

- **Compare** renewable and nonrenewable resources.
- **List** uses of fossil fuels.
- **Identify** alternatives to fossil fuel use.

Why It's Important

Wise use of natural resources is important for the health of all life on Earth.

Review Vocabulary
geyser: a spring that emits intermittent jets of heated water and steam

New Vocabulary
- natural resource
- renewable resource
- nonrenewable resource
- petroleum
- fossil fuel
- hydroelectric power
- nuclear energy
- geothermal energy

Natural Resources

An earthworm burrowing in moist soil eats decaying plant material. A robin catches the worm and flies to a tree. The leaves of the tree use sunlight during photosynthesis. Leaves fall to the ground, decay, and perhaps become an earthworm's meal. What do these living things have in common? They rely on Earth's **natural resources**—the parts of the environment that are useful or necessary for the survival of living organisms.

What kinds of natural resources do you use? Like other organisms, you need food, air, and water. You also use resources that are needed to make everything from clothes to cars. Natural resources supply energy for automobiles and power plants. Although some natural resources are plentiful, others are not.

Renewable Resources The Sun, an inexhaustible resource, provides a constant supply of heat and light. Rain fills lakes and streams with water. When plants carry out photosynthesis, they add oxygen to the air. Sunlight, water, air, and the crops shown in **Figure 1** are examples of renewable resources. A **renewable resource** is any natural resource that is recycled or replaced constantly by nature.

Figure 1 Cotton and wood are renewable resources. Cotton cloth is used for rugs, curtains, and clothing. A new crop of cotton can be grown every year. Wood is used for furniture, building materials, and paper. It will take 20 years for these young trees to grow large enough to harvest.

Supply and Demand Even though renewable resources are recycled or replaced, they are sometimes in short supply. Rain and melted snow replace the water in streams, lakes, and reservoirs. Sometimes, there may not be enough rain or snowmelt to meet all the needs of people, plants, and animals. In some parts of the world, especially desert regions, water and other resources usually are scarce. Other resources can be used instead, as shown in **Figure 2.**

Figure 2 In parts of Africa, firewood has become scarce. People in this village now use solar energy instead of wood for cooking.

Nonrenewable Resources Natural resources that are used up more quickly than they can be replaced by natural processes are **nonrenewable resources.** Earth's supply of nonrenewable resources is limited. You use nonrenewable resources when you take home groceries in a plastic bag, paint a wall, or travel by car. Plastics, paints, and gasoline are made from an important nonrenewable resource called petroleum, or oil. **Petroleum** is formed mostly from the remains of microscopic marine organisms buried in Earth's crust. It is nonrenewable because it takes hundreds of millions of years for it to form.

Reading Check *What are nonrenewable resources?*

Minerals and metals found in Earth's crust are nonrenewable resources. Petroleum is a mineral. So are diamonds and the graphite in pencil lead. The aluminum used to make soft-drink cans is a metal. Iron, copper, tin, gold, silver, tungsten, and uranium also are metals. Many manufactured items, like the car shown in **Figure 3,** are made from nonrenewable resources.

Figure 3 Iron, a nonrenewable resource, is the main ingredient in steel. Steel is used to make cars, trucks, appliances, buildings, bridges, and even tires.
Infer *what other nonrenewable resources are used to build a car.*

Observing Mineral Mining Effects

Procedure

1. Place a **chocolate-chip cookie** on a **paper plate.** Pretend the chips are mineral deposits and the rest of the cookie is Earth's crust.
2. Use a **toothpick** to locate and dig up the mineral deposits. Try to disturb the land as little as possible.
3. When mining is completed, try to restore the land to its original condition.

Analysis

1. How well were you able to restore the land?
2. Compare the difficulty of digging for mineral deposits found close to the surface with digging for those found deep in Earth's crust.
3. Describe environmental changes that might result from a mining operation.

Try at Home

Fossil Fuels

Coal, oil, and natural gas are nonrenewable resources that supply energy. Most of the energy you use comes from these fossil fuels, as the graph in **Figure 4** shows. **Fossil fuels** are fuels formed in Earth's crust over hundreds of millions of years. Cars, buses, trains, and airplanes are powered by gasoline, diesel fuel, and jet fuel, which are made from oil. Coal is used in many power plants to produce electricity. Natural gas is used in manufacturing, for heating and cooking, and sometimes as a vehicle fuel.

Fossil Fuel Conservation Billions of people all over the world use fossil fuels every day. Because fossil fuels are nonrenewable, Earth's supply of them is limited. In the future, they may become more expensive and difficult to obtain. Also, the use of fossil fuels can lead to environmental problems. For example, mining coal can require stripping away thick layers of soil and rock, as shown in **Figure 4,** which destroys ecosystems. Another problem is that fossil fuels must be burned to release the energy stored in them. The burning of fossil fuels produces waste gases that cause air pollution, including smog and acid rain. For these reasons, many people suggest reducing the use of fossil fuels and finding other sources of energy.

You can use simple conservation measures to help reduce fossil fuel use. Switch off the light when you leave a room and turn off the television when you're not watching it. These actions reduce your use of electricity, which often is produced in power plants that burn fossil fuels. Hundreds of millions of automobiles are in use in the United States. Riding in a car pool or taking public transportation uses fewer liters of gasoline than driving alone in a car. Walking or riding a bicycle uses even less fossil fuel. Reducing fossil fuel use has an added benefit—the less you use, the more money you save.

Figure 4 Coal is a fossil fuel. It often is obtained by strip mining, which removes all the soil above the coal deposit. The soil is replaced, but it takes many years for the ecosystem to recover.
Identify *the resource that provided 84 percent of the energy used in the United States in 1999.*

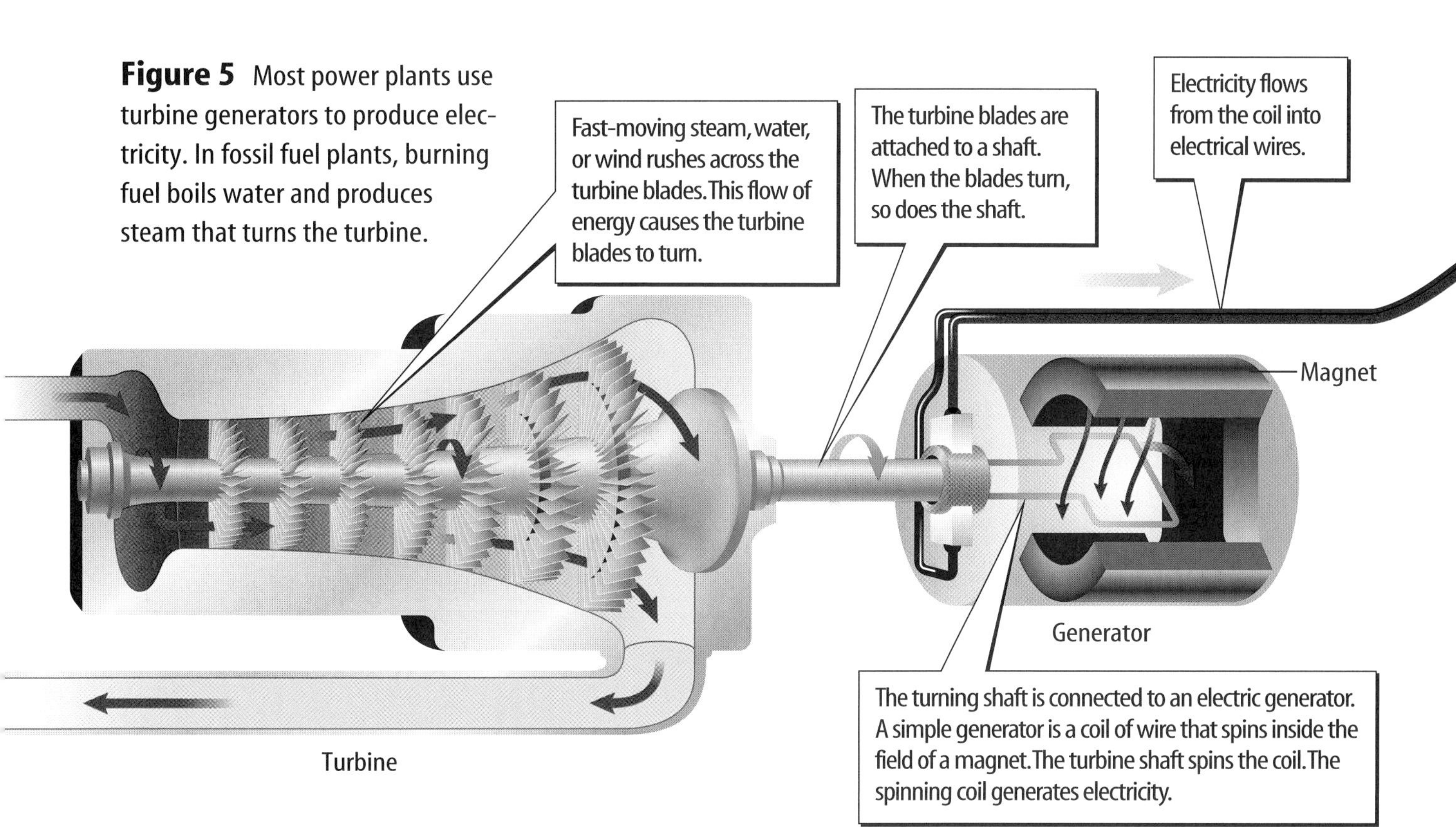

Fast-moving steam, water, or wind rushes across the turbine blades. This flow of energy causes the turbine blades to turn.
The turbine blades are attached to a shaft. When the blades turn, so does the shaft.
Electricity flows from the coil into electrical wires.
Magnet
Generator
The turning shaft is connected to an electric generator. A simple generator is a coil of wire that spins inside the field of a magnet. The turbine shaft spins the coil. The spinning coil generates electricity.
Turbine

Figure 5 Most power plants use turbine generators to produce electricity. In fossil fuel plants, burning fuel boils water and produces steam that turns the turbine.

Alternatives to Fossil Fuels

Another approach to reducing fossil fuel use is to develop other sources of energy. Much of the electricity used today comes from power plants that burn fossil fuels. As **Figure 5** shows, electricity is generated when a rotating turbine turns a coil of wires in the magnetic field of an electric generator. Fossil-fuel power plants boil water to produce steam that turns the turbine. Alternative energy sources, including water, wind, and atomic energy can be used instead of fossil fuels to turn turbines. Also, solar cells can produce electricity using only sunlight, with no turbines at all. Some of these alternative energy sources—particularly wind and solar energy—are so plentiful they could be considered inexhaustible resources.

Water Power Water is a renewable energy source that can be used to generate electricity. **Hydroelectric power** is electricity that is produced when the energy of falling water is used to turn the turbines of an electric generator. Hydroelectric power does not contribute to air pollution because no fuel is burned. However, it does present environmental concerns. Building a hydroelectric plant usually involves constructing a dam across a river. The dam raises the water level high enough to produce the energy required for electricity generation. Many acres behind the dam are flooded, destroying land habitats and changing part of the river into a lake.

Energy Oil and natural gas are used to produce over 60 percent of the energy supply in the United States. Over half of the oil used is imported from other countries. Many scientists suggest that emissions from the burning of fossil fuels are principally responsible for global warming. In your Science Journal, write what you might do to persuade utility companies to increase their use of water, wind, and solar power.

Wind Power Wind power is another renewable energy source that can be used for electricity production. Wind turns the blades of a turbine, which powers an electric generator. When winds blow at least 32 km/h, energy is produced. Wind power does not cause air pollution, but electricity can be produced only when the wind is blowing. So far, wind power accounts for only a small percentage of the electricity used worldwide.

Nuclear Power Another alternative to fossil fuels makes use of the huge amounts of energy in the nuclei of atoms, as shown in **Figure 6. Nuclear energy** is released when billions of atomic nuclei from uranium, a radioactive element, are split apart in a nuclear fission reaction. This energy is used to produce steam that rotates the turbine blades of an electric generator.

Nuclear power does not contribute to air pollution. However, uranium is a nonrenewable resource, and mining it can disrupt ecosystems. Nuclear power plants also produce radioactive wastes that can seriously harm living organisms. Some of these wastes remain radioactive for thousands of years, and their safe disposal is a problem that has not yet been solved. Accidents also are a danger.

Figure 6 Nuclear power plants are designed to withstand the high energy produced by nuclear reactions.
Describe *how heat is produced in a nuclear reactor.*

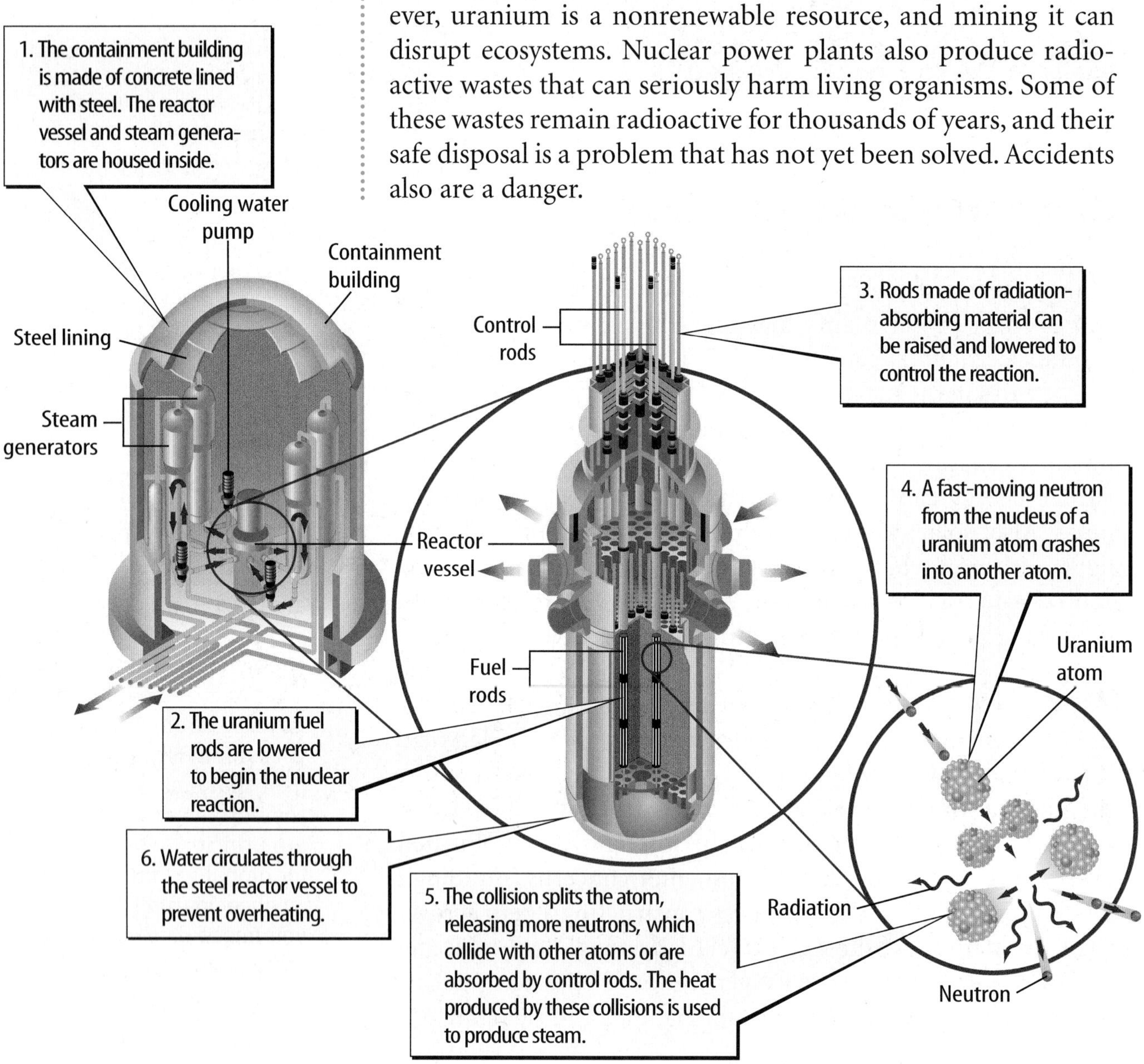

1. The containment building is made of concrete lined with steel. The reactor vessel and steam generators are housed inside.

Cooling water pump
Containment building
Steel lining
Steam generators
Control rods
Reactor vessel
Fuel rods

2. The uranium fuel rods are lowered to begin the nuclear reaction.

3. Rods made of radiation-absorbing material can be raised and lowered to control the reaction.

4. A fast-moving neutron from the nucleus of a uranium atom crashes into another atom.

Uranium atom
Radiation
Neutron

5. The collision splits the atom, releasing more neutrons, which collide with other atoms or are absorbed by control rods. The heat produced by these collisions is used to produce steam.

6. Water circulates through the steel reactor vessel to prevent overheating.

Geothermal Energy The hot, molten rock that lies deep beneath Earth's surface is also a source of energy. You see the effects of this energy when lava and hot gases escape from an erupting volcano or when hot water spews from a geyser. The heat energy contained in Earth's crust is called **geothermal energy.** Most geothermal power plants use this energy to produce steam to generate electricity.

Geothermal energy for power plants is available only where natural geysers or volcanoes are found. A geothermal power plant in California uses steam produced by geysers. The island nation of Iceland was formed by volcanoes, and geothermal energy is plentiful there. Geothermal power plants supply heat and electricity to about 90 percent of the homes in Iceland. Outdoor swimming areas also are heated with geothermal energy, as shown in **Figure 7.**

Figure 7 In Iceland, a geothermal power plant pumps hot water out of the ground to heat buildings and generate electricity. Leftover hot water goes into this lake, making it warm enough for swimming even when the ground is covered with snow.

Reading Check *Where does geothermal energy come from?*

Solar Energy The most inexhaustible source of energy for all life on Earth is the Sun. Solar energy is an alternative to fossil fuels. One use of solar energy is in solar-heated buildings. During winter in the northern hemisphere, the parts of a building that face south receive the most sunlight. Large windows placed on the south side of a building help heat it by allowing warm sunshine into the building during the day. Floors and walls of most solar-heated buildings are made of materials that absorb heat during the day. During the night, the stored heat is released slowly, keeping the building warm. **Figure 8** shows how solar energy can be used.

Figure 8 The Zion National Park Visitor Center in Utah is a solar-heated building designed to save energy. The roof holds solar panels that are used to generate electricity. High windows can be opened to circulate air and help cool the building on hot days. The overhanging roof shades the windows during summer.

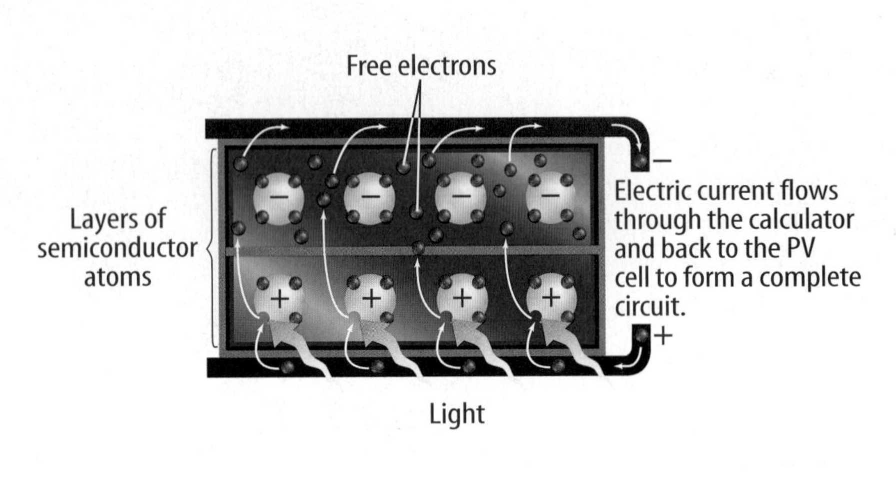

Figure 9 Light energy from the Sun travels in tiny packets of energy called photons. Photons crash into the atoms of PV cells, knocking electrons loose. These electrons create an electric current.

Solar Cells Do you know how a solar-powered calculator works? How do spacecraft use sunlight to generate electricity? These devices use photovoltaic (foh toh vohl TAY ihk) cells to turn sunlight into electric current, as shown in **Figure 9.** Photovoltaic (PV) cells are small and easy to use. However, they produce electricity only in sunlight, so batteries are needed to store electricity for use at night or on cloudy days. Also, PV cells presently are too expensive to use for generating large amounts of electricity. Improvements in this technology continue to be made, and prices probably will go down in the future. As **Figure 10** shows, solar buildings and PV cells are just two of the many ways solar energy can be used to replace fossil fuels.

section 1 review

Summary

Natural Resources

- All living things depend on natural resources to survive.
- Some resources are renewable, while other resources, such as petroleum, are nonrenewable.

Fossil Fuels

- Most of the energy that humans use comes from fossil fuels.
- Fossil fuels must be burned to release the energy stored in them, which causes air pollution.

Alternatives to Fossil Fuels

- Alternatives to fossil fuels include water power, wind power, nuclear power, geothermal energy, and solar energy.
- The Sun provides the most inexhaustible supply of energy for all life on Earth.

Self Check

1. **Summarize** What are natural resources?
2. **Compare and contrast** renewable and nonrenewable resources. Give five examples of each.
3. **Describe** the advantages and disadvantages of using nuclear power.
4. **Describe** two ways solar energy can be used to reduce fossil fuel use.
5. **Think Critically** Explain why the water that is used to cool the reactor vessel of a nuclear power plant is kept separate from the water that is heated to produce steam for the turbine generators.

Applying Math

6. **Solve One-Step Equations** Most cars in the U.S. are driven about 10,000 miles each year. If a car can travel 30 miles on one gallon of gasoline, how many gallons will it use in a year?

Science Online life.msscience.com/self_check_quiz

NATIONAL GEOGRAPHIC

VISUALIZING SOLAR ENERGY

Figure 10

Sunlight is a renewable energy source that provides an alternative to fossil fuels. Solar technologies use the Sun's energy in many ways—from heating to electricity generation.

▲ POWER PLANTS In California's Mojave Desert, an experimental solar power plant used hundreds of mirrors to focus sunlight on a water-filled tower. The steam produced by this system generates enough electricity to power 2,400 homes.

▼ ELECTRICITY Photovoltaic (PV) cells turn sunlight into electric current. They are commonly used to power small devices, such as calculators. Panels that combine many PV cells provide enough electricity for a home—or an orbiting satellite, such as the International Space Station, below.

▲ WATER HEATING Water is heated as it flows through small pipes in this roof-mounted solar heat collector. The hot water then flows into an insulated tank for storage.

▼ COOKING In hot, sunny weather, a solar oven or panel cooker can be used to cook a pot of rice or heat water. The powerful solar cooker shown below reaches even higher temperatures. It is being used to fry food.

▼ INDOOR HEATING South-facing windows and heat-absorbing construction materials turn a room into a solar collector that can help heat an entire building, such as this Connecticut home.

section 2

Pollution

as you read

What You'll Learn

- **Describe** types of air pollution.
- **Identify** causes of water pollution.
- **Explain** methods that can be used to prevent erosion.

Why It's Important

By understanding the causes of pollution, you can help solve pollution problems.

Review Vocabulary
atmosphere: the whole mass of air surrounding Earth

New Vocabulary

- pollutant
- acid precipitation
- greenhouse effect
- ozone depletion
- erosion
- hazardous waste

Keeping the Environment Healthy

More than six billion people live on Earth. This large human population puts a strain on the environment, but each person can make a difference. You can help safeguard the environment by paying attention to how your use of natural resources affects air, land, and water.

Air Pollution

On a still, sunny day in almost any large city, you might see a dark haze in the air, like that in **Figure 11.** The haze comes from pollutants that form when wood or fuels are burned. A **pollutant** is a substance that contaminates the environment. Air pollutants include soot, smoke, ash, and gases such as carbon dioxide, carbon monoxide, nitrogen oxides, and sulfur oxides. Wherever cars, trucks, airplanes, factories, homes, or power plants are found, air pollution is likely. Air pollution also can be caused by volcanic eruptions, wind-blown dust and sand, forest fires, and the evaporation of paints and other chemicals.

Smog is a form of air pollution created when sunlight reacts with pollutants produced by burning fuels. It can irritate the eyes and make breathing difficult for people with asthma or other lung diseases. Smog can be reduced if people take buses or trains instead of driving or if they use vehicles, such as electric cars, that produce fewer pollutants than gasoline-powered vehicles.

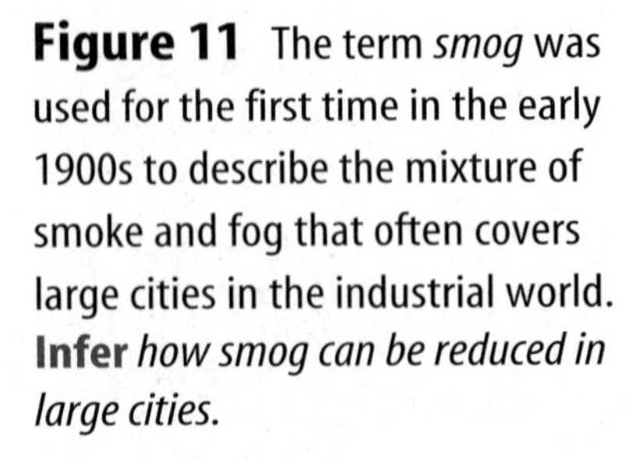

Figure 11 The term *smog* was used for the first time in the early 1900s to describe the mixture of smoke and fog that often covers large cities in the industrial world. **Infer** *how smog can be reduced in large cities.*

Figure 12 Compare these two photographs of the same statue. The photo on the left was taken before acid rain became a problem. The photo on the right shows acid rain damage. The pH scale, shown below, indicates whether a solution is acidic or basic.

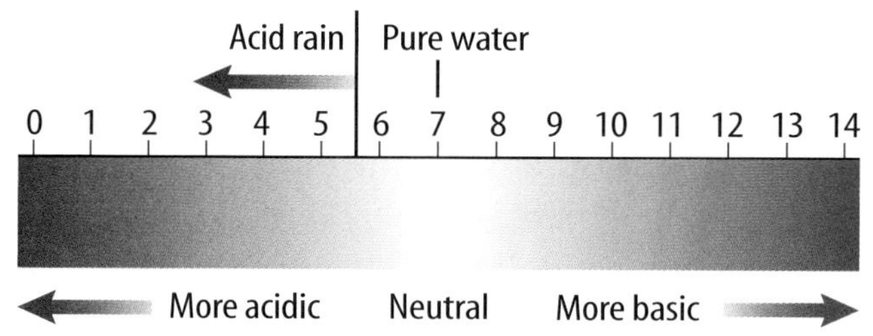

Acid Precipitation

INTEGRATE Chemistry

Water vapor condenses on dust particles in the air to form droplets that combine to create clouds. Eventually, the droplets become large enough to fall to the ground as precipitation—mist, rain, snow, sleet, or hail. Air pollutants from the burning of fossil fuels can react with water in the atmosphere to form strong acids. Acidity is measured by a value called pH, as shown in **Figure 12. Acid precipitation** has a pH below 5.6.

Effects of Acid Rain Acid precipitation washes nutrients from the soil, which can lead to the death of trees and other plants. Runoff from acid rain that flows into a lake or pond can lower the pH of the water. If algae and microscopic organisms cannot survive in the acidic water, fish and other organisms that depend on them for food also die.

Preventing Acid Rain Sulfur from burning coal and nitrogen oxides from vehicle exhaust are the pollutants primarily responsible for acid rain. Using low-sulfur fuels, such as natural gas or low-sulfur coal, can help reduce acid precipitation. However, these fuels are less plentiful and more expensive than high-sulfur coal. Smokestacks that remove sulfur dioxide before it enters the atmosphere also help. Reducing automobile use and keeping car engines properly tuned can reduce acid rain caused by nitrogen oxide pollution. The use of electric cars, or hybrid-fuel cars that can run on electricity as well as gasoline, also could help.

Measuring Acid Rain

Procedure

1. Collect **rainwater** by placing a clean **cup** outdoors. Do not collect rainwater that has been in contact with any object or organism.
2. Dip a piece of **pH indicator paper** into the sample.
3. Compare the color of the paper to the pH chart provided. Record the pH of the rainwater.
4. Use separate pieces of pH paper to test the pH of **tap water** and **distilled water.** Record these results.

Analysis

1. Is the rainwater acidic, basic, or neutral?
2. How does the pH of the rainwater compare with the pH of tap water? With the pH of distilled water?

Topic: Global Warming
Visit life.msscience.com for Web links to information about global warming.

Activity Describe three possible impacts of global warming. Provide one fact that supports global warming and one fact that does not.

Greenhouse Effect

Sunlight travels through the atmosphere to Earth's surface. Some of this sunlight normally is reflected back into space. The rest is trapped by certain atmospheric gases, as shown in **Figure 13.** This heat-trapping feature of the atmosphere is the **greenhouse effect.** Without it, temperatures on Earth probably would be too cold to support life.

Atmospheric gases that trap heat are called greenhouse gases. One of the most important greenhouse gases is carbon dioxide (CO_2). CO_2 is a normal part of the atmosphere. It is also a waste product that forms when fossil fuels are burned. Over the past century, more fossil fuels have been burned than ever before, which is increasing the percentage of CO_2 in the atmosphere. The atmosphere might be trapping more of the Sun's heat, making Earth warmer. A rise in Earth's average temperature, possibly caused by an increase in greenhouse gases, is known as global warming.

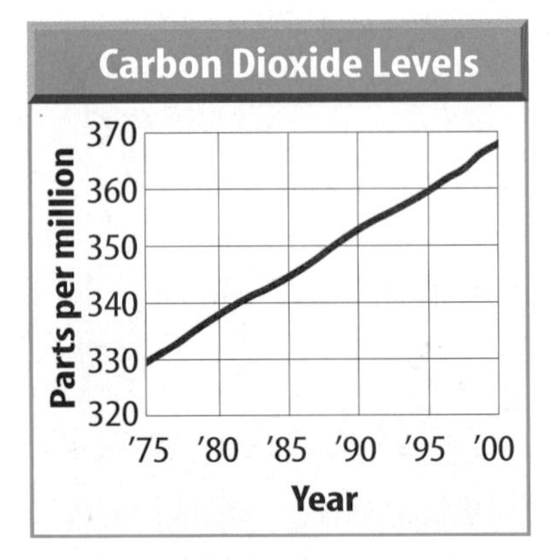

Global Warming Temperature data collected from 1895 through 1995 indicate that Earth's average temperature increased about 1°C during that 100-year period. No one is certain whether this rise was caused by human activities or is a natural part of Earth's weather cycle. What kinds of changes might be caused by global warming? Changing rainfall patterns could alter ecosystems and affect the kinds of crops that can be grown in different parts of the world. The number of storms and hurricanes might increase. The polar ice caps might begin to melt, raising sea levels and flooding coastal areas. Warmer weather might allow tropical diseases, such as malaria, to become more widespread. Many people feel that the possibility of global warming is a good reason to reduce fossil fuel use.

Figure 13 The moment you step inside a greenhouse, you feel the results of the greenhouse effect. Heat trapped by the glass walls warms the air inside. In a similar way, atmospheric greenhouse gases trap heat close to Earth's surface.

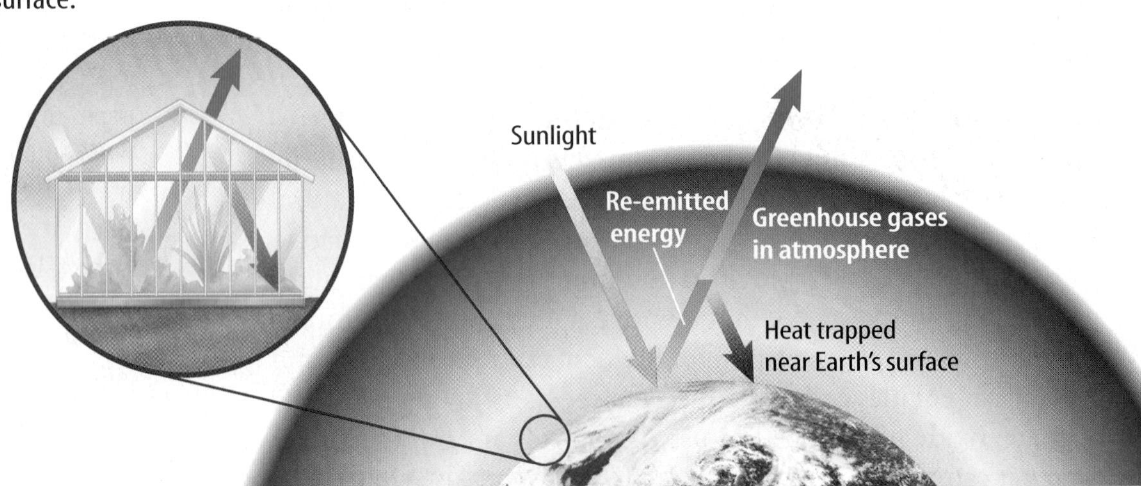

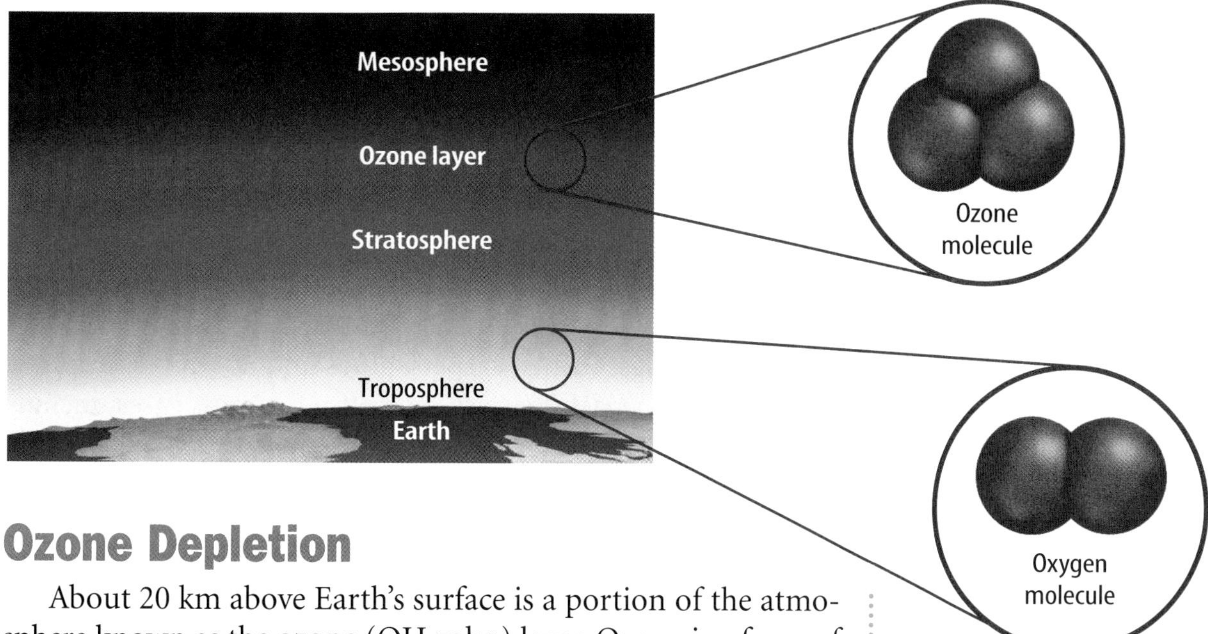

Figure 14 The atmosphere's ozone layer absorbs large amounts of UV radiation, preventing it from reaching Earth's surface. Ozone molecules are made of three oxygen atoms. They are formed in a chemical reaction between sunlight and oxygen. The oxygen you breathe has two oxygen atoms in each molecule.
Infer *what will happen if the ozone layer continues to thin.*

Ozone Depletion

About 20 km above Earth's surface is a portion of the atmosphere known as the ozone (OH zohn) layer. Ozone is a form of oxygen, as shown in **Figure 14.** The ozone layer absorbs some of the Sun's harmful ultraviolet (UV) radiation. UV radiation can damage living cells.

Every year, the ozone layer temporarily becomes thinner over each polar region during its spring season. The thinning of the ozone layer is called **ozone depletion.** This problem is caused by certain pollutant gases, especially chlorofluorocarbons (klor oh FLOR oh kar bunz) (CFCs). CFCs are used in the cooling systems of refrigerators, freezers, and air conditioners. When CFCs leak into the air, they slowly rise into the atmosphere until they arrive at the ozone layer. CFCs react chemically with ozone, breaking apart the ozone molecules.

UV Radiation Because of ozone depletion, the amount of UV radiation that reaches Earth's surface could be increasing. UV radiation could be causing a rise in the number of skin cancer cases in humans. It also might be harming other organisms. The ozone layer is so important to the survival of life on Earth that world governments and industries have agreed to stop making and using CFCs.

Ozone that is high in the upper atmosphere protects life on Earth. Near Earth's surface though, it can be harmful. Ozone is produced when fossil fuels are burned. This ozone stays in the lower atmosphere, where it pollutes the air. Ozone damages the lungs and other sensitive tissues of animals and plants. For example, it can cause the needles of a Ponderosa pine to drop, harming growth.

What is the difference between ozone in the upper atmosphere and ozone in the lower atmosphere?

Air Quality Carbon monoxide enters the body through the lungs. It attaches to red blood cells, preventing the cells from absorbing oxygen. In your Science Journal, explain why heaters and barbecues designed for outdoor use never should be used indoors.

Indoor Air Pollution

Air pollution can occur indoors. Today's buildings are better insulated to conserve energy. However, better insulation reduces the flow of air into and out of a building, so air pollutants can build up indoors. For example, burning cigarettes release hazardous particles and gases into the air. Even nonsmokers can suffer ill effects from secondhand cigarette smoke. As a result, smoking no longer is allowed in many public and private buildings. Paints, carpets, glues and adhesives, printers, and photocopy machines also give off dangerous gases, including formaldehyde. Like cigarette smoke, formaldehyde is a carcinogen, which means it can cause cancer.

Carbon Monoxide Carbon monoxide (CO) is a poisonous gas that is produced whenever charcoal, natural gas, kerosene, or other fuels are burned. CO poisoning can cause serious illness or death. Fuel-burning stoves and heaters must be designed to prevent CO from building up indoors. CO is colorless and odorless, so it is difficult to detect. Alarms that provide warning of a dangerous buildup of CO are being used in more and more homes.

Radon Radon is a naturally occurring, radioactive gas that is given off by some types of rock and soil, as shown in **Figure 15.** Radon has no color or odor. It can seep into basements and the lower floors of buildings. Radon exposure is the second leading cause of lung cancer in this country. A radon detector sounds an alarm when levels of the gas in indoor air become too high. If radon is present, increasing a building's ventilation can eliminate any damaging effects.

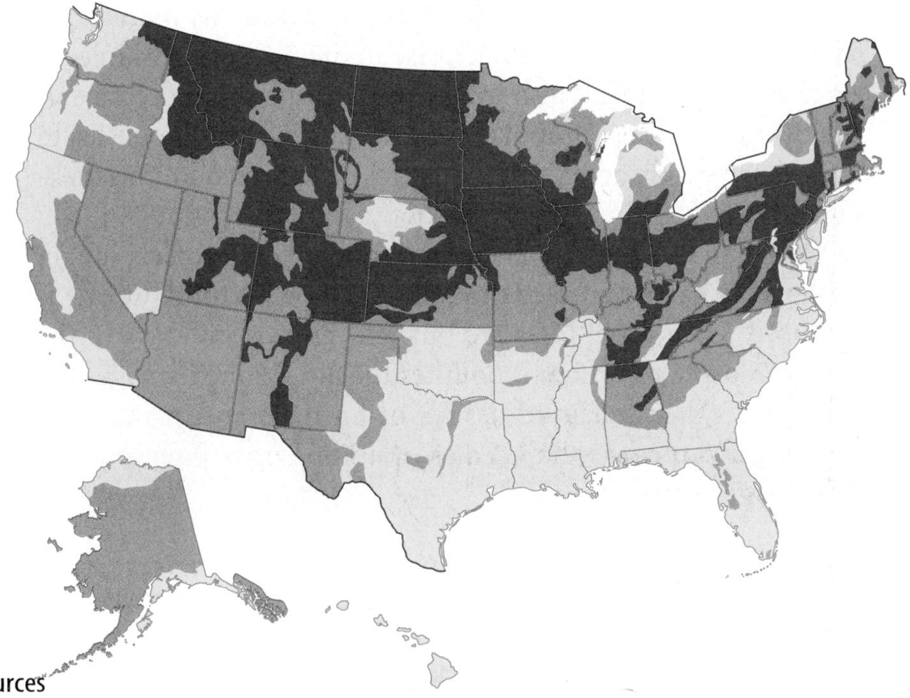

Figure 15 The map shows the potential for radon exposure in different parts of the United States. **Identify** *the area of the country with soils that produce the most radon gas.*

Geologic radon potential
Low Moderate High

When rain falls on roads and parking lots, it can wash oil and grease onto the soil and into nearby streams.

Rain can wash agricultural pesticides and fertilizers into lakes, streams, or oceans.

Industrial wastes are sometimes released directly into surface waters.

Figure 16 Pollution of surface waters can occur in several ways, as shown above.

Water Pollution

Pollutants enter water, too. Air pollutants can drift into water or be washed out of the sky by rain. Rain can wash land pollutants into waterways, as shown in **Figure 16.** Wastewater from factories and sewage-treatment plants often is released into waterways. In the United States and many other countries, laws require that wastewater be treated to remove pollutants before it is released. But, in many parts of the world, wastewater treatment is not always possible. Pollution also enters water when people dump litter or waste materials into rivers, lakes, and oceans.

Surface Water Some water pollutants poison fish and other wildlife, and can be harmful to people who swim in or drink the water. For example, chemical pesticides sprayed on farmland can wash into lakes and streams. These chemicals can harm the insects that fish, turtles, or frogs rely on for food. Shortages of food can lead to deaths among water-dwelling animals. Some pollutants, especially those containing mercury and other metals, can build up in the tissues of fish. Eating contaminated fish and shellfish can transfer these metals to people, birds, and other animals. In some areas, people are advised not to eat fish or shellfish taken from polluted waterways.

Algal blooms are another water pollution problem. Raw sewage and excess fertilizer contain large amounts of nitrogen. If they are washed into a lake or pond, they can cause the rapid growth of algae. When the algae die, they are decomposed by huge numbers of bacteria that use up much of the oxygen in the water. Fish and other organisms can die from a lack of oxygen in the water.

Figure 17 In 1996, the oil tanker *Sea Empress* spilled more than 72 million kg of oil into the sea along the coast of Wales. More than $40 million was spent on the cleanup effort, but thousands of ocean organisms were destroyed, including birds, fish, and shellfish.

Ocean Water Rivers and streams eventually flow into oceans, bringing their pollutants along. Also, polluted water can enter the ocean in coastal areas where factories, sewage-treatment plants, or shipping activities are located. Oil spills are a well-known ocean pollution problem. About 4 billion kg of oil are spilled into ocean waters every year. Much of that oil comes from ships that use ocean water to wash out their fuel tanks. Oil also can come from oil tanker wrecks, as shown in **Figure 17.**

Groundwater Pollution can affect water that seeps underground, as shown in **Figure 18.** Groundwater is water that collects between particles of soil and rock. It comes from precipitation and runoff that soaks into the soil. This water can flow slowly through permeable layers of rock called aquifers. If this water comes into contact with pollutants as it moves through the soil and into an aquifer, the aquifer could become polluted. Polluted groundwater is difficult—and sometimes impossible—to clean. In some parts of the country, chemicals leaking from underground storage tanks have created groundwater pollution problems.

Figure 18 Water from rainfall slowly filters through sand or soil until it is trapped in underground aquifers. Pollutants picked up by the water as it filters through the soil can contaminate water wells.

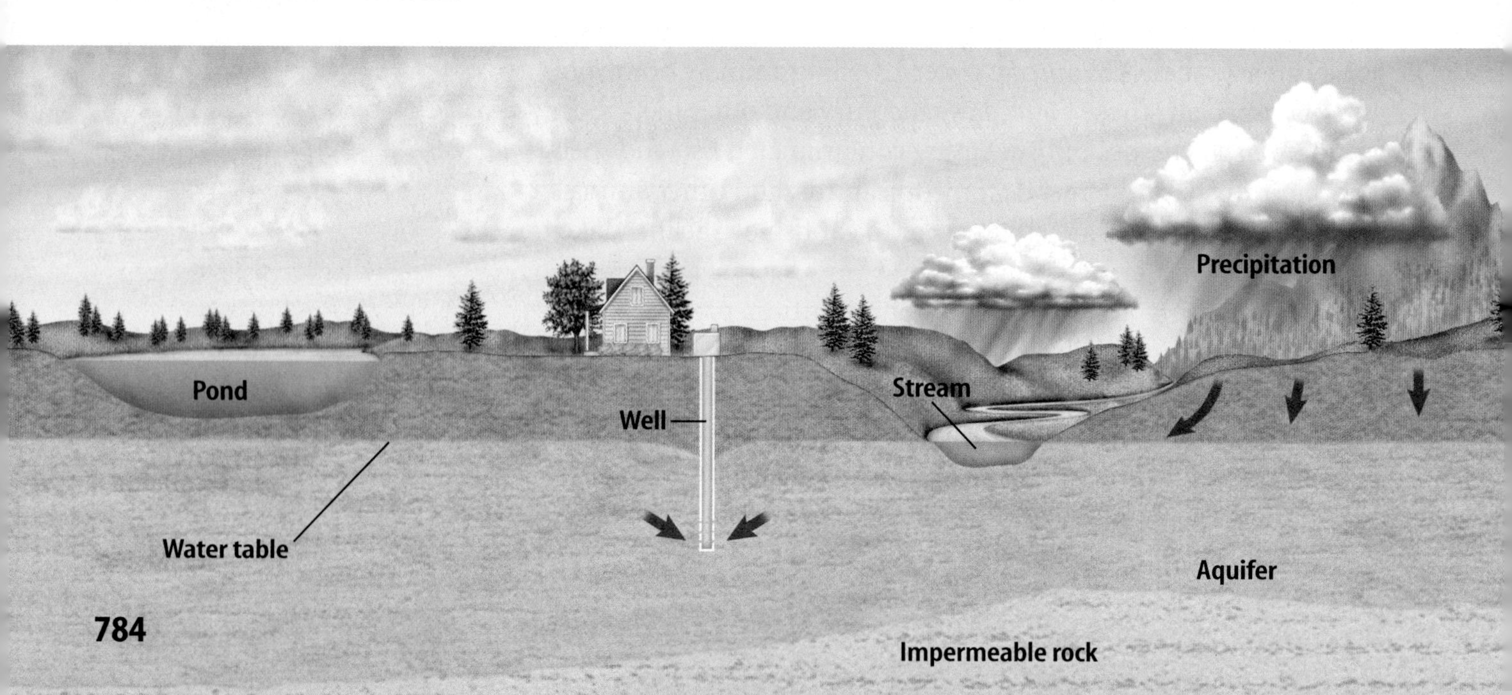

Contour plowing reduces the downhill flow of water.

Figure 19 The farming methods shown here help prevent soil erosion.
Infer *why soil erosion is a concern for farmers.*

On steep hillsides, flat areas called terraces reduce downhill flow.

In strip cropping, cover crops are planted between rows to reduce wind erosion.

In no-till farming, soil is never left bare.

Soil Loss

Fertile topsoil is important to plant growth. New topsoil takes hundreds or thousands of years to form. The Launch Lab at the beginning of this chapter shows that rain washes away loose topsoil. Wind also blows it away. The movement of soil from one place to another is called **erosion** (ih ROH zhun). Eroded soil that washes into a river or stream can block sunlight and slow photosynthesis. It also can harm fish, clams, and other organisms. Erosion is a natural process, but human activities increase it. When a farmer plows a field or a forest is cut down, soil is left bare. Bare soil is more easily carried away by rain and wind. **Figure 19** shows some methods farmers use to reduce soil erosion.

Soil Pollution

Soil can become polluted when air pollutants drift to the ground or when water leaves pollutants behind as it flows through the soil. Soil also can be polluted when people toss litter on the ground or dispose of trash in landfills.

Solid Wastes What happens to the trash you throw out every week? What do people do with old refrigerators, TVs, and toys? Most of this solid waste is dumped in landfills. Most landfills are designed to seal out air and water. This helps prevent pollutants from seeping into surrounding soil, but it slows normal decay processes. Even food scraps and paper, which usually break down quickly, can last for decades in a landfill. In populated areas, landfills fill up quickly. Reducing the amount of trash people generate can reduce the need for new landfills.

Figure 20 Leftover paints, batteries, pesticides, drain cleaners, and medicines are hazardous wastes that should not be discarded in the trash. They should never be poured down a drain, onto the ground, or into a storm sewer. Most communities have collection facilities where people can dispose of hazardous materials like these.

Hazardous Wastes Waste materials that are harmful to human health or poisonous to living organisms are **hazardous wastes.** They include dangerous chemicals, such as pesticides, oil, and petroleum-based solvents used in industry. They also include radioactive wastes from nuclear power plants, from hospitals that use radioactive materials to treat disease, and from nuclear weapons production. Many household items also are considered hazardous, such as those shown in **Figure 20.** If these materials are dumped into landfills, they could seep into the soil, surface water, or groundwater over time. Hazardous wastes usually are handled separately from trash. They are treated in ways that prevent environmental pollution.

Reading Check *What are hazardous wastes?*

section 2 review

Summary

Air Pollution and Acid Precipitation

- Vehicles, volcanoes, forest fires, and even wind-blown dust and sand can cause air pollution.
- Acid rain washes nutrients from the soil, which can harm plants.

Greenhouse Effect and Ozone Depletion

- CO_2 is a greenhouse gas that helps warm Earth.
- The ozone layer protects life on Earth.

Indoor Air Pollution, Water Pollution, Soil Loss, and Soil Pollution

- Pollutants can build up inside of buildings.
- There are many sources of water pollutants.
- Wind and rain can erode bare soil.
- Pollutants in soil decay more slowly than in air.

Self Check

1. **List** four ways that air pollution affects the environment.
2. **Explain** how an algal bloom can affect other pond organisms.
3. **Describe** possible causes and effects of ozone depletion.
4. **Think Critically** How could hazardous wastes in landfills eventually affect groundwater?

Applying Math

5. **Solve a One-Step Equation** A solution of pH 4 is 10 times more acidic than one of pH 5, and it is 10 times more acidic than a solution of pH 6. How many times more acidic is the solution of pH 4 than the one of pH 6?

The Greenhouse Effect

You can create models of Earth with and without heat-reflecting green-house gases. Then, experiment with the models to observe the greenhouse effect.

Real-World Question

How does the greenhouse effect influence temperatures on Earth?

Goals

- **Observe** the greenhouse effect.
- **Describe** the effect that a heat source has on an environment.

Materials

1-L clear-plastic, soft-drink bottles with tops cut off and labels removed (2)
thermometers (2)
**temperature probe*
potting soil
masking tape
plastic wrap
rubber band
lamp with 100-W lightbulb
watch or clock with second hand
**Alternate materials*

Safety Precautions

Procedure

1. Copy the data table and use it to record your temperature measurements.
2. Put an equal volume of potting soil in the bottom of each container.
3. Use masking tape to attach a thermometer to the inside of each container. Place each thermometer at the same height above the soil. Shield each thermometer bulb by putting a double layer of masking tape over it.

Changes in Temperature

Time (min)	Open Container Temperature (°C)	Closed Container Temperature (°C)
0		
2	Do not write in this book.	
4		
6		

4. Seal the top of one container with plastic wrap held in place with a rubber band.
5. Place the lamp with the exposed 100-W lightbulb between the two containers and exactly 1 cm away from each. Do not turn on the light.
6. Let the setup sit for 5 min, then record the temperature in each container.
7. Turn on the light. Record the temperature in each container every 2 min for 15 min to 20 min. Graph the results.

Conclude and Apply

1. **Compare and contrast** temperatures in each container at the end of the experiment.
2. **Infer** What does the lightbulb represent in this experimental model? What does the plastic wrap represent?

Communicating Your Data

Average the data obtained in the experiments conducted by all the groups in your class. Prepare a line graph of these data. **For more help, refer to the Science Skill Handbook.**

section 3

The Three Rs of Conservation

as you read

What You'll Learn

- **Recognize** ways you can reduce your use of natural resources.
- **Explain** how you can reuse resources to promote conservation.
- **Describe** how many materials can be recycled.

Why It's Important

Conservation preserves resources and reduces pollution.

Review Vocabulary
reprocessing: to subject to a special process or treatment in preparation for reuse

New Vocabulary
- recycling

Conservation

A teacher travels to school in a car pool. In the school cafeteria, students place glass bottles and cans in separate containers from the rest of the garbage. Conservation efforts like these can help prevent shortages of natural resources, slow growth of landfills, reduce pollution levels, and save people money. Every time a new landfill is created, an ecosystem is disturbed. Reducing the need for landfills is a major benefit of conservation. The three Rs of conservation are reduce, reuse, and recycle.

Reduce

You contribute to conservation whenever you reduce your use of natural resources. You use less fossil fuel when you walk or ride a bicycle instead of taking the bus or riding in a car. If you buy a carton of milk, reduce your use of petroleum by telling the clerk you don't need a plastic bag to carry it in.

You also can avoid buying things you don't need. For example, most of the paper, plastic, and cardboard used to package items for display on store shelves is thrown away as soon as the product is brought home. You can look for products with less packaging or with packaging made from recycled materials. What are some other ways you can reduce your use of natural resources?

Figure 21 Worn-out automobile tires can have other useful purposes.

Reuse

Another way to help conserve natural resources is to use items more than once. Reusing an item means using it again without changing it or reprocessing it, as shown in **Figure 21.** Bring reusable canvas bags to the grocery store to carry home your purchases. Donate clothes you've outgrown to charity so that others can reuse them. Take reusable plates and utensils on picnics instead of disposable paper items.

Recycle

If you can't avoid using an item, and if you can't reuse it, the next best thing is to recycle it. **Recycling** is a form of reuse that requires changing or reprocessing an item or natural resource. If your city or town has a curbside recycling program, you already separate recyclables from the rest of your garbage. Materials that can be recycled include glass, metals, paper, plastics, and yard and kitchen waste.

Reading Check *How is recycling different from reusing?*

Plastics Plastic is more difficult to recycle than other materials, mainly because several types of plastic are in use. A recycle code marked on every plastic container indicates the type of plastic it is made of. Plastic soft-drink bottles, like the one shown in **Figure 22,** are made of type 1 plastic and are the easiest to recycle. Most plastic bags are made of type 2 or type 4 plastic; they can be reused as well as recycled. Types 6 and 7 can't be recycled at all because they are made of a mixture of different plastics. Each type of plastic must be separated carefully before it is recycled because a single piece of a different type of plastic can ruin an entire batch.

Figure 22 Many soft-drink bottles are made of PETE, which is the most common type of recyclable plastic. It can be melted down and spun into fibers to make carpets, paintbrushes, rope, and clothing.
Identify *other products made out of recycled materials.*

Topic: Recycling

Visit life.msscience.com for Web links to information about recycling bottles and cans.

Activity Write one argument in support of a money deposit for bottles and cans and one argument against it. Provide data to support one of your arguments.

Metals The manufacturing industry has been recycling all kinds of metals, especially steel, for decades. At least 25 percent of the steel in cans, appliances, and automobiles is recycled steel. Up to 100 percent of the steel in plates and beams used to build skyscrapers is made from reprocessed steel. About one metric ton of recycled steel saves about 1.1 metric tons of iron ore and 0.5 metric ton of coal. Using recycled steel to make new steel products reduces energy use by 75 percent. Other metals, including iron, copper, aluminum, and lead also can be recycled.

You can conserve metals by recycling food cans, which are mostly steel, and aluminum cans. It takes less energy to make a can from recycled aluminum than from raw materials. Also, remember that recycled cans do not take up space in landfills.

Glass When sterilized, glass bottles and jars can be reused. They also can be melted and re-formed into new bottles, especially those made of clear glass. Most glass bottles already contain at least 25 percent recycled glass. Glass can be recycled again and again. It never needs to be thrown away. Recycling about one metric ton of glass saves more than one metric ton of mineral resources and reduces the energy used to make new glass by 25 percent or more.

Applying Science

What items are you recycling at home?

Many communities have recycling programs. Recyclable items may be picked up at the curbside, taken to a collection site, or the resident may hire a licensed recycling handler to pick them up. What do you recycle in your home?

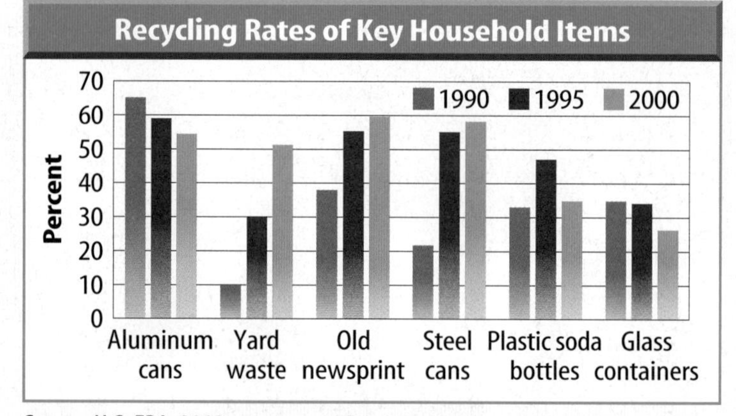

Source: U.S. EPA, 2003

Identifying the Problem

This bar graph shows the recycling rates in the U. S. of six types of household items for the years 1990, 1995, and 2000. What are you and your classmates' recycling rates?

Solving the Problem

For one week, list each glass, plastic, and aluminum item you use. Note which items you throw away and which ones you recycle. Calculate the percentage of glass, plastic, and aluminum you recycled. How do your percentages compare with those on the graph?

Paper Used paper is recycled into paper towels, insulation, newsprint, cardboard, and stationery. Ranchers and dairy farmers sometimes use shredded paper instead of straw for bedding in barns and stables. Used paper can be made into compost. Recycling about one metric ton of paper saves 17 trees, more than 26,000 L of water, close to 1,900 L of oil, and more than 4,000 kW of electric energy. You can do your part by recycling newspapers, notebook and printer paper, cardboard, and junk mail.

Figure 23 Composting is a way of turning plant material you would otherwise throw away into rich garden soil. Dry leaves and weeds, grass clippings, vegetable trimmings, and nonmeat food scraps can be composted.

Reading Check *What nonrenewable resource(s) do you conserve by recycling paper?*

Compost Grass clippings, leaves, and fruit and vegetable scraps that are discarded in a landfill can remain there for decades without breaking down. The same items can be turned into soil-enriching compost in just a few weeks, as shown in **Figure 23.** Many communities distribute compost bins to encourage residents to recycle fruit and vegetable scraps and yard waste.

Buy Recycled People have become so good at recycling that recyclable materials are piling up, just waiting to be put to use. You can help by reading labels when you shop and choosing products that contain recycled materials. What other ways of recycling natural resources can you think of?

section 3 review

Summary

Conservation
- The three Rs of conservation are reduce, reuse, and recycle.

Reduce
- You can contribute to conservation by reducing your use of natural resources.

Reuse
- Some items can be used more than once, such as reusable canvas bags for groceries.

Recycle
- Some items can be recycled, including some plastics, metal, glass, and paper.
- Grass clippings, leaves, and fruit and vegetable scraps can be composted into rich garden soil.

Self Check

1. **Describe** at least three actions you could take to reduce your use of natural resources.
2. **Describe** how you could reuse three items people usually throw away.
3. **Think Critically** Why is reusing something better than recycling it?

Applying Skills

4. **Make and Use Tables** Make a table of data of the number of aluminum cans thrown away in the United States: 2.7 billion in 1970; 11.1 billion in 1974; 21.3 billion in 1978; 22.7 billion in 1982; 35.0 billion in 1986; 33.8 billion in 1990; 38.5 billion in 1994; 45.5 billion in 1998; 50.7 billion in 2001.

Solar Cooking

Goals

- **Research** designs for solar panel cookers or box cookers.
- **Design** a solar cooker that can be used to cook food.
- **Plan** an experiment to measure the effectiveness of your solar cooker.

Possible Materials

poster board
cardboard boxes
aluminum foil
string
wire coat hangers
clear plastic sheets
oven bags
black cookware
thermometer
stopwatch
timer
glue
tape
scissors

Alternate materials

Safety Precautions

WARNING: *Be careful when cutting materials. Your solar cooker will get hot. Use insulated gloves or tongs to handle hot objects.*

Real-World Question

The disappearance of forests in some places on Earth has made firewood extremely difficult and expensive to obtain. People living in these regions often have to travel long distances or sell some of their food to get firewood. This can be a serious problem for people who may not have much food to begin with. Is there a way they could cook food without using firewood? How would you design and build a cooking device that uses the Sun's energy?

Make the Model

1. **Design** a solar cooker. In your Science Journal, explain why you chose this design and draw a picture of it.
2. **Write** a summary explaining how you will measure the effectiveness of your solar cooker. What will you measure? How will you collect and organize your data? How will you present your results?

3. **Compare** your solar cooker design to those of other students.
4. Share your experimental plan with students in your class. Discuss the reasoning behind your plan. Be specific about what you intend to test and how you are going to test it.
5. Make sure your teacher approves your plan before you start working on your model.
6. Using all of the information you have gathered, construct a solar cooker that follows your design.

Test the Model

1. **Test** your design to determine how well it works. Try out a classmate's design. How do the two compare?

Analyze Your Data

1. Combine the results for your entire class and decide which type of solar cooker was most effective. How could you design a more effective solar cooker, based on what you learned from this activity?
2. **Infer** Do you think your results might have been different if you tested your solar cooker on a different day? Explain. Why might a solar cooker be more useful in some regions of the world than in others?

Conclude and Apply

1. **Infer** Based on what you've read and the results obtained by you and your classmates, do you think that your solar cooker could boil water? Explain.
2. **Compare** the amount of time needed to cook food with a solar cooker and with more traditional cooking methods. Assuming plenty of sunlight is available, would you prefer to use a solar cooker or a traditional oven? Explain.

Prepare a demonstration showing how to use a solar cooker. Present your demonstration to another class of students or to a group of friends or relatives. **For more help, refer to the Science Skill Handbook.**

Beauty Plagiarized

by Amitabha Mukerjee

I wandered lonely as a cloud –
Except for a motorboat,
Nary a soul in sight.
Beside the lake beneath the trees,
Next to the barbed wire fence,
There was a picnic table
And beer bottle caps from many years.
A boat ramp to the left,
And the chimney from a power station on the
other side,
A summer haze hung in the air,
And the lazy drone of traffic far away.

Crimson autumn of mists and mellow fruitfulness
Blue plastic covers the swimming pools
The leaves fall so I can see
Dark glass reflections in the building
That came up
where the pine cones crunched underfoot . . .

And then it is snow
White lining on trees and rooftops . . .
And through my windshield wipers
The snow is piled dark and grey . . .
Next to my driveway where I check my mail
Little footprints on fresh snow —
A visiting rabbit.

I knew a bank where the wild thyme blew
Over-canopied with luscious woodbine
It is now a landfill —
Fermentation of civilization
Flowers on TV
Hyacinth rose tulip chrysanthemum
Acres of colour
Wind up wrapped in decorous plastic,
In this landfill where oxlips grew. . .

Understanding Literature

Cause and Effect Recognizing cause-and-effect relationships can help you make sense out of what you read. One event causes another event. The second event is the effect of the first event. In the poem, the author describes the causes and effects of pollution and waste. What effects do pollution and the use of nonrenewable resources have on nature in the poem?

Respond to the Reading

1. To plagiarize is to copy without giving credit to the source. In this poem, who or what has plagiarized beauty?
2. What do the four verses in the poem correspond to?
3. **Linking Science and Writing** Write a poem that shows how conservation methods could restore the beauty in nature.

The poet makes a connection between the four seasons of the year and the pollution and waste products created by human activity, or civilization. For example, in the spring, a landfill for dumping garbage replaces a field of wildflowers. Describing four seasons instead of one reinforces the poet's message that the beauty of nature has been stolen, or plagiarized.

Reviewing Main Ideas

Section 1 Resources

1. Natural resources are the parts of the environment that supply materials needed for the survival of living organisms.
2. Renewable resources are being replaced continually by natural processes.
3. Nonrenewable resources cannot be replaced or are replaced very slowly.
4. Energy sources include fossil fuels, wind, solar energy, geothermal energy, hydroelectric power, and nuclear power.

Section 2 Pollution

1. Most air pollution is made up of waste products from the burning of fossil fuels.
2. The greenhouse effect is the warming of Earth by a blanket of heat-reflecting gases in the atmosphere.
3. Water can be polluted by acid rain and by the spilling of oil or other wastes into waterways.
4. Solid wastes and hazardous wastes dumped on land or disposed of in landfills can pollute the soil. Erosion can cause the loss of fertile topsoil.

Section 3 The Three Rs of Conservation

1. You can reduce your use of natural resources in many ways.
2. Reusing items is an excellent way to practice conservation.
3. In recycling, materials are changed in some way so that they can be used again.
4. Materials that can be recycled include paper, metals, glass, plastics, yard waste, and nonmeat kitchen scraps.

Visualizing Main Ideas

Copy and complete the following concept map using the terms smog, acid precipitation, *and* ozone depletion.

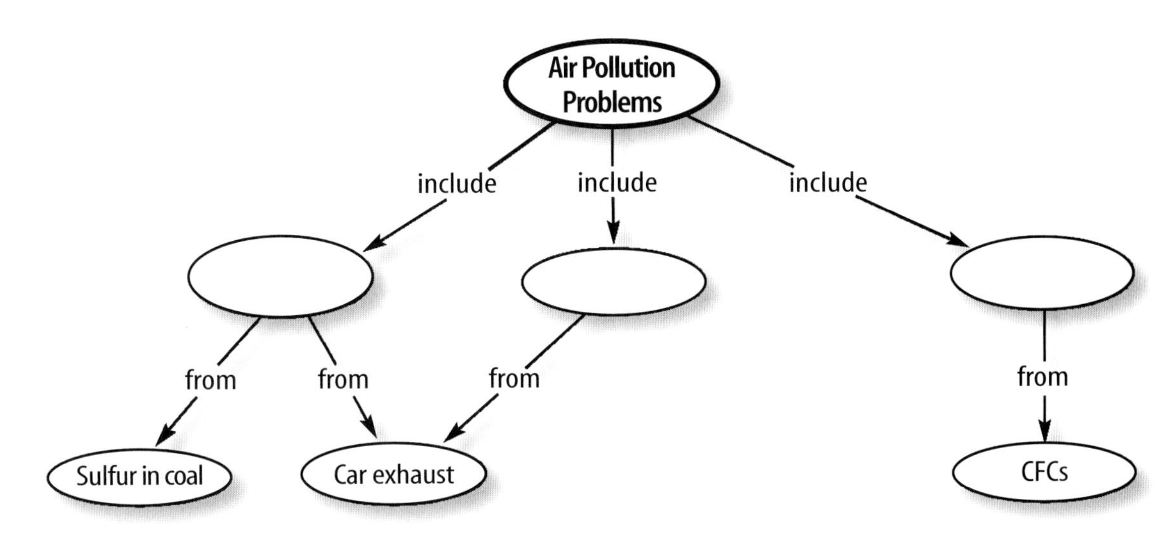

chapter 27 Review

Using Vocabulary

acid precipitation p. 779
erosion p. 785
fossil fuel p. 772
geothermal energy p. 775
greenhouse effect p. 780
hazardous waste p. 786
hydroelectric power p. 773
natural resource p. 770
nonrenewable resource p. 771
nuclear energy p. 774
ozone depletion p. 781
petroleum p. 771
pollutant p. 778
recycling p. 789
renewable resource p. 770

Explain the differences in the vocabulary words given below. Then explain how the words are related. Use complete sentences in your answers.

1. fossil fuel—petroleum
2. erosion—pollutant
3. ozone depletion—acid precipitation
4. greenhouse effect—fossil fuels
5. hazardous wastes—nuclear energy
6. hydroelectric power—fossil fuels
7. acid precipitation—fossil fuels
8. ozone depletion—pollutant
9. recycle—nonrenewable resources
10. geothermal energy—fossil fuels

Checking Concepts

Choose the word or phrase that best answers the question.

11. An architect wants to design a solar house in the northern hemisphere. For maximum warmth, which side of the house should have the most windows?
    A) north **C)** east
    B) south **D)** west
12. Of the following, which is considered a renewable resource?
    A) coal **C)** sunlight
    B) oil **D)** aluminum

Use the photo below to answer question 13.

13. Which energy resource is shown in the photo?
    A) solar energy
    B) geothermal energy
    C) hydroelectric energy
    D) photovoltaic energy
14. Which of the following is a fossil fuel?
    A) wood **C)** nuclear power
    B) oil **D)** photovoltaic cell
15. Which of the following contributes to ozone depletion?
    A) carbon dioxide **C)** CFCs
    B) radon **D)** carbon monoxide
16. What is a substance that contaminates the environment called?
    A) acid rain **C)** pollutant
    B) pollution **D)** ozone
17. If there were no greenhouse effect in Earth's atmosphere, which of the following statements would be true?
    A) Earth would be much hotter.
    B) Earth would be much colder.
    C) The temperature of Earth would be the same.
    D) The polar ice caps would melt.
18. Which of the following can change solar energy into electricity?
    A) photovoltaic cells
    B) smog
    C) nuclear power plants
    D) geothermal power plants

Science Online life.msscience.com/vocabulary_puzzlemaker

Thinking Critically

19. **Explain** how geothermal energy is used to produce electricity.
20. **Infer** why burning wood and burning fossil fuels produce similar pollutants.

Use the photos below to answer question 21.

21. **Draw a Conclusion** Which would make a better location for a solar power plant—a polar region (left) or a desert region (right)? Why?
22. **Explain** why it is beneficial to grow a different crop on soil after the major crop has been harvested.
23. **Infer** Is garbage a renewable resource? Why or why not?
24. **Summarize** Solar, nuclear, wind, water, and geothermal energy are alternatives to fossil fuels. Are they all renewable? Why or why not?
25. **Draw Conclusions** Would you save more energy by recycling or reusing a plastic bag?
26. **Recognize Cause and Effect** Forests use large amounts of carbon dioxide during photosynthesis. How might cutting down a large percentage of Earth's forests affect the greenhouse effect?
27. **Form a hypothesis** about why Americans throw away more aluminum cans each year.
28. **Compare and contrast** contour farming, terracing, strip cropping, and no-till farming.

Performance Activities

29. **Poster** Create a poster to illustrate and describe three things students at your school can do to conserve natural resources.

Applying Math

Use the table below to answer questions 30 and 31.

Estimated Recycling Rates

Item	Percent Recycled
Aluminum cans	60
Glass beverage bottles	31
Plastic soft-drink containers	37
Newsprint	56
Magazines	23

30. **Recycling Rates** Make a bar graph of the data above.
31. **Bottle Recycling** For every 1,000 glass beverage bottles that are produced, how many are recycled?
32. **Nonrenewable Resources** 45.8 billion (45,800,000,000) cans were thrown away in 2000. If it takes 33.79 cans to equal one pound and the average scrap value is $0.58/lb, then what was the total dollar value of the discarded cans?
33. **Ozone Depletion** The thin ozone layer called the "ozone hole" over Antarctica reached nearly 27,000,000 km^2 in 1998. To conceptualize this, the United States has a geographical area of 9,363,130 km^2. How much larger is the "ozone hole" in comparison to the United States?
34. **Increased CO_2 Levels** To determine the effects of increased CO_2 levels in the atmosphere, scientists increased the CO_2 concentration by 70 percent in an enclosed rain forest environment. If the initial CO_2 concentration was 430 parts per million, what was it after the increase?

chapter 27 Standardized Test Practice

Part 1 Multiple Choice

Record your answers on the answer sheet provided by your teacher or on a sheet of paper.

1. From what natural resource are plastics, paints, and gasoline made?
   A. coal **C.** iron ore
   B. petroleum **D.** natural gas

Use the illustration below to answer questions 2–4.

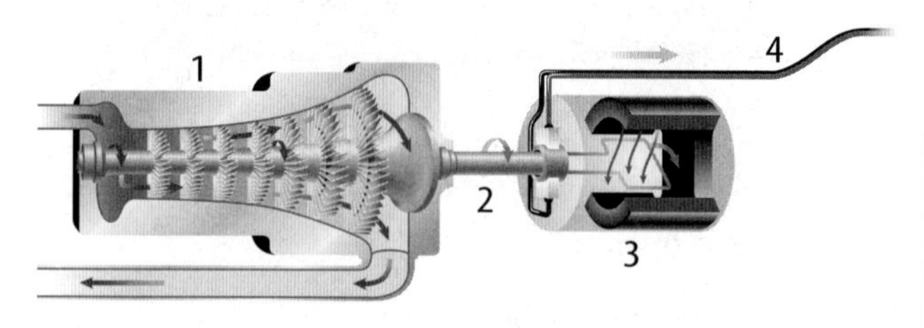

2. What is produced by the mechanism shown in the illustration?
   A. electricity **C.** petroleum
   B. coal **D.** plastic

3. In which section are the turbine blades found?
   A. 1 **C.** 3
   B. 2 **D.** 4

4. Which section represents the generator?
   A. 1 **C.** 3
   B. 2 **D.** 4

5. Which of the following is necessary for the production of hydroelectric power?
   A. wind
   B. access to a river
   C. exposure to sunlight
   D. heat from below Earth's crust

6. With which type of alternative energy are photovoltaic cells used?
   A. hydroelectric power
   B. geothermal energy
   C. nuclear energy
   D. solar energy

7. Which of the following is a type of air pollution that results when sunlight reacts with pollutants produced by burning fuels?
   A. ozones **C.** smog
   B. acid rain **D.** UV radiation

Use the photograph below to answer questions 8 and 9.

8. What is the name of the method of farming illustrated above?
   A. contour plowing **C.** terracing
   B. strip cropping **D.** no-till farming

9. What is the purpose of the method shown in the illustration?
   A. to decrease soil erosion from wind
   B. to decrease soil erosion from water flow
   C. to decrease acid rain production
   D. to increase the return of nutrients to the soil

Test-Taking Tip

Qualifying Terms Look for qualifiers in a question. Such questions are not looking for absolute answers. Qualifiers could be words such as most likely, most common, or least common.

Question 18 The qualifier in this question is *possible*. This indicates that there is uncertainty about the effects of global warming.

Part 2 | Short Response/Grid In

Record your answers on the answer sheet provided by your teacher or on a sheet of paper.

10. Give one example of a renewable source of energy and one example of a nonrenewable source of energy.

Use the illustration below to answer questions 11 and 12.

11. What type of alternative energy is the girl using in the diagram?

12. Name one benefit and one drawback to using this type of energy for cooking.

13. What are two ways that smog can be reduced?

14. A group of students collects rain outside their classroom, then tests the pH of the collected rain. The pH of the rain is 7.2. Can the students say that their rain is acid rain? Why or why not?

15. Why do we depend on the greenhouse effect for survival?

16. What is the cause of algal blooms in lakes and ponds?

Part 3 | Open Ended

Record your answers on a sheet of paper.

17. Are renewable resources always readily available? Explain.

18. What are the possible worldwide effects of global warming? What causes global warming? Why do some people think that using fossil fuels less will decrease global warming?

19. A family lives in a house that uses solar panels to heat the hot water, a wood-burning stove to heat the house, and a windmill for pumping water from a well into a tower where it is stored and then piped into the house as needed. What would be the result if there was no sunlight for two weeks?

Use the illustration below to answer question 20.

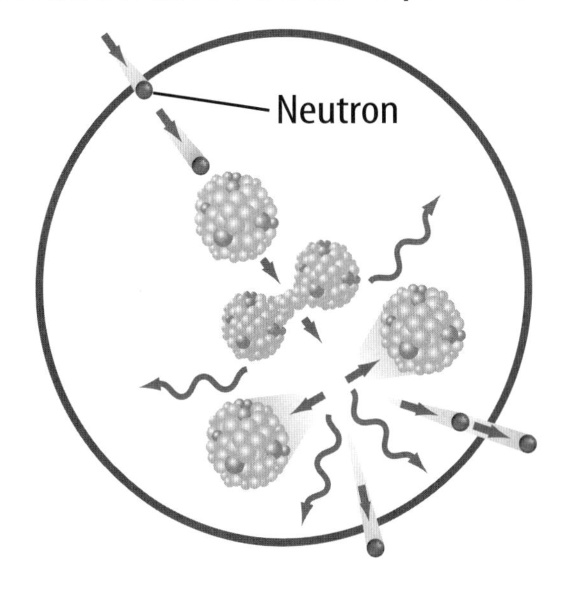

20. What does the diagram represent?

21. Explain how different kinds of plastics are recycled.

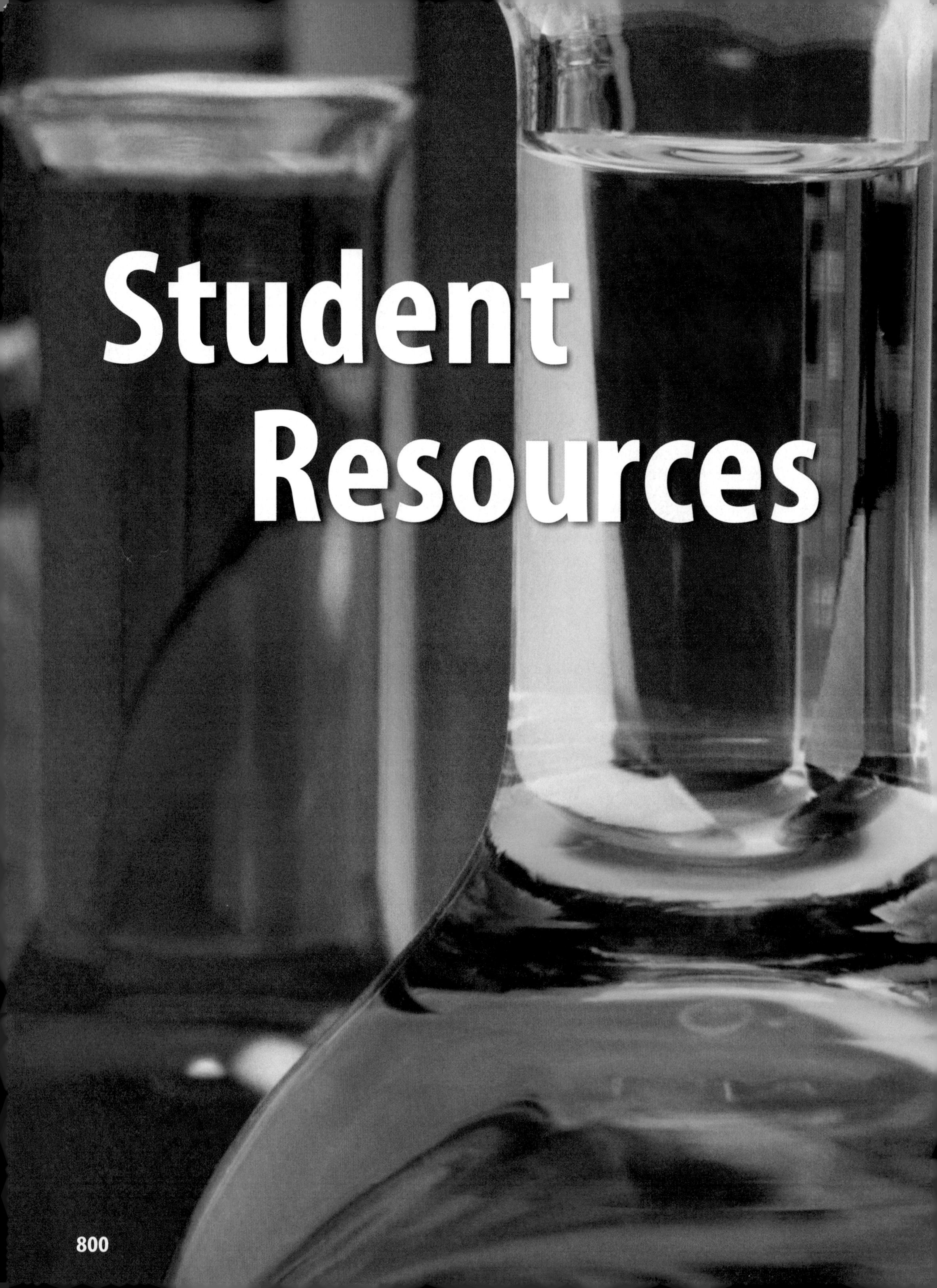

Student Resources

CONTENTS

Scientific Methods

Scientists use an orderly approach called the scientific method to solve problems. This includes organizing and recording data so others can understand them. Scientists use many variations in this method when they solve problems.

Identify a Question

The first step in a scientific investigation or experiment is to identify a question to be answered or a problem to be solved. For example, you might ask which gasoline is the most efficient.

Gather and Organize Information

After you have identified your question, begin gathering and organizing information. There are many ways to gather information, such as researching in a library, interviewing those knowledgeable about the subject, testing and working in the laboratory and field. Fieldwork is investigations and observations done outside of a laboratory.

Researching Information Before moving in a new direction, it is important to gather the information that already is known about the subject. Start by asking yourself questions to determine exactly what you need to know. Then you will look for the information in various reference sources, like the student is doing in **Figure 1.** Some sources may include textbooks, encyclopedias, government documents, professional journals, science magazines, and the Internet. Always list the sources of your information.

Figure 1 The Internet can be a valuable research tool.

Evaluate Sources of Information Not all sources of information are reliable. You should evaluate all of your sources of information, and use only those you know to be dependable. For example, if you are researching ways to make homes more energy efficient, a site written by the U.S. Department of Energy would be more reliable than a site written by a company that is trying to sell a new type of weatherproofing material. Also, remember that research always is changing. Consult the most current resources available to you. For example, a 1985 resource about saving energy would not reflect the most recent findings.

Sometimes scientists use data that they did not collect themselves, or conclusions drawn by other researchers. This data must be evaluated carefully. Ask questions about how the data were obtained, if the investigation was carried out properly, and if it has been duplicated exactly with the same results. Would you reach the same conclusion from the data? Only when you have confidence in the data can you believe it is true and feel comfortable using it.

Interpret Scientific Illustrations As you research a topic in science, you will see drawings, diagrams, and photographs to help you understand what you read. Some illustrations are included to help you understand an idea that you can't see easily by yourself, like the tiny particles in an atom in **Figure 2.** A drawing helps many people to remember details more easily and provides examples that clarify difficult concepts or give additional information about the topic you are studying. Most illustrations have labels or a caption to identify or to provide more information.

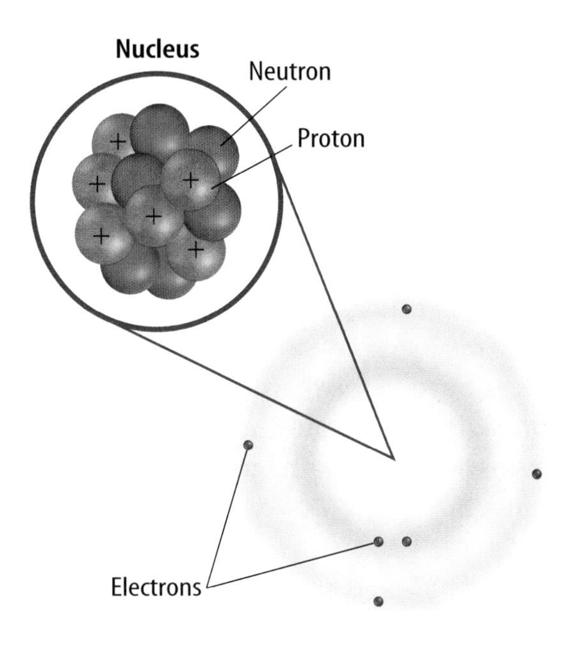

Figure 2 This drawing shows an atom of carbon with its six protons, six neutrons, and six electrons.

Concept Maps One way to organize data is to draw a diagram that shows relationships among ideas (or concepts). A concept map can help make the meanings of ideas and terms more clear, and help you understand and remember what you are studying. Concept maps are useful for breaking large concepts down into smaller parts, making learning easier.

Network Tree A type of concept map that not only shows a relationship, but how the concepts are related is a network tree, shown in **Figure 3.** In a network tree, the words are written in the ovals, while the description of the type of relationship is written across the connecting lines.

When constructing a network tree, write down the topic and all major topics on separate pieces of paper or notecards. Then arrange them in order from general to specific. Branch the related concepts from the major concept and describe the relationship on the connecting line. Continue to more specific concepts until finished.

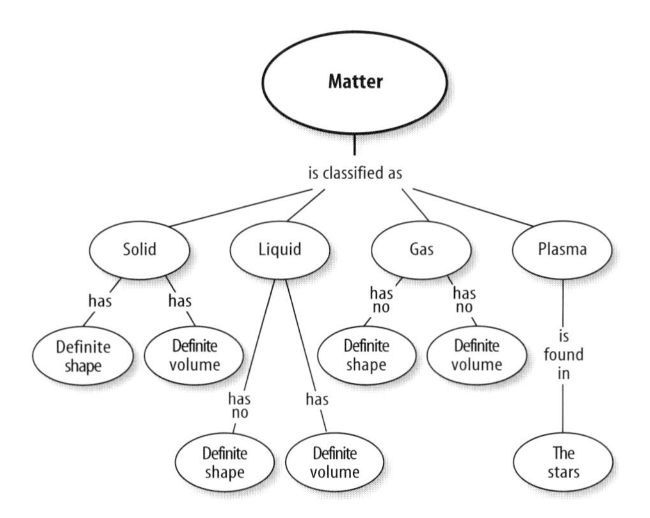

Figure 3 A network tree shows how concepts or objects are related.

Events Chain Another type of concept map is an events chain. Sometimes called a flow chart, it models the order or sequence of items. An events chain can be used to describe a sequence of events, the steps in a procedure, or the stages of a process.

When making an events chain, first find the one event that starts the chain. This event is called the initiating event. Then, find the next event and continue until the outcome is reached, as shown in **Figure 4.**

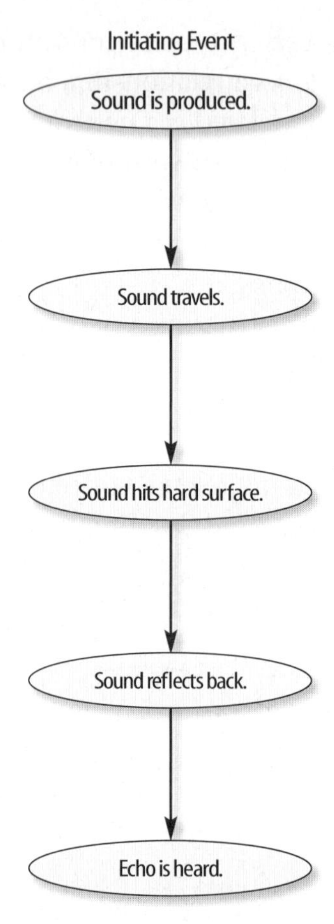

Figure 4 Events-chain concept maps show the order of steps in a process or event. This concept map shows how a sound makes an echo.

Cycle Map A specific type of events chain is a cycle map. It is used when the series of events do not produce a final outcome, but instead relate back to the beginning event, such as in **Figure 5.** Therefore, the cycle repeats itself.

To make a cycle map, first decide what event is the beginning event. This is also called the initiating event. Then list the next events in the order that they occur, with the last event relating back to the initiating event. Words can be written between the events that describe what happens from one event to the next. The number of events in a cycle map can vary, but usually contain three or more events.

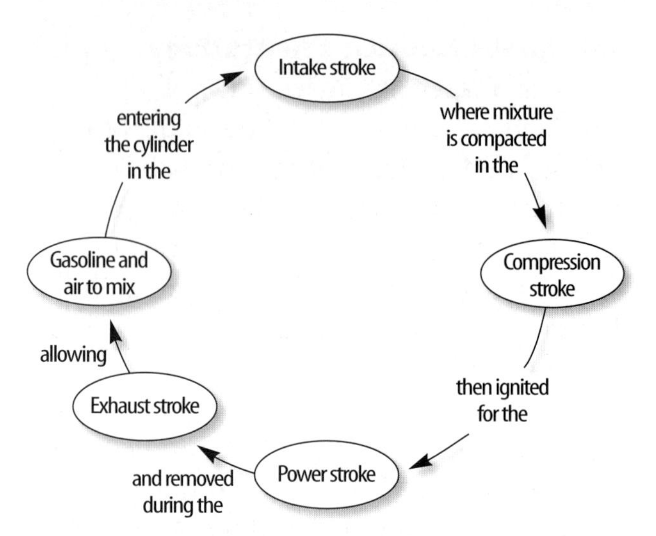

Figure 5 A cycle map shows events that occur in a cycle.

Spider Map A type of concept map that you can use for brainstorming is the spider map. When you have a central idea, you might find that you have a jumble of ideas that relate to it but are not necessarily clearly related to each other. The spider map on sound in **Figure 6** shows that if you write these ideas outside the main concept, then you can begin to separate and group unrelated terms so they become more useful.

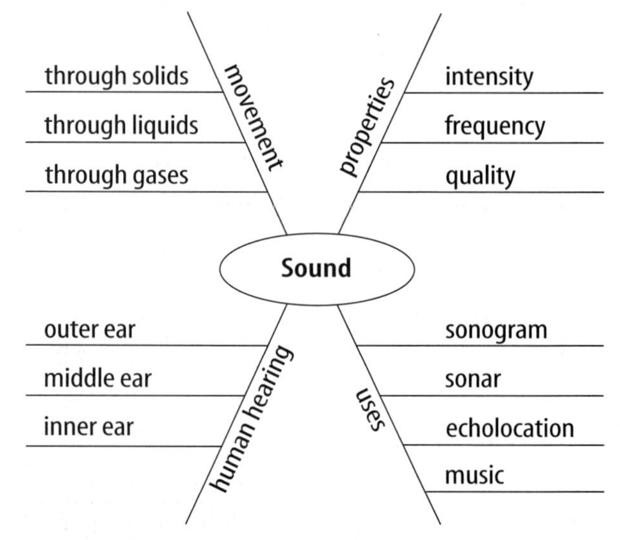

Figure 6 A spider map allows you to list ideas that relate to a central topic but not necessarily to one another.

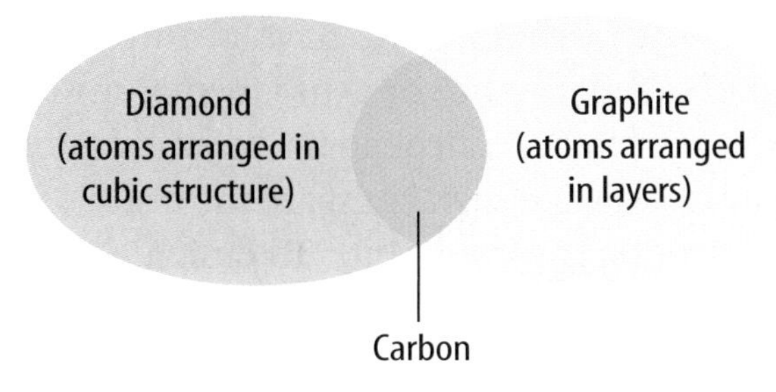

Figure 7 This Venn diagram compares and contrasts two substances made from carbon.

Venn Diagram To illustrate how two subjects compare and contrast you can use a Venn diagram. You can see the characteristics that the subjects have in common and those that they do not, shown in **Figure 7.**

To create a Venn diagram, draw two overlapping ovals that that are big enough to write in. List the characteristics unique to one subject in one oval, and the characteristics of the other subject in the other oval. The characteristics in common are listed in the overlapping section.

Make and Use Tables One way to organize information so it is easier to understand is to use a table. Tables can contain numbers, words, or both.

To make a table, list the items to be compared in the first column and the characteristics to be compared in the first row. The title should clearly indicate the content of the table, and the column or row heads should be clear. Notice that in **Table 1** the units are included.

Table 1 Recyclables Collected During Week

Day of Week	Paper (kg)	Aluminum (kg)	Glass (kg)
Monday	5.0	4.0	12.0
Wednesday	4.0	1.0	10.0
Friday	2.5	2.0	10.0

Make a Model One way to help you better understand the parts of a structure, the way a process works, or to show things too large or small for viewing is to make a model. For example, an atomic model made of a plastic-ball nucleus and pipe-cleaner electron shells can help you visualize how the parts of an atom relate to each other. Other types of models can by devised on a computer or represented by equations.

Form a Hypothesis

A possible explanation based on previous knowledge and observations is called a hypothesis. After researching gasoline types and recalling previous experiences in your family's car you form a hypothesis—our car runs more efficiently because we use premium gasoline. To be valid, a hypothesis has to be something you can test by using an investigation.

Predict When you apply a hypothesis to a specific situation, you predict something about that situation. A prediction makes a statement in advance, based on prior observation, experience, or scientific reasoning. People use predictions to make everyday decisions. Scientists test predictions by performing investigations. Based on previous observations and experiences, you might form a prediction that cars are more efficient with premium gasoline. The prediction can be tested in an investigation.

Design an Experiment A scientist needs to make many decisions before beginning an investigation. Some of these include: how to carry out the investigation, what steps to follow, how to record the data, and how the investigation will answer the question. It also is important to address any safety concerns.

Test the Hypothesis

Now that you have formed your hypothesis, you need to test it. Using an investigation, you will make observations and collect data, or information. This data might either support or not support your hypothesis. Scientists collect and organize data as numbers and descriptions.

Follow a Procedure In order to know what materials to use, as well as how and in what order to use them, you must follow a procedure. **Figure 8** shows a procedure you might follow to test your hypothesis.

Procedure

1. Use regular gasoline for two weeks.
2. Record the number of kilometers between fill-ups and the amount of gasoline used.
3. Switch to premium gasoline for two weeks.
4. Record the number of kilometers between fill-ups and the amount of gasoline used.

Figure 8 A procedure tells you what to do step by step.

Identify and Manipulate Variables and Controls In any experiment, it is important to keep everything the same except for the item you are testing. The one factor you change is called the independent variable. The change that results is the dependent variable. Make sure you have only one independent variable, to assure yourself of the cause of the changes you observe in the dependent variable. For example, in your gasoline experiment the type of fuel is the independent variable. The dependent variable is the efficiency.

Many experiments also have a control—an individual instance or experimental subject for which the independent variable is not changed. You can then compare the test results to the control results. To design a control you can have two cars of the same type. The control car uses regular gasoline for four weeks. After you are done with the test, you can compare the experimental results to the control results.

Collect Data

Whether you are carrying out an investigation or a short observational experiment, you will collect data, as shown in **Figure 9.** Scientists collect data as numbers and descriptions and organize it in specific ways.

Observe Scientists observe items and events, then record what they see. When they use only words to describe an observation, it is called qualitative data. Scientists' observations also can describe how much there is of something. These observations use numbers, as well as words, in the description and are called quantitative data. For example, if a sample of the element gold is described as being "shiny and very dense" the data are qualitative. Quantitative data on this sample of gold might include "a mass of 30 g and a density of 19.3 g/cm^3."

Figure 9 Collecting data is one way to gather information directly.

Figure 10 Record data neatly and clearly so it is easy to understand.

When you make observations you should examine the entire object or situation first, and then look carefully for details. It is important to record observations accurately and completely. Always record your notes immediately as you make them, so you do not miss details or make a mistake when recording results from memory. Never put unidentified observations on scraps of paper. Instead they should be recorded in a notebook, like the one in **Figure 10.** Write your data neatly so you can easily read it later. At each point in the experiment, record your observations and label them. That way, you will not have to determine what the figures mean when you look at your notes later. Set up any tables that you will need to use ahead of time, so you can record any observations right away. Remember to avoid bias when collecting data by not including personal thoughts when you record observations. Record only what you observe.

Estimate Scientific work also involves estimating. To estimate is to make a judgment about the size or the number of something without measuring or counting. This is important when the number or size of an object or population is too large or too difficult to accurately count or measure.

Sample Scientists may use a sample or a portion of the total number as a type of estimation. To sample is to take a small, representative portion of the objects or organisms of a population for research. By making careful observations or manipulating variables within that portion of the group, information is discovered and conclusions are drawn that might apply to the whole population. A poorly chosen sample can be unrepresentative of the whole. If you were trying to determine the rainfall in an area, it would not be best to take a rainfall sample from under a tree.

Measure You use measurements everyday. Scientists also take measurements when collecting data. When taking measurements, it is important to know how to use measuring tools properly. Accuracy also is important.

Length To measure length, the distance between two points, scientists use meters. Smaller measurements might be measured in centimeters or millimeters.

Length is measured using a metric ruler or meter stick. When using a metric ruler, line up the 0-cm mark with the end of the object being measured and read the number of the unit where the object ends. Look at the metric ruler shown in **Figure 11.** The centimeter lines are the long, numbered lines, and the shorter lines are millimeter lines. In this instance, the length would be 4.50 cm.

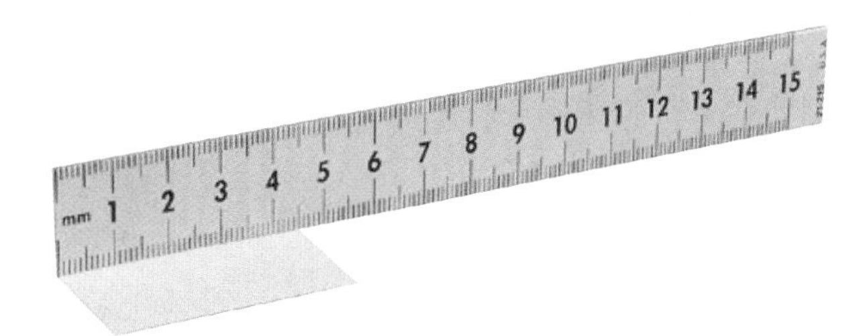

Figure 11 This metric ruler has centimeter and millimeter divisions.

Mass The SI unit for mass is the kilogram (kg). Scientists can measure mass using units formed by adding metric prefixes to the unit gram (g), such as milligram (mg). To measure mass, you might use a triple-beam balance similar to the one shown in **Figure 12.** The balance has a pan on one side and a set of beams on the other side. Each beam has a rider that slides on the beam.

When using a triple-beam balance, place an object on the pan. Slide the largest rider along its beam until the pointer drops below zero. Then move it back one notch. Repeat the process for each rider proceeding from the larger to smaller until the pointer swings an equal distance above and below the zero point. Sum the masses on each beam to find the mass of the object. Move all riders back to zero when finished.

Instead of putting materials directly on the balance, scientists often take a tare of a container. A tare is the mass of a container into which objects or substances are placed for measuring their masses. To mass objects or substances, find the mass of a clean container. Remove the container from the pan, and place the object or substances in the container. Find the mass of the container with the materials in it. Subtract the mass of the empty container from the mass of the filled container to find the mass of the materials you are using.

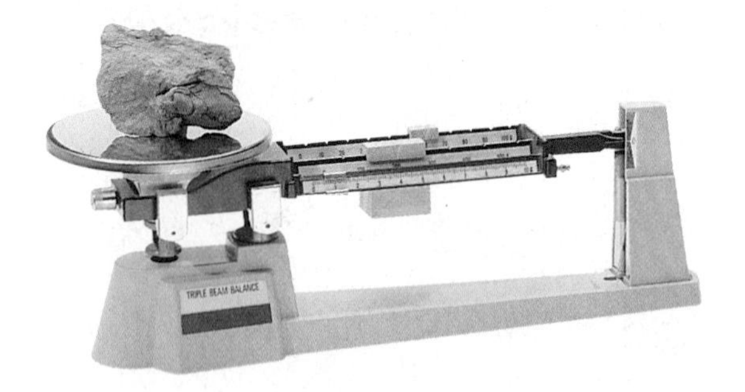

Figure 12 A triple-beam balance is used to determine the mass of an object.

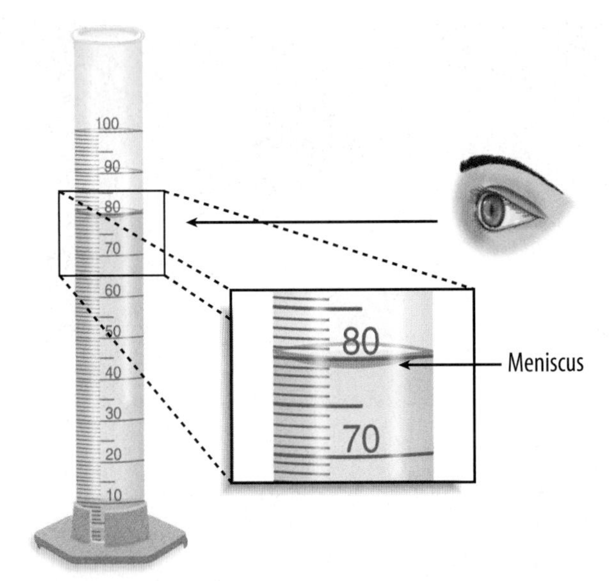

Figure 13 Graduated cylinders measure liquid volume.

Liquid Volume To measure liquids, the unit used is the liter. When a smaller unit is needed, scientists might use a milliliter. Because a milliliter takes up the volume of a cube measuring 1 cm on each side it also can be called a cubic centimeter ($cm^3 = cm \times cm \times cm$).

You can use beakers and graduated cylinders to measure liquid volume. A graduated cylinder, shown in **Figure 13,** is marked from bottom to top in milliliters. In lab, you might use a 10-mL graduated cylinder or a 100-mL graduated cylinder. When measuring liquids, notice that the liquid has a curved surface. Look at the surface at eye level, and measure the bottom of the curve. This is called the meniscus. The graduated cylinder in **Figure 13** contains 79.0 mL, or 79.0 cm^3, of a liquid.

Temperature Scientists often measure temperature using the Celsius scale. Pure water has a freezing point of 0°C and boiling point of 100°C. The unit of measurement is degrees Celsius. Two other scales often used are the Fahrenheit and Kelvin scales.

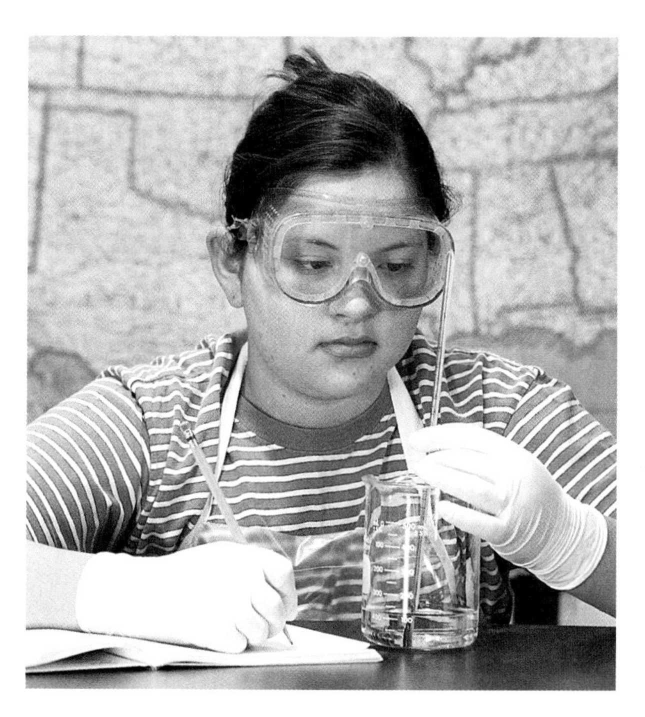

Figure 14 A thermometer measures the temperature of an object.

Scientists use a thermometer to measure temperature. Most thermometers in a laboratory are glass tubes with a bulb at the bottom end containing a liquid such as colored alcohol. The liquid rises or falls with a change in temperature. To read a glass thermometer like the thermometer in **Figure 14,** rotate it slowly until a red line appears. Read the temperature where the red line ends.

Form Operational Definitions An operational definition defines an object by how it functions, works, or behaves. For example, when you are playing hide and seek and a tree is home base, you have created an operational definition for a tree.

Objects can have more than one operational definition. For example, a ruler can be defined as a tool that measures the length of an object (how it is used). It can also be a tool with a series of marks used as a standard when measuring (how it works).

Analyze the Data

To determine the meaning of your observations and investigation results, you will need to look for patterns in the data. Then you must think critically to determine what the data mean. Scientists use several approaches when they analyze the data they have collected and recorded. Each approach is useful for identifying specific patterns.

Interpret Data The word *interpret* means "to explain the meaning of something." When analyzing data from an experiment, try to find out what the data show. Identify the control group and the test group to see whether or not changes in the independent variable have had an effect. Look for differences in the dependent variable between the control and test groups.

Classify Sorting objects or events into groups based on common features is called classifying. When classifying, first observe the objects or events to be classified. Then select one feature that is shared by some members in the group, but not by all. Place those members that share that feature in a subgroup. You can classify members into smaller and smaller subgroups based on characteristics. Remember that when you classify, you are grouping objects or events for a purpose. Keep your purpose in mind as you select the features to form groups and subgroups.

Compare and Contrast Observations can be analyzed by noting the similarities and differences between two more objects or events that you observe. When you look at objects or events to see how they are similar, you are comparing them. Contrasting is looking for differences in objects or events.

Science Skill Handbook

Recognize Cause and Effect A cause is a reason for an action or condition. The effect is that action or condition. When two events happen together, it is not necessarily true that one event caused the other. Scientists must design a controlled investigation to recognize the exact cause and effect.

Draw Conclusions

When scientists have analyzed the data they collected, they proceed to draw conclusions about the data. These conclusions are sometimes stated in words similar to the hypothesis that you formed earlier. They may confirm a hypothesis, or lead you to a new hypothesis.

Infer Scientists often make inferences based on their observations. An inference is an attempt to explain observations or to indicate a cause. An inference is not a fact, but a logical conclusion that needs further investigation. For example, you may infer that a fire has caused smoke. Until you investigate, however, you do not know for sure.

Apply When you draw a conclusion, you must apply those conclusions to determine whether the data supports the hypothesis. If your data do not support your hypothesis, it does not mean that the hypothesis is wrong. It means only that the result of the investigation did not support the hypothesis. Maybe the experiment needs to be redesigned, or some of the initial observations on which the hypothesis was based were incomplete or biased. Perhaps more observation or research is needed to refine your hypothesis. A successful investigation does not always come out the way you originally predicted.

Avoid Bias Sometimes a scientific investigation involves making judgments. When you make a judgment, you form an opinion. It is important to be honest and not to allow any expectations of results to bias your judgments. This is important throughout the entire investigation, from researching to collecting data to drawing conclusions.

Communicate

The communication of ideas is an important part of the work of scientists. A discovery that is not reported will not advance the scientific community's understanding or knowledge. Communication among scientists also is important as a way of improving their investigations.

Scientists communicate in many ways, from writing articles in journals and magazines that explain their investigations and experiments, to announcing important discoveries on television and radio. Scientists also share ideas with colleagues on the Internet or present them as lectures, like the student is doing in **Figure 15.**

Figure 15 A student communicates to his peers about his investigation.

SAFETY SYMBOLS

SAFETY SYMBOLS	HAZARD	EXAMPLES	PRECAUTION	REMEDY
DISPOSAL	Special disposal procedures need to be followed.	certain chemicals, living organisms	Do not dispose of these materials in the sink or trash can.	Dispose of wastes as directed by your teacher.
BIOLOGICAL	Organisms or other biological materials that might be harmful to humans	bacteria, fungi, blood, unpreserved tissues, plant materials	Avoid skin contact with these materials. Wear mask or gloves.	Notify your teacher if you suspect contact with material. Wash hands thoroughly.
EXTREME TEMPERATURE	Objects that can burn skin by being too cold or too hot	boiling liquids, hot plates, dry ice, liquid nitrogen	Use proper protection when handling.	Go to your teacher for first aid.
SHARP OBJECT	Use of tools or glassware that can easily puncture or slice skin	razor blades, pins, scalpels, pointed tools, dissecting probes, broken glass	Practice common-sense behavior and follow guidelines for use of the tool.	Go to your teacher for first aid.
FUME	Possible danger to respiratory tract from fumes	ammonia, acetone, nail polish remover, heated sulfur, moth balls	Make sure there is good ventilation. Never smell fumes directly. Wear a mask.	Leave foul area and notify your teacher immediately.
ELECTRICAL	Possible danger from electrical shock or burn	improper grounding, liquid spills, short circuits, exposed wires	Double-check setup with teacher. Check condition of wires and apparatus.	Do not attempt to fix electrical problems. Notify your teacher immediately.
IRRITANT	Substances that can irritate the skin or mucous membranes of the respiratory tract	pollen, moth balls, steel wool, fiberglass, potassium permanganate	Wear dust mask and gloves. Practice extra care when handling these materials.	Go to your teacher for first aid.
CHEMICAL	Chemicals can react with and destroy tissue and other materials	bleaches such as hydrogen peroxide; acids such as sulfuric acid, hydrochloric acid; bases such as ammonia, sodium hydroxide	Wear goggles, gloves, and an apron.	Immediately flush the affected area with water and notify your teacher.
TOXIC	Substance may be poisonous if touched, inhaled, or swallowed.	mercury, many metal compounds, iodine, poinsettia plant parts	Follow your teacher's instructions.	Always wash hands thoroughly after use. Go to your teacher for first aid.
FLAMMABLE	Flammable chemicals may be ignited by open flame, spark, or exposed heat.	alcohol, kerosene, potassium permanganate	Avoid open flames and heat when using flammable chemicals.	Notify your teacher immediately. Use fire safety equipment if applicable.
OPEN FLAME	Open flame in use, may cause fire.	hair, clothing, paper, synthetic materials	Tie back hair and loose clothing. Follow teacher's instruction on lighting and extinguishing flames.	Notify your teacher immediately. Use fire safety equipment if applicable.

Eye Safety
Proper eye protection should be worn at all times by anyone performing or observing science activities.

Clothing Protection
This symbol appears when substances could stain or burn clothing.

Animal Safety
This symbol appears when safety of animals and students must be ensured.

Handwashing
After the lab, wash hands with soap and water before removing goggles.

Safety in the Science Laboratory

The science laboratory is a safe place to work if you follow standard safety procedures. Being responsible for your own safety helps to make the entire laboratory a safer place for everyone. When performing any lab, read and apply the caution statements and safety symbol listed at the beginning of the lab.

General Safety Rules

1. Obtain your teacher's permission to begin all investigations and use laboratory equipment.
2. Study the procedure. Ask your teacher any questions. Be sure you understand safety symbols shown on the page.
3. Notify your teacher about allergies or other health conditions which can affect your participation in a lab.
4. Learn and follow use and safety procedures for your equipment. If unsure, ask your teacher.

5. Never eat, drink, chew gum, apply cosmetics, or do any personal grooming in the lab. Never use lab glassware as food or drink containers. Keep your hands away from your face and mouth.
6. Know the location and proper use of the safety shower, eye wash, fire blanket, and fire alarm.

Prevent Accidents

1. Use the safety equipment provided to you. Goggles and a safety apron should be worn during investigations.
2. Do NOT use hair spray, mousse, or other flammable hair products. Tie back long hair and tie down loose clothing.
3. Do NOT wear sandals or other open-toed shoes in the lab.
4. Remove jewelry on hands and wrists. Loose jewelry, such as chains and long necklaces, should be removed to prevent them from getting caught in equipment.
5. Do not taste any substances or draw any material into a tube with your mouth.
6. Proper behavior is expected in the lab. Practical jokes and fooling around can lead to accidents and injury.
7. Keep your work area uncluttered.

Laboratory Work

1. Collect and carry all equipment and materials to your work area before beginning a lab.
2. Remain in your own work area unless given permission by your teacher to leave it.

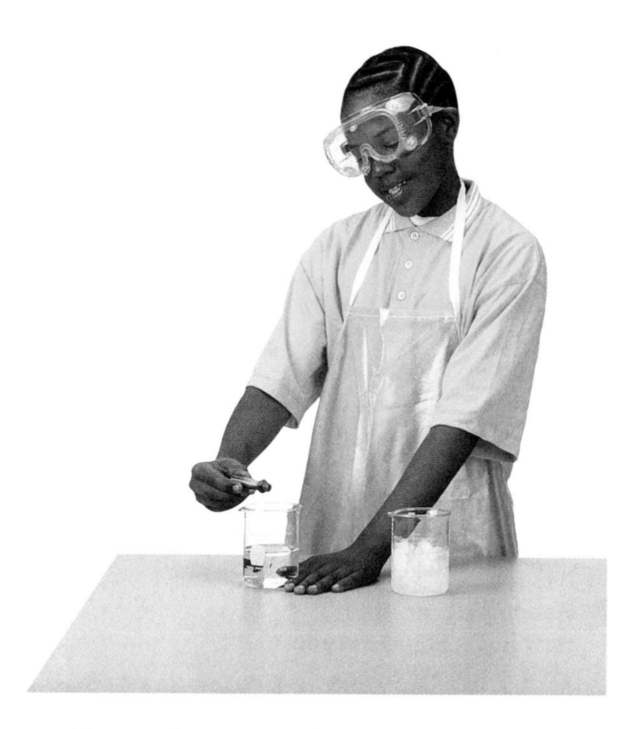

3. Always slant test tubes away from yourself and others when heating them, adding substances to them, or rinsing them.
4. If instructed to smell a substance in a container, hold the container a short distance away and fan vapors towards your nose.
5. Do NOT substitute other chemicals/substances for those in the materials list unless instructed to do so by your teacher.
6. Do NOT take any materials or chemicals outside of the laboratory.
7. Stay out of storage areas unless instructed to be there and supervised by your teacher.

Laboratory Cleanup

1. Turn off all burners, water, and gas, and disconnect all electrical devices.
2. Clean all pieces of equipment and return all materials to their proper places.
3. Dispose of chemicals and other materials as directed by your teacher. Place broken glass and solid substances in the proper containers. Never discard materials in the sink.
4. Clean your work area.
5. Wash your hands with soap and water thoroughly BEFORE removing your goggles.

Emergencies

1. Report any fire, electrical shock, glassware breakage, spill, or injury, no matter how small, to your teacher immediately. Follow his or her instructions.
2. If your clothing should catch fire, STOP, DROP, and ROLL. If possible, smother it with the fire blanket or get under a safety shower. NEVER RUN.
3. If a fire should occur, turn off all gas and leave the room according to established procedures.
4. In most instances, your teacher will clean up spills. Do NOT attempt to clean up spills unless you are given permission and instructions to do so.
5. If chemicals come into contact with your eyes or skin, notify your teacher immediately. Use the eyewash or flush your skin or eyes with large quantities of water.
6. The fire extinguisher and first-aid kit should only be used by your teacher unless it is an extreme emergency and you have been given permission.
7. If someone is injured or becomes ill, only a professional medical provider or someone certified in first aid should perform first-aid procedures.

From Your Kitchen, Junk Drawer, or Yard

1 Your Daily Drink

Real-World Question
How much do you drink in a day?

Possible Materials
- 500-mL measuring cup
- calculator

Procedure
1. When you drink a bottle or can of juice, soda, water, or other beverage, look on the label of the container to find out the volume of your drink in milliliters.
2. Record the volumes of all the can and bottle drinks you consume at home in one day in your Science Journal.
3. Use a measuring cup to measure the amount of liquids you drink during one day that you pour from a larger container. Record these volumes in your Science Journal.
4. Add up the volumes of all the drinks you consumed during the day.

Conclude and Apply
1. How much liquid did you drink during the day?
2. Infer how you would measure the mass of the foods you ate in one day.

2 Cell Sizes

Real-World Question
How do different cells compare in size?

Possible Materials
- meterstick
- metric ruler
- white paper
- pencil
- pen
- masking tape

Procedure
1. Make a dot on a white sheet of paper with a pencil.
2. Use a metric ruler to make a second dot 1 mm away from the first dot. This distance represents the average length of a bacteria cell.
3. Measure a distance 8 mm away from the first dot and make a third dot. This distance represents the average length of a red blood cell.
4. Mark a spot on the floor with a piece of tape and use the meterstick to measure a distance of 7 meters. Mark this distance with a second piece of tape. This distance represents the average length of an amoeba cell.

Conclude and Apply
1. The distance between the first and second dot is 1,000 times longer than the actual size of a bacterium cell. Calculate the length of an actual bacterium cell.
2. A large chicken egg is just one cell, and it is 100 times longer than an amoeba cell. Calculate the distance you would have to measure to represent the average length of a hen's egg.

Adult supervision required for all labs.

Extra Try at Home Labs

3 Many Mixtures

Real-World Question

What do different mixtures in the body look like?

Possible Materials

- cooking oil
- bottled water
- grape juice
- apple juice
- cornstarch
- flour
- clear drinking glasses (3)
- measuring cup
- spoon
- flashlight

Procedure

1. Pour 50 mL of water and 50 mL of cooking oil into a glass, stir the liquids together, and place the glass on a table to observe what type of mixture forms.
2. Pour 150 mL of water and 150 mL of grape juice into a glass and stir the liquids together. Darken the room and shine a flashlight into the mixture.
3. Pour 300 mL of water into a glass and add a spoonful of cornstarch. Stir the mixture, darken the room, and shine a flashlight into the mixture.

Conclude and Apply

1. Describe what happened to the water and cooking oil mixture.
2. Describe what happened to the light beam when you passed it through the water and grape juice mixture. What happened to the beam in the water and cornstarch mixture?
3. Identify the types of mixtures you made.

4 Putting Down Roots

Real-World Question

Can the stem cells of plants reproduce root cells for a new plant?

Possible Materials

- houseplant
- scissors
- metric ruler
- rooting hormone
- glasses or jars (3)
- water
- magnifying lens

Procedure

1. Examine the stems of a houseplant and locate a node on three different stems. A node looks like a small bump.
2. Cut 3 stems off the plant at a 45° angle about 3–4 mm below the node.
3. Remove all flowers and flower buds from the cut stems.
4. Dip the cut end of each stem into rooting hormone solution.
5. Place the end of each stem into a separate glass of water and observe them for a week.

Conclude and Apply

1. Describe what happened to the ends of the stems.
2. Infer why flowers and buds should be removed from the stems.
3. Infer how plant stem cells can produce root cells.

Adult supervision required for all labs.

5 Rolls Off the Tongue

Real-World Question

Is the ability to roll your tongue dominant or recessive?

Possible Materials

- Science Journal
- pencil
- pen
- calculator

Procedure

1. Ask your friends, family members, and other people you know at school to roll their tongues. To be able to roll their tongues, they must be able to stick out their tongues and curl the sides up so that they touch each other.
2. Try to interview as many people as possible to collect a large sample of data.
3. Record the number of people who can and who cannot roll their tongues in your Science Journal.

Conclude and Apply

1. Calculate the percentage of people who can roll their tongues and the percentage of people who cannot roll their tongues.
2. Infer whether or not the ability to roll your tongue is a dominant or recessive trait.
3. Infer how many children in a family could roll their tongue if neither of their parents could roll their tongues.

6 Frozen Fossils

Real-World Question

How can we model the formation of an amber fossil?

Possible Materials

- small glass jar with lid
- honey
- ruler
- dead insect or spider
- small rubber insect or spider
- freezer

Procedure

1. Thoroughly wash and dry a small glass jar and its lid.
2. Pour 3 cm of honey into the jar. Do not pour honey down the sides of the jar.
3. Search for a dead insect or spider around your home or school and drop it into the center of the honey's surface.
4. Pour another 3 cm of honey into the jar to cover the organism.
5. Place the jar in the freezer overnight.

Conclude and Apply

1. Explain how you modeled the formation of an amber fossil.
2. Infer how amber fossils help scientists observe adaptations of organisms over time.

7 Beating Bacteria

Real-World Question
How do we protect our foods from bacteria?

Possible Materials
- heavy cream or whole milk
- plastic drinking glasses (3)
- refrigerator

Procedure
1. Pour milk into three identical glasses until each glass is three-quarters full.
2. Place one glass in the back of your refrigerator where it will not be disturbed.
3. Place a second glass in a dark, cool place such as inside a kitchen cabinet.
4. Place the third glass on a windowsill in direct sunlight.
5. Observe the milk in each glass every day for several days. In your Science Journal, write down your observations about what is happening to the milk each day.

Conclude and Apply
1. Describe what happened to the milk in each of the glasses.
2. Infer what caused the change in the milk on the windowsill.
3. Infer how a refrigerator protects food from bacteria.

8 Fungus Foods

Real-World Question
How do we protect our foods from mold?

Possible Materials
- unopened jars of baby food or some other type of food such as spaghetti sauce or salsa (4)
- plate
- spoon
- magnifying lens

Procedure
1. Open one of the jars of food and spread the contents out over a plate.
2. Place the plate in a cool place away from direct sunlight where it will not be disturbed for several days.
3. Open a second jar of food and set it next to the plate without the lid on.
4. Break the seal of the third jar of food and set it next to the second jar with the lid on.
5. Do not open the fourth jar but place it next to the other two jars.
6. Observe the food in the jars and on the plate each day for the next several days. Record your observations in your Science Journal.

Conclude and Apply
1. Describe what happened to the food in each container.
2. Explain how food companies protect the foods you buy from fungi such as mold.

Adult supervision required for all labs.

Extra Try at Home Labs

9 Prickly Plants

Real-World Question
Why does a cactus have needles?

Possible Materials
- toilet paper roll or paper towel roll (cut in half)
- transparent tape
- toothpicks (15)
- metric ruler
- oven mitt
- self-sealing bag
- water
- measuring cup

Procedure
1. Pour 50 mL of water into a small self-sealing bag.
2. Stuff the bag of water into the toilet paper roll so that the bag is just inside the roll's rim.
3. Stand the roll on a table and hold it firmly with one hand. Place the oven mitt on your other hand and try to take the bag out of the roll.
4. If needed, place the bag back into the roll.
5. Securely tape toothpicks around the lip of the roll about 1 cm apart. About 4 cm of each toothpick should stick up above the rim.
6. Hold the roll on the table, put the oven mitt on, and try to take the bag out of the roll without breaking the toothpicks.

Conclude and Apply
1. Compare how easy it was to remove the water bag from the toilet paper roll with and without the toothpicks protecting the water.
2. Infer why desert cacti have needles.

10 Feed the Birds

Real-World Question
How can you attract hummingbirds to your yard?

Possible Materials
- plastic water bottle (20 ounce) with bent tube or straw (10 cm length)
- sugar
- water
- pot
- wooden spoon
- red plastic flowers
- red tape or red plastic wrap
- narrow net bag
- strong string or wire

Procedure
1. Wrap red tape or fasten red plastic wrap around the bottle. Fasten red flowers to the bottle.
2. Boil 800 mL of water and gradually stir in 200 mL of sugar until all the sugar is dissolved. Let the solution cool and fill the bottle with the solution.
3. Hang the bottle upside down in the net bag and hang the feeder by a string in a shady location. Hang it above bright red or pink flowers if possible.
4. Observe your feeder daily and watch for hummingbirds.
5. Clean your bottle and refill it with fresh sugar solution every three days.

Conclude and Apply
1. Explain why hummingbirds, bees, bats, and other animals are important to plant reproduction.
2. Infer why flowers are usually brightly colored.

Adult supervision required for all labs.

11 Breathing Plants

Real-World Question
How do plants breathe?

Possible Materials
- houseplant
- petroleum jelly
- paper towel
- soap
- water

Procedure
1. Scoop some petroleum jelly out of the jar with your fingertips and coat the top of three or four leaves of the houseplant with jelly. Cover the entire top surface of the leaves only.
2. Coat the bottom of three or four different leaves with a layer of jelly.
3. Choose two or three stems not connected to the leaves you have covered with jelly. Coat these stems from top to bottom with a layer of jelly. Cover the entire stems but not their leaves.
4. Wash your hands with soap and water.
5. Observe the houseplant for three days.

Conclude and Apply
1. Describe what happened to the leaves and stems covered with jelly.
2. Infer how plants breathe.

12 A Sponge of Water

Real-World Question
How much water does a sponge hold?

Possible Materials
- large natural sponge
- large artificial sponge (without soap)
- bucket
- shallow basin or pan
- water
- measuring cups (2)
- towel

Procedure
1. Submerge a large, natural sponge in a bucket of water for about 2 minutes so that it fills up with water. Wait until bubbles stop rising up from the sponge.
2. Place the two measuring cups next to the bucket.
3. Carefully remove the sponge from the water and hold it over a measuring cup. Squeeze all the water out of the sponge into the measuring cup. If you reach the top mark on the measuring cup, use the second measuring cup.

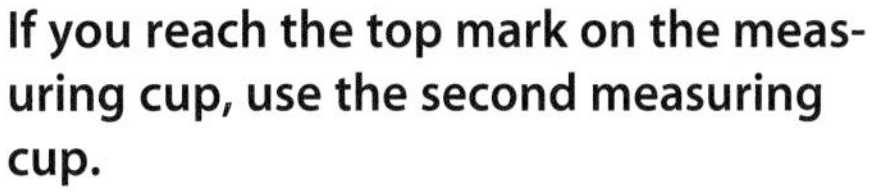

4. Repeat steps 1–3 using the artificial sponge. Compare your results.

Conclude and Apply
1. How much water did each sponge hold?
2. Infer how Roman soldiers used sponges as canteens.

13 Invertebrate Groceries

Real-World Question

What types of invertebrates can you find in your local grocery store?

Possible Materials

- access to local grocery store
- guidebook to ocean invertebrates

Procedure

1. Go with an adult to the largest grocery store in your area.
2. Search the seafood section in the store for invertebrates sold as food.
3. Search the grocery aisles selling canned meat for invertebrates sold in cans.
4. Record all the invertebrates you find in the store and identify each invertebrate as a mollusk, worm, arthropod, or echinoderm.
5. If you find more than one type of an organism such as crabs, ask a grocery store employee to identify the different species for you.

Conclude and Apply

1. Identify the types of invertebrates you found.
2. What types of invertebrates were not in your grocery store?
3. Infer why we should be concerned about ocean pollution.

14 Frog Lives

Real-World Question

What do the stages of frog metamorphosis look like?

Possible Materials

- gravel
- small aquarium or bucket
- water
- clean plastic milk jug
- *Anacharis* plants
- tadpole

Procedure

1. Pour a 2-cm layer of aquarium gravel or small pebbles into a small aquarium.
2. Fill the aquarium with water and allow the water to sit for 7 days.
3. Fill a clean, plastic milk jug with tap water and allow it to sit for 7 days.
4. Plant several stalks of *Anacharis* plants in the gravel.
5. Purchase or catch a tadpole and place it in the aquarium along with some of the water from the place where you collected it.
6. Replace 10 percent of the aquarium's water every other day with water from the milk jug.
7. Observe your tadpole each day for 2–3 weeks as it completes its metamorphosis into a frog.

Conclude and Apply

1. Describe the metamorphosis of your frog.
2. Infer why you added the plant *Anacharis* to the aquarium.

Adult supervision required for all labs.

15 Fly Like a Bird

Real-World Question

How does a bird's wings keep it in the air?

Possible Materials

- tennis ball or racquetball
- softball
- flying disc
- stopwatch or watch with second hand

Procedure

1. Throw a ball in a straight line and parallel to the ground as hard as you can.
2. Have a partner use a stopwatch to time how long the ball stays in the air.
3. Throw a flying disc in a straight line and parallel to the ground so that it hovers.
4. Have a partner use a stopwatch to time how long the flying disc stays in the air.
5. Repeat steps 1–4 several times and record your times in your Science Journal.

Conclude and Apply

1. Compare the maximum time the tennis ball stayed in the air to the longest time the flying disc stayed up.
2. Compare the shape of the flying disc when you hold it flat in front of you to the shape of a bird wing. How are they similar?
3. Infer how a bird's wings allow it to fly.

16 Fighting Fish

Real-World Question

How will a Siamese fighting fish react to a mirror?

Possible Materials

- 2 small fish bowls, glass bowls, or small jars
- water (aquarium, purified)
- 1 male Siamese fighting fish (*Betta splendens*)
- 1 female Siamese fighting fish (*Betta splendens*)
- mirror

Procedure

1. Place a male Siamese fighting fish in a small fish bowl with water and a female Siamese fighting fish into a second bowl with water.
2. Place a mirror up to the bowl with the male fish so that he sees his reflection and observe his reaction.
3. Place a mirror up to the bowl with the female fish so that she sees her reflection and observe her reaction.

Conclude and Apply

1. Describe how the male and the female reacted to their reflections.
2. What type of behavior did the male Siamese fighting fish display?
3. Infer why the male fish displays this type of behavior.

Adult supervision required for all labs.

17 Measuring Melanin

Real-World Question
How does the amount of melanin vary from person to person?

Possible Materials
- several plastic bandages
- color swatches from a paint or hardware store

Procedure
1. Obtain color swatches from a paint store or hardware store.
2. Identify the color swatch that most closely matches your skin color. Record the color in your Science Journal.
3. Place a plastic bandage around the end of one of your fingers and leave the bandage on for two days.
4. Remove the bandage and observe the change in your skin color. Identify the color swatch that most closely matches the faded skin color. Record the color in your Science Journal.
5. Follow steps 2–4 to measure the amount of melanin in the skin of several other people.

Conclude and Apply
1. Identify how many shades of color your skin and the skin of the other people you tested changed.
2. Infer why your skin color changed.

18 Vitamin Search

Real-World Question
How many vitamins and minerals are in the foods you eat?

Possible Materials
- labels from packaged foods and drinks
- nutrition guidebook or cookbook

Procedure
1. Create a data table to record the "% Daily Value" of important vitamins and minerals for a variety of foods.
2. Collect packages from a variety of packaged foods and check the Nutrition Facts chart for the "% Daily Value" of all the vitamins and minerals it contains. These values are listed at the bottom of the chart.
3. Use cookbooks or nutrition guidebooks to research the "% Daily Value" of vitamins and minerals found in several fresh fruits and vegetables such as strawberries, spinach, oranges, and lentils.

Conclude and Apply
1. Infer why a healthy diet includes fresh fruits and vegetables.
2. Infer why a healthy diet includes a wide variety of healthy foods.

Nutrition Facts
Serving Size 1 Meal

Amount Per Serving	% Daily Value*
Calories 330 Calories from Fat 60	
Total Fat 7g	10%
Saturated Fat 3.5g	17%
Polyunsaturated Fat 1g	
Monounsaturated Fat 2.5g	
Cholesterol 35mg	12%
Sodium 460mg	19%
Total Carbohydrate 52g	18%
Dietary Fiber 6g	24%
Sugars 17g	
Protein 15g	

Vitamin A 15% • Vitamin C 70%
Calcium 4% • Iron 10%

* Percent Daily Values are based on a 2,000 calorie diet. Your daily values may be higher or lower depending on your calorie needs.

	Calories	2,000	2,500
Total Fat	Less than	65g	80g
Sat Fat	Less than	20g	25g
Cholesterol	Less than	300mg	300mg
Sodium	Less than	2,400mg	2,400mg
Total Carbohydrate		300g	375g
Dietary Fiber		25g	30g

Adult supervision required for all labs.

19 What's in blood?

Real-World Question

What are the proportions of the components of your blood?

Possible Materials 

- bag of brown rice
- white rice
- small bag of wild rice
- measuring cup
- large bowl or large cooking tray

Procedure

1. Measure 1.25 liters of brown rice and pour the rice into a large bowl or on a cooking tray.
2. Measure 100 mL of wild rice and pour it into the bowl or on the tray.
3. Count out 50 grains of white rice and place them in the bowl or on the tray.
4. Mix the three types of rice thoroughly and observe the proportions of the three major components of your blood.

Conclude and Apply

1. Infer what type of rice represents red blood cells.
2. Infer what type of rice represents white blood cells.
3. Infer what type of rice represents platelets.

20 Modeling Glucose

Real-World Question

What does a molecule of glucose used in respiration look like?

Possible Materials

- red polystyrene balls (7)
- blue polystyrene balls (11)
- yellow polystyrene balls (14)
- box of toothpicks

Procedure

1. Use toothpicks to connect the 6 red polystyrene balls together in a line. These balls represent carbon atoms.
2. Attach a blue ball and a yellow ball to the first carbon atom. The blue ball goes to the left and the yellow ball to the right. The blue ball represents an oxygen atom, and the yellow ball represents a hydrogen atom.
3. Attach a hydrogen atom and oxygen atom in a line to the right of the second, fourth and fifth carbon atoms. Attach a hydrogen atom to the left of each of these carbon atoms.
4. Attach a hydrogen atom and oxygen atom in a line to the left of the third carbon atom and a hydrogen atom to the right.
5. Attach two hydrogen atoms and one oxygen atom to the sixth carbon atom. Attach a hydrogen atom to the oxygen atom.

Conclude and Apply

1. Construct models of the other molecules used during respiration.
2. Infer why oxygen is needed during respiration.

Adult supervision required for all labs.

21 Pupil Power

Real-World Question

Why do your eyes have pupils?

Possible Materials

- mirror
- flashlight

Procedure

1. Examine your eyes in a mirror in a brightly lit room. Observe the size of the pupil and iris of each eye.
2. Darken the room so that there is no light.
3. Turn on a flashlight and cover its light with the palm of your hand.
4. Hold the flashlight about 10–15 cm in front of your mouth with the beam facing straight up.
5. Quickly remove your hand and immediately observe the pupils of your eyes.

Conclude and Apply

1. Describe what happened to the pupils of your eyes.
2. Infer why your pupils responded to the light in this way.

22 Identifying Iodine

Real-World Question

How many iodine rich foods are in your home?

Possible Materials

- Science Journal
- pen or pencil

Procedure

1. The element iodine is needed in a person's diet for the thyroid gland to work properly. Search your kitchen for the foods richest in iodine including fish, shellfish (clams, oysters, mussels, scallops), seaweed, and kelp.
2. Record all the iodine rich foods you find in your Science Journal.
3. Search your kitchen for foods that may contain iodine if they are grown in rich soil such as onions, mushrooms, lettuce, spinach, green peppers, pineapple, peanuts, and whole wheat bread.
4. Record all of these foods that you find in your Science Journal.

Conclude and Apply

1. Infer whether or not your family eats a diet rich in iodine.
2. Infer why iodine is often added to table salt.
3. Research two diseases caused by a lack of iodine in the diet.

Adult supervision required for all labs.

23 Acid Defense

Real-World Question
How is stomach acid your internal first line of defense?

Possible Materials
- drinking glasses (2)
- milk
- cola or lemon juice
- masking tape
- marker
- measuring cup

Procedure
1. Pour 100 mL of milk into each glass.
2. Pour 20 mL of cola into the second glass.
3. Using the masking tape and marker, label the first glass *No Acid* and the second glass *Acid*.
4. Place the glasses in direct sunlight and observe the mixture each day for several days.

Conclude and Apply
1. Compare the odor of the mixture in both glasses after one or two days.
2. Infer how this experiment modeled one of your internal defenses against disease.

24 Rock Creatures

Real-World Question
What types of organisms live under stream rocks?

Possible Materials
- waterproof boots
- ice cube tray (white)
- aquarium net
- bucket
- collecting jars
- guidebook to pond life

Procedure
1. Search under the rocks of a local stream. Look for aquatic organisms under the rocks and leaves of the stream. Compare what you find in fast- and slow-moving water.
2. Carefully pull the organisms you find off the rocks and put them into separate compartments of your ice cube tray. Take care not to injure the creatures you find.
3. Use your net and bucket to collect larger organisms.
4. Use your guidebook to pond life to identify the organisms you find.
5. Release the organisms back into the stream once you identify them.

Conclude and Apply
1. Identify and list the organisms you found under the stream rocks.
2. Infer why so many aquatic organisms make their habitats beneath stream rocks.

25 A Light in the Forest

Real-World Question
Does the amount of sunlight vary in a forest?

Possible Materials
- empty toilet paper or paper towel roll
- Science Journal

Procedure:
1. Copy the data table into your Science Journal.
2. Go with an adult to a nearby forest or large patch of trees.
3. Stand near the edge of the forest and look straight up through your cardboard tube. Estimate the percentage of blue sky and clouds you can see in the circle. This percentage is the amount of sunlight reaching the forest floor.
4. Record your location and estimated percentage of sunlight in your data table.
5. Test several other locations in the forest. Choose places where the trees completely cover the forest floor and where sunlight is partially coming through.

Data Table

Location	% of Sunlight

Conclude and Apply
1. Explain how the amount of sunlight reaching the forest floor changed from place to place.
2. Infer why it is important for leaves and branches to stop sunlight from reaching much of the forest floor.

26 Immovable Mollusks

Real-World Question
How do mollusks living in intertidal ecosystems hold on to rocks?

Possible Materials
- plastic suction cup
- water
- paper towel or sponge

Procedure
1. Moisten a paper towel or sponge with water.
2. Press a plastic suction cup on the moist towel or sponge until the entire bottom surface of the cup is wet.
3. Firmly press the suction cup down on a kitchen counter for 10 seconds.
4. Grab the top handle of the suction cup and try removing the cup from the counter by pulling it straight up.

Conclude and Apply
1. Describe what happened when you tried to remove the cup from the counter.
2. Infer how mollusks living in intertidal ecosystems withstand the constant pull of ocean waves and currents.

Adult supervision required for all labs.

27 UV Watch

Real-World Question
How can you find out about the risks of ultraviolet radiation each day?

Possible Materials
- daily newspaper with weekly weather forecasts
- graph paper

Procedure
1. Use the local newspaper or another resource to get the weather forecast for the day.
2. Check the UV (ultraviolet light) index for the day. If it provides an hourly UV index level, record the level for 1:00 P.M.
3. Find a legend or do research to discover what the numbers of the UV index mean.
4. Record the UV index everyday for 10 days and graph your results on graph paper.

Conclude and Apply
1. Explain how the UV index system works.
2. Research several ways you can protect yourself from too much ultraviolet light exposure.

Computer Skills

People who study science rely on computers, like the one in **Figure 16,** to record and store data and to analyze results from investigations. Whether you work in a laboratory or just need to write a lab report with tables, good computer skills are a necessity.

Using the computer comes with responsibility. Issues of ownership, security, and privacy can arise. Remember, if you did not author the information you are using, you must provide a source for your information. Also, anything on a computer can be accessed by others. Do not put anything on the computer that you would not want everyone to know. To add more security to your work, use a password.

Use a Word Processing Program

A computer program that allows you to type your information, change it as many times as you need to, and then print it out is called a word processing program. Word processing programs also can be used to make tables.

Figure 16 A computer will make reports neater and more professional looking.

Learn the Skill To start your word processing program, a blank document, sometimes called "Document 1," appears on the screen. To begin, start typing. To create a new document, click the *New* button on the standard tool bar. These tips will help you format the document.

- The program will automatically move to the next line; press *Enter* if you wish to start a new paragraph.
- Symbols, called non-printing characters, can be hidden by clicking the *Show/Hide* button on your toolbar.
- To insert text, move the cursor to the point where you want the insertion to go, click on the mouse once, and type the text.
- To move several lines of text, select the text and click the *Cut* button on your toolbar. Then position your cursor in the location that you want to move the cut text and click *Paste.* If you move to the wrong place, click *Undo.*
- The spell check feature does not catch words that are misspelled to look like other words, like "cold" instead of "gold." Always reread your document to catch all spelling mistakes.
- To learn about other word processing methods, read the user's manual or click on thc *Help* button.
- You can integrate databases, graphics, and spreadsheets into documents by copying from another program and pasting it into your document, or by using desktop publishing (DTP). DTP software allows you to put text and graphics together to finish your document with a professional look. This software varies in how it is used and its capabilities.

Use a Database

A collection of facts stored in a computer and sorted into different fields is called a database. A database can be reorganized in any way that suits your needs.

Learn the Skill A computer program that allows you to create your own database is a database management system (DBMS). It allows you to add, delete, or change information. Take time to get to know the features of your database software.

- Determine what facts you would like to include and research to collect your information.
- Determine how you want to organize the information.
- Follow the instructions for your particular DBMS to set up fields. Then enter each item of data in the appropriate field.
- Follow the instructions to sort the information in order of importance.
- Evaluate the information in your database, and add, delete, or change as necessary.

Use the Internet

The Internet is a global network of computers where information is stored and shared. To use the Internet, like the students in **Figure 17,** you need a modem to connect your computer to a phone line and an Internet Service Provider account.

Learn the Skill To access internet sites and information, use a "Web browser," which lets you view and explore pages on the World Wide Web. Each page is its own site, and each site has its own address, called a URL. Once you have found a Web browser, follow these steps for a search (this also is how you search a database).

Figure 17 The Internet allows you to search a global network for a variety of information.

- Be as specific as possible. If you know you want to research "gold," don't type in "elements." Keep narrowing your search until you find what you want.
- Web sites that end in *.com* are commercial Web sites; *.org, .edu,* and *.gov* are non-profit, educational, or government Web sites.
- Electronic encyclopedias, almanacs, indexes, and catalogs will help locate and select relevant information.
- Develop a "home page" with relative ease. When developing a Web site, NEVER post pictures or disclose personal information such as location, names, or phone numbers. Your school or community usually can host your Web site. A basic understanding of HTML (hypertext mark-up language), the language of Web sites, is necessary. Software that creates HTML code is called authoring software, and can be downloaded free from many Web sites. This software allows text and pictures to be arranged as the software is writing the HTML code.

Use a Spreadsheet

A spreadsheet, shown in **Figure 18,** can perform mathematical functions with any data arranged in columns and rows. By entering a simple equation into a cell, the program can perform operations in specific cells, rows, or columns.

Learn the Skill Each column (vertical) is assigned a letter, and each row (horizontal) is assigned a number. Each point where a row and column intersect is called a cell, and is labeled according to where it is located—Column A, Row 1 (A1).

- Decide how to organize the data, and enter it in the correct row or column.
- Spreadsheets can use standard formulas or formulas can be customized to calculate cells.
- To make a change, click on a cell to make it activate, and enter the edited data or formula.
- Spreadsheets also can display your results in graphs. Choose the style of graph that best represents the data.

EM-016C-MSS02.excl

	A	B	C	D	E
1	Test Runs	Time	Distance	Speed	
2	Car 1	5 mins	5 miles	60 mph	
3	Car 2	10 mins	4 miles	24 mph	
4	Car 3	6 mins	3 miles	30 mph	

Figure 18 A spreadsheet allows you to perform mathematical operations on your data.

Use Graphics Software

Adding pictures, called graphics, to your documents is one way to make your documents more meaningful and exciting. This software adds, edits, and even constructs graphics. There is a variety of graphics software programs. The tools used for drawing can be a mouse, keyboard, or other specialized devices. Some graphics programs are simple. Others are complicated, called computer-aided design (CAD) software.

Learn the Skill It is important to have an understanding of the graphics software being used before starting. The better the software is understood, the better the results. The graphics can be placed in a word-processing document.

- Clip art can be found on a variety of internet sites, and on CDs. These images can be copied and pasted into your document.
- When beginning, try editing existing drawings, then work up to creating drawings.
- The images are made of tiny rectangles of color called pixels. Each pixel can be altered.
- Digital photography is another way to add images. The photographs in the memory of a digital camera can be downloaded into a computer, then edited and added to the document.
- Graphics software also can allow animation. The software allows drawings to have the appearance of movement by connecting basic drawings automatically. This is called in-betweening, or tweening.
- Remember to save often.

Presentation Skills

Develop Multimedia Presentations

Most presentations are more dynamic if they include diagrams, photographs, videos, or sound recordings, like the one shown in **Figure 19.** A multimedia presentation involves using stereos, overhead projectors, televisions, computers, and more.

Learn the Skill Decide the main points of your presentation, and what types of media would best illustrate those points.

- Make sure you know how to use the equipment you are working with.
- Practice the presentation using the equipment several times.
- Enlist the help of a classmate to push play or turn lights out for you. Be sure to practice your presentation with him or her.
- If possible, set up all of the equipment ahead of time, and make sure everything is working properly.

Figure 19 These students are engaging the audience using a variety of tools.

Computer Presentations

There are many different interactive computer programs that you can use to enhance your presentation. Most computers have a compact disc (CD) drive that can play both CDs and digital video discs (DVDs). Also, there is hardware to connect a regular CD, DVD, or VCR. These tools will enhance your presentation.

Another method of using the computer to aid in your presentation is to develop a slide show using a computer program. This can allow movement of visuals at the presenter's pace, and can allow for visuals to build on one another.

Learn the Skill In order to create multimedia presentations on a computer, you need to have certain tools. These may include traditional graphic tools and drawing programs, animation programs, and authoring systems that tie everything together. Your computer will tell you which tools it supports. The most important step is to learn about the tools that you will be using.

- Often, color and strong images will convey a point better than words alone. Use the best methods available to convey your point.
- As with other presentations, practice many times.
- Practice your presentation with the tools you and any assistants will be using.
- Maintain eye contact with the audience. The purpose of using the computer is not to prompt the presenter, but to help the audience understand the points of the presentation.

Math Review

Use Fractions

A fraction compares a part to a whole. In the fraction $\frac{2}{3}$, the 2 represents the part and is the numerator. The 3 represents the whole and is the denominator.

Reduce Fractions To reduce a fraction, you must find the largest factor that is common to both the numerator and the denominator, the greatest common factor (GCF). Divide both numbers by the GCF. The fraction has then been reduced, or it is in its simplest form.

Example Twelve of the 20 chemicals in the science lab are in powder form. What fraction of the chemicals used in the lab are in powder form?

Step 1 Write the fraction.

$$\frac{\text{part}}{\text{whole}} = \frac{12}{20}$$

Step 2 To find the GCF of the numerator and denominator, list all of the factors of each number.

Factors of 12: 1, 2, 3, 4, 6, 12 (the numbers that divide evenly into 12)

Factors of 20: 1, 2, 4, 5, 10, 20 (the numbers that divide evenly into 20)

Step 3 List the common factors.

1, 2, 4.

Step 4 Choose the greatest factor in the list.

The GCF of 12 and 20 is 4.

Step 5 Divide the numerator and denominator by the GCF.

$$\frac{12 \div 4}{20 \div 4} = \frac{3}{5}$$

In the lab, $\frac{3}{5}$ of the chemicals are in powder form.

Practice Problem At an amusement park, 66 of 90 rides have a height restriction. What fraction of the rides, in its simplest form, has a height restriction?

Add and Subtract Fractions To add or subtract fractions with the same denominator, add or subtract the numerators and write the sum or difference over the denominator. After finding the sum or difference, find the simplest form for your fraction.

Example 1 In the forest outside your house, $\frac{1}{8}$ of the animals are rabbits, $\frac{3}{8}$ are squirrels, and the remainder are birds and insects. How many are mammals?

Step 1 Add the numerators.

$$\frac{1}{8} + \frac{3}{8} = \frac{(1 + 3)}{8} = \frac{4}{8}$$

Step 2 Find the GCF.

$$\frac{4}{8} \text{ (GCF, 4)}$$

Step 3 Divide the numerator and denominator by the GCF.

$$\frac{4}{4} = 1, \ \frac{8}{4} = 2$$

$\frac{1}{2}$ of the animals are mammals.

Example 2 If $\frac{7}{16}$ of the Earth is covered by freshwater, and $\frac{1}{16}$ of that is in glaciers, how much freshwater is not frozen?

Step 1 Subtract the numerators.

$$\frac{7}{16} - \frac{1}{16} = \frac{(7 - 1)}{16} = \frac{6}{16}$$

Step 2 Find the GCF.

$$\frac{6}{16} \text{ (GCF, 2)}$$

Step 3 Divide the numerator and denominator by the GCF.

$$\frac{6}{2} = 3, \ \frac{16}{2} = 8$$

$\frac{3}{8}$ of the freshwater is not frozen.

Practice Problem A bicycle rider is going 15 km/h for $\frac{4}{9}$ of his ride, 10 km/h for $\frac{2}{9}$ of his ride, and 8 km/h for the remainder of the ride. How much of his ride is he going over 8 km/h?

Unlike Denominators To add or subtract fractions with unlike denominators, first find the least common denominator (LCD). This is the smallest number that is a common multiple of both denominators. Rename each fraction with the LCD, and then add or subtract. Find the simplest form if necessary.

Example 1 A chemist makes a paste that is $\frac{1}{2}$ table salt (NaCl), $\frac{1}{3}$ sugar ($C_6H_{12}O_6$), and the rest water (H_2O). How much of the paste is a solid?

Step 1 Find the LCD of the fractions.

$\frac{1}{2} + \frac{1}{3}$ (LCD, 6)

Step 2 Rename each numerator and each denominator with the LCD.

$1 \times 3 = 3, \ 2 \times 3 = 6$

$1 \times 2 = 2, \ 3 \times 2 = 6$

Step 3 Add the numerators.

$\frac{3}{6} + \frac{2}{6} = \frac{(3 + 2)}{6} = \frac{5}{6}$

$\frac{5}{6}$ of the paste is a solid.

Example 2 The average precipitation in Grand Junction, CO, is $\frac{7}{10}$ inch in November, and $\frac{3}{5}$ inch in December. What is the total average precipitation?

Step 1 Find the LCD of the fractions.

$\frac{7}{10} + \frac{3}{5}$ (LCD, 10)

Step 2 Rename each numerator and each denominator with the LCD.

$7 \times 1 = 7, \ 10 \times 1 = 10$

$3 \times 2 = 6, \ 5 \times 2 = 10$

Step 3 Add the numerators.

$\frac{7}{10} + \frac{6}{10} = \frac{(7 + 6)}{10} = \frac{13}{10}$

$\frac{13}{10}$ inches total precipitation, or $1\frac{3}{10}$ inches.

Practice Problem On an electric bill, about $\frac{1}{8}$ of the energy is from solar energy and about $\frac{1}{10}$ is from wind power. How much of the total bill is from solar energy and wind power combined?

Example 3 In your body, $\frac{7}{10}$ of your muscle contractions are involuntary (cardiac and smooth muscle tissue). Smooth muscle makes up $\frac{3}{15}$ of your muscle contractions. How many of your muscle contractions are made by cardiac muscle?

Step 1 Find the LCD of the fractions.

$\frac{7}{10} - \frac{3}{15}$ (LCD, 30)

Step 2 Rename each numerator and each denominator with the LCD.

$7 \times 3 = 21, \ 10 \times 3 = 30$

$3 \times 2 = 6, \ 15 \times 2 = 30$

Step 3 Add the numerators.

$\frac{21}{30} - \frac{6}{30} = \frac{(21 - 6)}{30} = \frac{15}{30}$

Step 4 Find the GCF.

$\frac{15}{30}$ (GCF, 15)

$\frac{1}{2}$

$\frac{1}{2}$ of all muscle contractions are cardiac muscle.

Example 4 Tony wants to make cookies that call for $\frac{3}{4}$ of a cup of flour, but he only has $\frac{1}{3}$ of a cup. How much more flour does he need?

Step 1 Find the LCD of the fractions.

$\frac{3}{4} - \frac{1}{3}$ (LCD, 12)

Step 2 Rename each numerator and each denominator with the LCD.

$3 \times 3 = 9, \ 4 \times 3 = 12$

$1 \times 4 = 4, \ 3 \times 4 = 12$

Step 3 Add the numerators.

$\frac{9}{12} - \frac{4}{12} = \frac{(9 - 4)}{12} = \frac{5}{12}$

$\frac{5}{12}$ of a cup of flour.

Practice Problem Using the information provided to you in Example 3 above, determine how many muscle contractions are voluntary (skeletal muscle).

Multiply Fractions To multiply with fractions, multiply the numerators and multiply the denominators. Find the simplest form if necessary.

Example Multiply $\frac{3}{5}$ by $\frac{1}{3}$.

Step 1 Multiply the numerators and denominators.

$$\frac{3}{5} \times \frac{1}{3} = \frac{(3 \times 1)}{(5 \times 3)} = \frac{3}{15}$$

Step 2 Find the GCF.

$\frac{3}{15}$ (GCF, 3)

Step 3 Divide the numerator and denominator by the GCF.

$\frac{3}{3} = 1$, $\frac{15}{3} = 5$

$\frac{1}{5}$

$\frac{3}{5}$ multiplied by $\frac{1}{3}$ is $\frac{1}{5}$.

Practice Problem Multiply $\frac{3}{14}$ by $\frac{5}{16}$.

Find a Reciprocal Two numbers whose product is 1 are called multiplicative inverses, or reciprocals.

Example Find the reciprocal of $\frac{3}{8}$.

Step 1 Inverse the fraction by putting the denominator on top and the numerator on the bottom.

$\frac{8}{3}$

The reciprocal of $\frac{3}{8}$ is $\frac{8}{3}$.

Practice Problem Find the reciprocal of $\frac{4}{9}$.

Divide Fractions To divide one fraction by another fraction, multiply the dividend by the reciprocal of the divisor. Find the simplest form if necessary.

Example 1 Divide $\frac{1}{9}$ by $\frac{1}{3}$.

Step 1 Find the reciprocal of the divisor.

The reciprocal of $\frac{1}{3}$ is $\frac{3}{1}$.

Step 2 Multiply the dividend by the reciprocal of the divisor.

$$\frac{\frac{1}{9}}{\frac{1}{3}} = \frac{1}{9} \times \frac{3}{1} = \frac{(1 \times 3)}{(9 \times 1)} = \frac{3}{9}$$

Step 3 Find the GCF.

$\frac{3}{9}$ (GCF, 3)

Step 4 Divide the numerator and denominator by the GCF.

$\frac{3}{3} = 1$, $\frac{9}{3} = 3$

$\frac{1}{3}$

$\frac{1}{9}$ divided by $\frac{1}{3}$ is $\frac{1}{3}$.

Example 2 Divide $\frac{3}{5}$ by $\frac{1}{4}$.

Step 1 Find the reciprocal of the divisor.

The reciprocal of $\frac{1}{4}$ is $\frac{4}{1}$.

Step 2 Multiply the dividend by the reciprocal of the divisor.

$$\frac{\frac{3}{5}}{\frac{1}{4}} = \frac{3}{5} \times \frac{4}{1} = \frac{(3 \times 4)}{(5 \times 1)} = \frac{12}{5}$$

$\frac{3}{5}$ divided by $\frac{1}{4}$ is $\frac{12}{5}$ or $2\frac{2}{5}$.

Practice Problem Divide $\frac{3}{11}$ by $\frac{7}{10}$.

Use Ratios

When you compare two numbers by division, you are using a ratio. Ratios can be written 3 to 5, 3:5, or $\frac{3}{5}$. Ratios, like fractions, also can be written in simplest form.

Ratios can represent probabilities, also called odds. This is a ratio that compares the number of ways a certain outcome occurs to the number of outcomes. For example, if you flip a coin 100 times, what are the odds that it will come up heads? There are two possible outcomes, heads or tails, so the odds of coming up heads are 50:100. Another way to say this is that 50 out of 100 times the coin will come up heads. In its simplest form, the ratio is 1:2.

Example 1 A chemical solution contains 40 g of salt and 64 g of baking soda. What is the ratio of salt to baking soda as a fraction in simplest form?

Step 1 Write the ratio as a fraction.

$$\frac{\text{salt}}{\text{baking soda}} = \frac{40}{64}$$

Step 2 Express the fraction in simplest form.

The GCF of 40 and 64 is 8.

$$\frac{40}{64} = \frac{40 \div 8}{64 \div 8} = \frac{5}{8}$$

The ratio of salt to baking soda in the sample is 5:8.

Example 2 Sean rolls a 6-sided die 6 times. What are the odds that the side with a 3 will show?

Step 1 Write the ratio as a fraction.

$$\frac{\text{number of sides with a 3}}{\text{number of sides}} = \frac{1}{6}$$

Step 2 Multiply by the number of attempts.

$$\frac{1}{6} \times 6 \text{ attempts} = \frac{6}{6} \text{ attempts} = 1 \text{ attempt}$$

1 attempt out of 6 will show a 3.

Practice Problem Two metal rods measure 100 cm and 144 cm in length. What is the ratio of their lengths in simplest form?

Use Decimals

A fraction with a denominator that is a power of ten can be written as a decimal. For example, 0.27 means $\frac{27}{100}$. The decimal point separates the ones place from the tenths place.

Any fraction can be written as a decimal using division. For example, the fraction $\frac{5}{8}$ can be written as a decimal by dividing 5 by 8. Written as a decimal, it is 0.625.

Add or Subtract Decimals When adding and subtracting decimals, line up the decimal points before carrying out the operation.

Example 1 Find the sum of 47.68 and 7.80.

Step 1 Line up the decimal places when you write the numbers.

$$\begin{array}{r} 47.68 \\ +\ 7.80 \\ \hline \end{array}$$

Step 2 Add the decimals.

$$\begin{array}{r} 47.68 \\ +\ 7.80 \\ \hline 55.48 \end{array}$$

The sum of 47.68 and 7.80 is 55.48.

Example 2 Find the difference of 42.17 and 15.85.

Step 1 Line up the decimal places when you write the number.

$$\begin{array}{r} 42.17 \\ -15.85 \\ \hline \end{array}$$

Step 2 Subtract the decimals.

$$\begin{array}{r} 42.17 \\ -15.85 \\ \hline 26.32 \end{array}$$

The difference of 42.17 and 15.85 is 26.32.

Practice Problem Find the sum of 1.245 and 3.842.

Multiply Decimals To multiply decimals, multiply the numbers like any other number, ignoring the decimal point. Count the decimal places in each factor. The product will have the same number of decimal places as the sum of the decimal places in the factors.

Example Multiply 2.4 by 5.9.

Step 1 Multiply the factors like two whole numbers.
$24 \times 59 = 1416$

Step 2 Find the sum of the number of decimal places in the factors. Each factor has one decimal place, for a sum of two decimal places.

Step 3 The product will have two decimal places.
14.16

The product of 2.4 and 5.9 is 14.16.

Practice Problem Multiply 4.6 by 2.2.

Divide Decimals When dividing decimals, change the divisor to a whole number. To do this, multiply both the divisor and the dividend by the same power of ten. Then place the decimal point in the quotient directly above the decimal point in the dividend. Then divide as you do with whole numbers.

Example Divide 8.84 by 3.4.

Step 1 Multiply both factors by 10.
$3.4 \times 10 - 34, 8.84 \times 10 = 88.4$

Step 2 Divide 88.4 by 34.

$$\begin{array}{r} 2.6 \\ 34\overline{)88.4} \\ -\underline{68} \\ 204 \\ -\underline{204} \\ 0 \end{array}$$

8.84 divided by 3.4 is 2.6.

Practice Problem Divide 75.6 by 3.6.

Use Proportions

An equation that shows that two ratios are equivalent is a proportion. The ratios $\frac{2}{4}$ and $\frac{5}{10}$ are equivalent, so they can be written as $\frac{2}{4} = \frac{5}{10}$. This equation is a proportion.

When two ratios form a proportion, the cross products are equal. To find the cross products in the proportion $\frac{2}{4} = \frac{5}{10}$, multiply the 2 and the 10, and the 4 and the 5. Therefore $2 \times 10 = 4 \times 5$, or $20 = 20$.

Because you know that both proportions are equal, you can use cross products to find a missing term in a proportion. This is known as solving the proportion.

Example The heights of a tree and a pole are proportional to the lengths of their shadows. The tree casts a shadow of 24 m when a 6-m pole casts a shadow of 4 m. What is the height of the tree?

Step 1 Write a proportion.

$$\frac{\text{height of tree}}{\text{height of pole}} = \frac{\text{length of tree's shadow}}{\text{length of pole's shadow}}$$

Step 2 Substitute the known values into the proportion. Let h represent the unknown value, the height of the tree.

$$\frac{h}{6} = \frac{24}{4}$$

Step 3 Find the cross products.
$h \times 4 = 6 \times 24$

Step 4 Simplify the equation.
$4h - 144$

Step 5 Divide each side by 4.

$$\frac{4h}{4} = \frac{144}{4}$$
$$h = 36$$

The height of the tree is 36 m.

Practice Problem The ratios of the weights of two objects on the Moon and on Earth are in proportion. A rock weighing 3 N on the Moon weighs 18 N on Earth. How much would a rock that weighs 5 N on the Moon weigh on Earth?

Use Percentages

The word *percent* means "out of one hundred." It is a ratio that compares a number to 100. Suppose you read that 77 percent of the Earth's surface is covered by water. That is the same as reading that the fraction of the Earth's surface covered by water is $\frac{77}{100}$. To express a fraction as a percent, first find the equivalent decimal for the fraction. Then, multiply the decimal by 100 and add the percent symbol.

Example Express $\frac{13}{20}$ as a percent.

Step 1 Find the equivalent decimal for the fraction.

$$\begin{array}{r} 0.65 \\ 20\overline{)13.00} \\ \underline{12\ 0} \\ 1\ 00 \\ \underline{1\ 00} \\ 0 \end{array}$$

Step 2 Rewrite the fraction $\frac{13}{20}$ as 0.65.

Step 3 Multiply 0.65 by 100 and add the % sign.

$0.65 \times 100 = 65 = 65\%$

So, $\frac{13}{20} = 65\%$.

This also can be solved as a proportion.

Example Express $\frac{13}{20}$ as a percent.

Step 1 Write a proportion.

$\frac{13}{20} = \frac{x}{100}$

Step 2 Find the cross products.

$1300 = 20x$

Step 3 Divide each side by 20.

$\frac{1300}{20} = \frac{20x}{20}$

$65\% = x$

Practice Problem In one year, 73 of 365 days were rainy in one city. What percent of the days in that city were rainy?

Solve One-Step Equations

A statement that two things are equal is an equation. For example, $A = B$ is an equation that states that A is equal to B.

An equation is solved when a variable is replaced with a value that makes both sides of the equation equal. To make both sides equal the inverse operation is used. Addition and subtraction are inverses, and multiplication and division are inverses.

Example 1 Solve the equation $x - 10 = 35$.

Step 1 Find the solution by adding 10 to each side of the equation.

$x - 10 = 35$

$x - 10 + 10 = 35 + 10$

$x = 45$

Step 2 Check the solution.

$x - 10 = 35$

$45 - 10 = 35$

$35 = 35$

Both sides of the equation are equal, so $x = 45$.

Example 2 In the formula $a = bc$, find the value of c if $a = 20$ and $b = 2$.

Step 1 Rearrange the formula so the unknown value is by itself on one side of the equation by dividing both sides by b.

$a = bc$

$\frac{a}{b} = \frac{bc}{b}$

$\frac{a}{b} = c$

Step 2 Replace the variables a and b with the values that are given.

$\frac{a}{b} = c$

$\frac{20}{2} = c$

$10 = c$

Step 3 Check the solution.

$a = bc$

$20 = 2 \times 10$

$20 = 20$

Both sides of the equation are equal, so $c = 10$ is the solution when $a = 20$ and $b = 2$.

Practice Problem In the formula $h = gd$, find the value of d if $g = 12.3$ and $h = 17.4$.

Use Statistics

The branch of mathematics that deals with collecting, analyzing, and presenting data is statistics. In statistics, there are three common ways to summarize data with a single number—the mean, the median, and the mode.

The **mean** of a set of data is the arithmetic average. It is found by adding the numbers in the data set and dividing by the number of items in the set.

The **median** is the middle number in a set of data when the data are arranged in numerical order. If there were an even number of data points, the median would be the mean of the two middle numbers.

The **mode** of a set of data is the number or item that appears most often.

Another number that often is used to describe a set of data is the range. The **range** is the difference between the largest number and the smallest number in a set of data.

A **frequency table** shows how many times each piece of data occurs, usually in a survey. **Table 2** below shows the results of a student survey on favorite color.

Table 2 Student Color Choice

Color	Tally	Frequency
red	||||	4
blue	~~||||~~	5
black	||	2
green	|||	3
purple	~~||||~~ ||	7
yellow	~~||||~~ |	6

Based on the frequency table data, which color is the favorite?

Example The speeds (in m/s) for a race car during five different time trials are 39, 37, 44, 36, and 44.

To find the mean:

Step 1 Find the sum of the numbers.
$39 + 37 + 44 + 36 + 44 = 200$

Step 2 Divide the sum by the number of items, which is 5.
$200 \div 5 = 40$

The mean is 40 m/s.

To find the median:

Step 1 Arrange the measures from least to greatest.
36, 37, 39, 44, 44

Step 2 Determine the middle measure.
36, 37, <u>39</u>, 44, 44

The median is 39 m/s.

To find the mode:

Step 1 Group the numbers that are the same together.
44, 44, 36, 37, 39

Step 2 Determine the number that occurs most in the set.
<u>44, 44</u>, 36, 37, 39

The mode is 44 m/s.

To find the range:

Step 1 Arrange the measures from largest to smallest.
44, 44, 39, 37, 36

Step 2 Determine the largest and smallest measures in the set.
<u>44</u>, 44, 39, 37, <u>36</u>

Step 3 Find the difference between the largest and smallest measures.
$44 - 36 = 8$

The range is 8 m/s.

Practice Problem Find the mean, median, mode, and range for the data set 8, 4, 12, 8, 11, 14, 16.

Use Geometry

The branch of mathematics that deals with the measurement, properties, and relationships of points, lines, angles, surfaces, and solids is called geometry.

Perimeter The **perimeter** (P) is the distance around a geometric figure. To find the perimeter of a rectangle, add the length and width and multiply that sum by two, or $2(l + w)$. Find perimeters of irregular figures by adding the length of the sides.

Example 1 Find the perimeter of a rectangle that is 3 m long and 5 m wide.

Step 1 You know that the perimeter is 2 times the sum of the width and length.
$P = 2(3 \text{ m} + 5 \text{ m})$

Step 2 Find the sum of the width and length.
$P = 2(8 \text{ m})$

Step 3 Multiply by 2.
$P = 16 \text{ m}$

The perimeter is 16 m.

Example 2 Find the perimeter of a shape with sides measuring 2 cm, 5 cm, 6 cm, 3 cm.

Step 1 You know that the perimeter is the sum of all the sides.
$P = 2 + 5 + 6 + 3$

Step 2 Find the sum of the sides.
$P = 2 + 5 + 6 + 3$
$P = 16$

The perimeter is 16 cm.

Practice Problem Find the perimeter of a rectangle with a length of 18 m and a width of 7 m.

Practice Problem Find the perimeter of a triangle measuring 1.6 cm by 2.4 cm by 2.4 cm.

Area of a Rectangle The **area** (A) is the number of square units needed to cover a surface. To find the area of a rectangle, multiply the length times the width, or $l \times w$. When finding area, the units also are multiplied. Area is given in square units.

Example Find the area of a rectangle with a length of 1 cm and a width of 10 cm.

Step 1 You know that the area is the length multiplied by the width.
$A = (1 \text{ cm} \times 10 \text{ cm})$

Step 2 Multiply the length by the width. Also multiply the units.
$A = 10 \text{ cm}^2$

The area is 10 cm^2.

Practice Problem Find the area of a square whose sides measure 4 m.

Area of a Triangle To find the area of a triangle, use the formula:

$$A = \frac{1}{2}(\text{base} \times \text{height})$$

The base of a triangle can be any of its sides. The height is the perpendicular distance from a base to the opposite endpoint, or vertex.

Example Find the area of a triangle with a base of 18 m and a height of 7 m.

Step 1 You know that the area is $\frac{1}{2}$ the base times the height.
$A = \frac{1}{2}(18 \text{ m} \times 7 \text{ m})$

Step 2 Multiply $\frac{1}{2}$ by the product of 18×7. Multiply the units.
$A = \frac{1}{2}(126 \text{ m}^2)$
$A = 63 \text{ m}^2$

The area is 63 m^2.

Practice Problem Find the area of a triangle with a base of 27 cm and a height of 17 cm.

Circumference of a Circle The **diameter** (d) of a circle is the distance across the circle through its center, and the **radius** (r) is the distance from the center to any point on the circle. The radius is half of the diameter. The distance around the circle is called the **circumference** (C). The formula for finding the circumference is:

$$C = 2\pi r \text{ or } C = \pi d$$

The circumference divided by the diameter is always equal to 3.1415926... This nonterminating and nonrepeating number is represented by the Greek letter π (pi). An approximation often used for π is 3.14.

Example 1 Find the circumference of a circle with a radius of 3 m.

Step 1 You know the formula for the circumference is 2 times the radius times π.
$C = 2\pi(3)$

Step 2 Multiply 2 times the radius.
$C = 6\pi$

Step 3 Multiply by π.
$C = 19$ m

The circumference is 19 m.

Example 2 Find the circumference of a circle with a diameter of 24.0 cm.

Step 1 You know the formula for the circumference is the diameter times π.
$C = \pi(24.0)$

Step 2 Multiply the diameter by π.
$C = 75.4$ cm

The circumference is 75.4 cm.

Practice Problem Find the circumference of a circle with a radius of 19 cm.

Area of a Circle The formula for the area of a circle is:
$A = \pi r^2$

Example 1 Find the area of a circle with a radius of 4.0 cm.

Step 1 $A = \pi(4.0)^2$

Step 2 Find the square of the radius.
$A = 16\pi$

Step 3 Multiply the square of the radius by π.
$A = 50 \text{ cm}^2$

The area of the circle is 50 cm^2.

Example 2 Find the area of a circle with a radius of 225 m.

Step 1 $A = \pi(225)^2$

Step 2 Find the square of the radius.
$A = 50625\pi$

Step 3 Multiply the square of the radius by π.
$A = 158962.5$

The area of the circle is 158,962 m^2.

Example 3 Find the area of a circle whose diameter is 20.0 mm.

Step 1 You know the formula for the area of a circle is the square of the radius times π, and that the radius is half of the diameter.
$A = \pi\left(\frac{20.0}{2}\right)^2$

Step 2 Find the radius.
$A = \pi(10.0)^2$

Step 3 Find the square of the radius.
$A = 100\pi$

Step 4 Multiply the square of the radius by π.
$A = 314 \text{ mm}^2$

The area is 314 mm^2.

Practice Problem Find the area of a circle with a radius of 16 m.

Volume The measure of space occupied by a solid is the **volume** (V). To find the volume of a rectangular solid multiply the length times width times height, or $V = l \times w \times h$. It is measured in cubic units, such as cubic centimeters (cm^3).

Example Find the volume of a rectangular solid with a length of 2.0 m, a width of 4.0 m, and a height of 3.0 m.

Step 1 You know the formula for volume is the length times the width times the height.

$V = 2.0 \text{ m} \times 4.0 \text{ m} \times 3.0 \text{ m}$

Step 2 Multiply the length times the width times the height.

$V = 24 \text{ m}^3$

The volume is 24 m^3.

Practice Problem Find the volume of a rectangular solid that is 8 m long, 4 m wide, and 4 m high.

To find the volume of other solids, multiply the area of the base times the height.

Example 1 Find the volume of a solid that has a triangular base with a length of 8.0 m and a height of 7.0 m. The height of the entire solid is 15.0 m.

Step 1 You know that the base is a triangle, and the area of a triangle is $\frac{1}{2}$ the base times the height, and the volume is the area of the base times the height.

$V = \left[\frac{1}{2}(b \times h)\right] \times 15$

Step 2 Find the area of the base.

$V = \left[\frac{1}{2}(8 \times 7)\right] \times 15$

$V = \left(\frac{1}{2} \times 56\right) \times 15$

Step 3 Multiply the area of the base by the height of the solid.

$V = 28 \times 15$

$V = 420 \text{ m}^3$

The volume is 420 m^3.

Example 2 Find the volume of a cylinder that has a base with a radius of 12.0 cm, and a height of 21.0 cm.

Step 1 You know that the base is a circle, and the area of a circle is the square of the radius times π, and the volume is the area of the base times the height.

$V = (\pi r^2) \times 21$

$V = (\pi 12^2) \times 21$

Step 2 Find the area of the base.

$V = 144\pi \times 21$

$V = 452 \times 21$

Step 3 Multiply the area of the base by the height of the solid.

$V = 9490 \text{ cm}^3$

The volume is 9490 cm^3.

Example 3 Find the volume of a cylinder that has a diameter of 15 mm and a height of 4.8 mm.

Step 1 You know that the base is a circle with an area equal to the square of the radius times π. The radius is one-half the diameter. The volume is the area of the base times the height.

$V = (\pi r^2) \times 4.8$

$V = \left[\pi\left(\frac{1}{2} \times 15\right)^2\right] \times 4.8$

$V = (\pi 7.5^2) \times 4.8$

Step 2 Find the area of the base.

$V = 56.25\pi \times 4.8$

$V = 176.63 \times 4.8$

Step 3 Multiply the area of the base by the height of the solid.

$V = 847.8$

The volume is 847.8 mm^3.

Practice Problem Find the volume of a cylinder with a diameter of 7 cm in the base and a height of 16 cm.

Science Applications

Measure in SI

The metric system of measurement was developed in 1795. A modern form of the metric system, called the International System (SI), was adopted in 1960 and provides the standard measurements that all scientists around the world can understand.

The SI system is convenient because unit sizes vary by powers of 10. Prefixes are used to name units. Look at **Table 3** for some common metric prefixes and their meanings.

Table 3 Common SI Prefixes

Prefix	Symbol	Meaning	
kilo-	k	1,000	thousand
hecto-	h	100	hundred
deka-	da	10	ten
deci-	d	0.1	tenth
centi-	c	0.01	hundredth
milli-	m	0.001	thousandth

Example How many grams equal one kilogram?

Step 1 Find the prefix *kilo* in **Table 3.**

Step 2 Using **Table 3,** determine the meaning of *kilo.* According to the table, it means 1,000. When the prefix *kilo* is added to a unit, it means that there are 1,000 of the units in a "*kilo*unit."

Step 3 Apply the prefix to the units in the question. The units in the question are grams. There are 1,000 grams in a kilogram.

Practice Problem Is a milligram larger or smaller than a gram? How many of the smaller units equal one larger unit? What fraction of the larger unit does one smaller unit represent?

Dimensional Analysis

Convert SI Units In science, quantities such as length, mass, and time sometimes are measured using different units. A process called dimensional analysis can be used to change one unit of measure to another. This process involves multiplying your starting quantity and units by one or more conversion factors. A conversion factor is a ratio equal to one and can be made from any two equal quantities with different units. If 1,000 mL equal 1 L then two ratios can be made.

$$\frac{1{,}000 \text{ mL}}{1 \text{ L}} = \frac{1 \text{ L}}{1{,}000 \text{ mL}} = 1$$

One can covert between units in the SI system by using the equivalents in **Table 3** to make conversion factors.

Example 1 How many cm are in 4 m?

Step 1 Write conversion factors for the units given. From **Table 3,** you know that 100 cm = 1 m. The conversion factors are

$$\frac{100 \text{ cm}}{1 \text{ m}} \text{ and } \frac{1 \text{ m}}{100 \text{ cm}}$$

Step 2 Decide which conversion factor to use. Select the factor that has the units you are converting from (m) in the denominator and the units you are converting to (cm) in the numerator.

$$\frac{100 \text{ cm}}{1 \text{ m}}$$

Step 3 Multiply the starting quantity and units by the conversion factor. Cancel the starting units with the units in the denominator. There are 400 cm in 4 m.

$$4 \cancel{\text{m}} \times \frac{100 \text{ cm}}{1 \cancel{\text{m}}} = 400 \text{ cm}$$

Practice Problem How many milligrams are in one kilogram? (Hint: You will need to use two conversion factors from **Table 3.**)

Table 4 Unit System Equivalents

Type of Measurement	Equivalent
Length	1 in = 2.54 cm 1 yd = 0.91 m 1 mi = 1.61 km
Mass and Weight*	1 oz = 28.35 g 1 lb = 0.45 kg 1 ton (short) = 0.91 tonnes (metric tons) 1 lb = 4.45 N
Volume	1 in^3 = 16.39 cm^3 1 qt = 0.95 L 1 gal = 3.78 L
Area	1 in^2 = 6.45 cm^2 1 yd^2 = 0.83 m^2 1 mi^2 = 2.59 km^2 1 acre = 0.40 hectares
Temperature	$°C = \frac{(°F - 32)}{1.8}$ K = °C + 273

*Weight is measured in standard Earth gravity.

Convert Between Unit Systems **Table 4** gives a list of equivalents that can be used to convert between English and SI units.

Example If a meterstick has a length of 100 cm, how long is the meterstick in inches?

Step 1 Write the conversion factors for the units given. From **Table 4,** 1 in = 2.54 cm.

$$\frac{1\text{ in}}{2.54\text{ cm}} \quad \textit{and} \quad \frac{2.54\text{ cm}}{1\text{ in}}$$

Step 2 Determine which conversion factor to use. You are converting from cm to in. Use the conversion factor with cm on the bottom.

$$\frac{1\text{ in}}{2.54\text{ cm}}$$

Step 3 Multiply the starting quantity and units by the conversion factor. Cancel the starting units with the units in the denominator. Round your answer based on the number of significant figures in the conversion factor.

$$100\ \cancel{\text{cm}} \times \frac{1\text{ in}}{2.54\ \cancel{\text{cm}}} = 39.37\text{ cm}$$

The meterstick is 39.4 in long.

Practice Problem A book has a mass of 5 lbs. What is the mass of the book in kg?

Practice Problem Use the equivalent for in and cm (1 in = 2.54 cm) to show how 1 in^3 = 16.39 cm^3.

Precision and Significant Digits

When you make a measurement, the value you record depends on the precision of the measuring instrument. This precision is represented by the number of significant digits recorded in the measurement. When counting the number of significant digits, all digits are counted except zeros at the end of a number with no decimal point such as 2,050, and zeros at the beginning of a decimal such as 0.03020. When adding or subtracting numbers with different precision, round the answer to the smallest number of decimal places of any number in the sum or difference. When multiplying or dividing, the answer is rounded to the smallest number of significant digits of any number being multiplied or divided.

Example The lengths 5.28 and 5.2 are measured in meters. Find the sum of these lengths and record your answer using the correct number of significant digits.

Step 1 Find the sum.

5.28 m	2 digits after the decimal
+ 5.2 m	1 digit after the decimal
10.48 m	

Step 2 Round to one digit after the decimal because the least number of digits after the decimal of the numbers being added is 1.

The sum is 10.5 m.

Practice Problem How many significant digits are in the measurement 7,071,301 m? How many significant digits are in the measurement 0.003010 g?

Practice Problem Multiply 5.28 and 5.2 using the rule for multiplying and dividing. Record the answer using the correct number of significant digits.

Scientific Notation

Many times numbers used in science are very small or very large. Because these numbers are difficult to work with scientists use scientific notation. To write numbers in scientific notation, move the decimal point until only one non-zero digit remains on the left. Then count the number of places you moved the decimal point and use that number as a power of ten. For example, the average distance from the Sun to Mars is 227,800,000,000 m. In scientific notation, this distance is 2.278×10^{11} m. Because you moved the decimal point to the left, the number is a positive power of ten.

The mass of an electron is about 0.000 000 000 000 000 000 000 000 000 000 911 kg. Expressed in scientific notation, this mass is 9.11×10^{-31} kg. Because the decimal point was moved to the right, the number is a negative power of ten.

Example Earth is 149,600,000 km from the Sun. Express this in scientific notation.

Step 1 Move the decimal point until one non-zero digit remains on the left.

1.496 000 00

Step 2 Count the number of decimal places you have moved. In this case, eight.

Step 3 Show that number as a power of ten, 10^8.

The Earth is 1.496×10^8 km from the Sun.

Practice Problem How many significant digits are in 149,600,000 km? How many significant digits are in 1.496×10^8 km?

Practice Problem Parts used in a high performance car must be measured to 7×10^{-6} m. Express this number as a decimal.

Practice Problem A CD is spinning at 539 revolutions per minute. Express this number in scientific notation.

Make and Use Graphs

Data in tables can be displayed in a graph—a visual representation of data. Common graph types include line graphs, bar graphs, and circle graphs.

Line Graph A line graph shows a relationship between two variables that change continuously. The independent variable is changed and is plotted on the x-axis. The dependent variable is observed, and is plotted on the y-axis.

Example Draw a line graph of the data below from a cyclist in a long-distance race.

Table 5 Bicycle Race Data

Time (h)	Distance (km)
0	0
1	8
2	16
3	24
4	32
5	40

Step 1 Determine the x-axis and y-axis variables. Time varies independently of distance and is plotted on the x-axis. Distance is dependent on time and is plotted on the y-axis.

Step 2 Determine the scale of each axis. The x-axis data ranges from 0 to 5. The y-axis data ranges from 0 to 40.

Step 3 Using graph paper, draw and label the axes. Include units in the labels.

Step 4 Draw a point at the intersection of the time value on the x-axis and corresponding distance value on the y-axis. Connect the points and label the graph with a title, as shown in **Figure 20.**

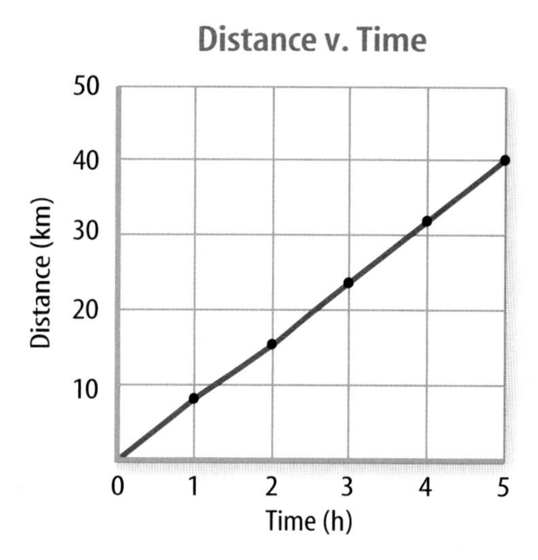

Figure 20 This line graph shows the relationship between distance and time during a bicycle ride.

Practice Problem A puppy's shoulder height is measured during the first year of her life. The following measurements were collected: (3 mo, 52 cm), (6 mo, 72 cm), (9 mo, 83 cm), (12 mo, 86 cm). Graph this data.

Find a Slope The slope of a straight line is the ratio of the vertical change, rise, to the horizontal change, run.

$$\text{Slope} = \frac{\text{vertical change (rise)}}{\text{horizontal change (run)}} = \frac{\text{change in } y}{\text{change in } x}$$

Example Find the slope of the graph in **Figure 20.**

Step 1 You know that the slope is the change in y divided by the change in x.

$$\text{Slope} = \frac{\text{change in } y}{\text{change in } x}$$

Step 2 Determine the data points you will be using. For a straight line, choose the two sets of points that are the farthest apart.

$$\text{Slope} = \frac{(40-0) \text{ km}}{(5-0) \text{ hr}}$$

Step 3 Find the change in y and x.

$$\text{Slope} = \frac{40 \text{ km}}{5\text{h}}$$

Step 4 Divide the change in y by the change in x.

$$\text{Slope} = \frac{8 \text{ km}}{\text{h}}$$

The slope of the graph is 8 km/h.

Bar Graph To compare data that does not change continuously you might choose a bar graph. A bar graph uses bars to show the relationships between variables. The x-axis variable is divided into parts. The parts can be numbers such as years, or a category such as a type of animal. The y-axis is a number and increases continuously along the axis.

Example A recycling center collects 4.0 kg of aluminum on Monday, 1.0 kg on Wednesday, and 2.0 kg on Friday. Create a bar graph of this data.

Step 1 Select the x-axis and y-axis variables. The measured numbers (the masses of aluminum) should be placed on the y-axis. The variable divided into parts (collection days) is placed on the x-axis.

Step 2 Create a graph grid like you would for a line graph. Include labels and units.

Step 3 For each measured number, draw a vertical bar above the x-axis value up to the y-axis value. For the first data point, draw a vertical bar above Monday up to 4.0 kg.

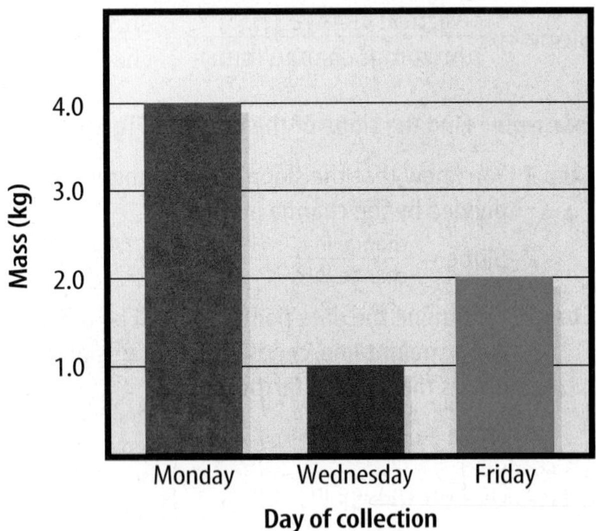

Practice Problem Draw a bar graph of the gases in air: 78% nitrogen, 21% oxygen, 1% other gases.

Circle Graph To display data as parts of a whole, you might use a circle graph. A circle graph is a circle divided into sections that represent the relative size of each piece of data. The entire circle represents 100%, half represents 50%, and so on.

Example Air is made up of 78% nitrogen, 21% oxygen, and 1% other gases. Display the composition of air in a circle graph.

Step 1 Multiply each percent by 360° and divide by 100 to find the angle of each section in the circle.

$$78\% \times \frac{360^\circ}{100} = 280.8^\circ$$

$$21\% \times \frac{360^\circ}{100} = 75.6^\circ$$

$$1\% \times \frac{360^\circ}{100} = 3.6^\circ$$

Step 2 Use a compass to draw a circle and to mark the center of the circle. Draw a straight line from the center to the edge of the circle.

Step 3 Use a protractor and the angles you calculated to divide the circle into parts. Place the center of the protractor over the center of the circle and line the base of the protractor over the straight line.

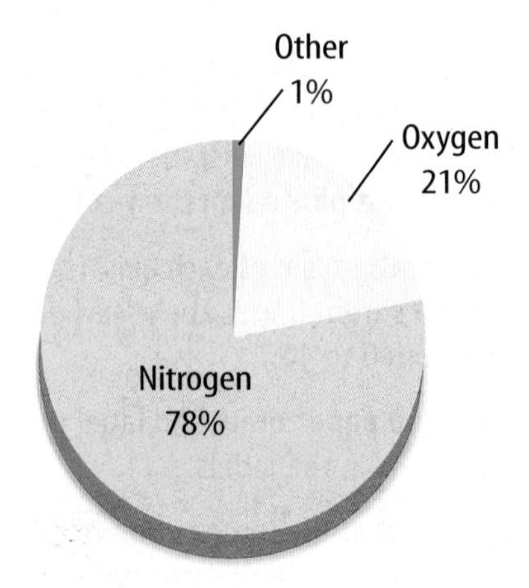

Practice Problem Draw a circle graph to represent the amount of aluminum collected during the week shown in the bar graph to the left.

Use and Care of a Microscope

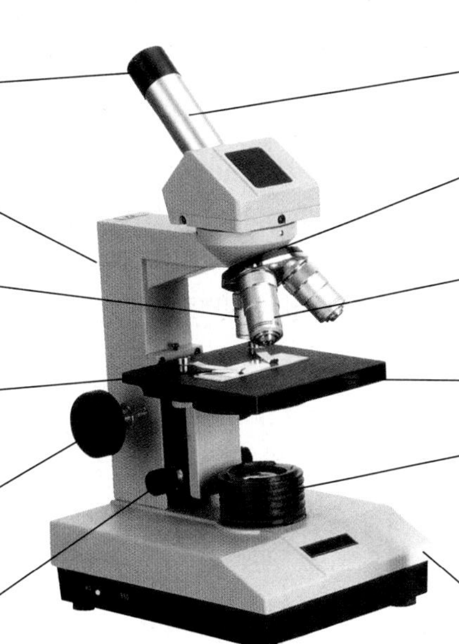
Eyepiece Contains magnifying lenses you look through.
Arm Supports the body tube.
Low-power objective Contains the lens with the lowest power magnification.
Stage clips Hold the microscope slide in place.
Coarse adjustment Focuses the image under low power.
Fine adjustment Sharpens the image under high magnification.
Body tube Connects the eyepiece to the revolving nosepiece.
Revolving nosepiece Holds and turns the objectives into viewing position.
High-power objective Contains the lens with the highest magnification.
Stage Supports the microscope slide.
Light source Provides light that passes upward through the diaphragm, the specimen, and the lenses.
Base Provides support for the microscope.

Caring for a Microscope

1. Always carry the microscope holding the arm with one hand and supporting the base with the other hand.
2. Don't touch the lenses with your fingers.
3. The coarse adjustment knob is used only when looking through the lowest-power objective lens. The fine adjustment knob is used when the high-power objective is in place.
4. Cover the microscope when you store it.

Using a Microscope

1. Place the microscope on a flat surface that is clear of objects. The arm should be toward you.
2. Look through the eyepiece. Adjust the diaphragm so light comes through the opening in the stage.
3. Place a slide on the stage so the specimen is in the field of view. Hold it firmly in place by using the stage clips.
4. Always focus with the coarse adjustment and the low-power objective lens first. After the object is in focus on low power, turn the nosepiece until the high-power objective is in place. Use ONLY the fine adjustment to focus with the high-power objective lens.

Making a Wet-Mount Slide

1. Carefully place the item you want to look at in the center of a clean, glass slide. Make sure the sample is thin enough for light to pass through.
2. Use a dropper to place one or two drops of water on the sample.
3. Hold a clean coverslip by the edges and place it at one edge of the water. Slowly lower the coverslip onto the water until it lies flat.
4. If you have too much water or a lot of air bubbles, touch the edge of a paper towel to the edge of the coverslip to draw off extra water and draw out unwanted air.

Diversity of Life: Classification of Living Organisms

A six-kingdom system of classification of organisms is used today. Two kingdoms—Kingdom Archaebacteria and Kingdom Eubacteria—contain organisms that do not have a nucleus and that lack membrane-bound structures in the cytoplasm of their cells. The members of the other four kingdoms have a cell or cells that contain a nucleus and structures in the cytoplasm, some of which are surrounded by membranes. These kingdoms are Kingdom Protista, Kingdom Fungi, Kingdom Plantae, and Kingdom Animalia.

Kingdom Archaebacteria

one-celled; some absorb food from their surroundings; some are photosynthetic; some are chemosynthetic; many are found in extremely harsh environments including salt ponds, hot springs, swamps, and deep-sea hydrothermal vents

Kingdom Eubacteria

one-celled; most absorb food from their surroundings; some are photosynthetic; some are chemosynthetic; many are parasites; many are round, spiral, or rod-shaped; some form colonies

Kingdom Protista

Phylum Euglenophyta one-celled; photosynthetic or take in food; most have one flagellum; euglenoids

Phylum Bacillariophyta one-celled; photosynthetic; have unique double shells made of silica; diatoms

Phylum Dinoflagellata one-celled; photosynthetic; contain red pigments; have two flagella; dinoflagellates

Phylum Chlorophyta one-celled, many-celled, or colonies; photosynthetic; contain chlorophyll; live on land, in freshwater, or salt water; green algae

Phylum Rhodophyta most are many-celled; photosynthetic; contain red pigments; most live in deep, saltwater environments; red algae

Phylum Phaeophyta most are many-celled; photosynthetic; contain brown pigments; most live in saltwater environments; brown algae

Phylum Rhizopoda one-celled; take in food; are free-living or parasitic; move by means of pseudopods; amoebas

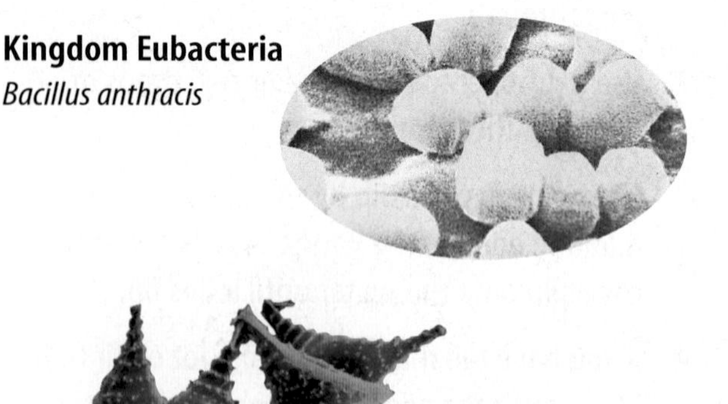

Kingdom Eubacteria
Bacillus anthracis

Phylum Chlorophyta
Desmids

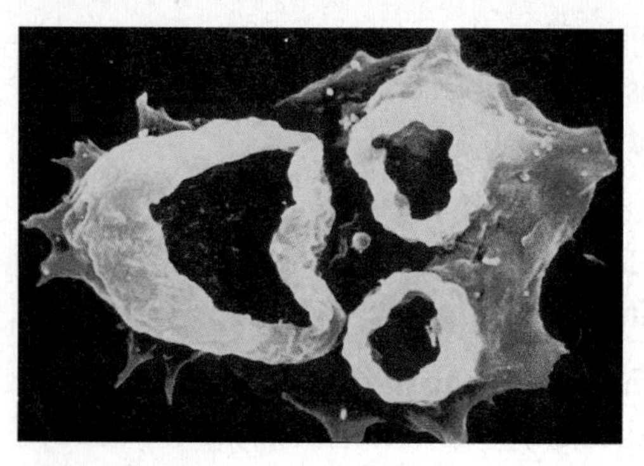

Amoeba

Phylum Zoomastigina one-celled; take in food; free-living or parasitic; have one or more flagella; zoomastigotes

Phylum Ciliophora one-celled; take in food; have large numbers of cilia; ciliates

Phylum Sporozoa one-celled; take in food; have no means of movement; are parasites in animals; sporozoans

Phylum Myxomycota
Slime mold

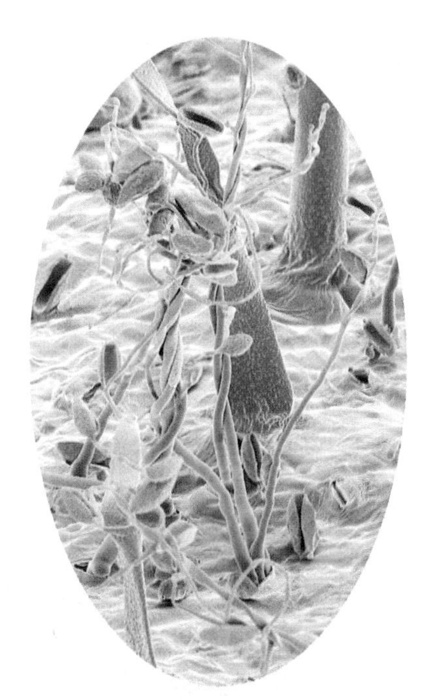

Phylum Oomycota
Phytophthora infestans

Phyla Myxomycota and Acrasiomycota one- or many-celled; absorb food; change form during life cycle; cellular and plasmodial slime molds

Phylum Oomycota many-celled; are either parasites or decomposers; live in freshwater or salt water; water molds, rusts and downy mildews

Kingdom Fungi

Phylum Zygomycota many-celled; absorb food; spores are produced in sporangia; zygote fungi; bread mold

Phylum Ascomycota one- and many-celled; absorb food; spores produced in asci; sac fungi; yeast

Phylum Basidiomycota many-celled; absorb food; spores produced in basidia; club fungi; mushrooms

Phylum Deuteromycota members with unknown reproductive structures; imperfect fungi; *Penicillium*

Phylum Mycophycota organisms formed by symbiotic relationship between an ascomycote or a basidiomycote and green alga or cyanobacterium; lichens

Lichens

Kingdom Plantae

Divisions Bryophyta (mosses), **Anthocerophyta** (hornworts), **Hepaticophyta** (liverworts), **Psilophyta** (whisk ferns) many-celled nonvascular plants; reproduce by spores produced in capsules; green; grow in moist, land environments

Division Lycophyta many-celled vascular plants; spores are produced in conelike structures; live on land; are photosynthetic; club mosses

Division Arthrophyta vascular plants; ribbed and jointed stems; scalelike leaves; spores produced in conelike structures; horsetails

Division Pterophyta vascular plants; leaves called fronds; spores produced in clusters of sporangia called sori; live on land or in water; ferns

Division Ginkgophyta deciduous trees; only one living species; have fan-shaped leaves with branching veins and fleshy cones with seeds; ginkgoes

Division Cycadophyta palmlike plants; have large, featherlike leaves; produces seeds in cones; cycads

Division Coniferophyta deciduous or evergreen; trees or shrubs; have needlelike or scale-like leaves; seeds produced in cones; conifers

Division Gnetophyta shrubs or woody vines; seeds are produced in cones; division contains only three genera; gnetum

Division Anthophyta dominant group of plants; flowering plants; have fruits with seeds

Kingdom Animalia

Phylum Porifera aquatic organisms that lack true tissues and organs; are asymmetrical and sessile; sponges

Phylum Cnidaria radially symmetrical organisms; have a digestive cavity with one opening; most have tentacles armed with stinging cells; live in aquatic environments singly or in colonies; includes jellyfish, corals, hydra, and sea anemones

Phylum Platyhelminthes bilaterally symmetrical worms; have flattened bodies; digestive system has one opening; parasitic and free-living species; flatworms

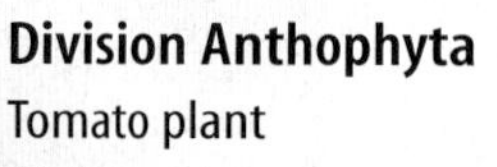

Division Anthophyta
Tomato plant

Division Bryophyta
Liverwort

Phylum Platyhelminthes
Flatworm

Phylum Chordata

Phylum Nematoda round, bilaterally symmetrical body; have digestive system with two openings; free-living forms and parasitic forms; roundworms

Phylum Mollusca soft-bodied animals, many with a hard shell and soft foot or footlike appendage; a mantle covers the soft body; aquatic and terrestrial species; includes clams, snails, squid, and octopuses

Phylum Annelida bilaterally symmetrical worms; have round, segmented bodies; terrestrial and aquatic species; includes earthworms, leeches, and marine polychaetes

Phylum Arthropoda largest animal group; have hard exoskeletons, segmented bodies, and pairs of jointed appendages; land and aquatic species; includes insects, crustaceans, and spiders

Phylum Echinodermata marine organisms; have spiny or leathery skin and a water-vascular system with tube feet; are radially symmetrical; includes sea stars, sand dollars, and sea urchins

Phylum Chordata organisms with internal skeletons and specialized body systems; most have paired appendages; all at some time have a notochord, nerve cord, gill slits, and a post-anal tail; include fish, amphibians, reptiles, birds, and mammals

PERIODIC TABLE OF THE ELEMENTS

Columns of elements are called groups. Elements in the same group have similar chemical properties.

Key: Element — Hydrogen; Atomic number — 1; Symbol — H; Atomic mass — 1.008; State of matter (Gas symbol)

Gas / Liquid / Solid / Synthetic

The first three symbols tell you the state of matter of the element at room temperature. The fourth symbol identifies elements that are not present in significant amounts on Earth. Useful amounts are made synthetically.

Period	1	2	3	4	5	6	7	8	9
1	Hydrogen 1 H 1.008 (Gas)								
2	Lithium 3 Li 6.941 (Solid)	Beryllium 4 Be 9.012 (Solid)							
3	Sodium 11 Na 22.990 (Solid)	Magnesium 12 Mg 24.305 (Solid)							
4	Potassium 19 K 39.098 (Solid)	Calcium 20 Ca 40.078 (Solid)	Scandium 21 Sc 44.956 (Solid)	Titanium 22 Ti 47.867 (Solid)	Vanadium 23 V 50.942 (Solid)	Chromium 24 Cr 51.996 (Solid)	Manganese 25 Mn 54.938 (Solid)	Iron 26 Fe 55.845 (Solid)	Cobalt 27 Co 58.933 (Solid)
5	Rubidium 37 Rb 85.468 (Solid)	Strontium 38 Sr 87.62 (Solid)	Yttrium 39 Y 88.906 (Solid)	Zirconium 40 Zr 91.224 (Solid)	Niobium 41 Nb 92.906 (Solid)	Molybdenum 42 Mo 95.94 (Solid)	Technetium 43 Tc (98) (Synthetic)	Ruthenium 44 Ru 101.07 (Solid)	Rhodium 45 Rh 102.906 (Solid)
6	Cesium 55 Cs 132.905 (Solid)	Barium 56 Ba 137.327 (Solid)	Lanthanum 57 La 138.906 (Solid)	Hafnium 72 Hf 178.49 (Solid)	Tantalum 73 Ta 180.948 (Solid)	Tungsten 74 W 183.84 (Solid)	Rhenium 75 Re 186.207 (Solid)	Osmium 76 Os 190.23 (Solid)	Iridium 77 Ir 192.217 (Solid)
7	Francium 87 Fr (223) (Solid)	Radium 88 Ra (226) (Solid)	Actinium 89 Ac (227) (Solid)	Rutherfordium 104 Rf (261) (Synthetic)	Dubnium 105 Db (262) (Synthetic)	Seaborgium 106 Sg (266) (Synthetic)	Bohrium 107 Bh (264) (Synthetic)	Hassium 108 Hs (277) (Synthetic)	Meitnerium 109 Mt (268) (Synthetic)

The number in parentheses is the mass number of the longest-lived isotope for that element.

Rows of elements are called periods. Atomic number increases across a period.

The arrow shows where these elements would fit into the periodic table. They are moved to the bottom of the table to save space.

Lanthanide series	Cerium 58 Ce 140.116 (Solid)	Praseodymium 59 Pr 140.908 (Solid)	Neodymium 60 Nd 144.24 (Solid)	Promethium 61 Pm (145) (Synthetic)	Samarium 62 Sm 150.36 (Solid)
Actinide series	Thorium 90 Th 232.038 (Solid)	Protactinium 91 Pa 231.036 (Solid)	Uranium 92 U 238.029 (Solid)	Neptunium 93 Np (237) (Synthetic)	Plutonium 94 Pu (244) (Synthetic)

Metal

Metalloid

Nonmetal

The color of an element's block tells you if the element is a metal, nonmetal, or metalloid.

Science Online
Visit life.msscience.com for updates to the periodic table.

10	11	12	13	14	15	16	17	18
								Helium 2 **He** 4.003
			Boron 5 **B** 10.811	Carbon 6 **C** 12.011	Nitrogen 7 **N** 14.007	Oxygen 8 **O** 15.999	Fluorine 9 **F** 18.998	Neon 10 **Ne** 20.180
			Aluminum 13 **Al** 26.982	Silicon 14 **Si** 28.086	Phosphorus 15 **P** 30.974	Sulfur 16 **S** 32.065	Chlorine 17 **Cl** 35.453	Argon 18 **Ar** 39.948
Nickel 28 **Ni** 58.693	Copper 29 **Cu** 63.546	Zinc 30 **Zn** 65.409	Gallium 31 **Ga** 69.723	Germanium 32 **Ge** 72.64	Arsenic 33 **As** 74.922	Selenium 34 **Se** 78.96	Bromine 35 **Br** 79.904	Krypton 36 **Kr** 83.798
Palladium 46 **Pd** 106.42	Silver 47 **Ag** 107.868	Cadmium 48 **Cd** 112.411	Indium 49 **In** 114.818	Tin 50 **Sn** 118.710	Antimony 51 **Sb** 121.760	Tellurium 52 **Te** 127.60	Iodine 53 **I** 126.904	Xenon 54 **Xe** 131.293
Platinum 78 **Pt** 195.078	Gold 79 **Au** 196.967	Mercury 80 **Hg** 200.59	Thallium 81 **Tl** 204.383	Lead 82 **Pb** 207.2	Bismuth 83 **Bi** 208.980	Polonium 84 **Po** (209)	Astatine 85 **At** (210)	Radon 86 **Rn** (222)
Darmstadtium 110 **Ds** (281)	Roentgenium 111 **Rg** (272)	Ununbium * 112 **Uub** (285)		Ununquadium * 114 **Uuq** (289)				

* The names and symbols for elements 112 and 114 are temporary. Final names will be selected when the elements' discoveries are verified.

Europium 63 **Eu** 151.964	Gadolinium 64 **Gd** 157.25	Terbium 65 **Tb** 158.925	Dysprosium 66 **Dy** 162.500	Holmium 67 **Ho** 164.930	Erbium 68 **Er** 167.259	Thulium 69 **Tm** 168.934	Ytterbium 70 **Yb** 173.04	Lutetium 71 **Lu** 174.967
Americium 95 **Am** (243)	Curium 96 **Cm** (247)	Berkelium 97 **Bk** (247)	Californium 98 **Cf** (251)	Einsteinium 99 **Es** (252)	Fermium 100 **Fm** (257)	Mendelevium 101 **Md** (258)	Nobelium 102 **No** (259)	Lawrencium 103 **Lr** (262)

Cómo usar el glosario en español:
1. Busca el término en inglés que desees encontrar.
2. El término en español, junto con la definición, se encuentran en la columna de la derecha.

Pronunciation Key

Use the following key to help you sound out words in the glossary.

a	**b**ack (BAK)	**ew**	f**oo**d (FEWD)
ay	d**ay** (DAY)	**yoo**	p**ure** (PYOOR)
ah	f**a**ther (FAH thur)	**yew**	f**ew** (FYEW)
ow	fl**ow**er (FLOW ur)	**uh**	comm**a** (CAH muh)
ar	c**ar** (CAR)	**u (+ con)**	r**u**b (RUB)
e	l**e**ss (LES)	**sh**	**sh**elf (SHELF)
ee	l**ea**f (LEEF)	**ch**	na**t**ure (NAY chur)
ih	tr**i**p (TRIHP)	**g**	**g**ift (GIHFT)
i (i + con + e)	**i**d**e**a (i DEE uh)	**j**	**g**em (JEM)
oh	g**o** (GOH)	**ing**	s**ing** (SING)
aw	s**o**ft (SAWFT)	**zh**	vi**s**ion (VIH zhun)
or	**or**bit (OR buht)	**k**	**c**a**k**e (KAYK)
oy	c**oi**n (COYN)	**s**	**s**eed, **c**ent (SEED, SENT)
oo	f**oo**t (FOOT)	**z**	**z**one, rai**s**e (ZOHN, RAYZ)

English — A — Español

abiotic: nonliving, physical features of the environment, including air, water, sunlight, soil, temperature, and climate. (p. 712)

abiótico: características inertes y físicas del medio ambiente, incluyendo el aire, el agua, la luz solar, el suelo, la temperatura y el clima. (p. 712)

acid precipitation: precipitation with a pH below 5.6—which occurs when air pollutants from the burning of fossil fuels react with water in the atmosphere to form strong acids—that can pollute water, kill fish and plants, and damage soils. (p. 779)

lluvia ácida: precipitación con un pH menor de 5.6—lo cual ocurre cuando los contaminantes del aire provenientes de la quema de combustibles fósiles reaccionan con el agua en la atmósfera para formar ácidos fuertes—que puede contaminar el agua, matar peces y plantas, y dañar los suelos. (p. 779)

active immunity: long-lasting immunity that results when the body makes its own antibodies in response to a specific antigen. (p. 655)

inmunidad activa: inmunidad duradera que resulta cuando el cuerpo produce sus propios anticuerpos en respuesta a un antígeno específico. (p. 655)

active transport: energy-requiring process in which transport proteins bind with particles and move them through a cell membrane. (p. 77)

transporte activo: proceso que requiere energía y en el cual las proteínas de transporte se unen con partículas y las trasladan a través de la membrana celular. (p. 77)

adaptation: any variation that makes an organism better suited to its environment. (p. 158)

adaptación: cualquier variación que haga que un organismo se adapte mejor a su medio ambiente. (p. 158)

aerobe (AY rohb): any organism that uses oxygen for respiration. (p. 188)

aerobio: cualquier organismo que utiliza oxígeno para respirar. (p. 188)

aggression: forceful behavior, such as fighting, used by an animal to control or dominate another animal in order to protect their young, defend territory, or get food. (p. 464)

agresión: comportamiento violento, como la lucha, manifestado por un animal para controlar o dominar a otro animal con el fin de proteger a sus crías, defender su territorio o conseguir alimento. (p. 464)

algae (AL jee): chlorophyll-containing, plantlike protists that produce oxygen as a result of photosynthesis. (p. 211)

algas: protistas parecidos a las plantas; contienen clorofila y producen oxígeno como resultado de la fotosíntesis. (p. 211)

allele (uh LEEL): an alternate form that a gene may have for a single trait; can be dominant or recessive. (p. 126)

allergen: substance that causes an allergic reaction. (p. 667)

allergy: overly strong reaction of the immune system to a foreign substance. (p. 666)

alveoli (al VEE uh li): tiny, thin-walled, grapelike clusters at the end of each bronchiole that are surrounded by capillaries; carbon dioxide and oxygen exchange takes place. (p. 571)

amino acid: building block of protein. (p. 513)

amniotic egg: egg covered with a shell that provides a complete environment for the embryo's development; for reptiles, a major adaptation for living on land. (p. 413)

amniotic (am nee AH tihk) sac: thin, liquid-filled, protective membrane that forms around the embryo. (p. 635)

anaerobe (AN uh rohb): any organism that is able to live without oxygen. (p. 188)

angiosperms: flowering vascular plants that produce fruits containing one or more seeds; monocots and dicots. (p. 257)

antibiotics: chemicals produced by some bacteria that are used to limit the growth of other bacteria. (p. 193)

antibody: a protein made in response to a specific antigen that can attach to the antigen and cause it to be useless. (p. 654)

antigen (AN tih jun): any complex molecule that is foreign to your body. (p. 654)

anus: opening at the end of the digestive tract through which wastes leave the body. (p. 347)

appendages: jointed structures of arthropods, such as legs, wings, or antennae. (p. 370)

artery: blood vessel that carries blood away from the heart, and has thick, elastic walls made of connective tissue and smooth muscle tissue. (p. 544)

ascus (AS kus): saclike, spore-producing structure of sac fungi. (p. 224)

asexual reproduction: a type of reproduction—fission, budding, and regeneration—in which a new organism is produced from one organism and has DNA identical to the parent organism. (p. 101)

asthma: lung disorder in which the bronchial tubes contract quickly and cause shortness of breath, wheezing, or coughing; may occur as an allergic reaction. (p. 576)

alelo: forma alternativa que un gen puede tener para un rasgo único; puede ser dominante o recesivo. (p. 126)

alergeno: sustancia que causa una reacción alérgica. (p. 667)

alergia: reacción exagerada del sistema inmune a una sustancia extraña. (p. 666)

alvéolos: pequeños racimos de pared delgada que se encuentran al final de cada bronquíolo y que están rodeados por capilares; aquí tiene lugar el intercambio de oxígeno y dióxido de carbono. (p. 571)

aminoácido: bloque de construcción de las proteínas. (p. 513)

huevo amniótico: huevo cubierto por un cascarón coriáceo que proporciona un ambiente completo para el desarrollo del embrión; para los reptiles, una gran adaptación para vivir en la tierra. (p. 413)

saco amniótico: membrana protectora delgada y llena de líquido que se forma alrededor del embrión. (p. 635)

anaerobio: cualquier organismo capaz de vivir sin oxígeno. (p. 188)

angiospermas: plantas vasculares que producen flores y frutos que contienen una o más semillas; pueden ser monocotiledóneas o dicotiledóneas. (p. 257)

antibióticos: químicos producidos por algunas bacterias, utilizados para limitar el crecimiento de otras bacterias. (p. 193)

anticuerpo: proteína formada en respuesta a un antígeno específico y que puede unirse al antígeno para provocar que éste sea inutilizado. (p. 654)

antígeno: molécula compleja que es extraña para el cuerpo. (p. 654)

ano: apertura al final del tracto digestivo a través de la cual los desechos salen del cuerpo. (p. 347)

apéndices: estructuras articuladas de los artrópodos, como las patas, alas o antenas. (p. 370)

arteria: vaso sanguíneo que transporta sangre desde el corazón y tiene paredes gruesas y elásticas hechas de tejido conectivo y tejido muscular liso. (p. 544)

asca: estructura en forma de saco en donde los ascomicetos producen esporas. (p. 224)

reproducción asexual: tipo de reproducción—fisión, gemación y regeneración—en el que un organismo da origen a uno nuevo de ADN idéntico al organismo progenitor. (p. 101)

asma: desorden pulmonar en el que los tubos bronquiales se contraen rápidamente y causan dificultad para respirar, silbido o tos; puede ocurrir como una reacción alérgica. (p. 576)

atmosphere: air surrounding Earth; is made up of gases, including 78 percent nitrogen, 21 percent oxygen, and 0.03 percent carbon dioxide. (p. 713)

atriums (AY tree umz): two upper chambers of the heart that contract at the same time during a heartbeat. (p. 541)

auxin (AWK sun): plant hormone that causes plant leaves and stems to exhibit positive phototropisms. (p. 314)

axon (AK sahn): neuron structure that carries messages away from the cell body. (p. 595)

atmósfera: aire que rodea a la Tierra; está compuesta de gases, incluyendo 78% de nitrógeno, 21% de oxígeno y 0.03% de dióxido de carbono. (p. 713)

aurículas: las dos cámaras superiores del corazón que se contraen al mismo tiempo durante el latido cardiaco. (p. 541)

auxina: hormona vegetal que causa que las hojas y tallos de las plantas desarrollen un fototropismo positivo. (p. 314)

axón: estructura de la neurona que transmite los mensajes desde el cuerpo de la célula. (p. 595)

B

basidium (buh SIH dee uhm): club-shaped, reproductive structure in which club fungi produce spores. (p. 224)

behavior: the way in which an organism interacts with other organisms and its environment; can be innate or learned. (p. 456)

bilateral symmetry: body parts arranged in a similar way on both sides of the body, with each half being nearly a mirror image of the other half. (p. 335)

binomial nomenclature (bi NOH mee ul • NOH mun klay chur): two-word naming system that gives all organisms their scientific name. (p. 24)

biogenesis (bi oh JEH nuh sus): theory that living things come only from other living things. (p. 19)

biological vector: disease-carrying organism, such as a rat, mosquito, or fly, that spreads infectious disease. (p. 661)

biomes (BI ohmz): large geographic areas with similar climates and ecosystems; includes tundra, taiga, desert, temperate deciduous forest, temperate rain forest, tropical rain forest, and grassland. (p. 744)

biosphere: part of Earth that supports life, including the top portion of Earth's crust, the atmosphere, and all the water on Earth's surface. (p. 684)

biotic (bi AH tihk): features of the environment that are alive or were once alive. (p. 712)

bladder: elastic, muscular organ that holds urine until it leaves the body through the urethra. (p. 580)

brain stem: connects the brain to the spinal cord and is made up of the midbrain, the pons, and the medulla. (p. 598)

bronchi (BRAHN ki): two short tubes that branch off the lower end of the trachea and carry air into the lungs. (p. 571)

budding: form of asexual reproduction in which a new, genetically-identical organism forms on the side of its parent. (p. 224)

basidio: estructura reproductora en forma de mazo en la que los basidiomicetos producen esporas. (p. 224)

comportamiento: forma en la que un organismo interactúa con otros organismos y su entorno; puede ser innato o aprendido. (p. 456)

simetría bilateral: disposición de las partes del cuerpo de manera similar a ambos lados de éste, de tal forma que cada mitad es una imagen especular de la otra. (p. 335)

nomenclatura binomial: sistema de denominación de dos palabras que da a todos los organismos su nombre científico. (p. 24)

biogénesis: teoría que sostiene que los seres vivos sólo provienen de otros seres vivos. (p. 19)

vector biológico: organismo portador de enfermedades, como las ratas, mosquitos y moscas, que propagan enfermedades infecciosas. (p. 661)

biomas: grandes áreas geográficas con climas y ecosistemas similares; incluyen la tundra, la taiga, el desierto, el bosque caducifolio templado, el bosque lluvioso templado, la selva húmeda tropical y los pastizales. (p. 744)

biosfera: capa de la Tierra que alberga la vida, incluyendo la porción superior de la corteza terrestre, la atmósfera y toda el agua de la superficie terrestre. (p. 684)

biótico: características del ambiente que tienen o alguna vez tuvieron vida. (p. 712)

vejiga: órgano muscular elástico que retiene la orina hasta que ésta sale del cuerpo por la uretra. (p. 580)

tronco cerebral: conecta al cerebro con la médula espinal y está compuesto por el mesencéfalo, el puente de Varolio y la médula. (p. 598)

bronquios: dos tubos cortos que se ramifican en la parte inferior de la tráquea y llevan el aire a los pulmones. (p. 571)

gemación: forma de reproducción asexual en la que se forma un organismo nuevo y genéticamente idéntico al lado de su progenitor. (p. 224)

C

cambium (KAM bee um): vascular tissue that produces xylem and phloem cells as a plant grows. (p. 255)

capillary: microscopic blood vessel that connects arteries and veins; has walls one cell thick, through which nutrients and oxygen diffuse into body cells, and waste materials and carbon dioxide diffuse out of body cells. (p. 545)

carbohydrate (kar boh HI drayt): nutrient that usually is the body's main source of energy. (p. 514)

carbon cycle: model describing how carbon molecules move between the living and nonliving world. (p. 725)

cardiac muscle: striated, involuntary muscle found only in the heart. (p. 493)

carnivore: animal that eats only other animals or the remains of other animals; mammals having large, sharp canine teeth and strong jaw muscles for eating flesh. (pp. 331, 437)

carrying capacity: largest number of individuals of a particular species that an ecosystem can support over time. (p. 691)

cartilage: tough, flexible tissue that joins vertebrae and makes up all or part of the vertebrate endoskeleton; in humans, thick, smooth, flexible, and slippery tissue layer that covers the ends of bones, makes movement easier by reducing friction, and absorbs shocks. (pp. 395, 486)

cell: smallest unit of an organism that can carry on life functions. (p. 14)

cell membrane: protective outer covering of all cells that regulates the interaction between the cell and the environment. (p. 38)

cell theory: states that all organisms are made up of one or more cells, the cell is the basic unit of life, and all cells come from other cells. (p. 51)

cellulose (SEL yuh lohs): chemical compound made out of sugar; forms tangled fibers in the cell walls of many plants and provides structure and support. (p. 242)

cell wall: rigid structure that encloses, supports, and protects the cells of plants, algae, fungi, and most bacteria. (p. 39)

central nervous system: division of the nervous system, made up of the brain and spinal cord. (p. 597)

cerebellum (sur uh BEH lum): part of the brain that controls voluntary muscle movements, maintains muscle tone, and helps maintain balance. (p. 598)

cámbium: tejido vascular que produce las células del xilema y floema conforme crece la planta. (p. 255)

capilar: vaso sanguíneo microscópico que conecta las arterias con las venas; su pared tiene el grosor de una célula y los nutrientes y el oxígeno se difunden a través de ella hacia las células del cuerpo y los materiales de desecho y el dióxido de carbono hacia afuera de éstas. (p. 545)

carbohidrato: nutriente que generalmente es la principal fuente de energía para el cuerpo. (p. 514)

ciclo del carbono: modelo que describe cómo se mueven las moléculas de carbono entre el mundo vivo y el mundo inerte. (p. 725)

músculo cardiaco: músculo estriado involuntario que sólo se encuentra en el corazón. (p. 493)

carnívoro: animal que se alimenta exclusivamente de otros animales o de los restos de otros animales; mamífero con caninos largos y afilados y músculos fuertes en la mandíbula que le sirven para alimentarse de carne. (pp. 331, 437)

capacidad de carga: el mayor número de individuos de una especie en particular que un ecosistema puede albergar en un periodo de tiempo. (p. 691)

cartílago: tejido resistente y flexible que conecta a las vértebras y constituye todo o parte del endoesqueleto de los vertebrados; en los humanos, capa gruesa y lisa de tejido resbaladizo y flexible que cubre los extremos de los huesos, facilita el movimiento reduciendo la fricción y absorbe los impactos. (pp. 395, 486)

célula: la unidad más pequeña de un organismo que puede continuar con sus funciones vitales. (p. 14)

membrana celular: capa externa protectora de todas las células y reguladora de la interacción entre la célula y su entorno. (p. 38)

teoría celular: establece que todos los organismos están formados por una o más células, que la célula es la unidad básica de la vida y que las células provienen de otras células. (p. 51)

celulosa: compuesto químico formado por azúcares y que forma fibras intrincadas en la pared celular de muchas plantas proporcionando estructura y soporte. (p. 242)

pared celular: estructura rígida que envuelve, sostiene y protege a las células de las plantas, algas, hongos y de la mayoría de las bacterias. (p. 39)

sistema nervioso central: parte del sistema nervioso, compuesto por el cerebro y la médula espinal. (p. 597)

cerebelo: parte del cerebro que controla los movimientos de los músculos voluntarios, mantiene el tono muscular y ayuda a mantener el equilibrio. (p. 598)

cerebrum (suh REE brum): largest part of the brain, where memory is stored, movements are controlled, and impulses from the senses are interpreted. (p. 598)

cerebro: la parte más grande del encéfalo, donde se almacena la memoria, se controlan los movimientos y se interpretan los impulsos provenientes de los sentidos. (p. 598)

chemical digestion: occurs when enzymes and other chemicals break down large food molecules into smaller ones. (p. 523)

digestión química: ocurre cuando las enzimas y otros químicos desintegran las moléculas grandes de los alimentos en otras más pequeñas. (p. 523)

chemosynthesis (kee moh SIN thuh sus): process in which producers make energy-rich nutrient molecules from chemicals. (p. 727)

quimiosíntesis: proceso a través del cual los productores fabrican moléculas ricas en energía a partir de agentes químicos. (p. 727)

chemotherapy (kee moh THAYR uh pee): use of chemicals to destroy cancer cells. (p. 670)

quimioterapia: uso de sustancias químicas para destruir las células cancerosas. (p. 670)

chlorophyll (KLOR uh fihl): green, light-trapping pigment in plant chloroplasts that is important in photosynthesis. (p. 304)

clorofila: pigmento verde que absorbe luz y que se encuentra en los cloroplastos de las plantas; es importante para la fotosíntesis. (p. 304)

chloroplast: green, chlorophyll-containing, plant-cell organelle that uses light energy to produce sugar from carbon dioxide and water. (p. 42)

cloroplasto: organelo de las células vegetales, de color verde, que contiene clorofila y que usa la luz solar para convertir el dióxido de carbono y el agua en azúcar. (p. 42)

chordate: animal that has a notochord, a nerve cord, pharyngeal pouches, and a postanal tail present at some stage in its development. (p. 394)

cordado: animal que posee notocordio, un cordón nervioso, bolsas faríngeas y que presenta cola postnatal en alguna etapa de su desarrollo. (p. 394)

chromosome: structure in a cell's nucleus that contains hereditary material. (p. 98)

cromosoma: estructura en el núcleo celular que contiene el material hereditario. (p. 98)

chyme (KIME): liquid product of digestion. (p. 527)

quimo: producto líquido de la digestión. (p. 527)

cilia (SIH lee uh): in protists, short, threadlike structures that extend from the cell membrane of a ciliate and enable the organism to move quickly. (p. 215)

cilio: en los protistas, estructuras cortas en forma de hilo que se extienden desde la membrana celular de un ciliado y permiten al organismo moverse rápidamente. (p. 215)

climate: average weather conditions of an area over time, including wind, temperature, and rainfall or other types of precipitation such as snow or sleet. (p. 717)

clima: condiciones meteorológicas promedio de un área durante un periodo de tiempo; incluye viento, temperatura y precipitación pluvial u otros tipos de precipitación como la nieve o el granizo. (p. 717)

climax community: stable, end stage of ecological succession in which balance is in the absence of disturbance. (p. 743)

clímax comunitario: etapa final estable de la sucesión ecológica en la cual se da un equilibrio en ausencia de alteraciones. (p. 743)

closed circulatory system: blood circulation system in which blood moves through the body in closed vessels. (p. 362)

sistema circulatorio cerrado: sistema circulatorio sanguíneo en el cual la sangre se mueve a través del cuerpo en vasos cerrados. (p. 362)

cochlea (KOH klee uh): fluid-filled structure in the inner ear in which sound vibrations are converted into nerve impulses that are sent to the brain. (p. 608)

cóclea: estructura del oído interno llena de líquido en la que las vibraciones sonoras se convierten en impulsos nerviosos que son enviados al cerebro. (p. 608)

commensalism: a type of symbiotic relationship in which one organism benefits and the other organism is not affected. (p. 698)

comensalismo: tipo de relación simbiótica en la que un organismo se beneficia sin afectar al otro. (p. 698)

community: all the populations of different species that live in an ecosystem. (p. 686)

comunidad: todas las poblaciones de diferentes especies que viven en un mismo ecosistema. (p. 686)

condensation: process that takes place when a gas changes to a liquid. (p. 721)

condensación: proceso que tiene lugar cuando un gas cambia a estado líquido. (p. 721)

conditioning: occurs when the response to a stimulus becomes associated with another stimulus. (p. 460)

condicionamiento: ocurre cuando la respuesta a un estímulo llega a estar asociada con otro estímulo. (p. 460)

consumer: organism that cannot create energy-rich molecules but obtains its food by eating other organisms. (p. 697)

consumidor: organismo que no puede fabricar moléculas ricas en energía por lo que debe obtener su alimento ingiriendo otros organismos. (p. 697)

contour feathers: strong, lightweight feathers that give birds their coloring and shape and that are used for flight. (p. 430)

plumas de contorno: plumas fuertes y ligeras que dan a las aves su colorido y forma y que son usadas para volar. (p. 430)

control: standard to which the outcome of a test is compared. (p. 9)

control: estándar contra el que se compara el resultado de una prueba. (p. 9)

coral reef: diverse ecosystem formed from the calcium carbonate shells secreted by corals. (p. 757)

arrecife de coral: ecosistema diverso conformado de caparazones de carbonato de calcio secretados por los corales. (p. 757)

coronary (KOR uh ner ee) circulation: flow of blood to and from the tissues of the heart. (p. 541)

circulación coronaria: flujo sanguíneo desde y hacia los tejidos del corazón. (p. 541)

courtship behavior: behavior that allows males and females of the same species to recognize each other and prepare to mate. (p. 465)

comportamiento de cortejo: comportamiento que permite que los machos y hembras de la misma especie se reconozcan entre sí y se preparen para el apareamiento. (p. 465)

crop: digestive system sac in which earthworms store ingested soil. (p. 366)

buche: saco del sistema digestivo en el que los gusanos de tierra almacenan el suelo ingerido. (p. 366)

cuticle (KYEW tih kul): waxy, protective layer that covers the stems, leaves, and flowers of many plants and helps prevent water loss. (p. 242)

cutícula: capa cerosa protectora que recubre el tronco, hojas y flores de muchas plantas y ayuda a prevenir la pérdida de agua. (p. 242)

cyclic behavior: behavior that occurs in repeated patterns. (p. 468)

comportamiento cíclico: comportamiento que ocurre en patrones repetidos. (p. 468)

cytoplasm: constantly moving gel-like mixture inside the cell membrane that contains heredity material and is the location of most of a cell's life processes. (p. 38)

citoplasma: mezcla parecida al gel y que está en constante movimiento dentro de la membrana celular, contiene material hereditario y es en donde tiene lugar la mayor parte de los procesos vitales de la célula. (p. 38)

D

day-neutral plant: plant that doesn't require a specific photoperiod and can begin the flowering process over a range of night lengths. (p. 316)

planta de día neutro: planta que no requiere de un fotoperiodo específico y que puede comenzar su periodo de floración basándose en un rango de duración de las noches. (p. 316)

dendrite: neuron structure that receives messages and sends them to the cell body. (p. 595)

dendrita: estructura de la neurona que recibe mensajes y los envía al cuerpo de la célula. (p. 595)

dermis: skin layer below the epidermis that contains blood vessels, nerves, oil and sweat glands, and other structures. (p. 497)

dermis: capa de la piel debajo de la epidermis que contiene vasos sanguíneos, nervios, glándulas sudoríparas, glándulas sebáceas y otras estructuras. (p. 497)

desert: driest biome on Earth with less than 25 cm of rain each year; has dunes or thin soil with little organic matter, where plants and animals are adapted to survive extreme conditions. (p. 750)

desierto: el bioma más seco sobre la Tierra con menos de 25 centímetros cúbicos de lluvia al año; tiene dunas o un suelo delgado con muy poca materia orgánica y aquí las plantas y animales están adaptados para sobrevivir en condiciones extremosas. (p. 750)

diaphragm (DI uh fram): muscle beneath the lungs that contracts and relaxes to move gases in and out of the body. (p. 572)

diafragma: músculo que está debajo de los pulmones y que se contrae y relaja para mover gases hacia dentro y fuera del cuerpo. (p. 572)

dicot: angiosperm with two cotyledons inside its seed, flower parts in multiples of four or five, and vascular bundles in rings. (p. 258)

diffusion: a type of passive transport in cells in which molecules move from areas where there are more of them to areas where there are fewer of them. (p. 75)

digestion: mechanical and chemical breakdown of food into small molecules that cells can absorb and use. (p. 523)

diploid (DIHP loyd): cell whose similar chromosomes occur in pairs. (p. 104)

DNA: deoxyribonucleic acid; the genetic material of all organisms; made up of two twisted strands of sugar-phosphate molecules and nitrogen bases. (p. 110)

dominant (DAH muh nunt): describes a trait that covers over, or dominates, another form of that trait. (p. 128)

down feathers: soft, fluffy feathers that provide an insulating layer next to the skin of adult birds and that cover the bodies of young birds. (p. 430)

dicotiledónea: angiosperma con dos cotiledones dentro de su semilla, partes florales en múltiplos de cuatro o cinco y haces vasculares distribuidos en anillos. (p. 258)

difusión: tipo de transporte pasivo en las células en el que las moléculas se mueven de áreas de mayor concentración de éstas hacia áreas de menor concentración. (p. 75)

digestión: desintegración mecánica y química de los alimentos en moléculas pequeñas que las células pueden absorber y utilizar. (p. 523)

diploide: célula cuyos cromosomas similares están en pares. (p. 104)

ADN: ácido desoxirribonucleico; material genético de todos los organismos constituido por dos cadenas trenzadas de moléculas de azúcar-fosfato y bases de nitrógeno (p. 110)

dominante: describe un rasgo que encubre o domina a otra forma de ese rasgo. (p. 128)

plumón: plumas suaves y esponjadas que proporcionan una capa aislante junto a la piel de las aves adultas y que cubren los cuerpos de las aves jóvenes. (p. 430)

E

ecology: study of the interactions that take place among organisms and their environment. (p. 685)

ecosystem: all the living organisms that live in an area and the nonliving features of their environment. (p. 685)

ectotherm: vertebrate animal whose internal temperature changes when the temperature of its environment changes. (p. 397)

egg: haploid sex cell formed in the female reproductive organs. (p. 104)

embryo: fertilized egg that has attached to the wall of the uterus. (p. 635)

embryology (em bree AH luh jee): study of embryos and their development. (p. 167)

emphysema (em fuh SEE muh): lung disease in which the alveoli enlarge. (p. 575)

endocytosis (en duh si TOH sus): process by which a cell takes in a substance by surrounding it with the cell membrane. (p. 78)

endoplasmic reticulum (ER): cytoplasmic organelle that moves materials around in a cell and is made up of a complex series of folded membranes; can be rough (with attached ribosomes) or smooth (without attached ribosomes). (p. 43)

ecología: estudio de las interacciones que se dan entre los organismos y su medio ambiente. (p. 685)

ecosistema: conjunto de organismos vivos que habitan en un área y las características de su medio ambiente. (p. 685)

ectotérmico: animal vertebrado cuya temperatura interna cambia cuando cambia la temperatura de su ambiente. (p. 397)

óvulo: célula sexual haploide que se forma en los órganos reproductivos femeninos. (p. 104)

embrión: óvulo fertilizado que se ha adherido a la pared del útero. (p. 635)

embriología: el estudio de los embriones y su desarrollo. (p. 167)

enfisema: enfermedad pulmonar en la cual se dilatan los alvéolos. (p. 575)

endocitosis: proceso mediante el cual una célula capta una sustancia rodeándola con su membrana celular. (p. 78)

retículo endoplásmático (RE): organelo citoplasmático que transporta materiales dentro de una célula y está formado por una serie compleja de membranas plegadas; puede ser rugoso (con ribosomas adosados) o liso (sin ribosomas adosados). (p. 43)

endoskeleton: supportive framework of bone and/or cartilage that provides an internal place for muscle attachment and protects a vertebrate's internal organs. (p. 395)

endoesqueleto: estructura ósea y/o cartilaginosa de soporte que proporciona un medio interno para la fijación de los músculos y que protege a los órganos internos de los vertebrados. (p. 395)

endospore: thick-walled, protective structure produced by some bacteria when conditions are unfavorable for survival. (p. 197)

endospora: estructura protectora de pared gruesa que es producida por algunas bacterias cuando las condiciones son desfavorables para su supervivencia. (p. 197)

endotherm: vertebrate animal with a nearly constant internal temperature. (pp. 397, 430)

endotérmico: animal vertebrado con una temperatura interna casi constante. (pp. 397, 430)

energy pyramid: model that shows the amount of energy available at each feeding level in an ecosystem. (p. 729)

pirámide de energía: modelo que muestra la cantidad de energía disponible en cada nivel alimenticio de un ecosistema. (p. 729)

enzyme: a type of protein that regulates nearly all chemical reactions in cells; a type of protein that speeds up chemical reactions in the body without being changed or used up itself. (pp. 71, 524)

enzima: tipo de proteína que regula casi todas las clases de reacciones químicas en las células; tipo de proteína que acelera las reacciones químicas en el cuerpo sin ser utilizada o consumida. (pp. 71, 524)

epidermis: outer, thinnest skin layer that constantly produces new cells to replace the dead cells rubbed off its surface. (p. 496)

epidermis: la capa más delgada y externa de la piel que produce constantemente células nuevas para reemplazar las células muertas que se pierden por fricción de su superficie. (p. 496)

equilibrium: occurs when molecules of one substance are spread evenly throughout another substance. (p. 75)

equilibrio: ocurre cuando las moléculas de una sustancia están diseminadas completa y uniformemente a lo largo de otra sustancia. (p. 75)

erosion: movement of soil from one place to another. (p. 785)

erosión: movimiento del suelo de un lugar a otro. (p. 785)

estivation: inactivity in hot, dry months. (p. 407)

estivación: inactividad durante los meses cálidos y secos. (p. 407)

estuary: extremely fertile area where a river meets an ocean; contains a mixture of freshwater and saltwater and serves as a nursery for many species of fish. (p. 758)

estuario: área extremadamente fértil donde un río desemboca en el océano; contiene una mezcla de agua dulce y salada y sirve como vivero para muchas especies de peces. (p. 758)

evaporation: process that takes place when a liquid changes to a gas. (p. 720)

evaporación: proceso que tiene lugar cuando un líquido cambia a estado gaseoso. (p. 720)

evolution: change in inherited characteristics over time. (p. 154)

evolución: cambio en las características heredadas a través del tiempo. (p. 154)

exocytosis (ek soh si TOH sus): process by which vesicles release their contents outside the cell. (p. 78)

exocitosis: proceso mediante el cual las vesículas liberan su contenido fuera de la célula. (p. 78)

exoskeleton: thick, hard, outer covering that protects and supports arthropod bodies and provides places for muscles to attach. (p. 370)

exoesqueleto: cubierta externa, dura y gruesa que protege y soporta el cuerpo de los artrópodos y proporciona lugares para que los músculos se fijen. (p. 370)

F

fat: nutrient that stores energy, cushions organs, and helps the body absorb vitamins. (p. 515)

grasa: nutriente que almacena energía, amortigua a los órganos y ayuda al cuerpo a absorber vitaminas. (p. 515)

fermentation: process by which oxygen-lacking cells and some one-celled organisms release small amounts of energy from glucose molecules and produce wastes such as alcohol, carbon dioxide, and lactic acid. (p. 84)

fermentación: proceso mediante el cual las células carentes de oxígeno y algunos organismos unicelulares liberan pequeñas cantidades de energía a partir de moléculas de glucosa y producen desechos como alcohol, dióxido de carbono y ácido láctico. (p. 84)

fertilization: in sexual reproduction, the joining of a sperm and egg. (p. 104)

fertilización: en la reproducción sexual, la unión de un óvulo y un espermatozoide. (p. 104)

fetal stress: can occur during the birth process or after birth as an infant adjusts from a watery, dark, constant-temperature environment to its new environment. (p. 638)

estrés fetal: puede ocurrir durante el proceso del nacimiento o luego del mismo mientras un nuevo ser humano se adapta de un ambiente acuoso, oscuro y de temperatura constante a su nuevo ambiente. (p. 638)

fetus: in humans, a developing baby after the first two months of pregnancy until birth. (p. 636)

feto: en los humanos, bebé en desarrollo desde los primeros dos meses de embarazo hasta el nacimiento. (p. 636)

fin: structure used by fish for steering, balancing, and movement. (p. 399)

aleta: estructura parecida a un abanico usada por los peces para mantener la dirección, equilibrio y movimiento. (p. 399)

fission: simplest form of asexual reproduction in which two new cells are produced with genetic material identical to each other and identical to the previous cell. (p. 188)

fisión: la forma más simple de reproducción asexual en la que se producen dos nuevas células cuyo material genético es idéntico entre sí y al de la célula antecesora. (p. 188)

flagellum: long, thin whiplike structure that helps organisms move through moist or wet surroundings. (pp. 187, 212)

flagelo: estructura alargada en forma de látigo que ayuda a algunos organismos a desplazarse de un lugar a otro en medios acuosos. (pp. 187, 212)

food group: group of foods—such as bread, cereal, rice, and pasta—containing the same type of nutrients. (p. 520)

grupo alimenticio: grupo de alimentos—como el pan, el cereal, el arroz y la pasta—que contiene el mismo tipo de nutrientes. (p. 520)

food web: model that shows the complex feeding relationships among organisms in a community. (p. 728)

cadena alimenticia: modelo que muestra las complejas relaciones alimenticias entre los organismos de una comunidad. (p. 728)

fossil fuels: nonrenewable energy resources—coal, oil, and natural gas—that formed in Earth's crust over hundreds of millions of years. (p. 772)

combustibles fósiles: recursos energéticos no renovables—carbón, petróleo y gas natural—que se formaron en la corteza terrestre durante cientos de millones de años. (p. 772)

free-living organism: organism that does not depend on another organism for food or a place to live. (p. 344)

organismo de vida libre: organismo que no depende de otro para alimentarse o para tener un lugar en donde vivir. (p. 344)

frond: leaf of a fern that grows from the rhizome. (p. 278)

fronda: hoja de un helecho que crece a partir del rizoma. (p. 278)

G

gametophyte (guh MEE tuh fite) stage: plant life cycle stage that begins when cells in reproductive organs undergo meiosis and produce haploid cells (spores). (p. 275)

etapa de gametofito: etapa del ciclo de vida de las plantas que comienza cuando las células en los órganos reproductores, a través de la meiosis, producen células haploides (esporas). (p. 275)

gene: section of DNA on a chromosome that contains instructions for making specific proteins. (p. 112)

gen: sección de ADN en un cromosoma, el cual contiene instrucciones para la formación de proteínas específicas. (p. 112)

genetic engineering: biological and chemical methods to change the arrangement of a gene's DNA to improve crop production, produce large volumes of medicine, and change how cells perform their normal functions. (p. 141)

ingeniería genética: métodos biológicos y químicos para cambiar la disposición del ADN de un gen y así mejorar la producción de cosechas, producir grandes volúmenes de un medicamento, o cambiar la forma en que las células realizan sus funciones normales. (p. 141)

genetics (juh NEH tihks): the study of how traits are inherited through the actions of alleles. (p. 126)

genética: estudio de la forma como se heredan los rasgos a través de las acciones de los alelos. (p. 126)

genotype (JEE nuh tipe): the genetic makeup of an organism. (p. 130)

genotipo: composición genética de un organismo. (p. 130)

genus: first word of the two-word scientific name used to identify a group of similar species. (p. 24)

género: primera palabra, de las dos palabras del nombre científico, que se usa para identificar a un grupo de especies similares. (p. 24)

geothermal energy: heat energy within Earth's crust, available only where natural geysers or volcanoes are located. (p. 775)

energía geotérmica: energía calórica en el interior de la corteza terrestre disponible sólo donde existen géiseres o volcanes. (p. 775)

germination: series of events that results in the growth of a plant from a seed. (p. 290)

germinación: serie de eventos que resultan en el crecimiento de una planta a partir de una semilla. (p. 290)

gestation period: period during which an embryo develops in the uterus; the length of time varies among species. (p. 441)

periodo de gestación: periodo durante el cual un embrión se desarrolla en el útero; este periodo varía de una especie a otra. (p. 441)

gills: organs that exchange carbon dioxide for oxygen in the water. (p. 360)

agallas: órganos que intercambian dióxido de carbono y oxígeno en el agua. (p. 360)

gizzard: muscular digestive system structure in which earthworms grind soil and organic matter. (p. 366)

molleja: estructura muscular del sistema digestivo en la que los gusanos de tierra muelen el suelo y materia orgánica. (p. 366)

Golgi bodies: organelles that package cellular materials and transport them within the cell or out of the cell. (p. 43)

aparato de Golgi: organelo que concentra los materiales celulares y los transporta hacia adentro o afuera de la célula. (p. 43)

gradualism: model describing evolution as a slow process by which one species changes into a new species through a continuing series of mutations and variations over time. (p. 160)

gradualismo: modelo que describe la evolución como un proceso lento mediante el cual una especie existente se convierte en una especie nueva a través de series continuas de mutaciones y variaciones a través del tiempo. (p. 160)

grasslands: temperate and tropical regions with 25 cm to 75 cm of precipitation each year that are dominated by climax communities of grasses; ideal for growing crops and raising cattle and sheep. (p. 751)

pastizales: regiones tropicales y templadas con 25 a 75 centímetros cúbicos de lluvia al año; son dominadas por el clímax comunitario de los pastos e ideales para la cría de ganado y ovejas. (p. 751)

greenhouse effect: heat-trapping feature of the atmosphere that keeps Earth warm enough to support life. (p. 780)

efecto de invernadero: característica de la atmósfera que le permite atrapar calor y mantener la Tierra lo suficientemente caliente para favorecer la vida. (p. 780)

guard cells: pairs of cells that surround stomata and control their opening and closing. (p. 253)

células oclusoras: pares de células que rodean al estoma y que controlan su cierre y apertura. (p. 253)

gymnosperms: vascular plants that do not flower, generally have needlelike or scalelike leaves, and produce seeds that are not protected by fruit; conifers, cycads, ginkgoes, and gnetophytes. (p. 256)

gimnospermas: plantas vasculares que no florecen, generalmente tienen hojas en forma de aguja o de escama y producen semillas que no están protegidas por el fruto; se clasifican en coníferas, cicadáceas, ginkgoales y gnetofitas. (p. 256)

H

habitat: place where an organism lives and that provides the types of food, shelter, moisture, and temperature needed for survival. (p. 687)

hábitat: lugar donde vive un organismo y que le proporciona los tipos de alimento, refugio, humedad y temperatura necesarios para su supervivencia. (p. 687)

haploid (HAP loyd): cell that has half the number of chromosomes as body cells. (p. 105)

haploide: célula que posee la mitad del número de cromosomas que tienen las células somáticas. (p. 105)

hazardous wastes: waste materials, such as pesticides and leftover paints, that are harmful to human health or poisonous to living organisms. (p. 786)

desperdicios peligrosos: materiales de desecho como los pesticidas y residuos de pintura nocivos para la salud humana o dañinos para los organismos vivos. (p. 786)

hemoglobin (HEE muh gloh bun): chemical in red blood cells that carries oxygen from the lungs to body cells, and carries some carbon dioxide from body cells back to the lungs. (p. 551)

hemoglobina: sustancia química de los glóbulos rojos que transporta oxígeno de los pulmones a las células del cuerpo y parte del dióxido de carbono de las células del cuerpo a los pulmones. (p. 551)

herbivore: animal that eats only plants or parts of plants; mammals with large premolars and molars for eating only plants. (pp. 331, 437)

herbívoro: animal que se alimenta exclusivamente de plantas o de partes de las plantas; mamífero con premolares y molares grandes que se alimenta exclusivamente de plantas. (pp. 331, 437)

heredity (huh REH duh tee): the passing of traits from parent to offspring. (p. 126)

herencia: transferencia de rasgos de un progenitor a su descendencia. (p. 126)

hermaphrodite (hur MA fruh dite): animal that produces both sperm and eggs in the same body. (p. 338)

hermafrodita: animal que produce óvulos y espermatozoides en el mismo cuerpo. (p. 338)

heterozygous (heh tuh roh ZI gus): describes an organism with two different alleles for a trait. (p. 130)

heterocigoto: describe a un organismo con dos alelos diferentes para un rasgo. (p. 130)

hibernation: inactivity in cold weather; cyclic response of inactivity and slowed metabolism that occurs during periods of cold temperatures and limited food supplies. (pp. 407, 469)

hibernación: inactividad durante periodos de clima frío; respuesta cíclica de inactividad y disminución del metabolismo que ocurre durante periodos de bajas temperaturas y suministro limitado de alimento. (pp. 407, 469)

homeostasis: regulation of an organism's internal, life-maintaining conditions. (pp. 15, 595)

homeostasis: control de las condiciones internas que mantienen la vida de un organismo. (pp. 15, 595)

hominid: humanlike primate that appeared about 4 million to 6 million years ago, ate both plants and meat, and walked upright on two legs. (p. 171)

homínido: primate con forma de humano que apareció entre 4 y 6 millones de años atrás, se alimentaba de plantas y carne, y caminaba erguido sobre sus dos pies. (p. 171)

homologous (huh MAH luh gus): body parts that are similar in structure and origin and can be similar in function. (p. 168)

homólogos: partes del cuerpo que son similares en estructura y origen y que pueden tener funciones similares. (p. 168)

homozygous (hoh muh ZI gus): describes an organism with two alleles that are the same for a trait. (p. 130)

homocigoto: describe a un organismo con dos alelos iguales para un rasgo. (p. 130)

Homo sapiens: early humans that likely evolved from Cro-Magnons. (p. 173)

Homo sapiens: humanos primitivos que probablemente evolucionaron a partir de los CroMagnon. (p. 173)

hormone (HOR mohn): in humans, chemical produced by the endocrine system, released directly into the bloodstream by ductless glands; affects specific target tissues, and can speed up or slow down cellular activities. (p. 622)

hormona: en los humanos, sustancia química producida por el sistema endocrino, liberada directamente al torrente sanguíneo mediante glándulas sin conductos; afecta a tejidos que constituyen blancos específicos y puede acelerar o frenar actividades celulares. (p. 622)

host cell: living cell in which a virus can actively multiply or in which a virus can hide until activated by environmental stimuli. (p. 52)

célula huésped: célula viva en la que un virus puede reproducirse activamente o en la que un virus puede ocultarse hasta que es activado por estímulos del medio ambiente. (p. 52)

hybrid (HI brud): an offspring that was given different genetic information for a trait from each parent. (p. 128)

híbrido: un descendiente que recibe de cada progenitor información genética diferente para un rasgo. (p. 128)

hydroelectric power: electricity produced when the energy of falling water turns the blades of a generator turbine. (p. 773)

energía hidroeléctrica: electricidad producida cuando la energía generada por la caída del agua hace girar las aspas de una turbina generadora. (p. 773)

hyphae (HI fee): mass of many-celled, threadlike tubes forming the body of a fungus. (p. 222)

hifa: masa de tubos multicelulares en forma de hilos formando el cuerpo de los hongos. (p. 222)

hypothesis: prediction that can be tested. (p. 8)

hipótesis: predicción que puede probarse. (p. 8)

I

immune system: complex group of defenses that protects the body against pathogens—includes the skin and respiratory, digestive, and circulatory systems. (p. 652)

sistema inmune: grupo complejo de defensas que protege al cuerpo contra agentes patógenos –incluye la piel y los sistemas respiratorio, digestivo y circulatorio. (p. 652)

imprinting: occurs when an animal forms a social attachment to another organism during a specific period following birth or hatching. (p. 459)

impronta: ocurre cuando un animal forma un vínculo social con otro organismo durante un periodo específico después del nacimiento o eclosión. (p. 459)

incomplete dominance: production of a phenotype that is intermediate between the two homozygous parents. (p. 134)

dominancia incompleta: producción de un fenotipo intermedio entre dos progenitores homocigotos. (p. 134)

infectious disease: disease caused by a virus, bacterium, fungus, or protist that is spread from an infected organism or the environment to another organism. (p. 661)

enfermedad infecciosa: enfermedad causada por virus, bacterias, hongos o protistas, propagada por un organismo infectado o del medio ambiente hacia otro organismo. (p. 661)

innate behavior: behavior that an organism is born with and does not have to be learned, such as a reflex or instinct. (p. 457)

comportamiento innato: comportamiento con el que nace un organismo y que no necesita ser aprendido, tal como los reflejos o los instintos. (p. 457)

inorganic compound: compound, such as H_2O, that is made from elements other than carbon and whose atoms usually can be arranged in only one structure. (p. 71)

compuesto inorgánico: compuesto, como H2O, formado por elementos distintos al carbono y cuyos átomos generalmente pueden estar organizados en sólo una estructura. (p. 71)

insight: form of reasoning that allows animals to use past experiences to solve new problems. (p. 461)

comprensión: forma de razonamiento que permite a los animales usar experiencias pasadas para solucionar problemas nuevos. (p. 461)

instinct: complex pattern of innate behavior, such as spinning a web, that can take weeks to complete. (p. 458)

instinto: patrón complejo de comportamiento innato, como tejer una telaraña, que puede durar semanas para completarse. (p. 458)

intertidal zone: part of the shoreline that is under water at high tide and exposed to the air at low tide. (p. 758)

zona litoral: parte de la línea costera que está bajo el agua durante la marea alta y expuesta al aire durante la marea baja. (p. 758)

invertebrate: animal without a backbone. (p. 334)

invertebrado: animal que no posee columna vertebral. (p. 334)

involuntary muscle: muscle, such as heart muscle, that cannot be consciously controlled. (p. 491)

músculo involuntario: músculo, como el músculo cardiaco, que no puede controlarse conscientemente. (p. 491)

J

joint: any place where two or more bones come together; can be movable or immovable. (p. 487)

articulación: cualquier lugar en donde se unen dos o más huesos, pudiendo ser fija o flexible. (p. 487)

K

kidney: bean-shaped urinary system organ that is made up of about 1 million nephrons and filters blood, producing urine. (p. 578)

kingdom: first and largest category used to classify organisms. (p. 23)

riñón: órgano del sistema urinario en forma de fríjol, compuesto por cerca de un millón de nefronas; filtra la sangre y produce la orina. (p. 578)

reino: la primera y más grande categoría utilizada para clasificar a los organismos. (p. 23)

L

larynx: airway to which the vocal cords are attached. (p. 571)

law: statement about how things work in nature that seems to be true consistently. (p. 10)

lichen (LI kun): organism made up of a fungus and a green alga or a cyanobacterium. (p. 226)

ligament: tough band of tissue that holds bones together at joints. (p. 487)

limiting factor: anything that can restrict the size of a population, including living and nonliving features of an ecosystem, such as predators or drought. (p. 690)

long-day plant: plant that generally requires short nights—less than ten to 12 hours of darkness—to begin the flowering process. (p. 316)

lymph (LIHMF): tissue fluid that has diffused into lymphatic capillaries. (p. 556)

lymphatic system: carries lymph through a network of lymph capillaries and vessels, and drains it into large veins near the heart; helps fight infections and diseases. (p. 556)

lymph nodes: bean-shaped organs found throughout the body that filter out microorganisms and foreign materials taken up by the lymphocytes. (p. 556)

lymphocyte (LIHM fuh site): a type of white blood cell that fights infection. (p. 556)

laringe: vía respiratoria que contiene las cuerdas vocales. (p. 571)

ley: enunciado acerca de cómo funciona todo en la naturaleza y que constantemente parece ser verdadero. (p. 10)

liquen: organismo formado por un hongo y un alga verde o una cianobacteria. (p. 226)

ligamento: banda de tejido resistente que mantiene unidos a los huesos de las articulaciones. (p. 487)

factor limitante: cualquier factor que pueda restringir el tamaño de una población, incluyendo las características biológicas y no biológicas de un ecosistema, tales como los depredadores o las sequías. (p. 690)

planta de día largo: planta que generalmente requiere de noches cortas—menos de 12 horas de oscuridad—para comenzar su proceso de floración. (p. 316)

linfa: fluido tisular que se ha difundido hacia los capilares linfáticos. (p. 556)

sistema linfático: sistema que transporta la linfa a través de una red de vasos y capilares linfáticos y la vierte en venas grandes cerca del corazón; ayuda a combatir enfermedades e infecciones. (p. 556)

ganglio linfático: órganos en forma de fríjol que se encuentran en todo el cuerpo; filtran y extraen microorganismos y materiales extraños captados por los linfocitos. (p. 556)

linfocito: tipo de glóbulo blanco que combate las infecciones. (p. 556)

M

mammals: endothermic vertebrates that have hair, teeth specialized for eating certain foods, and mammary glands; in females, mammary glands produce milk for feeding their young. (p. 436)

mamíferos: vertebrados endotérmicos que poseen pelo y dientes especializados para comer cierto tipo de alimentos y cuyas hembras tienen glándulas mamarias que producen leche para alimentar a sus crías. (p. 436)

mammary glands: glands of mammals; in females, produce milk to feed their young. (p. 436)

glándulas mamarias: glándulas productoras de leche que las hembras de los mamíferos usan para alimentar a sus crías. (p. 436)

mantle: thin layer of tissue that covers a mollusk's body organs; secretes the shell or protects the body of mollusks without shells. (p. 360)

manto: capa delgada de tejido que recubre los órganos corporales de los moluscos; secreta el caparazón o protege el cuerpo de los moluscos sin caparazón. (p. 360)

marsupial: a mammal with an external pouch for the development of its immature young. (p. 440)

marsupial: mamífero con una bolsa externa para el desarrollo de sus crías inmaduras. (p. 440)

mechanical digestion: breakdown of food through chewing, mixing, and churning. (p. 523)

digestión mecánica: desdoblamiento del alimento a través de la masticación, mezcla y agitación. (p. 523)

medusa (mih DEW suh): cnidarian body type that is bell-shaped and free-swimming. (p. 339)

medusa: tipo corporal de los cnidarios con forma de campana y nado libre. (p. 339)

meiosis (mi OH sus): reproductive process that produces four haploid sex cells from one diploid cell and ensures offspring will have the same number of chromosomes as the parent organisms. (p. 105)

meiosis: proceso reproductivo que produce cuatro células sexuales haploides a partir de una célula diploide y asegura que la descendencia tendrá el mismo número de cromosomas que los organismos progenitores. (p. 105)

melanin: pigment produced by the epidermis that protects skin from sun damage and gives skin and eyes their color. (p. 497)

melanina: pigmento producido por la epidermis que protege a la piel del daño producido por la luz solar y le da a la piel y a los ojos su color. (p. 497)

menstrual cycle: hormone-controlled monthly cycle of changes in the female reproductive system that includes the maturation of an egg and preparation of the uterus for possible pregnancy. (p. 630)

ciclo menstrual: ciclo mensual de cambios en el sistema reproductor femenino, el cual es controlado por hormonas e incluye la maduración de un óvulo y la preparación del útero para un posible embarazo. (p. 630)

menstruation (men STRAY shun): monthly flow of blood and tissue cells that occurs when the lining of the uterus breaks down and is shed. (p. 630)

menstruación: flujo mensual de sangre y células tisulares que ocurre cuando el endometrio uterino se rompe y se desprende. (p. 630)

metabolism: the total of all chemical reactions in an organism. (p. 81)

metabolismo: el conjunto de todas las reacciones químicas en un organismo. (p. 81)

metamorphosis: process in which many insect species change their body form to become adults; can be complete (egg, larva, pupa, adult) or incomplete (egg, nymph, adult). (p. 372)

metamorfosis: proceso a través del cual muchas especies de insectos cambian su forma corporal para convertirse en adultos; puede ser completa (huevo, larva, pupa, adulto) o incompleta (huevo, ninfa, adulto). (p. 372)

migration: instinctive seasonal movement of animals to find food or to reproduce in better conditions. (p. 470)

migración: movimiento estacional instintivo de los animales para encontrar alimento o para reproducirse en mejores condiciones. (p. 470)

mineral: inorganic nutrient that regulates many chemical reactions in the body. (p. 518)

mineral: nutriente inorgánico que regula una gran cantidad de reacciones químicas en el cuerpo. (p. 518)

Glossary/Glosario

mitochondrion: cell organelle that breaks down food and releases energy. (p. 42)

mitocondria: organelo celular que degrada nutrientes y libera energía. (p. 42)

mitosis (mi TOH sus): cell process in which the nucleus divides to form two nuclei identical to each other, and identical to the original nucleus, in a series of steps (prophase, metaphase, anaphase, and telophase). (p. 98)

mitosis: proceso celular en el que el núcleo se divide para formar dos núcleos idénticos entre sí e idénticos al núcleo original, a través de varias etapas (profase, metafase, anafase y telofase). (p. 98)

mixture: a combination of substances in which the individual substances do not change or combine chemically but instead retain their own individual properties; can be gases, solids, liquids, or any combination of them. (p. 69)

mezcla: una combinación de sustancias en la que las sustancias individuales no cambian ni se combinan químicamente pero mantienen sus propiedades individuales; pueden ser gases, sólidos, líquidos o una combinación de ellos. (p. 69)

molting: shedding and replacing of an arthropod's exoskeleton. (p. 370)

muda: muda y reemplazo del exoesqueleto de un artrópodo. (p. 370)

monocot: angiosperm with one cotyledon inside its seed, flower parts in multiples of three, and vascular tissues in bundles scattered throughout the stem. (p. 258)

monocotiledóneas: angiospermas con un solo cotiledón dentro de la semilla, partes florales dispuestas en múltiplos de tres y tejidos vasculares distribuidos en haces diseminados por todo el tallo. (p. 258)

monotreme: a mammal that lays eggs with tough, leathery shells and whose mammary glands do not have nipples. (p. 440)

monotrema: mamífero que pone huevos con cascarón coriáceo y resistente y cuyas glándulas mamarias carecen de pezones. (p. 440)

muscle: organ that can relax, contract, and provide the force to move bones and body parts. (p. 490)

músculo: órgano que puede relajarse, contraerse y proporcionar la fuerza para mover los huesos y las partes del cuerpo. (p. 490)

mutation: any permanent change in a gene or chromosome of a cell; may be beneficial, harmful, or have little effect on an organism. (p. 114)

mutación: cualquier cambio permanente en un gen o cromosoma de una célula; puede ser benéfica, perjudicial o tener un pequeño efecto sobre un organismo. (p. 114)

mutualism: a type of symbiotic relationship in which both organisms benefit. (p. 698)

mutualismo: tipo de relación simbiótica en la que ambos organismos se benefician. (p. 698)

mycorrhizae (mi kuh RI zee): network of hyphae and plant roots that helps plants absorb water and minerals from soil. (p. 226)

micorriza: estructura formada por una hifa y las raíces de una planta y que ayuda a las plantas a absorber agua y minerales del suelo. (p. 226)

N

natural resources: parts of Earth's environment that supply materials useful or necessary for the survival of living organisms. (p. 770)

recursos naturales: partes del medio ambiente terrestre que proporcionan materiales útiles o necesarios para la supervivencia de los organismos vivos. (p. 770)

natural selection: a process by which organisms with traits best suited to their environment are more likely to survive and reproduce; includes concepts of variation, overproduction, and competition. (p. 156)

selección natural: proceso mediante el cual los organismos con rasgos mejor adaptados a su ambiente tienen mayor probabilidad de sobrevivir y reproducirse; incluye los conceptos de variación, sobreproducción y competencia. (p. 156)

nephron (NEF rahn): tiny filtering unit of the kidney. (p. 579)

nefrona: pequeña unidad de filtrado del riñón. (p. 579)

nerve cord: tubelike structure above the notochord that in most chordates develops into the brain and spinal cord. (p. 395)

cordón nervioso: estructura en forma de tubo sobre el notocordio que en la mayoría de los cordados se desarrolla en el cerebro y en la médula espinal. (p. 395)

neuron (NOO rahn): basic functioning unit of the nervous system, made up of a cell body, dendrites, and axons. (p. 595)

neurona: unidad básica de funcionamiento del sistema nervioso, formada por un cuerpo celular, dendritas y axones. (p. 595)

niche: in an ecosystem, refers to the unique ways an organism survives, obtains food and shelter, and avoids danger. (p. 699)

nicho: en un ecosistema, se refiere a las formas únicas en las que un organismo sobrevive, obtiene alimento, refugio y evita el peligro. (p. 699)

nitrogen cycle: model describing how nitrogen moves from the atmosphere to the soil, to living organisms, and then back to the atmosphere. (p. 722)

ciclo del nitrógeno: modelo que describe cómo se mueve el nitrógeno de la atmósfera al suelo, a los organismos vivos y de nuevo a la atmósfera. (p. 722)

nitrogen fixation: process in which some types of bacteria in the soil change nitrogen gas into a form of nitrogen that plants can use. (p. 722)

fijación del nitrógeno: proceso en el cual algunos tipos de bacterias en el suelo transforman el nitrógeno gaseoso en una forma de nitrógeno que las plantas pueden usar. (p. 722)

nitrogen-fixing bacteria: bacteria that convert nitrogen in the air into forms that can be used by plants and animals. (p. 194)

bacterias fijadoras de nitrógeno: bacterias que convierten el nitrógeno presente en el aire en formas que pueden ser usadas por plantas y animales. (p. 194)

noninfectious disease: disease, such as cancer, diabetes, or asthma, that is not spread from one person to another. (p. 666)

enfermedad no infecciosa: enfermedad que no se trasmite de una persona a otra, como el cáncer, la diabetes o el asma. (p. 666)

nonrenewable resources: natural resources, such as petroleum, minerals, and metals, that are used more quickly than they can be replaced by natural processes. (p. 771)

recursos no renovables: recursos naturales, como el petróleo, los minerales y los metales, que son utilizados más rápidamente de lo que pueden ser reemplazados mediante procesos naturales. (p. 771)

nonvascular plant: plant that absorbs water and other substances directly through its cell walls instead of through tubelike structures. (p. 245)

planta no vascular: planta que absorbe agua y otras sustancias directamente a través de sus paredes celulares en vez de utilizar estructuras tubulares. (p. 245)

notochord: firm but flexible structure that extends along the upper part of a chordate's body. (p. 394)

notocordio: estructura firme pero flexible que se extiende a lo largo de la parte superior del cuerpo de un cordado. (p. 394)

nuclear energy: energy produced from the splitting apart of billions of uranium nuclei by a nuclear fission reaction. (p. 774)

energía nuclear: energía producida a partir del fraccionamiento de billones de núcleos de uranio mediante una reacción de fisión nuclear. (p. 774)

nucleus: organelle that controls all the activities of a cell and contains hereditary material made of proteins and DNA. (p. 40)

núcleo: organelo que controla todas las actividades de una célula y que contiene el material hereditario formado por proteínas y ADN. (p. 40)

nutrients (NEW tree unts): substances in foods—proteins, carbohydrates, fats, vitamins, minerals, and water—that provide energy and materials for cell development, growth, and repair. (p. 512)

nutrientes: sustancias de los alimentos—proteínas, carbohidratos, grasas, vitaminas, minerales y agua—que proporcionan energía y materiales para el desarrollo, crecimiento y reparación de las células. (p. 512)

O

olfactory (ohl FAK tree) cell: nasal nerve cell that becomes stimulated by molecules in the air and sends impulses to the brain for interpretation of odors. (p. 609)

célula olfatoria: célula nerviosa nasal que al ser estimulada por moléculas del aire envía impulsos al cerebro para la interpretación de los olores. (p. 609)

omnivore: animal that eats plants and animals or animal flesh; mammals with incisors, canine teeth, and flat molars for eating plants and other animals. (pp. 331, 437)

omnívoro: animal que se alimenta de plantas y animales; mamífero con incisivos, caninos y molares planos que se alimenta de plantas y otros animales. (pp. 331, 437)

open circulatory system: blood circulation system in which blood moves through vessels and into open spaces around the body organs. (p. 360)

organ: structure, such as the heart, made up of different types of tissues that all work together. (p. 45)

organelle: structure in the cytoplasm of a eukaryotic cell that can act as a storage site, process energy, move materials, or manufacture substances. (p. 40)

organic compounds: compounds that always contain hydrogen and carbon; carbohydrates, lipids, proteins, and nucleic acids are organic compounds found in living things. (p. 70)

organism: any living thing. (p. 14)

osmosis: a type of passive transport that occurs when water diffuses through a cell membrane. (p. 76)

ovary: in plants, swollen base of an angiosperm's pistil, where egg-producing ovules are found; in humans, female reproductive organ that produces eggs and is located in the lower part of the body. (pp. 285, 629)

ovulation (ahv yuh LAY shun): monthly process in which an egg is released from an ovary and enters the oviduct, where it can become fertilized by sperm. (p. 629)

ovule: in seed plants, the female reproductive part that produces eggs. (p. 283)

ozone depletion: thinning of Earth's ozone layer caused by chlorofluorocarbons (CFCs) leaking into the air and reacting chemically with ozone, breaking the ozone molecules apart. (p. 781)

sistema circulatorio abierto: sistema circulatorio sanguíneo en el que la sangre se mueve a través de vasos y entra en espacios abiertos alrededor de los órganos corporales. (p. 360)

órgano: estructura, como el corazón, que consiste en diferentes tipos de tejidos que trabajan conjuntamente. (p. 45)

organelo: estructura del citoplasma de una célula eucariota que puede actuar como sitio de almacenamiento, procesamiento de energía, movimiento de materiales o elaboración de sustancias. (p. 40)

compuestos orgánicos: compuestos que siempre contienen hidrógeno y carbono; los carbohidratos, lípidos, proteínas y ácidos nucleicos son compuestos orgánicos que se encuentran en los seres vivos. (p. 70)

organismo: cualquier ser vivo. (p. 14)

ósmosis: tipo de transporte pasivo que ocurre cuando el agua se difunde a través de una membrana celular. (p. 76)

ovario: en las plantas, base abultada del pistilo de una angiosperma donde se encuentran los óvulos productores de huevos; en los humanos, órgano reproductor femenino que produce óvulos y está localizado en la parte inferior del cuerpo. (pp. 285, 629)

ovulación: proceso mensual en el que un óvulo es liberado de un ovario y entra al oviducto, donde puede ser fertilizado por los espermatozoides. (p. 629)

óvulo: en las gimnospermas, la parte reproductiva femenina que produce huevos. (p. 283)

agotamiento del ozono: adelgazamiento de la capa de ozono de la Tierra causado por los clorofluorocarbonos (CFCs) que escapan al aire y reaccionan químicamente con el ozono rompiendo sus moléculas. (p. 781)

P

parasitism: a type of symbiotic relationship in which one organism benefits and the other organism is harmed. (p. 698)

passive immunity: immunity that results when antibodies produced in one animal are introduced into another's body; does not last as long as active immunity. (p. 655)

passive transport: movement of substances through a cell membrane without the use of cellular energy; includes diffusion, osmosis, and facilitated diffusion. (p. 74)

parasitismo: tipo de relación simbiótica en la que un organismo se beneficia y el otro es perjudicado. (p. 698)

inmunidad pasiva: inmunidad que resulta cuando los anticuerpos producidos en un animal son introducidos en el cuerpo de otro; no es tan duradera como la inmunidad activa. (p. 655)

transporte pasivo: movimiento de sustancias a través de la membrana celular sin usar energía celular; incluye difusión, ósmosis y difusión facilitada. (p. 74)

pasteurization (pas chur ruh ZAY shun): process in which a liquid is heated to a temperature that kills most bacteria. (p. 658)

pasterización: proceso mediante el cual un líquido es calentado a una temperatura que mata a la mayoría de las bacterias. (p. 658)

pathogen: disease-producing organism. (p. 197)

patógeno: organismo que produce enfermedades. (p. 197)

periosteum (pur ee AHS tee um): tough, tight-fitting membrane that covers a bone's surface and contains blood vessels that transport nutrients into the bone. (p. 485)

periostio: membrana fuertemente adherida y resistente que cubre la superficie de los huesos, contiene vasos sanguíneos y transporta nutrientes al interior del hueso. (p. 485)

peripheral nervous system: division of the nervous system, made up of all the nerves outside the CNS; connects the brain and spinal cord to other body parts. (p. 597)

sistema nervioso periférico: parte del sistema nervioso, compuesto por todos los nervios fuera del sistema nervioso central; conecta al cerebro y a la médula espinal con las otras partes del cuerpo. (p. 597)

peristalsis (per uh STAHL sus): waves of muscular contractions that move food through the digestive tract. (p. 526)

peristalsis: contracciones musculares ondulantes que mueven el alimento a través del tracto digestivo. (p. 526)

petroleum: nonrenewable resource formed over hundreds of millions of years mostly from the remains of microscopic marine organisms buried in Earth's crust. (p. 771)

petróleo: recurso no renovable formado durante cientos de millones de años, en su mayoría a partir de los restos de organismos marinos microscópicos sepultados en la corteza terrestre. (p. 771)

pharyngeal pouches: in developing chordates, the paired openings found in the area between the mouth and digestive tube. (p. 395)

bolsas faríngeas: en los cordados en desarrollo, las aperturas pareadas que se encuentran en el área entre la boca y el tubo digestivo. (p. 395)

pharynx (FER ingks): tubelike passageway for food, liquid, and air. (p. 570)

faringe: pasaje en forma de tubo por donde circulan alimentos, líquidos y aire. (p. 570)

phenotype (FEE nuh tipe): outward physical appearance and behavior of an organism as a result of its genotype. (p. 130)

fenotipo: apariencia física externa y comportamiento de un organismo como resultado de su genotipo. (p. 130)

pheromone (FER uh mohn): powerful chemical produced by an animal to influence the behavior of another animal of the same species. (p. 465)

feromona: químico potente producido por un animal para influir en el comportamiento de otro animal de la misma especie. (p. 465)

phloem (FLOH em): vascular tissue that forms tubes that transport dissolved sugar throughout a plant. (p. 255)

floema: tejido vascular que forma tubos que transportan azúcares disueltos a toda la planta. (p. 255)

photoperiodism: a plant's response to the lengths of daylight and darkness each day. (p. 316)

fotoperiodicidad: la respuesta de una planta a la duración de la luz y de la oscuridad cada día. (p. 316)

photosynthesis (foh toh SIHN thuh sihs): process by which plants and many other producers use light energy to produce a simple sugar from carbon dioxide and water and give off oxygen. (pp. 82, 305)

fotosíntesis: proceso mediante el cual las plantas y muchos otros organismos productores utilizan la energía luminosa para producir azúcares simples a partir de dióxido de carbono y agua y desprender oxígeno. (pp. 82, 305)

phylogeny (fi LAH juh nee): evolutionary history of an organism; used today to group organisms into six kingdoms. (p. 23)

filogenia: historia evolutiva de un organismo; usada hoy para agrupar a los organismos en seis reinos. (p. 23)

pioneer species: species that break down rock and build up decaying plant material so that other plants can grow; first organisms to grow in new or disturbed areas. (pp. 247, 740)

especies pioneras: especies que descomponen la roca y acumulan material vegetal en descomposición para que otras plantas puedan crecer; los primeros organismos que crecen en áreas nuevas o alteradas. (pp. 247, 740)

pistil: female reproductive organ inside the flower of an angiosperm; consists of a sticky stigma, where pollen grains land, and an ovary. (p. 285)

pistilo: órgano reproductivo femenino en la flor de una angiosperma; consiste en un ovario y un estigma pegajoso donde caen los granos de polen. (p. 285)

placenta: an organ that develops from tissues of the embryo and tissues that line the inside of the uterus and that absorbs oxygen and food from the mother's blood. (p. 441)

placental: a mammal whose offspring develop inside the female's uterus. (p. 441)

platelet: irregularly shaped cell fragment that helps clot blood and releases chemicals, that help form fibrin. (p. 551)

plasma: liquid part of blood, made mostly of water, in which oxygen, nutrients, and minerals are dissolved. (p. 550)

pollen grain: small structure produced by the male reproductive organs of a seed plant; has a water-resistant coat, can develop from a spore, and contains gametophyte parts that will produce sperm. (p. 281)

pollination: transfer of pollen grains to the female part of a seed plant by agents such as gravity, water, wind, and animals. (p. 281)

pollutant: substance that contaminates any part of the environment. (p. 778)

polygenic (pah lih JEH nihk) inheritance: occurs when a group of gene pairs acts together and produces a specific trait, such as human eye color, skin color, or height. (p. 136)

polyp (PAH lup): cnidarian body type that is vase-shaped and is usually sessile. (p. 339)

population: all the organisms that belong to the same species living in a community. (p. 686)

postanal tail: muscular structure at the end of a developing chordate. (p. 394)

preening: process in which a bird rubs oil from an oil gland over its feathers to condition them. (p. 430)

pregnancy: period of development—usually about 38 or 39 weeks in female humans—from fertilized egg until birth. (p. 634)

primates: group of mammals including humans, monkeys, and apes that share characteristics such as opposable thumbs, binocular vision, and flexible shoulders. (p. 170)

producer: organism, such as a green plant or alga, that uses an outside source of energy like the Sun to create energy-rich food molecules. (p. 696)

placenta: órgano que se desarrolla a partir de tejido embrionario y de los tejidos que cubren la pared interna del útero y que absorbe oxígeno y alimentos de la sangre de la madre. (p. 441)

placentario: mamífero cuyas crías se desarrollan en el útero de la hembra. (p. 441)

plaqueta: fragmento celular de forma irregular que ayuda a coagular la sangre y libera químicos que ayudan a formar fibrina. (p. 551)

plasma: parte líquida de la sangre compuesta principalmente por agua y en la que se encuentran disueltos oxígeno, nutrientes y minerales. (p. 550)

grano de polen: estructura pequeña producida por los órganos reproductivos masculinos de los espermatófitos; tiene una cubierta resistente al agua, puede desarrollarse a partir de una espora y contiene partes del gametofito que producirán esperma. (p. 281)

polinización: transferencia de los granos de polen a la parte femenina de un espermatófito a través de agentes como la gravedad, el agua, el viento y los animales. (p. 281)

contaminante: sustancia que contamina cualquier parte del medio ambiente. (p. 778)

herencia poligénica: ocurre cuando un grupo de pares de genes actúa conjuntamente y produce un rasgo específico, tal como el color de los ojos, el color de la piel, o la estatura en los humanos. (p. 136)

pólipo: tipo corporal de los cnidarios con forma de jarro y usualmente sésil. (p. 339)

población: todos los organismos que pertenecen a la misma especie dentro de una comunidad. (p. 686)

cola postnatal: estructura muscular en el extremo de un cordado en desarrollo. (p. 394)

acicalamiento: proceso mediante el cual las aves acondicionan sus alas frotándoles grasa producida por una glándula sebácea. (p. 430)

embarazo: período del desarrollo—generalmente unas 38 o 39 semanas en las hembras humanas—que va desde el óvulo fertilizado hasta el nacimiento. (p. 634)

primates: grupo de mamíferos que incluye a los humanos, monos y simios, los cuales comparten características como pulgares opuestos, visión binocular y hombros flexibles. (p. 170)

productor: organismo, como una planta o un alga verde, que utiliza una fuente externa de energía, como la luz solar, para producir moléculas de nutrientes ricas en energía. (p. 696)

protein: large molecule that contains carbon, hydrogen, oxygen, nitrogen, and sometimes sulfur and is made up of amino acids; used by the body for growth and for replacement and repair of body cells. (p. 513)

prothallus (proh THA lus): small, green, heart-shaped gametophyte plant form of a fern that can make its own food and absorb water and nutrients from the soil. (p. 278)

protist: one- or many-celled eukaryotic organism that can be plantlike, animal-like, or funguslike. (p. 210)

protozoan: one-celled, animal-like protist that can live in water, soil, and living and dead organisms. (p. 215)

pseudopods (SEW duh pahdz): temporary cytoplasmic extensions used by some protists to move about and trap food. (p. 216)

pulmonary circulation: flow of blood through the heart to the lungs and back to the heart. (p. 542)

punctuated equilibrium: model describing the rapid evolution that occurs when mutation of a few genes results in a species suddenly changing into a new species. (p. 160)

Punnett (PUN ut) square: a tool to predict the probability of certain traits in offspring that shows the different ways alleles can combine. (p. 130)

proteína: molécula grande que contiene carbono, hidrógeno, oxígeno, nitrógeno y algunas veces azufre, constituida por aminoácidos y usada por el cuerpo para el crecimiento y reemplazo o reparación de las células del cuerpo. (p. 513)

prótalo: gametofito pequeño de color verde en forma de corazón, de un helecho, que puede producir su propio alimento y absorber agua y nutrientes del suelo. (p. 278)

protista: organismo eucariota unicelular o pluricelular que puede parecerse a las plantas, a los animales o a los hongos. (p. 210)

protozoario: protista unicelular similar a los animales y que puede vivir en el agua, en el suelo y en los organismos vivos o muertos. (p. 215)

pseudópodos: extensiones citoplasmáticas temporales usadas por algunos protistas para moverse y atrapar alimento. (p. 216)

circulación pulmonar: flujo sanguíneo del corazón hacia los pulmones y de regreso al corazón. (p. 542)

equilibrio punteado: modelo que describe la evolución rápida que ocurre cuando la mutación de unos pocos genes resulta en que una especie cambie rápidamente para convertirse en otra especie. (p. 160)

Cuadrado de Punnett: herramienta para predecir la probabilidad de ciertos rasgos en la descendencia mostrando las diferentes formas en que los alelos pueden combinarse. (p. 130)

R

radial symmetry: body parts arranged in a circle around a central point. (p. 335)

radioactive element: element that gives off a steady amount of radiation as it slowly changes to a nonradioactive element. (p. 165)

radula (RA juh luh): in gastropods, the tonguelike organ with rows of teeth used to scrape and tear food. (p. 361)

recessive (rih SE sihv): describes a trait that is covered over, or dominated, by another form of that trait and seems to disappear. (p. 128)

recycling: conservation method that is a form of reuse and requires changing or reprocessing an item or natural resource. (p. 789)

reflex: simple innate behavior, such as yawning or blinking, that is an automatic response and does not involve a message to the brain; automatic, involuntary response to a stimulus; controlled by the spinal cord. (pp. 475, 601)

simetría radial: disposición de las partes del cuerpo circularmente alrededor de un punto central. (p. 335)

elemento radiactivo: elemento que emite una cantidad estable de radiación mientras se convierte lentamente en un elemento no radiactivo. (p. 165)

rádula: órgano de los gasterópodos en forma de lengua con filas de dientecillos usado para raspar y desgarrar alimentos. (p. 361)

recesivo: describe un rasgo que está encubierto, o que es dominado, por otra forma del mismo rasgo y que parece no estar presente. (p. 128)

reciclaje: método de conservación como una forma de reutilización y que requiere del cambio o reprocesamiento del producto o recurso natural. (p. 789)

reflejo: comportamiento innato simple, como bostezar o parpadear, que constituye una respuesta automática y no requiere el envío de un mensaje al cerebro; respuesta automática e involuntaria a un estímulo controlada por la médula espinal. (pp. 475, 601)

renewable resources: natural resources, such as water, sunlight, and crops, that are constantly being recycled or replaced by nature. (p. 771)

recursos renovables: recursos naturales, como el agua, la luz solar y los cultivos, que son reciclados o reemplazados constantemente por la naturaleza. (p. 771)

respiration: process by which producers and consumers release stored energy from food molecules; series of chemical reactions used to release energy stored in food molecules. (pp. 83, 307)

respiración: proceso mediante el cual los organismos productores y consumidores liberan la energía almacenada en las moléculas de los alimentos; serie de reacciones químicas usadas para liberar la energía almacenada en las moléculas de los alimentos. (pp. 83, 307)

retina: light-sensitive tissue at the back of the eye; contains rods and cones. (p. 605)

retina: tejido sensible a la luz situado en la parte posterior del ojo; contiene conos y bastones. (p. 605)

rhizoids (RI zoydz): threadlike structures that anchor nonvascular plants to the ground. (p. 246)

rizoides: estructuras en forma de hilos que anclan las plantas no vasculares al suelo. (p. 246)

rhizome: underground stem. (p. 278)

rizoma: tallo subterráneo. (p. 278)

ribosome: small cytoplasmic structure on which cells make their own proteins. (p. 42)

ribosoma: estructura citoplasmática pequeña en la que las células producen sus propias proteínas. (p. 42)

RNA: ribonucleic acid; a type of nucleic acid that carries codes for making proteins from the nucleus to the ribosomes. (p. 112)

ARN: ácido ribonucleico; tipo de ácido nucleico que transporta los códigos para la formación de proteínas del núcleo a los ribosomas. (p. 112)

S

saprophyte: organism that uses dead organisms as a food source and helps recycle nutrients so they are available for use by other organisms. (pp. 194, 222)

saprófito: organismo que usa a los organismos muertos como una fuente de alimento y ayuda a reciclar los nutrientes de tal forma que estén disponibles para ser usados por otros organismos. (pp. 194, 222)

scales: thin, hard plates that cover a fish's skin and protect its body. (p. 399)

escamas: placas duras y delgadas que cubren la piel de los peces y protegen su cuerpo. (p. 399)

scientific methods: procedures used to solve problems and answer questions that can include stating the problem, gathering information, forming a hypothesis, testing the hypothesis with an experiment, analyzing data, and drawing conclusions. (p. 7)

métodos científicos: procedimientos utilizados para solucionar problemas y responder a preguntas; puede incluir el establecimiento de un problema, recopilación de información, formulación de una hipótesis, comprobación de la hipótesis con un experimento, análisis de la información y presentación de conclusiones. (p. 7)

sedimentary rock: a type of rock, such as limestone, that is most likely to contain fossils and is formed when layers of sand, silt, clay, or mud are cemented and compacted together or when minerals are deposited from a solution. (p. 164)

roca sedimentaria: tipo de roca, como la piedra caliza, con alta probabilidad de contener fósiles y que se forma cuando las capas de arena, sedimento, arcilla o lodo son cementadas y compactadas o cuando los minerales de una solución son depositados. (p. 164)

semen (SEE mun): mixture of sperm and a fluid that helps sperm move and supplies them with an energy source. (p. 628)

semen: mezcla de espermatozoides y un fluido que ayuda a la movilización de los espermatozoides y les suministra una fuente de energía. (p. 628)

sessile (SE sile): describes an organism that remains attached to one place during most of its lifetime. (p. 337)

sésil: organismo que permanece adherido a un lugar durante la mayor parte de su vida. (p. 337)

setae (SEE tee): bristlelike structures on the outside of each body segment that helps segmented worms move. (p. 365)

cerdas: estructuras en forma de cilios presentes en la parte externa de cada segmento corporal y que ayudan a los gusanos segmentados a moverse. (p. 365)

sex-linked gene: an allele inherited on a sex chromosome and that can cause human genetic disorders such as color blindness and hemophilia. (p. 139)

sexual reproduction: a type of reproduction in which two sex cells, usually an egg and a sperm, join to form a zygote, which will develop into a new organism with a unique identity. (p. 104)

sexually transmitted disease (STD): infectious disease, such as chlamydia, AIDS, or genital herpes, that is passed from one person to another during sexual contact. (p. 662)

short-day plant: plant that generally requires long nights—12 or more hours of darkness—to begin the flowering process. (p. 316)

skeletal muscle: voluntary, striated muscle that moves bones, works in pairs, and is attached to bones by tendons. (p. 493)

skeletal system: all the bones in the body; forms an internal, living framework that provides shape and support, protects internal organs, moves bones, forms blood cells, and stores calcium and phosphorus compounds for later use. (p. 484)

smooth muscle: involuntary, nonstriated muscle that controls movement of internal organs. (p. 493)

social behavior: interactions among members of the same species, including courtship and mating, getting food, caring for young, and protecting each other. (p. 462)

society: a group of animals of the same species that live and work together in an organized way, with each member doing a specific job. (p. 463)

soil: mixture of mineral and rock particles, the remains of dead organisms, air, and water that forms the topmost layer of Earth's crust and supports plant growth. (p. 714)

sori: fern structures in which spores are produced. (p. 278)

species: group of organisms that share similar characteristics and can reproduce among themselves producing fertile offspring. (p. 154)

sperm: haploid sex cell formed in the male reproductive organs; in humans, male reproductive cells produced in the testes. (pp. 104, 628)

spiracles (SPIHR ih kulz): openings in the abdomen and thorax of insects through which air enters and waste gases leave. (p. 371)

spontaneous generation: idea that living things come from nonliving things. (p. 19)

gen ligado al sexo: un alelo heredado en un cromosoma sexual y que puede causar desórdenes genéticos humanos como daltonismo y hemofilia. (p. 139)

reproducción sexual: tipo de reproducción en la que dos células sexuales, generalmente un óvulo y un espermatozoide, se unen para formar un zigoto, el cual se desarrollará para formar un nuevo organismo con identidad única. (p. 104)

enfermedad de transmisión sexual (ETS): enfermedad infecciosa como la clamidiasis, SIDA y herpes genital, transmitida de una persona a otra mediante contacto sexual. (p. 662)

planta de día corto: planta que generalmente requiere de noches largas—12 horas o más de oscuridad—para comenzar su proceso de floración. (p. 316)

músculo esquelético: músculo estriado voluntario que mueve los huesos, trabaja en pares y se fija a los huesos por medio de los tendones. (p. 493)

sistema esquelético: todos los huesos en el cuerpo forman una estructura viva interna que proporciona forma y soporte, protege a los órganos internos, mueve a los huesos, forma células sanguíneas y almacena compuestos de calcio y fósforo para uso posterior. (p. 484)

músculo liso: músculo no estriado involuntario que controla el movimiento de los órganos internos. (p. 493)

comportamiento social: interacciones entre los miembros de la misma especie, incluyendo el cortejo y el apareamiento, la obtención de alimento, el cuidado de las crías y la protección de unos a otros. (p. 462)

sociedad: grupo de animales de la misma especie que vive y trabaja conjuntamente de forma organizada, con cada miembro realizando una tarea específica. (p. 463)

suelo: mezcla de partículas minerales y rocas, restos de organismos muertos, aire y del agua que forma la capa superior de la corteza terrestre y favorece el crecimiento de las plantas. (p. 714)

soros: estructuras de los helechos en donde se producen las esporas. (p. 278)

especie: grupo de organismos que comparten características similares entre sí y que pueden reproducirse entre ellos dando lugar a una descendencia fértil. (p. 154)

espermatozoides: células sexuales haploides que se forman en los órganos reproductores masculinos; en los humanos, células reproductoras masculinas producidas por los testículos. (pp. 104, 628)

espiráculos: aperturas del abdomen y tórax de los insectos a través de las cuales entra aire y salen gases de desecho. (p. 371)

generación espontánea: idea que sostiene que los seres vivos proceden de seres inertes. (p. 19)

sporangium (spuh RAN jee uhm): round spore case of a zygote fungus. (p. 225)

spores: waterproof reproductive cell of a fungus that can grow into a new organism; in plants, haploid cells produced in the gametophyte stage that can divide by mitosis to form plant structures or an entire new plant or can develop into sex cells. (pp. 223, 275)

sporophyte (SPOR uh fite) stage: plant life-cycle stage that begins when an egg is fertilized by a sperm. (p. 275)

stamen: male reproductive organ inside the flower of an angiosperm; consists of an anther, where pollen grains form, and a filament. (p. 285)

stinging cells: capsules with coiled triggerlike structures that help cnidarians capture food. (p. 340)

stomata (STOH muh tuh): tiny openings in a plant's epidermis through which carbon dioxide, water vapor, and oxygen enter and exit. (pp. 253, 303)

succession: natural, gradual changes in the types of species that live in an area; can be primary or secondary. (p. 740)

symbiosis: any close relationship between species, including mutualism, commensalism, and parasitism. (p. 698)

synapse (SIHN aps): small space across which an impulse moves from an axon to the dendrites or cell body of another neuron. (p. 597)

systemic circulation: largest part of the circulatory system, in which oxygen-rich blood flows to all the organs and body tissues, except the heart and lungs, and oxygen-poor blood is returned to the heart. (p. 543)

esporangio: estructura redondeada que contiene las esporas de un zigomiceto. (p. 225)

esporas: célula reproductora impermeable de un hongo, la cual puede convertirse en un nuevo organismo; en las plantas, las células haploides producidas en la etapa de gametofito que pueden dividirse por mitosis para formar las estructuras de la planta o una planta nueva, o que pueden convertirse en células sexuales. (pp. 223, 275)

etapa de esporofito: etapa del ciclo de vida de una planta que comienza cuando un huevo es fertilizado por un esperma. (p. 275)

estambre: órgano reproductor masculino dentro de la flor de una angiosperma, que consiste en un filamento y una antera donde se forman los granos de polen. (p. 285)

cnidocitos: cápsulas con estructuras enrolladas en forma de gatillo y que ayudan a los cnidarios a capturar su alimento. (p. 340)

estomas: aperturas pequeñas en la superficie de la mayoría de las hojas de las plantas, las cuales permiten que entre y salga dióxido de carbono, agua y oxígeno. (pp. 253, 303)

sucesión: cambios graduales y naturales en los tipos de especies que viven en un área; puede ser primaria o secundaria. (p. 740)

simbiosis: cualquier relación estrecha entre especies, incluyendo mutualismo, comensalismo y parasitismo. (p. 698)

sinapsis: espacio pequeño a través del cual un impulso se mueve del axón a las dendritas o al cuerpo celular de otra neurona. (p. 597)

circulación sistémica: la parte más grande del sistema circulatorio en la que la sangre rica en oxígeno fluye hacia todos los órganos y tejidos corporales excepto el corazón y los pulmones, y la sangre pobre en oxígeno regresa al corazón. (p. 543)

taiga (TI guh): world's largest biome, located south of the tundra between 50°N and 60°N latitude; has long, cold winters, precipitation between 35 cm and 100 cm each year, cone-bearing evergreen trees, and dense forests. (p. 746)

taste bud: major sensory receptor on the tongue; contains taste hairs that send impulses to the brain for interpretation of tastes. (p. 610)

taiga: el bioma más grande del mundo, localizado al sur de la tundra entre 50° y 60° de latitud norte; tiene inviernos prolongados y fríos, una precipitación que alcanza entre 35 y 100 centímetros cúbicos al año, coníferas perennifolias y bosques espesos. (p. 746)

papila gustativa: receptor sensorial principal de la lengua que contiene cilios gustativos que envían impulsos al cerebro para interpretación de los sabores. (p. 610)

temperate deciduous forest: biome usually having four distinct seasons, annual precipitation between 75 cm and 150 cm, and climax communities of deciduous trees. (p. 747)

temperate rain forest: biome with 200 cm to 400 cm of precipitation each year, average temperatures between 9°C and 12°C, and forests dominated by trees with needlelike leaves. (p. 747)

tendon: thick band of tissue that attaches bones to muscles. (p. 493)

tentacles (TEN tih kulz): armlike structures that have stinging cells and surround the mouths of most cnidarians. (p. 340)

testis: male organ that produces sperm and testosterone. (p. 628)

theory: explanation of things or events based on scientific knowledge resulting from many observations and experiments. (p. 10)

tissue: group of similar cells that work together to do one job. (p. 45)

toxin: poisonous substance produced by some pathogens. (p. 197)

trachea (TRAY kee uh): air-conducting tube that connects the larynx with the bronchi; is lined with mucous membranes and cilia, and contains strong cartilage rings. (p. 571)

tropical rain forest: most biologically diverse biome; has an average temperature of 25°C and receives between 200 cm and 600 cm of precipitation each year. (p. 748)

tropism: positive or negative plant response to an external stimulus such as touch, light, or gravity. (p. 312)

tube feet: hydraulic, hollow, thin-walled tubes that end in suction cups and enable echinoderms to move. (p. 380)

tundra: cold, dry, treeless biome with less than 25 cm of precipitation each year, a short growing season, permafrost, and winters that can be six to nine months long. Tundra is separated into two types: arctic tundra and alpine tundra. (p. 745)

bosque caducifolio templado: bioma que generalmente tiene cuatro estaciones distintas, con una precipitación anual entre 75 y 150 centímetros cúbicos y un clímax comunitario de árboles caducifolios. (p. 747)

bosque lluvioso templado: bioma con 200 a 400 centímetros cúbicos de precipitación al año; tiene una temperatura promedio entre 9 y 12°C y bosques dominados por árboles de hojas aciculares. (p. 747)

tendón: banda gruesa de tejido que une los músculos a los huesos. (p. 493)

tentáculos: estructuras en forma de brazo, que poseen cnidocitos y rodean la boca de la mayoría de los cnidarios. (p. 340)

testículos: órganos masculinos que producen espermatozoides y testosterona. (p. 628)

teoría: explicación de cosas o eventos basándose en el conocimiento científico resultante de muchas observaciones y experimentos. (p. 10)

tejido: grupo de células similares que trabajan conjuntamente para hacer una tarea. (p. 45)

toxina: sustancia venenosa producida por algunos patógenos. (p. 197)

tráquea: tubo conductor de aire que conecta a la laringe con los bronquios y que está recubierta por una membrana mucosa y cilios; está formada por anillos cartilaginosos resistentes. (p. 571)

selva húmeda tropical: el bioma más diverso biológicamente; tiene una temperatura promedio de 25°C y recibe entre 200 y 600 centímetros cúbicos de precipitación al año. (p. 748)

tropismo: respuesta positiva o negativa de una planta a un estímulo externo como el rozamiento, la luz o la gravedad. (p. 312)

pie tubular: tubos hidráulicos de pared delgada que terminan en copas de succión y que permiten moverse a los equinodermos. (p. 380)

tundra: bioma sin árboles, frío y seco, con menos de 25 centímetros cúbicos de precipitación al año; tiene una estación corta de crecimiento y permafrost e inviernos que pueden durar entre 6 y 9 meses. La tundra se divide en dos tipos: tundra ártica y tundra alpina. (p. 745)

U

umbilical cord: connects the embryo to the placenta; moves food and oxygen from the placenta to the embryo and removes the embryo's waste products. (p. 441)

cordón umbilical: conecta al embrión con la placenta, lleva nutrientes y oxígeno de la placenta al embrión y retira los productos de desecho de éste. (p. 441)

ureter: tube that carries urine from each kidney to the bladder. (p. 580)
urethra (yoo REE thruh): tube that carries urine from the bladder to the outside of the body. (p. 580)
uterus: in female humans, hollow, muscular, pear-shaped organ where a fertilized egg develops into a baby. (p. 629)
urinary system: system of excretory organs that rids the blood of wastes, controls blood volume by removing excess water, and balances concentrations of salts and water. (p. 577)
urine: wastewater that contains excess water, salts, and other wastes that are not reabsorbed by the body. (p. 578)

uréter: tubo que conduce a la orina de cada riñón hacia la vejiga. (p. 580)
uretra: tubo que conduce a la orina de la vejiga al exterior del cuerpo. (p. 580)
útero: en seres hermanos femeninos, órgano en forma de pera, hueco y musculoso, en el que un óvulo fertilizado se desarrolla en bebé. (p. 629)
sistema urinario: sistema de órganos excretores que elimina los desechos de la sangre, controla el volumen de sangre eliminando el exceso de agua y balancea las concentraciones de sales y agua. (p. 577)
orina: líquido de desecho que contiene el exceso de agua, sales y otros desechos que no son reabsorbidos por el cuerpo. (p. 578)

V

vaccination: process of giving a vaccine by mouth or by injection to provide active immunity against a disease. (p. 655)
vaccine: preparation made from killed bacteria or damaged particles from bacterial cell walls or viruses that can prevent some bacterial and viral diseases. (p. 199)
vagina (vuh JI nuh): muscular tube that connects the lower end of a female's uterus to the outside of the body; the birth canal through which a baby travels when being born. (p. 629)
variable: something in an experiment that can change. (p. 9)
variation: inherited trait that makes an individual different from other members of the same species and results from a mutation in the organism's genes. (p. 158)
vascular plant: plant with tubelike structures that move minerals, water, and other substances throughout the plant. (p. 245)
vein: blood vessel that carries blood back to the heart, and has one-way valves that keep blood moving toward the heart. (p. 544)
ventricles (VEN trih kulz): two lower chambers of the heart, that contract at the same time, during a heartbeat. (p. 542)
vertebrae: backbones that are joined by flexible cartilage and protect a vertebrate's spinal nerve cord. (p. 395)
vertebrate: animal with a backbone. (p. 334)

vacunación: proceso de aplicar una vacuna por vía oral o mediante una inyección para proporcionar inmunidad activa contra una enfermedad. (p. 655)
vacuna: preparación fabricada a partir de bacterias muertas o partículas dañadas de las paredes celulares bacterianas o virus y que puede prevenir algunas enfermedades bacterianas y virales. (p. 199)
vagina: tubo musculoso que conecta el extremo inferior del útero de una hembra con el exterior del cuerpo; el canal del nacimiento a través del cual sale un bebé al nacer. (p. 629)
variable: condición que puede cambiar en un experimento. (p. 9)
variación: rasgo heredado que hace que un individuo sea diferente a otros miembros de su misma especie como resultado de una mutación de sus genes. (p. 158)
planta vascular: planta con estructuras semejantes a tubos, las cuales sirven para movilizar minerales, agua y otras sustancias a toda la planta. (p. 245)
vena: vaso sanguíneo que lleva sangre de regreso al corazón y tiene válvulas unidireccionales que mantienen a la sangre en movimiento hacia el corazón. (p. 544)
ventrículos: las dos cámaras inferiores del corazón que se contraen al mismo tiempo durante el latido cardiaco. (p. 542)
vértebra: huesos de la espalda unidos por cartílago flexible y que protegen la médula espinal de los vertebrados. (p. 395)
vertebrado: animal que posee columna vertebral. (p. 334)

vestigial (veh STIHJ ee ul) structure: structure, such as the human appendix, that doesn't seem to have a function and may once have functioned in the body of an ancestor. (p. 168)

villi (VIH li): fingerlike projections covering the wall of the small intestine that increase the surface area for food absorption. (p. 528)

virus: a strand of hereditary material surrounded by a protein coating that can infect and multiply in a host cell. (pp. 52, 658)

vitamin: water-soluble or fat-soluble organic nutrient needed in small quantities for growth, for preventing some diseases, and for regulating body functions. (p. 516)

voluntary muscle: muscle, such as a leg or arm muscle, that can be consciously controlled. (p. 491)

estructura vestigial: estructura, como el apéndice humano, que no parece tener alguna función pero que pudo haber funcionado en el cuerpo de un antepasado. (p. 168)

vellosidades: proyecciones en forma de dedos que cubren la pared del intestino delgado y que incrementan la superficie de absorción de nutrientes. (p. 528)

virus: pieza de material hereditario rodeado de una capa de proteína que infecta y se multiplica en las células huéspedes. (pp. 52, 658)

vitamina: nutriente orgánico soluble al agua y al aceite, necesario en pequeñas cantidades para el crecimiento, prevención de algunas enfermedades y regulación de las funciones del cuerpo. (p. 516)

músculo voluntario: músculo, como el músculo de una pierna o un brazo, que puede controlarse conscientemente. (p. 491)

W

water cycle: model describing how water moves from Earth's surface to the atmosphere and back to the surface again through evaporation, condensation, and precipitation. (p. 721)

water-vascular system: network of water-filled canals that allows echinoderms to move, capture food, give off wastes, and exchange carbon dioxide and oxygen. (p. 380)

wetland: a land region that is wet most or all of the year. (p. 755)

ciclo del agua: modelo que describe cómo se mueve el agua de la superficie de la Tierra hacia la atmósfera y nuevamente hacia la superficie terrestre a través de la evaporación, la condensación y la precipitación. (p. 721)

sistema vascular acuoso: red de canales llenos de agua que permiten a los equinodermos moverse, capturar alimento, eliminar sustancias de desecho e intercambiar dióxido de carbono y oxígeno. (p. 380)

zona húmeda: región lluviosa la mayor parte del año. (p. 755)

X

xylem (ZI lum): vascular tissue that forms hollow vessels that transport substances, other than sugar, throughout a plant. (p. 256)

xilema: tejido vascular que forma vasos ahuecados que trasportan todo tipo de sustancias, excepto azúcares, en toda la planta. (p. 256)

Z

zygote: new diploid cell formed when a sperm fertilizes an egg; will divide by mitosis and develop into a new organism. (p. 104)

zigoto: célula diploide nueva formada cuando un espermatozoide fertiliza a un óvulo; se dividirá por mitosis y se desarrollará para formar un nuevo organismo. (p. 104)

Italic numbers = illustration/photo **Bold numbers = vocabulary term**
lab *= indicates a page on which the entry is used in a lab*

A

B

Index

C

Index

Index

Index

Index

S

Index

U

V

Index

Index

Magnification Key:
Magnifications listed are the magnifications at which images were originally photographed.
LM—Light Microscope
SEM—Scanning Electron Microscope
TEM—Transmission Electron Microscope

Acknowledgments: Glencoe would like to acknowledge the artists and agencies who participated in illustrating this program: Absolute Science Illustration; Andrew Evansen; Argosy; Articulate Graphics; Craig Attebery represented by Frank & Jeff Lavaty; CHK America; John Edwards and Associates; Gagliano Graphics; Pedro Julio Gonzalez represented by Melissa Turk & The Artist Network; Robert Hynes represented by Mendola Ltd.; Morgan Cain & Associates; JTH Illustration; Laurie O'Keefe; Matthew Pippin represented by Beranbaum Artist's Representative; Precision Graphics; Publisher's Art; Rolin Graphics, Inc.; Wendy Smith represented by Melissa Turk & The Artist Network; Kevin Torline represented by Berendsen and Associates, Inc.; WILDlife ART; Phil Wilson represented by Cliff Knecht Artist Representative; Zoo Botanica.

Photo Credits

Cover Klaus Nigge/National Geographic Image Collection/ Getty Images; **i ii** Klaus Nigge/National Geographic Image Collection/Getty Images; **vii** Aaron Haupt; **viii** John Evans; **ix** (t)PhotoDisc, (b)John Evans; **x** (l)John Evans, (r)Geoff Butler; **xi** (l)John Evans, (r)PhotoDisc; **xii** PhotoDisc; **xiii** Doug Perrine/Innerspace Visions; **xiv** (t)Mark Burnett, (b)Frans Lanting/Minden Pictures; **xvi** Dr. Jeremy Burgess/ Science Photo Library/Photo Researchers; **xvii** Fateh Singh Rathore/Peter Arnold, Inc.; **xix** Ariel Skelley/The Stock Market/CORBIS; **xx** Richard Thatcher/David R. Frazier Photolibrary; **xxi** Mark E. Gibson/Visuals Unlimited; **xxii** Rod Planck/Photo Researchers; **xxvi** Geoff Butler; **xxvii** Matt Meadows; **xxviii** M. Schliwa/Visuals Unlimited; **1** CORBIS; **2–3** Diane Scullion Littler; **3** (l)Jonathan Eisenback/ PhotoTake, NYC/PictureQuest, (r)Janice M. Sheldon/Picture 20-20/PictureQuest; **4–5** A. Witte/C. Mahaney/Getty Images; **6** Kjell B. Sandved/Visuals Unlimited; **8 9** Mark Burnett; **11** Tek Image/Science Photo Library/Photo Researchers; **12** Mark Burnett, (label)no credit needed; **13** Mark Burnett; **14** (t)Michael Abbey/Science Source/Photo Researchers, (bl)Aaron Haupt, (br)Michael Delannoy/Visuals Unlimited; **15** Mark Burnett; **16** (tr)Mark Burnett, (tcr)A. Glauberman/ Photo Researchers, (bl bcl br)Runk/Schoenberger from Grant Heilman, (others)Dwight Kuhn; **17** (t)Bill Beaty/Animals Animals, (bl)Tom & Therisa Stack/Tom Stack & Assoc., (br)Michael Fogden/Earth Scenes; **18** Aaron Haupt; **19** Geoff Butler; **22** (t)Arthur C. Smith III From Grant Heilman, (bl)Hal Beral/Visuals Unlimited, (br)Larry L. Miller/Photo Researchers; **23** Doug Perrine/Innerspace Visions; **24** (l)Brandon D. Cole, (r)Gregory Ochocki/Photo Researchers; **25** (l)Zig Leszczynski/Animals Animals, (r)R. Andrew Odum/Peter Arnold, Inc.; **26** Alvin E. Staffan; **27** Geoff Butler; **28** (t)Jan Hinsch/Science Photo Library/Photo Researchers, (b)Mark Burnett; **29** Mark Burnett; **30** Marc Von Roosmalen/AP; **31** (l)Mark Burnett, (r)Will & Deni McIntyre/Photo Researchers; **32** KS Studios/Mullenix; **33** Jeff Greenberg/Rainbow; **34** Dwight Kuhn; **35** Dave Spier/Visuals Unlimited; **36–37** Nancy Kedersha/Science Photo Library/Photo Researchers; **39** David M. Phillips/Visuals Unlimited; **40** (t)Don Fawcett/Photo Researchers, (b)M. Schliwa/Visuals Unlimited; **42** (t)George B. hapman/ Visuals Unlimited, (b)P. Motta & T. Naguro/Science Photo Library/Photo Researchers; **43** (t)Don Fawcett/Photo Researchers, (b)Biophoto Associates/Photo Researchers; **47** (l)Biophoto Associates/Photo Researchers, (r)Matt Meadows; **48** (bkgd)David M. Phillips/Visuals Unlimited, (cw from top)courtesy Nikon Instruments Inc., David M. Phillips/ Visuals Unlimited, Kathy Talaro/Visuals Unlimited, Michael Abbey/Visuals Unlimited, Michael Gabridge/Visuals Unlimited; **49** (tl)James W. Evarts, (tr)Bob Krist/CORBIS, (cl)courtesy Olympus Corporation, (cr)Michael Abbey/ Visuals Unlimited, (bl)Karl Aufderheide/Visuals Unlimited, (br)Lawrence Migdale/Stock Boston/PictureQuest; **52** (l)Richard J. Green/Photo Researchers, (c)Dr. J.F.J.M. van der Heuvel, (r)Gelderblom/Eye of Science/Photo Researchers; **55** Pam Wilson/Texas Dept. of Health; **56 57** Matt Meadows; **58** (t)Quest/Science Photo Library/Photo Researchers, (b)courtesy California University; **59** (l)Keith Porter/Photo Researchers, (r)NIBSC/Science Photo Library/Photo Researchers; **61** Biophoto Associates/Science Source/Photo Researchers; **62** P. Motta & T. Naguro/Science Photo Library/ Photo Researchers; **64–65** Jane Grushow/Grant Heilman Photography; **67** Bob Daemmrich; **69** (t)Runk/ Schoenberger from Grant Heilman, (b)Klaus Guldbrandsen/ Science Photo Library/Photo Researchers; **74** (l)John Fowler, (r)Richard Hamilton Smith/CORBIS; **75** KS Studios; **76** Aaron Haupt; **77** Visuals Unlimited; **78** Biophoto Associates/Science Source/Photo Researchers; **80** Matt Meadows; **82** Craig Lovell/ CORBIS; **83** John Fowler; **84** David M. Phillips/Visuals Unlimited; **85** (l)Grant Heilman Photography, (r)Bios (Klein/ Hubert)/Peter Arnold; **86** (t)Runk/Schoenberger from Grant Heilman, (b)Matt Meadows; **87** Matt Meadows; **88** Lappa/ Marquart; **89** CNRI/Science Photo Library/Photo Researchers; **90** Biophoto Associates/Science Source/Photo Researchers; **94–95** Zig Leszcynski/Animals Animals; **96** (l)Dave B. Fleetham/Tom Stack & Assoc., (r)Cabisco/Visuals Unlimited; **98** Cabisco/Visuals Unlimited; **99** (tl)Michael Abbey/Visuals Unlimited, (others)John D. Cunningham/Visuals Unlimited; **100** (l)Matt Meadows, (r)Nigel Cattlin/Photo Researchers; **101** (l)Barry L. Runk from Grant Heilman, (r)Runk/ Schoenberger from Grant Heilman; **102** (l)Walker England/ Photo Researchers, (r)Tom Stack & Assoc.; **103** Runk/ Schoenberger from Grant Heilman; **104** Dr. Dennis Kunkel/ PhotoTake NYC; **105** (tl)Gerald & Buff Corsi/Visuals Unlimited, (r)Fred Bruenner/Peter Arnold, Inc., (bl)Susan McCartney/Photo Researchers; **107** (l)John D. Cunningham/ Visuals Unlimited, (c)Jen & Des Bartlett/Bruce Coleman, Inc., (r)Breck P. Kent; **108** (tl)Artville, (tr)Tim Fehr, (c)Bob Daemmrich/Stock Boston/PictureQuest, (bl)Troy Mary Parlee/Index Stock/PictureQuest, (br)Jeffery Myers/Southern Stock/PictureQuest; **114** Stewart Cohen/Stone/Getty Images; **116** (t)Tom McHugh/Photo Researchers, (b)file photo; **117** Monica Dalmasso/Stone/Getty Images; **118** (t)Philip Lee Harvey/Stone, (b)Lester V. Bergman/CORBIS; **120** Walker England/Photo Researchers; **122** Barry L. Runk from Grant Heilman; **123** Cabisco/Visuals Unlimited; **124–125** Ashley Cooper/CORBIS; **125** (l r)Geoff Butler; **126** Stewart Cohen/Stone/Getty Images; **129** (bkgd)Jane Grushow from Grant Heilman, (others)Special Collections, National Agriculture Library; **130** Barry L. Runk From Grant Heilman; **132** Richard Hutchings/Photo Researchers; **134** (l)Grant Heilman Photograhy, (r)Gemma Giannini from Grant Heilman; **135** Raymond Gehman/CORBIS; **136** Dan McCoy from Rainbow; **137** (l)Phil Roach/Ipol, Inc., (r)CNRI/Science

Photo Library/Photo Researchers; **138** Gopal Murti/PhotoTake, NYC; **139** Tim Davis/Photo Researchers; **140** (t)Renee Stockdale/Animals Animals, (b)Alan & Sandy Carey/Photo Researchers; **143** Tom Meyers/Photo Researchers; **144** (t)Runk/Schoenberger from Grant Heilman, (b)Mark Burnett; **145** Richard Hutchings; **146** KS Studios; **151** CNRI/Science Photo Library/Photo Researchers; **152–153** B.G. Thomson/Photo Researchers; **155** Barbera Cushman/DRK Photo; **156** (l c)Tui De Roy/Minden Pictures, (r)Tim Davis/Photo Researchers; **158** (l)Gregory G. Dimijian, M.D./Photo Researchers, (r)Patti Murray/Animals Animals; **159** (l)George McCarthy/CORBIS, (r) Arthur Morris/Visuals Unlimited; **160** (l)Joe McDonald/Animals Animals, (c)Tom McHugh/Photo Researchers, (r)Tim Davis/Photo Researchers; **161** James Richardson/Visuals Unlimited; **162** Frans Lanting/Minden Pictures; **163** (l)Dominique Braud/Earth Scenes, (c)Carr Clifton/Minden Pictures, (r)John Cancalosi/DRK Photo; **164** (cw from top)Ken Lucas/Visuals Unlimited, John Cancalosi/DRK Photo, Larry Ulrich/DRK Photo, John Cancalosi/Peter Arnold, Inc., Sinclair Stammers/Science Photo Library/Photo Researchers; **165** John Kieffer/Peter Arnold, Inc.; **168** (l)Kees Van Den Berg/Photo Researchers, (r)Doug Martin; **169** Peter Veit/DRK Photo; **170** (l)Mark E. Gibson, (r)Gerard Lacz/Animals Animals; **171** (l)Michael Dick/Animals Animals, (r)Carolyn A. McKeone/Photo Researchers; **172** (l)John Reader/Science Photo Library/Photo Researchers, (r)Archivo Iconografico, S.A./CORBIS; **173** Francois Ducasse/Rapho/Photo Researchers; **174** (b)Kenneth W. Fink/Photo Researchers; **174–175** Aaron Haupt; **176** (t)Oliver Meckes/E.O.S/Gelderblom/Photo Researchers, (b)Lara Jo Regan/Saba; **177** (l)Koster-Survival/Animals Animals, (r)Stephen J. Keasemann/DRK Photo; **179** Tom Tietz/Stone/Getty Images; **180** Gregory G. Dimijian, M.D./Photo Researchers; **181** Patti Murray/Animals Animals; **182–183** Richard Hamilton Smith/CORBIS; **183** (l)Courtesy Broan-NuTone, (r)CORBIS; **184–185** Scimat/Photo Researchers; **186** (l)Oliver Meckes/Photo Researchers, (c r)CNRI/Science Photo Library/Photo Researchers; **188** Dr. L. Caro/Science Photo Library/Photo Researchers; **189** (l to r)Dr. Dennis Kunkel/PhotoTake NYC, David M. Phillips/Visuals Unlimited, R. Kessel/G. Shih/Visuals Unlimited, Ann Siegleman/Visuals Unlimited, SCIMAT/Photo Researchers; **190** (t)T.E. Adams/Visuals Unlimited, (b)Frederick Skavara/Visuals Unlimited; **191** R. Kessel/G. Shih/Visuals Unlimited; **192** T.E. Adams/Visuals Unlimited; **193** (tl)M. Abbey Photo/Photo Researchers, (tr)Oliver Meckes/Eye of Science/Photo Researchers, (bl)S. Lowry/University of Ulster/Stone/Getty Images, (br)A.B. Dowsett/Science Photo Library/Photo Researchers; **194** Ray Pfortner/Peter Arnold, Inc.; **195** (bkgd tl)Jeremy Burgess/Science Photo Library/Photo Researchers, (tr)Ann M. Hirsch/UCLA, (bl)John D. Cunningham/Visuals Unlimited, (br)Astrid & Hanns-Frieder Michler/Science Photo Library/Photo Researchers; **196** (l)Paul Almasy/CORBIS, (r)Joe Munroe/Photo Researchers; **197** (t)Terry Wild Studio, (b)George Wilder/Visuals Unlimited; **198** Amanita Pictures; **199** John Durham/Science Photo Library/Photo Researchers; **200** (t)KS Studios, (b)John Evans; **201** John Evans; **202** (t)P. Canumette/Visuals Unlimited, (c)Dr. Philippa Uwins, The University of Queensland, (bl)Dan Hoh/AP/Wide World Photos, (br)Reuters NewMedia Inc./CORBIS; **204** Carolina Biological /Visuals Unlimited; **206** (l)R. Kessel/G. Shih/Visuals Unlimited, (r)A. B. Dowsett/Science Photo Library/Photo Researchers; **207** (l)Breck P. Kent/Earth Scenes, (r)Ray Pfortner/Peter Arnold, Inc.; **208–209** Steve Austin; Papilio/CORBIS; **211** (l)Jean Claude Revy/PhotoTake, NYC, (r)Anne Hubbard/Photo Researchers; **212** (tl)NHMPL/Stone/Getty Images, (tr)Microfield Scienctific Ltd/Science Photo Library/Photo Researchers, (bl)David M. Phillips/Photo Researchers, (br)Dr. David Phillips/Visuals Unlimited; **213** (l)Pat & Tom Leeson/Photo Researchers, (r)Jeffrey L. Rotman/Peter Arnold, Inc.; **214** Walter H. Hodge/Peter Arnold, Inc.; **215** Eric V. Grave/Photo Researchers; **216** (t)Kerry B. Clark, (b)Astrid & Hanns-Frieder Michler/Science Photo Library/Photo Researchers; **217** Lennart Nilsson/Albert Bonniers Forlag AB; **218** (l)Ray Simons/Photo Researchers, (c)Matt Meadows/Peter Arnold, Inc., (r)Gregory G. Dimijian/Photo Researchers; **219** (t)Dwight Kuhn, (b)Mark Steinmetz; **220** Richard Calentine/Visuals Unlimited; **221** Biophoto Associates/Science Source/Photo Researchers; **222** James W. Richardson/Visuals Unlimited; **223** Carolina Biological Supply/Phototake, NYC; **224** (tl)Mark Steinmetz, (tr)Ken Wagner/Visuals Unlimited, (b)Dennis Kunkel; **225** (l)Science VU/Visuals Unlimited, (r)J.W. Richardson/Visuals Unlimited; **226** (tl)Bill Bachman/Photo Researchers, (tc)Frank Orel/Stone/Getty Images, (tr)Charles Kingery/PhotoTake, NYC, (b)Nancy Rotenberg/Earth Scenes; **227** (tl tc)Stephen Sharnoff, (tr)Biophoto Associates/Photo Researchers, (bl)L. West/Photo Researchers, (br)Larry Lee Photography/CORBIS; **228** (l)Nigel Cattlin/Holt Studios International/Photo Researchers, (r)Michael Fogden/Earth Scenes; **229** Ray Elliott; **230** Mark Steinmetz; **231** (tr)Mark Steinmetz, (b)Ken Wagner/Visuals Unlimited; **232** (t)Alvarode Leiva/Liaison, (b)courtesy Beltsville Agricultural Research Center-West/USDA; **233** (l)Michael Delaney/Visuals Unlimited, (r)Mark Steinmetz; **237** Mark Thayer Photography, Inc., Robert Calentine/Visuals Unlimited, Henry C. Wolcott III/Getty Images; **238–239** Peter Adams /Getty Images; **240** Tom Stack & Assoc.; **241** Laat-Siluur; **242** (t)Kim Taylor/Bruce Coleman, Inc., (b)William E. Ferguson; **243** (tl)Amanita Pictures, (tr)Ken Eward/Photo Researchers, (bl)Photo Researchers, (br)Amanita Pictures; **244** (cw from top)Dan McCoy from Rainbow, Philip Dowell/DK Images, Kevin & Betty Collins/Visuals Unlimited, David Sieren/Visuals Unlimited, Steve Callaham/Visuals Unlimited, Gerald & Buff Corsi/Visuals Unlimited, Mack Henley/Visuals Unlimited, Edward S. Ross, Douglas Peebles/CORBIS, Gerald & Buff Corsi/Visuals Unlimited, Martha McBride/Unicorn Stock Photos; **245** (t)Gail Jankus/Photo Researchers, (b)Michael P. Fogden/Bruce Coleman, Inc.; **246** (l)Larry West/Bruce Coleman, Inc., (c)Scott Camazine/Photo Researchers, (r)Kathy Merrifield/Photo Researchers; **247** Michael P. Gadomski/Photo Researchers; **249** (t)Farrell Grehan/Photo Researchers, (bl)Steve Solum/Bruce Coleman, Inc., (bc)R. Van Nostrand/Photo Researchers, (br)Inga Spence/Visuals Unlimited; **250** (t)Joy Spurr/Bruce Coleman, Inc., (b)W.H. Black/Bruce Coleman, Inc.; **251** Farrell Grehan/Photo Researchers; **252** Amanita Pictures; **253** (l)Nigel Cattlin/Photo Researchers, Inc., (c)Doug Sokel/Tom Stack & Assoc., (r)Charles D. Winters/Photo Researchers; **254** Bill Beatty/Visuals Unlimited; **256** (l)Doug Sokell/Tom Stack & Assoc., (tc)Robert C. Hermes/Photo Researchers, (r)Bill Beatty/Visuals Unlimited, (bc)David M. Schleser/Photo Researchers; **257** (cw from top)E. Valentin/Photo Researchers, Dia Lein/Photo Researchers, Eva Wallander, Tom Stack & Assoc., Joy Spurr/Photo Researchers; **259** (l)Dwight Kuhn, (c)Joy Spurr/Bruce Coleman, Inc., (r)John D. Cunningham/Visuals Unlimited; **260** (l)J. Lotter/Tom Stack & Assoc., (r)J.C. Carton/Bruce Coleman, Inc.; **262** (t)Inga Spence/Visuals Unlimited, (b)David Sieren/Visuals Unlimited; **263** Jim Steinberg/Photo Researchers; **264** (t)Michael Rose/Frank Lane Picture Agency/ CORBIS, (b)Dr. Jeremy Burgess/Science Photo Library/Photo Researchers; **266** Stephen P. Parker/Photo Researchers; **270–271** Massimo Mastrorillo/CORBIS; **272** (l)Stephen Dalton/Photo Researchers, (r)Matt Meadows; **273** (l)Holt Studios/Nigel Cattlin/Photo Researchers, (r)Inga Spence/ Visuals Unlimited; **274** (l)H. Reinhard/OKAPIA/Photo Researchers, (c)John W. Bova/Photo Researchers, (r)John D. Cunningham/Visuals Unlimited; **276** (l)Biology Media/Photo Researchers, (c)Andrew Syred/Science Photo Library/Photo Researchers, (r)Runk/Schoenberger from Grant Heilman; **278 279** Kathy Merrifield 2000/Photo Researchers; **280** Matt Meadows; **281** (l)John Kaprielian/Photo Researchers, (r)Scott Camazine/Sue Trainor/Photo Researchers; **282** Dr. WM. H. Harlow/Photo Researchers; **283** Christian Grzimek/OKAPIA/ Photo Researchers; **284** (l)Stephen P. Parker/Photo Researchers, (c)M. J. Griffith/Photo Researchers, (r)Dan Suzio/Photo Researchers; **285** (tl)Gustav Verderber/Visuals Unlimited, (tr)Rob Simpson/Visuals Unlimited, (b)Alvin E. Staffan/Photo Researchers; **286** (tl)C. Nuridsany & M. Perennou/Science Photo Library/Photo Researchers, (tr)Merlin D. Tuttle/Photo Researchers, (bl)Anthony Mercreca Photo/Photo Researchers, (bc)Kjell B. Sandved/Photo Researchers, (br)Holt Studios LTD/Photo Researchers; **287** William J. Weber/Visuals Unlimited; **289** (tl)Kevin Shafer/ CORBIS, (bcl)Tom & Pat Leeson, (bcr)Darryl Torckler/Stone/ Getty Images, (others)Dwight Kuhn; **290 292** Doug Martin; **293** Matt Meadows; **294** (t)Kevin Laubacher/FPG., (b)Michael Black/Bruce Coleman, Inc.; **295** Oliver Meckes/Photo Researchers; **296** Ed Reschke/Peter Arnold, Inc.; **298** Andrew Syred/Science Photo Library/Photo Researchers; **299** (l)Dan Suzio/Photo Researchers, (r)Alvin E. Staffan/Photo Researchers; **300–301** Terry Thompson/Panoramic Images; **301** Matt Meadows; **303** Dr. Jeremy Burgess/Science Photo Library/Photo Researchers; **304** (l)John Kieffer/Peter Arnold, Inc., (r)Runk/Schoenberger from Grant Heilman; **305** M. Eichelberger/Visuals Unlimited; **307** (t)Jacques Jangoux/Peter Arnold, Inc., (b)Jeff Lepore/Photo Researchers; **308** Michael P. Gadomski/Photo Researchers; **312** (l)Scott Camazine/Photo Researchers, (c r)Matt Meadows; **315** (t)Artville, (cl)Runk/ Schoenberger from Grant Heilman, (c cr)Prof. Malcolm B. Wilkins/University of Glasgow, (bl)Eric Brennan, (br)John Sohlden/Visuals Unlimited; **316** Jim Metzger; **318** (t)Ed Reschke/Peter Arnold, Inc., (b)Matt Meadows; **319** Matt Meadows; **320** Greg Vaughn/Getty Images; **321** (l)Norm Thomas/Photo Researchers, (r)S.R. Maglione/Photo Researchers; **322** Runk/|Schoenberger from Grant Heilman; **324** Matt Meadows; **325** Scott Camazine/Photo Researchers; **326–327** Glenn W. Elison; **326** (inset)Stephen St. John/ National Geographic Image Collection; **328–329** N. Sefton/ Photo Researchers; **329** Icon Images; **330** Zig Leszczynski/ Animals Animals; **331** (l)Jeff Foott/DRK Photo, (c)Leonard Lee Rue III/DRK Photo, (r)Hal Beral/Visuals Unlimited; **332** (t)Ken Lucas/Visuals Unlimited, (bl)Joe McDonald/ Visuals Unlimited, (br)Zig Leszczynski/Animals Animals; **333** (tl)Tom J. Ulrich/Visuals Unlimited, (tc)Peter & Beverly Pickford/DRK Photo, (tr)Fred McConnaughey/Photo Researchers, (b)Stuart Westmoreland/Mo Yung Productions/ Norbert Wu Productions; **335** (l)Stephen J. Krasemann/DRK Photo, (r)Ford Kristo/DRK Photo; **336** (l)Glenn Oliver/ Visuals Unlimited, (r)Andrew J. Martinez/Photo Researchers; **339** (t)Norbert WU/DRK Photo, (bl)Fred Bavendam/Minden Pictures, (br)H. Hall/OSF/Animals Animals;

340 Gerry Ellis/GLOBIO.org; **342** David B. Fleetham/Visuals Unlimited; **343** Larry Stepanowicz/Visuals Unlimited; **345** (tl)T.E. Adams/Visuals Unlimited, (tr)Science VU/ Visuals Unlimited, (b)Oliver Meckes/Eye of Science/Photo Researchers; **346** Triarch/Visuals Unlimited; **347** Oliver Meckes/Ottawa/Photo Researchers; **348** (tr)NIBSC/Science Photo Library/Photo Researchers, (cl)Sinclair Stammers/ Science Photo Library/Photo Researchers, (cr)Arthur M. Siegelman/Visuals Unlimited, (bl)Oliver Meckes/Photo Researchers, (bcl)Andrew Syred/Science Photo Library/Photo Researchers, (bcr)Eric V. Grave/Photo Researchers, (br)Cabisco/Visuals Unlimited; **349** R. Calentine/Visuals Unlimited; **350** (t)T.E. Adams/Visuals Unlimited, (b)Bob Daemmrich; **351** Matt Meadows; **352** (t)PhotoDisc, (b)Shirley Vanderbilt/Index Stock; **353** James H. Robinson/Animals Animals; **354** R. Calentine/Visuals Unlimited; **355** Donald Specker/Animals Animals; **356** (l)Fred McConnaughey/Photo Researchers, (r)Fred Bavendam/Minden Pictures; **357** (l)Stephen J. Krasemann/DRK Photo, (r)Glenn Oliver/ Visuals Unlimited; **358–359** Michael & Patricia Fogden/ CORBIS; **360** Wayne Lynch/DRK Photo; **361** (l)Jeff Rotman Photography, (r)James H. Robinson/Animals Animals; **362** (t)David S. Addison/Visuals Unlimited, (b)Joyce & Frank Burek/Animals Animals; **363** Clay Wiseman/Animals Animals; **364** Bates Littlehales/Animals Animals; **365** Beverly Van Pragh/ Museum Victoria; **366** Donald Specker/Animals Animals; **367** (t)Charles Fisher, Penn State University, (bl)Mary Beth Angelo/Photo Researchers, (br)Kjell B Sandved/Visuals Unlimited; **368** St. Bartholomew's Hospital/Science Photo Library/Photo Researchers; **370** Tom McHugh/Photo Researchers; **371** (t)Ted Clutter/Photo Researchers, (b)Kjell B. Sandved/Visuals Unlimited; **374** Lynn Stone; **375** (l)Bill Beatty/Animals Animals, (r)Patti Murray/Animals Animals; **376** (tl)Bill Beatty/Wild & Natural, (tc)Robert F. Sisson, (tr)Index Stock, (cl)Brian Gordon Green, (c)Joseph H. Bailey/National Geographic Image Collection, (cr)Jeffrey L. Rotman/CORBIS, (b)Timothy G. Laman/National Geographic Image Collection; **377** (t)James P. Rowan/DRK Photo, (b)Leonard Lee Rue/Photo Researchers; **378** Ken Lucas/Visuals Unlimited; **379** Tom Stack & Assoc.; **380** Scott Smith/Animals Animals; **381** Clay Wiseman/Animals Animals; **382** (tl)Andrew J. Martinez/Photo Researchers, (tr)David Wrobel/Visuals Unlimited, (b)Gerald & Buff Corsi/Visuals Unlimited; **383** Ken Lucas/Visuals Unlimited; **384 385** Matt Meadows; **386** (t)David M. Dennis, (b)Harry Rogers/Photo Researchers; **387** (l)Charles McRae/Visuals Unlimited, (r)Mark Moffet/Minden Pictures; **388** Leroy Simon/Visuals Unlimited; **389** William Leonard/DRK Photo; **390** Joyce & Frank Burek/Animals Animals; **391** (l)Clay Wiseman/Animals Animals, (r)Scott Smith/Animals Animals; **392–393** Robert Lubeck/Animals Animals; **394** Fred Bavendam/Minden Pictures; **395** Omni-Photo Communications; **396** (t to b)H. W. Robison/Visuals Unlimited, Flip Nicklin/Minden Pictures, John M. Burnley/Photo Researchers, George Grall/National Geographic Image Collection, M. P. Kahl/DRK Photo, Grace Davies/Omni-Photo Communications; **397** T. A. Wiewandt/ DRK Photo; **398** Icon Images; **399** (l to r)Meckes/Ottawa/ Photo Researchers, Rick Gillis/University of Wisconsin-La Crosse, Runk/Schoenberger from Grant Heilman; **400** Ken Lucas/Visuals Unlimited; **401** (tl)James Watt/Animals Animals, (tc)Norbert Wu/DRK Photo, (tr)Fred Bavendam/ Minden Pictures, (br)Richard T. Nowitz/Photo Researchers; **402 404** Tom McHugh/Photo Researchers; **405** (t)Tom McHugh/Steinhart Aquarium/Photo Researchers, (bl)Bill

Credits

Kamin/Visuals Unlimited, (bc)Norbert Wu/DRK Photo, (br)Michael Durham/GLOBIO.org; **406** Runk/Schoenberger from Grant Heilman; **407** Fred Habegger/Grant Heilman; **408** (t)David Northcott/DRK Photo, (bl br)Runk/Schoenberger from Grant Heilman; **409** (l)Runk/Schoenberger from Grant Heilman, (r)George H. Harrison from Grant Heilman; **410** (l)Mark Moffett/Minden Pictures, (r)Michael Fogden/DRK Photo; **411** Rob and Ann Simpson/Visuals Unlimited; **412** Joe McDonald/Visuals Unlimited; **414** (tl)Klaus Uhlenhut/Animals Animals, (tr)Rob & Ann Simpson/Visuals Unlimited, (b)G and C Merker/Visuals Unlimited; **415** (t)Mitsuaki Iwago/Minden Pictures, (b)Belinda Wright/DRK Photo; **416** (tl)John Sibbick, (tr)Karen Carr, (c)Chris Butler/Science Photo Library/Photo Researchers, (bl)Jerome Connolly, courtesy The Science Museum of Minnesota, (br)Chris Butler; **418** (t)Steve Maslowski/Visuals Unlimited, (b)Michael Newman/PhotoEdit, Inc.; **419** KS Studios; **420** (tl)Hemera Technologies, Inc., (tc)Michael Fogden/Animals Animals, (tr)Tim Flach/Stone/Getty Images, (b)R. Rotolo/Liaison Agency; **422** John Cancalosi/DRK Photo; **424** (tl tr)Runk/Schoenberger from Grant Heilman, (b)Tom McHugh/Photo Researchers; **426–427** Theo Allofs/CORBIS; **428** Michael Habicht/Animals Animals; **430** (l)Crown Studios, (r)KS Studios; **431** (l)Lynn Stone/Animals Animals, (r)Arthur R. Hill/Visuals Unlimited; **433** (l)Zefa Germany/The Stock Market/CORBIS, (r)Sid & Shirley Rucker/DRK Photo; **434** (l)Kennan Ward/CORBIS, (t to b)Wayne Lankinen/DRK Photo, Ron Spomer/Visuals Unlimited, M. Philip Kahl/Gallo Images/CORBIS, Rod Planck/Photo Researchers, (bkgd)Steve Maslowski; **436** Stephen J. Krasemann/DRK Photo; **437** (t)Gerard Lacz/Animals Animals, (bl)Tom Brakefield/DRK Photo, (br)John David Fleck/Liaison Agency/Getty Images; **438** Bob Gurr/DRK Photo; **439** Amos Nachoum/The Stock Market/CORBIS; **440** (t)Jean-Paul Ferrero/AUSCAPE, (bl)Phyllis Greenberg/Animals Animals, (br)John Cancalosi/DRK Photo; **441** (t)Carolina Biological Supply/PhotoTake, NYC, (b)Doug Perine/DRK Photo; **442** (t to b)Stephen J. Krasemann/DRK Photo, David Northcott/DRK Photo, Zig Leszczynski/Animals Animals, Ralph Reinhold/Animals Animals, Anup Shah/Animals Animals, Mickey Gibson/Animals Animals; **443** (t to b)Fred Felleman/Stone/Getty Images, Robert Maier/Animals Animals, Tom Bledsoe/DRK Photo, Wayne Lynch/DRK Photo, Joe McDonald/Animals Animals, Kim Heacox/DRK Photo; **446** (t)David Welling/Animals Animals, (b)Wayne Lankinen/DRK Photo; **447** (t)Richard Day/Animals Animals, (b)Maslowski Photo; **448** (tr)Mark Burnett, (bl)Jeff Fott/DRK Photo, (br)Joe McDonald/Animals Animals; **449** (l r)Tom & Pat Leeson/DRK Photo, (b)Tom Brakefield/DRK Photo; **450** Hans & Judy Beste/Animals Animals; **452** Bob Gurr/DRK Photo; **453** (l)Crown Studios, (r)KS Studios; **454–455** D. Robert & Lorri Franz/CORBIS; **456** (l)Michel Denis-Huot/Jacana/Photo Researchers, (r)Zig Lesczynski/Animals Animals; **457** (l)Jack Ballard/Visuals Unlimited, (c)Anthony Mercieca/Photo Researchers, (r)Joe McDonald/Visuals Unlimited; **458** (t)Stephen J. Krasemann/Peter Arnold, Inc., (b)Leonard Lee Rue/Photo Researchers; **459** (t)The Zoological Society of San Diego, (b)Margret Miller/Photo Researchers; **462** Michael Fairchild/Peter Arnold, Inc.; **463** (t)Bill Bachman/Photo Researchers, (b)Fateh Singh Rathore/Peter Arnold, Inc.; **464** Jim Brandenburg/Minden Pictures; **465** Michael Dick/Animals Animals; **466** (l)Richard Thorn/Visuals Unlimited, (c)Arthur Morris/Visuals Unlimited, (r)Jacana/Photo Researchers; **467** (starfish)Peter J. Herring, (krill)T. Frank/Harbor Branch Oceanographic Institution, (others) Edith Widder/Harbor Branch Oceanographic Institution; **468** Stephen Dalton/Animals Animals; **469** Richard Packwood/Animals Animals; **470** Ken Lucas/Visuals Unlimited; **472** (t)The Zoological Society of San Diego, (b)Gary Carter/Visuals Unlimited; **473** Dave B. Fleetham/Tom Stack & Assoc.; **474** (t)Walter Smith/CORBIS, (b)Bios (Klein/Hubert)/Peter Arnold, Inc.; **475** (l)Valerie Giles/Photo Researchers, (r)J & B Photographers/Animals Animals; **476** not available; **478** Jim Brandenburg/Minden Pictures; **479** PhotoDisc; **480–481** Birgid Allig/Stone/Getty Images; **481** (inset)Don Mason/The Stock Market/CORBIS; **482–483** Charles O' Rear/CORBIS; **483** Matt Meadows; **484** John Serro/Visuals Unlimited; **488** Geoff Butler; **489** Photo Researchers; **490** Digital Stock; **491** Aaron Haupt; **492** M. McCarron, (t)C Squared Studios/PhotoDisc; **493** (l)Breck P. Kent, (c)Runk/Schoenberger from Grant Heilman, (r)PhotoTake, NYC/Carolina Biological Supply Company; **497** (tl)Clyde H. Smith/Peter Arnold, Inc., (tcl)Erik Sampers/Photo Researchers, (tcr)Dean Conger/CORBIS, (tr)Michael A. Keller/The Stock Market/CORBIS, (bl)Ed Bock/The Stock Market/CORBIS, (bcl)Joe McDonald/Visuals Unlimited, (bcr)Art Stein/Photo Researchers, (br)Peter Turnley/CORBIS; **499** Jim Grace/Photo Researchers; **500** Photo Researchers; **501** Mark Burnett; **504** Sara Davis/The Herald-Sun; **506** Breck P. Kent; **509** (l)Royalty-Free/CORBIS, (r)Jim Grace/Photo Researchers; **510–511** Meckes/Ottawa/Photo Researchers; **512 513 514** KS Studios; **515** (l)KS Studios, (r)Visuals Unlimited; **517** (cabbages, avocado) Artville, (liver)DK Images, (doctor)Michael W. Thomas, (girls)Digital Vision/PictureQuest, (Blood Cells, Blood Clot)David M. Philips/Visuals Unlimited, (others)Digital Stock; **518** Gary Kreyer from Grant Heilman; **519** Larry Stepanowicz/Visuals Unlimited; **520 521** KS Studios; **523** (l)KS Studios, (r)Tom McHugh/Photo Researchers; **525** Geoff Butler; **527** (l)Benjamin/Custom Medical Stock Photo, (r)Dr. K.F.R. Schiller/Photo Researchers; **528** Biophoto Associates/Photo Researchers; **530** KS Studios; **531** Matt Meadows; **532** Goldwater/Network/Saba Press Photos; **533 535** KS Studios; **537** (l)Jose Luis Pelaez, Inc./CORBIS, (r)Dean Berry/Index Stock Imagery; **538–539** Steve Allen/Getty Images; **540 543** Aaron Haupt; **545** Matt Meadows; **546** Martin M. Rotker; **547** (t)StudiOhio, (b)Matt Meadows; **549** First Image; **551** National Cancer Institute/Science Photo Library/Photo Researchers; **555** Meckes/Ottawa/Photo Researchers; **557** Aaron Haupt; **558** (t)Matt Meadows/Peter Arnold, Inc., (b)Matt Meadows; **560** no credit; **561** (l)Manfred Kage/Peter Arnold, Inc., (r)K.G. Murti/Visuals Unlimited; **566–567** The Image Bank/Getty Images; **568** Randy Lincks/CORBIS; **569** Dominic Oldershaw; **570** Bob Daemmrich; **573** Richard T. Nowitz; **575** (l c)SIU/Photo Researchers, (r)Geoff Butler; **576** Renee Lynn/Photo Researchers; **578** (l)Science Pictures Ltd./Science Photo Library/Photo Researchers, (r)SIU/Photo Researchers; **580** Paul Barton/The Stock Market/CORBIS; **581** (bkgd)Gunther/Explorer/Photo Researchers, (l r)Mark Burnett; **582** Richard Hutchings/Photo Researchers; **583** Biophoto Associates/Science Source/Photo Researchers; **584** (t)Larry Mulvehill/Photo Researchers, (b)Matt Meadows; **585** Matt Meadows; **587** (bkgd)Science Photo Library/CORBIS, (l)Ed Beck/The Stock Market/CORBIS, (t)Lane Medical Library, (tr)Gregg Ozzo/Visuals Unlimited, (b)Custom Medical Stock Photo, (br)Tom & DeeAnn McCarthy/The Stock Market/CORBIS;

592–593 John Terrance Turner/FPG/Getty Images; **594** KS Studios; **601** KS Studios; **602** Michael Newman/PhotoEdit, Inc.; **603** KS Studios; **607** Aaron Haupt; **611** Mark Burnett; **612** (t)Jeff Greenberg/PhotoEdit, Inc., (b)Amanita Pictures; **613** Amanita Pictures; **614** Toni Morrison; **615** (l)David R. Frazier/Photo Researchers, (r)Michael Brennan/CORBIS; **619** Eamonn McNulty/Science Photo Library/Photo Researchers; **620–621** Lawerence Manning/CORBIS; **621** John Evans; **622** David Young-Wolff/PhotoEdit, Inc.; **631** Ariel Skelley/The Stock Market/CORBIS; **633** David M. Phillips/Photo Researchers; **634** (l)Tim Davis/Photo Researchers, (r)Chris Sorensen/The Stock Market/CORBIS; **635** Science Pictures Ltd/Science Photo Library/Photo Researchers; **636** Petit Format/Nestle/Science Source/Photo Researchers; **638** (l)Jeffery W. Myers/Stock Boston, (r)Ruth Dixon; **639** (tl b)Mark Burnett, (tr)Aaron Haupt; **640** KS Studios; **641** (l)NASA/Roger Ressmeyer/CORBIS, (r)AFP/CORBIS; **642** (t)Chris Carroll/CORBIS, (b)Richard Hutchings; **643** Matt Meadows; **644** (t)Nancy Sheehan/PhotoEdit, Inc., (b)Martin B. Withers/Frank Lane Picture Agency/CORBIS; **645** (l)Bob Daemmrich, (r)Maria Taglienti/The Image Bank/Getty Images; **650–651** S. Lowry/University of Ulster/Stone/Getty Images; **652** Dr. P. Marazzi/Science Photo Library/Photo Researchers; **653** (l)Michael A. Keller/The Stock Market/CORBIS, (r)Runk/Schoenberger from Grant Heilman, (b)NIBSC/Science Photo Library/Photo Researchers; **656** CC Studio/Science Photo Library/Photo Researchers; **659** (t)Jack Bostrack/Visuals Unlimited, (cl)Cytographics Inc./Visuals Unlimited, (cr)Cabisco/Visuals Unlimited, (b)Visuals Unlimited; **660** MM/Michelle Del Guercio/Photo Researchers; **661** Holt Studios International/Nigel Cattlin/Photo Researchers; **662** (t)Oliver Meckes/Eye of Science/Photo Researchers, (b)Visuals Unlimited; **663** Oliver Meckes/Eye Of Science/Gelderblom/Photo Researchers; **664** Mark Burnett; **666** (l)Caliendo/Custom Medical Stock Photo, (r)Amanita Pictures; **667** (t)Andrew Syred/Science Photo Library/Photo Researchers, (b)Custom Medical Stock Photo; **668** (l)Jan Stromme/Bruce Coleman, Inc., (c)Mug Shots/The Stock Market/CORBIS, (r)J.Chiasson-Liats/Liaison Agency/Getty Images; **670** KS Studios; **672** (t)Tim Courlas, (b)Matt Meadows; **673** Matt Meadows; **674** Layne Kennedy/CORBIS; **675** (l)Gelderblom/Eye of Science/Photo Researchers, (r)Garry T. Cole/BPS/Stone/Getty Images; **680–681** (bkgd)Jodi Jacobson, Andrew A. Wagner; **681** (inset)L. Fritz/H. Armstrong Roberts; **682–683** Joe McDonald/Visuals Unlimited; **684** (l)Adam Jones/Photo Researchers, (tr)Richard Kolar/Animals Animals, (c)Tom Van Sant/Geosphere Project, Santa Monica/Science Photo Library/Photo Researchers, (br)G. Carleton Ray/Photo Researchers; **685** (t)John W. Bova/Photo Researchers, (b)David Young/Tom Stack & Assoc.; **687** (l)Zig Leszczynski/Animals Animals, (r)Gary W. Carter/Visuals Unlimited; **690** Mitsuaki Iwago/Minden Pictures; **691** Joel Sartore from Grant Heilman; **693** (t)Norm Thomas/Photo Researchers, (b)Maresa Pryor/Earth Scenes; **694** (r)Bud Neilson/Words & Pictures/PictureQuest, (others)Wyman P. Meinzer; **696** (l)Michael Abbey/Photo Researchers, (r)OSF/Animals Animals; (b)Michael P. Gadomski/Photo Researchers; **697** (tcr)Lynn M. Stone, (bl)Larry Kimball/Visuals Unlimited, (bcl)George D. Lepp/Photo Researchers, (bcr)Stephen J. Krasemann/Peter Arnold, Inc., (br)Mark Steinmetz, (others) William J. Weber; **698** (t)Milton Rand/Tom Stack & Assoc., (c)Marian Bacon/Animals Animals, (b)Sinclair Stammers/Science Photo Library/Photo Researchers; **699** (tl)Raymond A. Mendez/Animals Animals, (bl)Donald Specker/Animals Animals, (br)Joe McDonald/Animals Animals; **700** Ted Levin/Animals Animals; **701** Richard L. Carlton/Photo Researchers; **702** (t)Jean Claude Revy/PhotoTake, NYC, (b)OSF/Animals Animals; **703** Runk/Schoenberger from Grant Heilman; **704** Eric Larravadieu/Stone/Getty Images; **705** (l)C.K. Lorenz/Photo Researchers, (r)Hans Pfletschinger/Peter Arnold, Inc.; **706** CORBIS; **708** (l)Michael P. Gadomski/Photo Researchers, (r)William J. Weber; **710–711** Ron Thomas/Getty Images; **712** Kenneth Murray/Photo Researchers; **713** (t)Jerry L. Ferrara/Photo Researchers, (b)Art Wolfe/Photo Researchers; **714** (t)Telegraph Colour Library/FPG/Getty Images, (b)Hal Beral/Visuals Unlimited; **715** (l)Fritz Polking/Visuals Unlimited, (r)R. Arndt/Visuals Unlimited; **716** Tom Uhlman/Visuals Unlimited; **720** Jim Grattan; **723** (l)Runk/Schoenberger from Grant Heilman, (r)Rob & Ann Simpson/Visuals Unlimited, Runk/Schoenberger from Grant Heilman; **726** WHOI/Visuals Unlimited; **730** Gerald and Buff Corsi/Visuals Unlimited; **731** Jeff J. Daly/Visuals Unlimited; **732** Gordon Wiltsie/Peter Arnold, Inc.; **733** (l)Soames Summerhay/Photo Researchers, (r)Tom Uhlman/Visuals Unlimited; **738–739** William Campbell/CORBIS Sygma; **740** Jeff Greenberg/Visuals Unlimited; **741** Larry Ulrich/DRK Photo; **742** (bkgd)Craig Fujii/Seattle Times, (l)Kevin R. Morris/CORBIS, (tr br)Jeff Henry; **743** Rod Planck/Photo Researchers; **745** (t)Steve McCutcheon/Visuals Unlimited, (bl)Pat O'Hara/DRK Photo, (br)Erwin & Peggy Bauer/Tom Stack & Assoc.; **746** (tl)Peter Ziminski/Visuals Unlimited, (c)Leonard Rue III/Visuals Unlimited, (bl)C.C. Lockwood/DRK Photo, (br)Larry Ulrich/DRK Photo; **747** (t)Fritz Polking/Visuals Unlimited, (b)William Grenfell/Visuals Unlimited; **748** Lynn M. Stone/DRK Photo; **750** (l)Joe McDonald/DRK Photo, (r)Steve Solum/Bruce Coleman, Inc.; **751** Kevin Schafer; **753** W. Banaszewski/Visuals Unlimited; **754** (l)Dwight Kuhn, (r)Mark E. Gibson/Visuals Unlimited; **755** James R. Fisher/DRK Photo; **756** D. Foster/WHOI/Visuals Unlimited; **757** (l)C.C. Lockwood/Bruce Coleman, Inc., (r)Steve Wolper/DRK Photo; **758** (tl)Dwight Kuhn, (tr)Glenn Oliver/Visuals Unlimited, (b)Stephen J. Krasemann/DRK Photo; **759** (l)John Kaprielian/Photo Researchers, (r)Jerry Sarapochiello/Bruce Coleman, Inc.; **760** (t)Dwight Kuhn, (b)John Gerlach/DRK Photo; **761** Fritz Polking/Bruce Coleman, Inc.; **762** courtesy Albuquerque Public Schools; **763** (l)James P. Rowan/DRK Photo, (r)John Shaw/Tom Stack & Assoc.; **767** (l)Leonard Rue III/Visuals Unlimited, (r)Joe McDonald/DRK Photo; **768–769** Grant Heilman Photography; **770** (l)Keith Lanpher/Liaison Agency/Getty Images, (r)Richard Thatcher/David R. Frazier Photolibrary; **771** (t)Solar Cookers International, (bl)Brian F. Peterson/The Stock Market/CORBIS, (br)Ron Kimball Photography; **772** Larry Mayer/Liaison Agency/Getty Images; **775** (tr)Torleif Svenson/The Stock Market/CORBIS, (bl)Rob Williamson, (br)Les Gibbon/Cordaiy Photo Library Ltd./CORBIS; **776** Sean Justice; **777** (t)Lowell Georgia/Science Source/Photo Researchers, (cl)NASA, (c)CORBIS, (cr)Sean Sprague/Impact Visuals/PictureQuest, (bl)Lee Foster/Bruce Coleman, Inc., (br)Robert Perron; **778** Philippe Renault/Liaison Agency/Getty Images; **779** (l)NYC Parks Photo Archive/Fundamental Photographs, (r)Kristen Brochmann/Fundamental Photographs; **783** (l)Jeremy Walker/Science Photo Library/Photo Researchers, (c)John Colwell from Grant Heilman, (r)Telegraph Colour Library/FPG/Getty Images; **784** Wilford Haven/Liaison Agency/Getty Images; **785** (tl)Larry Mayer/Liaison Agency/Getty Images, (tr)ChromoSohm/The Stock Market/CORBIS, (cr)David R. Frazier Photolibrary, (br)Inga Spence/Visuals Unlimited; **786** (r)Andrew Holbrooke/The Stock Market/CORBIS, (Paint Cans)Amanita Pictures, (Turpantine, Paint thinner, epoxy)Icon Images, (Batteries) Aaron Haupt; **788** Paul A. Souders/CORBIS; **789** Icon Images; **791** Larry Lefever from Grant Heilman; **792** (t)Howard Buffett from Grant Heilman, (b)Solar Cookers International; **793** John D. Cunningham/Visuals Unlimited; **794** Frank Cezus/FPG/Getty Images; **796** Robert Cameron/Stone/Getty Images; **797** (l)Steve McCutcheon/Visuals Unlimited, (r)James N. Westwater; **798** David R. Frazier Photolibrary, Lebrum/Liaison Agency/Getty Images, Spencer Grant/PhotoEdit, Inc.; **800** PhotoDisc; **802** Tom Pantages; **806** Michell D. Bridwell/PhotoEdit, Inc.; **807** (t)Mark Burnett, (b)Dominic Oldershaw; **808** StudiOhio; **809** Timothy Fuller; **810** Aaron Haupt; **812** KS Studios; **813** Matt Meadows; **814 815** Aaron Haupt; **816** Milton Rand/Tom Stack & Assoc.; **817** Matt Meadows; **819** (t)Aaron Haupt, (b)Andrew J. Martinez/Photo Researchers; **820** Runk/Schoenberger from Grant Heilman; **821** Rod Joslin; **822** KS Studios; **824** David S. Addison/Visuals Unlimited; **826** Rod Planck/Photo Researchers; **828** Amanita Pictures; **829** Bob Daemmrich; **831** Davis Barber/PhotoEdit, Inc.; **847** Matt Meadows; **848** (l)Dr. Richard Kessel, (c)NIBSC/Science Photo Library/Photo Researchers, (r)David John/Visuals Unlimited; **849** (t)Runk/Schoenberger from Grant Heilman, (bl)Andrew Syred/Science Photo Library/Photo Researchers, (br)Rich Brommer; **850** (tr)G.R. Roberts, (l)Ralph Reinhold/Earth Scenes, (br)Scott Johnson/Animals Animals; **851** Martin Harvey/DRK Photo.

PERIODIC TABLE OF THE ELEMENTS

Columns of elements are called groups. Elements in the same group have similar chemical properties.

Gas

Liquid

Solid

Synthetic

The first three symbols tell you the state of matter of the element at room temperature. The fourth symbol identifies elements that are not present in significant amounts on Earth. Useful amounts are made synthetically.

Element	Hydrogen
Atomic number	1
Symbol	H
Atomic mass	1.008
State of matter	

	1	2	3	4	5	6	7	8	9
1	Hydrogen 1 **H** 1.008								
2	Lithium 3 **Li** 6.941	Beryllium 4 **Be** 9.012							
3	Sodium 11 **Na** 22.990	Magnesium 12 **Mg** 24.305							
4	Potassium 19 **K** 39.098	Calcium 20 **Ca** 40.078	Scandium 21 **Sc** 44.956	Titanium 22 **Ti** 47.867	Vanadium 23 **V** 50.942	Chromium 24 **Cr** 51.996	Manganese 25 **Mn** 54.938	Iron 26 **Fe** 55.845	Cobalt 27 **Co** 58.933
5	Rubidium 37 **Rb** 85.468	Strontium 38 **Sr** 87.62	Yttrium 39 **Y** 88.906	Zirconium 40 **Zr** 91.224	Niobium 41 **Nb** 92.906	Molybdenum 42 **Mo** 95.94	Technetium 43 **Tc** (98)	Ruthenium 44 **Ru** 101.07	Rhodium 45 **Rh** 102.906
6	Cesium 55 **Cs** 132.905	Barium 56 **Ba** 137.327	Lanthanum 57 **La** 138.906	Hafnium 72 **Hf** 178.49	Tantalum 73 **Ta** 180.948	Tungsten 74 **W** 183.84	Rhenium 75 **Re** 186.207	Osmium 76 **Os** 190.23	Iridium 77 **Ir** 192.217
7	Francium 87 **Fr** (223)	Radium 88 **Ra** (226)	Actinium 89 **Ac** (227)	Rutherfordium 104 **Rf** (261)	Dubnium 105 **Db** (262)	Seaborgium 106 **Sg** (266)	Bohrium 107 **Bh** (264)	Hassium 108 **Hs** (277)	Meitnerium 109 **Mt** (268)

The number in parentheses is the mass number of the longest-lived isotope for that element.

Rows of elements are called periods. Atomic number increases across a period.

The arrow shows where these elements would fit into the periodic table. They are moved to the bottom of the table to save space.

Lanthanide series	Cerium 58 **Ce** 140.116	Praseodymium 59 **Pr** 140.908	Neodymium 60 **Nd** 144.24	Promethium 61 **Pm** (145)	Samarium 62 **Sm** 150.36
Actinide series	Thorium 90 **Th** 232.038	Protactinium 91 **Pa** 231.036	Uranium 92 **U** 238.029	Neptunium 93 **Np** (237)	Plutonium 94 **Pu** (244)